AF712690

Mechanical Support for Heart Failure

Jamshid H. Karimov
Kiyotaka Fukamachi • Randall C. Starling
Editors

Mechanical Support for Heart Failure

Current Solutions and New Technologies

Springer

Editors
Jamshid H. Karimov
Lerner Research Institute, Cleveland Clinic
Cleveland Clinic Lerner College of Medicine of
Case Western Reserve University
Cleveland Clinic
Cleveland, OH
USA

Kiyotaka Fukamachi
Lerner Research Institute, Cleveland Clinic
Cleveland Clinic Lerner College of Medicine of
Case Western Reserve University
Cleveland Clinic
Cleveland, OH
USA

Randall C. Starling
Kaufman Center for Heart Failure
Heart, Thoracic, and Vascular Institute
Cleveland Clinic Lerner College of Medicine of
Case Western Reserve University
Cleveland Clinic
Cleveland, OH
USA

ISBN 978-3-030-47808-7 ISBN 978-3-030-47809-4 (eBook)
https://doi.org/10.1007/978-3-030-47809-4

© Springer Nature Switzerland AG 2020
This work is subject to copyright. All rights are reserved by the Publisher, whether the whole or part of the material is concerned, specifically the rights of translation, reprinting, reuse of illustrations, recitation, broadcasting, reproduction on microfilms or in any other physical way, and transmission or information storage and retrieval, electronic adaptation, computer software, or by similar or dissimilar methodology now known or hereafter developed.
The use of general descriptive names, registered names, trademarks, service marks, etc. in this publication does not imply, even in the absence of a specific statement, that such names are exempt from the relevant protective laws and regulations and therefore free for general use.
The publisher, the authors and the editors are safe to assume that the advice and information in this book are believed to be true and accurate at the date of publication. Neither the publisher nor the authors or the editors give a warranty, expressed or implied, with respect to the material contained herein or for any errors or omissions that may have been made. The publisher remains neutral with regard to jurisdictional claims in published maps and institutional affiliations.

This Springer imprint is published by the registered company Springer Nature Switzerland AG
The registered company address is: Gewerbestrasse 11, 6330 Cham, Switzerland

To all the forward thinkers and courageous creators of the past, present and future.

Foreword

More than 50 years have passed since the development and initial clinical testing of temporary mechanical cardiac support systems for blood pumps. The National Heart, Lung and Blood Institute (NHLBI) funded those early, ambitious efforts and clinical trials, with the collaboration of several multidisciplinary groups. Since then, research on the artificial heart was geared toward replacing the diseased organ and securing normal physiological functionality with a heart prosthesis. Many significant efforts toward this end have been pursued in the USA and in leading centers around the world.

In 1988, NHLBI funding for the total artificial heart was temporarily suspended pending a review by the Institutes of Medicine. Funding eventually resumed, but the NHLBI also began to fund other approaches, such as continuous flow devices, which offered the advantages of size, simplicity, and durability. Continuous flow devices now constitute the majority of pump implants, with excellent outcomes—more than 85 percent of patients on these devices are alive after 1 year, results that rival those of cardiac transplantation. Thrombus formation, which had posed major limitations to use, has been significantly reduced, but other post-implant complications continue to present challenges. The multidisciplinary approach and methodology that the artificial heart programs were built on continue to contribute extensively to the greater understanding of the metabolism of all organs and systems affected by cardiac circulatory devices. The advancement in science and technology has enabled today's innovative device-based treatment options for end-stage heart failure.

This book documents the current status of mechanical cardiac support, with chapters from the leaders of the many disciplines that have contributed to the increasing successes in the field. The research on artificial circulation continues, with the clinical experts and outstanding engineering teams remaining closely connected in efforts to bring better technologies to the patient's bedside. The search for the perfect pump—the goal of these groups—goes on.

Cleveland, OH — Leonard A. Golding, MB, BS, FRACS

Foreword

It was in 1957 that some will remember Soviet Union launched Sputnik, the first artificial orbiter that beeped around the earth during its three-week spin before its batteries quieted and the craft plunged from space. It was the beginning of the Space Race and the height of the Cold War. It was a triumph of engineering. That was also the year Kolff and Akutsu first replaced a canine heart with a rudimentary total artificial circulatory pump in research labs at the Cleveland Clinic. It successfully supported the animal for several hours. Kolff, and others (particularly DeBakey and Kantrowitz), had speculated in 1966 that the human heart would be replaced by a mechanical pump in due time. And so, 50 years ago, in a controversial operation in Houston, Texas, the first native heart extirpation with replacement by a total artificial heart in a fateful attempt to "bridge" a critically ill patient to heart transplant occurred. The operation sparked intense media, clinical, and scientific interest as well as a malpractice lawsuit. About 15 years later, the first "destination therapy" total artificial system was implanted in another controversial operation in Salt Lake City, Utah. Barney Clark was a terminally ill heart failure patient and not a heart transplant candidate. It was a desperate attempt to introduce a therapy that would achieve the goals of early mechanical circulatory support – permanent substitution of the human heart with an artificial pump. That procedure was also controversial but sparked even greater intensification of efforts to develop mechanical options for human circulatory support. These included development of short-term temporary pumps meant to be removed after circulatory recovery, more permanent but still shorter term bridges to a definitive heart transplant procedure, and so called destination (being home) devices meant to be permanent implants. As we look at the profession today, we are reaping the benefits of extraordinary work, passion, and bravery. Indeed, there was a courage to fail, a term used by social scientists Fox and Swazey while studying organ transplantation. Researchers, clinicians, patients, and families soldiered on to make mechanical circulatory support or replacement a clinical reality. The efforts have been grand, intense, expensive, controversial, and filled with failures. But in the end, they've been largely successful.

Thus appears the seminal compendium of *Mechanical Support for Heart Failure – Current Solutions and New Technologies* edited by Karimov, Fukamachi, and Starling. The effort sets status of this enterprise straight. The editors have assembled a multidisciplinary listing of notables in the field who represent the extraordinary diversity of scientific and clinical fields that have been necessary to make significant discoveries. It is a team sport. This enabled development of an entirely new and quite different professional enterprise that now saves lives previously doomed with death the destiny. No more nuclear powered total artificial hearts and ventricular assist devices, or flawed and discarded pumping power tools, but instead a sophisticated and growing armamentarium of circulatory assist pumps that save lives in dramatic fashion. Indeed, the future is now because this volume's editors and its contributors have, in large part, created the future. We in this field are not done. Studying this volume will assist with our journey.

The text begins with a review of the history of device development and ends with perspectives on future effort. In between are comprehensive essays on patient selection, physiologic and non-physiologic circulation, available technology, the challenge of pediatric patients, evolving technology, total artificial hearts, patient management and implantation techniques, and complications with their management. Study of this work by all invested or interested in the field of mechanical circulatory support is essential. Our patients are depending on our continued efforts in this endeavor.

James B. Young, MD
Chief Academic Officer, Professor of Medicine and Academic Dean
Cleveland Clinic Lerner College of Medicine of Case Western Reserve University, Cleveland Clinic
Cleveland, OH, USA

Preface

The development of heart failure treatment options based on mechanical circulatory support devices is ongoing. Despite tremendous technological breakthroughs, and reinforcing the clinical knowledge supporting the use of this lifesaving, device-oriented therapeutic option, hurdles and complications relating to extended clinical use remain. These challenges will spur us on toward more research to bring these technologies to the next level.

With overwhelming evidence in several areas—biocompatibility requirements, portability and durability of the ventricular assist devices, catheter-based technologies to provide hemodynamic support, artificial hearts in development as total cardiac replacement therapy, and advanced patient monitoring platforms—understanding the fundamental basis of device development is imperative. It is driven by clinical need, and acknowledges the inevitability that patient quality of life and device-related outcomes and performance are interconnected. This knowledge will define the overall course and long-term scenarios post-implant.

This book is a comprehensive overview of mechanical circulatory support. It covers many avenues—the history of mechanical circulatory device development; the reminiscences of individuals who paved the way; fundamentals of hemodynamics support; evolving technologies and innovation; advanced and novel devices in the development pipeline; clinical insights into some of the most devastating complications; and opportunities for advancement in the field.

Right from the start of this book project, it has been an incredible experience, for us as editors, to realize that in addressing various perspectives of this amazing technology, we would also be contributing to a broad state-of-the-art understanding of the clinical complications and the basic and advanced features of engineering, design, and industry perspectives of mechanical circulatory support devices.

In doing so, we acknowledge that we may have missed some important contributions. Our purpose was not to provide an exhaustive overview of what is available to most of us, but rather to focus on what will contribute significantly to advancing the field, through the further substantial, mutual efforts of engineers and clinicians behind the development of new technologies to contribute to patient quality of life.

It has been a privilege and an honor to have such outstanding contributors to this book. As editors, we would like to acknowledge the excellent work they have provided us. This knowledge is expected to withstand the test of time, and will remain practical for years to come.

We are indebted to all of the authors for their assistance and efforts in preparing this book for publication.

We hope this book will be helpful to everyone who is interested in the field of mechanical circulatory support, whether they are just starting their career or at a more established stage, and that readers will benefit from this collaborative effort.

Cleveland, OH, USA

Jamshid H. Karimov
Kiyotaka Fukamachi
Randall C. Starling

Contents

Part I Overview of the Mechanical Circulatory Support

1 **A History of Mechanical Circulatory Support** ... 3
Shelley McKellar

2 **A Personalized History of Rotary Blood Pumps** ... 19
Richard Wampler

3 **History of Pediatric Devices for Mechanical Circulatory Support** ... 37
Kurt A. Dasse and Priscilla C. Petit

4 **Overview of Mechanical Circulatory Support Devices and Concepts** ... 51
Juan Marcano, Aladdein Mattar, and Jeffrey A. Morgan

5 **Physiology of Blood Pump Circulation in Heart Failure** ... 63
Abhinav Saxena, Nir Uriel, and Daniel Burkhoff

6 **Engineering Requirements for Mechanical Circulatory Support Devices** ... 83
J. Timothy Baldwin

7 **Clinical Requirements for Mechanical Circulatory Support Devices** ... 91
Neel K. Ranganath, Katherine G. Phillips, and Nader Moazami

8 **Engineering Perspectives for Mechanical Circulatory Support Devices** ... 109
Kevin Bourque, Christopher Cotter, and Charles Dague

Part II Indications and Selection for Mechanical Circulatory Support

9 **Patient Population and Selection Criteria for Mechanical Circulatory Support** ... 131
Pavan Bhat and Randall C. Starling

10 **Mechanical Circulatory Support Therapies: Right Timing and Prognosis Considerations** 141
Felix Schoenrath and Evgenij Potapov

11 **Perioperative Considerations in Left Ventricular Assist Device Placement** 151
Ranjani Venkataramani, Michael Zhen-Yu Tong, and Shiva Sale

12 **Left Ventricular Assist Devices: Management and Surgical Techniques** 171
Robert J. Steffen and Benjamin C. Sun

13 **Mechanical Circulatory Support for Biventricular Failure: Patient Selection and Management Options** 177
Kimberly N. Hong, Hao A. Tran, Victor Pretorius, and Eric D. Adler

14 **Status and Availability of a Total Artificial Heart** 191
Katherine G. Phillips, Neel K. Ranganath, and Nader Moazami

15 **Extracorporeal Membrane Oxygenation Techniques and Modern Considerations** 207
Aly El Banayosy

16 **Less Invasive Extra-pericardial Placement (LIEPP) of LVAD** 221
Areo Saffarzadeh and Pramod Bonde

Part III Parameters of Physiologic and Non-physiologic Blood Flow

17 **Pulsatile Mechanical Circulation, Physiology, and Pump Technology** 231
Jack Copeland and Hannah Copeland

18 **Pathophysiological Determinants Relevant in Blood Pump Control** 253
Marianne Schmid Daners and Seraina Anne Dual

19 **End-Organ Physiology Under Continuous-Flow Mechanical Circulatory Support** 279
Egemen Tuzun

20 **Quantification of Pulsatility During Mechanical Circulatory Support** 301
Shigang Wang, Morgan K. Moroi, and Akif Ündar

21 **Pulsatility as an Option with Continuous-Flow Mechanical Circulatory Support Devices** 317
Chelsea Lancaster, Michael A. Sobieski, Mark S. Slaughter, and Steven Koenig

Part IV Current Technology and Methods

22 **Implantable Continuous-Flow Blood Pump Technology and Features** ... 337
Matthew L. Goodwin, Peter H. U. Lee, and Nahush A. Mokadam

23 **Partial Support with Mechanical Circulatory Support Devices** ... 359
Jordan R. H. Hoffman and Joseph C. Cleveland Jr

24 **Percutaneous Mechanical Circulatory Support Technologies** ... 379
Jerry D. Estep

25 **Computational Fluid Dynamics for Mechanical Circulatory Support Device Development** ... 399
Roland Graefe and Lutz Pauli

26 **Hemodynamic Modelling and Simulations for Mechanical Circulatory Support** ... 429
Libera Fresiello and Krzysztof Zieliński

27 **Options for Modeling and Simulations Used in Mechanical Circulatory Support Development** ... 449
David J. Horvath, Kiyotaka Fukamachi, and Jamshid H. Karimov

Part V Evolving Technologies and Device Concepts

28 **Emerging Continuous-Flow Blood Pump Technologies** ... 469
Harveen K. Lamba and Jeffrey A. Morgan

29 **The Development of Advanced Ventricular Assist Device as a Next Generation Ventricular Assist Device** ... 481
Takuma Miyamoto, Jamshid H. Karimov, David J. Horvath, Nicole Byram, and Kiyotaka Fukamachi

30 **Cleveland Clinic Total Artificial Heart** ... 493
Jamshid H. Karimov, David J. Horvath, and Kiyotaka Fukamachi

31 **The ReinVAD LVAD: Smart Technology to Enhance Long-Term Circulatory Support Therapy** ... 505
Roland Graefe, Erika Minguez, Andreas Henseler, and Reiner Körfer

32 **EVAHEART 2 Left Ventricular Assist System: A Hemocompatible Centrifugal Pump with Physiological Pulsatility** ... 535
Tadashi Motomura

33 **A Novel Rotary Total Artificial Heart Using a Single Shuttling Impeller** ... 545
Jeremy Glynn and Richard Wampler

34 **New Mechanical Circulatory Device: TORVAD** ... 555
Jeffrey Gohean and Richard Smalling

35 **BiVACOR Total Artificial Heart** ... 563
Daniel L. Timms and Frank Nestler

36 **The Penn State Pediatric Total Artificial Heart** ... 577
William J. Weiss, Raymond Newswanger, J. Brian Clark, and Jenelle M. Izer

37 **CorWave LVAD: Insight into Device Concept and Stage of Development** ... 587
Carl Botterbusch, Trevor Snyder, Pier-Paolo Monticone, Louis de Lillers, Alexandra Schmidt, and Charlotte Rasser

38 **Progress on Total Artificial Heart for Pediatric Patients** ... 599
Kiyotaka Fukamachi, Jamshid H. Karimov, and Takuma Miyamoto

39 **Progress on Wireless LVAD and Energy Sources for Mechanical Circulatory Systems** ... 609
John Valdovinos, Jiheum Park, Joshua Smith, and Pramod Bonde

Part VI Management Strategies and Limitations of MCS

40 **Risk Factors for Mechanical Circulatory Support Use and Risk Assessment** ... 623
Rajakrishnan Vijayakrishnan and Emma J. Birks

41 **Gastrointestinal Bleeding in Mechanical Circulatory Support Patients** ... 641
Katherine M. Klein and Vigneshwar Kasirajan

42 **Postoperative Management Strategies in Mechanical Circulatory Support Patients** ... 647
Tiffany Buda, Kimberly Miracle, and Marjorie Urban

43 **Palliative Care in MCS Patients: Insights into Current Practice and Outcomes** ... 671
Krista Dobbie and Kyle Neale

Part VII Future Perspectives

44 **Clinical Demands and Challenges for Future Mechanical Circulatory Support Technologies** ... 693
Adam D. DeVore and Joseph G. Rogers

45 Smart Mechanical Circulatory Support Devices: Requirements and Perspectives 701
Joshua P. Cysyk and Gerson Rosenberg

46 Novel Solutions for Patient Monitoring and Mechanical Circulatory Support Device Control 707
Martin Maw, Francesco Moscato, Christoph Gross, Thomas Schlöglhofer, and Heinrich Schima

Index 729

Contributors

Eric D. Adler, MD University of California San Diego Medical Center, San Diego, CA, USA

J. Timothy Baldwin, MS, PhD National Heart Lung and Blood Institute, Bethesda, MD, USA

Pavan Bhat, MD Kaufman Center for Heart Failure, Heart & Vascular Institute, Cleveland Clinic, Cleveland, OH, USA

Emma J. Birks, MD, PhD Division of Cardiovascular Medicine, Department of Medicine, University of Kentucky, Gill Heart and Vascular Institute, Louisville, KY, USA

Pramod Bonde, MD Yale School of Medicine, New Haven, CT, USA

Carl Botterbusch, MBA CorWave SA, Clichy, France

Kevin Bourque, MSME Abbott, Burlington, MA, USA

Tiffany Buda, MSN, RN, NE Cleveland Clinic, Cleveland, OH, USA

Daniel Burkhoff, MD, PhD Cardiovascular Research Foundation, New York, NY, USA

J. Brian Clark, MD The Pennsylvania State University, College of Medicine, Department of Pediatrics, Hershey, PA, USA

Nicole Byram, BS Department of Biomedical Engineering, Lerner Research Institute, Cleveland Clinic, Cleveland, OH, USA

Joseph C. Cleveland Jr, MD University of Colorado School of Medicine, Division of Cardiothoracic Surgery, Academic Office One, Aurora, CO, USA

Hannah Copeland, MD University of Mississippi Medical Center, Division of Cardiothoracic Surgery, Heart Transplantation and Mechanical Circulatory Support, Jackson, MS, USA

Jack Copeland, MD Department of Surgery, University of Arizona, Tucson, AZ, USA

Christopher Cotter, MSME Abbott, Burlington, MA, USA

Joshua P. Cysyk, PhD The Pennsylvania State University, College of Medicine, Department of Surgery, Division of Applied Biomedical Engineering, Hershey, PA, USA

Charles Dague, MSEE Abbott, Burlington, MA, USA

Marianne Schmid Daners, PhD Product Development Group Zurich, Department of Mechanical and Process Engineering, Zurich, Switzerland

Kurt A. Dasse, PhD University of Louisville Medical School, Inspired Therapeutics LLC, Cocoa, FL, USA

Louis de Lillers, Master in Management CorWave SA, Clichy, France

Adam D. DeVore, MD, MHS Department of Medicine and Duke Clinical Research Institute, Duke University School of Medicine, Durham, NC, USA

Krista Dobbie, MD Cleveland Clinic, Cleveland, OH, USA

Seraina Anna Dual, MSc Product Development Group Zurich, Department of Mechanical and Process Engineering, Zurich, Switzerland

Aly El Banayosy, MD INTEGRIS Baptist Medical Center, Oklahoma City, OK, USA

Jerry D. Estep, MD Department of Cardiovascular Medicine and Heart and Vascular Institute, Kaufman Center for Heart Failure, Cleveland Clinic, Cleveland, OH, USA

Libera Fresiello, MSc, PhD Department of Cardiovascular Sciences, Cardiac Surgery, Katholieke Universiteit Leuven, Leuven, Belgium

Institute of Clinical Physiology, National Research Council, Pisa, Italy

Kiyotaka Fukamachi, MD, PhD Lerner Research Institute, Cleveland Clinic, Cleveland Clinic Lerner College of Medicine of Case Western Reserve University, Cleveland Clinic, Cleveland, OH, USA

Jeremy Glynn, PhD Abbott, St. Paul, MN, USA

Jeffrey Gohean, MSME Windmill Cardiovascular Systems, Inc., Austin, TX, USA

Matthew L. Goodwin, MD The Ohio State University Wexner Medical Center, Columbus, OH, USA

Roland Graefe, Dr.-Ing ReinVAD GmbH, Aachen, Germany

Christoph Gross, MSc Center for Medical Physics and Biomedical Engineering and Deparment of Cardiac Surgery, Medical University of Vienna, Wien, Austria

Ludwig Boltzmann Institute for Cardiovascular Research, Vienna, Austria

Andreas Henseler, Dr.-Ing ReinVAD GmbH, Aachen, Germany

Jordan R. H. Hoffman, MPH, MD Vanderbilt University School of Medicine; Section of Surgical Sciences, Department of Cardiac Surgery, Nashville, TN, USA

Kimberly N. Hong, MD Department of Cardiology, University of California San Diego, La Jolla, CA, USA

David J. Horvath, MSME Department of Biomedical Engineering, Lerner Research Institute, Cleveland Clinic, Cleveland, OH, USA

R1 Engineering LLC, Euclid, OH, USA

Jenelle M. Izer, DVM The Pennsylvania State University, College of Medicine, Department of Comparative Medicine, Hershey, PA, USA

G. S. Kakos, MD Heart and Vascular Center, The Ohio State University Wexner Medical Center, Columbus, OH, USA

Jamshid H. Karimov, MD, PhD Lerner Research Institute, Cleveland Clinic, Cleveland Clinic Lerner College of Medicine of Case Western Reserve University, Cleveland Clinic, Cleveland, OH, USA

Vigneshwar Kasirajan, MD Virginia Commonwealth University, Pauley Heart Center, Richmond, VA, USA

Division of Cardiothoracic Surgery, Richmond, VA, USA

Katherine M. Klein, MD Virginia Commonwealth University, Pauley Heart Center, Richmond, VA, USA

Steven Koenig, PhD Departments of Bioengineering and Cardiothoracic Surgery, University of Louisville, Louisville, KY, USA

Reiner Körfer, Prof. Dr. med. Dr. h. c. ReinVAD GmbH, Aachen, Germany

Harveen K. Lamba, MD, MSc Division of Cardiothoracic Transplantation and Circulatory Support, Michael E. DeBakey Department of Surgery, Baylor College of Medicine / Texas Heart Institute, Houston, TX, USA

Chelsea Lancaster, BEng Department of Bioengineering, University of Louisville, Louisville, KY, USA

Peter H. U. Lee, MD, PhD, MPH The Ohio State University Wexner Medical Center, Columbus, OH, USA

Juan Marcano, MD Baylor College of Medicine – Texas Heart Institute, Houston, TX, USA

Aladdein Mattar, MD Baylor College of Medicine – Texas Heart Institute, Houston, TX, USA

Martin Maw, MSc Center for Medical Physics and Biomedical Engineering and Deparment of Cardiac Surgery, Medical University of Vienna, Wien, Austria

Ludwig Boltzmann Institute for Cardiovascular Research, Vienna, Austria

Shelley McKellar, PhD Department of Surgery, Schulich School of Medicine and Dentistry, University of Western Ontario, London, Ontario, Canada

Erika Minguez, MSc ReinVAD GmbH, Aachen, Germany

Kimberly Miracle, MSN, ACNP Cleveland Clinic, Cleveland, OH, USA

Takuma Miyamoto, MD, PhD Department of Biomedical Engineering, Lerner Research Institute, Cleveland Clinic, Cleveland, OH, USA

Nader Moazami, MD Department of Cardiothoracic Surgery, NYU Langone Health, New York, NY, USA

Nahush A. Mokadam, MD Division of Cardiac Surgery, The Ohio State University Wexner Medical Center, Columbus, OH, USA

Pier-Paolo Monticone, PhD Geneva, Switzerland

Jeffrey A. Morgan, MD Division of Congestive Heart Failure, Transplant and Mechanical Circulatory Support, Department of Cardiothoracic Surgery, Baylor College of Medicine – Texas Heart Institute, Houston, TX, USA

Morgan K. Moroi, BS Penn State Health Pediatric Cardiovascular Research Center, Department of Pediatrics, Hershey, PA, USA

Francesco Moscato, PhD Center for Medical Physics and Biomedical Engineering and Deparment of Cardiac Surgery, Medical University of Vienna, Wien, Austria

Ludwig Boltzmann Institute for Cardiovascular Research, Vienna, Austria

Tadashi Motomura, MD, PhD Evaheart, Inc., Houston, TX, USA

Kyle Neale, DO Cleveland Clinic, Cleveland, OH, USA

Frank Nestler, PhD BiVACOR, Inc., Houston, TX, USA

Raymond Newswanger, PhD The Pennsylvania State University, College of Medicine, Department of Surgery, Division of Applied Biomedical Engineering, Hershey, PA, USA

Jiheum Park, PhD Bonde Artificial Heart Laboratory, Yale School of Medicine, New Haven, CT, USA

Lutz Pauli, Dr.-Ing MAGMA Gießereitechnologie GmbH, Aachen, Germany

Priscilla C. Petit, PhD Quality and Regulatory Affairs, Inspired Therapeutics LLC, Cocoa, FL, USA

Katherine G. Phillips, BS NYU Langone Health, Department of Cardiothoracic Surgery, New York, NY, USA

Evgenij Potapov, MD Department of Cardiothoracic and Vascular Surgery, German Heart Center Berlin, DZHK (German Centre for Cardiovascular Research), partner site Berlin, Berlin, Germany

Victor Pretorius, MD Department of Cardiothoracic Surgery, University of California San Diego, La Jolla, CA, USA

Neel K. Ranganath, MD NYU Langone Health, Department of Cardiothoracic Surgery, New York, NY, USA

Charlotte Rasser, PhD CorWave SA, Clichy, France

Joseph G. Rogers, MD Department of Medicine and Duke Clinical Research Institute, Duke University School of Medicine, Durham, NC, USA

Division of Cardiology, Duke University School of Medicine, Durham, NC, USA

Gerson Rosenberg, PhD The Pennsylvania State University, College of Medicine, Department of Surgery, Division of Applied Biomedical Engineering, Hershey, PA, USA

Areo Saffarzadeh, MS, MD Yale School of Medicine, New Haven, CT, USA

Shiva Sale, MD Heart Transplantation and Mechanical Circulatory Support, CCLCM, Case Western Reserve University, Department of Cardiothoracic Anesthesiology, Cleveland Clinic Foundation, Cardiothoracic Anesthesia, Cleveland, OH, USA

Abhinav Saxena, MBBS Maimonides Medical Center, Brooklyn, NY, USA

Heinrich Schima, PhD Center for Medical Physics and Biomedical Engineering and Deparment of Cardiac Surgery, Medical University of Vienna, Vienna, Austria

Ludwig Boltzmann Institute for Cardiovascular Research, Vienna, Austria

Thomas Schlöglhofer, MSc Center for Medical Physics and Biomedical Engineering and Deparment of Cardiac Surgery, Medical University of Vienna, Wien, Austria

Ludwig Boltzmann Institute for Cardiovascular Research, Vienna, Austria

Alexandra Schmidt, MSc CorWave SA, Clichy, France

Felix Schoenrath, MD Department of Cardiothoracic and Vascular Surgery, German Heart Center Berlin, DZHK (German Centre for Cardiovascular Research), partner site Berlin, Berlin, Germany

Mark S. Slaughter, MD Department of Cardiothoracic Surgery, University of Louisville, Louisville, KY, USA

Richard Smalling, MD, PhD UTHealth/McGovern Medical School, Memorial Hermann Heart and Vascular Institute, Houston, TX, USA

Windmill Cardiovascular Systems, Inc, Austin, TX, USA

Joshua Smith, PhD Department of Electrical Engineering, Department of Computer Science and Engineering, University of Washington, Seattle, WA, USA

Trevor Snyder, PhD CorWave SA, Clichy, France

Michael A. Sobieski, RN, CCP Department of Cardiothoracic Surgery, University of Louisville, Louisville, KY, USA

Randall C. Starling, MD, MPH, FACC, FAHA, FESC, FHFSA Kaufman Center for Heart Failure, Cleveland Clinic, Cleveland, OH, USA

Robert J. Steffen, MD Minneapolis Heart Institute, Department of Cardiovascular Surgery, Minneapolis, MN, USA

Benjamin C. Sun, MD Minneapolis Heart Institute, Department of Cardiovascular Surgery, Minneapolis, MN, USA

Daniel J. Timms, PhD BiVACOR, Inc., Houston, TX, USA

Hao A. Tran, MD Department of Cardiology, University of California San Diego, La Jolla, CA, USA

Egemen Tuzun, MD, PhD Texas A&M University, Institute for Preclinical Studies (TIPS), College Station, TX, USA

Akif Ündar, PhD Penn State Health Pediatric Cardiovascular Research Center, Department of Pediatrics, Hershey, PA, USA

Penn State Health Pediatric Cardiovascular Research Center, Department of Pediatrics, and Department of Surgery and Bioengineering, Penn State Milton S. Hershey Medical Center, Penn State College of Medicine, Penn State Health Children's Hospital, Hershey, PA, USA

Penn State College of Medicine, Department of Pediatrics, Hershey, PA, USA

Marjorie Urban, MSN, ACNP Cleveland Clinic, Cleveland, OH, USA

Nir Uriel, MD, MSc Columbia and Cornell Universities, New York Presbyterian, New York, NY, USA

John Valdovinos, PhD Formerly Bonde Artificial Heart Lab, Yale School of Medicine, New Haven, CT, USA

Ranjani Venkataramani, MD Department of Anesthesiology and Critical Care, University of New Mexico, Albuquerque, NM, USA

Rajakrishnan Vijayakrishnan, MD Advanced Heart Failure Therapeutics and Mechanical Circulatory Support, Norton Thoracic Institute/Creighton University School of Medicine, St Joseph Hospital and Medical Center, Phoenix, AZ, USA

Richard Wampler, MD Oregon Health & Science University, Portland, OR, USA

Shigang Wang, MD Penn State Health Pediatric Cardiovascular Research Center, Department of Pediatrics, Hershey, PA, USA

William J. Weiss, PhD The Pennsylvania State University, College of Medicine, Department of Surgery, Division of Applied Biomedical Engineering, Hershey, PA, USA

T. E. Williams Jr, MD, PhD Heart and Vascular Center, The Ohio State University Wexner Medical Center, Columbus, OH, USA

Michael Zhen-Yu Tong, MD, MBA Mechanical Circulatory Support, CCLCM, Case Western Reserve University, Department of Thoracic and Cardiovascular Surgery, Cleveland, OH, USA

Krzysztof Zieliński, MSc, PhD Nalecz Institute of Biocybernetics and Biomedical Engineering, Polish Academy of Sciences, Warsaw, Poland

Part I

Overview of the Mechanical Circulatory Support

1 A History of Mechanical Circulatory Support

Shelley McKellar

During the first decades of the twentieth century, heart disease rose to be a leading cause of death in the United States, bypassing tuberculosis and pneumonia as the top disease killer of Americans. By 1950, almost 40% of all deaths were due to heart disease, which claimed more lives than the next three leading causes of death combined [1]. In 1948, the US Congress passed the National Heart Act, which established the National Heart Institute (later renamed the National Heart, Lung, and Blood Institute) to conduct research and training into the causes, prevention, diagnosis, and treatment of diseases of the heart and circulation. This stimulated a variety of basic and applied research projects, most notably the Framingham Heart Study, a long-term, ongoing cardiovascular study that established the concept of risk factors, specifically hypertension, high cholesterol, and cigarette smoking, as major contributors to the problem of heart disease in America [2–5]. In 1955, President Dwight D. Eisenhower suffered a heart attack, requiring a 7-week stay in the hospital for recovery, and the seriousness of heart disease was made ever more apparent. As Paul Dudley White, a leading cardiologist and one of Eisenhower's physicians, wrote at mid-century in his classic textbook *Heart Disease,* "heart disease has become the chief public health problem of our day [6]."

Since the 1950s, the mortality rates from cardiovascular disease have declined as physicians provided better treatments for patients with coronary artery disease, cardiac arrhythmias, cardiomyopathy, heart valve problems of stenosis as well as regurgitation and prolapse, heart infections, and congenital heart defects. However, there emerged an increased incidence of congestive heart failure due to cardiac damage in patients battling cardiovascular diseases [7]. Heart failure is a chronic, progressive condition, and it is life-threatening, with poor long-term prognosis. One physician commented, "Heart failure is a lethal disease with a worse life expectancy than many types of cancer" [8]. In 2014 an estimated 26 million people were living with heart failure worldwide—including 5.8 million Americans—and the incidence and prevalence of this condition were on the rise [9]. With approximately 3% of the world's population affected by symptomatic heart failure, the condition is emerging as a global health and financial burden. The gold standard treatment for heart failure is cardiac transplantation. But the limited availability of donor hearts and the sizeable number of patients who are ineligible for heart transplant support the development and clinical use of mechanical circulatory support technology in the treatment of heart failure.

S. McKellar, PhD (✉)
Department of Surgery, Schulich School of Medicine and Dentistry, University of Western Ontario, London, ON, Canada
e-mail: smckell@uwo.ca

© Springer Nature Switzerland AG 2020
J. H. Karimov et al. (eds.), *Mechanical Support for Heart Failure*,
https://doi.org/10.1007/978-3-030-47809-4_1

The Possibility of Mechanical Circulatory Support Revealed

During the 1930s, Alexis Carrel and Charles Lindbergh produced a glass perfusion pump, erroneously dubbed an "artificial heart," that successfully kept small explanted organs of birds or cats alive by circulating a fluid of oxygen, salts, and nutrient through the blood vessels to these organs [10]. William Laurence, a *New York Times* science writer, described Lindbergh's "chamber of artificial life" as a combined mechanical heart and lungs, an artificial blood supply, and even synthetic air that kept body parts alive for the first time outside the body [11]. Not long after, a photo of Carrel, Lindbergh, and the pump made the cover of *Time* magazine in 1938 [12]. It made for spectacular reporting, feeding the imagination of society for such a device.

In 1953, in Philadelphia, John Gibbon, Jr. corrected an atrial septal defect in 18-year-old Cecilia Bavolek, using his heart-lung machine. It was a historic operation, and after the hole in her heart was closed, Bavolek lived more than 40 years after the surgery [13]. Gibbon had demonstrated that mechanical circulatory support was possible. Other heart-lung machines, with different oxygenators, were also developed at this time as the technology continued to be refined. By the late 1950s, an increasing number of surgeons were performing cardiac bypass operations, combining hypothermia with extracorporeal circulation.

A momentous technological feat, the heart-lung machine established the feasibility of mechanical circulatory support. The pump-oxygenator served as an acceptable substitute for the patient's heart and lungs, but the technology was not without problems. The heart-lung machine could only be used temporarily; it was not a cure for the failing heart. Patients are kept on bypass only for the length of their operation, ideally not longer than 6 hours, to avoid risks of blood clotting, hemolysis, air embolism, or neurocognitive damage. When coming off bypass, some patients went into severe cardiopulmonary failure and died. There was a need for prolonged extracorporeal circulation to allow weakened hearts and lungs extra time to recover. The heart-lung machine allowed for greater corrective cardiac surgery, but it was not a therapeutic solution for the failing heart.

The US Artificial Heart Program as Key Catalyst

Key to launching the fledgling field of mechanical circulatory support was the establishment of the US Artificial Heart Program at the National Heart Institute, part of the National Institutes of Health in Bethesda, Maryland, in 1964. This federally sponsored, large-scale research and development program attracted both academic and industry investigators into this line of work. Federal funding was initially modest, but more significantly, it legitimated the research and framed how development of this technology would proceed. The government program served as an important catalyst for the advancement and evaluation of mechanical circulatory support systems, supporting research and development before handing device commercialization over to industry [14].

The Artificial Heart Program was the NHI's first targeted, extramural development program that adopted a systems development approach, as did many US space and defense science and technology projects, to plan, manage, and produce key deliverables toward the goal of building an artificial heart. This was a departure from the traditional funding structure of small grants and basic research. Instead, this was a structured program intended to solve problems and achieve set objectives. Industry and academic researchers vied for contract or grant funding for independent development of valves, biomaterials, blood interface studies, control mechanisms, or power systems, which, at a later date, the program intended to integrate into a workable device. Broadly encompassing, the Artificial Heart Program proposed the development of a range of mechanical circulatory support systems that included both assistive and total replacement devices for short-term and long-term use, different devices to be used in different clinical situations to battle heart failure [15].

Early Assist Devices: 1960s–1970s

Among a small group of surgical researchers building assist devices in the mid-1960s, Michael DeBakey and Adrian Kantrowitz reported clinical successes with their respective mechanical pumps. In Texas, DeBakey and his team experimented with several types of assist pumps that underscored the technological and biological complexities of building a mechanical pump. Problems of poor materials, blood interaction issues, and clunky power sources needed to be overcome in the development of this technology [16]. In 1966, DeBakey reported some positive clinical outcomes with his newest assist device: a hemispherical, pneumatically driven pump that rested outside the body, with two connecting tubes penetrating the patient's chest to attach to the left atrium and a systemic artery. It was a temporary device to be used for only a matter of days or weeks to provide short-term cardiac assistance after surgery. It sustained the life of 37-year-old Esperanza Del Valle Vasquez, whose weakened heart had failed to resume function after aortic and mitral valve replacement surgery. After 10 days, DeBakey removed the device, and the patient made a full recovery and was discharged from the hospital 1 month after her valve surgery [17, 18] (Fig. 1.1).

Thereafter, DeBakey reported more clinical cases, in which other patients also benefited from short-term pump assistance after heart surgery. In 1967, a 16-year-old girl recovered from mitral valve replacement surgery after using DeBakey's ventricular assist pump for 4 days. But despite patient success, DeBakey ruled out wide clinical application of this pump based on its prohibitive cost, the major surgical intervention required, and imperfect blood interface materials. The research team returned to the laboratory to review pump designs, to explore better biocompatible materials, and to evaluate the pump's performance based on bench tests and recent clinical use. DeBakey firmly stated that in its present stage of development, the use of cardiac assist devices must remain an investigative procedure [17].

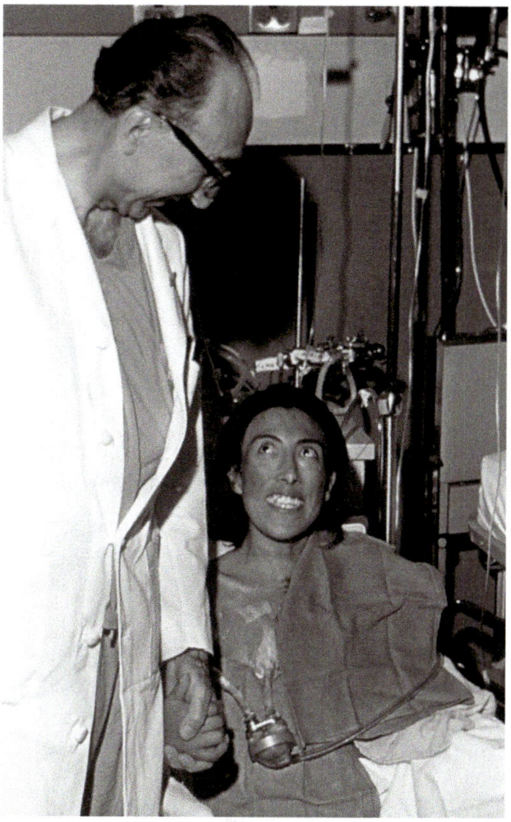

Fig. 1.1 Dr. Michael DeBakey with patient Esperanza Del Valle Vasquez, recovering from heart surgery with the support of an external cardiac assist pump, August 1966. (Courtesy of the Baylor College of Medicine Archives)

In New York, Kantrowitz developed counterpulsating assist devices, including a temporary intra-aortic balloon pump and a more permanent left ventricular assist device (VAD). The idea of the intra-aortic balloon pump was not new, as acknowledged by Kantrowitz [19]. It had originated in Willem Kolff's laboratory, where Spyridon Moulopoulos and others tested the device on dogs but never used it clinically [20]. Kantrowitz revisited their work to develop a practical system to offer temporary cardiac assist to patients in cardiogenic shock. He reconfigured the balloon to reduce occlusion and used helium instead of carbon dioxide as the driving gas. After three successful cases (from a total of five) in 1967, Kantrowitz claimed that the intra-aortic balloon pump constituted a simpler and more effective means of providing temporary mechanical circulatory support than DeBakey's left ventricular bypass system [21].

Kantrowitz also reported some success with his permanent left VAD, an auxiliary ventricle or patch with a connecting air tube that was sutured into the wall of the aorta. It was implanted in three patients in the early 1970s [22]. The longest surviving case was Haskell Shanks, a 63-year-old man in chronic congestive heart failure who had been bedridden for several years. He recovered from the operation, became ambulatory, and was discharged from hospital, returning home with different drive systems, including a portable battery unit. Shanks died 3 months after his implant from an infection at the site of the drive tube that connected his implanted device to the external power unit [23]. Whereas the acute heart failure patients had benefited from the balloon pump, Kantrowitz optimistically reported that the patch offered benefits to patients in severe, chronic left ventricular heart failure but that severe infectious complications thwarted its adoption [24].

Early Total Artificial Hearts: 1950s–1960s

Credited with the invention of the artificial kidney, Willem Kolff pursued the development of a total artificial heart (TAH) as a permanent mechanical replacement for a removed, failing human heart [25]. In the 1950s, while at the Cleveland Clinic, Kolff and Tetsuzo Akutsu experimented with plastic valves in the hearts of dogs before they boldly jumped ahead to build an entire plastic heart, composed of two polyvinyl chloride pumps. This plastic heart was a pneumatically driven device, built from a cast impression of the heart of a 60-pound dog. The mechanical heart was connected to an air compressor by plastic tubes. This air-driven heart was tested first in mock circulation on the bench (to ensure mechanical function), then implanted into a dead dog (to study surgical procedure and fit), and finally placed into a living dog in December 1957. The dog lived for 90 minutes with the implanted heart before device complications terminated the experiment. It was the first successful TAH implant in an animal in the Western world [26]. (Years later, Kolff and Akutsu learned that a Soviet scientist, Vladimir Demikhov, had performed a similar experiment in 1937 [27]) (Fig. 1.2).

For Kolff and Akutsu, it was proof of concept despite obvious problems. The efficacy of the device was questionable, given that the dog had hardly moved from the operating table. Obvious technological limitations were that the device was pneumatically driven, necessitating connection to an outside air compressor to power it, and that it had malfunctioned after 90 minutes. Equally problematic was the surgical procedure, which needed refinement. Heart removal was difficult and painstaking, and implanting the device required the tedious and time-consuming surgical connection of the mechanical heart to the major blood vessels. This was due to the design of this early device, which was modeled on the natural heart and the complexity of blood vessels surrounding it. Early artificial heart researchers grappled with issues of technical complexity, poor durability, and biomaterial problems in designing devices at this time.

Fig. 1.2 Dr. Willem Kolff, dubbed "the father of artificial organs," with multiple artificial heart prototypes developed in his laboratory during his career of more than 40 years. (Courtesy of Special Collections, J. Willard Marriott Library, University of Utah)

In December 1967, in Cape Town, South Africa, Christiaan Barnard performed the first transplantation of a human heart into a patient [28–30]. A tumultuous era in cardiac transplant surgery followed; most heart transplant patients died within weeks of the surgery owing to organ rejection or infection, and there were too few donor hearts. The challenges and uncertainties experienced in heart transplant surgery encouraged the work on, and perceived value of, mechanical circulatory support systems. Heart transplantation and artificial hearts became intertwined, emerging as complementary, and not necessarily competing, cardiac replacement strategies [14]. Mechanical hearts could now serve as a mechanical alternative to transplantation (a longer-term role) and as a bridge to transplantation (a shorter-term role).

In April 1969, Denton Cooley implanted a total artificial heart in heart failure patient, Haskell Karp, as a bridge to transplantation, in Houston, Texas. It was the first TAH implant in a patient. Built by Domingo Liotta, the Liotta-Cooley artificial heart was a two-ventricle, pneumatically driven device that sustained Karp for 64 hours. It worked well enough to bridge the patient to cardiac transplantation, but Karp died 32 hours thereafter. The device caused blood damage issues, and the patient was connected to a large, external power console that limited mobility. In the end, neither the device implant nor the heart transplant could reverse Karp's multiple problems, notable renal and respiratory complications that accompanied end-stage heart disease [31, 32]. The Liotta-Cooley artificial heart was never implanted in another human case. The implant operation stirred tremendous controversy within the medical community and society because of its experimental nature. It also severed the professional relationship of DeBakey and Cooley owing to allegations of device theft and lack of authorization to perform the implant procedure [14, 33] (Fig. 1.3).

Fig. 1.3 Dr. Denton Cooley, holding a Liotta-Cooley artificial heart, talks to reporters about the Haskell Karp case at a press conference at St. Luke's Hospital, Houston, Texas, April 7, 1969. (AP photo/Ed Kolenovsky; © 1969 The Associated Press)

In July 1981, Cooley performed a second mechanical heart implant, using the Akutsu III total artificial heart. This pneumatic, double-chambered device pumped blood by means of a reciprocating hemispherical diaphragm. Implanted as a bridge to transplantation, the device provided 39 hours of support for 36-year-old Willebrordus Meuffels, who could not be weaned from the heart-lung machine after his triple coronary bypass operation. A week after the transplant operation, Meuffels died as a result of massive infection and organ failure. The Akutsu III TAH was never used clinically again [34–37].

The Goal of Permanent Replacement: TAHs

Since the 1970s, numerous pneumatic, electromechanical, and electrohydraulic total artificial hearts have been designed and tested by various research teams, including Willem Kolff's team at the University of Utah, William Pierce's group at Pennsylvania State University, Yukihiko Nosé's unit at the Cleveland Clinic, Emil Bücherl's group at the Freie Universität Berlin, and Kazuhiko Atsumi's team at the University of Tokyo, among others. For these researchers, the goal was the permanent replacement of the diseased human heart with a viable total artificial heart. With their respective device, each group strove to improve device design, durability, performance, and long-term outcomes with extensive bench testing and implantation in animals. Only a few of these total artificial hearts were ever used clinically.

In December 1982, William DeVries implanted the pneumatic Jarvik-7 total artificial heart, as the first permanent TAH implantation, in Barney Clark, who was ineligible for a heart transplant. The Jarvik-7 TAH, developed in Willem Kolff's laboratory by Robert Jarvik, Don Olsen, and others, consisted of two ventricles with diaphragms and a novel, quick-connect system for inflow and outflow cuffs. Exiting the patient's body were drivelines, which connected the implanted pump to the external power and device monitoring console [38]. Clark lived 112 days with the Jarvik-7 TAH but he never left hospital. On occasion, Clark was able to get out of bed and walk with assistance, but his care was complicated. He battled many complications, including seizures, nosebleeds, a broken device valve, kidney and respiratory problems, and more. He ultimately died of severe pseudomembranous colitis and sepsis, which led to multi-organ failure [39, 40]. In 1984 and 1985, four more patients received Jarvik-7 hearts as permanent implants, with one recipient living 620 days with the artificial heart [41]. All patients struggled with issues of thromboembolism and sepsis. Its use attracted much public interest, but it was a mix of praise and disapproval. Device criticism peaked in 1988 when the implantable artificial heart was dubbed "The Dracula of Medical Technology" in a *New York Times* editorial [42] (Fig. 1.4).

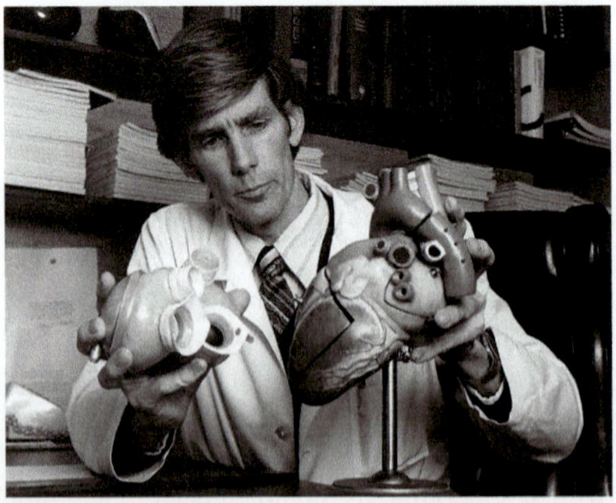

Fig. 1.4 Dr. William DeVries, holding a Jarvik-7 total artificial heart next to a model of the human heart, performed the first permanent TAH implantation in a patient, Barney Clark, in December 1982. (Courtesy of Special Collections, J. Willard Marriott Library, University of Utah)

The Jarvik-7 TAH (later renamed the Symbion TAH) and the availability of the smaller, improved Jarvik 7–70 TAH contributed to an international wave of TAH bridge implants during the late 1980s [43]. The new Jarvik 7–70 heart was 30% smaller than the original Jarvik-7 model and designed to fit into smaller patients without compromising performance or cardiac output. The smaller Jarvik-7 heart of 70 ml capacity was designed for a cardiac output of 7–8 liters (around 2 gallons) per minute, compared with the 11–12 liters per minute of the original 100 ml Jarvik-7 [44]. In 1990, the US Food and Drug Administration (FDA) recalled the Jarvik hearts, withdrawing its earlier investigational device exemption (IDE) approval, citing record-keeping and quality control problems by Symbion, Inc. In compliance with this ruling, Symbion representatives asked that hospital personnel destroy the devices or render them unusable [14].

The Penn State TAH was a fully implantable, orthotopically positioned, electromechanical heart system developed by William Pierce and his team of William Weiss, Gerson Rosenberg, James Donachy, and others at the Pennsylvania State University College of Medicine. Unlike the Jarvik-7 TAH, all components of the Penn State heart were implanted in the patient. The pump-drive unit, the compliance chamber, the battery, and the control system were situated in the chest and abdomen; no wires or tubes crossed the skin.

The pump-drive unit included the energy converter and two sac-type blood pumps. Pusher plates alternately compressed the right and left pump sacs, which filled passively. The compliance chamber maintained a constant pressure. A secondary energy transmission coil was implanted under the skin. In addition to the internal battery, an external battery pack (worn on a belt or in a shoulder bag) provided patient a backup power source. The Penn State heart underwent extensive preclinical testing, including some promising calf implant experiments [45–47]. In 2000, Abiomed, Inc. acquired the rights to the Penn State TAH, with the intention of incorporating this Penn State device technology, specifically its core energy converter system, into the AbioCor II device [14].

During the 1980s and 1990s, an Abiomed research team, under Robert Kung, developed the first-generation AbioCor Implantable Replacement Heart System, which was a self-contained, electrohydraulic total artificial heart. It did not require component lines to penetrate the patient's skin, which often caused infection; instead, it used a transcutaneous energy transfer (TET) system, thus promising improved patient mobility. The AbioCor TAH had four internal components: a plastic and titanium pump weighing about 2 pounds that was implanted orthotopically; an internal controller implanted in the patient's abdomen to regulate the movement of hydraulic fluid in the pump, its speed, and its pump rate; an internal lithium-ion battery, also placed abdominally; and an internal TET coil that received high-frequency power transmitted across the skin from an external TET coil. The AbioCor TAH also had four external components: the external TET coil, connected to either a bedside console or a portable TET module; a TET communication system; portable external batteries; and a bedside computer console that, via the radiofrequency communication system, recorded key information related to the function of the device, such as blood flow [48].

Between 2001 and 2004, 14 male patients received the AbioCor TAH as a permanent treatment for end-stage heart failure in an FDA-approved, multicenter clinical trial. The goal of the trial was to double the expected survival time of these patients from 30 to 60 days, with improved quality of life. Patient results ranged from immediate death to a 17-month survival case. The pump's performance was mixed. Thromboembolism remained a problem, and there were reported incidents of excessive bleeding, device failure, cerebrovascular events, and multisystem organ failure. Still device performance had been good enough to support ten patients beyond their 30 days of expected life. At Jewish Hospital-University of Louisville, Laman Gray and Robert Dowling performed the greatest number of these implant cases, including the first AbioCor TAH recipient Robert Tools, who lived 151 days with the artificial heart, as well as the longest-surviving AbioCor TAH recipient Tom Christerson, who was device-supported for nearly 17 months. The clinical trial yielded much device information that would be applied to an improved second-generation model, but no AbioCor II TAH was ever tested in humans [49–51].

The Symbion (formerly Jarvik) TAH technology reemerged as the SynCardia-CardioWest TAH, due to the determination of Jack Copeland, Richard Smith, and Marvin Slepian, during the 1990s and 2000s. It remained an air-driven mechanical heart system with three components: the pump unit of two prosthetic ventricles, the connecting drivelines, and an external pneumatic driver. Like earlier iterations, this device was a volume-displacement, pulsatile, diaphragm pump with two separate ventricles implanted orthotopically in the chest [52]. In 2004, the FDA approved the SynCardia TAH for American sales as a bridge device, based on the safety and efficacy data compiled during a 10-year multicenter clinical trial with this technology. SynCardia researchers developed a smaller 50 cc TAH that had a lower stroke value of 50 ml to deliver a cardiac output of approximately 6–8 liters per minute. It was intended for use in women and adolescents, for whom the 70 cc TAH was too large. SynCardia researchers significantly improved the driver technology, which became smaller, quieter, and more user-friendly over time. The Companion driver system was a new

55-pound hospital-based driver system with a touch screen interface, redundant compressors, and a continuous monitoring system. A portable Freedom driver system was developed for use outside the hospital [53, 54]. In 2018, the SynCardia TAH is the only TAH approved by the FDA, Health Canada, and the CE (Europe). Whereas the AbioCor TAH showcased a different, arguably more ideal technology platform as a fully implantable unit, it was the older, pneumatic technology that penetrated the market first for use as a BTT device and later for destination therapy (Table 1.1).

Table 1.1 Selected total artificial hearts and FDA regulatory status

Device	Description	FDA status
Liotta-Cooley heart, implanted at the Texas Heart Institute	A pneumatic, pulsatile TAH	n/a
Akutsu III heart, implanted at the Texas Heart Institute	A pneumatic, pulsatile TAH	n/a
Phoenix heart, implanted at the University of Arizona Medical Center	A pneumatic, pulsatile TAH	n/a
Jarvik-7 heart, by Kolff Medical-Symbion Inc.	A pneumatic, pulsatile TAH	1981 IDE approved 1990 IDE approval withdrawn
Penn State heart, implanted at the Penn State Hershey Medical Center	A pneumatic, pulsatile TAH	1985 IDE approved
SynCardia-CardioWest heart, by SynCardia Systems Inc.	A pneumatic, pulsatile TAH (an evolution of the Jarvik-7 heart technology)	1992 IDE approved 2004 PMA granted
AbioCor heart, by AbioMed Inc.	An electrohydraulic, pulsatile TAH	2001 IDE approved

McKellar [14], p. 254, Table 7.1. © 2018 Johns Hopkins University Press. Reprinted with permission of Johns Hopkins University Press

Note: FDA Food and Drug Administration, *TAH* total artificial heart, *IDE* investigational device exemption, *PMA* premarket approval

Wider Clinical Uptake: VADs

Ventricular assist devices made significant gains toward clinical acceptance during the latter part of the twentieth century. A VAD is a type of mechanical heart that unloads the pumping work of the native heart, assisting weakened ventricles to maintain blood circulation in the body. These devices are less complex in nature, both technologically and conceptually, than its more glamourous sibling, the TAH. Intuitively, it seemed to many stakeholders more doable and less drastic to develop a viable device that would not require removal of the native heart. Implanting a VAD—at a time when artificial heart valves and cardiac pacemakers were considered standard treatments—seemed much less radical than a TAH.

In the 1970s and 1980s, the standard device used for augmenting cardiac function was the intra-aortic balloon pump, developed a decade earlier by Adrian Kantrowitz. But balloon pumps worked only for limited and temporary assistance, and not in all heart failure cases. Extracorporeal membrane oxygenation (ECMO) technology, developed by Robert Bartlett, is a modified heart-lung bypass machine and provides temporary support of patients with reversible heart or lung failure, serving as a bridge to recovery or a bridge to a more robust support device [55]. During the 1980s, ECMO gained use as a mechanical circulatory support system for acute heart failure in roughly 50 specialized cardiac centers in the United States [56]. For about three decades, this technology provided the only cardiac support system for children, until the introduction of the Berlin Heart Pediatric VAD [57, 58]. In comparison to balloon pumps and ECMO, VADs offered short-term assistance and more, addressed left or right ventricular failure, and filled an important niche for those patients requiring longer-term cardiac assistance. VADs met the temporary circulatory support needs of the postoperative heart patient as a bridge-to-recovery device as well as provided longer cardiac assistance for patients battling chronic heart failure as bridge-to-transplantation devices and later destination therapy.

VADs are both similar to and different from TAHs. VADs may be implanted inside the chest or situated outside the body, may be used for short or long durations, and may be used as a bridge to recovery, a bridge to transplantation, or a destination therapy. Like TAH researchers, VAD investigators faced technological challenges of finding suitable biomaterials and power sources for their devices and reported operational difficulties of blood clotting, thrombus, and infection in animal models and patients. These shared device problems stimulated research crossover and exchange among the different artificial heart teams at various professional conferences. Not surprisingly, device projects overlapped in laboratories and among investigators.

First-generation VADs were large, pulsatile devices developed during the 1970s and 1980s. Four VAD research teams produced promising cardiac assist systems at this time. David Lederman at Avco-Everett Research Labs led the development of a sac-type assist device (a multisegment balloon pump) called the AVCO LVAD, which later became the ABIOMED BVS 5000 manufactured by ABIOMED Inc. At Pennsylvania State University College of Medicine, William Pierce and his team devised a pusher-plate diaphragm-type pump, the Pierce-Donachy pump, which was further developed and manufactured by the Thoratec Laboratories Corporation and renamed the Thoratec PVAD system. At Andros Inc., Peer Portner built a dual pusher-plate diaphragm-type pump referred to as the Andros solenoid-actuated LVAS system, later the Novacor LVAS distributed by Baxter HealthCare Corporation, Edwards Lifesciences LLC, and eventually the World Heart Corporation. John Norman teamed up with Thermo Electron Engineering Corporation's engineer Victor Poirier to develop a pusher-plate diaphragm-type pump, which evolved into the HeartMate assist device, manufactured by Thermo Cardiosystems Inc. [59–63] (Fig. 1.5).

Second-generation VADs secured therapeutic clout and greater clinical acceptance in the 1990s and the first decade of the 2000s. Two key events occurred that contributed to this. First, the

Fig. 1.5 Device inventor Peer Portner, PhD, talks to the press about the Novacor left ventricular assist system at the 3-year survival mark, a long-term VAD support record at that time, of Novacor implant patient Robert "Pete" Kenyon on August 9, 2001. Five months later, in January 2002, Kenyon received a donor heart at the Yale-New Haven Hospital, New Haven, Connecticut. (AP Photo/ Bob Child; © 2001 The Associated Press)

Randomized Evaluation of Mechanical Assistance for the Treatment of Congestive Heart Failure (REMATCH) study, conducted from May 1998 to July 2001, provided convincing, high-quality data supporting first-generation VAD device safety and clinical effectiveness over conventional medical treatment for end-stage heart failure patients. According to the REMATCH study results, patients implanted with Thoratec's HeartMate XVE LVAD lived more than twice as long as heart failure patients who were treated with medications only [64]. This data compelled heart specialists and their patients to consider mechanical pumps more seriously and secured FDA approval and third-party payer reimbursement for clinical use. The REMATCH study results prompted other device clinical trials, initiated by other device companies seeking entry into the medical marketplace [65].

Second, a key change in technology platforms shifted the design of VADs from volume displacement, pulsatile devices to continuous-flow, non-pulsatile pumps, resulting in second-generation assist devices. These devices were smaller, quieter, safer, and more effective than the clunky first-generation VADs. These new continuous-flow pumps, dubbed second-generation VADs, significantly reduced the problems of thromboembolism, infection, patient mobility, and device breakdowns, but the physiological implications of pulseless flow presented new, difficult clinical issues, such as dangerous gastrointestinal bleeding.

In the early 1980s, Richard Wampler and the Nimbus Corporation in Rancho Cordova, California, designed the Hemopump, a miniature axial-flow pump (about the size of a pencil eraser) that is inserted into the aorta, just above the aortic valve, and powered electrically by an external motor. This VAD uses pulsing electromagnetic fields to rotate a permanent magnet, spinning the pump blades rapidly to draw oxygenated blood from the left ventricle out to the body. The Hemopump was a simple and inexpensive device that was catheter-inserted through the femoral artery. Its significance is that it demonstrated the promise of continuous-flow technology. Its smaller and simpler design suggested greater mechanical durability, quieter operation, and fewer medical complications than the pulsatile VADs [66–68].

At the Texas Heart Institute, O.H. Frazier provided the laboratory resources and surgical expertise to implant investigational devices in calves and to translate experiment findings into device improvements, offering suggestions on how devices might work better in the human body. He performed the earliest clinical cases with many new devices, including the Hemopump. On April 17, 1988, Frazier implanted the Hemopump in Herb Kranich, a recent heart transplant recipient in severe organ rejection facing imminent death. The device worked, providing temporary mechanical support that allowed heart recovery, and the patient left hospital a few months later [69–71]. It prompted others to pay greater attention to continuous-flow technology, pump miniaturization, and device placement. During the 1990s and 2000s, several promising continuous-flow devices were in development, including the HeartMate II pump.

The HeartMate II pump is an axial-flow rotary VAD that consists of a blood pump, a percutaneous lead, an external power source, and a system driver. The pump has only one moving part, a cylindrical impeller or rotor with blades that propels blood along its axis from the left ventricle through the pump and into the circulation system. A magnetic field spins the rotor on its blood-lubricated bearings, and all inner surfaces of the device are covered with a textured, blood-coating system (also used in the older, pulsatile HeartMate device) to resist clot formation. The operating speed of the HeartMate II device is generally 8000 and 10,000 rpm, to generate an impressive flow rate of 10 liters (more than 2.6 gallons) per minute (stronger than pulsatile pumps' rate) [72, 73]. Former US Vice President Dick Cheney lived with his HeartMate II implant for 20 months before undergoing a heart transplant operation in 2012, and he credited the device for saving his life [74]. Other continuous-flow pumps under development in this period included the Jarvik 2000 Flowmaker Pump and the MicroMed DeBakey Noon VAD (later renamed the HeartAssist VAD), among others, toward providing both temporary and permanent support for heart failure patients [75–79] (Fig. 1.6).

The shift to this newer technology platform overcame some, but not all, former device problems. A newer problem with the second-generation devices related to the low pulsatility of these continuous-flow VADs. Patients implanted with the newer VADs had low, sometimes no, palpable pulse. The first-generation pulsatile VADs were audible pumps that mimicked the rhythmic contraction of the heart, reinforcing the pulse and the movement of blood through the heart and the body in sound and sensation for the patient. Continuous-flow devices were noticeably quieter, and no difference was felt physically by patients with the nonpulsatile movement of blood. One serious problem was the increased risk of gastrointestinal bleeding as a pathophysiological implication of near pulseless flow. In response, clinical researchers revised patient management and prevention strategies, such as setting a lower pump speed in the device and prescribing a lower dose of, if any, anticoagulants [80–82].

In 2005, the NHLBI sponsored the creation of the Interagency Registry for Mechanically Assisted Circulatory Support (INTERMACS) to

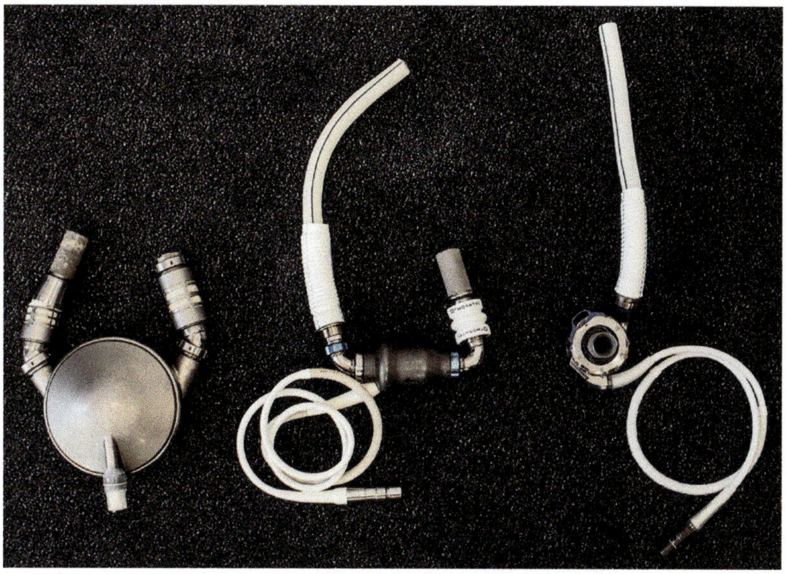

Fig. 1.6 HeartMate I, HeartMate II, and HeartMate III devices (l-r) lie on a table during a press conference at the Medical School in Hannover, Germany, on November 18, 2015. Note the decreasing size of the devices. The HeartMate II pump was one-seventh the size and one-fourth the weight of the older HeartMate pulsatile LVAD. (Photo by: Ole Spata/picture-alliance/dpa/AP Images; © 2015 The Associated Press)

track the ongoing use of long-term circulatory assist devices for the purpose of improving clinical practice. All patients receiving FDA-approved VAD implants for long-term support are subjects of INTERMACS, where details are collected regarding survival time, adverse events, device performance, and other issues, for scientific and clinical study of performance characteristics and direct comparisons of devices [83]. This national electronic database allowed the field to track and advance clinical practice with these devices. INTERMACS inspired the formation of an international registry—the International Society for Heart and Lung Transplantation Registry for Mechanically Assisted Circulatory Support (IMACS)—launched in January 2013, which is similarly managed and directly connected to the American database. Researchers can now compare how long and how well patients are living with different VADs in the United States and around the world [84].

Most recently, researchers are experimenting with a third-generation VAD, a second-generation design of the continuous-flow cardiac assist pumps. These devices are smaller in size and operate with frictionless movement of a rotor or diaphragm [85, 86]. The new design, based on hydrodynamic and magnetic bearings, increases device durability by eliminating heat from friction and the wear of components but without compromising blood flow rates and cardiac output. Another design advantage is the further miniaturization of the pump, which means less-invasive surgery and a smaller pocket (if any) in the chest cavity. These new, bearing-free, magnetically suspended pumps promise to be smaller, lighter, and easier to implant than older models [87, 88]. The HeartWare assist pump was the first device of this newer technology that becomes commercially available in the United States after FDA officials approved it in 2012. The HeartMate III, a centrifugal design that uses electromagnetic bearings, is currently under development [89, 90] (Table 1.2).

Conclusion

Historically, the effective development of TAHs and VADs occurred through iterative device improvement of benchwork, animal testing, and early clinical trials. It was a clinical need that prompted researchers to develop mechanical circulatory support systems, but it was never certain that this technology would be clinically realized. Throughout the history of mechanical circulatory support, device stakeholders faced tremendous technological challenges, and later political and

Table 1.2 Selected ventricular assist devices and FDA regulatory status

Device	Description	FDA status
HeartMate IP LVAS, by Thermo Cardiosystems Inc.	An implantable, pneumatic, pulsatile LVAD	1994 PMA granted
Thoratec PVAD, by Thoratec Laboratories Corp.	A paracorporeal, pneumatic, pulsatile BiVAD	1995 PMA granted 2003 PMA granted for home use with TLC-II Portable VAD Driver
Novacor LVAS, by Novacor Medical Corp.	An electrically powered, pulsatile, wearable assist system, with portable battery pack	1998 PMA granted
HeartMate VE LVAD, by Thermo Cardiosystems Inc.	An electrically powered, pulsatile, wearable assist system, with portable battery pack	1998 PMA granted
Thoratec IVAD, by Thoratec Corp.	An implantable version of the Thoratec PVAD	2004 PMA granted
Jarvik 2000 VAD, by Jarvik Heart Inc.	An implantable, axial-flow blood pump	2000 IDE approved
HeartMate II LVAD, by Thoratec Corp.	An implantable, axial-flow blood pump	2003 IDE approved 2008 PMA granted
HeartWare HVAD, by HeartWare International Inc.	An implantable, centrifugal LVAD	2008 IDE approved 2012 PMA granted

McKellar [14], p. 196, Table 5.1 and p. 231, Table 6.1. © 2018 Johns Hopkins University Press. Reprinted with permission of Johns Hopkins University Press
Note: FDA Food and Drug Administration, *LVAS* left ventricular assist system, *LVAD* left ventricular assist device, *IDE* investigational device exemption, *PMA* premarket approval, *PVAD* paracorporeal ventricular assist device, *BiVAD* biventricular assist device

socioeconomic queries, that seemed unsurmountable. Ultimately, it was the clinical success of VADs that convinced many in the cardiac medical field and society to adopt mechanical circulatory support systems as a viable life-extending therapy for heart failure patients. Their use expanded with market approval and device reimbursement, as well as with increased reporting of extended patient quality of life. For the foreseeable future, a technology-based approach to the treatment of end-stage heart failure will continue, thus assuring a place for mechanical pumps. Both TAHs and VADs will improve in efficiency, become smaller and more durable, and hopefully cause fewer adverse events in patients. There is also the hope that mechanical circulatory support systems will improve for biventricular heart failure to parallel heart transplantation as a standard treatment.

References

1. U.S. Department of Health and Human Services, Centers for Disease Control and Prevention, National Center for Health Statistics. Leading causes of death, 1900–98. [cited 2018 Oct 6]. Available from: https://www.cdc.gov/nchs/data/dvs/lead1900_98.pdf.
2. Dawber TR, Meadors GF, Moore FE Jr. Epidemiologic approaches to heart disease: the Framingham study. Am J Public Health Nations Health. 1951;41(3):279–81.
3. Kannel WB, Dawber TR, Kagan A, Revotskie N, Stokes J 3rd. Factors of risk in the development of coronary heart disease—six-year follow-up experience: the Framingham study. Ann Intern Med. 1961;55:33–50.
4. McKee PA, Castelli WP, McNamara PM, Kannel WB. The natural history of congestive heart failure: the Framingham study. N Engl J Med. 1971;285(26):14441–6.
5. Levy D, Brink S. A change of heart: how the people of Framingham, Massachusetts, helped unravel the mysteries of cardiovascular disease. New York: Knopf; 2005.
6. White PD. Heart disease. 3rd ed. New York: Macmillan; 1944.
7. Mann DL, Zipes DP, Libby P, Bonow RO, Braunwald E. Braunwald's heart disease: a textbook of cardiovascular medicine. 10th ed. Philadelphia: Elsevier Saunders; 2015.
8. American Heart Association. Heart failure patients living longer, but long-term survival still low. 15 May 2013. [cited 2018 Oct 6]. Available from: https://medicalxpress.com/news/2013-05-heart-failure-patients-longer-long-term.html.
9. Ponikowski P, Anker SD, AlHabib KF, Cowie MR, Force TL, Hu S, et al. Heart failure: prevent-

ing disease and death worldwide. ESC Heart Fail. 2014;1(1):4–25.
10. Hallowell C. Charles Lindbergh's artificial heart. Am Herit Invent Technol. 1985;1(2):58–62.
11. Laurence WL. Carrel, Lindbergh develop device to keep organs alive outside the body. NY Times. 21 June 1935; 1, 13.
12. Medicine: Men in Black. Time. 1938;31(24): cover story.
13. Gibbon JH Jr. Application of a mechanical heart and lung apparatus to cardiac surgery. Minn Med. 1954;37(3):171–85.
14. McKellar S. Artificial hearts: the allure and ambivalence of a controversial medical technology. Baltimore: Johns Hopkins University Press; 2018.
15. DHEW. Artificial heart program conference proceedings, 9–13 June 1969. Washington, DC: GPO; 1970.
16. DeBakey ME, Hall CW. Towards the artificial heart. New Sci. 1964;393:540–1.
17. DeBakey ME. Left ventricular bypass pump for cardiac assistance. Clinical experience. Am J Cardiol. 1971;27(1):3–11.
18. Mexican Heart Patient Operation by DeBakey and Baylor, 1966. Texas archive of the moving image. KHOU-TV Collection.
19. Kantrowitz A. Early mechanical left ventricular assistance. ASAIO J. 1993;39(4):834–9.
20. Moulopoulos SD, Topaz S, Kolff WJ. Diastolic balloon pumping (with carbon dioxide) in the aorta—a mechanical assistance to the failing circulation. Am Heart J. 1962;63:660–75.
21. Kantrowitz A, Tjonneland S, Freed PS, Phillips SJ, Butner AN, Sherman JL Jr. Initial clinical experience with intraaortic balloon pumping in cardiogenic shock. JAMA. 1968;203(2):113–8.
22. Sujansky E, Tjonneland S, Freed PS, Kantrowitz A. A dynamic aortic patch as a permanent mechanical auxiliary ventricle: experimental studies. Surgery. 1969;66(5):875–82.
23. U.S. National Library of Medicine, History of Medicine Division. Adrian Kantrowitz Papers. MS C 572. Clinical films: Haskell Shanks, 1971.
24. Kantrowitz A. Cardiac assist devices. 1972. J Extra Corpor Technol. 2007;39(2):119–24; discussion 117–8.
25. McKellar S. Chapter 15: Limitations exposed: Willem J. Kolff and his contentious pursuit of a mechanical heart. In: Li A, Heaman E, McKellar S, editors. Essays in honour of Michael Bliss: figuring the social. Toronto: University of Toronto Press; 2008.
26. Akutsu T, Kolff WJ. Permanent substitutes for valves and hearts. Trans Am Soc Artif Intern Organs. 1958;4(1):230–4.
27. Shumacker HB Jr. A surgeon to remember: notes about Vladimir Demikhov. Ann Thorac Surg. 1994;58:1196.
28. McKellar S. Clinical firsts – Christiaan Barnard's heart transplantations. N Engl J Med. 2017;377(23):2211–3.
29. Barnard CN. The operation: a human cardiac transplant: an interim report of a successful operation performed at Groote Schuur Hospital, Cape Town. S Afr Med J. 1967;41(48):1271–4.
30. Barnard CN. Human cardiac transplantation: an evaluation of the first two operations performed at the Groote Schuur Hospital, Cape Town. Am J Cardiol. 1968;22(4):584–96.
31. Hall CW, Liotta D, DeBakey ME. Bioengineering efforts in developing artificial hearts and assistors. Am J Surg. 1967;114(1):24–30.
32. Cooley DA, Liotta D, Hallman GL, Bloodwell RD, Leachman RD, Milam JD. Orthotopic cardiac prosthesis for two-staged cardiac replacement. Am J Cardiol. 1969;24(5):723–30.
33. DeBakey ME. The odyssey of the artificial heart. Artif Organs. 2000;24(6):405–11.
34. Cooley DA, Akutsu T, Norman JC, Serrato MA, Frazier OH. Total Artificial heart in two-staged cardiac transplantation. Cardiovasc Dis. 1981;8(3):305–19.
35. Cheng K, Meador JW, Serrato MA, Akutsu T. The design and fabrication of a new total artificial heart. Cardiovasc Dis. 1977;4(1):7–17.
36. Frazier OH, Akutsu T, Cooley DA. Total artificial heart (TAH) utilization in man. Trans Am Soc Artif Intern Organs. 1982;28:534–8.
37. Cooley DA. Staged cardiac transplantation: report of three cases. Heart Transplant. 1982;1:145–53.
38. Jarvik RK. The total artificial heart. Sci Am. 1981;244(1):74–80.
39. Joyce LD, DeVries WC, Hastings WL, Olsen DB, Jarvik RK, Kolff WJ. Response of the human body to the first permanent implant of the Jarvik-7 Total artificial heart. Trans Am Soc Artif Intern Organs. 1983;29:81–7.
40. DeVries WC, Anderson JL, Joyce LD, Anderson FL, Hammond EH, Jarvik RK, Kolff WJ. Clinical use of the total artificial heart. NEJM. 1984;310(5):273–8.
41. DeVries WC. The permanent artificial heart: four case reports. JAMA. 1988;259(6):849–59.
42. The Dracula of Medical Technology, NY Times. 16 May 1988; 16.
43. Johnson KE, Liska MB, Joyce LD, Emery RW. Registry report. Use of total artificial hearts: summary of world experience, 1969–1991. ASAIO J. 1992;38(3):M486–92.
44. Jarvik RK, DeVries WC, Semb BK, Koul B, Copeland JG, Levinson MM, Griffith BP, Joyce LD, Cooley DA, Frazier OH, et al. Surgical positioning of the Jarvik-7 artificial heart. J Heart Transplant. 1986;5(3):184–95.
45. Magovern JA, Rosenberg G, Pierce WS. Development and current status of a total artificial heart. Artif Organs. 1986;10(5):357–63.
46. Pierce WS, Rosenberg G, Snyder AJ, Pae WE Jr, Donachy JH, Waldhausen JA. An electric artificial heart for clinical use. Ann Surg. 1990;212(3):339–43.
47. Weiss WJ, Rosenberg G, Snyder AJ, Pierce WS, Pae WE, Kuroda H, et al. Steady state hemodynamic and energetic characterization of the Penn State/3M Health Care Total Artificial Heart. ASAIO J. 1999;45(3):189–93.

48. Kung RT, Yu LS, Ochs BD, Parnis SM, Macris MP, Frazier OH. Progress in the development of the ABIOMED total artificial heart. ASAIO J. 1995;41(3):M245–8.
49. Dowling RD, Gray LA, Etoch SW, Laks H, Marelli D, Samuels L, et al. The AbioCor implantable replacement heart. Ann Thorac Surg. 2003;75(6 Suppl):S93–9.
50. Dowling RD, Gray LA Jr, Etoch SW, Laks H, Marelli D, Samuels L, et al. Initial experience with the AbioCor implantable replacement heart system. J Thorac Cardiovasc Surg. 2004;127(1):131–41.
51. Frazier OH, Dowling RD, Gray LA Jr, Shah NA, Pool T, Gregoric I. The total artificial heart: where we stand. Cardiology. 2004;101(1–3):117–21.
52. Copeland JG, Smith RG, Arabia FA, Nolan PE, Banchy ME. The CardioWest total artificial heart as a bridge to transplantation. Semin Thorac Cardiovasc Surg. 2000;12(3):238–42.
53. Copeland JG, Copeland H, Gustafson M, Mineburg N, Covington D, Smith RG, Friedman M. Experience with more than 100 total artificial heart implants. J Thorac Cardiovasc Surg. 2012;143(3):727–34.
54. Jaroszewski DE, Anderson EM, Pierce CN, Arabia FA. The SynCardia freedom driver: a portable driver for discharge home with the total artificial heart. J Heart Lung Transplant. 2011;30(7):844–5.
55. Bartlett RH, Gazzaniga AB, Huxtable RF, Schippers HC, O'Connor MJ, Jefferies MR. Extracorporeal circulation (ECMO) in neonatal respiratory failure. J Thorac Cardiovasc Surg. 1977;74(6):826–33.
56. Allison PL, Kurusz M, Graves DF, Zwischenberger JB. Devices and monitoring during neonatal ECMO: survey results. Perfusion. 1990;5(3):193–201.
57. Almond CS, Buchholz H, Massicotte P, Ichord R, Rosenthal DN, Uzark K, et al. Berlin heart EXCOR pediatric ventricular assist device investigational device exemption study: study design and rationale. Am Heart J. 2011;162(3):425–35.
58. Almond CS, Morales DL, Blackstone EH, Turrentine MW, Imamura M, Massicotte MP, et al. Berlin Heart EXCOR pediatric ventricular assist device for bridge to heart transplantation in US children. Circulation. 2013;127(16):1702–11.
59. Champsaur G, Ninet J, Vigneron M, Cochet P, Neidecker J, Boissonnat P. Use of the Abiomed BVS System 5000 as a bridge to cardiac transplantation. J Thorac Cardiovasc Surg. 1990;100(1):122–8.
60. Pierce W, Parr GVS, Myers JL, Pae WE Jr, Bull AP, Waldhausen JA. Ventricular-assist pumping in patients with cardiogenic shock after cardiac operations. N Engl J Med. 1981;305(27):1606–10.
61. Portner PM, Oyer PE, Jassawalla JS, Miller PJ, Chen H, LaForge DH, et al. An implantable permanent left ventricular assist system for man. Trans Am Soc Artif Intern Organs. 1978;24:99–103.
62. Oyer PE, Stinson EB, Portner PM, Ream AK, Shumway NE. Development of a totally implantable, electrically actuated left ventricular assist system. Am J Surg. 1980;140(1):17–25.
63. Norman JC, McGee MG, Fuqua JM, Igo SR, Turner SA, Sterling R, et al. Development and evaluation of a long-term, implantable, electrically actuated left ventricular assist system: THI/Gould LVAS. Artif Organs. 1983;7(1):64–73.
64. Rose EA, Gelijns AC, Moskowitz AJ, Heitjan DF, Stevenson LW, Dembitsky W, et al. Long-term use of a left ventricular assist device for end-stage heart failure. N Engl J Med. 2001;345(20):1435–43.
65. McKellar S. Chapter 6: Disruptive potential: the 'landmark' REMATCH trial, left ventricular assist device (LVAD) technology and the surgical treatment of heart failure in the United States. In: Schlich T, Crenner C, editors. Technological change in modern surgery: historical perspectives on innovation. Rochester: University of Rochester Press; 2017.
66. Wampler RK, Moise JC, Frazier OH, Olsen DB. In vivo evaluation of a peripheral vascular access axial flow blood pump. ASAIO Trans. 1988;34(3):450–4.
67. Butler KC, Moise JC, Wampler RK. The Hemopump--a new cardiac prothesis device. IEEE Trans Biomed Eng. 1990;37(2):193–6.
68. Wampler RK, Baker BA, Wright WM. Circulatory support of cardiac interventional procedures with the Hemopump cardiac assist system. Cardiology. 1994;84(3):194–201.
69. Frazier OH, Nakatani T, Duncan JM, Parnis SM, Fuqua JM. Clinical experience with the Hemopump. ASAIO Trans. 1989;35(3):604–6.
70. Frazier OH, Wampler RK, Duncan JM, Dear WE, Macris MP, Parnis SM, et al. First human use of the Hemopump, a catheter-mounted ventricular assist device. Ann Thorac Surg. 1990;49(2):299–304.
71. Wampler RK, Frazier OH, Lansing AM, Smalling RW, Nicklas JM, Phillips SJ, et al. Treatment of cardiogenic shock with the Hemopump left ventricular assist device. Ann Thorac Surg. 1991;52(3):506–13.
72. Griffith BP, Kormos RL, Borovetz HS, Litwak K, Antaki JF, Poirier VL, et al. HeartMate II left ventricular assist system: from concept to first clinical use. Ann Thorac Surg. 2001;71(3 Suppl):S116–20; discussion S114–6.
73. Delgado R, Bergheim M. HeartMate II left ventricular assist device: a new device for advanced heart failure. Expert Rev Med Devices. 2005;2(5):529–32.
74. Cheney D, Reiner J. Heart: an American Medical Odyssey. New York: Scribner; 2013.
75. Westaby S, Banning AP, Jarvik R, Frazier OH, Pigott DW, Jin XY, et al. First permanent implant of the Jarvik 2000 Heart. Lancet. 2000;356(9233):900–3.
76. Frazier OH, Myers TJ, Jarvik RK, Westaby S, Pigott DW, Gregoric ID, et al. Research and development of an implantable, axial-flow left ventricular assist device: the Jarvik 2000 Heart. Ann Thorac Surg. 2001;71(3 Suppl):S125–32.
77. Jarvik R. Jarvik 2000 pump technology and miniaturization. Heart Fail Clin. 2014;10(1 Suppl):S27–38.
78. Noon GP, Morley D, Irwin S, Benkowski R. Development and clinical application of the MicroMed DeBakey VAD. Curr Opin Cardiol. 2000;15(3):166–71.

79. Noon GP, Morley DL, Irwin S, Abdelsayed SV, Benkowski RJ, Lynch BE. Clinical experience with the MicroMed DeBakey ventricular assist device. Ann Thorac Surg. 2001;71(3 Suppl):S133–8.
80. Morgan JA, Paone G, Nemeh HW, Henry SE, Patel R, Vavra J, et al. Gastrointestinal bleeding with the HeartMate II left ventricular assist device. J Heart Lung Transplant. 2012;31(7):715–8.
81. Amer S, Shah P, Hassan S. Gastrointestinal bleeding with continuous-flow left ventricular assist devices. Clin J Gastroenterol. 2015;8(2):63–7.
82. Joy PS, Kumar G, Guddati AK, Bhama JK, Cadaret LM. Risk factors and outcomes of gastrointestinal bleeding in left ventricular assist device recipients. Am J Cardiol. 2016;117(2):240–4.
83. Kirklin JK, Naftel DC, Stevenson LW, Kormos RL, Pagani FD, Miller MA, Ulisney K, Young JB. INTERMACS database for durable devices for circulatory support: first annual report. J Heart Lung Transplant. 2008;27(10):1065–72.
84. Kirklin JK, Cantor R, Mohacsi P, Gummert J, De By T, Hannan MM, Kormos RL, Schueler S, Lund LH, Nakatani T, Taylor R, Lannon J. First annual IMACS report: a global International Society for Heart and Lung Transplantation registry for mechanical circulatory support. J Heart Lung Transplant. 2016;35(4):407–12.
85. Hoshi H, Shinshi T, Takatani S, et al. Third-generation blood pumps with mechanical noncontact magnetic bearings. Artif Organs. 2006;30(5):324–38.
86. Lee JJ, Ahn CB, Choi J, Park JW, Song SJ, Sun K. Development of magnetic bearing system for a new third-generation blood pump. Artif Organs. 2011;35(11):1082–94.
87. Pagani FD. Continuous-flow rotary left ventricular assist devices with "3rd generation" design. Semin Thorac Cardiovasc Surg. 2008;20(3):255–63.
88. Moazami N, Steffen RJ, Naka Y, Jorde U, Bailey S, Murali S, et al. Lessons learned from the first fully magnetically levitated centrifugal LVAD trial in the United States: the DuraHeart trial. Ann Thorac Surg. 2014;98(2):541–7.
89. Farrar DJ, Bourque K, Dague CP, Cotter CJ, Poirier VL. Design features, developmental status, and experimental results with the HeartMate III centrifugal left ventricular assist system with a magnetically levitated rotor. ASAIO J. 2007;53(3):310–5.
90. Schmitto JD, Hanke JS, Rojas SV, Avsar M, Haverich A. First implantation in man of a new magnetically levitated left ventricular assist device (HeartMate III). J Heart Lung Transplant. 2015;34(6):858–60.

2 A Personalized History of Rotary Blood Pumps

Richard Wampler

There are in fact four very significant stumbling blocks in the way of grasping the truth which hinder every man however learned, and scarcely allow anyone to win a clear title to wisdom; namely, The Example of Weak and Unworthy Authority, Long Standing Custom, the Feeling of the Ignorant Crowd, and the Hiding of our Own Ignorance While Making a Display of Our Apparent Knowledge.

Every man is involved in these things, every rank is affected. For every person, in whatever walk of life both in application to study and in all forms of occupation arrives at the same conclusion by the worst arguments. Namely: This is a Pattern Set by our Elders, This the Custom, This is the Popular Belief: Therefore, it should be held.

Roger Bacon, "On the Causes of Error"
Opus Majus, Thirteenth Century

I am honored for the invitation to write this chapter as I was both a participant and witness to the difficult journey of bringing mechanical circulatory support devices (MCSDs), based on rotary blood pumps, to clinical readiness. As a pioneer in the field, I hope the reader will forgive my presumption in not only recounting the history of the technical development but also my personal experience of discovery, invention and development of a disruptive technology. During my career I have been a part of some amazing successes and soul-crushing failure. I share these stories in the hope that some of my experience and hard lessons will benefit those who choose to pursue the difficult path of innovative medical device development.

When I first began pursuing rotary blood pumps as senior medical student in 1974, there were many strongly held negative paradigms regarding the use of rotary blood pumps for mechanical circulatory support (MCS). It was widely and strongly believed that a pulsatile circulation was absolutely essential to sustainable human physiology. This belief was, largely, based on the fact that millions of years of evolution had resulted in creatures of all kinds that had pulsatile circulation, presumably, for compelling physiologic reasons. In addition, it was observed that the mortality of patients on cardiopulmonary bypass dramatically increased after even 2 hours of continuous flow bypass during heart surgery. Consequently, the notion of long-term pulseless support was deemed impossible if not absurd to contemplate.

There was also a widely held belief that rotary pumps would act like waring blenders for the blood producing levels of plasma-free hemoglobin (pf Hgb) that would result in severe renal injury and that loss of red blood cells (RBCs) would result in hemolytic anemia. These beliefs were strongly held in spite of the absence of experimental studies to support them. In fact, these beliefs were so widely and rigidly held that

R. Wampler, MD (✉)
Oregon Health & Science University,
Portland, OR, USA

© Springer Nature Switzerland AG 2020
J. H. Karimov et al. (eds.), *Mechanical Support for Heart Failure*,
https://doi.org/10.1007/978-3-030-47809-4_2

it was difficult to have a rational conversation on the subject. Stated succinctly, "If hemolysis did not kill the patient, non-pulsatile blood flow surely would." Finally, from an engineering perspective, it was argued that a pump with a shaft seal could not be durable due to accumulation of thrombus and denatured protein on the seal and ultimate seizure of the driveshaft.

Consequently, all efforts to create MCSDs, supported by the National Institute of Health, strived to emulate the pulse with biomimicry as a strategy. This was done in spite of the substantial technical challenges of large size, complexity, durability of flexing membranes, and the need for artificial valves.

I found it ironic that pump investigators didn't question their pulsatility strategy given the lesson of history developing heavier than air flight. Initially, inventors pursued a fruitless strategy of biomimicry by creating machines with flapping wings, but it took two bicycle mechanics from Dayton, Ohio, to demonstrate that a fixed wing device was a better strategy. Sometimes nature can be improved.

Epiphany

I remember very well the event in my life that beckoned me to pursue the path of rotary blood pumps for circulatory support. I was fortunate enough to have people in my youth that believed in me before I knew how to believe in myself. Most prominently, I remember my grandfather, Luster Foster. Luster was a dreamer who lived in a time of broken dreams during the great depression. He lost his home, struggled to support his young family, and lost his firstborn daughter to pneumonia. He taught me a lot about social justice, being human, how to dream, and how to create things with wood.

I was to lose him to heart disease during my junior year of medical school. I will never forget that feeling of desperation and helplessness as I watched his life slip through my fingers. I didn't realize, as I stood by his grave, that this event would radically alter the trajectory of my career to this day.

The Secret of my Success: Serendipity

I confess I would have accomplished very little if not for serendipity. Serendipity is a gift for discovering things, by accident or chance, while hunting for something else – described by some as "happy chance." Carl Jung coined the term synchronicity; some call it Karma or cosmic coincidence.

A well-known example of serendipity is Fleming's observation of a troublesome mold, penicillin, which was fouling up his experiments. This happy chance spawned an era of antibiotic development that has benefited countless patients with bacterial infections. Fleming had the following reflection on his discovery;

> *There are thousands of different molds and thousands of different bacteria, and that chance put that mold in the right spot at the right time was like winning the Irish Sweepstakes*
>
> Sir Alexander Fleming

For me, serendipity has often been the seed of inspiration.

The experience of my inventive journey is better described by a whimsical definition of serendipity: "serene stupidity." I liken my experience of serene stupidity to that of the Tom Hanks' character in the movie Forrest Gump. Forrest was a simple-minded person with good intentions who, by happy chance, ended up being a champion ping pong player, meeting three US presidents and founding a multimillion dollar restaurant franchise. In a similar way, a great deal of my accomplishments found me.

In addition, one of my other great assets was the fact that I kept my ignorance completely intact; it was my only defense. If I had sought out advice from experts in the field, I am sure they would have discouraged me from even trying. Finally, I trusted my crazy ideas and I was very persistent.

The Beginning of the Journey

During my senior year in medical school, all of my electives were with cardiovascular training centers around the country since, at that time, I thought I wanted to be a cardiac surgeon. Given my interest in heart surgery and being blessed with innate gifts of curiosity, creativity, good dexterity, resourcefulness, and a willingness to risk, I quickly began to imagine rotary artificial hearts. I immediately became convinced that some type of rotary pump was going to be the solution to the challenge of building an artificial heart. Looking back it is hard for me to say why I believed this nor where I got the audacity to think existing strategies of biomimicry by many others, more qualified than myself, were on the wrong track. At the time I had no background or education that would have supported my ideas. I can only say that I had the deepest intuition that my ideas, however far from the mainstream, were going to be the best way. Rotary blood pumps did have significant engineering advantages over pulsatile devices including (1) much smaller displaced and priming volume, (2) simplicity with one moving part, and (3) no artificial valves.

Fearful of criticism, I kept my ideas to myself. I have always been a doer, so I began making prototypes. I didn't let the fact that I didn't know what I was doing stop me. Figure 2.1 was my idea of a TAH with opposed centrifugal impellers produced by lost wax casting. I pumped expired bank blood with it using an electric drill for power. It pumped a lot but was very hemolytic. As far as I know, it was the first attempt to make an integrated total artificial heart from rotary blood pumps.

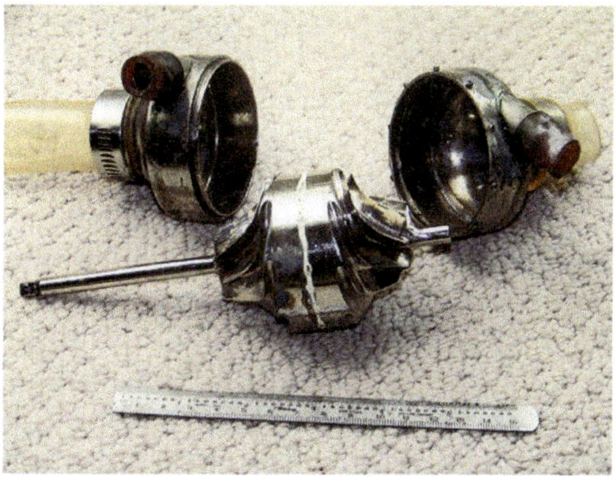

Fig. 2.1 The author's idea of a TAH built during residency in 1975 with opposed centrifugal impellers produced by lost wax casting

In Fig. 2.2 was my idea of an LVAD which used an axial flux gap motor prescient of the motors used in the HeartWare HVAD many years later. The motor ran once. I was not discouraged.

During my internship and residency, I spent a great deal of time researching work on artificial hearts and left ventricular assist devices. Although most research was focused on pumps capable of emulating a pulse, there was some research on rotary blood pumps. I was gratified to find that a few investigators, much smarter than me, had explored the idea of rotary blood pumps while I was still in high school.

The engineering advantages of centrifugal pumps were recognized by Saxton and Andrews [1] in a 1960 presentation at the annual meeting of the American Society of Artificial Internal Organs. Using commercial centrifugal pumps, Saxton demonstrated that their pumping behavior responded to inlet and outlet pressures in a way that emulated physiologic Starling autoregulatory behavior. Suspending disbelief regarding significant hemolysis, a number of canine experiments were performed using centrifugal pumps for biventricular assistance. All animals died within 15 minutes of severe pulmonary edema. No further work was reported. I did not become aware of Saxton's efforts until significant progress had been made in rotary blood pump devel-

Fig. 2.2 The author's idea of an LVAD built during residency in 1976 which used an axial flux gap motor prescient of the motors used in the HeartWare HVAD many years later

opment. It was, probably, fortuitous I was not aware of the failure of his animal studies as I might have been discouraged from pursuing my ideas.

Given prevailing concerns about hemolysis, I knew that the issue of tolerance to hemolysis was critical to establishing design criteria for a pump. I searched for studies which sought to quantify the tolerance to intravascular hemolysis. Though not related to pumps, there were already known intravascular hemolytic clinical conditions such as malaria and implanted mechanical valves. In spite of significant hemolysis, in many cases, these patients seemed able to compensate or tolerate the condition, but the toxicity and tolerance to hemolysis were not quantified and poorly understood. Nevertheless, it seemed to me that there was physiologic precedent for some degree of tolerance to chronic hemolysis and that rotary pumps might be designed that would have acceptable levels of blood injury.

In 1932 Ottenberg and Fox [2] injected 4–8 grams of their own free Hgb in medical students to measure the rate of excretion. Forty percent was excreted in 3 hours and total removal occurred at 20 hours. For plasma levels of 150 mg%, he found no urinary excretion of free Hgb. This study did not address the clinical situation wherein free hemoglobin would be continuously released.

Bernstein et al. [3], in 1965, conducted canine in vivo studies to determine the tolerance to continuous infusion of pf Hgb. Based on the known rates of hemolysis of existing pumps of 0.1 grams/100 liters pumped, the animals were proportionately infused with 0.1 mg/kg/min. These experiments ranged from 4 to 22 days. During infusion, except for episodes of sepsis, there was no elevation of BUN or urine hemoglobin spillage and with plasma-free hemoglobin elevation below 30 mg%.

Hemolysis due to continuous pumping with pumps available at the time was calculated to be 8.64 grams Hgb/day which could readily be replaced by the bone marrow. A couple of years after the Hemopump™ success, I made a back of the napkin calculation of RBC attrition per pass through a rotary pump with an index of hemolysis of 0.1 gms Hgb released/100 liters pumped. Assuming a blood volume of 5 liters and pump flow of 5 lpm, I got the astonishing result of only one RBC per 100,000 RBCs per pass. Even now it is hard to imagine it is so low and in the very early days I think I would have judged this number impossible to achieve. Sometimes ignorance can be a good thing.

Subsequently, in collaboration with Medtronic, Bernstein et al. developed a centrifugal pump in 1970, called the Hemodyne pump, intended for long-term left ventricular support [4]. This pump employed a straight vane pyrolytic carbon impeller and was driven with magnetic coupling. The seal was purged with saline. This pump demonstrated excellent hydraulic performance, and an index of hemolysis of, approximately, 50 mg/100 liters pumped, well below the level of existing pumps. The Hemodyne pump system incorporated a non-thoracotomy access described by Zwart [5] which used a peripheral inflow cannula passed retrograde across the aortic valve to gain ventricular inflow.

The Hemodyne pump was used in animal experiments of 24 hours with non-thoracotomy bypass [6]. Fifty-five calves underwent left ventricular pass using the Hemodyne pump via non-thoracotomy approach. Partial bypass flow of 3–6 lpm was achieved. Hematological changes were limited to moderate platelet depression and average plasma-free Hgb levels of 21 mg%.

This system was then used in limited clinical trials [7]. By March of 1970 two acute human cases were reported. These were patients in extremis and were supported for 9 and 5 and a half hours. The first patient had significant hemodynamic improvement with assistance and regained consciousness, but unidentified blood loss complicated management. Attempts to wean the patient were unsuccessful and the patient died. The second patient demonstrated improvement in hemodynamics and peripheral perfusion but developed massive gastrointestinal bleeding and died. The Hemodyne pump was ultimately abandoned because of limited seal life of approximately 1 month. However, the Hemodyne system did demonstrate the ability of a rotary pump to completely support systemic circulation with acceptable levels of blood injury and the viability of non-thoracotomy bypass. The accomplishment was largely forgotten. It would be 30 years later that I would apply the innovation of transvalvular ventricular access to the invention of the Hemopump.

In 1972 Kletschka and Rafferty [8] invented a blood pump based on the bladeless Tesla shear pump. Their pump used a set of nested bell-shaped bladeless cones and was initially marketed to provide blood flow during cardiopulmonary bypass and was introduced into the market as the Bio-Pump for support of cardiopulmonary bypass in 1976.

Although, approved for only 6 hours of use, it was adopted as an external pump for acute ventricular dysfunction following cardiopulmonary bypass by George Magovern beginning in 1980 [9]. The Bio-Medicus, asynchronous, vortex pump was used on 21 patients who failed to wean from cardiopulmonary bypass. They were supported for 2–9 days, ten patients were weaned from support, and five were discharged. This off-label use of the Bio-Medicus pump made the CEO of Bio-Medicus, Jim Lyons, very nervous, but it was a bold move and demonstrated that diminished pulsatility could provide effective physiologic function and paved the way for others.

I was encouraged and emboldened by these developments. It was reassuring to know that I wasn't the only one who saw the potential of rotary blood pumps and, more importantly, that evidence was mounting that hemolysis was not the barrier it was perceived to be. The issue of pulselessness remained.

Happy Chance

In 1976 fate took a hand. I went on a mission trip to a small village in Egypt called El Bayad. Life and farming were very primitive. One of the needs of the villagers was to help establish a safe water supply to mitigate infection from parasitic liver flukes and childhood diarrhea. Consequently, I read up on well designs and, particularly, on the concept of submersible pumps as depicted in Fig. 2.3. Submersible water pumps reside beneath the ground in the aquifer and "push" the water to the surface. While in El Bayad, I also witnessed the use of crude Archimedes screw pumps to lift water from irrigation canals to water fields. Figure 2.4 shows a villager turning a hand crank to power an Archimedes screw pump. This pump was

Fig. 2.3 Example of the concept of submersible pumps as depicted that the author consulted prior to mission trip to Egypt in 1976

Fig. 2.4 An Egyptian villager like one observed by author, turning a hand crank to power an Archimedes screw pump. (Robert Burch/Alamy Stock Photo)

ideal for lifting large volumes of water with small head rise. The Archimedes screw evolved with sophistication and modern materials into turbine pumps and jet engines. It would be 6 years later that I would connect these dots to come up with the invention of the Hemopump.

Leonard Golding adapted the Hemodyne pump to create a pulseless bovine model. Hemodyne pumps were placed to support the right and left heart, and the pulmonary artery was ligated. The heart was then fibrillated. Flows were balanced by manual adjustment. Survival from 34 to 100 days was reported [10]. Except for the immediate postoperative period, the liver functions were normal. Renal function was normal except for the 100 day animal which had a BUN of 29 at termination. These studies conclusively demonstrated that non-pulsatile flow was compatible with long-term sustainable and normal physiology. Clearly, pulsatile circulation was not as critical as had previously been thought.

After completing several years of surgical training at the University of Oregon in Portland, I moved to Sacramento in 1979 and worked in the trauma center at the University of California Davis Medical Center. I decided to pursue my crazy ideas in earnest and started to study engineering at California State University at Sacramento. Again fate took a hand. Alumni from the biomedical engineering program were aerospace engineers at Aerojet Liquid Rocket that were working on pulsatile LVAD actuators under National Institute of Health (NIH) contract. After completing a senior project with them, I joined the group when it splintered off to form Nimbus Medical. This effort was spear headed by John Moise and Kenneth Butler. The LVADs under development by Nimbus were funded by National Institute of Health and were intended to mimic the heart and were pulsatile with flexing membranes, reciprocating mechanisms and artificial valves. They were large and required major surgery on cardiopulmonary bypass to be implanted. I began to think of a way to simplify vascular access and minimize surgical risk.

In my inventive journey, I have found that it sometimes takes a while to connect the dots. I am sure, while in Egypt, that I did not once in the midst of that hot, desolate, primitive, wonderful place ever think of pumping blood with an Archimedes screw. Nor did I ever consider that the principle of a submersible pump could just as well be a Hemopump™ sustaining the flow of lifeblood to an injured heart. It was not until 1982, 6 years after my visit to Egypt that I suddenly found myself in a plane, a book in my hand opened to a diagram of a submersible pump and Archimedes screw. The final piece came to me when I recalled Zwart's transvalvular ventricular access [5]. It was then that serendipity and inspiration helped me connect the dots and the invention of an intravascular blood pump was born. Sometimes the problem has to be reframed and, in a metaphorical sense, rather than taking Mohammad to the mountain the mountain should be taken to Mohammad. My "eureka" moment imagined a miniature intravascular pump placed inside, not outside the body or heart. Whether or not such an imagined device was technically feasible would take several years to determine. Working with the Nimbus engineers, the system was reduced to practice and the Hemopump was born.

The Hemopump system was comprised of a disposable catheter-mounted pump assembly (Fig. 2.5) which included a 7 inch flexible inflow cannula connected to the inlet of a miniature (7 mm diameter) axial flow blood pump with a maximum rotational speed of 27,000 RPM. The flexible inflow cannula was placed retrograde across the aortic valve such that blood could be removed from the left ventricle and pumped into the descending aorta (Fig. 2.6). Torque was supplied to the pump rotor by a flexible drive cable within a drive shaft sheath. The flexible drive cable connected, proximally, to a permanent magnet within the housing. The pump housing was inserted into an external motor which electromagnetically applied torque to the permanent magnet. Sensorless commutation was achieved by back electromotive force sensing (emf). Blood was excluded from the pump bearings by means of a continuous flush of 100 cc/day of 40% dextrose infused at the pump console. The Hemopump was capable of producing a nominal

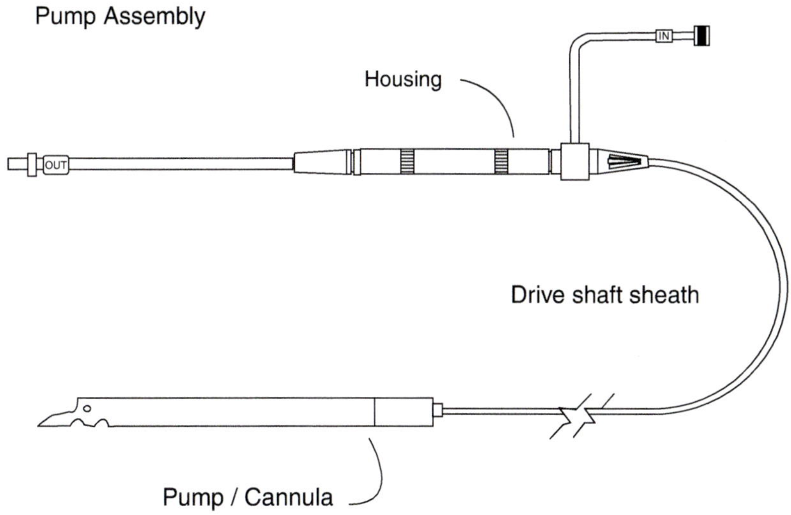

Fig. 2.5 The Hemopump system was comprised of a disposable catheter-mounted pump assembly, which included a 7 inch flexible inflow cannula connected to the inlet of a miniature (7 mm diameter) axial flow blood pump with a maximum rotational speed of 27,000 RPM. (Reproduced from Wampler and Frazier [13], with permission from Wolters Kluwer Health, Inc.)

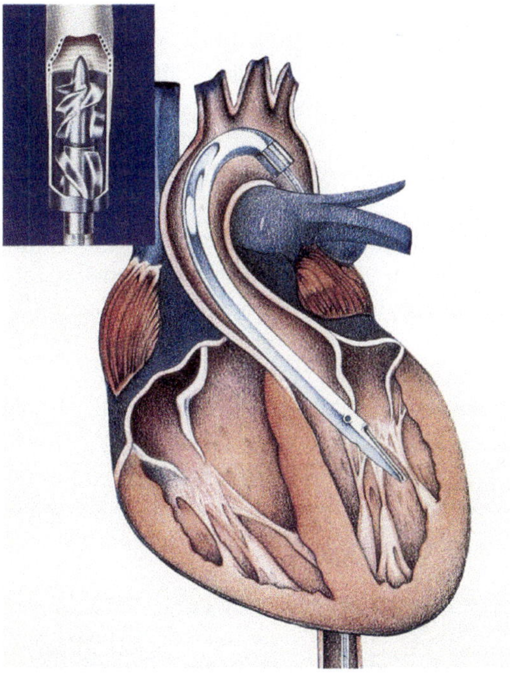

Fig. 2.6 The flexible inflow cannula was placed retrograde across the aortic valve such that blood could be removed from the left ventricle and pumped into the descending aorta. (Reproduced from Wampler and Frazier [13], with permission from Wolters Kluwer Health, Inc.)

flow of 3.5 lpm at a mean aortic pressure of 70 mmHg. A series of animal experiments up to 2 weeks duration were performed [11], and the first human implantation was performed on April 25, 1988 [12].

First Human Implantation of a Rotary Blood Pump: A Moment in History [13]

On April 23, 1988, I flew to the Texas Heart Institute in Houston Texas to perform the first human implantation of a Hemopump™ with our principle investigator, O. Howard (Bud) Frazier. Two Hemopumps™ were in my carry-on luggage – still crackling with the smell of ozone fresh from the radiation sterilizer. My mood ranged from great excitement and anticipation to stark terror, tormented by a nagging sense that I didn't really know what I was doing.

Would hemolysis be comparable to what we had seen in the animals?

How would circulatory physiology react to diminished pulsatility?

Would something catastrophic happen in humans that we did not see in the animals?

I spent a very restless night.

Patient Selection

I met with Bud early in the morning and we visited two potential candidates. We selected our first candidate, Mr. Herbert Kranich, a recent heart transplant recipient who was in the midst of a profound acute episode of rejection and cardiogenic shock with a cardiac index of 1.8 liters/m^2 and a pulmonary wedge pressure (PWP) of approximately

20 mmHg and impaired renal function. He was on respiratory support, intra-aortic balloon pump counter pulsation, and an array of pharmacologic agents. We had no doubt that Mr. Kranich was at the end of the therapeutic road and that he easily met the criteria for inclusion in the clinical trial. Bud confessed to me years later that he, in part, selected Herb because he was so sure he was going to die that his guilt would be eased if something catastrophic happened with the Hemopump™ that we had not seen in animals.

Case Report

We rushed to the operating room, and under general anesthesia Bud anastomosed an end-to-side 10 mm woven Dacron graft to the femoral artery. A heparinization dose of 2 mg/kg was given and the activated clotting time maintained at 2.0–2.5 control with continuous heparin infusion. The Hemopump™ was introduced into the abdominal aorta, and a silicone rubber plug with a center hole for the flexible drive cable sheath and a 9 mm outer diameter was secured to the proximal end of the graft with a ligature to prevent egress of blood. As soon as the pump was in the abdominal aorta, pumping was initiated to keep blood moving to prevent thrombus formation. Dr. Wayne Deer, a cardiologist, then advanced the pump up the aorta and readily introduced it retrograde across the aortic valve and into the left ventricle. The insertion into the heart took less than a minute. Acutely, the cardiac output rose from about 3–5.5 lpm, but by the time we left the operating room, the flow was back down to the low 3 lpm range. I was initially discouraged with this development since my simplistic surgical reasoning expected the flow of the Hemopump™ to arithmetically increase the cardiac output. This was the first of many lessons in the "new" physiology and clinical management of continuous flow pumps. Although Mr. Kranich's cardiac output was not significantly increased from pre-insertion levels of 3 lpm, his PWP fell from the mid-20s to 10 mmHg, and some remarkable, unexpected, events unfolded. Vasopressor infusions were reduced to maximize pump flows by decreasing the hydrostatic load on the pump to a mean aortic pressure of 55 mmHg. This maneuver resulted in peripheral warming, onset of urine production, and normal mentation despite a mean pressure of 55 mmHg. I began to speculate that the perfusion requirements of a working heart drive the pressure in the aorta and that the brain and kidney can function well at these non-physiologic pressures and with extremely diminished pulsatility. A number of deeply entrenched paradigms would fall in the years to come.

Over the next 46 hours, Mr. Kranich was treated with OKT3 (generic name muromonab-CD3), a new anti-rejection drug at the time, and he continued to improve with an increase in cardiac output and pulsatility of the arterial pressure. Lightning struck at 46 hours when the flexible drive cable broke with complete loss of support and the onset of some aortic insufficiency through the pump. Urgent removal of the Hemopump™ was performed at the bedside without anesthesia. The insertion graft was ligated close the anastomosis and the excess graft excised. The patient tolerated the short period of aortic insufficiency well. It was necessary to re-institute pharmacologic support, but, fortunately, there had been sufficient recovery of myocardial contractility that this set back was overcome.

In spite of all dark prognostications, Mr. Kranich did return home to Colorado with a normal ejection fraction and lived for another 2 and half years. Figure 2.7 is a photo of Mr. Kranich, Dr. Frazier and myself taken during his recovery.

A multicenter clinical trial was conducted in the United States under an investigation device

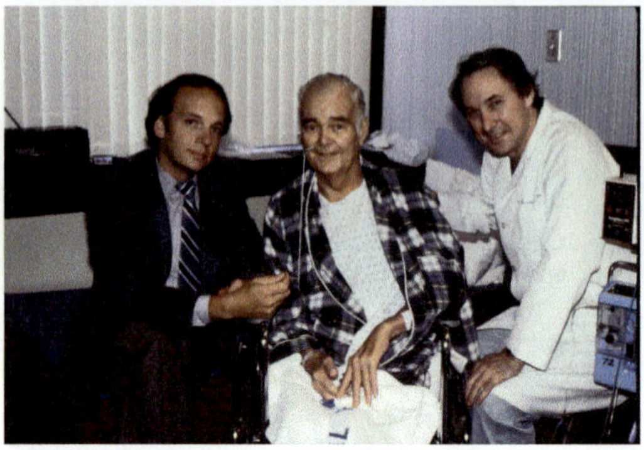

Fig. 2.7 Mr. Kranich, Dr. Frazier and the author taken during Mr. Kranich's recovery. (Photo credit: Texas Heart Institute)

exemption [14]. The trial was designed to measure the effectiveness of the Hemopump in the treatment of cardiogenic shock. Of the 53 patients enrolled, the Hemopump could be inserted in 41. Survival at 30 days was 31.7% compared to a survival of 16.6% in the non-insertion group. The mean pf Hgb was 26 mg%.

The practical demonstration that this radical idea worked, in spite of strong, widely held negative beliefs of experts captured a lot of attention in the field and popular press. We did not realize it at the time, but what Dr. Frazier calls a "Kitty Hawk" event for rotary blood pumps set loose a small explosion that radically altered the trajectory of MCSD development and foreshadowed the end of pulsatile devices. This was a very fortunate development for the field as the FDA-approved HeartMate vented electric LVAD had a limited life of 2 years.

Although the experience with the Hemopump demonstrated safety and efficacy for a temporary implantable blood pump, a couple of major technical barriers had to be overcome before long-term implantation could be accomplished. The shaft seal remained the weak link in pump life, due to stagnation on the back of the impeller thrombus and denatured protein accumulated on the seal and produced friction and ultimately seized the rotor shaft. Some innovation in shaft sealing would have to be developed to achieve durability, or a strategy for eliminating the shaft seal would have to be developed. Ultimately, the shaft seal was eliminated.

At the suggestion of OH Frazier, Nimbus explored the possibility of enlarging the Hemopump hydraulics and creating an implantable device. We determined that increasing the diameter of the Hemopump hydraulics by about 40% would provide 6 lpm of flow. Rather than transmitting torque to the impeller via a shaft penetrating the pump housing, torque was transmitted electromagnetically. In 1988 a patent application for a Chronic Ventricular Assist System was submitted [15]. Motor magnets were placed in the impeller hub, and a motor stator was positioned outside the pump housing adjacent to the rotor Fig. 2.8. The result was an integrated pump/electric motor to spin the hydraulic rotor. Commutation was performed using sensorless

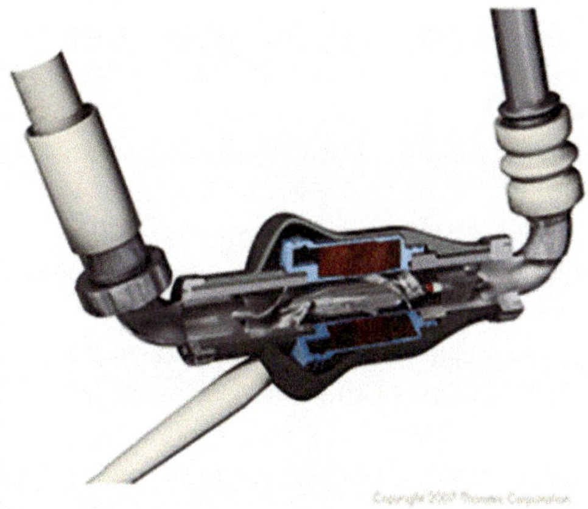

Fig. 2.8 Motor magnets were placed in the impeller hub, and a motor stator was positioned outside the pump housing adjacent to the rotor patent. (HeartMate II is a trademark of Abbott or its related companies. Reproduced with permission of Abbott, © 2019. All rights reserved)

back emf sensing which eliminated the need for implanting Hall sensors.

The first prototypes employed a purge seal like the Hemopump; however, this was considered impractical for a clinical device. The integrated motor pump design eliminated the need for a power shaft with a blood seal, but did not address the means of supporting the rotor. Fortunately, it was discovered that blood could serve as a lubricant and coolant. This idea, proposed by Robert Jarvik in 1995 [16] was radical as it ran contrary to conventional thinking which dictated that the bearing should be isolated from the blood with a seal. However, Jarvik showed that specialized bearing architecture and polishing and good washing of the bearing area to avoid stagnation and provide heat removal could produce a durable bearing without a seal. His blood-immersed bearings, adapted to what would become the Jarvik 2000, were demonstrated in the bovine model. In collaboration with OH Frazier at Texas Heart Institute, five intraventricular implants were performed with a maximum duration of 120 days [17]. This counterintuitive approach to rotor suspension with blood-immersed bearings was quickly embraced by many other pump developers, including Nimbus Medical.

Nimbus had confidence in their approach to use electromagnetic torque coupling for the hydraulic rotor, but did not believe that the purge seal design of the Hemopump could be durable. Funded by the NIH under an Innovative Research Grant, they began to explore various configurations of blood-immersed bearings. The final design of the hydraulic assembly Fig. 2.9 embodied spherical bearings Fig. 2.10 with a combination of sapphire and ceramic materials. These materials provided an excellent tribological combination and good heat transfer properties. Stationary bearing elements were supported by stator struts supported by the cylindrical pump housing. The first generation, called the AxiPump, was developed in a partnership with Bartley Griffith at the University of Pittsburg. In vivo experience [18] achieved a maximum of 120 days. The average pfHgb was 18.7 mg% ± 6.2 mg% with a range of 6.0–39.5 mg%. This device later evolved into what is now known as the HeartMate II. The first successful implantation of the final HeartMate II design was performed by OH Frazier in 2003 [19]. The patient survived for 2 and a half years.

Thoratec, Inc., gained an investigational device exemption to conduct a clinical trial of the HeartMate II as a bridge to cardiac transplantation (BTT) in patients awaiting heart transplantation [20]. The study period was from March 2005 to March 2006. One hundred thirty-three patients were enrolled without a concurrent control group. The principle outcomes were the proportions of patients who, at 180 days, had undergone transplantation, had cardiac recovery, or had ongoing mechanical support while remaining eligible for transplantation. Functional status and quality of life were also assessed.

Principle outcomes occurred in 100 patients (75%). Survival rate during support was 75% at 6 months and 68% at 12 months. Eleven patients (8%) had strokes. Bleeding requiring greater than two units of blood or packed cells occurred in 70 (53%) of patients. Evaluation of 78 patients at 3 months showed an improvement of NYHA functional class from 4.0 to 1.9 ± 0.7.

The HeartMate II achieved FDA PMA approval for bridge to cardiac transplantation in 2008. Subsequent approval for destination therapy was approved in 2010. At this writing over 26,000 HeartMate 2 LVADs have been implanted. The spherical bearing design has proven very

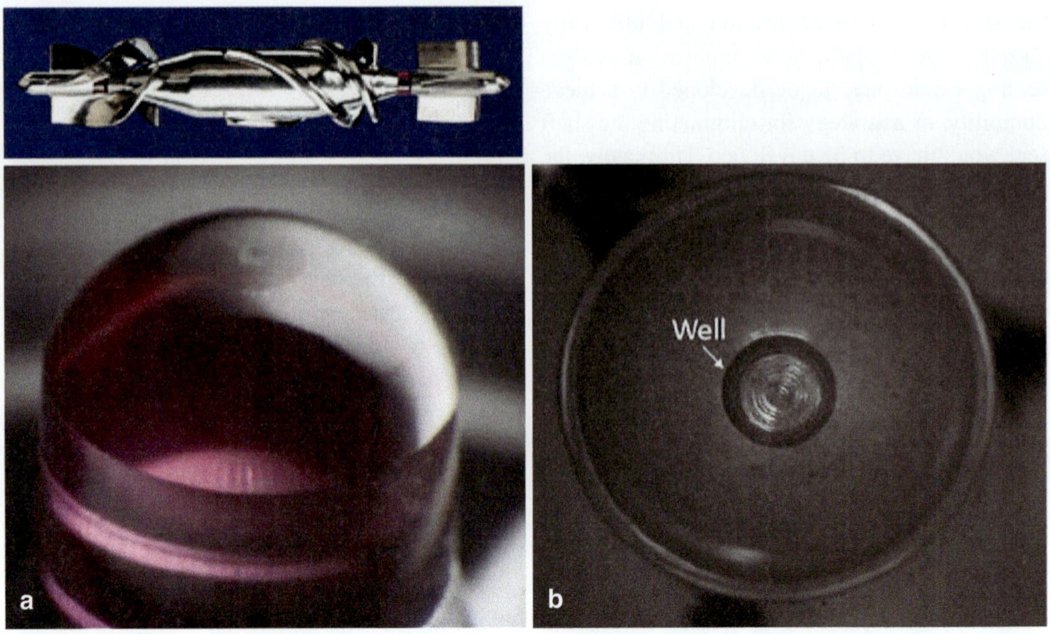

Figs. 2.9 and 2.10 The final design embodied spherical bearings with a combination of sapphire and ceramic materials. (**a**) Saphire sphere on rotating assembly (**b**) Alumina ceramic cup for stator support

durable, and, at this writing, there are over 30 patients that have gone for 13 years without a pump failure (conversation with David Farrar of Abbott Inc.). This was a great accomplishment, and I confess I would not have believed it possible since I was, initially, skeptical that mechanical bearings could be so durable.

Although the HeartMate II was the first rotary device to gain FDA PMA approval, it was not the first to be implanted in patients.

The first rotary pump LVAD to be placed in a human for long-term support was the DeBakey LVAD. The genesis for this device was the result of a collaboration initiated by a heart transplant recipient at NASA who had the vision to apply aerospace knowledge to design a rotary blood pump.

The DeBakey LVAD, MicroMed Houston Texas began clinical trials in 1999 [21]. Successful cardiac transplantation from the DeBakey LVAD was reported in 2002 [22]. This device placed small magnets in the blades rather than a single magnet in the hub. The rotor was supported by specially designed jeweled bearings. Most patients had very diminished pulsatility and non-palpable radial pulses. In spite of the grave concerns about the impact of pulseless circulation, there were no measurable ill effects. Metabolic metrics were normal and stable. Plasma-free hemoglobin was 3.3 mg% at 6 weeks. Patients reported no subjective difference between pulsatile and non-pulsatile flow. Considering that pulsatile circulation evolved over millions of years, it is a testament to the adaptability of human physiology that going from pulsatile to non-pulsatile flow, literally, with the flick of a switch, was tolerated extremely well. Although the DeBakey LVAD did not achieve PMA, it did demonstrate, definitively, that diminished and non-pulsatile circulation was tolerated very well.

CorAide LVAD

In 1990, Leonard Golding radically extended the lubricant properties of blood by demonstrating that blood could be used as a fluid to produce hydrodynamic support of a radial journal bearing in the CorAide [23]. This was an astonishing result since the minimum clearances on such a bearing was only about 0.002″ and was similar in geometry to test devices used to deliberately induce hemolysis in vitro. Consequently, one would have predicted the CorAide journal bearing to be very hemolytic. There is a lot to be said for just doing the experiment. Hydrodynamic bearings do not depend on surface lubrication and have a theoretical life of decades. A clinical BTT trial [24] was conducted from February 2005 to February 2006. Twenty-one patients were implanted with the CorAide LVAS. Seventeen patients 81% survived to 180 days or transplantation. Three patient deaths were attributed to pump polymer coating delamination which has since been mitigated. Unfortunately, Arrow abandoned the CorAide. Dr. Golding's successful use of blood as a hydraulic fluid for hydrodynamic bearings significantly extended the possibilities for blood pump rotor suspension. The basic architecture of the CorAide hydrodynamic journal bearing has been adapted to the SmartHeart total artificial heart. Based on Dr. Golding's success with blood hydrodynamic bearings, I later adapted the principle to hydrodynamic thrust bearings using blood as the fluid for what is now the HeartWare HVAD.

HeatWare HVAD

The axial flow LVADs, necessarily, have supporting struts in the form of stator blades that hold an element of the blood-immersed bearings such as the HM II and the Jarvik 2000. Unfortunately, this places the bearing in a place where the flow field tends to stagnate akin to the eye of a hurricane. This is a problem because stagnant blood tends to clot and does not remove bearing heat very well. Although, this bearing architecture usually performs well, thrombosis and rotor seizure do occur approximately 10% of the time. I had been thinking of ways to support an impeller without the use of mechanical bearing at the center of the axis of rotation while avoiding the complexity of active magnetic suspension. Serendipity led the way. I was visiting my mother in Indianapolis when I chanced by a kiosk in a shopping center that was displaying the magnetic top shown in Fig. 2.11. I watched it for a long time, bought one, and took it apart to reveal its secrets. The top was suspended by passive radial magnetic bearings which counteracted gravity. Earnshaw's law dictates that magnetic suspension requires active control of at least one range of motion. In the case of the top, stability was provided by a pointed pin at the axis of rotation constrained by a vertical glass surface. This was the key to the invention of what is now the HeartWare HVAD.

Using three dimensional magnetic analyses, it was possible to design a very powerful, miniaturized passive radial magnetic bearing. The axial constraint would be provided by hydrodynamic thrust bearings on the surface of wide blades also designed by FEA. The wide blades provided space for permanent magnets of a motor rotor, which were driven by axial magnetic slottless stators. This architecture avoided any struts or obstruction to the pump inlet (Fig. 2.12). In addition, the small size and discoid shape permit placement within the pericardial space to significantly simplify implantation.

A bridge to transplant and continued access trial for the HeartWare HVAD was reported by Mark Slaughter et al. [25]. A total of 332 patients completed the pivotal bridge to transplant and continued access protocol trial and completed 180-day primary endpoint assessment. Survival was 91% at 180 days and 84% at 360 days. Adverse events included GI bleeding 12.7%, driveline exit site infection 16.9%, hemolysis 1.2%, ischemic stroke 7.5%, and hemorrhagic stroke 7.8%.

The HeartWare HVAD received FDA approval for bridge to transplant on November of 2012 and destination on September 2017.

Although I did not intend or imagine that the HeartWare HVAD would be used in children, it has been adopted for this application because of its small size and has been used on children as young as 4 years old (ref. Fig. 2.13 is a photo of a 12 year old boy, Blake, who received a HeartWare HVAD following the onset of idiopathic cardiomyopathy). He was later trans-

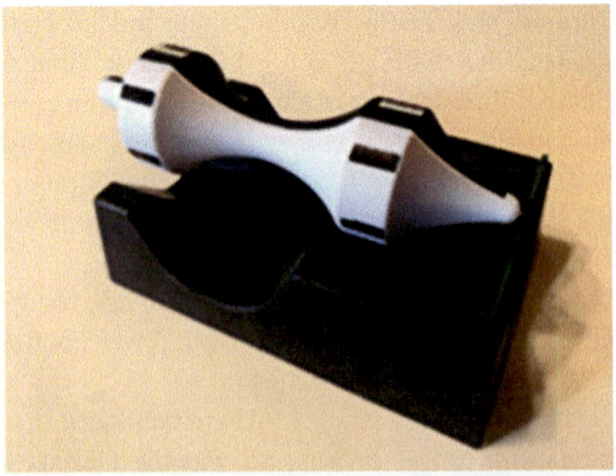

Fig. 2.11 A magnetic top like the one the author purchased and took apart to reveal its secrets. What he discovered was the key to the invention of what is now the HeartWare HVAD

Fig. 2.12 This architecture avoided any struts or obstruction to the pump inlet. (Photo courtesy of Medtronic)

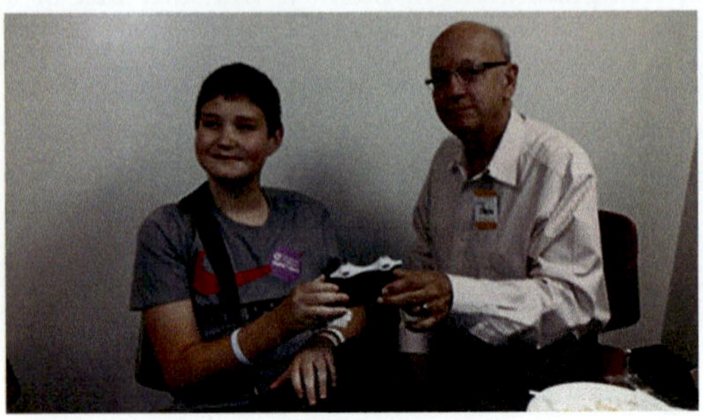

Fig. 2.13 Photo of Blake, a 12-year-old-boy who received a HeartWare HVAD following the onset of idiopathic cardiomyopathy. He was later transplanted

planted. A recent review of outcomes in the pediatric population [26] concluded that survival in children supported with the HeartWare HVAD is encouraging and comparable to young adults and there was no increased mortality in children weighing <20 kgs.

Complete Magnetic Suspension

Recently, Thoratec completed development of a completely magnetically suspended LVAD called the HM III. Based on perceived advantages of large gaps and magnetic suspension, it has enjoyed a very rapid recruitment during its clinical trial [27] and significant clinical usage. The MOMEMTUM trial compared 190 HeartMate III patients 176 HeartMate II patients. The composite primary endpoint was survival at 2 years free of disabling stroke or survival free of reoperation to replace or remove a malfunction device. Seventy-nine percent of HM III patients achieved the primary endpoint compared to 60.2% HM II patients. Pump removal for the HM III was 1.6% versus 17.0% for the HM II. The stroke rate was 10.1% vs. 19%. Gastrointestinal bleeding was equivalent, 27% for HM III vs 27.3% for HM II.

One theoretical advantage of magnetic suspension is that it avoids moving contact between the rotor and housing which has the potential benefit of unlimited mechanical life. It is interesting to note, however, there have not been reports of either the HeartMate II or the HeartWare HVAD wearing out. Pump thrombosis does occur in both the HeartMate II and the HeartWare HVAD but has not been associated with mechanical wear. Electrical failures in the form of broken wires and connectors can result in catastrophic failure since this results in abrupt loss of ventricular support. The electronics for magnetic suspension of the HM III are implanted within the pump and, consequently, would require reoperation if it were to fail and the patient was fortunate to survive long enough to reach the hospital.

The Future

Although, patients quickly adapt to non-pulsatility, there are complications. Gastrointestinal bleeding occurs in over 25% of patients on ventricular assistance with a rotary blood pump [28]. The high incidence of GI bleeding [29] was not seen as frequently with the pulsatile LVADs 6.8/100 patient years versus 63/100 patient years with the continuous flow LVADs. GI bleeding often requires hospitalization and is associated with significant additional cost. Arterio-venous malformations of the small intestine are observed on endoscopy, but the etiology remains controversial. Lyle Joyce described the occurrence of "acquired" von Willebrand syndrome [30] in patients with rotary pumps which may play a role in GI bleeding. Rotary pumps are not waring blenders for RBCs, but most of them do, seriously, deplete von Willebrand factor. von Willebrand factor was not even on the radar in the early days of rotary blood pump development. In fact, I had to look it up and refresh my medical school education when Dr. Joyce published his findings. Nor was the issue of platelet depletion and activation a part of the discussion in the early days of rotary blood pump development. We now know that alterations to these clotting elements do occur and can significantly alter the hemostatic behavior of the blood. von Willebrand factor depletion, as in the case of the clinical syndrome, can predispose to bleeding. Its role in contributing to GI bleeding is controversial. Depletion of platelets can also predispose to bleeding, and activation of platelets can predispose to hypercoagulability and thrombosis.

A great deal of progress has been made in adapting rotary blood pumps to left ventricular support devices, but significant improvements must be made. At this juncture, the adverse event profile, particularly, GI bleeding and stroke preclude use in less sick patients. It is essential that these adverse events are significantly decreased.

If GI bleeding is, in part, caused by diminished pulsatility, it could be that introducing speed changes to emulate pulsatility could mitigate this complication. Although it is simple to talk about introducing pulsatility, there are complexities in determining just what would constitute a clinically effective pulse contour. For instance, it is doubtful that a wavy pressure change will have much of a physiological effect. Joly et al. suggest that a pulsatility of less than 35 mmHg is associated with increased risk of GI bleeding [31]. Suspending disbelief about what would be a clinically effective pulse, it has not been shown that existing pumps are capable of producing a pulse that is close to a human physiologic pulse.

In addition, there is not agreement on the essential elements of the aortic pressure profile that determine its effectiveness.

Gastrointestinal bleeding, in general, is manageable, and not usually fatal or debilitating, but stroke is a major, debilitating, and expensive complication.

Stroke

Acharya et al reviewed 7,112 patients in the INTERMACS registry for stroke [32]. The mean follow-up was 9.79 months (range 0.02–34.96 months). Of all patients 752 (10.57%) had at least one stroke, with an incidence of 0.123 strokes per patient/year. Ischemic strokes account for 51.38% and hemorrhagic 48.62%. Thirty-day survival of ischemic strokes was 80.7% vs. 45.3% for hemorrhagic. Pre-implant predictors included female sex, heparin-induced thrombocytopenia, hypertension, and the intra-aortic balloon pump.

There is not agreement on the source of embolic strokes, but, misplaced hubris notwithstanding, I suspect that it is not originating from de novo pump thrombosis. There are two observations in support of this assertion: (1) pump thrombosis, even ones that are seized, are not significantly associated with strokes, and (2) the HeartMate III, which had only 1.7% incidence of thrombosis in the MOMENTUM trial, still had an incidence of 10.1% stroke. Although this was less than the HeartMate II control group, it is very close the INTERMACS stroke rate of 10.57%. This suggests that emboli are originating from inside the heart and passing through the large clearances of the HeartMate III. One possible intraventricular source is the outer surface of the inflow cannula and the interface of the myocardium with the inflow cannula. Various potential mitigations have been tried, most commonly, sintered titanium spheres. This treatment does appear to improve the interface with the myocardium, but has not been entirely effective.

Fig. 2.14 Evaheart proposed an inflow that does, indeed, remove the inflow cannula with sewing ring reminiscent of a mitral valve sewing ring. (Reprinted with permission from Evaheart, Inc., Houston, Texas)

From my surgical training, I retreat to the dictum "When in doubt, take it out". Evaheart proposed an inflow that does, indeed, remove the inflow cannula (Fig. 2.14) with a sewing ring reminiscent of a mitral valve sewing ring. Clinical trials have been approved, and I am anxious to see the results. If effective in reducing strokes, it would be a huge improvement and could expand the potential market for LVADs to class three patients.

Percutaneous Site Infection

The last big complication of LVADs is infection of the percutaneous exit site of the electrical driveline. The incidence of infections has improved as the diameter of the percutaneous wire has decreased, particularly when compared to the early pneumatic hoses, but it is still a significant problem, 23.8% for HM III, and results in hospital re-admissions.

It has been known and demonstrated that wireless transcutaneous power transmission has the potential to mitigate complications of percutaneous wires. These systems were first demonstrated in the early days of the NIHLB funding in 1977. Huge strides in implantable electronic devices such as implantable defibrillators and CRTs could be leveraged to significantly miniaturize a wireless system. Clinicians are very excited about the possibility, but I do not share their unbridled enthusiasm. There is a lot to consider.

Transcutaneous power would require additional implantable components including a secondary transformer, on board controller and battery. Additional conductors and connectors would need to be implanted. The battery would have to be surgically changed periodically. When existing external controllers fail, it is an easy thing to replace it at home. Implanted electronic failure would require operation if the patient was stable enough to make it to the hospital. Implanted components will increase the cost significantly, at least the order of magnitude of what is the going rate for IAD or CRT, approximately, $20,000. I believe there is an intermediate option.

We need only look to the past. A lot of research was done on developing a percutaneous skin button funded by NHLBI starting in 1980. The strategy of the skin button was to increase the area of the interface to distribute skin loading to prevent evulsion and tearing of the interface and to create a scaffold that encourage good ingrowth of skin into the skin button. A recent concept, Viaderm, is under development by Hbeat and shown in Fig. 2.15. A skin button could dramatically reduce the incidence of infection and avoid the increased complexity and cost of transcutaneous power transmission and preserve the advantage of an external controller.

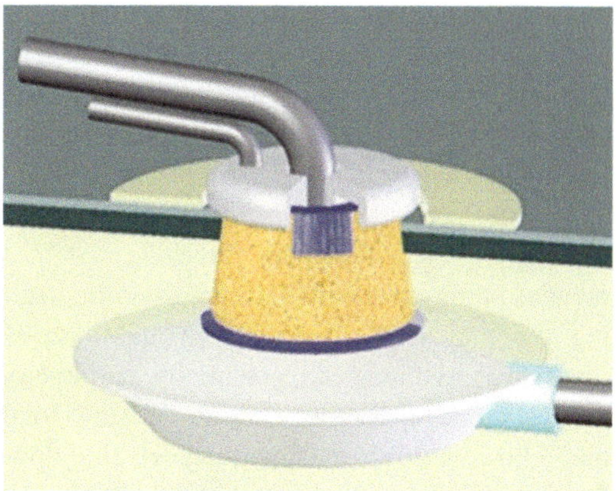

Fig. 2.15 A recent concept, Viaderm, is under development by Hbeat. (Reproduced with permission of Cardiac Assist Holding, LLC)

Pump Intelligence

Left ventricular assist devices (LVADs) have proven effective in patients with severe congestive heart failure (CHF) as bridge to cardiac transplantation and destination therapy. However, the hope for LVAD support as an effective myocardial recovery therapy, for which the diseased heart has healed with restored function enabling the device to be removed, rarely, occurs. The limited number of patients that have demonstrated myocardial recovery during LVAD support (less than 5%) raises the following questions:

Q1 – Are end-stage ventricles so damaged that most are just not recoverable?

Q2 – Are current medical management and/or LVAD control strategies ineffective?

Q3 – Do clinicians even attempt to wean patients?

Q4 – Could myocardial recovery and weaning be increased with intelligently controlled assistance?

Existing rotary pump LVADs employ speed control which can be adjusted by the clinician. However, there is no active control of pump flow or ventricular unloading. Pump flow is, passively, determined only by the pressure difference between the pump outlet and inlet. This means there is significantly more flow during systole when the gradient is near zero and much less during diastole when the gradient is approximately 80 mmHg. This may not be the optimal way to assist the heart during the various phases of assistance and remodeling, particularly, when attempting to wean a patient from support. The present strategy for weaning, if it is even attempted, is to lower the pump speed during a clinic visit using end-diastolic dimensions with echo as a metric. The patient is then sent home and followed at intervals, reevaluated, and the pump speed changed, down or up based on how well reduced assistance is tolerated. One problem with this strategy is the unintended consequence that reduced pump speed will decrease diastolic loading and may lead to increase in end-diastolic pressure and volume and, potentially, pulmonary congestion. If it were possible to decrease total pump flow while also maintaining diastolic flow. the risk of pulmonary congestion could be mitigated. By precisely and intentionally controlling the percentage of assistance occurring during diastole and systole of the heart, it should be possible to optimize assistance to each patient and each phase of assistance and possible recovery.

Reflections I am very fortunate to have been blessed with an extraordinary career – I never expected to surprise myself. Only my mother imagined I would accomplish the things of which I have been a part. I had a lot of help. Although there were many unknown heroes who made significant contributions during this journey, I would like to name a few.

Doctor O.H. (Bud) Frazier has been my colleague and dear friend during this long endeavor. He has been a believer in rotary blood pumps and worked with me and Robert Jarvik for many years. The importance of his clinical role in bringing rotary blood pumps to acceptance in the treatment of late stage heart failure cannot be overstated. The use of rotary blood pumps for MCS would have happened much later, or not at all, without him.

Leonard Golding was an early "true believer" whom I first met when I went to the Cleveland Clinic working on a pulsatile LVAD. His publications on the safety of non-pulsatile blood flow toppled a deeply held paradigm. He also did a great deal of work using non-pulsatile pumps for extracorporeal support and would have been involved with implantable LVADs if not for a stroke that cut his surgical career short. Not to be stopped, he became more active in research with the artificial organs laboratory resulting in the development of the CorAide LVAD that proved that blood lubricated hydrodynamic bearings could be safe and durable. I adapted the principle of hydrodynamic blood bearings the thrust bearings of the HeartWare HVAD.

Finally, I owe a lot to John Moise and Ken Butler of Nimbus Medical. They believed in my ideas on rotary pumps and the Hemopump and its potential back when most people thought it was crazy. They and I endured the bitter disappointment of the commercial failure of the Hemopump. In spite of this, they soldiered on against great financial and technical barriers, which I thought insurmountable, to develop the very clinically successful HeartMate II. The field of MCS would not be where it is today except for their vision and perseverance in the midst of a very difficult time.

Lessons Learned

To those continuing the journey to improve rotary blood pump MCSD, I offer the following thoughts and unsolicited advice:

1. A lot of what you have been told is not true. Always question your assumptions, particularly, the unconscious ones. I have wasted a lot of time going down blind alleys.
2. Trust your crazy ideas. However, be warned, ideas are a lot of trouble and are risky.
3. The invention or intellectual property is, at most, 20% of the formula of success. Without good management, finance, and regulatory, there is little hope of bringing even the best ideas to success. This is humbling for inventors and engineers that have the myopic view that the "thing" is everything.
4. "If you did not under estimate the work ahead of you, you'd never embark on any new journey." Henry Branson, former Chair of Surgery at the University of Pittsburg.
5. Fail fast, fail often, fail cheap.
6. Better is the enemy of good enough.
7. "Never, never, never give up." Winston Churchill

8. The following quote has been my mantra all these years;

   Whatever you can do,
   Or dream you can,
   Begin it,
   Boldness has genius, power and magic in it.
   Goethe

I am grateful for the honor and privilege to have been part of the journey to bring rotary blood pumps to clinical use. I am also grateful to have met a number of recipients of my inventions along the way. For me, it is deeply emotional and spiritual experience to see the impact this work has made on the lives of so many. As I visit with them, I often remember that long ago day, walking on a dusty road along the Nile River when it all began.

I wait to see what will happen next.
Richard Wampler, AKA Capt'n Hemo
Loomis California 2019

References

1. Saxton G, Andrews C. An ideal heart pump with hydrodynamic characteristics analogous to the mammalian heart. Trans Am Soc Artif Intern Organs. 1960;6:288–91.
2. Ottenberg R, Fox CL Jr. Rate of removal of hemoglobin from circulation and its renal threshold in human beings. Am J Phys. 1938;123:516.
3. Bernstein EF, Dorman FD, Blackshear PL Jr, Scott DR. Prolonged mechanical circulatory support: analysis of certain physical and physiologic considerations. Surgery. 1965;57(1):103–22.
4. Bernstein EF, Dorman FD, Blackshear PL Jr, Scott DR. An efficient, compact blood pump for assisted circulation. Surgery. 1970;68(1):105–15.
5. Zwart HH, Kralios A, Collan R, Kolff WJ. Transarterial closed chest left ventricular (TaCLV) bypass. Trans Am Soc Artif Intern Organs. 1969;15:386.
6. Bernstein EF, DeLaria GA, Johansen KH, Shuman RL, Stasz P, Reich S. Twenty-four hour left ventricular bypass with a centrifugal blood pump. Ann Surg. 1975;181(4):412–7.
7. Zwardt HH, Kralios A, Kwan-Gett CS, Backman DK, Foote JL, Andrade JD, Calton FM, Schoonmaker F, Kolff WJ. First clinical application of transarterial closed-chest left ventricular (TaCLV) bypass. Trans Am Soc Artif Intern Organs. 1970;16:386–91.
8. Rafferty EH, Kletschka HD, Inventors. Electrically driven pumps capable of use as heart pump. United States patent US 3,647,324. 1972 Mar 7.
9. Magovern GJ, Park SB, Maher TD. Use of centrifugal pump without anticoagulants for postoperative left ventricular assist. World J Surg. 1985;9(1):25–36.
10. Yada I, Golding LR, Harasaki H, Jacobs G, Koike S, Yozu R, Sato N, Fujimoto LK, Snow J, Olsen E, Murabayashi S, Venkatesen VS, Kiraly R, Nose Y. Physiopathological studies of nonpulsatile blood flow in chronic model. Trans Am Soc Artif Intern Organs. 1983;29:520–5.
11. Wampler RK, Moise JC, Frazier OH, Olsen DB. In vivo evaluation of a peripheral vascular access axial flow blood pump. Trans Am Soc Artif Intern Organs. 1988;23:450–4.
12. Frazier OH, Wampler RK, Duncan MJ, Dear WE, Macris MP, Parnis SM, Fuqua JM. First human use of the hemopump, a catheter mounted ventricular assist device. Ann Thorac Surg. 1990;49:299–304.
13. Wampler RK, Frazier O. The Hemopump™, the first intravascular ventricular assist device. ASAIO J. 2019;65(3):297–300.
14. Wampler RK, Frazier OH, Lansing AM, Smalling RW, Nicklas JM, Phillips SJ, Guyton RA, Golding AR. Treatment of cardiogenic shock with the Hemopump left ventricular assist device. Ann Thorac Surg. 1991;52:506–13.
15. Moise JC, Wampler RK, Butler KC, Inventors. Chronic ventricular assist system. United States patent US 4,908,012. 1990 Mar 13.
16. Jarvik RK. System considerations favoring rotary artificial hearts with blood-immersed bearings. Artif Organs. 1995;19(7):565–70.
17. Macris MP, Parnis SM, Frazier OH, Fuqua JM, Jarvik RK. Development of an implantable ventricular assist system. Ann Thorac Surg. 1997;63:367–70.
18. Butler KC, Maher TR, Borovetz HS, Kormos RL, Antaki JF, Kameneva M, Griffith BP, Zerbe T, Schaffer FD. Development of an axial flow blood pump LVAS. Am Soc Artif Intern Organs. 1992;38:m296–300.
19. Frazier OH, Delgado RM III, Kar BK, Patel V, Gregoric ID, Myers TJ. First clinical use of the redesigned HeartMate II left ventricular assist system in the United States. Tex Heart Inst J. 2004;31:157–9.
20. Miller LW, Pagani FD, Russel SD, John R, Boyle AJ, Aaronson KD, Conte JV, Naka Y, Mancici D, Delgado RM, MacGillivray TE, Farrar DJ, Frazier OH. Use of a continuous-flow device in patients awaiting heart transplantation. N Engl J Med. 2007;357(9):885–96.
21. Wieselthaler GM, Schima H, Hiesmayr M, Pacher R, Laufer G, Noon GP, DeBakey M, Wolner E. First clinical experience with the DeBakey VAD continuous-axial-flow pump for bridge to transplant. Circulation. 2000;101(4):356–9.
22. Christiansen S, Van Aken H, Breithardt G, Sheld H-H, Hammel D. Successful cardiac transplantation after 4 cases of DeBakey left ventricular assist device failure. J Heart Lung Transplant. 2002;21:706–9.
23. Golding LAR, Smith WA, Bodmann DR. Cleveland clinic rotodynamic pump program. Artif Organs. 1996;20(5):481–4.

24. Saeed D, Arusoglu L, Gazzoli F, Hetzer R, Morshius M, Alloni A, Vigano M, Koefer R, Golding LA, El Banayosy A. Results of the European clinical trial of Arrow CorAide left ventricular assist system. Artif Organs. 2013;37(2):121–7.
25. Slaughter MS, Pagani FD, McGee EC, Birks EJ, Cotts WG, Gregoric I, Frazier OH, Icenogle T, Najjar SS, Boyce SW, Acker MA, John R, Hathaway DR, Najarian KB, Aaronson KD. HeartWare ventricular assist system for bridge to transplant: combined results of the bridge to transplant and continued access protocol trial. J Heart Lung Transplant. 2013;13: 675–83.
26. VanderPluym CJ, Adachi IA, Niebler R, Griffiths E, Fynn-Thompson F, Chen S, O'Connor MJ, Machado D, Hawkins B, Bleiweis MS, Koehl DA, Cantor RS, Morales D, Lorts A. Outcomes of children supported with an intracorporeal continuous-flow left ventricular assist device. J Heart Lung Transplant. 2019;38(4):385–93.
27. Mehra MR, Goldstein DJ, Uriel N, Cleveland JC, Yuzefpolskaya M, Salerno C, Walsh MN, Milano CA, Patel CB, Ewald GA, Itoh A, Dean D, Krishnamoorthy A, Cotts WG, Tatooles AJ, Jorde UP, Bruckner BA, Estep JD, Jeevanandam V, Sayer G, Horstmanshof JW, Long JW, Gulati S, Skipper ER, O'Connell JB, Heatley G, Sood P, Naka Y. Two-year outcome with a magnetically levitated cardiac pump in heart failure. N Engl J Med. 2018;378(15):1386–95.
28. Kirklin JK, Xie R, Cowger J, Nakatani T, Schueler S, Taylor R, Lannon J, Mohacsi P, Gummert J, Goldstein D, Caliskan K, Hannan M. Second annual report from ISHLT mechanically assisted circulatory support registry. J Heart Lung Transplant. 2018;37:685–91.
29. Crow S, John R, Boyle A, Shumway S, Liao K, Colvin-Adams M, Toninato C, Missov E, Pritzker M, Martin C, Garry D, Thomas W, Joyce L. Gastrointestinal bleeding rates in recipients of nonpulsatile and pulsatile left ventricular assist systems. J Thorac Cardiov Surg. 2009;137(1):208–15.
30. Crow S, Chen D, Milano C, Thomas W, Joyce L, Piacentino V, Sharma R, Wu J, Arepally G, Bowles D, Rogers J, Villamizar-Ortiz N. Acquired von Willebrands syndrome in continuous flow ventricular assist device recipients. Ann Thorac Surg. 2010;90:1263–9.
31. Joly JM, el-Dabh A, Kirklin JK, Marshell R, Smith MG, Acharya D, Rajapreyar IN, Tallaj JA, Tresler M, Pamboukian SV. High atrial pressure and low pulse pressure predict gastrointestinal bleeding in patients with left ventricular assist device. J Card Fail. 2018;24(8):487–93.
32. Acharya K, Loyaga-Rendon R, Morgan CJ, Sands KA, Panboukian SV, Rajapreyar I, Holman WL, Kirklin JK, Jallaj JA. INTERMACS analysis of stroke during support with continuous-flow left ventricular assist devices: risk factors and outcomes. JACC Heart Fail. 2017;5(10):703–11.

History of Pediatric Devices for Mechanical Circulatory Support

3

Kurt A. Dasse and Priscilla C. Petit

Introduction

The evolution of mechanical circulatory support (MCS) systems over the past century reflects a magnificent series of historic breakthroughs in pediatric surgery and cardiology by courageous and innovative clinicians, scientists, and engineers. The road to current pediatric MCS systems required incredible teamwork, pushing the limits, and most importantly brilliant clinicians with the guts to trust their instincts to advance the field while often facing criticism, doubt, and resistance.

These early pioneers were obsessed with finding a means to perform surgery on the pediatric heart. They often suffered when they lost a child while advancing new technology and surgical solutions. These brilliant surgical pioneers made major advances, yet some stopped performing surgery forever after creating new procedures and technology due to the unbearable toll in terms of loss of lives along the way.

Nevertheless, their early efforts ended up saving the lives of thousands of pediatric patients and ultimately led to the devices and technology we share today. The history of the development of MCS devices is filled with passionate pursuits by these pioneers with many lessons learned. The purpose of this chapter is to share the evolution of breakthroughs that led to the mechanical circulatory support devices we have today for the surgical treatment of pediatric patients thanks to our forefathers and their courageous accomplishments.

Early History Leading to the Need for Pediatric Mechanical Circulatory Devices

The concept of creating an artificial heart for patients in end-stage heart failure was initially conceived in 1929 when Alexis Carrel teamed up with the famous aviator and engineer, Charles Lindbergh. Together, they demonstrated it was possible to keep an organ alive outside of the body using a perfusion machine [1]. Alexis Carrel was awarded the Nobel Prize for his contributions to vascular surgery and transplantation in 1912.

The path to developing mechanical circulatory support (MCS) systems began with the desperate need to somehow perform surgery on pediatric patients with congenital heart disorders. The discoveries that set the stage for performing pediatric cardiac surgery first required understanding

K. A. Dasse, PhD (✉)
University of Louisville Medical School, Inspired Therapeutics LLC, Cocoa, FL, USA
e-mail: kdasse@inspiredtherapeuticsllc.com

P. C. Petit, PhD
Quality and Regulatory Affairs, Inspired Therapeutics LLC, Cocoa, FL, USA
e-mail: ppetit@inspiredtherapeuticsllc.com

© Springer Nature Switzerland AG 2020
J. H. Karimov et al. (eds.), *Mechanical Support for Heart Failure*,
https://doi.org/10.1007/978-3-030-47809-4_3

the anatomic and physiologic basis for congenital cardiac anomalies. For example, Neils Stensen of Copenhagen, in 1671, described the pathology of a stillborn fetus with multiple congenital anomalies, which is now recognized as tetralogy of Fallot initially characterized by Etienne-Louis Fallot in 1888 [2]. Around the same time, Eduard Sandifort described the symptoms of a young child whom he called "blue boy." These, and many other early discoveries, served to create the basis for the emergence of cardiology and ultimately pediatric cardiac surgery.

Pediatric cardiology and surgery became a reality after 33-year-old Robert Gross ligated a patent ductus in 1938 [3]. He operated on a 7 years old girl to close a ductus. The surgery was performed at Brigham and Boston Children's Hospital without the knowledge of the chief surgeon who was opposed to performing the surgery. Gross trusted his instincts and successfully performed the surgery despite the protest.

Dr. Helen Taussig, a brilliant and determined cardiologist, believed if Dr. Gross could ligate a ductus, then it might be possible to construct a ductus. She asked Dr. Gross to consider the operation, but he declined. Alternatively, she found Dr. Alfred Blalock at Johns Hopkins and asked him to perform the operation. In November 1944, with the help of skilled surgical technician Vivien Thomas, Blalock amazed the surgical community and forever changed the life of a severely ill child with tetralogy of Fallot by anastomosing the left subclavian artery to the pulmonary artery. This procedure, known as the Blalock-Taussig operation [4], resulted in a total repair of tetralogy of Fallot in the young girl. Credit must go to Taussig because the procedure became reality due to her incredible instincts, insight, and persistence.

Dr. Denton Cooley was a 24-year-old surgical intern at Johns Hopkins at the time and was asked to assist in the operation [5]. His role was to administer the intravenous fluids. After the surgery was completed, he recalled seeing the baby turn from blue to pink. Unfortunately, the baby died within a few days. Nevertheless, Dr. Cooley believed the Blalock-Taussig procedure marked the dawn of heart surgery (Fig. 3.1).

Fig. 3.1 Dr. Denton Cooley. (Photo Credit: Texas Heart Institute)

Now that the anatomy of pediatric cardiac disorders was becoming clearer and surgery was becoming an option, the need for mechanical circulatory support became imperative. Dr. Cooley was the first to report the use of cardiopulmonary bypass to repair an aorticopulmonary septal defect [6].

Dr. O. Wangensteen, chief of surgery at the University of Minnesota, encouraged an Associate Professor of Surgery, Dr. Clarence Dennis, to develop an artificial heart-lung machine that would allow open-heart surgeries. Dr. Wangensteen was known to encourage the members of his department to do research. "They could not just be intellectual parasites using ideas and methods developed by others. Residents should contribute to the 'heritage' of the institution" [7].

A key breakthrough came about in 1952 when Dr. F. John Lewis, assisted by Lillehei, used hypothermia to perform the first successful pediatric open-heart surgery for correction of an atrial septal defect in a 5-year-old girl at the University of Minnesota [8, 9]. Before hypothermia, surgeons had approximately 5 minutes to perform a congenital cardiac surgical procedure until the patient might suffer brain damage. Lillehei and Lewis were keenly aware that every minute counted once they opened the heart.

Lillehei believed hypothermia was a major contribution, but he also believed cooling only provided the surgeon another 5 minutes for a total of 10 minutes to perform the operative procedure. Not all of Dr. Lewis's cases were successful, and he was known to go home after he lost a child and declined to see anyone except his wife who brought him cold compresses to help him with his pounding headaches [10].

Lillehei went on to create the cross-circulation procedure whereby he used a matching candidate such as a parent to supply the oxygenated blood directly to the pediatric patient undergoing surgery [11]. This pump technology used for the cross-circulation was also used in conjunction with the Lillehei-DeWall oxygenator [12]. He also advanced the "azygos flow concept" using the azygos vein to provide 10% of the flow to the heart. The first clinical use of cross-circulation was used for closure of a VSD in an 11-year-old boy in 1954 by Lillehei [13].

DeBakey is recognized for developing the roller pump while in medical school in 1934 [14]. Another faculty member wanted to find a pump for his laboratory that could be used to modify the pulse wave during laboratory experiments [15]. DeBakey was asked to help find a pump but could not find any references in the medical school library. One of his closest former classmates was studying engineering who advised him to do his research at the engineering library. DeBakey discovered that in the early nineteenth century, an article had been written about the use of rubber tubing being compressed to pump fluid. This article is what gave DeBakey the idea for the roller pump. Eventually he got two rollers to work by using adjustable rollers. If the rollers were adjustable, one could control the pressure exerted on blood, thus preventing hemolysis. DeBakey intended his original idea for the roller pump to be used for blood transfusions (Fig. 3.2).

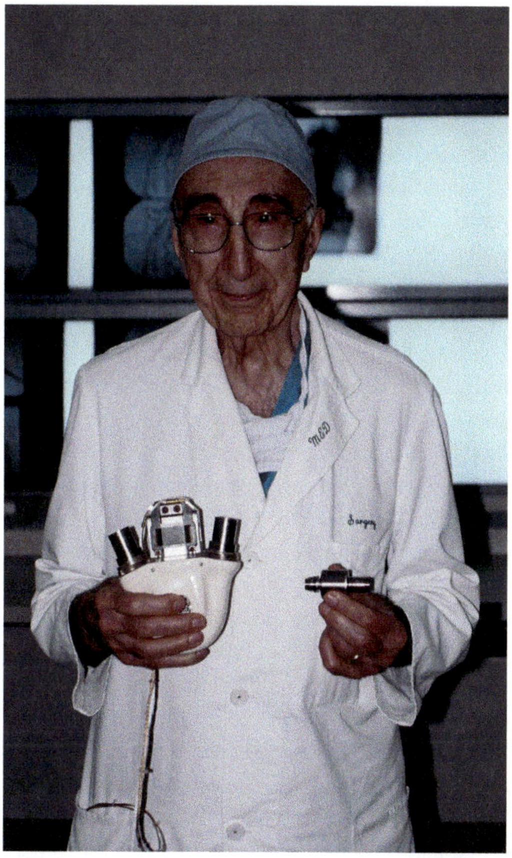

Fig. 3.2 Dr. Michael DeBakey. (Photo Credit: NASA/public domain: https://commons.wikimedia.org/wiki/File:Michael_DeBakey.jpg)

During the early 1940s before World War II, surgical pioneers like Dr. John Gibbon, Jr. had been working on a heart-lung machine but needed a pump. DeBakey met Dr. Gibbon at a medical conference in the 1930s where they discussed the concept of a heart-lung machine. DeBakey told Dr. Gibbon about the idea for his roller pump and suggested that he had a pump that he might be able to use. This meeting with Dr. Gibbon led to the implementation of the roller pump for cardiopulmonary bypass machines that are used to this day. In 1953, while at the Jefferson Medical School in Philadelphia, Dr. Gibbon reported the first successful use of extracorporeal circulation using an oxygenator, but he abandoned the method after five subsequent failures [16, 17]. He placed a moratorium on the use of the cardiopulmonary bypass machine and never performed another case.

Walt Lillehei went on to report the results from the first 32 patients he treated with ventricular septal defects, tetralogy of Fallot, and atrioventricular defects using the cross-circulation technique [11]. Despite encouraging results, he was criticized for this technique because it placed two individuals at risk instead of one. The risks associated with the procedure appeared to outweigh the benefits, so Lillehei continued his quest for a better means of supporting the pediatric patient during heart surgery.

Meanwhile, Dr. Clarence Dennis had been conducting research to develop a "rolling screen oxygenator" at the University Minnesota since 1946. He had been observing the progressive deterioration of young patient whom he suspected suffered from an atrial septal defect. He elected to use his oxygenator to perform surgery on the patient in 1951 [18]. Upon opening the heart, he was surprised by the amount of blood filling the heart and that she had multiple defects. He was unable to save her. Nevertheless, he persisted and was able to use the oxygenator to perform successful surgery in 1955.

Lillehei had been following a young female patient when she died before he could perform the surgery [10]. After her death, he went to see the pathologist and asked to examine her heart. When he opened her heart, he saw the defect and repaired it in a simple manner. This experience appeared to strengthen his resolve to create new surgical techniques to operate on pediatric patients.

Early open-heart surgery was fraught with challenges in the 1940s. There were inaccurate diagnoses, a lack of knowledge regarding the pathology, and limitations for how to rid the heart of enough blood to perform the surgery.

Lillehei was intent upon operating on the heart to repair congenital defects. He traveled to Boston to meet with Dr. Gross. Gross had become a legend by this time because of his innovative closed-heart surgery. He was reportedly guarded when Lillehei came to visit and believed a cardiopulmonary bypass machine would have limited usefulness.

Despite these concerns and the known potential for air embolism, the DeWall-Lillehei helix bubble oxygenator was developed and used clinically beginning in May 1955. The oxygenator was evaluated in a large series of patients and ultimately became the device of choice for open-heart operations.

There is an interesting myth about how the concept of the bubble oxygenator came to into existence. Lillehei was well-known to go to a local bar after a long day of surgery. One evening he was out with DeWall and a few colleagues, and according to lore, one of the folks came up with the idea for the bubble oxygenator after seeing bubbles in his beer [10].

Lillehei was intrigued with the research done by his colleague, Clarence Dennis, and the concept of a cardiopulmonary bypass machine. However, he was struck by the complexity of Dennis's machine and the fact that it took so many clinicians to operate it. Lillehei continued his pursuit of finding the ideal CPB machine.

Dr. William Norwood developed a stage-one procedure for transposition of the great arteries that allowed many of these children to survive [19]. Glenn created the second phase of the procedure in 1958 by creating the Glenn shunt [20]. Fontan from Paris developed an operation to bypass the right side of the heart, which became a successful palliative procedure for tricuspid atresia [21]. Transposition of the great arteries, previously a highly lethal condition, became treatable in the newborn with hope for long-term survival.

In 1966, William Rashkind of Philadelphia introduced balloon atrial septostomy [22]. This therapeutic intervention was carried out in newborn infants with transposition of the great arteries allowing surgery to be performed on stable infants. Dr. Rashkind has been called the father of interventional pediatric cardiology due to his many significant contributions.

Diagnostic advances needed to be developed to successfully perform pediatric surgery. Edler and Hertz described the use of ultrasound for imaging of the heart in 1954 [23]. Additional advances included the creation of angiography, cardiac ultrasound, color flow Doppler, and magnetic resonance imaging. Clinical teams comprised of pediatric surgeons, cardiologists, pathologists, physiologists, and intensivists were organized and trained to successfully operate and medically manage children with cardiac problems. The stage was set for technological advances that would enable longer-term support of pediatric patients in need of mechanical circulatory support.

Key Events Leading Up to Pediatric ECMO

Perhaps the earliest pioneer to think about performing gas exchange in the human was Robert Hooke who lived from 1635 to 1703 who conceptualized the notion of an oxygenator [24]. The first extracorporeal perfusion of a dog was carried out by Brukhonenko and Tchetchulin in 1928 [25].

Willem Kolff noted that blood became oxygenated when passed through the cellophane chambers of an artificial kidney in 1944 raising awareness that artificial gas exchange was a possibility [26]. In 1955, at the Mayo Clinic, Kirklin improved upon Gibbon's device and treated eight patients with cardiopulmonary bypass [27]. The development of the first membrane oxygenator, by Clowes, enabled prolonged cardiopulmonary bypass to become feasible [28].

In 1965, Rashkind and coworkers were the first to use a bubble oxygenator to support a neonate dying of respiratory failure [29]. Dorson and colleagues also reported the use of a membrane oxygenator for cardiopulmonary bypass in infants [30]. In 1970, Baffes et al. reported the successful use of extracorporeal membrane oxygenation to support infants with congenital heart defects who were undergoing cardiac surgery [31].

Dr. Donald Hill reported the first adult survivor on ECMO in 1972 after saving a man who was dying from acute respiratory insufficiency [32]. The patient was a young man suffering from the adult respiratory distress syndrome (ARDS). A multicenter clinical trial of extracorporeal circulation for adults with ARDS was organized by the National Institutes of Health in 1975.

A milestone in the history of pediatric ECMO was reached in 1975 when Bartlett and associates pioneered the treatment of infants in acute, reversible respiratory failure [33]. A child was admitted to Dr. Bartlett's service who had aspirated a large quantity of meconium and developed a severe chemical pneumonia. The respiratory dysfunction was considered so hopeless that there appeared to be nothing to lose but to try ECMO. Dr. Bartlett brought in a machine from his laboratory and explained to the mother through an interpreter what they were going to attempt. The mother disappeared abandoning the baby unsure of the outcome and afraid she might be arrested and possibly deported [34]. The nurses named the baby Esperanza, which is a Spanish feminine name meaning "hope" or "expectation." After 3 days of bypass, the child completely recovered. After Esperanza recovered, Dr. Bartlett continued to treat infants with ECMO who were dying of respiratory insufficiency. Unlike the high mortality observed in adults, 75% of the babies recovered.

In 1978 Kolobow and Gattinoni described using extracorporeal circulation to remove carbon dioxide [35]. The extracorporeal circulation was used to decrease the risk of barotrauma due to overventilation with a ventilator.

In 1979, a randomized controlled trial was conducted on adult patients with severe acute respiratory failure. A 90% mortality rate was reported for patients on ECMO. For several years, ECMO was used mostly for neonatal and pediatric patients with only a small number of highly specialized centers pursuing ECMO in adult patients. Several other clinical trials were conducted, but all of them were criticized for flawed designs. Excellent reviews of these clinical trials are available [36, 37].

Starting in 1989, treatment with ECMO was advanced by the Extracorporeal Life Support Organization (ELSO) [38], an International Consortium of Health Care Centers. Participating centers voluntarily contribute detailed data to the ELSO registry. The ELSO organization creates guidelines for the participating centers. While the original indication for use was for patients in acute respiratory failure, the use of ECMO expanded to include support of patients with chronic respiratory failure as a bridge to transplant, to support the failing heart, as well as for cardiopulmonary support for organ donation after determination of death.

Technological advances in ECMO over the past three decades have improved the safety and effectiveness of the therapy decreasing morbidity and improving survival of patients requiring ECMO support. Improvements in ECMO have been made with superior pumps, cannulae, and by knowledge gained for how to better anticoagulate the patient and extend the useful life of the circuit.

Evolution of Pediatric Ventricular Assist Devices

The NHLBI artificial heart program was established in 1964. The program called for the development of short- and long-term circulatory assist devices as well as cardiac replacement pumps.

The first clinical application of a pneumatically driven ventricular assist device (VAD) is attributed to DeBakey in 1966, in which a 37-year-old woman was successfully supported for 10 days with a paracorporeal circuit following complex cardiac surgery [39]. In 1967, DeBakey and associates used a left atrial auxiliary ventricular assist device in a 16-year-old girl after postoperative cardiac failure after a mitral valve replacement [40]. Long-term mechanical circulatory support was attempted with a pulsatile membrane pump by Cooley in 1969 [41].

In June of 1966 at Maimonides Hospital in Brooklyn, New York, Dr. Adrian Kantrowitz had the opportunity to perform the first human heart transplant in the world which would have been before Christiaan Barnard and Norman Shumway [42]. Dr. Kantrowitz had performed over 100 animal experiments in preparation for performing this historic transplant. The donor was an anencephalic newborn with a normal beating heart. However, a senior pediatrician and an anesthesiologist in the operating room were opposed to harvesting the heart until the heart of the newborn stopped beating. Dr. Kantrowitz argued that a newborn without a brain had no life. Despite his plea to harvest the heart before it stopped, the hospital insisted the heart could only be removed after the clinical death of the donor. Kantrowitz could only watch the baby die. And, when the heart was ultimately examined, it was blue and unable to be revived. He elected to abort the transplant and wait for another opportunity to perform world's first heart transplant (Fig. 3.3).

Meanwhile, on December 3, 1967 in Cape Town, Africa, Dr. Christiaan Barnard performed the world's first human heart transplant. The donor was a young brain-dead female that had suffered two skull fractures. The recipient was a male with severe end-stage heart disease. The recipient lived 18 days after the transplant.

Dr. Adrian Kantrowitz went on to perform the first pediatric heart transplant on December 6, 1967, 3 days after Barnard's transplant. This was the second heart transplant in the world and the first to be performed in the United States. He had the conviction and courage to overcome the skepticism and resistance of those around him when he performed the surgery. He transplanted

Fig. 3.3 Dr. Adrian Kantrowitz. (Photo Courtesy of Jean Kantrowitz)

the heart from a 2-day-old anencephalic baby to a boy dying from congestive heart failure. The recipient died 6 hours after the transplant. Nevertheless, this marked the beginning of heart transplantation in the United States.

Dr. Kantrowitz is credited along with his brother, Arthur Kantrowitz, a physicist, for inventing the intra-aortic balloon pump (IABP) and for performing the first clinical use the balloon pump in 1968 [43]. Dr. Kantrowitz also performed one of the earliest LVAD cases in the United States [44].

The first centrifugal pump used for cardiopulmonary bypass was produced by Bio-Medicus Inc. in 1976 (Medtronic Bio-Medicus, Inc., Eden Prairie, MN) [45]. Medtronic agreed to buy Bio-Medicus Inc. for $190 million in a bid to diversify its line of cardiovascular medical devices [46].

The MEDOS/HIA DeltaStream was an extracorporeal rotary blood pump that was also used for pediatric support [47]. MEDOS products are made in Germany and date back to 1987. The DeltaStream has been used as both a cardiac assist device and for ECMO.

Dr. William S. Pierce, James Donachy, and Dr. Gerson Rosenberg from Penn State Milton S. Hershey Medical Center developed the first pneumatic, sac-type heart-assist pump in the early 1970s [48]. The Pierce-Donachy ventricular assist device, more recently known as the Thoratec pneumatic ventricular assist device (Thoratec PVAD, Pleasanton, CA), was used to support nearly 4000 patients for right, left, and biventricular support including pediatric patients. The initial testing of the Pierce-Donachy VAD started in 1973. Penn State performed the first implant in 1976. Food and Drug Administration approval was granted in 1980, and, by 1985, the Penn State heart-assist pump was in widespread use.

Pioneers including Dr. Michael DeBakey and Clarence Dennis played a key role in working with the National Institutes of Health (NIH) to create the artificial heart program. NIH invited proposals to develop an implantable, integrated, electrically powered left heart-assist system in 1980. The goal was to develop a viable system that could allow patient mobility.

David Lederman, an aerospace engineer, founded Applied Biomedical Corporation in 1981, now called Abiomed. The Abiomed BVS 5000 was the first extracorporeal VAD approved by FDA in 1992. The BVS 5000 was successfully used to support larger children [49].

Peer Portner, Ph.D., born in Mombasa, Kenya, began his career as a senior physicist at Andros Inc., a research and development firm headquartered in Berkeley, California. He began experimenting to develop an LVAD during the 1970s. By 1984, he developed the Novacor LVAD [50]. The first recipient of the Novacor LVAD was 51-year-old male who was supported with the pump for 8 days until he underwent cardiac transplantation. Novacor devices had been implanted in approximately 1800 patients by the time of Portner's death in 2009.

Dr. O. Howard (Bud) Frazier implanted one of the first LVADs in a pediatric patient in 1987 [51]. He reported using an extracorporeal Bio-Medicus BP-80 device for 12 hours to support a 9-year-old boy as a bridge to transplant. Dr. Frazier also discussed the concept of using non-pulsatile devices for ventricular assistance. Dr. Frazier and his team blazed the trail for the majority of the future ventricular assist devices used for adult and pediatric support (Fig. 3.4).

Thermo Cardiosystems Incorporated (TCI), a publicly traded subsidiary of Thermo Electron Corporation, was founded in 1988. Victor Poirier served as the President and CEO of TCI with

Fig. 3.4 Dr. O. Howard "Bud" Frazier. (Photo Credit: Texas Heart Institute)

Dasse serving as Senior Vice President. The first-generation positive displacement pump was the HeartMate Implantable Pneumatic (IP) LVAS. The pump had a diaphragm and porcine valves, both of which were designed to flex approximately 40 million times per year. The HeartMate IP LVAS was approved via a Premarket Approval (PMA) in 1993 after decades of research and development.

TCI went on to develop the HeartMate XVE LVAS, a vented electric (VE) version of the device. The XVE LVAD incorporated a set of follower bearings that rotated against a helical cam, which converted rotary motion into a linear motion like a piston. The actuator flexed a diaphragm like the IP device, and the pump also had valves. The devices were implanted in patients with a body surface area (BSA) as low as 1.5 m^2 which included larger pediatric patients and small women.

In addition to the developments in the United States, the Toyobo and Zeon pumps were developed in Japan in 1990 [52]. They were successfully used to support pediatric patients as well as adults.

Meanwhile, Dr. Richard Wampler along with Ken Butler and the Nimbus team of engineers developed the Hemopump [53]. Wampler described conceiving of the concept for the Hemopump after observing an Archimedes screw-type pump on a trip to Egypt. This development represented a radical departure from the positive displacement pumps that were in use clinically and became the first step toward demonstrating clinical utility of continuous flow designs.

There were great concerns regarding the risk of hemolysis with a small rotating impeller racing at 27,000 rpm. There were other concerns regarding the lack of pulse compared to the pulsatile pumps. This sparked great debate about the need for a pulse. Many debates centered around pulsatile versus non-pulsatile pumps. The term non-pulsatile was eventually replaced with continuous flow when the MCS community later learned that continuous flow pumps allow the native pulse to exist providing the speed of the device is kept low enough to allow the aortic valve to open and shut. Future developments included operating the continuous flow pumps in a pulsatile manner.

The Hemopump was first used in 1988 at the Texas Heart Institute by Dr. O. Howard Frazier to support a 61 years old man, a heart transplant recipient who experienced an acute episode of rejection and cardiogenic shock. The pump successfully sustained him for 46 hours. The device was evaluated in seven additional patients by Dr. Frazier in the 1990s [54]. The Hemopump was later shown to be successful for short-term circulatory support for the pediatric population. Dr. Wampler is credited as being the father of the rotary blood pump revolution. The development of the Hemopump set the stage for the next generation of continuous flow pumps that followed.

Wampler, Butler, and John Moise went on to create the second-generation VAD, HeartMate II, through a collaborative effort between Nimbus and Pittsburgh's McGowan Center for Organ Engineering and the Department of Biomedical Engineering at the University of Pittsburgh. Dr. Harvey Borovetz, the lead biomedical engineer at the University of Pittsburgh, along with Drs. Bartley Griffith and Robert Kormos, cardiac surgeons, worked together with Nimbus to improve the bearing design and oversee the animal studies.

Nimbus was acquired by Thermo Cardiosystems (TCI) in 1996. TCI brought the expertise to take the Nimbus device and finalize the design for commercialization. The device was branded as HeartMate II by TCI. Poirier initially acquired HeartMate II and the rights to the intellectual property for HeartMate III as an insurance policy in the event that one of the devices failed to be proven safe and effective. Both devices went on to be successfully commercialized.

This second generation of continuous flow blood pumps, rotary pumps with mechanical bearings and/or seals (but no diaphragm or valves), entered the clinical arena with promising results. Additional examples of second-generation pumps included the Microbe DeBakey (Dr. Michael DeBakey), Terumo DuraHeart™ (Dr. Chisato Nojiri), Jarvik 2000™ (Dr. Robert Jarvik), CorAide (Dr. Leonard Golding), and the

Berlin Heart Incor™ VADs (Robert Kroslowitz). Among these devices, the MicroMed DeBakey, Jarvik 2000, and HeartMate II were evaluated for pediatric VAD support.

A landmark clinical trial, the Randomized Evaluation of Mechanical Assistance for The Treatment of Congestive Heart Failure (REMATCH), was conducted which provided the scientific data for FDA to approve the HeartMate Vented Electric assist device for destination therapy in 2002 [55]. Medicare approved the device for reimbursement in 2003 as a long-term implantable device. The stage was set for multiple clinical trials of devices to evaluate long-term durable mechanical circulatory support for both adult and pediatric applications.

Third-generation continuous flow pumps, rotary pumps without mechanical bearings, include both axial flow and centrifugal flow devices. Third-generation centrifugal devices included extracorporeal devices such as the CentriMag and PediMag devices developed by Levitronix, implantable devices such as HeartMate III which was co-developed by Levitronix and Thoratec, and the HVAD LVAS developed by Dr. Richard Wampler for HeartWare. Additional devices included HeartQuest™ (Dr. James Long) and the VentrAssistTM (Dr. John Woodward).

Levitronix LLC was founded by Kurt Dasse and George Hatsopoulos from the United States. Additional founders included Drs. Reto Schoeb, Natalie Barletta, and Thomas Gempp from Switzerland. Whereas TCI had acquired the intellectual property for the implantable magnetically levitated technology for ventricular assistance from Sulzer in Switzerland, Levitronix acquired the rights to the extracorporeal applications. The Levitronix team developed the first extracorporeal magnetically levitated centrifugal pump to enter the market. The pump was designed to provide hemodynamic support as a bridge to decision for up to 30 days until the patient recovered, underwent transplantation, or required to a longer-term implantable pump.

The CentriMag® pump was designed to support adult and pediatric patients with weights ranging from 25 to 40 kg. Levitronix further developed the PediMag® pump, branded as PediVAS outside the United States, designed to support pediatric patients weighing 25 kg or less. Both devices continue to be widely used throughout the world for ventricular assistance and ECMO and have been recognized to be cost-effective when used for these therapies [56].

Berlin Heart GmbH was originally founded as Mediport Kardiotechnik GmbH by the German Heart Institute Berlin in 1996. Berlin Heart AG was created after merging with Mediport Kardiotechnik GmbH in 2000. The US subsidiary, Berlin Heart Inc., was established in Texas in 2005 with Robert Kroslowitz serving as the president and chief executive officer. Berlin Heart developed the implantable INCOR VAD and the pneumatically driven pulsatile extracorporeal EXCOR VAD in the early 1990s. The pediatric-size EXCOR VAD was introduced into clinical practice by the German Heart Institute in Berlin by Roland Hetzer, MD, Ph.D., in 1992 [57]. The first child in the United States was supported by EXCOR Pediatric in 2000. To date, Berlin Heart has the only PMA-approved device in the United States and is the only device of its kind available for infants and children with severe heart failure. EXCOR Pediatric received FDA approval for the US market in 2011.

Abiomed purchased AG of Aachen, Germany, in 2005. The Impella 2.5 heart pump developed by Impella CardioSystems and marketed by Abiomed has been used to support pediatric patients [58, 59]. Abiomed received FDA 510(k) clearance for its Impella 2.5 heart pump for partial circulatory support for periods of up to 6 hours during cardiac procedures not requiring cardiopulmonary bypass in 2008. In March 2015, the Impella 2.5 heart pump received FDA Premarket Approval (PMA) for elective and urgent high-risk PCI procedures.

Another extracorporeal centrifugal device, TandemHeart (CardiacAssist, Inc., Pittsburgh, PA), was first used at the Texas Heart Institute in 2005 [60]. The device is used as a short-term VAD that employs peripheral cannulation using a 21 Fr transseptal venous cannula (62 or 72 cm in length). The cannula is too large to use in smaller children (neonates and pediatric patients). Therefore its use is limited to larger adolescents.

Adachi and Jaquiss recently provided an excellent overview of the experience with currently used pediatric VAD and ECMO devices [61]. The interest in pediatric MCS systems has significantly increased this past decade largely due to the number of children with end-stage heart failure and the lack of donors. Approximately 400 pediatric heart transplants are performed worldwide per year, and the length of stay in the hospital has increased along with the waiting time for transplantation. There is a growing need for improved pediatric MCS technology.

Almond et al. reported that a third of the pediatric patients that were transplanted after being bridged with ECMO died during the same admission [62]. Despite the substantial research leading to safer and more effective ECMO therapy, there remains a need for alternative MCS strategies for bridging children to heart transplantation.

Dr. Timothy Baldwin has led the effort at the National Institutes of Health to establish the PumpKIN (Pumps for Kids, Infants, and Neonates) Program to facilitate the development of novel pediatric MCS devices [63]. Excellent devices were developed under the PumpKIN program including the PediaFlow™ VAD (developed by The University of Pittsburgh, LaunchPoint Technologies Inc. and World Heart), the Penn State Pediatric VAD, the PediPL extracorporeal device (a pediatric pump/lung developed by Levitronix and University of Maryland), and Ension's ECLS device. These devices were developed under the PumpKIN program but were not commercialized largely due to lack of corporate financing.

One device continues to be funded by the PumpKIN Program which is the Jarvik 2015 implantable LVAD. Preclinical results for the Jarvik 2015 LVAS were published in 2015 [64]. The Jarvik 2015 LVAS received Investigational Device Exemption (IDE) approval by the Food and Drug Administration in 2016. Clinical sites have been established, and a pilot trial is underway. Meanwhile, the pediatric results with the Berlin Heart EXCOR have steadily improved due to advances in anticoagulation, and optimal medical management has been implemented.

Promising new devices are under development designed to support pediatric patients. Dr. Mark Rodefeld is developing the viscous impeller pump (VIP), a dynamic pump with kinetic energy continuously imparted to the fluid [65]. The device is intended for use in failed Fontan patients. The pump is optimized for low-pressure augmentation over a broad range of flow conditions and does not require rate adjustment to change performance. Vadovations is also developing a small (size of a triple AAA battery) mixed flow pump that may ultimately be used for pediatric patients [66]. These and other devices under development may serve as the next generation of devices that advance the field of MCS one more step toward safer and more effective devices specifically designed for pediatric patients.

Conclusions

The passion to learn how to surgically operate on pediatric patients with congenital anomalies paved the way toward creating MCS devices as we know them today. The early pioneers, like Gross, Lillehei, Dennis, DeBakey, Gibbon, Cooley, Kantrowitz, Blalock, and Taussig, made history by being bold and committed to seeing into the heart, making more accurate diagnoses and developing surgical techniques to repair the tiny cardiac defects that threatened the life of their pediatric patients.

Each decade has brought new, innovative solutions for oxygenators with better materials, optimized surface area, active mixing, and know-how to enhance gas exchange. We have witnessed the evolution from positive displacement to continuous flow pumps with elimination of valves, diaphragms, and mechanical bearings. Each decade has brought us closer to viable technology and intelligence for how to adequately support the pediatric patient. However, the ideal device remains to be developed. Further improvements in the materials, hemocompatibility, hydrodynamics, and control schemes for the pumps must be developed to reduce the risk of thrombosis, bleeding, and infection, adverse effects that still plague the industry.

Recent advances with implantable sensors may lead to optimal control schemes. Alternative means for powering the devices through innovative percutaneous or transcutaneous energy transmission systems may reduce infection. Improvements for how to better anticoagulate pediatric patients are being made. New hemocompatible coatings are under development, and basic research continues to uncover the molecular basis for risk factors for thrombosis.

Enhancements in imaging continue to be made to aid in the visualization of defects and treatment options for pediatric patients. Studies are in progress to help visualize optimal fit and fixation of a pediatric VAD before attempting to implant a device. Novel pump designs continue to emerge, and when combined with the tools described above, it is inevitable better devices will become available. It will not only require better devices, it will also require optimized patient selection criteria, medical therapy, and standardized patient and caretaker training and education.

There are significant challenges in developing acceptable ventricular assist devices for the pediatric population that differ from those facing adults. Anticoagulation poses a particularly difficult challenge for pediatric patients treated with a VAD. Cannulation and the orientation of the inflow and outflow conduits of implantable VADs are critical when interfacing with the different anatomic anomalies. How do implantable devices accommodate growth and weight changes of the pediatric patient? How well will implantable devices hold up to the rigorous activity of young patients? Will pediatric patients perceive a reasonable quality of life living with an implantable VAD? One fact we know for sure is that for a pediatric device to be considered a success, it will be necessary to reduce the adverse event rates for these patients.

Perhaps the next generation of percutaneous, endovascularly placed VADs will prove suitable for the larger pediatric patients. New devices specifically designed to support patients with single ventricle disease (Fontan failure) may extend survival and quality of life for these patients. A scaled-down extracorporeal pump with a portable/wearable controller may prove suitable for pediatric transplant candidates with relatively short waiting times providing the risk of decannulation is mitigated. New devices like these are under development with the hope that each innovation associated with their development will advance the field and improve the quality of life, safety, and survival of pediatric patients in need of MCS.

References

1. Dutkowski P, de Rougemont O, Clavien PA. Alexis Carrel: genius, innovator and ideologist. Am J Transplant. 2008;8(10):1998–2003.
2. Dabbagh A, Conte AH, Lubin L. Congenital heart disease in pediatric and adult patients: anesthetic and perioperative management. Cham: Springer; 2017.
3. Randolph JG, Robert E. Gross lecture notes on the early development of pediatric surgery in the United States. J Pediatr Surg. 2012;47:10–6.
4. Gott VL. And it happened during our lifetime. Ann Thorac Surg. 1993;55(5):1057–64.
5. Cooley AC. 100,000 hearts: a surgeon's memoir. 1st ed. Austin: Dolph Briscoe Center for American History, University of Texas at Austin; 2012.
6. Cooley DA, McNamara DR, Latson JR. Aorticopulmonary septal defect: diagnosis and surgical treatment. Surgery (St. Louis). 1957; 42:101.
7. Braile PM, de Godoy F. Paths to Cardiology. Braz Arch Cardiol. 1996;66:1.
8. Lewis FJ, Taufic M. Closure of atrial septal defects with the aid of hypothermia: experimental accomplishments and the report of one successful case. Surgery (St. Louis). 1953;33:52.
9. Lillehei CW. The Society Lecture. European Society for Cardiovascular Surgery Meeting, Montpellier, France, September 1992. The birth of open-heart surgery: then the golden years. Cardiovasc Surg. 1994;2(3):308–17.
10. Miller GW. King of hearts. The true story of a maverick who pioneered open heart surgery. New York: Crown Publishers; 2000.
11. Lillehei CW, Cohen M, Warden HE, Varco RL. The direct vision intracardiac correction of congenital anomalies by controlled cross circulation. Results in thirty-two patients with ventricular septal defects, tetralogy of Fallot, and atrioventricularis communis defects. Surgery. 1955;38:11–29.
12. DeWall RA, Warden H, Read RC, Gott V, Ziegler R, Varco R, Lillehei CW. A simple, expendable, artificial oxygenator for open heart surgery. Surg Clin N Am. 1956;4(36):1025–34.
13. Coher M, Lillehei CW. A quantitative study of the azygos factor during vena caval occlusion in the dog. Surg Gynecol Obstet. 1954;98:225–32.

14. DeBakey M. A simple continuous-flow blood transfusion instrument. New Orleans Med Surg J. 1934;87:386–9.
15. DeBakey Archives at the National Library of Medicine: Interview by Dan Schanche, Tape #5 – Side A, Pierre Hotel, New York.
16. Gibbon JH. Application of a mechanical heart and lung apparatus to cardiac surgery. Minn Med. 1954;37:171–85.
17. Murtra M. The adventure of cardiac surgery. Eur J Cardiothorac Surg. 2002;21:167–80.
18. Dennis C, Spreng DS, Nelson GE, et al. Development of a pump oxygenator to replace the heart and lungs: an apparatus applicable to human patients, and application to one case. Ann Surg. 1951;134:709–21.
19. Norwood WI, Lang P, Hansen DD. Physiologic repair of aortic atresia hypoplastic left heart syndrome. N Engl J Med. 1983;308:23–6.
20. Glenn WWL. Circulatory bypass of the right side of the heart: IV. Shunt between superior vena cava and distal right pulmonary artery: report of clinical application. N Engl J Med. 1958;259:117.
21. Fontan F, Baudet E. Surgical repair of tricuspid atresia. Thorax. 1971;26:240–8.
22. Rashkind WJ, Miller WW. Creation of an atrial septal defect without thoracotomy. A palliative approach to complete transposition of the arteries. JAMA. 1966;196:991–2.
23. Edler I, Hertz CH. Use of ultrasonic reflectoscope for continuous recording of movement of heart walls. Kungl Fysiogr Sallsk Lund Forhandl. 1954;24:5–9.
24. Lim MW. The history of extracorporeal oxygenators. Anesthesia. 2006 Oct;61(10):984–95.
25. Brukhonenko SS, Tchechulin SI. Experiments on isolation of dog's head. Trudi Nauchnogo Khimiko-Pharm Inst. 1928;20:7–43.
26. Kolff WJ, Berk TJ. Artificial kidney: dialyzer with great area. Acta Med Scand. 1944;117:121–34.
27. Kirklin JW, DuShane JW, Patrick RT, Donald DE, Hetzel PS, Harshbarger HG, Wood HE. Intracardiac surgery with the aid of a mechanical pumpoxygenator system (gibbon type): report of eight cases. Mayo Clin Proc. 1955;30:201–6.
28. Clowes GHA Jr, Neville WE. Further development of a blood oxygenator dependent upon the diffusion of gases through plastic membranes. Trans Am Soc Artif Intern Organs. 1957;3:52–8.
29. Rashkind WJ, Freeman A, Klein D, Toft RW. Evaluation of a disposable plastic, low volume, pumpless oxygenator as a lung substitute. J Pediatr. 1965;66:94–102.
30. Dorson W Jr, Baker E, Cohen ML, Meyer B, Molthan M, Trump D. A perfusion system for infants. Trans Am Soc Artif Intern Organs. 1969;15:155–60.
31. Baffes T, Patel K, Jegathesan S. Total cardiopulmonary bypass with the Lan-dé-Edwards membrane oxygenator. Am J Cardiol. 1972;29:672–7.
32. Hill JD, O'Brien TG, Murray JJ, et al. Extracorporeal oxygenation for acute post-traumatic respiratory failure (shock-lung syndrome): use of the Bramson Membrane Lung. N Engl J Med. 1972;286:629–34.
33. Bartlett RH, Gazzaniga AB, Jefferies R, et al. Extracorporeal membrane oxygenation (ECMO) cardiopulmonary support in infancy. Trans Am Soc Artif Intern Organs. 1976;22:80–8.
34. Wolfson P. The development and use of extracorporeal membrane oxygenation in neonates. Presented at the symposium, "Gibbon & His Heart-Lung Machine: 50 Years & Beyond;" 2003 May 2; Philadelphia. 2003 Dec; 76(6). p. S2224–9.
35. Kolobow T, Gattinoni L, Tomlinson T, White D, Pierce J, Iapichino G. The carbon dioxide membrane lung (CDML): a new concept. Trans Am Soc Artif Intern Organs. 1977;23:17–21.
36. ELSO Registry Publications. Available from: https://www.elso.org/Publications.aspx.
37. Sidebotham D. Extracorporeal membrane oxygenation—understanding the evidence: CESAR and beyond. J Extra Corpor Technol. 2011;43(1):23–6.
38. Combes A, Pesenti A, Ranieri VM. Fifty years of research in ARDS. Is extracorporeal circulation the future of acute respiratory distress syndrome management? Am J Respir Crit Care Med. 2017;195(9):1161–70.
39. De Bakey ME, Liotta D, Hall CW. Left-heart bypass using an implantable blood pump. Mechanical devices to assist the failing heart, vol. 1283. Washington, DC: National Academy of Sciences–National Research Council; 1966. p. 223–9.
40. DeBakey ME. Left ventricular bypass pump for cardiac assistance. Am J Cardiol. 1971;27:3–11.
41. Cooley DA, Liotta D, Hallman GL, Bloodwell RD, Leachman RD, Milam JD. Orthotopic cardiac prosthesis for two-staged cardiac replacement. Am J Cardiol. 1969;24:723–30.
42. McRae D. Every second counts. The race to transplant the first human heart. New York: Berkley Books; 2006.
43. Kantrowitz A, Tjønneland S, Freed PS, et al. Initial clinical experience with intra-aortic balloon pumping in cardiogenic shock. JAMA. 1968;203(2):113–8.
44. Phillips S, Jaron D, Freed P, Zorzi G, Aris A, Kantrowitz A. Hemodynamic studies with a permanently implanted left ventricular assist device. Am J Cardiol. 1972;29(2):285–6.
45. Lynch MF, Peterson D, Baker V. Centrifugal blood pumping for open-heart surgery. Minn Med. 1978;61:536–7.
46. Los Angeles Times. "Medtronic Agrees to Acquire Bio-Medicus". 1990 June 9; Associated Press.
47. Huang SC, Chi NH, Chen CA, Chen YS, Chou NK, Ko WJ, Wang SS. Left ventricular assist for pediatric patients with dilated cardiomyopathy using the Medos VAD cannula and a centrifugal pump. Artificial organs. Abstract from the 5th international conference on pediatric mechanical circulatory support systems & pediatric cardiopulmonary perfusion; 2009 May 27–30. Dallas.

48. Pierce WS, Donachy JH, Landis DL, Brighton JA, Rosenberg G, Migliore JJ, Prophet GA, White WJ, Waldhausen JA. Prolonged mechanical support of the left ventricle. Circulation. 1978;58(3 Pt 2):133–46.
49. Throckmorton AL, Allaire PE, Gutgesell HP, Matherne GP, Olsen DB, Wood HG, Allaire JH, Patel SM. Pediatric circulatory support systems. ASAIO J. 2002;48(3):216–21.
50. Jassawalla JS, Daniel MA, Chen H, Lee J, LaForge D, Billich J, Ramasamy N, Miller PJ, Oyer PE, Portner PM. In vitro and in vivo testing of a totally implantable left ventricular assist system. ASAIO Trans. 1988;34(3):470–5.
51. Frazier OH, Bricker JT, Macris M, Cooley DA. Use of a left ventricular assist device as a bridge to transplant in a pediatric patient. Tex Heart Inst J. 1989;16:46–50.
52. Takano H, Nakatani T. Ventricular assist systems: experience in Japan with Toyobo pump and Zeon pump. Ann Thorac Surg. 1996;61(1):317–22.
53. Butler KC, Moise JC, Wampler RK. The Hemopump--a new cardiac prothesis device. IEEE Trans Biomed Eng. 1990;37(2):193–6.
54. Frazier OH, Wampler RK, Duncan JM, Dear WE, Macris MP, Parnis SM, Fuqua JM. First human use of the Hemopump, a catheter-mounted ventricular assist device. Ann Thorac Surg. 1990;49(2):299–304.
55. Rose EA, Moskowitz M, Packer M, et al. The REMATCH trial: rationale, design, and end points. Ann Thorac Surg. 1999;67:723–30.
56. Borisenko O, Wylie G, Payne J, et al. The cost impact of short-term ventricular assist devices and extracorporeal life support systems therapies on the National Health Service in the UK. Interact Cardiovasc Thorac Surg. 2014;19:141–8.
57. Hetzer R, Alexi-Meskishvili V, Weng Y, Hübler M, Potapov E, Drews T, Hennig E, Kaufmann F, Stiller B. Mechanical cardiac support in the young with the Berlin Heart EXCOR pulsatile ventricular assist device: 15 years' experience. Semin Thorac Cardiovasc Surg Pediatr Card Surg Annu. 2006;9:99–108.
58. Dimas VV, Murthy R, Guleserian KJ. Utilization of the Impella 2.5 micro-axial pump in children for acute circulatory support. Catheter Cardiovasc Interv. 2014;83(2):261–2.
59. Murthy R, Brenes J, Dimas VV, Guleserian KJ. Ringed polytetrafluoroethylene (Gore-Tex) tunneled "chimney" graft for pediatric use of Impella 2.5 axial flow pump. J Thorac Cardiovasc Surg. 2014;147(4):1421–2.
60. Kar B, Adkins L, Civitello A, Loyalka P, Palanichamy N, Gemmato C, Myers T, Gregoric I, Delgado R. Clinical experience with the TandemHeart® percutaneous ventricular assist device. Tex Heart Inst J. 2006;33(2):111–5.
61. Adachi I, Jaquiss R. Mechanical circulatory support in children. Curr Cardiol Rev. 2016;12(2):132–40.
62. Almond CS, Singh TP, Gauvreau K, Piercey GE, Fynn-Thompson F, Rycus PT, Bartlett RH, Thiagarajan RR. Extracorporeal membrane oxygenation for bridge to heart transplantation among children in the United States: analysis of data from the Organ Procurement and Transplant Network and Extracorporeal Life Support Organization Registry. Circulation. 2011;123(25):2975–84.
63. Baldwin JT, Borovetz HS, Duncan BW, Gartner MJ, Jarvik RK, Weiss WJ, Hoke TR. The national heart, lung, and blood institute pediatric circulatory support program. Circulation. 2006;113(1):147–55.
64. Baldwin JT, Adachi I, Teal J, Almond CA, Jaquiss RD, Massicotte MP, Dasse K, Siami FS, Zak V, Kaltman JR, Mahle WT, Jarvik R. Closing in on the PumpKIN trial of the Jarvik 2015 ventricular assist device. Semin Thorac Cardiovasc Surg Pediatr Card Surg Annu. 2017;20:9–15.
65. Rodefeld M, Frankel SH, Giridharan G. Cavopulmonary assist: empowering the univentricular Fontan circulation. Semin Thorac Cardiovasc Surg Pediatr Card Surg Annu. 2011;14(1):45–54.
66. Coghill PA, Kanchi S, Azartash-Namin Z, Long JW, Snyder TA. Benchtop von Willebrand factor testing: comparison of commercially available ventricular assist devices and evaluation of variables for a standardized test method. ASAIO J. 2018. https://doi.org/10.1097/MAT.0000000000000849. [Epub ahead of print].

4 Overview of Mechanical Circulatory Support Devices and Concepts

Juan Marcano, Aladdein Mattar, and Jeffrey A. Morgan

Abbreviations

CHF	Congestive heart failure
ECMO	Extracorporeal membrane oxygenation
EKG	Electrocardiography
FDA	Food and Drug Administration
IABP	Intra-aortic balloon pump
LV	Left ventricle
LVAD	Left ventricular assist device
LVEF	Left ventricular ejection fraction
MCS	Mechanical circulatory support
PAPi	Pulmonary Artery Pulsatility Index
PCI	Percutaneous coronary intervention
VA ECMO	Venous-Arterial Extracorporeal membrane oxygenation

J. Marcano, MD · A. Mattar, MD
Baylor College of Medicine – Texas Heart Institute, Houston, TX, USA
e-mail: juan.marcano@bcm.edu; aladdein.mattar@bcm.edu

J. A. Morgan, MD (✉)
Division of Congestive Heart Failure, Transplant and Mechanical Circulatory Support, Department of Cardiothoracic Surgery, Baylor College of Medicine – Texas Heart Institute, Houston, TX, USA

Introduction

Mechanical circulatory support (MCS) is not a new concept but instead an evolving technology since Dr. Michael DeBakey implanted the first left ventricular assist device in 1966; the device was developed by Dr. Domingo Liotta in the Research Labs of the Texas Heart Institute at St. Luke's Episcopal Hospital in Houston.

The significant advances in anticoagulation therapy, surgical technique, device designs, imaging, and postoperative care resulted in better, safer, and more accessible devices that sometimes can even be placed at patient's bedside with minimal amount of equipment and considerable low risk of complications.

Today, the general indications for MCS are ischemic or nonischemic acute decompensated ventricular failure, acute on chronic congestive heart failure, refractory hypotension after cardiac surgery, acute dilated myocardiopathy, right ventricular failure, after heart or lung transplantation due to refractory hypotension, and right ventricular dysfunction.

There are different types of mechanical circulatory support:

Non-durable, percutaneous.

- *Intracorporeal.*
  - Pulsatile.
- Intra-aortic balloon pump.

© Springer Nature Switzerland AG 2020
J. H. Karimov et al. (eds.), *Mechanical Support for Heart Failure*,
https://doi.org/10.1007/978-3-030-47809-4_4

- Short-term ventricular support, less than 4 days.
  – Non-pulsatile, axial flow. Micro-axial flow devices (Impella pump).
- Short-term and medium-term support, 4 to 6 days.
- *Extracorporeal.*
- Non-pulsatile, centrifugal flow.
- TandemHeart, extracorporeal membrane oxygenation (ECMO).
- Short-term and medium-term support for tandem and prolonged times for ECMO.

Durable, implantable.

- Non-pulsatile flow. HeartWare, HeartMate III. Long-term ventricular support.

The Intra-aortic Balloon Pump (IABP)

(Maquet Holding B.V & Co. KG, Germany)

The IABP is a percutaneous device that constitutes the simplest and more widely available form of mechanical circulatory support (MCS) and consists of an intra-aortic counterpulsation balloon capable of augmenting the luminal pressure during diastole increasing the coronary (by increasing the aortic root pressure), cerebral, and systemic circulation. When the balloon is deflated, the luminal pressure decreases or afterload which therefore decrease the left ventricle work.

The presystolic deflation lowers the resistance to the systolic ejection, therefore decreasing the myocardial work and oxygen demand with a more favorable supply/demand ratio. The expected increase in the cardiac output is between 0.5 and 1.0 L per minute [1].

The concept of the counterpulsation balloon is not new, and it was proposed by Moulopoulos et al. in 1962 [2], and the first reported implantation goes back to June 29, 1967, when Kantrowitz et al. were called to evaluate a 45-year-old female patient who was "comatose, anuric, cold, and cyanotic" and whose blood pressure was unobtainable. The IABP was placed in and the patient was supported for about 7 hours, her condition stabilized, and the device was removed. That patient completely recovered and was discharged from hospital [3].

The device its designed to the balloon to be inflated during diastole and its synchronized to the EKG, also has a sensor that can trigger the balloon based in the blood pressure. Although the initial studies used carbon dioxide, currently helium is used as shuttle gas for sufficient speed and to assure appropriate timing [3]. Additionally, helium is rapidly absorbed in case of accidental balloon rupture.

The IABP can be used as support in patients with cardiogenic shock (defined as cardiac index <2.2 liters/min/m^2 and systolic blood pressure less than 90 mmHg) [4] secondary to reversible cause like acute coronary syndrome, right after percutaneous coronary intervention to increase coronary perfusion [5], as well as bridge therapy from patients with acute on chronic heart failure in cardiogenic shock to other forms of more durable left ventricular support and ultimately heart transplant.

Sintek et al. suggested that the more severe the left ventricular dysfunction is, the less benefit from IABP support [6], and Fried et al. in a recent publication found that in a selected group of chronic CHF with cardiogenic shock the IABP was associated with high likelihood of bridge to durable left ventricular assist device of heart transplant or discharge without escalation and that poor right ventricular function as measured by the Pulmonary Artery Pulsatility Index (PAPi) as well as ischemic cardiomyopathy can be associated with clinical deterioration with IABP therapy [7].

The most common contraindications for IABP placement are the presence of moderate to severe aortic insufficiency, severe peripheral vascular disease, large aortic aneurysm, and the presence of aortic stents. The most common complications are migration, bleeding, limb ischemia with need for either exchange or removal and vascular repair, as well as balloon rupture [8].

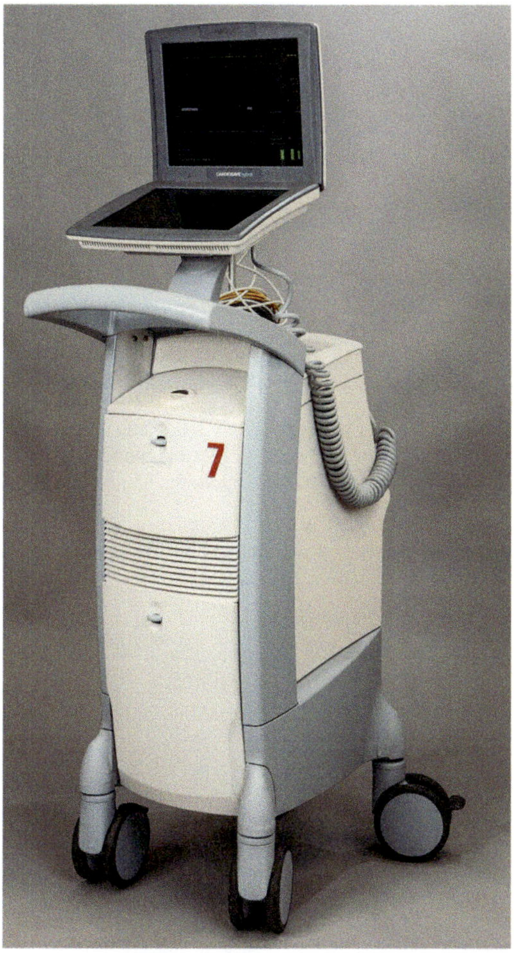

Fig. 4.1 Intra-aortic balloon pump cart. (Maquet Holding B.V & Co. KG, Germany)

The equipment consists of two parts: the catheter and the console. The flexible catheter is equipped with the balloon and two ports that allow pressure monitoring and the delivery of the gas to the balloon (Fig. 4.1). The command console has the monitor and computer system, the helium reservoir, and the cycling pumping system. The console is mobile which makes transport easier, and in cases like axillary access, the patient can ambulate with the device in place (Fig. 4.2).

Removal of the device can be safely performed at bedside if no vascular complications were noted. Direct pressure is recommended for at least 25 min. The use of temporary compression devices is adequate as long as there is appropriate control of the bleeding at the insertion site.

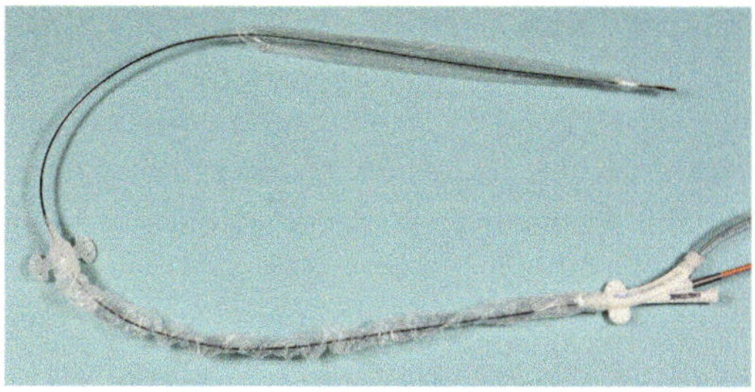

Fig. 4.2 Intra-aortic balloon pump catheter. The counterpulsation balloon can be seen in the distal end of the catheter. (Maquet Holding B.V & Co. KG, Germany)

The Impella Pump

(Abiomed INC, Denver, Massachusetts, United States)

The Impella is a percutaneous trans-valvular micro-axial pump that generates continuous flow based on the Archimedes screw pump. This is a very ingenious and ancient concept that according to historians is attributed to Archimedes (287–212 B.C.) and seen during one of his visits to Egypt where it was used for irrigation and to remove water from the ships. Later on, the idea was introduced in the ancient Greece for multiple purposes [9].

The mechanism consists in a helicoid screw, and with each turn, the screw propels a constant volume determined by the angle of inclination, diameter, and lead or pitch of the screw. The main advantage of this design is that it can flow large amount of volume with relatively low rotational speed (Fig. 4.1).

The main advantages of the device are the unloading of the left ventricle with improvement of the hemodynamics and myocardial oxygen consumption and subsequent improvement of the end-organ perfusion [10].

In the market there are different types of devices available [11].

Impella 2.5

The 2.5 pump is approved for temporary ventricular support and indicated during high-risk percutaneous intervention (<6 hours) and short-term support for cardiogenic shock (4 days or less). The catheter's diameter is 9 Fr with flow up to 2.5 L, and it has a 12 Fr pump motor; it is normally inserted through femoral approach.

Impella CP

This is FDA approved for short-term ventricular support in cardiogenic shock (4 days or less). The catheter's diameter is 9 Fr with flow up to 4.5 L, and it has a 14 Fr pump motor. It is normally inserted through femoral approach.

Impella 5.0

This is approved for mid-term ventricular support in cardiogenic shock (6 days or less). The catheter's diameter is 9 Fr with flow up to 5 L, and it has a 21 Fr pump motor. It can be inserted in the femoral or axillary artery. For the axillary artery, the placement is surgical and requires the use of an 8–10 mm vascular graft, usually to the right axillary artery. Main advantage of the axillary access is the ability of the patient to ambulate which improves recovery and decreases the risk of deconditioning.

The system consists of the catheter with micro-axial pump and the control console. The distal end of the catheter is placed in the left ventricle, away from the mitral valve apparatus where the inflow will be located in the ventricle and the outflow in the ascending aorta (Figs. 4.3, 4.4, 4.5, and 4.6).

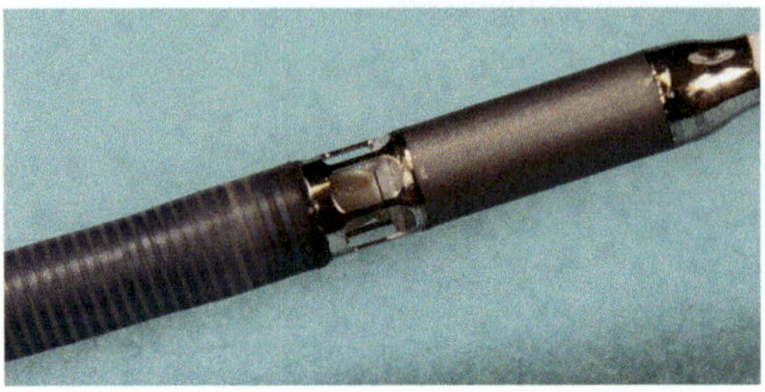

Fig. 4.3 Impella pump catheter. Distal end shows the inlet with the propulsion mechanism based on an Archimedes screw. (Abiomed INC, Denver, Massachusetts, United States)

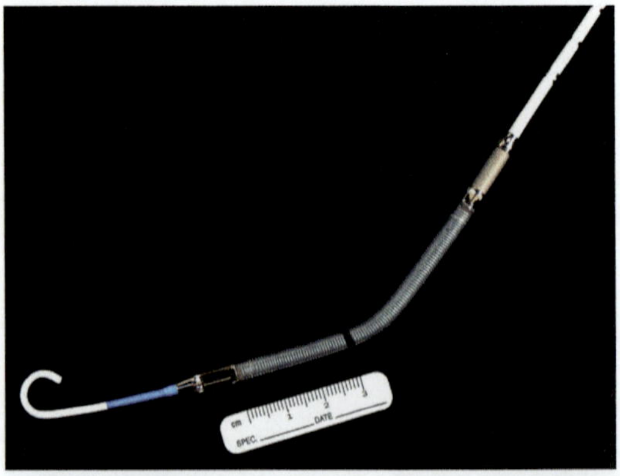

Fig. 4.4 Impella pump catheter 2.5 L. Catheter has three main parts: inflow (1), outflow (2), and pump motor (3). (Abiomed INC, Denver, Massachusetts, United States)

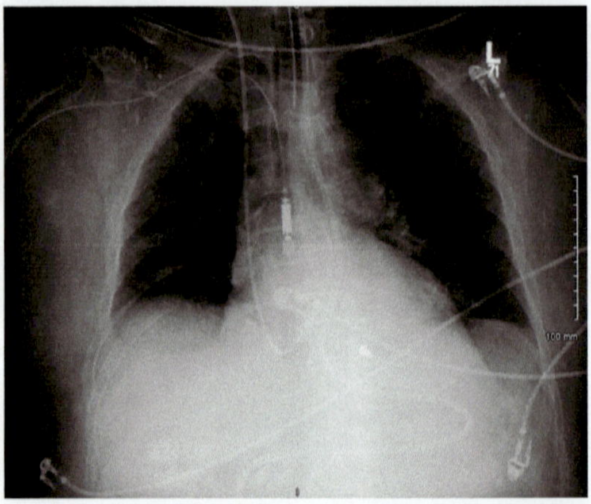

Fig. 4.5 Chest X-ray showing Impella 5.0 in place, right axillary approach. Catheter has three main parts: inflow (1), outflow (2), and pump motor (3). (Abiomed INC, Denver, Massachusetts, United States)

The PROTECT II trial (448 patients, prospective, randomized trial of hemodynamic support with Impella 2.5 versus IABP in patient undergoing high-risk PCI) showed that the 30-day incidence of major adverse events was not different for patients with IABP or Impella 2.5 hemodynamic support. However, trends for improved outcomes were observed for Impella 2.5-supported patients at 90 days [12]. Additional data coming from the same study and recently published by Goldstein and colleagues reported preserved aortic and mitral valve function with the Impella 2.5 during transthoracic echocardio-

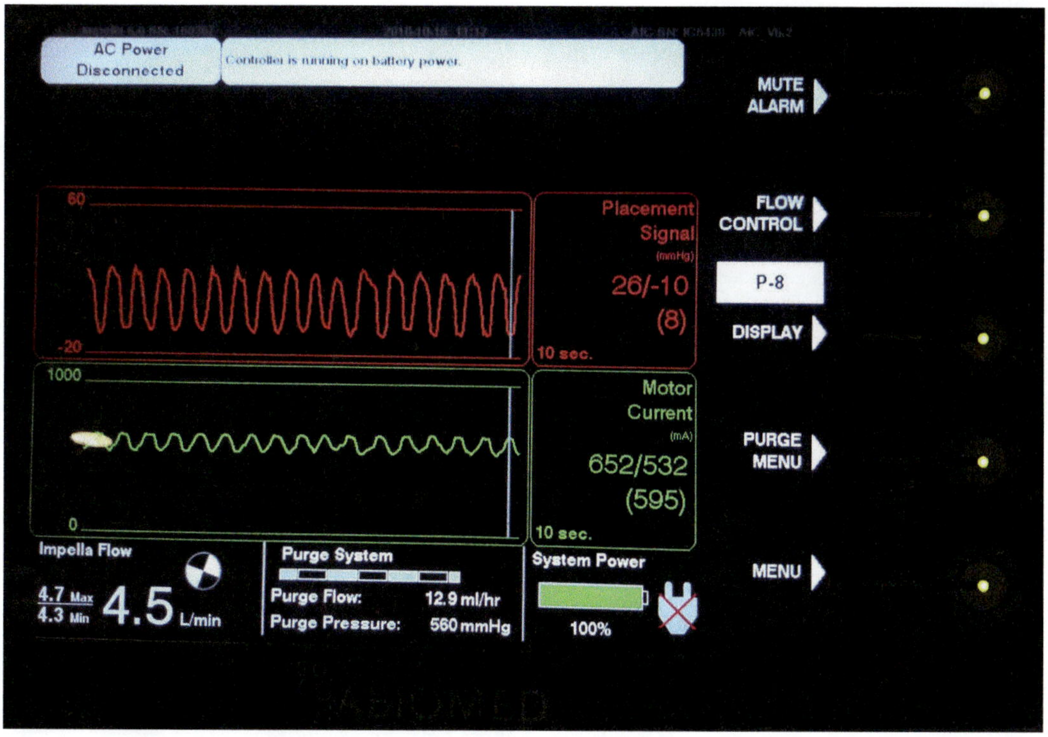

Fig. 4.6 Impella pump console screen. Main screen shows information related to system pressure, motor current, pump flow, and power. In this case, power setting is 8 (P-8) with 4.5 L of flow. (Abiomed INC, Denver, Massachusetts, United States)

gram performed at baseline, 1 month, and 3 months and also found an average of 22% relative increase in the LVEF compared with baseline [13].

Recent animal studies showed that the LV unloading with the Impella can improve coronary flow and infarct zone perfusion in ischemic heart by decreasing the end-diastolic wall stress and improving microvascular flow [14].

The Impella is considered a percutaneous left ventricular assist device (LVAD) and can be used for temporary support as well as bridge for destination therapy (transition to LVAD implantation) in patient with end-stage heart failure.

The main contraindications for placement are the presence of mechanical aortic prosthesis, left ventricular thrombus, and severe peripheral vascular disease, and the main complications are extremity ischemia, insertion site bleeding, and hemolysis.

TandemHeart

(Tandem-Life, Pittsburgh, Pennsylvania, United States)

The TandemHeart device is a non-durable, percutaneous centrifugal flow pump that is designed to flow blood from the left atrium and transfer pressurized blood to the aorta via the femoral artery; this allows to decompress the left ventricle and improve the systemic blood pressure. The pump can unload up to 90% of the left ventricle with flows up to 5 LPM, keeping pressures of 90 mmHg through the entire cardiac cycle, and produces 0.8 to 0.9 watts of cardiac power. The device requires the percutaneous catheter to be introduced to the left atrium through the atrial septum which is technically more demanding than other modalities of percutaneous mechanical support [15] (Fig. 4.7).

This device works independently of the heart rate and native cardiac function, but severe right

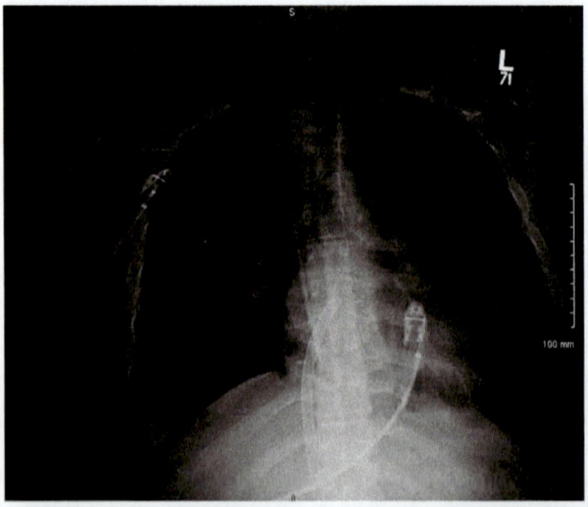

Fig. 4.7 Chest X-ray showing TandemHeart transseptal outflow cannula in place (arrow). (Tandem-Life, Pittsburgh, United States)

ventricular failure is a contraindication for left-sided heart placement due to the poor left atrial filling.

The system consists of four parts: the pump, the transseptal cannula (has a curved distal tip with 14 holes for blood drainage), the outflow arterial cannula, and the main console where parameters like flow, speed, and line pressure can be monitored [15] (Figs. 4.7, 4.8, and 4.9).

Main indications for use are cardiogenic shock and high-risk PCI. The device is commonly implanted in the Cath Lab and usually requires a large-sized cannula as the flow can be limited by the size of the outflow cannula (up to 19 Fr).

Kar et al. demonstrated in a cohort of 117 patient with both ischemic and nonischemic severe refractory cardiogenic shock already in either pressor and/or support with IABP improvement of cardiac index, systolic blood pressure and mixed venous saturation as well metabolic perfusion markers as creatinine and lactic acid with after TandemHeart implantation with average support of 5.8 ± 4.75 days [16]. Additionally, there is small case series reporting improving survival in patients with acute coronary syndrome complicated with ventricular septal defects in cardiogenic shock, especially with early implantation [17].

Main contraindications for use are severe peripheral vascular disease, ventricular septal defect, the presence of thrombus in any of the atria, aortic insufficiency, and right ventricular failure as stated before. Severe bleeding is a contraindication as well, as anticoagulation is required to avoid pump thrombosis.

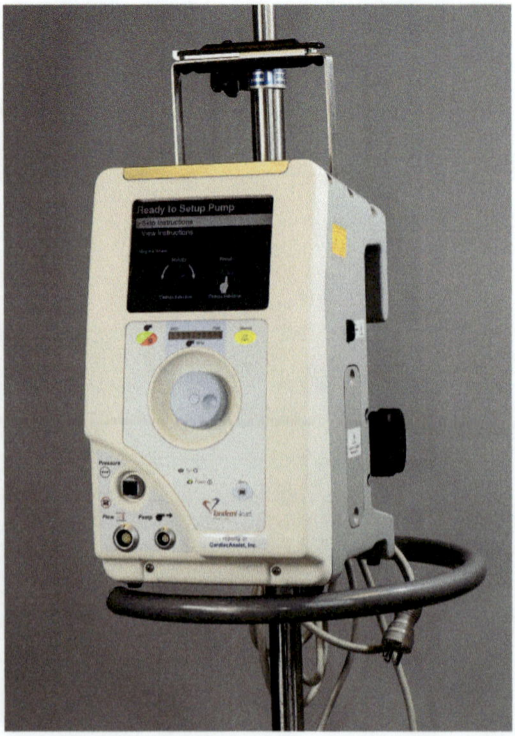

Fig. 4.8 TandemHeart console. (Tandem-Life, Pittsburgh, United States)

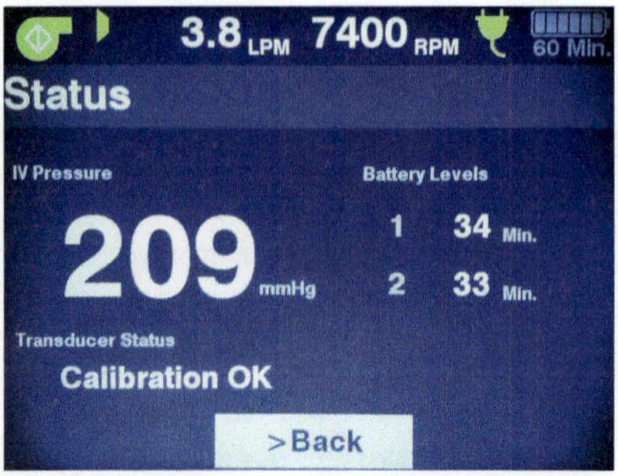

Fig. 4.9 TandemHeart console screen (Tandem-Life, Pittsburgh, United States) main view that shows IV pressure, pump speed, and pump flow. In this case, pump speed is 7400 RPM with flow of 3.8 LPM

Extracorporeal Membrane Oxygenation (ECMO)

The extracorporeal membrane oxygenation is a very powerful modality of mechanical support that depending on the configuration used (venous-arterial versus venous-venous) can support critical ill patient with cardiac and/or respiratory failure.

The concept of ECMO is not new and its efficacy is readily documented [18–21].

ECMO is a technology that involves taking deoxygenated blood from the patient through an outflow cannula, pumping it through a membrane oxygenator, and then returning it to the patient through an inflow cannula. Typical configurations are veno-venous (VV) and veno-arterial (VA), which provide respiratory and combined respiratory and circulatory support, respectively. In VV ECMO, blood is withdrawn from the central venous system (CVS) and then passed through a pump and oxygenator system to oxygenate the blood. The oxygenated blood is returned to the right atrium via CVS. This can be achieved with either a dual-site or a single-site configuration. In VA ECMO, blood is withdrawn from a central vein and returned to a large-caliber peripheral or central artery, providing both hemodynamic and respiratory support. The critical distinction is the status of the right ventricle and heart and if cardiac support is necessary.

VA ECMO support uses an artery to deliver oxygenated blood (inflow) to the body, bypassing the pulmonary circulation, while a vein is used to deliver deoxygenated blood (outflow) to the ECMO oxygenator. In a recently described mobile ECMO model, a shorter percutaneous venous outflow cannula (22–24 French) is placed in the internal jugular or subclavian vein to achieve the so-called sport model. The arterial cannula is placed directly into the axillary artery or the innominate artery usually with a tube graft [22].

The system is composed of the main console, the oxygenator, and the inflow/outflow cannulas (Fig. 4.10).

The most common indication for ECMO support is cardiogenic shock, either ischemic or nonischemic. In cases of nonischemic, fulminant myocarditis and sepsis-associated cardiomyopathy are strong indications.

Recently, pulmonary hypertension and right-sided heart failure have become other indications, due to the capacity to unload the right ventricle. In cardiac surgery, is not unusual in in patients who are unable to immediately be weaned from cardiopulmonary bypass secondary to post-cardiotomy shock.

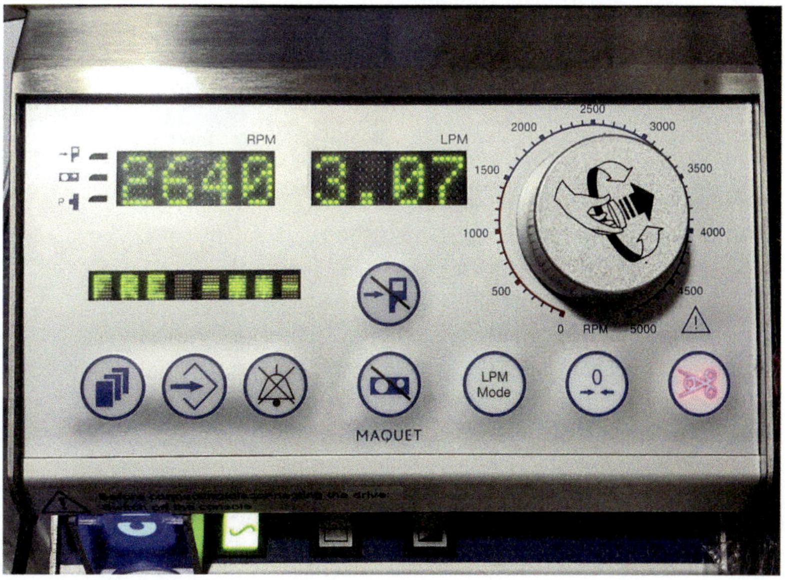

Fig. 4.10 ECMO console. The roller pump speed and subsequent flows can be controlled using a hand knob. In this patient speed is 2640 RPM with flow of 3 liters. (Maquet Holding B.V & Co. KG, Germany)

Finally, VA ECMO is used with a great success as a bridge to left ventricular assist device (LVAD) implantation or cardiac transplantation in patients with terminal heart failure [23, 24].

The most common complications are bleeding and circuit thrombosis that always carries a risk for embolism. Adequate anticoagulation is a key element to prevent thrombotic events and can be associated with increased risk of disseminated intravascular coagulation and heparin-induced thrombocytopenia. On the other hand, the risk of infection is a frequent hazard associated with VA ECMO. Groin cannulation is associated with increased risk of limb ischemia, and antegrade reperfusion cannula is strongly recommended [25, 26].

HeartMate III

(Thoratec Corporation, Pleasanton, California)

The HeartMate 3 (HM3) is the next generation of continuous flow and a significant change in the technology compared with the predecessor the HeartMate 2 (HM2).

The evolution of the implantable flow pumps goes from very large-sized and pulsatile devices with many bearing and complex mechanisms to progressively smaller axial continuous flow with more simple and elegant designs combining hydraulic and electromagnetic forces to totally magnetically levitated centrifugal continuous flow pump as the HeartMate 3.

Multiple studies already showed the improved survival and decreased incidence of adverse effects in patients when LVAD are used as bridge for recovery or transplantation [27].

This particular pump design allows the HM3 to decrease hemolysis due to consistent flow pathway and intrinsic pump pulsatility secondary to the levitational movement and the lack of bearing or other mechanical parts, in theory, should increase durability as decreased the mechanical wear [28].

Also, the pump allows to measure the pulsatility index (PI), whose calculation according to the manufacturer [29] represents cardiac pulsatility and is defined as the magnitude of the flow pulse when the left ventricle contracts that causes an increase in pump flow during cardiac systole. PI values typically range from 1 to 10. In general, the magnitude of the PI value is related to the amount of assistance provided by the pump. Higher values indicate more ventricular filling and higher pulsatility (i.e., the pump is providing less support to the left ventricle). Lower values indicate less ventricular filling and lower pulsatility (i.e., the pump is providing greater support and further unloading the ventricle).

The device system has four parts: the pump, the driveline, the pocket controller, and the batteries. The axial pump is magnetically levitated with no internal bearings. The pocket controller is the bridge between the pump and the batteries and allows to control and monitor the LVAD function and is also equipped with a backup battery. The driveline brings power to the pump and in the HM3 is designed in a modular fashion, so the externalized portion can be easily replaced if needed. The batteries are connected to the pocket controller with protected cables and are designed to provide up to 17 hours of autonomy (Figs. 4.11 and 4.12).

The HeartMate 3 is indicated for patients with heart failure refractory to guideline-directed medical therapy requiring short term (bridge to transplant or recovery) and recently FDA

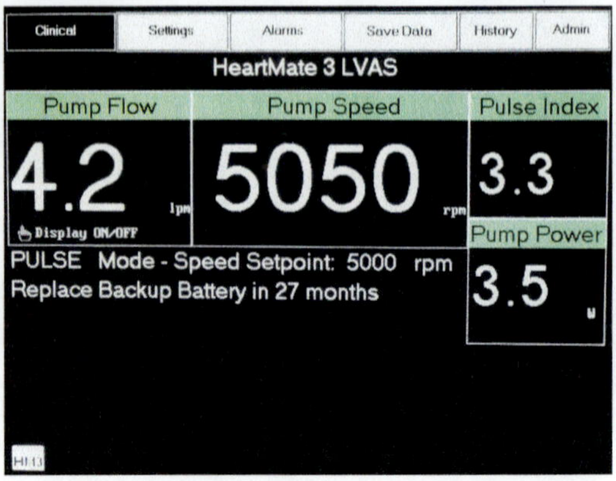

Fig. 4.11 HeartMate 3 continuous flow pump. The pump is totally magnetically levitated with no bearings, in consequence more efficient and in theory more durable. (Thoratec Corporation, Pleasanton, California)

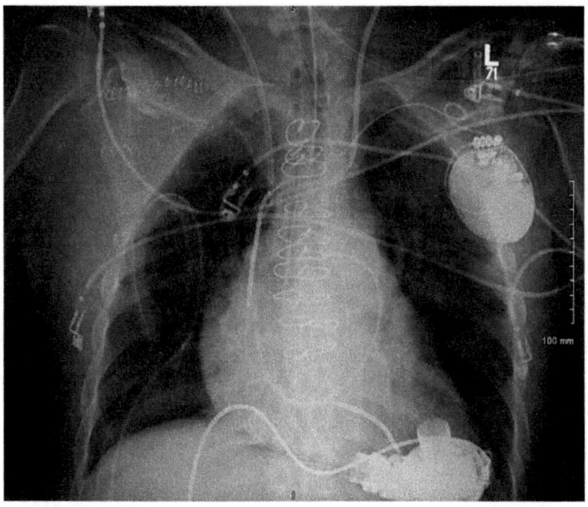

Fig. 4.12 Chest X-ray showing implanted LVAD device. Enlarged cardiac silhouette secondary to end-stage heart failure can be noted

approved as destination therapy in patients not candidates for transplantation.

The most common complications after LVAD implantation are pump thrombosis, hemolysis, driveline infections, GI bleeding, and stroke. Severe right heart failure can be a complication and is a contraindication for placement unless biventricular support is planned or immediately available during the procedure. Of note, there is early but growing experience in the use of HM3 for biventricular support with a small case series recently reported by Lavee et al. (8/9 with BiVAD support for 95 to 636 with mean of 266 days; 7 discharged home and 1 transplanted after 98 days of support) [30].

Morales et al. recently reported the use of HM3 in a congenital patient with failing Fontan circulation [31]. The decreased size of the device compared with prior generation has potential for future use in the pediatric population, where size is a very important and limiting factor for implantation [32].

The implantation is surgical and has four main steps:

1. *Pericardium access.* Subcostal, median sternotomy, and minimally invasive access through small thoracotomy have been reported. The most commonly used access is median sternotomy.
2. *Pump implantation.* The pump is introduced in the left ventricle using a special coring device. The most common placement sites are the apex and the diaphragmatic surface. The idea is that the inflow is located in the middle of the ventricular cavity, between the septum and the free wall and away from the mitral valve apparatus. The device is fixed to the ventricle using nonabsorbable sutures in a circular fashion.
3. *Outflow graft placement.* The pump outflow graft is placed around the apex, towards the right atrium, and implanted to the ascending aorta with nonabsorbable sutures.
4. *Driveline exteriorization.* The drive is passed through and tunneled in the subcutaneous tissue in a subcostal position, right or left, depending on the surgeon preference.

The MOMENTUM study published in 2016 which was a multicenter study comparing the outcomes of patients not eligible for heart transplant or requiring prolonged therapy undergoing HM2 placement versus HM3 placement (mechanical bearing axial flow versus fully levitated centrifugal pump) in which preliminary analysis (6-month outcomes) reported significant improvement in the rate of strokes and pump thrombosis. Mehra et al. recently published the 2-year follow-up results of 366 patients (190 HM3, 176 HM2) and showed that although the rates of death and disabling stroke were similar, the overall rate of stroke was significantly lower in the HM3 (10.1%) compared with HM2 (19.2%) with a $P = 0.02$, less need for pump exchange or removal (1.6% versus 17%), and less pump thrombosis (1.1% versus 15.7%) [33, 34].

Additionally, economic analysis from the same data showed decreased rate of readmission, hospital days, and cost, irrespective of the intended goal of therapy [35].

Krabatsch et al. in a prospective, non-randomized trial in Europe reported 12% gastrointestinal bleeding, 16% driveline infections, 18% strokes, and 2% outflow graft thrombosis with no hemolysis, pump thrombosis, or pump malfunction through 1 year. The 6-minute walk test distance increased from a mean of 273 m to 371 m

($P<0.0001$). EQ-5D quality-of-life score increased from a mean of 52.7 to 70.8 ($P=0.0006$) [36].

References

1. Freedman RJ. The intra-aortic balloon pump system: current roles and future directions. J Appl Cardiol. 1991;6:313.
2. Moulopoulos SD, Topaz S, Kolff WJ. Diastolic balloon pumping (with carbon dioxide) in the aorta- a mechanical assistance to the failing heart. Am Heart J. 1962;63:669–75.
3. Kantrowitz. Origins of intraaortic balloon pumping. Ann Thorac Surg 1990;50:672–4.
4. van Diepen S, Katz JN, Albert NM, et al. Contemporary management of cardiogenic shock: a scientific statement from the American heart association. Circulation. 2017;136(16):e232–68. https://doi.org/10.1161/CIR.0000000000000525.
5. Ohman EM, George BS, White CJ, Kern MJ, Gurbel PA, Freedman RJ, et al. Use of aortic counterpulsation to improve sustained coronary artery patency during acute myocardial infarction. Results of a randomized trial. The randomized IABP study group. Circulation. 1994;90(2):792–9.
6. Sintek MA, Gdowski M, Lindman BR, Nassif M, Lavine KJ, Novak E, et al. Intra-aortic balloon Counterpulsation in patients with chronic heart failure and cardiogenic shock: clinical response and predictors of stabilization. J Card Fail. 2015;21(11):868–76.
7. Fried JA, Nair A, Takeda K, Clerkin K, Topkara VK, Masoumi A, et al. Clinical and hemodynamic effects of intra-aortic balloon pump therapy in chronic heart failure patients with cardiogenic shock. J Heart Lung Transplant. 2018.
8. Elahi MM, Chetty GK, Kirke R, Azeem T, Hartshorne R, Spyt TJ. Complications related to intra-aortic balloon pump in cardiac surgery: a decade later. Eur J Vascul Endovas Surg. 2005;29(6):591–4.
9. Archimedes screw. Encyclopedia Britannica on-line. Accessed 6 Sep 2018. https://www.britannica.com/technology/Archimedes-screw.
10. Miller PE, Solomon MA, McAreavey D. Advanced percutaneous mechanical circulatory support devices for cardiogenic shock. Crit Care Med. 2017;45(11):1922–9.
11. Abiomed INC. Accessed 1 Sep 2018. http://www.abiomed.com/impella.
12. O'Neill WW, Kleiman NS, Moses J, Henriques JP, Dixon S, Massaro J, et al. A prospective, randomized clinical trial of hemodynamic support with Impella 2.5 versus intra-aortic balloon pump in patients undergoing high-risk percutaneous coronary intervention: the PROTECT II study. Circulation. 2012;126(14):1717–27.
13. Goldstein JA, Dixon SR, Douglas PS, Ohman EM, Moses J, Popma JJ, et al. Maintenance of valvular integrity with Impella left heart support: results from the multicenter PROTECT II randomized study. Catheter Cardiovasc Interv. 2018;92(4):813–7.
14. Watanabe S, Fish K, Kovacic JC, Bikou O, Leonardson L, Nomoto K, et al. Left ventricular unloading using an impella cp improves coronary flow and infarct zone perfusion in ischemic heart failure. J Am Heart Assoc. 2018;7(6). https://doi.org/10.1161/JAHA.117.006462.
15. CardiacAssist, INC. Accessed 23 Sep 2018. http://www.tandemlife.com/tandemheart-kit/.
16. Kar B, Gregoric ID, Basra SS, Idelchik GM, Loyalka P. The percutaneous ventricular assist device in severe refractory cardiogenic shock. J Am Coll Cardiol. 2011;57(6):688–96.
17. Gregoric ID, Jacob LP, La Francesca S, Bruckner BA, Cohn WE, Loyalka P, et al. The Tandem Heart as a bridge to a long-term axial-flow left ventricular assist device (bridge to bridge). Tex Heart Inst J. 2008;35(2):125–9.
18. Makdisi G, Wang IW. Extra corporeal membrane oxygenation (ECMO) review of a lifesaving technology. J Thorac Dis. 2015;7(7):E166–76.
19. Rihal CS, Naidu SS, Givertz MM, et al. 2015 SCAI/ACC/HFSA/STS clinical expert consensus statement on the use of percutaneous mechanical circulatory support devices in cardiovascular care: endorsed by the American Heart Association, the Cardiological Society of India, and Sociedad Latino Americana de Cardiologia Intervencion; affirmation of value by the Canadian Association of Interventional Cardiology-Association Canadienne de Cardiologie d'intervention. J Am Coll Cardiol. 2015;65:e1–20.
20. Abrams D, Combes A, Brodie D, et al. Extracorporeal membrane oxygenation in cardiopulmonary disease in adults. J Am Coll Cardiol. 2014;63:2769–78.
21. Takayama H, et al. Clinical outcome of mechanical circulatory support for refractory cardiogenic shock in the current era. J Heart Lung Transplant. 2013;32:106–11.
22. Biscotti M, Bacchetta M. The "sport model": extracorporeal membrane oxygenation using the subclavian artery. Ann Thorac Surg. 2014;98(4):1487–9.
23. Chung JC, Tsai PR, Chou NK, Chi NH, Wang SS, Ko WJ. Extracorporeal membrane oxygenation bridge to adult heart transplantation. Clin Transpl. 2010;24:375–80.
24. Barth E, Durand M, Heylbroeck C, et al. Extracorporeal life support as a bridge to high-urgency heart transplantation. Clin Transpl. 2012;26:484–8.
25. Sun HY, Ko WJ, Tsai PR, et al. Infections occurring during extracorporeal membrane oxygenation use in adult patients. J Thorac Cardiovasc Surg. 2010;140:1125–32.
26. Jackson KW, Timpa J, McIlwain RB, et al. Side-arm grafts for femoral extracorporeal membrane oxygenation cannulation. Ann Thorac Surg. 2012;94:e111–2.
27. John R, Pagani FD, Naka Y, Boyle A, Conte JV, Russell SD, et al. Post-cardiac transplant survival after support with a continuous-flow left ventricular

assist device: impact of duration of left ventricular assist device support and other variables. J Thorac Cardiovasc Surg. 2010;140(1):174–81.

28. Bourque K, Cotter C, Dague C, et al. Design rationale and preclinical evaluation of the HeartMate 3 left ventricular assist system for hemocompatibility. Am Soc Artificial Int Organs. 2016;62:375–83.
29. Thoratec Corporation, HeartMate 3 left ventricular assist system. Instructions for use. http://www.thoratec.com/_assets/download-tracker/HM3/100168999/100168999.A_HM3.IFU.PMA.Clip.LT.FinalPDF.2018.1012.pdf.
30. Lavee J, Mulzer J, Krabatsch T, Marasco S, McGiffin D, Garbade J, et al. An international multicenter experience of biventricular support with HeartMate 3 ventricular assist systems. J Heart Lung Transplant. 2018;17.
31. Lorts A, Villa C, Riggs KW, Broderick J, Morales DLS. First use of HeartMate 3 in a failing fontan circulation. Ann Thorac Surg. 2018;106(5):e233–4.
32. Burki S, Adachi I. Pediatric ventricular assist devices: current challenges and future prospects. Vasc Health Risk Manag. 2017;13:177–85.
33. Mehra MR, Goldstein DJ, Uriel N, Cleveland JC, Jr, Yuzefpolskaya M, Salerno C, et al. Two-year outcomes with a magnetically levitated cardiac pump in heart failure. N Engl J Med. 2018;378(15):1386–95.
34. Heatley G, Sood P, Goldstein D, et al. Clinical trial design and rationale of the multicenter study of MagLev Technology in Patients Undergoing Mechanical Circulatory Support Therapy with HeartMate 3 (MOMENTUM 3) investigational device exemption clinical study protocol. J Heart Lung Transplant. 2016;35:528–36.
35. Mehra MR, Salerno C, Cleveland JC, Pinney S, Yuzefpolskaya M, Milano CA, et al. Healthcare resource use and cost implications in the MOMENTUM 3 long-term outcome study. Circulation. 2018;138(18):1923–34.
36. Krabatsch T, Netuka I, Schmitto JD, Zimpfer D, Garbade J, Rao V, et al. Heartmate 3 fully magnetically levitated left ventricular assist device for the treatment of advanced heart failure −1-year results from the Ce mark trial. J Cardiothorac Surg. 2017;12(1):23-017-0587-3.

Physiology of Blood Pump Circulation in Heart Failure

5

Abhinav Saxena, Nir Uriel, and Daniel Burkhoff

Introduction

Left ventricular assist devices (LVADs) play a crucial role in providing hemodynamic support in patients with end-stage chronic heart failure. LVADs are used in patients awaiting cardiac transplantation by acting as a bridge to transplant (BTT) and as destination therapy (DT) in patients not eligible for heart transplant [1–3]. More recently, LVADs have been used as bridge to decision (BTD) in patients with uncertain eligibility for transplantation and as bridge to recovery (BTR) in critically ill patients expected to sufficiently or totally recover without the need for a transplant. Their use for all the above applications is expected to increase in the future as devices become more compact and safer to use.

LVADs actively interact with the native heart and circulation to effectively improve end-organ perfusion while unloading the left ventricle. These factors each contribute independently the profound reverse ventricular remodeling observed during prolonged LVAD support [4]. Understanding the physiology of LVAD hemodynamics is vital for clinicians to improve patient care especially since there will be an increasing number of LVAD patients in the coming years. We therefore aim to present a clinically relevant review of the physiology of LVADs as applied to patients with chronic heart failure.

A. Saxena, MBBS
Maimonides Medical Center, Brooklyn, NY, USA

N. Uriel, MD
Columbia and Cornell Universities, New York Presbyterian, New York, NY, USA
e-mail: nu2126@cumc.columbia.edu

D. Burkhoff, MD, PhD (✉)
Cardiovascular Research Foundation,
New York, NY, USA
e-mail: dburkhoff@crf.org

Types of Ventricular Assist Device

The human heart is a complex volume displacement, pulsatile pump. First generation of LVAD's mimicked this concept. Due to their large size, high rates of adverse events, and device failures [5], their use was supplanted entirely as soon as smaller, continuous-flow devices became available.

Second-generation axial-flow pumps utilize a rotational pump design with ceramic contact bearings. The blood enters and is pushed forward by a screwing motion eventually exiting the pump coaxially [6]. Despite smaller size and higher long-term reliability due to only one moving part [7], contact bearings are prone to frictional wear overtime, incomplete bearing wash, potential for stasis, and thrombus formation at the rotor-bearing interface [8]. Further iterations resulted in the current third generation of pumps which are centrifugal in design with noncontact

© Springer Nature Switzerland AG 2020
J. H. Karimov et al. (eds.), *Mechanical Support for Heart Failure*,
https://doi.org/10.1007/978-3-030-47809-4_5

bearings. In these pumps, the blood enters the pump, is rotated by the impeller, and ejected at 90° to the inlet flow. The use of noncontact bearings facilitates increased blood flow around the impeller and better washing of the impeller surface. This is expected to increase pump longevity by reducing mechanical wear [9].

Continuous-Flow Left Ventricular Assist Device (cf-LVAD) Hemodynamics

Cf-LVADs, including both axial and centrifugal pumps, impart kinetic energy and accelerate the blood by impeller rotation [10]. Pump function of cf-LVADs is characterized by pump speed (rpm), electrical power consumption (watts), flow (L/min), and the degree of pressure pulsatility during operation (via the pulsatility index, PI). The operator sets speed; power is a measure of current drawn by and voltage applied to the pump and relates for blood flow. In clinical practice, flow is not measured directly but is estimated from rotational speed (rpm) and power consumption in a majority of contemporary cf-LVDs; HeartAssist 5 and aVAD pumps use a transit-time ultrasonic probe to measure flow directly. The flow through the cf-LVADs is dictated by pump rotational speed, blood viscosity (related to hematocrit), preload pressure at the pump inlet, and afterload pressure at pump outlet according to the pump's unique pressure-flow characteristics, the HQ curve [11].

Pressure-Flow Relationship (HQ Curve)

Pump pressure head (H), or the pressure gradient (ΔP) across the pump, is the pressure difference between the inlet and outlet ports of the pump. In the case of LVADs that pump from the LV to the aorta, ΔP = aortic pressure − LV pressure + combined pressure loss across the inlet cannula and outlet graft [12]. At fixed operating speed, ΔP dictates flow according to the pressure-flow relationship, the so-called HQ curve, which is unique to each pump. Clinically, the relevant HQ curve is that of the entire system, which includes the pump, the inflow cannula, and the outflow graft [6].

HQ curves are typically generated in mock loops by measuring the pressure difference between the system inlet and outlet while gradually increasing resistance to outflow to the point of pump shutoff. Different curves are generated at different operating pump speeds and plotting ΔP on the *y*-axis and pump flow on the *x*-axis [12]. However, from a physiological and clinical perspective, it is more appropriate to plot ΔP on the *x*-axis (since this is the clinically independent parameter) and pump flow (the dependent parameter) on the *y*-axis. In general, cf-LVAD flow is inversely proportional to ΔP as depicted for an axial-flow pump (HeartMate 2) in Fig. 5.1a [13] and for a centrifugal flow pump (HVAD) in Fig. 5.1b [9]. Axial-flow pumps tend to have relatively linear HQ curves, whereas the HQ curve of centrifugal pumps is more nonlinear.

During normal operation, ΔP changes during the cardiac cycle, mainly due to cyclic variations of ventricular pressure during contraction (Fig. 5.2a). In systole, as the LV contracts, there is a reduction in ΔP, and flow is at a maximum. During diastole, with LV relaxation, ΔP increases, and pump flow decreases (Fig. 5.2b). Thus, due to the time-varying pressure gradient, pump flow also varies along the HQ curve with each cardiac cycle, even with a closed aortic valve [14, 15] (Fig. 5.2c, d). Accordingly, while flow from these pumps is continuous, it is not generally speaking, constant.

Significant differences can exist between centrifugal and axial-flow pumps that impact on the relative degrees of flow pulsatility for a given change in pressure, sensitivity to afterload resistances, and responses to suction [16, 10]. Importantly, differences between pumps do not result in major differences in clinical effectiveness since, in practice, RPMs are adjusted to provide the degree of support needed based on individual patient needs.

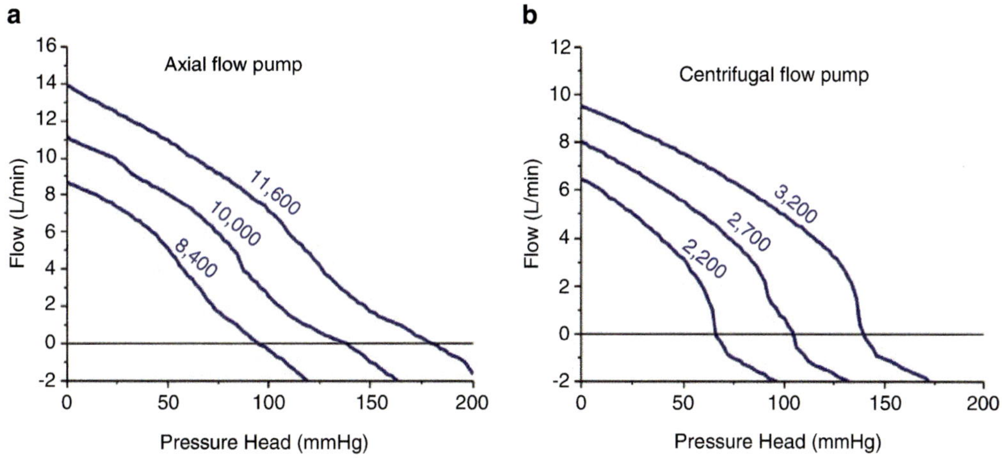

Fig. 5.1 (**a**) HQ relationship representative of a HeartMate II axial-flow pump at three specified RPMs. (**b**) HQ relationship representative of an HVAD centrifugal flow pump at three specified RPMs. (Created with Harvi-Online http://harvi.online)

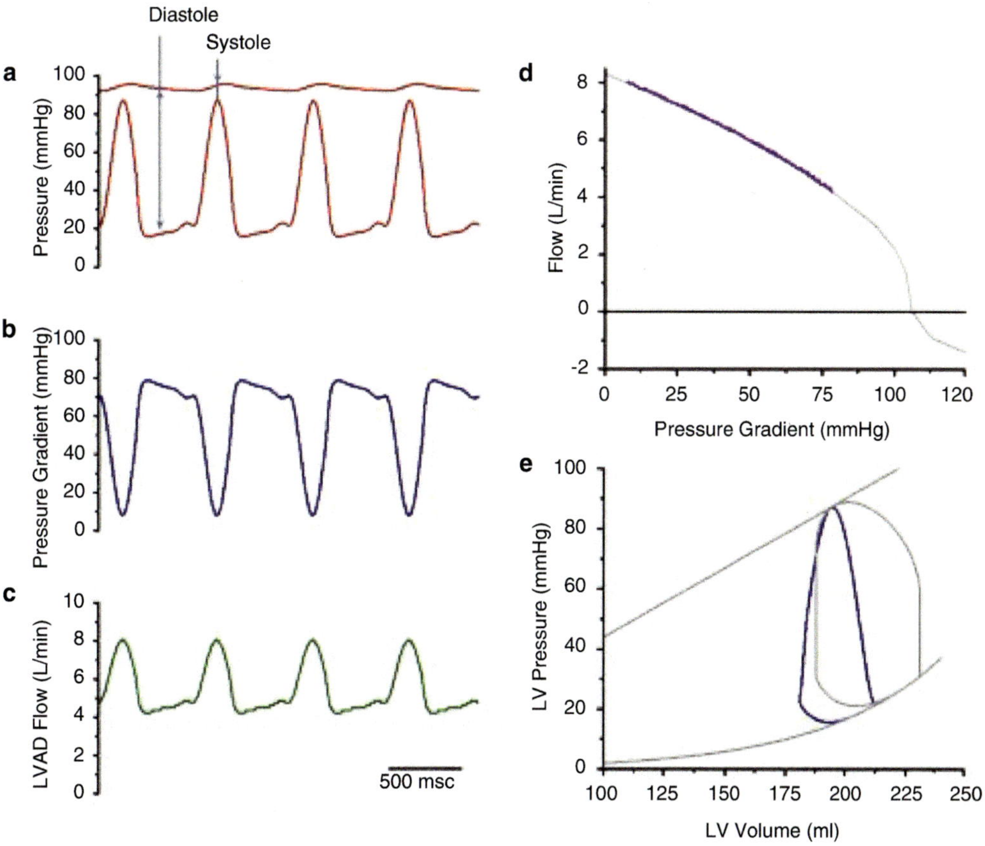

Fig. 5.2 Variations of aortic and ventricular pressures during the cardiac cycle (**a**), the resulting time-varying pressure gradient across the pump inlet and outlet during the cardiac cycle (**b**), and resulting flow waveform (**c**). The flow waveform is determined by the time-varying pressure gradient as it projects onto the HQ curve at the specified RPM (**d**). LVAD flow impacts on LV filling and mechanics as depicted on the pressure-volume diagram (**e**) which shows in comparison with pre-LVAD conditions (dark gray loop) a leftward shift and transition from rectangular to a triangular loop (blue) [19, 23]. (Created with Harvi-Online http://harvi.online)

Impact of LVAD Pumping on Ventricular Mechanics and Energetics

Physiology of Myocardial Energetics

Myocardial oxygen consumption is affected by multiple factors which include preload, afterload, muscle mass, heart rate, and contractility [17, 18]. Coronary blood flow (CBF) on the other hand is driven by a difference between the mean arterial pressure in diastole and downstream pressure related to the mean right atrial pressure as well as the left ventricular end-diastolic pressure (LVEDP) [17].

Pressure-volume analysis unifies the complex interactions listed above. Left ventricular pressure-volume area (PVA) is defined as the area on the pressure-volume diagram bounded by the end-systolic and end-diastolic pressure-volume relationships and the systolic portion of the pressure-volume curve (Fig. 5.3) [19]. It is equal to the sum of external stroke work (SW) plus the residual potential energy (PE) stored inside the myocardium at end systole: PVA = SW + PE. PVA is equal to the total mechanical work performed by the heart on each heartbeat and provides a load-independent index of oxygen consumption per beat [20–22].

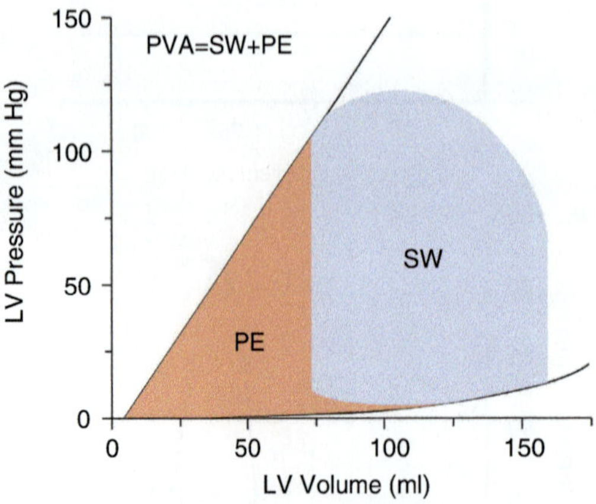

Fig. 5.3 Pressure-volume area (PVA). The PVA, composed of the external stroke work (SW) and the mechanical potential energy (PE) stored in the myocardium at end diastole, represents the total mechanical work performed by the heart and correlates closely with total myocardial oxygen consumption per beat [19, 20]. (Reprinted from Burkhoff et al. [19], with permission from Elsevier)

Ideal VAD-assisted hemodynamic effects, therefore, should minimize left ventricular end-diastolic pressure and PVA (leftward shift of PV loop) while improving systemic perfusion by providing normal cardiac output and blood pressure [20].

LVAD Impact on Ventricular Mechanics

Blood flow due to cf-LVADs impacts ventricular mechanics in several ways, which are readily appreciated on the ventricular pressure-volume diagram (Fig. 5.2e). First, pumping the blood directly from the LV reduces LV volume and diastolic pressure. Chronic unloading by LVADs leads to reverse ventricular remodeling that underlies, in large part, promotes recovery of myocardial function [4, 23, 24]. Second, the shape of the ventricular pressure-volume loop transitions from a rectangular shape to a triangular shape; this is because with continuous flow from the LV, ventricular volume is always decreasing, and there is a loss of the isovolumic phases of contraction and relaxation [19]. Both the degree of unloading and the degree of triangulation of the pressure-volume loop are pump RPM-dependent as illustrated in Fig. 5.4 in a preclinical model in which both HVAD and HeartMate II were studied.

Impact of Blood Inertia on the Instantaneous Pressure-Flow Relationship

In reality, the steady-state HQ relationships depicted in Fig. 5.1a, b do not adequately describe the dynamics of pump flow due to the inertia of the blood which causes instantaneous pressure-flow relationships to deviate from the curves that are measured under steady-state conditions.

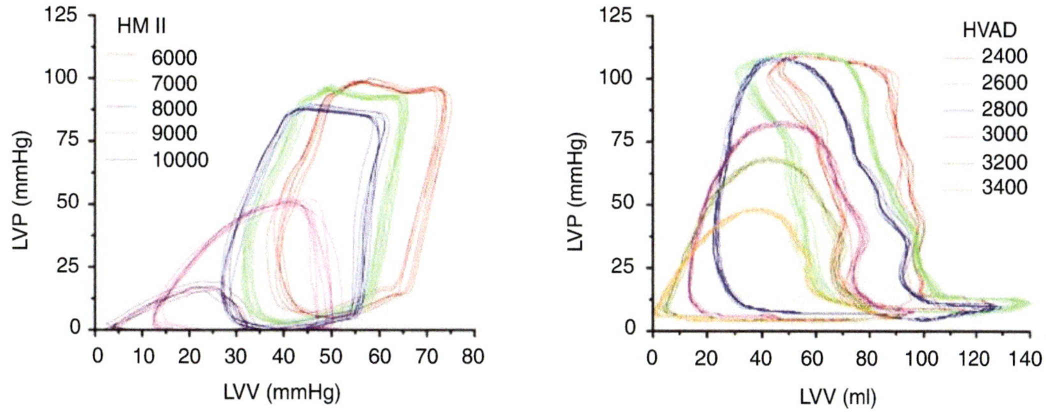

Fig. 5.4 Pressure-volume loops from a preclinical model in which LVAD speed is gradually ramped. Baseline loops (without LVAD pumping) shown in red. As RPMs are increased for both HeartMate II and HVAD, the loops shift progressively leftward toward lower volumes (dose-dependence of unloading) and become increasingly triangular. (Created with Harvi-Online http://harvi.online)

Using mock circulatory loops coupled with a pneumatic mock ventricle, several studies have demonstrated that the instantaneous LVAD pressure-flow relationship deviates from the steady-state HQ curves, demonstrating hysteresis as illustrated in Fig. 5.5 [25, 26]. We can further define the impact of inertia and hysteresis on overall function using in silico modeling [27] as illustrated in Fig. 5.6. The presence of inertia (and thus hysteresis around the steady-state curve, (Fig. 5.6a, b) decreases peaks and troughs of the flow waveform (Fig. 5.6c), thus decreasing intrinsic VAD flow pulsatility, but, interestingly, does not substantially impact on the average flow and insignificantly impacts on the LV pressure-volume loop (Fig. 5.6d).

Pump Power and Flow Relationship

For centrifugal pumps, there is a reasonably linear relationship between the electrical power drawn and the flow generated by the pump. Accordingly, flow estimates provided by pumps such as the HVAD which are based on established lookup tables which relate RPMs, blood viscosity (related to hematocrit), and electrical power to flow are considered reasonably reliable [28, 29]. In contrast, axial-flow pumps exhibit a nonlinear, U-shaped relationship between electrical current and flow [6]. Under this circumstance, flow estimates are considered less reliable.

Artificial Flow Pulsatility

As detailed above, flow from cf-LVADs varies with ventricular contraction and therefore can introduce a degree of arterial pressure pulsatility even if the aortic valve does not open. Based on the explanations above, intrinsic pulsatility due to ventricular contraction depends on the slope of the HQ curve [15]. The degree of pulsatility has been indexed clinically by the pulsatility index (PI) which is calculated and displayed in different ways for different pumps. For HeartMate II and III, PI is calculated as beat-to-beat amplitude between the maximal flows and minimal flows averaged over 10–15 seconds and divided by the average flow according to the formula: (maximum flow – minimum flow)/average flow. In HVADs pulsatility is displayed as real-time waveforms (Fig. 5.7). PI is inversely related to speed under conditions of constant preload and afterload [11].

Due to the potential physiological importance of pulsatility, several devices incorporate algorithms to vary RPMs to artificially introduce additional pulsatility; the clinical benefits

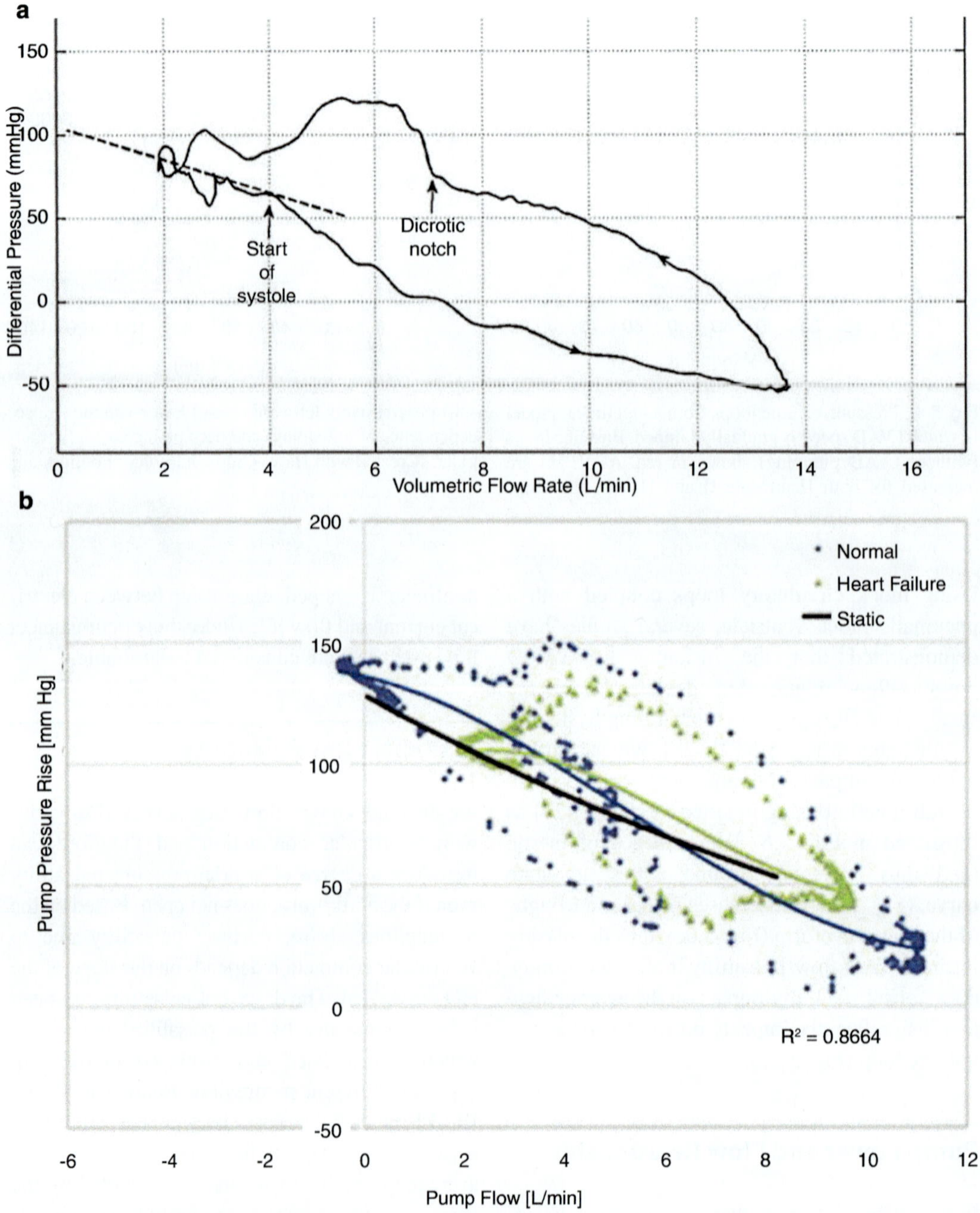

Fig. 5.5 HeartMate II instantaneous HQ (flow-pressure) loops measured in mock loops showing hysteresis around the steady-state relationship. (**a**) Study by Noor et al. [26]. (Reprinted from Noor et al. [26], with permission from John Wiley and Sons). (**b**) Study by Sunagawa et al. [25]. (Reprinted from Sunagawa et al. [25], Copyright (2015), with permission from Elsevier)

of such artificial pulsatility remain controversial [30, 31]. Purported advantages of pulsatile circulation in VAD patients include decreased blood stasis in the ventricle, intermittent aortic valve opening, decreased risk of ventricular suction by allowing intermittent LV filling, and

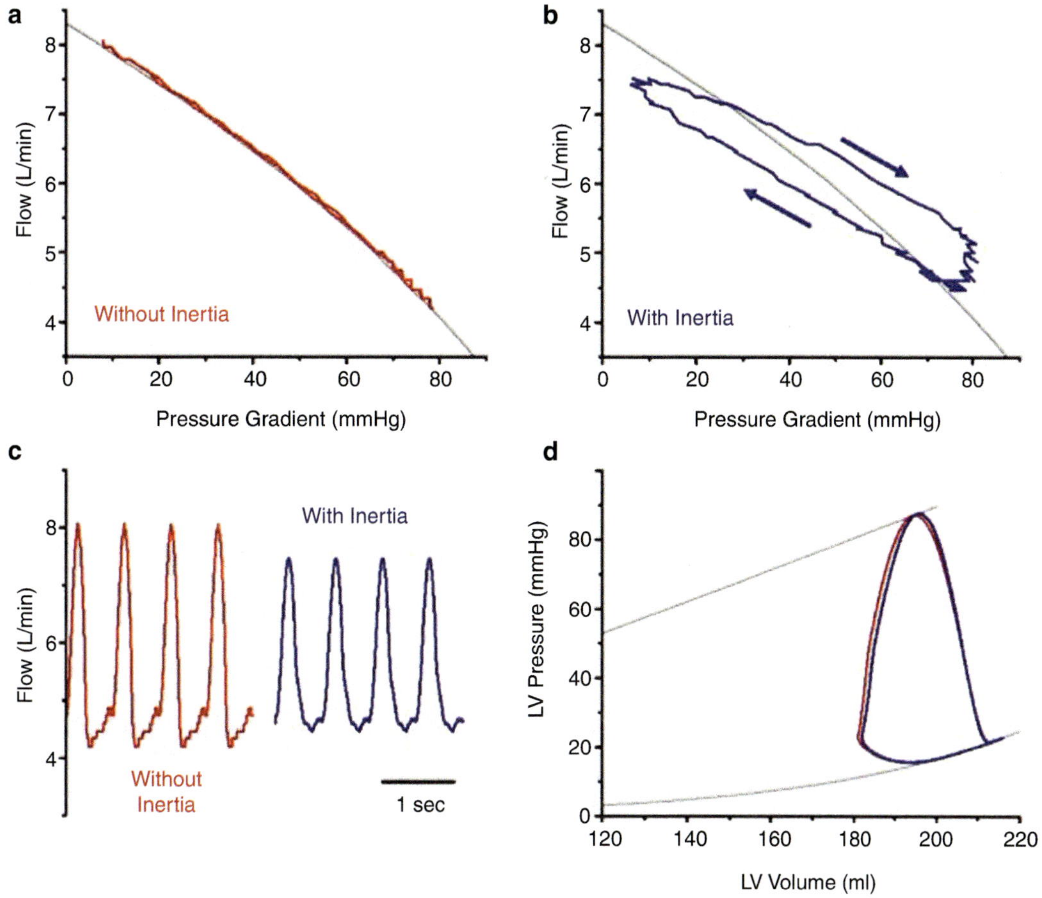

Fig. 5.6 Simulation model [27] showing the impact of inertia on the instantaneous LVAD pressure-flow relationship. (**a**) Without inertia, the instantaneous LVAD pressure-flow relationship follows the steady-state curve (gray). (**b**) With inertia added to the model, the loop deviates from the steady-state curve in a manner similar to that observed experimentally (as in Fig. 5.5). (**c**) The instantaneous flow signal pulsatility is decreased by the presence of inertia, but the mean flow is not significantly impacted. (**d**) Despite the impact of inertia on the flow signal, the impact on the ventricular pressure-volume loop is insignificant. (Created with Harvi-Online http://harvi.online)

potentially beneficial effects on end-organ function [32]. Some studies indicate pulsatile flow maintains lymphatic flow, decreases systemic vascular resistance [33], and improves autonomic function [34].

Accordingly, some cf-LVAD (e.g., HVAD and HM III) have an artificial pulse mode which induces pulsatile flow by transiently and rapidly varying pump speed [9]. HVAD has a so-called Lavare cycle that has been in place outside of the US for some time and more recently introduced in the US [35]. Benefits of this algorithm and pulsatility, in general, have not been definitively established.

Impact of RPMs on LVAD Flow and Total Flow to the Body

Ramp Test

One of the challenges faced in the care of LVAD patients is understanding how to optimally set the RPMs. Ideally, speed would be adjusted to simultaneously achieve normal values of filling pressures (both CVP and PCWP), arterial pressure, and total blood flow to the body. Unfortunately, this is not always possible simply by adjusting speeds, and medical management is required for adjustment of arterial resistance and

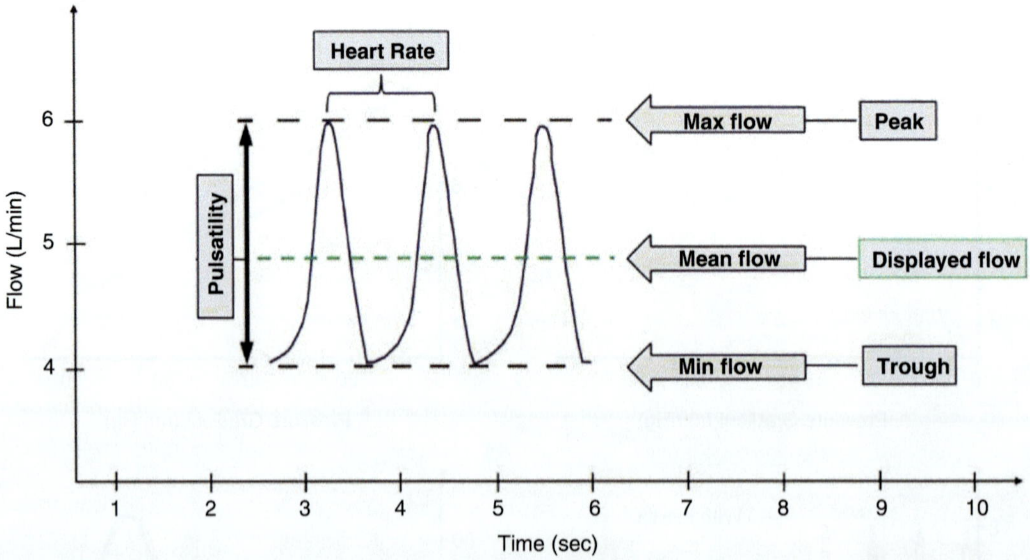

Fig. 5.7 A typical HVAD waveform [28]. (Reprinted from Rich and Burkhoff [28], https://journals.lww.com/asaio-journal/pages/default.aspx with permission from Wolters Kluwer Health, Inc)

Table 5.1 Characteristics of axial and centrifugal-flow pump ramp tests

pump	Lower rate	Upper rate	Speed increment	Parameters to stop the test	Device optimization goals	Features of pump thrombosis	Parameter slopes for device malfunction
HMII [38]	8000 RPM	12,000 RPM	400 RPM	Suction event (and/or) LVEDD <3.0 cm	Intermittent AV opening MAP >65 mm Hg MR not more than mild in severity	Minimal change in LVEDD with an increase in pump speed Clinical parameters: ↑LDH	LVEDD slope > −0.16 rpm/increment (regardless of AV closure)
HVAD [39]	2300 RPM	3200 RPM	100 RPM				LVEDD slope (varies with aortic valve closure) Open AV valve LVEDD slope > −0.09 rpm/increment Closed AV valve LVEDD slope > −0.15 rpm/increment

Abbreviations: *LVEDD* left ventricular end-diastolic dimension, *AV* aortic valve, *RPM* revolutions per minute, *MAP* mean arterial pressure, *MR* mitral regurgitation, *LDH* lactate dehydrogenase

fluid status. Understanding of the complex interactions between the LVAD and body properties has been enhanced through the use of speed ramp tests.

More specifically, hemodynamic speed ramp tests in cf-LVADs can be used to assess the dynamic interactions between device speed, left and right ventricular filling pressures (PCWP and CVP, respectively), as well as valve function using invasive pulmonary artery and echocardiographic methods. Ramp tests are used in the initial postoperative care of cf-LVAD patients to determine the appropriate LVAD speed [36] and also in stable LVAD patients to optimize hemodynamic conditions through speed and medication adjustments [37] as well as diagnose device malfunction and need for surgical or conservative interventions. Table 5.1

describes the characteristics of ramp tests for HeartMate II and HVAD devices, devices for which most information is currently available.

HeartMate II Echocardiographic Ramp Test

Uriel et al. defined a systematic approach to perform and analyze hemodynamic ramp tests [38]. The ramp test protocol for axial-flow pump (HM II) is performed by reducing the RPM of the pump to 8000 RPM and measuring left ventricular end-diastolic dimension (LVEDD), left ventricular end-systolic dimension (LVESD), frequency of AV opening, degree of AR, degree of mitral regurgitation (MR), right ventricular systolic pressure (RVSP), Doppler blood pressure, heart rate, pump power, pulsatility index (PI), and pump flow. Subsequently, the speed is increased by 400 RPM at 2-minute intervals, and all measurements are repeated until the pump reaches 12,000 RPM or the maximum tolerable speed. The ramp test is stopped if there is a suction event or if the LVEDD decreases to less than 3 cm. Pump thrombosis can be diagnosed when there is minimal change in the LVEDD with an increase in pump speed. The slope of RPM-LVEDD correlates with pump thrombosis or severe outflow obstruction due to an uncoupled relationship between increases of pump speeds and decreases in LVEDD. For HM II, an RPM-LVEDD slope > −0.16 rpm/increment is significant for device malfunction.

HVAD Echocardiographic Ramp Test

Ramp studies performed in a centrifugal pump (e.g., HVAD) is similar to that of an axial-flow pump ramp-study protocol. However, given differences in pump operating speeds, the ramp test (HVAD protocol) is started at 2300 rpm. Subsequently, the speed is increased in steps of 100 rpm to a max of 3200 rpm. Criteria for stopping the ramp studies are the same as detailed above [37]. The parameter slopes for HVAD are significantly different from HMII and vary with the aortic valve (AV) status. With AV valve open, the RPM-LVEDD slope for device malfunction was > −0.09 rpm/increment and with closed AV > −0.15 rpm/increment [39].

Invasive ramp tests with the use of a pulmonary artery catheter to assess CVP and PCWP can provide a more detailed assessment of the underlying hemodynamic state [37, 40]. Doppler-TTE-derived variables from the LVAD outflow graft were also recently shown to predict PCWP, CO, and SVR reliably and could potentially reduce the need for invasive testing [41]. Ramp tests in stable LVAD patients are reproducible and may represent a hemodynamic fingerprint for a patient. Changes in the ramp test can be used to assess device malfunction or alterations in volume status, peripheral vascular resistance and offer an opportunity to optimize medication doses and device settings [40].

Impact of RPM on Left Ventricle

Cf-LVADs pump blood continuously from the LV to aorta independent of the cardiac cycle. As a result, there is a loss of normal isovolumetric periods, and the PV loop morphology changes to a triangular shape from the normal rectangular or trapezoidal shape. With further increases in RPMs, the LV becomes progressively unloaded, and the PV loop shifts to the left (Fig. 5.8a). The leftward shift signifies a reduction in peak LV pressure generation and marked reduction in PVA and MVO2. As the degree of unloading increases, there is an increasing dissociation between aortic and left ventricular pressures (Fig. 5.8b–e).

Impact of RPM on Total Body Flow

While LVAD flow increases with an increase in RPMs, increases in RPM do not always result in increased overall flow to the body as detailed in Fig. 5.9. With the initiation of LVAD flow, LVAD

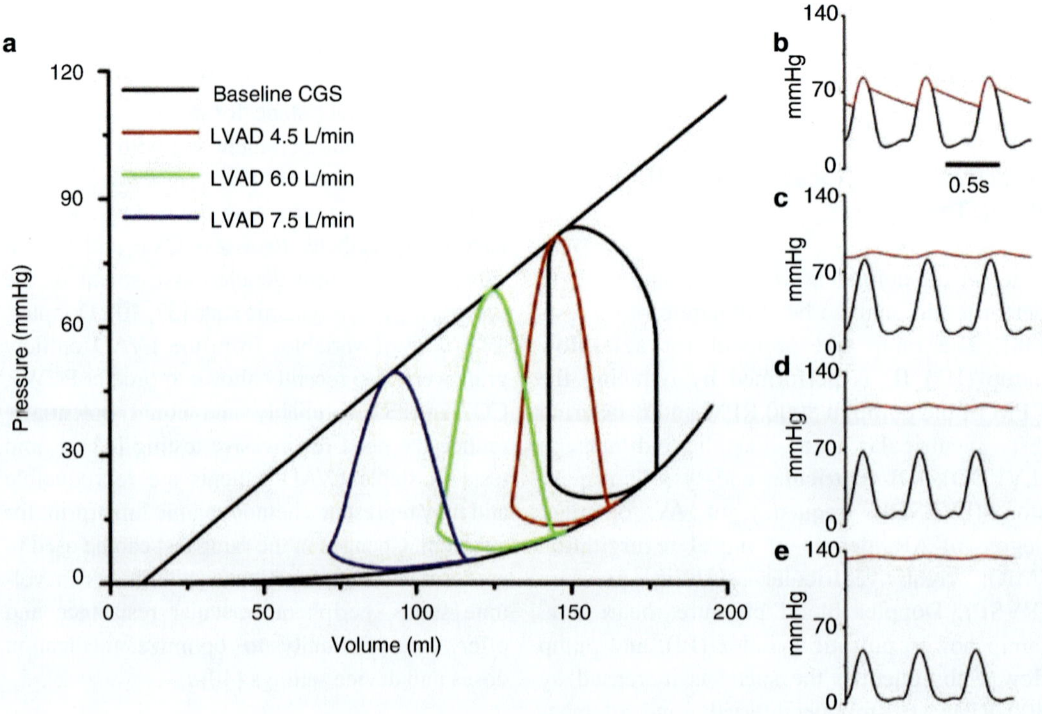

Fig. 5.8 Impact of RPM on the left ventricle. (**a**) Progressive unloading of the left ventricle with leftward shift of the PV loop. (**b–e**) Increasing dissociation can be noted between LV and aortic pressures with a progressive increase in the degree of LV unloading [19]. (Reprinted from Burkhoff et al. [19], Copyright (2015), with permission from Elsevier)

flow will unload the LV and increase afterload pressure, thus reducing intrinsic CO from the heart. As RPMs are increased, LVAD flow progressively increases, and intrinsic CO decreases. While RPMs are within a range where part of the flow is from the heart and part of the flow is from the LVAD is said to be providing *partial support*. Thus, within this range, while LVAD flow is increasing significantly with RPM increments, total flow seen by the body increases, but by a smaller amount. At some point, aortic pressure increases such that the LV no longer ejects and the aortic valve remains closed. After this point, total flow seen by the body is only provided by the pump; this is *full support* condition, and the slope of the curve relating RPMs to total flow increases.

From a terminology perspective, it is also important to distinguish between the degree of support and the degree of LV unloading (Fig. 5.10). As detailed in the next section, full support and full unloading are not synonymous.

Defining and Quantifying Ventricular Unloading

Promotion of ventricular reverse remodeling depends on adequate LV unloading and achieving VAD-assisted ideal hemodynamic state [42]. *LV unloading* has been defined as the reduction of total mechanical power expenditure (PVA·HR) of the ventricle which correlates with reductions in myocardial oxygen consumption and hemodynamic forces that lead to ventricular remodeling [19, 23]. Full unloading, therefore, only occurs when PVA has reached a minimal value (Fig. 5.10, rightmost panel). In contrast, as detailed above, *full support* occurs when there is uncoupling of arterial pressure and left

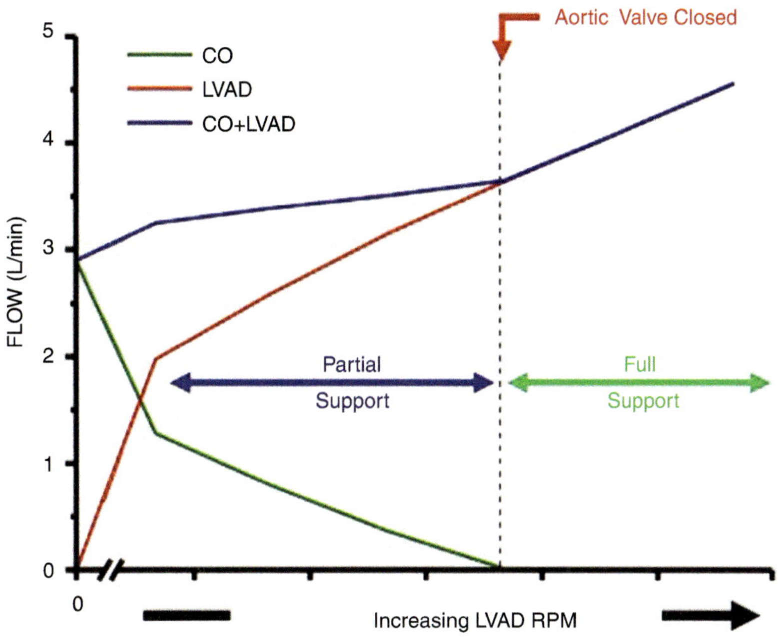

Fig. 5.9 Impact of LVAD speed on LVAD flow, intrinsic cardiac, and total flow to the body (=CO+LVAD). With the initiation of LVAD flow, LVAD flow will unload the LV and increase afterload pressure, thus reducing intrinsic CO from the heart. As RPMs are increased, LVAD flow progressively increases, and intrinsic CO decreases. This is a *partial support* condition in which part of the flow is from the heart and part is from the LVAD. Total flow seen by the body increases, but by a smaller amount. At some point, aortic pressure increases such that the LV no longer ejects and the aortic valve remains closed. After this point, total flow seen by the body is only provided by the pump. This is *full support* condition. (Created with Harvi-Online http://harvi.online)

ventricular systolic pressure resulting in a closed aortic valve [9]. Thus, full support can be achieved, while the LV is only minimally unloaded (middle panels, Fig. 5.10). Clinically, significant mitral regurgitation can also signify inadequate unloading. Pulsatile first-generation pumps running on the eject-on-full mode offered profound LV unloading [43]. There is conflicting evidence about the equivalence of LV unloading with pulsatile VADs vs. cf-LVADs, although both can offer adequate hemodynamic support [7, 43, 44].

Impact of RPM on CVP and PCWP

Assessment of complex VAD-ventricular interactions based on physical examination can be challenging, in particular as it relates to assessing an LVAD patient's volume status. In a study evaluating the use of invasive hemodynamic ramp test for device optimization in clinically stable cf-LVAD patients, at baseline, only 43% of the patients had normal CVP and PCWP at their original RPM settings (Fig. 5.11). During the ramp test, with an increase in speed, cardiac output increased, and PCWP decreased with no significant change in the CVP and SBP. 56% of the patients required adjustment of their pump speed from its original setting to achieve CVP and PCWP close to normal range [37]. Finally, in a large percentage of the patients studied, CVP and PCWP could not be optimized, signifying the need for altered medical therapy (diuretics and/or afterload reduction). The finding that there was no rpm-dependent change in CVP suggested the beneficial impact of LV unloading on RV function [37]. The findings were similar in patients supported by HM2 and HVAD. The conclusion of the study is that clinical assessment of volume status in LVAD patients is very challenging, and the hemody-

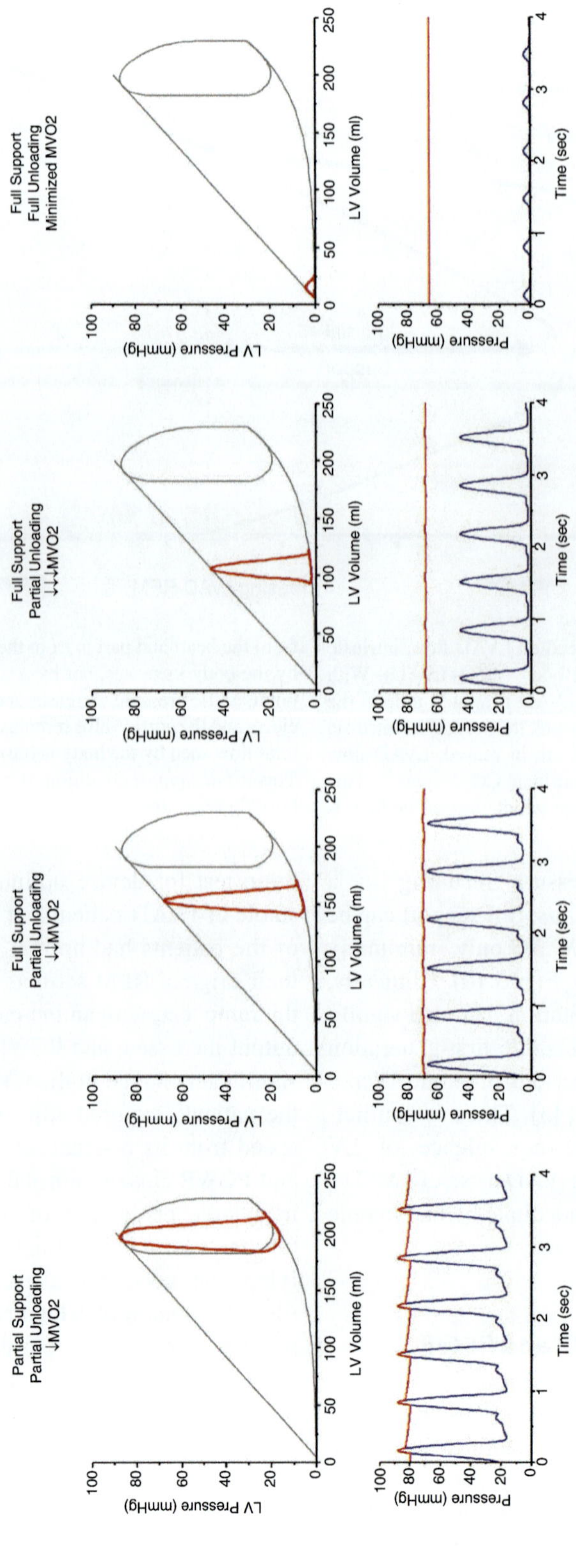

Fig. 5.10 Degree of support versus degree of unloading. (Created with Harvi-Online http://harvi.online)

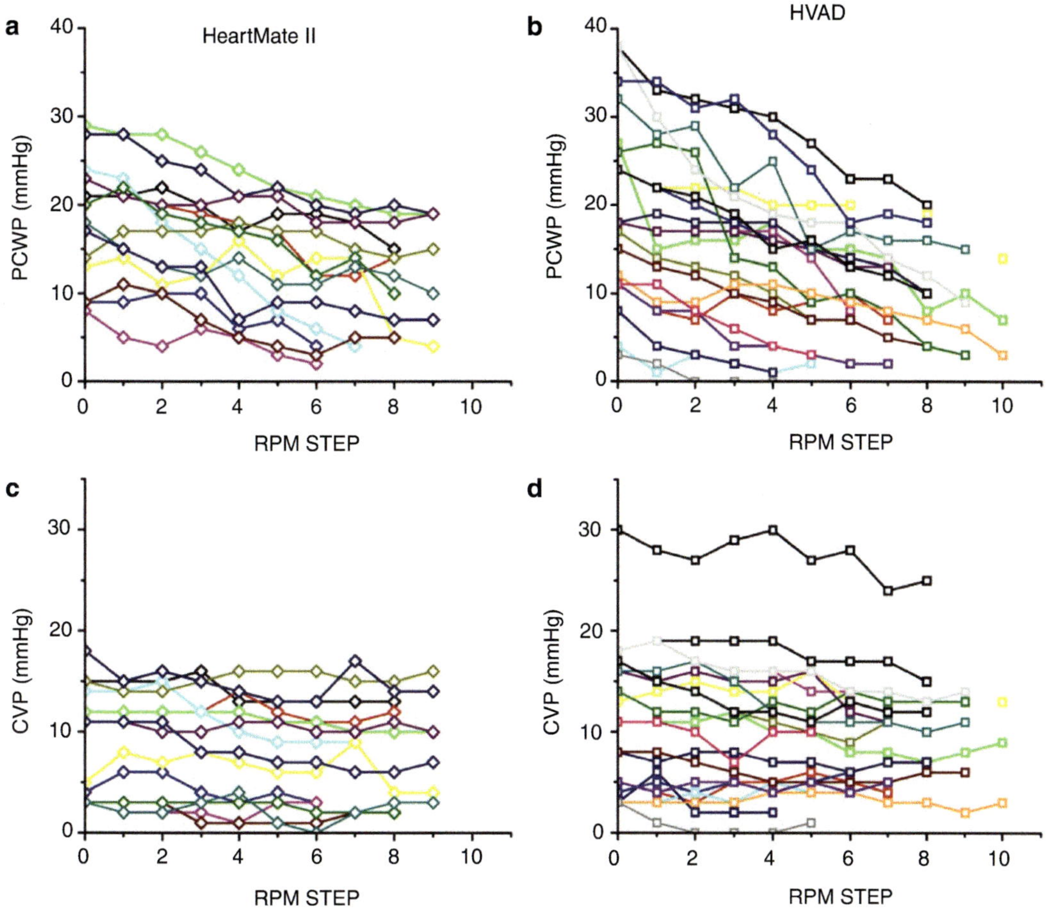

Fig. 5.11 Impact of LVAD RPMs on pulmonary capillary wedge pressure (PCWP) and central venous pressure (CVP) in patients supported by either HeartMate II or HVAD [37]. (Reprinted from Uriel et al. [37], Copyright (2016), with permission from Elsevier)

namic ramp tests can be helpful not only in determining optimal speed setting but perhaps more importantly for adjusting medical therapy, particularly diuretic dosing.

Special Considerations

RV Failure with LVAD

Right heart failure is an important problem encountered in patients undergoing LVAD support. It occurs with a frequency of ~20–25% in the perioperative period, with ~1–5% of patients requiring at least temporary mechanical right ventricular support. In recent studies, approximately 30% of patients experience right heart failure during chronic LVAD support. Hemodynamic and mechanical interaction between right and left heart occurs due to their connection in series and anatomical coupling from a shared interventricular septum. LVADs can have either a beneficial or a detrimental effect on RV function. Beneficial effects can include a decrease in RV afterload (i.e., reduction of PCWP), favorable alteration of the RV geometry due to a reduction in impingement of the RV by the interventricular septum, as well as improved coronary flow as a result of increased mean arterial pressure. Deleterious effects on RV function can result from several factors. LVADs increase venous return to the right heart, thereby increasing the preload which can potentially overwhelm the RV. An increase in RV afterload pressure can also occur when increased flow passes through a fixed pulmonary vascular resistance. In addition, reduced LV pressure generation due to LV unloading reduces LV contribution to RV pressure generation, and the effect is referred to as interventricular dependence that is mediated mainly by the interventricular septum [45].

Pulmonary Hypertension

Pulmonary hypertension (PH) due to chronic vascular remodeling can occur in chronic heart failure. Decoupling of pulmonary artery diastolic pressure (PADP) and PCWP defined as >5 mmHg difference between the two pressures has been observed in PH. A study reports of 43–48% heart failure patients with LVAD had decoupling of PADP and PCWP at baseline speeds, and it was a significant predictor for the composite endpoint of death and heart failure readmissions. 30% of the patients with decoupling could be normalized after a combined invasive hemodynamic measurement and ramp test-based change in the device setting. Normalization was significantly associated with 1 year heart failure readmission-free survival compared to non-normalized group [46].

Clinical LVAD Physiology

Understanding ventricle-VAD interaction enables identification of clinical pathology especially when flow waveforms are available as detailed by Drs. Rich and Burkhoff [28]. Table 5.2 summarizes common pathologic conditions and VAD responses.

The concept is illustrated by one of these conditions: the detection of hypertension is of particular relevance.

Hypertension

Elevated blood pressure is associated with increased risk of stroke [47], aortic regurgitation [48], and pump thrombosis [49]. It is recommending that mean arterial pressures (MAP) be maintained within the range of 70–80 mm Hg [36], and MAPs >90 mm Hg are not recommended. Due to high afterload sensitivity, hypertension affects the amount of cardiac support and unloading provided by the cf-LVAD. In this regard, the centrifugal pumps, with their flatter H-Q curves, have a higher afterload sensitivity than axial-flow pumps.

As detailed previously, the cf-LVAD operate on the rpm-dependent HQ curve during the cardiac cycle, and, as a result, the flow waveform can be characterized by its peak flow, mean flow, and trough flow during each cycle (Fig. 5.7). Increased arterial pressure is mediated by an increased SVR and can alter the magnitude of all components of the flow to varying degrees. During hypertensive periods, the pressure gradient between the aorta and the LV (ΔP) increases particularly during diastole. As a result, while peak flow (during systole) may be little affected, trough flow (during diastole), and as a result, the mean flow, can be markedly reduced depending on the degree of elevation of the arterial pressure; this is associated with a significant increase in pulsatility (Fig. 5.12a). In certain circumstances, either no diastolic flow and even negative flow or flow reversal can occur [25, 50].

The opposite situation occurs in a patient who is hypotensive (Fig. 5.12b). In such a case, ΔP is lower than normal, and there is less variability of pressure gradient. Accordingly, mean flow is increased, and flow pulsatility is decreased.

VAD-Exercise Physiology

Aerobic exercise in healthy individuals results in increased heart rate, increased stroke volume, decrease in systemic vascular resistance, and mild to moderate elevation in blood pressure without elevation in intracardiac filling pressures. Whereas in heart failure, stroke volume is

Table 5.2 Common pathologic conditions and VAD responses [28]

Condition	Loading conditions for the pump	ΔP systole	ΔP diastole	Speed	Power	Flow	Pulsatility	Hemodynamic parameters
Hypovolemia	Decreased preload	↑	↑	Constant	Low	Low	Low	CVP ↓ PCWP↓ CO↓
Hypervolemia	Increased preload	↓	↓	Constant	High	High	High	CVP↑ PCWP↑ CO →
RV failure	Decreased preload	↑	↑	Constant	Low	Low	Low	CVP ↑ PCWP↓ CO↓
Moderate hypertension	Increased afterload	↑	↑↑	Constant	Low	Low	High (systolic predominant flow)	–
Severe hypertension	Increased afterload	↑↑	↑↑	Constant	Low	Low	Low (reduced flow in systole and diastole)	PCWP↑
Outflow obstruction	Increased afterload	↑	↑	Constant	Low	Low	Variable	PCWP↑ CO↓
Inflow obstruction	Decreased preload	↑	↑	Constant	Low	Low	Variable	PCWP↑ CO↓
Relative Low pump speed	–	–	–	Low	Low	Low	High	PCWP↑ or → CO↓
Relative high pump speed	–	–	–	High	High	High	Low	PCWP↓ CO↑
Sepsis/vasodilation	Decreased afterload	↓	↓	Constant	High	High	Low	PCWP↓ CO↑
LV recovery	–	↓	↑	Constant	High	High	High	PCWP↓ CO↑
Pump thrombosis	–	–	–	Constant	High (power spikes)	High (overestimation)	Variable	CO↓
Aortic insufficiency	Increased preload, Decreased afterload	↓	↓	Constant	High	High	Low	PCWP↑ CO↓
Tamponade	–	↑	→	Constant	Low	Low	Low	CO↓

Abbreviations: *CVP* central venous pressure, *PCWP* pulmonary capillary wedge pressure, *CO* cardiac output, *LV* left ventricle, *RV* right ventricle

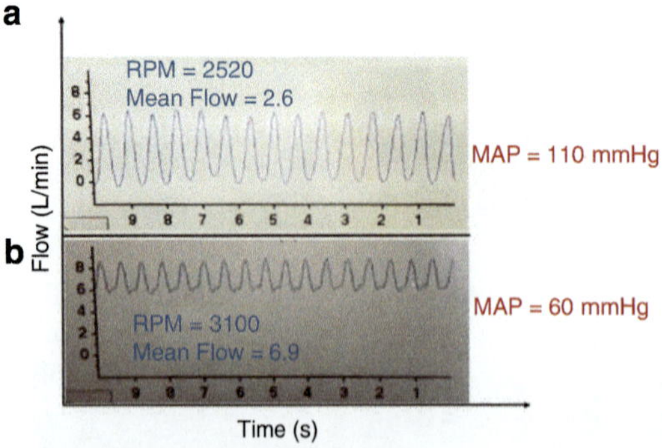

Fig. 5.12 (**a**) Typical waveforms for a hypertensive patient (low flow, high pulsatility). (**b**) Flow waveform for a hypotensive patient (high flow, low pulsatility) [28]. (Reprinted from Rich and Burkhoff [28], https://journals.lww.com/asaiojournal/pages/default.aspx, with permission from Wolters Kluwer Health, Inc)

increased at the expense of LVEDV [51, 52]. Hemodynamic responses to exercise in cf-LVADs are yet to be thoroughly characterized. At rest, LVADs improve hemodynamics, functional capacity, and ventilatory function. However, during exercise, despite hemodynamic support provided by LVAD, patients are unable to reach age- and sex-predicted normal aerobic capacity, and exercise performance is reduced by half the expected value [53, 54]. The reduction in functional capacity during exercise is multifactorial [53, 54]. No difference in hemodynamic support and exercise capacity has been noted between pulsatile and cf-LVADs [7].

Hemodynamic Factors Affecting VAD Flow During Exercise

The optimal setting for cf-LVADs (described in ramp tests) permits intermittent AV opening. As a result, at rest, majority of the CO is maintained by the VAD. During exercise, an increase in heart rate, total cardiac output, mean systemic arterial pressure, mean PA pressure, wedge pressure, right atrial pressure, minute ventilation, and LVEDV has been reported [52].

As the native LV and VAD work as two pumps in parallel, any change in the filling pressure has a simultaneous effect on flows through both the pumps. Jerson et al. showed that pump flow during exercise increases initially and later stabilizes at a moderate level despite a gradual increase in total cardiac output until maximum exercise suggesting an increased contribution from the native ventricle [13]. Physiologically this can be explained as follows: an increase in LV filling pressures causes a reduction in pump ΔP, and pump flow increases through the VAD. At the same time, increase in preload and end-systolic elastance invokes the Frank-Starling relationship and facilitates opening of the aortic valve and ejection of the blood from the LV. On PV loops the effect of exercise at constant RPM results in a change from triangular shape of the loop at rest to a more trapezoidal shape with exercise (Fig. 5.13) [52]. With both pumps working in parallel, an initial increase in CO along with MAP is noted. The elevated MAP results in an increased ΔP across the pump and prevents further increase in pump flow [13].

Chronotropic Incompetence and Preload

The ability to augment heart rate plays a major role in augmenting stroke volume in healthy patients and has been associated with decreased exercise performance in heart failure [55]. The importance of chronotropic incompetence in patients with cf-LVAD is controversial. Early studies in calves showed that at fixed pump speeds, LVAD flow occurred predominantly in systole in exercise and was caused by an increase in the heart rate [56]. However, recently, Muthiah et al. [57] have demonstrated that changes in maximum and minimum heart rates by adjusting the pacemaker function did not show a significant change in cardiac output through the VAD in patients with closed AV valves. In contrast, changes in preload was associated with significant changes in the LVAD flow in the same set of patients as evaluated by tilt table testing. This was further supported in a study by Hu et al. [58] wherein recumbent position and not heart rate was associated with increased pump flow suggesting that preload plays a key role during exercise.

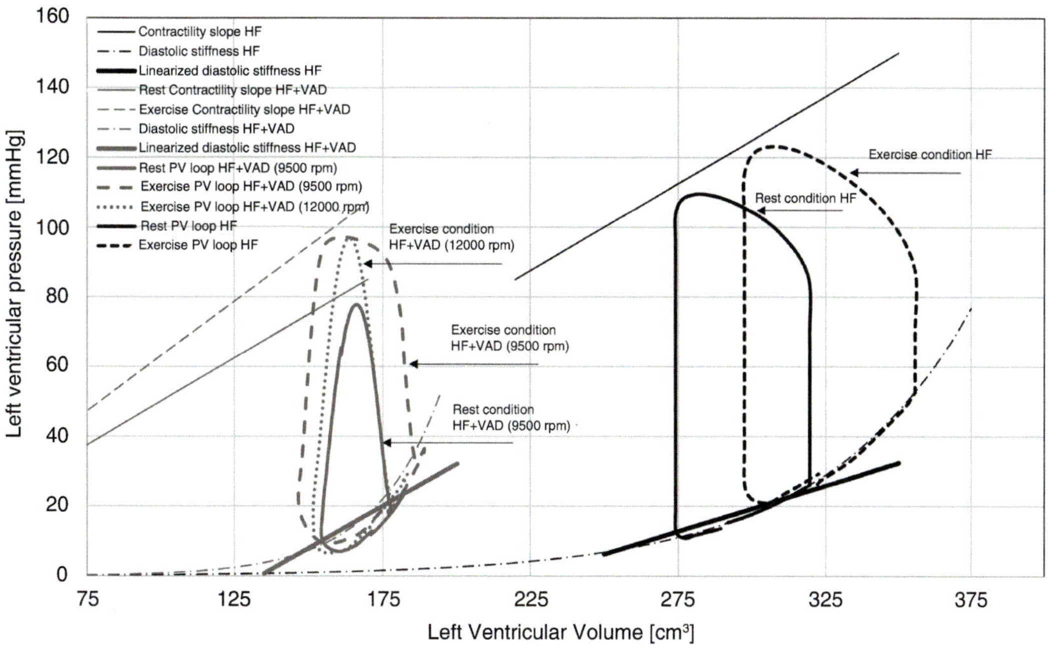

Fig. 5.13 Physiology of exercise in a heart failure patient versus an LVAD patient at different RPMs. (Reproduced under an open access license (CC BY) [52])

Effect of Pump Speed Adjustment During Exercise

In current practice, cf-LVADs are run at fixed RPMs irrespective of the level of activity. Several studies have assessed the role of active pump modulation with mixed results [59, 60]. No significant benefit of increasing pump speed during exercise in terms of total cardiac output has been reported in recent literature [61, 62]. On the contrary, increased in PCWP and RAP during exercise were noted, even with increased speed [52, 62, 63]. The likely explanation for this phenomenon is that, in contrast to the failed unassisted ventricle, the assisted ventricle has stiffer elastic properties and works on the nonlinear portion of diastolic stiffness resulting in increased filling pressures with minimal changes in the LVEDD. Exercise-adjusted pump speeds may still have a beneficial effect on the PV loop morphology as evidenced by return of the PV loop to a more triangular shape when LVAD speed is increased during exercise. At present, unloading provided by Cf-LVADs remains suboptimal. Ideal algorithms for pump speed during exercise at present are not available, and LV suction remains a pitfall of increasing pump speed.

Summary

In addition to restoring systemic blood flow and blood pressure, LVADs play a crucial role in unloading the left ventricle and facilitating reverse remodeling and recovery. We have provided an in-depth description of ventricular mechanics, vascular properties, and pump hemodynamics, thus providing a strong foundation for understanding the complex nature of their interactions that determine blood flow to the body, blood pressure, and LV unloading. In practice, these concepts are fundamental to the understanding and interpretation of hemodynamic ramp tests. In addition, we have detailed how factors intrinsic to the pump, the heart, and the vasculature (systemic and pulmonary) impact LVAD-assisted circulation and lead to either suboptimal hemodynamic support or, in extreme cases, complications such as RV dysfunction and pump thrombosis resulting in hemodynamic collapse.

References

1. Slaughter MS, Rogers JG, Milano CA, Russell SD, Conte JV, Feldman D, et al. Advanced heart failure treated with continuous-flow left ventricular assist device. N Engl J Med. 2009;361(23):2241–51.
2. Miller LW, Pagani FD, Russell SD, John R, Boyle AJ, Aaronson KD, et al. Use of a continuous-flow device in patients awaiting heart transplantation. N Engl J Med. 2007;357(9):885–96.
3. Pagani FD, Miller LW, Russell SD, Aaronson KD, John R, Boyle AJ, et al. Extended mechanical circulatory support with a continuous-flow rotary left ventricular assist device. J Am Coll Cardiol. 2009;54(4):312–21.
4. Kim G, Uriel N, Burkhoff D. Reverse remodeling and myocardial recovery in heart failure. Nat Rev Cardiol. 2018;15(2):83.
5. Hunt SA, Frazier OH. Mechanical circulatory support and cardiac transplantation. Circulation. 1998;97(20):2079–90.
6. Moazami N, Fukamachi K, Kobayashi M, Smedira NG, Hoercher KJ, Massiello A, et al. Axial and centrifugal continuous-flow rotary pumps: a translation from pump mechanics to clinical practice. J Heart Lung Transplant. 2013;32(1):1–11.
7. Haft J, Armstrong W, Dyke DB, Aaronson KD, Koelling TM, Farrar DJ, et al. Hemodynamic and exercise performance with pulsatile and continuous-flow left ventricular assist devices. Circulation. 2007;116(11 Suppl):I8–15.
8. Pagani FD. Continuous-flow rotary left ventricular assist devices with "3rd generation" design. Semin Thorac Cardiovasc Surg. 2008;20(3):255–63.
9. Farrar DJ, Bourque K, Dague CP, Cotter CJ, Poirier VL. Design features, developmental status, and experimental results with the Heartmate III centrifugal left ventricular assist system with a magnetically levitated rotor. ASAIO J. 2007;53(3):310–5.
10. Capoccia M. Mechanical circulatory support for advanced heart failure: are we about to witness a new "Gold Standard"? J Cardiovasc Dev Dis. 2016;3(4):35.
11. Tchoukina I, Smallfield MC, Shah KB. Device management and flow optimization on left ventricular assist device support. Crit Care Clin. 2018;34(3):453–63.
12. Griffith BP, Kormos RL, Borovetz HS, Litwak K, Antaki JF, Poirier VL, et al. HeartMate II left ventricular assist system: from concept to first clinical use. Ann Thorac Surg. 2001;71(3 Suppl):S116–20; discussion S4–6.
13. Martina J, de Jonge N, Rutten M, Kirkels JH, Klopping C, Rodermans B, et al. Exercise hemodynamics during extended continuous flow left ventricular assist device support: the response of systemic cardiovascular parameters and pump performance. Artif Organs. 2013;37(9):754–62.
14. Potapov EV, Loebe M, Nasseri BA, Sinawski H, Koster A, Kuppe H, et al. Pulsatile flow in patients with a novel nonpulsatile implantable ventricular assist device. Circulation. 2000;102(19 Suppl 3):III183–7.
15. Tagusari O, Yamazaki K, Litwak P, Antaki JF, Watach M, Gordon LM, et al. Effect of pressure-flow relationship of centrifugal pump on in vivo hemodynamics: a consideration for design. Artif Organs. 1998;22(5):399–404.
16. Griffith K, Jenkins E, Pagani FD. First American experience with the Terumo DuraHeart left ventricular assist system. Perfusion. 2009;24(2): 83–9.
17. Chatterjee K. Coronary hemodynamics in heart failure and effects of therapeutic interventions. J Card Fail. 2009;15(2):116–23.
18. Rodbard S, Williams CB, Rodbard D, Berglung E. Myocardial tension and oxygen uptake. Circ Res. 1964;14:139–49.
19. Burkhoff D, Sayer G, Doshi D, Uriel N. Hemodynamics of mechanical circulatory support. J Am Coll Cardiol. 2015;66(23):2663–74.
20. Burkhoff D, Naidu SS. The science behind percutaneous hemodynamic support: a review and comparison of support strategies. Catheter Cardiovasc Interv. 2012;80(5):816–29.
21. Suga H. Total mechanical energy of a ventricle model and cardiac oxygen consumption. Am J Phys. 1979;236(3):H498–505.
22. Takaoka H, Takeuchi M, Odake M, Hayashi Y, Hata K, Mori M, et al. Comparison of hemodynamic determinants for myocardial oxygen consumption under different contractile states in human ventricle. Circulation. 1993;87(1):59–69.
23. Uriel N, Sayer G, Annamalai S, Kapur NK, Burkhoff D. Mechanical unloading in heart failure. J Am Coll Cardiol. 2018;72(5):569–80.
24. Hall JL, Fermin DR, Birks EJ, Barton PJ, Slaughter M, Eckman P, et al. Clinical, molecular, and genomic changes in response to a left ventricular assist device. J Am Coll Cardiol. 2011;57(6):641–52.
25. Sunagawa G, Byram N, Karimov JH, Horvath DJ, Moazami N, Starling RC, et al. In vitro hemodynamic characterization of HeartMate II at 6000 rpm: implications for weaning and recovery. J Thorac Cardiovasc Surg. 2015;150(2):343–8.
26. Noor MR, Ho CH, Parker KH, Simon AR, Banner NR, Bowles CT. Investigation of the characteristics of HeartWare HVAD and Thoratec HeartMate II under steady and pulsatile flow conditions. Artif Organs. 2016;40(6):549–60.
27. Burkhoff D, Dickstein ML, Schleicher T. Harvi – Online. Retrieved from https://harvi.online/ 2017 updated 4/29/2017. Available from: http://harvi.online/.

28. Rich JD, Burkhoff D. HVAD flow waveform morphologies: theoretical foundation and implications for clinical practice. ASAIO J. 2017;63(5):526–35.
29. Ayre PJ, Vidakovic SS, Tansley GD, Watterson PA, Lovell NH. Sensorless flow and head estimation in the VentrAssist rotary blood pump. Artif Organs. 2000;24(8):585–8.
30. Hornick P, Taylor K. Pulsatile and nonpulsatile perfusion: the continuing controversy. J Cardiothorac Vasc Anesth. 1997;11(3):310–5.
31. Russell SD, Rogers JG, Milano CA, Dyke DB, Pagani FD, Aranda JM, et al. Renal and hepatic function improve in advanced heart failure patients during continuous-flow support with the HeartMate II left ventricular assist device. Circulation. 2009;120(23):2352–7.
32. Bourque K, Dague C, Farrar D, Harms K, Tamez D, Cohn W, et al. In vivo assessment of a rotary left ventricular assist device-induced artificial pulse in the proximal and distal aorta. Artif Organs. 2006;30(8):638–42.
33. Guan Y, Karkhanis T, Wang S, Rider A, Koenig SC, Slaughter MS, et al. Physiologic benefits of pulsatile perfusion during mechanical circulatory support for the treatment of acute and chronic heart failure in adults. Artif Organs. 2010;34(7):529–36.
34. Cornwell WK 3rd, Tarumi T, Stickford A, Lawley J, Roberts M, Parker R, et al. Restoration of pulsatile flow reduces sympathetic nerve activity among individuals with continuous-flow left ventricular assist devices. Circulation. 2015;132(24):2316–22.
35. Zimpfer D, Strueber M, Aigner P, Schmitto JD, Fiane AE, Larbalestier R, et al. Evaluation of the HeartWare ventricular assist device Lavare cycle in a particle image velocimetry model and in clinical practice. Eur J Cardiothorac Surg. 2016;50(5):839–48.
36. Slaughter MS, Pagani FD, Rogers JG, Miller LW, Sun B, Russell SD, et al. Clinical management of continuous-flow left ventricular assist devices in advanced heart failure. J Heart Lung Transplant. 2010;29(4 Suppl):S1–39.
37. Uriel N, Sayer G, Addetia K, Fedson S, Kim GH, Rodgers D, et al. Hemodynamic ramp tests in patients with left ventricular assist devices. JACC Heart Fail. 2016;4(3):208–17.
38. Uriel N, Morrison KA, Garan AR, Kato TS, Yuzefpolskaya M, Latif F, et al. Development of a novel echocardiography ramp test for speed optimization and diagnosis of device thrombosis in continuous-flow left ventricular assist devices: the Columbia ramp study. J Am Coll Cardiol. 2012;60(18):1764–75.
39. Uriel N, Levin AP, Sayer GT, Mody KP, Thomas SS, Adatya S, et al. Left ventricular decompression during speed optimization ramps in patients supported by continuous-flow left ventricular assist devices: device-specific performance characteristics and impact on diagnostic algorithms. J Card Fail. 2015;21(10):785–91.
40. Imamura T, Burkhoff D, Rodgers D, Adatya S, Sarswat N, Kim G, et al. Repeated ramp tests on stable LVAD patients reveal patient-specific hemodynamic fingerprint. ASAIO J. 2018;64(6):701–7.
41. Grinstein J, Imamura T, Kruse E, Kalantari S, Rodgers D, Adatya S, et al. Echocardiographic predictors of hemodynamics in patients supported with left ventricular assist devices. J Card Fail. 2018;24(9):561–7.
42. Pennings KA, Martina JR, Rodermans BF, Lahpor JR, van de Vosse FN, de Mol BA, et al. Pump flow estimation from pressure head and power uptake for the HeartAssist5, HeartMate II, and HeartWare VADs. ASAIO J. 2013;59(4):420–6.
43. Klotz S, Deng MC, Stypmann J, Roetker J, Wilhelm MJ, Hammel D, et al. Left ventricular pressure and volume unloading during pulsatile versus nonpulsatile left ventricular assist device support. Ann Thorac Surg. 2004;77(1):143–9; discussion 9–50.
44. Garcia S, Kandar F, Boyle A, Colvin-Adams M, Lliao K, Joyce L, et al. Effects of pulsatile- and continuous-flow left ventricular assist devices on left ventricular unloading. J Heart Lung Transplant. 2008;27(3):261–7.
45. Farrar DJ, Compton PG, Hershon JJ, Fonger JD, Hill JD. Right heart interaction with the mechanically assisted left heart. World J Surg. 1985;9(1):89–102.
46. Imamura T, Chung B, Nguyen A, Rodgers D, Sayer G, Adatya S, et al. Decoupling between diastolic pulmonary artery pressure and pulmonary capillary wedge pressure as a prognostic factor after continuous flow ventricular assist device implantation. Circ Heart Fail. 2017;10(9):e003882.
47. Nassif ME, Tibrewala A, Raymer DS, Andruska A, Novak E, Vader JM, et al. Systolic blood pressure on discharge after left ventricular assist device insertion is associated with subsequent stroke. J Heart Lung Transplant. 2015;34(4):503–8.
48. Patil NP, Mohite PN, Sabashnikov A, Dhar D, Weymann A, Zeriouh M, et al. Does postoperative blood pressure influence development of aortic regurgitation following continuous-flow left ventricular assist device implantation? Eur J Cardiothorac Surg. 2016;49(3):788–94.
49. Najjar SS, Slaughter MS, Pagani FD, Starling RC, McGee EC, Eckman P, et al. An analysis of pump thrombus events in patients in the HeartWare ADVANCE bridge to transplant and continued access protocol trial. J Heart Lung Transplant. 2014;33(1):23–34.
50. Akimoto T, Yamazaki K, Litwak P, Litwak KN, Tagusari O, Mori T, et al. Relationship of blood pressure and pump flow in an implantable centrifugal blood pump during hypertension. ASAIO J. 2000;46(5):596–9.
51. Fagard R. Athlete's heart. Circulation. 2001;103(6):E28–9.
52. Fresiello L, Rademakers F, Claus P, Ferrari G, Di Molfetta A, Meyns B. Exercise physiology with a left ventricular assist device: analysis of heart-pump interaction with a computational simulator. PLoS One. 2017;12(7):e0181879.

53. Loyaga-Rendon RY, Plaisance EP, Arena R, Shah K. Exercise physiology, testing, and training in patients supported by a left ventricular assist device. J Heart Lung Transplant. 2015;34(8):1005–16.
54. Hayward CS, Fresiello L, Meyns B. Exercise physiology in chronic mechanical circulatory support patients: vascular function and beyond. Curr Opin Cardiol. 2016;31(3):292–8.
55. Pina IL, Apstein CS, Balady GJ, Belardinelli R, Chaitman BR, Duscha BD, et al. Exercise and heart failure: a statement from the American Heart Association Committee on exercise, rehabilitation, and prevention. Circulation. 2003;107(8):1210–25.
56. Akimoto T, Yamazaki K, Litwak P, Litwak KN, Tagusari O, Mori T, et al. Rotary blood pump flow spontaneously increases during exercise under constant pump speed: results of a chronic study. Artif Organs. 1999;23(8):797–801.
57. Muthiah K, Gupta S, Otton J, Robson D, Walker R, Tay A, et al. Body position and activity, but not heart rate, affect pump flows in patients with continuous-flow left ventricular assist devices. JACC Heart Fail. 2014;2(4):323–30.
58. Hu SX, Keogh AM, Macdonald PS, Kotlyar E, Robson D, Harkess M, et al. Interaction between physical activity and continuous-flow left ventricular assist device function in outpatients. J Card Fail. 2013;19(3):169–75.
59. Jacquet L, Vancaenegem O, Pasquet A, Matte P, Poncelet A, Price J, et al. Exercise capacity in patients supported with rotary blood pumps is improved by a spontaneous increase of pump flow at constant pump speed and by a rise in native cardiac output. Artif Organs. 2011;35(7):682–90.
60. Schima H, Vollkron M, Jantsch U, Crevenna R, Roethy W, Benkowski R, et al. First clinical experience with an automatic control system for rotary blood pumps during ergometry and right-heart catheterization. J Heart Lung Transplant. 2006;25(2):167–73.
61. Brassard P, Jensen AS, Nordsborg N, Gustafsson F, Moller JE, Hassager C, et al. Central and peripheral blood flow during exercise with a continuous-flow left ventricular assist device: constant versus increasing pump speed: a pilot study. Circ Heart Fail. 2011;4(5):554–60.
62. Muthiah K, Robson D, Prichard R, Walker R, Gupta S, Keogh AM, et al. Effect of exercise and pump speed modulation on invasive hemodynamics in patients with centrifugal continuous-flow left ventricular assist devices. J Heart Lung Transplant. 2015;34(4):522–9.
63. Burkhoff D, Dickstein ML, Schleicher T. Harvi – Online. Retrieved from http://harvi.online/. 2017.

Engineering Requirements for Mechanical Circulatory Support Devices

J. Timothy Baldwin

Introduction

The medical device industry heavily relies upon engineering principles to design, analyze, develop, test, and eventually manufacture their devices. Because mechanical circulatory support (MCS) devices incorporate complex features into their designs and provide critical life-sustaining function, successful development of these devices involves many engineering disciplines and requires well-organized, thorough, and thoughtful application of those disciplines. Over the past five decades, the technology and designs of MCS devices, particularly ventricular assist devices (VADs), have evolved, spanning from the early versions of implantable pneumatic positive displacement pumps (which led to the start of the NIH Artificial Heart Program) to the current generation of technologically advanced continuous flow VADs [1–5]. Evolution of the devices is evident from their designs. The early positive displacement VADs such as the Thoratec PVAD (Fig. 6.1) used air compressors, simple analog controllers, pneumatic hoses, and polymer blood sacs. These were followed by electro-mechanical positive displacement VADs which used mechanized pusher plates operating on polymer blood sacs (e.g., Arrow LionHeart LVAS and Novacor VAS) or a flexible diaphragm coupled to a rigid housing (HeartMate XVE, Fig. 6.2) [6–8]. These systems involved motors, batteries, cables, more sophisticated controllers, and, in the case of the HeartMate XVE, a texturized metal blood-contacting surface. These, in turn, were followed by smaller and more reliable rotary blood pumps that exposed blood to high shear for such short periods of time that low levels of hemolysis and platelet activation could be achieved. While the first clinical success of a rotary blood pump was realized through the use of the Hemopump, the clinical field is now dominated by the current generation of continuous flow VADs such as the Jarvik 2000, HeartMate II (Fig. 6.2), HVAD, and HeartMate 3 (Fig. 6.3) [2–4, 9, 10]. The last three have all been successfully evaluated in clinical

Fig. 6.1 The Thoratec Paracorporeal Ventricular Assist Device or PVAD™. The PVAD™ was developed in the 1970s to provide left, right, or bi-ventricular pulsatile support for weeks to months. The PVAD™ system is composed of the blood pump (shown in image), cannulae, and a pneumatic driver. (Image provided by St. Jude Medical, Inc. CentriMag, PediMag, and PVAD are trademarks of Abbott or its related companies. Reproduced with permission of Abbott, © 2019. All rights reserved)

J. T. Baldwin, MS, PhD (✉)
National Heart Lung and Blood Institute, Bethesda, MD, USA
e-mail: Tim.Baldwin@fda.hhs.gov

© Springer Nature Switzerland AG 2020
J. H. Karimov et al. (eds.), *Mechanical Support for Heart Failure*,
https://doi.org/10.1007/978-3-030-47809-4_6

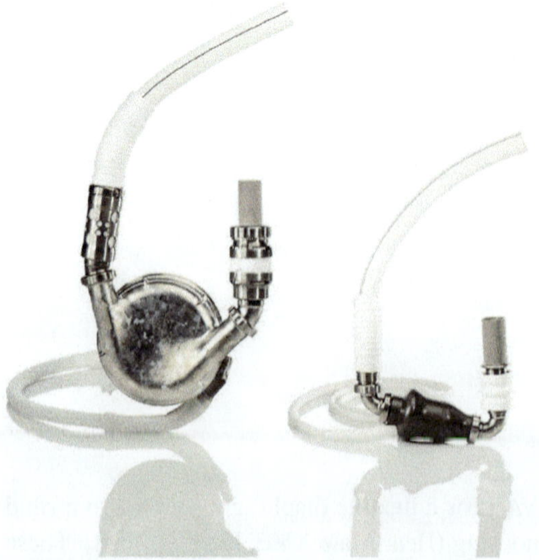

Fig. 6.2 The Thoratec HeartMate XVE™ (left) and HeartMate II™ (right) LVADs. The HM XVE™ was a widely used electro-mechanical pulsatile flow VAD with a textured surface that promoted endothelialization. The HM II™ was the first FDA-approved continuous flow VAD for bridge to transplant and destination therapy indications. It provided superior durability and, due to its smaller size, ease of implantation over the HM XVE™. (Images provided by St. Jude Medical, Inc. HeartMate II, HeartMate 3, and HeartMate are trademarks of Abbott or its related companies. Reproduced with permission of Abbott, © 2019. All rights reserved)

Fig. 6.3 The centrifugal pump used in the HeartMate 3™ system has fully magnetically levitated rotor and delivers up to 10 liters of blood/minute. The system, which is designed and used as a bridge to transplant or recovery and for destination therapy, can be used in patients ranging from large children to adults. The system pulses the flow to reduce the potential for thrombosis and has a wearable external battery pack and a small controller. (Image provided by St. Jude Medical, Inc. HeartMate II, HeartMate 3, and HeartMate are trademarks of Abbott or its related companies. Reproduced with permission of Abbott, © 2019. All rights reserved)

trials in the USA which led to FDA approvals over the past decade [11–13]. The various MCS devices from the latest generation resulted from different and more advanced technologies than their predecessors. These included technologies to design and manufacture the devices (such as processes to produce highly polished blood-contacting titanium surfaces and computational fluid dynamics used in the design process) as well as those incorporated into the designs (such as digital controllers, various bearing designs including magnetically levitated bearings, and reliable and rechargeable batteries). Now, other technologies are being considered and developed for future use in MCS devices. One is transcutaneous energy transmission systems (TETs) which were successfully used in the LionHeart VAD and the AbioCor TAH [14, 15]. Other technologies waiting in the wings for MCS systems include a system to wirelessly power the devices, while another uses magnetic coupling of an actuating piston to a motor to create a valveless, pulsatile, rotary, dual-piston, positive displacement pump [16, 17].

Engineering MCS Devices

With the diverse sizes, modes of operation, intended populations, and intended durations of use of MCS devices, the technologies incorporated into their designs and the engineering involved in developing the devices necessarily vary to a large degree. And, with more advanced MCS devices being developed for the future, including those for small pediatric patients, the engineering and technology used in these devices will continue to vary and evolve [18–20].

The engineering involved to develop MCS devices over the past five decades has been needed to address the materials used in, loads imposed by, fluid mechanics created within, power required by, and the control of the devices to result in acceptable biocompatibility, durability, and reliability. Clearly, the engineering required for the development of the early generation of positive displacement pneumatic MCS devices is not the same as it was for the latest generation of continuous flow blood pumps. The early pneumatic devices were fabricated using polymer blood sacs, polycarbonate housings, and various types of prosthetic heart valves, were driven using large air compressors to produce pulsatile flows, and were controlled by analog circuits. Developing these features for these devices involved some basic mechanical, electrical, materials, and chemical engineering and some of the emerging field of biomedical engineering to address the issue of biocompatibility. The latest generation of MCS devices involves implantable batteries, magnetically levitated bearings, highly polished titanium surfaces, pressure and flow sensors, sophisticated digital controllers, passivated surfaces, precisely machined components, and various engineering tools to design, develop, and analyze device performance. As a result, the development of MCS devices today involves advanced types of mechanical, electrical, materials, and chemical engineering as well as other sub-specialties such as computer, biomedical, reliability, human factors, and software engineering. And the engineering needed for the next generation of MCS devices will continue to evolve to fit the needs of those future designs.

Differing MCS Designs and Their Corresponding Engineering Requirements

The specific engineering requirements for a mechanical circulatory support device heavily depend on the design and specifications for the device. This can be illustrated by considering two devices with clear differences in their indications for use which lie on opposite ends of the expected duration-of-use spectrum and differ substantially in expected patient size and etiology of heart failure: the Abbott PediMag Blood Pump (Fig. 6.4) and the SynCardia temporary TAH (Fig. 6.5). The PediMag (also known as the PediVAS outside the USA) is a magnetically levitated continuous flow pump designed to provide partial or full extracorporeal circulatory support for hours to weeks in children. In contrast, the SynCardia TAH is a pneumatically actuated positive displacement designed to replace the failing heart in order to bridge patients to a heart transplant. The PediMag, being designed for short-term support in small children, is used extracorporeally and is designed to only deliver the needed maximum flow rate of less than 2 LPM [21]. In contrast, the SynCardia TAH is necessarily fully implantable and can provide more than 9 LPM because it is designed and used to replace the heart and provide the chronic required systemic and pulmonary circulation in a range of patients including larger adults [22]. To meet the indications for use and also due to their basic designs, each of these devices has significantly different engineering requirements. For example, SynCardia TAH's flexible polymer blood sacs (and other system components such as the pneumatic tubing, compressor, controller, and heart valves) needed to be designed and engineered to be durable enough to withstand the chronic cyclic loading by the pneumatic driver to pump the full requirements of the systemic and pulmonary systems at 110 or more beats/minute for months or years. Apparently, these goals were met. According to the FDA's Summary of Safety and Effectiveness for the SynCardia TAH, the device has a 1-year reliability of 88% with 90% confidence [23]. This level of reliability has been demonstrated through clinical reports of device use of up to 3.75 years and with many patients who have survived longer than 1 year of support on the device [22].

In contrast, because the PediMag/PediVAS pump uses a magnetically levitated rotor for acute support, the durability for any one use of the device's disposable pump head and cannulae needs only be approximately 30 days, the longest duration expected when used as part of an ECMO system. Per the published Summary of Safety and Probable Benefit for the CentriMag right ventricular assist system, which is the same as the PediMag system except for the pump head, the system has a 30-day 90% reliability with 90% confidence [24]. With no bearings and no wearing parts in the pump head and being used for such a short period of time, the reliability of the PediMag system is primarily a function of the motor, console, monitor, and flow probe which are intended to be used repeatedly. Therefore, the reliability of the system at any time of use must be based on the system's robustness to function

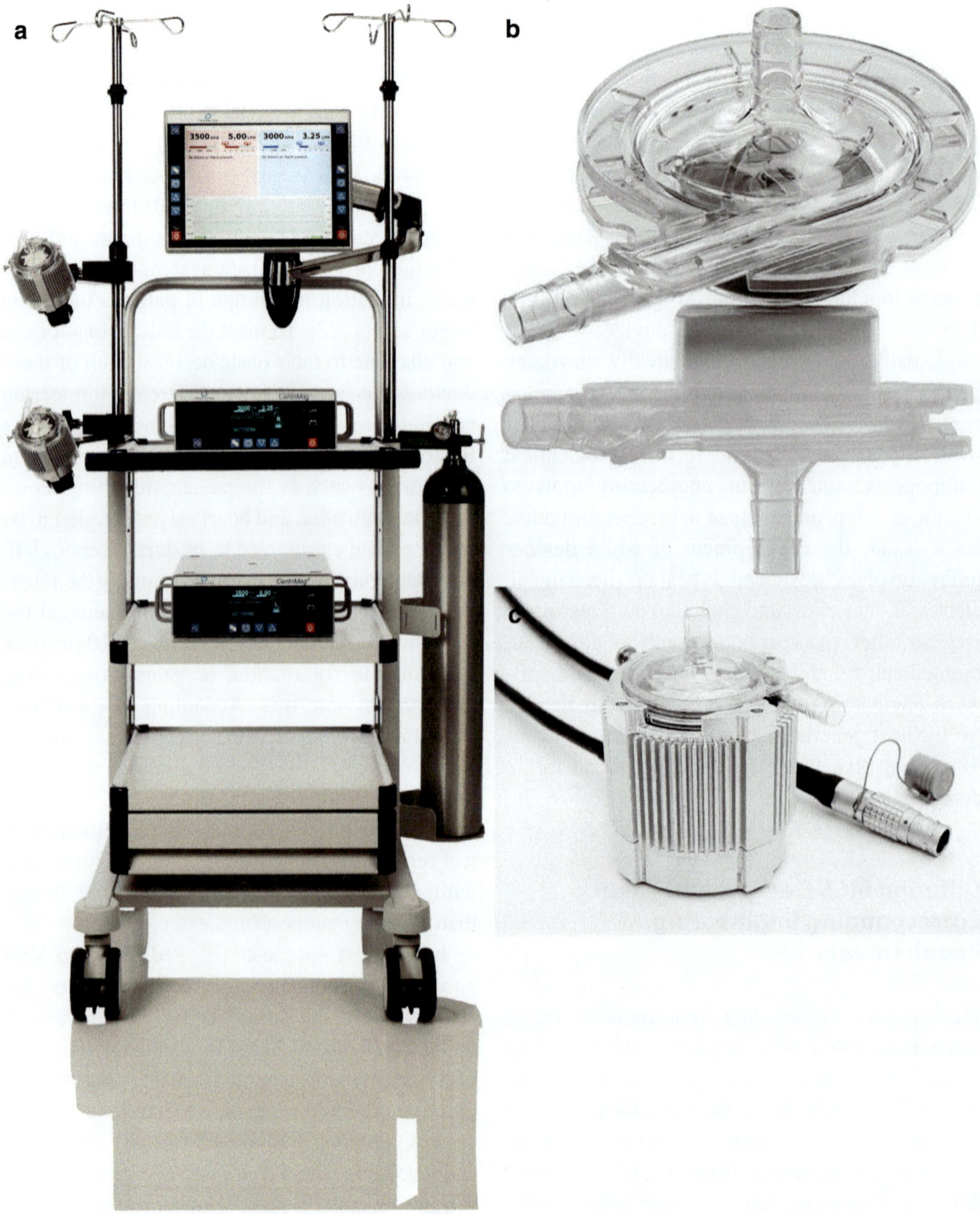

Fig. 6.4 The PediMag™ blood pump was designed to be used as part of the CentriMag™ acute circulatory support system (**a**) to provide short-term support of up to 1.5 L/min in children. System components include a single-use PediMag™ pump head which operates using a magnetically levitated rotor (**b**), a reusable motor to which the pump head is coupled (**c**), and other reusable system components. (Images provided by St. Jude Medical, Inc. CentriMag, PediMag, and PVAD are trademarks of Abbott or its related companies. Reproduced with permission of Abbott, © 2019. All rights reserved)

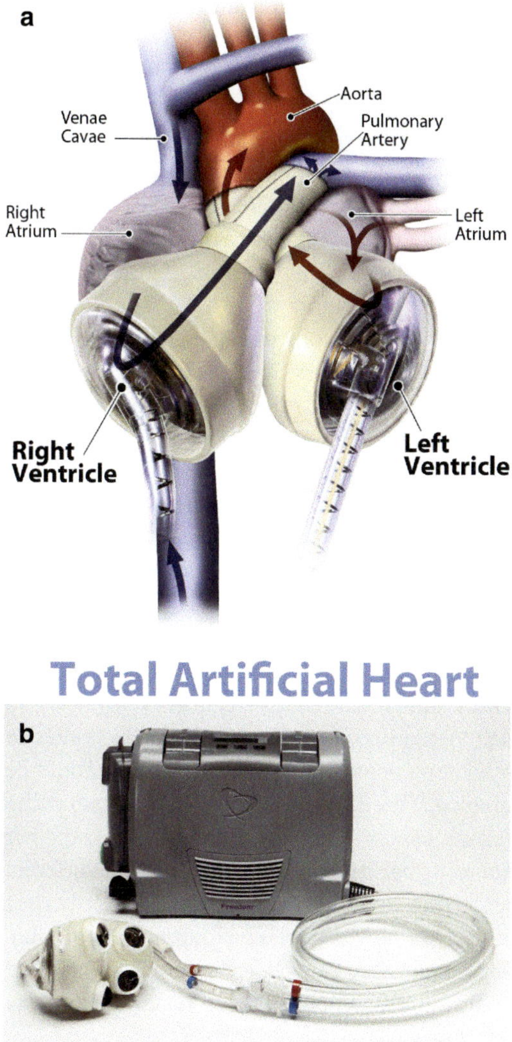

Fig. 6.5 The SynCardia temporary Total Artificial Heart illustrated as implanted (**a**) and with the Freedom® Portable Driver (**b**). In contrast to the PediMag™ blood pump (Fig. 6.4), the Syncardia TAH completely replaces the heart, and its 50 cc and 70 cc sizes provide the ability for its use in adolescent and adult patients. Photo credit: SynCardia. (Reproduced with permission pf SynCardia Systems, © 2019)

properly over repeated use of the reusable components. Fortunately, because the system is extracorporeal, backup components can be readily exchanged for any component failures, and this is, indeed, the procedure for malfunction of any of these components [25].

Clearly the durability performance requirements for these two devices are drastically different, and, as a result, the engineering and designs needed to meet those requirements are, accordingly, drastically different. Similarly, the related engineering and designs vary for both devices to meet the specified device requirements for other device characteristics such as functional performance, anatomic fit, user interface, controller performance, and others. The same is true for any device: the engineering requirements for any MCS device vary with its performance requirements and indications for use. As such, there is no standard set of engineering requirements for MCS devices. Instead, MCS device engineering requirements are determined using the same tools used throughout the medical device field and in other high-tech industries such as the aerospace and automotive businesses, where safety is paramount: risk analysis.

Risk Analysis to Engineer MCS Devices

Because MCS devices are high risk, Class III Medical Devices, it is imperative for the companies who develop and manufacture them to analyze and manage the risk presented by them. Risk analysis is a part of the risk management process which is used to minimize and control risks associated with the use of a product, including medical devices such as MCS devices [26]. The details of risk analysis as they are particularly applied to MCS devices are beyond the scope of this chapter. However, it is important to understand the basics of risk analysis to appreciate how they are related to and dictate the engineering requirements of MCS devices. In a nutshell, risk analysis for any MCS device involves identifying hazards that could be experienced by users of the device and determining the severity or consequences of all identified hazards and the probability of each occurring. For MCS devices and medical devices in general, many hazards are commonly referred to as adverse events and range from highly severe (e.g., ischemic stroke and pump failure) to less severe (e.g., temporary reduced perfusion and percutaneous driveline infection). The risk of any hazard is estimated by incorporating both the severity of the hazard and probability of the hazard occurring. Severity can range from negligible (e.g., simple temporary discomfort) to catastrophic (e.g., patient death) and can be considered qualitative or quantitative using a numerical scale [26]. Likewise, the probability of a hazard occurring can range from frequent to improbable and can be considered qualitative or quantitative using a scale or actual estimates of the probability [26]. The risks determined by the risk analysis are then evaluated to determine if they are acceptable or not and, of those that are not, which need to be reduced.

Engineering of MCS devices is a key part to the risk analysis because it is generally focused on minimizing risks by incorporating engineering principles into the design to meet goals for structural and system reliability, biocompatibility, hemocompatibility, and others. As indicated previously, the engineering required for each MCS device will vary and will be assessed and depends on the risk analysis. To illustrate this, consider the previous comparison of the durability of the PediMag and SynCardia TAH. Many adverse events/hazards (stroke, pump failure, etc.) are similar for the two devices as are the assessment of severity of those hazards. However, the probability of each type of hazard occurring will be different, especially when considering the expected maximum duration of use. For an adverse event such as pump failure, the severity for both devices would be death unless the device could be successfully replaced or restarted in time. But, for the PediMag device, the duration over which pump failure needs to be minimized is 30 days, and components that are critical to its durability are the motor, controller, pump head, and cannulae. As such, the engineering requirements would be to minimize the possibility of pump failure for the pump head and cannulae for 30 days and the motor and other reusable equipment of the specified lifespan of those devices. In contrast, in the SynCardia TAH the duration over which pump failure needs to be minimized is a year or more, and some of the components that are critical to its durability are the polymer bloods sacs, the prosthetic valves, the compressor, and the controller. And with these, the replaceable components, such as the compressor and controller, may be serviceable and replaceable, so they may also be exchanged. As a result, the acceptable probability of failure of those components can be higher than of other non-replaceable components, and the engineering of those components may not require as high a reliability as that of other critical components such as the blood sacs and heart valves in this particular device for this particular hazard/adverse event. However, the engineering of the blood sacs, the heart valves for use in an artificial heart (as opposed for use to replace natural heart valves), the compressor, the controller, and other system components needs to be performed so that the risks are not only minimized but are acceptable.

For MCS devices, including these two examples, risk assessments are made by the manufacturers and reviewed by regulatory agencies to help determine if approvals of clinical studies or marketing applications for the devices should be granted. Those risk assessments go beyond pump failure, covering hazards such as stroke and other neurological dysfunctions, all types of different infections, arrhythmias, and others as well. They also cover how they can be minimized by design and engineering as well as the implementation of quality management systems for manufacturing of the devices.

This chapter on engineering requirements for MCS devices has demonstrated that due to their specific uses and requirements, no specific engineering requirements for MCS devices exist. Instead, engineering requirements are determined and used to design MCS devices based on the specific performance requirements for the device. How well an MCS device meets its engineering requirements is determined through pre-clinical and clinical analyses and evaluations and the ensuing risk analysis. The ISO Standard for Circulatory Support Devices (ISO 14708-5:2010E) provides guidance on the types of testing that should be performed which help to inform the risk analysis [27]. Testing specified in that standard includes that for (1) pump performance, fluid dynamics, vibration, and cavitation of internal components; (2) both internal and external control and drive units, including programming; (3) batteries and electrical connections; (4) grafts, conduits, and cuffs; (5) prosthetic heart valves and transcutaneous energy transmission systems; and (5) electromagnetic compatibility, materials and biocompatibility, and others. So, while no specific engineering requirements exist for MCS devices, with standards such as this which cover the range of technologies involved in MCS devices and the international standard for risk management (which is referenced in ISO 14708), manufacturers and developers have standard ways to determine, assess, refine, and, hopefully, meet the engineering requirements for their MCS devices.

Disclaimer

The views expressed are those of the author and not necessarily those of the National Heart, Lung, and Blood Institute or the National Institutes of Health.

References

1. Joyce LD, Noon GP, Joyce DL, DeBakey ME. Mechanical circulatory support–a historical review. ASAIO J. 2004;50(6):x–xii.
2. Delgado R, Bergheim M. HeartMate II left ventricular assist device: a new device for advanced heart failure. Expert Rev Med Devices. 2005;2(5):529–32.
3. Larose JA, Tamez D, Ashenuga M, Reyes C. Design concepts and principle of operation of the HeartWare ventricular assist system. ASAIO J. 2010;56(4):285–9.
4. Bourque K, Cotter C, Dague C, Harjes D, Dur O, Duhamel J, et al. Design rationale and preclinical evaluation of the HeartMate 3 left ventricular assist system for hemocompatibility. ASAIO J. 2016;62(4):375–83.
5. Englert JA 3rd, Davis JA, Krim SR. Mechanical circulatory support for the failing heart: continuous-flow left ventricular assist devices. Ochsner J. 2016;16(3):263–9.
6. Weiss WJ, Rosenberg G, Snyder AJ, Donachy J Sr, Reibson J, Kawaguchi O, et al. A completely implanted left ventricular assist device. Chronic in vivo testing. ASAIO J. 1993;39(3):M427–32.
7. Wheeldon DR, Jansen PG, Portner PM. The Novacor electrical implantable left ventricular assist system. Perfusion. 2000;15(4):355–61.
8. Dowling RD, Park SJ, Pagani FD, Tector AJ, Naka Y, Icenogle TB, et al. HeartMate VE LVAS design enhancements and its impact on device reliability. Eur J Cardiothorac Surg. 2004;25(6):958–63.
9. Frazier OH, Wampler RK, Duncan JM, Dear WE, Macris MP, Parnis SM, et al. First human use of the Hemopump, a catheter-mounted ventricular assist device. Ann Thorac Surg. 1990;49(2):299–304.
10. Selzman CH, Koliopoulou A, Glotzbach JP, McKellar SH. Evolutionary Improvements in the Jarvik 2000 left ventricular assist device. ASAIO J. 2018;64(6):827–30.
11. Mehra MR, Goldstein DJ, Uriel N, Cleveland JC Jr, Yuzefpolskaya M, Salerno C, et al. Two-year outcomes with a magnetically levitated cardiac pump in heart failure. N Engl J Med. 2018;378(15):1386–95.
12. Slaughter MS, Rogers JG, Milano CA, Russell SD, Conte JV, Feldman D, et al. Advanced heart failure treated with continuous-flow left ventricular assist device. N Engl J Med. 2009;361(23):2241–51.
13. Rogers JG, Pagani FD, Tatooles AJ, Bhat G, Slaughter MS, Birks EJ, et al. Intrapericardial left ventricular assist device for advanced heart failure. N Engl J Med. 2017;376(5):451–60.
14. Dowling RD, Gray LA Jr, Etoch SW, Laks H, Marelli D, Samuels L, et al. The AbioCor implantable replacement heart. Ann Thorac Surg. 2003;75(6 Suppl):S93–9.
15. Mehta SM, Pae WE Jr, Rosenberg G, Snyder AJ, Weiss WJ, Lewis JP, et al. The LionHeart LVD-2000: a completely implanted left ventricular assist device for chronic circulatory support. Ann Thorac Surg. 2001;71(3 Suppl):S156–61; discussion S83–4.
16. Waters BH, Park J, Bouwmeester JC, Valdovinos J, Geirsson A, Sample AP, et al. Electrical power to run ventricular assist devices using the free-range resonant electrical energy delivery system. J Heart Lung Transplant. 2018;37(12):1467–74.
17. Letsou GV, Pate TD, Gohean JR, Kurusz M, Longoria RG, Kaiser L, et al. Improved left ventricular unloading and circulatory support with synchronized pulsatile left ventricular assistance compared with continuous-flow left ventricular assistance in an acute porcine left ventricular failure model. J Thorac Cardiovasc Surg. 2010;140(5):1181–8.
18. Burki S, Adachi I. Pediatric ventricular assist devices: current challenges and future prospects. Vasc Health Risk Manag. 2017;13:177–85.
19. Olia SE, Wearden PD, Maul TM, Shankarraman V, Kocyildirim E, Snyder ST, et al. Preclinical performance of a pediatric mechanical circulatory support device: the PediaFlow ventricular assist device. J Thorac Cardiovasc Surg. 2018;156(4):1643–51, e7.
20. Baldwin JT, Adachi I, Teal J, Almond CA, Jaquiss RD, Massicotte MP, et al. Closing in on the PumpKIN trial of the Jarvik 2015 ventricular assist device.

Semin Thorac Cardiovasc Surg Pediatr Card Surg Annu. 2017;20:9–15.
21. Lorts A, Zafar F, Adachi I, Morales DL. Mechanical assist devices in neonates and infants. Semin Thorac Cardiovasc Surg Pediatr Card Surg Annu. 2014;17(1):91–5.
22. Slepian MJ, Alemu Y, Girdhar G, Soares JS, Smith RG, Einav S, et al. The Syncardia() total artificial heart: in vivo, in vitro, and computational modeling studies. J Biomech. 2013;46(2):266–75.
23. Accessdata.fda. PMA summary of safety and effectiveness data for the CardioWest temporary total artificial heart (TAH-t) 2004. Available from: https://www.accessdata.fda.gov/cdrh_docs/pdf3/P030011b.pdf.
24. Accessdata.fda. HDE summary of safety and probable benefit of the CentriMag RVAS 2008. Available from: https://www.accessdata.fda.gov/cdrh_docs/pdf7/H070004b.pdf.
25. Thoratec.com. 2018. Available from: http://www.thoratec.com/_assets/download-tracker/centrimag/0047/PL-0047_Rev_08_2nd_Generation_CentriMag_System_Operating_Manual_(US).pdf.
26. Organization IS. ISO 14971: Medical devices-application of risk management to medical devices 2007.
27. Organization IS. ISO 14708-5: Implants for surgery-active implantable medical devices. Circulatory support devices: ISO; 2010.

7 Clinical Requirements for Mechanical Circulatory Support Devices

Neel K. Ranganath, Katherine G. Phillips, and Nader Moazami

Introduction

In the last 20 years, the field of MCS has made tremendous progress in both temporary and long-term support devices, owing largely to the development of continuous-flow rotary blood pumps [1]. The previously utilized pulsatile volume-displacement pumps [2–4] were encumbered by inherent design limitations, most notably that pulsatile pumps require a chamber large enough to accommodate an adequate stroke volume. Thus, pulsatile pumps are relatively large and can only be implanted in patients with a large body habitus, to the exclusion of most women and adolescents, and often required extensive surgical dissection for appropriate placement. Additionally, limited long-term mechanical durability of cyclic pumps frequently required reoperations for device exchanges. Pulsatile pumps are also audible during normal operation and require larger caliber percutaneous lines, which are more susceptible to infection, in order to vent air. Though still effective in treating medically refractory heart failure, these design factors and their significant associated morbidity limited the widespread use of pulsatile MCS devices [5, 6].

N. K. Ranganath, MD · K. G. Phillips, BS
N. Moazami, MD (✉)
NYU Langone Health, Department of Cardiothoracic Surgery, New York, NY, USA
e-mail: neel.ranganath@nyulangone.org; nader.moazami@nyulangone.org

Continuous-flow rotary blood pumps allowed for a significant reduction in pump size, lower caliber percutaneous lines, and quiet operation [1, 7]. Following approval of the HeartMate II for destination therapy in 2010, the rate of durable MCS device placement increased dramatically as durable MCS devices could then be offered to women, adolescents, and older patients who are not transplant candidates [8]. In response to this rapid increase in clinical application, the International Society of Heart and Lung Transplantation (ISHLT) released an executive summary for MCS, including guidelines for patient selection [9]. However, considering the excellent clinical outcomes of currently approved durable MCS devices [10] and the newer pump technologies on the horizon, the role of MCS continues to expand, and patient selection is oftentimes a subjective, risk-benefit consideration: Do the potential survival and quality-of-life advantages of durable MCS device implantation outweigh the significant perioperative morbidity and mortality for this patient?

Temporary MCS has also seen greatly expanded clinical role in recent years. However, while long-term MCS owed its expanded role to a breakthrough technology, the technology necessary for temporary MCS has been available and utilized by clinicians in the form of cardiopulmonary bypass (CPB) since 1953 [11]. Early forays into temporary MCS involved prolonged weaning of patients with

© Springer Nature Switzerland AG 2020
J. H. Karimov et al. (eds.), *Mechanical Support for Heart Failure*,
https://doi.org/10.1007/978-3-030-47809-4_7

postcardiotomy cardiogenic shock using peripheral CPB, which is essentially analogous to extracorporeal membrane oxygenation (ECMO). However, continuous-flow rotary blood pump technology has also benefited temporary MCS by allowing the development of microaxial devices, many of which can be percutaneously implanted [12]. Additionally, many clinicians are advocating for earlier institution of temporary MCS in cardiogenic shock, inferring that earlier intervention would halt the pathological process and result in superior outcomes. Regardless of the anticipated duration of support, the role of MCS is expanding, and appropriate patient and device selection is critical in the face of rapidly changing technology.

Long-Term MCS and End-Stage Heart Failure

Long-term MCS is dedicated to the treatment of medically refractory, end-stage heart failure. The potential etiologies of congestive heart failure are extensive, but ultimately a decrease in ventricular performance leads to reduced end-organ perfusion and several deleterious compensatory mechanisms (Fig. 7.1). Reduced renal blood flow signals salt and water retention by activating the renin-angiotensin-aldosterone system [13–17]. Increased autonomic nervous system stimulation results in increased systemic vascular resistance, increased arterial blood pressure, and increased left ventricular filling pressures [18–21]. The

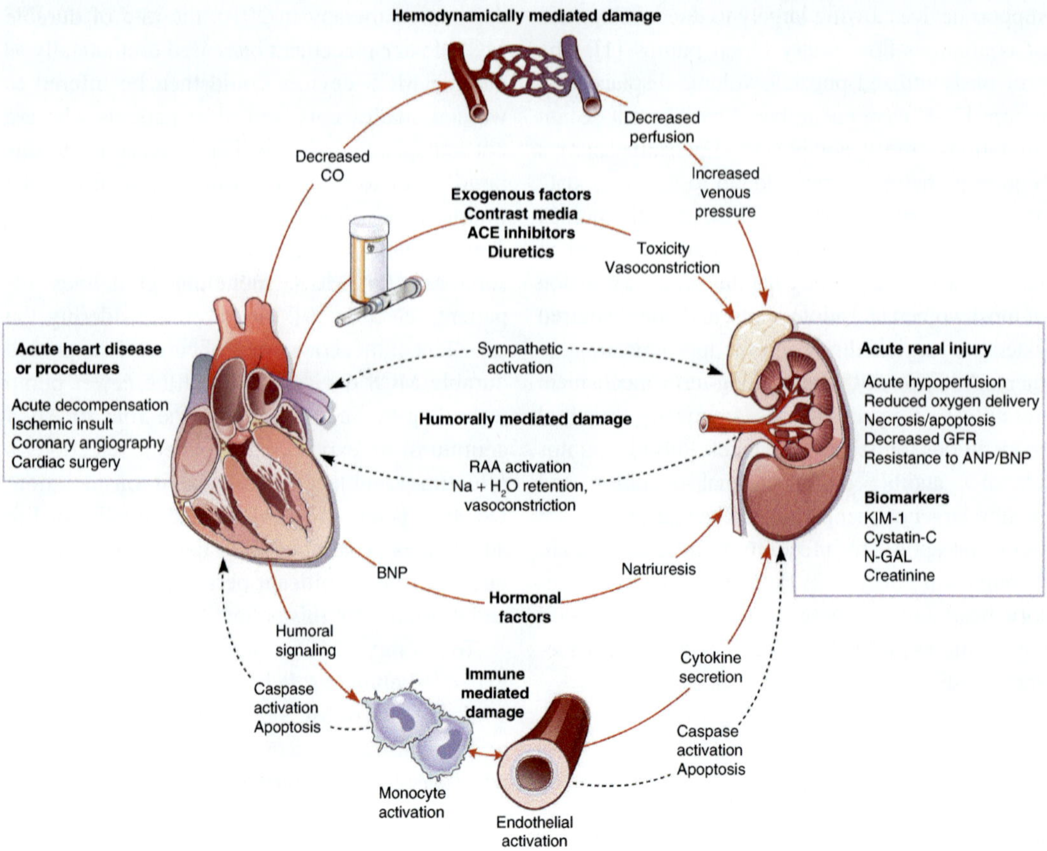

Fig. 7.1 Heart failure pathophysiology. ACE angiotensin-converting enzyme, ANP atrial natriuretic peptide, BNP B-type natriuretic peptide, CO cardiac output, GFR glomerular filtration rate, KIM kidney injury molecule, N-GAL neutrophil gelatinase-associated lipocalin, RAA renin angiotensin aldosterone. (Reproduced from Ronco et al. [71])

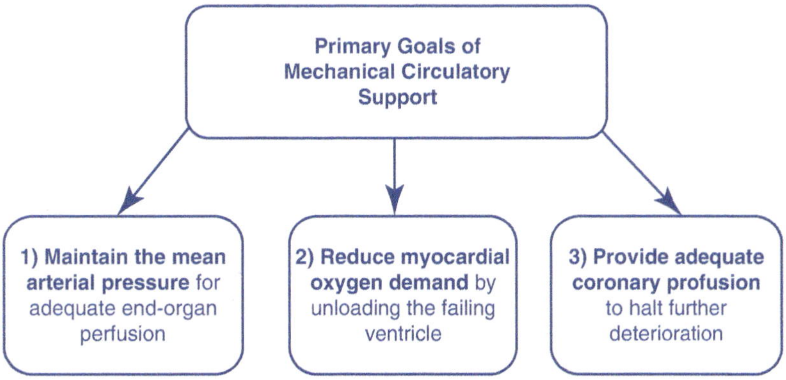

Fig. 7.2 Three primary goals of mechanical circulatory support

body's predominantly parasympathetic tone shifts to a more sympathetic state, causing an increase in circulating catecholamines, which may precipitate arrhythmias such as atrial fibrillation or ventricular tachycardia [22]. Fluid retention and progressive volume overload lead to pulmonary edema with subsequent mesenteric congestion, reducing the function of the liver, kidneys, and gastrointestinal tract. Cytokines and other inflammatory mediators, released in response to progressive volume overload, also have important effects in the pathophysiology of heart failure [23, 24]. TNF alpha, in particular, is thought to mediate the cachexia seen in many end-stage heart failure patients [25]. Atrial natriuretic peptide and B-type natriuretic peptide, released in response to increased blood volume stretching the atrial and ventricular walls, act to decrease systemic vascular resistance and induce natriuresis [26, 27]. However, the theoretical counterbalancing effects of the natriuretic peptides are overwhelmed by the profound activation of the RAS system and increased sympathetic tone [28]. The goal of long-term MCS is to halt or delay the progressive clinical deterioration described above, palliate the associated symptoms, and improve survival in medically refractory heart failure.

General Goals of MCS Therapy

Whether being utilized temporarily for acute cardiogenic shock to allow recovery from a reversible cardiovascular pathology or long term to improve survival and quality of life by mitigating the deleterious physiological effects of end-stage heart failure, the general goals of MCS therapy are the same: (1) maintain mean arterial pressure for adequate end-organ perfusion; (2) reduce myocardial oxygen demand by unloading the failing ventricle; and (3) provide adequate coronary perfusion to halt further deterioration and maximize the potential for myocardial recovery (Fig. 7.2) [29].

Maintain End-Organ Perfusion The primary goal of MCS is to compensate for the low output state of cardiogenic shock by maintaining an adequate mean arterial pressure (MAP) and blood flow to vital organs. Adequate perfusion of these organs is evidenced by increased urine output, resolution of prerenal azotemia, decreased serum bilirubin, and decreased serum lactate.

Unloading the Failing Ventricle Another goal of MCS is to reduce myocardial oxygen demand by unloading the failing ventricle. Based on Laplace's law, the wall tension of myocardium increases proportionally with the diameter of the ventricle (Fig. 7.3) [30], and myocardial oxygen consumption increases to tolerate this increased wall tension. By maintaining the ventricle in a decompressed state, we reduce the diameter of the ventricle and minimize tension on the myofibrils. Reducing myocardial oxygen demand is particularly important when ischemic cardiomyopathy is the etiology.

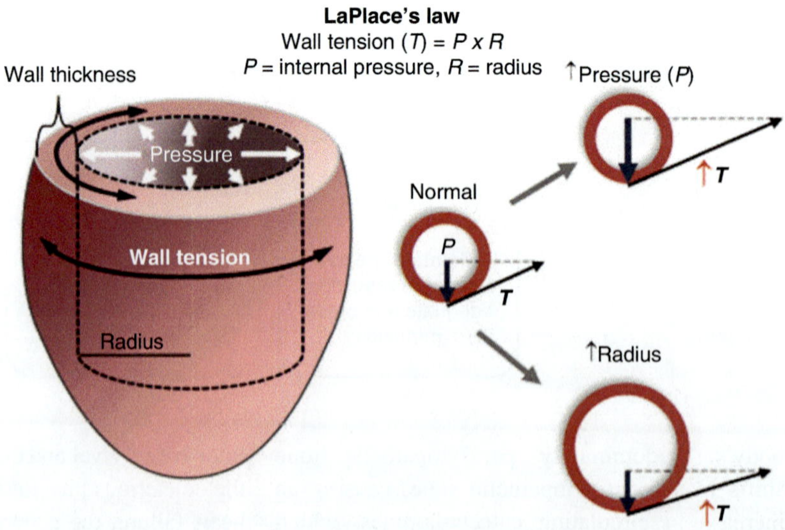

Fig. 7.3 Laplace's law. (Reproduced from Berthiaume et al. [72])

Coronary Perfusion To maximize the myocardium's potential for recovery and prevent further deterioration of the ventricle, adequate coronary perfusion must be maintained during MCS. However, coronary perfusion rarely requires additional consideration beyond maintaining an adequate MAP in the aortic root. Additionally, decompression of the ventricle decreases wall stress, which increases transmyocardial perfusion while reducing myocardial oxygen demand. Thus, ventricular unloading both increases myocardial oxygen delivery and minimizes myocardial oxygen demand.

Major Indications for MCS

There are four major indications for MCS: bridge to transplant (BTT), bridge to candidacy (BTC), bridge to recovery (BTR), and destination therapy (DT) (Fig. 7.4) [31]. Determining the goals of therapy for each individual patient is critical, because device strategy has profound implications on the anticipated duration of MCS, which is a major factor in device selection.

Bridge to Transplant BTT is indicated for patients who are actively listed for transplant but will likely not survive or will develop progressive end-organ dysfunction secondary to impaired perfusion prior to organ availability. BTT was the most prevalent device strategy for an implanted MCS device prior to 2010, when the HeartMate II was approved for DT. The target population for BTT are patients who (1) have worsening end-organ dysfunction, as evidenced by progressive prerenal azotemia or liver failure; (2) are anticipated to have long wait-list times, such as those with blood group O or who are highly sensitized; and/or (3) desire improved quality of life while waiting for an organ to become available. BTT patients are usually considered for durable MCS devices as their duration of support is on the order of months to years.

Bridge to Candidacy BTC is indicated for patients not currently listed for transplant but who do not have an absolute, permanent contraindication to solid-organ transplant. The target population for BTC are patients who may become eligible for transplant after a period of MCS resulting in improved end-organ function, resolution of a comorbid condition such as completing cancer treatment, or institution of necessary lifestyle changes such as discontinuing substance abuse. The optimal outcome for a BTC patient is ultimately to receive a transplant, so they are typically evaluated in the same manner as BTT patients and are considered for a durable MCS device. Once a BTC patient is actively listed for transplant, he or she is redesignated as BTT. If a

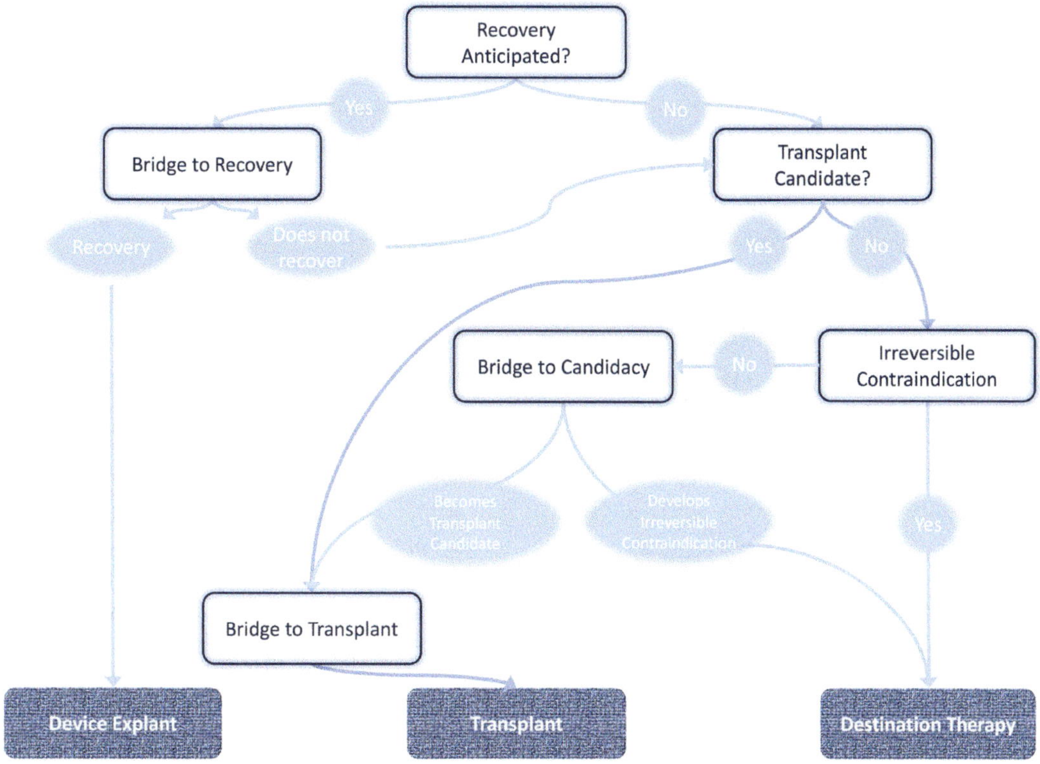

Fig. 7.4 Algorithm for decisions in mechanical circulatory support: bridge-to-transplantation and destination therapy

BTC patient develops an absolute, irreversible contraindication to transplant, they are redesignated as DT.

Bridge to Recovery BTR is indicated for patients who require temporary circulatory support for a reversible cardiovascular injury. The expectation in BTR is that the patient will recover cardiac function and MCS will be discontinued without requiring consideration for a transplant. Potential etiologies include postcardiotomy cardiogenic shock, fulminant myocarditis, or peripartum cardiomyopathy. BTR has the shortest anticipated duration of support of all strategies and typically utilizes a temporary MCS device, which may be implanted centrally or percutaneously. Central implantation of a temporary MCS device is most often in the setting of a patient who is undergoing cardiac surgery for another indication; the device placement may have been planned preoperatively, such as in very high-risk coronary artery bypass patients, or decided upon intraoperatively, such as for a patient failing to wean from bypass. BTR patients may be redesignated as BTT, BTC, or DT if they do not recover from their cardiac injury; in this scenario, their temporary MCS strategy may be transitioned to a long-term durable MCS device.

Destination Therapy DT is indicated for patients who require long-term MCS but are not eligible for transplant due to one or more irreversible contraindications. Older patients (>70 years) with end-stage heart failure, who are very unlikely to be considered for a transplant, comprise the majority of DT patients. The dramatic increase in the rate of patients receiving durable MCS devices temporally correlates with the approval of the HeartMate II for DT. Currently available long-term MCS devices are usually durable enough for BTT patients, as they typically remain on the wait list for months to a few years prior to receiving a transplant. However, due to the extremely limited supply of suitable cardiac allograft donors, many clinicians view DT as the device strategy that should receive the most attention from manufacturers.

Evaluation of Candidates for MCS

All patients should have any reversible causes of heart failure addressed prior to being considered for durable MCS device implantation, and patients should have their transplant candidacy assessed prior to receiving a device implantation. Additionally, New York Heart Association functional classification and Interagency Registry for Mechanically Assisted Support (INTERMACS) profile [32] should be determined in all patients being considered for MCS [9]. Many criteria are considered to determine whether the survival and quality-of-life advantages of long-term MCS device implantation outweigh the significant perioperative morbidity and mortality; these criteria include hemodynamic factors, clinical examination, laboratory studies, and psychosocial assessment. Factors supporting the need for MCS include cool and clammy extremities indicative of poor peripheral perfusion; low systemic blood pressure; tachycardia; rales or jugular venous distention; reduced urine output even in response to diuretics; laboratory evidence of hepatic dysfunction; and laboratory evidence of prerenal azotemia.

Hemodynamic Evaluation There is no specific hemodynamic criteria the define candidacy for long-term MCS, but a cardiac index less than 2.0 L/min or a stroke volume less than 25 cc/beat despite the use of vasoactive drugs is usually considered a requirement. Stroke volume should be routinely evaluated as some patients may maintain a cardiac index greater than 2.0 L/min, but only because tachycardia is compensating for a low stroke volume; these patients are actually in an advanced phase of decompensation and should be considered for long-term MCS. Mixed venous saturation should be evaluated in patients with an adequate measured cardiac output but who appear otherwise compromised.

Evaluation of the right ventricle is necessary to determine high-risk patients as right heart failure is one of the most important causes of morbidity and mortality after left-sided MCS device placement, usually occurring in the setting of pre-existing right ventricular failure [33]. Contributing factors include altered right ventricle geometry, particularly leftward shift of the interventricular septum caused by acute left ventricular unloading, and increased venous return overloading the disease right ventricle (Fig. 7.5). Whether or not a patient develops right heart failure depends on the diseased right ventricle's ability to respond to these acutely changing hemodynamic loads. Though there is a protective reduction in right ventricle afterload, overloading of the right ventricle caused by increased preload is often the dominant pathophysiologic process in the immediate post-implant period. Many hemodynamic factors and risk scores have been evaluated to predict which patients will go on to develop right ventricular failure after left ventricular assist device placement [34–41]; among those, pulmonary artery pulsatility index is emerging as a promising independent predictor and is especially useful because it is more predictive in patients receiving inotropes (Fig. 7.6) [42, 43].

Other hemodynamic criteria include the requirement for adequate intracardiac filling pressures to distinguish low cardiac output due to hypovolemia from overdiuresis. Right atrial pressure >10 mm Hg and pulmonary capillary wedge pressure >15 mm Hg reflect an appropriate volume status.

Evaluation of Renal Function Renal dysfunction has consistently been shown to portend a higher risk of morbidity and mortality in patients after durable MCS device implantation [44]. A creatinine value of 3.5 mg/dL was used as the cutoff point for exclusion in the REMATCH trial [45], and this value has been reduced to 2.5 mg/dL in more recent trials. However, using high serum creatinine to define renal dysfunction can drastically underestimate the severity of renal disease in patients with lower muscle mass, including many women and elderly patients. Estimated

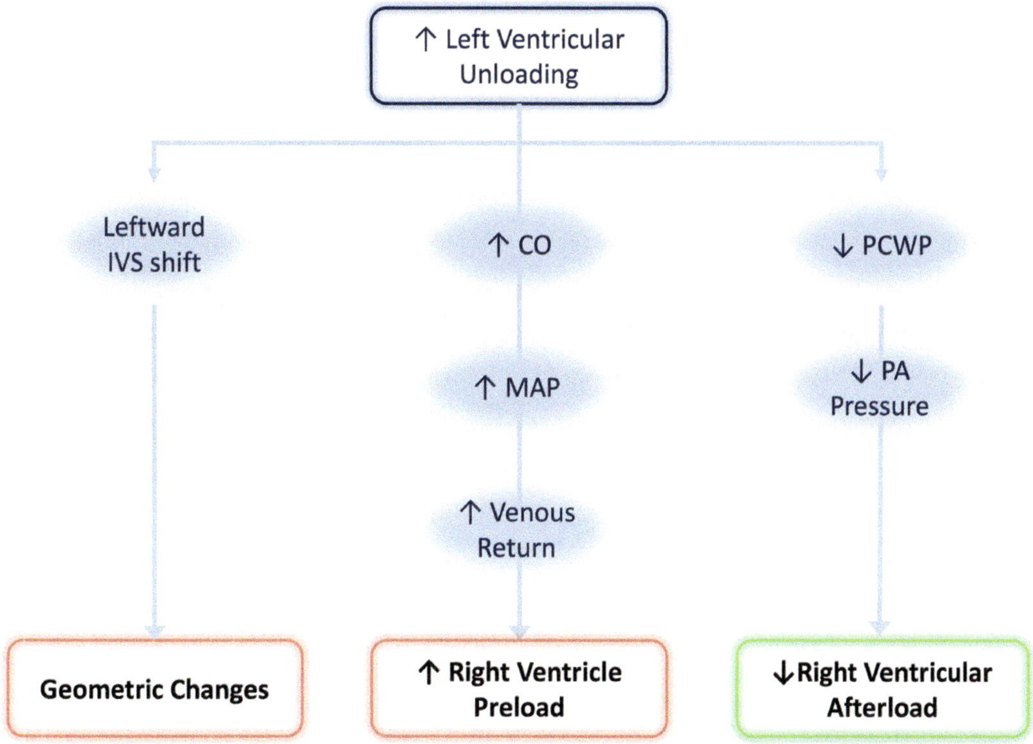

Fig. 7.5 Pathophysiology of right ventricular failure

Pulmonary Artery Pulsatility Index

$$PaPi = \frac{PA\ Systolic\ Pressure - PA\ Diastolic\ Pressure}{Right\ Atrial\ Pressure}$$

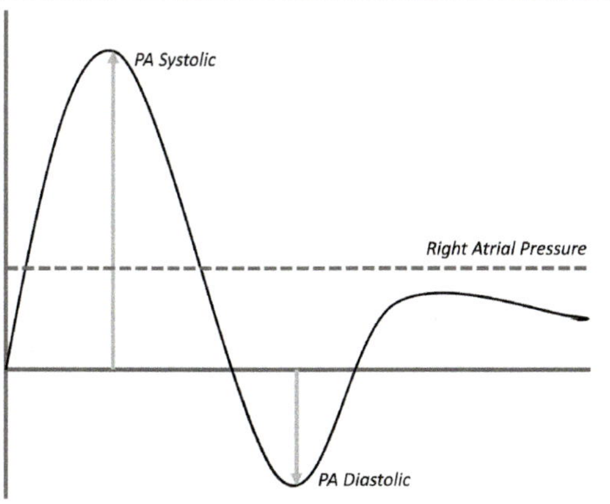

Fig. 7.6 Pulmonary artery pulsatility index

glomerular filtration rate less than 0.5 mL/kg/min is an appropriate alternative to serum creatinine for assessing renal dysfunction in most patients. Additionally, many centers do not measure creatinine clearance until the patient has been on an intravenous drug regimen associated with a cardiac index greater than 2.4 L/min for 24–48 hours. Renal dysfunction is likely reversible if it improves on such a regimen. Twenty-four hour urine studies in addition to standard urinalysis should be utilized looking for the presence of inflammatory cells or eosinophils. High urine protein excretion or the presence of cells in the urine may warrant a renal biopsy to exclude primary kidney disease, rather than heart failure, as the cause of renal dysfunction.

Patients on long-term hemodialysis support are considered extremely high risk for long-term MCS due to increased risk of complications, most often related to infection or vascular access. It may be difficult to assess the reversibility of renal dysfunction following long-term MCS device implantation [36]. Renal dysfunction is potentially reversible if it is secondary to decreased perfusion associated with heart failure but is unlikely to recover if due to primary kidney disease or the direct nephrotoxic effects of commonly prescribed angiotensin-converting enzyme inhibitors, angiotensin II receptor blockers, or nonsteroidal anti-inflammatory drugs. Many patients who were not deemed transplant candidates secondary to their poor renal function subsequently underwent transplantation because their renal dysfunction reversed after a prolonged period of MCS; in the HeartMate II BTT trial, patients with elevated serum creatinine prior to device implant had an average decrease of 1.8 mg/dL to 1.4 mg/dL [46].

Evaluation of Hepatic Function Hepatic dysfunction has also been associated with poor outcomes following durable MCS device implantation [47]. Previous studies have reported and suggested that total serum bilirubin, aspartate aminotransferase, and alanine aminotransferase levels greater than three times normal are independent risk factors for adverse outcomes [48]. All patients should be screened for a history of severe alcohol abuse and tested for previous infection with viral hepatitis. Simple ultrasound of the liver is an adequate screening test for infiltrative masses warranting biopsy. The most common cause of hepatic dysfunction in this population is right heart failure, which may improve following an appropriate duration of MCS. In the HeartMate II BTT trial, total serum bilirubin, aspartate aminotransferase, and alanine aminotransferase levels returned to normal after 6 months of support in patients with abnormal baseline levels [46]. Cirrhosis is predictive of poor outcome following device implantation, and many centers screen for cirrhosis with hepatic ultrasound even in the presence of significant right ventricular failure.

Evaluation of Gastrointestinal Bleeding Long-term MCS is associated with an increased risk of gastrointestinal bleeding secondary to the development of arteriovenous malformations. Additionally, durable MCS devices require systemic anticoagulation to prevent pump thrombosis, exacerbating the risk of gastrointestinal bleeding. Patients with a history of gastrointestinal bleeding should be carefully screened with both upper and lower endoscopy prior to durable MCS device implantation to assess the etiology and rule out malignancy.

Evaluation of Pulmonary Function There are no absolute pulmonary function criteria or thresholds for exclusion prior to durable MCS implantation. However, patients with severe pulmonary dysfunction are at increased risk perioperatively, especially related to prolonged mechanical ventilation and increased duration of stay in the intensive care unit [44]. Severe cardiac dysfunction may impede accurate pulmonary function testing, but if forced vital capacity, forced expiratory volume at one second, and diffusion capacity of carbon monoxide are all less than 50% predicted, long-term MCS may not be an appropriate option. Additionally, mechanical ventilation prior to device implantation is one of the highest risk factors associated with adverse outcomes [48].

Patients with low oxygen saturation, less than 92% on room air, should undergo a sonicated saline injection during echocardiography to rule out a right-to-left shunt caused by an atrial septal defect or patent foramen ovale. If no shunt is found, these patients should undergo spiral computed tomography to rule out chronic venous thromboembolic disease. Sleep-disordered breathing, including obstructive sleep apnea, is an underappreciated cause of hypoxemia, pulmonary hypertension, and right heart failure. If screening tests such as nocturnal oximetry are suggestive of a sleep-disordered breathing, patients should undergo a formal sleep study. Patients with a proven sleep-disordered breathing should receive appropriate therapy and have their durable MCS device implantation delayed for several months, possibly allowing recovery of right ventricular function and reducing the risk of requiring a right ventricular assist device.

Evaluation of Abnormal Coagulation Thrombocytopenia, anemia, and coagulopathy are associated with poor outcomes following long-term MCS, and attempts should be made to identify the cause and correct any abnormalities prior to device implantation [49]. Preoperative platelet counts less than 149,000/uL increase the risk of mortality prior to hospital discharge sevenfold. Prolonged international normalized ratio should also be normalized prior to durable MCS device implantation, if possible, though this may prove to be difficult in patients with significant hepatic dysfunction. Some centers screen for hereditary and autoimmune coagulation disorders by measuring von Willebrand factor, protein C, protein S, anticardiolipin antibodies, and antithrombin III levels, in addition to other factors.

Screening for Malignancy Preoperative screening for malignancy should be performed in accordance with national guidelines [50] or directed by abnormal clinical findings. In the generally older population of long-term MCS candidates, prognosis is a critical determination when evaluating for device implantation. Patients with a previous malignancy associated with significantly decreased survival should have an appropriate disease-free interval prior to durable MCS device implantation. However, patients with a malignancy associated with a good prognosis may be reasonable candidates for long-term MCS.

Peripheral Vascular Disease Patients with significant peripheral vascular disease are often not considered candidates for long-term MCS as their atherosclerotic disease burden limits postoperative ambulation and recovery. Abdominal ultrasound and ankle-brachial indices are appropriate screening tools in selected patients, including all patients with diabetes. Ankle-brachial indices may be falsely elevated in diabetes due to noncompliant arteries, so pulse volume recordings may be more accurate in diagnosing peripheral vascular disease in that population. Carotid stenosis should be evaluated in any high-risk patient, especially those with an audible carotid bruit or a history suggestive of transient ischemic attacks. Patients with a prior stroke and permanent residual neurological deficit are not generally considered good candidates for durable MCS device implantation.

Evaluation for Infection Infectious complications, especially sepsis, are a major cause of morbidity and mortality in the long-term MCS population owing to a staggering number of risk factors: prolonged mechanical ventilation, indwelling catheters, and skin flora changes associated with prolonged hospitalization. Correcting as many of these factors as possible prior to device implantation minimizes the risk of postoperative infectious complications. Patients with active systemic infection should not be considered for durable MCS device implantation, and device insertion should not be delayed in the setting of potentially treatable localized infections.

Evaluation of Nutrition Malnutrition is extremely common in heart failure patients and has significant implications on postoperative infectious complications [51]. In the post-REMATCH study, almost all patients who expired from infection after durable MCS device implantation had a preoperative albumin less than 3.2 mg/dL [49]. The etiology of malnutrition in heart failure patients is multifactorial and includes poor appetite secondary to inflammatory cytokines, particularly TNF alpha [52]. Additionally, cachexia is associated with increased perioperative mortality, and malnutrition has profound implications on immune function, especially relating to wound healing and impaired T-lymphocyte function [53].

As has been described previously in cardiac surgery, obesity is not associated with increased mortality in durable MCS device implantation [54]. However, obese patients have a higher prevalence of postoperative infectious complications, especially wound infection. Obesity is a growing national epidemic and is very prevalent in heart failure patients, and recent reports have shown no deleterious effects of obesity on outcomes in patients following durable MCS device implantation. Sometimes, long-term MCS may be utilized as BTC to support obese patients as they attempt to lose sufficient weight to become eligible for transplant; however, very few patients achieve this weight loss goal.

Psychosocial Evaluation Guidelines for cardiac transplantation require that all candidates receive a formal mental health evaluation by a trained healthcare professional [55], and this requirement is equally critical for long-term MCS candidates. In addition to screening for psychiatric illness, this evaluation should assess neurocognitive function as having a durable MCS device requires a significant capacity for self-care. Preoperative screening should include assessment compliance with medications and medical therapies, compliance with physician appointments, and any history of substance abuse or dependence [56]. Perhaps the most crucial psychosocial consideration when assessing a candidate for MCS is the presence of an adequate social support system, which may include a spouse, significant other, family members, or friends. Patients are usually not discharged to home without the availability of someone who has passed the device education test and would be able to respond to the various device alarms for at least the first few months after implantation. These alarms may be alerting to a mechanical failure resulting in significant hemodynamic compromise and decreased mental functioning for the patient, who may not be able to effectively respond to the alarms.

A relatively common situation is a patient who presents in acute cardiogenic shock and are intubated shortly after presentation to the healthcare facility. These patients are often too compromised to participate in their own care, and a thorough psychosocial assessment may not be possible. In this scenario, every effort should be made to obtain as much background information as possible from medical records and the patients support system, as these patients may have a history of significant substance abuse or other social considerations that may complicate their candidacy for long-term MCS.

Patient Selection and Eligibility

For the use of a durable MCS device as destination therapy, the Centers for Medicare and Medicaid Services uses the same criteria for eligibility for payment as the entry criteria used in the REMATCH study: (1) not a transplant candidate; (2) New York Heart Association class IV heart failure on maximal medical therapy >60–90 days; (3) ejection fraction <25%; and (4) peak oxygen consumption (VO_2 max) <12 mL/kg/min [45]. Clinicians must consider that using percent of predicted peak VO_2 may be more reliable than absolute peak VO_2 values in older candidates because the predicted peak VO_2 is derived from nomograms stratified by age, gender, and body surface area. BTT patients are evaluated for eligibility by the same criteria as DT patients, with the exception that they are by definition transplant candidates. Due to the considerable donor-recipient mismatch in heart transplantation, the criteria for transplant candidacy are very exclusive (Table 7.1).

Long-term MCS for patients in acute cardiogenic shock is generally not indicated unless the following criteria are met: (1) ventricular function is not anticipated to recover without long-term MCS support; (2) cannot maintain normal hemodynamics and vital organ function with temporary MCS devices or who cannot be weaned from temporary MCS devices or inotropic support; (3) potential for recovery of end-organ function and improved quality of life; and (4) do not have irreversible end-organ damage. Patients who are inotrope-dependent, even if they are not in acute cardiogenic shock, should be considered as candidates for MCS as they have very high mortality with medical management alone [9].

Device Characteristics Required for Left Ventricular Support

As mentioned previously, the goals of a left-sided MCS device are to unload the left ventricle, maintain adequate mean arterial pressure, and maintain coronary perfusion. Functionally, this requires an inflow from the left ventricle, a pump to propel blood into the high-pressure systemic circulation, and an outflow into the aortic root. Configurations will vary among different devices and especially between temporary and durable devices, but those three components are univer-

Table 7.1 Recipient selection for heart transplantation. Based on Shanewise, 2005 [70]

Indications

1. Systolic heart failure
   A. Included etiology
      1. Ischemic
      2. Dilated
      3. Valvular
      4. Hypertensive
      5. Other
   B. Excluded etiology
      1. Amyloid (controversial)
      2. HIV infection
      3. Cardiac sarcoma
2. Ischemic heart disease with intractable angina
   A. Ineffective maximal tolerated medical therapy
   B. Not a candidate for direct myocardial revascularization, percutaneous revascularization, or transmyocardial revascularization procedure
   C. Unsuccessful myocardial revascularization
3. Intractable arrhythmia
   A. Uncontrolled with pacing cardioverter defibrillator
      1. Not amenable to electrophysiology-guided single or combination medical therapy
      2. Not a candidate for ablative therapy
4. Hypertrophic cardiomyopathy
   A. Class IV symptoms persist despite interventional therapies
      1. Alcohol injection of septal artery
      2. Myotomy or myomectomy
      3. Mitral valve replacement
      4. Maximal medical therapy
      5. Pacemaker therapy
5. Congenital heart disease in which severe fixed pulmonary hypertension is not a complication
6. Cardiac tumor
   A. Confined to the myocardium
   B. No evidence of distant disease revealed by metastatic workup

Absolute contraindications

1. Age >70 years (may vary at different centers)
2. Fixed pulmonary hypertension (unresponsive to pharmacological intervention)
   A. Pulmonary vascular resistance >5 woods units
   B Transpulmonary gradient >15 mmHg
3. Systemic illness that will limit survival despite transplant
   A. Neoplasm other than skin cancer (<5 years disease-free survival)
   B. HIV/AIDS (CDC definition CD4 count <200 cells/mm^3)
   C. Systemic lupus erythematosus (SLE) or sarcoid that has multisystem involvement and is currently active
   D. Any systemic process with a high probability of recurrence in the transplanted heart
   E. Irreversible renal or hepatic dysfunction

sally present. Ideal pump performance characteristics for a durable MCS device are shown in Table 7.2. A long-term MCS device should be able to maintain flow over a range of 2–6 L/min; the lower range of 2 L/min represents a minimum flow rate for clinically significant MCS; and the upper range of 6 L/min represents an adequate cardiac output for a large patient who is unlikely to be performing excessively strenuous exercise. Ideally, the pump should be able to provide that range of flow while maintaining a pressure differential of 20–60 mm Hg. Normalized power vs. flow rate over the operating speed range is also critical to enable automatic control of the flow rate. Availability of a pulsatility mode is also preferred.

Other non-hemodynamic factors also play a critical role in the feasibility of a durable MCS device. We mentioned earlier when discussing the limitations of the previous generation of pulsatile devices that small device size is necessary for implantation in patients with a smaller body

Table 7.2 Assumed ideal pumping performance and procedural characteristics

Hemodynamic characteristics
Range of nominal flows (excluding pulsatile transients): 2–6 L/min
Range of nominal pump differential pressures: 20–60 mm Hg
Pulsatility mode available
Normalized power vs. flow over operating speed range (to enable automatic control)
Procedural characteristics
Small size and light weight
Versatility of device insertion (right atrium or right ventricle)
Shorter inflow cannula
Adjustable sewing ring height
Bloodless and off-pump device insertion
Option for minimally invasive implantation techniques
Single controller for automatic control in biventricular assist device configuration

habitus, including most women and adolescents. Similarly, an ideal device would be as lightweight as possible to place minimal force on surrounding structures. Due to variations in patient anatomy, an especially important consideration for patients who had prior chest surgery, implantation should be possible in a variety of locations; for left-sided devices, these sites include the left atrium, left ventricle, or subdiaphragmatic. Ideally, small device dimensions and having multiple sites for implantation should provide minimally invasive options for device insertion with a bloodless, off-pump technique.

Special Considerations for Right Ventricular Support

Despite the therapeutic success of left ventricular assist devices, concomitant right ventricular failure and the lack of appropriate long-term durable right-sided devices have limited the full potential that MCS can offer to patients with biventricular failure. Despite the relatively low rate of temporary right ventricular assist device support in the United States, a significant number of patients who have been successfully discharged following LVAD placement remain symptomatic from right ventricular failure and undergo recurrent readmissions for management of heart failure [57, 58]. Additionally, delayed right ventricular failure is increasingly being recognized in a subset of patients receiving long-term LVAD support [59, 60]. It is imperative that newer devices designed or adapted for right ventricular support be developed to meet this growing clinical requirement.

The common working definition of right ventricular failure according to the Interagency Registry for Mechanically Assisted Circulatory Support (INTERMACS) is a persistence of signs and symptoms of right ventricular dysfunction evidence by elevated central venous pressure >18 mm Hg with a cardiac index <2.0 L/min/m^2 in the absence of elevated left atrial filling pressure (pulmonary capillary wedge pressure) >18 mm Hg, cardiac tamponade, ventricular arrhythmias, and/or pneumothorax requiring placement of an RVAD, or the use of intravenous inotropes or inhaled nitric oxide for greater than 14 days following LVAD implantation. The development of right ventricular failure requiring RVAD support in LVAD recipients is associated with a high mortality regardless of the timing of device insertion, with patients requiring RVAD support having higher inhospital mortality, 48% vs 9.5%, and lower 1-year survival, 40% vs 82%, compared to LVAD recipients [61]. These data suggest that there is limited benefit in prophylactic RVAD placement in patients at high risk of right ventricular failure after LVAD insertion.

The right ventricle has approximately one sixth the muscle mass of the left ventricle and reflects characteristics of a high-volume/low-pressure pump [62]. Anatomically, the right ventricle forms a crescent-shaped changer that wraps around the cone-shaped left ventricle [63]. Another unique anatomic characteristic of the right ventricle is its course inner surface due to the high level of "reinforcement" with interweaving strands of muscle, the trabeculae carneae, which allows efficient generation of power without excessively thickening the wall of the chamber [64]. The right ventricle is normally 3–5 mm in thickness. With current pump design, several factors make proper device placement challenging: protrusion of the inflow into the highly trabeculated right ventricle, relative proximity of the tricuspid chordae, distance of the inflow from the interventricular septum, and proper device inflow port orientation and fixation. With the HeartWare ventricular assist device adapted as an RVAD (Fig. 7.7), some surgeons have placed extra epicardial spacers to prevent the device

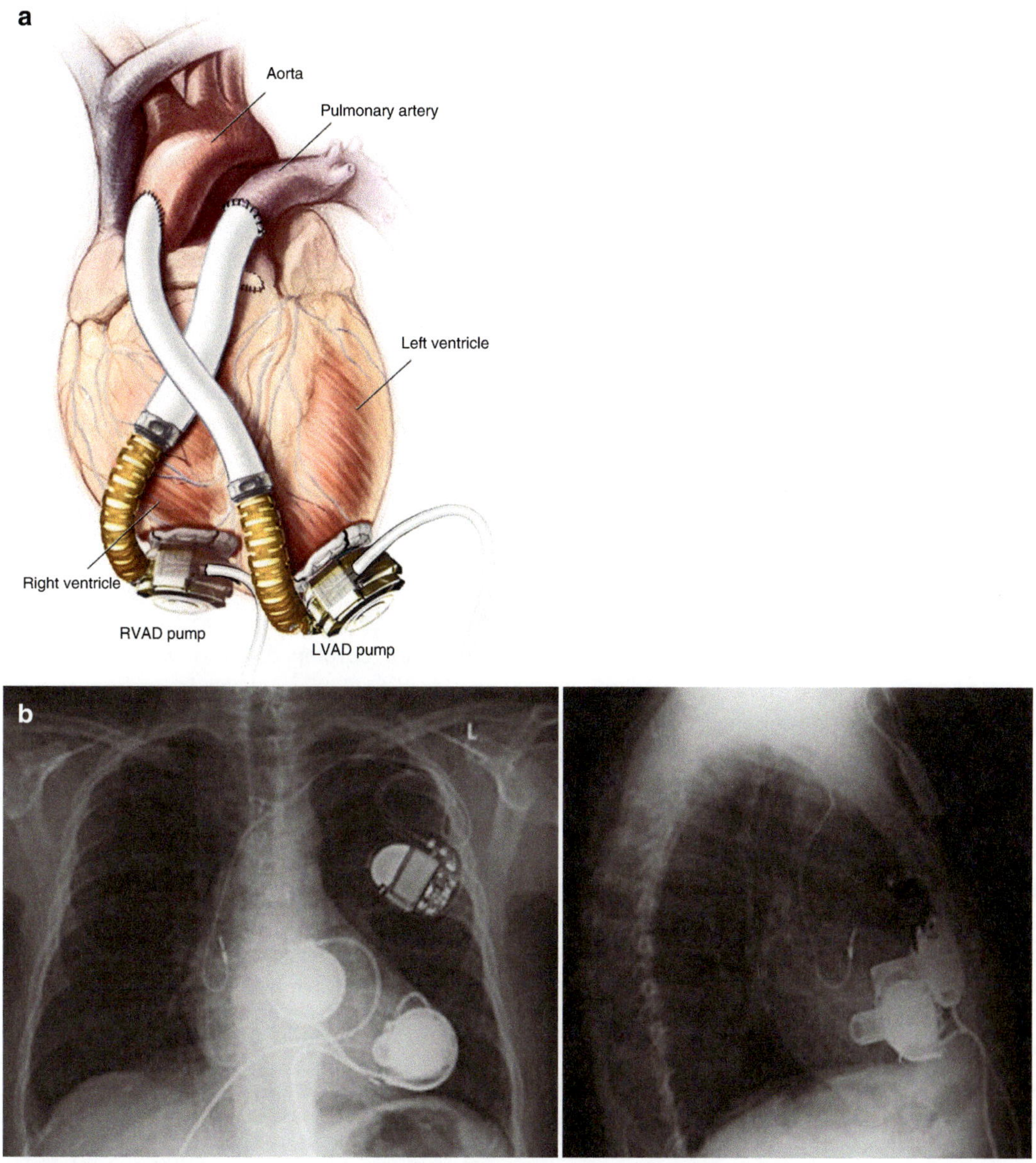

Fig. 7.7 (a) Schematic of HVAD as RAVD (Used with permission of Mayo Foundation for Medical Education and Research, all rights reserved.) (b) (Reproduced from Krabatsch et al. [65], with permission from Wolters Kluwer Health, Inc)

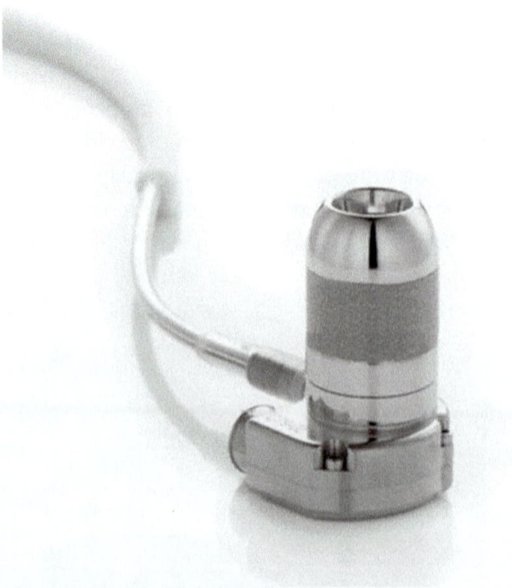

Fig. 7.8 MVAD. (Reproduced with permission of Medtronic, Inc)

inlet from projecting too far into the right ventricle cavity [65, 66]. The HVAD has also been placed in the right atrium, but similar concerns exist in addition to the increased risk of suction events due to the thin-walled nature of the right atrium [67]. Furthermore, the relatively large size of these pumps creates difficulty in closing the sternum if the device is implanted on the free anterior surface of the right ventricle, so diaphragmatic placement is often preferred.

Physiologically, RVADs pump into the pulmonary circulation, which is usually lower pressure than the systemic circulation; this results in significantly lower hydraulic loads and power consumption rates compared to LVADs. To account for this discrepancy causing overload of the pulmonary circulation, RVADs must be configured at very low pump speeds, or the outflow graft must be restricted in cross-sectional area [68]. Currently, no durable MCS devices have been designed specifically for RVAD application, and the lack of reliable devices limits the MCS options for patients with biventricular failure or isolated right ventricular failure.

The HeartWare MVAD is a miniature wide-blade, continuous-flow pump featuring a magnetic axial rotor and a combination of passive permanent magnets and hydrodynamic bearings that totally suspend the rotor in operation [69]. This low-profile device requires volume displacement of only 15 mL, operates in the 11,000–18,000 RPM range, and can provide flow up to 7 L/min (Fig. 7.8). The MVAD has provided full right ventricular support in a porcine biventricular assist model, but is not yet available for clinical use.

Conclusion

Though heart transplantation remains the gold standard treatment for end-stage heart failure, short supply of suitable donors has limited its availability as an option for the majority of heart failure patients. In medically refractory patients, long-term MCS has emerged as an efficacious and relatively safe treatment strategy. Previous pulsatile devices were hindered by their large profile and poor mechanical durability, but the newer generation of continuous-flow devices have shown great promise as both bridge to transplantation and destination therapy. Additionally, temporary MCS devices are being utilized very effectively as bridge to recovery. As more advanced technology is incorporated and newer devices are developed, MCS will continue to grow as a field to meet the increasing clinical demand.

References

1. Miller LW, Pagani FD, Russell SD, John R, Boyle AJ, Aaronson KD, et al. Use of a continuous-flow device in patients awaiting heart transplantation. N Engl J Med. 2007;357(9):885–96.
2. Frazier OH, Rose EA, Oz MC, Dembitsky W, McCarthy P, Radovancevic B, et al. Multicenter clinical evaluation of the HeartMate vented electric left ventricular assist system in patients awaiting heart transplantation. J Thorac Cardiovasc Surg. 2001;122(6):1186–95.
3. Deng MC, Edwards LB, Hertz MI, Rowe AW, Keck BM, Kormos R, et al. Mechanical circulatory support device database of the International Society for Heart and Lung Transplantation: third annual report–2005. J Heart Lung Transplant. 2005;24(9):1182–7.

4. Slaughter MS, Tsui SS, El-Banayosy A, Sun BC, Kormos RL, Mueller DK, et al. Results of a multicenter clinical trial with the thoratec implantable ventricular assist device. J Thorac Cardiovasc Surg. 2007;133(6):1573–80.
5. Baldwin JT, Robbins RC, National Heart L, Blood Institute Working Group. Executive summary for the National Heart, Lung, and Blood Institute Working Group on next generation ventricular assist devices for destination therapy. Semin Thorac Cardiovasc Surg. 2005;17(4):369–71.
6. Frazier OH, Kirklin JK. Mechanical circulatory support. Oxford: Elsevier; 2006.
7. Griffith BP, Kormos RL, Borovetz HS, Litwak K, Antaki JF, Poirier VL, et al. HeartMate II left ventricular assist system: from concept to first clinical use. Ann Thorac Surg. 2001;71(3 Suppl):S116–20; discussion S4–6.
8. Kirklin JK, Naftel DC, Kormos RL, Stevenson LW, Pagani FD, Miller MA, et al. Second INTERMACS annual report: more than 1,000 primary left ventricular assist device implants. J Heart Lung Transplant. 2010;29(1):1–10.
9. Feldman D, Pamboukian SV, Teuteberg JJ, Birks E, Lietz K, Moore SA, et al. The 2013 International Society for Heart and Lung Transplantation Guidelines for mechanical circulatory support: executive summary. J Heart Lung Transplant. 2013;32(2):157–87.
10. Kirklin JK, Pagani FD, Kormos RL, Stevenson LW, Blume ED, Myers SL, et al. Eighth annual INTERMACS report: special focus on framing the impact of adverse events. J Heart Lung Transplant. 2017;36(10):1080–6.
11. Gibbon JH Jr. Application of a mechanical heart and lung apparatus to cardiac surgery. Minn Med. 1954;37(3):171–85; passim.
12. Miller PE, Solomon MA, McAreavey D. Advanced percutaneous mechanical circulatory support devices for cardiogenic shock. Crit Care Med. 2017;45(11):1922–9.
13. Merrill AJ, Morrison JL, Branno ES. Concentration of renin in renal venous blood in patients with chronic heart failure. Am J Med. 1946;1(5):468.
14. Zhang J, Pfaffendorf M, van Zwieten PA. Hemodynamic effects of angiotensin II and the influence of angiotensin receptor antagonists in pithed rabbits. J Cardiovasc Pharmacol. 1995;25(5):724–31.
15. Hall JE, Coleman TG, Guyton AC, Balfe JW, Salgado HC. Intrarenal role of angiotensin II and [des-Asp1]angiotensin II. Am J Phys. 1979;236(3):F252–9.
16. Pratt JH. Role of angiotensin II in potassium-mediated stimulation of aldosterone secretion in the dog. J Clin Invest. 1982;70(3):667–72.
17. Ichikawa I, Pfeffer JM, Pfeffer MA, Hostetter TH, Brenner BM. Role of angiotensin II in the altered renal function of congestive heart failure. Circ Res. 1984;55(5):669–75.
18. Leimbach WN Jr, Wallin BG, Victor RG, Aylward PE, Sundlof G, Mark AL. Direct evidence from intraneural recordings for increased central sympathetic outflow in patients with heart failure. Circulation. 1986;73(5):913–9.
19. Swedberg K, Viquerat C, Rouleau JL, Roizen M, Atherton B, Parmley WW, et al. Comparison of myocardial catecholamine balance in chronic congestive heart failure and in angina pectoris without failure. Am J Cardiol. 1984;54(7):783–6.
20. Bristow MR, Ginsburg R, Minobe W, Cubicciotti RS, Sageman WS, Lurie K, et al. Decreased catecholamine sensitivity and beta-adrenergic-receptor density in failing human hearts. N Engl J Med. 1982;307(4):205–11.
21. Bell-Reuss E, Trevino DL, Gottschalk CW. Effect of renal sympathetic nerve stimulation on proximal water and sodium reabsorption. J Clin Invest. 1976;57(4):1104–7.
22. Cohn JN, Levine TB, Olivari MT, Garberg V, Lura D, Francis GS, et al. Plasma norepinephrine as a guide to prognosis in patients with chronic congestive heart failure. N Engl J Med. 1984;311(13):819–23.
23. Hedayat M, Mahmoudi MJ, Rose NR, Rezaei N. Proinflammatory cytokines in heart failure: double-edged swords. Heart Fail Rev. 2010;15(6):543–62.
24. Katz AM. Pathophysiology of heart failure: identifying targets for pharmacotherapy. Med Clin North Am. 2003;87(2):303–16.
25. Levine B, Kalman J, Mayer L, Fillit HM, Packer M. Elevated circulating levels of tumor necrosis factor in severe chronic heart failure. N Engl J Med. 1990;323(4):236–41.
26. Brunner-La Rocca HP, Kaye DM, Woods RL, Hastings J, Esler MD. Effects of intravenous brain natriuretic peptide on regional sympathetic activity in patients with chronic heart failure as compared with healthy control subjects. J Am Coll Cardiol. 2001;37(5):1221–7.
27. Jensen KT, Eiskjaer H, Carstens J, Pedersen EB. Renal effects of brain natriuretic peptide in patients with congestive heart failure. Clin Sci (Lond). 1999;96(1):5–15.
28. Greenberg BH. Congestive heart failure textbook. 2nd ed. Philadelphia: Lippincott Williams & Williams; 2000.
29. Kapur N, Esposito M. use of acute mechanical circulatory support devices in the setting of cardiogenic shock: pump fiction or an emerging reality? Latest in cardiology [Internet]. 2016. Available from: https://www.acc.org/latest-in-cardiology/articles/2016/08/31/08/01/use-of-acute-mechanical-circulatory-support-devices.
30. Mirsky I. Left ventricular stresses in the intact human heart. Biophys J. 1969;9(2):189–208.
31. Stewart GC, Givertz MM. Mechanical circulatory support for advanced heart failure: patients and technology in evolution. Circulation. 2012;125(10):1304–15.
32. Stevenson LW, Pagani FD, Young JB, Jessup M, Miller L, Kormos RL, et al. INTERMACS profiles of advanced heart failure: the current picture. J Heart Lung Transplant. 2009;28(6):535–41.

33. Ranganath NK, Smith DE, Moazami N. The Achilles' heel of left ventricular assist device therapy: right ventricle. Curr Opin Organ Transplant. 2018;23(3):295–300.
34. Cowger J, Sundareswaran K, Rogers JG, Park SJ, Pagani FD, Bhat G, et al. Predicting survival in patients receiving continuous flow left ventricular assist devices: the HeartMate II risk score. J Am Coll Cardiol. 2013;61(3):313–21.
35. Atluri P, Goldstone AB, Fairman AS, MacArthur JW, Shudo Y, Cohen JE, et al. Predicting right ventricular failure in the modern, continuous flow left ventricular assist device era. Ann Thorac Surg. 2013;96(3):857–63; discussion 63–4.
36. Kormos RL, Teuteberg JJ, Pagani FD, Russell SD, John R, Miller LW, et al. Right ventricular failure in patients with the HeartMate II continuous-flow left ventricular assist device: incidence, risk factors, and effect on outcomes. J Thorac Cardiovasc Surg. 2010;139(5):1316–24.
37. Loghmanpour NA, Kormos RL, Kanwar MK, Teuteberg JJ, Murali S, Antaki JF. A Bayesian model to predict right ventricular failure following left ventricular assist device therapy. JACC Heart Fail. 2016;4(9):711–21.
38. Puwanant S, Hamilton KK, Klodell CT, Hill JA, Schofield RS, Cleeton TS, et al. Tricuspid annular motion as a predictor of severe right ventricular failure after left ventricular assist device implantation. J Heart Lung Transplant. 2008;27(10):1102–7.
39. Vivo RP, Cordero-Reyes AM, Qamar U, Garikipati S, Trevino AR, Aldeiri M, et al. Increased right-to-left ventricle diameter ratio is a strong predictor of right ventricular failure after left ventricular assist device. J Heart Lung Transplant. 2013;32(8):792–9.
40. Boegershausen N, Zayat R, Aljalloud A, Musetti G, Goetzenich A, Tewarie L, et al. Risk factors for the development of right ventricular failure after left ventricular assist device implantation-a single-centre retrospective with focus on deformation imaging. Eur J Cardiothorac Surg. 2017;52(6):1069–76.
41. Kashiyama N, Toda K, Nakamura T, Miyagawa S, Nishi H, Yoshikawa Y, et al. Evaluation of right ventricular function using liver stiffness in patients with left ventricular assist device. Eur J Cardiothorac Surg. 2017;51(4):715–21.
42. Kang G, Ha R, Banerjee D. Pulmonary artery pulsatility index predicts right ventricular failure after left ventricular assist device implantation. J Heart Lung Transplant. 2016;35(1):67–73.
43. Morine KJ, Kiernan MS, Pham DT, Paruchuri V, Denofrio D, Kapur NK. Pulmonary artery pulsatility index is associated with right ventricular failure after left ventricular assist device surgery. J Card Fail. 2016;22(2):110–6.
44. Slaughter MS, Pagani FD, Rogers JG, Miller LW, Sun B, Russell SD, et al. Clinical management of continuous-flow left ventricular assist devices in advanced heart failure. J Heart Lung Transplant. 2010;29(4 Suppl):S1–39.
45. Rose EA, Moskowitz AJ, Packer M, Sollano JA, Williams DL, Tierney AR, et al. The REMATCH trial: rationale, design, and end points. Randomized evaluation of mechanical assistance for the treatment of congestive heart failure. Ann Thorac Surg. 1999;67(3):723–30.
46. Russell SD, Rogers JG, Milano CA, Dyke DB, Pagani FD, Aranda JM, et al. Renal and hepatic function improve in advanced heart failure patients during continuous-flow support with the HeartMate II left ventricular assist device. Circulation. 2009;120(23):2352–7.
47. Reinhartz O, Farrar DJ, Hershon JH, Avery GJ Jr, Haeusslein EA, Hill JD. Importance of preoperative liver function as a predictor of survival in patients supported with Thoratec ventricular assist devices as a bridge to transplantation. J Thorac Cardiovasc Surg. 1998;116(4):633–40.
48. Aaronson KD, Patel H, Pagani FD. Patient selection for left ventricular assist device therapy. Ann Thorac Surg. 2003;75(6 Suppl):S29–35.
49. Lietz K, Long JW, Kfoury AG, Slaughter MS, Silver MA, Milano CA, et al. Outcomes of left ventricular assist device implantation as destination therapy in the post-REMATCH era: implications for patient selection. Circulation. 2007;116(5):497–505.
50. Smith RA, Cokkinides V, Brooks D, Saslow D, Brawley OW. Cancer screening in the United States, 2010: a review of current American Cancer Society guidelines and issues in cancer screening. CA Cancer J Clin. 2010;60(2):99–119.
51. Holdy K, Dembitsky W, Eaton LL, Chillcott S, Stahovich M, Rasmusson B, et al. Nutrition assessment and management of left ventricular assist device patients. J Heart Lung Transplant. 2005;24(10):1690–6.
52. Sharma R, Anker SD. Cytokines, apoptosis and cachexia: the potential for TNF antagonism. Int J Cardiol. 2002;85(1):161–71.
53. Mano A, Fujita K, Uenomachi K, Kazama K, Katabuchi M, Wada K, et al. Body mass index is a useful predictor of prognosis after left ventricular assist system implantation. J Heart Lung Transplant. 2009;28(5):428–33.
54. Coyle LA, Ising MS, Gallagher C, Bhat G, Kurien S, Sobieski MA, et al. Destination therapy: one-year outcomes in patients with a body mass index greater than 30. Artif Organs. 2010;34(2):93–7.
55. Miller LW. Listing criteria for cardiac transplantation: results of an American Society of Transplant Physicians-National Institutes of Health conference. Transplantation. 1998;66(7):947–51.
56. Cogswell R, Smith E, Hamel A, Bauman L, Herr A, Duval S, et al. Substance abuse at the time of left ventricular assist device implantation is associated with increased mortality. J Heart Lung Transplant. 2014;33(10):1048–55.
57. Kavarana MN, Pessin-Minsley MS, Urtecho J, Catanese KA, Flannery M, Oz MC, et al. Right ven-

tricular dysfunction and organ failure in left ventricular assist device recipients: a continuing problem. Ann Thorac Surg. 2002;73(3):745–50.
58. Dang NC, Topkara VK, Mercando M, Kay J, Kruger KH, Aboodi MS, et al. Right heart failure after left ventricular assist device implantation in patients with chronic congestive heart failure. J Heart Lung Transplant. 2006;25(1):1–6.
59. Takeda K, Takayama H, Colombo PC, Yuzefpolskaya M, Fukuhara S, Han J, et al. Incidence and clinical significance of late right heart failure during continuous-flow left ventricular assist device support. J Heart Lung Transplant. 2015;34(8):1024–32.
60. Kapelios CJ, Charitos C, Kaldara E, Malliaras K, Nana E, Pantsios C, et al. Late-onset right ventricular dysfunction after mechanical support by a continuous-flow left ventricular assist device. J Heart Lung Transplant. 2015;34(12):1604–10.
61. Takeda K, Naka Y, Yang JA, Uriel N, Colombo PC, Jorde UP, et al. Timing of temporary right ventricular assist device insertion for severe right heart failure after left ventricular assist device implantation. ASAIO J. 2013;59(6):564–9.
62. Dell'Italia LJ. Anatomy and physiology of the right ventricle. Cardiol Clin. 2012;30(2):167–87.
63. Ryan JJ, Archer SL. The right ventricle in pulmonary arterial hypertension: disorders of metabolism, angiogenesis and adrenergic signaling in right ventricular failure. Circ Res. 2014;115(1):176–88.
64. Driessen MM, Baggen VJ, Freling HG, Pieper PG, van Dijk AP, Doevendans PA, et al. Pressure overloaded right ventricles: a multicenter study on the importance of trabeculae in RV function measured by CMR. Int J Cardiovasc Imaging. 2014;30(3):599–608.
65. Krabatsch T, Potapov E, Stepanenko A, Schweiger M, Kukucka M, Huebler M, et al. Biventricular circulatory support with two miniaturized implantable assist devices. Circulation. 2011;124(11 Suppl):S179–86.
66. Deuse T, Schirmer J, Kubik M, Reichenspurner H. Isolated permanent right ventricular assistance using the HVAD continuous-flow pump. Ann Thorac Surg. 2013;95(4):1434–6.
67. Shehab S, Macdonald PS, Keogh AM, Kotlyar E, Jabbour A, Robson D, et al. Long-term biventricular HeartWare ventricular assist device support–case series of right atrial and right ventricular implantation outcomes. J Heart Lung Transplant. 2016;35(4):466–73.
68. Krabatsch T, Hennig E, Stepanenko A, Schweiger M, Kukucka M, Huebler M, et al. Evaluation of the HeartWare HVAD centrifugal pump for right ventricular assistance in an in vitro model. ASAIO J. 2011;57(3):183–7.
69. McGee E Jr, Chorpenning K, Brown MC, Breznock E, Larose JA, Tamez D. In vivo evaluation of the HeartWare MVAD pump. J Heart Lung Transplant. 2014;33(4):366–71.
70. Shanewise J. Cardiac transplantation. Anesthesiol Clin North Am. 2004;22(4):753–65.
71. Ronco C, Haapio M, House AA, Anavekar N, Bellomo R. Cardiorenal syndrome. J Am Coll Cardiol. 2008;52(19):1527–39.
72. Berthiaume JM, Kirk JA, Ranek MJ, Lyon RC, Sheikh F, Jensen BC, et al. Pathophysiology of heart failure and an overview of therapies. In: Buja LM, Butany J, editors. Cardiovascular pathology. 4th ed. New York: Elsevier; 2015.

8 Engineering Perspectives for Mechanical Circulatory Support Devices

Kevin Bourque, Christopher Cotter, and Charles Dague

Introduction

As soon as one decides to solve heart failure-related circulatory dysfunction with a medical device, he is conceding that previously unutilized physical principles, starkly non-human materials, and a cadre of professionals charged with shoehorning science into solutions will enter the scene. That is to say: engineers are now involved. In some ways, their approach will resemble those employed in other pursuits of progress, such as crossing a body of water (bridge, airplane), communicating across distances (printing press, cellular phone), and surviving cold weather (fireplace, oil burner). Recognize that there is a problem. Define it. Determine an achievable scope. Consider the solution space. Adapt existing technologies. Invent and innovate as needed. And so forth.

But in some ways, this task is different. It is generally true that engineering involves the exploitation of nature's raw materials – sunlight, oxygen, magnetism – and the wherewithal to recognize, combine, and adjust these resources to improve human existence. But for all the outward accomplishments of mankind to lengthen and improve every aspect of his earthly existence through technological advancement, there is something special, something *insular*, in applying technology to improve oneself. To *heal* oneself.

For millions of years, mankind's only protections against disease and injury were chance and genetic advantage, and expected survival remained nearly constant, increasing at a glacial rate. Technological advancements over the past few thousand years, geometrically accelerated by the Enlightenment and the Industrial Revolution, have changed the algebra. It is *volition*, not evolution, that determines life expectancy and quality of life today.

Among the wide range of modern medical devices available, MCS devices are among the most mission critical, literally life sustaining, and compensating for a fundamental physical deficiency that invariably progresses to patient death. Such devices are often implanted, intrinsically invasive (to varying degrees), and once an MCS patient's immunologic responses have adapted to the intrusion, it is often difficult to distinguish where the man ends and the machine begins. The invention, design, development, and production of those devices involve efforts rendered large by their criticality. The fundamentals of those efforts will be covered in this chapter. Bear in mind that however prosaic some of the details of implementation may appear, it is never too far from an MCS engineer's mind that he is striving for nothing less lofty than to postpone the shuffling of his mortal coil.

K. Bourque, MSME (✉) · C. Cotter, MSME
C. Dague, MSEE
Abbott, Burlington, MA, USA
e-mail: kevin.bourque@abbott.com; chris.cotter@abbott.com; chad.dague@abbott.com

© Springer Nature Switzerland AG 2020
J. H. Karimov et al. (eds.), *Mechanical Support for Heart Failure*,
https://doi.org/10.1007/978-3-030-47809-4_8

Fundamentals, First Principles, and Design

Native Circulation

It is important to recognize that, disregarding a variety of physiologic shunts and mass transfers that do not significantly corrupt the model, human circulation is a closed system of two pumps in series (Fig. 8.1a, b), which may be conceptualized via an electrical circuit analogy (Fig. 8.1c).

Within this simplified model and in the steady state, conservation of matter provides that blood flow is the same everywhere in the circuit. Whatever blood flow enters the systemic pump also enters the pulmonic pump. This can be temporarily untrue without violating conservation law if fluid accumulates somewhere in the circuit, a routine possibility in human physiology and pathophysiology.

Further observations may be made. It is interesting to consider two cases: one in which both pumps have identical capacity and the other in which one pump overmatches the other.

For the first case, it can be conceived how identical operation of two serial pumps would result in constant flow without accumulation on either side, but in practice neither the pumps nor their environments are likely to be identical, so a compensatory mechanism(s) is required to achieve balance. In humans, relatively simple mechanical (Laplace) and complex neurohormonal systems automatically respond to imbalances to restore *homeostasis* via adjustments in beat rate and contractility force. Details of physiologic compensatory mechanisms are outside of this chapter's scope, but they are mentioned to foreshadow similar effects in MCS pumps, discussed below. The milieu of automatically restoring a physical system to a "home," or "target," position is ubiquitous in engineering, and a simplified model of a negative feedback system is shown in Fig. 8.2. The great value of such a system is that when a desired output is not being attained, e.g., cardiac output is suddenly reduced due to an increase in arterial pressure, the system changes to attain that output, not only *directionally* but *in proportion* to the deviation. The

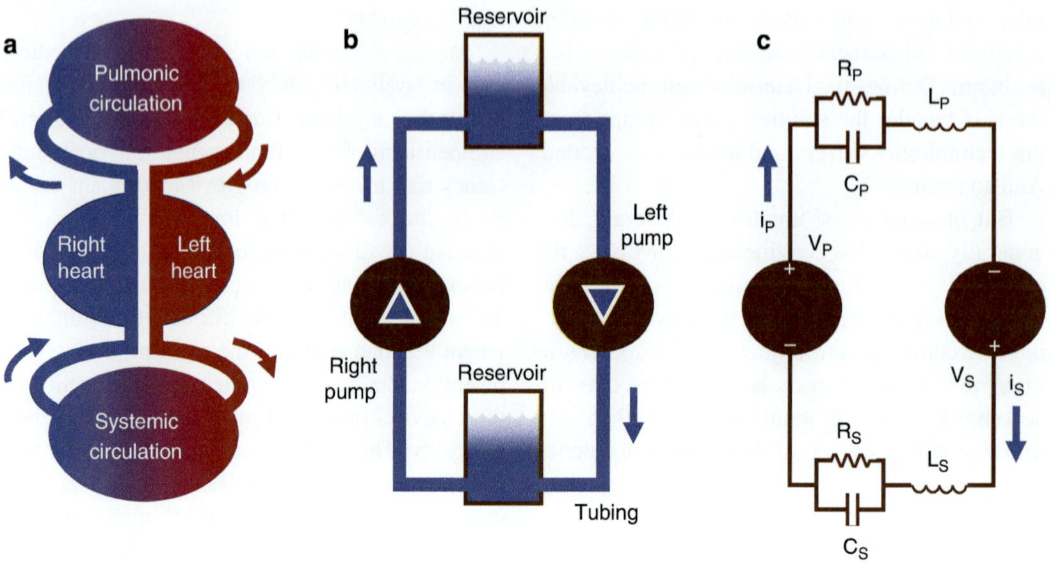

Fig. 8.1 Models of two pumps in series. (**a**) The human circulatory system. (**b**) The human circulatory system can be simplified and modeled as a closed system of two pumps in series. (**c**) This can be analogized with an electrical circuit (C). V_x = voltage, i_x = current, R_x = resistance, L_x = inductance, C_x = capacitance, x = P (pulmonic) or S (systemic)

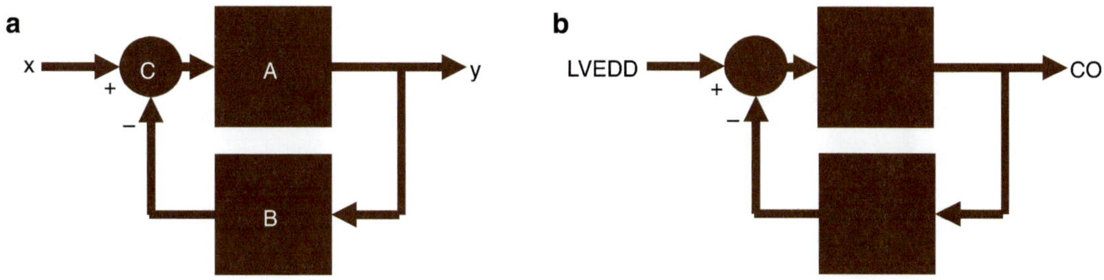

Fig. 8.2 Simplified negative feedback model. (**a**) A transformational system, A, may be described for which an output, y, is a function of the input, x, i.e., y = f(x). A measurement of the output is converted at B back into the transformational system units, x = f(y), and compared to the desired value at C. If identical, the input remains unchanged. If the output has exceeded the target, the new input is adjusted downward (hence the minus), in proportion to the difference. (**b**) For example, Laplace's law dictates that a certain LVEDD will produce a certain CO in a given environment, i.e., CO = f(LVEDD). If the actual CO is lower than expected, perhaps due to an environment change, the LVEDD will be increased in proportion to the difference. (Subtracting a negative value leads to an *increase* in input, hence *negative* feedback.) *LVEDD* = left ventricular end-diastolic diameter, *CO* = cardiac output

further the variable has deviated, the more forcefully it is nudged back to the desired value.

In the second case, in which one pump (ventricle) overmatches the other, flow persists provided the less productive pump permits fluid to pass. However, with one pump incapacitated, the setting is not normal. In pathophysiology, negative feedback mechanisms actuate as described above, and either the dominant pump increases its exertion (work) or the flow is reduced in comparison to the first case. In humans, this translates to increased metabolic demand in incipient heart failure, decreased cardiac output in advanced heart failure, or some combination of both. As might be expected, there are typically limits to the range over which such a negative feedback system is effective, and the erosion of the Frank-Starling mechanism as a cardiac output regulation system in end-stage heart failure is a complex example of a system departing from the homeostatic range [1].

Mechanical (Artificial) Circulation

Modern MCS devices include *intravascular devices* that augment circulation in an existing conduit, *assist devices* that create a bypass conduit around an existing conduit, and *replacement devices* that are substitutes for exiting conduits. The latter group includes the total artificial hearts (TAH) and the use of two ventricular assist devices in place of resected ventricles, invariably performed in pairs due to the dependence of the interventricular septum in contraction. Since ventriculectomy disables natural compensatory responses related to contractility force and beat rate, the mechanical pumps must substitute a means of negative feedback, and to an extent they can. Referring to Fig. 8.1b, it is apparent that the output of one pump provides the input to the other, so for these systems to balance, a pump that receives more than expected flow from the other must either increase its own output, influence the other pump to decrease its output, or both. For most pump types, this is provided by a monotonically inverse relationship between flow and pressure head.

The bypass configuration describes typical left ventricular assist device (LVAD) cannulation, for which some degree of native cardiac output is preserved (Fig. 8.3).

The relationship between flow and pressure head determines LVAD output as for the serial ventricular replacement case, except here the parallel native circulatory path participates in determining the pressure head and whichever physiologic compensatory systems have prevailed despite heart failure continue to contribute. Because the parallel configuration has an intrinsic redundancy that makes the failure of either system potentially survivable and retains some native cardiac physiology, LVADs are more often employed than TAHs except when ventricular

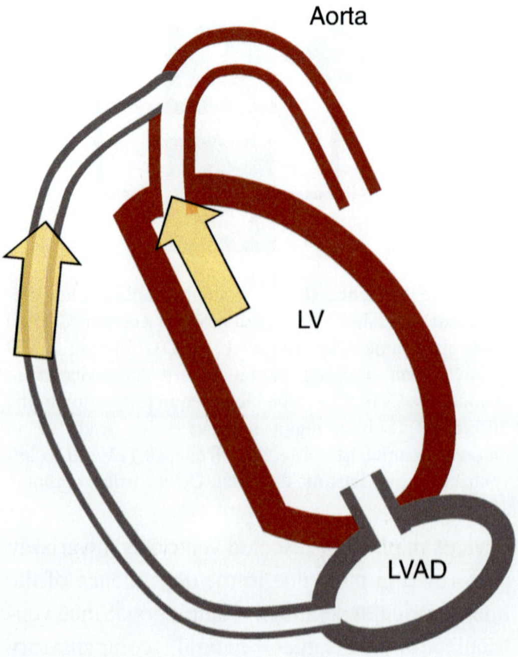

Fig. 8.3 Typical LVAD (bypass) configuration. Parallel flow paths (yellow arrows) provide perfusion from the LV to the aorta. *LVAD* = left ventricular assist device, *LV* = left ventricle

resection is indicated. Exceptional LVAD conditions, such as oversewn aortic valve and aortic stenosis or insufficiency, convert the configuration to a serial one, and severe dependence upon the LVAD in end-stage HF attenuates device failure mitigation.

Intravascular pumps are better described by the parallel case, as are both veno-venous and veno-arterial cannulation techniques applied in extracorporeal life support (ECLS) applications, although context is important because what is being bypassed differs immensely among those examples. Intra-aortic balloon pumps (IABPs) are excluded from this chapter because they achieve cardiac output increase via afterload reduction, which is not pumping in the sense of mechanically impelling flow [2].

Pump Types

MCS pumps can be broadly categorized as *positive displacement* or *continuous flow*. That other types have not been successfully applied to pumping blood should not discourage future consideration, but they will not be reviewed here. Positive displacement pumps (Table 8.1) in some ways mimic the human cardiac cycle but are relative large (because they only pump part of the time, requiring a filling period); loud (because the means of switching between filling and pumping stages typically generate acoustic noise); and unreliable (because they typically have solenoids, ball bearings, valves, polymeric sacs, or other components that wear out).

Table 8.1 A partial list of MCS devices used clinically. (Note that regulatory approval of these devices is limited to various geographies)

Positive displacement	Thoratec® PVAD™
	World Heart Novacor
	Arrow LionHeart LVAD
	Thoratec HeartMate® XVE LVAD
	Berlin Heart EXCOR®
	Abiomed AbioCor®
	SynCardia TAH
Continuous flow	Jarvik Heart Jarvik 2000®
	ReliantHeart HeartAssist5®
	Abbott HeartMate II™ LVAD
	MicroMed Debakey VAD®
	Ventracor VentrAssist™
	Terumo Duraheart™
	Medtronic HeartWare™ HVAD™
	Abbott HeartMate 3™ LVAD

Continuous flow pumps (Table 8.1) running at a constant speed have an intrinsically time-invariant output that does not mimic the human cardiac cycle. Although this initially generated consternation as a clinical option, there is no longer controversy that long-term survival with improved quality of life can be achieved with continuous flow pumps, even for entirely pump-dependent patients. Because they do not switch between filling and pumping modes, they are generally smaller, quieter, and more reliable than positive displacement pumps [3].

Continuous flow pump components include a stationary housing (stator) that contains a motor, a rotor that is induced to turn by the motor and bears a means of impelling fluid, and a bearing(s) that permits the rotor to turn relative to the stator.

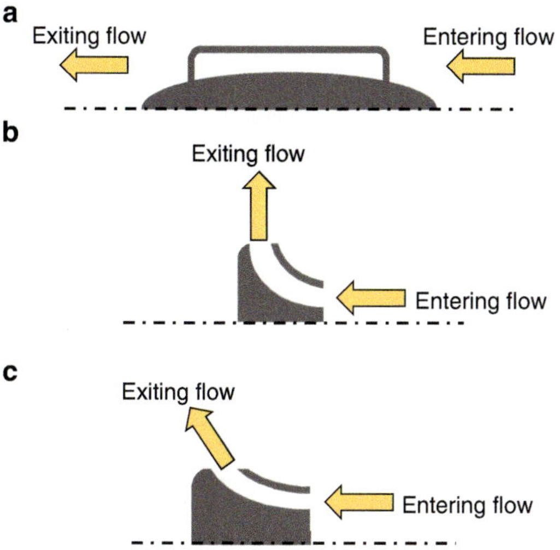

Fig. 8.4 Continuous flow pumps characterized by flow direction. The schematic illustrates cross sections through various rotor types. (**a**) The output of an *axial* pump is in the same direction as the input. (**b**) The output of a *centrifugal* pump is perpendicular to the input. (**c**) The output of a *mixed* flow pump falls somewhere in between

Continuous flow pumps can be categorized by the relationship between their input and output directions, worthwhile because that factor permits generalization of certain characteristics (Fig. 8.4).

In MCS, the working fluid is human blood, which produces markedly different challenges than for other pump applications, such as pumping water, petroleum, and harsh chemicals. As such, the type and nature of the bearing(s) used to support the rotor determine the design challenges, and categorizing continuous flow pumps by bearing type is worthwhile (Fig. 8.5).

Fore and aft mechanical bearings are normally axially spaced to permit a small amount of clearance (~1–10 micrometer) between the mating rotor and stator bearings to avert wear that would occur if the fore and aft bearings clamped the rotor. Examined at the macroscopic level, the term "mechanical" is a misnomer for blood-immersed mechanical bearings since in most operating conditions, a thin film of fluid will interrupt physical contact between mating bearings. They are distinct from hydrodynamic bearings in three ways: the bearing area is much smaller, restriction of the rotor to move in any

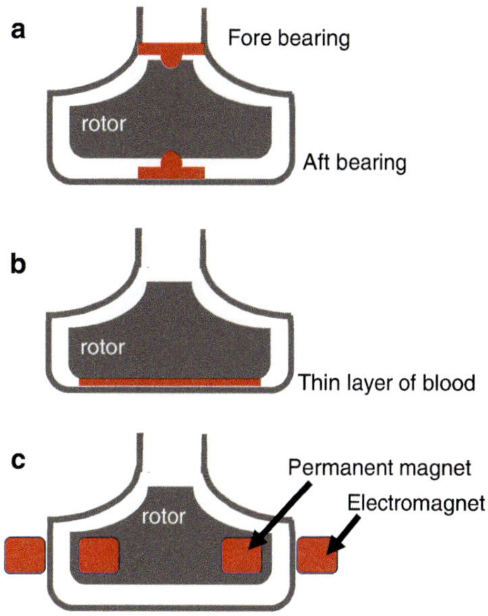

Fig. 8.5 Continuous flow pumps categorized by bearing type. The schematic illustrates cross sections through pumps with various bearing types. (**a**) *Mechanical bearings* support the rotor fore and aft.† (**b**) *Hydrodynamic bearings* support the rotor with a thin layer of blood. (**c**) *Magnetic bearings* support the rotor by levitating it with servo-controlled electromagnets that interact with permanent magnets. †Classical pump design involves mechanical bearings, e.g., ball bearings, isolated from the working fluid by seals. However, the consequence of toxic seal debris entering the bloodstream and corrosive blood entering the motor is so severe that such bearings are generally not employed in mechanical circulatory support

direction is much greater, and whatever fluid force is generated between mating bearings is not required for support [4, 5].

Hydrodynamic bearings exploit strategically shaped mating surfaces that entrain a wedge(s) of fluid between the rotor and stator that, once a critical relative speed is reached, levitates the rotor. The properties of the fluid and surface tension determine the rotor-stator gap (~10–100 micrometer) and levitation force, which balances electromagnetic forces related to the motor, hydraulic rotor forces such as thrust, gravity, and any other applied force, such as a deliberate magnetic bias to ensure the rotor is initially in the proper position to engage the bearing [6, 7].

An alternative means of levitation is with a magnetic bearing. In the axisymmetric configuration of a motor, a centered rotor with an enclosed permanent magnet is attracted to the motor's iron components equally in all directions. However, Earnshaw's theorem states that this is not a stable configuration, so electromagnets and rotor position measurements are used in a negative feedback control scheme to levitate the rotor. The resulting rotor-stator gap (~100–1000 micrometer) is determined by the electromagnet design and is unrelated to rotor speed; in fact, with a magnetic bearing, the rotor need not turn at all to accomplish levitation [8, 9].

Fundamental Pump Concepts

A pump is a machine that converts a readily available form of power, such as electrical, into the hydraulic power desired by accelerating a fluid via a rotor with impeller surfaces or blades. Hydraulic power is the product of the pressure and the volume flow rate of a fluid [10]. *Volume flow rate* is readily measured. Conceptually, one can use a stopwatch to measure how long it takes to fill a one-liter container, express the rate in liters per minute (L/min), or more practically utilize measurement options such as thermodilution or ultrasound. *Pressure* can be directly measured with transducers. Alternatively, pressure can be inferred with a manometer, a vertical, transparent tube in which a fluid of known density is exposed to pressure at both ends, and since pressure and gravity balance, the elevation of this manometer fluid is proportional to the pressure. A manometer is calibrated to its fluid so, for example, a pressure that is above atmospheric pressure may cause a column of mercury (Hg), a common manometer fluid, to rise a certain number of millimeters (mm). This measurement is not pressure but *pressure head*, measured in units of length, such as mmHg. The measurement technique is so ubiquitous that the terms pressure and pressure head are often imprecisely interchanged. The conversion shown in Fig. 8.6 is useful for common MCS units.

In normal adult physiology, resting cardiac output is approximately 5 L/min, and mean aortic and left ventricular pressure heads are 90 mmHg and 10 mmHg, respectively, so normal adult systemic cardiac output power is

$$\left(5.0\frac{\text{L}}{\text{min}}\right)\times(90\,\text{mmHg}-10\,\text{mmHg})\times 0.0022\frac{\text{W}}{\text{mmHg}\times\frac{\text{L}}{\text{min}}}$$

= 0.88 W, or about a Watt. Because the difference between pulmonary arterial and ventricular pressures is much lower than on the systemic side, pulmonic cardiac output power is proportionally lower. In exercise, the demand is potentially several times higher [11].

A fundamentally important characterization of continuous flow pumps relates volume flow rate (Q) to pressure head (H, Fig. 8.7).

This relationship offers clinical insight. For example, the LVAD shown in Fig. 8.3 has its inflow exposed to the left ventricle and its outflow exposed to the aorta. Because the heart is beating, the pump is exposed to widely time-variant pressures throughout the cardiac cycle. In Fig. 8.8, it is apparent that during cardiac systole, there is relatively little difference between aortic and ventricular pressure; con-

$$\frac{101{,}325\,Pa}{760\,mmHg}\times\frac{\left[\frac{N}{m^2}\right]}{Pa}\times\frac{1000\,cm^3}{L}\times\left[\frac{m}{100\,cm}\right]^3\times\frac{min}{60\,s}\times\frac{W}{\left[\frac{N\cdot m}{s}\right]}$$
$$=0.0022\frac{W}{mmHg\times\frac{L}{min}}$$

Fig. 8.6 Convenient conversion factor to calculate hydraulic power from the product of pressure head and volume flow rate. If H [mmHg] and Q [L/min] are known, the conversion factor is 0.0022 W/mmHg/(L/min). H = pressure head, mmHg = millimeter of mercury, Q = volume flow rate, L/min = liter per minute, Pa = Pascal, N = Newton, m = meter, cm = centimeter, L = liter, min = minute, s = second

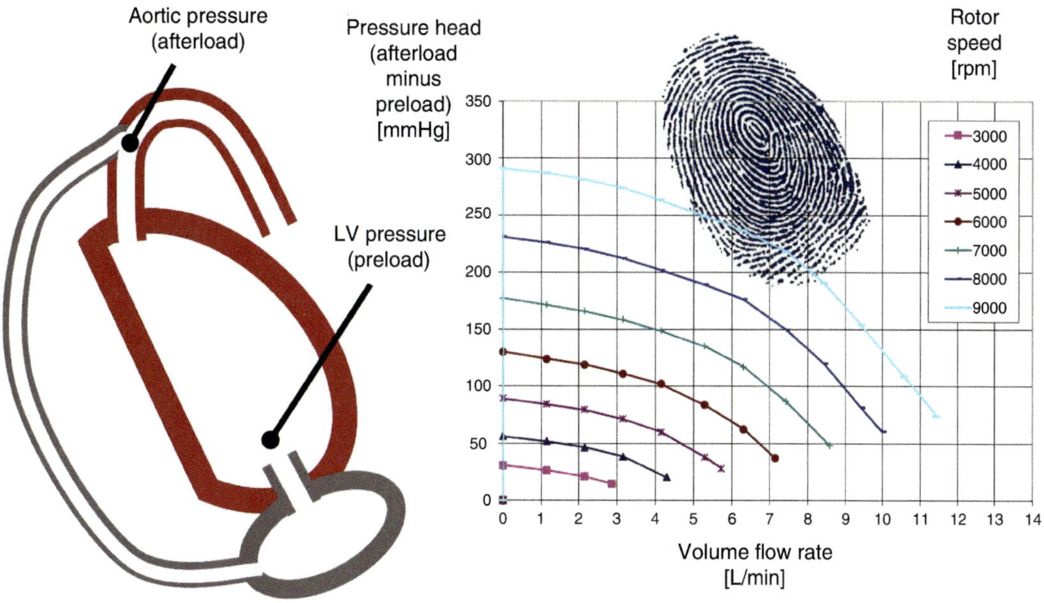

Fig. 8.7 The H-Q characteristic. A continuous flow pump of a certain design, set to a certain rotor speed, exposed to a certain preload and afterload, will always produce the same Q, other factors (including fluid properties) being equal in proportion to the difference between afterload and preload, H. In effect, it is a "fingerprint" of a pump design. Identical pumps in an identical setting will obey this relationship. H = pressure head, Q = volume flow rate, LV = left ventricle

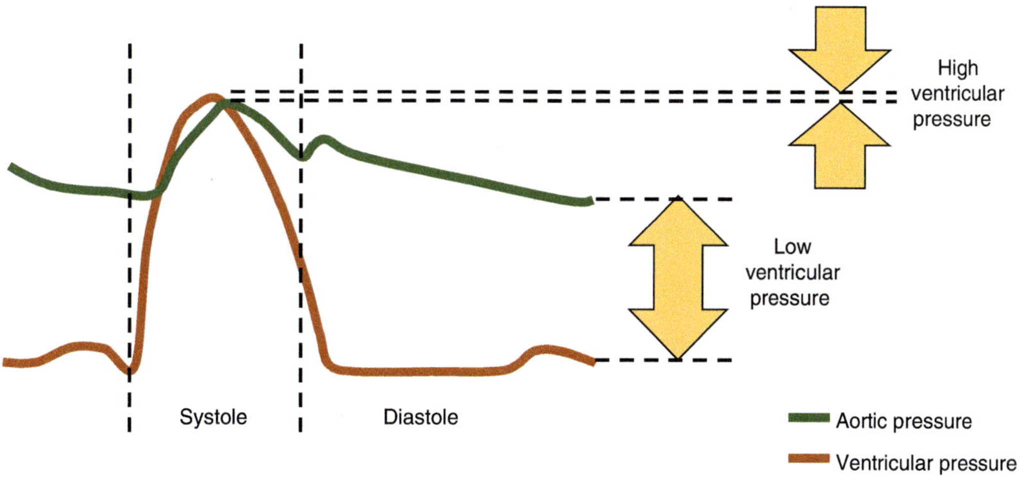

Fig. 8.8 Variation in the difference between aortic and ventricular pressure during the cardiac cycle

versely, during diastole there is a greater difference. It follows that pump flow is greater during systole (larger difference between aortic and ventricular pressure) than during diastole, to an extent determined by the H-Q characteristic.

The relationship between pressure and flow is inverse to an extent that varies by pump type. In Fig. 8.9, axial and centrifugal (radial) pump H-Q characteristics are generalized. Since pressure head is the difference between aortic and LV pressure head, considerably different clinical

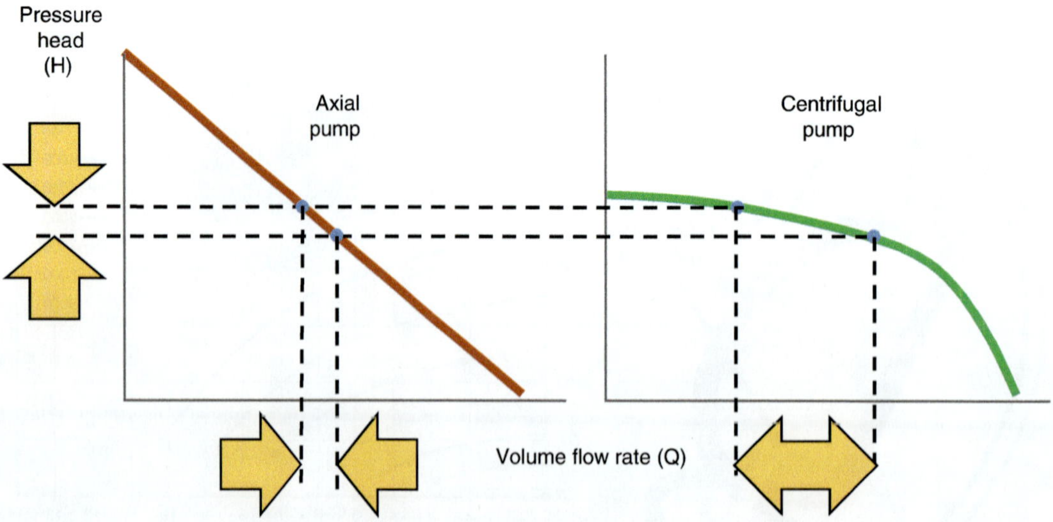

Fig. 8.9 Generalized H-Q characteristics. H = pressure head, Q = volume flow rate

conditions, e.g., aortic and LV pressure head of 75 and 10 mmHg, respectively, and 100 and 35 mmHg, respectively, would result in the same flow rate. It is often heard that one pump is "afterload sensitive" and another is "preload sensitive," but it should be clear that all pumps are sensitive to both afterload and preload. Context is required, so, for example, if the topic is low flow in systemic hypertension, where the centrifugal pump characteristic is flatter than that of the axial pump, the conclusion may be that a centrifugal pump is more afterload sensitive. One should appreciate, however, that at high flow rates, a centrifugal pump is markedly *less* afterload sensitive.

Another important characterization of continuous flow pumps relates volume flow rate to motor power (Fig. 8.10).

The relationship between power and flow is direct, a consequence of conservation of energy. A continuous flow pump of a certain design, set to a certain rotor speed, will always consume the same amount of power to produce a certain flow, other factors being equal. There is a common misconception surrounding this immutable principle; for example, that elevated arterial pressure in hypertension that creates an impediment to flow will cause motor power to *increase*. In fact, the opposite can readily be demonstrated with the Power-Q characteristic, as hypertension causes a *reduction* in flow, thus a *reduction* in motor power. The paradox likely stems from experiences in which a new impediment triggers an increase in effort to compensate, but a pump left at the same rotor speed facing an increase in arterial pressure does not compensate. It does *less* work, not more [12].

The *efficiency* of converting available electrical power to desired hydraulic power involves not only the pump but the other power-consuming components of the system (Fig. 8.11).

Efficiency is generally defined as the ratio of "what you get" to "what you paid for it," in this case the ratio of hydraulic power and electrical power delivered to the system. There are intrinsic limits to how efficiently hydraulic energy can be produced, and 25% overall efficiency is a reasonable MCS target. Accordingly, about 4 W is typically required to meet the 1 W demand derived previously.

Pump Design

In systolic heart failure, the native ventricle is unable to produce the cardiac output demanded. A host of sociologic adaptations and pharmacologic options are available as initial therapies

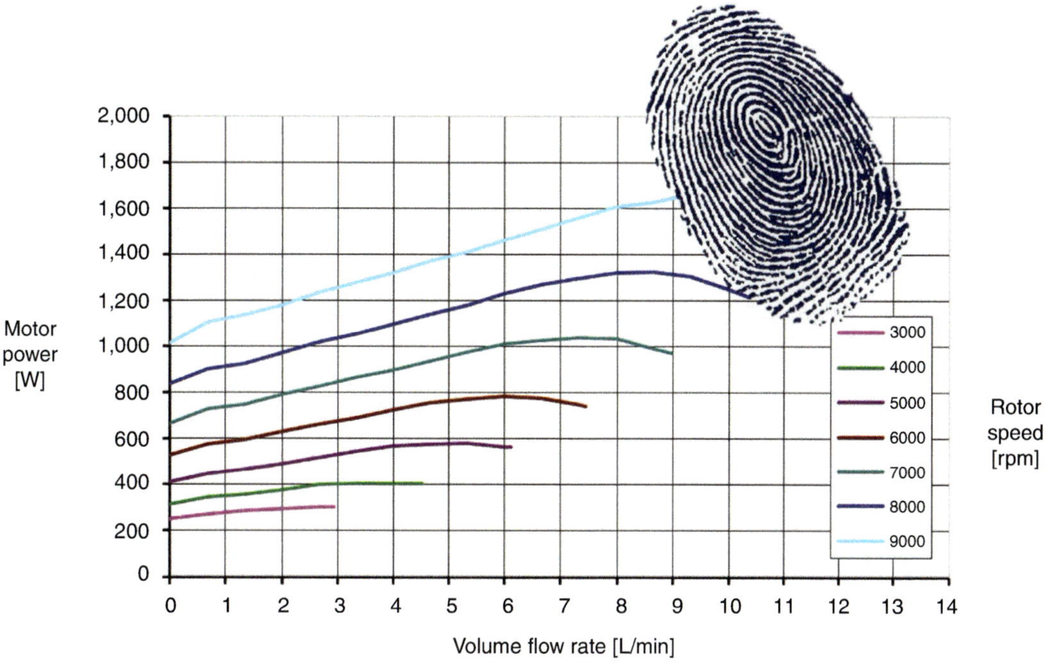

Fig. 8.10 The Power-Q characteristic. A continuous flow pump of a certain design, set to a certain rotor speed, will always consume the same amount of power to produce a certain Q, other factors being equal. In effect, it is a "fingerprint" of a pump design. Identical pumps in an identical setting will obey this relationship. Q = volume flow rate

for what is, with the possibility of an oscillating course, an ultimately progressive disease that culminates in death. In advanced stages of this progression, the tolerance for risk, invasiveness, and side effect typically increases as the number and effectiveness of therapeutic options dwindles. The modern application of chronic MCS devices is generally confined to these latter stages because devices bear intrinsic risk, such as with the initial surgery, adverse events related to the therapy, device malfunction, and use and misuse in an expanding array of activities of daily living. Adoption of chronic MCS therapy is driven by the trade-off between mortality benefit and associated morbidity [13]. That acute MCS devices are used for shorter durations, often in settings in which medical support is available, is mitigating.

To reconcile the nature of modern era MCS devices with the present clinical threshold for the therapy, it is helpful to make this sweeping generalization: if a pump is made smaller, it will be less efficient, less hemocompatible, of lower capacity, less invasive, and less costly, other factors being equal. Trade-offs among these factors, as well as between reliability and required duration, have led to an invisible barrier that has precluded MCS therapy from further penetration into earlier stages of heart failure progression. For example, one may start with a pump that is ideally efficacious in the setting of chronic, full support, and hypothesize that it could be made smaller, only to learn that the smaller pump now produces undesirable blood damage; to avoid that, pump capacity is reduced and is no longer sufficient for the application or, if a less ill population is identified for which the capacity is sufficient, surgical risk and adverse events are no longer acceptable. It should be apparent that the goal of the engineer ought to be negating or inverting antagonistic tradeoffs or, if not possible, optimizing the combination. Technological progress that reduces adverse events clearly moves the invisible barrier favorably.

Designing a continuous flow pump to move blood has some similarities to and differences

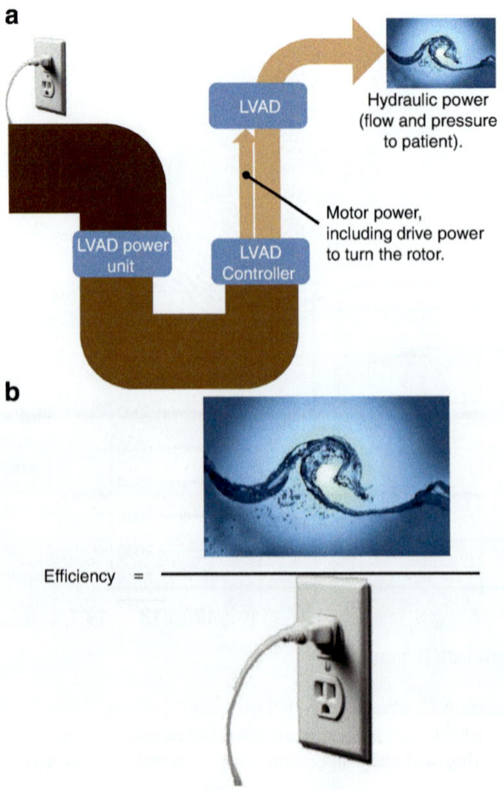

Fig. 8.11 Power dissipation in an LVAD system. (**a**) Starting at the wall outlet that represents the overall power delivered to the system, the path narrows at each system component, symbolizing power dissipation. The power delivered to the LVAD is already significantly less than at the wall outlet, and the final conversion to hydraulic power by the pump decreases it further. (**b**) Efficiency is the ratio of "what you get" and "what you paid for it." LVAD = left ventricular assist device

from designing for other applications. It starts with defining attributes such as pressure head and volume flow rate. The numerous references that guide the pump designer in this task combine analytical physics and empirically established closed form expressions such as affinity laws and dimensionless quantities such as Reynolds number, derived via dimensional analysis [14]. The challenge in determining an optimal design lies in careful consideration of requirements, acknowledging that favoring one attribute often comes at the expense of another.

A traditional pump design process starts with identifying the pump specific speed, n_s. Specific speed for pumps, like Reynolds number for pipe flow, identifies performance similarities for geometrically different designs. Specific speed is defined as:

$$n_s = \frac{n\sqrt{Q}}{H^{3/4}}$$

where n is pump speed in revolutions per minute (rpm), Q is pump flow rate in gallons per minute (gpm), and H is pump head in feet (ft). When used as a type number, Q and H are defined at the best efficiency point for any pump speed; thus, specific speed is a constant. Specific speed is the speed at which a pump must rotate to generate 1 gpm of pump flow at 1 ft of pump head, and it can be suggestive of a pump type – axial, centrifugal, or mixed – that optimizes efficiency. However, in MCS, matching form factor to anatomical and hemocompatibility targets is at least as likely to influence type selection.

Pump size and shape are driven by a few critical aspects. Pump size is proportional to capacity (Q and H). Pump shape, particularly if implantable, is driven by patient habitus, which can be quite diverse, and elements of the implant procedure and desired performance over the intended support duration. For example, axial flow pumps tend to be long and cylindrical, while centrifugal pumps tend to be flatter and disc shaped. In intravascular, intraventricular, and surgical pocket applications, the long, cylindrical shape may be preferred. Where protrusion is to be minimized, the flatter disc shape may be preferred (Fig. 8.12). Other considerations include the types of incisions and anatomical manipulation required, cannulation options, and the pump circuit de-airing process.

Many industrial pumps designed for a specific purpose operate within a narrow, well-controlled pump head range, as do MCS pumps, spanning only about 100–200 mmHg in the extreme. The H-Q characteristics introduced earlier, which vary with pump type, play a dominant role in performance attributes. As an example, if a centrifugal pump design results in a characteristic that is too flat, especially if normal operation is predominantly in the flat region, a relatively subtle increase in arterial pressure could result in dramatically lower, *or zero*, flow (Fig. 8.13). This "pump shutoff" condition could be detrimental to a pump-dependent patient and increase risk of blood stasis and

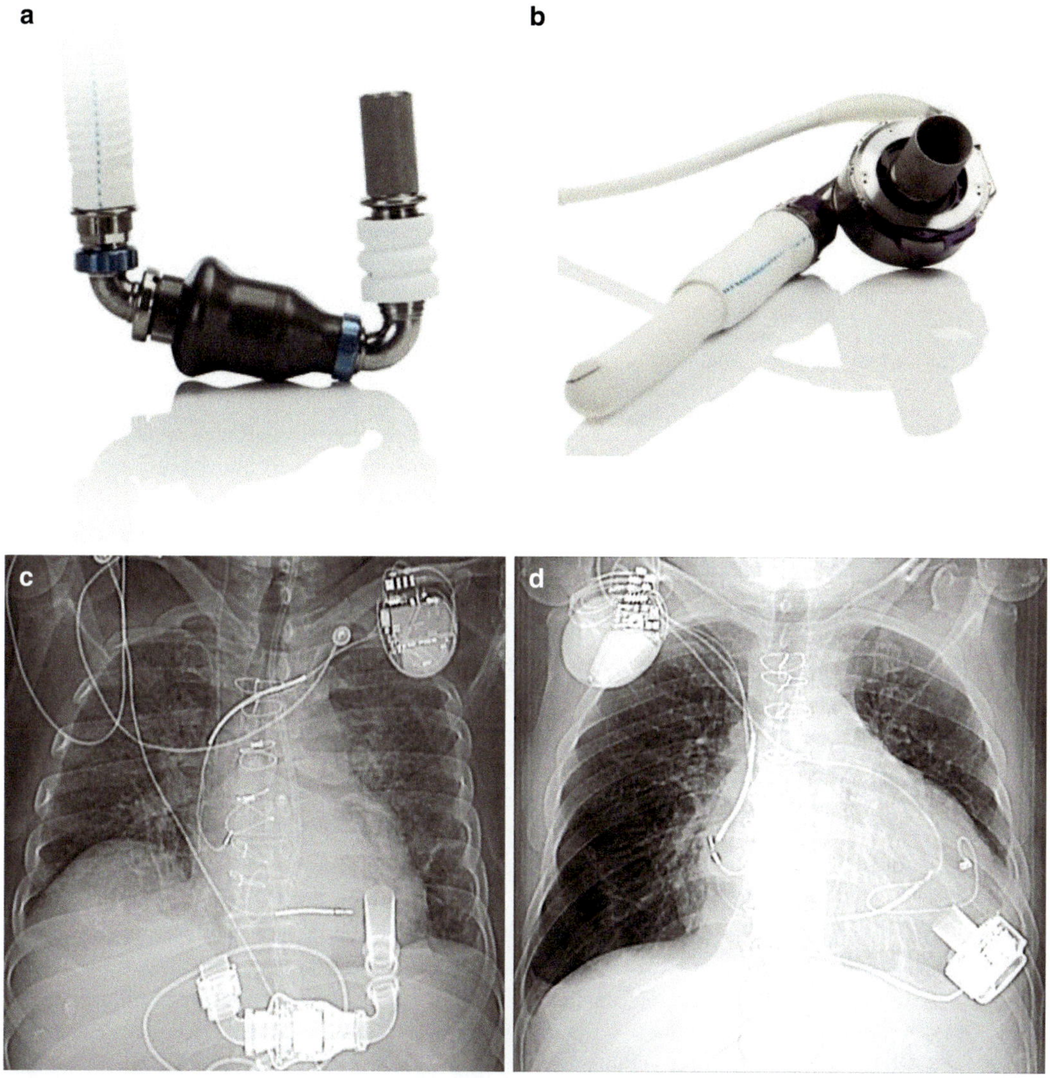

Fig. 8.12 Example of implant location influencing pump type selection. (**a**) The Abbott HMII LVAD, an axial flow pump. (**b**) The Abbott HM3, a centrifugal flow pump. (**c**) Chest radiograph of the HMII in a subdiaphragmatic pocket. (**d**) Chest radiograph of the HM3 positioned intrathoracically directly on the ventricular epicardium. LVAD = left ventricular assist device, HMII = HeartMate II™ LVAD, HM3 = HeartMate 3™ LVAD. (8.12 a, b, and c reprinted with permission from Abbott © 2018, all rights reserved; HeartMate II and HeartMate 3 are trademarks of Abbott or its related companies. 8.12d reprinted from Schmitto et al. [27])

pump thrombosis. However, if normal operation is closer to the steep region, the shutoff pressure head may be suitably distant. H-Q characteristics are not inherently good or bad; rather, the pump designer may appropriately leverage them based on desired pump performance.

The pump designer can, to some extent, tune the H-Q characteristic through careful consideration of the impeller and casing design. Impeller diameter, blade shape and number, and the presence or absence of a shroud can profoundly affect pump performance.

Impeller diameter is directly related to pump capacity. The number of impeller blades is related to pump size and pressure head with smaller, lower pressure head pumps requiring fewer blades. Impeller blades are typically curved backwards, but blades can have either

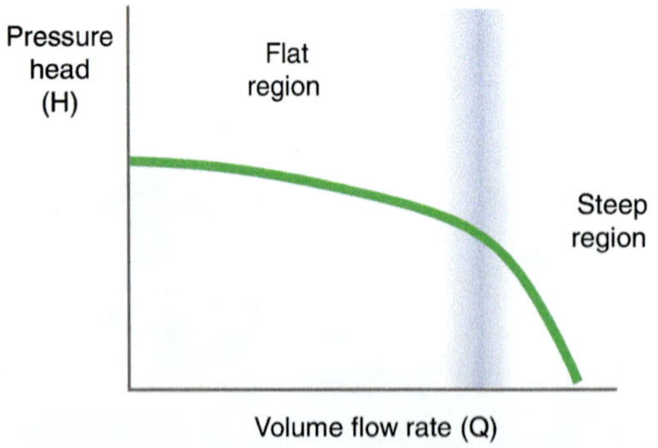

Fig. 8.13 The variable slope of a centrifugal pump H-Q characteristic curve. H = pressure head, Q = volume flow rate

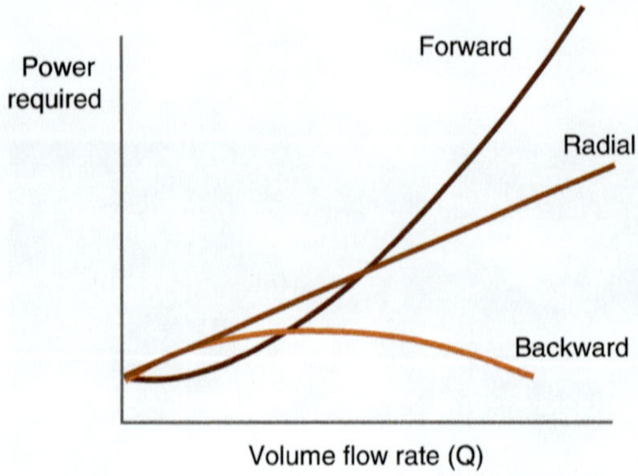

Fig. 8.14 The effect of impeller blade sweep direction on the power required to produce flow

no curvature (radial) or even forward curvature, affecting the power-capacity relationship (Fig. 8.14). A critical aspect of impeller blade shape is the discharge angle, defined as the angle between a line tangent to the trailing edge of the impeller blade and a line tangent to the impeller diameter at the same point. The way the discharge angle is defined results in an acute angle for the backward curved blade, a right angle for a radial blade, and an obtuse angle for a forward curved blade. Capacity generally increases with increasing discharge angle, so this parameter is related to maximum output and the steepness of the H-Q curve. Discharge angle can also influence linearity of the relationship between motor power and volume flow rate, which can increase utility of the relationship [14].

A routine pump design goal is to minimize required motor power, maximizing efficiency. Leakage is the pump capacity lost to fluid that does not discharge from the pump to the patient, including fluid that shunts from the pressure side of a blade to the suction side and that which recirculates from the trailing blade edge to the leading edge in the space between the rotor and stator. When leakage occurs, pump capacity is less than impeller capacity; the ratio of pump and impeller capacity is called *volumetric efficiency*. Most industrial pumps have very high volumetric efficiency, but when blood is the working fluid, there are unique requirements to avoid low flow, stasis, high shear stress, and long residence time, and sacrificing volumetric efficiency to avoid prothrombotic conditions is common.

Designing a blood pump for compatibility with blood extends beyond volumetric efficiency to a constellation of competing requirements holistically termed *hemocompatibility*, which covers thrombosis, bleeding, and stroke [15]. It is helpful to take liberties with the classical Virchow's triad, adapted for MCS, to appreciate the interaction (Fig. 8.15). To some degree, all three factors can be influenced by design, although antagonistic requirements, unclear or poorly understood physiologic mechanisms, and patient variability make optimization extremely difficult. Foreign body insult can be minimized by reducing blood-contacting area and carefully selecting materials for biocompatibility and favorable interaction. Stasis can be minimized by ensuring ample flow in all regions and avoiding flow irregularities such as boundary layer separation and eddies. Hypercoagulability can be minimized by reducing hemolysis, avoiding blood exposure to high shear stress, and minimizing the required level of antiplatelet and anticoagulation therapy by avoiding pro-thrombotic design features. All of the tactics to avoid thrombosis also favorably impact the risk of gastrointestinal and mucosal bleeding, and a primary design objective is to reduce the sensitivity of MCS therapy to patient phenotype, including the often unknown propensity to clot or bleed.

The blood pump designer must identify blood contacting surface materials and flow patterns that minimize adverse interactions with blood stemming from stasis, shear, or inflammation.

Basic biocompatibility is well defined by international standards. Surface geometry, smoothness, and coatings can be tuned to foster a desired reaction with blood. Computational tools and empirical test platforms are available to predict and observe flow irregularities, respectively.

The pump characteristics described earlier are routinely exploited to offer clinically meaningful product features. Because motor power is routinely measured during operation, and the speed is also known by the system, the Power-Q relationship can be utilized to predict the volume flow rate (Fig. 8.16). This is termed *sensorless flow estimation*, since a sensor is not needed. Accuracy is enhanced by monotonic linearity of the relationship and isolation of rotor drive power measurement, since power consumption unrelated to turning the rotor unnecessarily contributes measurement error.

Because flow, thus pump power, varies throughout the cardiac cycle in direct proportion to preload and afterload, it is possible to assess the *degree of pulsatility* in the pump output (Fig. 8.17). This parameter can be displayed to offer a clinician potentially useful corroboration or refutation of other clinical indicators. The parameter is a figure of merit, sensitive to many factors related to the uniqueness of a pump

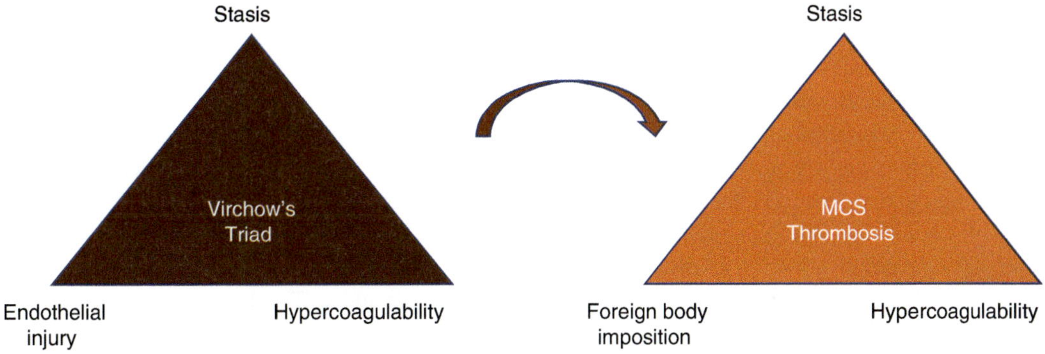

Fig. 8.15 Virchow's Triad adapted for mechanical circulatory support. Virchow's endothelial injury is substituted with foreign body (device) insult. To some degree, all factors can be influenced by design. MCS = mechanical circulatory support

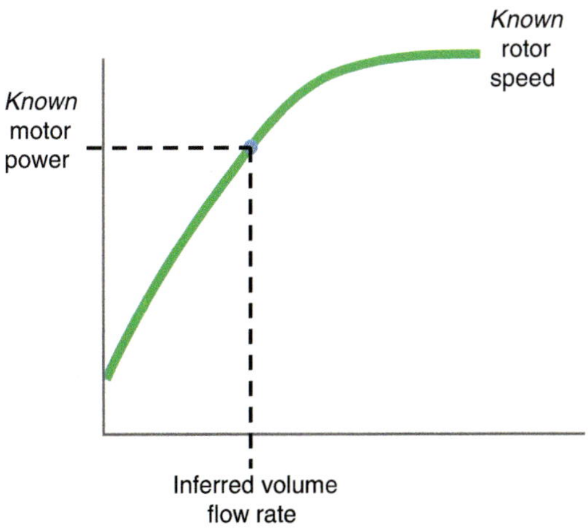

Fig. 8.16 Sensorless flow estimation. Internal to the pump, knowledge of rotor speed and motor power permits inference of volume flow rate

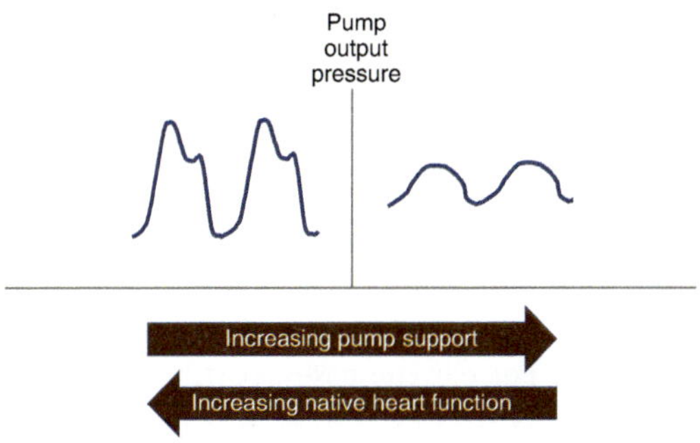

Fig. 8.17 Two examples of pump output through two cardiac cycles each to illustrate how the degree of pulsatility can be quantified. The difference between minimum and maximum motor power is proportional to the ratio of native cardiac and pump output power. Decreased pulsatility in the pump output, on the right side in the illustration (decreased variation in motor power), indicates increasing pump support relative to native heart function and vice versa

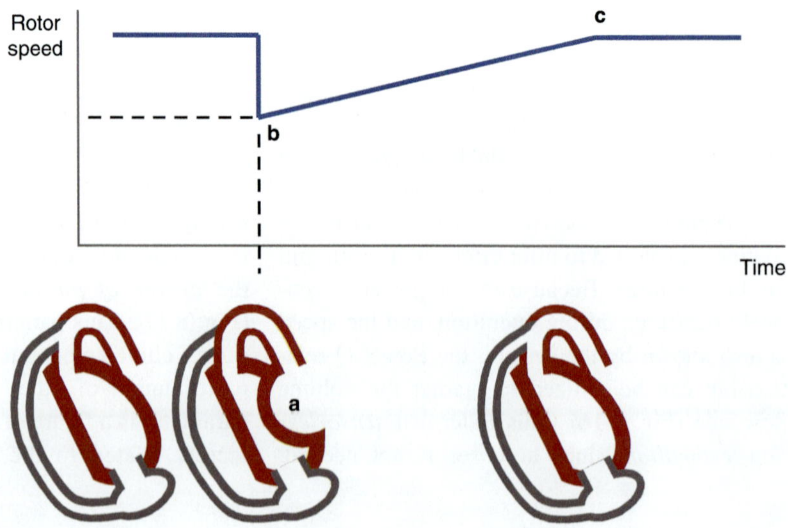

Fig. 8.18 Suction detection algorithm. If a sudden change in the degree of pulsatility is detected, ventricular suction may be suspected (**a**), and the speed may be quickly lowered a pre-set amount (**b**), in principle permitting the ventricle to refill. The speed can then be gradually increased to the prior therapeutic level (**c**)

design, confounding comparison among different pumps. The degree of pulsatility can be monitored for sudden changes, which could indicate *ventricular suction*. Algorithms have been employed to respond to such events by rapidly decreasing the speed by a pre-set amount, in principle allowing an evacuated ventricle to refill, then gradually returning to the original speed to restore the intended level of support (Fig. 8.18). Appropriately tuned, false positives are benign such that false negatives can be avoided by making the feature somewhat oversensitive.

Controller and Power System Design

MCS support may be acute, for hours or days, or chronic, spanning months or years. For acute cases, only clinicians interact with the device and peripheral equipment, so patient training is not necessary. For chronic cases, patients generally must interact with the system to ambulate. This places additional burden on designers to ensure that patients and "lay user" caregivers of widely varied health, skill, and dexterity can safely interact with the system to maintain proper device function and comprehend how to respond to messages and alarms. A careful balance must be struck between too much information (e.g., unnecessary alarms) and too little information (e.g., neglecting to alert a patient of a life-threatening condition).

A typical chronic MCS system consists of the VAD, a patient interface device for communicating the state of the system to the patient, a clinician interface device, and a power supply. Power supply options generally include a portable battery(ies) and utility line power available as alternating current (AC). Modern batteries usually employ lithium-ion chemistry for its high power density. Although carrying batteries is an inconvenience, this configuration offers the patient mobility untethered from utility power, enabling many of the routine activities of daily living. The utility power option is typically used when the patient is sedentary or sleeping, as this alleviates the concern of sleeping through battery end-of-charge.

A patient interface device is required to communicate system operating status and to alert the patient to conditions needing resolution. Typically referred to as the System Controller, this device contains hardware and software that determines power management, user interface, and the alarm system. A robustly validated alarm system is a critical component of patient safety. It is utilized by patients, caregivers, nurses, and physicians to confirm conditions are normal or, if conditions are not normal, to provide troubleshooting information, produce audio and visual alarms if appropriate, and to indicate palliative actions. Per IEC 60601-1-8, alarms are separated by priority. High-priority alarms require immediate attention and are indicated by flashing red light-emitting diodes (LEDs) and a high urgency audio tone. Medium priority alarms require prompt but not immediate attention, indicated by yellow LEDs (flashing or non-flashing) and a lower urgency audio tone. Low priority alarms require attention that can appropriately be delayed and may or may not have associated audio or visual signals.

A clinician interface device enables adjustment of LVAD rotor speed and system parameters as clinically required, and a convenient graphical interface for the clinician to view historical data stored by the System Controller. Since the patient does not use this device, it represents a way to provide diagnostically useful information to the clinician while limiting access by the patient, who gets his information from the System Controller.

In general, chronic LVADs are designed for those of adolescent to geriatric age, varying degrees of illness, with potential physical or psychological handicaps and with different social and financial supports. Designers aim to accommodate the great majority of potential patients and other users by defining user profiles covering the patient, caregiver, nurse, physicians, operating room support staff, VAD coordinators, paramedics and firefighters in the local community, and others. The system must be easy to understand and use for this heterogeneous group whether they are trained or untrained people, a significant design challenge. Designing intuitively manipulated connectors for joining components has been historically difficult, as the scenario for breaking and making connections can be stressful due to a performance problem, a loud alarm, a physical limitation, or technical incomprehension.

Engineers, trained to evaluate "what if" scenarios in the event of a failure, must resist over-complicating the system with ostensible mitigations. Fail-safe features meant to mitigate remotely likely events can instead create new risks, potentially reducing system reliability. It is important to use a design Failure Mode and Effects Analysis (dFMEA) to evaluate these probabilities to determine if the risk warrants mitigation. For example, a non-actionable, low-severity system problem should be properly classified as lower priority than a high-priority alarm, which could lead to undue anxiety or inappropriate actions that trigger an event of much higher severity. It is critical that any alarm presented to the patient must be well understood, actionable, and necessary.

Development, Verification, and Validation

An MCS product design and development effort is typically initiated either to meet an unmet customer (patient or clinician) need, sometimes accompanied by a complementary technological advance, or to respond to a newly appreciated shortcoming in an existing product. Accordingly, we conceptually separate *new product development* (NPD) from *product sustaining* efforts. NPD programs may involve next-generation,"game-changing" advances of a scope typically more comprehensive than product sustaining scope, which usually addresses only one or few features at a time to adapt or improve in response to customer feedback or field observations. All programs require cross-functional participation, including program management, marketing, research and development, manufacturing, supply chain, quality, post-market analysis, regulatory, and clinical, but NPD programs usually have larger teams. All programs are governed by the same quality and compliance standards and regulatory pathways, but larger scope often makes NPD programs take longer to complete. The application and prioritization of resources are driven by optimizing patient benefit. An incremental product change may be relatively quickly employed, but a more time-consuming NPD project often offers more definitive solutions to shortcomings in existing products, so there is a need for discipline in neither under-pursuing nor over-pursuing individual field observations.

New Product Development

NPD in MCS is typically characterized by relatively long durations, driven by the intrinsic challenges of designing a reliable system with implanted parts. The human homeostatic environment is hostile to evolutionarily unfamiliar metal and plastic devices, and among an increasingly active patient population with longer support durations, human interaction with the non-implanted parts is often *more* hostile. The relatively long duration of mission-critical verification and validation testing (which includes some tests that must run for extended periods), the general need to acquire extensive product performance experience internally prior to product launch, and the most stringent regulatory requirements for active, implantable medical devices also protract development times. This produces opportunity pressure to include as much as possible within the scope of an NPD project. Organizations typically devote a portion of their engineering and science resources to invention and innovation strategically targeting anticipated NPD needs by establishing technical feasibility and practicality. Discoveries and breakthroughs so attained have led to some of the most meaningful leaps forward in reducing MCS-related adverse events and enhancing patient quality of life.

Sustaining of Existing Products

Many of the components comprising MCS commercial products are especially designed and fabricated for the application, but customization is frequently neither practical nor justified for commonplace components such as fasteners and electrical components. Because consumer products such as cellular phones and computers generate the demand for many components utilized in medical devices, and those markets have far shorter lifecycles, frequent revisions are commonplace. Component obsolescence is a constant source of supply chain risk, and anticipating looming unavailability of a component, identifying a substitute, adapting the device for the new component, and verifying functionality are a continuous effort.

A formal, structured complaint system is required of MCS organizations to capture and manage remarks, dissatisfactions, mishaps with the medical device, injuries to patients, and adverse events. These experiences are assessed to evaluate imminent or future hazards and potential corrective actions, and lessons learned are fed forward into projects for potential design and development improvements.

Verification and Validation

Product development is governed by *design control* principles that simultaneously satisfy business objectives and regulations, such as FDA 21 CFR Part 820 and EU Medical Device Regulation [16, 17], both of which depend upon patient safety and product quality and efficacy. Formal *design reviews* conducted by responsible staff at defined project phases provide opportunity to scrutinize work already accomplished for quality and rigor and to evaluate whether the existing plan should be continued, adjusted, or cancelled.

Formal user needs must be established and *validated*, confirming that the right product has been produced to meet needs. Formal inputs (requirements) and outputs must be established, with clear traceability between input(s) and output(s). Inputs may directly address specific user needs or be derived from system architecture, international standards, or risk management activities. Each input requirement must be *verified*, confirming that the product has been made properly to meet requirements.

To overcome the intractability of testing medical devices in all possible use environments, international standards are used as benchmarks for comparing and classifying widely varying devices. Conceptually, passing a certain test standard gives some assurance that a device will adequately perform in the vagaries of real-world conditions. Compliance with some high-level standards is mandated; others of narrower scope are chosen for appropriateness using engineering judgment in view of the clinical indication, habitus, and environment. Many aspects of VAD systems are not adequately covered by existing standards, so custom test protocols and apparatuses must be designed for verification. It is the responsibility of the engineering team to ensure the adequacy of test approach.

MCS verification testing often includes assuring long-term reliability and durability. With a goal of establishing in a short period of time that a device will endure a long period of clinical use, accelerated tests are devised. For example, if it is anticipated that a connector will be used as many as ten times per day for 5 years, an apparatus may be designed to exercise the connector by repeatedly connecting and disconnecting it rapidly, such that 20,000 cycles can be imposed in several days instead of several years. Such acceleration techniques range widely from cycling batteries through charge and discharge, bending cables back and forth to elicit fatigue, and many others. It is the engineer's responsibility to appropriately account for how results differ from real-world experience by virtue of acceleration techniques. Applying a factor of safety is common. Continuing a test beyond requirements to learn where failure occurs offers insight into the margin from expected failure, whether the test failure mode matches expected clinical failure mode, and whether an endurance limit exists.

The timeliness of accelerated tests has obvious value, but a real-time life test of MCS systems, for which operating conditions and loads are not accelerated, has been historically important in eliciting clinically relevant failure modes. Even in the continuous flow pump era in which fewer parts that wear out are employed, exercising the device in simulated real-world conditions is universally done and expected by regulatory bodies. Interim reports at important intervals of a test that may run indefinitely, perhaps for 10 years or longer, are generally acceptable to initiate clinical trials and support regulatory approvals. At some point, the direct relevance and increasing number of clinical patients eclipse the value of the real-time life test in determining reliability and durability.

Highly-accelerated life tests (HALT) are often utilized early in development and are distinct from the real-time life test. In HALT, unnaturally aggressive loads and conditions are applied to deliberately cause failure to elicit weaknesses and identify areas for increasing design robustness. As such, they contribute to design for product reliability and durability but are not used in verification.

In previously describing controller and power system design elements, the importance of user interaction was highlighted. The real-time life test, normally automated and relatively unattended, does nothing to validate the usability of the MCS system. The field of *human factors and usability engineering* (HFUE) techniques has emerged as a structured way to characterize the users and their environments to avert real-world hazards. HFUE is addressed from the moment the device is picked up from hospital inventory, through surgery and convalescence, and after the patient has returned home and resumed relatively normal activities. For MCS, AAMI/ANSI HE75:2009 and ISO/IEC 62366:2007 are the applicable standards [18, 19]. A useful guidance document, *Applying Human Factors and Usability Engineering to Optimize Medical Device Design*, reflects the FDA's present outlook on the topic [20].

A ubiquitous concept in MCS device development is *risk management*, which is so critical that it extends beyond mere process to cultural norm. A number of tools are available, including Failure Mode and Effect Analysis (FMEA), that ensure that product risk is minimized at every stage of design, development, and production. Risk management considerations can determine statistical methods and sample sizes, lead to identification of new requirements to mitigate risk, and help arbitrate among design trade-offs. ANSI/AAMI/ISO 14971:2007 is the applicable standard for MCS [21].

Future Directions

Many of the principles covered in this chapter have governed MCS throughout the decades of its existence, even as experience, insight, and technological advances have led to devices that eclipse earlier-generation products in safety, reliability, usability, and therapeutic efficacy. Morbidity, mortality, and patient quality of life with today's commercial devices have improved tremendously [22, 23]. However, it is not complacence that such progress should inspire, but confident diligence in further advances that minimize or eradicate adverse events related to the device, revolutionize the ease and peace of mind with which patients accept their therapy, and even contribute to reversal of the disease itself.

Among the dominant MCS-related adverse events, thrombosis, stroke, and bleeding are interrelated, often antagonistically, and engineers should strive to minimize aggravating pre-existing coagulopathies or creating new inflammatory and pro-thrombotic scenarios by applying previously discussed hemocompatibility considerations. A critical aspect is averting device vulnerability to incipient thrombosis. As modern pumps routinely operate at a single, clinician-applied setting, the term "Smart Pump" has been coined to describe enhancements that advantageously alter therapy. Automatically changing rotor speed to produce an artificial pulse or to occasionally disrupt steady flow conditions has been implemented [24, 25], and optimizing such algorithms, e.g., by synchronizing or counter-synchronizing with the cardiac cycle, is possible. Implantable sensors have been applied to characterize activity and measure attributes such as pressure [26], and these can conceivably be integrated in devices for use in algorithms that match therapy to physiologic demand.

Obviating the percutaneous driveline in MCS devices by implanting a power source and recharging it via a transcutaneous energy transmission system (TETS) would eliminate driveline exteriorization site infection, a prevalent adverse event that persists throughout support duration. TETS technology is readily available and has been successfully commercialized in other applications, but the relatively high power consumption, implanted setting, safety requirements, and strict regulations produce formidable engineering challenges. Nevertheless, a fully implanted left ventricular assist system (FILVAS) will eventually emerge because, in addition to the morbidity benefit, the quality of life benefit is profound: extended periods untethered from peripheral equipment and the freedom to shower, to swim, and to outwardly appear as though unafflicted by heart failure.

Finally, the ultimate goal should not be forgotten, and engineers have a potential role. Heart failure patients wish to be cured as much as cancer patients do, and although that may not be a realistic goal for all afflicted, it is conceivable that *cardiac remission* is as feasible as cancer remission, with future similar odds of success. Adjunctive disease-modifying therapies – pharmacologic, gene-altering, or otherwise – are likely necessary, but MCS could contribute more than merely stabilizing perfusion. Advanced "Smart Pump" techniques could be applied to enhance disease-modifying therapies, assess candidacy for or progress toward remission, and prepare the cardiovascular system for device removal. Although the psychological benefit of hope for meaningful post-intervention life is obvious, as for patients who endure chemotherapy and radiation for cancer-free survival, it should also be appreciated that eventual removal of the MCS device removes associated risks as well.

Disclosures The authors are employees of Abbott.

References

1. Katz AM. Heart failure pathophysiology, molecular biology, and clinical management. Philadelphia: Lippincott Williams & Wilkins; 2000. p. 41–4.
2. Schampaert S, Rutten MC, et al. Modeling the interaction between the intra-aortic balloon pump and the cardiovascular system: the effect of timing. ASAIO J. 2013;59(1):30–6. https://doi.org/10.1097/MAT.0b013e3182768ba9.
3. Slaughter MS, Rogers JG, Milano CA, Russell SD, Conte JV, Feldman D, et al. Advanced heart failure treated with continuous-flow left ventricular assist device. N Engl J Med. 2009;361(23):2241–51. https://doi.org/10.1056/NEJMoa0909938.
4. Griffith BP, Kormos RL, Borovetz HS, Litwak K, Antaki JF, Poirier VL, et al. HeartMate II left ventricular assist system: from concept to first clinical use. Ann Thorac Surg. 2001;71(3 Suppl):S116–20; discussion S114–6.

5. Selzman CH, Koliopoulou A, Glotzbach JP, McKellar SH. Evolutionary improvements in the Jarvik 2000 left ventricular assist device. ASAIO J. 2018;64(6):827–30. https://doi.org/10.1097/MAT.0000000000000743.
6. Chung MK, Zhang N, Tansley GD, Woodard JC. Impeller behavior and displacement of the VentrAssist implantable rotary blood pump. Artif Organs. 2004;28(3):287–97.
7. Larose JA, Tamez D, Ashenuga M, Reyes C. Design concepts and principle of operation of the HeartWare ventricular assist system. ASAIO J. 2010;56(4):285–9. https://doi.org/10.1097/MAT.0b013e3181dfbab5.
8. Morshuis M, Schoenbrodt M, Nojiri C, Roefe D, Schulte-Eistrup S, Boergermann J, et al. DuraHeart magnetically levitated centrifugal left ventricular assist system for advanced heart failure patients. Expert Rev Med Devices. 2010;7(2):173–83. https://doi.org/10.1586/erd.09.68.
9. Bourque K, Cotter C, Dague C, Harjes D, Dur O, Duhamel J, et al. Design rationale and preclinical evaluation of the HeartMate 3 left ventricular assist system for hemocompatibilitY. ASAIO J. 2016;62(4):375–83. https://doi.org/10.1097/MAT.0000000000000388.
10. Mironer A. Engineering fluid mechanics. New York: McGraw-Hill; 1979. p. 213.
11. Guyton AC, Hall JE. Textbook of medical physiology. 11th ed. Philadelphia: Elsevier Saunders; 2006. p. 232–3.
12. Arnold WS, Bourque K. The engineer and the clinician: understanding the work output and troubleshooting of the HeartMate II rotary flow pump. J Thorac Cardiovasc Surg. 2013;145(1):32–6. https://doi.org/10.1016/j.jtcvs.2012.07.113.
13. Maltais S, Kilic A, Nathan S, Keebler M, Emani S, Ransom J, et al. PREVENtion of HeartMate II pump thrombosis through clinical management: the PREVENT multi-center study. J Heart Lung Transplant. 2017;36(1):1–12. https://doi.org/10.1016/j.healun.2016.10.001.
14. Stepanoff AJ. Centrifugal and axial flow pumps. 2nd ed. Florida: Krieger; 1957.
15. Uriel N, Colombo PC, Cleveland JC, Long JW, Salerno C, Goldstein DJ, et al. Hemocompatibility-related outcomes in the MOMENTUM 3 trial at 6 months: a randomized controlled study of a fully magnetically levitated pump in advanced heart failure. Circulation. 2017;135(21):2003–12. https://doi.org/10.1161/CIRCULATIONAHA.117.028303.
16. Quality System (QS) Regulation/Medical Device Good Manufacturing Practices, 21 C.F.R. Part 820 (1978).
17. Regulation (EU) 2017/745 of the European Parliament and of the Council of 5 April 2017 on medical devices, amending Directive 2001/83/EC, Regulation (EC) No 178/2002 and Regulation (EC) No 1223/2009 and repealing Council Directives 90/385/EEC and 93/42/EEC (2017).
18. American National Standards Institute. Human factors engineering – Design of medical devices. Virginia: ANSI/AAMI/HE75:2009 (r2013).
19. International Organization for Standardization. Medical Devices – Application of usability engineering to medical devices. Switzerland: ISO/IEC 62366:2007.
20. U.S. Food and Drug Administration. Guidance for Industry: Applying human factors and usability engineering to optimize medical device design. Maryland: The Administration; 2016.
21. American National Standards Institute. Medical Devices – Application of risk management to medical devices. Virginia: ANSI/AAMI/ISO 14971:2007 (r2010).
22. Rogers JG, Pagani FD, Tatooles AJ, Bhat G, Slaughter MS, Birks EJ, et al. Intrapericardial left ventricular assist device for advanced heart failure. N Engl J Med. 2017;376(5):451–60. https://doi.org/10.1056/NEJMoa1602954.
23. Mehra MR, Goldstein DJ, Uriel N, Cleveland JC Jr, Yuzefpolskaya M, Salerno C, et al. Two-year outcomes with a magnetically levitated cardiac pump in heart failure. N Engl J Med. 2018;378(15):1386–95. https://doi.org/10.1056/NEJMoa1800866.
24. Zimpfer D, Strueber M, Aigner P, Schmitto JD, Fiane AE, Larbalestier R, et al. Evaluation of the HeartWare ventricular assist device Lavare cycle in a particle image velocimetry model and in clinical practice. Eur J Cardiothorac Surg. 2016;50(5):839–48.
25. Bourque K, Dague C, Farrar D, Harms K, Tamez D, Cohn W, Tuzun E, Poirier V, Frazier OH. In vivo assessment of a rotary left ventricular assist device-induced artificial pulse in the proximal and distal aorta. Artif Organs. 2006;30(8):638–42.
26. Desai AS, Bhimaraj A, Bharmi R, Jermyn R, Bhatt K, Shavelle D, et al. Ambulatory hemodynamic monitoring reduces heart failure hospitalizations in "Real-World" clinical practice. J Am Coll Cardiol. 2017;69(19):2357–65. https://doi.org/10.1016/j.jacc.2017.03.009.
27. Schmitto JD, Hanke JS, Rojas SV, Avsar M, Haverich A, et al. First implantation in man of a new magnetically levitated left ventricular assist device (HeartMate III). J Heart Lung Transplant. 2015;34(6):858–60.

Part II

Indications and Selection for Mechanical Circulatory Support

9 Patient Population and Selection Criteria for Mechanical Circulatory Support

Pavan Bhat and Randall C. Starling

Advances in the treatment for heart failure have progressed significantly over the last 50 years which has led to an increasing worldwide population of patients living with heart failure. Despite significant advances, advanced heart failure is still associated with significant mortality, estimated to be 33% at 1 year [1]. During this time cardiac transplantation remains the definitive therapy for end-stage heart failure to reduce mortality and improve quality of life. However, despite the numerous advances in other aspects of heart failure, there hasn't been any significant increase in the number of donor hearts until recently. The surge in donor hearts has primarily been attributed to an increase related to the opioid epidemic and the use of hepatitis C donors [2]. However, the number of patients living with end-stage heart failure has increased yielding a supply-demand mismatch, which mechanical circulatory support (MCS) has helped address. More than 5000 cardiac transplants occur each year around the world although it is estimated that up to 50,000 people are candidates for transplantation [3]. This mismatch leads to a growing number of transplant eligible patients that are awaiting transplants and an ever-expanding pool of patients that are considered ineligible for transplant due to age or comorbidities. Prior to mid-1990s, these transplant ineligible patients had few options, but fortunately improving outcomes with MCS has changed that paradigm.

Almost 30 years later, we see the left ventricular assist device (LVAD) as the dominant durable mechanical circulatory support for patients with end-stage heart failure (Stage D). The LVAD has evolved from short-term extracorporeal and pulsatile devices containing valves and mechanical bearing into the smaller intracorporeal devices currently in use. The HeartMate III® is comprised of a centrifugal flow pump with a magnetically levitated rotor, programmed with intermittent speed reduction to reduce stasis, and provides "wash out" to reduce adverse events. Future advances will focus on miniaturization of the device and internalization of the power source hence eliminating the percutaneous driveline.

The improvement in LVAD technology has allowed its use in three basic patients' subsets: bridge-to-transplant (BTT), destination therapy (DT), and bridge to decision (BTD). BTT patients are awaiting transplant but are too sick to continue to wait with medical therapy alone. DT patients include patients with non-reversible comorbidities that preclude them from transplant or age and are using LVAD as lifelong support. BTD is a heterogeneous group of patients with potentially reversible

P. Bhat, MD
Kaufman Center for Heart Failure, Heart & Vascular Institute, Cleveland Clinic, Cleveland, OH, USA
e-mail: BHATP@ccf.org

R. C. Starling, MD, MPH, FACC, FAHA, FESC, FHFSA (✉)
Kaufman Center for Heart Failure, Cleveland Clinic, Cleveland, OH, USA
e-mail: STARLIR@ccf.org

© Springer Nature Switzerland AG 2020

J. H. Karimov et al. (eds.), *Mechanical Support for Heart Failure*,
https://doi.org/10.1007/978-3-030-47809-4_9

contraindications that could be potential transplant candidates in the future. Given that organ transplantation rates are inadequate to meet demand, DT LVADs have become the major area of expansion for MCS. There has been a tenfold increase in DT LVAD implantation rates since 2010 [4]. Determining which patients are ideal candidates for LVADs and the timing on when to implant these devices will improve as the heart failure specialists continue to gain experience with these patients.

Patient Selection

Looking at the inclusion criteria for the REMATCH trial, the HeartMate II trial, and MOMENTUM 3 trial provides insight into the scope of suitable patients for MCS. The REMATCH trial was the first randomized control trial to assess the benefit of LVADs in 129 patients who were ineligible for transplant using the HeartMate XVE, an early pulsatile LVAD [5]. Inclusion criteria in that trial were patients with NYHA class IV heart failure for at least 90 days despite guideline directed therapy, an EF <25%, and a peak oxygen consumption of <12 ml per kilogram per minute or continued need for intravenous inotropic therapy. The inclusion criteria were later broadened to include patients who had been on at least 14 days of support of an intra-aortic balloon pump. In the final enrollment of 129 patients, the majority (73%) was inotrope-dependent. The results showed a 48 percent reduction in the risk of death from any cause in the patients that received the LVAD; however this was matched by a high rate of serious adverse events, 2.35 times the medical group. Also, these early LVADs typically required replacement after only approximately 18 months. Given these above limitations, early LVADS were not widely adopted as it was felt the adverse event burden was prohibitively high.

The next major generational change in LVAD technology occurred after the HeartMate II trial [6]. Compared to the HeartMate XVE which was pulsatile, the HeartMate II LVAD (HMII) was a continuous axial flow device. The HeartMate II trial was designed to include significantly less ill patients than the REMATCH population. The inclusion criteria included patients with NYHA IIIB or IV symptoms, dependence on an intra-aortic balloon pump for at least 7 days, or dependence on inotropes for at least 14 days. Patients were excluded based on irreversible end organ damage: renal, liver, or pulmonary disease. Two hundred patients were enrolled; the mean age was younger than REMATCH (62 vs 68 years). Survival in the HeartMate II trial was 68% at 1 year and 58% at 2 years which was significantly improved compared to REMATCH.

The next major advance in LVAD technology occurred with the MOMENTUM 3 trial with HeartMate III [7]. The HeartMate III was engineered to improve on the HeartMate II by moving to a fully magnetically levitated centrifugal pump. The HeartMate III also varied the flow speed to create an artificial pulse despite it being a continuous flow device. The inclusion was very broad and included patients with NYHA class III and IV symptoms, EF <25%, and either inotrope-dependent or cardiac index <2.2 L/min/m2 while not on inotropes. MOMENTUM 3 enrolled 294 patients, and 152 were assigned to the HeartMate III, and 142 were assigned to the HeartMate II. The outcomes at 6 months showed superior outcomes with the HeartMate III; a reduction in pump thrombosis led to lower rates of reoperation for pump malfunction. Subsequent publications with additional patients and 24 months follow-up clearly demonstrated the reduction in pump thrombosis, device exchange, stroke, and GI bleeding with the HeartMate III as compared with the HeartMate II.

Creation of the Interagency Registry for Mechanically Assisted Circulatory Support (INTERMACS) provided contemporary data to look at appropriate patient selection and outcomes [8]. Stevenson and colleagues created seven distinct INTERMACS patient profiles to classify patients regarding potential for further LVAD implantation [9] (Fig. 9.1). The INTERMACS patient profiles have created a standardized tool to discuss patient classification when considering advanced therapy. One of the key advances of the INTERMACS profiles is that it gives a general guideline about the timing/urgency for intervention needed depending on the severity of heart failure. The INTERMACS

Profile	Description	NYHA class	Modifiers	Timescale for intervention	Number of implants January-June 2012 Total n	%
1	Critical cardiogenic shock	IV	TCS/A	Hours	1307	19.7
2	Progressive decline during inotropie support	IV	TCS/A	Hours to days	2664	40.1
3	Stable but inotrope-dependent	IV	A/FF	Weeks	1515	22.8
4	Resting symptoms	IV	A/FF	Weeks to months	791	11.9
5	Exertion intolerant	IIIb/IV	A/FF	Variable	184	2.7
6	Exertion limited	IIIb	A/FF	Variable	104	1.5
7	Advanced NYHA III	III/IIIb	A	Not indicated	61	
Total					6633 (7 unspecified)	100

A, Frequent arrhythmia; FF, 'frequent flyer' (frequent hospitalizations); INTERMACS, Interagency Registry for Mechanically Assisted Circulatory Support; NYHA, New York Heart Association; TCS, temporary circulatory support.
Modified from Stevenson et al.[11] and from Table 4 of the 2013 Annual INTERMACS Report.[7]

Fig. 9.1 INTERMACS profiles. (Adapted from Kirklin et al. [4], Copyright 2013 with permission from Elsevier, and from Stevenson et al. [9], Copyright 2009 with permission from Elsevier)

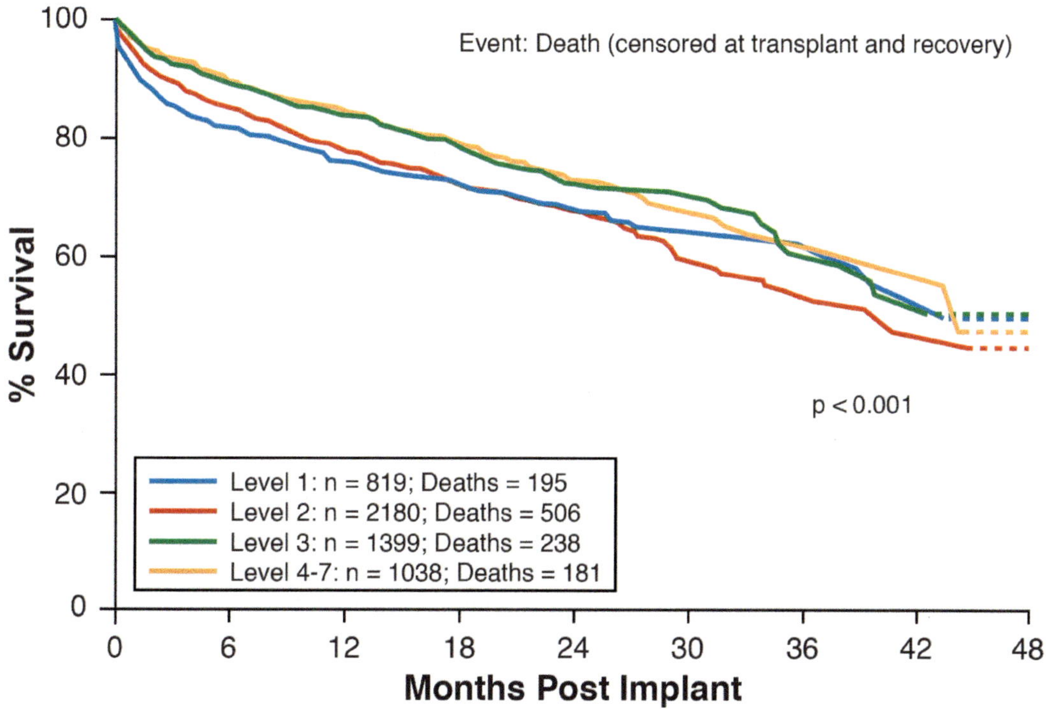

Fig. 9.2 Survival by INTERMACS level: adult Primary continuous flow LVADS and BIVADs, DT, and BTT (June 2006–June 2012). (Reproduced from Kirklin et al. [4], Copyright 2013, with permission from Elsevier)

profiles have also been closely correlated to mortality (Fig. 9.2).

The INTERMACS profiles allow risk stratifying patients with advanced heart failure and picking the appropriate time for LVAD implantation. A study by Boyle and colleagues looked at 101 BTT and DT LVAD patients broken into three groups by their INTERMACS profiles [10]. The study showed that survival at 36 months was better in the INTERMACS profiles that were less acutely ill (INTERMACS profile 2 and higher) compared to the patients with critical cardiogenic shock (INTERMACS profile 1). This study demonstrates the importance of INTERMACS profile risk stratification in selecting patients for MCS. Specifically, inferior survival is seen with

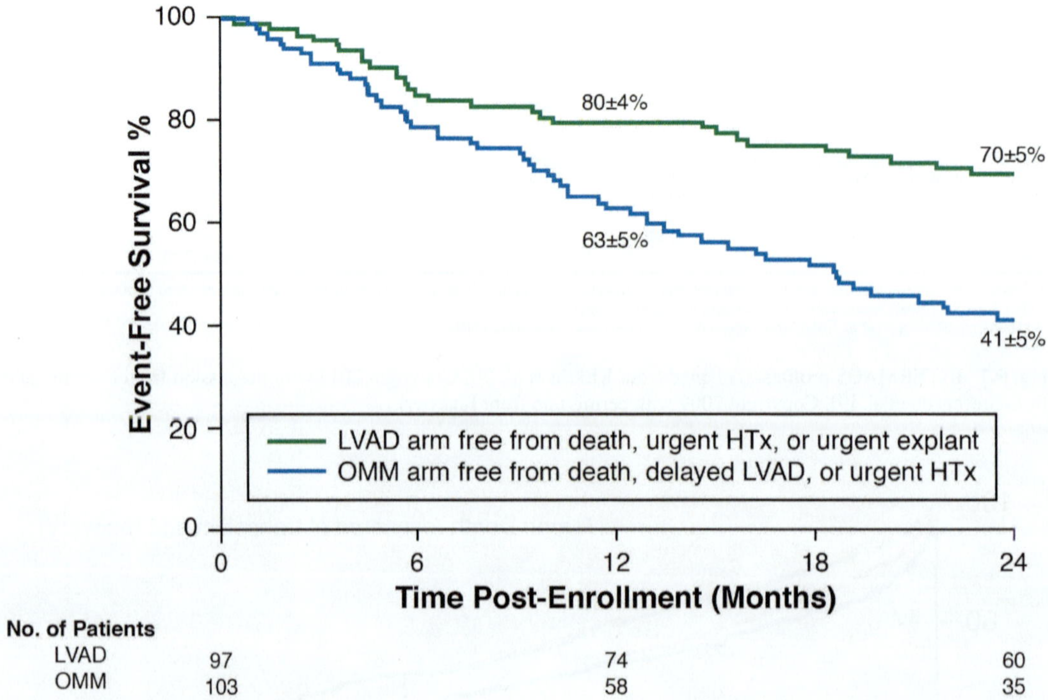

Fig. 9.3 Survival as treated: optimal medical management versus LVAD. (Adapted from Estep et al. [11], Copyright 2015,with permission from Elsevier)

patients with INTERMACS profiles 1 and 2, with potentially more favorable outcomes with implantation of durable LVADs in INTERMACS profiles 3–6. The ROADMAP study investigated patients that were non-inotrope-dependent/less ill with higher INTERMACS profiles [4–7]. The ROADMAP trial was a prospective non-randomized observational study of 200 ambulatory NYHA class IIIb/IV patients with half receiving LVAD and half receiving optimal medical management (OMM) at enrollment. The study showed that survival with improved functional status at 1 year was improved in the LVAD group as opposed to the OMM group [11] (Fig. 9.3). The 2-year study results of ROADMAP showed survival with improvement in 6-min walk distance was superior with LVAD compared OMM at 2 years [12]. Additionally, a reduction in adverse events related to gastrointestinal bleeding and arrhythmias beyond 1 year was observed in the LVAD group. One caveat with the ROADMAP study results is that there were some differences in baseline characteristics between the two groups (non-randomized study). The patients receiving LVADs had lower baseline health-related quality of life, reduced Seattle Heart Failure Model 1-year survival, and were primarily INTERMACS profile 4 versus profiles 5 to 7. The data from the ROADMAP trial provides important information that elective LVAD implantation in ambulatory advanced heart failure patients can improve event-free survival and quality of life.

Recognizing "Red Flags": Identifying Signs of Deterioration

The data from the prior studies reviewed suggests the early recognition of clinical deterioration of advanced heart failure patients is key for successful LVAD implantation with improved survival and quality of life. Selecting the right patient and timing is the crux of treating advanced heart failure with MCS. Identifying the red flags, signs of early clinical deterioration, and proceeding with

MCS strategically help identify the patients that would benefit the most.

Repeat Hospitalization

A heart failure patient with repeated hospitalization should prompt referral to an advanced heart failure specialist. The analysis of the Candesartan in Heart Failure: Assessment of Reduction in Mortality and morbidity (CHARM) trials by Solomon and CHARM investigators showed that in patients with chronic heart failure, the highest risk of death is in the early period after discharge after a hospitalization for a heart failure exacerbation [13]. Additionally they showed that mortality is directly related to duration and frequency of heart failure hospitalizations. These results point toward the patients with repeated hospitalizations being a vulnerable population that may benefit from earlier initiation of MCS.

Medication Intolerance

Angiotensin-converting enzyme (ACE) inhibition or angiotensin receptor-neprilysin inhibitor (ARNI) and β-blockade remain the cornerstones of optimal medical therapy. Patients that cannot tolerate ACEI/ARNI or β-blocker therapy should be referred for evaluation for MCS or advanced heart failure therapy as these patients represent a patient population with more severe disease. Analysis of the Acute Study of Clinical Effectiveness of Nesiritide in Decompensated Heart Failure (ASCEND-HF) by Patel et al. showed that hypotension with acute decompensated heart failure was an independent risk factor for increased 30-day mortality and heart failure readmission [14]. As regards to β-blockade, prior studies by Fonarow et al. have shown that patients who have had their β-blocker withdrawn during a hospitalization for acute decompensated heart failure had a 2.3 higher risk of mortality compared to similar patients that continued the β-blocker [15]. Overall, the intolerance of optimal medical management for heart failure should warn the clinician that their patient is at higher risk and should be evaluated for advanced heart failure therapies.

Exercise Cardiopulmonary Stress Testing

Formal exercise cardiopulmonary stress testing allows a standardized and objective measurement of the aerobic capacity of patients with heart failure which has important prognostic information. Mancini and colleagues conducted a study in 114 ambulatory patients with heart failure that established a peak VO_2 cut off of <14 mL/kg/minute as the criterion for which 1-year survival was significantly lower than transplantation, approximately 70% lower [16]. Contrasting, individuals with peak VO_2 >14 mL/kg/minute had a 6% 1-year mortality. This suggests that peak VO_2 can be reliably used to risk stratify ambulatory patients that would benefit from MCS.

Cardiac Hemodynamics

Hemodynamic criteria for cardiogenic shock as measured by a pulmonary artery catheter include a reduced cardiac index <2.2 L/min/m^2. These individuals typically require temporary support with inotropes or mechanical support in the form of intra-aortic balloon pump or percutaneous LVAD. The inclusion criteria for MOMENTUM 3 discussed earlier in this chapter included patients with cardiac index <2.2 L/min/m^2; however this is not a strict criterion for LVAD implantation. Typically, heart failure patients who present in cardiogenic shock as evidenced by invasive hemodynamic monitoring should be evaluated for MCS.

Inotrope Dependence

The inclusion criteria from the previously discussed REMATCH trial, the HeartMate II trial, and MOMENTUM 3 trial all included patients that were inotrope-dependent due to the fact that chronic inotrope use is known to be associated with a high mortality. A retrospective analysis by Hauptman et al. has shown that chronic milrinone use is associated with excessive all-cause mortality at 12 months (58.8%) and also associated with significant healthcare expenditure. Given the high mortality rate associated with optimal medical management and inotrope therapy in the REMATCH study, patients that are inotrope-dependent should be considered for MCS if candidates. A previous study by Gorodeski et al. has also shown high mortality (~50% at 6 months) in patients discharged from the hospital on dobutamine or milrinone and recommended advanced therapies or palliative care and inactivation of ICD if not candidates for MCS [17].

End Organ Dysfunction

One of the goals of early initiation of MCS in appropriate candidates is to prevent long-term injury to other organs. Renal and hepatic dysfunctions are two organ systems that are critical to assess when determining if a patient is an acceptable candidate for MCS. End-stage renal disease (ESRD) is a poor prognostic marker and will often exclude a patient from being accepted as a MCS candidate. Similarly, patients that have poor baseline renal function that is felt not to be cardiorenal and or reversible may progress to dialysis after MCS, which is associated with high mortality and morbidity. In a retrospective analysis done by Bansal and colleagues, 81.9% of patients with ESRD prior LVAD implantation died during the follow-up, and the median time of death was 16 days after LVAD placement for patients with ESRD [18]. Similarly, hepatic dysfunction can occur from congestive hepatopathy. Ascites, elevated liver enzymes, or elevated bilirubin should raise the concern for congestive hepatopathy. The model for end-stage liver disease (MELD) score incorporates the patient's creatinine, total bilirubin, and international normalized ratio (INR) into a scoring system that has been used to prioritize patients for liver transplantation. Work by Kim and colleagues has shown that MELD scores also are excellent predictors for 1-year endpoints in patients undergoing heart failure evaluation [19]. Overall, early end organ dysfunction should be an indicator that the patient needs evaluation for MCS or transplantation.

Timing

A wide variety of factors are involved in the decision to define the appropriate timing for LVAD implantation. The difficulty lies in the variable nature of the prognosis of heart failure in individual patients. Previous studies have shown that NYHA functional class remains a powerful predictor of survival in the ambulatory heart failure population [20]; however further studies have shown significant interoperator variability in classification of NYHA class II and class III [21]. The optimal timing is balancing act between the burden of adverse effects associated with LVAD therapy and the long-term mortality and morbidity associated with progressive heart failure.

The data from the ROADMAP study suggests that waiting until a patient is INTERMACS profile 1–2 may result worse outcomes. The Seattle Heart Failure Model (SHFM) is a predictive model in the pre-LVAD era that assists in prognostication for heart failure patients. The analysis of ROADMAP data using the SHFM by Lanfear and colleagues has shown that the SHFM in non-inotrope-dependent advanced heart failure patients was predictive of overall survival but underestimated the risk of clinical worsening and LVAD implantation [22]. In the same study, another risk stratification that models the HeartMate II Risk Score (HMRS) had marginal discrimination and underestimated survival post-LVAD implantation. Cowger et al. analyzed the HeartMate risk score (HMRS) for LVAD mortality risk prediction and showed the HMRS discrimination was good for predicting mortality with receiver-operating curve: 0.71, 95% CI (0.65–0.75) [20].

Other exclusion criteria that need to be accounted for when considering timing of MCS are age, prior cardiac surgeries, renal failure, and right ventricular (RV) failure. Specifically, RV dysfunction when characterized by ascites increased right atrial (RA) pressures, or bilirubin >2 mg predicts early mortality after LVAD, so a careful assessment of RV function is necessary. RV assessment can be performed by qualitative assessment by echocardiography or quantitative measures such as tricuspid annular plane systolic excursion (TAPSE), right ventricular work index (RVSWI), and invasive PA catheter measurement of hemodynamics. Overall several factors are integrated into the decision for MCS, and they must be balanced. Assessing adequacy of RV function is extremely challenging and must be carefully evaluated with surgical collaboration. Emerging approaches include preoperative percutaneous LVADs to unload the RV and assess recovery of RV function. The additional decision that is crucial is a preoperative decision with the cardiac surgeon to place a temporary RVAD at the time of LVAD implant that can be removed early in 5–7 days post-LVAD implant. The goal is to avoid the scenario where the patient requires prolonged biventricular support and cannot be discharged and heart transplant becomes the only viable option [23, 24].

I-NEED-HELP

I - IV inotropes

N - NYHA IIIB/IV persistently elevated natriuretic peptides

E - End-organ dysfunction

E - Ejection fraction ≤35%

D - Defibrillator shocks

H - Hospitalization >1

E - Edema despite escalating diuretics

L - Low blood pressure, high heart rate

P - Prognostic medication - progressive intolerance or down-titration of GDMT

Fig. 9.4 "I NEED HELP." (Adapted from Yancy et al. [25], Copyright 2018, with permission from Elsevier)

Conclusions

The patient selection criteria for MCS are complex as it involves integrating data from INTERMACS profile, clinical variables, invasive hemodynamic assessment, and recent hospitalization history. Identifying patients before incipient end organ failure ensues is key. A simple mnemonic "I NEED HELP" provides the essential checklist of variables that alert the clinician to advanced heart failure [25] (Fig. 9.4). Recently it has been demonstrated that the event rate is surprisingly high in non-inotrope-dependent, "stable" ambulatory heart failure patients [25].

Our experience with the ROADMAP data has reinforced that earlier initiation of MCS with ambulatory patients with INTERMACS scores 4–6 as opposed to 1–2 can lead to better outcomes. Risk models such as SHFM and HMRS alone are insufficient to appropriate risk stratify these patients. Appropriate risk stratification requires a multidisciplinary approach with input of heart failure specialists, cardiac surgeons, case managers, and social workers.

Bibliography

1. Lloyd-Jones D, Adams RJ, Brown TM, Carnethon M, Dai S, De Simone G, et al. Heart Disease and Stroke Statistics-2010 Update A Report From the American Heart Association WRITING GROUP MEMBERS. 2010 [cited 2018 Nov 14]; Available from: http://www.americanheart.org/presenter.jhtml.
2. Mehra MR, Jarcho JA, Cherikh W, Vaduganathan M, Lehman RR, Smits J, et al. The Drug-Intoxication Epidemic and Solid-Organ Transplantation. N Engl J Med [Internet]. 2018 May 17 [cited 2019 Feb 12];378(20):1943–5. Available from: http://www.nejm.org/doi/10.1056/NEJMc1802706.
3. Taylor DO, Edwards LB, Boucek MM, Trulock EP, Aurora P, Christie J, et al. Registry of the International Society for Heart and Lung Transplantation: twenty-fourth official adult heart transplant report--2007. J Heart Lung Transplant [Internet]. 2007 Aug 1 [cited 2018 Nov 14];26(8):769–81. Available from: http://www.ncbi.nlm.nih.gov/pubmed/17692781.

4. Kirklin JK, Naftel DC, Kormos RL, Stevenson LW, Pagani FD, Miller MA, et al. SPECIAL FEATURE Fifth INTERMACS annual report: Risk factor analysis from more than 6,000 mechanical circulatory support patients Site and patient enrollment. J Hear Lung Transplant [Internet]. 2013 [cited 2018 Nov 14];32:141–56. Available from: https://doi.org/10.1016/j.healun.2012.12.004.
5. Ric R Ose EA, Nnetine G Elijns AC, Lan M Oskowitz AJ, Aniel H Eitjan DF, Ynne S Tevenson LW, Alter Embitsky WD, et al. The New England Journal of Medicine. Long-term use of a left ventricular assist device for end-stage heart failure for the r andomized e valuation of m echanical a ssistance for the t reatment of c ongestive h eart failure [Internet]. N Engl J Med. 2001 [cited 2018 Nov 14]. 345. Available from: www.nejm.org.
6. Slaughter MS, Rogers JG, Milano CA, Russell SD, Conte J V, Feldman D, et al. Advanced heart failure treated with continuous-flow left ventricular assist device abstract. N Engl J Med [Internet]. 2009 [cited 2018 Nov 14];17:2241–51. Available from: https://www.nejm.org/doi/pdf/10.1056/NEJMoa0909938.
7. Mehra MR, Naka Y, Uriel N, Goldstein DJ, Cleveland JC, Colombo PC, et al. A fully magnetically levitated circulatory pump for advanced heart failure. N Engl J Med [Internet]. 2017 Feb 2 [cited 2018 Nov 14];376(5):440–50. Available from: http://www.nejm.org/doi/10.1056/NEJMoa1610426.
8. Kirklin JK, Naftel DC, Kormos RL, Stevenson LW, Pagani FD, Miller MA, et al. Fifth INTERMACS annual report: risk factor analysis from more than 6,000 mechanical circulatory support patients. J Hear Lung Transplant [Internet]. 2013 Feb 1 [cited 2018 Nov 14];32(2):141–56. Available from: https://www.sciencedirect.com/science/article/pii/S105324981201457X.
9. Stevenson LW, Pagani FD, Young JB, Jessup M, Miller L, Kormos RL, et al. INTERMACS profiles of advanced heart failure: the current picture. J Hear Lung Transplant [Internet]. 2009 Jun 1 [cited 2018 Nov 14];28(6):535–41. Available from: https://www.sciencedirect.com/science/article/pii/S1053249809001910.
10. Boyle AJ, Ascheim DD, Russo MJ, Kormos RL, John R, Naka Y, et al. Clinical outcomes for continuous-flow left ventricular assist device patients stratified by pre-operative INTERMACS classification. J Heart Lung Transplant [Internet]. 2011 Apr 1 [cited 2018 Dec 3];30(4):402–7. Available from: http://www.ncbi.nlm.nih.gov/pubmed/21168346.
11. Estep JD, Starling RC, Horstmanshof DA, Milano CA, Selzman CH, Shah KB, et al. Risk assessment and comparative effectiveness of left ventricular assist device and medical management in ambulatory heart failure patients. J Am Coll Cardiol [Internet]. 2015 Oct 20 [cited 2019 Feb 12];66(16):1747–61. Available from: http://www.ncbi.nlm.nih.gov/pubmed/26483097.
12. Starling RC, Estep JD, Horstmanshof DA, Milano CA, Stehlik J, Shah KB, et al. Risk assessment and comparative effectiveness of left ventricular assist device and medical management in ambulatory heart failure patients: the ROADMAP Study 2-year results. JACC Hear Fail [Internet]. 2017 Jul 1 [cited 2018 Dec 3];5(7):518–27. Available from: https://www.sciencedirect.com/science/article/pii/S2213177917301646?via%3Dihub.
13. Solomon SD, Dobson J, Pocock S, Skali H, Mcmurray JJ V, Granger CB, et al. Influence of nonfatal hospitalization for heart failure on subsequent mortality in patients with chronic heart failure. 2007 [cited 2018 Dec 3]; Available from: http://circ.ahajournals.org.
14. Patel PA, Heizer G, O'Connor CM, Schulte PJ, Dickstein K, Ezekowitz JA, et al. Hypotension during hospitalization for acute heart failure is independently associated with 30-day mortality findings from ASCEND-HF. Circ Hear Fail. 2014.
15. Fonarow GC, Abraham WT, Albert NM, Stough WG, Gheorghiade M, Greenberg BH, et al. Heart failure influence of beta-blocker continuation or withdrawal on outcomes in patients hospitalized with heart failure findings from the OPTIMIZE-HF program. 2008 [cited 2018 Dec 3]; Available from: https://ac.els-cdn.com/S0735109708015039/1-s2.0-S0735109708015039-main.pdf?_tid=2cc15f30-aac6-4369-b273-66a24ee7b87b&acdnat=1543829242_fb20cc53d8df1395bc4a5533426b7ef7.
16. Mancini DM, Eisen H, Kussmaul W, Mull R, Edmunds LH, Wilson JR. Value of peak exercise oxygen consumption for optimal timing of cardiac transplantation in ambulatory patients with heart failure. Circulation [Internet]. 1991 Mar [cited 2018 Dec 3];83(3):778–86. Available from: http://www.ncbi.nlm.nih.gov/pubmed/1999029.
17. Gorodeski EZ, Chu EC, Reese JR, Shishehbor MH, Hsich E, Starling RC. Prognosis on chronic dobutamine or milrinone infusions for Stage D Heart Failure. 2009 [cited 2019 Feb 12]; Available from: http://www2.sas.com/proceedings/sugi26/p214-26.pdf.
18. Bansal N, Hailpern SM, Katz R, Hall YN, Tamura MK, Kreuter W, et al. Outcomes associated with left ventricular assist devices among recipients with and without end-stage renal disease. JAMA Intern Med [Internet]. 2018 Feb 1 [cited 2018 Dec 3];178(2):204. Available from: http://archinte.jamanetwork.com/article.aspx?doi=10.1001/jamainternmed.2017.4831.
19. Kim MS, Kato TS, Farr M, Wu C, Givens RC, Collado E, et al. Hepatic dysfunction in ambulatory patients with heart failure: application of the MELD scoring system for outcome prediction. J Am Coll Cardiol [Internet]. 2013 Jun 4 [cited 2018 Dec 3];61(22):2253–61. Available from: http://www.ncbi.nlm.nih.gov/pubmed/23563127.
20. Cowger J, Sundareswaran K, Rogers JG, Park SJ, Pagani FD, Bhat G, et al. Predicting survival in patients receiving continuous flow left ventricular assist devices: the heartmate II risk score. J Am Coll Cardiol [Internet]. 2013 Jan 22 [cited 2019 Feb

12];61(3):313–21. Available from: https://www.sciencedirect.com/science/article/pii/S0735109712055933?via%3Dihub.

21. Raphael C, Briscoe C, Davies J, Whinnett ZI, Manisty C, Sutton R, et al. Heart failure and cardiomyopathy. Limitations of the New York Heart Association functional classification system and self-reported walking distances in chronic heart failure. Heart [Internet]. 2007 [cited 2019 Feb 12];93:476–82. Available from: www.heartjnl.com.
22. Lanfear DE, Levy WC, Stehlik J, Estep JD, Rogers JG, Shah KB, et al. Accuracy of seattle heart failure model and heartmate II risk score in non–inotrope-dependent advanced heart failure patients. Circ Hear Fail [Internet]. 2017 May [cited 2018 Dec 3];10(5). Available from: https://www.ahajournals.org/doi/10.1161/CIRCHEARTFAILURE.116.003745.
23. Bellavia D, Iacovoni A, Scardulla C, Moja L, Pilato M, Kushwaha SS, et al. Prediction of right ventricular failure after ventricular assist device implant: systematic review and meta-analysis of observational studies. Eur J Heart Fail [Internet]. 2017 Jul 1 [cited 2019 Feb 12];19(7):926–46. Available from: http://doi.wiley.com/10.1002/ejhf.733.
24. Konstam MA, Kiernan MS, Bernstein D, Bozkurt B, Jacob M, Kapur NK, et al. On behalf of the American Heart Association Coun-cil on Clinical Cardiology; Council on Cardiovascular Disease in the Young; and Council on cardiovascular surgery and anesthesia. Circulation [Internet]. 2018 [cited 2019 Feb 12];137:578–622. Available from: http://circ.ahajournals.org.
25. Yancy CW, Januzzi JL, Allen LA, Butler J, Davis LL, Fonarow GC, et al. 2017 ACC expert consensus decision pathway for optimization of heart failure treatment: answers to 10 pivotal issues about heart failure with reduced ejection fraction. J Am Coll Cardiol [Internet]. 2018 Jan [cited 2019 Feb 12];71(2):201–30. Available from: https://linkinghub.elsevier.com/retrieve/pii/S073510971741641X.

Mechanical Circulatory Support Therapies: Right Timing and Prognosis Considerations

10

Felix Schoenrath and Evgenij Potapov

Short-Term MCS

Cardiogenic shock (CS) is a clinical condition of inadequate (end-organ) tissue perfusion due to cardiac dysfunction irrespective of the injurious mechanism. Between 2003 and 2010, the incidence of CS caused by acute myocardial infarction (AMI) rose from 6.5% to 10.1%, but in-hospital mortality decreased significantly from 44.6% to 33.8% in the USA [1]. Among patients with congestive heart failure, the incidence of CS rose from 0.5% to 1%, and the mortality decreased from 44.2% to 26.1% in the USA [2]. The definition of CS includes reduced systemic perfusion and increased residual volume of the left or right or both ventricles and therefore increased filling pressures within the affected cardiac chambers [3]. If these hemodynamic derangements persist, multi-organ ischemia, increased lactate accumulation, and hepatic and venous congestion are the possible consequences, and these factors transmit the initially reversible hemodynamic problem into a multi-organ metabolic problem that cannot be treated by hemodynamic support alone. This perpetuating phenomenon is similar in different forms of shock, and to overcome it, early and goal-directed therapy is needed, i.e., in hemorrhagic or septic shock as well as in cardiogenic shock, as evidenced by the term "the golden hour of shock." Acute myocardial infarction is the most important underlying disease for CS. Other important etiologies of CS are decompensated advanced heart failure due to a variety of chronic diseases such as ischemic, valvular heart disease or dilated cardiomyopathy and other acute settings: toxic cardiomyopathy, takotsubo cardiomyopathy, myocarditis, or peripartum cardiomyopathy [4]. To identify potential acute CS etiologies, the current ESC guidelines use the CHAMP acronym (acute Coronary syndrome, Hypertension emergency, Arrhythmia, Mechanical cause, Pulmonary embolism). Important triggers for acute decompensation in chronic disease are infections, arrhythmias, and pharmacological or dietary non-adherence; further potential triggers are mentioned in these guidelines [5]. Theoretically, after specific treatment and when possible and less-invasive strategies, especially inotropic support, have been exhausted without achieving sustained stabilization of patients underlying disease and before the onset of significant end-organ dysfunction, mechanical circulatory support should be evaluated (in a minority of patients and in some countries, an emergency heart transplantation may be an option) in refractory cardiogenic shock. Therefore, referral to a tertiary hub center with short- and long-term MCS expertise needs

F. Schoenrath, MD (✉) · E. Potapov, MD
Department of Cardiothoracic and Vascular Surgery, German Heart Center Berlin, DZHK (German Centre for Cardiovascular Research), partner site Berlin, Berlin, Germany
e-mail: schoenrath@dhzb.de; potapov@dhzb.de

© Springer Nature Switzerland AG 2020
J. H. Karimov et al. (eds.), *Mechanical Support for Heart Failure*,
https://doi.org/10.1007/978-3-030-47809-4_10

to be secured as soon as possible. Even more as compared to chronic heart failure, time is of the essence in decision-making in these cases [6]. Regional heart failure networks and cardiac arrest centers should aim to reduce the time before support can be commenced. After referral, swift evaluation of patients is necessary, and many options, including short- or long-term MCS, heart transplantation, or in some cases even palliation can stand at the end of this process. Inadequate perfusion, prior myocardial infarction, and oliguria as predictors of impaired 30-day survival were identified in the GUSTO-I database [7] and therefore further indicate advanced treatment strategies like MCS. Especially in situations of potential recovery of end-organ as well as myocardial function (including acute coronary syndrome with early appropriate revascularization, fulminant myocarditis, stress cardiomyopathy, or drug intoxication), with a positive risk/benefit balance, short-term MCS is a potent and useful treatment possibility [8]. However, refractory cardiac arrest and a cummulation of further clinical situations (see Table 10.1 Part 7 Chap. 39) might be inappropriate for MCS, since long-term prognosis is lacking. Only the intra-aortic balloon pump (IABP) (Class III level of evidence (LOE) B) and MCS in general (Class IIb, LOE C) are mentioned, and no comparison is made between the two, nor any preference is presented within the guidelines [5].

Table 10.1 Clinical situations that can influence decision-making about short-term mechanical circulatory support [adapted from 7]

Contraindication	Specific considerations
Age	>75 years
Frailty	
Psychiatric disorders	Uncontrolled
Addiction habits	Uncontrolled
Preexisting factors	
Severe kidney disease	
Liver cirrhosis	
Cerebral stroke	With residual neurological findings
Neoplasia	
Dementia	
Under therapy	
Multi-organ failure	Ongoing
Severe acidosis	pH < 7,0
Cardiac arrest	With no flow >5 minutes or lack of witness
Severe sepsis	
Neurological impairment	After CPR

CPR Cardiopulmonary resuscitation

According to this and taking into account the previously mentioned pathophysiological considerations, patient selection for MCS is crucial. It is difficult in the acute setting to accurately assess patient prognosis and/or response to medical treatment on admission and consequently define the best approach in each individual case. In some cases, the initial prognosis is unknown, and short-term MCS is considered, but long-term strategies have to be reconsidered after an appropriate time period (bridge to decision concept). Evaluation of the bridge to long-term MCS concept was performed by [9] in a single-center cohort retrospectively. Survival rates after 31 months were comparable (approximately 52%) to primarily implanted long-term MCS outcomes, although the BTB cohort was initially more severe ill. Lactate and impaired liver function were predictors for impaired survival, but overall Lebreton et al. showed that the BTB concept allows the implementation of long-term MCS in severely ill with adequate survival [9]. Further assessment of risk factors for impaired postoperative outcome was performed by Tsyganenko et al. in an all-comers cohort at a tertiary center. After evaluation of 618 patients, primarily hospitalized in refractory cardiogenic shock and supported by ECLS, liver dysfunction, inflammatory status, and obesity were seen to increase the risk for mid-term mortality after the implantation of long-term mechanical circulatory support [10]. After patient selection for short-term MCS in general, device selection is of key importance; since the multiple etiologies causing cardiogenic shock and the different patterns of myocardial involvement (i.e., left ventricle, right ventricle, biventricular), there is no "one device fits all" strategy. Current modes of short-term MCS can be characterized by one of four circuit configurations: (1) LV to aorta assist devices (CentriMag and Impella); (2) left atrium (LA) to systemic artery assist devices (TandemHeart); (3) right atrium (RA) to systemic artery assist devices (venous–arterial extracorporeal life support [V-A ECLS]); and (4) RA to pulmonary artery assist devices (Impella RP, Levitronix, TandemHeart). All available devices improve cardiac output and blood pressure to a certain amount, but their specific features result in different hemodynamic profiles. So far, the implications of these on clinical outcomes have not yet been studied in adequately powered randomized trials [11]. A practical approach favors in a scenarios of combined shock or if need for oxygenation or decarboxylation exists, ECLS, since it provides the advantages of full circulatory and ventilator support, rapidly, if required, at the bedside. Its installation is also less expensive than that of other devices, and it offers the possibility of extension or de-escalation therapy. It is therefore broadly used. Furthermore, in patients with hypertrophic cardiomyopathy and other restrictive forms with preserved ejection fraction, an LV suction device do not properly work and should be avoided. In patients with primarily left ventricular impairment, Impella devices (Impella CP delivers maximally 3.7 L/min and the Impella 5.0/5.5 deliver 5–5.5 L/min of Cardiac output support), TandemHeart (4 l/min), or again ECLS are possible treatment options. The Impella (Abiomed, Danvers, MA) is a continuous, nonpulsatile, axial-flow Archimedes screw pump that provides active support by expelling aspirated blood from the LV into the ascending aorta. Importantly, the Impella facilitates stability even in the setting of tachyarrhythmias or electromechanical disassociation. However, careful patient selection regarding RV function and meticulous attention to good technique are critical given reported complications, which include device dislocation and device malfunction because of thrombosis, causing hemolysis. Further complications are bleeding requiring transfusion, arrhythmias, limb ischemia, tamponade, aortic or mitral valve injury, and stroke [11].

Despite the seemingly logical conclusion that a pumping device (Impella) should be superior to an augmented one (IABP), no proof for this concept has been achieved with the smaller Impella 2.5. The ISAR Shock trial compared IABP and Impella 2.5 in acute myocardial infarction (AMI) complicated by cardiogenic shock. Overall mortality at 30 days was 46% in both groups [12]. Even more effective Impella CP device does not performed better. Ouweneel et al. compared in a similar setting Impella CP and IABP and found no differences in mortality at 30 days (46% vs. 50%) and 6 months (50% vs. 50%) [13]. A similar situation occurs if TandemHeart was compared to IABP [14–16]. The RECOVER I feasibility study [17] showed that hemodynamics improved in a postcardiotomy setting after Impella 5.0 was inserted, but since it was a single-arm design, no comparison to other devices could be drawn. Most of the trials presented and published so far comparing different short-term MCS devices in acute cardiogenic shock suffer from major methodological limitations, including small sample size, non-randomization, and selective selection. Therefore, the actual benefit of short-term cardiac assist devices should be balanced with associated costs, and the risk of complications so far is not completely understood. Furthermore, in one of the few large prospective randomized trials in cardiogenic shock so far, the IABP shock II trial, mortality at 30 days, was approximately 40% and similar among patients in the IABP group and those in the control group (without device support) [18]. Since there is no single ideal device, their use should be primarily guided by clinical judgment and local experience [19] A potential selection algorithm, used in our institution, is displayed in Fig. 10.1 (Part 7; Chap. 39).

Long-Term Mechanical Circulatory Support

Finding the right time for long-term MCS implantation is a balance between the positive effect of mechanical support of the failing heart on survival and/or on quality of life (QoL) and the potential adverse effects inherited by this therapy. In patients in NYHA class II–III, the risk of complications associated with the surgery per se, the anticoagulation required, and infection is

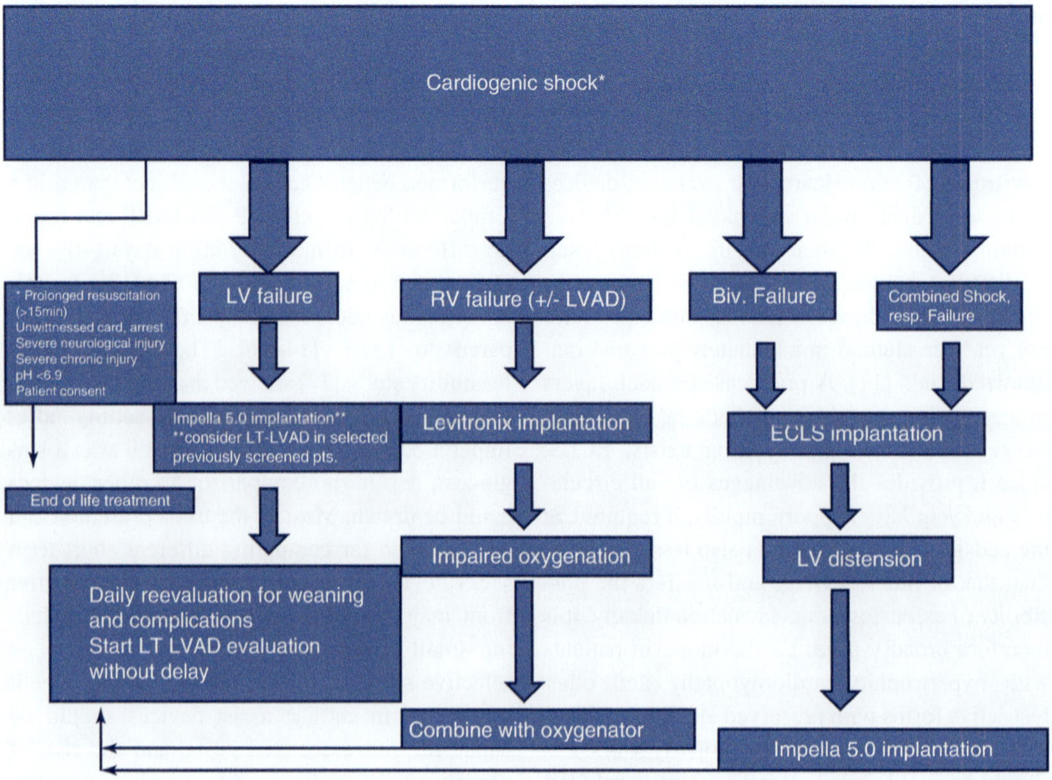

Fig. 10.1 A potential selection algorithm, for short-term mechanical circulatory support *LV* left ventricle, *RV* right ventricle, *BIV* biventricular, *RESP* respiratory, *LT-LVAD* long-term left ventricular assist device, *PTS* patients, *ECLS* extracorporeal life support

currently probably unacceptably higher than the benefit in terms of survival and improvement of QoL. On the other hand, in a patient dying from cardiogenic shock, the benefit is absolutely clear [20], but the outcome might be impaired nevertheless mechanical circulatory support is provided or not. The goal of this chapter is to find an individual pathway to determine the right time point for every patient for implantation of a long-term MCS. It should be remembered that, with new technology such as transcutaneous energy transfer, eliminating a major source for infection and improving QoL, this timepoint is likely to shift toward earlier implantation. Durable MCS could follow short-term MCS in appropriate patients, and durable MCS is often implanted in somewhat more stable patients, since thoughtful evaluation is needed. The European and American Heart Failure guidelines indicate that if patients with acute heart failure cannot be stabilized with medical therapy, MCS systems can be used to unload the failing ventricle and maintain sufficient end-organ perfusion [6, 21]. To more accurately define this situation, the INTERMACS classification is used. The INTERMACS profiles were developed in 2009 with the first INTERMACS report, since the widely used New York Heart Association classification did not provide proper differentiation especially in Class III and IV. In 2009, when the seen profiles were introduced, 80% of long-term device implantations were performed in patients in profile I and II [22].Due to worse survival in IM profile 1 (52%) and IM profile II (63%), the ESC guidelines state that IM profile I has a clear preference for ECLS or short-term devices, and even in IM profile II, ECLS has its place within the guidelines (together with LVAD therapy). For IM profile III to V the guidelines state a clear preference for LVAD therapy with survival rates ranging from 78% to 93%, and in selected cases in advanced NYHA III patients (INTERMACS profile VI and VII), LVAD implantation is an option as well. If compared to current implant rates and rates over the last decade, the situation has shifted swiftly. In 2009, 80% of all implants were in IM profile I and II. The percentage of implants in IM profile I decreased from 41% in 2006 to 14% in 2011 and has remained constant at this level since then, but a further decrease was seen in IM level II during the recent years [23, 24]. In 2012, 20% of patients underwent pulsatile device implantation, whereas nowadays only 5% of implants are pulsatile devices as recorded in the INTERMACS registry. Since these two aspects (pulsatile devices and low INTERMACS profile) are relevant risk factors in many models or registries, the decrease should have led to an increase in survival, but in fact higher individual patient risk profiles over the years led to a relatively constant survival rate of 80% at 1 year [23]. In comparison, in the European registry (EUROMACS), implantation numbers of pulsatile VAD are lower (3.5%), while percentages within the different IM classes are quite similar (14% IM I; 30% IM II) [25]. With these findings both large registries stand in contrast to current guidelines, since every sixth to seventh implantation is performed within INTERMACS profile I. In selected patients, i.e., those on a transplant list or already scheduled for LVAD implantation, previously stable on inotropes, with rapid decline during the waiting period, implantation in IM level I might be reasonable, but in other cases initial short-term MCS support has to be favored. After having discussed the potential limitation in too sick patients, the same focus has to be given to patients maybe too healthy for implantation. A non-randomized trial to study this question is the ROADMAP study [26]. This study found improved survival with improved functional status after HeartMate II LVAD implantation compared to optimal medical therapy in patients in INTERMACS profile IV to VII. LVAD patients experienced more frequent adverse events but improved more in quality of life and in a depression scale. In 2017 the ROADMAP investigators reevaluated the previously enrolled patients and compared 2-year results of patients in IM level IV to those of patients in IM profile V–VII [27]. After studying more balanced cohorts (INTERMACS profile IV and INTERMACS profile V–VII), event-free survival with LVAD was superior to medical therapy in both groups, but the composite endpoint of survival with improved quality of life only met significance for LVAD therapy in patients in IM profile IV. Since the cohort was enrolled from 2011 to 2013 and the study group received a HM II pump, in the light of contemporary left ventricular assist systems (HM 3, HeartWare), these results might be newly discussed. To close this gap, the currently enrolling *Early* Versus *Emergency Left Ventricular Assist Device Implantation in Patients Awaiting Cardiac Transplantation* trial [ClinicalTrials.gov Identifier: NCT02387112] was conducted, to prospectively compare outcome after LVAD implantation in IM profile IV–VII in patients on a transplant list under optimal medical care to patients who receive emergent LVAD implantation or high urgent transplant listing during further deterioration. A further trial within a similar context, the REVIVE-IT study, published in 2016 was conducted to investigate whether LVAD implantation in a destination therapy setting might improve outcome compared to medical therapy in IM level IV to VII but was discontinued in 2015 after the FDA placed the trial on hold following a series of consecutive pump thrombus events observed in patients receiving a LVAD [28]. Another clinical highly relevant patient cohort is the group of transplant and LVAD ineligible patients. These patients were studied within the MEDAMACS registry [29]. These patients had a high mortality of 23% in 10 months, suggesting them to be in a very high-risk group. Despite their poor prognosis, only 30% had undergone proper evaluation for advanced therapies. This is of special interest, since other risk factors, so far not addressed that adversely affect outcome as judged by clinicians, led to ineligibility, notably non-adherence, and limited social support. These factors warrant further investigation, as they are currently not part of classic LVAD selection models but clearly seem to be associated with poorer prognosis. Parameters, which are currently used to individualize patient risk profile, are age, (female) sex, chronic pulmonary disease, peripheral vascular disease, and renal dysfunction as well as overall nutritional state [24]. The biggest contributors to early mortality are the need for a right ventricular assist device and need for chronic renal replacement therapy prior to implantation [24]. As expected, increased surgical complexity, either through previous cardiac surgery or through concomitant cardiac procedures to the implant operation, increases early risk [24]. Although these aspects are clear, no widely used specific risk score for contemporary devices exists. Therefore, an interesting approach is a Bayesian model to predict survival, such as was published by Kanwar et al. using the INTERMACS data [30]. Such a model might be of particular interest, since it has the great advantage of showing how clinical variables together affect predicted mortality without every variable being known. Furthermore a reasonable prediction can be assessed even when only key variables are known. In this work, renal and hepatic dysfunction, lower INTERMACS profile, and temporary MCS were predictors for short-term outcome, whereas frailty, age, and chronic kidney disease were important predictors for 12-month outcome. A recently published risk score from a single-center cohort is the PENN-Columbia risk score which includes again age, creatinine, bilirubin, and body mass index as well as right ventricular dysfunction and grade of aortic valve insufficiency but shows only acceptable discrimination of low- and intermediate-risk cohort compared to the high-risk cohort and is therefore of limited benefit [31]. Two simple to assess scores are the MELD and the MELD XI score, using creatinine and bilirubin as risk markers, the MELD score together with INR, the Meld XI score without. They have been used in the LVAD population again only in small trials and have not been evaluated in large registries but suggest that creatinine and bilirubin values could predict outcome after LVAD implantation [32, 33]. Furthermore the HeartMate II risk score, which was evaluated on the INTERMACS registry, showed predictive value for 90-day mortality using age, creatinine, albumin, and INR [34]. It is important to notice that two parameters that are used in all risk models, creatinine and bilirubin, or more generally renal and hepatic end-organ impairment, are both risk factors and risk makers. Both become more and more affected over time, with more severe heart failure and with biventricular involvement. Therefore, impaired survival in patients with elevated creatinine and bilirubin is to some extent only because implantation of a VAD is conducted (too) late. Heart failure leads to reduced renal perfusion, activation of the renin-angiotensin-aldosterone system, and inflammatory response via cardiorenal syndrome type II and venous congestion [35, 36]. Due to heart failure-mediated inflammation and congestion, nonalcoholic liver disease can develop and can lead to liver cirrhosis in a later stage [37, 38]. Due to the ISHLT guidelines [39], there is a class I recommendation to perform ultrasound of the liver, if liver function tests are abnormal, to exclude primary liver disease. This is of utmost importance for patients with Fontan physiology (level of evidence: C). For patients with suspected cirrhosis, further radiologic and tissue confirmations in conjunction with a hepatology consultation are indicated (level of evidence: C). So far, literature on liver biopsy and its predictive value for reversibility is scarce [40, 41], and no recommendation is given for noninvasive methods such as Fibroscan and ultrasound-based scan (vibration measurement) of liver stiffness or the LiMAx test, a real-time breathing test, with measurement of CYP 1A2 and therefore of the maximal liver function capacity. Therefore, prevention of acute kidney injury and hepatic impairment is to some extent synonymous with prevention of right heart failure. In a clinical scenario, although there are no randomized controlled trials, maintaining end-organ perfusion pressure, reducing CVP, and maintaining right ventricular output or even right ventricular mechanical support can lead to end-organ improvement, and it might have prognostic implications, since reversibility of the end-organ involvement improves prognosis. It might be concluded that right heart function plays a crucial role, since postoperative short- and long-term mortality is dependent on it as well as the possible need for biventricular support, but the identification is still challenging since LVADs support only the left ventricle, and immediately after device implantation, the right ventricle is exposed to the risk of developing severe dysfunction, i.e., right ventricular failure (RVF) [42]. The LVAD decompresses the left ventricle causing a leftward shift of the interventricular septum, which results in a more spherical shape of the right ventricle, thus reducing the mechanical contractile properties of the right-sided chambers. In addition, right ventricular (RV) hemodynamics change after LVAD implantation because RV output must match that generated by the device. Right ventricular failure (RVF) can be a fatal complication usually presenting within 2 weeks after device implantation [43]. Unfortunately, several aspects of RVF following LVAD implantation are still uncertain, and even the reported prevalence of RVF in this population has been variable (ranging between 13% and 51%) [44]. Temporary mechanical right ventricular support is needed in approximately 5–10% in the current era [24]. Bellavia et al. reviewed currently available body of literature about right heart failure and concluded that a few clinical characteristics, in particular need for renal replacement therapy or mechanical ventilation, show adequate specificity for the occurrence of RVF but that they have very low sensitivity. Among humoral markers, INR (even slightly increased over 1.1), NT-proBNP, and white blood cell count may be helpful for the identification of patients at risk. Invasive hemodynamic assessment should be focused on central venous pressure, right ventricular stroke work index, and mean arterial pressure. Finally, qualitative assessment of RV performance, RV/LV diameter ratio, and RV free wall longitudinal two-dimensional strain could be adopted in routine echocardiographic evaluation of these patients [45]. The issue with all these findings is how to appropriately apply them to the patient in front of you, especially since no adequate investigated medical treatment options exist. So far some of the markers could be measured or estimated, but if a patient exhibits a high risk for right heart failure, it remains unclear what strategy is the best (direct long-term biventricular support vs. left ventricular support with intra- and early postoperative decision-making on temporary and potentially long-term right ventricular support at a later stage). This aspect becomes even more difficult since currently available continuous-flow (cf) devices are not approved for right ventricular support and reimbursement differs from country to country. Two recently published studies by Eulert-Grehn et al. and Vierecke et al. target this concern [46, 47]. Eulert-Grehn et al. suggest that biventricular support with two continuous-flow pumps is feasible, and Vierecke et al. did not find any difference between BIVAD implantation and LVAD +/− temporary RVAD implantation regarding outcome. Considering the aspect of higher flexibility with LVAD +/− temporary RVAD implantation (intraoperatively and during the potential weaning phase), this concept, previously also described by [48], might be favorable, as long as no outcome data is available that shows a preference for one of the two approaches. With respect to the currently used continuous-flow VAD type for left ventricular support, the MOMENTUM III trial showed that the HeartMate 3 device, a fully magnetically levitated centrifugal-flow pump, was superior after 2 years to the HeartMate II device, a mechanical-bearing axial-flow pump, with regard to reoperation to replace or remove a malfunctioning device [49]. So far, the IM reports have not stated device type (axial vs. centrifugal flow) as a risk factor for early or later mortality [24], nor did Mueller et al. find any difference with regard to 1-year survival and complication profile in a large single-center cohort comparing HM 3 and HeartWare HVAD for left ventricular support in 200 patients [50]. The endurance trial [51] investigated transplant ineligible patients with regard to implantation of HM II or HVAD. HVAD in this study was found to be noninferior with respect to a composite of survival free from disabling stroke or device removal for malfunction or failure. The high rate of strokes in the study cohort (HVAD vs. HM II – 29.7% vs. 12.1%) was statistically significant, and increased mean arterial blood pressure was found to be an independent risk factor. Therefore, the ENDURANCE supplemental trial was designed, again comparing HM II and HVAD with strict blood pressure control, and confirmed that adequate blood pressure control was associated with a reduced risk for stroke in HVAD subjects. Additionally, with regard to the composite endpoint of survival free from disabling stroke and need for device exchange or urgent transplantation or death, the HVAD system was superior to the control [52].

References

1. Kolte D, Khera S, Aronow WS, Mujib M, Palaniswamy C, Sule S, et al. Trends in incidence, management, and outcomes of cardiogenic shock complicating ST-elevation myocardial infarction in the United States. J Am Heart Assoc. 2014;3(1):e000590. https://doi.org/10.1161/JAHA.113.000590.
2. Gupta N, Aggarwal S, Gaglianello N, Bangalore S, Cinquegrani MP. Trends in incidence, management and outcomes of heart failure patients with cardiogenic shock. Circulation. 2018;130:A13061.
3. Reynolds HR, Hochman JS. Cardiogenic shock: current concepts and improving outcomes. Circulation. 2008;117(5):686–97. https://doi.org/10.1161/CIRCULATIONAHA.106.613596.
4. Crespo-Leiro MG, Metra M, Lund LH, Milicic D, Costanzo MR, Filippatos G, et al. Advanced heart failure: a position statement of the heart failure Association of the European Society of cardiology. Eur J Heart Fail. 2018; https://doi.org/10.1002/ejhf.1236. [Epub ahead of print].
5. Ponikowski P, Voors AA, Anker SD, Bueno H, Cleland JGF, Coats AJS, et al. ESC guidelines for the diagnosis and treatment of acute and chronic heart failure: the task force for the diagnosis and treatment of acute and chronic heart failure of the European Society of Cardiology (ESC)developed with the special contribution of the heart failure association (HFA) of the ESC. Eur Heart J. 2016;37(27):2129–200. https://doi.org/10.1093/eurheartj/ehw128.
6. Lund LH, Trochu JN, Meyns B, Caliskan K, Shaw S, Schmitto JD, et al. Screening for heart transplantation and left ventricular assist system: results from the ScrEEning for advanced heart failure treatment (SEE-HF) study. Eur J Heart Fail. 2018 Jan;20(1):152–60. https://doi.org/10.1002/ejhf.975.
7. Hasdai D, Holmes DR Jr, Califf RM, Thompson TD, Hochman JS, Pfisterer M, et al. Cardiogenic shock complicating acute myocardial infarction: predictors of death. GUSTO investigators. Global utilization of streptokinase and tissue-plasminogen activator for occluded coronary arteries. Am Heart J. 1999;138(1 Pt 1):21–31.
8. Bonello L, Delmas C, Schurtz G, Leurent G, Bonnefoy E, Aissaoui N, et al. Mechanical circulatory support in patients with cardiogenic shock in intensive care units: A position paper of the "Unité de Soins Intensifs de Cardiologie" group of the French Society of Cardiology, endorsed by the "Groupe Athérome et Cardiologie Interventionnelle" of the French Society of Cardiology. Arch Cardiovasc Dis. 2018:S1875–2136(18)30075–5. https://doi.org/10.1016/j.acvd.2018.03.008.
9. Lebreton G, Pozzi M, Mastroianni C, Léger P, Pavie A, Leprince P. Extracorporeal life support as a bridge to bridge: a strategy to optimize ventricular assist device results. Eur J Cardiothorac Surg. 2015;48(5):785–91. https://doi.org/10.1093/ejcts/ezu516.
10. Tsyganenko D, Gromann TW, Schoenrath F, Mueller M, Mulzer J, Starck C, et al. Predictors of mid-term outcome in patients undergoing implantation of a ventricular assist device directly after extracorporeal life support in a large single center cohort. Accepted EJCTS. https://doi.org/10.1093/ejcts/ezy351.
11. Mandawat A, Rao SV. Percutaneous Mechanical Circulatory Support Devices in Cardiogenic Shock. Circ Cardiovasc Interv. 2017;10(5):e004337. https://doi.org/10.1161/CIRCINTERVENTIONS.116.004337.
12. Seyfarth M, Sibbing D, Bauer I, Fröhlich G, Bott-Flügel L, Byrne R, et al. A randomized clinical trial to evaluate the safety and efficacy of a percutaneous left ventricular assist device versus intra-aortic balloon pumping for treatment of cardiogenic shock caused by myocardial infarction. J Am Coll Cardiol. 2008;52:1584–1588. https://doi.org/10.1016/j.jacc.2008.05.065.
13. Ouweneel DM, Eriksen E, Sjauw KD, van Dongen IM, Hirsch A, Packer EJ, et al. Percutaneous mechanical circulatory support versus intra-aortic balloon pump in cardiogenic shock after acute myocardial infarction. J Am Coll Cardiol. 2017;69:278–87. https://doi.org/10.1016/j.jacc.2016.10.022.
14. Thiele H, Sick P, Boudriot E, Diederich KW, Hambrecht R, Niebauer J, et al. Randomized comparison of intra-aortic balloon support with a percutaneous left ventricular assist device in patients with revascularized acute myocardial infarction complicated by cardiogenic shock. Eur Heart J. 2005;26:1276–83. https://doi.org/10.1093/eurheartj/ehi161.
15. Burkhoff D, Cohen H, Brunckhorst C, O'Neill WW. TandemHeart Investigators Group. A randomized multicenter clinical study to evaluate the safety and efficacy of the TandemHeart percutaneous ventricular assist device versus conventional therapy with intraaortic balloon pumping for treatment of cardiogenic shock. Am Heart J. 2006;152:469.e1–469. e8. https://doi.org/10.1016/j.ahj.2006.05.031.
16. Kar B, Gregoric ID, Basra SS, Idelchik GM, Loyalka P. The percutaneous ventricular assist device in severe refractory cardiogenic shock. J Am Coll

Cardiol. 2011;57:688–96. https://doi.org/10.1016/j.jacc.2010.08.613.

17. Griffith BP, Anderson MB, Samuels LE, Pae WE Jr, Naka Y, Frazier OH. The recover I: A multicenter prospective study of Impella 5.0/ LD for postcardiotomy circulatory support. J Thorac Cardiovasc Surg. 2013;145:548–54. https://doi.org/10.1016/j.jtcvs.2012.01.067.
18. Thiele H, Zeymer U, Neumann FJ, Ferenc M, Olbrich HG, Hausleiter J, et al. Intraaortic balloon support for myocardial infarction with cardiogenic shock. N Engl J Med. 2012;367(14):1287–96. https://doi.org/10.1056/NEJMoa1208410.
19. Rihal CS, Naidu SS, Givertz MM, Szeto WY, Burke JA, Kapur NK, et al. 2015 SCAI/ACC/HFSA/STS Clinical expert consensus statement on the use of percutaneous mechanical circulatory support devices in cardiovascular care: endorsed by the American Heart Assocation, the Cardiological Society of India, and Sociedad Latino Americana de Cardiologia Intervencion; Affirmation of Value by the Canadian Association of Interventional Cardiology-Association Canadienne de Cardiologie d'Intervention. J Am Coll Cardiol 2015;65:e7–26.
20. Lietz K, Long JW, Kfoury AG, Slaughter MS, Silver MA, Milano CA, et al. Outcomes of left ventricular assist device implantation as destination therapy in the post-REMATCH era: implications for patient selection. Circulation. 2007;116(5):497–505.
21. Yancy CW, Jessup M, Bozkurt B, Butler J, Casey DE Jr, Colvin MM, et al. 2017 ACC/AHA/HFSA focused update of the 2013 ACCF/AHA guideline for the Management of Heart Failure: A report of the American College of Cardiology/American Heart Association task force on clinical practice guidelines and the Heart Failure Society of America. J Am Coll Cardiol. 2017;70(6):776–803. https://doi.org/10.1016/j.jacc.2017.04.025.
22. Stevenson LW, Pagani FD, Young JB, Jessup M, Miller L, Kormos RL, et al. INTERMACS profiles of advanced heart failure: the current picture. J Heart Lung Transplant. 2009;28(6):535–41. https://doi.org/10.1016/j.healun.2009.02.015.
23. Kirklin JK, Naftel DC, Kormos RL, Stevenson LW, Pagani FD, Miller MA, et al. The fourth INTERMACS annual report: 4,000 implants and counting. J Heart Lung Transplant. 2012;31(2):117–26. https://doi.org/10.1016/j.healun.2011.12.001.
24. Kirklin JK, Pagani FD, Kormos RL, Stevenson LW, Blume ED, Myers SL, et al. Eighth annual INTERMACS report: special focus on framing the impact of adverse events. J Heart Lung Transplant. 2017;36(10):1080–6. https://doi.org/10.1016/j.healun.2017.07.005.
25. de By TMMH, Mohacsi P, Gahl B, Zittermann A, Krabatsch T, Gustafsson F, et al. The European registry for patients with mechanical circulatory support (EUROMACS) of the European Association for Cardio-Thoracic Surgery (EACTS): second report. Eur J Cardiothorac Surg. 2017; https://doi.org/10.1093/ejcts/ezx320.
26. Estep JD, Starling RC, Horstmanshof DA, Milano CA, Selzman CH, Shah KB, et al. Risk assessment and comparative effectiveness of left ventricular assist device and medical Management in Ambulatory Heart Failure Patients: results from the ROADMAP study. J Am Coll Cardiol. 2015;66(16):1747–61. https://doi.org/10.1016/j.jacc.2015.07.075.
27. Shah KB, Starling RC, Rogers JG, Horstmanshof DA, Long JW, Kasirajan V, et al. Left ventricular assist devices versus medical management in ambulatory heart failure patients: an analysis of INTERMACS profiles 4 and 5 to 7 from the ROADMAP study. J Heart Lung Transplant. 2018;37(6):706–14. https://doi.org/10.1016/j.healun.2017.12.003.
28. Pagani FD, Aaronson KD, Kormos R, Mann DL, Spino C, Jeffries N, et al. The NHLBI REVIVE-IT study: understanding its discontinuation in the context of current left ventricular assist device therapy. J Heart Lung Transplant. 2016;35(11):1277–83. https://doi.org/10.1016/j.healun.2016.09.002.
29. Ambardekar AV, Forde-McLean RC, Kittleson MM, Stewart GC, Palardy M, Thibodeau JT, et al. High early event rates in patients with questionable eligibility for advanced heart failure therapies: results from the medical arm of mechanically assisted circulatory support (Medamacs) registry. J Heart Lung Transplant. 2016;35(6):722–30. https://doi.org/10.1016/j.healun.2016.01.014.
30. Kanwar MK, Lohmueller LC, Kormos RL, Teuteberg JJ, Rogers JG, Lindenfeld J, et al. A Bayesian model to predict survival after left ventricular assist device implantation. JACC Heart Fail. 2018;6(9):771–9. https://doi.org/10.1016/j.jchf.2018.03.016.
31. Birati EY, Hanff TC, Maldonado D, Grandin EW, Kennel PJ, Mazurek JA, et al. Predicting long term outcome in patients treated with continuous flow left ventricular assist device: the penn-columbia risk score. J Am Heart Assoc. 2018;7(6): e006408. https://doi.org/10.1161/JAHA.117.006408.
32. Deo SV, Daly RC, Altarabsheh SE, Hasin T, Zhao Y, Shah IK, et al. Predictive value of the model for end-stage liver disease score in patients undergoing left ventricular assist device implantation. ASAIO J. 2013;59(1):57–62. https://doi.org/10.1097/MAT.0b013e31827c0c77.
33. Yang JA, Kato TS, Shulman BP, Takayama H, Farr M, Jorde UP, et al. Liver dysfunction as a predictor of outcomes in patients with advanced heart failure requiring ventricular assist device support: use of the model of end-stage liver disease (MELD) and MELD eXcluding INR (MELD-XI) scoring system. J Heart Lung Transplant. 2012;31(6):601–10. https://doi.org/10.1016/j.healun.2012.02.027.
34. Cowger JA, Castle L, Aaronson KD, Slaughter MS, Moainie S, Walsh M, et al. The HeartMate II risk score: an adjusted score for evaluation of all continuous-flow left ventricular assist devices.

ASAIO J. 2016;62(3):281–5. https://doi.org/10.1097/MAT.0000000000000362.

35. Ross DW, Stevens GR, Wanchoo R, Majure DT, Jauhar S, Fernandez HA, et al. Left ventricular assist devices and the kidney. Clin J Am Soc Nephrol. 2018;13(2):348–55. https://doi.org/10.2215/CJN.04670417.
36. Tromp TR, de Jonge JA. Joles. Left ventricular assist devices: a kidney's perspective. Heart Fail Rev. 2015;20(4):519–32. https://doi.org/10.1007/s10741-015-9481-z.
37. Ballestri S, Lonardo A, Bonapace S, Byrne CD, Loria P, Targher G. Risk of cardiovascular, cardiac and arrhythmic complications in patients with non-alcoholic fatty liver disease. World J Gastroenterol. 2014;20(7):1724–45. https://doi.org/10.3748/wjg.v20.i7.1724.
38. Nikolaou M, Parissis J, Yilmaz MB, Seronde MF, Kivikko M, Laribi S, et al. Liver function abnormalities, clinical profile, and outcome in acute decompensated heart failure. Eur Heart J. 2013;34(10):742–9. https://doi.org/10.1093/eurheartj/ehs332.
39. Feldman D, Pamboukian SV, Teuteberg JJ, Birks E, Lietz K, Moore SA, et al. The 2013 International Society for Heart and Lung Transplantation guidelines for mechanical circulatory support: executive summary. J Heart Lung Transplant. 2013;32(2):157–87. https://doi.org/10.1016/j.healun.2012.09.013.
40. Xuereb L, Go PH, Kaur B, Akrawe S, Borgi J, Paone G, et al. Should patients with hepatic fibrosis undergo LVAD implantation: A comparative analysis. ASAIO J. 2016;62(4):498–500. https://doi.org/10.1097/MAT.0000000000000375.
41. Potthoff A, Schettler A, Attia D, Schlue J, Schmitto JD, Fegbeutel C, et al. Liver stiffness measurements and short-term survival after left ventricular assist device implantation: A pilot study. J Heart Lung Transplant. 2015;34(12):1586–94. https://doi.org/10.1016/j.healun.2015.05.022.
42. Harjola VP, Mebazaa A, Celutkiene J, Bettex D, Bueno H, Chioncel O, et al. Contemporary management of acute right ventricularfailure: a statement from the heart failure association and the working group on pulmonary circulation and right ventricular function of the European Society of Cardiology. Eur J Heart Fail. 2016;18:226–41.
43. Frea S, Pidello S, Bovolo V, Iacovino C, Franco E, Pinneri F, et al. Prognostic incremental role of right ventricular function in acute decompensation of advanced chronic heart failure. Eur J Heart Fail. 2016;18:564–72.
44. Slaughter MS, Rogers JG, Milano CA, Russell SD, Conte JV, Feldman D, et al. Advanced heart failure treated with continuous flow left ventricular assist device. N Engl J Med. 2009;361:2241–51.
45. Bellavia D, Iacovoni A, Scardulla C, Moja L, Pilato M, Kushwaha SS, et al. Prediction of right ventricular failure after ventricular assist device implant: systematic review and meta-analysis of observational studies. Eur J Heart Fail. 2017;19(7):926–46. https://doi.org/10.1002/ejhf.733. Review.
46. Eulert-Grehn JJ, Lanmüller P, Schönrath F, Solowjowa N, Müller M, Mulzer J, et al. Two implantable continuous-flow ventricular assist devices in a biventricular configuration: technique and results. Interact Cardiovasc Thorac Surg. 2018; https://doi.org/10.1093/icvts/ivy228. [Epub ahead of print].
47. Vierecke J, Gahl B, de By TMMH, Antretter H, Beyersdorf F, Caliskan K, et al. Results of Primary Biventricular Ventricular Support: An Analysis of data from the EUROMACS Registry. Submitted to EJCTS.
48. Loforte A, Stepanenko A, Potapov EV, Musumeci F, Dranishnikov N, Schweiger M, et al. Temporary right ventricular mechanical support in high-risk left ventricular assist device recipients versus permanent biventricular or total artificial heart support. Artif Organs. 2013;37(6):523–30. https://doi.org/10.1111/aor.12038.
49. Mehra MR, Goldstein DJ, Uriel N, Cleveland JC Jr, Yuzefpolskaya M, Salerno C, et al. Two-year outcomes with a magnetically levitated cardiac pump in heart failure. N Engl J Med. 2018;378(15):1386–95. https://doi.org/10.1056/NEJMoa1800866.
50. Mueller M, Hoermandinger C, Richter G, Mulzer J, Tsyganenko D, Krabatsch T, et al. Retrospective 1-year outcome follow-up in 200 patients supported with HeartMate 3 and HeartWare left ventricular assist devices in a single centre. E.Eur J Cardiothorac Surg. 2020;57(6):1160–5. https://doi.org/10.1093/ejcts/ezaa017.
51. Rogers JG, Pagani FD, Tatooles AJ, Bhat G, Slaughter MS, Birks EJ, et al. Intrapericardial left ventricular assist device for advanced heart failure. N Engl J Med. 2017;376(5):451–60. https://doi.org/10.1056/NEJMoa1602954.
52. Milano CA, Rogers JG, Tatooles AJ, Bhat G, Slaughter MS, Birks EJ, et al. HVAD: The ENDURANCE Supplemental trial. JACC Heart Fail. 2018;6(9):792–802. https://doi.org/10.1016/j.jchf.2018.05.012.

Perioperative Considerations in Left Ventricular Assist Device Placement

11

Ranjani Venkataramani, Michael Zhen-Yu Tong, and Shiva Sale

Mechanical circulatory support has emerged as a viable treatment option for patients with end-stage heart failure. Even though heart transplantation is considered to be the gold standard, timely availability of a donor heart has been a limiting factor in its universal application. Moreover, not all patients are eligible for transplantation, and the heart failure state can be improved from the circulatory support provided by the durable modern assist devices. Continuous flow left ventricular assist devices (CF-LVAD) are the commonly used durable support systems in patients with advanced heart failure and are known to improve survival and quality of life [1]. In this chapter, we are going to focus on perioperative anesthetic care of these patients during the implantation of an assist device.

Contemporary rotary CF-LVAD pumps are classified into either axial pumps such as HeartMate II (Abbott, Abbott Park IL) or centrifugal pumps consisting of HVAD (Medtronic, Minneapolis, MN) and HeartMate 3 (Abbott, Abbott Park IL). These devices have variable flow capacities and are implanted in the body using different techniques. They are governed by the same principle of generating continuous flow of blood out of the left ventricle (LV) into the aorta throughout the cardiac cycle.

Pre-operative Anesthetic Considerations

Patients presenting for assist device placement are in end-stage heart failure with varying levels of circulatory decompensation. There are no commonly practiced defined criteria for patient selection for LVAD implantation. Quite often the FDA approval requirement for destination therapy is used as selection criteria for advanced device therapy (Table 11.1).

In practice, we are going to see assist device implanted in individuals outside these strictly defined criteria.

Electronic Supplementary Material The online version of this chapter (https://doi.org/10.1007/978-3-030-47809-4_11) contains supplementary material, which is available to authorized users.

R. Venkataramani, MD
Department of Anesthesiology and Critical Care, University of New Mexico, Albuquerque, NM, USA
e-mail: Rvenkataramani@salud.unm.edu

M. Z.-Y. Tong, MD, MBA
Mechanical Circulatory Support, CCLCM, Case Western Reserve University, Department of Thoracic and Cardiovascular Surgery, Cleveland, OH, USA
e-mail: tongz@ccf.org

S. Sale, MD (✉)
Heart Transplantation and Mechanical Circulatory Support, CCLCM, Case Western Reserve University, Department of Cardiothoracic Anesthesiology, Cleveland Clinic Foundation, Cardiothoracic Anesthesia,, Cleveland, OH, USA
e-mail: sales@ccf.org

© Springer Nature Switzerland AG 2020
J. H. Karimov et al. (eds.), *Mechanical Support for Heart Failure*,
https://doi.org/10.1007/978-3-030-47809-4_11

Table 11.1 FDA approval requirement for destination therapy (HeartMate2-DT trial)

1.NYHA IIIb-IV symptoms for at least 45 of the last 60 days
2. Heart failure symptoms failed to respond to optimal medical management
3. LVEF<25%
4. Peak VO2 < 14 ml/kg/min or continued need for i.v. inotropic therapy owing to symptomatic hypotension, decreasing renal function, or worsening pulmonary congestion
5. i.v. inotropic medications for ≥14 days
6. Intra-aortic balloon pump support ≥7 days

Table 11.2 INTERMACS profile of advanced heart failure state and profiles @device implant[2]

Level	Definition	Description	@device implant* [Jan 12,013-Dec31 2016]
1	Critical cardiogenic shock	"Crash and burn"	17%
2	Progressive decline	"Sliding fast"	34%
3	Stable but inotrope dependent	"Stable but dependent"	32%
4	Recurrent advanced heart failure	Frequent flyer	13%
5	Exertion intolerant	House bound	2%
6	Exertion limited	Walking wounded	0.6%
7	Advanced NYHA class III	Advanced NYHA class III	0.5%

Many of these patients are in a rapidly progressing heart failure state, and decision to advance therapy is based on projected survival benefits with the device. Treatment decisions in these patients can be challenging considering the bleak outlook of the untreated heart failure state. INTERMACS has developed a classification [2] (Table 11.2) of these advanced heart failure patients in order to stratify the patients according to severity of the heart failure. This classification also helps in the decision to institute advanced therapy in these acutely ill patients. Severely ill patients, in cardiogenic shock on temporary circulatory assist devices, are classified as INTERMACS 1 profile. These patients have the worse outcomes because of the acuity of their illness [2]. Therefore many of these patients are hemodynamically stabilized on temporary assist devices before being considered for durable devices. INTERMACS 4 and above are considered as ambulatory patients. The majority of the patients, considered for durable MCS assist device, are in profiles 2–3. There appears to be a survival benefits for device implantation in less severely ill patients [3].

Table 11.3 Device strategy @implantation (IMACS Jan 1, 2013-Dec 312,016 n = 14,062)

Device strategy	%
Listed for transplant	28
Bridge to candidacy	29
Destination therapy	41
Other	2

IMACS International Society for Heart and Lung Transplantation Mechanically Assisted Circulatory Support

According to the second IMACS (ISHLT Mechanically Assisted Circulatory Support) report, the global database for cardiac assist devices, destination therapy remains the common strategy (Table 11.3).

Absolute contraindications for LVAD therapy include systemic illness with a life expectancy of less than 2 years or malignancy within 5 years, irreversible renal or hepatic dysfunction, severe obstructive pulmonary disease, active severe systemic infection, or other systemic diseases with multiorgan involvement. All patients have to undergo thorough psychosocial evaluation to determine the candidacy, and appropriate remedial measures are considered before the device placement in identified patients. Poor nutritional status is an important predictor of poor outcome and should be addressed before considering for the device implantation surgery. However, LVAD may be an acceptable option for a patient with recent cancer, which might theoretically be cured but who is unable to survive the 5-year disease-free interval typically required for heart transplantation. Similarly, controlled infection with human immunodeficiency virus [4] or advanced end-organ function secondary to heart failure may not preclude patients from LVAD implantation.

The circulatory disturbance induced by acute deterioration of heart function results in some end-organ hypoperfusion. This can result in organ dysfunction depending on the autoregulatory reserves of the organs. In organ failure state due to cardiogenic shock, it is important to observe for any improvement of organ function with circulatory support and rule out intrinsic disease before being considered for assist device placement.

Hepatic Dysfunction

Passive hepatic congestion from heart failure or tricuspid regurgitation often results in twofold increase in transaminases, sometimes with bilirubin rise. In uncertain scenarios, it is advisable to check portal pressures and proceed with liver biopsy to rule out primary liver dysfunction before being considered for LVAD.

Renal Dysfunction

Renal dysfunction seen in advanced heart failure is typically due to a low cardiac output state with reduced renal perfusion, cardio renal syndrome (high venous pressures), and intrinsic renal disease from high-dose diuretics to treat volume overload. Therefore in patients being considered for LVAD therapy, some degree of renal function impairment has to be expected considering the advanced heart failure. LVAD should optimize the renal perfusion, and therefore it is observed that renal function tend to improve after LVAD implantation. After the initial improvement in the renal function, 1 to 6 months post implant, it is not uncommon to see a deterioration before a stable "plateau" is reached [5]. In patients with renal dysfunction receiving durable LVAD, the pre-op eGFR is identified as a prognostic indicator. In a contemporary retrospective analysis, authors noted that at 12 months, the risk-adjusted survival was comparable between patients with eGFR more or less than 60 ml/1.73m^2 [6]. A GFR >30 ml/min is usually acceptable, and in the absence of significant intrinsic renal disease, the GFR usually improves post LVAD implantation. Even though many centers consider eGFR less than 30 ml/1.73m^2 as a relative contraindication, the notion is being challenged as too conservative considering the recovery of renal function with improved perfusion. Device placement can be considered if intrinsic renal disease can be excluded and with evidence of recovery with optimization of heart failure therapy. ESRD with dialysis dependence remains a contraindication to LVAD placement given the complexity in managing these patients in outpatient dialysis centers along with the increase risk of infection. Preexisting renal dysfunction is associated with increased risk of RV failure after device implantation [7].

Medications

An end-stage heart failure patient usually receives a combination of afterload-reducing agents such as angiotensin converting enzyme (ACE) inhibitors, angiotensin receptor blocker (ARB), and vasodilators such as nitroprusside to aid to assist forward flow. Occasionally, patients on inotropic drips such as dobutamine and milrinone may be encountered. ACEI/ARBs may be associated with refractory hypotension, and the timing needs consideration while scheduled for LVAD surgery. Vasodilators are continued till the patient reaches the operating room at which point they are turned off or dosage reduced in anticipation of induction of anesthesia, thus avoiding precipitous decrease in blood pressure which can be aggravated by the accompanying positive pressure ventilation. Any preexisting inotropes are continued through the procedure until the patient is on cardiopulmonary bypass (CPB).

Anticoagulation Status

Many of these patients presenting for LVAD placement are on some anticoagulants or antiplatelet agents. Acutely ill patients are commonly started on heparin, which is continued until surgery if there are no concerns for sternal reentry.

Oral anticoagulants are withdrawn prior to nonemergency surgery with a goal to return to normal hemostatic function. The return of near normal hemostatic function depends on the degree of altered pharmacokinetics in these patients, and this could be less predictable with patients with renal or hepatic dysfunction. Coagulation and platelet function tests should be within normal limits for ambulatory patients and redo-sternotomy patients. The risk of bleeding should be weighed against the higher risk of right heart failure in patients receiving multiple transfusions.

Prior Surgeries

With the advances in cardiac surgery, many patients would have had prior cardiac surgeries. Implications of these surgeries, including valve replacements, repairs, and bypass grafts, must be taken into account with regard to surgical planning. A CT scan of the chest helps look for structures such as the RV, RA, aorta, lung, or great veins that may be adherent or near the sternum during sternotomy. Presence of such findings might help guide the surgeon to consider alternate cannulating sites such as femoral vessels and pre-place wires or cannulas to be able to initiate CPB in the event of catastrophic hemorrhage during sternotomy. Plan for sternal reentry should be discussed during surgical planning. These patients are at risk of blood transfusions considering the hemorrhagic risks from mediastinal adhesions and difficult sternal entry. In patients with previous LVAD, if CPB is needed prior to opening the chest, the RPM should be reduced to prevent suction of the LV but high enough to overcome the AI.

Pre-op RV Risk Stratification:

The right heart failure (RHF) is one of the important determinants of prognosis after LVAD implantation. It is consistently associated with higher perioperative morbidity and mortality [8, 9]. Patients with evidence of right heart failure had a lower chance of survival to transplantation than those without in bridge to transplantation population [10]. Therefore, it is important to understand the pathophysiology of right heart failure after left heart mechanical support with LVAD. The etiology of RVF is multifactorial with complex interplay of the physiological alteration induced by the device, underlying altered loading conditions and primary myocardial dysfunction (Table 11.4).

Table 11.4 Perioperative factors adversely affecting RV function during LVAD implantation

Increased venous return
Variable effect of device on RV afterload
Septal shift after LV unloading
Preexisting RV contractile dysfunction
Peri-operative factors
Air embolism
Untreated coronary artery disease
Massive transfusion
Tamponade
Positive pressure ventilation

Perioperative RV function is essentially dependent on the functional ability of the right ventricle in adapting to the adverse conditions during the perioperative course. According to INTERMACS, RHF after LVAD implantation is defined as presence of signs and symptoms of persistent RV dysfunction. RVF is further classified into mild (≤7 days), moderate(8–14 days), and severe(˃14 days) depending on the duration of the circulatory failure state [8]. Most widely used definition is based on hemodynamics and sustained need of high degree of support as below:

- *Elevated CVP (>18 mmHg) with depressed Cardiac index <2 liters/min/m²*
- *Absence of elevated pulmonary capillary wedge pressure(PCWP <18 mmHg)*
- *RVAD implantation or requirement of prolonged nitric oxide or inotropic therapy(>1 week)*

With the wide experience with LVAD implanted over the years, it is well known that the majority of the RV failure that ensues after LVAD implantation are temporary. Right ventricular performance improves over time secondary to RV adaptation and amelioration of the adverse

Table 11.5 Factors associated with right heart failure after LVAD implantation

Patient characteristics	Laboratory	Hemodynamics	Imaging
Female	Hepatic dysfunction	CI <2.2 l/min/m^2	Pre-op RV dysfunction
Small BSA	Renal dysfunction [S Cr ≥2 mg/dl]	RVSWI<250	Severe TR
Non-ischemic etiology		CVP/PCWP>0.64	RV SAX/LAX>0.6
Redo-sternotomies		Intra-op CVP >18	
Preoperative inotropes		Intra-op low PAP	
Mechanical ventilation		PAPI<2	
Pre-op MCS			

physiological conditions [11]. Therefore, identifying patients who are at risk of postoperative RHF, timely recognition of failing right heart and implementing appropriate support to improve the ventricular performance are important to stall the relentless progression of RHF.

Risk Stratification for Postoperative RVF

Prediction of RVF in patients undergoing LVAD implantation is not well defined considering the complex interplay of the various factors involved. It is important to consider the clinical, hemodynamic, imaging, and laboratory factors in the risk stratification of these patients (Tables 11.5 and 11.6).

Several scoring systems [12–14] have been designed to address this issue of preoperative risk stratification of RHF. Practically, it is difficult to apply these scores in day-to-day practice because of:

1. Change in device type over the years
2. Single institution derivation
3. Lack of uniform definition of RVF

Many of these risk scores perform only modestly when applied outside the validation cohort. Soliman et al. [15] designed the EUROMACS right-sided heart failure risk score in continuous flow devices to predict right sided failure. It is derived and validated in comparably larger cohort of contemporary patient population and has an advantage of being simple to apply. The prediction rate of right heart failure was similar in both prediction and validation cohorts with significant an increase noted from 12.5% for a score of 0 to 2 to a 42.4% for a score of >4.

Table 11.6 EUROMACS risk score for right heart failure: preoperative risk model

Parameter	Points
Severe RV dysfunction	2
Ratio of RA/PCWP ≥0.54	2
Advanced INTERMACS class 1 through 3	2
Need for ≥3 intravenous inotropes	2.5
Hemoglobin ≤10 g/dL	1
Low risk	0–2
Intermediate risk	2.5–4
High risk	>4

Risk stratification of these patients should assist the physician to identify highest risk patients and apply aggressive treatment strategies early to improve the outcomes.

Intraoperative Management

All patients with late stages of heart failure are at high risk of hemodynamic instability with changes in the loading conditions that are commonly observed while undergoing an invasive procedure. Therefore, large bore intravenous access for fluid administration and arterial line for invasive blood pressure monitoring must be placed before induction. Oftentimes patients are critically sick and optimized in the intensive care unit with inotropes and vasodilator infusions before surgery. Preexisting inotropic medication infusions should be continued to avoid hemodynamic collapse. However, infusion dose of vasodilators should be decreased or stopped to avoid deleterious hypotension with the induction medications. Occasionally temporary ventricular assist devices are placed to optimize the circulatory state as a bridge to stabilize the patient before LVAD implantation. Transportation of such patients should involve personnel who are trained in operating these devices and requires appropriate planning to avoid accidental decannulation or disconnection.

Induction

Anesthetic induction can be associated with negative inotropic effect of anesthetic agents, associated vasodilation, decrease in sympathetic tone, and positive pressure ventilation. This can result in further decrease in cardiac performance and should be monitored. Low cardiac output can result in longer circulation time and can result in longer time to the effect of medications administered. All the available induction agents have negative inotropic effect and should be titrated to effect. Large doses of opioids should be avoided in opioid naïve patients which may decrease heart rate and avoid blunting sympathetic tone. Adjuncts such as magnesium can be used for its salient features such as vasodilation and decrease in dose of induction agents required.

Despite abovementioned measures, hypotension is common at induction due to vasodilation and loss of sympathetic tone. Patients are also intravascular volume depleted due to aggressive diuresis, fluid restriction, and being NPO for surgery. Abrupt transition from spontaneous to positive pressure ventilation can lead to significant preload reduction. Inhalation induction if not contraindicated can be used for gradual transition to positive pressure ventilation. Attention should be paid to avoid hypoventilation and resulting hypercarbia. If required, intravenous fluids should be judiciously administered to maintain preload which is rarely required.

Many patients should be considered full stomach in the setting of significant ascites, and modified rapid sequence induction is preferred in this scenario. Sternotomy is the standard approach for LVAD placement; however, left thoracotomy can be sometimes used for device placement, and lung isolation might be required for adequate exposure.

Due to risk of postoperative bleeding, additional large bore peripheral intravenous access and central venous access are obtained after induction. Even though there is lack of evidence of time-based change of invasive lines, it is judicious to change long-standing invasive lines prior to device implantation. It is also our practice to replace any central lines and swans older than 7 days to avoid potential contamination of the newly implanted pump.

Swan Ganz catheters are routinely placed as it provides valuable information regarding the PA pressure and LV filling pressures. TEE guidance might be required for Swan Ganz catheter placement as RV dilation and tricuspid regurgitation can make the placement challenging.

Intraoperative Echocardiography

Pre-cardiopulmonary Bypass Assessment

The American Society of Echocardiography (ASE) has published recommendations for echocardiographic assessment for an LVAD [16] including intraoperative imaging during implantation, and an exam should be performed by either an experienced cardiologist or cardiac anesthesiologist with advanced perioperative TEE expertise. After induction of anesthesia and placement of TEE probe, detailed comprehensive TEE exam should be performed to confirm preoperative findings and any missed diagnosis and identify any additional new pathologies or any interval change in right heart function from last ECHO exam and surgery that might alter the surgical or treatment plan. The key points in the TEE examination during LVAD implantation are highlighted (Table 11.7).

Right ventricular dysfunction is a well-known complication after LVAD placement and therefore evaluation of RV function in of paramount importance during LVAD placement. TEE evaluation is going to help in risk assessment and instituting appropriate measures to support right ventricular function. Right ventricle must be evaluated thoroughly using multiple TEE views. Qualitative assessment, ventricular shape, fractional area change, TAPSE, right ventricle tissue Doppler velocity, and RV strain are used to access RV function. RV tissue Doppler velocities

Table 11.7 Key points in the transesophageal echocardiographic (TEE) examination during left ventricular assist device (LVAD) implantation

LVAD-specific TEE examination
Cardiac chambers:
1. Right ventricular function
2. Indwelling thrombus
3. Shunts – patent foramen ovale (PFO), atrial septal defect or ventricular septal defect
Valvular function
1. More than trivial aortic insufficiency
2. More than mild mitral stenosis
3. More than moderate mitral insufficiency
4. More than mild tricuspid regurgitation
5. Any pulmonary regurgitation if RVAD is planned
Aorta
1. Mobile atheroma in ascending aorta and arch

<10 cm/s and RV free wall strain <−14% are sensitive predictors of post-LVAD implant RV failure [15]. RV free wall longitudinal strain −9.0 or less is associated with increased risk of RV failure after LVAD placement [16]. With the available evidence and the constraints in the measurement, it is difficult to come up with any cut-off values for RV strain and tissue Doppler measurements for accurate prediction of post-op RHF.

Hemodynamic Factors Predicting Post-implant Right Heart Failure

Many preoperative hemodynamic factors such as CVP, PA pressures, ratio of CVP/PCWP, right ventricular stroke volume index (RVSWI), and pulmonary artery pulsatility index (PAPi) have been proposed to assess and predict post-LVAD severe right heart failure. All these parameters offer an insight into preexisting right ventricular function and a spot measurement in the course of heart failure treatment (Table 11.4). Apart from contractility and loading conditions, ventriculo-arterial coupling is well recognized as an important factor determining the right heart output under adversely changing working conditions. These factors are particularly helpful in titrating treatment to assess the functional improvement to optimize the patient preoperatively. There is variability in the performance of these factors as predictor of post-implant right heart failure in the current literature. Therefore, these hemodynamic factors should be used with other clinical and imaging parameters to identify the patients at risk of RVF and stratify the posed risk.

Pulmonary Artery Pressure Index (PAPi)

The pulmonary artery pulsatility index (PAPi) has evolved as a useful dimensionless parameter in identifying and monitoring the right ventricle at risk. PAPi [(systolic pulmonary artery pressure – diastolic pulmonary artery pressure)/central venous pressure] is a hemodynamic index that predicts severe RVF [17]. A PAPi <2 was associated with increased odds of right ventricular failure following LVAD placement. The clinical value of PAPi is further strengthened by its independence from stroke volume, which avoids the need for accurate measures of cardiac output.

PAPi represents a pure right-sided heart measurement that gauges RV systolic effectiveness and is hemodynamically more removed from the influence of the left side of the heart. The authors noted PAPi to represent a better predictor of RVF after LVAD patients in patients receiving inotropes and that the CVP/PCWP was less predictive in patients receiving inotropes. This is attributed to the score being not affected by the decrease in LA pressure observed secondary to improved left ventricular performance by inotropes. This response is of practical significance in the operating room to gauge the right ventricular reserves. An increase in PAPi to a value above 2 with inotropic support pre-CPB may be suggestive of adaptive right ventricle with adequate reserves and can assist in making therapeutic decisions to support the right ventricular function (Table 11.5).

A comprehensive valvular examination on TEE is necessary to identify any coexisting lesions requiring intervention during the device implantation surgery. Preoperative aortic insufficiency needs to be graded for severity using standard available scoring system. Aortic valve needs to be repaired or replaced if there is moderate or severe aortic regurgitation irrespective of device implantation indication. This is to prevent ineffective circulation from aortic insufficiency as the blood returns to the pump. This closed loop circulation results in inadequate LV decompression and increase wall stress. The aortic valve should be evaluated in multiple views mid-esophageal long axis, mid-esophageal aortic valve short axis, and the deep trans-gastric views. The degree of AI should be assessed using new ASE guidelines with special emphasis on vena contracta, ratio of regurgitant jet width/LVOT width and area, and diastolic flow reversal in descending aorta. However, despite best efforts, AI might be underestimated as increased LV diastolic pressure due to heart failure and decreased systemic vascular resistance, and low mean arterial pressure under general anesthesia will lead to a decrease in regurgitant volume. If the aortic valve needs replacement, then bioprosthetic valves are used in favor to mechanical valves secondary to thromboembolic implications. Any preexisting mechanical valves in any position will need replacement with a bioprosthetic valve at the time of LVAD implantation. Continuous flow devices (CFD) with decreased flow across the AV can result in structural changes in AV resulting in new onset or worsening of preexisting aortic insufficiency [18]. The prevalence of aortic insufficiency increases with the duration of support and is up to 10% after 1 year to 50% after 18 months of circulatory support [19, 20]. It is notable that the progression to significant regurgitation is higher in patients with AR prior to implantation [19, 21]. Therefore, in cases where duration of therapy with CFDs is longer as in destination therapy, mild AR or more should be repaired to improve the efficacy of the therapy. Other patient factors that might contribute to the decision-making include persistently closed AV, short body surface area, higher pump speed resulting in excessive unloading, old age, systemic hypertension, and increased aortic size [19, 21, 22].

Any direct communication between right and left side of the heart can result in right to left shunt on activating the device and should be identified prepump. Patent foramen ovale (PFO) being the most common cause of intracardiac shunt, interatrial septum should be closely interrogated for the patency of foramen ovale. Due to elevated left sided pressures, flow through the PFO may not be easy to identify. To improve the odds of detecting a PFO, modified Valsalva technique could be used along with agitated saline. However, it might not be possible to identify a PFO despite best efforts until the left side pressure decreases and should be looked at for immediately post CPB. Any detected PFO must be surgically closed during implantation.

Low blood flow and arrhythmias associated with heart failure predispose for intracardiac thrombus. Although thrombus could be formed in any cardiac chamber, left atrial appendage, and LV apex are especially susceptible for clot formation. Systemic embolization of an undetected thrombus would have adverse consequences. In patients on temporary circulatory support, thrombi are oftentimes formed within and around cannulas due to stasis and thrombogenic surfaces; however, visualization of thrombus could be extremely challenging due to ECHO artifact. Cardiac chambers should be examined in multiple views both before and after removal of the device or cannula.

Tricuspid regurgitation in heart failure is often secondary to annular dilatation and intracardiac leads across the valve. Unlike mitral regurgitation, tricuspid regurgitation does not improve after left ventricular unloading with LVAD implantation. Patients with preexisting TR tends to require longer inotropic support, longer hospital stays, and increased mortality after LVAD placement [17]. TR should be quantified preoperatively, and any moderate to severe TR should be considered for repair during LVAD placement [18]. Tricuspid valve procedures were also shown to be associated with early postoperative decrease in end-systolic and end-diastolic RV area in patients receiving continuous flow LVAD [23]. These positive results of tricuspid procedures on regurgitation and RV function have not consistently resulted in improved RV function or survival [18, 24].

It is common for patients to have MR due to left ventricular dilation and tethering. MR secondary to ventricular tethering tend to improve after LV unloading with device placement. There is no consensus on surgical mitral valve repair for significant MR during LVAD implantation. In a multivariate analysis of the INTERMACS database, authors found that concomitant mitral valve procedures did not affect the early or late survival after continuous flow LVAD implantation [23]. Authors also noted a trend toward improved survival in patients receiving concomitant MVPs. The visual analogue score (VAS) for quality of life was significantly better and also the probability of readmission decreased with MV procedures. The findings are suggestive of the fact that in selective patients, concomitant MVP can have improved outcomes than isolated LVAD implantation. It remains to be determined the characteristics of these selective patients in whom surgical mitral valve intervention can be beneficial. Primary mitral valve pathology resulting in moderate to severe regurgitation should be identified and repaired. Mitral valve should be assessed for mitral stenosis as moderate/severe MS would limit the blood flow to the device. If detected, the mitral valve needs to be replaced during LVAD implantation. With mitral clip being increasingly used for treating mitral regurgitation, the mitral valve has to be evaluated for resulting stenosis. In selective cases, it is feasible to explant the clip without damaging the leaflets through the apical ventriculotomy site. Similarly, the pulmonary valve should be evaluated for evidence of significant PI, which can interfere with RVAD function if planned. Additionally, any prosthetic valve should be evaluated for structural degeneration.

The aorta needs to be scanned for significant atherosclerotic plaques in the ascending aorta and arch, which result in changes in cannulation and outflow graft placement strategies. TEE is limited in ability in scanning ascending aorta, and Epi aortic scanning can be used in case of non-confirmatory findings.

In minimally invasive device placement, left thoracotomy approach is used to place the inflow cannula. In a remodeled failing heart, the normal relationship between intercostal space and cardiac apex is expected to change. Transthoracic echocardiography can be used to identify the intercostal space and the favorable incision site on the chest to access the apex of the heart. TEE is also useful in identifying the apical coring site and the direction by introducing an echogenic needle in the apex and identifying the position of the needle in the ventricle in relation to the ventricular walls (Video 11.1, Fig. 11.1).

Separation from CPB

Intraoperative echocardiography is used while separating from CPB to ascertain the proper device placement and assist in weaning from CPB according to the interaction between the native heart and the device. The rotational speed is increased gradually with volume loading the heart. Left ventricular size should decrease, while the interventricular septum remains in the neutral position without deviating into either ventricle. The right ventricle should be monitored for any dilation and change in shape while increasing the intravascular volume. Right ventricular dysfunction is common during weaning, and stroke volume response to preload augmentation may not be normal. Deviation of the IVS to the left can be the sign of high relative pump speed for the loading conditions or worsening RV function. With IVS shifting to left, there should be no further increase in pump speed until loading conditions are changed by volume administration as tolerated or gradual increase in afterload in cases of vasodilation.

RV failure should be aggressively treated with inotropes, selective pulmonary vasodilators and avoiding persistent vasodilation. In resistant cases, RVAD should be considered early before chest closure. In severe RV failure on high pharmacological support, patients should be closely monitored for any further deterioration dictating the need for mechanical support. Interatrial septum should be rechecked for any patent foramen ovale that is evident after the decrease in the left-sided pressures.

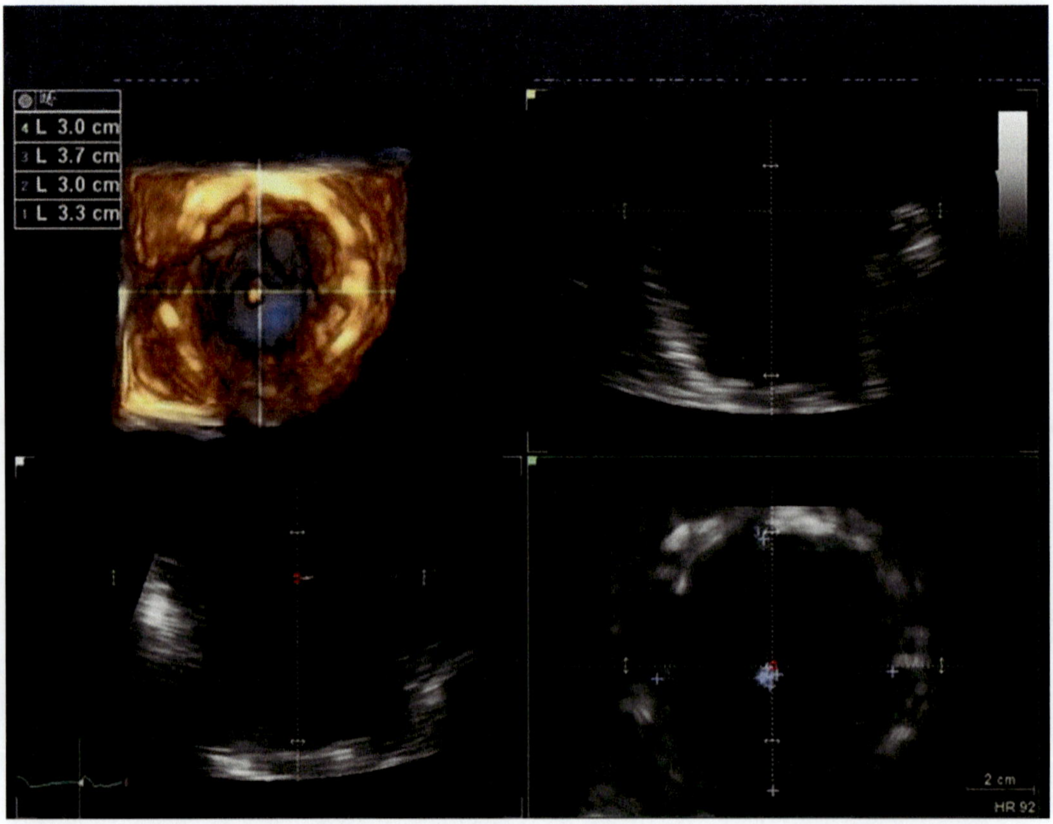

Fig. 11.1 TEE spatial and directional guidance for apical coring during inflow cannula implantation

Inflow and Outflow Cannula Interrogation

The Inflow Cannula

The inflow cannula is inserted in the LV apex with the goal to be coaxial with the mitral inflow. Biplane imaging in orthogonal axis, with the current generation of 3D probes, can identify any malalignment of the cannula (Videos 11.2 and 11.3). Additionally enface 3D view of the inflow cannula can easily provide both spatial and directional information in relation to the ventricular walls (Video 11.4). Ideally, the inflow cannula has to be equidistant from the LV walls at normal operational speed; color Doppler of the inflow cannula should show laminar flow from the mitral valve to the cannula with no evidence of flow acceleration or regurgitation [25]. It is not uncommon to observe brief regurgitation at lower pump speeds and in some pumps which has speed modulation program to induce flow variation (Fig. 11.2).

Spectral Doppler imaging should show a blood velocity in the inflow cannula <1.5 m/s, and the flow is pulsatile depending on the contractile state of the LV.

The centrifugal pump inflow cannula may be hard to interrogate using spectral Doppler due to the artifact. This artifact, previously described as the waterfall artifact [26], is responsible for the inability to accurately measure velocities due to signal interference imposed by the pump itself. The effect of this interference can be minimized by increasing the travel time of the Doppler signal by increasing the color Doppler sector size and also by decreasing the aliasing velocity. (Videos 11.5 and 11.6) (Fig. 11.3).

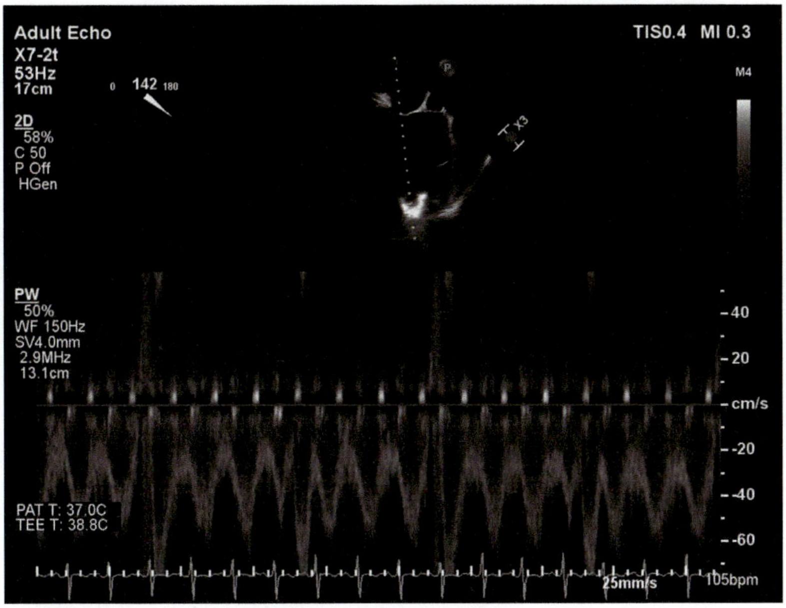

Fig. 11.2 A brief reversal of flow observed during the programmed decrease in speed in Heartmate III

Outflow Cannula

The outflow cannula can be visualized in the mid-esophageal long axis view of the ascending aorta.

To measure flow velocity in the outflow graft, the PW sample volume should be 1 cm proximal to the aortic anastomosis. The peak velocity in the outflow graft in axial flow pumps usually ranges from 1.0 to 2.0 m/s. In the event of outflow cannula obstruction, color Doppler shows high velocity flow with acceleration in the ascending aorta with a velocity that is greater than 2 m/s. We do not have any normative values for the cannula velocities in modern centrifugal pumps (Figs. 11.4 and 11.5).

Post CPB and ICU Management

Suction Event

Suction events can be common during immediate post-CPB phase in the presence of predisposing factors. These factors are high relative pump rotational speed, ongoing bleeding, RV failure, inappropriate vasodilation, and repeated surgical manipulation of the heart. The suction event is a progressive phenomenon (Fig. 11.6) and rate of progression can vary depending on the severity of precipitating factors. In a severe suction event, the left ventricular cavity is obliterated and mitral valve is stented open (Video 11.7). It is important to identify the event in the early stages which usually responds to decrease in pump speed and reversing the precipitating factors (Table 11.8). Recovery from a suction event depends on how well the right ventricle is performing under the adverse conditions. Severe recalcitrant events may necessitate transient cardiopulmonary extracorporeal support. The majority of the suction events results in escalation of pharmacological support for RV function.

Right Ventricular Failure

Acute right ventricular RV failure after LVAD insertion occurs in approximately 10%–30% of patients [8] and is associated with a high mortality. The cardinal features are low LVAD flows in the setting of elevated CVP and a decompressed LV. In patients requiring LVAD support, unloading of the LV and increasing forward flow can have adverse consequences on RV function:

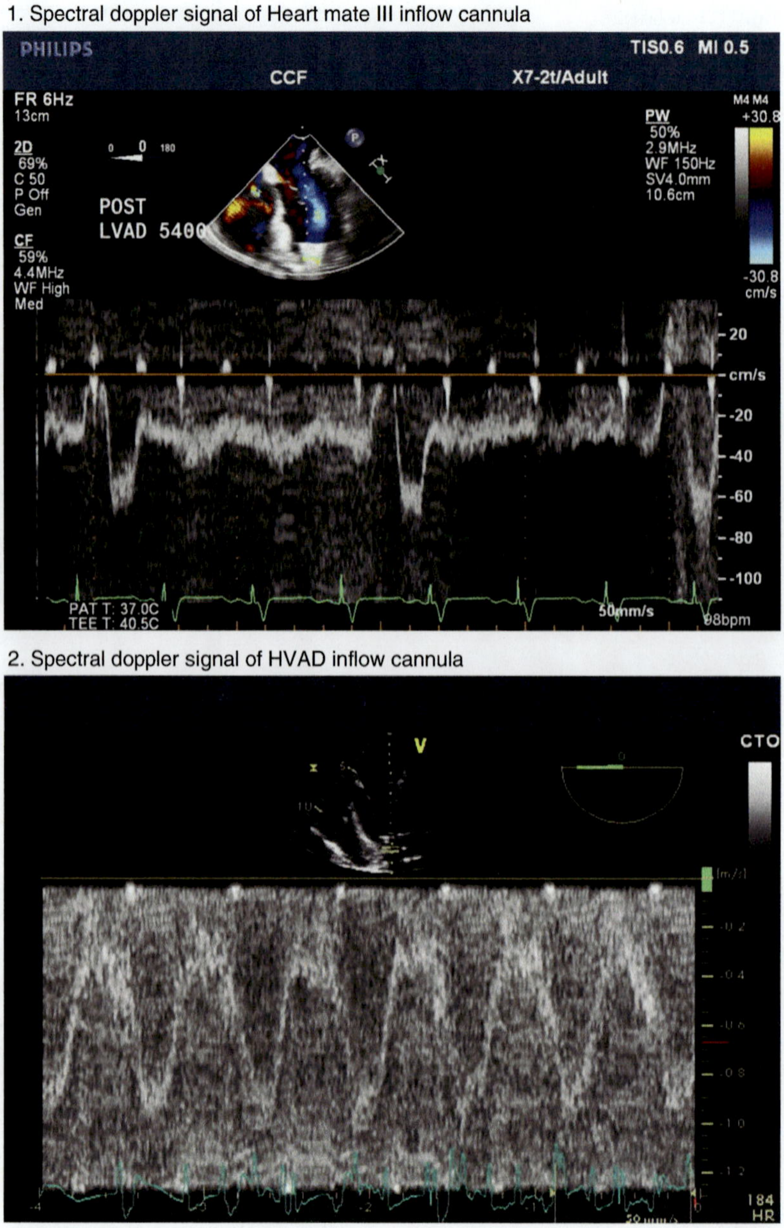

Fig. 11.3 Spectral Doppler signal of the inflow cannula in (1) Heartmate III and (2) HVAD with velocities under 1.5 m/s

1. The dysfunctional RV may not be able to handle the increased load from blood return generated by the newly placed LVAD to the right heart.
2. Excessive decompression of the left ventricle with associated leftward shift of the interventricular septum alters RV size and shape. This can cause impairment of effective RV contraction and may also decrease the significant contribution of the septal contraction to RV output. In addition, septal shift may worsen tricuspid valve regurgitation.

Other factors which may contribute to postoperative RV failure include:

- Suboptimal myocardial protection during surgery, particularly if the right coronary artery (RCA) territory has been grafted.
- Injury of the LAD from the inflow cannula insertion in patients with large wrap around LAD.
- Long CPB time.
- RV myocardial ischemia caused by coronary air or thrombotic embolism.

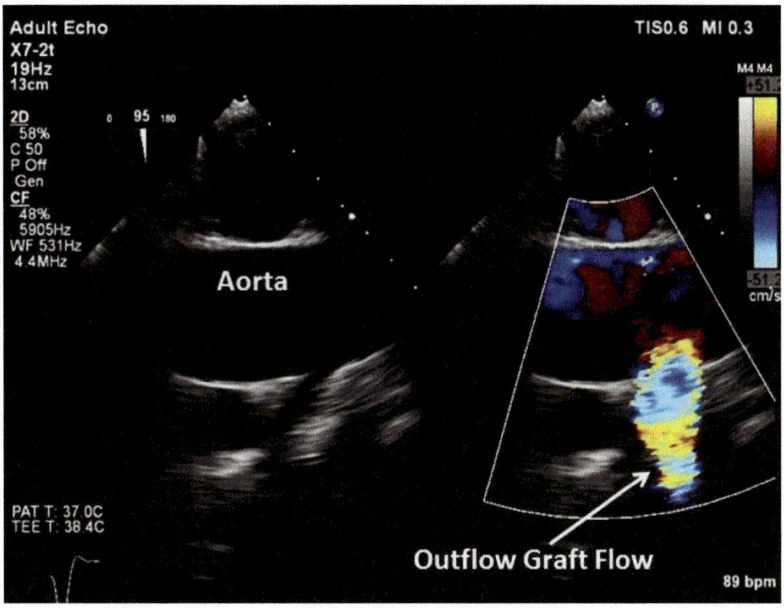

Fig. 11.4 Ascending aorta long axis view showing color flow Doppler across the outflow cannula

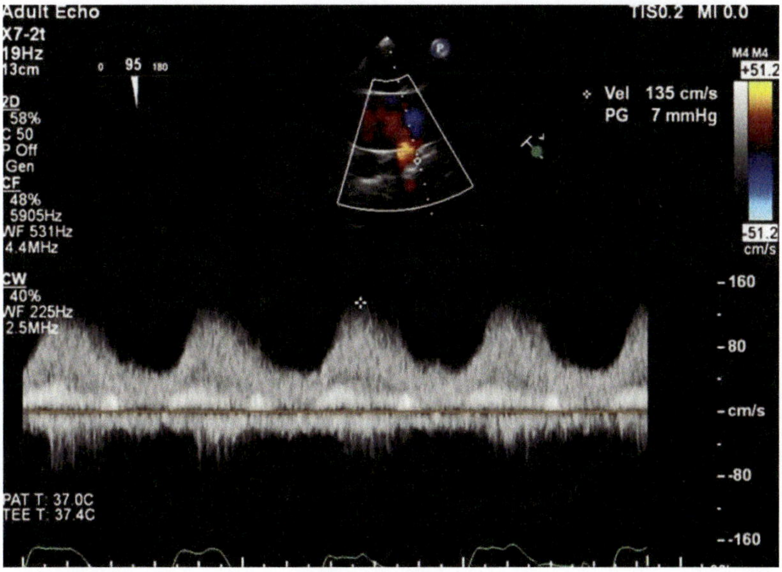

Fig. 11.5 CW Doppler across the outflow cannula depicting velocities under 1.5 m/s

- Atrial arrhythmias or loss of atrioventricular synchrony.
- Hypotension secondary to vasodilation is not well tolerated by RV.
- Worsening pulmonary hypertension due to:

1. The inflammatory response to CPB
2. Excessive administration of blood and blood products
3. Protamine reaction
4. Suboptimal milieu related to hypoxia, hypercarbia, acidosis, high airway pressures, and presence of large pleural effusions

Management of RVF in the LVAD patient relies on preoperative risk stratification and planning the appropriate circulatory support (Table 11.9):

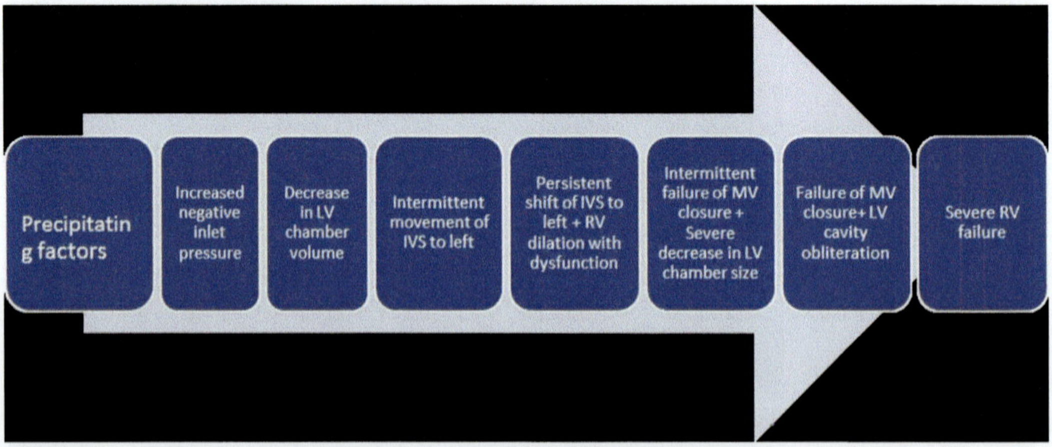

Fig. 11.6 Progression of a suction event-disproportionate unloading of the left ventricle to the given preload

Table 11.8 Treatment of a suction event

Decrease in pump speed by at least 200 rpm in early stages or lowest allowable in severe cases
Treat the precipitating factors
Consider increasing right ventricular support to improve RV output
Guide volume administration to RV shape and function
ACLS if PA pressures are non-pulsatile
Extracorporeal support in case of persistent RV failure

- Maintaining sinus rhythm and contractility
- Optimizing preload
- Maintaining adequate RV perfusion pressure (MAP)
- Avoiding factors leading to change in ventricular shape
- Avoiding sudden changes in right ventricular afterload

Pharmacological Support

During the weaning from CPB, pharmacologic inotropic support is the first line of treatment (Table 11.10) after ensuring that all the physiological variables that affect the RV function are optimized. Epinephrine is a potent beta stimulant and is used before starting inodilators. However, it can be arrhythmogenic due to its beta agonist properties. The milrinone and/or dobutamine provides effective inotropic support while also effectively reducing PVR. Milrinone is less arrhythmogenic and has a salutary effect of pulmonary vasodilation in patients with pulmonary hypertension. Patient has to be closely monitored for hypotension with its systemic vasodilatory effect and longer half-life. It is not uncommon to start vasopressors such as norepinephrine or vasopressin to counter the variable systemic vasodilation that results from the inodilators.

In situations requiring further reductions in PVR to improve right heart performance, selective pulmonary vasodilators such as inhaled nitric oxide or inhaled epoprostenol are likely to be the most effective therapy. The avid binding of nitric oxide by hemoglobin permits pulmonary vasodilation without negative inotropic effects or dilation of the systemic vasculature. The expected hemodynamic responses in patients that benefit from this therapy include an increase in LVAD flow and a reduction in central venous pressure (CVP). Pulmonary artery (PA) pressures may however decrease [27, 28] or remain unchanged because of improved transpulmonary flow. The timing of starting the selective pulmonary vasodilators in the operating room is controversial, and there is some early evidence suggesting better response with drug initiation before CPB in comparison to its initiation after termination of bypass [27]. Association of epoprostenol with increased bleeding is certainly of concern in these surgeries [27].There is a need for further evidence to identify the effect of induced platelet dysfunction on transfusion requirements.

Table 11.9 Stratifying RV support based on preoperative RV function

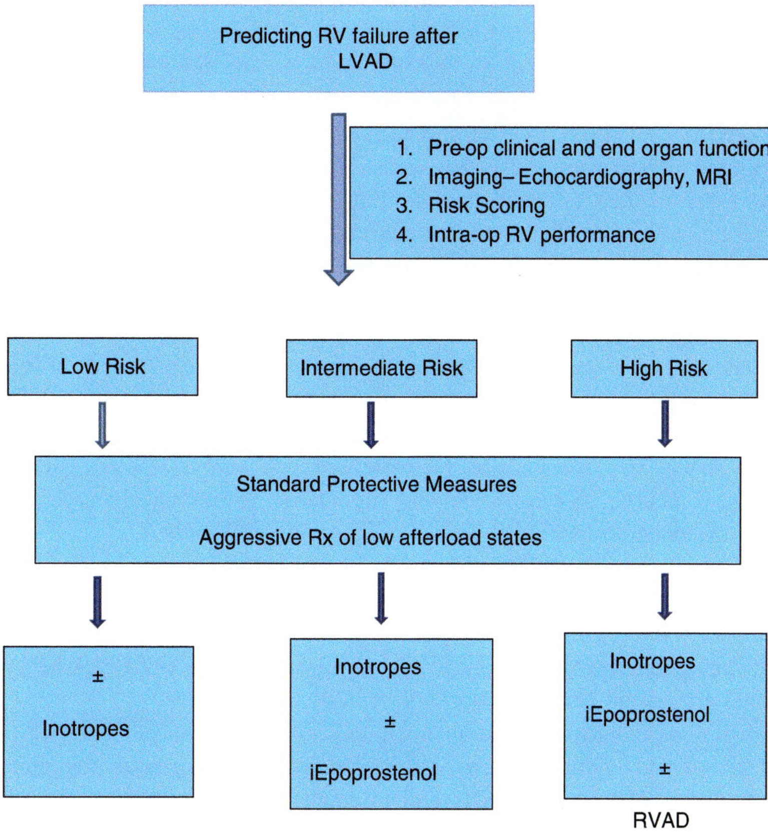

Table 11.10 Common medications used in pharmacological support of failing right heart

Medication	Pharmacological Action	Dosing range
Epinephrine	β agonist, increase contractility	0.05–0.15mcg/kg/min
Milrinone	PDE inhibitor, increase contractility, decrease afterload	0.175–0.5mcg/kg/min
Norepinephrine	α and β agonist, increase contractility and vasoconstriction	0.05–0.15mcg/kg/min
Vasopressin	Activates V1 receptors, vasoconstriction	0.02–0.1mcg/min
Dobutamine	β agonist, increase contractility	5–15mcg/kg/min
Inhaled Epoprostenol	Activates adenylate cyclase, pulmonary vasodilation	0.01–0.05 mcg/kg/min
Inhaled nitric oxide	Activates Guanylyl cyclase, pulmonary vasodilation	4–40 ppm

Fluid Management

Avoid volume overload in pre-implant state: CVP should be maintained <15 mm Hg prior to surgery.

Post-implant intravascular volume management: The goal is to optimize the intravascular volume for the given RV function to maintain an adequate device preload to achieve target systemic output. This could be a challenge in post-CPB bleeding scenarios because of continuous blood loss and repeated surgical handling of the heart to contain the bleeding sites. It is a common practice to target a CVP of 10–15 mmHg with chest closed. A low device preload state puts the ventricle at risk of suck down especially in left ventricles with a smaller cavity size. Therefore, it is advisable to run the device at a lower allowable speed to decrease the effect of excessive adverse unloading.

If LVAD flows continue to be low, patient is hemodynamically unstable with evidence of significant RV dysfunction, and requiring high doses of inotropes and vasopressors, mechanical support of the RV is considered. There is no specific device designed and approved for durable RV support. Therefore, mechanical support of RV is to be considered as a temporary measure and planned early institution of right ventricular assist device (RVAD) is favored for better outcomes [29].

If in spite of pharmacological support the cardiac index is <2.0 liters/min/m^2 with the CVP >15 mm Hg and a echocardiographic evidence of severe RV dysfunction, a temporary RVAD should be considered before leaving the operating room [30]. Patients at high risk for further deterioration of RV function before transfer to ICU should be identified, to facilitate further decision making on mechanical RV support in the immediate postoperative period or a high for mechanical support should be identified with any further deterioration of circulatory status. The temporary RVAD should assist in the right ventricular remodeling and acclimatize the dysfunctional RV to the changing working conditions. The RV mechanical support is instituted with extracorporeal continuous flow pump with the cannulation of femoral vessels and pulmonary artery. A graft is sutured on to main pulmonary artery and is exteriorized via a thoracotomy which is cannulated directly. This approach first described by Strauch et al. [31] allows the chest to be closed, and the RVAD can be weaned without the need to open the chest. A dual-lumen RVAD cannula (Protek Duo, Livanova PLC, UK) is designed so that a single catheter can be introduced through right internal jugular vein (RA to PA assist) thus assisting early ambulation. Some centers prefer percutaneously placed catheter, IMPELLA RP (Abiomed Inc., Danvers, USA) from femoral artery with the advantage of being a minimally invasive approach. This catheter placement requires fluoroscopy in addition to the TEE for guiding the placement. This microaxial pump drains the blood from IVC and delivers it to the pulmonary artery. The pump position needs to be closely monitored, and the patient cannot ambulate with the catheter introduced from the femoral vessels.

The RVAD flows are usually maintained between 2 and 3.5 L/min depending on the native RV function. The flows are adjusted by monitoring the right and left filling pressures and echocardiography. A left atrial line is particularly valuable with RVAD for monitoring the device preload. The CVP should be less than 15 mmHg, and right ventricle should appear decompressed on echocardiography. Weaning from RVAD is usually not attempted for 48–72 hours of biventricular support [11]. Recovery of the RV depends on the extent of intrinsic myocardial dysfunction, systemic factors, and LVAD management strategies. The degree of recovery is variable, and it is prudent to wean the mechanical support when the function is sufficient to maintain the systemic output. It is not uncommon to escalate the pharmacological support in order to attain the target RV output.

Utility of Left Atrial Pressure Monitoring (LAP)

The preload to the LV may vary depending on the intravascular volume status, right ventricular function, and pump speed. It is not always possible to make a confident interpretation of clinical condition with PA catheter given the gradient between PA diastolic pressure and LA pressure. An LA line with pressure monitoring is useful in directly measuring device preload and assessing the LV unloading. A rough schematic is depicted in Table 11.11. Monitoring both right and left atrial pressures simultaneously should enable early detection of right heart failure or suction event before hemodynamic compromise.

The LA line is usually inserted via the right upper pulmonary vein or through Sondergaard's groove by the surgeon and the position of which should be confirmed by echocardiography. The LA line might read erroneously high values if the

Table 11.11 Left atrial pressure monitoring

LAP (mm Hg)	Implication
<8	Low preload, early suction event
8–15	Optimal preload delivery
>15	Suggests increased preload, inadequate unloading

tip rests very close to the mitral valve wherein the MR jet causes the catheter to record higher than normal LA pressures.

The LAP is beneficial in circumstances such as:

- Titrate RVAD flows to a LAP (12–15 mm Hg)
- Presence of small LV cavity(High risk of suction event)
- High risk of RV failure with expectant management to guide need for RVAD

Bleeding

It is a very common cause of "takebacks" after LVAD surgeries. Early postoperative bleeding is more commonly multifactorial resulting from long bypass runs and hemostatic dysfunction, preexisting multiorgan dysfunction, preoperative anticoagulation, prior cardiac surgeries with extensive mediastinal dissection, and surgical misadventures. Bleeding into the mediastinum can have dramatic effect on device filling from hypovolemia and compressive effect on the heart. Even a localized pericardial collection can result in adverse hemodynamic effects. Filling pressures tend to increase if there is a compressive effect on the right side, but a localized left sided collection can result in hypotension by decreasing the device filling before increasing the central venous pressure. Traditional hemodynamic sign of pressure equalization may not be seen with the device in situ. A left atrial line (LA) if present may sometimes help in this situation along with prompt bedside echo/TEE to help identify the situation. Prompt re-exploration and decompression of the chest cavity are key in mitigating the instability. High index of suspicion is necessary and early TEE is indicated is transthoracic echocardiographic windows are unsatisfactory. Unidentified significant pericardial collection can result in suction events, and prompt surgical evacuation of the collection is indicated.

Hypovolemia from bleeding also places the patient at risk of suck down effect especially at higher rotational speed. Decreasing the pump speed to the given decrease in preload is an important initial measure before replacing the volume and monitoring for subsequent development of right ventricular failure.

High LVAD Flows, Low Pulsatility, and Hypotension: Vasodilatation, Sepsis

Vasodilation may be a combination of SIRS response or from TRALI with massive transfusion. Low threshold to treat with antibiotics after cultures in the event of a strong suspicion for sepsis induced afterload reduction.

Hypertension

ISHLT recommendations are to keep MAP below 80 mm Hg. Hypertension with high MAP is evident as low flows, high pulsatility on the LVAD console. Occurrence of hypertension is quite common in scenarios where the blood loss has abated, and the patients generate intrinsic sympathetic response during sedation hold and extubation. Vasodilators such as nitroprusside or nitroglycerine may serve as rapidly titratable medications to assess response to vasodilation. Uncontrolled systemic hypertension is associated with increased neurological complications [32].

Arrhythmias

Arrhythmias are fairly common affecting nearly 50% of LVAD populations with ventricular arrhythmias being reported in as many as 22–59% of patients [33]. Most patients with LVAD and low EF have an ICD or CRT-D device that needs to be re-interrogated or turned back on after deactivation during the intra-operative period. Ventricular arrhythmias may be triggered by a "suck down" of the LV cavity especially in states of low preload. This must be recognized and treated with fluids and/or RV support based on the mechanism stated above. Cardioversion may be performed in the routine fashion except placing pads in an area away from the LVAD pump.

References

1. Breathett K, Allen LA, Ambardekar AV. Patient-centered care for left ventricular assist device therapy: current challenges and future directions. Curr Opin Cardiol. 2016;31(3):313–20.
2. Boyle AJ, Ascheim DD, Russo MJ, Kormos RL, John R, Naka Y, et al. Clinical outcomes for continuous-flow left ventricular assist device patients stratified by pre-operative INTERMACS classification. J Heart Lung Transplant. 2011;30(4):402–7.
3. Kirklin JK, Xie R, Cowger J, de By T, Nakatani T, Schueler S, et al. Second annual report from the ISHLT mechanically assisted circulatory support registry. J Heart Lung Transplant. 2018;37(6):685–91.
4. Sims DB, Uriel N, Gonzalez-Costello J, Deng MC, Restaino SW, Farr MA, et al. Human immunodeficiency virus infection and left ventricular assist devices: a case series. J Heart Lung Transplant. 2011;30(9):1060–4.
5. Franz DD, Stedman MR, Myers SL, Naftel DC, Silver SA, Banerjee D, et al. Long-term changes in kidney function after left ventricular assist device implant: an analysis of the STS Intermacs database. J Heart Lung Transplant. 2019;38(4):S89–90.
6. Kilic A, Chen CW, Gaffey AC, Wald JW, Acker MA, Atluri P. Preoperative renal dysfunction does not affect outcomes of left ventricular assist device implantation. J Thoracic Cardiovas Surg. 2018;156(3):1093–101.e1.
7. Ross DW, Stevens GR, Wanchoo R, Majure DT, Jauhar S, Fernandez HA, et al. Left ventricular assist devices and the kidney. Clin J Am Soc Nephrol. 2018;13(2):348–55.
8. Kormos RL, Teuteberg JJ, Pagani FD, Russell SD, John R, Miller LW, et al. Right ventricular failure in patients with the HeartMate II continuous-flow left ventricular assist device: incidence, risk factors, and effect on outcomes. J Thorac Cardiovasc Surg. 2010;139(5):1316–24.
9. Dang NC, Topkara VK, Mercando M, Kay J, Kruger KH, Aboodi MS, et al. Right heart failure after left ventricular assist device implantation in patients with chronic congestive heart failure. J Heart Lung Transplant. 2006;25(1):1–6.
10. Baumwol J, Macdonald PS, Keogh AM, Kotlyar E, Spratt P, Jansz P, et al. Right heart failure and “failure to thrive” after left ventricular assist device: Clinical predictors and outcomes. J Heart Lung Transplant. 2011;30(8):888–95.
11. Dandel M, Hetzer R. Temporary assist device support for the right ventricle: pre-implant and post-implant challenges. Heart Fail Rev. 2018;23(2):157–71.
12. Atluri P, Goldstone AB, Fairman AS, MacArthur JW, Shudo Y, Cohen JE, et al. Predicting right ventricular failure in the modern, continuous flow left ventricular assist device era. Ann Thorac Surg. 2013;96(3):857–63.. discussion 63-4
13. Drakos SG, Janicki L, Horne BD, Kfoury AG, Reid BB, Clayson S, et al. Risk factors predictive of right ventricular failure after left ventricular assist device implantation. Am J Cardiol. 2010;105(7):1030–5.
14. Fitzpatrick JR III, Frederick JR, Hsu VM, Kozin ED, O'Hara ML, Howell E, et al. 97: preoperative parameters at the time of left ventricular assist device placement predict the need for biventricular mechanical support. J Heart Lung Transplant. 2008;27(2):S93–S4.
15. Soliman OII, Akin S, Muslem R, Boersma E, Manintveld OC, Krabatsch T, et al. Derivation and validation of a novel right-sided heart failure model after implantation of continuous flow left ventricular assist devices: the EUROMACS (European registry for patients with mechanical circulatory support) right-sided heart failure risk score. Circulation. 2018;137(9):891–906.
16. Stainback RF, Estep JD, Agler DA, Birks EJ, Bremer M, Hung J, et al. Echocardiography in the Management of Patients with left ventricular assist devices: recommendations from the American Society of Echocardiography. J Am Soc Echocardiogr. 2015;28(8):853–909.
17. Kang G, Ha R, Banerjee D. Pulmonary artery pulsatility index predicts right ventricular failure after left ventricular assist device implantation. J Heart Lung Transplant. 2016;35(1):67–73.
18. John R, Mantz K, Eckman P, Rose A, May-Newman K. Aortic valve pathophysiology during left ventricular assist device support. J Heart Lung Transplant. 2010;29(12):1321–9.
19. Cowger J, Pagani FD, Haft JW, Romano MA, Aaronson KD, Kolias TJ. The development of aortic insufficiency in left ventricular assist device-supported patients. Circ Heart Fail. 2010;3(6):668–74.
20. Cowger J, Rao V, Massey T, Sun B, May-Newman K, Jorde U, et al. Comprehensive review and suggested strategies for the detection and management of aortic insufficiency in patients with a continuous-flow left ventricular assist device. J Heart Lung Transplant. 2015;34(2):149–57.
21. Toda K, Fujita T, Domae K, Shimahara Y, Kobayashi J, Nakatani T. Late aortic insufficiency related to poor prognosis during left ventricular assist device support. Ann Thorac Surg. 2011;92(3):929–34.
22. Patil NP, Mohite PN, Sabashnikov A, Dhar D, Weymann A, Zeriouh M, et al. Does postoperative blood pressure influence development of aortic regurgitation following continuous-flow left ventricular assist device implantation?Dagger. European journal of cardio-thoracic surgery: official journal of the European Association for Cardio-thoracic Surgery. 2016;49(3):788–94.
23. Maltais S, Topilsky Y, Tchantchaleishvili V, McKellar SH, Durham LA, Joyce LD, et al. Surgical treatment of tricuspid valve insufficiency promotes early reverse remodeling in patients with axial-flow left ventricular assist devices. J Thorac Cardiovasc Surg. 2012;143(6):1370–6.

24. Robertson JO, Grau-Sepulveda MV, Okada S, O'Brien SM, Matthew Brennan J, Shah AS, et al. Concomitant tricuspid valve surgery during implantation of continuous-flow left ventricular assist devices: a Society of Thoracic Surgeons database analysis. J Heart Lung Transplant. 2014;33(6):609–17.
25. Ferns J, Dowling R, Bhat G. Evaluation of a patient with left ventricular assist device dysfunction. ASAIO journal (American Society for Artificial Internal Organs: 1992). 2001;47(6):696–8.
26. Shah NR, Cevik C, Hernandez A, Gregoric ID, Frazier OH, Stainback RF. Transthoracic echocardiography of the HeartWare left ventricular assist device. J Card Fail. 2012;18(9):745–8.
27. Groves DS, Blum FE, Huffmyer JL, Kennedy JL, Ahmad HB, Durieux ME, et al. Effects of early inhaled epoprostenol therapy on pulmonary artery pressure and blood loss during LVAD placement. J Cardiothorac Vasc Anesth. 2014;28(3):652–60.
28. Argenziano M, Choudhri AF, Moazami N, Rose EA, Smith CR, Levin HR, et al. Randomized, double-blind trial of inhaled nitric oxide in LVAD recipients with pulmonary hypertension. Ann Thorac Surg. 1998;65(2):340–5.
29. Fitzpatrick JR 3rd, Frederick JR, Hiesinger W, Hsu VM, McCormick RC, Kozin ED, et al. Early planned institution of biventricular mechanical circulatory support results in improved outcomes compared with delayed conversion of a left ventricular assist device to a biventricular assist device. J Thorac Cardiovasc Surg. 2009;137(4):971–7.
30. Deschka H, Holthaus AJ, Sindermann JR, Welp H, Schlarb D, Monsefi N, et al. Can perioperative right ventricular support prevent postoperative right heart failure in patients with biventricular dysfunction undergoing left ventricular assist device implantation? J Cardiothorac Vasc Anesth. 2016;30(3):619–26.
31. Strauch JT, Franke UF, Madershahian N, Wahlers T. Right ventricular assist device implantation--a new transcutaneous approach. Thorac Cardiovasc Surg. 2004;52(6):378–9.
32. Teuteberg JJ, Slaughter MS, Rogers JG, McGee EC, Pagani FD, Gordon R, et al. The HVAD left ventricular assist device: risk factors for neurological events and risk mitigation strategies. JACC Heart Fail. 2015;3(10):818–28.
33. Nakahara S, Chien C, Gelow J, Dalouk K, Henrikson CA, Mudd J, et al. Ventricular arrhythmias after left ventricular assist device. Circ Arrhythm Electrophysiol. 2013;6(3):648–54.

12 Left Ventricular Assist Devices: Management and Surgical Techniques

Robert J. Steffen and Benjamin C. Sun

Successful implantation of a left ventricular assist device requires planning and execution in the presurgical, intraoperative, and postoperative phases of the procedure. This chapter will detail current knowledge about best practices in all phases of the implantation using guideline statements, current research, and operative photographs from our institution.

Preoperative Optimization

Preoperative assessment of coronary artery anatomy, valvular function, presence of a patent foramen ovale or atrial septal defect, optimization of right ventricular function, pulmonary vascular resistance, volume status, nutritional state, and infectious state are all key in operative success.

Bilateral heart catheterizations should be performed on all pre-LVAD patients. Right heart catheterization yields right ventricular preload and afterload. Central venous pressure (CVP) prior to implant should be less than 15 mmHg. This may require admission to the intensive care unit (ICU) for administration of inotropes and diuretics, intra-aortic balloon pump, or hemodialysis. Achieving euvolemia prior to surgery is critical. Pulmonary arterial pressure and its reactivity are also assessed. Mean pulmonary artery pressure near the CVP is concerning for end-stage right heart failure. Pulmonary hypertension may induce right heart failure postoperatively. Either of these will limit the right heart's ability to fill the left heart and yield a postoperative low-flow state.

Left heart catheterization is used to evaluate coronary anatomy and position and patency of previous bypass grafts. Significant lesions of a dominant right coronary artery may cause right heart ischemia and subsequent failure. Although there is no consensus, significant right-sided lesions are often bypassed.

Aortic insufficiency (AI) is poorly tolerated in a decompressed left ventricle and will likely worsen over time. AI greater than mild in severity should be addressed at the time of surgery. As the majority of forward flow goes through the LVAD, aortic stenosis (AS) is generally well tolerated. Severe AS without AI can be addressed at the surgeon's discretion. Mitral regurgitation is almost always functional and is rarely treated. Mitral stenosis (MS), however, limits ventricular filling and moderate MS or greater should be treated as well. Tricuspid regurgitation greater than moderate merits fixing.

Atrial septal defects and patent foramen ovale should be assessed for and, if found, fixed at the time of surgery. LVADs decompress the left side

R. J. Steffen, MD (✉) · B. C. Sun, MD
Minneapolis Heart Institute, Department of Cardiovascular Surgery, Minneapolis, MN, USA
e-mail: Robert.steffen@allina.com; Benjamin.sun@allina.com

© Springer Nature Switzerland AG 2020
J. H. Karimov et al. (eds.), *Mechanical Support for Heart Failure*,
https://doi.org/10.1007/978-3-030-47809-4_12

of the heart, so shunt flow will increase right to left after the device is started.

Valve replacement is almost uniformly performed with tissue valves. The aortic valve in a VAD patient will be relatively static compared with traditional ventricular unloading. If a mechanical prosthesis is present in the aortic position, it should be replaced with a tissue valve. Previously implanted, normally function tissue prosthesis can otherwise be left in place.

Any known infections should be treated prior to implanting an LVAD.

Nutritional status is addressed and treated as indicated.

Cross-sectional imaging with CT or MR is useful for planning in patients with prior cardiac surgery or known aneurysmal disease.

The exit site of the driveline should be discussed with the patient prior to surgery. Patients who are right-handed often prefer to have it on their left for ease with dressing changes.

Operation

The patient is brought to the operating room, and monitoring lines including an arterial line, central line, and Swan-Ganz catheter are placed. A transesophageal echocardiography probe is placed in the esophagus. Antimicrobial prep is used to prep from the chin down to the ankles. Prophylactic intravenous (i.v.) antibiotics with Gram-positive coverage are used, typically a cephalosporin. The nares should be swabbed for MRSA, and, if present, nasal bactroban is administered, and i.v. vancomycin is added to the cephalosporin.

Surgical Access

There are multiple surgical approaches to implant current-generation LVADs. Access is needed to the LV apex and ascending aorta. Traditionally, the operation is performed through a median sternotomy. The LV can also easily be accessed via a left thoracotomy (usually at the 4th interspace). The ascending aorta can be accessed from a right thoracotomy, hemi-sternotomy, or full sternotomy. A potential advantage of a non-sternotomy LVAD implant is that patients subsequently undergoing heart transplant do not require a redo sternotomy.

Some older-generation pumps require the creation of a pocket abutting the diaphragm for the LVAD to sit in. The pump pocket is then created by transecting the left-sided attachments of the chest wall to the diaphragm either with a stapler or electrocautery. Meticulous attention to hemostasis of the pump pocket is key as the patients are often coagulopathic and the muscle edges bleed.

Tunneling of the Driveline

Access to the pericardium is obtained via one of the approaches listed above. The pump is brought into the pericardium, and the driveline is tunneled through the abdominal wall (Fig. 12.1). The majority of the tunneling is performed through the subcutaneous tissue, and the last part of the driveline is brought through the rectus sheath at the edge of the pericardium. Patients with prior abdominal surgery or hernias may require an endoscope to assist with the tunneling to prevent visceral injury. On the externalized portion of the driveline, the entire velour should be under the skin edge to prevent local infection tracking to the device.

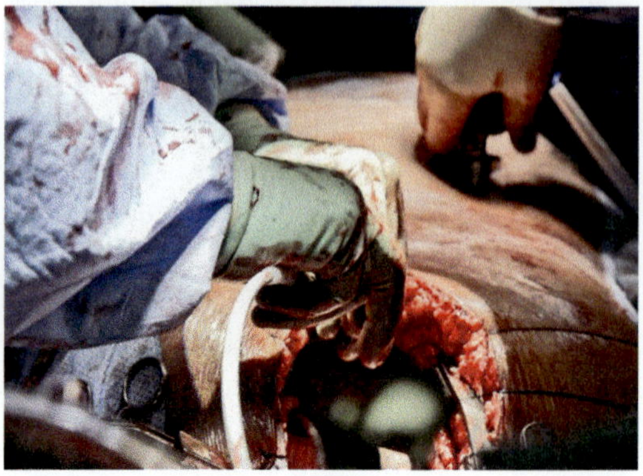

Fig. 12.1 The driveline is brought through the rectus muscle as it exits the pericardium and is tunneled through the subcutaneous tissue to the exit site

Pump Implantation

At this point the pericardium is opened, and access to the left ventricle and ascending aorta has been obtained. Cardiopulmonary bypass with central cannulation and a beating heart is often initiated at this time to support the implant. Replacing the aortic valve or insufficient space on the aorta to place a side-biting clamp would require arresting the heart. Patent foramen ovale, atrial septal defect, or severe tricuspid regurgitation requires bicaval cannulation and opening of the right atrium to repair. Peripheral cannulation is an alternative in patients with large enough vasculature and no aortic calcification. Some centers implant LVADs without the use of cardiopulmonary bypass.

Sewing Cuff

Visual and echocardiographic inspections are both used to determine the site of the sewing cuff. The pump should sit at the apex of the heart, face the mitral valve, and be away from the interventricular septum. With the LV apex in the surgical field of view, pushing on the LV apex in conjunction with continuous echo will show you where the pump will sit on the heart. This is often around the site of the dimple and traditionally 1–2 cm lateral to the left anterior descending artery.

The sewing cuff is then sewn on to the left ventricle (Fig. 12.2). Different techniques have been described, and they vary based on which pump is being implanted. Most include interrupted, plegetted, 2-0 horizontal mattress sutures to hold the cuff in place and a subsequent running prolene suture for apposition of epicardium to the sewing cuff. Ensuring hemostasis of the suture line at this point is key because placing repair sutures becomes more difficult once the pump is in place.

Coring

The apical muscle core is then removed. The technique varies depending on the device implanted which may include a coring device

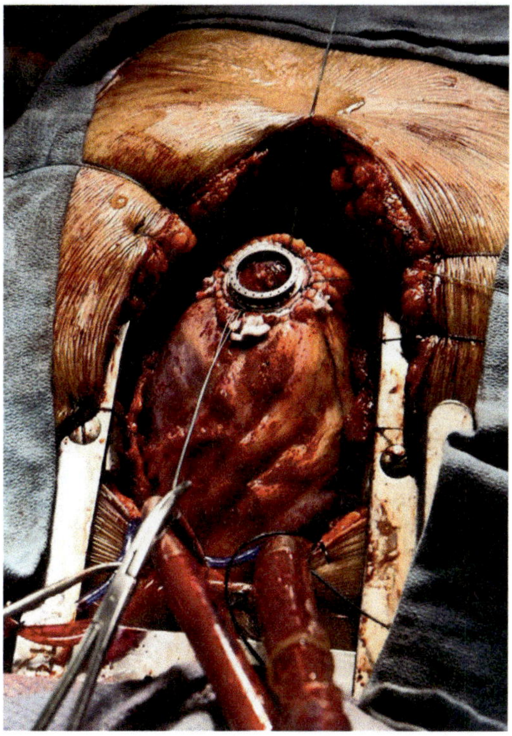

Fig. 12.2 The sewing cuff is sutured to the left ventricular apex with a both plegetted, horizontal mattress sutures and a running, prolene suture

(Fig. 12.3). If not, one method is to make a slit made into the muscle in the center of sewing cuff. A Foley catheter is placed through the slit, into the LV, and the balloon is inflated. The coring tool is placed over the Foley, through the muscle, and into the left ventricle. Assessing the coring tool to ensure the core has been removed and hasn't fallen into the LV is critical in avoiding embolus (Fig. 12.4). The LV should be assessed for thrombus and any found should be removed. Trabeculae that may obstruct the pump inflow can be trimmed as well (Fig. 12.5).

Implanting LVAD

The pump is then placed into the left ventricular apex through the sewing cuff. Each device has its unique way of securing the pump to the sewing cuff (Fig. 12.6). Ensuring the device is securely locked to the sewing cuff is essential.

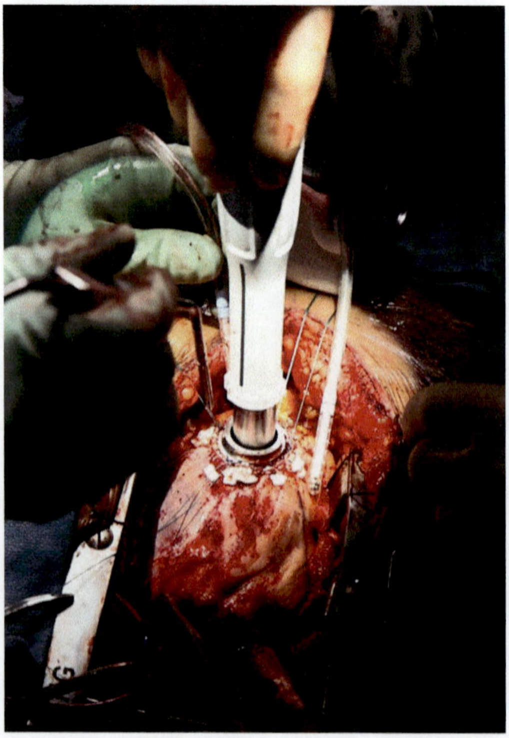

Fig. 12.3 An apical coring device is used to remove the muscular core within the sewing cuff

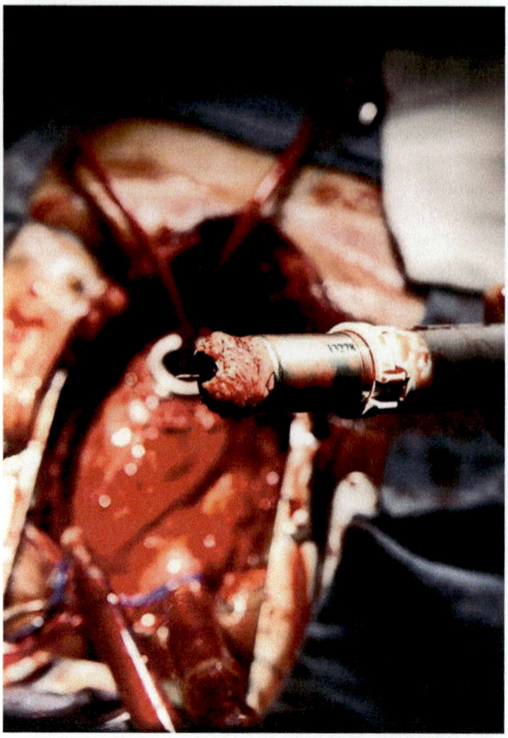

Fig. 12.4 Apical core specimen in the coring tool

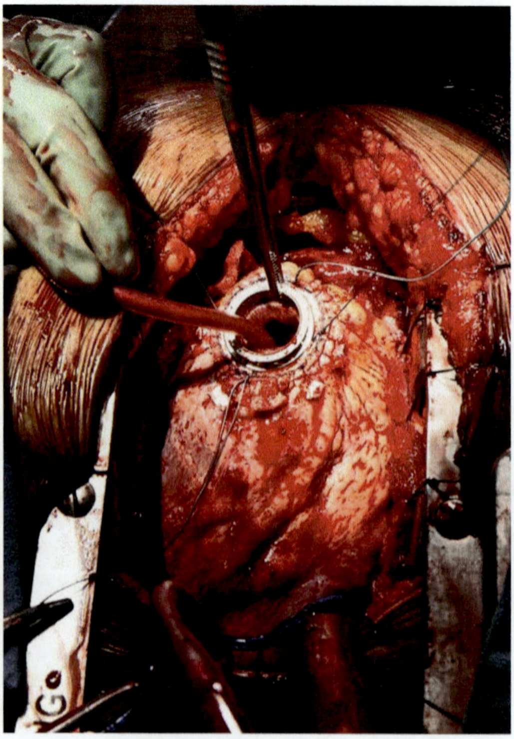

Fig. 12.5 With the apex core, the LV cavity is assessed for thrombus and obstructing trabeculae

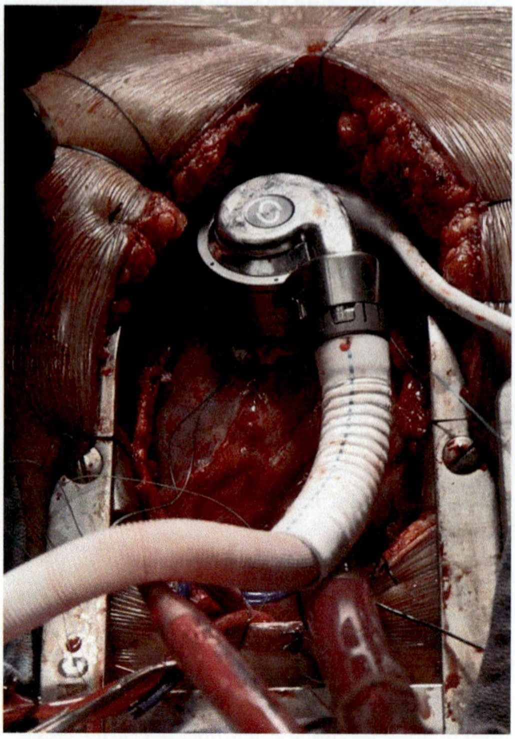

Fig. 12.6 The pump is placed in the LV apex and secured to the sewing cuff

The implanted pump and outflow graft need to be de-aired. This can be done by raising the outflow graft and, if on cardiopulmonary bypass, filling the heart.

Outflow Graft

The outflow graft is then sewn to the ascending aorta (Fig. 12.7). This can be done using a side-biting clamp. The length of the outflow graft should be such that it runs along the diaphragm and up the right side of the heart, not crossing directly underneath the sternum (Fig. 12.8). The LVAD is now implanted and ready to function. The pump and the graft should be meticulously de-aired.

Prior to weaning cardiopulmonary bypass and starting the LVAD, the right ventricle must be ready to support left ventricular filling. Any metabolic abnormality leading to acidosis or hypercarbia should be corrected. Inotropic drips and inhaled pulmonary vasodilators may be necessary in diseased right ventricles or patients with pulmonary hypertension.

Once ready, the patient can be weaned from cardiopulmonary bypass, and the pump started. Continuous assessment of interventricular septal position, right ventricular function, and arterial pressure while increasing pump speed is critical.

Most pumps are afterload sensitive, meaning the pump flow is indirectly proportional to the mean arterial pressure. Fluid shifts and vasoplegia typical in cardiac surgery patients can lead to significant pump flow fluctuations from hypotension. As pump flow increases, the right ventricle is forced to push more blood to the left side to maintain balanced flow. If it can't keep up, the septum moves into the LV cavity and obliterates the lumen. As such, pump speed must be constantly monitored and adjusted dur-

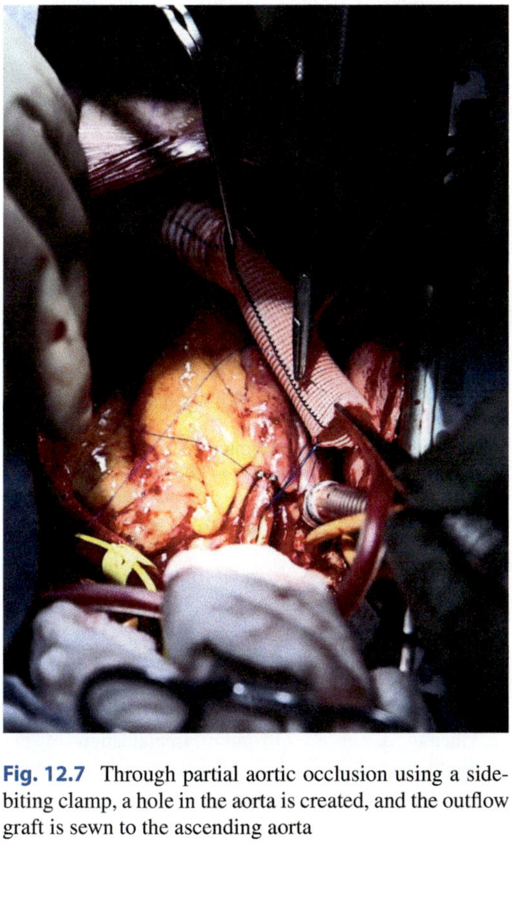

Fig. 12.7 Through partial aortic occlusion using a side-biting clamp, a hole in the aorta is created, and the outflow graft is sewn to the ascending aorta

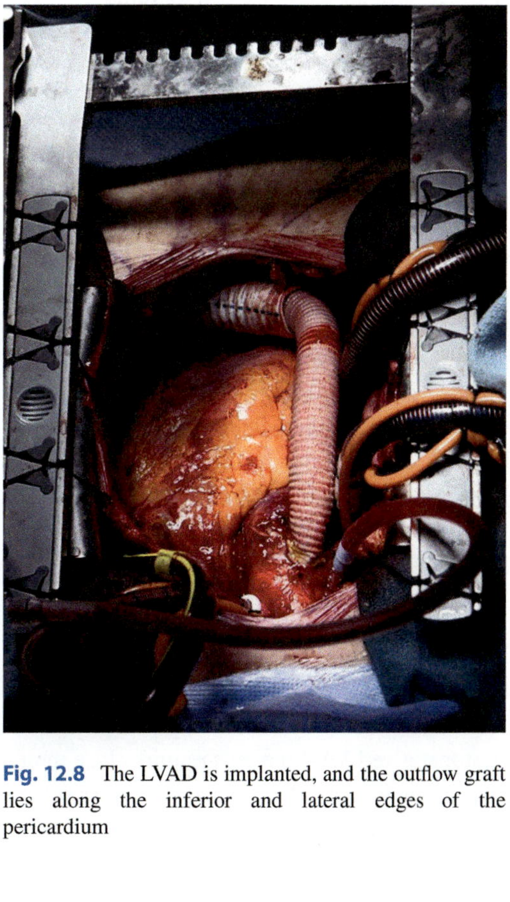

Fig. 12.8 The LVAD is implanted, and the outflow graft lies along the inferior and lateral edges of the pericardium

ing this time to prevent a "suction event." A suction event happens when LV muscle is entrapped into the pump and inhibits flow. Should a suction event occur, the pump should be immediately shut off, the LV allowed to decompress, and the pump slowly restarted. Meticulous attention to right ventricular function during this period is key to preventing RV failure and need of a temporary RVAD.

Temporary RVAD

Preoperative concern for significant RV dysfunction or intraoperative RV failure can be treated by a temporary RVAD. Surgically, a 7 or 8 mm graft is sewn to the distal main PA and tunneled out of the chest. An arterial cannula can be placed through the graft and used as the RVAD outflow. A venous cannula placed through the femoral vein can serve as the inflow. This configuration allows for a slow wean of support, and once the RV recovers, the cannulas can be removed without having to reopen the chest. If there is significant concern for the need of a temporary RVAD, it is easier to sew the graft on the PA prior to implanting the LVAD as clamping the PA can cause a decrease in RV output.

Chest Closure

Prior to leaving the operating room, pump flow and pulsatility index should be stable, the interventricular septum should be midline on echo, and hemostasis should be achieved. In patients being bridged to transplant, some have advocated for leaving a Gore-Tex membrane around the pump to minimize adhesions. The driveline and outflow graft should be kept away from the posterior sternum to avoid injury in the case of reentry.

Postoperative VAD Management

Postoperative LVAD patients require close monitoring immediately after surgery. A chest x-ray in the ICU is performed for routine assessment as well as checking the lie of the VAD. Pump parameters, arterial blood pressure, and volume status are constantly measured and corrected. Drops in the pulsatility index or pump flows merit transesophageal echo assessment, and inability to correct them in the ICU should prompt consideration to return for operative exploration.

References

1. Sun BC, et al. Placement of long-term implantable ventricular assist devices without the use of cardiopulmonary bypass. J Heart Lung Transplant. 2008;27(7):718–21.
2. Awad H, Sun BC, et al. Initial experience with off-pump left ventricular assist device implantation in single center: retrospective analysis. J Card Surg. 2010;5(1):123.
3. Feldman D, et al. The 2013 International Society for Heart and Lung Transplantation guidelines for mechanical circulatory support: executive summary. J Heart Lung Transplant. 2013;32:157–87.
4. Morgan JA, et al. The 2013 International Society for Heart and Lung Transplantation guidelines for mechanical circulatory support: intraoperative and immediate postoperative management. J Heart Lung Transplant. 2013;32:1–21.
5. Danter MR, et al. A prospective, controlled, unblinded, multi-center clinical trial to evaluate the thoracotomy implant technique of the HVAD system in patients with advanced heart failure: results of the LATERAL trial. J Heart Lung Transplant. 2017; 36(4).
6. Popov, et al. HeartWare left ventricular assist device implantation through bilateral anterior thoracotomy. Ann Thorac Surg. 2012;93:674–6.
7. Schmitto JD, et al. Implantation of a centrifugal pump as a left ventricular assist device through a novel, minimized approach: upper hemisternotomy combined with anterolateral thoracotomy. J Thorac Cardiovasc Surg. 2012;143(2):511–3.
8. Sileshi B, et al. In-hospital outcomes of a minimally invasive off-pump left thoracotomy approach using a centrifugal continuous-flow left ventricular assist device. J Heart Lung Transplant. 2015;34(1): 107–12.
9. Pierson RN, et al. Left ventricular assist device implantation via left thoracotomy: alternative to repeat Sternotomy. Ann Thorac Surg. 2002;73(3):997–9.
10. Cheung A, et al. Minimally invasive, off-pump explant of a continuous-flow left ventricular assist device. J Heart Lung Transplant. 2010;29(7):808–10.
11. Cheung A, et al. Off-pump implantation of the HeartWare HVAD left ventricular assist device through minimally invasive incisions. Ann Thorac Surg. 2011;91(4):1294–6.

13 Mechanical Circulatory Support for Biventricular Failure: Patient Selection and Management Options

Kimberly N. Hong, Hao A. Tran, Victor Pretorius, and Eric D. Adler

Abbreviations

BiVAD	Biventricular assist device
CVP	Central venous pressure
EF	Ejection fraction
LVAD	Left ventricular assist device
iNO	Inhaled nitric oxide
INTERMACS	Interagency Registry for Mechanically Assisted Circulatory Support
ISHLT	International Society for Heart and Lung Transplantation
PAPi	Pulmonary artery pulsatility index
PCWP	Pulmonary capillary wedge pressure
PVR	Pulmonary vascular resistance
MAP	Mean arterial pressure
MR	Mitral regurgitation
MVS	Mitral valve surgery
RAP	Right atrial pressure
RHC	Right heart catheterization
RVAD	Right ventricular assist device
RVF	Right ventricular failure
RVSWI	Right ventricular stroke work index
TR	Tricuspid regurgitation

K. N. Hong, MD · H. A. Tran, MD
Department of Cardiology, University of California San Diego, La Jolla, CA, USA
e-mail: knhong@ucsd.edu; hatran@ucsd.edu

V. Pretorius, MD
Department of Cardiothoracic Surgery, University of California San Diego, La Jolla, CA, USA
e-mail: gpretorius@ucsd.edu

E. D. Adler, MD (✉)
University of California San Diego Medical Center, San Diego, CA, USA
e-mail: eradler@ucsd.edu

Introduction

Biventricular failure is a heterogeneous clinical entity that includes primary restrictive and/or dilated cardiomyopathies involving both ventricles as well as right ventricular failure due to chronic left ventricular dysfunction. Regardless of etiology, it is associated with increased morbidity and mortality, and its management depends on the reversibility of right ventricular remodeling after preload, afterload, and contractility are optimized, either medically or by left ventricular support alone. In primary biventricular dysfunction, particularly in non-dilated restrictive cardiomyopathies, anatomical constraints may limit the use of left ventricular assist devices, and instead transplantation or TAH is required. This chapter will focus on the patient selection and management of individuals with pre-existing biventricular failure either from chronic left ventricular failure or dilated cardiomyopathies, as well as those who develop right ventricular failure (RVF)

© Springer Nature Switzerland AG 2020
J. H. Karimov et al. (eds.), *Mechanical Support for Heart Failure*,
https://doi.org/10.1007/978-3-030-47809-4_13

after left ventricular assist devices (LVAD) have been placed [1, 2]. We will discuss methods for risk stratifying individuals with biventricular dysfunction and pre-, peri-, and postoperative management strategies for improving RV recovery and survival to transplant in these patients.

Anatomically, the right ventricle (RV) hugs the left ventricle (LV). Transverse muscle fibers comprise the thinner free wall, and longitudinal fibers interlace in the interventricular septal wall with LV muscle fibers. These muscle fibers in the septal wall reach to the tricuspid annulus and free wall via the crista supraventricularis [3–6]. RV ejection is therefore largely achieved passively via longitudinal contraction of the LV and interventricular septum. Thus, circumstances which compromise the geometric relationship between the LV and RV disproportionately affect RV function. Additionally, under ordinary conditions the pulmonary vascular bed is a highly compliant low-pressure system that can recruit additional vasculature to reduce circuit resistance when changes in loading conditions and oxygenation occur. Because it faces a low-pressure system, unlike the LV, the RV is primarily a volume pump. It has no isovolumic contraction and relaxation phases and is perfused in both systole and diastole. For these reasons, under preserved physiological conditions, less energy is required by the RV to generate an equivalent LV stroke volume, but this output is more sensitive to afterload because perfusion can be compromised. Understanding the normal physiological conditions that govern RV function provides a framework for how pre-existing left-sided heart failure, acute changes in loading conditions, and LVAD support can result in RV failure.

Epidemiology

The incidence of RV failure in individuals with heart failure with reduced ejection fraction is estimated at 48%, with some series reporting rates higher than 60% in dilated nonischemic populations [7]. By unloading the LV and decreasing left atrial pressures, pulmonary arterial compliance is augmented and RV afterload is reduced. Thus, while there is a reasonable chance for RV recovery post-LVAD, depending on degree of RV remodeling and fibrosis, the opposite is also true where RV failure can be unmasked, develop, or worsen because of a mismatch between LV and RV outputs.

Historically, the definition of RV failure post-LVAD implantation has varied [8]. The Interagency Registry for Mechanically Assisted Circulatory Support (INTERMACS) defines RV failure after LVAD as a collection of symptoms (e.g., edema) and signs (e.g., elevated bilirubin) requiring inotropic or mechanical support [9]. Overall rates of RV failure after left ventricular assist device (LVAD) are estimated to range between 20% and 35% [3, 10–12], with ~20% occurring in the early postoperative period (within 30 days of surgery) compared to ~10% at a later time [6, 13, 14]. Although the incident rate for RV failure may vary depending on etiology, its presence is associated with increased mortality and decreased survival to transplant [15–19]. In a cohort of patients undergoing LVAD placement who went on to require unplanned RV support with an extracorporeal RVAD, survival at 3 months was 59% compared to 95% in those who did not require RVAD [20]. Additionally, based on the International Society for Heart and Lung Transplantation (ISHLT) Mechanically Assisted Circulatory Support Registry which combines data from 35 countries, 1-year survival for individuals undergoing biventricular support as their initial treatment strategy was only 53% [17].

Improving outcomes in patients with pre-existing biventricular failure involves distinguishing individuals who will benefit from biventricular support as their initial bridging strategy from those who can be managed with just LVAD alone. Additionally, appropriate hemodynamic and echocardiographic monitoring will be important for managing patients who go on to develop late RV failure after LVAD implant.

Risk Stratification

Risk stratification has been attempted by combining echocardiography (Table 13.1) and invasive hemodynamics (Table 13.2) with other clinical parameters (Table 13.3). Echocardiography is

Table 13.1 Echocardiographic parameters used to evaluate right ventricular failure

Parameter	Details	Threshold	Positive evidence	Negative evidence	Limitations [21]
TAPSE	4-chamber view M-mode of lateral annulus	<7.5 mm	23–25	26–34	Transducer angle dependent Affected by prior CTS
RV E/E’	Pulsed Doppler of tricuspid inflow/TDI of lateral annulus Obtain at end expiration	>10	23		Transducer angle dependent Affected by prior CTS and TR
RV strain		−9.6 to −14%	23,25, 28, 35–37	30	Platform dependent
FAC	4-chamber view Trabeculations should be excluded	<31%	23, 32	24, 29, 30, 35, 38	Asymmetric RV shape
S′	TDI of basal segment of lateral annulus Must be aligned with Doppler	<4.4	38	23, 28, 29	Transducer angle dependent Affected by prior CTS
Sphericity index	4-chamber view Ratio between RV mid-ventricular and longitudinal measurements	>0.6 or increase over 1 year	38, 39		Asymmetric RV shape
3D RV strain		>−10.1	26		Asymmetric RV shape
3D RVEF			26		
TR		Grade III/IV	39	30	Load dependent
RVEDD	4-chamber Basal or mid-ventricular	>50 mm	30		Asymmetric RV shape
RV:LV	Mid-ventricular short axis during systole Ratio of RV diameter to LV diameter	>0.75	29, 40	31, 32, 34, 35	

CTS cardiothoracic surgery, *EF* ejection fraction, *FAC* fractional area change, *LV* left ventricle, *RVEDD* right ventricular end diastolic dimension, *RV* right ventricle, *TAPSE* tricuspid annular plane systolic excursion, *TDI* tissue Doppler index, *TR* tricuspid regurgitation

commonly used to quantify pre-existing RV dysfunction and predict RV failure after LVAD. While it is a noninvasive and relatively inexpensive imaging modality, the RV’s location behind the sternum and its asymmetric shape and varying contractile mechanics can pose challenges to accurate measurements by echocardiogram. The echocardiographic parameters used to estimate RV function can be categorized into measures estimating contractile function and volumes that estimate severity of physiological remodeling (Table 13.1). Variables estimating contractility include tricuspid annular plane systolic excursion, systolic velocity with tissue Doppler imaging, RVEF, RV fractional change, and myocardial deformation by tissue Doppler imaging [22–25, 27–29, 31, 34–39]. These parameters have been criticized for their dependence on the angle of the transducer as well as their sensitivity to loading conditions [21]. A promising alternative for measuring myocardial deformation/strain that is independent of the transducer angle is speckle

Table 13.2 Hemodynamic parameters used to evaluate RV failure

Parameter	Details	Threshold	Positive evidence	Negative evidence
RAP		>15 mmHg	19, 31, 33	41, 42
RAP:PCWP	Ratio between RAP and PCWP	>0.63	19, 31–33, 43	
RVSWI	(mPAP-RAP)x SVI; SVI=CI/HR	<250–402 mmHg*mL/m2	19, 31, 33, 34, 41, 42, 44	
PVR	(mPAP-PCWP)/CO		33	31, 41, 42
PAPI	(PASP-PADP)/CVP	<2	31–33	
TPG	mPAP-PCWP		33, 45	31
mPAP			41	42
dPAP			41	31
CO/CI			44	31, 33, 42
PCWP				31

CI cardiac index, *CO* cardiac output, *CVP* central venous pressure, *dPAP* diastolic pulmonary artery pressure, *mPAP* mean pulmonary artery pressure, *PAPI* pulmonary artery pulsatility index, *PADP* pulmonary artery diastolic pressure, *PASP* pulmonary artery systolic pressure, *PCWP* pulmonary capillary wedge pressure, *PVR* pulmonary vascular resistance, *RAP* right atrial pressure, *RVSWI* right ventricular stroke work index, *TPG* transpulmonary gradient

imaging [27, 35, 37]. Volumetric variables include right-sided chamber volumes by 2D and 3D, RV sphericity index, and tricuspid regurgitation [21, 25, 36, 37, 48, 49]. Importantly, none of the aforementioned variables have been shown in observational datasets to consistently predict RV dysfunction after LVAD [36].

Thus, due to a lack of evidence, current guidelines do not recommend the use of any single echocardiographic parameter for predicting RV failure after LVAD [50]. Though MRI remains the gold standard for quantifying RVEF and volumes, its use is limited by incompatibilities with older automatic implantable cardioverter-defibrillators and LVAD after implantation [4]. Computed tomography has also been shown to correlate well with RV fractional area change and is another potential modality for assessing function and volumes in individuals without renal dysfunction [51]. And more recently, there is evidence of good correlation between RV ejection fraction assessments by computed tomography and radionuclide ventriculography [52].

Invasive Hemodynamics

Right heart catheterization (RHC) provides important invasive hemodynamic data including cardiac output, right- and left-sided filling pressures, and afterload via the calculated resistances across the pulmonary and systemic vascular beds. Interestingly, none of these variables by themselves have been consistently associated with RVF post-LVAD (Table 13.2). This is likely due to the nonlinear relationship that these variables have with RVF and the importance of these values as they relate to one another. For example, because pulmonary vascular resistance (PVR) is dependent on both the pressure gradient between the pulmonary artery and left atrium and cardiac output, the interpretation of a high PVR driven by an elevated pressure gradient is very different from one driven by a low cardiac output. Similarly, a high central venous pressure (CVP) in the presence of a high pulmonary capillary wedge pressure (PCWP) is less ominous than an isolated one paired with a low PCWP. As a result, variables such as CVP:PCWP which captures the relationship between left- and right-sided filling pressures and pulmonary artery pulsatility index (PAPi) and right ventricular stroke work index (RVSWI), which are measures of RV work capacity, have been shown to be better predictors of RV recovery after LVAD implantation [18, 30, 33, 40–43]. Lastly, hemodynamic maneuvers using milrinone, nitroprusside, or inhaled nitric oxide (iNO) to either reduce left-sided filling pressures or preferentially vasodilate the pulmonary bed can help to identify fixed PVR which impact RV recovery after LVAD, as well as discriminate between an RV with contractile reserve and one that cannot be optimized by changes in loading conditions.

Table 13.3 Risk scores used to evaluate RV failure after LVAD implantation

Risk score	*N* (%CVAD[a])	Score variables	Outcome	C-statistic study	C-statistic validation	Reference
Fitzpatrick et al.	266 (2.2%)	RVSWI, CI, RVD, Cr, prior CTS, SBP	RVAD	–	0.63	[43, 45]
Michigan Score	197 (15%)	Vasopressor, AST, bilirubin, Cr	ECMO/RVAD, iNOx48h or inotropes x14d or on discharge, iNO >=48 h	0.73	0.50	[26, 44, 45]
Drakos et al.	175 (14%)	PVR, IABP, destination therapy, inotrope dependency, obesity, preoperative ACEi or ARB, preoperative beta-blocker	RVAD, iNO >=48 h, inotropes > 14d	0.743	0.55	[45, 46]
Pittsburgh Decision Tree	183 (22%)	TPG, age, RA pressure, HR, INR, ALT, white blood count, inotropes	RVAD	0.87	–	[44]
CRITT	218 (36%)	CVP, RVD, intubation, TR, tachycardia	RVAD	0.8	0.6	[41, 47]
EUROMACS	2000 (100%)	RA/PCWP, hgb, >1 inotrope, INTERMACS 1–3, severe RVD	ECMO/RVAD, iNO > =48 h, inotropes x14d or on discharge with/in 30d	0.7	0.67	[32]
Heartmate II	484 (100%)	CVP:PCWP, BUN, intubated	RVAD, inotropes > 14d, inotropes after 14d	0.68	0.58	[18, 47]
INTERMACS	9976 (100%)	INTERMACS, prior CTS, Cr, dialysis, ECMO, INR, total bilirubin, WBC, PA pulse pressure, RAP, SV, LVEDD, TR, concomitant procedure	RVAD w/in 14d	0.78	–	[15]

ACEi angiotensin-converting enzyme inhibitor, *ARB* angiotensin receptor blocker, *CI* cardiac index, Cr creatinine, *CTS* cardiothoracic surgery, *CVP* central venous pressure, *ECMO* extracorporeal membrane oxygenator, *HR* heart rate, *iNO* inhaled nitric oxide, *INR* international normalized ratio, *LVEDD* left ventricular end diastolic dimension, *PA* pulmonary artery, *PCWP* pulmonary capillary wedge pressure, *PVR* pulmonary vascular resistance, *RA* right atrial, *RVSWI* right ventricular stroke work index, *RVD* right ventricular dilation, *RVAD* right ventricular assist device, *SBP* systolic blood pressure, *SV* stroke volume, *TPG* transpulmonary gradient, TR tricuspid regurgitation

[a]% of continuous flow devices in study population

Risk Scores

There are several risk scores for predicting RV failure after LVAD placement that combine imaging and hemodynamic parameters with demographics, clinical characteristics, and markers of end-organ function (Table 13.3). All of these risk scores were derived from retrospective datasets, and unfortunately none have been successfully validated outside of their original datasets, with c-statistics in external validation cohorts ranging from 0.5 to 0.67 [44, 45, 47]. The moderate performance of these models is likely due to differences between study populations, i.e., device indication and type of VAD, as well as how RV failure was defined [47]. Thus, while these risk scores can help with risk stratifying LVAD candidates, they should be used concurrently with imaging and hemodynamic data.

Management

Preoperative Optimization

All individuals with biventricular dysfunction prior to LVAD placement should have their preload, afterload, and contractility optimized prior to their index operation regardless of the device strategy (BiVAD versus LVAD alone). As mentioned previously, the RV's thin-walled anatomy allows it to accommodate higher volumes; however, at some point the myofibrils will be overstretched and TR will develop, both of which will lead to a drop in cardiac output. Although the optimal filling pressure is determined by multiple factors, a CVP < 15 mmHg is generally targeted either by aggressive IV diuresis augmented by inotropes or even ultrafiltration or CVVH if needed [11, 12, 53, 54]. In terms of afterload reduction, appropriate oxygenation strategies to increase recruitment of the pulmonary bed and reduce vasoconstriction can be trialed. Similarly, inotropes can be used for the dual effect of augmenting RV function as well as for their pulmonary vasodilative effect [53]. The role of direct pulmonary vasodilators is unclear, even with high pulmonary arterial pressures and elevated PVR, particularly because left-sided filling pressures are likely very elevated [11]. Nonetheless, despite the lack of evidence, inhaled pulmonary vasodilators are frequently used in patients with failing RVs after LVAD [55–57]. And lastly, maintaining adequate perfusion to the RV is important, thus higher MAPs with a goal that is dependent on RV filling and pulmonary arterial pressures. There is evidence that utilization of intra-aortic balloon pumps in individuals with biventricular failure may help optimize right ventricular function by augmenting perfusion pressures while simultaneously unloading the left ventricle and improving right ventricular afterload [58, 59].

Surgical and Intraoperative Management Strategies

Once the preoperative variables have been optimized, the patient's status can be reassessed. If there is evidence of persistent poor biventricular function, then consideration for biventricular support as the index surgery should be made. There are several studies which show either improved or a trend towards improved outcomes in patients who undergo a biventricular support strategy as their index surgery [60–62]. In the Fitzpatrick et al. study, which utilized both durable and temporary paracorporeal RVADs, only heart rate and diastolic blood pressure were different between the planned and delayed BiVAD groups [62]; however, survival to discharge was higher in the planned group compared to the unplanned group (51% versus 29%, $p = 0.046$). In another study by Takeda et al., although there was a trend towards improved survival to discharge between the planned and unplanned groups, the difference did not reach statistical significance (50% versus 30%, respectively, $p = 0.064$) [61]. In this study, which included both durable and the temporary devices as well (with central cannulation utilized in all devices), there were more preoperative differences between those with a planned and an unplanned BiVAD, including age, body mass index, diabetes, hypertension, etiology of heart failure, albumin, and hematocrit. Similarly, a multicenter registry study by Shah et al. showed a trend towards improved 1-year survival between planned and unplanned BiVADs (74% vs 40%, $p = 0.11$) [60]. Interestingly in this study, although the staged BiVADs were older, the patients had lower RAP, PAPI, and RAP:PCWP, suggesting a better preload and RV reserve profile compared to those who underwent planned BiVAD [60]. Although these data are retrospective and limited by small sample size, they suggest that a more aggressive support strategy may be beneficial in appropriately selected candidates.

In addition to deciding between a biventricular support strategy and left-sided support alone, surgical planning includes consideration of concomitant procedures as well as surgical approach. Significant mitral regurgitation (MR) and tricuspid regurgitation (TR) are often seen in these patients, whether the etiology is functional, ischemic, or even primary from pacemaker lead impingement. As a result, consideration needs to be made regarding concomitant valve procedures. With LVAD implantation and unloading of the LV, functional MR is expected to decrease, with one observational study showing a 58% decrease in moderate-to-severe MR to less than moderate MR without any mitral intervention [63]. However, based on observational data, the incidence of RV failure is higher in patients with greater than moderate residual MR after LVAD implantation [64]. In a different retrospective study comparing individuals who underwent concomitant mitral valve surgery (MVS) to those who did not but had a spontaneous reduction in MR immediately post-implantation, actuarial survival at 1 year was higher in the MVS group (69% vs 59%, $p = 0.028$) [65]. In terms of preoperative variables, while only BTT indication for LVAD was significantly higher in the MVS compared to non-MVS group ($p = 0.017$), higher rates of residual MR in postoperative echocardiograms were noted in the non-MVS group [65].

In regard to tricuspid regurgitation, a systematic review of LVAD trials published in 2016 found one study that reported reduced risk of RVAD support after LVAD with concomitant tricuspid valve surgery (TVS) [66]. In the pooled analysis of six studies, however, there was no difference in RVAD placement post-LVAD between those who underwent TVS and those who did not (RR 1.42, 95% CI 0.54–3.76) [66]. In subsequent observational studies comparing outcomes in LVAD recipients with at least moderate TR who underwent TVS compared to those who did not, although reductions in TR were noted postoperatively, there were no differences in mortality or RVAD support [67, 68]. In a different study by Nakanishi et al., tricuspid annular dilation with residual TR after LVAD was associated with increased mortality [48, 49]; however, TVS was not evaluated in this study. Although it seems reasonable to repair or replace the mitral or tricuspid valve if there is a primary lesion, in the absence of good randomized controlled trials evaluating MV or TV interventions in LVADs, it is difficult to recommend any valvular interventions based on functional insufficiencies alone. Particularly when the addition of concomitant procedures adds cardiopulmonary bypass time and increases the risk for myocardial ischemia which can also impact postoperative RV function [65, 66].

Lastly, surgical approach, i.e., sternotomy versus a lateral thoracotomy, is another consideration. The U.S. Food and Drug Administration recently approved a lateral approach for the HeartWare HVAD (Medtronic, Framingham, MA) durable device. This lateral approach has been proposed to protect against right ventricular dysfunction by minimizing the anterior manipulation of the heart needed to expose the apex for inflow cannulation, as well as preserving the anterior aspect of the pericardium. The frequency of right heart dysfunction and RVAD requirement at 30 days was 22.1% and 0.7%, respectively, in the lateral approach trial and 23.3% and 2.6% in the pivotal trial conducted via a median sternotomy [69, 70]. Because the lateral approach trial was designed as a non-inferiority trial on survival alone, while this data is hypothesis generating, it does not currently support a lateral thoracotomy approach for reducing RV dysfunction.

Intraoperative transesophageal echocardiogram should be performed to ensure that the inflow cannula is parallel to the intraventricular septum so that the LV can be maximally without increasing the risk of suction events [54]. Additionally, the LVAD speed should be adjusted to maintain the septum at midline to preserve the RV's longitudinal contractility and not upset the interventricular dependence between the left and right ventricles. Additionally, although minimizing MR while allowing for the aortic valve to open intermittently is recommended, if there is concern for RV dysfunction, LVAD speeds can be set lower to minimize the risk of overloading the RV [39]. Lastly, prior to coming off bypass, the filling pressures and PVR should be monitored closely as cardiopulmonary bypass, mechanical ventilation, and volume shifts can all mediate pulmonary vasoconstriction. Inotropes and inhaled pulmonary vasodilators such as iNO should be added to assist with weaning from the cardiopulmonary bypass machine if needed [71].

Postoperative Optimization

Surveillance

The management of patients who develop RV failure after LVAD depends on the etiology of their right heart failure as well as their initial indication for LVAD support, i.e., destination therapy versus bridge-to-transplant. In those with pre-existing biventricular failure with progressive RV dysfunction who are not candidates for transplant, medical optimization may be the only option as durable biventricular support options are limited.

Supporting the heterogeneous makeup of patients developing RV failure after LVAD implantation is the study from Rich et al. where individuals with RV failure after LVAD as DT therapy were stratified into early, defined as within 30 days of discharge, and late RV failure, after 30 days from discharge [13]. Of note, all patients had to have survived to 30 days post-discharge to have been included in the study population; thus, the median time to late RV failure in this patient population was 480 days. Compared to individuals who did not develop RV failure, those who developed late RV failure had higher body mass index, blood urea nitrogen, and CVP:PCWP ($p < 0.05$). Interestingly, their HMII RV and Michigan risk scores were also higher ($p < 0.05$) [13], and yet their median time to RV failure was 480 days. This suggests that these risk scores, which were developed to predict RV failure during index hospitalization, can help to discriminate those at higher risk of developing post-LVAD RV failure, but they are not good at prognosticating the acuity and time course of it. Also notable was that early RV failure was not predictive of late RV failure in this study. Although preoperative predictors of early RV failure were not provided in this study for comparison, this suggests that early and late RV failures do not exist on a continuum and that with late RV failure, management strategies should focus on preventing RV dysfunction from progressing to RV failure. Resultantly, routine surveillance RAMP studies, both invasively and by echocardiogram, should be conducted and LVAD settings optimized for a protective RV strategy that balances LV unloading with RV preload while keeping the septum midline [72]. Routine monitoring will also allow for early identification of increasing PVR and institution of pulmonary vasodilator therapy if indicated.

Medical Management

As mentioned earlier, because RV output is afterload dependent, reducing pulmonary pressures and PVR is one strategy to augment RV function. Preoperatively, the greatest concern for using pulmonary vasodilators in WHO II pulmonary hypertension is the risk of pulmonary edema when pulmonary resistance is lowered and increased flow overloads the left heart. Once an LVAD has been implanted and the left side of the heart appropriately unloaded, this risk is reduced. Postoperatively, risk for worsening right ventricular function can be considered to occur in two phases. In the early phase, pulmonary vasoconstriction mediated by factors directly related to the surgery, including cardiopulmonary bypass, ventilation strategies, and blood products, can cause RV failure. In the later phase, pre-existing vascular dysfunction from high postcapillary pressures or chronic remodeling from increased right ventricular loading conditions post-VAD [56, 57] is the likelier cause. There is significant data suggesting utility in iNO and prostacyclin analogs in this immediate postoperative phase for reducing RV afterload [55–57, 73, 74]. Between these options the data for iNO is most robust, although prostacyclin analogs are favored because of reduced cost [56]. iNO has been shown in both observational and randomized controlled studies to decrease PVR in the immediate postoperative period [75, 76]. Another randomized controlled trial using right ventricular dysfunction as the primary clinical outcome found no difference in RVD in those bridged with iNO in the first 48 hours postoperatively compared to those who were not bridged (9.6% versus 15.6%, $p = 0.33$). A major limitation of this study which confounded the intention-to-treat analysis is that more than 25% of the placebo group crossed over into the iNO group [73]. It is also important to note that when titrating iNO off, there can be a rebound increase in pulmonary vasoconstriction, so close invasive hemodynamic monitoring may be appropriate depending on the severity of RV dysfunction being managed [77]. Thus, in the immediate postoperative phase, it is reasonable to consider bridging the patient with either inhaled NO or prostacyclin analogs while the patient's volume status is optimized and the patient is extubated.

After the immediate postoperative period, there is some data to suggest that use of phosphodiesterase-5 inhibitors, specifically sildenafil, can reduce PVR after LVAD [74, 78, 79]. In one study which prospectively selected individuals with elevated PVR even after a PCWP of <15 mmHg was achieved, sildenafil was given and PA pressures and PVR were compared to those from historical controls [79]. In this study, lower PVR was seen at 2–4 weeks after LVAD implantation and persisted until the last follow-up RHC 12 to 15 weeks postoperatively [79]. Interestingly, none of the studies saw a change in MAP with sildenafil, except for one which noted an increase in MAP post-LVAD on sildenafil [74, 78]. Similarly, no studies showed an increase in PCWP which suggests appropriate unloading by the LVAD. Although all of these studies were observational and additional studies are needed, considering that it is an oral medication with a relatively safe side effect profile, it is reasonable to trial sildenafil as a means of lowering PVR and augmenting RV function in hemodynamically stable patients post-LVAD [74, 78, 79].

Mechanical Circulatory Support Options

Although there is increasing experience using the HVAD device off-label as a right ventricular support device, options for durable biventricular devices outside of SynCardia's total artificial heart are limited. Additionally, the morbidity associated with these devices is significant. In one multicenter registry which looked only at HVADs implanted as RVADs, 37% of those implanted devices developed device thrombosis. The median time to thrombosis in this study was just 47 days [60]. As a result, the majority of patients who undergo biventricular assist device support are appropriately bridge-to-recovery or bridge-to-transplant [60, 80–82]. For temporary support, the options include extracorporeal, paracorporeal, and percutaneous devices [60, 82–91]. Because of this, in situations of biventricular failure or right ventricular failure post-LVAD that are refractory to medical therapy, the most important decisions to make are (1) the reversibility of the RV failure and if it is irreversible (2) the candidacy of the individual for transplant.

Predicting RV recovery after RVAD is difficult, with retrospective analyses estimating successful RVAD weaning to range between 40% and 75% [61, 87, 88]. Again, because these studies are not randomized, these rates are highly dependent on patient selection. In the Takeda et al. study, higher white blood cell count and lower creatinine were associated with successful weaning, compared to the Saito et al. study where CVP and RVSWI was. These studies suggest that successful RV recovery depends on perioperative preservation of end-organ function, as well as the ability to successfully optimize RV loading conditions and augment RVSWI.

Conclusions

The first step of managing biventricular failure involves appropriately risk stratifying individuals so that appropriate medical or surgical treatment options can be initiated. This is important not just for discriminating between individuals who may require right-sided mechanical support but also for determining severity and thus reversibility of RV failure. Once an individual has been selected for BiVAD support, pre-, peri-, and postoperative optimization of RV loading conditions and monitoring are critical. Although future research should focus on new biomarkers and imaging modalities to improve on current parameters and risk scores, as the technical design of RVADs improves and allows them to be implanted for longer, advances in managing their side effect profile will also be important.

References

1. Felker GM, Thompson RE, Hare JM, et al. Underlying causes and long-term survival in patients with initially unexplained cardiomyopathy. N Engl J Med. 2000;342:1077–84.
2. McKenna WJ, Maron BJ, Thiene G. Classification, epidemiology, and global burden of cardiomyopathies. Circ Res. 2017;121:722–30.
3. Konstam MA, Kiernan MS, Bernstein D, et al., on behalf of the American Heart Association Council on Clinical Cardiology; Council on Cardiovascular Disease in the Young; and Council on Cardiovascular Surgery and Anesthesia. Evaluation and management of right-sided heart failure: a scientific statement from the American Heart Association. Circulation 2018;Apr 19.
4. Houston BA, Shah KB, Mehra MR, Tedford RJ. A new "twist" on right heart failure with left ventricular assist systems. J Heart Lung Transplant. 2017;36(7):701–7.
5. Hayek S, Sims DB, Markham DW, Butler J, Kalogeropoulos AP. Assessment of right ventricular function in left ventricular assist device candidates. Circ Cardiovasc Imaging. 2014;7:379–89.
6. Kimmaliardjuk DM, Ruel M. Cardiac passive-aggressive behavior? The right ventricle in patients with a left ventricular assist device. Expert Rev Cardiovasc Ther. 2017;15(4):267–76.
7. Iglesias-Garriz I, Olalla-Gómez C, Garrote C, López-Benito M, Martín J, Alonso D, Rodríguez MA. Contribution of right ventricular dysfunction to heart failure mortality: a meta-analysis. Rev Cardiovasc Med. 2012;13:e62–9.
8. Amsallem M, Mercier O, Kobayashi Y, Moneghetti K, Haddad F. Forgotten no more: a focused update on the right ventricle in cardiovascular disease. JACC Heart Fail. 2018;6(11):891–903.1.
9. Interagency Registry for Mechanically Assisted Circulatory Support. Appendix A - adverse event definitions. UAB School of Medicine.http://www.uab.edu/medicine/intermacs/appendices/app-a-5-0. Accessed 9/6/2018.
10. Bellavia D, Iacovoni A, Scardulla C, Moja L, Pilato M, Kushwaha SS, Senni M, Clemenza F, Agnese V, Falletta C, Romano G, Maalouf J, Dandel M. Prediction of right ventricular failure after ventricular assist device implant: systematic review and meta-analysis of observational studies. Eur J Heart Fail. 2017;19(7):926–46.
11. Lampert BC, Teuteberg JJ. Right ventricular failure after left ventricular assist devices. J Heart Lung Transplant. 2015;34(9):1123–30.
12. Dell'Aquila AM, Schneider SR, Stypmann J, Ellger B, Redwan B, Schlarb D, et al. Survival results after implantation of intrapericardial third-generation centrifugal assist device: an INTERMACS-matched comparison analysis. Artif Organs. 2014;38(5):383–90.
13. Rich JD, Gosev I, Patel CB, Joseph S, Katz JN, Eckman PM, et al. The incidence, risk factors, and outcomes associated with late right-sided heart failure in patients supported with an axial-flow left ventricular assist device. J Heart Lung Transplant. 2017;36(1):50–8.
14. MacGowan GA, Schueler S. Right heart failure after left ventricular assist device implantation: early and late. Curr Opin Cardiol. 2012;27(3):296–300.
15. Kiernan MS, Grandin EW, Brinkley M Jr, Kapur NK, Pham DT, Ruthazer R et al. Early right ventricular assist device use in patients undergoing continuous-flow left ventricular assist device implantation: incidence and risk factors from the interagency registry for mechanically assisted circulatory support. Circ Heart Fail. 2017;10(10).
16. Cleveland JC, Naftel DC, Reece TB, Murray M, Antaki J, Pagani FD, et al. Survival after biventricular assist device implantation: an analysis of the interagency registry for mechanically assisted circulatory support database. J Heart Lung Transplant. 2011;30(8):862–9.
17. Kirklin JK, Xie R, Cowger J, Nakatani T, Schueler S, Taylor R, et al. Second annual report from the ISHLT mechanically assisted circulatory support registry. J Heart Lung Transplant. 2018;37(6):685–91.
18. Kormos RL, Teuteberg JJ, Pagani FD, Russell SD, John R, Miller LW, et al. Right ventricular failure in patients with the HeartMate II continuous-flow left ventricular assist device: incidence, risk factors, and effect on outcomes. J Thorac Cardiovasc Surg. 2010;139(5):1316–24.
19. LaRue SJ, Raymer DS, Pierce BR, Nassif ME, Sparrow CT, Vader JM. Clinical outcomes associated with INTERMACS-defined right heart failure after left ventricular assist device implantation. J Heart Lung Transplant. 2017;36(4):475–7.
20. Yoshioka D, Takayama H, Garan RA, Topkara VK, Han J, Kurlansky P, et al. Contemporary outcome of unplanned right ventricular assist device for severe right heart failure after continuous-flow left ventricular assist device insertion. Interact Cardiovasc Thorac Surg. 2017;24(6):828–34.
21. Cohen DG, Thomas JD, Freed BH, Rich JD, Sauer AJ. Echocardiography and continuous-flow left ventricular assist devices: evidence and limitations. JACC Heart Fail. 2015;3(7):554–64.
22. Kato TS, Jiang J, Schulze PC, Jorde U, Uriel N, Kitada S, et al. Serial echocardiography using tissue Doppler

and speckle tracking imaging to monitor right ventricular failure before and after left ventricular assist device surgery. JACC Heart Fail. 2013;1(3):216–22.
23. Puwanant S, Hamilton KK, Klodell CT, Hill JA, Schofield RS, Cleeton TS, et al. Tricuspid annular motion as a predictor of severe right ventricular failure after left ventricular assist device implantation. J Heart Lung Transplant. 2008;27(10):1102–7.
24. Dandel M, Potapov E, Krabatsch T, Stepanenko A, Löw A, Vierecke J, et al. Load dependency of right ventricular performance is a major factor to be considered in decision making before ventricular assist device implantation. Circulation. 2013;128(11 Suppl 1):S14–23.
25. Magunia H, Dietrich C, Langer HF, Schibilsky D, Schlensak C, Rosenberger P et al. 3D echocardiography derived right ventricular function is associated with right ventricular failure and mid-term survival after left ventricular assist device implantation. Int J Cardiol. 2018. pii: S0167-5273.
26. Matthews JC, Koelling TM, Pagani FD, Aaronson KD. The right ventricular failure risk score a pre-operative tool for assessing the risk of right ventricular failure in left ventricular assist device candidates. J Am Coll Cardiol. 2008;51(22):2163–72.
27. Cameli M, Lisi M, Righini FM, Focardi M, Lunghetti S, Bernazzali S, et al. Speckle tracking echocardiography as a new technique to evaluate right ventricular function in patients with left ventricular assist device therapy. J Heart Lung Transplant. 2013;32(4):424–30.
28. Vivo RP, Cordero-Reyes AM, Qamar U, Garikipati S, Trevino AR, Aldeiri M, et al. Increased right-to-left ventricle diameter ratio is a strong predictor of right ventricular failure after left ventricular assist device. J Heart Lung Transplant. 2013;32(8):792–9.
29. Aissaoui N, Salem JE, Paluszkiewicz L, Morshuis M, Guerot E, Gorria GM, et al. Assessment of right ventricular dysfunction predictors before the implantation of a left ventricular assist device in end-stage heart failure patients using echocardiographic measures (ARVADE): combination of left and right ventricular echocardiographic variables. Arch Cardiovasc Dis. 2015;108(5):300–9.
30. Kang G, Ha R. Banerjee D3. Pulmonary artery pulsatility index predicts right ventricular failure after left ventricular assist device implantation. J Heart Lung Transplant. 2016;35(1):67–73.
31. Raina A, Seetha Rammohan HR, Gertz ZM, Rame JE, Woo YJ, Kirkpatrick JN. Postoperative right ventricular failure after left ventricular assist device placement is predicted by preoperative echocardiographic structural, hemodynamic, and functional parameters. J Card Fail. 2013;19(1):16–24.
32. Soliman OII, Akin S, Muslem R, Boersma E, Manintveld OC, Krabatsch T, et al. EUROMACS Investigators. Derivation and validation of a novel right-sided heart failure model after implantation of continuous flow left ventricular assist devices: the EUROMACS (European registry for patients with mechanical circulatory support) right-sided heart failure risk score. Circulation. 2018;137(9):891–906.
33. Kato TS, Farr M, Schulze PC, Maurer M, Shahzad K, Iwata S, et al. Usefulness of two-dimensional echocardiographic parameters of the left side of the heart to predict right ventricular failure after left ventricular assist device implantation. Am J Cardiol. 2012;109(2):246–51.
34. Grant AD, Smedira NG, Starling RC, Marwick TH. Independent and incremental role of quantitative right ventricular evaluation for the prediction of right ventricular failure after left ventricular assist device implantation. J Am Coll Cardiol. 2012;60(6):521–8.
35. Beck DR, Foley L, Rowe JR, Moss AF, Weitzel NS, Reece TB, et al. Right ventricular longitudinal strain in left ventricular assist device surgery-a retrospective cohort study. J Cardiothorac Vasc Anesth. 2017;31(6):2096–102.
36. Dandel M, Krabatsch T, Falk V. Left ventricular vs. biventricular mechanical support: decision making and strategies for avoidance of right heart failure after left ventricular assist device implantation. Int J Cardiol. 2015;198:241–50.
37. Cameli M, Loiacono F, Sparla S, Solari M, Iardino E, Mandoli GE, et al. Systematic left ventricular assist device implant eligibility with non-invasive assessment: the SIENA protocol. J Cardiovasc Ultrasound. 2017;25(2):39–46.
38. Potapov EV, Stepanenko A, Dandel M, Kukucka M, Lehmkuhl HB, Weng Y, et al. Tricuspid incompetence and geometry of the right ventricle as predictors of right ventricular function after implantation of a left ventricular assist device. J Heart Lung Transplant. 2008;27(12):1275–81.
39. Kukucka M, Stepanenko A, Potapov E, Krabatsch T, Redlin M, Mladenow A, et al. Right-to-left ventricular end-diastolic diameter ratio and prediction of right ventricular failure with continuous-flow left ventricular assist devices. J Heart Lung Transplant. 2011;30(1):64–9.
40. Ochiai Y, McCarthy PM, Smedira NG, Banbury MK, Navia JL, Feng J, et al. Predictors of severe right ventricular failure after implantable left ventricular assist device insertion: analysis of 245 patients. Circulation. 2002;106(12 Suppl 1):I198–202.
41. Atluri P, Goldstone AB, Fairman AS, MacArthur JW, Shudo Y, Cohen JE, et al. Predicting right ventricular failure in the modern, continuous flow left ventricular assist device era. Ann Thorac Surg. 2013;96(3):857–63; discussion 863-4.
42. Kinugawa K, Imamura T, Kato N, Endo M, Inaba T, Maki H, et al. Combination evaluation of preoperative risk indices predicts requirement of biventricular assist device. Circ J. 2012;76(12):2785–91.
43. Fitzpatrick JR, Frederick JR, Hsu VM, Kozin ED, O'Hara ML, Howell E, et al. Risk score derived from pre-operative data analysis predicts the need for biventricular mechanical circulatory support. J Heart Lung Transplant. 2008;27(12):1286–92.

44. Wang Y, Simon MA, Bonde P, Harris BU, Teuteberg JJ, Kormos RL, et al. Decision tree for adjuvant right ventricular support in patients receiving a left ventricular assist device. J Heart Lung Transplant. 2012;31(2):140–9.
45. Pettinari M, Jacobs S, Rega F, Verbelen T, Droogne W, Meyns B. Are right ventricular risk scores useful? Eur J Cardiothorac Surg. 2012;42(4):621–6.
46. Drakos SG, Janicki L, Horne BD, Kfoury AG, Reid BB, Clayson S, et al. Risk factors predictive of right ventricular failure after left ventricular assist device implantation. Am J Cardiol. 2010;105(7):1030–5.
47. Kalogeropoulos AP, Kelkar A, Weinberger JF, Morris AA, Georgiopoulou VV, Markham DW, et al. Validation of clinical scores for right ventricular failure prediction after implantation of continuous-flow left ventricular assist devices. J Heart Lung Transplant. 2015;34(12):1595–603.
48. Nakanishi K, Homma S, Han J, Takayama H, Colombo PC, Yuzefpolskaya M, et al. Prevalence, predictors, and prognostic value of residual tricuspid regurgitation in patients with left ventricular assist device. J Am Heart Assoc. 2018;24:7(13).
49. Nakanishi K, Homma S, Han J, Takayama H, Colombo PC, Yuzefpolskaya M, et al. Usefulness of tricuspid annular diameter to predict late right sided heart failure in patients with left ventricular assist device. Am J Cardiol. 2018;122(1):115–20.
50. Stainback RF, Estep JD, Agler DA, Birks EJ, Bremer M, Hung J, et al. American Society of Echocardiography. Echocardiography in the management of patients with left ventricular assist devices: recommendations from the American Society of Echocardiography. J Am Soc Echocardiogr. 2015;28(8):853–909.
51. Garcia-Alvarez A, Fernandez-Friera L, Lau JF, Sawit ST, Mirelis JG, Castillo JG, et al. Evaluation of right ventricular function and post-operative findings using cardiac computed tomography in patients with left ventricular assist devices. J Heart Lung Transplant. 2011;30(8):896–903.
52. Lairez O, Delmas C, Fournier P, Cassol E, Méjean S, Pascal P, et al. Feasibility and accuracy of gated blood pool SPECT equilibrium radionuclide ventriculography for the assessment of left and right ventricular volumes and function in patients with left ventricular assist devices. J Nucl Cardiol. 2018;25(2):625–34.
53. de Asua I, Rosenberg A. On the right side of the heart: medical and mechanical support of the failing right ventricle. J Intensive Care Soc. 2017;18(2):113–20.
54. Marzec LN, Ambardekar AV. Preoperative evaluation and perioperative management of right ventricular failure after left ventricular assist device implantation. Semin Cardiothorac Vasc Anesth. 2013;17(4):249–61.
55. Antoniou T, Prokakis C, Athanasopoulos G, Thanopoulos A, Rellia P, Zarkalis D, et al. Inhaled nitric oxide plus iloprost in the setting of post-left assist device right heart dysfunction. Ann Thorac Surg. 2012;94(3):792–8.
56. Sabato LA, Salerno DM, Moretz JD, Jennings DL. Inhaled pulmonary vasodilator therapy for management of right ventricular dysfunction after left ventricular assist device placement and cardiac transplantation. Pharmacotherapy. 2017;37(8):944–55.
57. Critoph C, Green G, Hayes H, Baumwol J, Lam K, Larbalestier R, et al. Clinical outcomes of patients treated with pulmonary vasodilators early and in high dose after left ventricular assist device implantation. Artif Organs. 2016;40(1):106–14.
58. Ntalianis A, Kapelios CJ, Kanakakis J, Repasos E, Pantsios C, Nana E, et al. Prolonged intra-aortic balloon pump support in biventricular heart failure induces right ventricular reverse remodeling. Int J Cardiol. 2015;192:3–8.
59. Kapelios C, Terrovitis I, Ntalianis A, Kaldara E, Repasos E, Siskas P, et al. Intra-aortic balloon pump improves right ventricular function in patients with end-stage congestive heart failure. Circulation. 2018;128:A18182.
60. Shah P, Ha R, Singh R, Cotts W, Adler E, Kiernan M, et al. Multicenter experience with durable biventricular assist devices. J Heart Lung Transplant. 2018;37(9):1093–101.
61. Takeda K, Naka Y, Yang JA, Uriel N, Colombo PC, Jorde UP, et al. Outcome of unplanned right ventricular assist device support for severe right heart failure after implantable left ventricular assist device insertion. J Heart Lung Transplant. 2014;33(2):141–8.
62. Fitzpatrick JR, Frederick JR, Hiesinger W, Hsu VM, McCormick RC, Kozin ED, et al. Early planned institution of biventricular mechanical circulatory support results in improved outcomes compared with delayed conversion of a left ventricular assist device to a biventricular assist device. J Thorac Cardiovasc Surg. 2009;137(4):971–7.
63. Kitada S, Kato TS, Thomas SS, Conwell SD, Russo C, Di Tullio MR, et al. Pre-operative echocardiographic features associated with persistent mitral regurgitation after left ventricular assist device implantation. J Heart Lung Transplant. 2013;32(9):897–904.
64. Ertugay S, Kemal HS, Kahraman U, Engin C, Nalbantgil S, Yagdi T, et al. Impact of residual mitral regurgitation on right ventricular systolic function after left ventricular assist device implantation. Artif Organs. 2017;41(7):622–7.
65. Tanaka A, Onsager D, Song T, Cozadd D, Kim G, Sarswat N, et al. Surgically corrected mitral regurgitation during left ventricular assist device implantation is associated with low recurrence rate and improved midterm survival. Ann Thorac Surg. 2017;103(3):725–33.
66. Dunlay SM, Deo SV, Park SJ. Impact of tricuspid valve surgery at the time of left ventricular assist device insertion on postoperative outcomes. ASAIO J. 2015;61(1):15–20.
67. Han J, Takeda K, Takayama H, Kurlansky PA, Mauro CM, Colombo PC, et al. Durability and clinical impact of tricuspid valve procedures in patients receiving a continuous-flow left ventricular assist device. J Thorac Cardiovasc Surg. 2016;151(2):520–7.

68. Song HK, Gelow JM, Mudd J, Chien C, Tibayan FA, Hollifield K, et al. Limited utility of tricuspid valve repair at the time of left ventricular assist device implantation. Ann Thorac Surg. 2016;101(6):2168–74.
69. Danter MR, McGee EC, Strueber M, Maltais S, Mokadam NA, Weisenthaler GM, et al. A prospective, controlled, un-blinded, multi-center clinical trial to evaluate the thoracotomy implant technique of the HVAD system in patients with advanced heart failure: results of the LATERAL trial. J Heart Lung Transplant. 2017;36(4):S66.
70. Slaughter MS, Pagani FD, McGee EC, Birks EJ, Cotts WG, Gregoric I, et al. Use of the HeartWare ventricular assist system for bridge to transplant: combined results of the ADVANCE and CAP trial. J Heart Lung Transplant. 2013;32:675–83.
71. Morgan JA, Paone G, Nemeh HW, Murthy R, Williams CT, Lanfear DE, et al. Impact of continuous-flow left ventricular assist device support on right ventricular function. J Heart Lung Transplant. 2013;32(4):398–403.
72. Imamura T, Chung B, Nguyen A, Sayer G, Uriel N. Clinical implications of hemodynamic assessment during left ventricular assist device therapy. J Cardiol. 2018;71(4):352–8.
73. Potapov E, Meyer D, Swaminathan M, Ramsay M, El Banayosy A, Diehl C, et al. Inhaled nitric oxide after left ventricular assist device implantation: a prospective, randomized, double-blind, multicenter, placebo-controlled trial. J Heart Lung Transplant. 2011;30(8):870–8.
74. Hamdan R, Mansour H, Nassar P, Saab M. Prevention of right heart failure after left ventricular assist device implantation by phosphodiesterase 5 inhibitor. Artif Organs. 2014;38(11):963–7.
75. Wagner F, Dandel M, Günther G, Loebe M, Schulze-Neick I, Laucke U, et al. Nitric oxide inhalation in the treatment of right ventricular dysfunction following left ventricular assist device implantation. Circulation. 1997;96(9 Suppl):II-291-6.
76. Argenziano M, Choudhri AF, Moazami N, Rose EA, Smith CR, Levin HR, et al. Randomized, double-blind trial of inhaled nitric oxide in LVAD recipients with pulmonary hypertension. Ann Thorac Surg. 1998;65(2):340–5.
77. Macdonald PS, Keogh A, Mundy J, Rogers P, Nicholson A, Harrison G, et al. Adjunctive use of inhaled nitric oxide during implantation of a left ventricular assist device. J Heart Lung Transplant. 1998;17(3):312–6.
78. Baker WL, Radojevic J, Gluck JA. Systematic review of phosphodiesterase-5 inhibitor use in right ventricular failure following left ventricular assist device implantation. Artif Organs. 2016;40(2):123–8.
79. Tedford RJ, Hemnes AR, Russell SD, Wittstein IS, Mahmud M, Zaiman AL, et al. PDE5A inhibitor treatment of persistent pulmonary hypertension after mechanical circulatory support. Circ Heart Fail. 2008;1(4):213–9.
80. Tran HA, Pollema TL, Silva Enciso J, Greenberg BH, Barnard DD, Adler ED, et al. Durable biventricular support using right atrial placement of the HeartWare HVAD. ASAIO J. 2018;64(3):323–7.
81. Cork DP, Tran HA, Silva J, Barnard D, Greenberg B, Adler ED, et al. A case series of biventricular circulatory support using two ventricular assist devices: a novel operative approach. Ann Thorac Surg. 2015;100(4):e75–7.
82. Levin AP, Jaramillo N, Garan AR, Takeda K, Takayama H, Yuzefpolskaya M, et al. Outcomes of contemporary mechanical circulatory support device configurations in patients with severe biventricular failure. J Thorac Cardiovasc Surg. 2016;151(2):530–5.e2.
83. Scherer M, Sirat AS, Moritz A, Martens S. Extracorporeal membrane oxygenation as perioperative right ventricular support in patients with biventricular failure undergoing left ventricular assist device implantation. Eur J Cardiothorac Surg. 2011;39(6):939–44.
84. De Silva RJ, Soto C, Spratt P. Extra corporeal membrane oxygenation as right heart support following left ventricular assist device placement: a new cannulation technique. Heart Lung Circ. 2012;21(4):218–20.
85. Leidenfrost J, Prasad S, Itoh A, Lawrance CP, Bell JM, Silvestry SC. Right ventricular assist device with membrane oxygenator support for right ventricular failure following implantable left ventricular assist device placement. Eur J Cardiothorac Surg. 2016;49(1):73–7.
86. Schmack B, Weymann A, Popov AF, Patil NP, Sabashnikov A, Kremer J, et al. Concurrent Left Ventricular Assist Device (LVAD) implantation and percutaneous temporary RVAD support via CardiacAssist Protek-Duo TandemHeart to preempt right heart failure. Med Sci Monit Basic Res. 2016;22:53–7.
87. Saito S, Sakaguchi T, Miyagawa S, Nishi H, Yoshikawa Y, Fukushima S, et al. Recovery of right heart function with temporary right ventricular assist using a centrifugal pump in patients with severe biventricular failure. J Heart Lung Transplant. 2012;31(8):858–64.
88. Haneya A, Philipp A, Puehler T, Rupprecht L, Kobuch R, Hilker M, et al. Temporary percutaneous right ventricular support using a centrifugal pump in patients with postoperative acute refractory right ventricular failure after left ventricular assist device implantation. Eur J Cardiothorac Surg. 2012;41(1):219–23.
89. Sultan I, Kilic A. Short-term circulatory and right ventricle support in cardiogenic shock: extracorporeal membrane oxygenation, Tandem Heart, CentriMag, and Impella. Heart Fail Clin. 2018;14(4):579–83.
90. Shehab S, Newton PJ, Allida SM, Jansz PC, Hayward CS. Biventricular mechanical support devices--clinical perspectives. Expert Rev Med Devices. 2016;13(4):353–65.
91. Gregory SD, Timms D, Gaddum N, Mason DG, Fraser JF. Biventricular assist devices: a technical review. Ann Biomed Eng. 2011;39(9):2313–28.

14 Status and Availability of a Total Artificial Heart

Katherine G. Phillips, Neel K. Ranganath, and Nader Moazami

Introduction

The invention of the total artificial heart (TAH) is one of the greatest symbolic advances in modern medicine. Though technological strides have been made since the first implantation of a permanent TAH, the Jarvik-7, in 1982, these devices still only supported 1.4% of all adult heart recipients in 2016 [1]. Heart disease is the leading cause of death in most of the world, and its prevalence is expected to increase 46% from 2012 to 2030 [2, 3]. Heart transplantation has provided exceptional short- and long-term outcomes in carefully selected patients, with a median survival of 10.7 years [4, 5], but has major limitations, the most notable of which is donor organ supply.

Close to half of all patients (41%) awaiting heart transplant in 2012 were receiving mechanical circulatory support, indicating the escalating medical urgency to develop more physiologic and durable circulatory support devices [5]. Current mechanical circulatory support devices not only increase the likelihood of survival while awaiting transplantation, but they also improve hemodynamics, which reduces other organ dysfunction, increases post-transplant survival, and optimizes the use of donor hearts [6] (Fig. 14.1). Due to the relative ease of assisting the failing heart rather than replacing it, most clinically used circulatory support devices for congestive heart failure are left ventricular assist devices (LVADs). The TAH is predominantly use as a bridge to transplantation reserved for patients with biventricular heart failure in whom an LVAD would be inadequate. However, the TAH also solves other issues not adequately addressed by LVAD such as cardiac arrhythmias, severe multi-valvular disease, ventricular clots, large post-infarct ventricular septal defects, or left ventricular rupture and even certain localized tumors of the heart. Additionally, future devices may be most valuable as destination therapy for patients who are not suitable candidates for transplant.

From its inception the TAH was designed to mimic the native physiology of the human heart through a two-ventricle system, with unidirectional inlet and outlet valves that connect to the atria and provide pulsatile output through volume displacement. Limited durability, large size, complex anticoagulation regimens, and significant post-implantation complications have limited the widespread applicability of these devices [7]. While many TAHs have been developed, only the SynCardia temporary TAH has been clinically utilized in a large number of patients (Fig. 14.2).

In the past 10 years, a paradigm shift in LVADs toward smaller and simpler continuous-flow

K. G. Phillips, BS · N. K. Ranganath, MD
N. Moazami, MD (✉)
NYU Langone Health, Department of Cardiothoracic Surgery, New York, NY, USA
e-mail: neel.ranganath@nyulangone.org; nader.moazami@nyulangone.org

© Springer Nature Switzerland AG 2020
J. H. Karimov et al. (eds.), *Mechanical Support for Heart Failure*,
https://doi.org/10.1007/978-3-030-47809-4_14

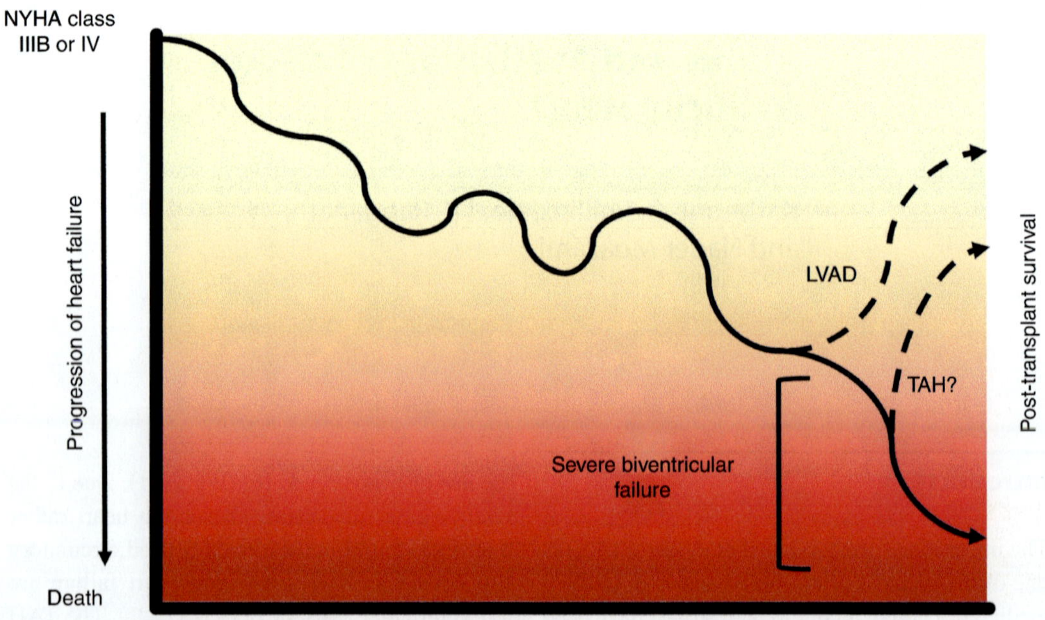

Fig. 14.1 Role of mechanical support devices in management of progressive heart failure. These devices not only increase the likelihood of survival while awaiting transplantation, but they also improve the hemodynamics of those awaiting transplant and thereby increase the chance of post-transplant survival, thus improving the optimal use of donor hearts. (Based on prior model by Renlund et al., 2004 [6])

	AbioCor	ReinHeart	SynCardia	CARMAT	BiVACOR	Cleveland CFTAH
Flow type	Pulsatile	Pulsatile	Pulsatile	Pulsatile	Continuous	Continuous
Therapeutic goal	Destination Therapy	Bridge-to-transplantation	Both	Both	Bridge-to-transplantation	Bridge-to-transplantation
Flow (L/min)	4-8	4.8-7.5	70cc: 7±1 50cc: 5±1	2-9	Up to 16L/min with no hemolysis	<9
Smaller size			√50cc: BSA≥1.2m²	No, BSA≥1.7m²	✓	√P-CFTAH BSA≥0.3m²
Portable driver		√2kg, 12hr	√6kg, 12hr	√3kg, 5hr	✓	
Entirely implantable	✓	✓		No, 8mm driveline		
Special features				Biocompatible Auto-regulates	Durability	Efficient Durability Remotely monitor
Human Implantation	✓ 14 Patients	Animal Studies, 2 day survival	✓ >1,700 Patients	✓ 10 patients	Animal studies, 30 day survival	Animal studies, 90 day survival

Fig. 14.2 Characteristics of current clinical and investigational total artificial hearts

rotary blood pumps which create their own unique non-pulsatile, non-laminar blood flow profile has renewed interest in TAHs which employ the same technology. The smaller size, energy efficiency, and durability inherent to continuous-flow pumps make them a promising avenue for future TAHs design. Nevertheless, these devices have their own set of complications, potentially linked to this hemodynamic departure from normal pulsatile physiology [8]. Further research into this relationship may provide valuable insights to decrease post-implantation complications of TAHs. This chapter will examine the history, current status, and potential future directions of the total artificial heart by examining previous and current TAH technology.

Historical Perspective

Initial Inspirations

The history of the total artificial heart is a culmination of scientific, philosophical, ethical, and political values. In the fifteenth century, Leonardo da Vinci was the first to describe the concept of the total artificial heart as an inspiration [9]. This was expanded upon by M. Le Gallois who believed "…if in the place of the heart could be… artificially formed…then life might be indefinitely maintained" [10].

However, ideas of such a device preceeded its existance by several centuries. It was not until 1957 that Dr. Tetsuzo Akutso and Dr. Willem Kloff implanted their first TAH into a dog at the Cleveland Clinic [11]. This monumental implantation inspired an increase in research of TAH, leading to Dr. Michael DeBakey's successful request of President Lyndon B. Johnson for the creation of a national research program with an equally ambitious goal: to produce a fully functional artificial heart by the time that humans set foot on the moon [9]. In April 1969, three months prior to humankind's first steps on the moon, Dr. Dennis Cooley (Texis Heart Institute) became the first surgeon to implant a TAH in a human patient for 64 hours of support prior to heart transplantation. Thirteen years later, there was significant excitement when Dr. Barney Clark recieved the first permanent TAH in 1982, a Jarvik-7, named after one of it's inventors, Dr. Robert Jarvik. The primary surgeon, Dr. DeVries, believed, "in a very short period of time we had seen something that was bigger than each of us. A lot of people, even those who were not religious, were reverent and attributed the success to God. As we saw the artificial heart beat in Dr. Clark, the feeling was not aren't we great, but aren't we small."

AbioCor®

On July 2, 2001, Dr. Gray and Dr. Dowling (University of Louisville Health and Science Center) made history implanting the AbioCor® implantable replacement heart (IRH), the first completely internal TAH to be placed in a human. An implantable TAH had been the focus of several previous designs such as the Sarns-3M TAH (3M Health Care, USA, in conjunction with the Pennsylvania State University, University Park, PA, USA) and the Nimbus TAH (Nimbus, USA, in conjunction with the Cleveland Clinic, Cleveland, OH, USA). The Sarns-3M TAH was tested in 14 calves in 1993 and had a 150-day survival, yet it was never adapted to human use due to concerns of size, reliability, and durability.

Since the conception of the device in 1981, ABIOMED worked toward the unique goal of designing the device as a permanent therapy. Departing from previous TAHs designed to bridge patients to transplantation, the AbioCor® was intended for patients with end-stage biventricular heart failure in whom heart transplantation was contraindicated.

Robert Tools, the first patient to receive the AbioCor® and survive without a living heart, was the perfect candidate. Prior to TAH implantation, he was rapidly approaching organ failure and was determined to have less than 1 month to live. He hardly had the energy to raise his head from his pillow. For Tools, the AbioCor® was a tremendous improvement, however short, to his last months of life. Although he eventually succumbed to stroke and abdominal bleeding, he survived 151 days, well beyond ABIOMED's initial survival goal of 60 days post-implantation [12]. His level of activity was improved to the point where he took 20 trips out of the hospital. In interviews, he said he was capable of walking a city block without stopping and took trips to the ice cream store and barbeque joint. The device also afforded him precious time with his grandchildren [13, 14]. Robert Tools' story is exemplary in both the successes and limitations of the AbioCor®.

After initial results over a 12-year period on 120 calves showed promise, with many of the animals surviving past the designed 90-day study term and even capable of exercising on a treadmill, the US Food and Drug Administration (FDA) subsequently approved the AbioCor® for a multicenter clinical trial in January of 2001 (the University of Louisville, the University of

Arizona Medical Center, Hahnemann University, Massachusetts General Hospital/Brigham and Women's Hospital, and the Texas Heart Institute) [9]. The device was approved for patients with end-stage biventricular heart failure ineligible for heart transplantation with a 30-day predicted mortality of <70%. Due to the severely ill state of the recipients prior to transplantation, the design of the device emphasized improved quality of life in addition to prolongation of life [14, 15].

Technical Specifications

As a totally implantable TAH, the AbioCor® consisted of entirely separate internal and external units which did not require transcutaneous wires or connection, thus eliminating substantial risk of bacterial infection. The internal units consisted of a thoracic unit, battery, controller, and transcutaneous energy transfer wire. Approximately the size of a grapefruit, the thoracic unit consisted of two alternating pumping chambers, designed to function as pseudo ventricles, with a high-efficiency pump (4000–8000 rpm) situated between them capable of adjusting to different resistances of systemic and pulmonary vascular systems and systolic ejection durations. A switching valve allowed alternation of flow of a hydraulic fluid between the two separate chambers, making the TAH an active flow device. When this fluid entered one of the sides, it created a negative pressure, filling the displaced volume with an equal amount of blood in the opposing chamber and resulting in alternating left and right systole. The rate of this switching valve determined the heart rate, which was between 75 and 100 beats/min, and resulted in flows ranging from 4 to 8 L/min.

In order to compensate for differential stroke volumes on the right and left side, an atrial balance chamber absorbed some of the hydraulic fluid whenever left atrial pressure reached a particular level, resulting in displaced right chamber volume, filling, and subsequent stroke volume. Consequently, venous return to the left atria was then diminished, and left atrial pressure decreased as pulmonary and systemic balance was restored. This intuitive system allowed for basic autoregulation such that the amount of fluid in the AbioCor® did not need to be adjusted over time like several of its gas-dependent predecessors [9].

All internal facets of the device were seamless and made entirely of polyurethane (Angioflex) ensuring smooth blood flow. However, at a hefty 2 lbs and close to the size of a grapefruit, the thoracic unit's size restricted its use to men who had adequate mediastinal space.

Most revolutionary was AbioCor's power system, which utilized a novel method of transcutaneous energy transfer. This system consisted of both internal and external lithium ion-based batteries, which allowed the device to be entirely implanted. Energy from the external battery, which was powered either by a bedside console or a portable patient-carried module, was transmitted wirelessly to the internal battery by converting electricity to electromagnetic waves. The internal battery was able to power the device for up to 20 minutes in the event of temporary disconnection.

Clinical Outcomes

This novel design was put to the test in 14 patients over the next 2 years, and, unfortunately, few had better outcomes than Tools. All of the patients were male with significant comorbidities. All were bedridden prior to implantation. Six of seven patients in an initial cohort reported by Dowley and colleagues had a predicted 30-day mortality that was as high as 89%. Of the 14 patients total, 2 died during surgery and 2 others did not survive the goal "60-day" period, due to hemorrhage (n = 2), air embolism (n = 1), and stroke (n = 1). Of the ten patients who did survive beyond 60 days, all had improvement of various degrees to quality of life and four were able to attend activities outside of the hospital.

The devices provided support for an average of 4.5 months, with five patients surviving past 9 months. Cerebrovascular accidents (CVA) were common (n = 9) and in six patients led to the withdrawal of support. Three of these CVA were attributed to a complication with the left atrial inflow struts, which had been developed to fit the animal model without complications and were subsequently removed. Multisystem organ failure and sepsis were other common mechanisms of death ($n = 4$), and device failure occurred in two patients. One of these patients survived 512 days, and although his device failure was predicted, he opted not to have a second device implanted [16].

While the AbioCor® may not have achieved its goal of becoming a truly permanent treatment for heart failure, it was able to extend life by several months and provide significantly improved quality of life and time with families for patients in whom death was imminent and no other treatment options existed. It was approved for compassionate use in September 2006, achieving the final goal of the ambitious artificial heart program that had been initiated by Lyndon B. Johnson 42 years prior [17]. ABIOMED, however, found further development of the TAH to be too expensive, and this device is no longer being manufactured or in clinical use.

Clinical

SynCardia Temporary TAH

The SynCardia temporary TAH is the only artificial heart to receive widespread clinical use, with an estimated >1700 implantations in over 35 years of clinical use, and is the only TAH to reach commercial use approval. Unlike the AbioCor®, the SynCardia temporary TAH was originally designed as a bridge to transplantation.

Evolution of the SynCardia Temporary TAH

The SynCardia temporary TAH has a rich engineering history beginning with design of the Jarvik hearts by veterinarian Dr. Don Olsen; medical engineer Dr. Robert Jarvik; chair of the division for artificial organs at the University of Utah, Dr. Willem Kolff; and surgeon Dr. William DeVries. After a 268-day survival in a calf implanted with the Jarvik 7, the team was awarded FDA approval for implantation in humans. On December 1, 1982, Barney Clark, who suffered from end-stage ischemic heart failure, was the first human to receive the device and subsequently survived 112 days despite no expectation of surviving the immediate postoperative period. He agreed to participate in the study out of a passion for science. A second patient to receive the Jarvik survived 620 days, the longest of any patient ever receiving a TAH at the time [18].

Despite these initial promising results, infection proved to be a major limitation of prolonged support in the initial cohort of six patients to receive the Jarvik 7, with four patients dying of septic complications. Simultaneously, cyclosporine was approved as the first immunosuppressant drug, leading to dramatic improvements in survival after transplantation and with it an increasing focus on bridging moribund heart failure patients to heart transplantation [9]. Since then, the Jarvik technology changed hands and names several times, due to noncompliance with FDA guidelines and regulations. It was renamed the Symbion Artificial Heart, then the CardioWest TAH, and, for the past 15 years, the SynCardia temporary TAH.

Technical Specifications

The SynCardia temporary TAH is designed as a bridge-to-transplantation device for individuals with end-stage biventricular heart failure. It is a biventricular, pneumatically driven blood pump providing pulsatile flow. It displaces 400 mL of volume and weighs 160 g, about the weight of a hockey puck. It consists of two ventricles and four SynHall mechanical valves (SynCardia Systems Inc.), allowing for unidirectional flow during pump systole and diastole.

Inside of each of these ventricles lies a four-layer polyurethane diaphragm that partitions the chamber between air and blood. This balloon-

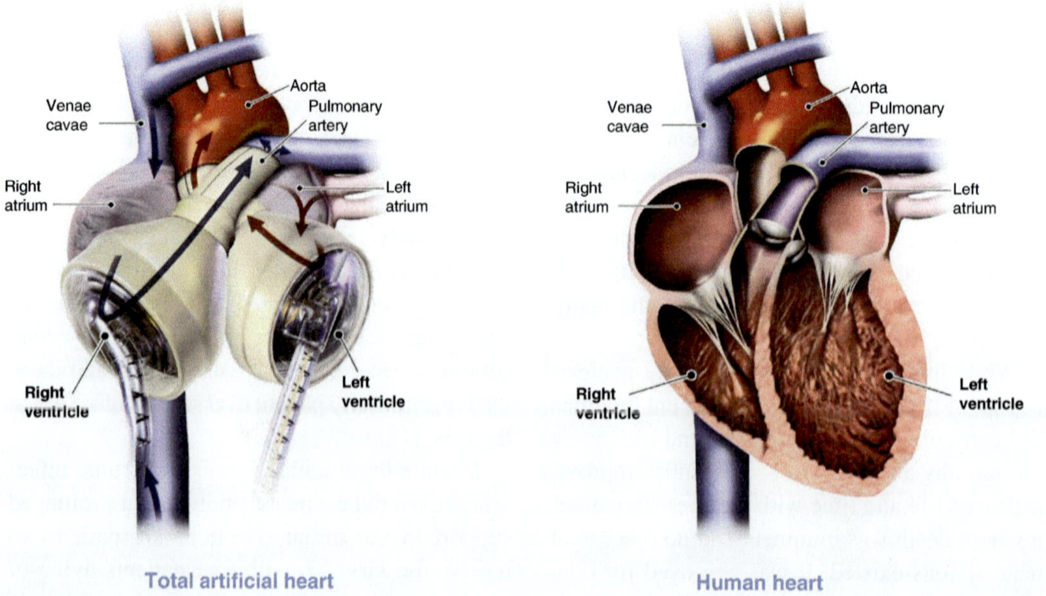

Fig. 14.3 SynCardia is a pneumatically driven pump that alters between pump systole and diastole. (Image courtesy of SynCardia, © SynCardia)

like diaphragm is what allows for the creation of pulsatile flow; it deflates to allow blood to enter the ventricle through the inflow valve and subsequently re-inflates to expel blood from the chamber (Fig. 14.3). Cannulas which connect to the two air-filled spaces and exit the body are what allow for this alternating pattern. Unlike the AbioCor®, which allowed for totally transcutaneous energy delivery, these cannulas exit the body cavity through the abdominal wall where they attach to drivelines which power the device.

Despite the increased risk of infection from this continuous connection, there are advantages to this configuration. In having this direct connection with the external components, the elements of the pump that needed to reside inside of the body could be simplified and minimized. Additionally, the components that were most likely to fail from cyclic fatigue are external to the body, allowing for easier monitoring and maintenance. These external parts are changed preemptively every 6–12 weeks.

The device is connected to either an external drive console that weighs as much as 400 pounds and maintains drive pressure to control the heart rate and duration of pump systole or, more recently, the Freedom Driver, developed as a portable alternative to this external console which allowed for outpatient use. The Freedom Driver has a battery life of 2 hours and weighs 13.11 pounds and able to be worn in a backpack or shoulder bag (Fig. 14.4). Future drivers are in development which will extend battery life and limit weight even further. Despite this advance, most patients require continued hospitalization due to inadequate social support, hemodynamic instability, or concurrent comorbidities. Concerns of driver safety have clouded both the Freedom Driver, which was recalled in 29 patients due to pump failure without any of the traditional alarms sounding, and the second-generation non-portable Companion 2 Driver System (C2), which had increased 1-year mortality and risk of stroke (31% vs. 16%) when compared to patients with the first-generation system.

One Size Does Not Fit All

The design of the SynCardia temporary TAH has extended application of the TAH to patients with thoracic body cavities that were tradition-

Fig. 14.4 Freedom Portable Driver. (Image courtesy of SynCardia, © SynCardia)

ally too small for older models. The use of external drivelines rather than a totally internal system minimizes the electronic hardware that needs to reside inside the body cavity. Additionally, the two ventricles are connected by Velcro, allowing surgeons to tailor the shape of the device to the patient's anatomy. It is also available in two sizes: a 70 cc model and a smaller 50 cc model, which has been used in children and adolescents with a body surface area range between 1.2 and 1.85m^2 (Fig. 14.5). Heart rate and drive pressure are regulated to maintain partial fill of each ventricle, thus reducing risk of thrombus formation while maintaining complete ejection of blood in the ventricle. The 70 cc model is capable of producing a cardiac output as high as >9L/min [19].

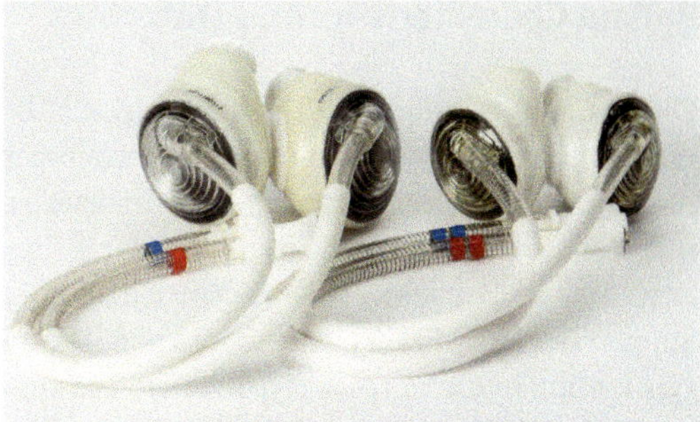

Fig. 14.5 The SynCardia temporary TAH is available in two sizes, 70 cc (left) and more recently 50 cc (right), which has allowed for expansion of use of the device in women and children who were previously unable to receive a TAH due to size limitations. (Image courtesy of SynCardia, © SynCardia)

Clinical Implementation of SynCardia Temporary TAH

In April 2018, 10-year-old Gabriel Gonzalez became the youngest patient to receive a TAH. Diagnosed with left ventricular non-compaction cardiomyopathy, the 50 cc TAH served as a vital bridge to his transplantation 3 months later. Gabriel is one of many children who has been supported by the SynCardia temporary TAH while awaiting an ideal match. Since its FDA approval in 2004 for patients with irreversible biventricular heart failure, >75 individuals have received the smaller 50 cc model and >1700 have received the 70 cc model. Greater than 80% of patients are ambulatory within 2 weeks of surgery, and several patients have survived on the device for greater than 4.5 years.

Perhaps the greatest accomplishment of the 50 cc model is its expansion of the population eligible for TAH therapy to many patients with smaller thoracic cavity size. Unlike the 70 cc TAH where only 11% of implants were in women, 61% of the 50 cc device implants have been in women. Additionally, 18% of the 50 cc TAH implants have been in pediatric patients (< 18 years old).

Patient Characteristics and Outcomes

Patients who receive the TAH are among the sickest patients, complicating most outcomes comparisons to other mechanical support devices. Based on data provided by the Interagency Registry for Mechanically Assisted Circulatory Support (INTERMACS), the predominant etiologies leading to use of the TAH are dilated cardiomyopathy (50%) and ischemic cardiomyopathy (20%), the majority of which had characteristics of right ventricular failure (82%).

One year after surgery, more than half of patients receiving a TAH are transplanted (53.3%), 32.3% die before transplant, and 14.4% are alive with the device still in place at 12 months. One quarter (24%) of those patients were able to be discharged from the hospital with device support. The primary cause of death was multisystem organ failure (36.4%), followed by neurological dysfunction (17.9%) and withdrawal of support (11.7%). Major infection (70%), stroke (22.7%), and gastrointestinal hemorrhage (20%) remain common in the first 6 months following implantation. Major device malfunction during this time period occurred in 7.1% of patients. Implantation at a low-volume center (≤10 TAHs) was greatly associated with increased risk of death, with 71% one-year survival at high-volume centers compared with 57% at low-volume centers [20].

Patients receiving biventricular support are more critically ill preoperatively [20] and have consistently shown lower survival rates post-implantation when compared to patients receiving isolated left ventricular support (Fig. 14.6). In addition to concerns of device failure, severe comorbidities such as pulmonary hypertension

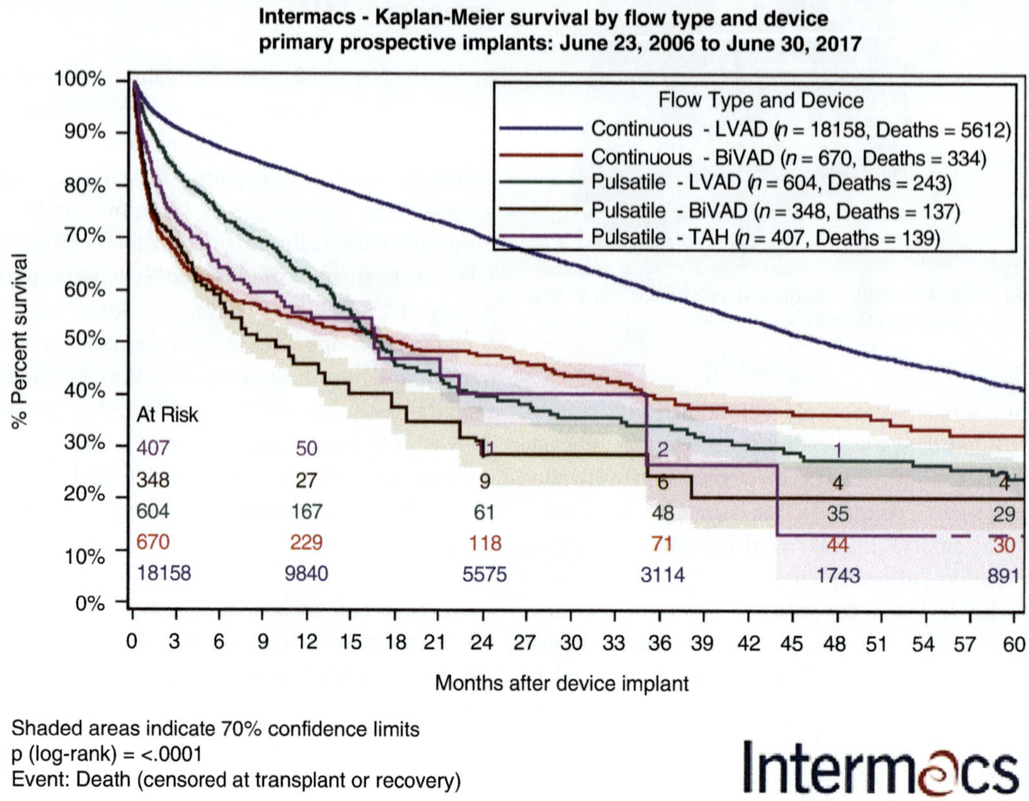

Fig. 14.6 INTERMACS survival data comparing forms of mechanical support devices. Reprinted with permission from INTERMACS

and liver failure are inherently more prevalent in patients requiring biventricular support. Thus, comparing outcomes between those requiring biventricular support and those receiving isolated left ventricular support may not be appropriate given the discrepancy in patient condition and indications for support prior to implantation. Among patients requiring biventricular support, a series of 383 patients showed no difference in survival while on support or after heart transplant between paracorporeal BiVAD, implantable BiVAD, and the TAH. However, survival after TAH was superior in patients requiring prolonged support (>90 days), likely due to the lower incidence of neurological events [15].

Investigational

Continuous-Flow TAH

Most initial TAHs attempted to mimic the heart's inherent pulsatility by using ventricles that alternated configuration between pump systole and diastole. However, pulsatile flow pumps require a chamber as large as the stroke volume to generate this flow, which limits their use to individuals with large thoracic cavities. On the other hand, continuous-flow pumps do not require a large chamber and thus allow for much smaller devices. In the 1980s, Golding and colleagues were the first to show in a mammalian model that non-pulsatile flow could support perfusion for prolonged periods of time [21]. By 1987, Qian and colleagues were developing a single-device rotary TAH but encountered issues balancing flow between the right and left sides [22].

After a stall in this development, several groups began using dual rotary LVADs for the treatment of patients who were ineligible for TAH or traditional LVAD therapy. In 2012, Frazier and Cohn implanted a dual HeartMate II® with customized cuffs (Fig. 14.7a, b) in a 55-year-old patient with systemic amyloidosis, a contraindication for the SynCardia temporary TAH due to technical difficulties. The patient was hemodynamically stable in the postoperative period but eventually died when support was removed 5 weeks later due to progression of amyloidosis in the liver and lungs [23]. However, Frazier and Cohn were able to demonstrate the feasibility of totally non-pulsatile circulatory support.

Currently, two investigational continuous-flow totally artificial hearts (CFTAH), the BiVACOR (Houston, TX, USA) [24] and the Cleveland Heart (Cleveland, OH, USA) [25], now named the SmartHeart, have been designed with unique mechanisms to balance left and right heart flow. Both of the devices contain a single rotating pump with right and left impellers on opposing sides.

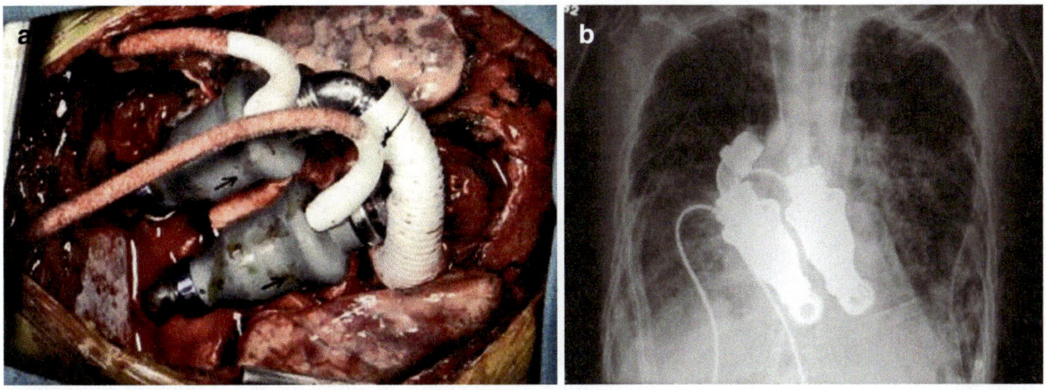

Fig. 14.7 (**a**, **b**) Placement of dual rotary HeartMate II® devices in an individual with amyloidosis who was ineligible for other treatment. This was the first human to be supported by a continuous-flow biventricular support. (Reprinted with permission from Frazier et al. [23])

Despite a constant speed of the impeller due to a singular rotating pump, pulmonary and systemic flows are delivered differently due to differences in impeller size and geometry. CFTAH are substantially simpler than their pulsatile counterparts, eliminating the need for various mechanical bearings and valves, which are subject to wear. Additionally, a singular pump allows for a more compact, energy-efficient device.

Cleveland Clinic SmartHeart CFTAH

Cleveland Heart and Cleveland Clinic are developing the SmartHeart TAH, a CFTAH that is based on many of the same design principles as their SmartHeart LVAD. The singular continuous motor is supported by a blood-lubricated fluid-film bearing. The device contains no sensors or diaphragms. This single motor drives two centrifugal impellers on both ends of the cylindrical shaft to allow for continuous flow. Lack of preload sensitivity due to systemic-pulmonary circulation imbalances, a challenge with CFTAH, is addressed through the use of a passive and self-balancing flow system. When an imbalance in pulmonary-systemic circulation occurs, the axis of rotation is shifted toward the side with lower atrial pressure, thus adjusting the output of that side. During exercise, for example, the physiological rise in systemic vascular resistance will cause left pump flow to decrease and left atrial pressure to increase as a result. In response, the pump's passive hydraulic feedback mechanism will respond to the atrial pressure gradient by shifting the rotor toward the lower-pressured right atrium, thus limiting pulmonary flow until left atrial pressure is decreased and the pressure gradient restored. This same mechanism protects against atrial suctioning [25, 26]. This innovative design allows for minimization of elements like valves and sensors that may cause additional thrombus and limit durability. Additionally, the compact size will continue to allow for implantation in smaller patients who are not candidates for the AbioCor® or SynCardia temporary TAH. Additional features, including remote tracking, will allow for monitoring of key patient parameters by clinical teams.

Though the device remains to be tested in humans, animal testing of the device is ongoing and promising. In vivo testing on 17 calves demonstrated reliable self-regulation and hemodynamic performance with no pump failures or chronic hemolysis to date. In three recent implantations, recovery to planned in vivo durations up to 90 days was achieved without post-operative anti-coagulation, all with good biocompatibility and no evidence of organ thromboembolism [27].

An even smaller device, the pediatric CFTAH (P-CFTAH) [28] has been developed with 30% reductions in all dimensions which will allow for implantation in infants and children with BSA $\geq$ 0.3m^2. Initial testing in four lambs revealed favorable anatomical fit and hemodynamics. Pump speed modulation was performed to generate an arterial pulsation [28]. These devices may contribute significantly as a bridge to transplantation for children with congenital heart defects.

BiVACOR

The BiVACOR continuous-flow pump, originated in 2008 in Australia, is now one of the predominant TAHs in preclinical development. While it is similar in design to the Cleveland Clinic SmartHeart, its central rotating disc is suspended through magnetic levitation in order to maintain pulmonary-systemic balance [24]. The BiVACOR uses positive feedback sensors which act on the electromagnetic bearing system to make minor adjustments in axial position, which can substantially change efficiency of the pumps. When a pressure gradient is sensed, the rotor is pulled toward the underperforming side, increasing the relative efficiency of this side until homeostasis is regained. The design is compact, efficient, and likely to be durable.

During in vitro mock circulation loop tests, the pump was capable of generating 16 L/min at a constant pressure of 100 mmHg. Additionally, in a preliminary in vivo test on a calf, different pump speed profiles were tested for 30 seconds each to determine their ability to generate pulsatility. Through modulation of speed, a physiological blood pressure ranging from 86 to 115 mmHg was generated at 30 bpm [29]. Seven additional calves have since been implanted with a maximum duration of 30 days. The average outflow was 12 L/min to maintain a pressure of 100 mmHg and consumed 13W, but even flows as high as 16L/min were generated with no evidence of hemolysis [9].

CARMAT

The CARMAT bioprosthetic artificial heart (C-TAH) was created as a collaboration in 2008 between technology firm Matra Défense (Airbus Group) and cardiac surgeon Alain Carpentier. The primary innovation of the device is the use of entirely bioprosthetic materials for all blood-facing surfaces, which have demonstrated reduced risk of thrombus formation in biological valves (partnership with Edwards since 2011). It maintains pulsatility through an electric-hydraulic power system, which minimizes noise heard by the patient when compared with pneumatically driven pumps like the SynCardia temporary TAH. Additionally, better mimicry of human physiology was accomplished through the creation of algorithms that incorporate reactions to blood pressure regulation by cardiac muscle and postural changes [30]. These features were all developed around the goal of C-TAH as a destination therapy.

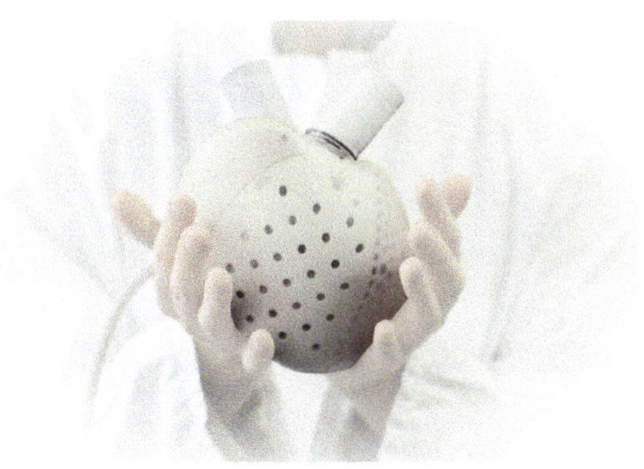

Fig. 14.8 CARMAT total artificial heart. (Reproduced with permission from Carmat)

All components of the device except for the battery reside in one unit, and a single 8 mm driveline exits the patient's skin. This unit contains right and left ventricles, each containing a compartment for blood and a compartment for silicone oil, separated by a hybrid membrane (Fig. 14.8). Pulsatility is created when the membrane is pushed back and forth based on force from the silicone oil which is acted on by a rotary pump. The hybrid membrane consists of a layer of bovine pericardial tissue chemically treated with glutaraldehyde, which is the blood-facing layer, and a bottom layer of polyurethane which contacts the silicone oil. Four Carpentier-Edwards bioprosthetic valves (Edwards Lifesciences; Irvine, CA, USA) and atrial connections are located at the inflow and outflow orifices [31].

Viscoelastic contraction of the device allows for regulation of pressure in a similar manner to the native heart, thus preserving valve durability. Internal ultrasound pressure sensors located inside of the liquid compartment of the ventricles continuously monitor device position, using an internal accelerometer, and pressure gradients to calculate preload and afterload. In order to obtain physician-specified inlet and outlet pressures, the device adjusts heart rate (35–150 beats per minute, uniform between sides) and stroke volume (30–65 mL, varies between sides) in response to sensor feedback, resulting in cardiac outputs between 2 and 9 L/min. This is in contrast to the SynCardia temporary TAH which only adjusts stroke volume in response to increased preload, which has an upper limit of 70 cc [32]. The C-TAH's unique feedback mechanism is also designed with a "fill-to-empty" pumping mode which ensures complete ejection and therefore prevents stasis and clot formation. Similar to the SynCardia temporary TAH, the pump has a 3 kg control that has a battery life of approximately 5 hours. One of the major limitations of the device is its large size, limiting its use to approximately 84% of men and 16% of women based on computed tomography (CT) scan analysis [31].

Clinical

After successful support in animal studies for up to 10 days, an initial feasibility study was done in the European Union in four patients, with three patients surviving for more than 30 days (74, 270, 254 days) [30, 33, 34]. These patients were all in the hospital at imminent risk of death from irreversible biventricular failure, and most adverse events post-implantation were related to the multiple comorbidities of the cohort. Two of the patients were able to return home on the wearable system and had modest improvements in physical capacity on 6-minute walk tests. Two patients died from device-related complications, one died from respiratory failure, and another from multi-organ failure. The causes for the two instances of device failure, failure of an electrical sensor and sudden low device cardiac output, have been identified and corrected in subsequent models. All patients developed renal dysfunction [30].

One of the primary advantages of the C-TAH design is the use of bovine pericardial tissue, which has been shown to acquire hemocompatibility through endogenous protein and fibrin deposition in bioprosthetic valves, becoming essentially thromboresistant after a period of several weeks. Anticoagulation therapy after implantation remains to be optimized for the C-TAH since it has a surface area approximately 24 times the size of a biological valve. The potential advantage of this design is evidenced by no visible evidence of thromboembolic depositions in major organs and device cavities. However, the first patient to receive the device did develop thrombocytopenia not controlled by platelet transfusions and increased fibrin monomers and D-dimers, suggestive of disseminated intravascular coagulation (DIC). Heparin therapy was used in the subsequent patients for early surface anticoagulation until a fibrin cap could form around the device surface [30].

In June 2018, the first patient was bridged from a C-TAH to heart transplantation. Due to his pulmonary hypertension, he was not initially a candidate for heart transplantation, but the C-TAH was able to support him for 8 months and he recovered from his pulmonary hypertension, thus making him eligible for transplantation. He is one of ten patients enrolled in the PIVOTAL clinical trial in the Czech Republic, Denmark, France, and Kazakhstan. Two separate cohorts of ten patients each will be implanted with the CARMAT with a primary end point of 180 days post-transplant survival or transplantation if prior to the 180-day mark. Surgical success has been 100%, and after initially promising results in the first ten patients, implantation began in the second cohort of patients [35].

ReinHeart® TAH

Despite its ultimate discontinued production, the AbioCor® showed that an entirely transcutaneous energy source was possible. However, these devices have been limited based on concerns about the durability of these complex systems. Since its inception in 2009, the ReinHeart® TAH (Aachen University, Germany) has the goal of being totally implantable yet relatively mainte-

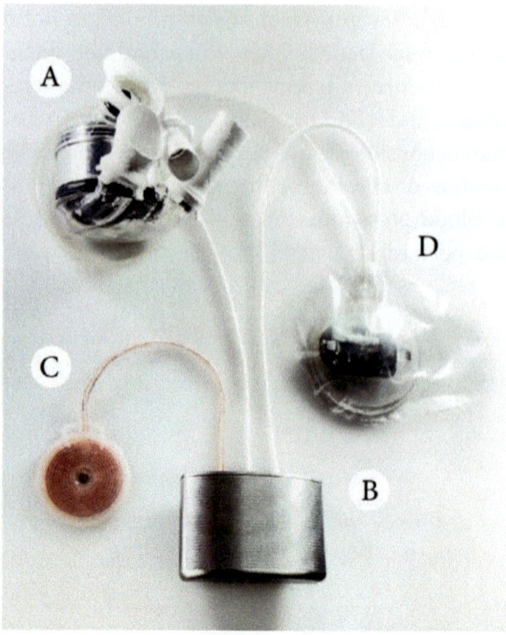

Fig. 14.9 ReinHeart® TAH implantable components: (**a**) pump unit, (**b**) internal controller, (**c**) internal coil, (**d**) compliance chamber. (Reprinted Pelletier et al. [36], with permission from Oxford University Press)

nance free and small enough to fit inside most body cavities [36].

The implantable components consist of a pump unit, internal controller, internal coil, and compliance chamber (Fig. 14.9). Similar to other pulsatile TAHs, the pump consists of two ventricles, both of which have a polyurethane membrane that separates them from a centrally located driver with disk-shaped pusher plates. This drive alternates its central axial location to displace fluid in one ventricle at a time, thus creating an alternating pulsatility with a maximal stroke volume of 60 cc. The only moving components to this linear drive are a bobbin and springs, which depend upon input from position and temperature sensors to drive their movement. The polyurethane membrane that separates the ventricle is not directly connected to the driver's disks and thus allows for sensitivity to preload and mitigation of suctioning. Additionally, an implanted compliance chamber helps regulate the balance between left and right ventricular filling through direction supplied by an implantable controller. The controller provides power supply to the pump through its connection with a TET energy transfer system analogous to that previously described for the AbioCor®. In the event of a disconnection between the TET system and the externally worn power supply, the internal battery can support the individual for 45 minutes. Two external batteries weighing <2kg total can provide energy to the TET system for up to 12 hours, or an additional third battery (1kg) can be added to provide 18 hours between charging. Each of the ventricles has four unidirectional bileaflet mechanical valves and attaches in a novel snap manner to prosthesis which can be tailored to the individual's anatomy.

The ReinHeart® is still early in its clinical development. Mock circulation loops displayed appropriate preload sensitivity, a 5L/min cardiac output when the pump cycled 120 times per minute, and a 99.4% washout of blood in the ventricles every 3 cycles, which should be high enough to prevent against clot formation. In vivo testing in bovine models is ongoing, and as of 2015, 12 animal trials reported complete circulatory support for up to 2 days. The placement of these devices proved to be more complex than expected and will likely be more complicated than that of the currently used SynCardia temporary TAH. Bleeding, pulmonary complications, and technical problems were the main reasons of termination in these early animal models. Next steps in the development of this model include expansion to a 90-day animal trial, durability testing of the linear drive for prolonged periods of time (4–7 years), and reduction of device weight to 800g [36].

Future Strategies

Expanding Patient Populations

While the TAH is predominantly used in patients with biventricular failure, this population appears to be expanding. In the past few years, acute and chronic heart transplant failure, cardiac malignancies, restrictive and infiltrative cardiomyopathies, partial ventricular thromboses, and other diagnoses have become increasingly common prior to TAH placement [20]. Additionally, these devices have successfully bridged patients experiencing LVAD and BiVAD failure [37].

These devices are predominantly used as a bridge to transplant or as a bridge to candidacy (59.3 and 37.6%, respectively); however, destination therapy is the fastest-growing indication for mechanical support and is the goal of many TAHs in development, such as the CARMAT [38]. Additional clinical trials are currently investigating the use of the SynCardia 70 cc temporary TAH as destination therapy. Should the future iterations of TAH technology prove viable as permanent therapy, this would have profound implications on the treatment of end-stage heart failure.

The development of the SynCardia 50 cc temporary TAH currently under clinical investigation as bridge-to-transplantation therapy has allowed for preliminary expansion to the pediatric population [39]. Improved surgical and medical management for congenital heart disease (CHD) substantially increased the number of patients surviving with CHD, and these patients are being successfully bridged to transplantation [40]. CHD patients are increasingly being bridged to transplantation with VAD, but this remains challenging due to the complexity in anatomic variation of CHD; TAH has successfully bridged select patients to transplantation and may be preferential in this population [41–45]. In particular, the TAH is uniquely capable of providing support for the multiple levels of failure leading to Fontan failure and can simultaneously increase cardiac output and lower CVP to address restrictive heart failure [40]. In addition to the pneumatically driven pump in clinical trials, other TAH in development, such as the P-CFTAH, may also be uniquely suited for this population [28].

Importance of Pulsatility

Despite the improvements in durability, miniaturization, and patient quality of life that continuous-flow devices have accomplished, they have also been associated with a unique side effect profile compared to previous generation devices including gastrointestinal bleeding, aortic valve insufficiency, impaired renal function, and thromboembolic complications [46]. While we know that natural selection has preserved pulsatile blood flow for 750 million years in our species, we do not fully understand the implications of attenuating or eliminating this pulsatility, such as occurs with continuous-flow pumps. While continuous-flow devices present much of the promise in TAH investigation, further understanding of the vascular changes associated with continuous-flow physiology may allow for improvement in patient outcomes and device design [8]. The CF-TAH was the first to show that speed modulation could be used to generate variations in systemic arterial pressure, potentially mitigating some of the adverse effects of continuous-flow devices.

Conclusion

Over the past half-century, significant milestones in the development of a replacement for the dying heart have been achieved. The only current clinically used TAH uses a pneumatically driven pump to generate pulsatile blood flow. TAH design has shifted toward continuous-flow devices which are smaller and more durable than their pulsatile counterparts. However, a deeper understanding of the consequences associated with losing or attenuating pulsatility is needed, and issues surrounding biocompatibility remain. While use of the TAH is currently limited to individuals with end-stage biventricular heart failure who are not candidates for other therapy, newer TAH designs show promise as destination therapy, which would significantly alleviate the increasing clinical burden of heart failure. Despite the innovation in this field, the development of a viable TAH with improved survival free of major adverse events and quality of life comparable to heart transplant recipients remains a lofty goal. Future changes in regulation may expand the role of heart transplantation in patients with severe biventricular failure. In that scenario, TAH technologies will be reserved for patients in extremis in whom a transplantation may not be viable.

References

1. Colvin M, Smith JM, Hadley N, Skeans MA, Carrico R, Uccellini K, et al. OPTN/SRTR 2016 annual data report: heart. Am J Transplant. 2018;18(Suppl 1):291–362.
2. Benjamin EJ, Blaha MJ, Chiuve SE, Cushman M, Das SR, Deo R, et al. Heart disease and stroke statistics-2017 update: a report from the American Heart Association. Circulation. 2017;135(10):e146–603.
3. Heidenreich PA, Albert NM, Allen LA, Bluemke DA, Butler J, Fonarow GC, et al. Forecasting the impact of heart failure in the United States: a policy statement from the American Heart Association. Circ Heart Fail. 2013;6(3):606–19.
4. Mancini D, Colombo PC. Left ventricular assist devices: a rapidly evolving alternative to transplant. J Am Coll Cardiol. 2015;65(23):2542–55.
5. Stehlik J, Edwards LB, Kucheryavaya AY, Benden C, Christie JD, Dobbels F, et al. The registry of the International Society for Heart and Lung Transplantation: twenty-eighth adult heart transplant report--2011. J Heart Lung Transplant. 2011;30(10):1078–94.
6. Renlund DG. Building a bridge to heart transplantation. N Engl J Med. 2004;351(9):849–51.
7. Goerlich CE, Frazier OH, Cohn WE. Previous challenges and current progress-the use of total artificial hearts in patients with end-stage heart failure. Expert Rev Cardiovasc Ther. 2016;14(10):1095–8.

8. Purohit SN, Cornwell WK 3rd, Pal JD, Lindenfeld J, Ambardekar AV. Living without a pulse: the vascular implications of continuous-flow left ventricular assist devices. Circ Heart Fail. 2018;11(6):e004670.
9. Cohn WE, Timms DL, Frazier OH. Total artificial hearts: past, present, and future. Nat Rev Cardiol. 2015;12(10):609–17.
10. LeGallois J. Experiments on the principles of life. Paris: D'Hautel; 1812.
11. Akutsu T, Kolff WJ. Permanent substitutes for valves and hearts. Am Soc Art Int Org. 1959;4:230–5.
12. SoRelle R. First AbioCor trial patient dies. Circulation. 2001;104(24):E9050–60.
13. Hamilton A. Abiocor Artificial Heart2001 March 7, 2018. Available from: http://content.time.com/time/specials/packages/article/0,28804,1936165_1936238_1936258,00.html.
14. Dowling RD, Gray LA Jr, Etoch SW, Laks H, Marelli D, Samuels L, et al. Initial experience with the AbioCor implantable replacement heart system. J Thorac Cardiovasc Surg. 2004;127(1):131–41.
15. Samuels L, Entwistle J, Holmes E, Fitzpatrick J, Wechsler A. Use of the AbioCor replacement heart as destination therapy for end-stage heart failure with irreversible pulmonary hypertension. J Thorac Cardiovasc Surg. 2004;128(4):643–5.
16. AbioCor implantable replaceable heart. Summary of safety and probable benefit - H04006. In: Administration FaD, editor, 2006.
17. SoRelle R. Third abiocor artificial heart implanted in Houston. Circulation. 2001;104(15):E9033–4.
18. DeVries WC. The permanent artificial heart. Four case reports. JAMA. 1988;259(6):849–59.
19. Sunagawa G, Horvath DJ, Karimov JH, Moazami N, Fukamachi K. Future prospects for the total artificial heart. Expert Rev Med Devices. 2016;13(2): 191–201.
20. Arabia FA, Cantor RS, Koehl DA, Kasirajan V, Gregoric I, Moriguchi JD, et al. Interagency registry for mechanically assisted circulatory support report on the total artificial heart. J Heart Lung Transplant. 2018;37:1304.
21. Golding LR, Jacobs G, Murakami T, Takatani S, Valdes F, Harasaki H, et al. Chronic nonpulsatile blood flow in an alive, awake animal 34-day survival. Trans Am Soc Artif Intern Organs. 1980;26:251–5.
22. Qian KX, Pi KD, Wang YP, Zhao MJ. Toward an implantable impeller total heart. ASAIO Trans. 1987;33(3):704–7.
23. Frazier OH, Cohn WE. Continuous-flow total heart replacement device implanted in a 55-year-old man with end-stage heart failure and severe amyloidosis. Tex Heart Inst J. 2012;39(4):542–6.
24. Timms D, Fraser J, Hayne M, Dunning J, McNeil K, Pearcy M. The BiVACOR rotary biventricular assist device: concept and in vitro investigation. Artif Organs. 2008;32(10):816–9.
25. Fumoto H, Horvath DJ, Rao S, Massiello AL, Horai T, Takaseya T, et al. In vivo acute performance of the Cleveland Clinic self-regulating, continuous-flow total artificial heart. J Heart Lung Transplant. 2010;29(1):21–6.
26. Fukamachi K, Horvath DJ, Massiello AL, Fumoto H, Horai T, Rao S, et al. An innovative, sensorless, pulsatile, continuous-flow total artificial heart: device design and initial in vitro study. J Heart Lung Transplant. 2010;29(1):13–20.
27. Karimov JH, Moazami N, Kobayashi M, Sale S, Such K, Byram N, et al. First report of 90-day support of 2 calves with a continuous-flow total artificial heart. J Thorac Cardiovasc Surg. 2015;150(3):687–93 e1.
28. Karimov JH, Horvath DJ, Byram N, Sunagawa G, Kuban BD, Gao S, et al. Early in vivo experience with the pediatric continuous-flow total artificial heart. J Heart Lung Transplant. 2018;37(8):1029–34.
29. Kleinheyer M, Timms DL, Greatrex NA, Masuzawa T, Frazier OH, Cohn WE. Pulsatile operation of the BiVACOR TAH - motor design, control and hemodynamics. Conf Proc IEEE Eng Med Biol Soc. 2014;2014:5659–62.
30. Latremouille C, Carpentier A, Leprince P, Roussel JC, Cholley B, Boissier E, et al. A bioprosthetic total artificial heart for end-stage heart failure: results from a pilot study. J Heart Lung Transplant. 2018;37(1):33–7.
31. Carpentier A, Latremouille C, Cholley B, Smadja DM, Roussel JC, Boissier E, et al. First clinical use of a bioprosthetic total artificial heart: report of two cases. Lancet. 2015;386(10003):1556–63.
32. Bizouarn P, Roussel JC, Trochu JN, Perles JC, Latremouille C. Effects of pre-load variations on hemodynamic parameters with a pulsatile autoregulated artificial heart during the early post-operative period. J Heart Lung Transplant. 2018;37(1):161–3.
33. Latremouille C, Duveau D, Cholley B, Zilberstein L, Belbis G, Boughenou MF, et al. Animal studies with the Carmat bioprosthetic total artificial heart. Eur J Cardiothorac Surg. 2015;47(5):e172–8; discussion e8-9.
34. Smadja DM, Susen S, Rauch A, Cholley B, Latremouille C, Duveau D, et al. The Carmat bioprosthetic total artificial heart is associated with early hemostatic recovery and no acquired von Willebrand syndrome in calves. J Cardiothorac Vasc Anesth. 2017;31(5):1595–602.
35. CARMAT completes patient enrollment in the first part of the PIVOTAL study in line with the objective of obtaining CE marking in 2019 [press release]. July 11th, 2018.
36. Pelletier B, Spiliopoulos S, Finocchiaro T, Graef F, Kuipers K, Laumen M, et al. System overview of the fully implantable destination therapy--ReinHeart-total artificial heart. Eur J Cardiothorac Surg. 2015;47(1):80–6.
37. Arabia FA, Moriguchi JD. Machines versus medication for biventricular heart failure: focus on the total artificial heart. Futur Cardiol. 2014;10(5):593–609.
38. Miller LW, Rogers JG. Evolution of left ventricular assist device therapy for advanced heart failure: a review. JAMA Cardiol. 2018;3(7):650–8.

39. Wells D, Villa CR, Simon Morales DL. The 50/50 cc total artificial heart trial: extending the benefits of the total artificial heart to underserved populations. Semin Thorac Cardiovasc Surg Pediatr Card Surg Annu. 2017;20:16–9.
40. Villa CR, Morales DLS. The total artificial heart in end-stage congenital heart disease. Front Physiol. 2017;8:131.
41. Morales DL, Khan MS, Gottlieb EA, Krishnamurthy R, Dreyer WJ, Adachi I. Implantation of total artificial heart in congenital heart disease. Semin Thorac Cardiovasc Surg. 2012;24(2):142–3.
42. Kirsch ME, Nguyen A, Mastroianni C, Pozzi M, Leger P, Nicolescu M, et al. SynCardia temporary total artificial heart as bridge to transplantation: current results at la pitie hospital. Ann Thorac Surg. 2013;95(5):1640–6.
43. Kirsch M, Mazzucotelli JP, Roussel JC, Bouchot O, N'Loga J, Leprince P, et al. Survival after biventricular mechanical circulatory support: does the type of device matter? J Heart Lung Transplant. 2012;31(5):501–8.
44. Rossano JW, Goldberg DJ, Fuller S, Ravishankar C, Montenegro LM, Gaynor JW. Successful use of the total artificial heart in the failing Fontan circulation. Ann Thorac Surg. 2014;97(4):1438–40.
45. Morales DLS, Lorts A, Rizwan R, Zafar F, Arabia FA, Villa CR. Worldwide experience with the syncardia total artificial heart in the pediatric population. ASAIO J. 2017;63(4):518–9.
46. Moazami N, Dembitsky WP, Adamson R, Steffen RJ, Soltesz EG, Starling RC, et al. Does pulsatility matter in the era of continuous-flow blood pumps? J Heart Lung Transplant. 2015;34(8):999–1004.

15 Extracorporeal Membrane Oxygenation Techniques and Modern Considerations

Aly El Banayosy

Extracorporeal life support (ECLS) encompasses various techniques that support patients with severe cardiac and/or pulmonary failure. In recent years, there has been a substantial increase in the use of ECLS in a broad population of patients with cardiopulmonary failure [1]. Extracorporeal membrane oxygenation (ECMO) may be configured as venoarterial ECMO (VA-ECMO) for cardiac support, veno-venous ECMO (VV-ECMO) for respiratory support, veno-arterial-veno ECMO (VAV-ECMO) for cardiac and respiratory failure, or extracorporeal carbon dioxide removal (ECCO2R) for lung protection [2]. ECMO to restore circulation during cardiac arrest is referred to as extracorporeal cardiopulmonary resuscitation (ECPR). In recent years, there has been a dramatic increase in the use of VA-ECMO for cardiogenic shock. This variation of ECMO is the primary focus of discussion in this chapter.

ECMO support is provided for stabilization of salvageable patients when death is imminent and other treatment options are not available. ECMO techniques are a modification of the standard cardiopulmonary bypass technique and allow for support durations (typically days to weeks) longer than usual cardiac surgery. ECMO support allows time for recovery of cardiac or pulmonary function or bridges patients to other definitive therapy, such as transplantation or durable left ventricular assist device (LVAD) support.

ECMO is a high-risk and complex therapy that requires substantial resources for initiation and continuing support. Effective ECMO requires an infrastructure or system of care with medical and technical expertise in all phases of advanced cardiopulmonary care. Ideally, a network consisting of a tertiary care center that provides all facets of advanced care is partnered with community healthcare providers to allow treatment of a broad range of patients in various communities [3–5]. Patient access to advanced cardiopulmonary care (including ECMO, transplant, durable LVAD, and other devices) with optimal medical therapy is an important challenge today.

Since the introduction of ECMO in the mid-1970s, its use has varied widely, with application predominantly in pediatrics until the 2009 H1N1 influenza epidemic, which affected a large number of adults [6–8]. Advancements in ECMO technology has improved its safety and portability, which has led to a dramatic increase in its use [9]. The ECMO circuit is smaller than the usual cardiopulmonary bypass, cannulae designed for ECMO are readily available, and the entire system is portable. These advances have allowed expansion into multiple hospital environments, such as the intensive care unit, cardiac catheterization laboratory, emergency department, and operating room, but also have allowed ECMO therapy to be instituted in community hospitals,

A. El Banayosy, MD (✉)
INTEGRIS Baptist Medical Center,
Oklahoma City, OK, USA

© Springer Nature Switzerland AG 2020
J. H. Karimov et al. (eds.), *Mechanical Support for Heart Failure*,
https://doi.org/10.1007/978-3-030-47809-4_15

permitting stabilization and safe transport to a tertiary care center. The application of VA-ECMO for cardiogenic shock has increased considerably in recent years. The ability to implement support in a variety of settings is important for the rapid restoration of circulation with adequate oxygenation. Prompt access to ECMO support avoids multiple organ failure, which is key to successful outcomes of the therapy [10–12]. ECMO support stabilizes the patient's condition, allowing time for necessary transport, accurate diagnosis of underlying pathology, and definitive therapy in a controlled environment.

The current growth rate of ECMO is outpacing the evidence supporting the current clinical applications and management practices [13, 14]. Standards of care for ECMO have not been established, and most practices are not supported by randomized, controlled clinical trials [15]. The goal of this chapter is to provide a contemporary review of the use of ECMO for support during profound cardiogenic shock and pulmonary failure.

ECMO Indications and Contraindications

ECMO may be indicated for support of patients with severe pulmonary or cardiac failure, or a combination of both. Because ECMO is not intended to resolve the primary etiology of cardiac or pulmonary failure, it is a temporary bridge to recovery or other definitive therapy [16]. Indications for VA-ECMO in cardiac failure include cardiogenic shock, refractory ventricular arrhythmia, ongoing cardiopulmonary resuscitation for cardiac arrest, and acute decompensated heart failure (Table 15.1) [15]. The leading causes of cardiogenic shock are acute myocardial infarction (70% to 80%) and acute decompensated chronic heart failure (up to 30%), and these conditions comprise the majority of patients needing ECMO support. It is about 50% in our multi-institutional high-volume shock network [17–31]. Patients with witnessed cardiac arrest with cardiopulmonary resuscitation (CPR) in progress may be candidates for VA-ECMO if there is a reasonable possibility that neurologic function is intact. In many patients, ECMO support as a bridge to decision is necessary to determine viability for other treatments. Patients with irreversible end-organ function may not be suitable candidates for transplantation or durable mechanical circulatory support. Patients may need a period of stabilization and assessment of neurologic function. Bridging to recovery for cardiac failure may result in successful weaning without further need for mechanical assistance, and patients are treated medically. A bridge-to-bridge approach may be appropriate for patients who cannot be weaned from support and are candidates for a durable LVAD or artificial heart. Bridge to transplant is an option for a few patients, but donor availability limits this treatment to a small number of patients.

Table 15.1 Indications for ECMO

Indications for ECMO
Indications for VA-ECMO
Severe refractory cardiogenic shock from
Myocardial infarction
Postcardiotomy/failure to wean cardiopulmonary bypass
Myocarditis
Acute decompensated chronic heart failure (cardiomyopathies and congenital heart disease)
Acute allograft failure/rejection
Refractory ventricular tachycardia
Cardiotoxicity/drug overdose
Severe right ventricular failure following LVAD or transplant
Septic shock
Pulmonary embolism
Cardiac arrest with CPR in progress
Indications for VV-ECMO
ARDS of any etiology, with $PaO_2/FiO_2 < 100$ on $FiO_2 > 90\%$
CO_2 retention on mechanical ventilation with a peak pressure > 30 cm H_2O
Severe air leak syndromes
Need for mechanical ventilation in patients on the lung transplant waiting list
Rapid respiratory and/or cardiac decompensation refractory to standard care (e.g., pulmonary embolism, airway obstruction)

Abbreviations: *VA-ECMO* venoarterial extracorporeal membrane oxygenation, *CPR* cardiopulmonary resuscitation, *LVAD* left ventricular assist device, PaO_2 arterial partial pressure of oxygen, FiO_2 fraction of inspired oxygen

Use of VV-ECMO for primary pulmonary failure with adequate cardiac function is almost always considered a bridge to recovery while the lungs recover from infection or other cause of severe adult respiratory distress syndrome (ARDS) (Table 15.1). Etiologies for severe respiratory failure that may require VV-ECMO include viral, bacterial, or fungal lung infections, primary lung disease (e.g., cystic fibrosis, primary pulmonary hypertension, idiopathic fibrosis), chest trauma, and bronchiolitis obliterans. Patients with severe air leak from any cause, when adequate oxygenation and CO_2 removal cannot be maintained, should be promptly supported with VV-ECMO. Acute severe respiratory failure in asthma or chronic obstructive pulmonary disease (COPD) may also benefit from this therapy.

ECMO Contraindications

Most contraindications to ECMO support are relative, while the risks are considered together with the potential benefits for individual patients. Relative contraindications include irreversible organ function, particularly the brain; conditions decreasing quality of life (neurologic damage, malignancy, severe risk of bleeding with anticoagulation); older age; patients who have been on conventional therapy for long duration; and aortic dissection. Although older age is associated with higher mortality, it should not be considered an absolute contraindication [20]. Lack of exit strategies should be considered as a contraindication.

The ECMO System

The basic ECMO system has a centrifugal blood pump, a membrane oxygenator, heat exchanger, inflow and outflow cannula, and tubing to establish a circuit between these components and the patient's circulatory system (Fig. 15.1). Pumps most commonly used include the CentriMag (Abbott, Chicago, Illinois, USA), Rotaflow (Maquet, Rastatt, Germany), or TandemHeart (TandemLife, Pittsburgh, Pennsylvania, USA). Blood is drained from the inflow cannula and is pumped through the oxygenator and back to the patient through the outflow cannula. The gas flow rate and the fraction of inspired oxygen (F_iO_2) through the oxygenator control the exchange of oxygen and carbon dioxide with the blood. VA-ECMO draws blood from the venous circulation and returns oxygenated blood to the arterial circulation; it is used primarily for cardiac support while providing pulmonary assist. VV-ECMO differs by returning oxygenated blood to the pulmonary circulation and is used for severe pulmonary failure in the presence of adequate native cardiac function (Fig. 15.2).

Management of ECMO

Cannulation

Initiation of VA-ECMO support with open chest cannulation may take place in the operating room for cases of postcardiotomy shock; however, the majority of cases are initiated outside of the operating room [21]. VA-ECMO support can be initiated quickly using percutaneous cannulation in multiple hospital environments, such as the cardiac catheterization laboratory, operating room, emergency department, and intensive care unit. Appropriate cannulation for ECMO is critical for successful outcomes. Central cannulation usually is performed for postcardiotomy support when the chest is open; peripheral cannulation is most often employed in nonsurgical cases. Cannulation is initiated only after all team members and system components are at the bedside. A bolus of heparin, 50–100 units per kilogram, is given just before cannula placement. Cannulation for VA-ECMO is most commonly via a femoral vein (inflow) and a femoral artery (outflow). Venous cannulas are 19–25 Fr in diameter with a length up to 60 cm and may be of single- or multistage configuration. The outflow cannula placed in the arterial system is 15–19 Fr in diameter and 20–25 cm long. Cannula sizes vary with the size of the patient, and the proper cannula diameter can be determined by assessment of the vessel diameter using vascular ultrasound. The Fr gauge

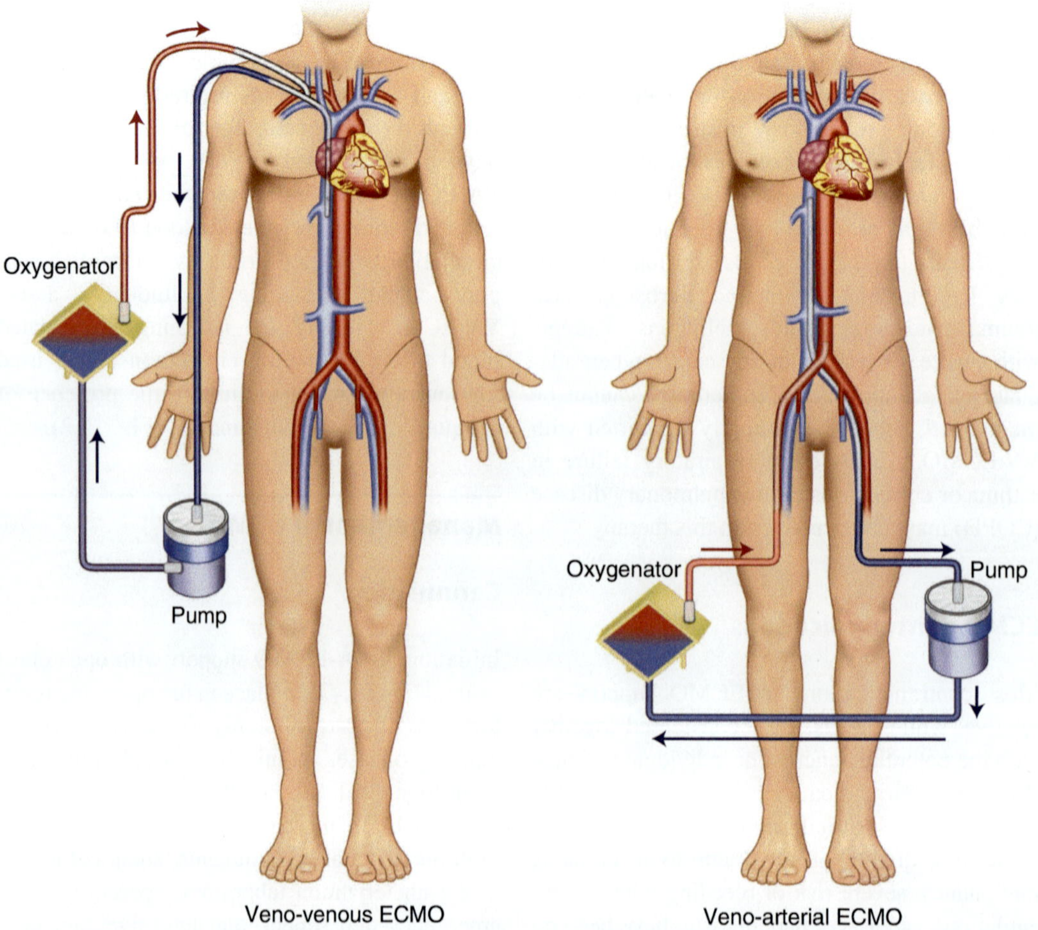

Fig. 15.1 Veno-venous (VV) ECMO (left) receives deoxygenated blood from the vena cava and returns oxygenated blood to the pulmonary artery. Venoarterial (VA) ECMO receives deoxygenated blood from the vena cava and returns oxygenated blood to the descending aorta

size used should be 3 times the size of the target vessel in mm [22]. Vascular ultrasound is useful to locate the target vessel and to assess for anatomic problems. The femoral arterial cannula ideally is at least more than 2 mm smaller than the vessel diameter to allow flow to the lower extremity. A properly selected cannula size allows enough blood flow to the leg and avoids ischemia. Near-infrared spectroscopy provides an assessment of tissue oxygenation and should be used to assess for leg ischemia [23].

Except when ECPR has been initiated, cannulation to provide blood flow distal to the cannulation site should be performed. Distal cannulation is completed by insertion of a 5–7 Fr sheath connected to the arterial outflow. Retrograde and antegrade cannulation can be accomplished by placement of a T-shaped Dacron graft on the femoral artery. This should be used when large cannula sizes are inserted or when femoral unilateral arterial and venous cannulation is performed. When distal arterial blood flow to the leg is inad-

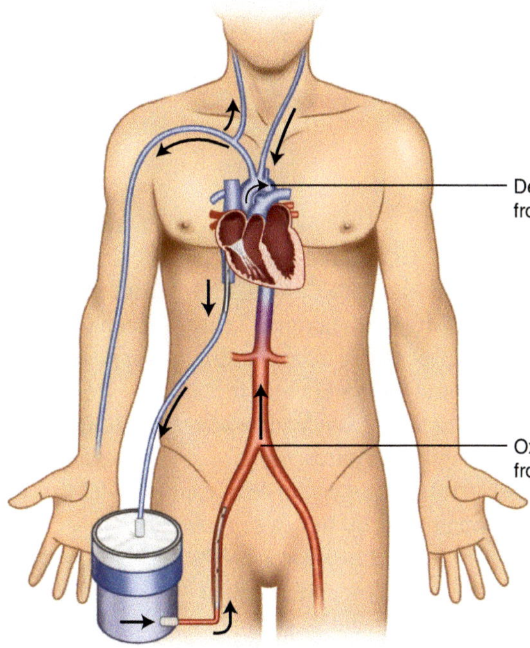

Fig. 15.2 Harlequin syndrome during ECMO support occurs when the left ventricle pumps deoxygenated blood to the head and upper extremities and oxygenated blood only reaches the lower extremities

equate, a separate perfusion cannula should be placed in the distal superficial femoral artery by cutdown or to the posterior tibial artery for retrograde perfusion.

When venous drainage is inadequate, an additional venous drainage cannula through a different vein may be necessary. It may be possible to change the cannula to a larger size, although removal and replacement of the cannula is sometimes very challenging. A damaged, kinked, or clotted cannula must be exchanged with a new cannula.

Initiating ECMO Support

The implementation of VA-ECMO for cardiac failure is usually more urgent than implementing VV-ECMO for respiratory failure [24, 25]. After a thorough assessment of the patient's condition is completed by the ECMO team, the risk and benefits are carefully considered to avoid fruitless use of resources. Once the decision to proceed with ECMO support has been made, a lead physician coordinates cannulation and initiation of support. The initiation of ECMO therapy is extremely time sensitive, and numerous activities take place concurrently or in rapid succession until the patient's condition is stabilized [26]. Perfusionists prepare the ECMO circuit, and nurses provide medical therapy as directed. The implanting physician determines the cannulation locations and cannula sizes to be inserted. Appropriate selection of cannula size should allow for full-support flow rate of 50–70 ml/kg/min in adults. After initiating support, VA-ECMO flow rate should be maintained in appropriate range to allow for adequate tissue perfusion or oxygenation, which is usually 4–6 l/min.

Monitoring

Patients should be monitored with an arterial pressure line and a pulmonary artery catheter. Continuous monitoring of the mixed venous oxygen saturation provides a real-time assessment of system perfusion. Pulmonary artery pressures are useful in assessing for left ventricular distension and the need for venting of the ventricle. Liberal use of echocardiography is invaluable in assessing for ventricular distension and the need for left ventricular venting [27]. It is also important to assess valvular function and thrombus formation within the heart.

Blood Gases

Oxygen delivery from ECMO should provide an arterial saturation greater than 95% (VA-ECMO) or at least 80% (VV-ECMO) with nominal ventilatory support. The difference between arterial and venous oxygen saturation is normally in the range of 20% to 30% when there is adequate oxygen delivery and consumption. A hematocrit of at least 40% assures adequate oxygen delivery. Removal of CO_2 is controlled by the sweep gas flow rate through the ECMO membrane. The ratio of gas flow rate to blood flow rate is initially set at 1:1 and then titrated to maintain the partial pressure of CO_2 ($PaCO_2$) within the desired range. Carbogen (5% CO_2/95% O_2) may be used as the sweep gas to help maintain the outlet $PaCO_2$ at about 40 mmHg. If the $PaCO_2$ is greater than 70 mmHg, increasing the sweep gas gradually over several hours will help prevent rapid change in the arterial pH.

Anticoagulation

Most of the modern ECMO circuits require low level of systemic anticoagulation to prevent thrombosis within the circuit. Heparin is the most common anticoagulant and is monitored by activated thromboplastin time, (40–60 seconds), activated clotting time (1.5–2.5 times baseline), or indirect heparin concentration with anti-factor Xa (0.3–0.5 IU/ml) [28]. Thromboelastography (TEG) may also be useful to assess the time and density of clot formation in response to kaolin. Clot density is affected by clotting factors, platelets, and fibrinolysis, so TEG provides more information than activated clotting time. TEG can be performed with or without an agent that inactivates heparin to separate the anticoagulant effect of heparin from other factors. In patients with heparin-induced thrombocytopenia (HIT), direct thrombin inhibitors, such as parenteral bivalirudin and argatroban, may be used [29]. When alternative anticoagulants are used, the activated partial thromboplastin time should be maintained between 50 and 60 seconds.

Ventilator Management

Since ECMO provides full gas exchange, mechanical ventilation is less critical during full support. However, adequate respiratory function is necessary before weaning is begun. Endotracheal intubation is often necessary to maintain the airway. Mechanical ventilation with low tidal volume (3–5 ml/kg) with low airway pressure is desired to prevent lung injury. Positive end-expiratory pressure (PEEP) of 10–15 mmHg helps to maintain alveolar expansion. The FiO_2 should be kept below 0.40.

Complications of ECMO

Bleeding

Bleeding is a common complication of ECMO support due to the necessary anticoagulation therapy and hematologic abnormalities. Blood transfusion should be used to maintain a hemoglobin concentration of at least 10 mg/dl. Bleeding can be minimized by careful attention to hemostasis at the cannula insertion sites and careful monitoring of anticoagulation therapy. In some cases, anticoagulation therapy may be decreased or stopped to help control excessive bleeding [30, 31].

Stroke

The rate of ischemic and hemorrhagic stroke during ECMO is approximately 4% [32]. The causes of stroke are due to anticoagulation therapy, the artificial surfaces of the ECMO circuit, and hemodynamic instability. Maintaining adequate flow rates and careful monitoring of anticoagulation status with adjustments in therapy are vital.

Anticoagulation should be increased in low-flow states and during weaning, and excessive dosing of anticoagulants must be avoided. Cannulation of the femoral artery is associated with a much lower stroke rate than cannulation of the carotid artery.

Infection

Infection is common in patients being supported by ECMO, with over 50% of patients acquiring bacteremia, which has a mortality rate greater than 60% [16]. As much as possible, sterile techniques should be used during emergent cannulation, and antibiotic therapy should be provided throughout support. Usual surgical infection-control measures must be employed.

Limb Ischemia

A potential complication of femoral arterial cannulation occurs when the outer cannula diameter is near equal to the inner diameter of the blood vessel, resulting in very low or absent distal blood flow [31]. Although this complication occurs in less than 5% of cases, it can be effectively treated with proper recognition and treatment [23]. When time permits, an ultrasound procedure is performed on the cannulation target vessel to measure its diameter, permitting optimal cannula size selection. A proper cannula size will allow enough blood flow to the leg. Leg ischemia can be monitored using near-infrared spectroscopy on the leg, which provides an assessment of tissue oxygenation. Retrograde and antegrade cannulation is performed by placement of a T-shaped Dacron graft on the femoral artery, with cannula directed in both directions.

Left Ventricular Overload

In certain conditions, aortic retrograde blood flow during VA-ECMO may increase left ventricular afterload, which increases left ventricular end-diastolic pressure, left atrial pressure, and pulmonary wedge pressure [23, 33]. The elevated left-sided pressure can lead to pulmonary edema, hemoptysis, and poor gas exchange [34]. Hypoxemia may become severe, and poorly oxygenated blood from the left ventricle enters the cerebral and coronary circulation, causing neurologic dysfunction and worsening myocardial function. Decreased output from the left ventricle due to high afterload may inhibit aortic valve opening, increasing the potential for clot formation within the left ventricle or aortic root [35].

Venting of the left ventricle is used in 15%–20% of patients supported with VA-ECMO [36]. Patients should be closely monitored with a pulmonary artery catheter and a right radial arterial pressure line to assess for excessive afterload and prevent left ventricle overload. The pulmonary artery diastolic pressure should be maintained less than 22 mmHg. Left ventricular contractility, pulse pressure, and aortic valve opening are assessed with the right radial arterial pressure waveform. A low or absent pulse pressure and no dicrotic notch indicates that VA-ECMO flow and left ventricular afterload exceed the ability of the left ventricle to eject blood, and the aortic valve remains closed. Echocardiography should be used to assess left ventricle and left atrial size and to aid in intravascular volume management.

Minimizing VA-ECMO flow may help to avert left ventricular distension, but a flow rate should always be maintained at a level that achieves adequate systemic perfusion assessed by lactate level, arterial pH, and central venous oxygen saturation. Left ventricular afterload may also be controlled by adjusting VA-ECMO flow with cautious dosing of vasodilators and inotropes and appropriate balance of intravascular volume.

Various methods have been used for left heart decompression during peripheral VA-ECMO. Intra-aortic balloon pump (IABP) support is easy to implant and is effective in certain cases. Transseptal placement of an 8- to 15-Fr cannula into the left atrium with blood drained into the venous inflow has been effective [34]. Another percutaneous method to decompress the left ventricle uses a blade-balloon septostomy to create a left-to-right shunt [37]. Atrial stenting decreases left atrial volume but requires surgical closure following termination of VA-ECMO. Draining the pulmonary artery into the ECMO inflow through a percutaneously placed 15-Fr cannula has been effective in two reported cases [38]. Direct cannulation at the apex through a mini-thoracotomy or a subcostal approach with placement of a 21- to 23-Fr cannula in the left ventricle is another effective means [35, 39, 40]. The Impella 2.5 and 5.0 ventricular assist devices (Abiomed Inc., Danvers, MA, USA) may be used for either primary mechanical support in cardiogenic shock or for left ventricular unloading during VA-ECMO support. Single-center studies have shown improved hemodynamics; however, controlled multicenter trials evaluating this technique have not been conducted [41–44]. It is important to note that there are no studies comparing the various techniques for left ventricle decompression. The technique used should be based on the level of expertise and training at the individual center.

Harlequin Syndrome

Harlequin syndrome is a complication that may occur during VA-ECMO with peripheral cannulation when respiratory function is poor and deoxygenated blood from the left ventricle is pumped into the arterial circulation and is the primary blood flow source to the coronary and carotid arteries [23, 45]. In severe cases, myocardial recovery is deterred, and cerebral ischemia may cause neurologic deficit. The presence of Harlequin syndrome can be detected by monitoring the arterial oxygen saturation at the right radial artery, as this is the most distal point from the VA-ECMO blood flow. A finger pulse oximeter on the right hand may give an early indication of oxygen desaturation. In cases of Harlequin syndrome, mechanical ventilation with appropriate FiO_2 and PEEP may help to maintain an arterial oxygen saturation of at least 90%. When the right radial oxygen saturation is less than 88%, the VA-ECMO flow may be too low and can be increased. Decreasing inotropic medications may also be useful. Use of beta-blockers to decrease heart rate may help decrease left heart output. When these measures fail to resolve the low radial artery saturation, cannulation of the ascending aorta for the VA-ECMO outflow should be employed. Also, the VA-ECMO outflow can be split between the femoral artery and a new cannula in the superior vena cava (SVC). When splitting the outflow, the cannula size in the femoral artery should be 17–19 Fr and should be 15–17 Fr in the SVC location [(31].)

Recirculation

Recirculation may occur with VV-ECMO—oxygenated blood from the ECMO outflow enters the drainage cannula without passing through the systemic circulation [46]. Recirculated blood does not contribute to oxygen delivery and decreases the overall efficiency of ECMO support. The pump speed and location of the inflow and outflow cannulas need to be assessed when recirculation occurs. Increasing the distance between the two cannulas and lowering flow rate may minimize the recirculation effects. Also, larger cannulas may allow high flow rates with minimal recirculation. Adding a second drainage cannula in a different location may also resolve the problem.

Weaning ECMO

Weaning from VA-ECMO versus VV-ECMO is very different, and techniques for both vary widely among institutions [47]. The optimal method for ECMO weaning is yet to be determined and is usually based on experience. Assessment for weaning begins soon after the initial stabilization, and daily trials of weaning should be performed. When considering weaning in patients, the primary cause of cardiac or pulmonary failure must be reversible, and other organ failure must be resolved or resolving [48]. Patients who do not have recoverable etiologies must be considered for transplantation, durable mechanical circulatory support, or hospice. Weaning from VV-ECMO requires return of adequate pulmonary function, whereas removal of VA-ECMO requires adequate cardiac function. Generally, weaning from VV-ECMO is more gradual, and careful attention to the adequacy of mechanical ventilation is given [47]. VA-ECMO weaning requires that patients maintain adequate cardiac output and blood pressure throughout the weaning process.

Weaning from VA-ECMO requires patients to maintain satisfactory hemodynamics before and throughout the weaning process. Inotropic support should be minimal, and mechanical support with an IABP or pVAD should be weaned. A weaning trial most often is performed in the operating room, and decannulation with proper hemostasis is performed for successful weaning trials. Echocardiography is used to assess for any ventricular distension, and blood pressure is continuously assessed. VA-ECMO flow is decreased in increments of 25% to 33% of the baseline flow rate, while the mean arterial blood pressure is maintained at greater than 65 mmHg. The patient should maintain a mixed venous saturation of at least 65% and an arterial saturation greater than 90% when ECMO flow is less than 1.5 l/min. Some centers advocate a slow weaning period of a few hours or even days, whereas others will perform the weaning trial in just an hour or two [16, 48].

Weaning from VV-ECMO begins once the patient has been stabilized, and minimal blood flow and sweep gas are set according to the patient's condition. The level of VV-ECMO support is decreased as the patient's condition improves and respiratory function returns. When VV-ECMO support is less than 30% of the maximum or initial amount of support, cardiac and respiratory function may allow removal of support. With the patient spontaneously breathing at an FiO_2 of less than 50%, the ECMO flow is decreased in increments of 1 l/min to a minimum of 1 l/min. When the patient maintains a $PaCO_2$ less than 50 mmHg and an arterial oxygen saturation greater than 95% for at least 1 hour, support can be removed [49].

ECMO Program Organization

ECMO is a complex therapy that should be managed at centers with appropriate experience and resources to ensure it is used effectively [24, 50]. Patients requiring ECMO support need the highest level of intensive care from a multidisciplinary team. New developing programs need to partner with experienced programs to provide a state-of-the-art system of care. Comprehensive advanced heart failure care programs at tertiary care centers serve as hub ECMO centers that are associated with regional centers. ECMO programs organized regionally can provide the high quality of care to as many patients as possible. Centers participating in a hub-and-spoke system of care should adhere to standardized protocols that detail criteria for the initiation of ECMO support, contraindications, follow-up care, and exit strategies.

ECMO programs should have a board-certified director with expertise in critical care, advanced heart failure, thoracic, cardiac, vascular, or trauma surgery or other board-certified specialist with specific training and experience in ECMO [24, 51, 52]. An ECMO coordinator, perfusionists, critical care nurses, and respiratory therapists provide technical and medical care. Ideally, institutional protocols guide team organization and individual team member responsibilities. Team members should complete ECMO training and periodically demonstrate competency [53]. An ECMO team coordinator is an essential part of the organization and communications. ECMO programs may use physician specialists, nurse practitioners, or RN to coordinate team activities and to provide triage between hub-and-spoke centers. Specialties needed to support the ECMO program include interventional cardiology, cardiothoracic surgery, pulmonology, neurology, nephrology, radiology, infectious disease, social workers, chaplains, palliative care, financial counselor, and hospital administrators. The responsibilities of each team member must be defined and agreed upon [15].

Quality assurance of the ECMO program must be a priority for the institution. Before starting the ECMO program, essential training and proficiency testing for all team members must be completed. Routine meetings with team members for review of training and equipment needs, staffing levels, patient volume, and case reviews are important. Referring spoke centers should participate in these review meetings. Morbidity and mortality meetings should be held to review any serious complication or death during ECMO support. Participation in the Extracorporeal Life Support Organization (ELSO) registry is useful for data comparison between a center and other institutions. New ECMO programs should complete a thorough analysis of the potential patient volume to assure that an appropriate amount of support exists. Hospital administration should be committed to financial support of the program costs and be prepared to adjust resources as the volume of cases varies.

ECMO Transport

There is an increasing number of patients that are being treated with ECMO in community hospitals. Ideally, patients with severe refractory cardiogenic shock or pulmonary failure will have ECMO support initiated at a major center that offers comprehensive advanced heart failure care [24]. However, many patients in need of emergent cardiopulmonary support who present at community hospitals may be initiated on-site by the local specialist or the ECMO team from a hub center. Regardless of the location at which ECMO support is initiated, the patient's follow-up care should take place at a center with complete advanced heart failure care and/or pulmonary care.

The ELSO has published ECMO transport guidelines, which should be considered by centers participating in transport programs [54]. Efficient advanced cardiac care systems, including high-volume hub hospitals, emergency medical services, and community-based spoke centers, may impact the outcomes of patients with profound cardiac and/or pulmonary failure [50]. A hub-and-spoke regional network entails three levels of care (Fig. 15.3) [55]. Level 1 centers provide comprehensive care, including heart transplant, long-term LVAD, total artificial heart, and short-term circulatory support. Level 2 centers offer cardiac catheterization and surgery and

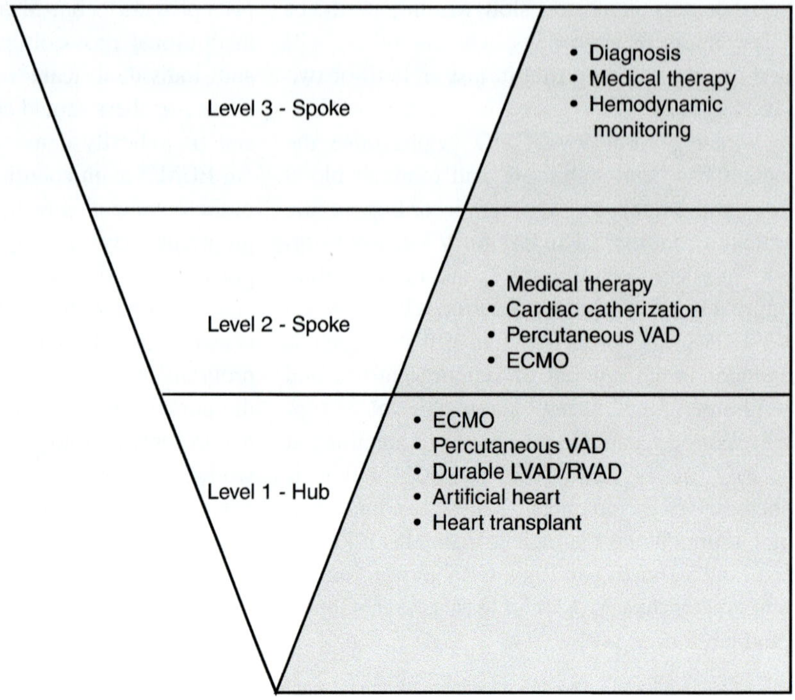

Fig. 15.3 Levels of care at hub-and-spoke centers in an advanced care network

percutaneous short-term mechanical support. Level 3 centers provide resuscitation and medical therapy to stabilize patients. Effective participation of this type of network requires commitment and communication from a variety of healthcare professionals. Protocols that outline communications, triage and patient selection, patient management, and the transport process must be established and strictly followed by all participating centers. Established patient selection criteria will help to minimize debate during the triage process. The Level 1 hospital is responsible for the overall coordination of the ECMO program.

A mobile ECMO team from the Level 1 center, with personnel trained in cannulation, transport of patients, initiation of support, and patient management, must be available 24 hours a day. Transport of critically ill patients with ECMO support, or those in need of support, must be prompt and efficient, as delays increase mortality [56]. The mobile ECMO team should consist of a combination of an ECMO specialist, ECMO coordinator, perfusionists, paramedics, and respiratory therapists. The team must be available for immediate transport to the patient's location. The ECMO coordinator at the hub center must have the ability and authority to choose the most appropriate means of transportation. Depending on distance and size of the metropolitan area, ground ambulance, helicopter, or fixed-wing aircraft are used as appropriate. Patients with severe cardiogenic shock or pulmonary failure at Level 3 centers should be promptly transported to the hub center if their condition is stable. For unstable cases, an ECMO transport team is dispatched to initiate therapy on-site. After the mobile ECMO team stabilizes the patient, safe transport to the hub hospital can take place.

Future Directions

ECMO support is unique in comparison to other healthcare technologies, as it was not developed specifically for current clinical applications. ECMO is an extension of cardiopulmonary bypass that has been routinely used in cardiac surgery. Today, ECMO is being applied in multiple different scenarios without randomized controlled clinical trials supporting the indications for use. Consequently, indications and contraindications need to be defined, and standards of care need to be established. Because patients with refractory cardiac or pulmonary failure present in extreme conditions, it is often difficult or impossible to discuss care with the patients or their next of kin. Ethical dilemmas related to this therapy need more thorough consideration in future research.

Extracorporeal cardiopulmonary resuscitation (ECPR) with ECMO may offer survival benefit for the large number of patients experiencing in-hospital and pre-hospital cardiac arrest [57, 58]. ECMO offers a means to rapidly restore circulation, which is the key component to achieving survival in patients with profound cardiogenic shock or cardiac arrest. Preliminary reports present the possible benefit of pre-hospital ECMO support initiated by emergency medical services [59]. Pre-hospital ECMO support is particularly challenging due to the absence of the multidisciplinary team present in the hospital environment.

As more patients with acute cardiopulmonary failure are being treated in community hospitals, more advanced care networks need to be established [4, 5, 60]. This includes the use of ECMO technology in use at more hospitals by more physician specialties. Presently, there are a large number of patients without access to this lifesaving therapy; this situation will improve as technology advances and more medical professionals are trained to provide the therapy. Outreach programs at tertiary care centers can initiate the development of networks; societies need to promote these activities as well.

Clinical research is necessary to better define populations of patients that will benefit from this therapy. Because patients are at high risk of death, randomized controlled trials are nearly impossible to conduct; however, studies that focus on assessing risks and benefits are needed to better define patient selection. Protocols for management and weaning from support need to be better defined to establish more uniform practices. The Extracorporeal Life Support Organization (ELSO) registry provides some useful information on practices and outcomes but needs to be expanded to provide more precise guidelines.

References

1. Stentz MJ, Kelley ME, Jabaley CS, et al. Trends in extracorporeal membrane oxygenation growth in the United States, 2011-2014. ASAIO J. 2018;65:712.
2. Conrad SA, Broman LM, Taccone FS, et al. The extracorporeal life support organization Maastricht treaty for nomenclature in extracorporeal life support. A position paper of the extracorporeal life support organization. Am J Respir Crit Care Med. 2018;198:447–51.
3. El-Banayosy A, Cobaugh D, Zittermann A, et al. A multidisciplinary network to save the lives of severe, persistent cardiogenic shock patients. Ann Thorac Surg. 2005;80:543–7.
4. Dini CS, Lazzeri C, Chiostri M, Gensini GF, Valente S. A local network for extracorporeal membrane oxygenation in refractory cardiogenic shock. Acute Card Care. 2015;17:49–54.
5. Aubin H, Petrov G, Dalyanoglu H, et al. Four-year experience of providing mobile extracorporeal life support to out-of-center patients within a suprainstitutional network-outcome of 160 consecutively treated patients. Resuscitation. 2017;121:151.
6. Davies A, Jones D, Bailey M, et al. Extracorporeal membrane oxygenation for 2009 influenza A(H1N1) acute respiratory distress syndrome. JAMA. 2009;302:1888–95.
7. Patroniti N, Zangrillo A, Pappalardo F, et al. The Italian ECMO network experience during the 2009 influenza A(H1N1) pandemic: preparation for severe respiratory emergency outbreaks. Intensive Care Med. 2011;37:1447–57.
8. Zangrillo A, Biondi-Zoccai G, Landoni G, et al. Extracorporeal membrane oxygenation (ECMO) in patients with H1N1 influenza infection: a systematic review and meta-analysis including 8 studies and 266 patients receiving ECMO. Crit Care. 2013;17:R30.
9. Stretch R, Sauer CM, Yuh DD, Bonde P. National trends in the utilization of short-term mechanical circulatory support: incidence, outcomes, and cost analysis. J Am Coll Cardiol. 2014;64:1407–15.
10. Steimer DA, Hernandez O, Mason DP, Schwartz GS. Timing of ECMO initiation impacts survival in influenza-associated ARDS. Thorac Cardiovasc Surg. 2018;67:212.
11. Chillcott S, Stahovich M, Earnhardt C, Dembitsky W. Portable rapid response extracorporeal life support: a center's 20-year experience with a registered nurse-run program. Crit Care Nurs Q. 2008;31:211–5.
12. Durinka JB, Bogar LJ, Hirose H, et al. End-organ recovery is key to success for extracorporeal membrane oxygenation as a bridge to implantable left ventricular assist device. ASAIO J. 2014;60:189–92.
13. Abrams D, Combes A, Brodie D. Extracorporeal membrane oxygenation in cardiopulmonary disease in adults. J Am Coll Cardiol. 2014;63:2769–78.
14. Morris AH. Exciting new ECMO technology awaits compelling scientific evidence for widespread use in adults with respiratory failure. Intensive Care Med. 2012;38:186–8.
15. Abrams D, Garan AR, Abdelbary A, et al. Position paper for the organization of ECMO programs for cardiac failure in adults. Intensive Care Med. 2018;44:717.
16. Guglin M, Zucker MJ, Bazan VM, et al. Venoarterial ECMO for Adults: JACC Scientific Expert Panel. J Am Coll Cardiol. 2019;73:698–716.
17. Thiele H, Ohman EM, Desch S, Eitel I, de Waha S. Management of cardiogenic shock. Eur Heart J. 2015;36:1223–30.
18. Harjola VP, Lassus J, Sionis A, et al. Clinical picture and risk prediction of short-term mortality in cardiogenic shock. Eur J Heart Fail. 2015;17:501–9.
19. Kar B, Gregoric ID, Basra SS, Idelchik GM, Loyalka P. The percutaneous ventricular assist device in severe refractory cardiogenic shock. J Am Coll Cardiol. 2011;57:688–96.
20. Lorusso R, Gelsomino S, Parise O, et al. Venoarterial extracorporeal membrane oxygenation for refractory cardiogenic shock in elderly patients: trends in application and outcome from the Extracorporeal Life Support Organization (ELSO) registry. Ann Thorac Surg. 2017;104:62–9.
21. McCarthy FH, McDermott KM, Kini V, et al. Trends in U.S. extracorporeal membrane oxygenation use and outcomes: 2002-2012. Semin Thorac Cardiovasc Surg. 2015;27:81–8.
22. Nanjayya VB, Murphy D. Ultrasound guidance for extra-corporeal membrane oxygenation: general guidelines. extracorporeal life support Organization; https://www.elso.org/Portals/0/Files/elso_Ultrasoundguideance_ecmogeneral_guidelines_May2015.pdf.
23. Meuwese CL, Ramjankhan FZ, Braithwaite SA, et al. Extracorporeal life support in cardiogenic shock: indications and management in current practice. Neth Heart J. 2018;26:58–66.
24. Combes A, Brodie D, Bartlett R, et al. Position paper for the organization of extracorporeal membrane oxygenation programs for acute respiratory failure in adult patients. Am J Respir Crit Care Med. 2014;190:488–96.
25. Kon ZN, Bittle GJ, Pasrija C, et al. Venovenous versus venoarterial extracorporeal membrane oxygenation for adult patients with acute respiratory distress syndrome requiring precannulation hemodynamic support: a review of the ELSO registry. Ann Thorac Surg. 2017;104:645–9.
26. Kilic A, Shukrallah BN, Kilic A, Whitson BA. Initiation and management of adult veno-arterial extracorporeal life support. Ann Transl Med. 2017;5:67.
27. Donker DW, Meuwese CL, Braithwaite SA, et al. Echocardiography in extracorporeal life support: a key player in procedural guidance, tailoring and monitoring. Perfusion. 2018;33:31–41.
28. Delmas C, Jacquemin A, Vardon-Bounes F et al. Anticoagulation monitoring under ECMO support: a

comparative study between the activated coagulation time and the anti-Xa activity assay. J Intensive Care Med. 2018:885066618776937.
29. Sanfilippo F, Asmussen S, Maybauer DM, et al. Bivalirudin for alternative anticoagulation in extracorporeal membrane oxygenation: a systematic review. J Intensive Care Med. 2016;32:312.
30. Chung YS, Cho DY, Sohn DS, et al. Is stopping heparin safe in patients on extracorporeal membrane oxygenation treatment? ASAIO J. 2017;63:32–6.
31. Koerner MM, Harper MD, Gordon CK, et al. Adult cardiac veno-arterial extracorporeal life support (VA-ECMO): prevention and management of acute complications. Ann Cardiothorac Surg. 2019;8: 66–75.
32. Nasr DM, Rabinstein AA. Neurologic complications of extracorporeal membrane oxygenation. J Clin Neurol. 2015;11:383–9.
33. Prasad A, Ghodsizad A, Brehm C, et al. Refractory pulmonary edema and upper body hypoxemia during veno-arterial extracorporeal membrane oxygenation-a case for atrial Septostomy. Artif Organs. 2018;42:664.
34. Aiyagari RM, Rocchini AP, Remenapp RT, Graziano JN. Decompression of the left atrium during extracorporeal membrane oxygenation using a transseptal cannula incorporated into the circuit. Crit Care Med. 2006;34:2603–6.
35. Centofanti P, Attisani M, La Torre M, et al. Left ventricular unloading during peripheral extracorporeal membrane oxygenator support: a bridge to life in profound cardiogenic shock. J Extra Corpor Technol. 2017;49:201–5.
36. Meani P, Gelsomino S, Natour E, et al. Modalities and effects of left ventricle unloading on extracorporeal life support: a review of the current literature. Eur J Heart Fail. 2017;19(Suppl 2):84–91.
37. Johnston TA, Jaggers J, McGovern JJ, O'Laughlin MP. Bedside transseptal balloon dilation atrial septostomy for decompression of the left heart during extracorporeal membrane oxygenation. Catheter Cardiovasc Interv. 1999;46:197–9.
38. Avalli L, Maggioni E, Sangalli F, Favini G, Formica F, Fumagalli R. Percutaneous left-heart decompression during extracorporeal membrane oxygenation: an alternative to surgical and transeptal venting in adult patients. ASAIO J. 2011;57:38–40.
39. Guirgis M, Kumar K, Menkis AH, Freed DH. Minimally invasive left-heart decompression during venoarterial extracorporeal membrane oxygenation: an alternative to a percutaneous approach. Interact Cardiovasc Thorac Surg. 2010;10:672–4.
40. Eudailey KW, Yi SY, Mongero LB, Wagener G, Guarrera JV, George I. Trans-diaphragmatic left ventricular venting during peripheral venous-arterial extracorporeal membrane oxygenation. Perfusion. 2015;30:701–3.
41. Moazzami K, Dolmatova EV, Cocke TP, et al. Left ventricular mechanical support with the Impella during extracorporeal membrane oxygenation. J Tehran Heart Cent. 2017;12:11–4.
42. Cheng A, Swartz MF, Massey HT. Impella to unload the left ventricle during peripheral extracorporeal membrane oxygenation. ASAIO J. 2013;59:533–6.
43. Pappalardo F, Schulte C, Pieri M, et al. Concomitant implantation of Impella(R) on top of veno-arterial extracorporeal membrane oxygenation may improve survival of patients with cardiogenic shock. Eur J Heart Fail. 2017;19:404–12.
44. Kawashima D, Gojo S, Nishimura T, et al. Left ventricular mechanical support with Impella provides more ventricular unloading in heart failure than extracorporeal membrane oxygenation. ASAIO J. 2011;57:169–76.
45. Tak VM, Holecek WF 3rd. Harlequin effect in angiography on extracorporeal membrane oxygenation mimicking aortic dissection. Ann Thorac Surg. 2018;106:e97.
46. Abrams D, Bacchetta M, Brodie D. Recirculation in venovenous extracorporeal membrane oxygenation. ASAIO J. 2015;61:115–21.
47. Grant AA, Hart VJ, Lineen EB, et al. A weaning protocol for venovenous extracorporeal membrane oxygenation with a review of the literature. Artif Organs. 2018;42:605.
48. Aissaoui N, El-Banayosy A, Combes A. How to wean a patient from veno-arterial extracorporeal membrane oxygenation. Intensive Care Med. 2015;41:902–5.
49. (ELSO) ELSO. Guidelines for adult respiratory failure. www.elso.org. 2017.
50. van Diepen S, Katz JN, Albert NM, et al. Contemporary Management of Cardiogenic Shock: a scientific statement from the American Heart Association. Circulation. 2017;136:e232–68.
51. Gutsche JT, Vernick WJ. Cardiac and critical care anesthesiologists may be ideal members of the Mobile ECMO team. J Cardiothorac Vasc Anesth. 2016;30:1439–40.
52. Lawson DS, Lawson AF, Walczak R, et al. North American neonatal extracorporeal membrane oxygenation (ECMO) devices and team roles: 2008 survey results of extracorporeal life support organization (ELSO) centers. J Extra Corpor Technol. 2008;40:166–74.
53. Salna M, Chicotka S, Biscotti M 3rd, et al. Management of Surge in extracorporeal membrane oxygenation transport. Ann Thorac Surg. 2018;105:528–34.
54. Dirnberger DR, Fiser R, Harvey C et al. Extracorporeal Life Support Organization (ELSO): Guidelines for ECMO transport. https://www.elso.org/2015.
55. Tchantchaleishvili V, Hallinan W, Massey HT. Call for organized statewide networks for management of acute myocardial infarction-related cardiogenic shock. JAMA Surg. 2015;150:1025–6.
56. Jaroszewski DE, Kleisli T, Staley L, et al. A traveling team concept to expedite the transfer and management of unstable patients in cardiopulmonary shock. J Heart Lung Transplant. 2011;30:618–23.
57. Tonna JE, Johnson NJ, Greenwood J, et al. Practice characteristics of emergency department extracorporeal cardiopulmonary resuscitation (eCPR) programs

in the United States: the current state of the art of emergency department extracorporeal membrane oxygenation (ED ECMO). Resuscitation. 2016;107: 38–46.
58. ELSO. Guidelines for ECPR cases. https://www.elso.org/2013..
59. Lamhaut L, Jouffroy R, Soldan M, et al. Safety and feasibility of prehospital extra corporeal life support implementation by non-surgeons for out-of-hospital refractory cardiac arrest. Resuscitation. 2013;84:1525–9.
60. Morshuis M, Bruenger F, Becker T, et al. Inter-hospital transfer of extracorporeal membrane oxygenation-assisted patients: the hub and spoke network. Ann Cardiothorac Surg. 2019;8:62–5.

16 Less Invasive Extra-pericardial Placement (LIEPP) of LVAD

Areo Saffarzadeh and Pramod Bonde

Introduction

The conventional operative approach for placement of a left ventricular assist device (LVAD) has been via median sternotomy [1]. Initially, sternal-sparing approaches were utilized predominately for high-risk cases [2–4]. However, each generation of LVADs has led to the progressive miniaturization of these devices enabling a variety of less invasive approaches for routine placement [5, 6]. The sternal-sparing or minimally invasive approach has been advocated for patients as a means to avoid the risks of redo sternotomy during placement of LVAD or during future transplant. While quality randomized studies comparing minimally invasive to traditional approaches is lacking, there is evidence to suggest that minimally invasive approaches may be associated with shorter cardiopulmonary bypass times, reduced blood transfusion requirements, and lower 30-day mortality [7, 8]. Recent evidence suggests that minimally invasive approach may also better protect right ventricular function [9, 10]. This chapter describes the preoperative planning and surgical technique used at our institution to safely accomplish a less invasive extra-pericardial placement (LIEPP) of LVAD.

A. Saffarzadeh, MS, MD · P. Bonde, MD (✉)
Yale School of Medicine, New Haven, CT, USA
e-mail: Areo.Saffarzadeh@yale.edu; pramod.bonde@yale.edu

Preoperative Planning

The preoperative CT and CXR are utilized to determine the optimal locations for incisions. We utilize a three-incision approach for LVAD placement as shown in Fig. 16.1. The approach includes a (1) left thoracotomy near the fifth or sixth intercostal space for access to the LV apex, (2) a subxiphoid incision to facilitate safe tunneling of the outflow graft, and (3) small right thoracotomy for outflow graft anastomosis to the ascending aorta near the second or third intercostal space. To determine the left thoracotomy site, the CT is examined to see where the LV apex meets the chest wall. For the right thoracotomy, the site of maximum contour of the aorta can be used to determine optimal interspace for aortic anastomosis.

We use single-lumen endotracheal intubation, a Swan-Ganz catheter, TEE, and CPB with femoral cannulation for most patients. This is communicated with the anesthesia team before starting the case. If IABP or Impella has been previously placed, we will often use the same side for cannulation. For positioning, the patient is placed supine, with the left arm outstretched and the right arm tucked. Pressure bags (500 cc in volume) are placed under the thorax on each side, which can be inflated during the case to elevate each side, if necessary, to improve exposure.

© Springer Nature Switzerland AG 2020

J. H. Karimov et al. (eds.), *Mechanical Support for Heart Failure*,
https://doi.org/10.1007/978-3-030-47809-4_16

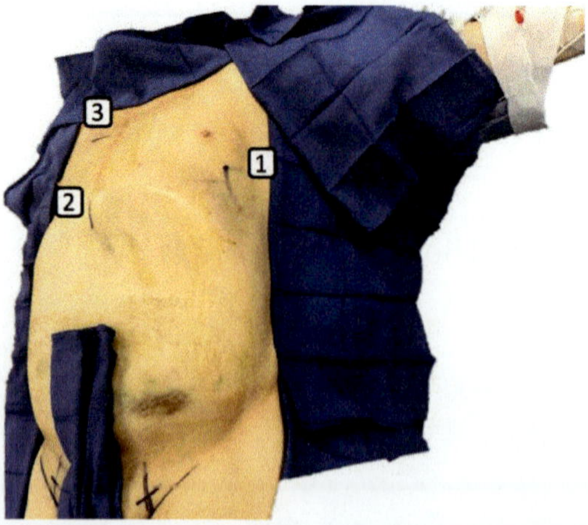

Fig. 16.1 Surgical approach for LVAD insertion. Skin markings show the incision sites for the less invasive extra-pericardial placement approach. (1) Left mini-thoracotomy at fifth intercostal space for exposure of the left ventricular apex to facilitate inflow cannulation anastomosis. Left ventricular apex is labeled with a blue dot. (2) Subxiphoid incision to tunnel the graft and drivelines. (3) Right thoracotomy at two-thirds intercostal space for outflow graft anastomosis to ascending aorta. Note: the groins are marked for cardiopulmonary bypass via the femoral arteries. Right arm is tucked, and left arm is out

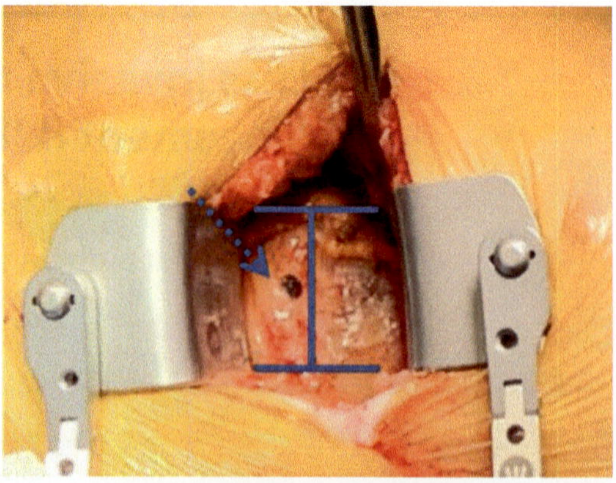

Fig. 16.2 Left ventricular apical exposure. This shows the 4–5 cm incision in the left fifth intercostal space, with rib retractors assisting in the exposure of the left ventricular apex. An H-shaped incision (oriented sideways) is used to open the pericardium. The center of the left ventricular apex is marked with a blue dot (dashed arrow). The patient in this scenario previously underwent a coronary artery bypass grafting, and there were multiple fibrous adhesions surrounding the pericardium

Surgical Steps

Exposure of LV Apex

We start with a small subxiphoid incision and enter the left and right pleural space. When entering the left pleural space, we confirm the interspace in which the LV apex contacts the chest wall for the planned left thoracotomy. This typically falls between the fifth and sixth intercostal space. To expose the LV apex via the left thoracotomy, it is important to open the interspace posteriorly to facilitate improved visualization of the intercostal space using a rib retractor.

Once the left thoracotomy is complete, an H-shaped incision is made on the pericardium to ensure proper exposure of the LV apex (Fig. 16.2). Using finger palpation with the assistance of TEE, we then determine the optimal center for the apical sewing ring and ventriculotomy site and mark the site with a pen.

Placement of Apical Sewing Ring

The apical sewing ring is used to secure the inflow cannula to the heart and should be centered at the LV apex. The sewing ring has a 38 mm diameter piece of felt transfixed to its rigid body to reinforce the inner aspect of the suture line. As shown in Fig. 16.3a, this piece of felt has a black 33 mm diameter circle drawn on it to mark the entry point for the needle. Sutures for the apical sewing ring are placed in a radial fashion circumferentially through the myocardium. To reinforce the exit point of the needle, we also create a second circular felt strip cut to be slightly wider than the sewing cuff as shown in Fig. 16.3b.

We attach the sewing ring to the LV apex using a “sew first and cut second” technique, meaning that the sutures are placed before removing the core of the apex. We attach the sewing ring using one 3-0 suture per quadrant (four sutures total). We start at the 12:00 o’clock position on the cuff and begin by placing the suture through the black line on the felt of the sewing

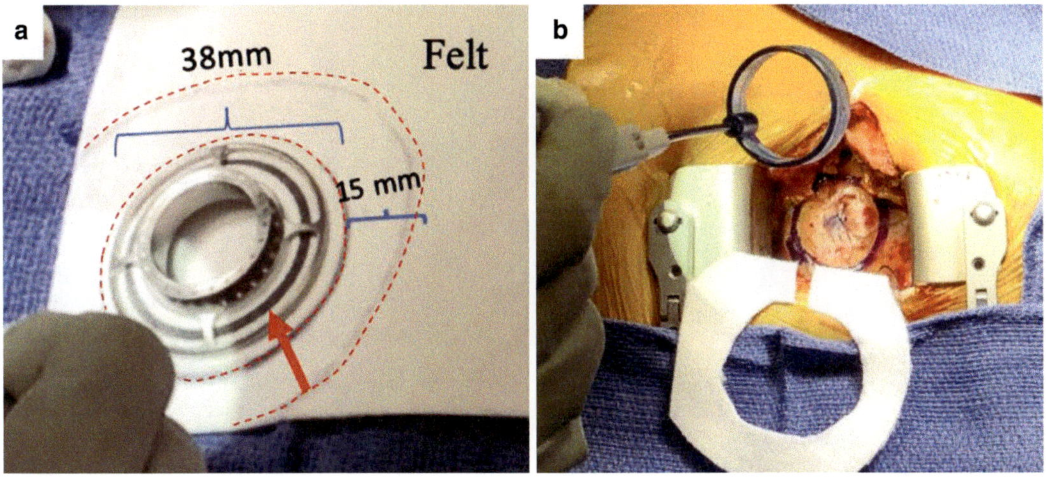

Fig. 16.3 Creation of felt strip to reinforce the outer suture line for the apical sewing ring. (**a**) The apical sewing ring has a black circle (red arrow) placed by the manufacturer to guide needle placement through the center of the felt portion apical sewing ring and to reinforce the inner suture line. The apical sewing ring is approximately 38 mm in diameter, and the black circle is 33 mm in diameter. We place the sewing ring on top of second piece of felt which is cut 15 mm wider (red dashed lines) to make another donut-shaped felt strip which will be used to reinforce the outer suture line. (**b**) A 33 mm St. Jude sizer, which is the same size as black circle on apical sewing ring, is used to draw a circle on the left ventricular apex to also help guide placement of the needle as it enters the heart (inner suture line). Also shown is the second felt strip which has been cut to be 15 mm wider than the sewing ring. This second strip will be used to reinforce the outer aspect of the suture line at the point where the needle leaves the heart

ring from superficial to deep (Fig. 16.4a). The suture is then used to take a full-thickness bite through the LV apex going radially outward (using the circle drawn on the LV apex with the 33 mm St. Jude sizer as a guide). The entrance point for the needle is shown with a blue dot in Fig. 16.4b and the exit point with a green dot. Finally, the suture is placed through the second felt strip from deep to superficial (Fig. 16.4c).

This three-step process is repeated at the 11:00, 10:00, and 9:00 o'clock position (Fig. 16.5a). The second suture is started where the first one ended at the 9:00 o'clock position, and a rubber shod is used to link the tail of the second suture with the needle of the first suture (Fig. 16.5a, yellow star). This three-step process is repeated with separate suture until reaching the 6:00 position. This is repeated from 6:00 to 3:00 position and then from 3:00 back to the 12:00 o'clock starting point. The apical sewing cuff is then parachuted down and should look similar to the drawing on Fig. 16.5b. A nerve hook is utilized to remove any redundant suture, and the loose ends of both sutures at the 12:00, 9:00, 6:00, and 3:00 o'clock position are snared together (Fig. 16.5c). With the sewing ring secured to the apex, we move on to tunneling the driveline and outflow graft before making the ventriculotomy and initiating cardiopulmonary bypass.

Tunneling the Drive Line

The driveline is the only element of the LVAD that exits the skin and contains cables that power and control the pump (Fig. 16.6a). Drive line infections are a significant cause of morbidity; therefore sterile management and technique is paramount as with the insertion of any prosthetic device. Using a tunneling device, the drive line is tunneled from the left thoracotomy site to the subxiphoid incision, and from the subxiphoid incision, it is tunneled subcutaneously to exit the skin in the right upper quadrant

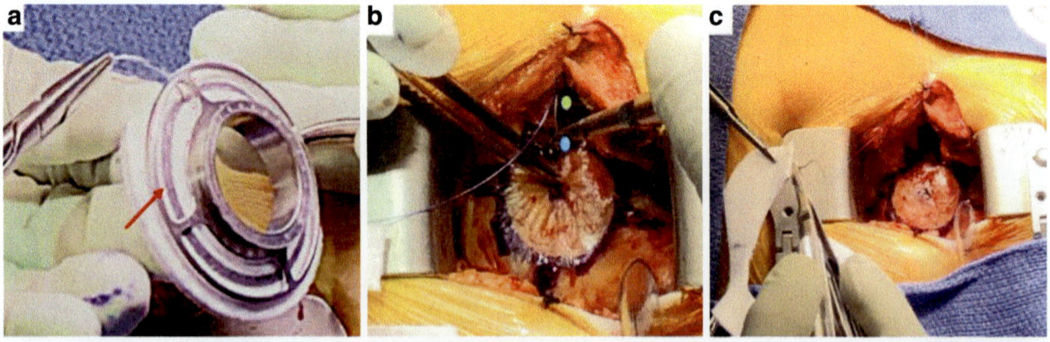

Fig. 16.4 Three-step process to attach apical sewing ring to left ventricular apex. (**a**) Step 1. The needle is placed from superficial-to-deep through the black line on the felt portion of the apical sewing ring (red arrow). (**b**) Step 2. The needle is then placed through the myocardium in an outward radial fashion using the circle drawn with the 33 mm St. Jude sizer on the left ventricular apex as the entrance point. The entrance point of the needle is shown in the blue dot and the exit point in the green dot. (**c**) Step 3. The needle is brought through the 15-mm-wide felt strip from deep to superficial (right image). This second felt strip reinforces to outer aspect of the suture line

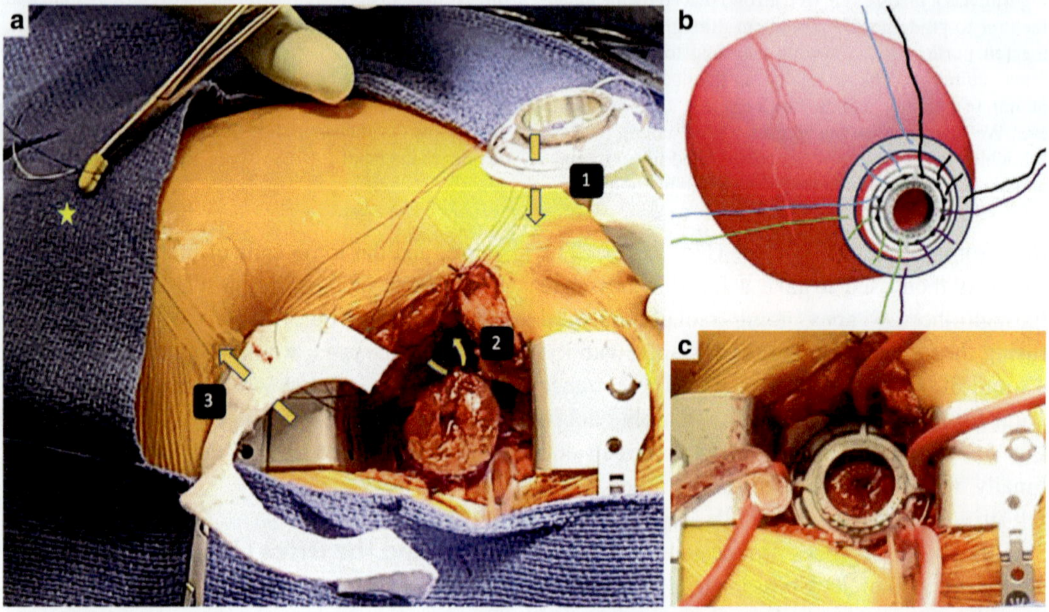

Fig. 16.5 One suture per quadrant technique: parachute apical cuff on the left ventricular apex. (**a**) Left: Shows the first suture which began at 12:00 o'clock and goes sequentially through (1) apical sewing ring, (2) myocardium, and (3) reinforcing felt strip. This has been repeated at the 11:00, 10:00, and 9:00 o'clock position. Note that the needle of the first suture has been snapped to the tail of the second suture (yellow star) which starts at the 9:00 o'clock position. (**b**) Top right: One suture is used per each quadrant denoted by the unique colors. The cuff has been parachuted down to the left ventricle here. Note that the end of one suture is paired with the beginning of the next suture, and these *separate* sutures (1 and 2, 2 and 3, 3 and 4, and 4 with 1) will be snared together. (**c**) Bottom right: Snares are placed to link sutures and to keep apical sewing ring in place

of the abdomen. Data from a registry used to evaluate driveline infections for the HeartMate II showed a driveline infection rate of 23% in the first year and 35% by the second year when the polyester velour-skin interface was used, which was decreased to 9% and 19%, respectively, when the interface was between skin and silicone [11].

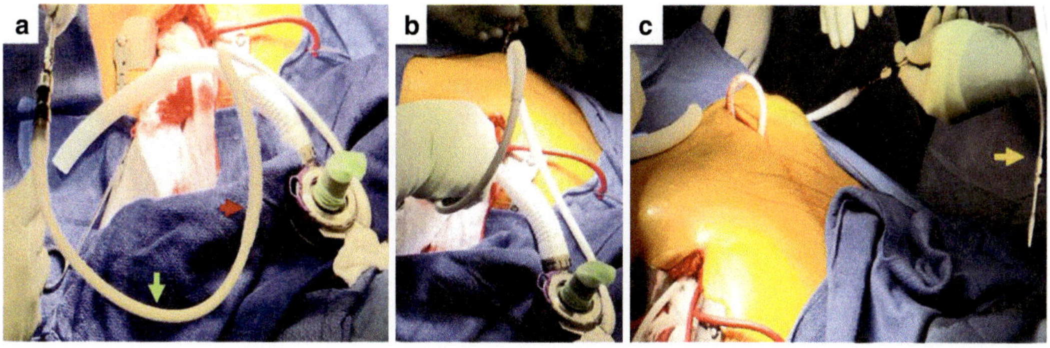

Fig. 16.6 Tunneling the driveline. (**a**) Shows the velour polyester covering of the driveline (red arrow) and the silicone portion of the driveline (green arrow). The velour covering is meant to remain intracorporeal. Note that the outflow graft is already connected to the pump and secured. (**b**) Tunneling begins from the left thoracotomy to the subxiphoid incision via the left hemithorax. (**c**) From the subxiphoid incision, the driveline is tunneled subcutaneously to exit the abdomen in the right upper quadrant. The tunneling device is shown with orange arrow

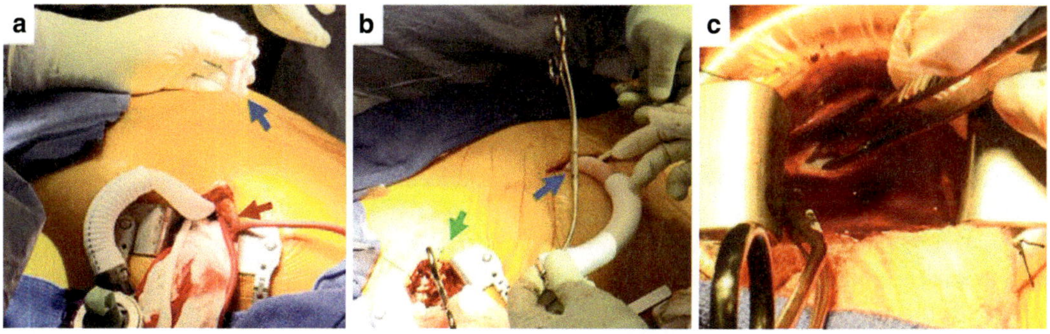

Fig. 16.7 Tunneling the outflow graft and aortic anastomosis. (**a**) This shows the outflow graft being tunneled from the left thoracotomy (red arrow) to the subxiphoid incision (blue arrow). (**b**) This shows the course the outflow graft will take in the right paramediastinal space between the subxiphoid incision (blue arrow) and right thoracotomy at two-thirds intercostal space (green arrow). (**c**) A side-biter is place on the aorta which has been incised prior to the anastomosis

To further minimize the risk of infection, it is recommended that the driveline be as long as possible in order to maximize the ingrowth on the velour covering. It is also important to avoid making too large of an incision at the skin as this can serve as an entry point for infection, which are notoriously difficult to treat. Considerations should be made to create an exit site that conforms to the patient's body habitus and to avoid any unnecessary kinking or wear and tear.

Tunneling the Outflow Graft and Aortic Anastomosis

The outflow graft connects the LVAD to the ascending aorta. Exposure of the ascending aorta involves a right mini-thoracotomy which is performed at the second or third intercostal space. This is often performed at the start of the case simultaneously with the left thoracotomy and subxiphoid incision. The superior aspect of the pericardium should be visible and is incised longitudinally with two stay sutures to improve exposure. The ideal exposure is where the right atrial appendage is inferior, the right brachiocephalic vein is superior, and neither structure obstructs the exposure of the ascending aorta.

To tunnel the outflow graft, electrocautery is first used to make a defect in the diaphragm, taking care to avoid injuring the phrenic nerve. As shown in Fig. 16.7a, the sealed outflow graft is brought from the left thoracotomy through this surgically created diaphragmatic defect and out of the subxiphoid incision. From here a second tunnel is made from the subxiphoid incision to the right thoracotomy (Fig. 16.7b) where it will eventually be anastomosed to the ascending aorta. The outflow graft will lay in this right paramediastinal space between the right lung and the right side of the pericardium. Prior to cutting the outflow graft to length, the proximal aspect of the graft (the screw ring) should be the correct anatomical position at the apex of the heart. The graft should be gently stretched to ensure there is no redundant material or kinking in the graft. To perform the anastomosis, a side-biter is placed on the aorta, and the anastomosis is performed in an end-to-side fashion with a 4-0 polypropylene suture in a running fashion.

Ventriculotomy and Attachment of Pump to the Apical Sewing Ring

To perform the ventriculotomy, cardiopulmonary bypass is first initiated via femoral cannulation to decompress the heart. At this point, the four sutures previously placed in each quadrant to secure to the apical sewing ring to the LV apex are tied to each other, starting at 3 o'clock position and then at 6, 9, and 12 o'clock. Our preference is to use an 11-blade to make the ventriculotomy (Fig. 16.8a), but a coring knife can also be used. The ventriculotomy is created in a manner so that the inflow canula is directed at the mitral valve and not at the interventricular septum. Once the core is removed, the LV is inspected for crossing trabecula or mural thrombi which should be excised (Fig. 16.8b). When attaching the LVAD to the apical sew ring, ensure that it is oriented so that the outflow graft and driveline cable are directed toward the midline to avoid unnecessary kinks (Fig. 16.8c).

Dealring

Once the pump is firmly attached to the apical sewing ring, we place a cross-clamp on the outflow graft near the xiphoid incision (Fig. 16.9). The patient is placed in a steep Trendelenburg position, and the outflow graft is held vertically to ensure there is an arch at the highest point. The side-biting clamp on the aorta is removed to backbleed. At this point, we place a 18g needle to vent the outflow graph between the pump and the cross-clamp. The flows of the LVAD are increased, and CPB is weaned to fill the heart and further deair. The cross-clamp is partially removed, and air bubbles are monitored with intraoperative TEE. Once no more air is no longer visible coming from the vent and no longer visible with TEE, the vent is removed, and a repair stitch is placed if needed.

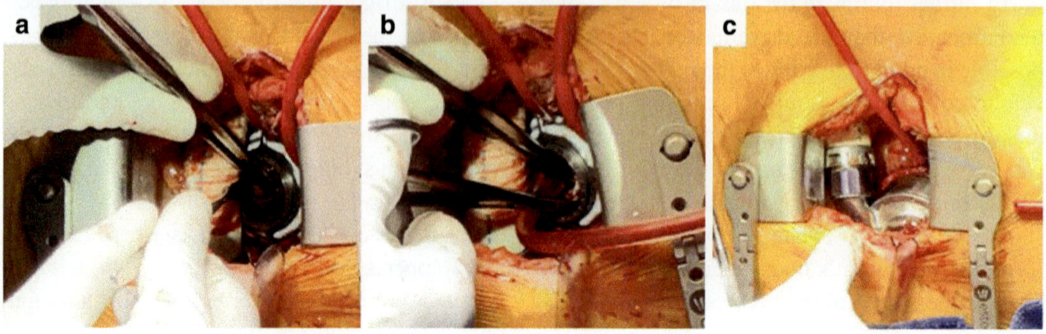

Fig. 16.8 Ventriculotomy and attachment of pump to the apical sewing ring. (**a**) Shows the ventriculotomy being performed with an #11 blade (a coring device can also be used here). (**b**) Crossing trabeculae which may obstruct the flow of blood to the pump are excised if present. (**c**) The left ventricular assist device (LVAD) is attached to the sewing ring; note that the outflow graft is oriented toward the midline prior to attaching to the left ventricular assist device to ensure there is no kinking

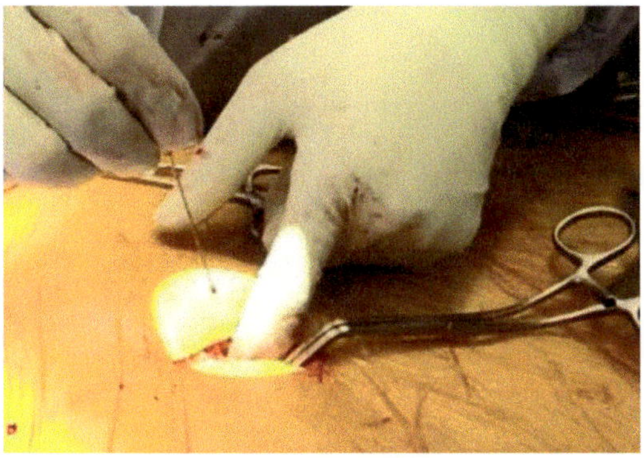

Fig. 16.9 Deairing of outflow graft. A cross-clamp is placed on the outflow graft at the subxiphoid incision. The patient is placed in Trendelenburg position, and the side-biting clamp on the aorta is removed. A 16g needle is being placed to vent the outflow graft between the pump and cross-clamp

Modifications of Technique for HMII

The technique described mainly applies to implantation of the HeartMate (HM) III and HVAD. For less invasive insertion of HM II, a similar approach is used; however the ventriculotomy is done first, and the entire ring is sutured in with interrupted pledgeted sutures ("cut first and sew second"). The driveline and outflow graft are tunneled and anastomosed in a similar fashion. Dearing is done via backbleeding when connecting the outflow graft to the pump.

References

1. Slaughter MS, Tsui SS, El-Banayosy A, Sun BC, Kormos RL, Mueller DK, et al. Results of a multicenter clinical trial with the Thoratec Implantable Ventricular Assist Device. J Thorac Cardiovasc Surg. 2007;133(6):1573–80. https://doi.org/10.1016/j.jtcvs.2006.11.050.
2. Tittle SL, Mandapati D, Kopf GS, Elefteriades JA. Alternate technique for implantation of left ventricular assist system: left thoracotomy for reoperative cases. Ann Thorac Surg. 2002;73(3):994–6. https://doi.org/10.1016/S0003-4975(01)03451-8.
3. Pierson RN, Howser R, Donaldson T, Merrill WH, Dignan RJ, Drinkwater DC, et al. Left ventricular assist device implantation via left thoracotomy: alternative to repeat sternotomy. Ann Thorac Surg. 2002;73(3):997–9. https://doi.org/10.1016/S0003-4975(01)03448-8.
4. Pfister R, Tozzi P, Hullin R, Yerly P, Jahns F-P, Prêtre R, et al. HeartMate 3 implantation via left anterolateral thoracotomy to avoid resternotomy in high risk patients. Multimedia Man Cardiothorac Surg. 2018;2018 https://doi.org/10.1510/mmcts.2018.026.
5. Wachter K, Franke U, Rustenbach C, Baumbach H. Minimally invasive versus conventional LVAD-implantation—an analysis of the literature. Thorac Cardiovasc Surg. 2018;67(03):156–63. https://doi.org/10.1055/s-0038-1627455.
6. Hanke J, Rojas S, Avsar M, Haverich A, Schmitto J. Minimally-invasive LVAD implantation: state of the art. Curr Cardiol Rev. 2015;11(3):246–51. https://doi.org/10.2174/1573403x1103150514151750.
7. Rojas SV, Avsar M, Hanke JS, Fischer L, Repges L, Meyer A, et al. Improved clinical outcome following a minimally invasive implantation technique for left ventricular assist devices implantation in adults with severe heart failure. J Am Coll Cardiol. 2013;61(10) https://doi.org/10.1016/s0735-1097(13)60717-2.
8. Cheung A, Soon J-L, Bashir J, Kaan A, Ignaszewski A. Minimal-access left ventricular assist device implantation. Innovations. 2014;9(4):281–5. https://doi.org/10.1097/IMI.0000000000000086.
9. Kazui T, Suryanarayana P, Singh S, Smith R, Avery R, Juneman E, et al. A minimally invasive left ventricular assist device surgical approach is more durable at protecting right ventricular function than a full sternotomy. J Heart Lung Transplant. 2017;36(4) https://doi.org/10.1016/j.healun.2017.01.1261.
10. Unsworth B, Casula RP, Yadav H, Baruah R, Hughes AD, Mayet J, et al. Contrasting effect of different cardiothoracic operations on echocardiographic right ventricular long axis velocities, and implications for interpretation of post-operative values. Int J Cardiol. 2013;165(1):151–60. https://doi.org/10.1016/J.IJCARD.2011.08.031.
11. Dean D, Kallel F, Ewald GA, Tatooles A, Sheridan BC, Brewer RJ, et al. Reduction in driveline infection rates: results from the HeartMate II Multicenter Driveline Silicone Skin Interface (SSI) Registry. J Heart Lung Transplant. 2015;34(6):781–9. https://doi.org/10.1016/J.HEALUN.2014.11.021.

Part III

Parameters of Physiologic and Non-physiologic Blood Flow

Pulsatile Mechanical Circulation, Physiology, and Pump Technology

17

Jack Copeland and Hannah Copeland

Introduction

For 500 million years, there have been hearts. The phylum Chordata (having a spine or notochord) is characterized as having a heart. The heart and lungs evolved into separate pulmonary and circulatory systems in amphibians about 350 million years ago. In the timeline of mammalian evolution, precursors to man appeared over 50 million years ago, and *Homo sapiens* (humans) the final descendant of the great apes appeared 250,000–500,000 years ago [1]. Evolution has created our incredibly complex biology. Yet, there is a consistency and simplicity. All higher animals have pulsatile, heart beating circulations that bear a striking resemblance. The embryonic development of the human heart in the early stages recapitulates the longitudinal two-chambered heart of fish. Our hearts and circulatory systems of vessels have evolved over a time. The physics, biology, anatomy, and physiology of this complex system provide 4–8 l of oxygenated blood flow every minute in adults. The resting beat rate of the heart is 70–100 per minute, and the volume of the pulse delivered is between 60 and 100 ml/beat. Arterial blood pressure is 120/70 mmHg, and the peak rate of pressure change as the blood is ejected from the left ventricle is over 1200 mmHg/s. About 80% of the resistance to flow is due to the microvasculature. Approximately 23% of the flow goes to the kidneys and 14% to the brain [2]. The remainder is distributed to other organs and tissues. Homeostasis of the blood in a normo-coagulable state has also evolved. The major factors in this complex and incompletely understood system are the platelets and the coagulation system. A fine balance prevents hypocoagulability that leads to bleeding and hypercoagulability that leads to clotting. The introduction of foreign materials such as those found in blood pumps to these systems causes activation as well as degradation of platelets and procoagulants. Antiplatelet and anticoagulant therapies are therefore necessary but may lead to bleeding. Bleeding also results from destruction of von Willebrand factor multimers by continuous-flow pumps [3].

The first circulatory assist for the open-heart surgery in 1954 [4] was biological. A small child could be supported by cross-circulation with his parent while the child's heart was emptied during surgical repair. This work of Dr. C.W. Lillehei triggered interest in open-heart surgery. The concept of using a "pump oxygenator" was pursued

J. Copeland, MD (✉)
Department of Surgery, University of Arizona, Tucson, AZ, USA

H. Copeland, MD
University of Mississippi Medical Center, Division of Cardiothoracic Surgery, Heart Transplantation and Mechanical Circulatory Support, Jackson, MS, USA
e-mail: hannahcopeland411@umc.edu

© Springer Nature Switzerland AG 2020
J. H. Karimov et al. (eds.), *Mechanical Support for Heart Failure*, https://doi.org/10.1007/978-3-030-47809-4_17

by many after the initial success of Gibbon using as an oxygenator a screen over which a thin layer of the blood passed allowing for gas exchange [5]. Continuous flow was then used with a series of different models of oxygenators including disc, bubble over "iron wool," and finally tiny hollow tubes. Cardiopulmonary bypass (CPB) was initially performed with roller pumps and later with centrifugal pumps that delivered a continuous flow. To compensate for the inadequacy of continuous flow, surgeons learned to operate quickly and to use hypothermia to slow total body metabolism. After an initial surge of interest in pulsatile cardiopulmonary bypass, it has nearly disappeared, but not because it was proven to be inferior. It was just more difficult to deliver a physiologic pulse compared to the simplicity of delivering continuous flow.

Mechanical circulatory support blood pumps in current clinical use and those that preceded them have provided life-saving blood flow at a price. The continuous-flow pumps, ventricular assist devices (VADs), came into use because they are smaller and thus easier to implant (Cheng et al.) and more durable than the pulsatile electric HeartMate XVE, a pump that had on average a 15- to 20-month durability (covered later in the chapter). The evaluation of these continuous-flow assist pumps is complicated by the function of the native heart. In some patients native heart pulsations are present and can be palpated and used for standard blood pressure measurement with a cuff. These patients' hearts open the aortic valve in native heart systole as often as every beat. Lumping together all continuous-flow patients, those that have retained more native heart pulsatility with those that have less call into question outcome reports particularly as they relate to adverse events. Some now call the flow with LVADs "low pulsatile flow" based upon the observation that a good proportion of the patients have some measurable pulse pressure [6]. In some reports [7] clinical complications with continuous-flow, non-pulsatile pumps have included pump thrombosis [8], ascending aortic thrombosis, hemolysis, aortic insufficiency, hemorrhagic and ischemic stroke, gastrointestinal bleeding, infection, right heart failure, aortic leaflet fusion, thinning of the peripheral blood vessel walls, endothelial damage, and sympathetic nervous system hyperactivity associated with hypertension [9, 10]. Other unanticipated, subtler acute, and chronic complications may be possible [11]. The intermediate-term changes created by continuous flow on the circulatory system and end organs are a pathophysiology that is poorly understood. However, improved short-term survival of heart failure patients with these devices compared to medical therapy has led to implantation of over 20,000 devices worldwide with the largest numbers in the USA and Germany [12]. Most patients do well for a limited time. Most seem to be neurologically intact and have normal end-organ function. However, looking at a composite end point of death, stroke, bleeding (primarily GI), device malfunction, and device-related infection [13], in over 12,000 patients from 2008 to 2014 with continuous-flow VADs, there was 50% occurrence at 3 months, 80% at 2 years, and 90% at 30 months. This end point omitted other complications that may have been concomitant including aortic insufficiency, right heart failure, leaflet fusion of the aortic valve, vascular wall pathology, endothelial damage, and hypertension. Also, those on continuous flow for bridge to transplantation (BTT) have only a 31% chance of transplantation after 1 year of continuous flow and just over 40% chance at 21 months. Those defined preimplantation as bridge to decision has only a 25% chance of transplantation at 2 years. Survival after a continuous-flow LVAD is 80%, 70%, 59%, and 48% at 1, 2, 3, and 4 years after implantation, respectively. Thus, the continued clinical experience with non-pulsatile devices has unveiled problems. Clinicians are now seeing and following patients, who have limited pulsation or no pulse at all. This is superimposed upon our complex and incompletely understood circulatory system. The question remains whether continuous flow is truly a good long-term intervention [14]. The experience with continuous-flow devices thus far has resulted in some [15–17] taking a renewed interest in pulsatile flow. Moazami and colleagues [15] concluded that "The true effects of chronic reduced pulsatility on hemostasis, the peripheral vasculature, and end organ function are largely unknown." They also reviewed continuous-flow device problems of hemostasis, hemolysis, ventricular recovery, vascular thinning, end-organ function, and characteristics of pulsatility and proposed methods of flow modulation. They concluded that adding pulsation to LVADs may improve problems in these areas.

The device complication profile was significantly different with pulsatile-flow devices. Most notable was the very low incidence of GI bleeding, hemorrhagic stroke, aortic insufficiency, aortic valve fusion, vascular pathology, and pump thrombosis. Cheng and colleagues [16] reviewed studies comparing continuous-flow devices and pulsatile-flow devices regarding left ventricular (LV) unloading, LV remodeling, bridge to transplantation, aortic fusion and regurgitation, gastrointestinal bleeding, and right heart failure. They concluded that the major advances of continuous-flow devices are smaller size and increased durability and that these were responsible for improvements in survival. Further, they propose modifying continuous-flow devices to make them pulsatile devices. To have a better grasp of pulsatile versus non-pulsatile, we may need to look beyond short- and intermediate-term clinical results. We may need to more critically evaluate the impact on the microcirculation. This chapter will focus on the effects of pulsatile devices on the circulatory system with consideration given to both macro- and microcirculation.

Basic Information

Microvascular Circulation

As blood flows through the vascular tree, arterial branches become progressively smaller until reaching the capillary bed (Table 17.1) [18]. Pulsation becomes blunted. Transmural capillary pressure varies from 4 mmHg in the liver to 40 mmHg in the kidney glomerulus. Capillaries are tiny vessels of endothelial cells one cell layer thick. At any moment, only 5% of the blood volume is in the capillaries, but the approximate cross-sectional area of the capillary bed is 4500 cm^2 [19], and the total surface area of the capillary walls in an adult is >70 m^2 [20]. This huge area facilitates exchange including gases, molecules, fluid, nutrients, wastes, and hormones. Capillary diameter is about 5 μm at the proximal end and 9 μm at the distal end [19]. The microvascular walls are lined with a 1–2 μm layer of glycocalyx or "endothelial surface layer" (Fig. 17.1) [21]. Red blood cells (RBCs) tend to migrate away from larger vessel walls and to aggregate. Red blood cells passing through capillaries move single file (Fig. 17.1). Red blood cells 8–10 μm in diameter deform and squeeze through capillaries (Fig. 17.2). Separation of RBC from plasma often occurs at branch points resulting in shunting and nonhomogeneous flow to the capillary bed. As blood passes through progressively smaller vessels, the viscosity drops until it reaches vessels that approximate and then become smaller than the size of the RBC. Then it rises. At this point, RBC deformation occurs into parachute shaped or asymmetric crepe-shaped cells and the viscosity increases. White blood cells including lymphocytes, monocytes, and neutrophils are slightly larger, have less flexible cell membranes, and have more viscous cytoplasm than RBCs [22]. In low flow or inflammatory conditions, they may develop pseudopods and attach to the precapillary endothelium obstructing proximal capillaries [22]. Under normal conditions, leukocytes deform denuding the endothelial surface layer and move more slowly through capillaries than red blood cells. When RBCs are following a leukocyte, they form a slow-moving "train." Mechanisms for capillary obstruction exist. Low energy, low flow, or low pressure could create microinfarctions and explain some observed clinical problems noted with continuous-flow device support. Venules may increase resistance in the system by RBC aggregation and slow rolling along walls by WBC.

Table 17.1 Relative sizes in microns

Capillary 4–9
Red blood cell 6–8
Lymphocyte 7–20
Monocyte 15–20
Neutrophil 12–15

Pulsatility may be necessary for optimal capillary bed perfusion. Inadequate microvascular perfusion in non-pulsatile support may be the

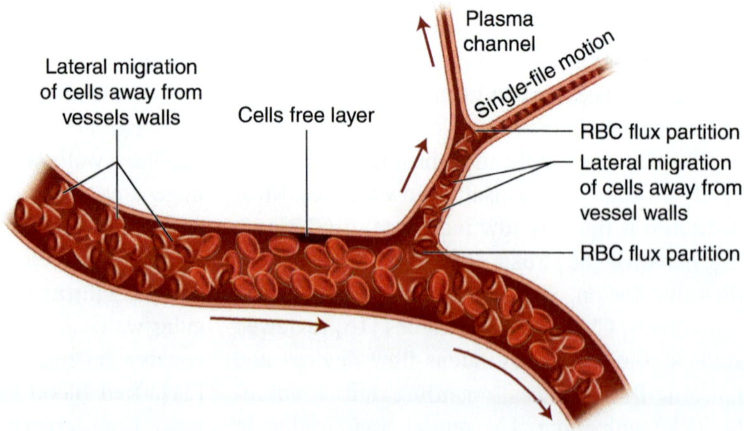

Fig. 17.1 Red blood cells changing shape and selectively flowing in one capillary branch. (Republished with permission of Annual Reviews, Inc., from Secomb [21]; permission conveyed through Copyright Clearance Center, Inc.)

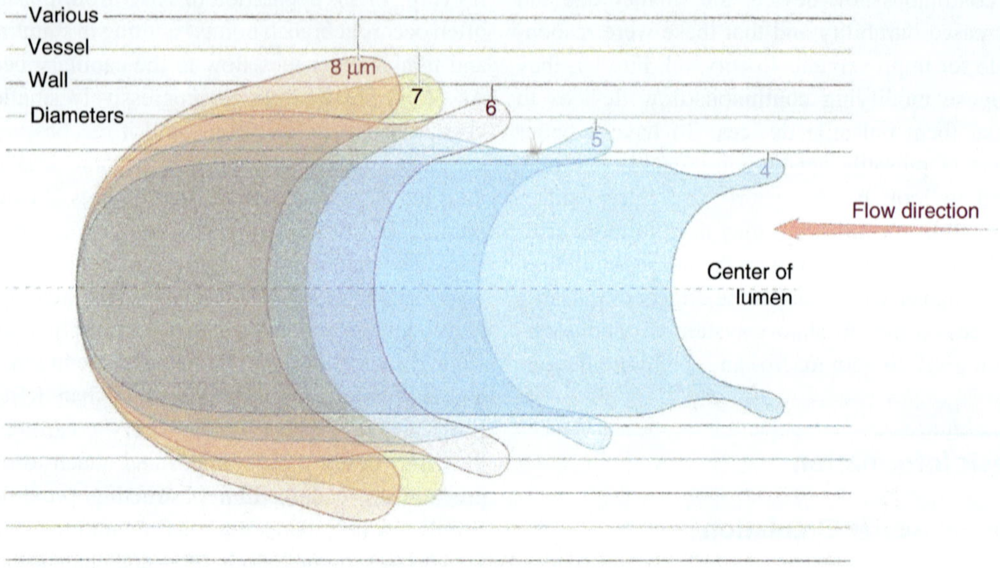

Fig. 17.2 RBC deformation varies with the capillary diameter. (Republished with permission of Annual Reviews, Inc., from Secomb [21]; permission conveyed through Copyright Clearance Center, Inc.)

cause of hemorrhagic stroke, AV malformation gastrointestinal bleeding, and other organ damage or dysfunction. The chronic effects of continuous-flow pumps in the microcirculation are not known. One wonders if simple non-pulsatile pressure is enough to make this system work optimally and chronically. And in the absence of pulsation, what increase in mean pressure must be necessary to drive blood cells through the capillary bed?

Besides the constraints of size of cellular elements and capillary diameter, arterioles, metarterioles, and precapillary sphincters may constrict in response to neural or humoral stimuli and thereby limit flow. Most of the resistance of the circulation is in the small arteries and arterioles.

They constrict when pressure increases and dilate when pressure decreases, thus regulating vascular tone [21]. Other pulsatility-sensitive mechanisms for regulating resistance include calcium regulation by integrin mechanical receptors and NO release of endothelial cells in response to shear stress. This autoregulation of flow maintains adequate brain and kidney blood flows. In the case of very low pressure at the capillary inflow, these tiny vessels are subject to a critical closing pressure. They spontaneously close due to surrounding interstitial pressure being greater than the pressure at the capillary entrance. Lymphatic flow is facilitated by pulsatility. Interstitial volume and pressure may increase in the absence of pulsatility [22].

Pulsatile Perfusion

The pathophysiology of chronic continuous-flow pumping is just becoming apparent. Paradoxically, patients who do better with continuous flow have stronger pulsatility from their native hearts. The microvascular circulation is a likely site of pathology. My (JC) impression is that pathologic lesions in this extensively redundant system may at first be difficult to detect and may need to reach a critical number over time before becoming clinically manifest. In contrast to continuous flow, pulsatile pumps produce pulses in humans and animals that are very similar to those from the heart when the stroke volume is delivered to the proximal aorta. Wave propagation and flows are in the normal range.

The native pulse starts with the heart, and the pulse wave is propagated along the aorta and other elastic arteries (Fig. 17.3). The speed of the wave increases with increasing rigidity of the vessel; thus, it is faster in stiffer and in older calcified systems [23]. Waves caused by pulsation become more prominent as they pass distally. This results from antegrade waves and reflected waves [23]. This complex wave interaction produces pressure-flow relationships that depend upon branching sites and reflected waves. This wave is associated with the blood that flows in a laminar fashion in the larger vessels and creates shear axial and radial stress and strain (Fig. 17.4). These forces, shear stress, and cyclic strain and stretch on the endothelium activate genes controlling production of signaling molecules

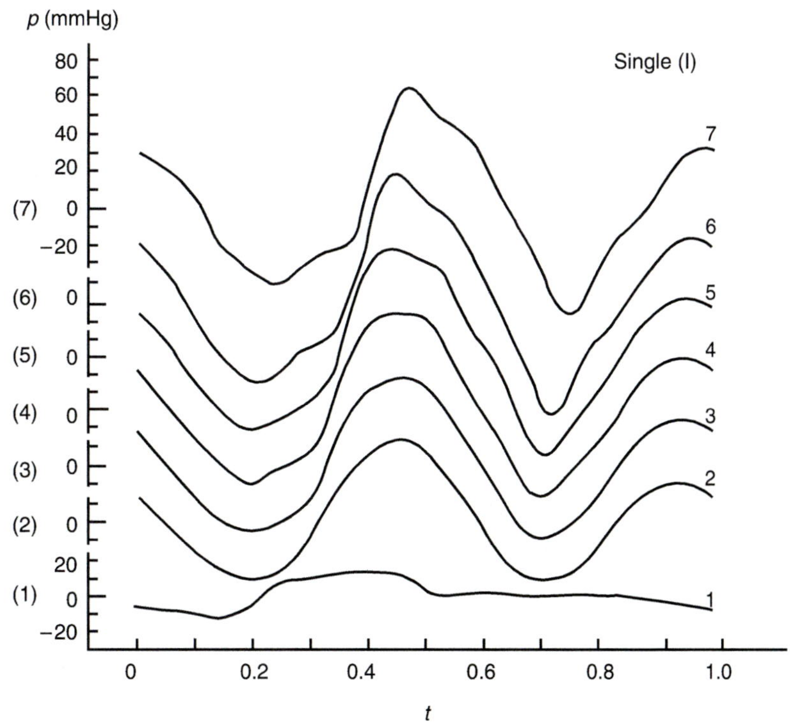

Fig. 17.3 Progression of a composite pressure wave along seven levels of an arterial tree. The wave changes shape and becomes amplified by wave reflections as it travels from the proximal aorta (1) to distal aorta (7). (Reprinted by permission from Springer Nature: "Wave reflections" in Zamir [23])

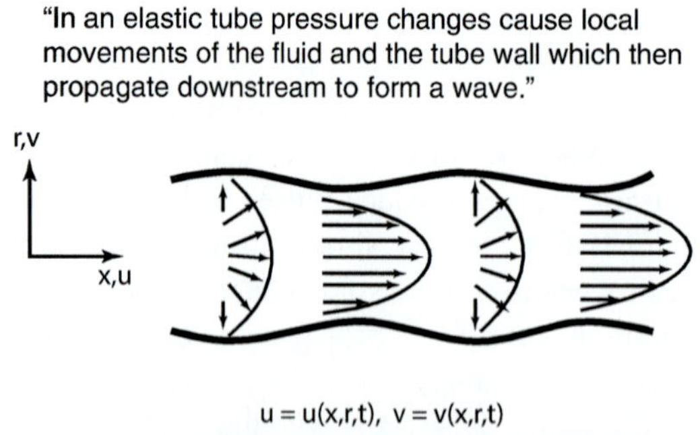

Fig. 17.4 The pulse wave propagation forces in an elastic vessel. (Reprinted by permission from Springer Nature: "Pulsatile flow in an elastic tube" in Zamir [23])

including growth factors that relate to endothelial apoptosis and integrins. Systemic vasorelaxation is NO production mediated and is related to vessel stretch [24]. Pulsatility is athero-protective. Pulsation also influences protein production and collagen synthesis in smooth muscle.

As the volume of the blood ejected from the heart or a pulsatile device reaches the larger vessels, the elastic lamina is stretched and causes local movements of the blood that leads to wave propagation. The aorta is "loaded" in systole and unloads in diastole, the Windkessel effect (Fig. 17.5).

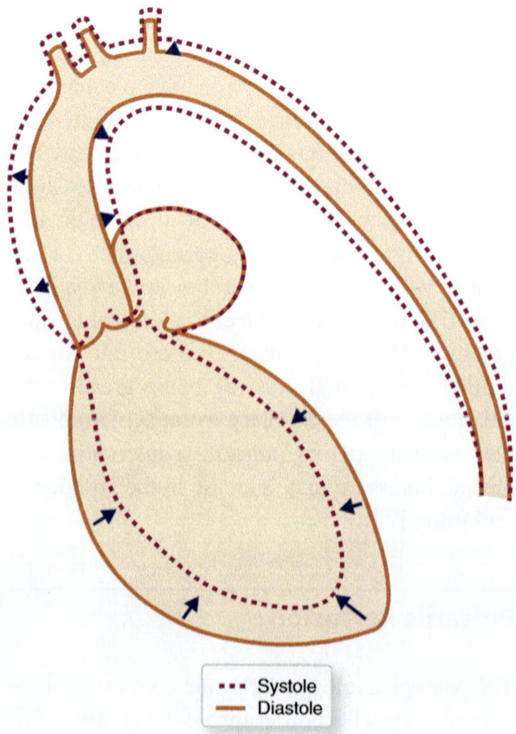

Fig. 17.5 Windkessel effect. The aorta enlarges in systole and decreases in size in diastole, thus pumping the blood during diastole when the heart is relaxed

As flow velocity increases within a vessel, the tendency to go from laminar to turbulent flow was described by Reynolds [25]. This is a phenomenon of parallel vectors of flow turning into flow with energy vibrating in random directions. The Reynolds number which relates fluid density and velocity with the diameter of the vessel and the fluid viscosity is used to assess this tendency. A number of >2000 poise is a conservative upper boundary of laminar flow. The aorta with an inner diameter of 2.5 cm under normal physiologic conditions with a flow of 5 l/min has a Reynolds number of 1063. Given the same conditions and flow, an inner diameter of 1.2 cm such as the HeartMate II outflow conduit would have a Reynolds number of 2211 poise.

Defining pulsatility clinically with current measurements available is most often done with pulse pressure (systolic/diastolic). Organ perfusion pressure is defined as mean arterial minus mean venous pressure. Recovery of end-organ function in critically ill patients depends on adequate organ perfusion. If the arterial pressure is adequate, but the central venous pressure is high (i.e., low organ perfusion pressure), perfusion may not be good enough to promote recovery. This concept seems to be most important for renal recovery. An example of this can be seen with the total artificial heart (TAH). With the TAH, the mean arterial pressure (MAP) is normal say 70 mmHg, and the CVP is usually around 10 or less giving a kidney perfusion pressure of 60 mmHg. If the MAP is 70 mmHg, but the CVP is 25 mmHg (from right heart failure after LVAD implantation), the kidney perfusion pressure is 45 mmHg. In patients with acute kidney injury (AKI), we nearly always see complete renal recovery with the TAH. In the multi-institutional TAH study (Ref. [26]), renal function returned to normal within 2 weeks. In a recent study from Virginia Commonwealth University [27] of patients with preimplantation renal insufficiency, 75% had renal function on TAH support improved from estimated glomerular filtration rate of 48 ml/min to 56–66 ml/min with a 6-month follow-up.

Pulsatility index (PI) has been used in HeartMate II continuous-flow LVAD patients to quantitate pulsatility (max. pump flow – minimum pump flow/average flow over 15 s × 10). Native heart pulsation can be strong enough to open the aortic valve or so weak that it just augments pump flow during systole, ejecting through the LVAD. For the adverse event of gastrointestinal bleeding, a lower pulsatility index (less pulsation in the circulation) correlates with more bleeding [28]. This is also true of aortic insufficiency. A lower pulse index is related to a greater incidence of aortic insufficiency.

Pulse Energy

Comparison of the Driving Force of Continuous and Pulsatile Pumps

In 1966, Shepard et al. described "energy equivalent pressure" (EEP) as the hemodynamic work per cc of blood pumped [29]. This measurement was based upon instantaneous measurements of pressure and flow. EEP is measured in mmHg. For continuous flow, the EEP equals the mean arterial pressure, a straight line over time. In a normal patient, one cardiac cycle, for pulsatile-flow EEP, is in the range of 40 mmHg greater than the mean. The magnitude of this difference over one systolic cycle is the area between the straight-line continuous flow curve and the pulsatile curve. No doubt, the energy of the pulse is much greater.

Another way of looking at this difference is surplus hemodynamic energy (SHE) which is equal to the EEP minus mean arterial pressure times 1332 (This converts the pressure to energy units (erg/cc)). Figure 17.6 shows a comparison of pulsatility of normal hearts, failing hearts, pulsatile VADs (PVAD), and continuous-flow VADs. There is almost no SHE for continuous flow. This subject is summarized nicely by Soucy, Slaughter et al. in a review of defining pulsatility in continuous-flow VADs [30].

The extra energy available with a pulse would appear to be an advantage. The microcirculation may need this extra energy to maintain adequate capillary flows as physiologic conditions change. Something simple like orthostatic change might be an example. Rising from a sitting to a standing position with continuous flow might limit cerebral perfusion.

Pulsatile Cardiopulmonary Bypass and ECMO

Pulsatility has been applied to cardiopulmonary bypass machines for support during routine operations as well as to extracorporeal membrane oxygenation (ECMO) when used for cardiac support. Damping of pulsations has been observed. This is most likely due to relatively small plastic

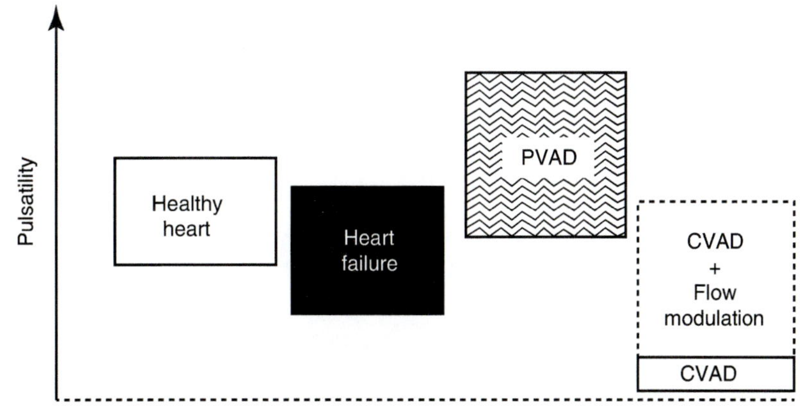

Fig. 17.6 SHE for the continuous-flow pumps is zero in the absence of native heart pulsation. PVAD is pulsatile ventricular assist device; CVAD is continuous-flow ventricular assist device. (Reprinted from Soucy et al. [30], with permission from Elsevier. This article supports the concept of modifying current continuous-flow pumps to give them some pulsatility)

cannulas placed in the aorta [31]. The better access options would seem to be either very large cannulas (24–28 Fr) or placement of a small graft (8–12 mm inner diameter). Peripheral grafting might simplify the procedure.

Speed changes in roller pumps create a pulse as does simultaneous use of intra-aortic balloon pumps. These are not physiologic pulses, and neither of these has become widely used; nevertheless many reports of improved results exist. A review of the literature by Sunagawa and colleagues covers the controversy [32].

Pulsatile flow on cardiopulmonary bypass has been found to optimize endothelial release of vasoactive substances, prevent activation of the renin-angiotensin system, decrease microcirculatory heterogeneity, reduce vasopressin and catecholamine levels, increase bronchial flow to the lungs, and improve subendocardial perfusion of the fibrillating heart. In clinical use, pulsatility has been found in some studies to decrease mortality [33], decrease the incidence of myocardial infarction, improve whole body flow, decrease pro-inflammatory cytokine release, improve postoperative renal function, reduce the length of hospital stay, and reduce the need for postoperative mechanical circulatory support. A meta-analysis reviewed over 200 papers comparing pulsatile bypass with continuous flow and concluded that there was "great variability among the techniques for creating pulsatility," but that with pulsatile flow, postoperative creatinine clearances were higher, and serum lactates were lower [34]. Improved results have also been reported in the pediatric population [35]. Pulsatility has also been used in ECMO cases with some good preliminary results [36].

Problems with adding pulsatility by speed control of the currently used centrifugal pumps include the requirement for a "minimal pump speed" to prevent backflow and to maintain pressure, the possibility of generating air microemboli with abrupt alternation from low to high speed, and the sensitivity to afterload. As afterload increases, centrifugal and axial pump output decreases. Current roller pumps can pump against high resistance and create high pressure. But, on the inflow side, they can create a vacuum. Both types of pumps do not seem to be well suited for duplicating a normal physiologic pulse. A cardiopulmonary bypass pump (VentriFlo True Pulse Pump [Design Mentor, Pelham, NH]) that works like the heart, displacing volume with each beat and producing pulses that appear physiologic, is currently undergoing animal testing [37]. It promises for the first time in routine cardiopulmonary bypass to deliver true pulsatile flow (Fig. 17.7). Animal studies compared capillary flow with the VentriFlo pulsatile pump and the Rotoflow centrifugal continuous flow pump at the same mean flow and pressure. At 6 h of normothermic bypass, a normal capillary flow pattern was maintained with the VentriFlo compared to an abnormal heterogeneous pattern with the Rotoflow that included high-flow shunting, slow flow, sluggish flow, and no-flow areas [38].

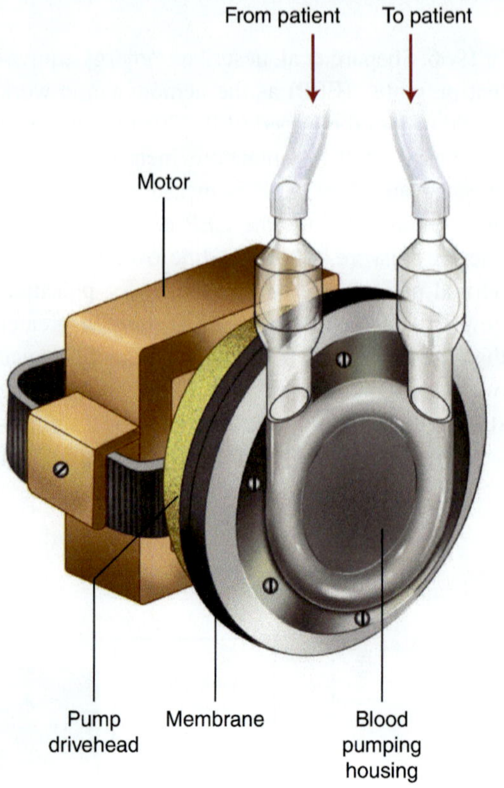

Fig. 17.7 VentriFlo True Pulse Pump. Inflow and outflow valves insure unidirectional flow. A diaphragm pushes the blood and is controlled by the motor. It delivers a biomimetic pulse. (Reproduced with permission from Design Mentor, Inc. © 2018 Design Mentor, Inc. All rights reserved)

Intra-Aortic Balloon Pumping

The intra-aortic balloon pump (IABP) has been widely used for over 50 years. This pneumatic technology has been very important in the treatment of severe ischemic heart disease before and after surgical intervention. An IABP catheter with balloon (usually 30–40 ml) is inserted via the femoral artery in the descending aorta. Pneumatic pulsation is timed (gated) with respect to the cardiac cycle as sensed using a pressure or an EKG tracing. It is used for counterpulsation meaning delivery of the pulse is timed to start just at the end of systole (aortic valve closure). Air is rapidly pumped into the balloon. Diastolic pressure and hence coronary artery flow is augmented. The second major effect is creating a presystolic pressure dip by deflating the balloon just before systole, thus reducing the afterload that the left ventricle must pump against. This is standard in patients with coronary artery disease, patients with heart failure and coronary artery disease, and some patients with heart failure alone. As a blood pump alone, it may augment the cardiac output by about 500–800 ml/min. This is not enough to make a significant difference in most heart failure patients. A laboratory comparison of ECMO, IABP, continuous-flow (CF) VAD, and pulsatile-flow (PF) VAD was done in a piglet model [39]. It demonstrated better left ventricular blood supply demand ratio during CFVAD, PFVAD, and ECMO but not IABP as well as improved global myocardial blood supply during PFVAD and CFVAD, but not ECMO or IABP.

As a treatment for severe heart failure that is not related to myocardial ischemia, the benefit of 500–800 ml/min additional flow with IABP is in most cases outweighed by the morbidity and mortality of the procedure. But, in patients with marginal coronary blood flow and heart failure, it can lead to significant improvement in cardiac function. The most common IABP morbidity is due to vascular injury or occlusion [40] from the intra-arterial catheter. Less commonly, we have seen bleeding problems when patients are heparinized for more than a few days, thrombocytopenia, aortic injury, distal embolism, and rarely perforation of the aorta (usually fatal).

Extracorporeal Pulsatile Ventricular Assist Devices

There have been two types of pulsatile VADs, extracorporeal and intracorporeal. Both have saved thousands of lives. The Pierce-Donachy (Thoratec) pumps developed by Donachy and associates were the most popular extracorporeal [41], first implanted by Hill and Farrar and associates [42] and then by many others [43] as well as our own center. The extracorporeal pumps sit outside the body. Inflow and outflow cannulas are connected to the native heart and the prosthetic ventricle. They are versatile in that they do not need to fit inside the body, and they may be connected to the native heart in a variety of ways. The blood is drained from the native heart by a transcutaneous tube. It enters the pumping chamber via an inflow one-way valve and is pumped out by air pressure on the outside of the sac. This creates a pulse as blood flows out the outflow valve and outflow cannula into the aorta or pulmonary artery. The pulse is not timed with the native heart pulsation (Fig. 17.8) [44]. Because implantation involves the simple placement of cannulas in the heart, it is technically easy for the surgeon and not too risky for the patient. Connection options include inflow from atrium or from the ventricle. Outflow is to the appropriate artery (aorta or pulmonary artery). It is also easy to use a biventricular configuration. In biventricular support, options for drainage from either the atrium or the ventricle exist. The RVAD

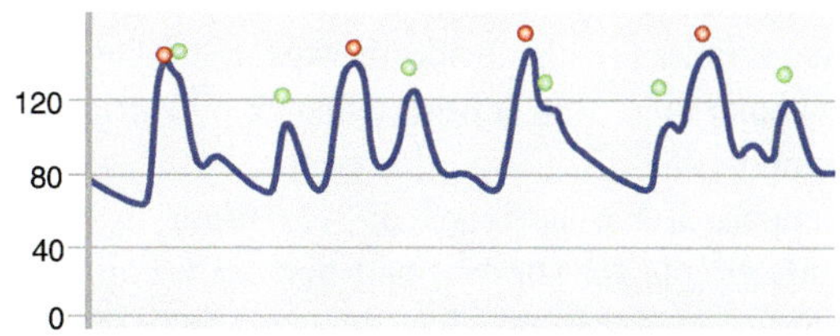

Fig. 17.8 Markers on asynchronous VAD. Arterial pressure tracing showing native heart pressure with peaks marked by green dots and LVAD pressure with peaks marked by red dots. The native heart rate is about 90 beats/min. The LVAD rate is about 60 beats/min. There is no coordination of the LVAD rate with the native heart rate (asynchronous). Simultaneous beats result in higher systolic pressure

may have a right atrial or right ventricular (body) inflow cannula filling the prosthetic ventricle. Likewise, the LVAD may have an inflow cannula in either the left atrium or the left ventricle (apex) filling the left VAD. The outflow cannulas are attached through a graft sewn directly to the pulmonary artery (RVAD) or aorta (LVAD). With direct ventricular drainage, ventricular decompression is superior, cardiac outputs from the device tend to be higher, and results with biventricular configuration are superior to atrial cannulation. They deliver adequate pulsatile flow to maintain life and give the adult-sized patient the option of a considerable amount of mobility with a portable driver that has worked best in the LVAD, not BIVAD, configuration.

In this era, choosing between biventricular support and left ventricular was not a problem. Identification of right heart failure was not difficult. Our program and many others used a biventricular configuration more often than univentricular (LVAD) [45, 46]. There was no hesitation to support the right ventricle when it was failing on maximal medical therapy as judged by high CVP (>15 mmHg) and low RV ejection fraction (<20% by Multigated Acquisition Scan (MUGA)).

These VADs delivered pulses that were asynchronous with respect to the native heartbeat. In adults, it was rare to see pump outputs of greater than 5 l/min in either the univentricular or biventricular configuration. Long and narrow cannulas may have been the limiting factor. One of the problems we encountered in large patients with body surface areas of >2.5 m^2 was continued heart failure with device therapy regardless of whether there was univentricular or biventricular support. We also found that once on mechanical support, a cardiac index of $\leq$2.0 l/min/m^2 was unfavorable for survival and that >2.5 l/min/m^2 cardiac index with device support in critically ill patients predicted survival [47, 48]. In our experience with the Thoratec VADs, this was achievable in most smaller patients (body surface area of <2 m^2) [49].

Extracorporeal Pulsatile LVADs in Children

We used the Thoratec PVAD in a small (17 kg) boy [50]. He was the smallest to have a Thoratec LVAD. The device has a 70 ml stroke volume. Pumping at full stroke volume caused severe systolic hypertension with peak pressures of 170 mmHg. To solve this problem, we converted from a full fill mode (blood pumped only when the prosthetic ventricle was full) to a partial fill mode (filled only to about 40 ml then ejected by controlling the beat rate and the % systole). This cured his hypertension by reducing the stroke volume. He went on to successful transplantation and long-term survival with the donor heart of a 35-year-old woman.

The basic function of the Thoratec pump was passive fullfilling of the prosthetic ventricle as detected by a Hall effect sensor followed by full ejection. The rate of the pumping was therefore dependent on the rate of device filling. In adults, this worked well, but not in small children because they require much smaller stroke volumes than 70 ml.

We were fortunate to be the first in the USA to use the Berlin Heart in the 7-year-old child in October 2000 [51]. He was transplanted after 1 month and is alive 18 years later. This device was made primarily for children in a variety of ventricular sizes with stroke volumes from 10 to 60 ml. It may be implanted in the same configurations as noted for the Thoratec device. It consists of a rigid case, polyurethane tricuspid inflow and outflow valves, and blood and air chambers separated by a polyurethane diaphragm. It has been used in hundreds of children with better results in patients with cardiomyopathy and dilated hearts than in children with complex congenital cardiac disease [52]. We used the Berlin Heart and the Medos Heart, both pulsatile, as well as the continuous Rotoflow pump in children and neonates to treat acutely decompensating congestive heart failure. In children <2 years old and infants with decompensating cardiomyopathies, LVAD decompression with a cannula in the left ventricular apex [53, 54] provided excellent LV decompression and nearly uniformly led to rapid (hours to days) myocardial recovery. Seventy percent had native heart recovery and device removal with no episodes of recurrent heart failure. The key to success appeared to be LV decompression regardless of the pumping mechanism. Once the LV size reduced, LV wall thickness increased, and the hearts began ejecting almost immediately. Continued LVAD support allowed time for nearly complete reverse remodeling. Weaning from support averaged 2 weeks. Heart failure did not recur.

Problems with small pulsatile devices in small children with cardiomyopathy have included bleeding (44%), hypertension (34.5%), major infection (25%), stroke (11–24%), end-organ failure (30–50%), and pump change-out in 30–100% of patients [55]. Survival to 1-year post-implant has been just over 80% for cardiomyopathy patients and 30% for congenital heart disease patients [47]. The large number of pump change-outs resulted from thrombosis. Causes include low flows, small cannula sizes, and polyurethane inflow and outflow valves. Our method of anticoagulation was consistent for over 15 years and has been reviewed in full for children [56] and for adults [57]. This protocol resulted in acceptable thromboembolic and bleeding rates, but children were more challenging than adults presumably from smaller tubes, lower flows, polyurethane valves, small devices, and less washing of the device by blood flow.

Implanted Pulsatile-Flow LVADs

Two implantable pulsatile LVADs were widely used, HeartMate (two models: HeartMate I and HeartMate XVE) and Novacor. Each was abandoned in favor of smaller, easier to implant, continuous-flow pumps: HeartMate II, HeartWare, and now HeartMate 3. The pulsatile HeartMate devices (HeartMate I and XVE) also had durability issues and the Novacor embolism issues. A clinical comparison of the two devices from the Bad Oeynhausen group showed a slight survival advantage with the Novacor and documented the adverse events that were specific to each device [58].

There was a third device, unique because it was the only totally implantable LVAD with no wires or tubes passing through the skin. It featured transcutaneous energy transmission by induction coils. It also required that a volume compensator be placed in the chest, a necessity for preventing vacuum as the pump diaphragm moved from full to empty. The compensator was a balloon that allowed air to flow in and out from pump to compensator and vice versa as the pump diaphragm moved back and forth. A complete implant included the pump, a controller (electronic brains of the device) with battery, a volume compensator placed in the left chest, and the energy receiver subcutaneously above the right breast. A brief history of this device named "Lion Heart" is well summarized by the principle investigator Dr. Walter Pae [59]. There was a significant reduction in infections with this totally implanted device as compared to all other devices that had transcutaneous "drive lines" [59].

Novacor Ventricular Assist System

The Novacor left ventricular assist system, the invention of Peer Portner, PhD, was first used successfully as a bridge to transplantation (BTT) by Dr. Philip Oyer in 1984 at nearly the same time as the first Thoratec BTT by Donald Hill [60, 61]. In a novel spring-loaded, passively filling device with a smooth surface polyurethane lining on dual pusher plates, bovine pericardial inflow and outflow valves, and Dacron tube inflow and outflow conduits (later the inflow conduit was PTFE), the Novacor was driven by a solenoid-based motor. The inflow conduit was placed in the apex of the left ventricle and the outflow conduit in the ascending aorta. It filled during native heart systole and pumped the blood forward in diastole (counterpulsation pumping mechanism). Outputs from the pump were in the 6 l/min range. The initially recommended implantation technique for the Novacor as well as for the HeartMate pulsatile pump was more difficult than extracorporeal VADs. It required creating a pocket in the left abdominal wall between the muscular and fascial layers that extended caudally to near the iliac crest. An alternative technique described by Icenogle (T. Icenogle, personal communication) was adopted in our program. It involved marsupial-like partitioning of the anterior abdominal cavity with a large thick perforated PTFE sheet that was stapled to the inside of the abdominal wall. The device was placed anterior to the sheet, between the sheet and the anterior abdominal wall muscles, leaving the abdominal contents posterior to the sheet. The cephalic end of the sheet was stapled or sewn to the diaphragm such that the device space was in direct communication with the pericardial space. This completely separated the pump from the abdominal contents and eliminated extensive abdominal wall surgery at implant and explant as well as resulting in much less bleeding. No infections were seen with this technique.

With the Novacor device, there were no issues of GI bleeding, aortic insufficiency, or pump thrombosis. The major issue in some programs was embolic stroke variously reported to occur in 20–47% of cases [62]. The source of the emboli was most likely the inflow conduit from the left ventricular apex to the pump. It was Dacron and later switched to PTFE and contained a tissue valve. It was 6–9 cm in length. Pseudomembrane formation was reported in this conduit and was implicated in the strokes. Of interest was a high inter-institutional variability in the reported incidence of stroke suggesting varied operative and management techniques may have had a major impact. Another component of this device that was successfully tried in sheep, but never in humans, was a transcutaneous belt energy transformer that would have made the device totally implantable. A review of the clinical experience was written by the Stanford group [63].

HeartMate Left Ventricular Assist Devices

The HeartMate LVAD (HeartMate I and HeartMate XVE models) was implantable and about the same size as the Novacor (a discus shape of 11 cm in diameter and 4 cm in thickness). The implantation technique and positioning were like that of the Novacor heart. It appeared in two forms. The first was pneumatic, a single pusher plate displaced by pulses of air pressure [64]. The other side of the pumping chamber was part of the rigid titanium case. Both the diaphragm and rigid wall inside the pumping chamber were microscopically "textured," sintered titanium on the rigid side, and flocked polyurethane on the pusher plate. Porcine xenograft valves on the inflow side and outflow side were part of their respective conduits. As time passed, the design was improved. Renamed the HeartMate XVE, it now had an electric motor [65]. The textured lining was envisioned to lead to an endothelial lining that would be non-thrombogenic. Rather it led to a coagulum of randomly distributed proteins and blood elements that resulted in a localized coagulopathy characterized by fibrinolysis and thrombin formation [66]. As a result of this, patients did not need any anticoagulant and were often just given aspirin alone or in combination with dipyridamole. The other issue created by this textured surface lining of the pump was immunological compromise. Opportunistic organisms infected the patients. Twenty-eight percent were infected with Candida within 3 months of implant [67]. It was found that CD4 T cells in patients with this device had increased susceptibility to cell death. This was a new finding not seen with smooth surface pump linings. Even so, the REMATCH trial [68] of 129 patients randomized between LVAD and medical treatment for very sick end-stage heart failure patients found significantly better survival at 1 year in the LVAD group (medical group survival 25%, LVAD group 52%). Complications with the LVAD included infection, bleeding, device malfunction, and neurologic dysfunction, and comparing the number for all complications, there were 2.35 times as many complications in the LVAD group as in the medical group. Right heart failure was noted in 17% of the implanted patients. By 2005, despite the wave of enthusiasm from the REMATCH trial, it was clear that a high percentage of HeartMate pumps were failing at 12–18 months following implantation primarily from inflow valve failure (regurgitation) and motor failure. This ended enthusiasm about this device as a long-term therapy. Of interest with this pump, GI bleeding and aortic insufficiency were rare, and there was no pump thrombosis.

The Total Artificial Heart

As of August 1, 2019, 1920 total artificial hearts had been implanted in humans since the Akutsu Heart by Cooley in 1969. The SynCardia (formerly called CardioWest and Jarvik-7) has accounted for 97% of implants and over 666 patient implant years with the longest being over 5 years [69]. This is an orthotopic biventricular (Fig. 17.9) pneumatic pulsatile pump lined with smooth segmented polyurethane, with tilting disk carbon pyrolite inflow and outflow valves (SynHall valves). A four-layered diaphragm separates the blood side from the air side (Fig. 17.10). Drive lines from the semirigid ventricles exit from the skin about 12 cm below the left costal margin and are connected to a driver that has adjustments for each ventricle for air pressure, beat rate, vacuum, and % systole (this determines the length of systole for one heart beat and is usually set at 50%). The flow rate of air exiting each ventricle is measured by pneumo-tachometers. Instantaneous measurement of the area under the air flow curve provides beat-to-beat stroke volume since the air volume exiting is equal to the blood volume entering the pump (fill volume). From this, instantaneous cardiac output is automatically calculated. Also, from this, the user sets the device so that the ventricle is only 70–80% filled (partial fill). Partial filling is a major advantage of pneumatic technology in total artificial hearts. It assures that the left and right ventricles are always in balance. It allows the patient to have an automatic "Starling" mechanism and allows the user to control for the effects of wide blood volume swings as might occur with volume transfusion. The ventricle maximal volume capacity is 70 ml. If we partially fill to "fill volume" of 50 ml (70% fill), at the end of the filling period, 20 ml of air remains on the other side of the four-layered diaphragm. If the right ventricle pumps more blood on the next beat, the increased volume simply displaces the left ventricle diaphragm a bit more, say 5 ml. And if less volume, say 10 ml, is pumped from the right on the next beat, the diaphragm is displaced by 10 ml less that the preceding beat. Similarly, if the patient goes for a walk and has increased venous return from his muscle pumps, more than 2 l/min increase in pump output can be automatically accommodated (20 ml reserve/beat times 120 beats/min provides 2400 ml/min increased flow). No sensor or feedback is required, and no adjustments need to be made. As it turns out, the fill volume is a surrogate for the CVP. An example of this is a patient on dialysis. Every 2- to 3-day dialysis, resulting in large volume shifts are not seen in patients with normal renal function. The clinician can watch the TAH fill volume each day. When it exceeds 60–65 ml, he sends the patient for hemodialysis and takes off enough fluid volume with the dialysis to drop the TAH fill volume to 50 ml. Thus, the mantra for the SynCardia TAH is "partial fill, full eject." In patients with normal renal function, it is rare to need any adjustment in the driver controls after the first few days following implantation.

There are three drivers available. The original console was very large containing an active console, a backup console, numerous alarms, and compressed air tanks. This has been an excellent driver but was phased out. The current drivers are the Companion which weighs about 50 pounds and is used in the hospital and the Freedom driver a 13.5-pound portable console that can be carried by the patient when he is discharged from the hospital.

Our first use of this device was in August of 1985 in a young man with biventricular failure

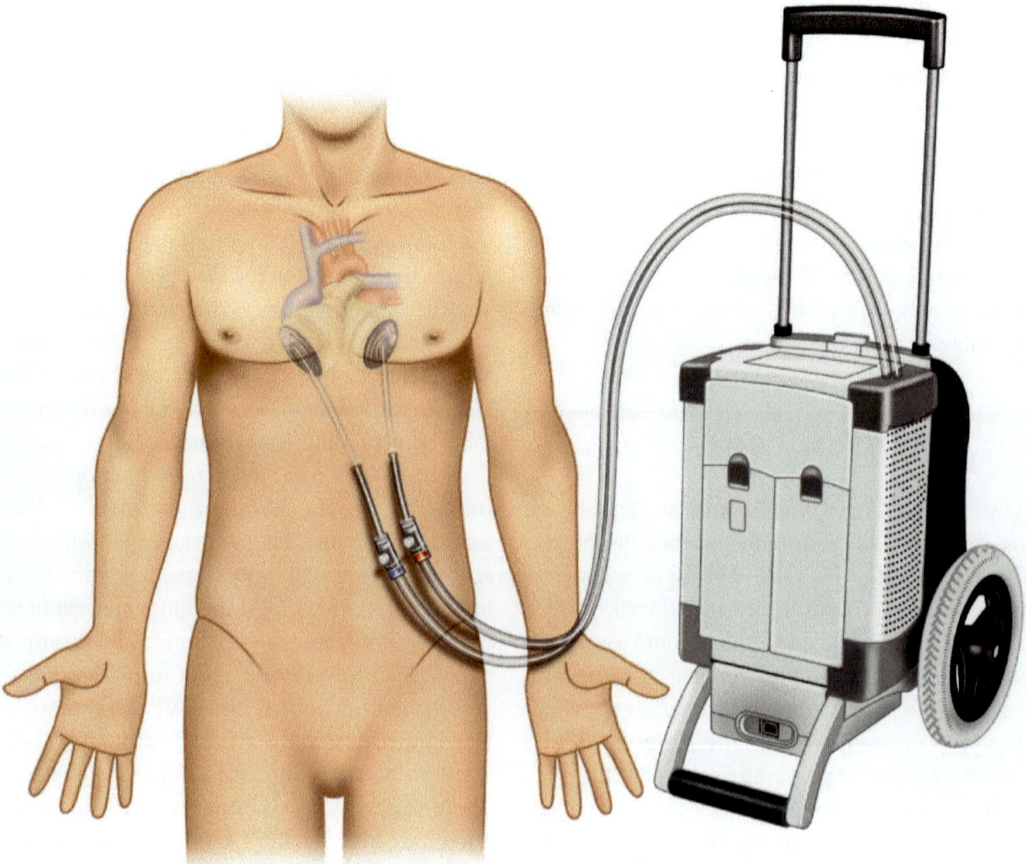

Fig. 17.9 SynCardia total artificial heart. The native ventricles are removed and replaced by prosthetic left and right ventricles that each have inflow and outflow valves (pyrolite carbon SynHall valves). The drive lines pass from the device in the mediastinum through layers of muscle, subcutaneous fat, and skin. They are attached in this drawing to the Companion Driver for inhospital use. Transition to a wearable driver (Freedom driver) is made prior to discharge. (Reprinted from "Total artificial heart" in Selke et al. [70], 2010 with permission from Elsevier)

[71]. The device replaced both ventricles and all four valves. It functioned well maintaining end-organ function in a critically ill person for whom there was no donor heart available. He was transplanted about 10 days later and lived for 5 years, the first successful bridge to transplant with a TAH. Since that time, we completed a multi-institutional trial from 1993 to 2002 [26]. In that trial 81 TAH-implanted patients were compared to matched controls. In the TAH group, 79% survived to transplantation, and 86% of those transplanted survived for over 1 year compared with the control group where 46% were transplanted and 69% of those survived for over 1 year after transplantation. More clinical data has been accumulated [72–75]. In these studies as in the multi-institutional study, patients were nearly all INTERMACs status 1 ("Crash and burn": expected survival hours to days). Only 15% of LVAD continuous-flow patients since 2008 have been INTERMACS profile 1 [76, 77]. For continuous-flow pumps, most recipients are less sick with 38% of recipients being INTERMACS profile 3 (inotrope dependent, may be at home). Another indication for using the TAH is the central venous pressure (CVP). This may be the

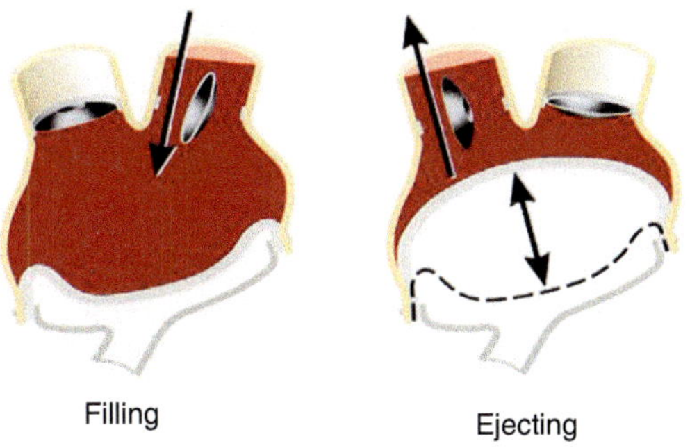

Fig. 17.10 Cutawaydrawing of SynCardia total artificial left ventricle showing partial filling (fill volume to 50–60 ml) and full ejection (full excursion of diaphragms ejects fully with each beat). (Reprinted from "Total artificial heart" in Selke et al. [70], 2010 with permission from Elsevier)

most important determinant of right heart failure in patients that are failing medical therapy indicating need for biventricular support. In our multi-institutional study, in patients on maximal medical therapy prior to implant, the mean CVP was 20 ± 7 mmHg. This stands in contrast to nearly every report on LVADs where the CVP before LVAD implantation has been around 12 mmHg and a CVP of >16 is a known risk factor for death after LVAD implantation [78]. One other factor that we considered pre-TAH implant was the RV ejection fraction on first-pass MUGA scan, where <20% ejection fraction was taken as right heart failure.

TAH Function

Function of the TAH is unique in current clinical usage in that it replaces orthotopically both ventricles giving the clinician complete control of the left- and right-sided circulations. There is no right heart failure with the TAH until the maximal pump flow of 9.8 l/min is exceeded (beat rate of 140 times full fill volume of 70 ml). Flow and output are not dependent on the shape and function of the native left and right ventricles. There are no native ventricles with the TAH. They have been removed to make room for the prosthetic ventricles. The criteria most often used to insure that the device will fit and not compress the inferior vena cava (IVC) or left pulmonary veins once the patient's chest is closed are BSA > 2 m^2, native heart LVEDD of ≥70 mm on echocardiogram, a "large heart" on chest X-ray with a CT ratio of >0.5, and on chest CT scan a distance of ≥10 cm from posterior sternum to anterior vertebral body at a level of T10 (tenth thoracic vertebra). There is also a recently introduced TAH with each ventricle 20 ml smaller, the "50 cc TAH." Seventy-five patients have been implanted [79] with 53% in women and 18% in children. Finally, when there is still doubt about size, "virtual implantation" has been done many times [80, 81] and provides the radiologist and surgeon with superimposed views of the prosthetic ventricles placed three dimensionally on the chest CT scan of the potential recipient of the TAH. As a frequent consultant on these studies, I believe that in almost all cases, this has been very helpful in making the "fit" determination.

The implantation technique is not difficult; it is just "new" to most transplant surgeons. It involves doing a cardiectomy on the ventricular side of the AV groove, trimming the ventricular muscle to within 1 cm of the AV groove, closing the orifice of the coronary sinus, ligating the left atrial appendage, sewing in atrial connectors and great vessel grafts, passing the transcutaneous drive lines, and attaching the two ventricles. A detailed description of the implantation technique was reported by Arabia [82]. Additional implant and management information have been reported by an author (JC) [83, 84].

Once implanted and functioning, the SynCardia TAH pulsatile flow to the lungs and the body is physiologic. In the multi-institutional study [26] involving mean values for 95 patients just after TAH implantation, the cardiac index rose from 1.9 to 3.2 l/min/m^2, the systolic pressure rose from 93 to 122 mmHg, the CVP fell from 20 to 14 mmHg, and the organ perfusion pressure (mean arterial pressure minus mean CVP) rose from 49 to 68 mmHg. Renal and hepatic function returned to normal from elevated pre-implant levels within 2 weeks. Most patients (75%) were out of bed within 1 week, and 60% were walking >100 ft within 2 weeks. A summary of 15 years of experience using the TAH with a consistent anticoagulation protocol [85] revealed that there were 8 strokes in 99 consecutive patients, most within 2–6 days after implantation and attributed to the operation. There were 2 strokes in chronically anticoagulated patients over a period of 24 patient years or 0.08 strokes per patient year. And there were two strokes in the two patients that developed prosthetic valvular infections. There were four GI bleeds, two known to be traumatic (esophagus and stomach) and two of unknown cause. There was no pump thrombosis. The anticoagulation protocol employed antiplatelet and anticoagulant therapy and monitored both. There was documented subclinical hemolysis attributed to the four mechanical valves. Hematocrit postimplantation was in the 20–25% range. Chronic anemia observed in these patients was felt to be due to low-grade hemolysis (free hemoglobin levels of <15–20 mg/dl, no obvious hemoglobinuria) and to inflammation caused by chronic contact of the blood with the TAH [86]. A moderate anemia may have some advantages. It decreases blood viscosity and by decreasing resistance and increasing pump output increases the "washing effect" of the blood passing through the ventricles. The high flows through the TAH, generally 6–8 l/min, are standard. Shear stress within the ventricles can be measured [87]. Thrombogenecity of various devices based upon platelet activation was found to be greater in continuous-flow devices [88] and much greater than that found in the pulsatile devices (M.J. Slepian, personal communication). Some antiplatelet drugs can control platelet activation in the TAH, but not in continuous-flow devices [89]. No cases of pump thrombosis have been reported with the SynCardia TAH.

While 70–80% of patients with the SynCardia device in experienced centers have been transplanted within 6 months [75] and the FDA device approval is limited to bridge to transplantation, the experience with long-term survival has been substantial. Over 395 patients have been implanted for more than 6 months, over 175 for 1 year, over 43 for 2 years, and over 6 for 4 years. An international experience with 47 SynCardia recipients that have had the device for at least 1 year has been published by Torregrossa and multiple institutions [90]. They addressed durability. There were 2 membrane perforations, 1 at 971 and the other at 801 days postimplantation in their group of 47 patients. Their search of the world literature is found including the two they reported a total of six reported incidents, three deaths. The earliest perforation was at 2 days and said to be related to an unauthorized resterilization at high temperature before implantation and perforation of one layer of the RV diaphragm. This patient survived with a reimplant of the RV. There were two events within 4–6 months of implantation with one death and three at over 693 days with two deaths. Thus, 0.4% of patients (approximately 1400 implants had been done at the time of their report) have had diaphragm failure, and 0.2% have subsequently died. An update from the company as of October 1, 2018, reports 8 perforations in 666 patient years or 0.01 events per patient year (S. Venkataramani, SynCardia Systems, Tucson, personal communication). The diaphragm has four layers, and there have not been any catastrophic leaks. This provides time to take remedial action.

Comparison of Pulsatile Devices: Extracorporeal VADs, LVAD, and TAH

From 1985 to 2000, at the University of Arizona, we had implanted 48 extracorporeal VADs (18 Symbion and 30 Thoratec), 37 implantable LVADs (Novacor), and 55 total artificial hearts (SynCardia, CardioWest). Patient selection was based upon failure of medical therapy, clinical condition, patient size, the presence or absence of right heart failure, and the presence or absence of renal failure. The study compared Thoratec extracorporeal (n = 26 [18 BIVAD and 8 LVAD]), Novacor LVAD (n = 23), and the TAH (n = 43) [91]. The TAH recipients were larger and had lower cardiac output and higher central venous pressures (CVPs) (20 ± 7 mmHg) than the other two groups. The percent surviving to transplantation were 75% TAH, 56% LVAD, and 38% BIVAD. Strokes were seen in 8% of TAH, 25% of LVAD, and 12% with the BIVADs. This unique single-center study comparing three devices had consistent selection and management protocols. Our conclusion was that biventricular support should be used in unstable patients with right heart failure. Patients too small for a TAH received a Thoratec. Novacor LVADs were reserved for more stable patients without right heart or renal failure.

Risk Factor Analysis

In the absence of studies directly comparing devices within the same institution, risk factor studies have been used to try to understand the impact of device types on the physiology of the patient. In 2008, we summarized ten pulsatile VAD and TAH risk factor studies [92]: two TAH, four Thoratec, two HeartMate, and two Novacor. For the TAH, significant multivariate risk factors for death included history of smoking and prothrombin time ≥16 s and had odds ratios of 10 and 4, respectively. For the Novacor, HeartMate, and Thoratec, CVP ≥ 16 mmHg, a diagnosis of right heart failure, renal failure markers (BUN > 40 mg/dl, creatinine elevation), reoperation, and mechanical ventilation were found to be significant predictors of mortality. These study results support our selection algorithm based upon our finding that biventricular failure in potential transplant candidates was best treated with the total artificial heart. If fitting the TAH appeared questionable, we used Thoratec BIVAD support.

Conclusions

It is difficult to think about modes of perfusion without comparing continuous and pulsatile flow. There is a huge experience with continuous-flow devices. Even the biggest enthusiasts admit there are many problems. Issues not seen with pulsatile devices have become apparent with continuous flow. Smaller size, ease of implantation, and greater durability favor continuous-flow devices. Disadvantages that lead to complications appear to be high shearing forces within the devices and, in many cases, decreased pulsatility or no pulsatility. Also, the LVAD does not treat right heart failure that in the previous era was routinely treated with preemptive pulsatile BIVAD or TAH placement. According to the INTERMACS registry, use of LVADs in the sickest patients and in patients with right heart failure has declined [13, 77]. Use in profile 1 patients has been stable at 15% for 10 years, for profile 2 patients use decreased from 40% to 36%, and in profile 3 use increased from 25% to 38%. At the same time in 17,000 implants, RV failure at the time of LVAD implantation was associated with a 3.76-fold increase in mortality [77]. We have learned that pulsation in LVAD patients is a good thing. There are less complications in patients with native heart pulsatility, and strategies for increasing native heart pulsatility in LVADs such as decreasing the rpm and allowing the aortic valve to be opened on every third beat are under consideration [12].

The physiology of pulsatile devices approaches and in the case of the TAH nearly duplicates the native heart. Shearing forces within the devices are much lower leading to less platelet activation that is controllable with available antiplatelet therapy. Size was not and is not a problem with the extracorporeal VADs that are still easily implanted. The implantable pulsatile LVADs required a bigger operation than current LVADs, but "not that big." Durability was a problem with the HeartMate pneumatic and electric models that became significant by postimplant months 12–18. But, on the positive side, device "auto-anticoagulation" was caused by a low-grade fibrinolytic state. Only minimal if any antiplatelet therapy was used. The Novacor electric LVAD was durable. It had a long inflow conduit that in some hands caused an unacceptable amount of thromboembolism. The TAH is not difficult to implant providing the fitting guidelines are met nor is it difficult to explant if the recommended technique is followed [77]. Orthotopic implantation provides the shortest device blood flow pathway and allows the highest pump outputs. It is useful for bridge to transplantation in sicker patients and those with right heart failure. The INTERMACS profile 1 patients, the sicker potential transplant recipients with higher CVPs, can be salvaged (Table 17.2 Indications for TAH). Destination therapy with the TAH is being actively studied.

By 12 months of implantation time, approximately twice the percentage of TAH recipients are transplanted compared with LVADs. TAH advantages include removal of diseased myocardium and complete circulatory control of right and left sides.

The pathologic consequences of continuous flow were mentioned in the introduction of this chapter. Concern for these and the high frequency of death and multiple end points found in the INTERMACS registry [9] leads to the question of developing new or reexamining older pulsatile systems. Addressing the issue of biventricular failure is simple – use a biventricular device.

Table 17.2 Indications for total artificial heart as a bridge to transplantation (BTT) in adults who have been selected as potential transplant candidates

Adequate size (for 70 ml ventricles)
BSA > 2 m^2
10 cm or > from posterior sternum to anterior vertebral body at T-10 on CT scan
LVEDD ≥ 70 mm on TTE or TEE
Cardiothoracic ratio > 0.5 on chest X-ray
Sudden death in transplant candidate
Transplant recipient with graft failure
Acute primary graft failure
Chronic: Coronary vasculopathy
Chronic rejection
INTERMACS I status
Right heart failure in face of maximal medical therapy (diuretics, inotropes)
CVP ≥ 20 mmHg in spite of maximal medical therapy
RVEF < 0.20 by first pass MUGA
Irreversible structural problems
Ischemic cardiomyopathy
Massive myocardial infarction
Postinfarction VSD
Surgical misadventures
Thrombosis of left ventricle
Selected myocardial tumors
Hypertrophic and infiltrative restrictive cardiomyopathy
LVAD failure

Compromising to find workable solutions has been the standard in mechanical perfusion since the days of Lillehei and Gibbon. Continuous flow has been the easier and more practical path to follow. Roller pumps and then centrifugal pumps led the way. Cardiopulmonary bypass with continuous-flow technology has been stagnant for decades. Some notable progress toward achieving true pulsatility for cardiopulmonary bypass has been mentioned [37, 38]. Axial flow pumps historically followed much later with the HeartMate II. Experience with continuous flow has led to great success but also to enough numbers of unanticipated and bad consequences that the question has been raised about the fundamental physiology. Does continuous flow provide enough energy to propel the blood through the circulatory system compared to pulsatile flow? Are we seeing inadequate capillary bed perfusion as a cause of GI bleeding in continuous-flow devices? If so, could this be present in other organs and causing problems such as hemorrhagic cerebral infarcts? Are we able to detect microcirculatory disease at an early stage? Beneficial effects of pulsatility on sympathetic nervous system function, neurohumoral signaling, and vascular and valvular function are also lost with continuous flow.

The other major difference between continuous-flow devices and pulsatile is trauma to the blood. Forces on platelets and red blood cells as well as proteins such as von Willebrand multimers are generated by rotary devices spinning at 2500–10,000 rpm. These are much greater than shear stress of pulsatile pumps that produce a true physiologic pulse. Consequently, with continuous flow pumps, platelet activation and hemolysis are seen and are procoagulant.

Durability is another consideration. If 6–10% of continuous-flow pumps thrombose within 1 year and need replacement and there is a significant mortality in the following year (30%) [93], durability is an issue. Granted some continuous-flow device, patients have survived for over 5 years. Most have not. Survival is 48%

at 5 years. The only pulsatile device to have durability issues was the HeartMate XVE.

Summarizing the pulsatile era, the results were good, and bridge to transplantation became successful. Pulsatility did not stand out as a problem. It simply mimicked the normal human circulation. Durability, size, and thromboembolism as we have outlined were problems with some of the devices. Durability in the broadest sense of the word means absence of device failure over some specified number of years. We have nothing mechanical that approaches the durability of the transplanted heart. Size reduction as seen with current continuous-flow pumps comes at the price of damage to the blood by obligatory continuous rotary flow, not to mention the problems inherent in lack of pulsatility and retention of diseased ventricles. A return to pulsatility in LVADs may now be in order.

The gold standard for pulsatile perfusion is volume displacement. Technologies may not necessarily need to be electrical and miniaturized. An example is left-right balance and an automatic Starling mechanism in the pneumatic TAH. Thus far, other implantable technologies have struggled with this balance that is vital for biventricular support and is simply accomplished with pneumatic technology and partial fill full eject pump management.

References

1. Bishopric NH. Evolution of the human heart from bacteria to man. Ann N Y Acad Sci. 2005;1047:13–29.
2. Barrett KE, Barman SM, Boitano S, Brooks HL. Ganong's review of medical physiology. 24th ed. New York: The McGraw Hill Companies, Inc.; 2012. p. 602.
3. Saeed O, Rangasamy S, Reyes M. Speed reduction does not restore high molecular weight von Willebrand multimers during HeartMate II support: an in vivo study. ASAIO J. 2018;64(5):e123–5.
4. Cooley DA. C. Walton Lillehei, the "Father of open heart surgery". Circulation. 1999;100(13):1364–5.
5. Hill JD. John H. Gibbon, Jr. Development of the first successful heart lung machine. Ann Thorac Surg. 1982;34:337–4.
6. Rangasamy S, Madan S, et al. Non-invasive measures of pulsatility and blood pressure during continuous flow left ventricular assist device. ASAIO J. 2019;65(3):241–6.
7. Cornwell WK, Urey M, et al. Continuous flow circulatory support: the Achilles heel of current left ventricular assist devices? Circ Heart Fail. 2015;8:850–2.
8. Starling R, Moazami N, Silvestry S, et al. Unexpected abrupt increase in left ventricular assist device thrombosis. NEJM. 2014;370:33–40.
9. Markham DW, Fu Q, Palmer MD, et al. Sympathetic neural and hemodynamic responses to upright tilt in patients with pulsatile and nonpulsatile left ventricular assist devices. Circ Heart Fail. 2013;6:293–9.
10. Cornwell WK, Tarumi T, et al. Restoration of pulsatile flow reduces sympathetic nerve activity among individuals with continuous flow left ventricular assist devices. Circulation. 2015;132:2316–22.
11. Purohit SN, Cornwell WK, et al. Living without a pulse. Circ Heart Fail. 2018;11:1–11. e004670. https://doi.org/10.1161/Circheartfailure.117.004670.
12. Aissaoui N, Morshuis M, Maoulida H, et al. Management of end-stage heart failure patients with or without ventricular assist device: an observational comparison of clinical and economic outcomes. Eur J Cardiothorac Surg. 2018;53(1):170–7. https://doi.org/10.1093/ejcts/ezx258. PMID:28950304ClinicalTrial
13. Kirklin JK, Naftel DC, et al. The seventh InterMacs annual report: 15,000 patients and counting. J Heart Lung Transplant. 2015;34:1495–504.
14. Miller L. We always need a pulse or do we? J Cardiovasc Trans Res. 2012;5:296–301.
15. Moazami N, Dembitsky WP, Adamson R, Steffen RJ, Soltesz EG, Starling RC, Fukamachi K. Does pulsatility matter in the era of continuous-flow blood pumps? J Heart Lung Transplant. 2015;34:999–1004.
16. Cheng A, Williamitis CA, Slaughter M. Comparison of continuous flow and pulsatile flow left ventricular assist devices: is there any advantage of pulsatility? Ann Cardiothorac Surg. 2014;3(6):573–81.
17. Guan Y, Karkhanis T, Wang S, et al. Physiologic benefits of pulsatile perfusion during mechanical circulatory support for the treatment of acute and chronic heart failure in adults. Artif Organs. 2010;34(7):529–36.
18. Barrett KE, Barman SM, Boitano S, Brooks HL. Ganong's review of medical physiology. 24th ed. New York: The McGraw Hill Companies, Inc.; 2012. p. 571.
19. Barrett KE, Barman SM, Biotano S, Brooks HL. Ganong's review of medical physiology. 24th ed. New Delhi: Tata McGraw-Hill; 2012. p. 571.
20. Popel AS, Johnson PC. Microcirculation and hemorheology. Annu Rev Fluid Mech. 2005;37:43–69.
21. Secomb TW. Blood flow in the microcirculation. Annu Rev Fluid Mech. 2017;49:443–61.
22. Schmid-Shoenbein GW. The damaging potential of leukocyte activation in microcirculation. Angiology. 1993;44(1):45–56.
23. Zamir M. The physics of pulsatile flow. New York: Springer-Verlag; 2000. p. 114–5.

24. Domanski MJ, Giddens DP. Optimization of pulsatile flow for mechanical circulatory support. J Heart Lung Transplant. 2013;32:577–8.
25. Zamir M. The physics of pulsatile flow. New York: Springer-Verlag; 2000. p. 18.
26. Copeland JG, Smith RG, Arabia FA, et al. Cardiac replacement with a total artificial heart as a bridge to transplantation. NEJM. 2004;351:859–67.
27. Quader MA, Goodreau AM, et al. Renal function recovery with total artificial heart support. ASAIO J. 2016;62:87–91.
28. Wever-Pinzon O, et al. Pulsatility and the risk of nonsurgical bleeding in patients supported with continuous flow ventricular assist device HeartMate II. Circ Heart Fail. 2013;6:517–26.
29. Shepard RB, Simpson BD, Sharp JF. Energy equivalent pressure. Arch Surg. 1966;93(5):30–40.
30. Soucy KG, Koenig SC, Giridharan GA, Sobieski MA, Slaughter MS. Defining pulsatility during continuous-flow ventricular assist device support. J Heart Lung Transplant. 2013;32:581–7.
31. Dinardo J. Anesthesia for cardiac surgery. Stamford: Appleton & Lange; 1998. p. 385.
32. Sunagawa G, Koprivanac M, Karimov JH, Mozami N, Fukamachi K. Is a pulse absolutely necessary during cardiopulmonary bypass? Expert Rev Med Devices. 2017;14(1):27–35. https://doi.org/10.1080/17434440.2017.1265445.
33. Ji B, Undar A. An evaluation of the benefits of pulsatile versus non-pulsatile perfusion during cardiopulmonary procedures in pediatric and adult patients. ASAIO J. 2006;52:357–61.
34. Sievert A, Sistino J. Meta-analysis of renal benefits of pulsatile perfusion in cardiac surgery. J Extra Corpor Technol. 2012;44:10–4.
35. Alkan-Bozkaya T, Akçevin A, Türkoğlu H, Ündar A. Impact of pulsatile perfusion on clinical outcomes of neonates and infants with complex pathologies undergoing cardiopulmonary bypass procedures. Artif Organs. 2013;37:82–6.
36. Agati S, Mignosa C, Ciccarello G, et al. Pulsatile ECMO in neonates and infants: first European clinical experience with a new device. ASAIO J. 2005;51:508–12.
37. Sunagawa G, Karimov JH. New technology to achieve true physiologic pulsatile flow during cardiopulmonary bypass. ASAIO annual conference, San Francisco, CA; June 15–18, 2016.
38. Sunagawa G, Karimov JH, et al. Impact of physiologic pulsatile flow on microcirculation during cardiopulmonary bypass. ASAIO poster presentation; 2016.
39. Bartoli CR, Koenig SC, et al. Extracorporeal membrane oxygenation vs counterpulsatile, pulsatile, and continuous flow left ventricular unloading for pediatric mechanical circulatory support. Pediatr Crit Care Med. 2013;14(9):e424–37.
40. Meco M, et al. Morbidity and mortality from intra-aortic balloon pumps. J Cardiovasc Surg. 2002;13:17–23.
41. Donachy JH, Landis DL, Rosenberg G, Prophet GA, Ferrar IO, Pierce WS. Design and evaluation of a left ventricular assist device: the angle port pump. In: Unger F, editor. Assisted circulation. Berlin: Springer-Verlag; 1979.
42. Farrar DJ, Hill JD, Gray LA, Pennington DG, McBride LR, Pierce WS, Pae WE, Glenville B, et al. Heterotopic prosthetic ventricles as a bridge to cardiac transplantation. N Engl J Med. 1988;318:333–40.
43. Pennington G, Kanter KR, et al. Seven years' experience with the Pierce-Donachy ventricular assist device. J Thorac Cardiovasc Surg. 1988;96:901–11.
44. Icenogle TB, Williams RJ, Smith RG, et al. Extracorporeal pulsatile biventricular support after cardiac transplantation. Ann Thorac Surg. 1989;47:614–6.
45. Farrar DJ, Hill JD, Pennington DG, et al. Preoperative and postoperative comparison of patients with univentricular and biventricular support with Thoratec ventricular assist device as a bridge to cardiac transplantation. J Thorac Cardiovasc Surg. 1997;113:202–9.
46. McBride LR, Naunheim KS, et al. Clinical experience with 111 Thoratec ventricular assist devices. Ann Thorac Surg. 1999;67:1233–9.
47. Mehta VK, Copeland JG, Arabia FA, Smith RG, Banchy ME. Analysis of premorbid comorbid factors associated with biventricular assist device and total artificial heart: a single center experience. J Heart Lung Transplant. 2000;19:65.
48. Mehta VK, Copeland JG, Arabia FA, Banchy ME, Smith RG. Mechanical ventricular support as a bridge to transplant: risk factors and selection. ASAIO J. 2000;46:192.
49. Copeland JG, Smith RG, Nolan P, et al. Comparison of the CardioWest total artificial heart, the Novacor left ventricular assist system, and the Thoratec ventricular assist system in bridge to transplantation. Ann Thorac Surg. 2001;71(3):S92–7.
50. Copeland JG, Arabia FA, Smith RG. Bridge to transplantation with a Thoratec LVAD in a 17 kg child. Ann Thorac Surg. 2001;71(3):1003–4.
51. Arabia FA, Tsau PH, Smith RG, Nolan P, Paramesh V, Bose R, Woolley DS, Sethi GK, Rhenman BE, Copeland JG. Pediatric bridge to heart transplantation: application of the Berlin Heart, Medos Heart, and Thoratec ventricular assist devices. J Heart Lung Transplant. 2006;25:16–21.
52. Morales D, Zafar F, Almond CS, et al. Berlin Heart EXCOR use in patients with congenital heart disease. J Heart Lung Transplant. 2017;36:1209–16.
53. Zimmerman H, Covington D, Smith RG, Inhat C, Barber B, Copeland JG. Recovery of dilated cardiomyopathies in infants and children using left ventricular assist devices. ASAIO J. 2010;56:364–8.
54. Ihnat C, Zimmerman H, Copeland JG, Meany FJ, Soboyna RE, Larsen BT, Blair D, Barber BJ. Left ventricular assist device support: a longitudinal echocardiographic follow-up and search for predictors of successful weaning in young chil-

dren with heart failure as a bridge to recovery. Congenit Heart Dis. 2011;6:234–40. https://doi.org/10.1111/j.1747-0803.2011.00494.x.
55. Maeda K, Almond C, et al. Refining of the pump exchange procedure in children supported with the Berlin Heart EXCOR ventricular assist device: 10 years experience at a single institution. J Heart Lung Transplant. 2015;34(4S):588.
56. Copeland H, Nolan P, Covington D, Gustafson M, Smith RG, Copeland JG. A method of anticoagulation for children on mechanical circulatory support. Artif Organs. 2011;35:1018–23.
57. Copeland J, Copeland H, Nolan P, Gustafson M, Slepian M, Smith R. Results with an anticoagulation protocol in 99 SynCardia total artificial heart patients. ASAIO J. 2013;59:216–20.
58. El-Banayosy A, Arusoglu L, et al. Novacor left ventricular assist system versus HeartMate vented electric left ventricular assist system as a long-term mechanical circulatory support device in bridging patients: a prospective study. J Thorac Cardiovasc Surg. 2000;119:581–7.
59. Pae W, Collell JM, et al. Does total implantability reduce infection with the use of a left ventricular assist device? The LionHeart experience in Europe. J Heart Lung Transplant. 2007;26:219–29.
60. Starnes V, Oyer P, Portner PM, et al. Isolated left ventricular assist as bridge to cardiac transplantation. J Thorac Cardiovasc Surg. 1988;96:62–71.
61. Portner PM, Oyer PE, Pennington G, et al. Implantable electrical left ventricular assist system: bridge to transplantation and the future. Ann Thorac Surg. 1989;47:142–50.
62. Schmid C, et al. Cerebral and systemic embolization during left ventricular support with the Novacor LVAS. Ann Thorac Surg. 1998;65(6):1703–10.
63. Robbins R, Kwon M, Portner P, Oyer PE. The totally implantable Novacor left ventricular assist system. Ann Thorac Surg. 2000;71(3):S162–5.
64. Frazier OH, Rose EA, et al. Multicenter clinical evaluation of the HeartMate 1000 IP left ventricular assist device. Ann Thorac Surg. 1992;53:1080–90.
65. Dowling RD, Park SJ, Pagani AJ, et al. HeartMate VE LVAS design enhancements and its impact on device reliability. Eur J Cardiothorac Surg. 2004;25(6):958–63.
66. Spanier T, Oz M, Levin H, et al. Activation of coagulation and fibrinolytic pathways in patients with left ventricular assist devices. J Thorac Cardiovasc Surg. 1996;112:1090–7.
67. Ankersmit HJ, Tugudea S, Spainier T, et al. Activation induced T-cell death and immune dysfunction after implantation of left ventricular assist device. Lancet. 1998;354(9178):550–5.
68. Rose E, Gelijns A, et al. Long term use of a left ventricular assist device for end stage heart failure. NEJM. 2001;345:1435–43.
69. Syncardia Systems Inc, 1992 E Silverlake Rd, Tucson Az, 85713.
70. Zimmerman H, Copeland JG, Aquila LA, Smith RG. Total Artificial Heart. In: Selke FW, DelNido PJ, Swanson SJ, editors. Sabiston and spencer surgery of the chest. Vol 2, Philadelphia: Saunders; 2010; pp 1525–32.
71. Copeland JG, Levinson MM, Smith RG, et al. The total artificial heart as a bridge to transplantation: a report of 2 cases. JAMA. 1986;256(21): 2191–5.
72. Copeland JG, Copeland H, Gustafson M, et al. Experience with over 100 total artificial heart implants. J Thorac Cardiovasc Surg. 2012;143:227–34.
73. Leprince P, Bonnet N, Rama A, et al. Bridge tot transplantation with the Jarvik-7 (CardioWest) total artificial heart: a single center 15 year experience. J Heart Lung Transplant. 2003;22:1296–303.
74. Torregrossa G, Anyanwu A, et al. SynCardia: the total artificial heart. Ann Cardiothorac Surg. 2014;3(6):612–20.
75. Arabia FA, Gregoric I, Kasarajan JD, Moriguchi JD, et al. Total artificial heart (TAH): survival, outcomes, risk factors, adverse events in InterMacs. J Heart Lung Transplant. 2016; abstract A 3190.
76. Stevenson LW, et al. INTERMACS profiles of advanced heart failure: the current picture. J Heart Lung Transplant. 2009;28:535–41.
77. Kirklin JK, et al. Eight annual INTERMACS report: special focus on framing the impact of adverse events. J Heart Lung Transplant. 2017;36:1080–6.
78. Leitz K, Long J, et al. Outcomes of left ventricular assist device implantation as destination therapy in the post-REMATCH era. Circulation. 2007;116:497–505.
79. Courtesy of SynCardia Systems Statistics as of August 1, 2018, 1992, Tucson, AZ.
80. Virtual sizing done by ASU/Phoenix Children's Hospital through SynCardia Systems, 1992, Tucson, AZ.
81. Moore R, Lorts A, et al. Virtual implantation of the 50 cc SynCardia total artificial heart. J Heart Lung Transplant. 2016;35:824.
82. Arabia FA, Copeland JG, Pavie A, Smith RG. Implantation technique for the CardioWest total artificial heart. Ann Thorac Surg. 1999;68:698–704.
83. Copeland JG, Arabia FA, Tsau P, et al. Total artificial hearts: bridge to transplantation. Cardiol Clin. 2003;21:101–13.
84. Copeland JG, Arabia FA, Smith RG, et al. Synthetic membrane neo-pericardium facilitates total heart explantation. J Heart Lung Transplant. 2001;20: 654–6.
85. Copeland J, Copeland H, Nolan P, Gustafson M, Slepian M, Smith R. Results with an anticoagulation protocol in 99 SynCardia total artificial heart recipients. ASAIO J. 2013;59:216–20.
86. Pierce CN, Larson DL, Arabia FA, Copeland JG. Inflammatory mediated chronic anemia in patients supported with a mechanical circulatory assistance device. J Extra Corpor Technol. 2004;36:10–5.

87. Slepian MJ, Alemu Y, et al. The SynCardia total artificial heart: *in vivo, in vitro, and computational modeling studies*. J Biomech. 2013;46:266–75.
88. Girdhar G, Xenos M, et al. Device thrombogenicity emulation: a novel method for optimizing mechanical circulatory support device thromboresistance. PLoS One. 2012;7(e32463):1–8.
89. Valerio L, Tran PL, et al. Aspirin has limited ability to modulate shear-mediated platelet activation associated with elevated shear stress of ventricular assist devices. Thromb Res. 2016;140:110–7.
90. Torregrossa G, Morshuis M, et al. Results with the SynCardia TAH beyond one year. ASAIO J. 2014;60(6):626–34.
91. Copeland JG, Smith RG, Arabia FA, et al. Comparison of the CardioWest total artificial heart, the Novacor left ventricular assist system, and the Thoratec ventricular assist system in bridge to transplantation. Ann Thorac Surg. 2001;71:S92–7.
92. Copeland JG, Smith RG, Bose RK, et al. Risk factor analysis for bridge to transplantation with the CardioWest TAH. Ann Thorac Surg. 2008;85:1639–44.
93. Moazami N, Milano C, et al. Pump replacement for left ventricular assist device failure can be done safely and is associated with a low mortality. Ann Thorac Surg. 2013;95:500–5.

18 Pathophysiological Determinants Relevant in Blood Pump Control

Marianne Schmid Daners and Seraina Anne Dual

Abbreviations

AoP	Aortic pressure
AoVF	Aortic valve flow
CO	Cardiac output
CVM	Cardiovascular model
CVP	Central venous pressure
CVS	Cardiovascular system
ECG	Electrocardiogram
EDP	End-diastolic pressure
EDPVR	End-diastolic pressure-volume relation
EDV	End-diastolic volume
EF	Ejection fraction
ESPVR	End-systolic pressure-volume relation
HIL	Hardware-in-the-loop
HMC	Hybrid mock circulation
HR	Heart rate
ITP	Intrathoracic pressure
LAP	Left atrial pressure
LV	Left ventricular, left ventricle
LVEDD	Left ventricular end-diastolic diameter
LVP	Left ventricular pressure
LVV	Left ventricular volume
MAP	Mean arterial pressure
mPAP	Mean pulmonary arterial pressure
MVM	Mean-value model
OCP	Optimal control problem
PF	Pump flow
PIP	Pump inlet pressure
PP	Pump power
PPLUT	Pump-power lookup table
PRS	Preload-responsive speed
PRSW	Preload recruitable stroke work
PS	Pump speed
PV	Pressure-volume
PVC	Premature ventricular contraction
PW	Pump work
SP	Systolic pressure
SW	Stroke work
TVEM	Time-varying elastance model
VAD	Ventricular assist device
VM	Valsalva maneuver

M. Schmid Daners, PhD (✉) · S. A. Dual, MSc
Product Development Group Zurich,
Department of Mechanical and Process Engineering,
Zurich, Switzerland
e-mail: marischm@ethz.ch;
seraina.dual@alumni.ethz.ch

In healthy subjects, the cardiac output corresponds to the pre- and afterload conditions as well as the contractility of the left ventricle (LV) and the heart rate (HR) [1]. In heart failure, ventricular assist devices (VADs) are implanted in parallel to the failing heart, as an alternative to heart transplantation [2], to maintain sufficient organ perfusion. Current VADs run at constant settings defined by the clinicians, i.e., at constant pump speed with superimposed pulsations by the residual native heart function [3] and intermittent pulsatile speed changes by the VAD [4, 5]. The pump speed is adjusted according to the assessment of the clinician at a specific point in time,

© Springer Nature Switzerland AG 2020
J. H. Karimov et al. (eds.), *Mechanical Support for Heart Failure*,
https://doi.org/10.1007/978-3-030-47809-4_18

starting immediately after implantation at the operating table. Further reevaluations and adjustments of the pump speed in the hospital aim at providing an estimated optimal cardiac output while avoiding adverse events. Until today, the pump speed does not automatically adjust to changes of perfusion requirements, e.g., when the patient is more or less active, during exercise or at rest, respectively. A physiologic controller would adapt the pump speed actively to the changing requirements. Such active adjustments need estimated or invasively measured signals to inform the controller in order to mimic the native feedback mechanism of the healthy heart. The most promising controllers rely on the Frank-Starling mechanism, and they need an invasively measured signal by a sensor [6]; however, the only attempt to integrate a sensor in a VAD was for monitoring only and never became established clinical practice [7]. The integration of sensors in VADs is a necessary condition to physiologically and dynamically adjust the pump speed and thereby improve clinical outcomes [8].

Pathophysiologic Determinants

Heart Conditions

In VAD patients, LV size, shape, anatomy, and function are more heterogeneous and susceptible to changes compared to the entire heart failure population. In heart failure, the native heart has suffered from multiple deficiencies. A total of 53% of all heart failure patients have a reduced ejection fraction, while 47% have a preserved ejection fraction [9]. A reduced ejection fraction is associated with a decreased left-ventricular (LV) contractility, while preserved ejection fraction is associated with a decreased LV elastance. In the first case, systolic dysfunction prevails as pathophysiologic mechanism, while diastolic dysfunction is most prevalent in the second case. The treatment with a VAD is most beneficial for patients, which have a reduced ejection fraction, as the pump provides the deficient blood flow. Indeed, 59% of VAD patients have an ejection fraction below 20%. The majority of patients implanted with a VAD are affected by cardiomyopathy, which comes hand in hand with a severely dilated LV. Prior to implantation, the mean LV inner diameter is 6 ± 1.4 cm [10] in this group of patients, which is estimated to represent an LV volume of 197–289 mL according to the Teichholz formula [11] and thus is above the clinical threshold for dilation. Normally, the dilated LV reduces in size after the implantation of a VAD, which can be associated with functional reverse remodeling [12].

In addition to the LV shape, also the mitral and aortic valves are affected by preexisting diseases and the implantation of a VAD. Mitral valve insufficiency occurs often together with or as a result of cardiac dilation following dilation of the valve ring. An insufficient valve causes mitral regurgitation upon contraction of the LV. Although the mitral valve insufficiency functions as a decompressing mechanism for the dilated and weak heart, the regurgitating blood volume eventually leads to overloading. An overloaded dilated LV is prone to impose further mechanical stress on the mitral annulus, which in turn favors mitral regurgitation. The VAD implantation usually improves mitral valve insufficiency by a reduction of the LV volume [13].

The aortic valve is affected negatively by the implantation of a VAD. The shunt from the LV apex to the aortic anastomosis, created by the VAD graft, exhibits unnatural flow patterns in the aortic root. In full support, the aortic valve remains closed, because the pressure in the LV is lower than the aortic pressure (AoP) throughout the cardiac cycle. If the aortic valve remains closed over a long period of time, it may cause calcification of the valve sails and aortic valve insufficiency [14]. In partial support, a part of the blood is still ejected through the aortic valve due to the native heart's function and the cardiac output (CO) is composed of a flow through the VAD and a flow through the aortic valve, hence regular openings of the aortic valve are ensured. The support condition largely depends on the residual contractility of the native LV as well as of the pump speed setting of the VAD. Directly after implantation of a VAD, full support is preferred, as it fully unloads the failing ventricle. When the ventricle is recovering, it could be favorable to apply partial support in order to allow aortic valve opening and LV training [15].

The elastance and the contractility of the heart are best explained using the relation between LV pressure and volume. In each cardiac cycle, the heart undergoes a period of relaxation (diastole) and contraction (systole). Plotting the pressures and volumes of one cardiac cycle yields the pressure-volume (PV) relation represented as a loop (see Fig. 18.1 left). The area inside the PV loop is the stroke work (SW). The PV loop is constrained by two curves: the end-diastolic pressure-volume relation (EDPVR) during diastole and the end-systolic pressure-volume relation (ESPVR) during systole. The EDPVR is determined by the elastance of the heart, which is constant in the short term. The relationship between the end-diastolic pressure (EDP) and the end-diastolic volume (EDV) has been described by Klotz et al. [16] with a simple exponential model viable for healthy and diseased ex vivo hearts:

$$EDV = a_1\left(\frac{EDP}{A}\right)^{1/B} + b_1, \quad (18.1)$$

with a_1, b_1 being patient-specific and A and B being 22.8 mmHg and 2.79, respectively. The ESPVR depicts the contractility of the heart. The contractility of the heart is influenced by the nervous system and may fluctuate in the short term.

The achievable contractility of the native heart depends on various factors. Prior to the implantation of a VAD, 40% of patients had an ischemic event [10]. Ischemic myocardium may turn into fibrotic, stiff scar tissue. On the contrary, the non-ischemic and overly dilated parts of the myocardium may be extremely distensible. In both cases, the contractility is likely more heterogeneous and reduced compared to that of healthy hearts. In the healthy heart, the contractility is estimated to be around 3.5 mmHg/mL, while with heart failure the decrease is estimated to go down to as little as 0.25 mmHg/mL [17–19].

The healthy heart has developed sophisticated mechanisms to react to pre- and afterload changes while continuously readjusting the CO to meet the changing requirements. Preload is equivalent to the pressure in the atria and afterload to the pressure in the aorta [1]. Preload increases during moderate to extreme exercise but also if the intravascular volume is increased for other reasons (increased volume status), such as high fluid intake. The Frank-Starling mechanism describes the built-in physiological adaptation of the myocardium to an increase in preload. Afterload increases if the resistance of the arterial vasculature increases. The baroreflex describes the feedback loop of sensing arterial afterload and reacting with an increase in heart rate and contractility of the heart. In the healthy heart, changes in preload augment the CO up to its triple value during extreme exercise [20], which is much more than changes induced by afterload. Yet, current VADs react only marginally to preload changes [17, 21–23].

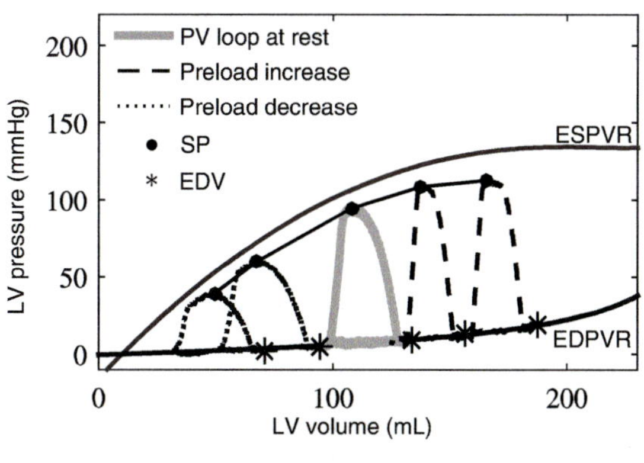

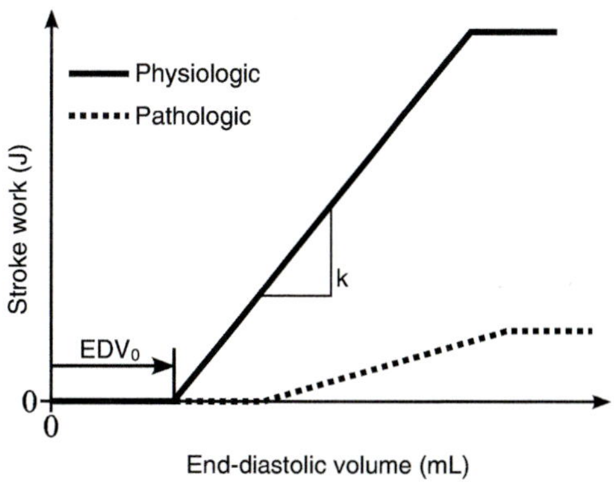

Fig. 18.1 *Left:* Pressure-volume (PV) relation that the physiologic controllers aim to restore. *Right:* Dependence of the stroke work on the preload condition. LV left ventricular, ESPVR end-systolic PV relation, EDPVR end-diastolic PV relation, SP systolic pressure, EDV end-diastolic volume, k gain of the stroke work to EDV relation

Preload Measures

The Frank-Starling mechanism elucidates the reason for the SW to increase with an increase in preload (see Fig. 18.1 right). The healthy heart's SW is the summation of the work of every single myocardial fiber. At high preload, the myocardial fibers are stretched more due to an increase in LV filling. At high stretch myocardial fibers operate at an energetically more optimal point such that their individual achievable work increases. This results in an overall higher SW. The PV loop depicts how an increase in preload increases EDP and EDV and how it shifts the line of the iso-volumetric contraction to the right (see Fig. 18.1 left). If everything else remains constant, the SW increases at the same time [1].

EDP and EDV reflect the SW increase associated with the Frank-Starling mechanism. Hence, the most direct surrogate of the LV preload are the EDP and EDV. As the EDPVR is relatively flat, the EDP undergoes smaller relative variations than the EDV during pre- and afterload changes. Hence, a measurement of EDP would require being more accurate than a measurement of EDV. The mean pulmonary arterial pressure (mPAP) could serve as an indirect measurement of preload as well, provided that the pulmonary system is not congested, the pulmonary resistance is not elevated, and the condition does not change over time. The CardioMEMS system measures mPAP continuously via an implantable pressure sensor and is used clinically [24].

As an alternative to the EDP and the mPAP, the systolic pressure (SP) undergoes larger variations, has a larger overall amplitude, and has been previously suggested to be used as an indirect measure of preload in the assisted pathological circulation even though it depends directly on afterload [25]. In the healthy heart, the SP depends on afterload and the contractility. However, once the LV is continuously unloaded by a VAD, the SP depends directly on preload as well, as simulations on a hybrid mock circulation have shown [21]. In contrast to the EDP and EDV, the SP is sensitive to the contractility of the native heart in the healthy and the assisted pathologic condition. In spite of the sensitivity to contractility, the SP should be considered as a well measurable signal for preload as it is more sensitive to preload changes than the EDP signal as it undergoes larger pressure variations.

Pump data of a VAD has been shown to provide indirect measures of LV preload. Preload signals derived from the pump data are advantageous as to date no invasive measurements of LV pressure and LV volume are available. Various approaches have been proposed for estimating pump flow [26] in order to use its indices as a preload surrogate or measure flow directly [27] with a flow probe [7, 28]. Pump flow increases with an increase in LV preload, and pump flow pulsatility increases with LV preload as well [29]. Maintaining a linear relationship between mean pump flow and arterial pulsatility has been suggested to imitate Frank Starling-like mechanisms [30]. Unfortunately, pump flow estimators still lack robustness and accuracy [26, 31].

Afterload Measures

Afterload is equivalent to the pressure in the aorta, which is composed of a mean arterial pressure (MAP) superimposed by the arterial pulsatility. Short-term fluctuations in AoP of up to 20 mmHg are common [32]. The healthy heart adapts immediately to an increase in AoP with an increase in contractility in order to maintain the CO. In the PV loop, the SP increases along the ESPVR to the right (see Fig. 18.1 left). In a VAD, however, pump flow is directly related to the pressure difference across the pump, namely, LV pressure and AoP. An increase in AoP immediately reduces pump flow due to the increased pressure difference compared to the healthy heart's function. Therefore, VADs are overproportionally afterload-sensitive, which motivates the implementation of a preload-sensitive control concept.

A sustained elevated stiffness of the arterial vessels in the periphery increases the resistance of the vasculature and hence of the AoP. A stiff vasculature is a common cause for heart failure and is thus to be expected in many VAD patients. Additionally, patients with full VAD support show increased aortic wall stiffness compared with matched HF patients without a VAD [33]. Both effects need to be considered for modeling the interaction of a VAD and the cardiovascular system (CVS) [19].

Arterial pulsatility is significantly reduced with current VADs, as evidenced by the fact that they are often referred to as continuous-flow pumps. Turbodynamic devices have been implanted for over 10 years now, and the organ function seems not to depend largely on the pulsatility of the flow. However, there are adverse events assigned to the non-pulsatile VAD flow, such as gastrointestinal bleeding that is related to the decrease of the von Willebrand factor [34], hemolysis, or device thrombosis [35]. Recently, newer VADs are programmed to generate superimposed pulsations [4, 5]. Many uncertainties remain about the advantage of having a pulsatile flow over a continuous flow, and its implementation continues to be debated [36]. However, if any additional risk was mitigated, an implementation of pulsatility with a continuous-flow VAD would result in a more physiological arterial flow pattern.

Pathophysiologic Events

As outlined above, an optimal physiologic controller would be sensitive to pre- and afterload changes to a similar degree as the native heart and should be tested accordingly. Isolated pre- and afterload variations by changing the intravascular volume (unstressed venous volume) and the systemic resistance are the first choices to test physiologic controllers. Furthermore, events in everyday life of a VAD patient are transferable to a combination of preload, afterload, and contractility changes. From rest to exercise, the pulmonary and systemic resistances are reduced, while the unstressed venous volume increases. From sleep to awake, the systemic resistance increases and it is accompanied by an increase in contractility. Finally, yet importantly, a physiologic controller should not be affected by a change in contractility of the native heart. The contractility is expected to increase to 51% (ejection fraction (EF) = 44%) in case of a functional reverse remolding of the LV and to decrease to 17% (EF = 22%) in case of progressive severe heart failure [6].

All preload, afterload, and contractility changes described so far occur in the range of 10 minutes, a couple of hours, or, in the case of the arterial stiffness or LV remodeling, over a lifetime. In contrast, there are a number of sudden events in everyday life of a VAD patient, which acutely affect her or his hemodynamic status [37]. Sudden events are related to hemodynamics (acute blood loss, head-up tilt, or suction), respiration (deep inspiration or expiration), physical strain (Valsalva or Mueller maneuver, coughing, heavy lifting, or bowel movement), or arrhythmia (atrial fibrillation or premature ventricular contraction) [38]. Elevated physical strain is accompanied by an increase in intrathoracic pressure. Even if the sudden events would not affect the native heart, they can potentially destabilize a feedback-based physiologic control system and lead to acute suction, overload, or pump speed oscillations [6, 38].

Not only hemodynamics but also estimated signals based on VAD motor current or sensor measurements are affected by a thrombus in the VAD [39] or by acute changes of the hematocrit. Currently, those signals are used for monitoring purposes, but they could influence the stability of physiologic controllers if, in the future, they were used as control inputs. The motor current used for the estimation of pump flow is strongly dependent on the hematocrit because the hematocrit affects the viscosity of the blood. The hematocrit can change rapidly, for instance, because of bleeding, and it is subject to variations from 15% (anemia) to 65% (polycythemia) [1, 40]. The effects of a thrombus on motor current are evident, as thrombogenetic events are identified in clinical practice by observing an increase in motor current [41].

The sudden events, which influence sensor measurements, depend on the underlying measurement principle applied. Pressure sensors are currently being integrated in the inlet cannula [42]. On the contrary, implantable volume sensors are not available yet [17, 43]. The most promising measurement principles for LV volume sensing are electric impedance [44] and ultrasonic distance measurements [45]. Drift is a common challenge in the development of implantable pressure sensors. Values of drift have been reported to range from 2 mmHg/year to 50 mmHg/year [46, 47]. Drift estimates for a volume sensor are not available, but based on volume estimation accuracies observed by echocardiography, they can be inferred to be 10–20% [48]. A volume estimation based on electric impedance is influenced by the hematocrit and by sudden changes in the position of the sensor. Intracardiac echocardiography is as sensitive to position of the sensor as electric impedance.

Adaptive Physiologic Pump Controllers

Motivation

Current pumps run at a predefined speed. If it is set in an optimal way, the LV is in a physiologic state (see Fig. 18.2a). Yet, because these pumps do not adjust the pump speed according to changes in preload in patients, they are susceptible to cause adverse events. Everyday situations such as rest, coughing, or bowel movements reduce the blood flow back to the heart resulting in a decreased preload. Fixed-speed pumps do not react to a decreased preload and hence they may run too high. This overpumping situation (see Fig. 18.2b) is observed clinically through low-flow alarms and a low LV volume. The non-physiologically high pump flow may reduce the LV volume to such a degree that suction events occur [50]. By contrast, exercise may increase backflow of blood to the heart and in turn increase preload. The increased blood volume dilates the heart chambers leading to volume and pressure overload and consequently to pulmonary congestion and even edema, as well as systemic venous congestion. This underpumping situation (see Fig. 18.2c) is clinically observed through an overly dilated ventricle and is characterized by a rightward septum shift [51].

Physiologic controllers should additionally account for different levels of assistance that the clinicians can set. During full support, the aortic valve remains closed because the LV pressure remains below the AoP, which is considered unfavorable for the valve's mechanical integrity [14]. During partial support, the failing heart might not be able to provide sufficient SW as it is less supported by the VAD. There is a trade-off between the opening of the aortic valve and the LV SW. During severe heart failure, full support guarantees sufficient organ perfusion and decompression of the cardiopulmonary system. The LV is unloaded completely, which results in a minimum LV SW and a closed aortic valve. During recovery, partial support allows to reload the LV and the aortic valve to open [52]. Depending on the volume status of the LV as well as the therapeutic goal, the VAD's operation needs to be adjusted.

Moreover, the continuous operation of VADs produces a non-physiologic and non-pulsatile arterial flow. VAD companies have recently introduced pump speed modulations, which are pulse profiles that superimpose an artificial, intermittent pulsatility to the mean pump speed (see Fig. 18.3 right). For the HeartMate 3 (Abbott, Abbott Park, IL, USA), an "artificial pulse" [4] is superimposed on the mean pump speed, and for the HVAD (Medtronic, Minneapolis, MN, USA), it is a "Lavare cycle" [5]. Both pulse profiles aim at having intermittent aortic valve-opening events during low pump speed and at enforcing pump washouts during the periods of maximum pump speed [53]. Yet, the pulse profiles are not synchronized with the cardiac cycle. The LV and AoP conditions are randomly superimposed by these pulse profiles, which may or may not be physiological. The waveform of the "artificial pulse" and the "Lavare cycle" are preset and do not respond to the hemodynamic state of the patient.

To effectively reduce adverse events associated with over- and underpumping events (see Fig. 18.2b, c), the VAD should react to the preload and impose a pulsatility synchronized with the native heart on the arterial flow. Yet, sensor measurements or estimated signals, which correspond to preload, are required such that the pump can adjust its speed accordingly.

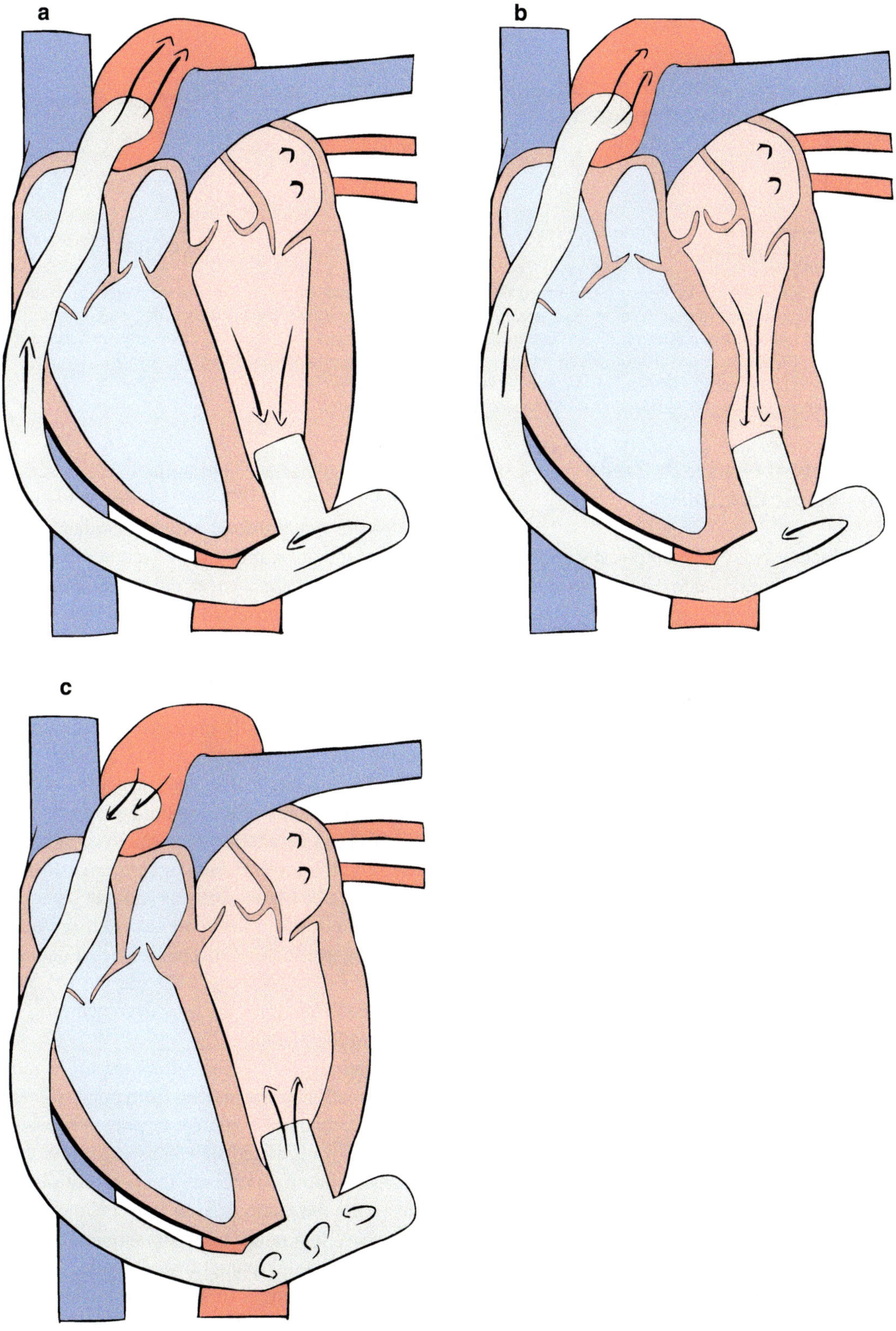

Fig. 18.2 Drawings of a turbodynamic ventricular assist device (VAD) implanted at the apex of the left ventricle and operated in parallel to the diseased heart. The outflow graft is connected to the ascending aorta. (**a**) Regular VAD operation. (**b**) Overpumping and (**c**) underpumping events due to preload changes during activity changes [49]. (Illustrations courtesy of Gregor Ochsner)

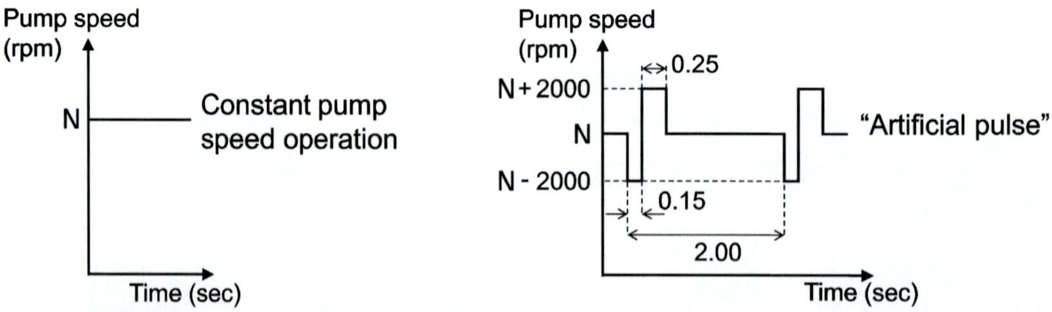

Fig. 18.3 *Left:* Predefined constant pump speed that is set by the clinician at a specific point in time. *Right:* The HeartMate 3 (Abbott, Abbott Park, IL, USA) has an "artificial pulse" [4] implemented that allows to "wash out" the pump and let the aortic valve open from time to time. For the same purpose, the HVAD (Medtronic, Minneapolis, MN, USA) has an additional "Lavare cycle" [5]. Both of these control strategies work such that they change the pump speed given by their respective software settings but without taking into account the hemodynamic state of the patient

Overview of Feedback-Based Physiologic Controllers

In contrast to state-of-the-art pump speed controllers that do not respond to the hemodynamic state of the patient, physiologic controllers make use of a measured or estimated signal as an input to the controller. The measured or estimated signal is compared to a reference signal, and the difference between the two is computed. Based on this difference, the controller computes the required change in pump speed and applies it to the pump (see Fig. 18.4). In this way, the feedback loop is closed and the pump speed can automatically respond to changes of the measured or estimated signal.

The majority of controllers, which have been proposed in literature for the physiologic control of VADs, rely on the input of a preload signal, such as the EDV [17] or the EDP [29]. The similarity to the healthy heart function motivates preload to be the favorable parameter to control the pump speed of a VAD. Preload-based controllers intend to mimic the Frank-Starling mechanism. The controllers mainly differ in the types of signals that are used as inputs, which are described in more detail below. Preload-responsive control can also be mixed with some afterload sensitivity, with the intention to maintain the absolute perfusion within a physiologic range [21, 23, 54–56].

Physiologic afterload conditions can be matched with pump-speed modulations of VADs to provide an increased pulsatility of the systemic arterial circulation [57–59]. Ideally, the pulse profiles are synchronized with the cardiac cycle such that the pulsatility can be increased and the LV unloading enhanced at the same time [52, 60]. During synchronized operation, the tradeoff between LV unloading and opening of aortic valve can be specifically accounted for the volume status of the LV as well as the therapeutic goal. By systematically changing the timing of the pulsation, the ventricular load and the pulsatility in the arteries can be adjusted as shown in an in vivo study with a volume-displacement VAD [61]. As an alternative to the synchronized pulsatility, the level of assistance (partial to full support) can be used directly as an input signal to a physiologic controller, provided that pressure signals are available [23]. This enables a defined load sharing between the VAD and the LV. The assistance is based on the time-averaged ratio of VAD flow and the total CO.

Preload-responsive controllers require a reliable estimation of the preload from pump signals or the placement of a volume or pressure sensor. Examples of estimated signals used for preload-responsive pump control are the pressure difference across the pump derived from the magnetic bearing of the pump [62] or signals from a combination of the pump flow, the pump speed, and the motor current [63]. A pressure sensor in the LV or the pump inlet can measure EDP and SP, which are related to the preload condition in a

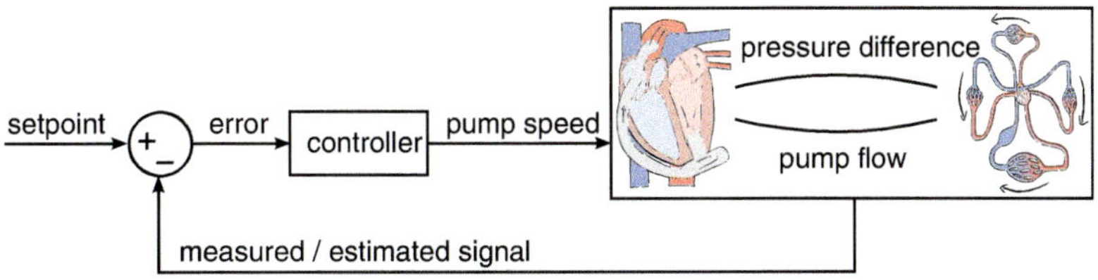

Fig. 18.4 Feedback-based control loop. Based on a measured and/or estimated signal the controller computes the desired pump speed (N_{des}) and applies it to the pump. (Illustrations courtesy of Gregor Ochsner)

VAD patient [25, 64]. The EDV maps preload as well and it can be measured by the use of an LV volume sensor. Yet, such sensors are not available in current VADs.

In the ideal case, sensors are integrated in the inflow cannula of a VAD such that the signal is recorded at the place of action and that the VAD implantation technique remains unchanged without any additional surgical steps being required. The integration of the sensor technology in a VAD remains a challenge. New technological developments are underway suggesting the integration of pressure sensors [8] and measurement principles for LV volume. The LV volume measurement could be obtained by ultrasonic distance measurement [45], by electrical impedance [44], by the correlation between the intra-cardiac depolarization amplitude and the LV volume [43] (using the Brody effect [65]), and finally by the correlation between EDP and LV volume [16]. Measuring LV volume based on ultrasonic distance measurements has proven to be a feasible approach.

In vitro results in heart phantoms obtained from heart failure patients achieved accuracies comparable to those obtained by 2D and 3D echocardiography based on a one-point calibration of the absolute LV volume [45]. Several pressure sensors are under investigation that can be placed in the LV [8, 46]. The integration of a commercially available pressure sensor into the inflow cannula is feasible with a Parylene-C diaphragm that has an adjusted surface interface topography [66]. This approach has proven short-term stability, which supports the realization of a long-term drift-free pressure recording [42]. A further approach to assess the LV preload could be the use of CardioMEMS (Abbott, Abbott Park, IL, USA), an already commercially available pressure sensor for short-term use that measures the pulmonary arterial pressure [67].

Implementation

Three preload-based physiologic controllers and an optimal control strategy for VADs are described. As a fundamental goal, the perfusion requirement of the patient shall be guaranteed at all times. Therefore, all controllers account for the Frank-Starling mechanism [1] in which the pump speed is to be adjusted to the preload condition of the LV; see Fig. 18.1. The pump speed is controlled, based either on LV volume or on LV pressure measurements. One of the controllers additionally imposes an artificial pulse profile synchronized with the cardiac cycle to increase the arterial pulse pressure. In addition, the optimized speed profile strategy allows to solve an arbitrary objective, e.g., to maximize the flow through the aortic valve and to minimize the LV SW.

The *preload-responsive speed (PRS) controller* takes the measured left ventricular volume (LVV) as an input and calculates the desired pump speed (N_{des}) [17]. The schematic of the controller is depicted in Fig. 18.5. It is based on the concept of the preload recruitable stroke work (PRSW) (see Fig. 18.1 right) [68]. This concept states that the SW depends linearly on the EDV in the healthy heart. Similarly, the pump work (PW) should increase with an increasing EDV in the assisted circulation. The PRS controller linearly maps the PW to the EDV as in the concept of the PRSW. The gradient of the PRSW k corresponds directly to the gain of the controller k_{prsw}. An increase in EDV

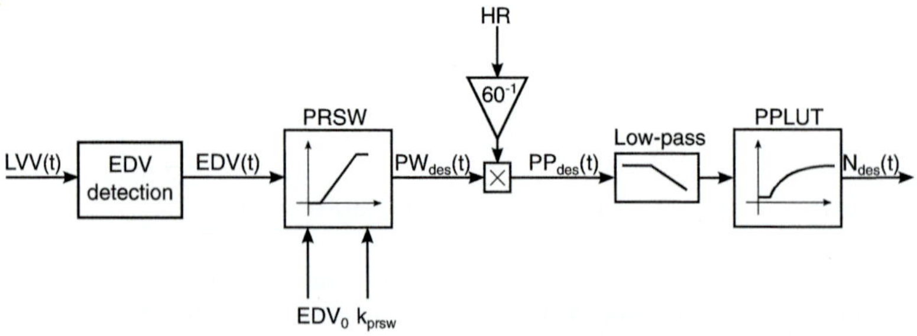

Fig. 18.5 Control scheme of the proportional preload-responsive speed controller that takes the left ventricular volume (LVV) as an input and computes the desired pump speed (N_{des}) as an output [17]

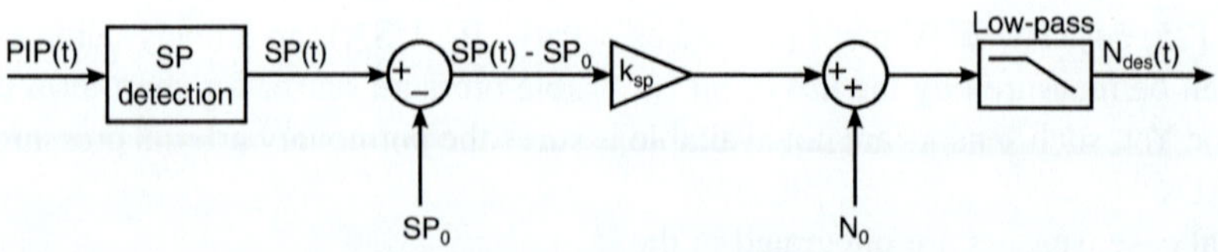

Fig. 18.6 Control scheme of the proportional systolic-pressure controller that takes the pump inlet pressure (PIP) as an input and computes the desired pump speed (N_{des}) as an output [21]

leads to an increase in PW. Finally, the N_{des} is computed from the filtered desired pump power (PP_{des}) applying the pump-power lookup table (PPLUT) that relates the hydraulic pump power to the pump speed.

The *systolic-pressure (SP) controller* is based on the pump inlet pressure [21]. The control scheme is depicted in Fig. 18.6. From the pump inlet pressure (PIP), the LV SP is detected and with a proportional gain k_{sp}, the desired pump speed N_{des} is computed. As the SP is related to preload in patients with a VAD, the SP controller can mimic the Frank-Starling mechanism.

The *multi-objective physiologic controller* [59] is based on the measurement of the PIP, and it also takes into account the current pump-speed (N) and pump-flow (PF) signals. The EDP, the SP, the HR, the mean arterial pressure (MAP), and the minimum PF (PF_{min}) are estimated from these three signals. The desired pump speed N_{des} is computed from all measured and estimated signals and a physiologic pump flow is generated. The EDP is extracted at every heartbeat and it is used as an input to the controller. Additionally, based on the PIP, the opening of the aortic valve is detected. The detection of the aortic valve opening allows a control that takes into account the level of assistance (mode) desired by the clinician. A change in mode enables LV unloading or loading (training) conditions. Furthermore, the controller provides a sinusoidal pulse profile to impose increased aortic pulse pressure that is synchronized with the cardiac cycle. The synchronization is based on the SP per heartbeat detection that applies the influence of the phase shift φ on the aortic pulse pressure described by Amacher et al. [52]. Moreover, the controller comprises safety algorithms, such as detecting LV suction, LV overload and pump backflow, and respective releases (control scheme see Fig. 18.7).

The *optimal control strategy* for VADs allows pulse-speed profiles of VADs beat-to-beat [69] (see Fig. 18.8) to be optimized with the same intention as the multi-objective controller does to account for various levels of assistance. In contrast to current implementations of artificial pulse profiles [4, 5], this strategy allows cycle-synchronized speed profiles to be optimized to arbitrary objectives. In this way an effective VAD interaction with the CVS can be achieved. The optimal control problem (OCP) considered the pulmonary arterial and the systemic venous pressure as well as the systemic arterial resistance. Based on this OCP, the blood flow through the aortic valve and the ventricular loading are adjusted and synchronized with the cardiac cycle, accounting for the tradeoff between aortic valve opening and LV SW. This particular

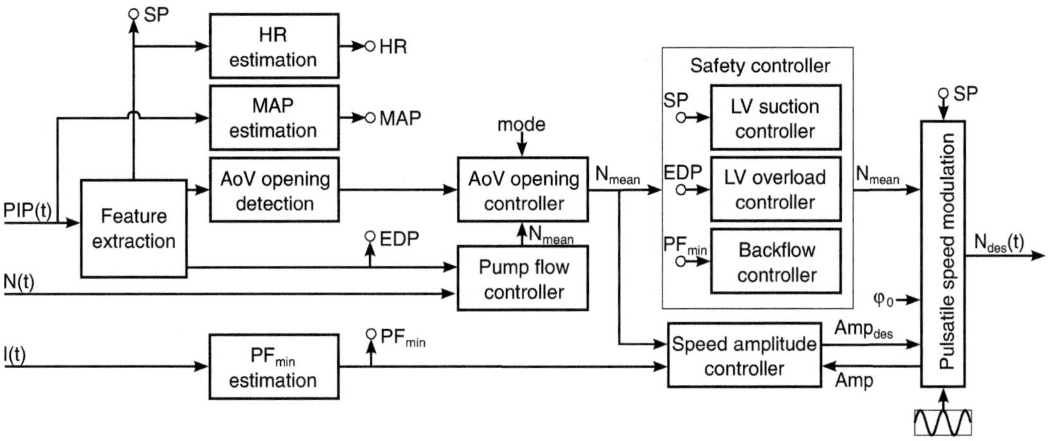

Fig. 18.7 Control scheme of the multi-objective physiologic controller that takes the pump inlet pressure (PIP), the current pump speed (*N*), and the pump current (*I*) as inputs and computes the desired pump speed (N_{des}) as an output. The controller provides a switch mode that the clinician can set to change the left ventricular loading. Additionally, a sinusoidal pulse profile is added to the controlled pump speed to increase the arterial pressure. Safety algorithms, such as left ventricular (LV) suction, the detection of overload and pump backflow, and release feature this physiologic controller [59]

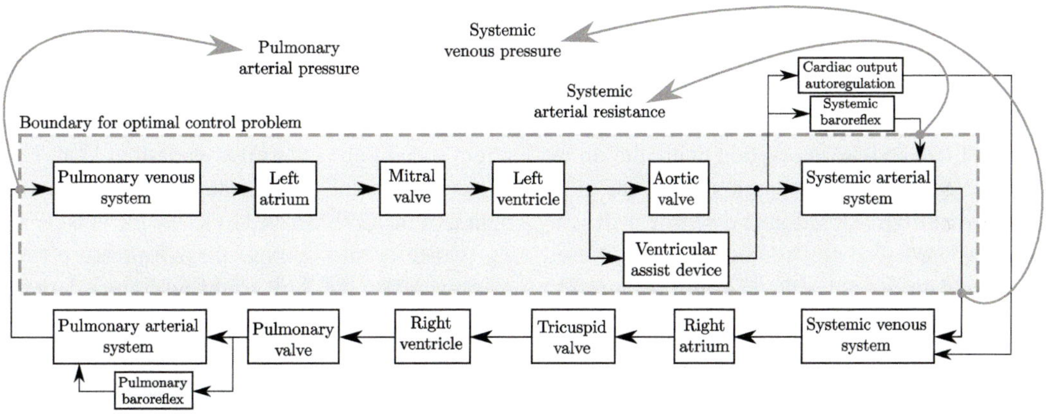

Fig. 18.8 Control scheme of the optimal control strategy for VADs. The optimal-control problem is based on the reduced model shown in the gray box. (Reprinted, from Amacher et al. [69] under CC BY 3.0, 2014 *Bioengineering*)

optimal-control problem is based on a mathematical model of the cardiovascular system (CVS) combined with a VAD, solved numerically and tested in vitro. The advantages of such a model are that the level of assistance is automatically adjusted according to the instantaneous patient requirement. The control strategy offers a powerful tool to develop patient-specific, optimized VAD actuation strategies, provided the patient-specific information of the cardiovascular system is available.

Validation of Physiologic Pump Controllers

Criteria

Novel control approaches for VADs need to adapt to varying pathophysiologic conditions applicable to the cohort of VAD patients and to accurately react to hemodynamic changes. In vitro and in vivo experiments are to be designed such that the sensitivity of the physiologic controllers to changes in hemodynamics can be tested alongside with their robustness towards sudden events. The sensitivity of the controllers is compared to the response of the healthy heart and the constant-speed operation of a VAD. The robustness is counted by the number of suction events and overload conditions occurring and the characteristics of pump speed changes provoked by the controller output.

An effective environment to test physiologic controllers in vitro must mimic pathophysiologic changes of the CVS like HR, EDV, and LV contractility variations. Preload and afterload need to be adjustable by controlling the unstressed venous volume and the systemic arterial resistance, respectively. Ideally, physiologic control mechanisms such as the CO autoregulation and the baroreflex are available. Conditions such as sleep-to-awake changes, premature ventricular contractions (PVC) that cause abrupt changes of the CO, or the Valsalva maneuver (VM) that increases the intrathoracic pressure (ITP) must be comprised in the numerical model of the CVS. Pump speed changes are to be applied in real time and in interaction with the simulated CVS. Hence, the sensor signals are to be recorded at high resolution and sensor drift is to be considered. The electronics of the VAD used to investigate the physiologic controller need to be accessible and controllable. Finally yet importantly, the working fluid should mimic the blood viscosity.

In vivo, the protocol must be designed as close as possible to the investigations performed in vitro to render the investigations comparable. The animal model used should imitate the anatomy and size of a human being as precisely as possible. Therefore, ovine, bovine, or pork models are valid animal choices. Optimally, the pathophysiologic state of the animal imitates baseline HR, EDV, and LV contractility of a patient with heart failure. Preload needs to be varied, for instance, with the infusion or draining of blood or blood-like solutions. Afterload needs to be increased, e.g., with a balloon catheter placed in the aorta or with drugs. These interventions affect short-term changes of HR and EDV. All electronic equipment like the VAD and sensors should be identical to those used for the in vitro testing, allowing for real-time interaction and recording. They should be positioned in the animal at places corresponding to those on the hybrid mock circulation, and furthermore, the sensor placement must be feasible for the surgeon.

Testing on Hybrid Mock Circulation

The hybrid mock circulation (HMC) maps individual pathophysiologies and allows the interaction with the physiologically controlled VAD in real time. The HMC shown in Fig. 18.9a is based on the hardware-in-the-loop (HIL) concept [70] and includes a ventricular suction emulation [71]. It consists of a hardware part and a numerical model of the CVS. The VAD is connected to the hydraulic interface that contains reservoirs with pneumatic actuation (hardware part) and interacts in real time with the numerical model of the CVS (software part). The pump that mimics the VAD (Deltastream DP2) is equipped with an encoder and an industrial motor controller to allow the pump speed to be controlled as desired [70]. The numerical model implemented is based on the model of the CVS derived by Colacino et al. [73].

Two reservoirs impose the LV pressure (left reservoir) and the AoP (right reservoir) derived from the numerical model and applied to the hardware flow loop. The pressures imposed to the two hydraulic interfaces provide testing conditions that a VAD implanted in a patient would also experience including the interaction with the circulation of the CVS. The pressures are measured and actively controlled instantaneously. This fast pressure control enables the replication of small pressure fluctuations as physiologically present due to the contraction of the heart. The flow through the VAD is measured with a transonic flow probe, and the water levels of the two reservoirs are equalized by a backflow pump. Additionally, the fluid viscosity is controlled [74] to account for a temperature-dependent viscosity of the glycerin-water mixture used. An exemplary representation of the LV pressure and the AoP imposed and measured are depicted in Fig. 18.9b.

The numerical model of the CVS allows the experimental settings to be changed arbitrarily, depending on the testing conditions required. The VAD can be operated at constant speed (CS) or with a physiologically controlled speed. The preload is varied by changing the settings of the unstressed venous volume of the systemic veins and adjusting the systemic arterial resistance by the baroreflex. The afterload is varied by a change of the resistance of the systemic arteries and adjusting the unstressed volume of the systemic veins by the CO autoregulation. The parameter settings of the HR and the LV contractility can be altered independently, as needed. In this way, individual conditions of the CVS allow a thorough testing of physiologic controllers for VADs.

The specific conditions for testing the physiologic controllers are as follows: During the preload changes, the unstressed venous volume of the systemic veins varies between 2200 and 2900 mL. The HR changes in the physiologic circulation from 60 to 80 bpm and in the pathologic circulation from 65 to 135 bpm, while the systemic arterial resistance is controlled by the baroreflex. During the afterload changes, the resistance of the systemic arteries changes from 0.5 to 1.8 mmHg·s/mL, the HR is kept constant at a physiologic 60 bpm and a pathologic 90 bpm, while the unstressed volume of the systemic veins is controlled by the CO autoregulation. The LV contractility for the PRS controller is varied with elastance settings of 3.5, 0.25, and 0.125 mmHg/mL and for the SP and multi-objective controller with LV contractility settings of 60%, 34%, and 10% and 51%, 34%, and 17%, respectively. All three controllers are analyzed for their stability and robustness towards sensor drift or LV contractility changes.

Furthermore, in a standardized comparison, several physiologic controllers developed for turbodynamic VADs were implemented and tested in vitro on the HMC [6]. The controllers investigated are based on the AoP and heart rate [56] (Con1), on the pump flow and EDP [29] (Con2), on the LV pressure [75] (Con3), on the differences between pump pressures and pump speeds [76] (Con4), on EDV [17] (Con5), and on SP [21] (Con6). The study considered pre- and afterload changes during daily activity, contractility variations that mimic the progression of heart failure or myocardium recovery as well as accounting for sensor drift due to material properties.

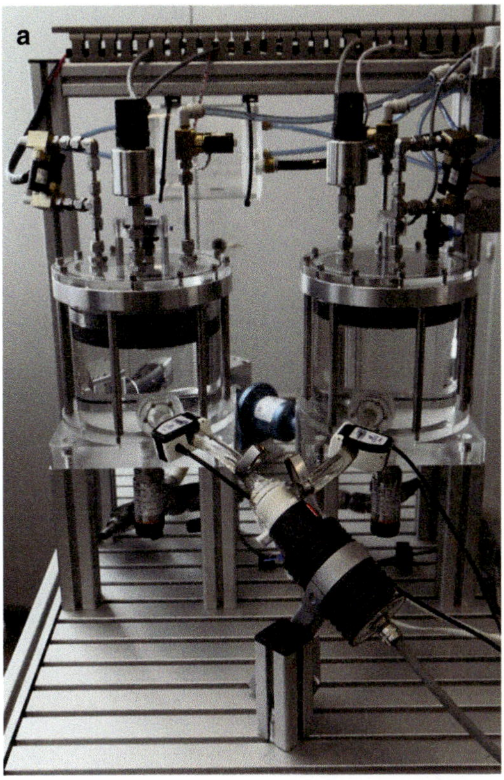

Fig. 18.9 (**a**) Hybrid mock circulation (HMC) allowing to dynamically test and validate new controllers developed in interaction with a mimicked cardiovascular system implemented in software. The pressures in the two reservoirs correspond to the left ventricular (LV) and aortic pressure (AoP) and are adjusted in real time [70]. In addition, ventricular suction is incorporated [71]. A movie of the HMC in action can be found on YouTube [72]. (**b**) Reference (bold gray line) and measured (black line) signals of pump speed, pump flow rate, aortic and LV pressure presenting the performance of the HMC. The HMC reacts to the increase in pump speed from partial support to full support resulting in a corresponding increase in pump flow rate. The mean LV pressure gradually decreases and the AoP increases by one step at the time of pump speed increase going along with a gradual pulsatility decrease in both compartments over time. In the bottom panel, the LV pressure-volume relation is depicted for a healthy heart (bold gray line) and a supported one (black line) investigated on the HMC. (Reprinted from Ochnser et al [70], with permission from IEEE, © 2013 IEEE)

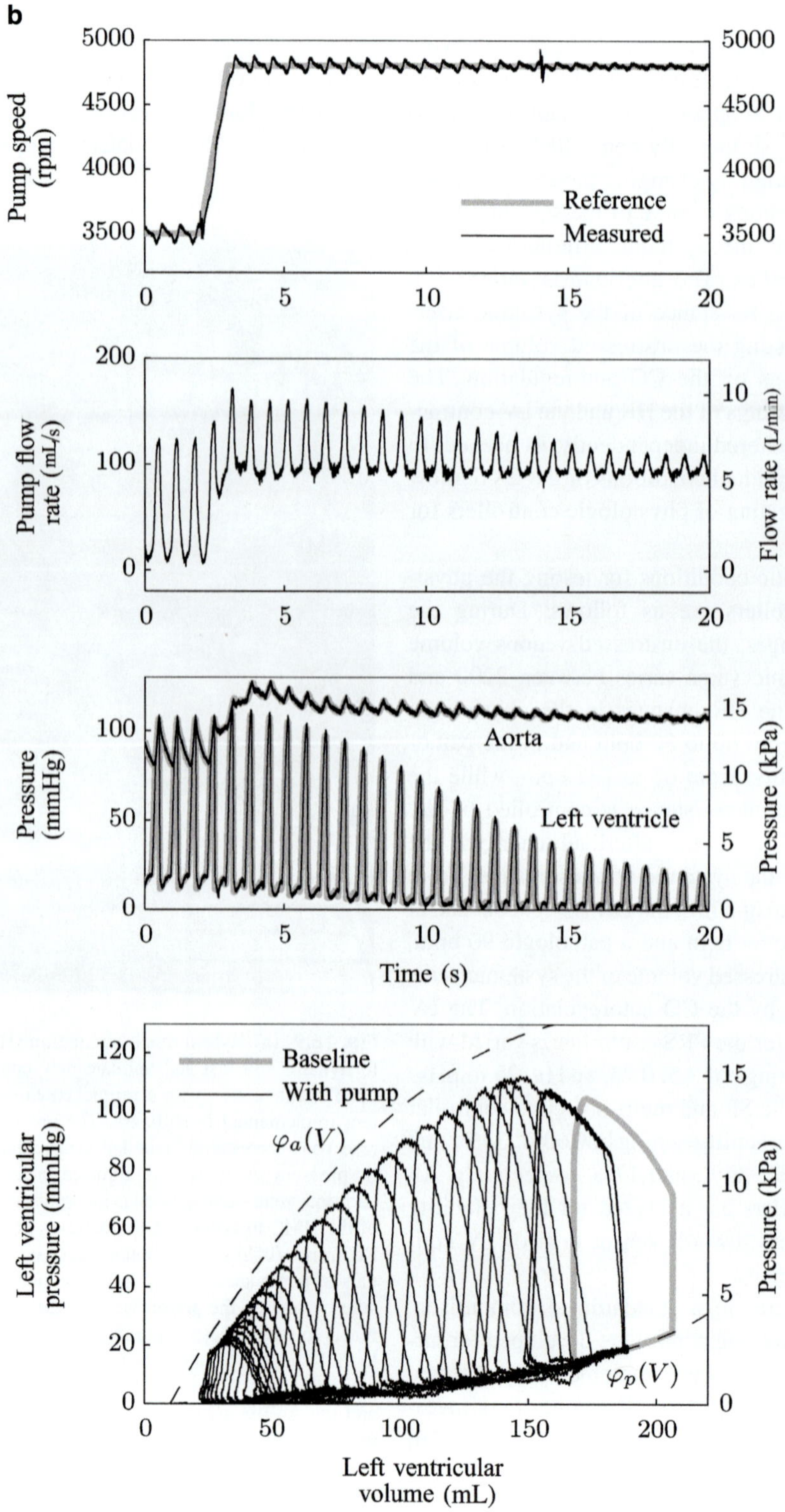

Fig. 18.9 (continued)

Testing In Vivo

In vivo, the hemodynamic testing conditions must be comparable to those on the HMC in order to ensure an accurate validation of the in vitro results [77]. With the same pump (Deltastream DP2) used in vitro as a VAD, the physiologic controllers are tested under pre- and afterload variations, and their performance is compared to the constant-pump-speed operation. Preload increase and decrease are imposed by infusing or draining ±500 mL of blood via the heart-lung machine. Afterload is increased by occluding the balloon catheter that is placed in the descending aorta.

The LV volume is measured continuously with sonomicrometry ultrasound crystals for the long and short axes of the LV. Three crystals are placed inside the ventricular wall and one on the inflow cannula (see Fig. 18.10). The PIP is measured on the inflow cannula and the LV pressure via the pigtail catheter. A transonic flow probe is positioned around the ascending aorta to measure the total CO. All signals are recorded at 500 Hz on a PC using an MF624 input/output card and MATLAB Real-Time Windows Target.

Performance on Hybrid Mock Circulation

All three preload-based physiologic controllers perform well in vitro during preload, afterload, and contractility changes and imitate the healthy heart function in terms of CO well.

The PRS controller was tested, and the results were compared to those obtained with a constant-speed VAD operation during pre- and afterload variations [17]. The comparison shows that the PRS controller provokes a VAD speed adjustment that is similar to the physiologic circulation (see Fig. 18.11). During a preload increase, with constant pump speed operation (C3), the CO is only marginally increased causing the LV volume to be markedly increased leading to LV overload, which can be prevented with the PRS controller. The PRS controller (C4) adjusts the pump speed such that the level of CO can be maintained according to a physiologic condition (C1). In contrast, during a preload decrease, the insufficient adjustment of the VAD with a constant pump speed provokes suction events, which can be prevented with the physiologic pump speed adjustment as well. During an afterload decrease or increase, the pump flow of the VAD with constant speed increased or decreased unphysiologically, whereas the pump flow controlled with the PRS controller is comparable to that of the physiologic heart functioning. The main result of the latter ability is that LV overload and suction are prevented. Because the PRS controller is directly related to the influence of preload changes, it is insensitive to contractility changes of the native heart.

Furthermore, the PRS controller is tested in vitro during acute pathophysiologic events [38] such as premature ventricular contraction (PVC)

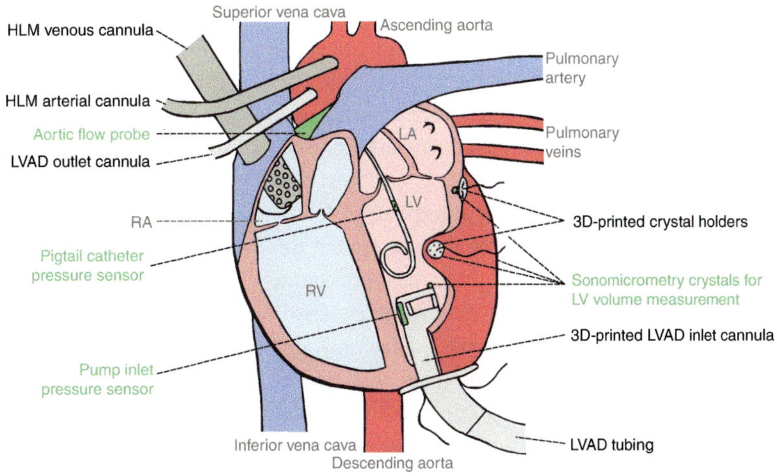

Fig. 18.10 Sketch of the inflow cannula and the sensor placement during the in vivo trial. HLM heart-lung machine, LVAD left ventricular assist device, RA right atrium, RV right ventricle, LA left atrium, LV left ventricle. (Reprinted from Ochsner et al. [77], with permission from Wolters Kluwer Health Inc.)

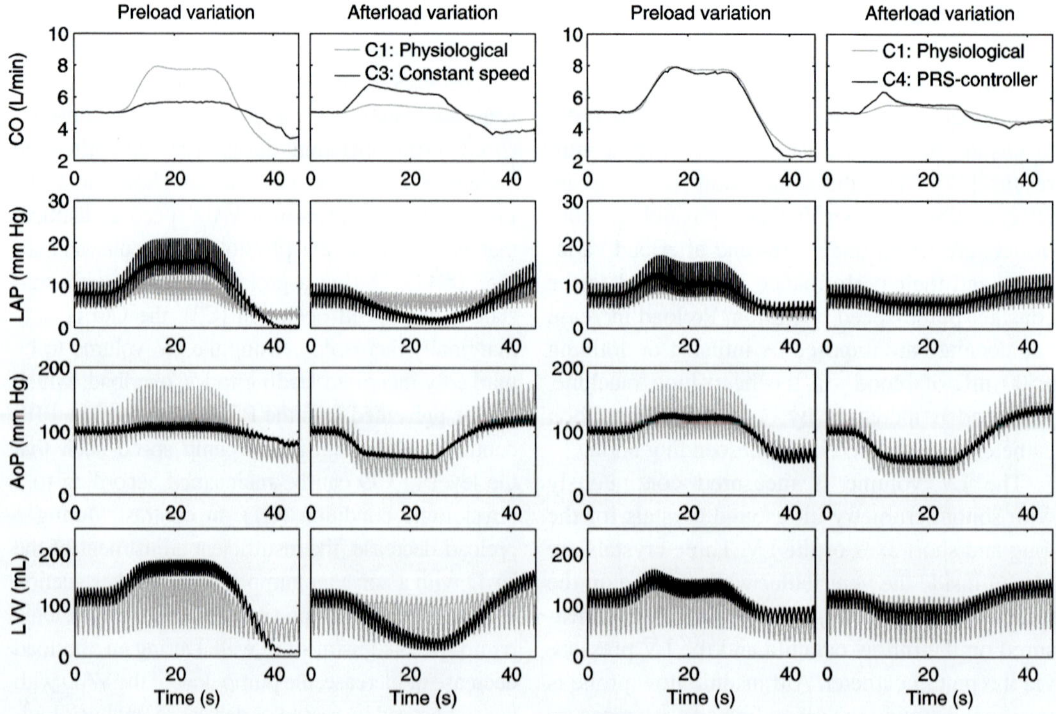

Fig. 18.11 Response of the preload-responsive speed (PRS) controller to preload and afterload variations. The performance of the PRS controller (C4) is compared with that of the constant-speed VAD operation (C3) under pathologic conditions and is evaluated based on the physiologic state with a healthy heart functioning (C1) without VAD support. CO cardiac output, LAP left atrial pressure, AoP aortic pressure, LVV left ventricular volume. (Reprinted from Ochsner et al. [17], with permission from John Wiley and Sons)

that cause abrupt changes of the CO or the VM that increases intrathoracic pressure (ITP). These results show that the PRS controller provides the ability to prevent suction during the VM, while with constant-speed operation, this is not the case. The PRS controller reduces the pump flow to the point that it leads to undesirably low AoP, while it remains physiologic with constant-speed operation. During PVCs, the PRS controller avoids sudden undesirable speed changes, but the pump backflow is increased as well. Overall, the PRS controller yields a fast, physiologic, and safe adaptation of the VAD's pump speed to the patient's perfusion requirement even under the hemodynamic influence of ITP variations and arrhythmia. Yet, a reliable LV volume sensor needs to be combined with the PRS controller in order to effectively prevent over- and underpumping, i.e., ventricular suction and backflow from the aorta to the left ventricle, respectively.

The SP controller was tested in vitro under conditions that are comparable to those of the PRS controller [21]. The results show that the response of the SP controller applied to the pathological circulation assisted by the VAD match the physiologic circulation well (see Fig. 18.12). With the SP controller during a preload increase, the CO is comparable to the physiologic CO, and suction can effectively be prevented, although during this period the resulting systolic pressure is similar to the reaction observed under constant-speed operation. During the afterload decrease, the CO is increased in accordance with the reaction of the healthy heart. In the case of a very low contractility, the SP controller was unable to sufficiently change the speed of the VAD such that suction events may occur. Because the SP controller is sensitive to afterload, relying on an estimate of the preload only, it is directly sensitive to contractility changes. Therefore, the SP controller reacts

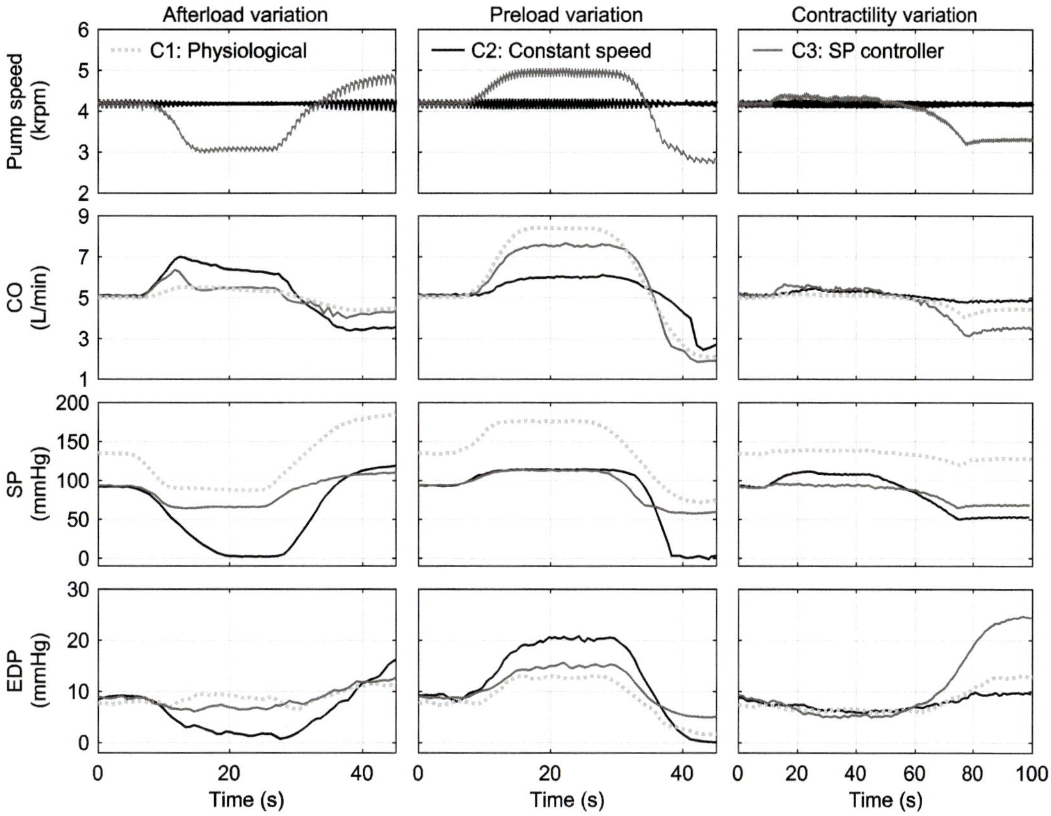

Fig. 18.12 Response of the systolic-pressure (SP) controller to after- and preload variations as well as contractility changes. The performance of the SP controller (C3) is compared with that of the constant-speed VAD operation (C2) under pathologic conditions and evaluated based on the physiologic state with a healthy heart functioning (C1) without VAD support. CO cardiac output, EDP end-diastolic pressure. (Reprinted from Petrou et al [21], with permission from John Wiley and Sons)

more aggressively and may provoke greater CO changes when the contractility changes.

The multi-objective physiologic controller allows the clinicians to decide whether the VAD is to unload the heart during severe heart failure conditions or reload the recovering heart as a kind of training function [59]. The controller is based on the PIP, the PS, as well as the estimated PF signal. The results are structured in order to show the interaction of the CVS and the controlled VAD for three testing modes, Modes 1–3. In Mode 1, the controller aims to regularly open and close the aortic valve regardless of the physiologic requirement of the circulation (results depicted in Fig. 18.13). In Mode 2, a predefined EDP is maintained, the VAD takes over most of the blood flow, the SW of the heart is minimized, and the aortic valve remains mostly closed. Mode 3 intends to support recovery through the myocardial training. The EDP gradually increases along with an increasing number of aortic valve openings. Additionally, pulsatile speed modulations provoke cyclic AoP increases. The results present a very ailing patient trying to exercise.

Furthermore, the safety algorithms for the prevention of suction, overload, and backflow guarantee a safe operation of the pump. The great advantage of the multi-objective controller is its ability to control the heart's load switching between unloading to training conditions and allowing for cyclic arterial pulse pressure increases.

The optimal control strategy for VADs has a physiologically motivated objective function

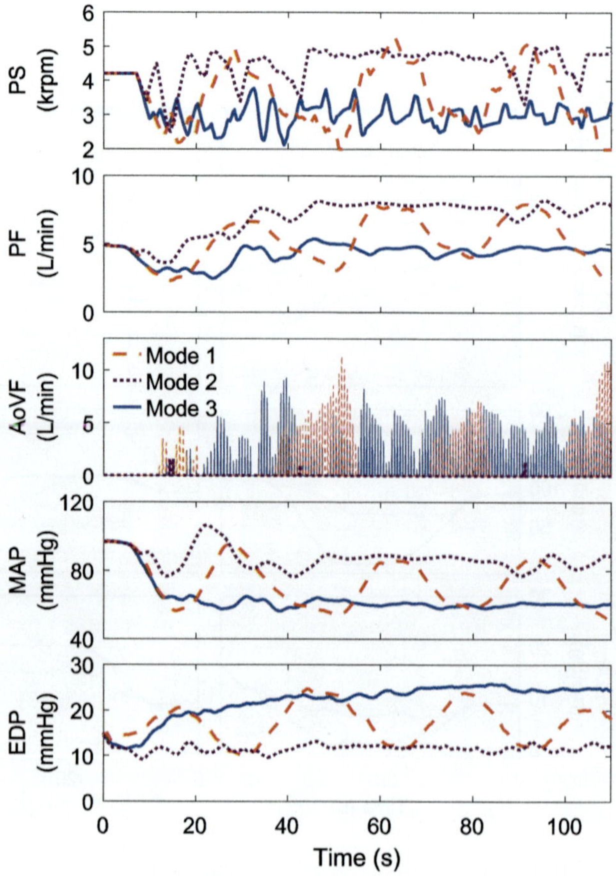

Fig. 18.13 Exemplary results of the multi-objective physiologic controller during exercise conditions with a contractility of 17%. The mean pump speed (PS), the mean pump flow (PF), the aortic valve flow (AoVF), the mean arterial pressure (MAP), and the end-diastolic pressure (EDP) are depicted under Mode 1–3 settings. (Reprinted by permission from Springer Nature: Petrou et al. [59] © 2017)

and allows pulsatile speed profiles of VADs to be optimized beat-to-beat [69]. Compared to the multi-objective controller that has a fixed sinusoidal speed profile, the level of assistance of the optimal control strategy is determined by the pulsatile speed profile derived by the optimal control problem (OCP). The results of the functional interaction between the VAD and the CVS tested in silico and validated on the HMC prove a better performance than those of constant or sinusoidal speed profiles. One drawback of the optimized speed profiles is the fact that the pump speed may change rapidly within one cardiac cycle. The hemocompatibility might be reduced due to the presence of high shear stress [53]. Further, these rapid speed changes might pose a challenge for the VAD with the fast de- and accelerations of the rotor due to the fluid inertia. With such an optimal control strategy for VADs, the pump performance can be adjusted further accounting for the intention to treat the individual patient, e.g., toward recovery of the native heart or a lifelong VAD support.

The HMC allows for a standardized comparison of various controllers and their interaction with the CVS. This is particularly valuable because a comparison with the constant-pump-speed operation alone is not sufficient to evaluate the best-suited control strategy. In general, during preload variations, the reactions of the six physiologic controllers tested (Con1–Con6) were similar to that of the healthy heart, and they all outperformed the constant-speed operation [6]. However, during afterload and contractility variations as well as drift, their performance differed, which led to LV suction events or overload. A high sensitivity of the controllers to drift and contractility changes was problematic thus required additional calibration. For example, controllers Con1 and Con2 required sensor calibrations during sensor drift, and controllers Con3 and Con6 needed offset corrections during contractility changes. The comparison study proves the power of exclusive preload-based controllers like Con1, Con2, and Con5, as already theoretically reported by Tchantchaleishvili et al. [8], and presents the challenges associated with such controllers for a robust clinical implementation.

Performance In Vivo

The operation of the PRS [17] and the SP controller [21] as well as the constant-pump-speed operation were evaluated in eight pigs [77]. With the cyclic increase of the preload and the afterload, the performances of the three control approaches were studied, and the in vitro results were validated. Additionally, the LV volume measurements based on sonomicrometry and the pressure sensor functioning integrated in the inflow cannula were tested as inputs to the respective controllers [42]. Moreover, from three pigs the intra-cardiac electrocardiogram (ECG) signal from the pigtail conductance catheter and the relation of the end-diastolic pressure to volume were monitored during all experiments. As reported in [43], any correlations between the depolarization amplitude (via the Brody effect [65]) and the EDP [16] to the LV volume were investigated as well.

The PRS controller as well as the SP controller revealed a robust performance and adapted the VAD flow according to the Frank-Starling mechanism of the heart (see Fig. 18.14). Ventricular suction and overload were mostly prevented; only one short period of suction over two cardiac cycles occurred with the PRS controller, none with the SP controller, and five with constant-speed operation. The two physiologic controllers reacted reliably to the hemodynamic changes. For example, in response to a preload reduction, the PRS and the SP controller reduced the pump speed effectively resulting in a reduced pump flow. A sensitivity analysis revealed the two controllers being comparably sensitive to a preload reduction as a healthy heart, but less sensitive to a preload increase. The latter can be explained by the high physiologic contractility of the healthy pig hearts that can compensate for the preload increase by increasing the flow through the aortic valve. Therefore, the VAD flow was not as effectively increased as expected.

During an afterload increase, the PRS controller increased the pump speed to compensate for the decreased pump flow. However, the reaction of the SP controller to the afterload increase was too strong, which may result in an excessive arterial pressure. In the pathological environment, the performance of the SP controller is expected to improve. In general, none of the experiments had to be stopped due to controller failure. The results were not affected negatively by the variability in sensor placement, for instance, the placement of the sonomicrometry ultrasound crystals. Moreover, the results were stable during arrhythmic periods, which were observed in two of the eight pigs.

The recordings based on the pressure sensor integrated in the inflow cannula [42] and the sonomicrometry crystals to measure LV volume were reliable inputs for the physiologic control of the pump. Yet, the packaging of the sensors and the overall integration in the inflow cannula need to be adjusted for long-term investigations. The assessment of the correlation between the EDV and the depolarization amplitude of the intra-cardiac ECG signal as well as the EDP was successful, and the respective accuracy of the LV volume estimation was quantified. The results reveal that in order to effectively estimate the EDV from both the depolarization amplitude and the EDP, repeated sensor calibration is needed [43]. The robustness of both signals to changes of hematocrit and LV contractility has yet to be evaluated. Ideally, a combination of the LV volume estimation based on both principles is used as a control input for a robust physiologic control.

Comparison of In Vitro and In Vivo Results

The effectiveness of the PRS and the SP controller compared to the constant-speed VAD operation was validated in vivo and their performance reflected the in vitro results. On the one hand,

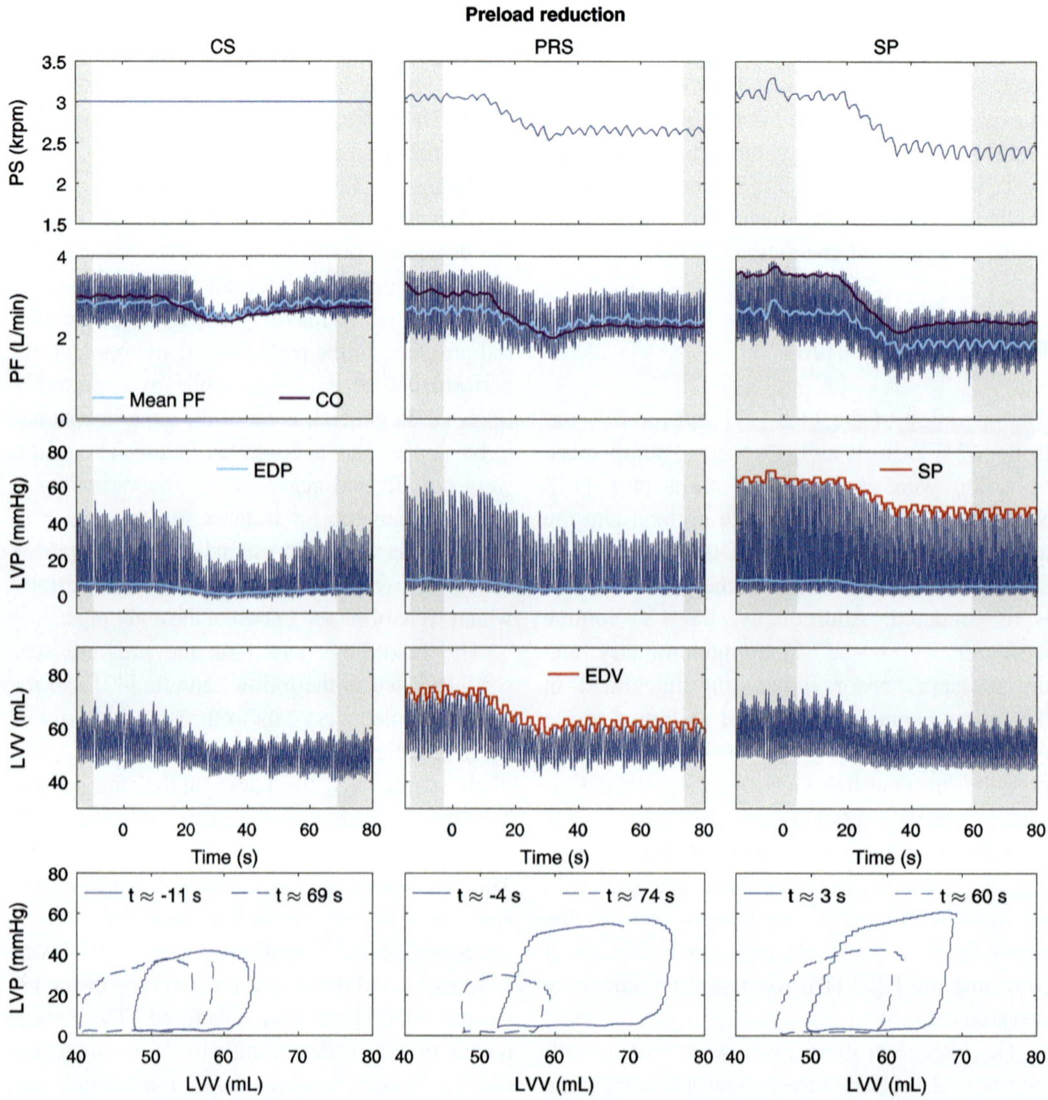

Fig. 18.14 Exemplary results of the in-vivo investigation [77]. The performance of the two physiologic controllers, the preload-responsive speed (PRS) controller based on a measurement of the left ventricular (LV) volume [17], and the systolic-pressure (SP) controller based on the systolic LV pressure [21] is compared to the constant-speed (CS) operation of the VAD. The signals of the pump speed (PS), the pump flow (PF), the left ventricular pressure (LVP), and volume (LVV) are depicted over a period of 80 s. Additionally, the mean PF, the cardiac output (CO) and the end-diastolic pressure (EDP) are shown. The end-diastolic volume (EDV) and SP are the input signals to the respective controllers. The pressure-volume loops show the steady-state conditions before and after the decrease of preload by the drainage of 500 mL of blood. The results shown prove the effectiveness of both the PRS and the SP controller. (Reprinted from Ochsner et al. [77], with permission from Wolters Kluwer Health Inc.)

the numerical cardiovascular model (CVM) has restrictions compared to the CVS of the pig with the entire physiologic feedback loops of the healthy organism active. Short periods of suction events that occurred in vivo were not predicted in vitro. Understanding the key dynamics that lead to the different results could potentially improve the CVM used in current in vitro studies. In the CVM the interaction of the right ventricular and the LV filling is not modelled, which could lead to inaccuracies described above. On the other hand, the HMC already allows physiologic controllers for VADs to be compared in a standardized and reproducible way [6].

Furthermore, due to the fact that the pig hearts were healthy, their contractility cannot be altered in vivo but can be mimicked easily on the HMC. Pathological hemodynamics are not easily modelled in vivo. The healthy heart's interaction with a VAD system was therefore more preload sensitive than that of heart failure patients such that the results of the in vitro and in vivo investigations differed in that respect. For instance, in vivo, the high contractility ability of the healthy heart may explain the strong reaction of the SP controller to the afterload increase which is different compared to the in vitro performance. In vivo, the physiologic controllers influenced the SW, the CO, and the EDP less than on the HMC.

In vitro, every arbitrary signal is available either computed by the numerical CVM or measured with a sensor that can be placed at the HMC where needed. In vivo, the access is limited and the sensors have to be placed carefully in the appropriate location. For example, LV volume on the HMC is computed numerically, while in the pig it is measured either by the sensor integrated in the pigtail catheter or by the sonomicrometry crystals. Moreover, due to animal-specific physiologic conditions, recalibrations of sensors and controller gain adjustments are mandatory. Again, this could explain the discrepancies of the number of suction events in vitro and in vivo.

The functioning of the pressure sensor integrated in the inflow cannula could be validated in vivo according to the short-term tests in vitro [42]. Yet, long-term stability and sensor drift could only be investigated in chronic animal trials. Moreover, the investigation of novel measurement principles for an LV volume sensor would not be possible on the current HMC. The only interface of the numerical model and the HMC are pressure-based and are not mimicking an integrated heart in a thorax. Therefore, additional testing environments are necessary, which additionally provide a physical model of the heart [45].

Outlook

The three preload-responsive physiologic controllers and the respective pressure and volume sensors are to be incorporated in state-of-the-art VADs, allowing the current pumps to adapt to the patients' perfusion requirement and guaranteeing a safe pump operation with the overall goal to significantly improve the patients' well-being and increase the survival rates. Redundant control approaches based on multiple measurements are favorable, e.g., combining the LV volume and LV pressure input. Ideally, these sensors are integrated in the inflow cannula of the VAD [42, 45]. Future work is necessary to develop these completely integrated, long-term stable, and hemo- and biocompatible sensor systems providing the controller with the required signal input.

The development and improvement of physiologic controllers can be advanced by the use of classic control theory with the advantage to conduct systematic stability and robustness analyses. To do so, a linear time-invariant plant model is required. This plant comprises a mean-value model of the cardiovascular system that is combined with a VAD. This mean-value model offers two advantages: (1) it is computationally less demanding than the widely used time-varying elastance model, and (2) it allows to extract the linear time-invariant plant for the model-based design of physiologic controllers [78].

Furthermore, the pathophysiology of the right ventricular dynamics [79] and its interaction with the septum needs to be studied in more depth as right ventricular failure can occur under left VAD support [80]. The cardiac output provided by the LV supported by a VAD needs to be handled by the right heart as well. The right ventricle may become overloaded due to the preload increase caused by the left VAD support, or the interventricular septum shifted due to a lack of pump speed adjustments. Therefore, the right and left ventricular filling must be well matched by the setting of the VAD speed providing an optimal cardiac output. Thus, the septal interaction between the two chambers should be studied in more detail. The challenging topic of right ventricular support with direct clinical implications can be addressed with the biventricular HMC, as it comprises a second interface to the right heart in the simulated CVM. Yet, to additionally test right ventricular support by a VAD, the extended HMC provides the respective interfaces to do so [81].

Ideally, testing facilities are standardized such that results can be compared more quantitatively among groups. The configuration of the HMC as well as the pathophysiologic testing criteria, such as the mimicked heart's contractility or the amount of pre- and afterload variations, should be standardized to effectively test novel sensor principles or physiologic controllers. In order to study the hemodynamic influences of the VAD support using speed modulations to improve the arterial pulse pressure, a synchronized VAD operation with the cardiac cycle is promising [52, 59]. Finally, there is an urgent need for the development of animal models of heart failure, which can mimic the pathologic heart function accurately to render the in vitro and in vivo results comparable.

References

1. Hall JE. Guyton and hall textbook of medical physiology. 13th ed. Philadelphia: Saunders; 2016.
2. Anand J, Singh SK, Antoun DG, Cohn WE, Frazier OHB, Mallidi HR. Durable mechanical circulatory support versus organ transplantation: past, present, and future. Biomed Res Int. 2015;2015:849571.
3. Patel SR, Jorde UP. Creating adequate pulsatility with a continuous flow left ventricular assist device: just do it! Curr Opin Cardiol. 2016;31(3):329–36.
4. Netuka I, Pya Y, Zimpfer D, Krabatsch T, Garbade J, Rao V, et al. HeartMate III, fully magnetically levitated left ventricular assist device for the treatment of advanced heart failure–results from the CE mark trial. J Card Fail. 2015;21(11):936.
5. Bhagra S, Bhagra C, Özalp F, Butt T, Ramesh B, Parry G, et al. Development of de novo aortic valve incompetence in patients with the continuous-flow HeartWare ventricular assist device. J Heart Lung Transplant. 2016;35(3):312–9.
6. Petrou A, Lee J, Dual S, Ochsner G, Meboldt M, Schmid Daners M. Standardized comparison of selected physiological controllers for rotary blood pumps: in vitro study. Artif Organs. 2018;42(3):E29–42.
7. Molina EJ, Boyce SW. Current status of left ventricular assist device technology. Semin Thorac Cardiovasc Surg. 2013;25(1):56–63.
8. Tchantchaleishvili V, Luc JGY, Cohan CM, Phan K, Hübbert L, Day SW, et al. Clinical implications of physiological flow adjustment in continuous-flow left ventricular assist devices. ASAIO J. 2017;63: 241–50.
9. Benjamin EJ, Blaha MJ, Chiuve SE, Cushman M, Das SR, Deo R, et al. Heart disease and stroke statistics—2017 update: a report from the American Heart Association. Circulation. 2017;135(10):e146.
10. Shah P, Birk S, Maltais S, Stulak J, Elmi A, Pagani FD, et al. Left ventricular assist device outcomes based on flow configuration and pre-operative left ventricular dimension: an interagency registry for mechanically assisted circulatory support analysis. J Heart Lung Transplant. 2017;36:640–9.
11. Teichholz LE, Kreulen T, Herman MV, Gorlin R. Problems in echocardiographic volume determinations: echocardiographic-angiographic correlations in the presence of absence of asynergy. Am J Cardiol. 1976;37:7–11.
12. Klotz S, Danser AHJ, Burkhoff D. Impact of left ventricular assist device (LVAD) support on the cardiac reverse remodeling process. Prog Biophys Mol Biol. 2008;97:479–96.
13. Wang TS, Hernandez AF, Felker GM, Milano CA, Rogers JG, Patel CB. Valvular heart disease in patients supported with left ventricular assist devices. Circ Heart Fail. 2014;7:215–22.
14. da Rocha e Silva JG, Meyer AL, Eifert S, Garbade J, Mohr FW, Strueber M. Influence of aortic valve opening in patients with aortic insufficiency after left ventricular assist device implantation. Eur J Cardiothorac Surg. 2015;49(3):784–7.
15. Burkhoff D, Klotz S, Mancini DM. LVAD-induced reverse remodeling: basic and clinical implications for myocardial recovery. J Card Fail. 2006;12(3):227–39.
16. Klotz S, Hay I, Dickstein ML, Yi GH, Wang J, Maurer MS, et al. Single-beat estimation of end-diastolic pressure-volume relationship: a novel method with potential for noninvasive application. Am J Physiol Heart Circ Physiol. 2006;291:H403–12.
17. Ochsner G, Amacher R, Wilhelm MJ, Vandenberghe S, Tevaearai H, Plass A, et al. A physiological controller for turbodynamic ventricular assist devices based on a measurement of the left ventricular volume. Artif Organs. 2014;38:527–83.
18. Takeuchi M, Odake M, Takaoka H, Hayashi Y, Yokoyama M. Comparison between preload recruitable stroke work and the end-systolic pressure-volume relationship in man. Eur Heart J. 1992;13(Suppl E):80–4.
19. Colacino FM, Moscato F, Piedimonte F, Arabia M, Danieli GA. Left ventricle load impedance control by apical VAD can help heart recovery and patient perfusion: a numerical study. ASAIO J. 2007;53(3):263–77.
20. Rodeheffer RJ, Gerstenblith G, Becker LC, Fleg JL, Weisfeldt ML, Lakatta EG. Exercise cardiac output is maintained with advancing age in healthy human subjects: cardiac dilatation and increased stroke volume compensate for a diminished heart rate. Circulation. 1984;69(2):203–13.
21. Petrou A, Ochsner G, Amacher R, Pergantis P, Rebholz M, Meboldt M, et al. A physiological controller for turbodynamic ventricular assist devices based on systolic left ventricular pressure. Artif Organs. 2016;40(9):842–55.
22. Arndt A, Nüsser P, Lampe B. Fully autonomous preload-sensitive control of implantable rotary blood pumps. Artif Organs. 2010;34:726–35.

23. Rüschen D, Opitz S, von Platen P, Korn L, Leonhardt S, Walter M. Robust physiological control of rotary blood pumps for heart failure therapy. at-Automatisierungstechnik. 2018;66(9):767–79.
24. Abraham WT, Perl L. Implantable hemodynamic monitoring for heart failure patients. J Am Coll Cardiol. 2017;70(3):389–98.
25. Moscato F, Vollkron M, Bergmeister H, Wieselthaler G, Leonard E, Schima H. Left ventricular pressure–volume loop analysis during continuous cardiac assist in acute animal trials. Artif Organs. 2007;31:369–76.
26. Petrou A, Kuster D, Lee J, Meboldt M, Schmid Daners M. Comparison of flow estimators for rotary blood pumps: an in vitro and in vivo study. Ann Biomed Eng. 2018;46(12): 2123-2134.
27. Pauls JP, Stevens MC, Bartnikowski N, Fraser JF, Gregory SD, Tansley G. Evaluation of physiological control systems for rotary left ventricular assist devices: an in-vitro study. Ann Biomed Eng. 2016;44(8):2377–87.
28. Schima H, Huber L, Schmallegger H, Drost CJ, Droudt A, Wieselthaler G, et al. Flow measurement at the pump head of centrifugal pumps: comparison of ultrasonic transit time and ultrasonic Doppler systems. Artif Organs. 1997;21(7):808–15.
29. Mansouri M, Salamonsen RF, Lim E, Akmeliawati R, Lovell NH. Preload-based starling-like control for rotary blood pumps: numerical comparison with pulsatility control and constant speed operation. PLoS One. 2015;10(4):e0121413.
30. Gaddum NR, Stevens M, Lim E, Fraser J, Lovell N, Mason D, et al. Starling-like flow control of a left ventricular assist device: in vitro validation. Artif Organs. 2014;38(3):E46–56.
31. AlOmari AHH, Savkin AV, Stevens M, Mason DG, Timms DL, Salamonsen RF, et al. Developments in control systems for rotary left ventricular assist devices for heart failure patients: a review. Physiol Meas. 2013;34:R1–R27.
32. Schima H, Trubel W, Moritz A, Wieselthaler G, Stohr HG, Thoma H, et al. Noninvasive monitoring of rotary blood pumps: necessity, possibilities, and limitations. Artif Organs. 1992;16(2):195–202.
33. Ambardekar AV, Hunter KS, Babu AN, Tuder RM, Dodson RB, Lindenfeld J. Changes in aortic wall structure, composition, and stiffness with continuous-flow left ventricular assist devices: a pilot study. Circ Heart Fail. 2015;8:944. CIRCHEARTFAILURE–114.
34. Crow S, John R, Boyle A, Shumway S, Liao K, Colvin-Adams M, et al. Gastrointestinal bleeding rates in recipients of nonpulsatile and pulsatile left ventricular assist devices. J Thorac Cardiovasc Surg. 2009;137:208–15.
35. Kirklin JK, Pagani FD, Kormos RL, Stevenson LW, Blume ED, Myers SL, et al. Eighth annual INTERMACS report: special focus on framing the impact of adverse events. J Heart Lung Transplant. 2017;36:1080–6.
36. Schima H, Dimitrov K, Zimpfer D. Debate: creating adequate pulse with a continuous flow ventricular assist device: can it be done and should it be done? Probably not, it may cause more problems than benefits! Curr Opin Cardiol. 2016;31(3):337–42.
37. Lim E, Salamonsen RF, Mansouri M, Gaddum N, Mason DG, Timms DL, et al. Hemodynamic response to exercise and head-up tilt of patients implanted with a rotary blood pump: a computational modeling study. Artif Organs. 2015;39(2):E24–35.
38. Petrou A, Pergantis P, Ochsner G, Amacher R, Krabatsch T, Falk V, et al. Response of a physiological controller for ventricular assist devices during acute pathophysiological events: an in vitro study. Biomed Eng/Biomedizinische Technik. 2017;62:623–33.
39. Kaufmann F, Krabatsch T. Using medical imaging for the detection of adverse events ("incidents") during the utilization of left ventricular assist devices in adult patients with advanced heart failure. Expert Rev Med Devices. 2016;13:463–74.
40. Petrou A, Kanakis M, Boës S, Pergantis P, Meboldt M, Schmid Daners M. Viscosity prediction in a physiologically controlled ventricular assist device. IEEE Trans Biomed Eng. 2018;65(10):2355–2364.
41. Jorde UP, Aaronson KD, Najjar SS, Pagani FD, Hayward C, Zimpfer D, et al. Identification and management of pump thrombus in the HeartWare left ventricular assist device system: a novel approach using log file analysis. JACC Heart Fail. 2015;3:849–56.
42. Staufert S, Hierold C. Novel sensor integration approach for blood pressure sensing in ventricular assist devices. Procedia Eng. 2016;168:71–5.
43. Dual SA, Ochsner G, Petrou A, Amacher R, Wilhelm MJ, Meboldt M, et al. R-wave magnitude: a control input for ventricular assist devices. In: Proceedings of the 8th international workshop on biosignal interpretation (BSI 2016); 2016. pp. 18–21. Available from: http://www.p.u-tokyo.ac.jp/~bsi2016/PDF/BSI2016_proceedings.pdf.
44. Cysyk J, Newswanger R, Popjes E, Pae W, Jhun CS, Izer J, et al. Cannula tip with integrated volume sensor for rotary blood pump control: early-stage development. ASAIO J. 2019;65(4):318-23.
45. Dual SA, Zimmermann JM, Neuenschwander J, Cohrs NH, Solowjowa N, Stark WJ, et al. Ultrasonic sensor concept to fit a ventricular assist device cannula evaluated using geometrically accurate heart phantoms. Artif Organs. 2019;43(5):467-77.
46. Brancato L, Keulemans G, Verbelen T, Meyns B, Puers R. An implantable intravascular pressure sensor for a ventricular assist device. Micromachines. 2016;7:135.
47. Troughton RW, Ritzema J, Eigler NL, Melton IC, Krum H, Adamson PB, et al. Direct left atrial pressure monitoring in severe heart failure: long-term sensor performance. J Cardiovasc Transl Res. 2011;4:3–13.
48. Dorosz JL, Lezotte DC, Weitzenkamp DA, Allen LA, Salcedo EE. Performance of 3-dimensional echocardiography in measuring left ventricular volumes and ejection fraction: a systematic review and meta-analysis. J Am Coll Cardiol. 2012;59:1799–808.

49. Ochsner G. Physiologic control of left ventricular assist devices. Diss. ETH No. 22061, ETH Zurich; 2014. Available from: https://doi.org/10.3929/ethz-a-010252528.
50. Mason DG, Hilton AK, Salamonsen RF. Reliable suction detection for patients with rotary blood pumps. ASAIO J. 2008;54(4):359–66.
51. Klima UP, Lee MY, Guerrero JL, LaRaia PJ, Levine RA, Vlahakes GJ. Determinants of maximal right ventricular function: role of septal shift. J Thorac Cardiovasc Surg. 2002;123:72–80.
52. Amacher R, Ochsner G, Schmid Daners M. Synchronized pulsatile speed control of turbodynamic left ventricular assis devices: review and prospects. Artif Organs. 2014;38(10):867–875.
53. Wiegmann L, Thamsen B, de Zélicourt D, Granegger M, Boës S, Schmid Daners M, et al. Fluid dynamics in the HeartMate 3: influence of the artificial pulse feature and residual cardiac pulsation. Artif Organs. 2018;43:363.
54. Casas F, Ahmed N, Reeves A. Minimal sensor count approach to Fuzzy Logic rotary blood pump flow control. ASAIO J. 2007;53:140–6.
55. Wang Y, Koenig S, Slaughter M, Giridharan G. Rotary blood pump control strategy for preventing left ventricular suction. ASAIO J. 2015;61(1):21–30.
56. Bullister E, Reich S, Sluetz J. Physiologic control algorithms for rotary blood pumps using pressure sensor input. Artif Organs. 2002;26:931–8.
57. Ising MS, Sobieski MA, Slaughter MS, Koenig SC, Giridharan GA. Feasibility of pump speed modulation for restoring vascular pulsatility with rotary blood pumps. ASAIO J. 2015;61(5):526–32.
58. Soucy KG, Koenig SC, Giridharan GA, Sobieski MA, Slaughter MS. Defining pulsatility during continuous-flow ventricular assist device support. J Heart Lung Transpl. 2013;32:581–7.
59. Petrou A, Monn M, Meboldt M, Schmid Daners M. A novel multi-objective physiological control system for rotary left ventricular assist devices. Ann Biomed Eng. 2017;45(12):2899–910.
60. Umeki A, Nishimura T, Takewa Y, Ando M, Arakawa M, Kishimoto Y, et al. Change in myocardial oxygen consumption employing continuous-flow LVAD with cardiac beat synchronizing system, in acute ischemic heart failure models. J Artif Organs. 2013;16:119–28.
61. Amacher R, Weber A, Brinks H, Axiak S, Ferreira A, Guzzella L, et al. Control of ventricular unloading using an electrocardiogram-synchronized Thoratec paracorporeal ventricular assist device. J Thorac Cardiovasc Surg. 2013;146(3):710–7.
62. Arndt A, Nüsser P, Lampe B. Fully autonomous preload-sensitive control of implantable rotary blood pumps. Artif Organs. 2010;34(9):726–35.
63. Schima H, Vollkron M, Jantsch U, Crevenna R, Roethy W, Benkowski R, et al. First clinical experience with an automatic control system for rotary blood pumps during ergometry and right-heart catheterization. J Heart Lung Transpl. 2006;25(2):167–73.
64. Klabunde R. Cardiovascular physiology concepts. Baltimore: Lippincott Williams & Wilkins; 2011.
65. Brody DA. A theoretical analysis of intracavitary blood mass influence on the heart-lead relationship. Circ Res. 1956 Nov;4:731–8.
66. Staufert S, Gutzwiller P, Mushtaq F, Hierold C. Surface nanostructuring of Ti6Al4 V surfaces for Parylene-C coatings with ultradurable adhesion. ACS Appl Nano Mater. 2018;1(4):1586–94.
67. Abraham WT, Adamson PB, Bourge RC, Aaron MF, Costanzo MR, Stevenson LW, et al. Wireless pulmonary artery haemodynamic monitoring in chronic heart failure: a randomised controlled trial. Lancet. 2011;377:658–66.
68. Takeuchi M, Odake M, Takaoka H, Hayashi Y, Yokoyama M. Comparison between preload recruitable stroke work and the end-systolic pressure-volume relationship in man. Eur Heart J. 1992;13:80–4.
69. Amacher R, Asprion J, Ochsner G, Tevaearai H, Wilhelm MJ, Plass A, et al. Numerical optimal control of turbo dynamic ventricular assist devices. Bioengineering. 2014;1:22–46.
70. Ochsner G, Amacher R, Amstutz A, Plass A, Schmid Daners M, Tevaearai H, et al. A novel interface for hybrid mock circulations to evaluate ventricular assist devices. IEEE Trans Biomed Eng. 2013;60:507–16.
71. Ochsner G, Amacher R, Schmid Daners M. Emulation of ventricular suction in a hybrid mock circulation. In: Proc. 2013 European control conference; 2013. p. 3108–12.
72. Hybrid mock circulation to evaluate VADs. Available from: youtube/_L1EaSJzERU.
73. Colacino FM, Moscato F, Piedimonte F, Danieli G, Nicosia S, Arabia M. A modified Elastance model to control mock ventricles in real-time: numerical and experimental validation. ASAIO J. 2008;54: 563–73.
74. Boës S, Ochsner G, Amacher R, Petrou A, Meboldt M, Schmid Daners M. Control of the fluid viscosity in a mock circulation. Artif Organs. 2018;42(1):68–77.
75. Moscato F, Arabia M, Colacino FM, Naiyanetr P, Danieli GA, Schima H. Left ventricle afterload impedance control by an axial flow ventricular assist device: a potential tool for ventricular recovery. Artif Organs. 2010;34(9):736–44.
76. Wang Y, Koenig SC, Slaughter MS, Giridharan GA. Rotary blood pump control strategy for preventing left ventricular suction. ASAIO J 2015;61(1):21–30.
77. Ochsner G, Wilhelm MJ, Amacher R, Petrou A, Cesarovic N, Röhrnbauer B, et al. In vivo evaluation of physiological control algorithms for LVADs based on left ventricular volume or pressure. ASAIO J. 2017;63:568–77.
78. Ochsner G, Amacher R, Schmid Daners M. A novel mean-value model of the cardiovascular system including a left ventricular assist device. Cardiovasc Eng Technol. 2017;8(2):120–30.
79. Greyson CR. Pathophysiology of right ventricular failure. Crit Care Med. 2008;36(1):S57–65.

80. Dandel M, Krabatsch T, Falk V. Left ventricular vs. biventricular mechanical support: decision making and strategies for avoidance of right heart failure after left ventricular assist device implantation. Int J Cardiol. 2015;198:241-50.
81. Petrou A, Granegger M, Meboldt M, Schmid Daners M. A versatile hybrid mock circulation for hydraulic investigations of active and passive cardiovascular implants. ASAIO J. 2019;65(5):495-502.

19 End-Organ Physiology Under Continuous-Flow Mechanical Circulatory Support

Egemen Tuzun

When you venture into the wilderness, do not expect to find a paved road.

C. Walton Lillehei

Introduction

Heart failure is one of the leading causes of death and hospitalization in the United States, and nearly 6.5 million Americans are currently living with congestive heart failure (CHF) [1]. Treatment of advanced heart failure often requires frequent hospitalization and prolonged intensive care, accounts for more hospitalizations annually than all forms of cancer combined, and is a major contributor to healthcare costs. Fifty percent of the patients with CHF either do not respond to medical therapy or become unresponsive to medical therapy within 5 years of diagnosis. While heart transplantation is the best treatment for heart failure, mechanical circulatory support has become an important alternative because only a limited number of donor hearts are available each year [2]. In addition, a growing number of people need cardiac support but are ineligible for heart transplantation. The clinical success of ventricular assist devices (VADs) offers an innovative strategy for increasing the life expectancy of patients with CHF. The landmark REMATCH trial established the left ventricular assist device (LVAD) as a successful alternative to medical therapy. LVAD recipients had 1- and 2-year survivals that were 2 times greater than patients treated with medical therapy alone [3].

While pulsatile flow conditions are considered the most physiologic, continuous-flow (nonpulsatile) ventricular assist devices (CFVADs) are also being developed to reduce some of the complications associated with pulsatile technology, such as infection, surgical bleeding, and mechanical device failure, and to address the needs of patients who are too small to be fitted with the current pulsatile devices. In a recently published study, CFVAD recipients demonstrated improved survival rates at 6-month, 1-year, and 2-year follow-up with $92 \pm 4\%$, $81 \pm 6\%$, and $74 \pm 6\%$, respectively [4].

When continuous-flow pumps are used as LVADs, the resulting aortic pressure waveform has reduced pulse amplitude and a rounded shape. Therefore, continuous-flow pumps are not really "nonpulsatile" pumps, although they do provide reduced pulse perfusion. The pressure and flow waveforms in the aorta have residual pulsatility because the flow output from a CFVAD is proportional to the pressure gradient across the pump. As long as the left ventricle is generating some pressure during each contraction, then the pressure gradient also will be pulsatile, and the resultant cyclic increase in pump flow will impart some degree of pulsatility to the arterial waveform [5] Fig. 19.1.

E. Tuzun, MD, PhD (✉)
Texas A&M University, Institute for Preclinical Studies (TIPS), College Station, TX, USA
e-mail: Egemen.tuzun@tamu.edu

© Springer Nature Switzerland AG 2020
J. H. Karimov et al. (eds.), *Mechanical Support for Heart Failure*,
https://doi.org/10.1007/978-3-030-47809-4_19

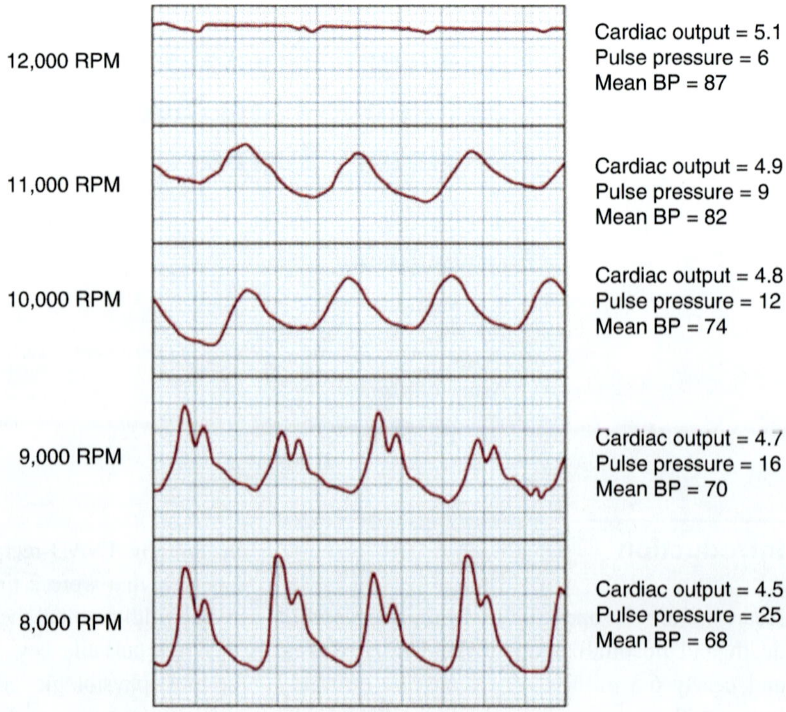

Fig. 19.1 Arterial pressure tracings after continuous flow left ventricular assist device implantation. With increasing speed, there is reduced pulse pressure as a larger volume of blood is removed from the left ventricular apex. Aortic valve opening is gradually reduced until it closes completely to provide continuous systemic blood flow. BP blood pressure, RPM revolutions per minute. (Republished with permission of McGraw-Hill Education, from Fuster et al. [5]; permission conveyed through Copyright Clearance Center, Inc.)

Recently, we have seen patients with advanced heart failure who because of poor unloading and severely impaired ventricular function have a weak or no detectable pulse, despite the absence of heart failure symptoms. As the worldwide experience with CFVADs is increasing, there is now considerable evidence that chronic support with continuous-flow and/or reduced-pulse circulatory assistance is clinically viable. Previous studies have demonstrated that nonpulsatile circulation can lead to certain vasomotor, metabolic, hemodynamic, hormonal, functional, and morphologic changes; however the topic still remains controversial [6]. The clinical relevance of pulsatility for VADs is unclear, especially since most evidence is largely developed to the context of cardiopulmonary bypass (CPB). To further confound matters, a uniform definition of what constitutes a physiologically meaningful pulse has hampered comparisons among studies. Indeed, unlike the CPB setting, wherein the heart is arrested, CFVAD circulation generally reflects a component of the native pulse generated in proportion to the patient's native ventricular function and in inverse proportion to the degree of mechanical circulatory support (MCS). The importance of pulsatility in long-term MCS is a relatively new topic. At the very least, it remains possible that in certain circumstances, it may be desirable to either temporarily or occasionally produce or enhance native pulsatility in CFVAD patients by production of an artificial pulse [7, 8].

Despite technological advances, prolonged use of CFVADs has raised concerns about their long-term effects upon flow properties and end-organ physiology, such as aortic valve leaflet fusion, increased cerebral and gastrointestinal bleeding complications, and/or lack of myocardial reverse remodeling. For this reason, surgeon's skill and expertise, which are obviously basic requirements to ensure a satisfactory long-term success rate of a procedure, should be supported by innovative predictive approaches provided by clinical and preclinical research studies. In this concept patient-specific simulations (in vitro, in vivo, or ex vivo models) are necessary to evaluate continuous-flow cardiac assist physiology, the efficacy of various possible treatments, and to plan and design the optimal surgical or device solution based on predictions of outcomes provided by these models. In order to achieve the above-mentioned goal, cardiac surgeons, cardiologists, engineers, basic scientists, radiologists, and veterinarians forming basic teams of academia and industry need to strictly collaborate to create accurate models for realistic simulations and solutions. Despite significantly improved survival, important adverse events of the CFVADs still affect patients' life quality, readmission rates to the hospital, and mortality/morbidity rates. To reduce and/or prevent those adverse effects, understanding the mechanism of continuous physiology at the microcirculatory and molecular level is undoubtedly crucial. Unfortunately, up to date, basic science studies investigating the mechanisms of the continuous-flow physiology are very limited. This chapter aims to summarize the existing literature comparing continuous- flow and pulsatile flow physiologies at the organ level and provide some published/pending preliminary data to shed a light on this complex, relatively unknown physiology.

Effects of CFVAD Flow Functional and Structural Vascular Adaptation

CFVADs unload the heart in parallel or in series. FDA-approved devices such as HeartMate II (axial), HeartMate III (centrifugal), and HeartWare (centrifugal) work parallel with the native heart via an alternate pathway (outflow graft) and provide blood flow to the aorta from the left ventricle or atrium depending on the inflow cannula location (Fig. 19.2). On the other hand, some short-term use percutaneously implantable CFVADs such as Impella 2.5 (axial) unload the heart in series. In both technologies, the pump differential pressure which is the difference between the left ventricular and aortic pressure is inversely related to the pump flow [9] (Fig. 19.3a, b). Since CFVADs continuously unload the ventricles during systole and diastole, the pump flows momentarily change during the cardiac cycle reaching out to maximum during systole and minimum during diastole. Considering that diastolic and mean arterial pressures increase at increasing pump supports, the aortic pulse pressure (AoPP), which is equal to the difference between aortic systolic (AoS) and diastolic (AoD) pressures (AoPP = AoS − AoD), diminishes at increasing pump speeds resulting in low or no pulse pressure in the arterial system along with a leftward shift on pressure-volume (PV) curves [9, 10] (Fig. 19.4a, b).

CFVADs are more preload- and afterload-sensitive pumps compared to healthy native hearts. In normal hearts cardiac output increases with increase in preload (by fluid replacement, increased venous return, etc.), but it is relatively insensitive to increase in afterload (by hypovolemia, hypertension, increased peripheral vascular resistance, etc.). In contrast, CFVAD flows fall quickly in response to preload and afterload reductions [9].

Aortic pulse pressure arises from the interaction of the heart, the systemic arterial system, and peripheral microcirculations. The complex

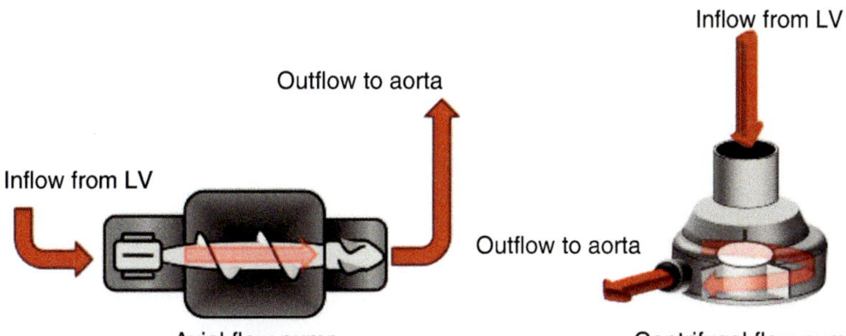

Fig. 19.2 Continuous-flow left ventricular assist devices may be classified as (left) axial or (right) centrifugal based on the direction of blood flow in relation to the axis of the device. (Republished with permission of Elsevier Science and Technology Journals, from Lim et al. [9]; permission conveyed through Copyright Clearance Center, Inc.)

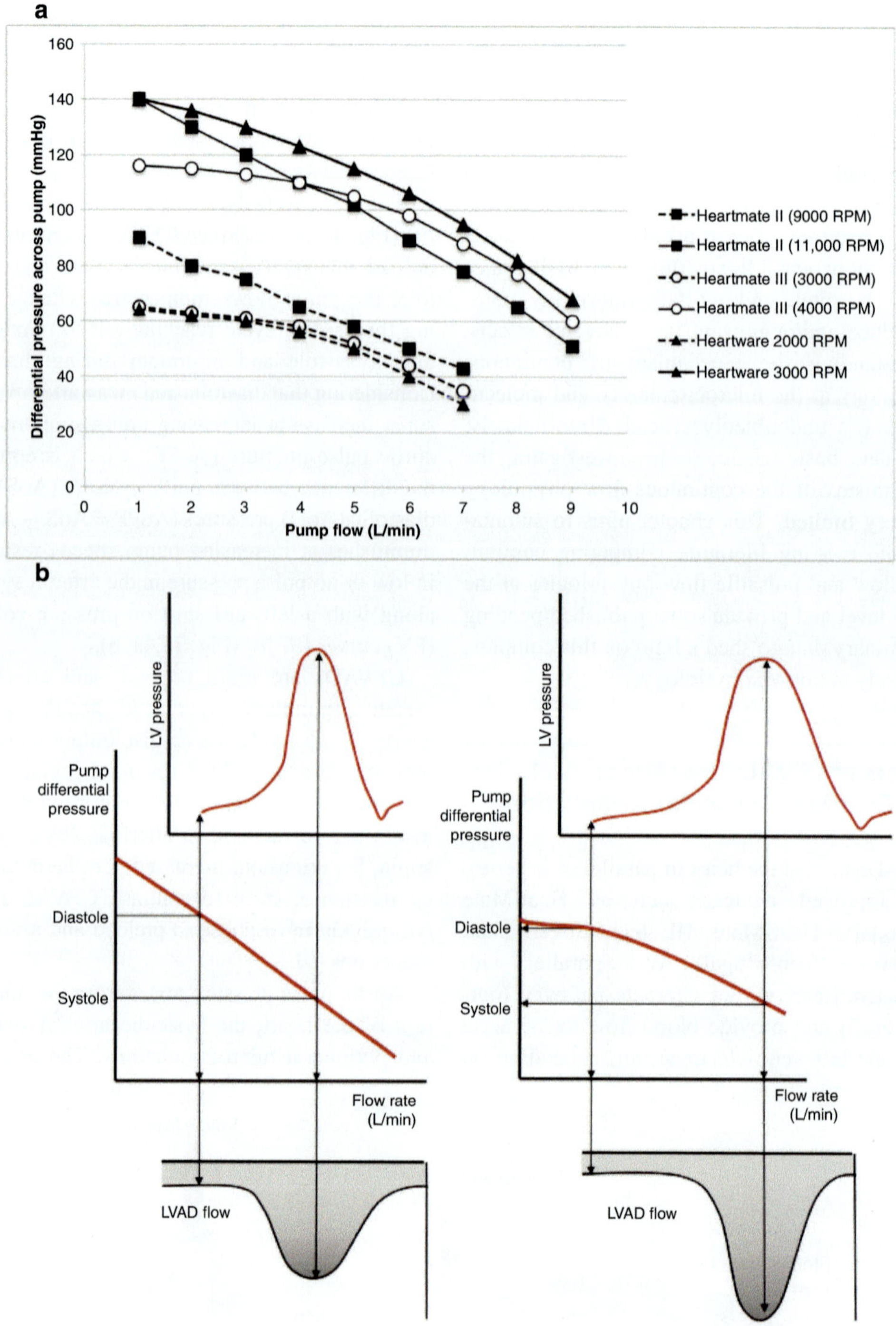

Fig. 19.3 (**a**) Pressure (pump differential pressure)–flow relationship for the axial flow HeartMate II and centrifugal-flow HeartMate III and HeartWare. Pump differential pressure is the absolute difference between aortic (systemic) blood pressure and ventricular pressure. (**b**) The flatter pressure-flow curves in (right) centrifugal pumps compared with (left) axial flow pumps result in (i) greater changes in flow between systole and diastole (assuming unchanged systemic arterial blood pressure) due to the higher pump differential pressure and (ii) smaller changes in pump differential pressure for a given change in pump flow, which may reduce the likelihood of suction events compared to axial flow pumps. LV, left ventricular; LVAD, left ventricular assist device. (Republished with permission of Elsevier Science and Technology Journals, from Lim et al. [9]; permission conveyed through Copyright Clearance Center, Inc.)

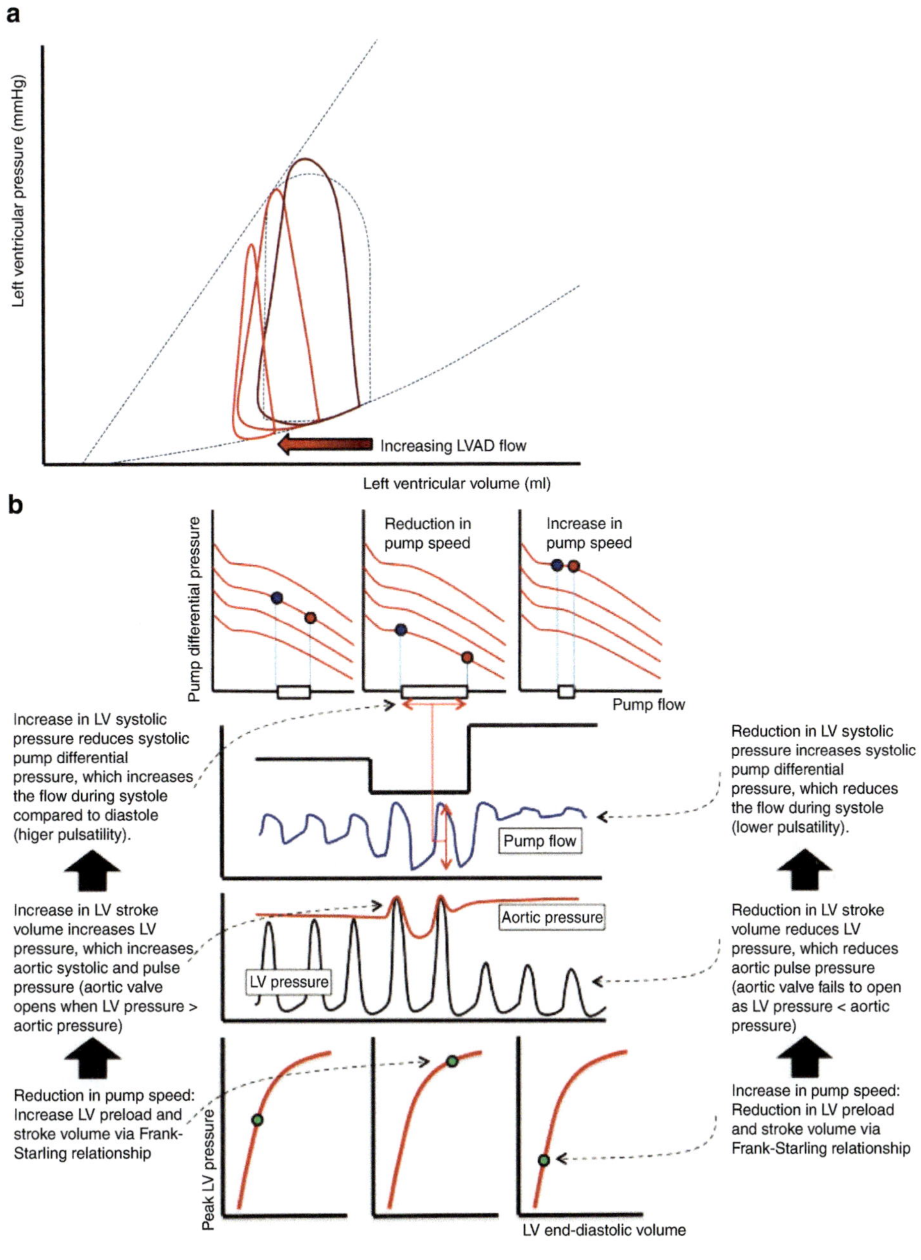

Fig. 19.4 (**a**) The effect of increasing pump speed/flow on the left ventricular pressure-volume loop. The unsupported left ventricular pressure-volume loop is shown (thin dotted line). Note the triangular shape of the loop due to loss of isovolumic phases. The left ventricular volume reduces at higher pump flow. Left ventricular diastolic filling improves at lower ventricular volume owing to the nonlinear end-diastolic pressure-volume relationship. LVAD, left ventricular assist device. (**b**) The effect of pump speed on left ventricular (LV) output and blood pressure. From top to bottom: pump differential pressure-flow curves (blue dots, diastole; red dots, systole), pump speed and flow, arterial and LV pressure, and native LV pressure–LV end-diastolic volume relationship. Reduction in pump speed, in the presence of native LV contractile function, increases the LV pressure (third panel) through the Frank-Starling effect (bottom panels), which increases the systolic-diastolic pump differential pressure and flow. This is reflected in the greater pulsatility in LVAD flow. LV end-diastolic volume drops at higher pump speed, reducing native LV ejection and pressure, thereby reducing pulsatility in LVAD flow. The arterial blood pressure increases at higher pump speed owing to an increase in diastolic pressure, but pulse pressure decreases, especially if the left ventricle fails to eject (aortic valve fails to open). (Republished with permission of Elsevier Science and Technology Journals, from Lim et al. [9]; permission conveyed through Copyright Clearance Center, Inc.)

interaction between hemodynamics and arterial remodeling precludes the ability to experimentally ascribe changes in aortic pulse pressure to particular adaptive responses. Chronic changes in blood pressure and flow (i.e., hypertension, heart failure) have been correlated with structural remodeling of the arterial wall both in the young and elderly, involving changes in lumen radius (primarily determined by shear stress), wall thickness, and stiffness of the arterial wall (primarily determined by circumferential wall stress) within days or weeks [11]. Naturally, the vascular adaptation is highly expected after the induction of reduced or nonpulsatile flow physiology in the arterial system based on CFVADs' distinct pressure and flow properties. Unfortunately, based on the difficulty in establishing a causal relationship between changes in a particular mechanical property to a particular stress in in vivo experiments, investors have relied on mathematical modeling to characterize the structural adaptation of the vasculature [12–15]. Such models have confirmed the critical role played by endothelial shear stress and circumferential wall stress, and they yielded the critical insight that a complex architecture can emerge from the identical adaptive rules universally applied to each vessel. Perhaps given the complexity arising from pulse wave propagation and reflection in arterial networks, most models have focused on nonpulsatile hemodynamics, and none have addressed adaptation of arterial stiffness [16–18]. Simultaneous prediction of the observed heterogeneous mechanical properties, pulsatile hemodynamics, and vascular stresses requires a set of adaptive rules be defined without assuming equilibrium a priori. Motivated by the observation that adaptation reduces the initial perturbation in vascular stresses in vivo [19], mathematical models purporting to explain the structural adaptation of vascular networks have universally assumed vascular adaptation to target stress "set points" [12, 20–22]. Cecchini et al. [23], however, questioned whether the set point is a distinct structural entity in biological control and is unnecessary to explain control of physiological systems. Instead, homeostasis can be achieved when a system with competing processes (providing negative feedback) yields an equilibrium "balance point." Based on this balance point approach, Nguyen et al. [11, 24] tested the hypothesis that pulse pressure homeostasis can emerge from physiological adaptation of systemic arteries to local mechanical stresses. In their work, they reported that a 50% reduction in inlet pulsatile flow had no acute impact on arterial compliances compared (Fig. 19.5a, silver bars), because resistances were adjusted to maintain arterial mean pressure. However, acute values of pulse pressures decreased in all arteries (Fig. 19.5b, silver bars), as well as values of endothelial shear stress and circumferential wall stress. Arterial adaptation caused radii and thicknesses to decrease and arterial elastic moduli to increase. Taken together, these adaptations caused all arterial compliances to decrease (Fig. 19.5a, black bars), which, in turn, increased pulse pressures back toward baseline values (Fig. 19.5b, black bars). The most significant restoration in pulse pressure occurred in the ascending and descending aorta (nearly 50% compared with acute values), and less significant restoration occurred in the iliac and femoral arteries. Arterial adaptation also increased shear stress toward baseline values in all vessels.

Unfortunately, despite advances in mathematical and in vivo models, long-term effects of CFVAD physiology on vessel structure are still unknown. Preliminary chronic animal experiments showed reduced wall thickness of the aorta and reduced volume ratio of vascular smooth muscle (VSM) myocytes as well as an increased proportion of VSM cells with low activity and low contractility in CFVAD supported animals [25, 26]. Some studies reported hypertrophy of VSM cells in the renal cortex arteries; however, no changes were observed in similar other studies [27, 28]. Among the limited number of clinical studies assessing vascular adaptation and structure in response to CFVADs, Potapov et al. [29, 30] showed that arterial pulsatility, which was low after the immediate postoperative phase, steadily increased in all patients over the course of CFVAD support. Although the authors attributed the increase of pulsatility to myocardial recovery, it is highly possible that increased stiffness of the

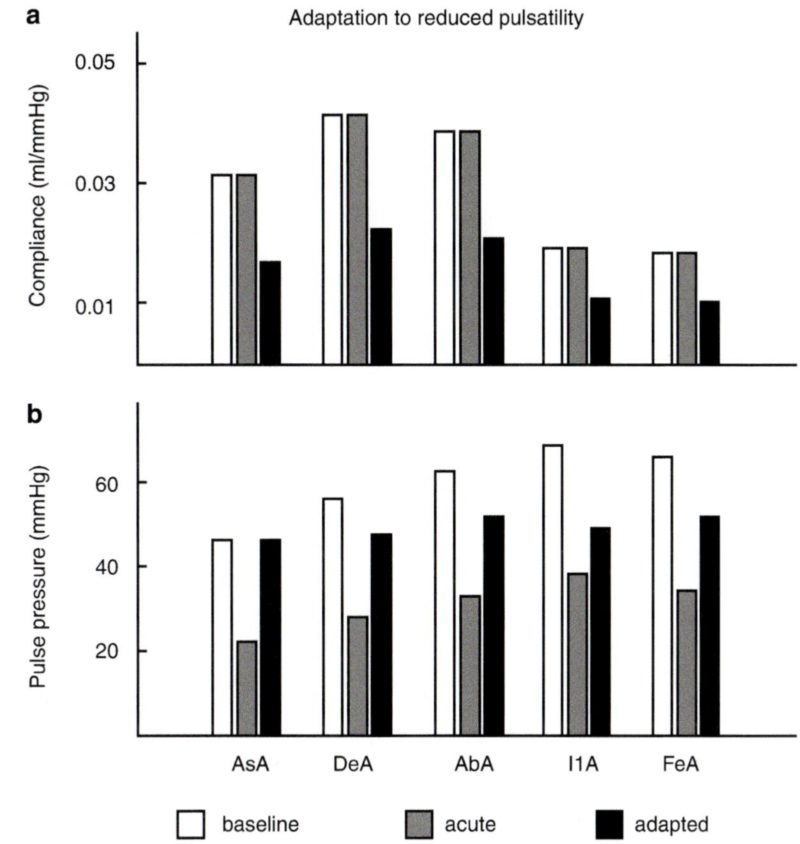

Fig. 19.5 Evidence of pulse pressure homeostasis in vessels along the aortic-femoral pathway in response to changes in input flow. AsA, ascending aorta; DeA, descending aorta; AbA, abdominal aorta; IlA, iliac artery; FeA, femoral artery. (**a**) A 50% decrease in peak and mean input flow had no acute effect on arterial compliance. Adaptation decreased arterial compliance in all vessels. (**b**) Decreased inlet pulsatile flow caused an acute decrease in pulse pressure, but the decrease in arterial compliance with adaptation caused pulse pressure to increase back toward baseline values. (From Nguyen et al. [11])

arterial system can cause an increase in arterial pulsatility as predicted by Nguyen et al. [11, 24].

Effects of CFVAD Flow on Left and Right Ventricular Hemodynamics, Geometry, and Microcirculation

Left ventricular (LV) unloading and restoration of the cardiac output are essential purposes for CFVAD implantation in end-stage heart failure patients. It has been demonstrated that CFVADs successfully reduce the left ventricular end-diastolic/end-systolic pressures and dimensions, myocardial workload, wall tension, and myocardial oxygen consumption under acute and chronic support [31–33]. As a result of mid- or long-term CFVAD support, reduction in left ventricular wall thickness and myocyte size along with increased collagen content and gene expression is also observed in patients with reverse remodeling [34, 35]. Considering that CFVAD preload is closely dependent to right heart output, pulmonary vascular resistance, and transpulmonary flow, excessive unloading of the left ventricle or poor transpulmonary blood flow may end up with ventricular suction and altered right ventricular geometry. While septal suction can create severe arrhythmias, it can also cause severe right ventricular (RV) dysfunction [36]. However, optimal CFVAD flows will result in decreased right ventricular afterload and improvement of right ventricular geometry and function [37, 38]. The increase in left ventricular output provided by CFVAD increases venous return and right ventricular preload which increases right ventricular output in adequate right ventricular contractile reserve. This effect will reduce pulmonary hypertension and pulmonary vascular resistance in chronic patient support [39, 40]. On the other hand, critical or poor right ventricular contractile reserve may cause increase wall tension and tricuspid regurgitation which may lead to right heart failure [41].

The terms coronary blood flow and regional myocardial perfusion are often used as synonyms, although they are not quite the same. Regional myocardial perfusion depends on coronary artery blood flow, using conductance vessels as epicardial coronary trunks and resistance vessels as coronary perforants, arterioles, and capillaries. Blood volume is distributed all through myocardial wall but is mostly present at subendocardial area, filled by approximately 80% of myocardial blood volume [8]. Eventually, the coronary blood drains through three types of channels: the subepicardial veins, the Thebesian veins, and the arteriosinusoidal vessels [42]. Although, the regional myocardial perfusion partially depends on the epicardial blood flow, other factors such as epicardial coronary artery stenosis, perfusion pressure, intraventricular pressures, flow mode (pulsatile vs continuous), myocardial geometry (wall shape and thickness), and intramural pressures (fractional shortening) may affect regional myocardial perfusion properties [8].

The major determinants of myocardial oxygen demand are wall tension, contractility, and heart rate [43]. The complex interaction between myocardial oxygen demand and coronary flow in healthy and failing hearts is regulated by changes in coronary vessel diameter and consequently the vascular resistance of the coronary arteries. The epicardial arteries are known to contribute to only 5% of the total coronary vascular resistance, while intramyocardial coronary arterioles are responsible for 95% of the total coronary vascular resistance [44]. Since myocardial perfusion occurs mainly during diastole, any change of the myocardial contractility pattern may affect epicardial and intramyocardial coronary flows [45]. In a study with an ex vivo beating heart loaded with an axial flow pump [46], significant decrease in dP/dtmax, myocardial oxygen consumption, left ventricular end-diastolic dimension, and pressure at higher speeds in healthy and failing hearts particularly observed at extensive left ventricular unloading which has been reported in in vivo healthy and/or heart failure animal models [31, 32, 47]. The decrease in dP/dtmax may be attributed to delayed LV relaxation and decreased time-tension index (TTI) secondary to extensive LV unloading. Unchanged tissue CO2 production (TCO2), an indirect indicator of myocardial cell metabolism, despite increasing epicardial coronary flows at increasing pump supports may be attributed to decreased coronary vascular resistance and TTI resulting in intramyocardial arteriovenous shunting along with well protected myocardial cell metabolism [42, 46, 48–50]. In induced heart failure scenarios created by major coronary artery ligation, partial unloading of the LV with lower pump speeds resulted in increased contractility noted by increased dP/dtmax, but extensive unloading resulted in delayed ventricular relaxation and decreased contractility as observed in healthy hearts. Although TCO2 was protected at increasing pump supports, significant increase in left coronary artery flow and decrease in arteriovenous oxygen difference suggested more pronounced intramyocardial arteriovenous shunting compared to healthy hearts possibly due to less stiff myocardium and lower intramyocardial pressure. Limited number of studies performed with microspheres or positron emission tomography techniques reported unchanged or decreased regional myocardial perfusion under CFVAD support in either healthy or diseased myocardiums, but they were not able to outline the mechanism of this phenomenon [31, 32, 51, 52]. This particular ex vivo beating heart study showed that higher CFVAD flows result in actively changing myocardial mechanical properties, and regional myocardial perfusion is closely dependent to several intrinsic and extrinsic factors such as degree of myocardial stiffness, coronary vascular resistance, TTI, and opening of the existing myocardial arteriovenous shunts. Knowing that two conditions, arteriovenous shunting and low oxygen utilization, may exist simultaneously as result of venous hyperoxia in the tissues, thus, there is a need to differentiate one condition from the other. It is well-known that under normal tissue perfusion conditions, one oxygen molecule is utilized as one carbon dioxide molecule is produced; however, low oxygen utilization is associated with low carbon dioxide production. Therefore, myocardial arteriovenous shunts may be associated with unchanged or increased carbon dioxide production. According to this study, it is likely that immediate occurrence of venous hyperoxia with unchanged myocardial cell metabolism despite increasing or unchanged coronary blood flows is strong evidence of intramyocardial arteriovenous shunt opening under continuous-flow physiology [46]. The effects of the existing or de novo arteriovenous shunt development on left ventricular recovery and reverse remodeling need to be investigated further studies.

It is also noteworthy that right coronary artery flow significantly increased at increasing pump supports along with a slight decrease or unchanged regional myocardial perfusion measured by microspheres similar to the left ventricle although the right ventricle is not unloaded by a CFVAD. In this two-chamber ex vivo mock loop study, the right ventricular wall function and tension was not affected by increased systemic venous return or pulmonary vascular resistance since right ventricular stroke volume was only limited with coronary sinus return. Therefore, one might speculate that increasing mean arterial pressure as a result of increasing reduced or nonpulsatile flow is solely responsible for this phenomenon since metabolic status and mechanical wall properties of the right ventricle have not changed compared to baseline conditions at increasing CFVAD supports. Considering 40% of the CFVAD implanted patients develop postoperative right heart failure, the possible occurrence of the intramyocardial arteriovenous shunts under CFVAD support and its mid- or long-term effects may be particularly important mostly in patients with limited preoperative right ventricular functional reserve, and this complex interaction needs further assessment [46].

Because of their small, compact design, CFVADs are optimal for minimally invasive and reoperative cardiac procedures because they can be implanted with an ascending, descending, abdominal aortic, or subclavian artery anastomosis through a small incision [53–58]. Although ascending aortic anastomosis is the primary choice for most of the clinicians, other alternative anastomosis locations preserve the sternum for subsequent cardiac transplantation and/or other cardiac procedures that require a median sternotomy. So far, however, controversial results from some animal and in vitro studies have precluded wide use of the alternative method, even for primary CFVAD implantations [59–61]. Kar and colleagues [59] reported that CFVADs implanted with an outflow graft anastomosis to the descending aorta have potential drawbacks such as aortic valve fusion and aortic root flow stagnation that may alter myocardial perfusion. Similarly, in their experimental study, Litwak and colleagues [61] reported that the mean aortic arch blood flow decreased when the CFVAD outflow graft anastomosis site was the descending aorta versus the ascending aorta. They also reported that, although mean aortic root and carotid-artery flow are independent of the CFVAD outflow graft location, competitive flow between the native heart and the CFVAD may result in turbulent mixing or stagnation in the aortic arch if the CFVAD outflow graft anastomosis is located in the descending aorta. Therefore, they suggested that anastomosis of the outflow graft to the ascending aorta may be optimal for a CFVAD. However, none of these investigators addressed the effect of stagnation on regional myocardial perfusion. Because of these controversies, the alternative anastomosis is mostly performed in patients who have undergone a previous sternotomy, coronary artery bypass patients with vein grafts, infected mediastinum, and/or patients with atherosclerotic disease of the ascending aorta [59, 62]. Only one in vivo study compared the effect of ascending vs descending outflow graft location on myocardial perfusion and showed no myocardial perfusion difference based on outflow graft location [63].

Effect of CFVAD Flow on Aortic Valve Structure and Function

The aortic valve regurgitation and fusion are common complications reported with an incidence of 11% at 6 months and 52% at 18 months after CFVAD implantations [64–69]. It is well-known that CFVADs operating at higher speeds and/or significantly reduced left ventricular contractility may cause extensive left ventricular unloading through the device outflow resulting in reduced wall stresses, myocardial contractility, and cardiac output [54, 70]. The resulting reduction of forward ejection across the aortic valve and increase of aortic diastolic pressure due to increased pump outflow into the aorta may impair aortic valve opening and cause aortic flow stagnation, which may in turn cause thrombosis, fusion, or regurgitation of the aortic valve during long-term support [71]. Based on limited number of preclinical and clinical studies, the mechanism of de novo aortic regurgitation and/or leaflet fusion is likely to be multifactorial. In a mock loop study, our group reported that even under healthy ventricular contraction conditions, the aortic valve duty cycle which is normally 30% dropped to 15% at increasing pump flows, with significant reduction in aortic leaflet flows [72]. In a study with an ex vivo beating heart loaded with a CFVAD, we also demonstrated that the pressure load on the aortic valve was almost 2.5 times higher than the baseline when the left ventricle was fully unloaded with a CFVAD in healthy heart experiments, whereas the pressure load was 7 times higher under failing heart conditions [73]. The echocardiographic data in the same study demonstrated that prior to the CFVAD support, failing ex vivo hearts had significantly smaller aortic root diameters compared to the healthy hearts which may be explained by the decreased aortic root pressure and wall tension secondary to the left ventricular failure. The initiation of CFVAD support resulted in pressure overload above the aortic valve which caused flattening of the aortic leaflets, increase in aortic root diameter during diastole, and retrograde flow through aortic leaflets due to altered leaflet coaptation in both healthy and failing hearts. The aortic root diameter decreased below baseline levels when the left ventricle is over unloaded at maximum CFVAD speeds due to collapsed left ventricular outflow tract and increased retrograde aortic flow [73]. Under physiological conditions, the radial length of the aortic leaflets does not change significantly during systolic aortic flow; however, when the leaflets close and coapt under increasing diastolic pressure, the radial length increases during diastole [74], and altered hemodynamic changes above and below the aortic valve may result in aortic root dilation, leaflet deformation, and aortic regurgitation immediately after continuous-flow VAD implantation.

Aortic valve commissural fusion and resultant insufficiency have begun to be considered complications of the prolonged use of the pulsatile or nonpulsatile left ventricular assist devices which may cause poor patient prognosis [75–77]. The aortic valve is a functional assembly composed of the three cusps, corresponding sinuses, and the sinotubular junction. It is characterized not only by morphological features but also its functional properties, which together create an environment that is optimal for distribution of diastolic pressure load and assures proper and timely valve opening and closure [78]. Under physiological conditions, a constant tissue renewal in aortic valve leaflets is demonstrated, while altered mechanical forces (i.e., hypertension) result in changes in their structural and biological properties [74, 79, 80]. There is a general tendency to speculate that similar changes occur in aortic leaflet structure in CFVAD implanted patients [65, 81–83].

Effect of CFVAD Flow on Hematological Parameters

The intravascular or intraventricular location of CFVADs leads to unavoidable continuous interaction with the hemostatic system, and bleeding and thrombosis continue to pose major challenges as source of morbidity, mortality, and device failure. It has been reported that CFVADs alter blood rheology and are recognized to cause increased shear stresses [84]. Increased shear levels occur in CFVAD patients as a result of the high-velocity blood flow generated by axial or centrifugal pumps [85]. Excessive shear stress is a major contributor to alterations in normal hemostatic mechanisms. The precise effect of altered mechanical forces is complex, and while these changes can contribute to thrombogenicity through platelet activation, they can also impair hemostasis through depletion of high-molecular-weight VWF multimers or by inducing platelet receptor shedding [86–88]. High shear forces have well-documented effects on the structure and function of VWF which in turn exerts effects on platelets by binding to platelet GPIbα of the GPIb-IX-V complex [89]. VWF circulates as large multimers that are cleaved to smaller multimers by ADAMTS-13 (a disintegrin and metalloproteinase with thrombospondin motifs-13). In high shear settings, these forces can alter the conformation of VWF's structural architecture, unfurl the molecules, and cause VWF deficiency or VWF syndrome in CFVAD patients. Receptor shedding induced by high shear stress causes a rapid and irreversible disturbance of normal platelet function which is also associated with an elevated clinical bleeding risk in CFVAD patients [90]. VWF levels are typically elevated in patients receiving CFVADs, reflecting endothelial activity and/or chronic inflammatory conditions [91, 92].

CFVADs may also be associated with changes in intracellular reactive oxygen species (ROS) in platelets and mitochondrial damage that could potentially contribute to increased nonsurgical bleeding observed in these patients [93, 94]. Numerous cells, particularly platelets, release small, submicron plasma membrane fragments which are known as microparticles. Those fragments perform a variety of functions such as stimulation of angiogenesis, expression of inflammatory mediators, and catalyzation of the coagulation cascade. In patients treated with CFVAD, there is a persistent rise in microparticles and consequent increase in cellular adhesion molecule expression and inflammatory cytokines potentially contributing to a perpetual cycle of vascular injury and altered hemostasis [85, 95, 96].

High shear stress, flow rates, and interaction with artificial surfaces created by rapidly spinning CFVADs can also cause damage to the circulating red blood cells (hemolysis) which is more pronounced during the first couple of days after CFVAD implantations. Despite laboratory evidence of high hemolysis levels earlier implant days, most patients are not clinically affected, and hemolysis rates undergo a relative decrease with ongoing CFVAD support. Elevated levels of prothrombin fragment 1.2 and D-dimer are reported to be the most common postoperatively which may persist up to 6–12 months after CFVAD implantation [97].

Effect of CFVAD Flow on Renal, Hepatic, and Pulmonary System

This was one of the most popular research topics in early years of CFVAD implantations. Several authors studied the effects of the continuous-flow physiology on kidneys in healthy vs failing animal models and compared nonpulsatile vs pulsatile blood flow physiology. Most of the preliminary studies reported detrimental effects to kidneys such as expansion of the proximal tubes, retention of blood cells and expansion of blood capillaries within the glomerulus, smooth muscle hypertrophy and periartritis in renal vessels, and upregulation in renin-angiotensin system [98–101]; in contrast recently published experimental and clinical studies showed that kidney microcirculation, function, and structure are not negatively affected after short- or long-term CFVAD support [26, 30–32, 102]. Data from INTERMACS registry reported an approximately 50% reduction in the rate of renal dysfunction in patients implanted with CFVADs compared to their pulsatile flow competitors [103]. Unfortunately, it is extremely difficult to assess pure chronic renal histopathological effects of the CFVADs in patients due to ethical concerns and preoperative risk factors such as age, hypertension, existing renal dysfunction, diabetes, right heart failure, and/or cachexia. Therefore currently most of our knowledge is based on long-term clinical studies monitoring renal functions with glomerular filtration rate, creatinine, and BUN levels [104].

Experimental and clinical studies of hepatic function in CFVAD patient populations have produced similar results to those of renal function. Several investigators reported normal liver function tests and histology after 3–7 months of CFVAD support in ovine and bovine models [26, 105, 106]. Although accurate preoperative liver function evaluation is difficult in heart failure patients, serum total bilirubin, serum glutamic-oxaloacetic transaminase (SGOT), serum glutamic-pyruvic transaminase (SGPT), alanine aminotransferase (ALT), and aspartate aminotransferase (AST) are useful to monitor liver functions. It has been shown that hepatic function is improved or maintained over the duration of continuous-flow LVAD support up to 3 years even in patients with periportal fibrosis [103, 107–110]. Direct comparisons of liver perfusion using continuous- flow and pulsatile flow LVADs have used the same liver function tests to measure long-term outcome but have not shown any difference between the two groups [111, 112]. Interestingly, unlike other end-organs, Yoshioka et al. [104] reported that causes of heart failure, comorbidities, and adverse events do not significantly alter hepatic functions possibly due to its dual blood supply from vena porta and hepatic artery and its significant reserve/regeneration capacity.

Pulmonary system and continuous-flow interaction is possibly the most underinvestigated topic in experimental and clinical CFVAD studies. In their 30-day bovine study, Ootaki et al. [113] demonstrated that CFVAD implanted as a right ventricular assist device (RVAD) caused wall thickening in the medium-sized intrapulmonary arteries with abundant mononuclear cells infiltrating the periarterial areas and endothelium. The artery structure was relatively conserved, with extensive hyperplasia of the smooth muscle layer of the vascular wall media and few inflammatory cells present. Angiotensin (Ang) II type 1 receptor (AT1R) was observed in the endothelial and inflammatory cells that infiltrated the pulmonary periarterial interstitial areas, and the serum ACE enzymatic activity significantly decreased after implantation of all animals at the study termination. The morphologic finding for the size of PAs demonstrating inflammatory responses parallels findings in primary pulmonary hypertension (PPH) patients, with decreased and increased biomechanical stimuli to the endothelium being the common thread. In contrast, in a 14-day caprine study, Sakaki et al. [114] reported no significant change was observed in either mean pulmonary arterial pressure or pulmonary vascular resistance index during the pulsatile pumping compared with that on the 14th day of nonpulsatile pumping. Blood gas data, extravascular lung water content, and serum level of angiotensin-converting enzyme were maintained within normal ranges, but histopathology was not performed. In Ootaki et al.'s abovementioned chronic (22–95 day) bovine study, similar changes were observed in the kidneys, but there was no change in the lung histology [113].

In patients implanted with RVAD and LVAD, it is very difficult to investigate continuous-flow and pulmonary structural interaction into details due to ethical concerns, concomitant right heart failure, pulmonary vascular disease, existing renal dysfunction, renal dysfunction, and other factors. Based on numerous clinical reports, CFVAD implanted as an LVAD decompresses the pulmonary circulation, reduces pulmonary artery pressures even in pulmonary hypertensive patients, and positively affects the patient's long-term outcome [115–118]. Single CFVAD support as an RVAD is extremely rare, and it is mostly combined with an oxygenator, as in ECMO form. RVAD is mostly used with an LVAD as a biventricular (BiVAD) support. The flow physiology and right-left circulation balance is very complex for this mechanical support technique and is not the topic of our chapter.

Effect of CFVAD Flow on Systemic Inflammatory Response and Endocrine System

The effects of low pulsatility on inflammatory mediators are still not completely understood in CFVAD recipients. It is well-known that inflammatory mediators contribute to the development and progression of heart failure [119, 120]. These include cytokines [Interleukin 6 (IL-6), Interleukin 1 beta (IL-1β), and tumor necrosis factor alpha (TNF-α)], which are cell-signaling protein molecules that regulate inflammatory responses, and chemokines [Interleukin 8 (IL-8), monocyte chemoattractant protein-1 (MCP-1), macrophage inflammatory protein-1β

(MIP-1β), interferon γ-induced protein-10 (IP-10), granulocyte-macrophage colony-stimulating factor (GM-CSF), and macrophage-derived chemokine (MDC)], which direct migration of leukocytes to the inflammation sites. From a clinical perspective, increased levels of inflammatory mediators are associated with deterioration of New York Heart Association (NYHA) functional classification [121–124].

Previous studies have shown that blood exposure to an artificial surface results in systemic inflammation, increasing IL-8, MCP-1, MIP-1β, IP-10, and GM-CSF, but not the traditional proinflammatory cytokines (TNF-α, IL-1β, IL-6) [125–128]. Compared with pulsatile devices, patients with CF-LVADs have been shown to have elevated levels of renin-angiotensin-aldosterone system (RAAS) neurohormones [129] which are known to increase the levels of inflammatory markers in the setting of heart failure, including TNF-α, IL-1β, IL-6, IL-8, and MCP-1, as well as the marker of systemic inflammation C-reactive protein (CRP) [130–135]. However, the augmented levels of inflammatory factors may also be related to de novo low pulsatile flow physiology, which potentially increases neurohormones. Several researchers already demonstrated that neurohormones have been implicated in the pathogenesis of inflammatory pathway activation in heart failure. In animal models, low pulsatile flow upregulates the renin-angiotensin-aldosterone system (RAAS) [26]. Furthermore, in another experimental models using comparing low pulsatile and pulsatile flow conditions, increased sympathetic renal nerve activity and systemic norepinephrine levels were detected in CFVAD supported animals [136, 137]. These findings suggest that in response to continuous flow, release of norepinephrine and epinephrine from the sympathetic neurons and adrenal medulla is increased, which may then activate the RAAS, thereby inducing an increase in inflammatory markers in patients with heart failure [138]. Therefore, elevated levels of neurohormones may also induce an increase in inflammation in CFVAD implanted recipients. Currently, inflammatory markers are used to monitor the prognosis of the CFVAD patient in terms of cardiac recovery and/or end-organ functions [139].

Small retrospective studies have reported improvements in hemoglobin A1c (A1C), fasting blood glucose level, circulating cortisol, and catecholamine levels after CFVAD implantation along with a reduction in insulin use [140–143]. Some other studies report decrease in brain natriuretic peptide (BNP) and endothelin-1 (ET-1) concentration after CFVAD implantation which is considered to be indicative of the recovery of native ventricular function [144–146]. No major down- or upregulations in adrenocorticotropic hormone (ACTH), thyroid-stimulating hormone (TSH), growth hormone (GH), follicle-stimulating hormone (FSH), and luteinizing hormone (LH) were observed in CFVAD patients in response to hypothalamic releasing hormone injection compared to healthy controls. The peripheral hormone levels (total T4, total T3, free T4, free T3, testosterone) were within normal ranges as well [147].

Effect of CFVAD Flow on Cerebrovascular System

Cerebral autoregulation which involves several vascular and neurogenic mechanisms regulates cerebral flow relatively constant despite constantly changing wide range of arterial blood pressures [148]. The vascular component is mainly characterized by constriction and dilation of vascular smooth muscle cells in response to changes in intraluminal pressures [149], as well as release of vasoactive substances (e.g., nitric oxide) from endothelial cells that regulates changes in arterial pressure [148]. Activation of sympathetic nerves may shift the autoregulatory plateau in response to fluctuations in arterial pressure [148]. Since all of these mechanisms are sensitive to changes in arterial pressure and pulsatility, it is extremely important to understand how this regulatory mechanism is responding to reduced pulse or nonpulsatile flow conditions in patients implanted with CFVADs.

Impaired cerebral autoregulation is a marker for adverse neurologic outcomes [150–152] which may lead to uncontrolled hypertension [5, 10] and cerebral hyperemia [153]. It was reported that nonpulsatile flow leads to endothelial dysfunction and reduced production of nitric oxide [154, 155] which is an important modulator of dynamic cerebral autoregulation [156]. It also has been shown that reduced pulsatility causes high levels of sympathetic activity [157] which attenuates increases in cerebral flow that would otherwise occur after abrupt increases in central arterial pressure in CFVAD patients, and it has also been suggested that increased sympathetic activity may be protective against cerebrovascular injury during acute elevations in mean arterial pressures [158, 159].

There are several preclinical and clinical studies in the literature assessing cerebral blood flow in CFVAD implanted subjects with contradictory results. In their acute animal experiments under healthy and heart failure conditions, Tuzun and Eya investigated cerebral microcirculation under pulsatile and nonpulsatile conditions with microsphere studies and couldn't find a difference between the study groups [31, 32]. Similar preclinical data were published by other investigators assessing CFVAD flow and cerebral perfusion interaction in short- and long-term studies [160–162]. In contrast, some investigators claimed that cerebral blood flow is better under pulsatile flow conditions [163], but interestingly in another study, the same group found no difference in perfusion under pulsatile and nonpulsatile flow conditions [164].

Recent clinical studies reported preserved cerebral blood flow autoregulation [165, 166] and neurocognitive functions [167] in patients implanted with CFVADs after long-term support. In a human autopsy study, no difference was found as well in cerebral arteriolar histology between patients supported with pulsatile and CF-LVADs [30]. Despite favorable results in preclinical and clinical studies, it is well-known that patients implanted with CFVADs experience 10–15% stroke rate in the first year which declines to 5–7% in the second year [168]. It was shown that nearly all CFVAD patients had at least one cerebral microbleed lesion in MRI controls and has been speculated that they are caused by fractures of fragile small vessels in the brain [169]. Tabit et al. [170] reported that CFVADs lead to altered angiogenesis and are associated with higher nonsurgical bleeding via an elevated level of angiopoietin-2. The exact mechanisms of abnormal neoangiogenesis along with increased vascular fragility in the brain of CFVAD implanted patients are still not well understood, and additional investigations are warranted to reduce the incidence of stroke.

Effect of CFVAD Flow on Gastrointestinal System

Gastrointestinal (GI) bleeding occurs in 15–30% of patients implanted with a CFVAD and is 10 times more frequent in patients supported with CFVADs than in patients with pulsatile pumps [171, 172]. This bleeding appears to be directly associated with reduced pulsatile blood flow, as similar bleeding is not seen in patients with mechanical valves and comparable levels of anticoagulation [173]. Unfortunately, the mechanisms underlying GI bleeding are still not well outlined. Several explanations for the association between GI bleeding and CFVAD interaction have been proposed [174] and include (1) acquired von Willebrand's disease due to increased shear stress in CFVAD-implanted patients [175–177], (2) increased intraluminal pressure coupled with muscular contraction which may result in dilated mucosal veins and the development of arteriovenous communications [171], (3) neurovascular etiology in which increased sympathetic tone results in smooth muscle relaxation and a subsequent propensity for angiodysplasia or hypoperfusion of the intestines [172], and (4) impaired platelet aggregation in response to shear stress and depleted VWF [178]. Whether the cause of GI bleeding is related to one or a combination of these factors, GI perfusion assessment is crucial to understand the complex interaction between GI bleeding and CFVAD physiology and to improve device design to prevent or reduce this serious complication.

There are very few preclinical studies to assess gastrointestinal perfusion under continuous-flow physiology. Tuzun et al. [179] demonstrated that gastrointestinal microperfusion assessed with PETCT (using Copper 62 radioisotope) and microsphere techniques is not compromised under continuous-flow or induced pulse conditions compared to baseline off-pump conditions in a healthy sheep model. The second important observation of this study was a significant increase in selectively cannulated cranial mesenteric vein partial oxygen pressures in both continuous and induced pulse modes immediately after the initiation of CFVAD support. Considering that they used a healthy heart model, under continuous or induced pulse supports, the total cardiac output and blood flow to the intestines theoretically should not change after the pump support which was proven by PETCT and microsphere studies. This study data revealed a significant increase in total carbon dioxide production at the intestinal region in both continuous and induced pulse support modes as calculated by the formula proposed by Rozin et al. [180]. This is a strong evidence of immediate intestinal AVS opening under continuous and/or induced pulse flow physiology without intestinal tissue perfusion defect as proved by microsphere and PET/CT measurement. Although one may speculate that the immediate occurrence of AVS at the gastrointestinal level may be linked to the development of intestinal arteriovenous malformations and intestinal bleeding complication, the authors refrain from making such predictions without chronic in vivo studies.

Conclusion

CFVAD treatments have become an established modality in the management of end-stage heart failure. Consequently, continuous-flow-related complications have emerged in time. Despite outstanding success of the long-term CFVAD therapy, gastrointestinal bleeding, hemorrhagic/ischemic stroke, and pump thrombosis remain as the major source of morbidity and mortality in this fragile population. As the mechanical circulatory support with CFVADs progresses, advances in the understanding of the physiopathological mechanisms underlying bleeding/thrombosis risk, patient risk stratification, potential use of novel anticoagulants, and induced pulse mode will play a major role to reduce adverse events and mortality.

References

1. Benjamin EJ, Blaha MJ, Chiuve SE, Cushman M, Das SR, Deo R, et al. Heart Disease and Stroke Statistics-2017 update: a report from the American Heart Association. Circulation. 2017;135(10):e146–603.
2. Slaughter MS, Pagani FD, McGee EC, Birks EJ, Cotts WG, Gregoric I, et al. HeartWare ventricular assist system for bridge to transplant: combined results of the bridge to transplant and continued access protocol trial. J Heart Lung Transplant. 2013;32:675–83.
3. Park SJ, Milano CA, Tatooles AJ, Rogers JG, Adamson RM, Steidley DE, HeartMate II Clinical Investigators, et al. Outcomes in advanced heart failure patients with left ventricular assist devices for destination therapy. Circ Heart Fail. 2012;5(2):241–8.
4. Schmitto JD, Pya Y, Zimpfer D, Krabatsch T, Garbade J, Rao V, et al. Long-term evaluation of a fully magnetically levitated circulatory support device for advanced heart failure-two-year results from the HeartMate 3CE Mark Study. Eur J Heart Fail. 2018;21:90.
5. Fuster V, Harrington RA, Narula J, Eapen ZJ. Hurst's the heart. 14th ed. New York: McGraw-Hill Education; 2017.
6. Soucy KG, Koenig SC, Giridharan GA, Sobieski MA, Slaughter MS. Rotary pumps and diminished pulsatility: do we need a pulse? ASAIO J. 2013;59(4):355–66.
7. Stewart GC, Givertz MM. Mechanical circulatory support for advanced heart failure: patients and technology in evolution. Circulation. 2012;125:1304–15.
8. Gewirtz H. Regulating myocardial blood flow in health and disease. Curr Cardiol Rep. 2009;11:117–24.
9. Lim HS, Howell N, Ranasinghe A. The physiology of continuous-flow left ventricular assist devices. J Card Fail. 2017;23(2):169–80.
10. Myers TJ, Bolmers M, Gregoric ID, Kar B, Frazier OH. Assessment of arterial blood pressure during support with an axial flow left ventricular assist device. J Heart Lung Transplant. 2009;28(5):423–7.
11. Nguyen PH, Tuzun E, Quick CM. Aortic pulse pressure homeostasis emerges from physiological adaptation of systemic arteries to local mechanical stresses. Am J Physiol Regul Integr Comp Physiol. 2016;311(3):R522–31.
12. Fridez P, Rachev A, Meister JJ, Hayashi K, Stergiopulos N. Model of geometrical and smooth muscle tone adaptation of carotid artery subject to step change in pressure. Am J Physiol Heart Circ Physiol. 2001;280:H2752–60.
13. Tsamis A, Stergiopulos N, Rachev A. A structure-based model of arterial remodeling in response

to sustained hypertension. J Biomech Eng. 2009;131:101004.
14. Pries AR, Secomb TW, Gaehtgens P. Structural adaptation and stability of microvascular networks: theory and simulations. Am J Physiol Heart Circ Physiol. 1998;275:H349–60.
15. Quick CM, Young WL, Leonard EF, Joshi S, Gao E, Hashimoto T. Model of structural and functional adaptation of small conductance vessels to arterial hypotension. Am J Physiol Heart Circ Physiol. 2000;279:H1645–53.
16. Noodergraaf A. Hemodynamics. In: Schwan HP, editor. Biological engineering. New York: McGraw-Hill; 1969. p. 556.
17. Reymond P, Bohraus Y, Perren F, Lazeyras F, Stergiopulos N. Validation of a patient-specific one-dimensional model of the systemic arterial tree. Am J Physiol Heart Circ Physiol. 2011;301: H1173–82.
18. Westerhof N, Bosman F, De Vries CJ, Noordergraaf A. Analog studies of the human systemic arterial tree. J Biomech. 1969;2:121–43.
19. Humphrey JD. Vascular adaptation and mechanical homeostasis at tissue, cellular, and sub-cellular levels. Cell Biochem Biophys. 2008;50:53–78.
20. Karsaj I, Humphrey JD. A multilayered wall model of arterial growth and remodeling. Mech Mater. 2012;44:110–9.
21. Manini S, Passera K, Huberts W, Botti L, Antiga L, Remuzzi A. Computational model for simulation of vascular adaptation following vascular access surgery in haemodialysis patients. Comput Methods Biomech Biomed Engin. 2014;17:1358–67.
22. Quick CM, Berger DS, Hettrick DA, Noordergraaf A. True arterial system compliance estimated from apparent arterial compliance. Ann Biomed Eng. 2000;28:291–301.
23. Cecchini AB, Melbin J, Noordergraaf A. Set-point: is it a distinct structural entity in biological control? J Theor Biol. 1981;93:387–94.
24. Nguyen PH, Coquis-Knezek SF, Mohiuddin MW, Tuzun E, Quick CM. The complex distribution of arterial system mechanical properties, pulsatile hemodynamics, and vascular stresses emerges from three simple adaptive rules. Am J Physiol Heart Circ Physiol. 2015;308(5):H407–15.
25. Nishimura T, Tatsumi E, Takaichi S, et al. Morphologic changes of the aortic wall due to reduced systemic pulse pressure in prolonged non pulsatile left heart bypass. ASAIO J. 1997;43:M691–5.
26. Saito S, Westaby S, Piggot D, et al. End-organ function during chronic nonpulsatile circulation. Ann Thorac Surg. 2002;74:1080–5.
27. Kihara S, Litwak KN, Nichols L, et al. Smooth muscle cell hypertrophy of renal cortex arteries with chronic continuous flow left ventricular assist. Ann Thorac Surg. 2003;75:17.
28. Nojiri C, Kijima T, Maekawa J, et al. Terumo implantable left ventricular assist system: results of long-term animal study. ASAIO J. 2000;46:117–22.
29. Potapov EV, Loebe M, Hennig E, Nasseri BA, Sinawski H, Kopitz M, et al. Pulsatile flow in patients with a novel non-pulsatile implantable LVAD. In: Swain JA, Utley JR, editors. 20th San Diego cardiothoracic surgery symposium: pathophysiology & techniques of cardiopulmonary bypass. San Diego; 2000. p. 61.
30. Potapov EV, Dranishnikov N, Morawietz L, Stepanenko A, Rezai S, Blechschmidt C, Lehmkuhl HB, Weng Y, Pasic M, Hübler M, Hetzer R, Krabatsch T. Arterial wall histology in chronic pulsatile-flow and continuous-flow device circulatory support. J Heart Lung Transplant. 2012;31(11):1171–6.
31. Tuzun E, Eya K, Chee HK, et al. Myocardial hemodynamics, physiology, and perfusion with an axial flow left ventricular assist device in the calf. ASAIO J. 2004;50:47–53.
32. Eya K, Tuzun E, Conger JL, et al. Effect of pump-flow mode of novel left ventricular assist device on end-organ perfusion in dogs with Doxorubicin-induced heart failure. ASAIO J. 2005;51:41–9.
33. McCarthy PM, Nakatani S, Vargo R, Kottke-Marchant K, Harasaki H, James KB, Savage RM, Thomas JD. Structural and left ventricular histologic changes after implantable LVAD insertion. Ann Thorac Surg. 1995;59(3):609–13.
34. Madigan JD, Barbone A, Choudhri AF, Morales DL, Cai B, Oz MC, Burkhoff D. Time course of reverse remodeling of the left ventricle during support with a left ventricular assist device. J Thorac Cardiovasc Surg. 2001;121(5):902–8.
35. Heerdt PM, Holmes JW, Cai B, Barbone A, Madigan JD, Reiken S, Lee DL, Oz MC, Marks AR, Burkhoff D. Chronic unloading by left ventricular assist device reverses contractile dysfunction and alters gene expression in end-stage heart failure. Circulation. 2000;102(22):2713–9.
36. Hayward CS, Salamonsen R, Keogh AM, Woodard J, Ayre P, Prichard R, Walker R, Kotlyar E, Macdonald PS, Jansz P, Spratt P. Effect of alteration in pump speed on pump output and left ventricular filling with continuous-flow left ventricular assist device. ASAIO J. 2011;57(6):495–500.
37. Tedford RJ, Hassoun PM, Mathai SC, Girgis RE, Russell SD, Thiemann DR, Cingolani OH, Mudd JO, Borlaug BA, Redfield MM, Lederer DJ, Kass DA. Pulmonary capillary wedge pressure augments right ventricular pulsatile loading. Circulation. 2012;125(2):289–97.
38. Kukucka M, Potapov E, Stepanenko A, Weller K, Mladenow A, Kuppe H, Habazettl H. Acute impact of left ventricular unloading by left ventricular assist device on the right ventricle geometry and function: effect of nitric oxide inhalation. J Thorac Cardiovasc Surg. 2011;141(4):1009–14.
39. Mikus E, Stepanenko A, Krabatsch T, Loforte A, Dandel M, Lehmkuhl HB, Hetzer R, Potapov EV. Reversibility of fixed pulmonary hypertension in left ventricular assist device support recipients. Eur J Cardiothorac Surg. 2011;40(4):971–7.

40. Kutty RS, Parameshwar J, Lewis C, Catarino PA, Sudarshan CD, Jenkins DP, Dunning JJ, Tsui SS. Use of centrifugal left ventricular assist device as a bridge to candidacy in severe heart failure with secondary pulmonary hypertension. Eur J Cardiothorac Surg. 2013;43(6):1237–42.
41. Holman WL, Bourge RC, Fan P, Kirklin JK, Pacifico AD, Nanda NC. Influence of longer term left ventricular assist device support on valvular regurgitation. ASAIO J. 1994;40(3):M454–9.
42. Addetia AM, Callaghan JC. Intramyocardial arteriovenous shunting of blood with nonpulsatile perfusion. J Thorac Cardiovasc Surg. 1981;81(2): 219–26.
43. Yang SS, Bentivoglio LG, Maranhao V, Goldberg H. Assessment of ventricular function. In: Yang SS, editor. From cardiac catheterization data to hemodynamic parameters. 3rd ed. Philadelphia: F.A. Davis Company; 1988. p. 207–8.
44. Kelley KO, Feigl EO. Segmental alpha-receptor-mediated vasoconstriction in the canine coronary circulation. Circ Res. 1978;43(6):908–17.
45. Losert U, Glogar D, Mayr H, et al. Regional myocardial blood flow during nonpulsatile left ventricular bypass in calves. Trans Am Soc Artif Intern Organs. 1982;28:86–92.
46. Tuzun E. In-vitro, ex-vivo and in-vivo modeling to assess continuous flow cardiac assist physiology and end-organ interactions. Eindhoven: Technische Universiteit Eindhoven; 2014. https://doi.org/10.6100/IR781404.
47. Saito A, Shiono M, Orime Y, et al. Effects of left ventricular assist device on cardiac function: experimental study of relationship between pump flow and left ventricular diastolic function. Artif Organs. 2001;25:728–32.
48. Griffith LD, Shoor PM, Dilley RB, Bernstein EF. Beneficial effects of nonpulsatile left ventricular bypass on myocardial energy utilization: interaction of oxygen demand and supply. Trans Am Soc Artif Intern Organs. 1978;24:298–304.
49. Laas J, Campbell CD, Takanashi Y, Replogle RL. Oxygen consumption of the left ventricle during transapical left ventricular bypass. J Thorac Cardiovasc Surg. 1980;80(2):280–8.
50. Pennock JL, Pierce WS, Prophet GA, Waldhausen JA. Myocardial oxygen utilization during left heart bypass; effect of varying percentage of bypass flow rate. Arch Surg. 1974;109(5):635–41.
51. Letsou GV, Sdringola S, Gregoric ID, Patel V, Myers TJ, Delgado RM, Frazier OH. Myocardial perfusion as assessed by positron emission tomography during long-term mechanical circulatory support. Congest Heart Fail. 2006;12(2):69–74.
52. Maybaum S, Epstein S, Beniaminovitz A, et al. Partial loading of the left ventricle during mechanical assist device support is associated with improved myocardial function, blood flow and metabolism and increased exercise capacity. J Heart Lung Transplant. 2002;21:446–54.
53. Frazier OH, Gregoric ID, Cohn WE. Initial experience with non-thoracic, extraperitoneal, off-pump insertion of the Jarvik 2000 Heart in patients with previous median sternotomy. J Heart Lung Transplant. 2006;25:499.
54. Gregoric ID, La Francesca S, Myers T, et al. A less invasive approach to axial flow pump insertion. J Heart Lung Transplant. 2008;27:423.
55. Hetzer R, Potapov EV, Weng Y, et al. Implantation of MicroMed DeBakey VAD through left thoracotomy after previous median sternotomy operations. Ann Thorac Surg. 2004;77:347.
56. Pierson RN 3rd, Howser R, Donaldson T, et al. Left ventricular assist device implantation via left thoracotomy: alternative to repeat sternotomy. Ann Thorac Surg. 2002;73:997.
57. Selzman CH, Sheridan BC. Off-pump insertion of continuous flow left ventricular assist devices. J Card Surg. 2007;22:320.
58. Neidlin M, Corsini C, Sonntag SJ, Schulte-Eistrup S, Schmitz-Rode T, Steinseifer U, Pennati G, Kaufmann TAS. Hemodynamic analysis of outflow grafting positions of a ventricular assist device using closed-loop multiscale CFD simulations: preliminary results. J Biomech. 2016;49(13):2718–25.
59. Kar B, Delgado RM 3rd, Frazier OH, et al. The effect of LVAD aortic outflow-graft placement on hemodynamics and flow: implantation technique and computer flow modeling. Tex Heart Inst J. 2005;32:294.
60. Kim HK, Son HS, Fang YH, et al. The effects of pulsatile flow upon renal tissue perfusion during cardiopulmonary bypass: a comparative study of pulsatile and nonpulsatile flow. ASAIO J. 2005;51:30.
61. Litwak KN, Koenig SC, Tsukui H, et al. Effects of left ventricular assist device support and outflow graft location upon aortic blood flow. ASAIO J. 2004;50:432.
62. Ahmed M, Aftab M, Singh S, Mallidi HR, Frazier OH. Left ventricular assist device outflow graft: alternative sites. Ann Cardiothorac Surg. 2014;3(5):541–5.
63. Tuzun E, Narin C, Gregoric ID, Cohn WE, Frazier OH. Ventricular assist device outflow-graft site: effect on myocardial blood flow. J Surg Res. 2011;171(1):71–5.
64. Cowger J, Pagani FD, Haft JW, Romano MA, Aaronson KD, Kolias TJ. The development of aortic insufficiency in LVAD-supported patients. Circ Heart Fail. 2010;3:668–74.
65. Mudd JO, Cuda JD, Halushka M, Soderlund KA, Conte JV, Russell SD. Fusion of aortic valve commissures in patients supported by a continuous axial flow left ventricular assist device. J Heart Lung Transplant. 2008;27(12):1269–74.
66. Aggarwal A, Raghuvir R, Eryazici P, Macaluso G, Sharma P, Blair C, et al. The development of aortic insufficiency in continuous-flow left ventricular assist device-supported patients. Ann Thorac Surg. 2013;95:493–8.
67. Pak SW, Uriel N, Takayama H, Cappleman S, Song R, Colombo PC, et al. Prevalence of de novo aor-

tic insufficiency during long-term support with left ventricular assist devices. J Heart Lung Transplant. 2010;29:1172–6.
68. Patil NP, Sabashnikov A, Mohite PN, Garcia D, Weymann A, Zych B, et al. De novo aortic regurgitation after continuous-flow left ventricular assist device implantation. Ann Thorac Surg. 2014;98:850–7.
69. Rose A, Park S, Bank A, Miller L. Partial aortic valve fusion induced by left ventricular assist device. Ann Thorac Surg. 2000;70:1270–4.
70. Gralla EJ, Fleischman RW, Luthra YK, Stadnicki SW. The dosing schedule dependent toxicities of adriamycin in beagle dogs and rhesus monkeys. Toxicology. 1979;13:263–73.
71. Frazier OH, Tuzun E, Cohn W, Tamez D, Kadipasaoglu KA, Mallia M. Total heart replacement with dual centrifugal ventricular assist devices. ASAIO J. 2005;51:224–9.
72. Tuzun E, Rutten M, Dat M, van de Vosse F, Kadipasaoglu C, de Mol B. Continuous-flow cardiac assistance: effects on aortic valve function in a mock loop. J Surg Res. 2011;171(2):443–7.
73. Tuzun E, Pennings K, van Tuijl S, de Hart J, Stijnen M, van de Vosse F, de Mol B, Rutten M. Assessment of aortic valve pressure overload and leaflet functions in an ex vivo beating heart loaded with a continuous flow cardiac assist device. Eur J Cardiothorac Surg. 2014;45(2):377–83.
74. Thubrikar MJ, Aouad J, Nolan SP. Comparison of the in vivo and in vitro mechanical properties of aortic valve leaflets. J Thorac Cardiovasc Surg. 1986;92(1):29–36.
75. Baradarian S, Dembitsky WP, Jaski B, et al. Left ventricular outflow tract obstruction associated with chronic ventricular assist device support. ASAIO J. 2002;48(6):665–7.
76. Letsou GV, Connelly JH, Delgado RM III, et al. Is native aortic valve commissural fusion in patients with long-term left ventricular assist devices associated with clinically important aortic insufficiency? J Heart Lung Transplant. 2006;25:395.
77. Toda K, Fujita T, Domae K, Shimahara Y, Kobayashi J, Nakatani T. Late aortic insufficiency related to poor prognosis during left ventricular assist device support. Ann Thorac Surg. 2011;92(3):929–34.
78. Robicsek F, Thubrikar MJ. Etiology of degenerative disease of the tri-leaflet aortic valve: a simple explanation for a complex problem. Z Kardiol. 2001;90(Suppl 6):35–8.
79. Schneider PJ, Deck JD. Tissue and cell renewal in the natural aortic valve of rats: an autoradiographic study. Cardiovasc Res. 1981;15(4):181–9.
80. Weston MW, Yoganathan AP. Biosynthetic activity in heart valve leaflets in response to in vitro flow environments. Ann Biomed Eng. 2001;29(9): 752–63.
81. Saito T, Wassilew K, Gorodetski B, Stein J, Falk V, Krabatsch T, Potapov E. Aortic valve pathology in patients supported by continuous-flow left ventricular assist device. Circ J. 2016;80(6):1371–7.
82. May-Newman K, Enriquez-Almaguer L, Posuwattanakul P, Dembitsky W. Biomechanics of the aortic valve in the continuous flow VAD-assisted heart. ASAIO J. 2010;56(4):301–8.
83. Hata H, Fujita T, Ishibashi-Ueda H, Nakatani T, Kobayashi J. Pathological analysis of the aortic valve after long-term left ventricular assist device support. Eur J Cardiothorac Surg. 2014;46(2): 193–7.
84. Fraser KH, Zhang T, Taskin ME, Griffith BP, Wu ZJ. A quantitative comparison of mechanical blood damage parameters in rotary ventricular assist devices: shear stress, exposure time and hemolysis index. J Biomech Eng. 2012;134(08):081002.
85. Casan J, Andrews RK, Gardiner EE, Davis AK. Mechanisms of platelet dysfunction in patients with implantable devices. Semin Thromb Hemost. 2018;44(1):12–9.
86. Nobili M, Sheriff J, Morbiducci U, Redaelli A, Bluestein D. Platelet activation due to hemodynamic shear stresses: damage accumulation model and comparison to in vitro measurements. ASAIO J. 2008;54(01):64–72.
87. Al-Tamimi M, Tan CW, Qiao J, et al. Pathologic shear triggers shedding of vascular receptors: a novel mechanism for downregulation of platelet glycoprotein VI in stenosed coronary vessels. Blood. 2012;119(18):4311–20.
88. Cheng H, Yan R, Li S, et al. Shear-induced interaction of platelets with von Willebrand factor results in glycoprotein Ibalpha shedding. Am J Physiol Heart Circ Physiol. 2009;297(06):H2128–35.
89. Tsai HM, Sussman II, Nagel RL. Shear stress enhances the proteolysis of von Willebrand factor in normal plasma. Blood. 1994;83(08):2171–9.
90. Hu J, Mondal NK, Sorensen EN, et al. Platelet glycoprotein Ibα ectodomain shedding and non-surgical bleeding in heart failure patients supported by continuous-flow left ventricular assist devices. J Heart Lung Transplant. 2014;33(01):71–9.
91. Meyer AL, Malehsa D, Bara C, et al. Acquired von Willebrand syndrome in patients with an axial flow left ventricular assist device. Circ Heart Fail. 2010;3(06):675–81.
92. Meyer AL, Malehsa D, Budde U, Bara C, Haverich A, Strueber M. Acquired von Willebrand syndrome in patients with a centrifugal or axial continuous flow left ventricular assist device. JACC Heart Fail. 2014;2(02):141–5.
93. Mondal NK, Sorensen EN, Pham SM, et al. Systemic inflammatory response syndrome in end-stage heart failure patients following continuous-flow left ventricular assist device implantation: differences in plasma redox status and leukocyte activation. Artif Organs. 2016;40(05):434–43.
94. Mondal NK, Li T, Chen Z, et al. Mechanistic insight of platelet apoptosis leading to non-surgical bleeding among heart failure patients supported by continuous-flow left ventricular assist devices. Mol Cell Biochem. 2017;433:125.

95. Diehl P, Aleker M, Helbing T, et al. Enhanced microparticles in ventricular assist device patients predict platelet, leukocyte and endothelial cell activation. Interact Cardiovasc Thorac Surg. 2010;11(02):133–7.
96. John R, Panch S, Hrabe J, et al. Activation of endothelial and coagulation systems in left ventricular assist device recipients. Ann Thorac Surg. 2009;88(04):1171–9.
97. Slaughter MS, Sobieski MA, Gallagher C, Graham J, Brandise J, Stein R. Fibrinolytic activation during long-term support with the HeartMate II left ventricular assist device. ASAIO J. 2008;54(1):115–9.
98. Sezai A, Shiono M, Orime Y, Nakata K, Sezai Y. Comparison studies of mayor organ microcirculations under pulsatile-and non-pulsatile-assisted circulations. Artif Organs. 1996;20:139–42.
99. Orime Y, Shiono M, Nakata K, Hata M, Sezai Y. The role of pulsatility in end-organ microcirculation after cardiogenic shock. ASAIO J. 1996;42:M724–9.
100. Nakata K, Shiono M, Orime Y, et al. Effect of pulsatile and non-pulsatile assist on heart and kidney microcirculation with cardiogenic shock. Artif Organs. 1996;20:681–4.
101. Ootaki C, Yamashita M, Ootaki Y, Kamohara K, Weber S, Klatte RS, Smith WA, Massiello AL, Emancipator SN, Golding LA, Fukamachi K. Reduced pulsatility induces periarteritis in kidney: role of the local renin-angiotensin system. J Thorac Cardiovasc Surg. 2008;136(1):150–8.
102. Sandner SE, Zimpfer D, Zrunek P, et al. Renal function and outcome after continuous flow left ventricular assist device implantation. Ann Thorac Surg. 2009;87:1072–8.
103. Kirklin JK, Naftel DC, Kormos RL, Stevenson LW, Pagani FD, Miller MA, Baldwin JT, Young JB. Fifth INTERMACS annual report: risk factor analysis from more than 6,000 mechanical circulatory support patients. J Heart Lung Transplant. 2013;32(2):141–56.
104. Yoshioka D, Takayama H, Colombo PC, Yuzefpolskaya M, Garan AR, Topkara VK, Han J, Kurlansky P, Naka Y, Takeda K. Changes in end-organ function in patients with prolonged continuous-flow left ventricular assist device support. Ann Thorac Surg. 2017;103(3):717–24.
105. Tuzun E, Roberts K, Cohn WE, Sargin M, Gemmato CJ, Radovancevic B, Frazier OH. In vivo evaluation of the HeartWare centrifugal ventricular assist device. Tex Heart Inst J. 2007;34(4):406–11.
106. Motomura T, Tuzun E, Yamazaki K, Tatsumi E, Benkowski R, Yamazaki S. Preclinical evaluation of the EVAHEART 2 centrifugal left ventricular assist device in bovines. ASAIO J. 2018;65:845. https://doi.org/10.1097/MAT.0000000000000869. [Epub ahead of print].
107. John R, Kamdar F, Liao K, Colvin-Adams M, Boyle A, Joyce L. Improved survival and decreasing incidence of adverse events with the HeartMate II left ventricular assist device as bridge-to-transplant therapy. Ann Thorac Surg. 2008;86(4):1227–34; discussion 1234–5.
108. Letsou GV, Myers TJ, Gregoric ID, Delgado R, Shah N, Robertson K, Radovancevic B, Frazier OH. Continuous axial-flow left ventricular assist device (Jarvik 2000) maintains kidney and liver perfusion for up to 6 months. Ann Thorac Surg. 2003;76(4):1167–70.
109. Deo SV, Sharma V, Altarabsheh SE, Hasin T, Dillon J, Shah IK, Durham LA 3rd, Stulak JM, Daly RC, Joyce LD, Park SJ. Hepatic and renal function with successful long-term support on a continuous flow left ventricular assist device. Heart Lung Circ. 2014;23(3):229–33.
110. Sargent JE, Dardas TF, Smith JW, Pal JD, Cheng RK, Masri SC, Shively KR, Colyer LM, Mahr C, Mokadam NA. Periportal fibrosis without cirrhosis does not affect outcomes after continuous flow ventricular assist device implantation. J Thorac Cardiovasc Surg. 2016;151(1):230–5.
111. Radovancevic B, Vrtovec B, de Kort E, Radovancevic R, Gregoric ID, Frazier OH. End-organ function in patients on long-term circulatory support with continuous- or pulsatile-flow assist devices. J Heart Lung Transplant. 2007;26(8):815–8.
112. Kamdar F, Boyle A, Liao K, Colvin-adams M, Joyce L, John R. Effects of centrifugal, axial, and pulsatile left ventricular assist device support on end-organ function in heart failure patients. J Heart Lung Transplant. 2009;28(4):352–9.
113. Ootaki C, Yamashita M, Ootaki Y, Saeed D, Horai T, Fumoto H, Massiello AL, Emancipator SN, Golding LA, Fukamachi K. Periarteritis in lung from a continuous-flow right ventricular assist device: role of the local renin-angiotensin system. Ann Thorac Surg. 2013;96(1):148–54.
114. Sakaki M, Taenaka Y, Tatsumi E, Nakatani T, Kinoshita M, Akagi H, Masuzawa T, Matsuo Y, Inoue K, Baba Y, et al. Pulmonary function in a non-pulsatile pulmonary circulation. ASAIO J. 1992;38(3):M366–9.
115. Atluri P, Fairman AS, MacArthur JW, Goldstone AB, Cohen JE, Howard JL, Zalewski CM, Shudo Y, Woo YJ. Continuous flow left ventricular assist device implant significantly improves pulmonary hypertension, right ventricular contractility, and tricuspid valve competence. J Card Surg. 2013;28(6):770–5.
116. Yuan N, Arnaoutakis GJ, George TJ, Allen JG, Ju DG, Schaffer JM, Russell SD, Shah AS, Conte JV. The spectrum of complications following left ventricular assist device placement. J Card Surg. 2012;27(5):630–8.
117. Slaughter MS, Rogers JG, Milano CA, Russell SD, Conte JV, Feldman D, Sun B, Tatooles AJ, Delgado RM 3rd, Long JW, Wozniak TC, Ghumman W, Farrar DJ, Frazier OH, HeartMate II Investigators. Advanced heart failure treated with continuous-flow left ventricular assist device. N Engl J Med. 2009;361(23):2241–51.

118. Zimpfer D, Zrunek P, Roethy W, Czerny M, Schima H, Huber L, Grimm M, Rajek A, Wolner E, Wieselthaler G. Left ventricular assist devices decrease fixed pulmonary hypertension in cardiac transplant candidates. J Thorac Cardiovasc Surg. 2007;133(3):689–95.
119. Anker SD, von Haehling S. Inflammatory mediators in chronic heart failure: an overview. Heart. 2004;90:464–70.
120. Mann DL. Inflammatory mediators and the failing heart: past, present, and the foreseeable future. Circ Res. 2002;91:988–98.
121. Stumpf C, Lehner C, Yilmaz A, Daniel WG, Garlichs CD. Decrease of serum levels of the anti-inflammatory cytokine interleukin-10 in patients with advanced chronic heart failure. Clin Sci (Lond). 2003;105:45–50.
122. Torre-Amione G, Kapadia S, Benedict C, et al. Proinflammatory cytokine levels in patients with depressed left ventricular ejection fraction: a report from the Studies of Left Ventricular Dysfunction (SOLVD). J Am Coll Cardiol. 1996;27:1201–6.
123. Aukrust P, Ueland T, Lien E, et al. Cytokine network in congestive heart failure secondary to ischemic or idiopathic dilated cardiomyopathy. Am J Cardiol. 1999;83:376–82.
124. Aukrust P, Ueland T, Muller F, et al. Elevated circulating levels of C-C chemokines in patients with congestive heart failure. Circulation. 1998;97:1136–43.
125. Lappegard KT, Bergseth G, Riesenfeld J, et al. The artificial surface-induced whole blood inflammatory reaction revealed by increases in a series of chemokines and growth factors is largely complement dependent. J Biomed Mater Res A. 2008;87:129–35.
126. Baufreton C, Intrator L, Jansen PG, et al. Inflammatory response to cardiopulmonary bypass using roller or centrifugal pumps. Ann Thorac Surg. 1999;67:972–7.
127. Paparella D, Yau TM, Young E. Cardiopulmonary bypass induced inflammation: pathophysiology and treatment. An update. Eur J Cardiothorac Surg. 2002;21:232–44.
128. Rossaint J, Berger C, Van Aken H, et al. Cardiopulmonary bypass during cardiac surgery modulates systemic inflammation by affecting different steps of the leukocyte recruitment cascade. PLoS One. 2012;7:e45738.
129. Welp H, Rukosujew A, Tjan TD, et al. Effect of pulsatile and non-pulsatile left ventricular assist devices on the renin-angiotensin system in patients with end-stage heart failure. Thorac Cardiovasc Surg. 2010;58(suppl 2):S185–8.
130. Caprio M, Newfell BG, la Sala A, et al. Functional mineralocorticoid receptors in human vascular endothelial cells regulate intercellular adhesion molecule-1 expression and promote leukocyte adhesion. Circ Res. 2008;102:1359–67.
131. Hashikabe Y, Suzuki K, Jojima T, Uchida K, Hattori Y. Aldosterone impairs vascular endothelial cell function. J Cardiovasc Pharmacol. 2006;47:609–13.
132. Luther JM, Gainer JV, Murphey LJ, et al. Angiotensin II induces interleukin-6 in humans through a mineralocorticoid receptor-dependent mechanism. Hypertension. 2006;48:1050–7.
133. Ozawa Y, Kobori H, Suzaki Y, Navar LG. Sustained renal interstitial macrophage infiltration following chronic angiotensin II infusion. Am J Physiol Renal Physiol. 2007;292:F330–9.
134. Ruiz-Ortega M, Ruperez M, Lorenzo O, et al. Angiotensin II regulates the synthesis of proinflammatory cytokines and chemokines in the kidney. Kidney Int Suppl. 2002;82:S12–22.
135. Sorescu D. Smad3 mediates angiotensin II- and TGF-beta1-induced vascular fibrosis: Smad3 thickens the plot. Circ Res. 2006;98:988–9.
136. Toda K, Tatsumi E, Taenaka Y, Masuzawa T, Takano H. Impact of systemic depulsation on tissue perfusion and sympathetic nerve activity. Ann Thorac Surg. 1996;62:1737–42.
137. Tatsumi E, Toda K, Taenaka Y, et al. Acute phase responses of vasoactive hormones to non pulsatile systemic circulation. ASAIO J. 1995;41:M460–5.
138. Triposkiadis F, Karayannis G, Giamouzis G, et al. The sympathetic nervous system in heart failure physiology, pathophysiology, and clinical implications. J Am Coll Cardiol. 2009;54:1747–62.
139. Kirklin JK, Naftel DC, Kormos RL, et al. The fourth INTERMACS annual report: 4,000 implants and counting. J Heart Lung Transplant. 2012;31:117–26.
140. Koerner M, El-Banayosy A, Eleuteri K, et al. Neurohormonal regulation and improvement in blood glucose control; reduction of insulin requirement in patients with a nonpulsatile ventricular assist device. Heart Surg Forum. 2014;2:E98–102.6.
141. Guglin M, Maguire K, Missimer T, Faber C, Caldeira C. Improvement in blood glucose control in patients with diabetes after implantation of left ventricular assist devices. ASAIO J. 2014;60:290–3.7.
142. Uriel N, Naka Y, Colombo P, et al. Improved diabetic control in advanced heart failure patients treated with left ventricular assist devices. Eur J Heart Fail. 2011;13:195–9.8.
143. Mohamedali B, Yost G, Bhat G. Mechanical circulatory sup-port improves diabetic control in patients with advanced heart failure. Eur J Heart Fail. 2014;16:1120–4.
144. Sato T, Seguchi O, Iwashima Y, Yanase M, Nakajima S, Hieda M, Watanabe T, Sunami H, Murata Y, Hata H, Fujita T, Kobayashi J, Nakatani T. Serum brain natriuretic peptide concentration 60 days after surgery as a predictor of long-term prognosis in patients implanted with a left ventricular assist device. ASAIO J. 2015;61(4):373–8.
145. Hellman Y, Malik AS, Lin H, Shen C, Wang IW, Wozniak TC, Hashmi ZA, Shaukat A, Pickrell J, Caccamo MA, Gradus-Pizlo I, Hadi A. B-type natriuretic peptide-guided therapy and length of hospital stay post left ventricular assist device implantation. ASAIO J. 2015;61(2):156–60.

146. Thompson LO, Skrabal CA, Loebe M, Lafuente JA, Roberts RR, Akgul A, Jones V, Bruckner BA, Thohan V, Noon GP, Youker KA. Plasma neurohormone levels correlate with left ventricular functional and morphological improvement in LVAD patients. J Surg Res. 2005;123(1):25–32.
147. Wieselthaler GM, Riedl M, Schima H, Wagner O, Waldhäusl W, Wolner E, Luger A, Clodi M. Endocrine function is not impaired in patients with a continuous. MicroMed-DeBakey axial flow pump. J Thorac Cardiovasc Surg. 2007;133(1):2–6.
148. Edvinsson L, Krause DN. Cerebral blood flow and metabolism. 2nd ed. Philadelphia: Lippincott Williams&Wilkins; 2002.
149. Folklow B. Description of the myogenic hypothesis. Circ Res. 1964;15:279–87.
150. Czosnyka M, Brady K, Reinhard M, et al. Monitoring of cerebrovascular autoregulation: facts, myths, and missing links. Neurocrit Care. 2009;10:373–86.
151. Ono M, Joshi B, Brady K, et al. Risks for impaired cerebral autoregulation during cardiopulmonary bypass and postoperative stroke. Br J Anaesth. 2012;109:391–8.
152. Roach GW, Kanchuger M, Mangano CM, et al. Adverse cerebral outcomes after coronary bypass surgery. N Engl J Med. 1996;335:1857–63.
153. Lietz K, Brown K, Ali SS, Colvin-Adams M, Boyle AJ, Anderson D, Weinberg AD, Miller LW, Park S, John R, Lazar RM. The role of cerebral hyperperfusion in postoperative neurologic dysfunction after left ventricular assist device implantation for end-stage heart failure. J Thorac Cardiovasc Surg. 2009;137(4):1012–9.
154. Lanzarone E, Gelmini F, Tessari M, et al. Preservation of endothelium nitric oxide release by pulsatile flow cardiopulmonary bypass when compared with continuous flow. Artif Organs. 2009;33:926–34.
155. Hutcheson IR, Griffith TM. Release of endothelium-derived relaxing factor is modulated by both frequency and amplitude of pulsatile flow. Am J Phys. 1991;261:H257–62.
156. White RP, Vallance P, Markus HS. Effect of inhibition of nitric oxide synthase on dynamic cerebral autoregulation in humans. Clin Sci. 2000;99: 555–60.
157. Markham DW, Fu Q, Palmer MD, et al. Sympathetic neural and hemodynamic responses to upright tilt in patients with pulsatile and nonpulsatile left ventricular assist devices. Circ Heart Fail. 2013;6:293–9.
158. Busija DW, Heistad DD, Marcus ML. Effects of sympathetic nerves on cerebral vessels during acute, moderate increases in arterial pressure in dogs and cats. Circ Res. 1980;46:696–702.
159. Heistad DD, Marcus ML. Effect of sympathetic stimulation on permeability of the blood-brain barrier to albumin during acute hypertension in cats. Circ Res. 1979;45:331–8.
160. Kashiwazaki S. Effects of artificial circulation by pulsatile and non pulsatile flow on brain tissues. Ann Thorac Cardiovasc Surg. 2000;6:389–96.
161. Murkin JM, Murphy DA, Finlayson DC, Waller JL. Cerebral autoregulation and flow/metabolism coupling during cardiopulmonary bypass: the influence of PaCO2. Anesth Analg. 1987;66:825–32.
162. Nishinaka T, Tatsumi E, Nishimura T, et al. Effects of reduced pulse pressure to the cerebral metabolism during prolonged non-pulsatile left heart bypass. Artif Organs. 2000;24:676–9.
163. Undar A, Masai T, Yang SQ, Goddard-Finegold J, Frazier OH, Fraser CD Jr. Effects of perfusion mode on regional and global organ blood flow in a neonatal piglet model. Ann Thorac Surg. 1999;68(4):1336–42; discussion 1342–3.
164. Undar A, Eichstaedt HC, Frazier OH, Fraser CD Jr. Monitoring regional cerebral oxygen saturation using near-infrared spectroscopy during pulsatile hypothermic cardiopulmonary bypass in a neonatal piglet model. ASAIO J. 2000;46(1):103–6.
165. Ono M, Joshi B, Brady K, Easley RB, Kibler K, Conte J, Shah A, Russell SD, Hogue CW. Cerebral blood flow autoregulation is preserved after continuous flow left ventricular assist device implantation. J Cardiothorac Vasc Anesth. 2012;26(6): 1022–8.
166. Cornwell WK 3rd, Tarumi T, Aengevaeren VL, Ayers C, Divanji P, Fu Q, Palmer D, Drazner MH, Meyer DM, Bethea BT, Hastings JL, Fujimoto N, Shibata S, Zhang R, Markham DW, Levine BD. Effect of pulsatile and nonpulsatile flow on cerebral perfusion in patients with left ventricular assist devices. J Heart Lung Transplant. 2014;33(12):1295–303.
167. Zimpfer D, Wieselthaler G, Czerny M, Fakin R, Haider D, Zrunek P, Roethy W, Schima H, Wolner E, Grimm M. Neurocognitive function in patients with ventricular assist devices: a comparison of pulsatile and continuous blood flow devices. ASAIO J. 2006;52(1):24–7.
168. Kirklin JK, Pagani FD, Kormos RL, Stevenson LW, Blume ED, Myers SL, Miller MA, Baldwin JT, Young JB, Naftel DC. Eighth annual INTERMACS report: special focus on framing the impact of adverse events. J Heart Lung Transplant. 2017;36(10):1080–6.
169. Yoshioka D, Okazaki S, Toda K, Murase S, Saito S, Domae K, Miyagawa S, Yoshikawa Y, Daimon T, Sakaguchi M, Sawa Y. Prevalence of cerebral microbleeds in patients with continuous-flow left ventricular assist devices. J Am Heart Assoc. 2017;6(9):e005955.
170. Tabit CE, Chen P, Kim GH, Fedson SE, Sayer G, Coplan MJ, Jeevanandam V, Uriel N, Liao JK. Elevated Angiopoietin-2 level in patients with continuous-flow left ventricular assist devices leads to altered angiogenesis and is associated with higher nonsurgical bleeding. Circulation. 2016;134(2):141–52.
171. Crow S, John R, Boyle A, et al. Gastrointestinal bleeding rates in recipients of nonpulsatile and pulsatile left ventricular assist devices. J Thorac Cardiovasc Surg. 2009;137(1):208–15.

172. Letsou GV, Shah N, Gregoric ID, Myers TJ, Delgado R, Frazier OH. Gastrointestinal bleeding from arteriovenous malformations in patients supported by the Jarvik 2000 axial-flow left ventricular assist device. J Heart Lung Transplant. 2005;24:105–9.
173. Slaughter MS. Hematologic effects of continuous flow left ventricular assist devices. J Cardiovasc Transl Res. 2010;6(8):815–8.
174. Suarez J, Patel CB, Felker GM, Becker R, Hernandez AF, Rogers JG. Mechanisms of bleeding and approach to patients with axial-flow left ventricular assist devices. Circ Heart Fail. 2011;4(6):779–84.
175. Uriel N, Pak SW, Jorde UP, Jude B, Susen S, Vincentelli A, Ennezat PV, Cappelman S, Naka Y, Mancini D. Acquired von Willebrand syndrome after continuous-flow mechanical device support contributes to a high prevalence of bleeding during long-term support and at the time of transplantation. J Am Coll Cardiol. 2010;56:1207–13.
176. Crow S, Chen D, Milano C, Thomas JW, Joyce L, Piacention V III, Sharma R, Wu J, Arepally G, Bowles D, Rogers J, Villamizar-Ortiz N. Acquired von Willebrand syndrome in continuous-flow ventricular assist device recipients. Ann Thorac Surg. 2010;90:1263–9.
177. Crow S, Milano C, Joyce L, Chen D, Arepally G, Bowles D, Thomas W, Ortiz NV. Comparative analysis of von Willebrand factor profiles in pulsatile and continuous left ventricular assist device recipients. ASAIO J. 2010;56:441–5.
178. Klovaite J, Gustafsson F, Mortensen SA, Sander K, Nielsen LB. Severely impaired von Willebrand factor-dependent platelet aggregation in patients with a continuous-flow left ventricular assist device (HeartMate II). J Am Coll Cardiol. 2009;53:2162–7.
179. Tuzun E, Chorpenning K, Liu MQ, Bonugli K, Tamez D, Lenox M, Miller MW, Fossum WF, Rutten M, Dat M, Van de Vosse F, Kadipasaoglu C, de Mol B. The effects of continuous and intermittent reduced speed modes on renal and intestinal perfusion in an ovine model. ASAIO J. 2014;60(1):19–24.
180. Rozin AP, Attias J, Presser D, Rosenberg H, Moscowitz M, Bentur Y. Alcohol poisoning and venous hyperoxia. Toxicol Mech Method. 2008;18:745–50.

20 Quantification of Pulsatility During Mechanical Circulatory Support

Shigang Wang, Morgan K. Moroi, and Akif Ündar

After millions of years of evolution, the human body still produces pulsatile blood flow. The beating heart generates pulsatile blood pressure and flow, affords high output under high pressure, and is able to tolerate high vascular resistance. The human heart appears well-adapted for us [1]. During systole, the left ventricle contracts and ejects blood into the aorta to produce a maximum systolic pressure. Meanwhile, the right ventricle simultaneously ejects blood into the lungs. During diastole, the left and right ventricles fill, and the aortic pressure decreases to a minimum diastolic pressure. The elastic recoil of the great vessels helps to force blood into the arterioles. Overall, the cardiovascular system is an ideal and highly efficient system capable of transporting blood in a pulsatile fashion to all of the body's tissues. Any attenuated or absence of blood flow pulsatility may lead to unclear clinical ramifications [2]. Acute or chronic heart failure is a condition in which the heart can't pump enough blood to meet the body's needs, resulting in multiple organ dysfunction because of reduced blood flow and pressure. Clinicians usually use simple hemodynamic data, such as cardiac output, mean arterial pressure, and systolic and diastolic pressures, to assess cardiovascular function and make quick cardiovascular diagnoses. However, hemodynamic pulsatility is often overlooked in clinical cardiovascular assessments.

Currently, mechanical circulatory support (MCS) is used to support the patient's heart and/or lungs to satisfy the metabolic demands of the human body as well as serve as a bridge to improvement of cardiac/lung function, permanent therapy, or heart transplant. Various MCS devices include intra-aortic balloon pumps (IABPs), percutaneous or implantable ventricular assist devices (VAD), and extracorporeal membrane oxygenation (ECMO) or extracorporeal life support (ECLS). All of these devices are used to improve cardiac and/or respiratory function; however, they utilize different mechanisms of action. For example, MCS devices can create continuous (non-pulsatile) flow as well as pulsatile flow. First-generation pneumatic VADs are powered by pressurized gas and selectively produce pulsatile flow, whereas

S. Wang, MD · M. K. Moroi, BS
Penn State Health Pediatric Cardiovascular Research Center, Department of Pediatrics, Hershey, PA, USA
e-mail: swang2@pennstatehealth.psu.edu; mmoroi@pennstatehealth.psu.edu

A. Ündar, PhD (✉)
Penn State Health Pediatric Cardiovascular Research Center, Department of Pediatrics, Hershey, PA, USA

Penn State Health Pediatric Cardiovascular Research Center, Department of Pediatrics, and Department of Surgery and Bioengineering, Penn State Milton S. Hershey Medical Center, Penn State College of Medicine, Penn State Health Children's Hospital, Hershey, PA, USA

Penn State College of Medicine, Department of Pediatrics, Hershey, PA, USA
e-mail: aundar@pennstatehealth.psu.edu

© Springer Nature Switzerland AG 2020
J. H. Karimov et al. (eds.), *Mechanical Support for Heart Failure*,
https://doi.org/10.1007/978-3-030-47809-4_20

some third-generation rotary pumps can generate non-pulsatile as well as pulsatile flow. The main physiologic difference between non-pulsatile and pulsatile MCS is that non-pulsatile MCS continuously unloads the left ventricle (LV) during the entire cardiac cycle, whereas pulsatile MCS unloads the ventricle intermittently and asynchronously. This difference may have an impact on cardiovascular pulsatility and possibly vital organ perfusion [3].

The debate on pulsatile flow during acute and chronic support has continued for more than half a century. During acute support, roller pumps are typically used as blood pumps, and most of them can create *diminished pulsatility* using the pulse width and frequency parameters in the pulsatile mode. The pressure and flow waveforms of this type of *pulsatility* are significantly different than waveforms of normal physiologic quality. Pulsatile MCS devices are commonly referred to as pneumatic assist devices for chronic support. Some investigators have suggested that pulsatile flow is more beneficial than conventional non-pulsatile perfusion during acute or chronic MCS due to improvements in organ perfusion, whereas some investigators claim that there are no differences in vital organ recovery between the perfusion modes. Therefore, it has become mandatory to precisely quantify pulsatility in heart failure patients and comparison of MCS devices.

Quantification of Pulsatile Flow

The fundamental problem of comparing different perfusion modes or different types of pulsatile flow is the lack of understanding regarding the precise quantification of pressure-flow waveforms generated by MCS [4]. The degree of pulsatility depends on pump type, pump speed, LV contractility, as well as preload and afterload pressures [3]. There are currently no standard criteria to describe the variations in pulsatile flow. Typically, systolic pressure, diastolic pressure and mean arterial pressure (MAP) are used to describe pulsatility. However, this approach disregards flow rate and the hemodynamic quality of the blood flow. Therefore, it is important to establish common parameters to describe pulsatility, such as pressure, flow, and energy levels. Without a precise quantification of pulsatility, it is impossible to make meaningful comparisons between different perfusion modes and devices for both research and clinical purposes [4]. The following sections outline various parameters available to evaluate the degree of pulsatility during CPB and MCS.

Systolic Pressure, Diastolic Pressure, and MAP

Each heartbeat consists of two periods: systole and diastole. Systole is the contraction phase of the heart, when blood is pumped out from the heart to the circulation. Diastole is a relaxation phase, when the ventricles are filled with blood. Systolic pressure is the maximum pressure generated during ventricular contraction; diastolic pressure is the minimum pressure produced during relaxation of the ventricles. They are measured in millimeters of mercury (mmHg) above the surrounding atmospheric pressure. MAP is the average pressure over one cardiac cycle and can be estimated from the systolic and diastolic pressure (Eq. 20.1).

$$\text{MAP}\left(\text{mmHg}\right) \approx \frac{\text{Systolic pressure} + 2\,\text{Diastolic pressure}}{3} \tag{20.1}$$

All of these pressures are associated with cardiac output, peripheral resistance, and elasticity. For example, in heart failure patients, patients experience a decrease in cardiac output, which can favor a decrease in systolic pressure and mean arterial pressure. Luckily, MCS is able to help increase cardiac output in these patients. Therefore, elevated systemic blood pressure may serve as a clinical indicator of hemodynamic improvement following implantation of an MCS device.

Pulse Pressure (PP)

Traditionally, pulsatility is quantified as the difference between the maximum systolic (P_{MAX}) and minimum diastolic aortic pressure (P_{MIN}). This difference is referred to as pulse pressure (PP) (see Fig. 20.1). Systemic and pulmonary pulse pressures are defined in Eqs. 20.2 and 20.3. As defined by PP, blood flow becomes continuous when PP has been diminished.

$$\text{Systemic PP}\left(\text{mmHg}\right) = \text{AoP}_{\text{MAX}} - \text{AoP}_{\text{MIN}} \quad (20.2)$$

$$\text{Pulmonary PP}\left(\text{mmHg}\right) = \text{PaP}_{\text{max}} - \text{PaP}_{\text{min}} \quad (20.3)$$

PP is a well-known hemodynamic parameter that correlates with cardiac output and is always available on real-time basis in the intensive care setting. The PP value can help indicate the strength of ventricular contraction and cardiac output. Low systemic PP (less than 25% of the aortic systolic pressure) is usually caused by low left ventricular stroke volume, such as in congestive heart failure, cardiogenic shock, or traumatic

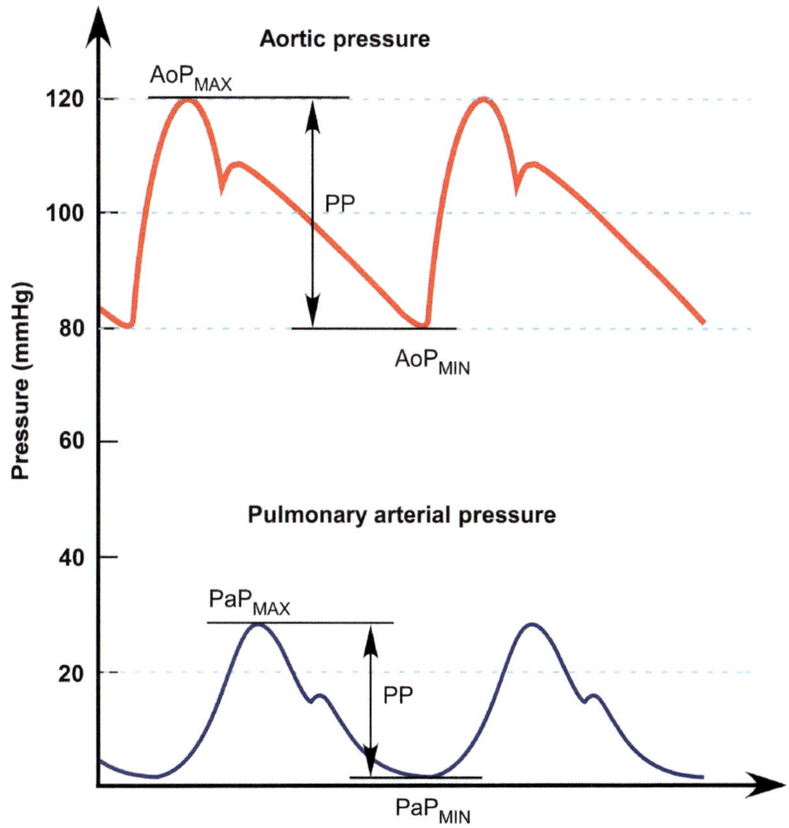

Fig. 20.1 Pulse pressure. AoP_{MAX} maximum aortic pressure, AoP_{MIN} minimum aortic pressure, PaP_{MAX} maximum pulmonary arterial pressure, PaP_{MIN} minimum pulmonary arterial pressure, *PP* pulse pressure

injury (i.e., significant blood loss). Low PP has been found to be a strong independent predictor of all-cause and cardiovascular death in patients with heart failure [5]. In patients with continuous MCS, high flow support can also result in low systemic PP [3]. Increased pump flow with decreased pulsatility may affect aortic valve opening times, aortic root flow, left atrial and left ventricular volumes, right ventricular function, and the amount of shear stress on blood components, increasing the risk of complications for continuous-flow LVAD patients [6]. The mean PP over the initial 6 hours in patients receiving ECMO is a sole independent hemodynamic predictor of in-hospital mortality and weaning failure with modest predictive accuracy [7].

High systemic PP (more than 100 mmHg during rest) may be caused by aortic stiffness, aortic valve regurgitation, and hyperthyroidism. With MCS, high PPs may also be generated if the artificial pulsatile flow meets an innate pulse wave generated by the patient's native heart. Still, PP is not the most efficient measure of pulsatility because it does not account for the hemodynamic properties of the flow.

Pulsatility Index (PI)

PI quantifies pulsatile circulation based on variations in flow velocities and has a direct correlation with cardiovascular resistance. It is the difference between maximum (V_{MAX}) and minimum (V_{MIN}) blood flow velocity, normalized to the average blood flow velocity (V_{MEAN}), as defined in Eq. 20.4 (see Fig. 20.2). PI is a unitless measurement of the pulsatile flow. In contrast to PP, the PI metric requires more advanced measurement equipment and techniques, such as ultrasound catheters or transonic flow probes.

$$\mathrm{PI} = \frac{\left(V_{MAX} - V_{MIN}\right)}{V_{MEAN}} \tag{20.4}$$

Lower PI indicates decreased pulsatility. Patients who have very poor left ventricular systolic function can have minimal pulsatility with $V_{MAX} - V_{MIN} \approx 0$. A low PI could also be generated with a more functional left ventricle if the pump flow were excessive and the ventricle driven to collapse (i.e., speed excessive for preload). The lower the PI, the greater the amount of support that can be provided by the pump. The greater the PI, the more the native heart is able to eject. Rising PIs in VA-ECMO patients can represent improving cardiac function [8].

The pulmonary artery pulsatility index (PAPi; Eq. 20.5) is a special hemodynamic index that predicts severe right ventricular failure (RVF) and serves as a key parameter to identify patients who will require right ventricular assistant device (RVAD) support before and after LVAD implantation [9].

$$\mathrm{PAPi} = \frac{\text{Systolic pulmonary artery pressure} - \text{Diastolic pulmonary artery pressure}}{\text{Central venous pressure}} \tag{20.5}$$

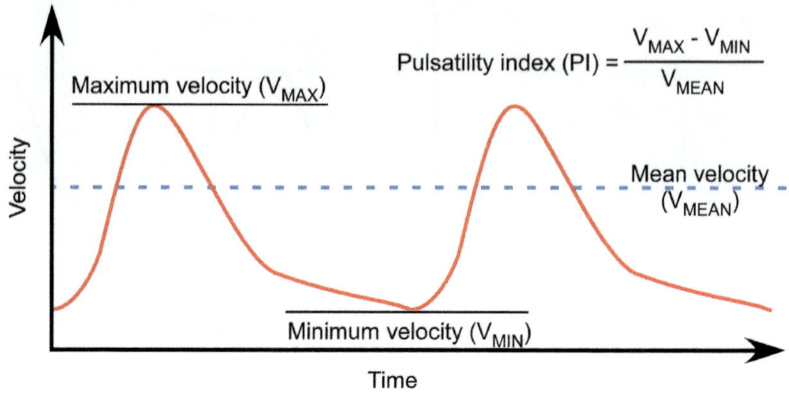

Fig. 20.2 Pulsatility index

Maximum Rate of Change of Pressure (dP/dt_{max})

dP/dt_{max} is an index derived from pressure that indicates the change in pressure ($P_2 - P_1$) over time ($t_2 - t_1$) (i.e., the maximal ascending slope of the pressure waveform; see Fig. 20.3 and Eq. 20.6 (v is velocity)). Usually, LV dP/dt_{max} is measured by echocardiography and used as an index of LV contractility. It is dependent on cardiac filling and heart rate. Higher LV dP/dt_{max} values may reflect better myocardial contractility, better pulsatility, great vessel resistance, or effect of inotropic agents. Aortic dP/dt_{max} is also used to assess ventricular contractility because it is closely correlated with the LV dP/dt_{max}.

$$\frac{dP}{dt_{max}}(\text{mmHg}/\text{s}) = \frac{P_2 - P_1}{t_2 - t_1} = \frac{4V_2^2 - 4V_1^2}{t_2 - t_1} \quad (20.6)$$

Heart failure patients have lower LV dP/dt_{max} [10]. Animal studies have shown that continuous-flow LVADs significantly decrease both pulse pressure and mean dP/dt_{max}. However, pulsatile LVADs can create higher dP/dt_{max} when compared to continuous-flow LVADs [11].

Pulse Power Index (PPI)

The PPI quantifies the relative power of a pulsatile waveform with respect to a non-pulsatile equivalent flow. It is based on transforming the pulsatile flow waveform into a series of simple sinusoidal waves using a fast Fourier transform (Eq. (20.7) [12]:

$$\text{PPI}\left[\left(\frac{\text{circles}}{\text{s}}\right)^2\right] = \sum_{i=0}^{n} \frac{A_i^2 \cdot \omega_i^2}{A_0^2} \quad (20.7)$$

where A_i is the amplitude of i-th harmonic of flow, A_0 is the amplitude of mean flow, and ω is frequency of flow (cycles/s).

PPI is calculated from the flow waveform rather than the pressure waveform to avoid the

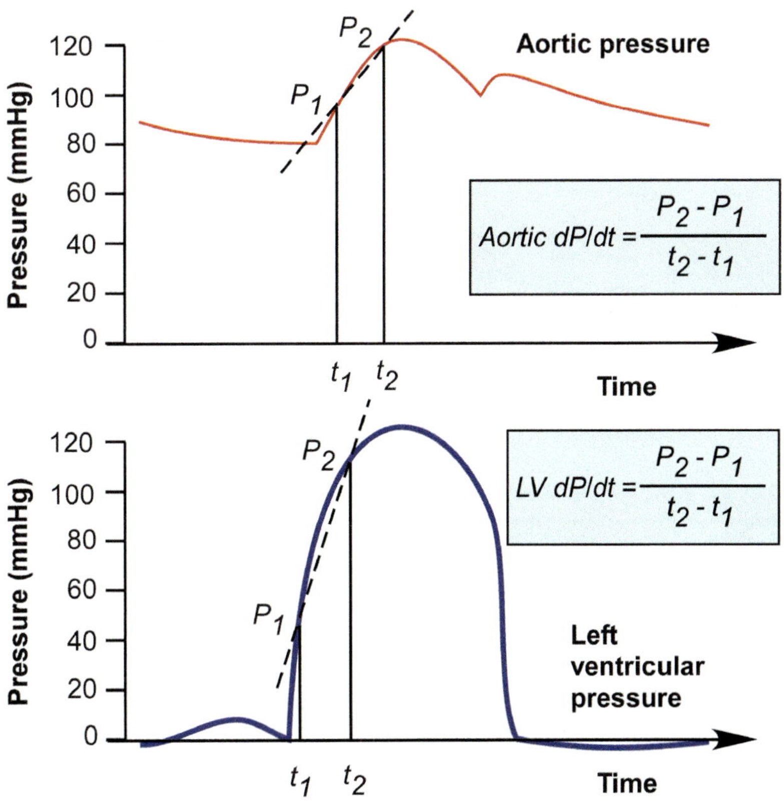

Fig. 20.3 Rates of changes of left ventricular (LV) and aortic pressures

effects of vascular tone and arterial wall stiffness. PPI uses two variables of pulse shape and pulse frequency, whereas PI only reflects the relative "sharpness" of pulse waveform, not its frequency. The PPI of continuous flow is 0. As the pulsatility or pulsatile frequency increases, the PPI increases (Fig. 20.4). In vitro and animal experiments have demonstrated that increasing the continuous-flow LVAD assist ratio significantly decreases the PI and PPI [13, 14]. PPI is a useful index to determine whether a continuous-flow LVAD achieves physiological support.

Energy Equivalent Pressure (EEP) and Surplus Hemodynamic Energy (SHE)

The above indices are typically used to clinically evaluate native ventricular pulsatility; however, they may not adequately quantify pulsatility in patients with MCS devices. This is because the combination of the patient's arterial pressure and the device's flow waveform produces waveforms that are significantly different (i.e., shape, amplitude, frequency, and phase) from those created by the native heart alone. More importantly, pulsatile flow is generated from energy gradients, not pressure gradients. In other words, both arterial pressure and flow waveforms must be used to quantify different perfusion modes [4].

The concept of EEP was introduced by Shepard et al. to more adequately quantify pulsatile flow [15]. EEP is used to quantitatively measure differences in steady-state and pulsatile blood flow, and it is defined in Eq. 20.8. EEP, the energy per unit blood volume pumped, is based on the ratio between the area beneath the hemodynamic power curve ($\int_{t1}^{t2} f p\,dt$) and the area beneath the pump flow curve ($\int_{t1}^{t2} f\,dt$) during each pulse cycle:

$$\mathrm{EEP}(\mathrm{mmHg}) = \int_{t1}^{t2} \frac{f \cdot p\,dt}{f\,dt} \qquad (20.8)$$

where f is the pump flow rate (ml/s), p is the arterial pressure (mmHg), and dt is the increment in time. EEP depends on the morphology of both pressure and flow waveforms and not solely the pressure waveforms. Under non-pulsatile flow, EEP equals mean aortic pressure (MAP), while

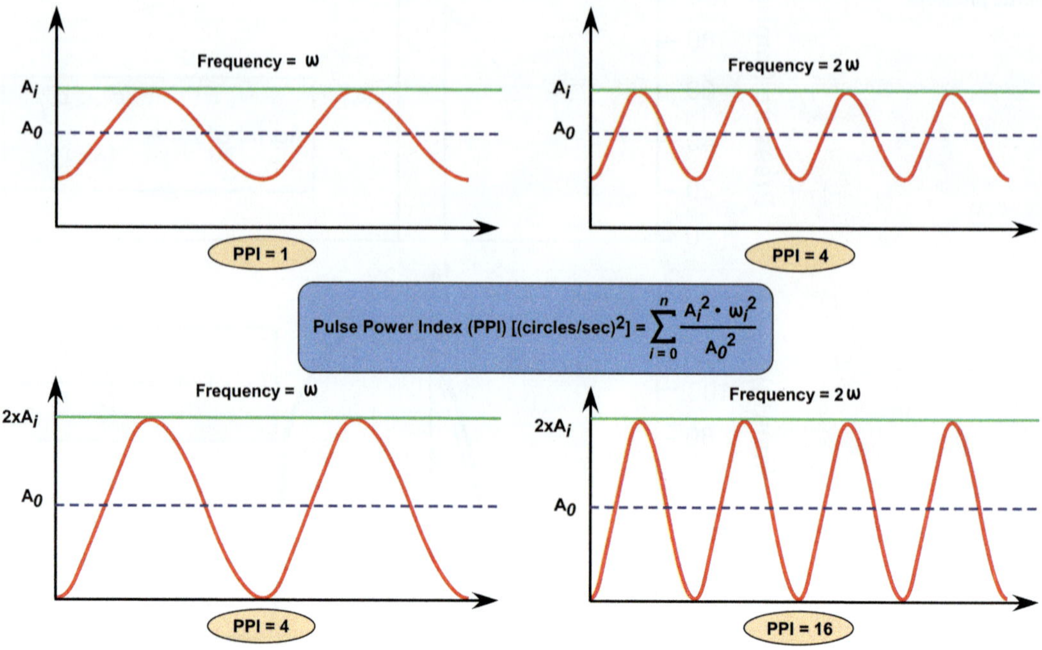

Fig. 20.4 Pulse power index (PPI). *A* amplitude, ω frequency

under pulsatile flow, EEP is always greater than MAP. The percentage pulsatile energy (PPE) is related to the ratio of the EEP and the MAP (Eq. 20.9) [16].

$$\mathrm{PPE}(\%) = \left[\frac{\mathrm{EEP}}{\mathrm{MAP}} - 1\right] \times 100 \qquad (20.9)$$

In the normal human heart, the difference between MAP and EEP is approximately 10–12% [17]. If PPE is more than 10% at the output of blood pump, the blood pump is considered to be capable of generating physiological-like pulsatile flow.

SHE is defined as the difference between EEP and mean arterial pressure (MAP), which is converted to the units of erg/cm^3 using a constant factor [15, 18] (Eq. 20.10; see Fig. 20.5). SHE is the energy created by the pulsatile flow minus the energy carried by steady flow at the same MAP and mean flow. SHE is a measure of how much additional energy is present due to the pulsatile nature of the waveform. It therefore corresponds to the additional hemodynamic energy generated from pulsatile flow alone. SHE is zero under 100% non-pulsatile flow [19] (see Fig. 20.6). The higher pulsatility, the greater SHE level.

$$\mathrm{SHE}\left(\frac{\mathrm{ergs}}{\mathrm{cm}^3}\right) = 1332\int_{t_1}^{t_2}\frac{f\cdot p dt}{f dt} - 1332 \quad \mathrm{MAP} = 1332 \quad (\mathrm{EEP} - \mathrm{MAP}) \qquad (20.10)$$

Pressure (p) and flow (f) must be instantaneous values and are expressed in mmHg and cm^3/s, respectively. Time in seconds is represented by t_1 and t_2. The constant 1332 converts pressure from units of mmHg to dynes/cm^2 (1 mmHg = 1332 dyn/cm^2). Ergs is identical to dyne × centimeter (i.e., work done by 1 dyne acting over 1 cm). The differences in SHE level and percent change between non-pulsatile and pulsatile VAD support are shown in Fig. 20.7 [20]. EEP and SHE are useful tools to quantify and compare different perfusion modes and types of pulsatility [18, 21, 22].

Total hemodynamic energy (THE) is the energy carried by blood, which is defined as the product of the force acting at a cross-sectional area (A) and the length of the blood column forced through that area in given time (Δt) [15]. The force equals the cross-sectional area A (cm^2) multiplied the instantaneous pressure P (mmHg). The length of the blood column is the length of blood passing through the cross-sectional area A in Δt seconds. Mathematically, the blood column is derived from the blood flow (F, cm^3/s) multiplied by the time Δt (s) divided by the cross-sectional area A (cm^2), expressed as a distance in centimeters. The net work done in accelerating the blood is $(PA) \times [(F\Delta t)/A]$, which simplifies to $PF\Delta t$. Thus, the total hemodynamic energy (THE) per cubic centimeter of blood passing through the cross-sectional area is the ratio of total work done in time Δt to blood volume passed through the cross-sectional area in the same period (Eq. 20.11). If the constant, 1332 ergs/cm^3/mmHg, is omitted, the total hemodynamic energy is expressed as an energy equivalent pressure in mmHg.

$$\mathrm{THE}\left(\frac{\mathrm{ergs}}{\mathrm{cm}^3}\right) = 1332\left[\int_{t_1}^{t_2}\frac{f\cdot p dt}{f dt}\right] \qquad (20.11)$$

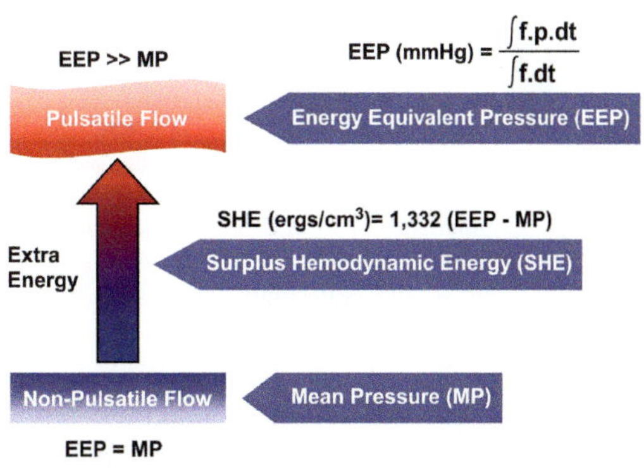

Fig. 20.5 Relationships among the mean pressure (MP), energy equivalent pressure (EEP), and the surplus hemodynamic energy (SHE)

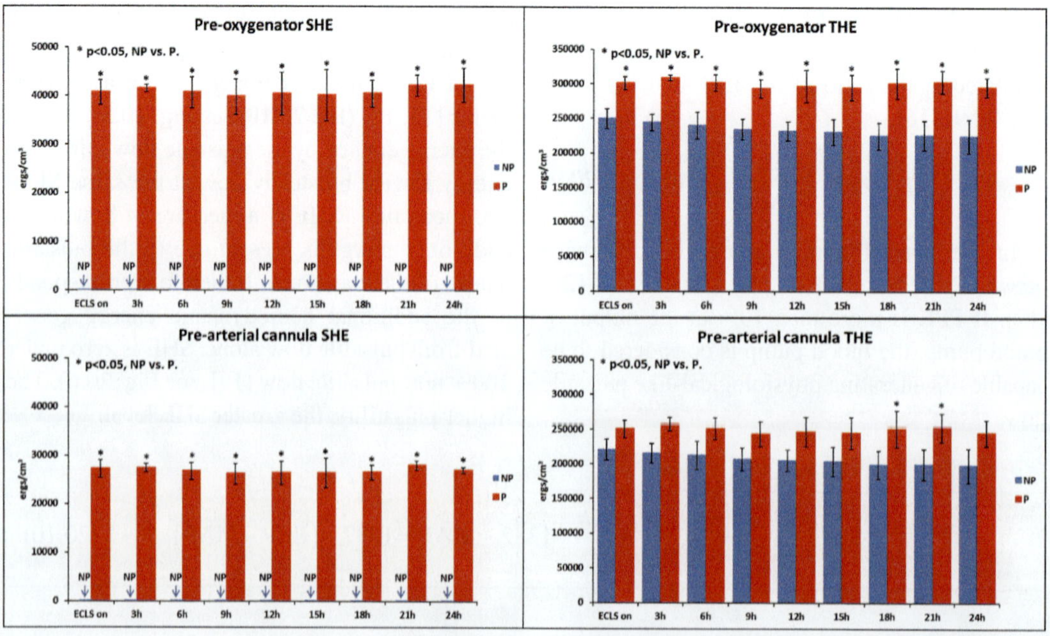

Fig. 20.6 Surplus hemodynamic energy (SHE) and total hemodynamic energy (THE) levels in non-pulsatile (NP) and pulsatile (P) groups during ECLS. (Reprinted from Wang et al. [19], with permission from John Wiley and Sons)

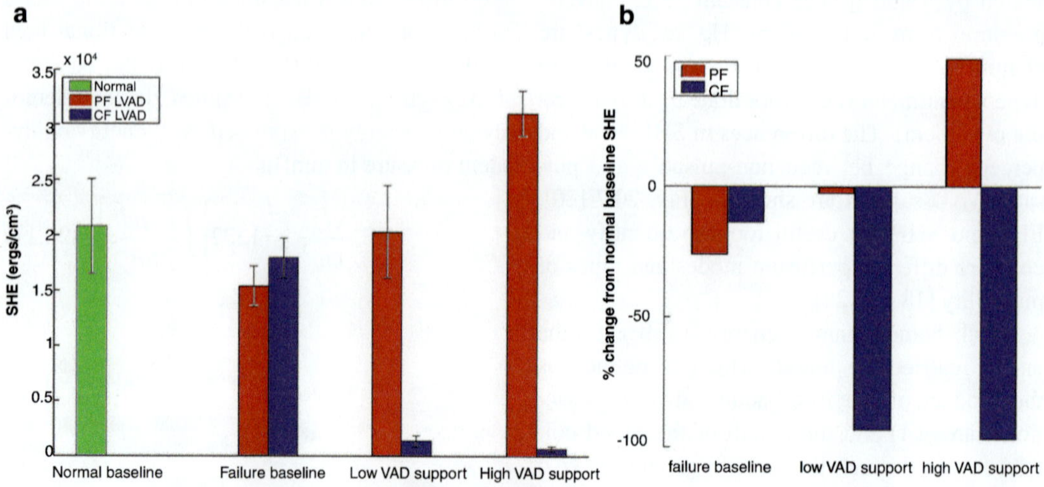

Fig. 20.7 Comparisons of surplus hemodynamic energy (SHE) (**a**) and SHE percentage change (**b**) with low and high pulsatile-flow (PF) and continuous-flow (CF) left ventricular assist device (VAD) support. (Reproduced from Travis et al. [20], with permission from Elsevier)

Pulsatile flow can deliver more THE and SHE to patients compared to non-pulsatile flow [19] (see Fig. 20.6). The extra hemodynamic energy may increase blood flow in microcirculation for capillary perfusion [23].

Current Pulsatile Devices for MCS

Intra-aortic balloon pumps (IABPs) are the most widely used MCS device for heart failure patients. A balloon is inserted percutaneously through a large blood vessel (typically, the femoral artery) and advanced directly into the ascending or thoracic aorta. Theoretically, balloon inflation at the onset of diastole will increase coronary perfusion, and balloon deflation at the end of diastole will result in reduction in aortic end-diastolic and systolic pressures, allowing for decreased LV afterload, decreased cardiac work, decreased myocardial oxygen consumption, and increased cardiac output. IABPs can be used alone or in combination with VA-ECMO. Currently, most of ECMO systems provide continuous flow. However, when combined with VA-ECMO, IABPs can generate some pulsatile support to help increase coronary perfusion. IABPs may also neutralize some adverse effects of VA-ECMO, such as the increased afterload [24]. The combination of IABP and VA-ECMO can be found to reduce mortality and successful weaning from VA-ECMO [25]. The use of concurrent IABP with ECMO may also reduce left ventricle end-diastolic diameter and pulmonary artery pressure [26].

Pneumatic pumps are known as first-generation devices and typically used for left, right, or bi-ventricular support. They are not utilized as an ECMO system because of their higher circuit resistance. These pumps are composed of inflow and outflow conduits, one-way valves, a pumping chamber, a battery pack, and a system controller and may be driven pneumatically by sending a predetermined air pressure through tubing to fill a sac cyclically (such as Berlin Heart Excor VAD, Thoratec PVAD and IVAD, SynCardia total artificial heart) or electrically using an electromagnetic pusher plate to drive blood flow (such as Thoratec HeartMate XVE). The inherent pulsatile flow generated by these pumps is suitable to long-term ventricular assistance for pediatric and adult heart failure patients [27].

Currently, roller pumps have been largely replaced by rotary pumps for use in MCS systems. Roller pumps can generate "ripple" pulses under non-pulsatile mode. They can also create pulsatile flow due to the acceleration and deceleration of the rollers. Despite this, they are seldom used to provide pulsatile flow during MCS because of their occlusive characteristics and safety issues.

Rotary pumps have three basic types – axial flow pumps, diagonal flow pumps, and radial centrifugal pumps. Because of their non-occlusiveness, rotary pumps can be used in all MCS systems. Axial pumps have efficient pumping capability and small size, supporting their typical use as VADs. Centrifugal pumps have better hydraulic efficiency than axial flow pumps. Technical improvements allow centrifugal pumps to undergo longer MCS durations with fewer bleeding and thromboembolic complications. Diagonal pumps are mixed flow pumps, combining the benefits of axial flow and radial centrifugal pumps. Rotary pumps generate true non-pulsatile flow. The non-pulsatile flow produced by these ECLS pumps mixes with the pulsatile flow generated by the patient's native heart, weakening the patient's circulatory pulsatility. While continuous ECLS flow is still capable of providing systemic perfusion, they also increase the afterload against which the left ventricle must work, and can even cause left ventricular distension and aortic insufficiency, depending on the level of support and the state of the heart [28, 29]. Still, continuous-flow rotary pumps have been successfully used for the treatment of end-stage heart failure in clinical practice. Fortunately, some rotary pumps can also create pulsatile flow by changing rotational speed. Pulsatile flow is expected to be superior to continuous flow during MCS. Further assessment of pulsatile MCS is required once pulsatile blood pumps become more available and used in clinical practice.

Pulsatile MCS Devices Under Development

Pulsatile Centrifugal Pump

A novel pump control system that was developed for use as a continuous-flow LVAD (EVAHEART; Sun Medical Technology Research Corporation, Nagano, Japan) is now being used to generate pulsatile flow in an in vivo model [30–32]. The controller can detect R waves of the ECG signal and instantly increase or decrease the rotational speed to the target speed after a configured delay time from the R wave to sync pulses with the patient's native heartbeat. The pulsatile settings of this pump include pulsatile rotational speed, the duration of systolic and diastolic phases, and pulsatile delay from the R wave. Animal experiments showed that this modified centrifugal pump can create ECG-synchronized pulsatile flow for MCS [30–32]. It displays the possibility of utilizing continuous-flow LVADs to provide physiological flow.

TORVAD

The TORVAD (Windmill Cardiovascular Systems, Inc., Austin, TX) LVAD is a valveless piston pump that has two pistons independently moving within a toroidal-shaped pumping chamber with inlet and outlet ports. The pump pushes blood by cyclically actuating a piston moving in the chamber while maintaining a second piston in a stationary position to occlude blood flow between the inlet and outlet ports. After each pump ejection, the pistons exchange roles. This special design allows it to deliver synchronous, pulsatile flow for pediatric (stroke volume, 15 ml; maximum flow rate, 4 L/min) and adult (stroke volume, 30 ml; maximum flow rate, 8 L/min) circulatory support. In vitro and animal experiments demonstrated that the TORVAD can generate synchronous pulsatile flow with low levels of shear stress, produce superior LV unloading, and preserve native aortic flow and pulsatility when compared to continuous-flow MCS devices [33–35].

The i-cor Diagonal Pump

The i-cor system (Xenios AG, Heilbronn, Germany) is the world's first electrocardiogram (ECG)-synchronized diagonal pump that changes rotational speed in order to achieve pulsatile flow. The i-cor diagonal pump has a separate pump head, a magnetic coupling of forces between the impeller and the driver, a high-tech ceramic bearing to improve durability and reduce blood trauma, a low priming volume (16 ml) to minimize hemodilution and reduce the blood-artificial surface interface, a large application spectrum (maximal flow rate, 8 L/min; revolution speed, 0–10,000 RPM) ranging from neonates to adults, a zero-flow mode to prevent backflow, a flow-measuring sensor with integrated bubble detector for safety, and an optional pulsatile flow mode to mimic physiological flow conditions. Although the i-cor system has not been approved by the Food and Drug Administration (FDA) in the United States, it has been used in clinical practice in Europe (CE approved for non-pulsatile flow).

The i-cor diagonal pump can generate non-pulsatile and pulsatile flow, and it can be easily switched between two modes. Pulsatile settings include pulsatile amplitude (100–5000 RPM in steps of 100 RPM), pulsatile width (default 200 ms), and pulsatile delay (−50 to 90 ms). Pulsatile settings have a direct impact on pulsatility. A greater pulsatile amplitude will generate higher pulsatility (see Fig. 20.8) [36]. A narrow pulsatile width will shorten duration of the pulse.

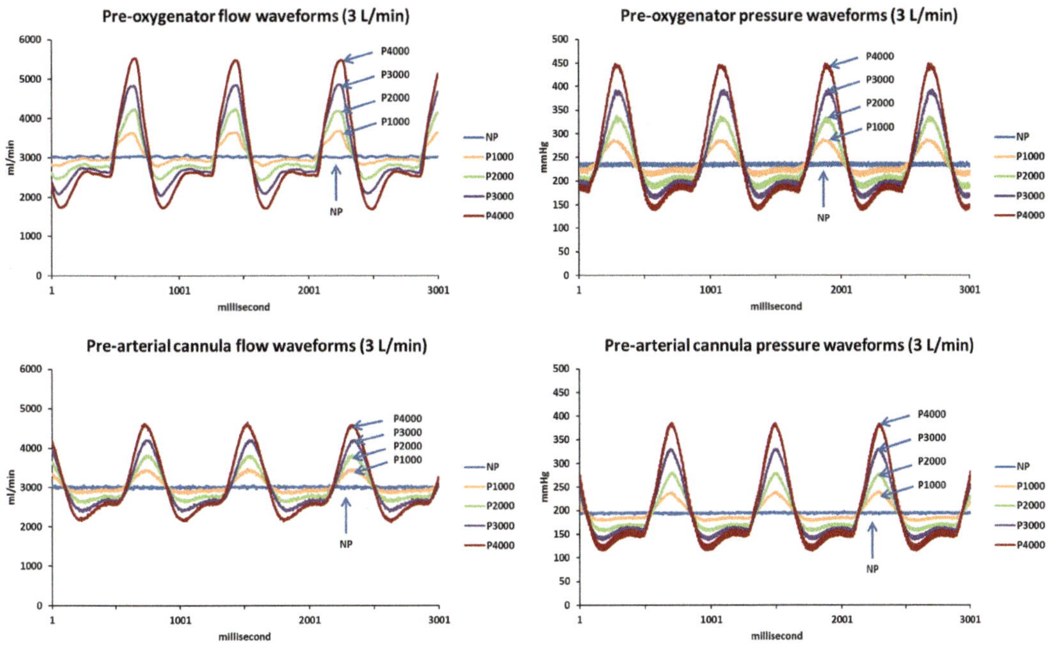

Fig. 20.8 Pre-oxygenator and pre-arterial cannula flow and pressure waveforms at 3 L/min under non-pulsatile and pulsatile modes with pulsatile amplitudes P1,000–P4,000 during ECLS. (Reprinted from Wang et al. [36], with permission from John Wiley and Sons)

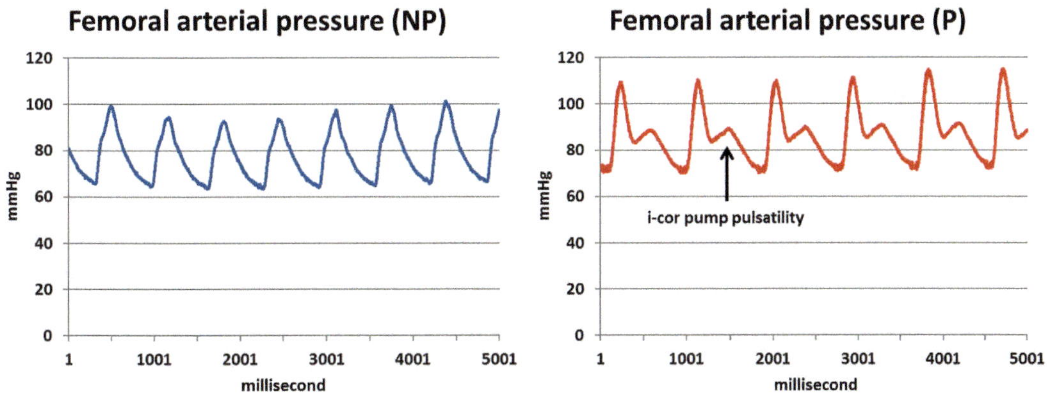

Fig. 20.9 Femoral artery pressures during non-pulsatile and pulsatile support during ECLS. (Reprinted from Wang et al. [19], with permission from John Wiley and Sons)

Adjusting the pulsatile delay will accurately match the pulses generated by the pump with the patient's native heartbeat. Most importantly, the i-cor diagonal pump can further synchronize its pulsatile flow with patient's intrinsic heart rhythm using ECG signal. The pump console can detect R-waves of the ECG signal to trigger the generation of pulses by the pump. Pulsatile flow can be synchronized to every detected R wave, every other R wave, or every third R wave, resulting in 1:1, 1:2, or 1:3 pulsatile assist ratios. If the i-cor console detects a higher heart rate, it can automatically adjust pulsatile assist ratio to decrease pulsatile frequency and allow for adequate filling of the ventricles. The i-cor system is set to deliver pulses during diastole and decrease extracorporeal blood flow during systole (see Fig. 20.9) [19]. This design permits delivery of

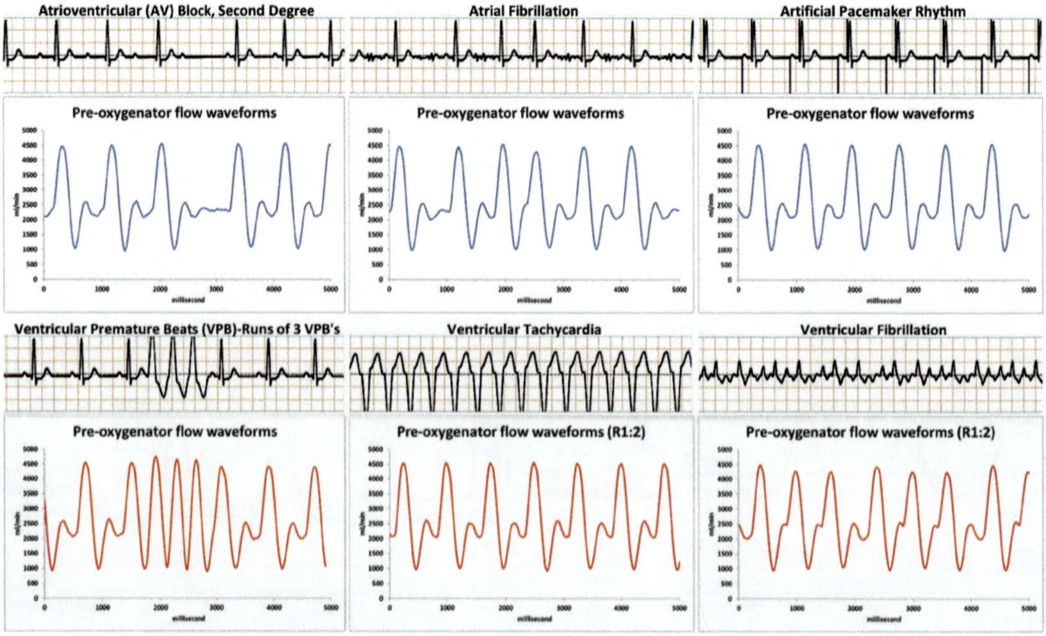

Fig. 20.10 ECG-synchronized pulsatile flow under atrial and ventricular arrhythmias in a simulated ECLS model. *R* pulsatile ratio. (Reprinted from Patel et al. [37], with permission from John Wiley and Sons)

pulsatile flow to improve coronary and peripheral perfusion without impeding systemic circulation. This also allows for decreasing LV afterload in the setting of ECLS. Animal and in vitro experiments have shown that the i-cor diagonal pump can generate ECG-synchronized, physiological quality pulsatile flow under normal sinus rhythm as well as various atrial and ventricular arrhythmias. The ECG-synchronized pump has also been shown to increase coronary blood flow, cardiac output, and mean arterial pressure during ventricular fibrillation when compared with standard continuous-flow ECMO pumps in pig models (see Fig. 20.10) [37–42].

Further modification of the i-cor pulsatile algorithm may improve pulsatility. The i-cor pump is modified from the Medos Deltastream DP3 pump. Both have identical pump heads, but the embedded control systems of the two pump consoles are quite different. The Medos console uses an internal pulsatile frequency (40–90 bpm) and pulsatile width, while the i-cor system can be triggered with the ECG. The pulsatile width of the Medos pump – a percentage of the RR interval in cardiac cycle (30–70%) – must be adjusted manually, whereas the i-cor pump is capable of automatically adjusting the pulsatile interval as the heart rate changes. In addition, the pulsatile amplitude is lower in the Medos pump than the i-cor pump.

The pulsatile waveforms are also significantly different between two pumps [43]. An in vitro study demonstrated that the i-cor pump performed better at low flow rates, whereas the Medos pump performed better at high flow rates with regard to hemodynamic energy generation and transmission. Therefore, further optimization of pulsatile control algorithms is necessary in order to achieve the most physiological-like pulsatile flow for clinical use.

For neonatal and pediatric MCS patients, high heart rate, low flow rate, and the use of small-caliber arterial cannulae may limit pulsatile flow delivery. Our in vitro study showed that the i-cor pump can automatically reduce the pulsatile frequency by increasing the assist ratio at higher heart rates under pulsatile mode [44]. Low rotational speeds at low flow rates give the i-cor pump more room to generate pulsatile flow. Based upon our in vitro evaluations, pulsatile amplitudes of

2000–3000 RPM are recommended for use in neonatal and pediatric MCS patients [45]. The i-cor pump can also serve as a non-pulsatile or pulsatile short-term cardiac assist device for neonatal and pediatric patients [46].

Currently, there are only limited data with i-cor pump and pulsatile flow in in-vivo ECLS models. It is documented that this pump at high rotational speeds with pulsatile flow, significantly increased the plasma free hemoglobin levels [43]. Before utilizing this ECLS system with the pulsatile flow in any clinical scenario, more animal experiments must be completed with the appropriate duration of ECLS support (>7 days). Additional hemolysis tests should also be completed after the necessary modifications of pulsatile flow algorithms with this system. Based on our current knowledge, it is not safe to utilize the i-cor diagonal pump for pediatric and adult ECLS patients with pulsatile flow.

Conclusions

MCS is a feasible option for the treatment of acute and chronic heart failure for pediatric and adult patients. The growing interest toward pulsatility during MCS creates a scientific need to accurately quantify hemodynamic pulsatility in the clinical setting. Although there are many parameters to be used to quantify pulsatility, the use of EEP and SHE indices is suggested to quantify hemodynamic energy and pulsatility. This will aid in comparing different MCS devices as well as patient outcomes. This may facilitate the selection of optimal MCS devices and further improve patient outcomes as well as quality of life.

References

1. Bettex DA, Prêtre R, Chassot PG. Is our heart a well-designed pump? The heart along animal evolution. Eur Heart J. 2014;35(34):2322–32.
2. Moazami N, Dembitsky WP, Adamson R, Steffen RJ, Soltesz EG, Starling RC, et al. Does pulsatility matter in the era of continuous-flow blood pumps? J Heart Lung Transplant. 2015;34(8):999–1004.
3. Soucy KG, Koenig SC, Giridharan GA, Sobieski MA, Slaughter MS. Defining pulsatility during continuous-flow ventricular assist device support. J Heart Lung Transplant. 2013;32(6):581–7.
4. Ündar A. Myths and truths of pulsatile and nonpulsatile perfusion during acute and chronic cardiac support. Artif Organs. 2004;28(5):439–43.
5. Schillaci G, Di Luzio S, Coluccini M, Gonzini L, Porcu M, Pozzar F, et al. A low pulse pressure is an independent predictor of mortality in heart failure: data from a large nationwide cardiology database (IN-CHF Registry). Ital Heart J. 2004;5(12):892–8.
6. Myers TJ, Bolmers M, Gregoric ID, Kar B, Frazier OH. Assessment of arterial blood pressure during support with an axial flow left ventricular assist device. J Heart Lung Transplant. 2009;28(5):423–7.
7. Park BW, Seo DC, Moon IK, Chung JW, Bang DW, Hyon MS, et al. Pulse pressure as a prognostic marker in patients receiving extracorporeal life support. Resuscitation. 2013;84(10):1404–8.
8. Kavi T, Esch M, Rinsky B, Rosengart A, Lahiri S, Lyden PD. Transcranial Doppler changes in patients treated with extracorporeal membrane oxygenation. J Stroke Cerebrovasc Dis. 2016;25(12):2882–5.
9. Kang G, Ha R, Banerjee D. Pulmonary artery pulsatility index predicts right ventricular failure after left ventricular assist device implantation. J Heart Lung Transplant. 2016;35(1):67–73.
10. Pandey AK, Penny WF, Bhargava V, Lai NC, Xu R, Hammond HK. Clinical evaluation of heart failure: agreement among tests. PLoS One. 2016;11(8):e0161536.
11. Ando M, Nishimura T, Takewa Y, Yamazaki K, Kyo S, Ono M, et al. Electrocardiogram-synchronized rotational speed change mode in rotary pumps could improve pulsatility. Artif Organs. 2011;35(10):941–7.
12. Grossi EA, Connolly MW, Krieger KH, Nathan IM, Hunter CE, Colvin SB, et al. Quantification of pulsatile flow during cardiopulmonary bypass to permit direct comparison of the effectiveness of various types of "pulsatile" and "nonpulsatile" flow. Surgery. 1985;98(3):547–54.
13. Kawahito S, Nakata K, Nonaka K, Sato T, Yoshikawa M, Takano T, et al. Analysis of the arterial blood pressure waveform using Fast Fourier Transform technique during left ventricular nonpulsatile assistance: in vitro study. Artif Organs. 2000;24(7):580–3.
14. Kawahito S, Takano T, Nakata K, Maeda T, Nonaka K, Linneweber J, et al. Analysis of the arterial blood pressure waveform during left ventricular nonpulsatile assistance in animal models. Artif Organs. 2000;24(10):816–20.
15. Shepard RB, Simpson DC, Sharp JF. Energy equivalent pressure. Arch Surg. 1966;93(5):730–40.
16. Teman NR, Mazur DE, Toomasian J, Jahangir E, Alghanem F, Goudie M, et al. A novel rotary pulsatile flow pump for cardiopulmonary bypass. ASAIO J. 2014;60(3):322–8.

17. Wright G. Hemodynamic analysis could resolve the pulsatile blood flow controversy. Ann Thorac Surg. 1994;58:1199–204.
18. Ündar A, Zapanta CM, Reibson JD, Souba M, Lukic B, Weiss WJ, et al. Precise quantification of pressure flow waveforms of a pulsatile ventricular assist device. ASAIO J. 2005;51(1):56–9.
19. Wang S, Izer JM, Clark JB, Patel S, Pauliks L, Kunselman AR, et al. In vivo hemodynamic performance evaluation of novel electrocardiogram-synchronized pulsatile and nonpulsatile extracorporeal life support systems in an adult swine model. Artif Organs. 2015;39(7):E90–E101.
20. Travis AR, Giridharan GA, Pantalos GM, Dowling RD, Prabhu SD, Slaughter MS, et al. Vascular pulsatility in patients with a pulsatile- or continuous-flow ventricular assist device. J Thorac Cardiovasc Surg. 2007;133(2):517–24.
21. Undar A, Frazier OH, Fraser CD Jr. Defining pulsatile perfusion: quantification in terms of energy equivalent pressure. Artif Organs. 1999;23(8):712–6.
22. Undar A, Masai T, Frazier OH, Fraser CD Jr. Pulsatile and nonpulsatile flows can be quantified in terms of energy equivalent pressure during cardiopulmonary bypass for direct comparisons. ASAIO J. 1999;45(6):610–4.
23. Krupičková P, Huptych M, Mormanová Z, Bouček T, Belza T, Šmíd O, et al. Effect of pulsatility on microcirculation in patients treated with extracorporeal cardiopulmonary resuscitation: a pilot study. ASAIO J. 2017;63(4):386–91.
24. Nuding S, Werdan K. IABP plus ECMO-is one and one more than two? J Thorac Dis. 2017;9(4):961–4.
25. Aso S, Matsui H, Fushimi K, Yasunaga H. The effect of intraaortic balloon pumping under venoarterial extracorporeal membrane oxygenation on mortality of cardiogenic patients: an analysis using a nationwide inpatient database. Crit Care Med. 2016;44(11):1974–9.
26. Petroni T, Harrois A, Amour J, Lebreton G, Brechot N, Tanaka S, et al. Intra-aortic balloon pump effects on macrocirculation and microcirculation in cardiogenic shock patients supported by venoarterial extracorporeal membrane oxygenation. Crit Care Med. 2014;42(9):2075–82.
27. Zafar F, Jefferies JL, Tjossem CJ, Bryant R 3rd, Jaquiss RD, Wearden PD, et al. Biventricular Berlin Heart EXCOR pediatric use across the United States. Ann Thorac Surg. 2015;99(4):1328–34.
28. Hong TH, Byun JH, Yoo BH, Hwang SW, Kim HY, Park JH. Successful left-heart decompression during extracorporeal membrane oxygenation in an adult patient by percutaneous transaortic catheter venting. Kor J Thorac Cardiovasc Surg. 2015;48(3):210–3.
29. Hatano M, Kinugawa K, Shiga T, Kato N, Endo M, Hisagi M, et al. Less frequent opening of the aortic valve and a continuous flow pump are risk factors for postoperative onset of aortic insufficiency in patients with a left ventricular assist device. Circ J. 2011;75(5):1147–55.
30. Date K, Nishimura T, Arakawa M, Takewa Y, Kishimoto S, Umeki A, et al. Changing pulsatility by delaying the rotational speed phasing of a rotary left ventricular assist device. J Artif Organs. 2017;20(1):18–25.
31. Date K, Nishimura T, Takewa Y, Kishimoto S, Arakawa M, Umeki A, et al. Shifting the pulsatility by increasing the change in rotational speed for a rotary LVAD using a native heart load control system. J Artif Organs. 2016;19(4):315–21.
32. Arakawa M, Nishimura T, Takewa Y, Umeki A, Ando M, Kishimoto Y, et al. Pulsatile support using a rotary left ventricular assist device with an electrocardiography-synchronized rotational speed control mode for tracking heart rate variability. J Artif Organs. 2016;19(2):204–7.
33. Gohean JR, Larson ER, Hsi BH, Kurusz M, Smalling RW, Longoria RG. Scaling the low-shear pulsatile TORVAD for pediatric heart failure. ASAIO J. 2017;63(2):198–206.
34. Gohean JR, George MJ, Chang KW, Larson ER, Pate TD, Kurusz M, et al. Preservation of native aortic valve flow and full hemodynamic support with the TORVAD using a computational model of the cardiovascular system. ASAIO J. 2015;61(3):259–65.
35. Letsou GV, Pate TD, Gohean JR, Kurusz M, Longoria RG, Kaiser L, et al. Improved left ventricular unloading and circulatory support with synchronized pulsatile left ventricular assistance compared with continuous-flow left ventricular assistance in an acute porcine left ventricular failure model. J Thorac Cardiovasc Surg. 2010;140(5):1181–8.
36. Wang S, Kunselman AR, Clark JB, Ündar A. In vitro hemodynamic evaluation of a novel pulsatile extracorporeal life support system: impact of perfusion modes and circuit components on energy loss. Artif Organs. 2015;39(1):59–66.
37. Wang S, Spencer SB, Kunselman AR, Ündar A. Novel ECG-synchronized pulsatile ECLS system with various heart rates and cardiac arrhythmias: an in vitro study. Artif Organs. 2017;41(1):55–65.
38. Patel S, Wang S, Pauliks L, Chang D, Clark JB, Kunselman AR, et al. Evaluation of a novel pulsatile extracorporeal life support system synchronized to the cardiac cycle: effect of rhythm changes on hemodynamic performance. Artif Organs. 2015;39(1):67–76.
39. Cremers B, Link A, Werner C, Gorhan H, Simundic I, Matheis G, et al. Pulsatile venoarterial perfusion using a novel synchronized cardiac assist device augments coronary artery blood flow during ventricular fibrillation. Artif Organs. 2015;39(1):77–82.
40. Ostadal P, Mlcek M, Gorhan H, Simundic I, Strunina S, Hrachovina M, et al. Electrocardiogram-synchronized pulsatile extracorporeal life support preserves left ventricular function and coronary flow in a porcine model of cardiogenic shock. PLoS One. 2018;13(4):e0196321.

41. Wang S, Patel S, Izer JM, Clark JB, Kunselman AR, Wilson RP, et al. Impact of different perfusion modalities on coronary and carotid blood flow velocities in an adult ECLS swine model. Artif Organs. 2018;42(9):918–32.
42. Wang S, Krawiec C, Patel S, Kunselman AR, Song J, Lei F, et al. Laboratory evaluation of hemolysis and systemic inflammatory response in neonatal nonpulsatile and pulsatile extracorporeal life support systems. Artif Organs. 2015;39(9):774–81.
43. Wang S, Force M, Moroi MK, Patel S, Kunselman AR, Ündar A. Effects of pulsatile control algorithms for diagonal pump on hemodynamic performance and hemolysis. Artif Organs. 2018;43:60; [Epub ahead of print]. https://doi.org/10.1111/aor.13284.
44. Wang S, Moroi MK, Force M, Kunselman AR, Ündar A. Impact of heart rate on pulsatile hemodynamic performance in a neonatal ECG-synchronized ECLS system. Artif Organs. 2018;43:81; [Epub ahead of print]. https://doi.org/10.1111/aor.13273.
45. Moroi M, Force M, Wang S, Kunselman AR, Ündar A. In vitro evaluation of ECG-synchronized pulsatile flow using i-cor diagonal pump in neonatal and pediatric ECLS systems. Artif Organs. 2018;42(7):E127–40.
46. Force M, Moroi M, Wang S, Kunselman AR, Ündar A. In vitro hemodynamic evaluation of ECG-synchronized pulsatile flow using i-cor pump as short-term cardiac assist device for neonatal and pediatric population. Artif Organs. 2018;42(8): E153–67.

21 Pulsatility as an Option with Continuous-Flow Mechanical Circulatory Support Devices

Chelsea Lancaster, Michael A. Sobieski, Mark S. Slaughter, and Steven Koenig

Advanced heart failure (HF) is estimated to affect over 6.5 million people with 250,000 to 300,000 individuals failing medical management in the United States alone [1, 2]. While heart transplantation remains the gold standard for the treatment of advanced HF, organ availability continues to be the limiting factor with less than 3500 patients worldwide receiving this treatment option annually. Additionally, the number of patients dying while on the heart transplant waiting list now exceeds 10% [3, 4]. Advances in medical and surgical treatment options, including durable mechanical circulatory support (MCS) devices, have improved survival and quality of life for this select population. Specifically, the technical and clinical management advances made in the MCS devices arena have significantly altered the clinical paradigm. The current data from the Risk Assessment and Comparative Effectiveness Management in Ambulatory Heart Failure Patients (ROADMAP) study demonstrated a clinically and statistically significant survival benefit to patients being supported by current MCS therapy compared to medically managed patients with advanced HF symptoms (New York Heart Association class III or IV) [4].

The original volume displacement pumps, although lifesaving, had significant issues with size compatibility, durability, and reliability. Each of these technology-limiting factors was addressed by the development and clinical introduction of continuous-flow pumps. By expanding the clinical use in smaller adults and some children and having more patients survive beyond 2 years due to improved durability and reliability [5], the long-term effect of diminished pulsatility on end-organ function and adverse events has become a renewed area of interest. Current CF-LVAD technology was not originally designed to produce vascular pulsatility or cyclic ventricular volume unloading, the absence of which may be associated with clinically significant adverse events, including gastrointestinal (GI) bleeding [6, 7], arteriovenous malformations (AVM) [8], hemorrhagic stroke and pump thrombus [9], aortic valve insufficiency (AI), vascular stiffness, and diastolic hypertension [10–15]. To mitigate these clinical risks and to promote therapeutic cardiovascular changes, algorithms for modulating rotary pump speed have been introduced clinically. In addition, research continues to try and develop novel pump speed modulation algorithms to optimize vascular pulsatility (hemo-

C. Lancaster, BEng
Department of Bioengineering, University of Louisville, Louisville, KY, USA
e-mail: Chelsea.Lancaster@louisville.edu

M. A. Sobieski, RN, CCP · M. S. Slaughter, MD
Department of Cardiothoracic Surgery, University of Louisville, Louisville, KY, USA
e-mail: m.sobieski@perfusion.ws; Mark.Slaughter@louisville.edu

S. Koenig, PhD (✉)
Departments of Bioengineering and Cardiothoracic Surgery, University of Louisville, Louisville, KY, USA
e-mail: Steven.Koenig@louisville.edu

© Springer Nature Switzerland AG 2020
J. H. Karimov et al. (eds.), *Mechanical Support for Heart Failure*,
https://doi.org/10.1007/978-3-030-47809-4_21

dynamic and histologic changes in blood vessels), volume unloading of the heart, and hemocompatibility (blood) related to end-organ perfusion to further improve adverse event profiles. In this chapter, we present a review of the development and potential clinical need for LVADs to behave physiologically, methods for defining pulsatility, and how pulsatility may be achieved with continuous-flow rotary blood pumps.

Case for Pulsatile Support

The potential benefits of and/or need for physiologic vascular pulsatile pressure and flow have been one of the most debated topics for both short-term (cardiopulmonary bypass, CPB) and long-term (LVAD) support in the field of mechanical circulatory support (MCS) devices. In the review article by Ji et al., many clinical benefits of *Pulsatile flow (PF) compared to continuous flow (CF) CPB support for pediatric and adult patients* were identified, including improved end-organ blood flow, reduced systemic inflammation, and lower post-operative mortality rate that was attributed to greater surplus hemodynamic energy (SHE) [16]. Similarly, in the review article by Soucy et al., *Rotary pumps and Diminished Pulsatility: Do we need a pulse?*, the authors speculate that diminished vascular pulsatility and ventricular volume unloading produced by rotary pumps lead to non-physiologic states that may be associated with clinically significant adverse events, including bleeding, stroke, and blood trauma, previously not reported with pulsatile devices [10]. A summary of clinically used pulsatile- and continuous-flow devices is presented for comparison in Table 21.1.

In addition, it has also been demonstrated that active blood pressure management is needed to maintain mean arterial pressures less than 80 mmHg with adequate anticoagulation regiments to reduce risk of thrombus and stroke as well as medication for BP management [17]. Active control of rotary pump speed has been introduced (Jarvik-2000, HVAD with Lavare, and HM3) to enable periodic opening of the aortic valve, prevent septal shift and ventricular suction, and avoid low flows that can lead to thrombogenesis and high shear stress that may induce blood trauma [13, 18].

Advancement in technology and novel use of these artificial blood pumps for the contemporary treatment of end-stage HF is a top priority of the National Heart, Lung, and Blood Institute (NHLBI). One focus area identified was the need to develop LVAD control strategies, includ-

Table 21.1 Summary of pulsatile- and continuous-flow left ventricular assist devices (LVAD) and their mode of operation, including key engineering specifications, benefits, and limitations associated with ability to achieve near-physiologic pulsatility and ventricular volume unloading [37, 53–56]

Device	Mode of operation	Pressure profiles	Engineering specs	References
HeartMate XVE (Abbott, Chicago, IL)	Electromechanically actuated		Stroke volume: 83 mL Up to 10 L/min Max 120 BPM	
HeartMate II (Abbott, Chicago, IL)	Axial (fixed speed)	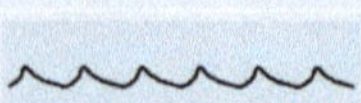	7000–12,000 rpm Up to 10 L/min	
HeartMate III (Abbott, Chicago, IL)	Centrifugal (+artificial pulse setting)	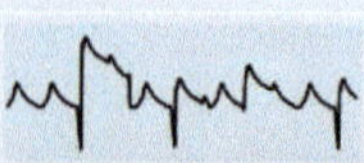	3000–9000 rpm Up to 10 L/min	
HVAD (Medtronic, Minneapolis, MN)	Centrifugal (fixed speed)		1800–4000 rpm Up to 10 L/min	
	Lavare cycle		1800–4000 rpm Up to 10 L/min	
Jarvik 2000 (Jarvik Heart, New York, NY)	Axial (fixed speed)	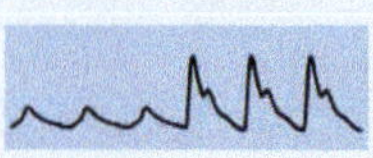	8000–12,000 rpm Up to 7 L/min	

PF pulsatile flow, *CF* continuous flow, *HVAD* HeartWare ventricular assist device, *BPM* beats per minute, *FDA* Food and Drug Administration, *BTT* bridge to transplantation ing pump speed modulation algorithms to mimic native healthy heart function with near-physiologic vascular pulsatility, phasic ventricular volume unloading, and minimal blood trauma that should help to reduce adverse events and promote myocardial remodeling and/or recovery. Specifically, these control strategies should generate pulsatile blood pressures and flows that eliminate intermittent low flow conditions due to high systemic vascular resistance and non-compliant vessels that may be associated with diminished, non-physiologic pulsatility, normalize left ventricular and atrial flow patterns, and reduce thrombus at the LV apex (inflow cannula) and in the aortic root (non-coronary cusp) while minimizing blood trauma. Toward achieving these design criteria, LVAD pump speed modulation therapy must also demonstrate hemodynamic, vascular, and blood component *efficacy*, *safety*, and *reliability*, which may also have the added benefit of the discovery of novel biomarker(s) and elucidation of physiologic mechanism(s) of favorable cardiovascular remodeling.

Methods and Metrics to Define Pulsatility

In healthy humans, the continuous time-varying arterial pressure and flow waveforms within a cardiac cycle vary in magnitude, period, and phase as a function of many physiologic factors, including preload (filling), afterload (vascular resistance, compliance, and inertance), contractility, volume, and heart rate. These pulsatile waveforms are characterized by peak-systolic (maximum) and end-diastolic (minimum) landmarks and morphology (shape), which in the clinical setting are routinely used for effective patient monitoring and treatment. However, in advanced HF patients supported by rotary blood pumps, these basic parameters may be insufficient for completely quantifying vascular pulsatility. For example, in a clinical study by Travis et al., aortic, ventricular, and LVAD pressures and flow waveforms were recorded from HF patients implanted with pulsatile (HeartMate XVE, PVAD) and rotary (HeartMate II, HVAD) devices operated clinically during low (aortic valve open) and high (aortic valve closed) pump speeds. The resulting hemodynamic waveforms may be interpreted as achieving pulsatility for both CF- and PF-LVADs at low and high support (Fig. 21.1). However, the authors applied the surplus hemodynamic energy (SHE) metric first proposed by Shepard et al. [19] and advocated by Undar et al. [14, 20, 21] for characterizing pulsatility with pediatric and adult in CPB devices. Similar to Undar's findings, Travis demonstrated PF-LVAD restored SHE to healthy physiologic values, while CF-LVAD substantially diminished SHE (non-physiologic) [22]. Subsequently, in the review article by Soucy et al., *Defining pulsatility during continuous-flow LVAD support*, the authors provided a detailed review and recommended a standardized metric for rigorously characterizing vascular pulsatility to enable comparison of future device development and speed modulation algorithms [23]. A brief review of four standard metrics (pulse pressure, pulsatility index, surplus hemodynamic energy, and input impedance) that have been reported in MCS device literature is presented, including the potential advantages and limitations of each technique.

Pulse pressure (*PP or ΔP*) is the most common metric used clinically for quantifying the magnitude of arterial pulsatility and is defined as the difference between peak aortic systolic pressure (maximum AoP) and aortic end-diastolic pressure (minimum AoP), as shown in Fig. 21.2a and Eq. 21.1. This metric can be easily obtained non-invasively (auscultatory, oscillometer, palpation, sphygmomanometer) or invasively intra- or post-operatively during LVAD support via an arterial line (fluid-filled transducer) with the continuous arterial waveform displayed on a medical monitor. However, these measurements may have limited accuracy due to signal conditioning (gain and phase distortion, low-pass filtering, sensor frequency response <40 Hz).

$$\mathrm{PP} = \left[\mathrm{AoP_{MAX}} - \mathrm{AoP_{MIN}}\right] \qquad (21.1)$$

Pulsatility index (*PI*) may be defined as the difference between the peak-systolic (V_{MAX}) and minimum diastolic (V_{MIN}) arterial blood flow velocity and normalized to the mean blood flow

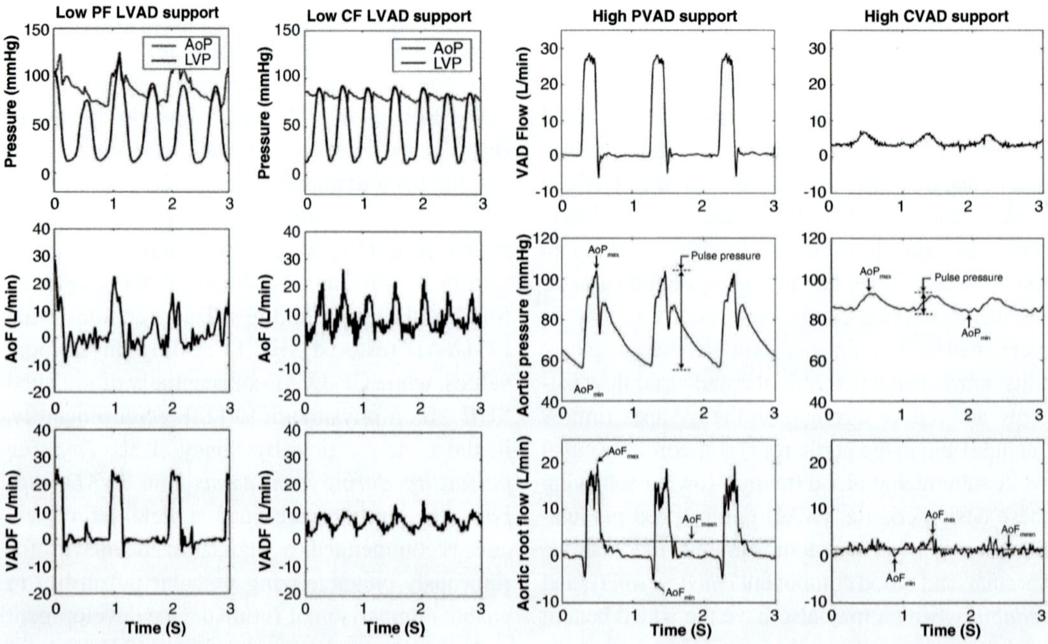

Fig. 21.1 Hemodynamic waveforms recorded intraoperatively in clinical heart failure patients implanted with continuous-flow (CF) and pulsatile-flow (PF) left ventricular assist devices (LVAD) during low (aortic valve open) and high (aortic valve closed) support. Aortic pressure (AoP), left ventricular pressure (LVP), aortic flow (AoF), and LVAD flow (VADF) were recorded with high-fidelity pressure catheters (Millar Instruments, Houston, TX) and flow probes (Transonic, Ithaca, NY) to quantify pressure and flow pulsatility using gradient (ΔP, ΔQ) and surplus hemodynamic energy (SHE). The waveform morphology appears to suggest varying levels of pulsatility may be achieved; however, SHE was non-physiologic with rotary pump (CF-LVAD) and near-physiologic with PF-LVAD for both low and high support. CVAD continuous ventricular assist device, PVAD pulsatile ventricular assist device. (Reproduced from Travis et al. [22] and Soucy et al. [23])

velocity (V_{MEAN}) over one cardiac cycle, as shown in Fig. 21.2b and Eq. 21.2. The metric may be obtained using non-invasive ultrasound instrumentation, but accurate identification of maximum and minimum landmarks may be difficult to achieve.

$$\mathrm{PI} = \left[\frac{V_{\mathrm{MAX}} - V_{\mathrm{MIN}}}{V_{\mathrm{MEAN}}}\right] \tag{21.2}$$

The HM II and HM III display a numeric calculation of pulsatility index (PI) based upon pump power consumption and the estimated pump flow quantified as the difference between the maximum and minimum flow divided by the average flow times 10 or [(max − min) ÷ average] × 10 (Fig. 21.3a, b). The HVAD displays an instantaneous, estimated time-varying flow waveform with distinct maximum and minimum landmarks to observe level of pulsatility (Fig. 21.3c).

Surplus hemodynamic energy (*SHE*) is a metric first introduced by Shepard et al. [19], to quantify the kinetic energy generated by the heart (source) and delivered to the vasculature and end organs (load). Graphically, the energy equivalent pressure (EEP) is represented by the area under pulsatile pressure (P) and flow (Q) waveforms over a cardiac cycle (Fig. 21.2c), and SHE is the difference between EEP and mean arterial pressure (MAP) as defined in Eq. 21.3. For a purely non-pulsatile waveform, the EEP will be equal to the MAP and does not produce SHE (0 erg/cm^3). For the case of rotary blood pumps, vascular pressure and flow waveforms will have substantially reduced pulsatility and altered waveform morphology (shape, ampli-

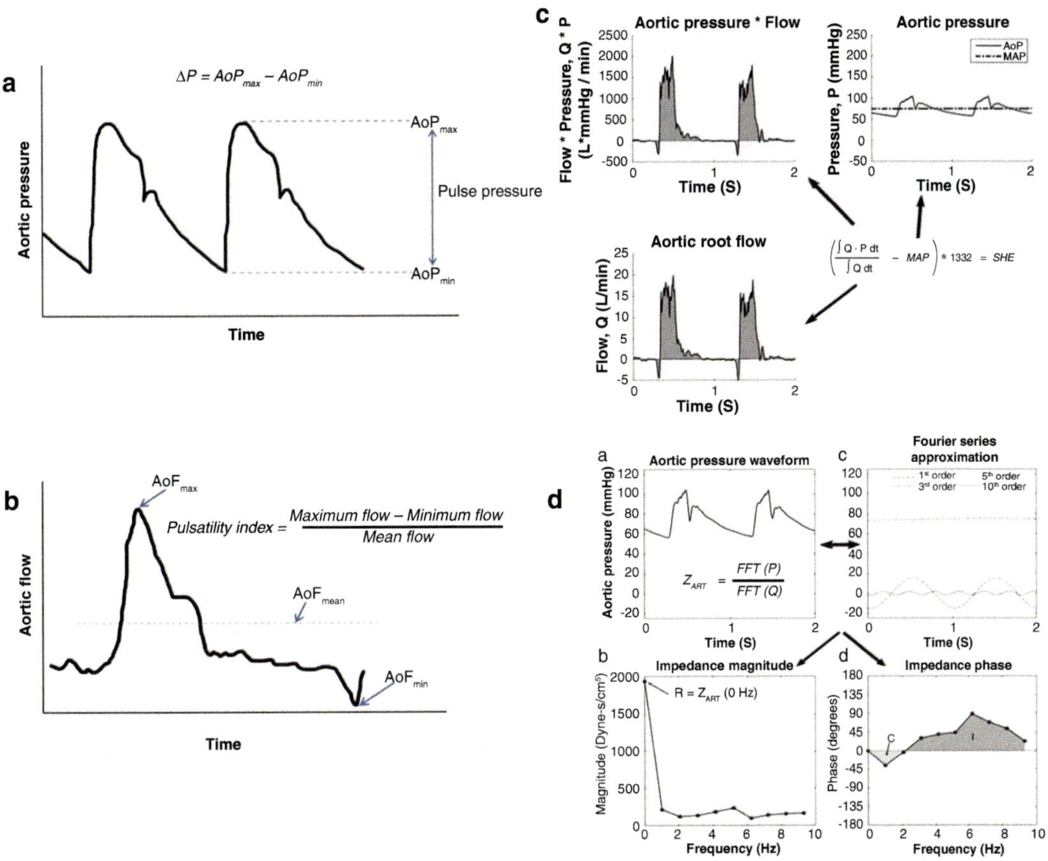

Fig. 21.2 Illustration of four methods used to quantify pressure and flow pulsatility: (**a**) pulse pressure (Δ*P*) calculated as the difference between the peak-systolic (max) and end-diastolic (min) aortic pressure (AoP); (**b**) pulsatility index (PI) calculated as the difference between the peak (max) and trough (min) aortic flow (AoF) divided by the mean AoF; (**c**) energy equivalent pressure (EEP) calculated as the area under the aortic pressure waveform times the area under the aortic flow waveform divided by the area under the aortic flow waveform. Surplus hemodynamic energy (SHE) is the EEP minus the mean aortic pressure (MAP), where EEP is derived from aortic pressure (*P*) and flow (*Q*) over cardiac cycle; and (**d**) vascular input impedance for systemic vasculature (Zart) calculated by taking the fast Fourier transform (FFT) of the time-varying aortic pressure and aortic flow (AoF) resulting in frequency-based magnitude and phase representations, where the 0th harmonic represents resistance (*R*), low harmonics compliance (*C*), and high harmonics inertance (*L*). (Reproduced from Soucy et al. [23])

tude, phase, and frequency) with increasing pump speed. Shepard, Undar, Soucy, and others have strongly advocated for using SHE as a standard metric for quantifying pulsatility and comparison of MCS devices as well as future pump speed modulation algorithms. High-fidelity pressure and flow sensors requiring invasive placement, signal conditioning, and data acquisition and analysis are required to accurately quantify EEP and SHE.

$$\text{SHE} = 1332 * [\text{EEP} - \text{MAP}], \quad \text{where EEP} = \frac{\int Q * P * dt}{\int Q * dt} \tag{21.3}$$

Input impedance (Z_{art}) is an engineering metric used to characterize the vascular load as a function of resistance, compliance, and inertance, which regulates the dissipation of hemodynamic energy. Vascular impedance is calculated using the fast Fourier transform (FFT) to transform continuous time-varying aortic pressure and flow waveforms to discrete magnitude and phase com-

Fig. 21.3 Left ventricular assist device (LVAD) control panels for (**a**) HeartMate II (Abbott, Abbott Park, IL), (**b**) HeartMate III (Abbott, Abbott Park, IL), and (**c**) HVAD (Medtronic, Minneapolis, MN) devices, illustrating pump flow (L/min), pump speed (rpm), power (W), and pulsatility index (PI) [57–59]. (Abbott and Medtronic Instructions for Use (IFU) documents)

ponents as a function of frequency (or harmonics) as defined in Eq. 21.4. Graphically, the FFT is a cumulative sum of sine waveforms that vary as a function of amplitude, phase, and frequency (Fig. 21.2d).

$$Z_{\text{ART}} = \frac{\text{FFT}(P)}{\text{FFT}(Q)} \tag{21.4}$$

In the time domain, vascular resistance component (R) can be calculated as the difference between mean arterial pressure (MAP) and mean central venous pressure (CVP) divided by the cardiac output (CO). In the frequency domain, R is the magnitude at the mean (DC or 0th frequency/harmonic) and has no phase (0 °). The vascular compliance (C) and inertance (L) can be derived from the frequency components, where C magnitudes and negative phase angles occur at lower frequencies and L magnitudes and positive phase angles dominate at higher frequencies. Maximum (high loss) and efficient (low loss) blood pressure and flow delivery are physiologically regulated by R, L, and C. In the case of *non-pulsatile* (or diminished pulsatility) rotary blood pumps, the vascular impedance will be dominated by resistance and may also have lower C (higher stiffness) due to changes in vascular structure, function, and/or phenotype. Calculating vascular input impedance also requires high-fidelity pressure and flow sensors, signal conditioning, and complex data acquisition and analysis techniques.

Rotary Blood Pump Speed Modulation

During the transition from first-generation volume displacement LVAD (HM XVE, PVAD, IVAD) to second-generation continuous-flow (CF) LVAD (Jarvik-2000, HM II, and HVAD), emphasis was placed on device reliability, safety, efficacy, and longevity. The CF-LVADs were operated clinically at fixed pump speeds (rpm) with device management focus on intrinsic pump parameters, such as estimated flow and speed (end-organ perfusion) and pulsatility index and power (early thrombus detection), and patient management with mean blood pressure, aortic valve function, ventricular septum alignment, and anticoagulation regiment. Despite clinical best practices for device and patient management as well as improvements to LVAD technology, clinically significant adverse events continued. Pump speed modulation was proposed as a potential solution(s) to address pump thrombus (washing) and bleeding (vascular pulsatility) and promote favorable myocardial remodeling (phasic volume unloading). Pump speed modulation may be achieved by varying LVAD speed (rpm) with the following engineering parameters defined: range (maximum and minimum rpm), profile/shape (sinusoidal, step response), time periods (fixed or physiologic, sec), and frequency (single or multiple cardiac cycles, Hz). However, there are potential engineering trade-offs and/or additional system requirements that may be needed to implement pump speed modulation, including device design (axial, centrifugal), integration of physical sensors and/or intrinsic pump parameters for real-time diagnostics and control of pump speed, increased pump power, and high time-varying (non-physiologic) shear stress. In the review article by Castagna et al., *Unique blood pressures and pulsatility of LVAD patients*, the authors conclude pump speed modulation, such as HM III artificial pulse, may require "accurate assessment and interpretation of (patho)physiologic consequences … to optimize device safety and efficacy" [24]. In this section, design criteria and engineering considerations (washing, device configuration, pulsatility/phasic volume unloading, and power/sensors/shear stress) toward achieving this objective are presented.

Design criteria for developing optimal pump speed modulation algorithms should be designed to address the clinically significant adverse events (thrombus, stroke, bleeding, and blood trauma) reported with rotary blood pumps. Pump speed modulation algorithms have been developed to improve pump washing, increase volume unloading, enable periodic opening of the aortic valve, and/or generate pulsatility while also eliminating low/stagnant flow minimizing shear stress. However, benchmarks for achieving physiologic (or near-physiologic) pulsatility and phasic volume unloading are not well-defined, and the requirements to preserve structure and function of the myocardial wall and arterial vessel tissue are unclear. Ideally, MCS devices would be designed to mimic native healthy heart and vascular function by producing physiologically equivalent ventricular and arterial hemodynamics, including pulsatility, phasic volume unloading, and blood shear stress. For example, normal arterial pressure and flow in a healthy human will have time-varying waveforms over a cardiac cycle (60–80 bpm) characterized by mean arterial pressure (60–85 mmHg), pulse pressure (ΔP 30–50 mmHg), and cardiac output (5–6 L/min) as well as the recommended kinetic energy metric (SHE = 15,000–20,000 erg/cm^3). In advanced heart failure patients with (or without) vascular disease (increased resistance and stiffness) supported by rotary blood pumps, these parameters are significantly altered, especially with increasing LVAD speed (rpm). It's reasonable to speculate whether pump speed modulation is required to restore hemodynamics to healthy physiologic range and/or replicate waveform morphology (magnitude, phase, shape, frequency) or if there is an acceptable near-physiologic range that minimizes the risk of adverse events. Similarly, rotary pumps continuously unload ventricular volume reducing the size of the left ventricle characterized by reduced peak-systolic and end-diastolic pressures and volumes that are different than in healthy humans. Subsequently, if peak-systolic pressure falls below aortic systolic pressure, the aortic valve may not open, and/or no ventricular blood volume will be ejected through the aortic valve. The absolute and difference in end-systolic and end-diastolic volumes, especially at higher pump speeds, are not representative of healthy human physiology. Currently, rotary blood pumps operating with pump speed modulation (HM III artificial pulse, HVAD Lavare cycle, Jarvik 2000) provide periodic cycles designed to improve pump washing, but do not achieve physiologic pulsatility or phasic volume unloading. The potential impact and/or continued risk of adverse events with long-term support are promising, but warrant further investigation.

Washing using speed modulation algorithms designed to reduce the risk of aortic valve and/or pump thrombus and promoting periodic aortic valve opening have been implemented for clinical use with the Jarvik 2000 (New York, NY), HeartMate III (Abbott, Abbott Park, IL), and HVAD (Medtronic, Minneapolis, MN). The Jarvik-2000, an axial rotary pump, was the first LVAD to employ cyclic speed reduction (1 cycle/min) to minimize the risk of thrombus formation in the non-coronary cusp by promoting periodic native heart ejection through the aortic valve. The HM III, a centrifugal rotary pump with a magnetically levitated rotor, features an *artificial pulse mode* algorithm that rapidly ramps the pump speed up and down simulating a pulse (Fig. 21.4) designed to reduce stasis and provide cyclic pump washing and increase pump pulsatility. During the artificial pulse mode, the pump speed is rapidly decreased by 2000 rpm (T = 150 ms) from the ini-

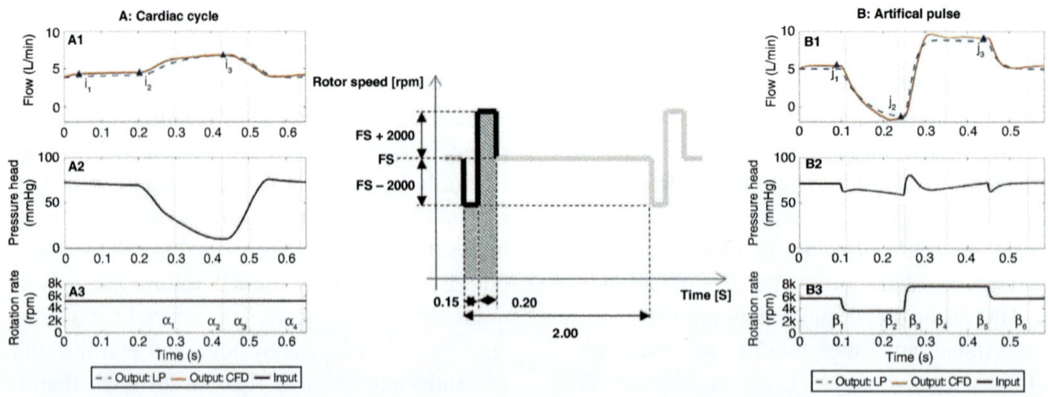

Fig. 21.4 Illustration of predicted pressure and flow waveforms achieved with HeartMate III centrifugal rotary pump (Abbott, Abbott Park, IL) operated at constant pump speed pulsatility (ΔP) of 5 mmHg (left) and with artificial pulse algorithm (middle) demonstrating ability to generate ΔP of 11 mmHg (right) [25, 60]. (Reproduced from Wiegmann et al. [25], with permission of John Wiley and Sons, and from Essandoh et al. [60])

tial set fixed speed, then is rapidly increased by 4000 rpm (T = 200 ms), and then returns to the fixed speed by decreasing 2000 rpm. The artificial pulse cycle is repeated every 2 s (30 cycles/min) asynchronous to the native heart, which results in an arterial pressure waveform with potential for four distinct landmarks (native heart systole and diastole, pump maximum and minimum pressures). The initial results are promising with the MOMENTUM3 clinical trial demonstrating no incidence of pump thrombus and trend toward a reduction in stroke rate [25]. The HVAD (Medtronic, Minneapolis, MN), a centrifugal rotary pump with hydrodynamic thrust bearings, integrated the Lavare cycle (Spanish translation washing) into its control strategy, to provide intermittent pump washing. The Lavare cycle algorithm is achieved by 200 rpm decrease (T = 2 s) followed by a 400 rpm increase (T = 1 s) and completed by a 200 rpm decrease returning to the initial set fixed speed (Fig. 21.5). The Lavare cycle is repeated once per minute asynchronous to the native heart, which is theorized to aid in washout during the ramp down phase and helps with ventricular volume unloading during the ramp up phase. Zimpfer et al. reported lower stroke rates and level of right heart failure in HF patients support by the HVAD with Lavare cycle [26].

The Jarvik-2000, HM III, and HVAD speed modulation algorithms operate asynchronously, which offers the advantage of not requiring a physical sensor and/or intrinsic pump parameters for synchronous timing with the native cardiac cycle. Future implementation of novel pump speed modulation algorithms may offer additional physiologic benefits to optimize pump performance and further reduce the risk of adverse events. For example, washing around the LV apex inflow may be further improved with algorithms designed to enable smaller end-diastolic volumes and augmented diastolic flow rates to reduce risk of thrombus formation and decrease incidence of embolic strokes. The ability to mimic near-physiologic flow fields in the left ventricle during diastole may also result in better cyclic washing around the apex minimizing the risk of thrombus formation to reduce the incidence of embolic stroke. Aortic valve opening may also be controlled to maximize washing of the leaflets to reduce the risk of thrombus in the aortic root (non-coronary cusp).

Pump Configuration (*Axial, Centrifugal*). Currently, LVAD volume unloading is directly related to the pressure gradient across the pump and diminished contractility generated by the native heart. Motor control is used to maintain a selected speed, but the flow delivered by the rotary blood pump at a given speed is determined by the pressure gradient. Rotary pumps have an inherent inefficiency related to afterload, where

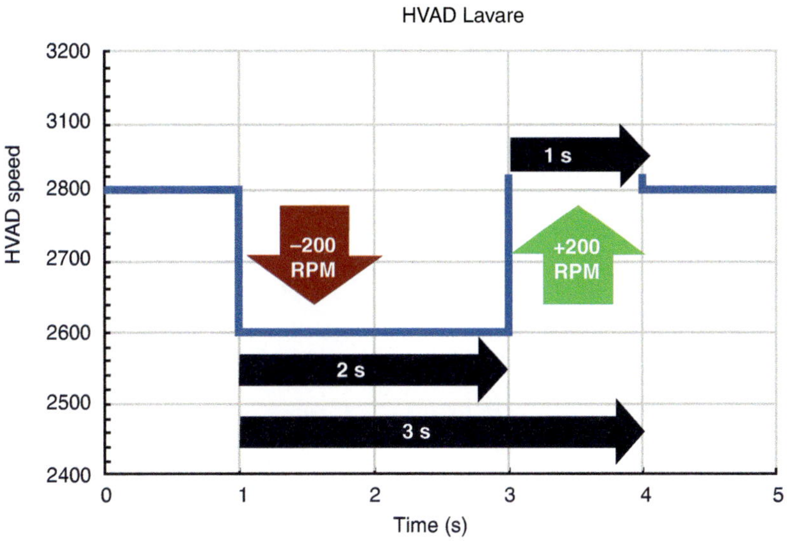

Fig. 21.5 Illustration demonstrating Lavare cycle algorithm for the HVAD centrifugal rotary pump (Medtronic, Minneapolis, MN). The Lavare algorithm is a 3-s cycle consisting of (1) decrease pump speed (200 rpm) for 2 s to increase preload and enhance mixing to promote better washing and reduce risk of thrombus, (2) increase pump speed (400 rpm) for 1 s to empty excess blood volume and enable pump blood washing, and (3) decrease pump speed (200 rpm) back to initial pump speed with the complete 3-s Lavare cycle repeated every minute. (Reproduced from Kumar et al. [26])

the flow produced at a higher pressure head significantly reduces forward flow in a direct relationship (i.e., as afterload increases, flow decreases). Specifically, without speed control, the distribution of assisted forward flow during the cardiac cycle varies. A CF-LVAD with the ability to modulate pump speed within the cardiac cycle will enable precise control (speed, magnitude, timing), such as higher pump speed during diastole and lower pump speed during systole for more efficient pump function and hemodynamic performance. The slew rate, defined as change in pump speed per unit time (rpm/second), is determined by motor design and as a general principle is dependent on the motor torque and physics. Axial pumps have significantly lower torque than centrifugal pumps, in part, because the motor magnets have smaller diameters due to size constraints. Centrifugal pumps have motor magnets with larger diameters that can generate greater torque to enable higher slew rates. For example, a centrifugal device operated at a pump speed of 2400 rpm can produce 2 L/min flow (*Q*) at a pressure head (*H*) of 100 mmHg, which could be increased to 6 L/min (*Q*) at 100 mmHg (*H*) for pump speed of 3000 rpm. Subsequently, the total time to change pump speed would be 600 rpm/2000 rpm/s or 300 ms, which may be dynamically achieved in patient with a heart rate of up to 100 bpm.

In the review article by Moazami et al. [27], *Axial and centrifugal continuous-flow rotary pumps …*, the authors provide a detailed summary of the engineering design and physiologic impact related to patient management. Axial devices (Archimedes screw principle) are in line with the native flow compared to centrifugal devices that are tangential (or perpendicular). The pressure gradient across these rotary pumps, as measured at the pump inlet (preload) and outlet (afterload), where the pressure head (*H*) is the instantaneous difference between the left ventricular and aortic pressure, and the resulting forward flow (*Q*) has an inverse relationship to the corresponding pressure head. Rotary pumps have one-third the preload sensitivity and three times the afterload sensitivity of the native heart [28]. Axial devices have a steeper *H-Q* curve that is less sensitive to afterload compared to centrifugal devices that have flat *H-Q* curves (Fig. 21.6a). Subsequently,

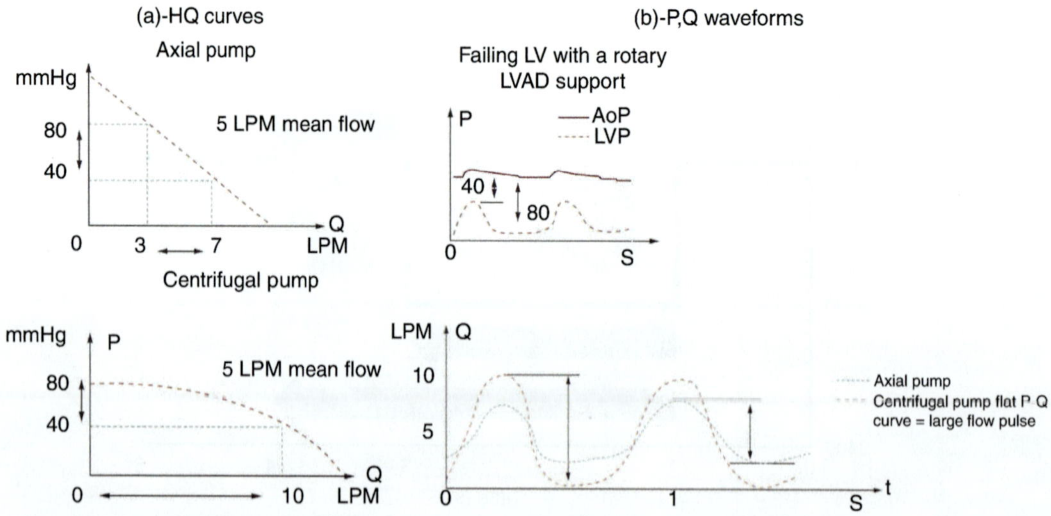

Fig. 21.6 (**a**) Illustration of differences in pressure gradient (*H*) and resulting flow (*Q*) relationship curves across each left ventricular assist device (LVAD) demonstrating higher preload and afterload sensitivity with axial (top left) compared to centrifugal (bottom left) blood pumps. (**b**) Illustration demonstrating centrifugal blood pumps are able to generate greater pulsatility (ΔP) than axial devices. AoP aortic pressure, LVP left ventricular pressure. (Reproduced from Mancini and Colombo [61])

centrifugal device design features may be more favorable toward pump speed modulation strategies. Specifically, centrifugal pumps can generate greater pulsatility (independent of native heart contractility, Fig. 21.6b) and have the ability to better control pump speed modulation (slew rate) and phasic ventricular volume unloading (vary pump speed for optimal pump flow during a single cardiac cycle) and are less prone to ventricular suction events at low preload conditions [27]. In a review article by Lim et al., *Physiology of continuous-flow LVADs*, the authors identify key physiological considerations in pump and patient management for second-generation axial (HeartMate II) and third-generation centrifugal (HeartMate III, HVAD) [29]. In addition to the challenges associated with diminishing pulsatility as pump speed increases, the ventricular pressure-volume relationship (PV loop) is also altered and characterized by changes in shape, phases, and position. Specifically, the ventricular pressures and leftward volume shift associated with increasing pump speed produce lower pressures and smaller volumes, but continuous flow produces triangular (loss of native heart isovolumic contraction and relaxation) and narrow (smaller native heart stroke volume) shape and lower peak-systolic pressure (non-ejection, aortic valve closed), conceptually transforming the ventricle from an ejecting pump to a blood reservoir. In addition, the intraventricular and aortic root flow fields are altered with the blood flow path primarily through the pump, including native heart ejection through the pump inflow conduit rather than the aortic valve. Collectively, the altered physiology of continuous-flow devices and flow path (diminished pulsatility and volume unloading) may contribute to vascular stiffness, increased pulmonary vascular resistance, reduced myocardial perfusion, increased inflammatory markers, and cardiac disuse atrophy [6, 10, 11, 14, 30].

Vascular pulsatility and phasic ventricular volume unloading using pump speed modulation has been proposed as a potential solution to the limitations of rotary pumps operated at fixed speeds or to further improve and optimize current modulation algorithms designed to provide washing. Pump speed modulation may be designed using the following parameters: shape (sinusoid, triangular, arbitrary), range (mean,

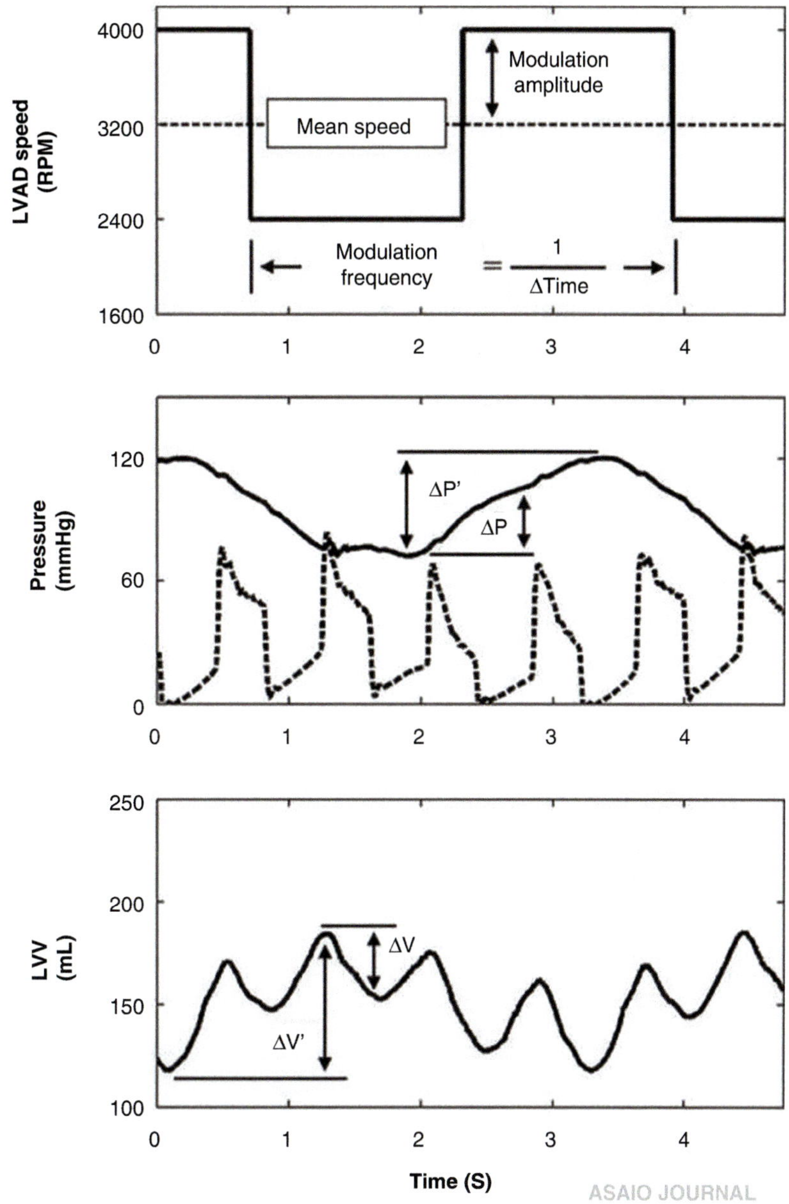

Fig. 21.7 Illustration of frequency (Hz) and amplitude (rpm) variables (top) for modulating left ventricular assist device (LVAD) to generate pulsatile pressure and flow (middle) and phasic volume unloading (bottom). (Reproduced from Ising et al. [62], with permission from Wolters Kluwer, Health Inc.)

maximum, minimum speeds, time period(s) within cardiac cycle, and frequency of profile (cycles per minute) to achieve a desired pulsatility and volume unloading, as shown in Fig. 21.7. In a pre-clinical study by Soucy et al. [31], *Rotary pump speed modulation for generating pulsatile flow and phasic left ventricular volume unloading*, the authors propose and demonstrate the integration and efficacy of fixed and modulated pump speed algorithms using synchronous co-pulsation, synchronous counter-pulsation, and asynchronous timing, which offer distinct potential physiologic benefits and limitation (Table 21.2). Synchronous co-pulsation, where the pump speed is increased during systole and decreased during diastole of the native heart, may be able to achieve a higher degree of pulsatility. Synchronous counter-pulsation, where the pump speed is decreased during systole and increased during diastole of the native heart, may provide

Table 21.2 Summary of left ventricular assist device (LVAD) operation modes that may be achieved by speed modulation with rotary pumps, sample hemodynamic waveforms illustrating principle of operation with level of pulsatility and ventricular volume unloading that may be achieved, and engineering and physiological benefits and limitations [31]

Pulsatile modality summary					
Pulse mode	Hemodynamic waveforms	PV loops	Benefits	Disadvantages	References
Fixed speed	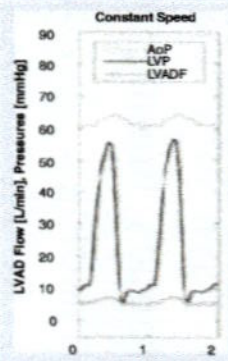	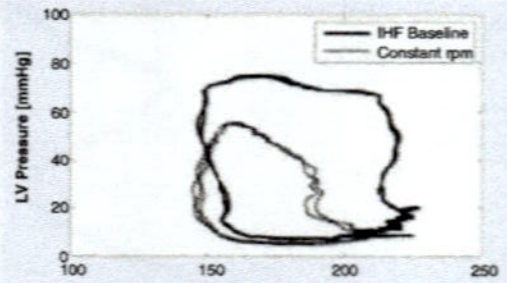		Diminished pulsatility Adverse events	
Synchronous (co-pulse)	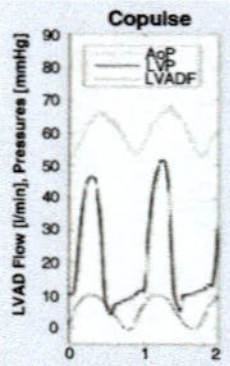	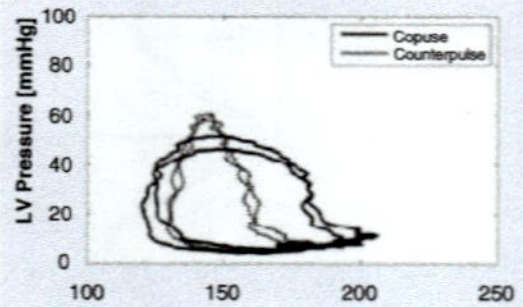	Increases pulsatility	May require sensor to trigger/time	
Synchronous (counter-pulse)	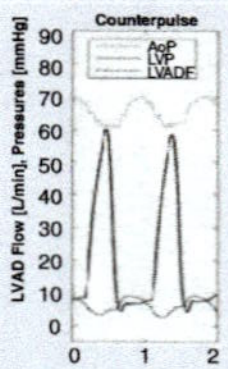	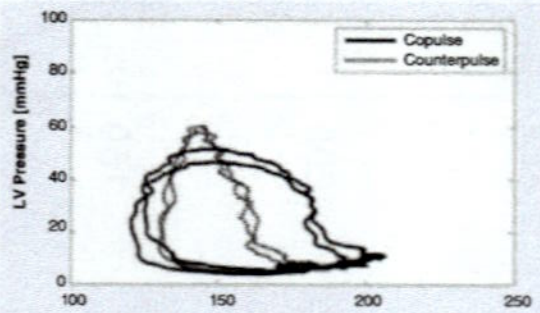	Reduction in external work Reduction of afterload Greater augmentation of myocardial perfusion	May require sensor to trigger/time	
Asynchronous	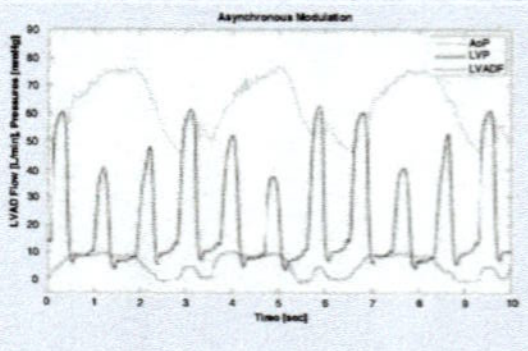	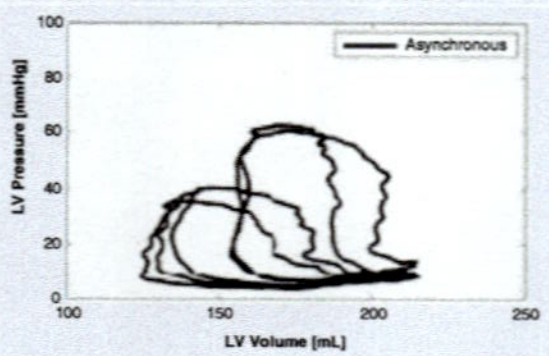	Sensor-less control Increases pulsatility Periodic opening of the aortic valve Increases washing capability		

PV loops pressure-volume loops

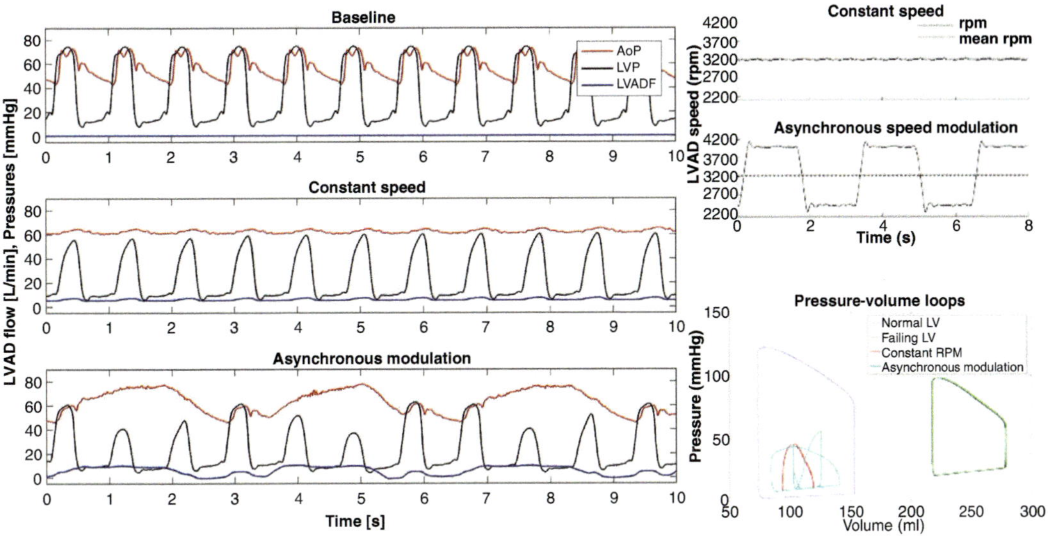

Fig. 21.8 Sample hemodynamic waveforms, including aortic pressure (AoP), left ventricular pressure (LVP), and left ventricular assist device flow (LVADF) (left) and resulting left ventricular pressure-volume relationship (bottom right) with LVAD off (baseline) and operated at constant pump speed and with asynchronous pump speed modulation (top right). These data demonstrated ability to demonstrate near-physiologic pulsatility and phasic volume unloading with asynchronous modulation compared to LVAD operated at fixed pump speed. (Reproduced from Soucy et al. [31])

a greater reduction in external work (small volume) and afterload as well as greater augmentation of myocardial perfusion. Asynchronous timing, where pump speed is cyclically increased and decreased independent of the native heart, provides the greatest increase in pulsatility (ΔP, SHE) and phasic volume unloading (ΔV), as shown in Fig. 21.8. Subsequently, asynchronous modulation periodically results in synchronous co-pulsation and counter-pulsation intermittently (in and out of phase) with the native heart over many cardiac cycles, resulting in the combined benefits of each, as well as periodic aortic valve opening and washing of the pump, inflow conduit, intraventricular chamber, and aortic root. For synchronized timing, a landmark trigger is required using physical sensor(s) and/or intrinsic pump parameter(s), which may be a challenge for reliability, whereas asynchronous modulation does not have this requirement.

In developing optimal rotary pump speed modulation algorithms, engineers and clinicians need to consider and weigh the potential benefits and trade-offs in controlling the diastolic-to-systolic flow ratio within the cardiac cycle by changing the pump speed, timing of speed changes, and the duration of speed changes. These approaches may be designed to optimize pump efficiency for better washing, achieve more flow at lower pump speeds, and/or optimize hemodynamics for maximum near-physiologic pulsatility (co-pulsation) or ventricular unloading (counter-pulsation). The ability to generate near-physiologic pulsatile blood pressure and flow may help to maintain normal vascular resistance and compliance, avoid low flow or suction conditions that can lead to thrombus formation (stroke), as well as maintain normal vascular structure and function to minimize risk of arteriovenous malformations (AVM) that can lead to gastrointestinal (GI) and/or neurological bleeding events.

Engineering trade-offs for implementing pump speed modulation may include the following requirements: physical sensor(s) and/or intrinsic pump parameter(s) for timing, increased power consumption, and potential for high (non-physiologic) shear stress. The search for accurate, reliable (gain, offset drift), and safe physical sensor(s) for integration with feedback control for MCS devices has been ongoing for decades [32]. The MicroMed DeBakey system (axial LVAD) included a transit-time flow probe on the outflow graft for continuous real-time measurement of pump flow, but the pump has undergone multiple design iterations and has not achieved widespread clinical use. The Intellect2 clinical trial is investigating the use of a pulmonary artery (PA) pressure sensor (CardioMEMS) in conjunction with the HeartMate III for patient management by monitoring right heart function [33]. In pre-clinical development, integration of potentially multiple sensors for monitoring pressure and flow is being considered [32]. Currently, non-invasive pressure measurement techniques, using pump display indices or Doppler ultrasound, are routinely used to assess pump pulsatility. However, the efficacy of Doppler ultrasound is unclear as potential issues with signal interference (noise), ability to reliably distinguish between peak-systolic and mean pressure landmarks, and inaccuracy due to tendency to overestimate mean pressure by up to 9 mmHg, especially in the presence of diminished pulsatility. Intrinsic pump parameters (power, speed, current, estimated flow, pulsatility index) are also being considered. Pump speed modulation algorithms have also been shown to require higher power requirements (20–40%) dependent upon head pressure (100 mmHg) compared to operation at fixed pump speeds [31]. Many studies have reported clinically insignificant levels of hemolysis in HF patients supported by axial and centrifugal blood pumps operated at fixed pump speeds despite higher (non-physiologic) shear stress [34–36]. In contrast, pump speed modulation may produce similar average elevated shear stress to fixed speed, but will have higher peak and lower minimum shear stresses for shorter time periods. Another important consideration is that hemolysis has been the primary metric used to assess blood trauma; however, acquired von Willebrand factor, platelet activation, thrombin formation, and cell deformation metrics are now being investigated to identify other potential clinically significant factors contributing to adverse events [12, 13, 37].

Pre-clinical Development of Pump Speed Modulation Algorithms

In vivo (*large animal models*) testing has been used to investigate candidate pump speed modulation algorithms as a function of shape, amplitude, and phase using a variety of CF-LVADs (CentriMag, EVAHEART, HeartMate II, HeartMate III, HVAD, and TORVAD) in pre-clinical development, proposed with large animal models (bovine, ovine, and porcine). Bourque et al. demonstrated the ability to achieve near-physiologic pulse pressures (ΔP 26–34 mmHg, dP/dt 139–238 mmHg/s) in the thoracic and abdominal aorta with mean arterial pressure greater than 30 mmHg by using the HeartMate III rapidly varying pump speed (1500–5500 rpm) while eliminating native heart participation (resecting ventricle) [38]. Pirbodaghi et al. investigated four select waveform shapes (sine, sawtooth, triangle, and square) for two separate phase conditions (0 °, 180 °) using the CentriMag synchronized to the native heart ECG, resulting in improved pulsatility and volume unloading. Differences in waveform morphology had no impact on the level of pulsatility achieved; however, differences in phase were observed as evidenced by increased pulsatility for co-pulsation and decreased end-diastolic volume and increased myocardial perfusion for counter-pulsation [39]. In subsequent studies confirmation of the physiologic benefits and limitations of synchronous (co-pulse and counter-pulse) and asynchronous pump speed modulation compared to fixed rate were demonstrated by multiple independent investigators [40]. In addition, Kishimoto et al. demonstrated the potential to minimize aortic insufficiency (AI) by adding a phase delay to pump speed modulation that allowed the aortic valve to open with high EVAHEART LVAD flow rates [41].

In vitro and in silico testing have also been used to evaluate pump speed modulation algorithms as a function of shape, phase, and univentricular or biventricular support using a variety of MCS devices in pre-clinical development (BiVACOR TAH, HeartMate II, HVAD, MicroMed DeBakey LVAD and TAH, and tVAD). Various of these algorithms have been proposed and tested in computer simulation and mock flow loop models. Kleinheyer et al. demonstrated the successful replication of a physiologic pulse morphology, with a pulse pressure of 10–35 mmHg, mean arterial pressure of 90 mmHg, and 5 L/min flow by implementing various pulse modulating strategies (sine, square 20% and 50%, saw tooth, LUT1, and LUT2) using a BiVACOR TAH (Houston, TX) [42]. The authors also indicated possible trade-offs between SHE levels and flow rate pulsatility related to the relative systolic duration in the speed profile shape. Tolpen et al. demonstrated periodic aortic valve opening by dropping pump speed from 10,000 to 7000 rpm for a period of 6 s without reduction in cardiac output for simulated low (CO 3.5 L/min, SV 50 ml, EF 28%) and moderate (CO 2.8 L/min, SV 40 ml, EF 22%) HF conditions [43]. Khalil et al. used a pair of MicroMed DeBakey pumps in series for TAH configuration and operated using four pulsatility modes (asynchronous left or right only, left and right co-pulsation, and left and right counter-pulsation) by changing pump speed (8000–11,000 rpm) to achieve near-physiologic pulse pressures (ΔP 16–61 mmHg) with corresponding decreases in left atrial pressure [44]. Shiose et al. created a pulse pressure by varying TAH pump speed (1500–4500 rpm) by 25% that required a 16% increase in pump power to maintain cardiac output. Bozkurt et al. used an isolated porcine heart model to investigate speed control algorithms with passive feedback control resulting in increased pulsatility and decreased end-diastolic volume [45]. Horobin et al. investigated blood trauma indices during HVAD operation at fixed and modulated pump speeds (3273 ± 430 rpm) with the following key findings: (1) no discernible differences in routine hemocompatibility biomarkers (hemolysis, platelet aggregation, vWF, ADAMTS13), (2) identified novel potential biomarker (RBC-NOS activity), and (3) greater potential risk for RBC deformability [46]. Use of computational models for evaluating pump speed modulation algorithms is a cost-effective preclinical method for theoretical analysis. A number of lumped parameter models have been used that have demonstrated similar hemodynamic benefits of pulsatility with pump speed modulation techniques [47–52].

Conclusion

With more than 15 years of clinical experience with continuous-flow LVAD support in HF patients, it is well proven that CF-LVADs provide significantly improved quality of life, survival rates, and functional status when compared to first-generation pulsatile-flow devices. Numerous clinical and pre-clinical studies have confirmed that there are no major differences between continuous and pulsatile-flow devices in terms of left ventricular unloading, systemic support, and end-organ function preservation. However, a significant number of LVAD therapy patients continue to suffer from major adverse events (stroke, bleeding, or thrombosis). The incidence of adverse events varies between device selection (HeartMate II, III, and HVAD) and indication for use (BTT or DT). Some of these difficulties may be attenuated by incorporating modulated pulsatility strategies into current and emerging continuous-flow devices.

References

1. Benjamin EJ, Blaha MJ, Chiuve SE, Cushman M, Das SR, Deo R, et al. Heart Disease and Stroke Statistics—2017 update: a report from the American Heart Association. Circulation. 2017;135:146–603.
2. Miller LW. Left ventricular assist devices are underutilized. Circulation. 2011;123(14):1552–8.
3. Colvin M, Smith JM, Hadley N, Skeans MA, Uccellini K, Lehman R, et al. OPTN/SRTR 2017 Annual Data Report: heart. Am J Transplant. 2017;17:1–72.
4. Estep JD, Starling RC, Horstmanshof DA, Milano CA, Selzman CH, Shah KB, et al. Risk assessment and comparative effectiveness of left ventricular assist device and medical management in ambulatory heart failure patients results from the ROADMAP study. J Am Coll Cardiol. 2015;66(16):1747–61.
5. Slaughter M, Rogers JG, Milano C, Russell SD, Conte JV, Feldman D, et al. Advanced heart failure treated with continuous-flow left ventricular assist device. N Engl J Med. 2009;361(23):2241–51.
6. Crow S, John R, Boyle A, Shumway S, Liao K, Colvin-adams M, et al. Gastrointestinal bleeding rates in recipients of nonpulsatile and pulsatile left ventricular assist devices. J Thorac Cardiovasc Surg. 2009;137(1):208–15.
7. Draper KV, Huang RJ, Gerson LB. GI bleeding in patients with continuous-flow left ventricular assist devices : a systematic review and meta-analysis. Gastrointest Endosc. Elsevier Ltd. 2014;80(3):435–46.

8. Demirozu ZT, Radovancevic R, Hochman LF, Gregoric ID, Letsou GV, Kar B, et al. Arteriovenous malformation and gastrointestinal bleeding in patients with the HeartMate II left ventricular assist device. J Hear Lung Transplant. Elsevier Inc. 2011;30(8):849–53.
9. Najjar SS. Late bleeding and neurological sequelae after HeartMate II left ventricular assist device. J Am Coll Cardiol. Elsevier. 2014;63(9):889–90.
10. Soucy K, Koenig S, Giridharan G, Sobieski M, Slaughter M. Rotary pumps and diminished pulsatility: do we need a pulse? ASAIO J. 2013;59:355–66.
11. Letsou GV, Shah N, Gregoric ID, Myers TJ, Delgado R, Frazier OH, et al. Gastrointestinal bleeding from arteriovenous malformations in patients supported by the Jarvik 2000 axial-flow left ventricular assist device. J Heart Lung Transplant. 2005;24:105–9.
12. Crow S, Chen D, Milano C, Thomas W, Joyce L, Iii VP, et al. Acquired von Willebrand syndrome in continuous-flow ventricular assist device recipients. Ann Thorac Surg. Elsevier Inc. 2010;90(4):1263–9.
13. Chen Z, Mondal NK, Ding J, Gao J, Griffith BP, Wu ZJ. Shear-induced platelet receptor shedding by non-physiological high shear stress with short exposure time: glycoprotein Ibα and glycoprotein VI. Thromb Res. 2015;135(4):692–8.
14. Undar A. Myths and truths of pulsatile and nonpulsatile perfusion during acute and chronic cardiac support. Artif Organs. 2004;28(5):439–43.
15. Krabatch T, Schweiger M, Dandal M, Stepanenko A, Drews T, Potapov E, et al. Is bridge to recovery more likely with pulsatile left ventricular assist devices than with nonpulsatile-flow systems? Ann Thorac Surg. 2011;91(5):1335–40.
16. Ji B, Ündar A. An evaluation of the benefits of pulsatile versus nonpulsatile perfusion during cardiopulmonary bypass procedures in pediatric and adult cardiac patients. ASAIO J. 2006;52(4):357–61.
17. Feldman CD, Pamboukian ÃSV, Teuteberg JJ, Moore SA, Morgan JA, Buchholz HW, et al. The 2013 International Society for Heart and Lung Transplantation Guidelines for mechanical circulatory support: executive summary. J Hear Lung Transplant. Elsevier. 2013;32(2):157–87.
18. Chen Z, Jena S, Giridharan G, Koenig S, Slaughter M, Griffith BP, et al. Flow features and device-induced blood trauma in CF-VADs under a pulsatile blood flow condition: a CFD comparative study. Int J Numer Method Biomed Eng. 2018;34(2):e2924.
19. Shepard RB, Simpson DC, Sharp JF. Energy equivalent pressure. Arch Surg. 1966;93:730–40.
20. Undar A, Zapanta C, Reibson J, Souba M, Lukic B, Weiss WJ, et al. Precise quantification of pressure flow waveforms of a pulsatile ventricular assist device. ASAIO J. 2005;51(3):56–9.
21. Bingyang J, Undar A. An evaluation of the benefits of pulsatile versus nonpulsatile perfusion during cardiopulmonary bypass procedures in pediatric and adult cardiac patients. ASAIO J. 2006;52(4):357–61.
22. Travis AR, Giridharan GA, Pantalos GM, Dowling RD, Prabhu SD, Slaughter MS, et al. Vascular pulsatility in patients with a pulsatile- or continuous-flow ventricular assist device. J Thorac Cardiovasc Surg. 2007;133(2):517–24.
23. Soucy K, Koenig S, Giridharan G, Sobieski M, Slaughter M. Defining pulsatility during continuous flow ventricular assist device support. J Heart Lung Transplant. 2013;32(6):581–7.
24. Castagna F, Stohr E, Pinsino A, Cockcroft J, Willey J, Reshad Garan A, et al. The unique blood pressures and pulsatility of LVAD patients: current challenges and future opportunities. Curr Hypertens Rep. 2017;19(10):85.
25. Wiegmann L, Thamsen B, De Zélicourt D, Granegger M, Boës S, Daners MS, et al. Fluid dynamics in the HeartMate 3: influence of the artificial pulse feature and residual cardiac pulsation. Artif Organs. 2018;43(4):363–76.
26. Kumar J, Elhassan A, Dimitrova G, Essandoh M. The Lavare cycle: a novel pulsatile feature of the HVAD continuous-flow left ventricular assist device. J Cardiothorac Vasc Anesth. Elsevier Inc. 2019;33(4):1170–1.
27. Moazami N, Fukamachi K, Kobayashi M, Smedira NG, Hoercher KJ, Massiello A, et al. Axial and centrifugal continuous-flow rotary pumps: a translation from pump mechanics to clinical practice. J Heart Lung Transplant. Elsevier. 2013;32(1):1–11.
28. Fukamachi K, Shiose A, Massiello A, David J, Starling C. Preload sensitivity in cardiac assist devices. Ann Thorac Surg. 2014;95(1):373–80.
29. Lim HS, Howell N, C-th F, Ranasinghe A, C-th F. The physiology of continuous-flow left ventricular. J Card Fail. Elsevier Inc. 2017;23(2):169–80.
30. Willey JZ, Boehme AK, Castagna F, Yuzefpolskaya M, Garan AR, Topkara V, et al. Hypertension and stroke in patients with left ventricular assist devices (LVADs). Curr Hypertens Rep. 2016;18(2):12.
31. Soucy K, Giridharan G, Choi Y, Sobieski M, Monreal G, Cheng A, et al. Rotary pump speed modulation for generating pulsatile flow and phasic left ventricular volume unloading in a bovine model of chronic ischemic heart failure. J Heart Lung Transplant. 2015;34:122.
32. Khare A. Estimation and control of the pump pressure rise and flow from intrinsic parameters for a magnetically-levitated axial blood pump. Thesis 2008.
33. Investigation to optimize hemodynamic management of left ventricular assist devices using the CardioMEMS™ (Intellect2). Abbott Medical Devices.
34. Krabatsch T, Potapov E, Stepanenko A, Schweiger M, Kukucka M, Huebler M, et al. Biventricular circulatory support with two miniaturized implantable assist devices. Circulation. 2011;124:S179.
35. Slaughter MS. Hematologic effects of continuous flow left ventricular assist devices. J Cardiovasc Transl Res. 2010;3:618–24.

36. Heilmann C, Geisen U, Benk C, Berchtold-herz M, Trummer G, Schlensak C, et al. Haemolysis in patients with ventricular assist devices: major differences between systems. Eur J Cardiothorac Surg. 2009;36:580–4.
37. Pirbodaghi T, Asgari S, Cotter C, Bourque K. Physiologic and hematologic concerns of rotary blood pumps: what needs to be improved? Heart Fail Rev. 2014;19:259–66.
38. Bourque K, Dague C, Farrar D, Harms K, Tamez D, Cohn W, et al. In vivo assessment of a rotary left ventricular assist device-induced artificial pulse in the proximal and distal aorta. Artif Organs. 2006;30(8):638.
39. Pirbodaghi T, Axiak S, Weber A, Gempp T, Vandenberghe S. Pulsatile control of rotary blood pumps: does the modulation waveform matter? J Thorac Cardiovasc Surg. 2012;144(4):970–7.
40. Pirbodaghi T, Weber A, Axiak S, Carrel T, Vandenberghe S. Asymmetric speed modulation of a rotary blood pump affects ventricular unloading. Eur J Cardiothorac Surg. 2013;43:383.
41. Kishimoto Y, Takewa Y, Arakawa M, Umeki A, Ando M, Nishimura T, et al. Development of a novel drive mode to prevent aortic insufficiency during continuous-flow LVAD support by synchronizing rotational speed with heartbeat. J Artif Organs. 2013;16(2):129–37.
42. Kleinheyer M, Timms D, Tansley G, Nestler F, Greatrex N, Frazier O, et al. Rapid speed modulation of a rotary total artificial heart impeller. Artif Organs. 2016;40:824.
43. Tolpen S, Janmaat J, Reider C, Kallel F, Farrar D, May-Newman K. Programmed speed reduction enables aortic valve opening and increased pulsatility in the LVAD-assisted heart. ASAIO J. 2015;61(5):540.
44. Khalil H, Kerr D, Schusterman MI, Cohn W, Frazier O, Radovancevic B. Induced pulsation of a continuous-flow total artificial heart in a mock circulatory system. J Heart Lung Transplant. 2010;29(5):568–73.
45. Shiose A, Nowak K, Horvath D, Massiello A, Golding L, Fukamachi K. Speed modulation of the continuous-flow total artificial heart to simulate a physiologic arterial pressure waveform. ASAIO J. 2010;56(5):403–9.
46. Horobin J, Simmonds M, Nandakumar D, Gregory S, Tansley G, Pauls J, et al. Speed modulation of the HeartWare HVAD to assess in vitro hemocompatibility of pulsatile and continuous flow regimes in a rotary blood pump. Artif Organs. 2018;42:879.
47. Arndt A, Nusser P, Graichen K, Muller J, Lampe B. Physiological control of a rotary blood pump with selectable therapeutic options: control of pulsatility gradient. Artif Organs. 2008;321(10):761–71.
48. Shi Y, Korakianitis T, Bowles C. Numerical simulation of cardiovascular dynamics with different types of VAD assistance. J Biomech. 2007;40(13):2919–33.
49. Cox L, Loerakker S, Rutten M, de Mol B, va de Vosse F. A mathematical model to evaluate control strategies for mechanical circulatory support. Artif Organs. 2009;33:593.
50. Choi S, Boston J, Antaki J. Hemodynamic controller for left ventricular assist device based on pulsatility ratio. Artif Organs. 2007;31(2):114–25.
51. Wu Y. Adaptive physiological speed/flow control of rotary blood pumps in permanent implantation using intrinsic pump parameters. ASAIO J. 2009;55(4):335–9.
52. Gao B, Chang Y, Gu K, Zeng Y, Liu Y. A pulsatile control algorithm of continuous-flow pump for heart recovery. ASAIO J. 2012;58:343.
53. Rich J, Burkhoff D. HVAD flow waveform morphologies: theoretical foundation and implications for clinical practice. ASAIO J. 2017;63:526–35.
54. Heartware products and technology: Heartware HVAD system.
55. HeartMate 3™ left ventricular assist system – P160054. US FDA. 2017.
56. Jarvik heart products: The Jarvik 2000. Jarvik Hear.
57. HEARTMATE 3™ left ventricular assist system instructions. 2009(September).
58. Thoratec. HeartMate II® LVAS – Left ventricular assist system: operating manual. 2013;1477:264.
59. HeartWare ® ventricular assist system instructions for use.
60. Essandoh M, Essandoh G, Stallkamp ED, Perez WJ. Spectral Doppler analysis of the HeartMate 3 left ventricular assist device inflow: new challenges presented by the artificial pulse technology. J Cardiothorac Vasc Anesth. Elsevier Inc. 2018;32(6):e4–5.
61. Mancini D, Colombo PC. Left ventricular assist devices. J Am Coll Cardiol. Elsevier Inc. 2015;65(23):2542–55.
62. Ising MS, Sobieski MA, Slaughter MS, Koenig SC, Giridharan GA. Feasibility of pump speed modulation for restoring vascular pulsatility with rotary blood pumps. ASAIO J. 2015;61(5):526–32.

Part IV

Current Technology and Methods

22 Implantable Continuous-Flow Blood Pump Technology and Features

Matthew L. Goodwin, Peter H. U. Lee, and Nahush A. Mokadam

Abbreviations

AC	alternating current
BTT	bridge to transplant
DC	direct current
DT	destination therapy
FDA	US Food and Drug Administration
IDE	investigational device exemption
ILS	intermittent low speed
LVAD(s)	left ventricular assist device(s)
LVAS	left ventricular assist system
MCS	mechanical circulatory support
NIH	National Institute of Health
PMDA	Pharmaceuticals and Medical Devices Agency
PTFE	polytetrafluoroethylene
VAD	ventricular assist device
VAS	ventricular assist system

M. L. Goodwin, MD · P. H. U. Lee, MD, PhD, MPH · N. A. Mokadam, MD (✉)
Division of Cardiac Surgery, The Ohio State University Wexner Medical Center, Columbus, OH, USA
e-mail: Matthew.Goodwin@osumc.edu; Peter.Lee@osumc.edu; Nahush.Mokadam@osumc.edu

Introduction

Durable mechanical circulatory support (MCS) has emerged as the standard of care for patients with advanced heart failure refractory to maximal medical therapy awaiting heart transplantation or as destination therapy. With the number of donor hearts for transplantation remaining relatively constant over the years [1], the demand for MCS devices as a treatment modality for advanced heart failure has driven considerable evolution in device design and technology during the last 40 years. The National Institutes of Health (NIH) prioritized circulatory support devices for heart failure with the inception of the artificial heart program in 1964 [2]. Left ventricular assist devices (LVADs) were first approved by the US Food and Drug Administration (FDA) as a bridge to transplantation in 1994 [3]. Initial LVAD technology utilized large, pneumatically driven, pulsatile pumps to provide adequate circulatory support. Long-term, pulsatile LVAD support was shown to improve survival and quality of life as destination therapy in heart failure patients [4]. However, the large size, limited durability, and complication profile of these early, pulsatile-flow devices hindered long-term success [5]. Continuous-flow LVAD technology is developed as an alternative to the early, pulsatile-flow devices.

Continuous-flow LVADs offer several advantages in pump design that aid in minimizing com-

© Springer Nature Switzerland AG 2020

J. H. Karimov et al. (eds.), *Mechanical Support for Heart Failure*, https://doi.org/10.1007/978-3-030-47809-4_22

plications associated with LVAD support [5]. Continuous-flow LVADs use axial-flow and centrifugal rotary pump designs, which eliminate the blood pumping chambers and volume compensation of earlier designs, making these devices substantially smaller and suitable for smaller patients [5]. These pumps are without internal valves and involve only the moving rotor, enhancing durability and producing silent pump operations [5]. When compared to early, pulsatile-flow devices, continuous-flow devices improved survival and decreased adverse events associated with MCS [6]. This chapter reviews continuous-flow LVADs and highlights the design features and technology of the axial-flow and centrifugal implantable rotary pumps approved for investigational or commercial use in the United States and Europe (Table 22.1). Clinical data will be reviewed, where available, to illustrate the effect of continuous-flow device design and technology on patient outcomes.

Table 22.1 Implantable continuous-flow VADs

Device	Manufacturer	Flow	Bearings	Size	Weight	Cardiac support	Approvals	
HeartMate II	Abbott	Axial	Mechanical, blood-immersed	8.1 cm × 4.3 cm	281 g	3–10 L/min	CE Mark 2005	FDA 2008 (BTT) FDA 2010 (DT)
Jarvik 2000	Jarvik Heart, Inc.	Axial	Mechanical cone, blood-immersed	7.8 cm × 2.6 cm	90 g	2–8.5 L/min	CE Mark 2005 PMDA 2013	FDA IDE
HeartAssist 5	ReliantHeart Inc.	Axial	Mechanical ball-in-cup, blood-immersed	7.1 cm × 3 cm	92 g	2–10 L/min	CE Mark 2010	FDA IDE
aVAD	ReliantHeart Inc.	Axial	Mechanical ball-in-cup, blood-immersed	2.48 cm outer diameter	84 g	2–7 L/min	CE Mark 2016	
INCOR	Berlin heart GmbH	Axial	Magnetic	12 cm × 3 cm	200 g	< 6 L/min	CE Mark 2003	
HVAD	Medtronic Inc.	Centrifugal	Magnetic levitation and hydrodynamic	6 cm × 2.8 cm	145 g	10 L/min	CE Mark 2009 (BTT) CE Mark 2012 (DT)	FDA 2012 (BTT) FDA 2017 (DT) FDA 2018 (Thoracotomy)
HeartMate 3	Abbott	Centrifugal	Magnetic levitation	5.03 cm × 3.4 cm	200 g	10 L/min	CE Mark 2015	FDA 2017 (BTT) FDA 2018 (DT)
EVAHEART 2	Sun Medical Technology Research Corp.	Centrifugal	Hydrodynamic levitation	5.1 cm × 67 cm	262 g	< 14 L/min	PMDA 2008	FDA IDE

VAD ventricular assist device, *CE Mark* certification mark within the European Economic Area, *FDA* Food and Drug Administration, *IDE* investigational device exemption, *PMDA* Japanese Pharmaceuticals and Medical Devices Agency, *BTT* bridge to transplant, *DT* destination therapy

Axial-Flow LVADs

HeartMate II

The HeartMate II Left Ventricular Assist System (LVAS) (Abbott, Abbott Park, IL) is an axial-flow, rotary, durable left ventricular assist device. The device project began in 1991 from a research collaboration between the Nimbus Company and the University of Pittsburgh's McGowan Center for Organ Engineering with the first human implantation in 2000 [7]. The HeartMate II received CE Mark approval in 2005, FDA approval as a bridge-to-transplant (BTT) device in 2008, and FDA destination therapy (DT) approval in 2010. The LVAS consists of the HeartMate II LVAD, pump assembly, a System Controller, Power Module, patient cable, batteries, battery clip, universal battery charger, and System Monitor with cable [8] (Fig. 22.1).

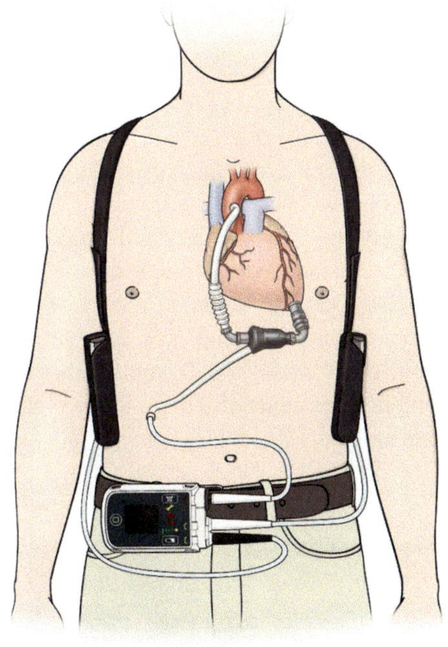

Fig. 22.1 The HeartMate II Left Ventricular Assist System (LVAS) with an axial-flow rotary pump. Blood enters through an inflow cannula in the left ventricular apex and is continuously pumped into the ascending aorta. The pump sits below the diaphragm and is connected to the System Controller via a percutaneous driveline cord. (HeartMate II and HeartMate 3 are trademarks of Abbott or its related companies. Reproduced with permission of Abbott, © 2019. All rights reserved)

The HeartMate II LVAD generates up to + 10 liters/minute (L/min) blood flow in parallel to the heart, by utilizing an electric motor that generates torque to drive the rotor. The motor creates a magnetic field that spins a permanent magnet located within the rotor. The magnetic rotor is located inside a thin-walled, 12 mm titanium duct that passes through the bore of the motor. An inlet stator straightens the blood flow field upon entering the rotor. The rotor uses three blades to impart kinetic energy to the flow field in the form of radial velocity [8]. As blood exits the rotor, an exit stator converts the radial velocity created by the rotor into an axial velocity [8]. The pump rotor apparatus is a smooth, polished titanium surface, which is washed directly by the main blood flow field to reduce the formation of thrombus. The pump apparatus weighs 281 grams (g) and measures 8.1 cm × 4.3 cm, with a gross volume of 63 ml (Fig. 22.2). Pump flow is dependent on the rotation speed of the rotor and the pressure difference between the inflow and outflow conduits.

Blood enters the pump via a 14 mm gelatin-impregnated knitted polyester with polypropylene reinforced inflow conduit [9]. The inflow conduit's proximal arm is a 19 mm textured titanium inlet extension that stimulates tissue ingrowth, creating a natural lining which has been attributed to a reduction in thromboembolic complications [10]. The inflow conduit connects to a flexible silicone sleeve and screw ring that attaches to the pump. The outflow graft consists of a 14 mm diameter sealed gelatin-impregnated woven polyester graft with an outer polytetrafluoroethylene bend relief sleeve [9]. Like the inflow conduit, the outflow graft attaches to the pump via a blue screw ring.

The HeartMate II LVAD contains a single cable driveline that extends from the LVAD through the skin to the System Controller. The driveline consists of three primary wires and three backup wires shielded in silicone sheaths and covered with woven polyester [9]. An electrical connector attaches the driveline to the System Controller. The driveline is tunneled through the right rectus sheath and exits the subcutaneous tissue in the right upper abdominal quadrant. The

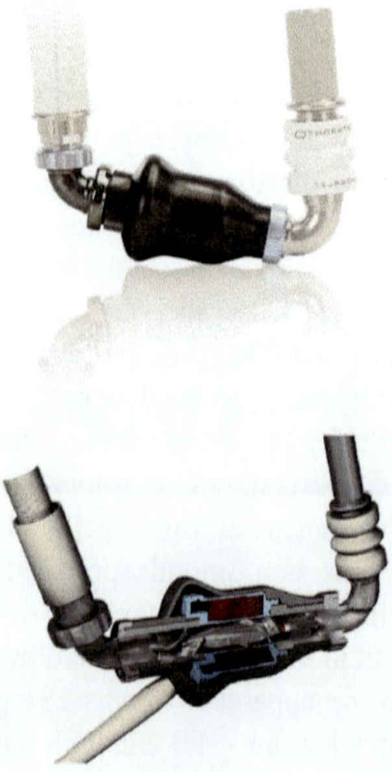

Fig. 22.2 HeartMate II axial-flow rotary pump with inlet stator, mechanical bearing system, three-bladed, magnetic rotor, and exit stator in a 12 mm diameter titanium duct. HeartMate II inflow cannula features a 19 mm textured titanium inlet extension, flexible silicone sleeve, and screw ring to attach to the pump. HeartMate II 14 mm diameter sealed gelatin-impregnated woven polyester graft with an outer polytetrafluoroethylene bend relief sleeve. (HeartMate II and HeartMate 3 are trademarks of Abbott or its related companies. Reproduced with permission of Abbott, © 2019. All rights reserved)

LVAD is powered through the System Controller by either the Power Module which connects to an alternating current (AC) electrical outlet or two HeartMate 12 Volt nickel metal hydride or 14 Volt lithium-ion direct current (DC) batteries. An emergency power pack (EPP) is a third, alternative power source available for use in emergency situations.

The HeartMate II LVAD is surgically implanted beneath the diaphragm in a preperitoneal or intra-abdominal position. The inflow cannula is inserted into the apex of the left ventricle and connected to the pump apparatus. A coring knife and apical sewing ring are used to prepare the left ventricular apex for inflow cannula insertion. The inflow cannula should avoid the left anterior descending artery and is directed towards the mitral valve orifice to optimize inflow in and away from the intraventricular septum and ventricular free wall with the visual aid of intraoperative transesophageal echocardiography. Inflow positioning is critical to LV unloading and reducing pump thrombus formation or hemolysis [5, 11, 12]. The outflow graft with bend relief is anastomosed to the ascending aorta in an end-to-side fashion. The woven polyester graft is stretched appropriately to prevent redundancy and placed in the right sternal pericardial well in an unkinked and untwisted orientation [5]. The outflow graft positioning is also important in maintaining pump function and limiting complications. The bend relief system had a propensity to migrate along the outflow graft which was modified by a snap ring design in 2010 [13, 14].

The System Controller serves as the main user interface for the HeartMate II LVAS (Fig. 22.3). It provides a variety of monitoring functions, implements selected operating modes, and notifies of system alarms. Monitoring functions include performance data processing and storage, battery life indicators, a bidirectional data link, driveline continuity check, and fault detection [9]. The System Controller is capable of implementing one of two operating modes. The primary operating mode is a fixed speed mode which is set by the medical team using the System Monitor or Display Module. The fixed speed mode maintains the pump at a constant speed between 6000 and 15,000 rpm and can be adjusted in increments of 200 rpm. A power save mode is entered when battery charge falls below

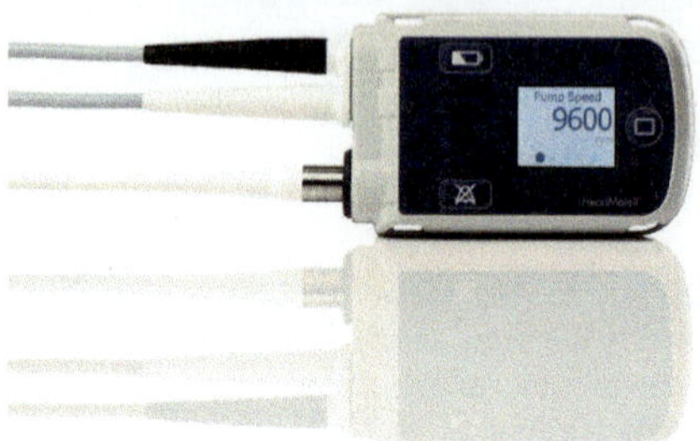

Fig. 22.3 HeartMate II System Controller. (HeartMate II and HeartMate 3 are trademarks of Abbott or its related companies. Reproduced with permission of Abbott, © 2019. All rights reserved)

a critical voltage. In addition, a backup operating system provides redundancy in the event that the primary operating mode becomes defective. An alarm system will notify the user of low flow or an inoperative pump, disconnection from power, low battery voltage, and low speed operation. The alarm notification system is categorized into hazard alarms and advisory alarms. Hazard alarms indicate hemodynamic support has been lost or is imminent.

The System Controller contains two leads for connection to a power source in addition to a connection for the percutaneous lead from the HeartMate II pump. The power options for the HeartMate II include the Power Module, a Mobile Power Unit, or battery operation. The Power Module is intended to provide power to the System Controller and pump in a clinical setting in which the System Monitor is applied. These clinical scenarios include pump start-up; changing the pump settings; viewing pump diagnostics, alarms, and history; and most inpatient situations. The System Monitor will display the pump flow, speed, power, and pulsatility index (PI).

Initial BTT trials for the HeartMate II device published by Miller et al. [15] and an extended enrollment protocol by Pagani et al. [16] showed durable and effective hemodynamic support with significant improvement in quality of life and functional status in patients awaiting transplant. A clinical trial for DT in nontransplant-eligible patients randomized patients 2:1 between the HeartMate II LVAD and the pulsatile-flow HeartMate XVE LVAD [6]. The HeartMate II device provided improved survival and survival free from stroke or reoperation at 2 years [6]. Similarly, nonrandomized, observational studies show improved event-free survival and functional status in advanced heart patients receiving the HeartMate II LVAD compared to optimal medical therapy with 2-year survival of 70 ± 5% [17, 18] (Fig. 22.4). However, the technology is not without complications. Bleeding remains the most commonly reported complication with the HeartMate II [6, 15, 16]. The axial-flow pump design with contact bearings necessitates anticoagulation use as well as affects platelet function [19]. The pump design also lends itself to thrombus formation. An increased incidence in pump thrombosis was observed with the HeartMate II in implants after 2011 [20, 21]. Risk factor analysis suggested that a number of patient-related factors and patient management strategies including anticoagulation therapy may have attributed to the increase in pump thrombosis [20]. Additional

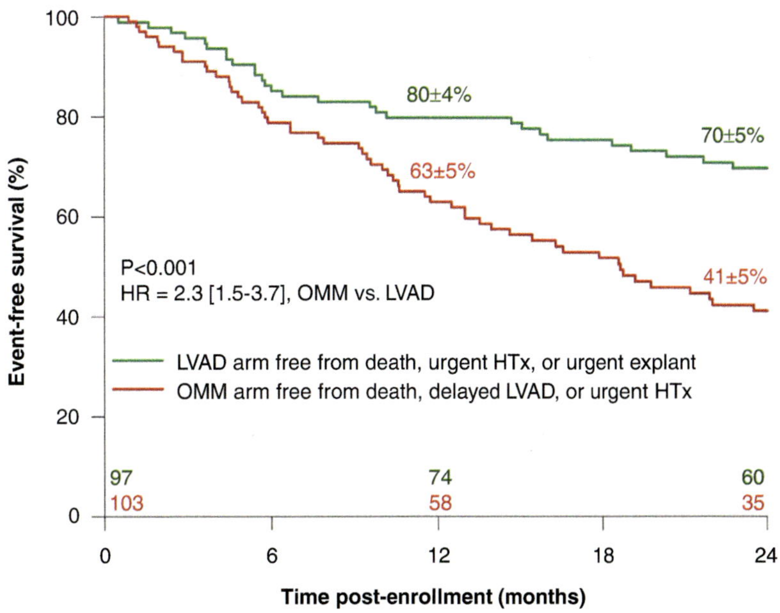

Fig. 22.4 A 24-month event-free survival on original therapy for nonrandomized patients receiving the HeartMate II LVAD or optimal medical therapy (OMM). (Reproduced from Starling et al. [18])

adverse events including infection (17%) and rehospitalization (86%) were significantly increased in HeartMate II patients at 2 years compared to patients maintained on medical therapy [18]. Patients undergoing LVAD implantation had an increased stroke risk (12%) in the first year after implantation which was not seen thereafter when compared to medical therapy [18].

Jarvik 2000

The Jarvik 2000 Ventricular Assist System (VAS) is a miniature axial-flow blood pump system (Jarvik Heart, Inc., New York, NY). Dr. Robert Jarvik began working with axial-flow blood pump technology in 1977 at the University of Utah, concurrent with the development of the Jarvik 7 heart [22]. The Jarvik 2000 VAS development began in 1987 and was supported in the early 1990s with research grants from the National Institutes of Health and New York State. The first human implant took place in 2000. The device received CE Mark certification in 2005 for BTT and DT. In 2000, the FDA granted an investigational device exemption (IDE) for BTT use as part of the Jarvik 2000 pilot study [23]. The FDA approved the destination trial named RELIVE for the evaluation of the Jarvik 2000 for permanent use in 2012. The system consists of the Jarvik 2000 VAD, an implanted power cable with post-auricular or abdominal exit points, differentiating the two Jarvik 2000 VAS models, outflow graft, the FlowMaker Controller, and rechargeable lithium-ion batteries with battery pack.

The Jarvik 2000 LVAD has undergone several stepwise, systematic improvements to enhance function and limit complications since its implementation [24] (Table 22.2). The current iteration has several unique design and technology features that will be described herein. The Jarvik LVAD pump is relatively small compared to other commercially available continuous-flow devices. The adult pump measures 7.8 cm × 2.6 cm and weighs 90 g, with a gross volume of 30 ml. The miniaturization of the pump is accomplished through its design. The low profile pump is placed into the ventricle within the pericardium,

Table 22.2 Milestone developments in the Jarvik 2000

2000	IDE approved by the FDA and pilot study initiated
2002	Introduced "Y cable" to permit battery changes without pump stoppage
2003	Began apical implantation via the left chest without cardiopulmonary bypass
2003	Developed intermittent low speed (ILS) controller to improve aortic washing
2004	Increased pull strength of external power cables from 35 to 200 pounds
2005	CE Mark approval
2005	FDA approval of pivotal trial in BTT patients
2005	Recovery case with intact apical pump, ligated aortic graft, and cable removal
2006	Radially reinforced PTFE grafts for bend relief and adhesion reduction
2008	Added titanium microsphere surface to reduce intraventricular thrombus
2009	Use of total ventricular assist with biventricular pumps
2010	Introduced cone bearings to reduce pump thrombus
2010	FDA-approved IDE supplement substituting cone for pin bearings
2010	Increased cable flex to driveline failure

IDE investigational device exemption, *FDA* Food and Drug Administration, *BTT* bridge to transplant, *PTFE* polytetrafluoroethylene. (Reproduced with permission from Selzman, CH[24])

eliminating the need for an inflow conduit or pump pocket. A Silastic sewing cuff is placed on the left ventricular apex and secures the pump within the myocardium [25]. Blood enters the pump by passing through an intraventricular inlet cage. The outer pump surface was modified with a titanium microsphere coating, preventing thrombus formation between the pump inflow orifice and endocardium [24]. A cone-bearing design supports the bladed, spinning titanium rotor. A stationary ring with streamlined posts matches the conical shape of the bearings and holds the rotor centered [22]. The cone-bearing design replaced a pin-bearing system which improved the hemodynamic profile of the pump and eliminated a thrombus-generating space in the bearing apparatus [24] (Fig. 22.5).

The blood-contacting surface of the pump is made of silicon carbide (Sic) and titanium alloy (Ti-6Al-4 V) [26]. Mechanical durability, ease of shaping, relatively low cost, and decent biocom-

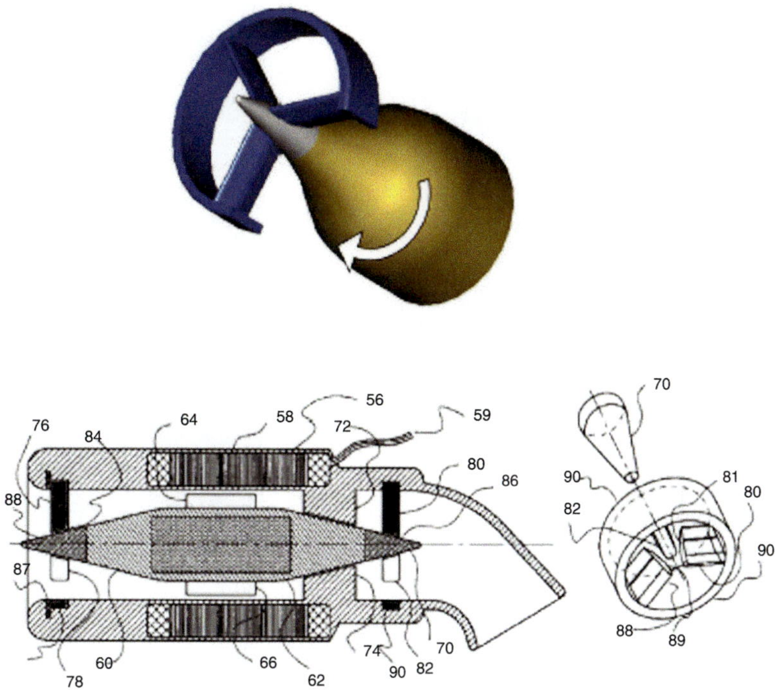

Fig. 22.5 The Jarvik 2000 cone-bearing design. The mechanical cone bearings replaced a pin bearing system and improved the hemodynamic profile of the pump and decreased thrombus generation. The cone bearings rotate in a stationary ring with streamlined posts and prevent radial deviation of the spinning magnetic rotor. (Reproduced from Jarvik [22])

patibility are characteristics of these materials [26]. The rotor is composed of a neodymium-iron-boron magnet, which under electromagnetic forces spins the rotor at 8000–12000 rpm [25]. The pump is capable of providing up to 8.5 L/minute of support.

Outflow to the Jarvik 2000 pump is provided by a pre-attached 16 mm Hemashield outflow graft. Blood enters the outflow graft after passing outflow stator blades downstream from the rotor. A radially supported polytetrafluoroethylene (PTFE) graft sutured to the flange of the outflow graft serves as a bend relief and adhesion reduction system for the outflow graft [24]. The outflow graft is anastomosed to either the ascending or descending aorta. Power to the device is provided by the percutaneous power cord. The cord uses a triple redundant cable insulated with polyurethane and partially covered with Dacron [25]. The Jarvik 2000 VAS is available in two models with either an abdominal exiting power cable or a novel, head-mounted post-auricular model. The post-auricular model, based on cochlear implant pedestals, is highly resistant to infection, allows normal showering and bathing, and is cosmetically acceptable [27, 28] (Fig. 22.6). Various implantation techniques including sternotomy and thoracotomy with and without cardiopulmonary bypass support have been described [29]. In addition, biventricular devices have been utilized to provide total ventricular support [30, 31].

The Jarvik 2000 is powered by an external FlowMaker Controller and battery pack. The controller is connected to the external cable from a post-auricular pedestal. The FlowMaker Controller is unique in that it features patient-adjusted speed settings and an intermittent low speed (ILS) mode. The pump speed is adjusted between five speed settings ranging from 8000 to 12,000 rpm. The controller contains a speed setting knob with which patients may adjust the pump speed based on exercise levels and symptoms. The ILS mode intermittently reduces the set speed of the pump to 7000 rpm for 8 seconds each minute. The pump will operate at the set speed for the remainder of the minute. The intermittent speed reduction permits increased preload and ventricular contractility, if present, thereby washing the aortic valve of stagnant blood and providing some systemic pulsatility

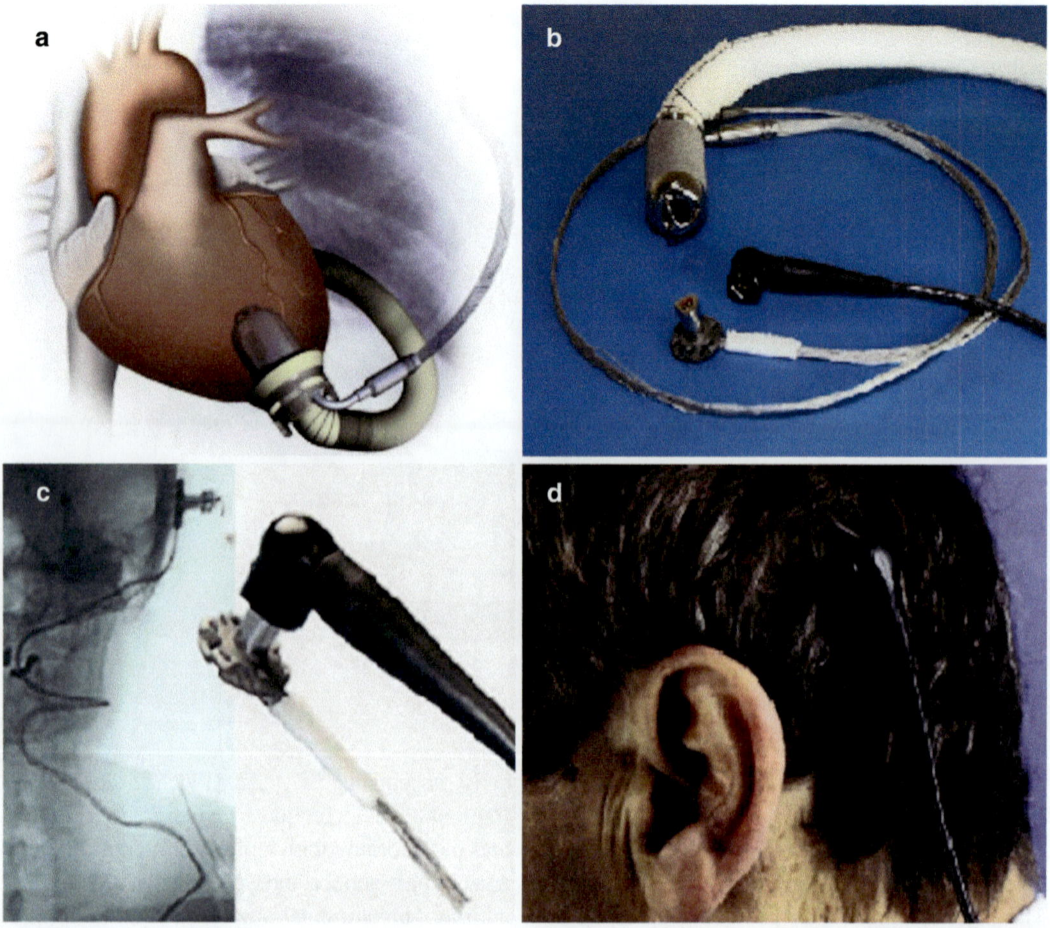

Fig. 22.6 (**a**) The Jarvik 2000 is placed in the pericardial space with the pump directed through the left ventricular apex, eliminating the need for inflow conduit. The outflow graft is anastomosed to the descending aorta. (**b**) 16 mm diameter outflow graft with proximal bend relief and percutaneous power cord with post-auricular pedestal. (**c**) Fluoroscopy showing the power cord tunneled to the post-auricular scalp. (**d**) Healed post-auricular pedestal with connected external power cable. (Reproduced from Jarvik [22])

[24] (Fig. 22.7). Lithium-ion batteries provide power to the Jarvik 2000 VAS for 7–12 hours under usual operating conditions. A Y cable connects the FlowMaker Controller to the battery and allows for battery changes without pump stoppage.

The unique technical and innovative advances of the Jarvik 2000 VAS have been mirrored with other implantable continuous-flow blood pumps. The Jarvik 2000 provides a small, durable intrapericardial device with a novel post-auricular power connector. The device features a patient-controlled adjustable speed setting with intermittent low speed mode. The pump can be implanted with a variety of surgical techniques and does not require pump removal in recovery patients. Additionally, the Jarvik 2000 technology is available in infant and child models. The child model, Jarvik 2015 VAD, is currently under investigation in a prospective, multicenter, single-arm feasibility study, the PumpKIN clinical trial, as a bridge to transplant or recovery in children.

Outcomes for the Jarvik 2000 VAS are more readily available in data obtained outside the United States, where the Jarvik 2000 VAS gained CE Mark approval in 2005 and Japanese Pharmaceuticals and Medical Devices Agency (PMDA) approval in 2013. Westaby et al.

a

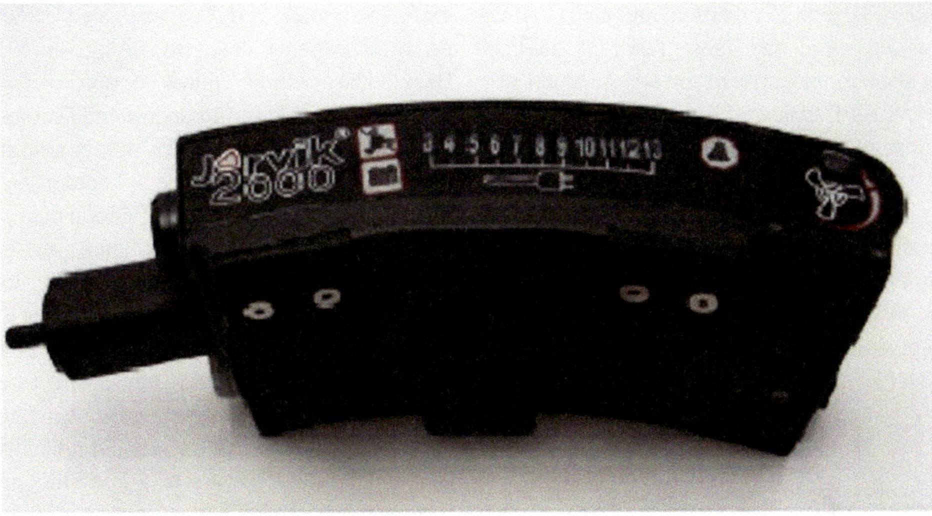

b

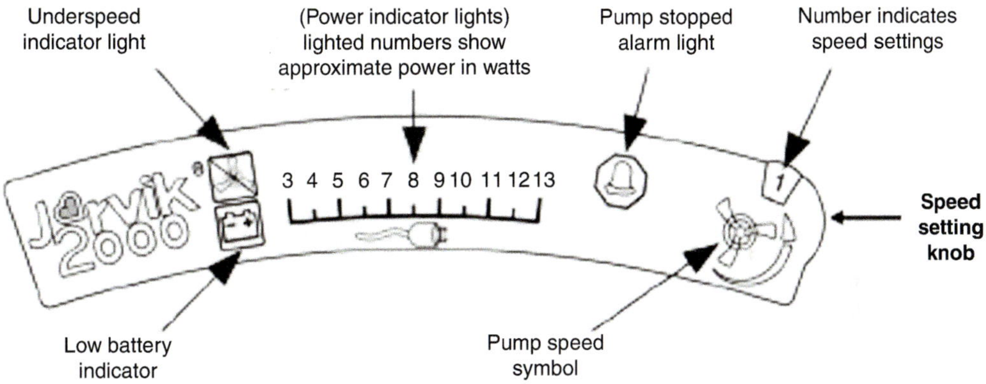

c

Setting number	Approximate Jarvik 2000 VAD speed	Approximate flow
1	8,000 RPM	1-2 L/min
2	9,000 RPM	2-3 L/min
3	10,000 RPM	4-5 L/min
4	11,000 RPM	5-7 L/min
5	12,000 RPM	7-8.5 L/min

Fig. 22.7 (**a**) The Jarvik 2000 Intermittent Low Speed (ILS) FlowMaker Controller provides power to the LVAD. (**b**) Diagram of the FlowMaker Controller top panel including the self-adjustable speed knob. (**c**) The five speed settings of the Jarvik 2000 with approximate rpms and flows associated with each speed setting. (Reproduced from Jarvik [22])

reported a 74% 2-year survival after CE Mark approval, an appreciable learning curve with the device [32]. In this series, 38/46 (83%) patients were without thromboembolism at 2 years, and freedom from LVAD failure, driveline infection, and skull pedestal infection was 100%, 100%, and 94%, respectively [32]. Similarly, the Japanese Registry for Mechanically Assisted Circulatory Support database reports overall 1- and 2-year survival as 85.0% and 79.3%, respectively [33]. Bleeding, infection, and neurological dysfunction were the most reported adverse events after implantation of the Jarvik 2000 [33]. In the US BTT trial, 150 patients had the Jarvik 2000 implanted from 2005 to 2011 with a primary outcome of successful transplantation or listing 180 days post-implant [34]. During the study period, considerable changes were made to the Jarvik 2000 pump design to limit adverse events. Overall, 67.3% of patients met the primary outcome [34]. This figure increased to 90.9% after the implementation of the cone-bearing system [34].

HeartAssist5

The HeartAssist5 (ReliantHeart Inc., Houston, TX) is a continuous axial-flow, miniaturized VAD. In 1988, the initial design was jointly developed by heart surgeons, Drs. Michael E. DeBakey and George P. Noon, in collaboration with the Baylor College of Medicine and engineers from NASA. The device, originally called the MicroMed DeBakey Noon VAD, was first implanted in Berlin, Germany, in 1998, marking the first continuous-flow device implanted in a human [35]. The first implantation in the United States took place in 2002 as a bridge to transplant [35]. The latest iteration of the VAD, the HeartAssist5, received CE Mark approval in 2010. In the United States, the HeartAssist5 device is limited to investigational use. The FDA granted IDE clinical trial approval in 2014, which has since terminated due to low enrollment. The pump size is miniaturized, measuring 7.1 cm × 3 cm, weighing 92 g, with a priming volume of 25 ml. The pump is capable of providing 2–10 L/min of support. The HeartAssist 5 offers several unique features to continuous-flow blood pump technology.

The HeartAssist5 VAD is comprised of an axial-flow pump system. Blood is directed in through the inlet cannula. A unique feature of the HeartAssist5 is the availability of two inflow cannula options to accommodate varying body sizes. The titanium inflow cannula is available in a 140-degree inflow angle with a 60 mm outflow graft for small body surface area (BSA) or pediatric patients down to 18 kg and 0.7 m^2 BSA. The second inflow option, for larger patients, is angled at 115 degrees and comes with a 90 mm outflow graft. The pump assembly is made of a titanium flow tube. It contains an inlet, three-bladed flow straightener. Blood then passes an inducer connected to the magnetic, six-bladed impeller. Additional guide vanes are located downstream to the impeller. The impeller is connected to silicon carbide ball-and-cup bearings and is powered by a three-phase direct current (DC) motor located in the stator housing.

The inflow cannula is secured to an apical sewing ring. The sewing ring is a silicone ring wrapped in polyester with a 0 Prolene drawstring suture used to attach the cannula to the apex of the left ventricle. The inflow cannula is textured with sintered titanium beads to improve tissue ingrowth and integration within the ventricle. The outflow graft, available in two sizes as detailed above, is a pre-clotted gelatin weave graft. It is attached to the distal pump with a wedge nut and is anastomosed to the ascending aorta. The proximal outflow graft is covered by a plastic graft protector that provides bend relief and attaches directly to a flow probe. This flow probe is another unique feature of the HeartAssist5 VAD. The probe is a custom, ultrasonic real-time flow sensor that encircles the outflow graft and can accurately measure blood flow. As opposed to other continuous-flow devices which estimate pump flow, direct measurement of blood flow eliminates external factors to the blood flow estimates and assists in patient management [36].

The HeartAssist5 is implanted in the pericardial space which eliminates diaphragm or abdominal wall dissection needed for larger pumps. The pump assembly contains a percutaneous power cord which is tunneled through the abdominal wall. The percutaneous power cord contains the optical flow probe cable, and both connect to the external Controller. The Controller provides power to the LVAD via two lithium-ion batteries and displays the operating parameters of the device on an LCD display. It contains the power management system, the motor controller, data acquisition memory, a microprocessor, and ultrasonic flow measurement system. A unique feature of the HeartAssist5 controller is an internal radio for one-way wireless GSM cell phone connectivity to transmit pump data for remote monitoring. The HeartAssst5 technology of remote monitoring provides physicians with secure access to real-time and historical VAD parameters and alarms from wireless devices. Clinicians can receive text messaging or email of pump alarms with links to live data. The remote monitoring and wireless data transmission provide helpful information to clinicians [37], but all pump parameters are controlled through the HeartAttendant console.

The HeartAttendant is a portable console used to program the Controller and set the LVAD speed and alarm levels. Device speeds are set between 7500 and 12,500 rpm to generate flows up to 10 L/min. The HeartAttendant displays LVAD performance data, battery charge levels, power usage, and alarm history. It is also capable of wireless transmission of HeartAsssist5 performance data. When not connected to the HeartAttendant, patients utilize an ergonomically designed, reinforced fabric organizer called the VADPAK to hold the Controller and batteries and enhance mobility.

At this time, prospective clinical data is limited for the HeartAssist5 device. The US BTT IDE trial has been terminated due to reduced enrollment. The technology of the HeartAssist5 has been miniaturized into the new aVAD device (ReliantHeart Inc., Houston, TX) (Table 22.1). This pump utilizes similar rotor and bearing design as its predecessor, the HeartAssist5, but it is positioned within the left ventricle through the apex with a 90-degree outflow elbow connecting to the outflow graft. The device received CE Mark approval in 2016 with the first human implant taking place that same year [38]. The aVAD is currently under regulatory review in the United States and not approved for investigation or commercial use.

INCOR

The Berlin Heart INCOR (Berlin Heart GmbH, Berlin, Germany) is an implantable, axial-flow rotary LVAD. The device was developed by the Berlin Heart in cooperation with the German Heart Institute, Berlin, Germany [39]. It received CE Mark approval in 2003 for BTT and DT. The pump assembly measures 12 cm × 3 cm with an approximate weight of 200 g and a blood volume of 60 ml. The pump speed is set between 5000 and 10,000 rpm and is capable of providing flow of 6 L/min. The first human implant took place in 2002 and has since been used to support over 670 heart failure patients. The INCOR LVAD shares several design features with previously described axial-flow devices with several unique technological features of its own.

The INCOR inflow cannula is positioned in the left ventricle apex. The cannula is made of medical grade silicone with a textured titanium intramyocardial head which offers several advantages. The silicone cannula material is leak proof and does not require pre-clotting prior to implantation. The cannula is flexible to compensate for the movements of the beating heart and limits inflow obstruction. In 2005, the inflow cannula was lengthened to reduce the rate of thromboembolic complications [40]. The cannula is coupled to the INCOR pump during implantation using snap-in connector.

The INCOR pump is an axial-flow blood pump made of biocompatible titanium with a heparin coating. Blood enters the pump tube through a stationary inlet guide vane which produces laminar flow to the rotor. The rotor is a magnetic levitating bladed impeller with magnetic bearings both of which have no contact with the pump tube. Electromagnets in the pump tube and hubs keep the axial rotor centered within the pump tube and deliberately transfer the rotor back and forth during operation to promote pump washing. Additionally, sensors in the pump apparatus control the magnetic bearings and measure the pressure differential and flow through the pump [41]. A stationary diffuser converts the rotational flow from the impeller to laminar flow. The impeller design and blades have been optimized to limit hemolysis [42]. A silicone outflow conduit connects the pump to the outflow graft with bend relief. The INCOR offers graft or silicone outflow cannula options that are anastomosed to the ascending aorta. The pump is connected to a percutaneous power cord coated with Dacron velour and silicone. Design modifications to the power cord shortened the Dacron velour length of the driveline to ensure the silicone portion exits the abdominal skin which decreases driveline infections [43, 44]. The size of the INCOR pump assembly makes subdiaphragmatic implantation necessary in an abdominal wall pocket.

The driveline is tunneled through the abdominal wall and connects to a power cord from the INCOR control unit. The INCOR control unit regulates the pump speed, the magnetic bearings, alarm systems, and power supply. The control unit has several features to improve pump function. Pulsatility control regulates left ventricle filling allowing ventricular contraction and aortic valve opening. Suction protection limits myocardial wall suction events. The active magnetic bearings enable periodic flow change which facilitates washing out of the ventricle and pump tube. The control unit stores pump performance data and transmits pump data to a laptop with the INCOR monitoring program that is used to optimize therapy. The main power and backup batteries are connected to the control unit to supply the LVAD with power.

The levitation system of the INCOR rotary pump accentuates an evolution in continuous-flow LVAD design and technology seen in pumps designed with a centrifugal blood flow path. Initial clinical experiences with the INCOR LVAD supported the efficacy of the pump with event-free survival, transplantation, and adverse event profiles similar to other axial-flow devices [39, 45]. In a retrospective review of the Italian heart failure registry, 42 INCOR LVADs were implanted from 2006 to 2012 [40]. The 1- and 2-year survival rates according to log rank tests and Kaplan-Meier curves were 74% and 60%, respectively [40]. In this series, there was not a single episode of GI bleeding or pump thrombosis with a median support duration of 372 days [40]. Commercial availability of the INCOR device ceased in 2018; however, the device design featuring a magnetically levitating axial rotor is worth noting as continuous-flow devices featuring this technology continued to evolve.

The continuous-flow rotary pumps with axial-flow design that have been discussed thus far in the chapter featured innovative technology and design features that improved survival, quality of life, and acceptance of LVAD therapy for long-term mechanical circulatory support in patients with advanced heart failure. However, limitations in axial-flow pump design contribute to the complications and shortcomings observed with MCS. With the exception of the INCOR LVAD, axial-flow pumps mostly share contact bearing systems and inlet and outlet stators in design which increase friction, pump wear and failure, heat generation, and thrombus formation. Furthermore, axial-flow rotary pump design creates significant pressure differentials across the pump which can result in ventricular collapse and suction events when filling pressures are abruptly reduced [46]. Many of the limitations of axial-flow rotary pumps are addressed with improvements in pump technology centrifugal pump design.

Centrifugal LVADs

Continuous-flow LVADs with a centrifugal configuration utilize a rotor or impeller suspended in the blood flow path with a noncontact bearing or bearingless system. Centrifugal pumps incorporate either magnetic or hydrodynamic levitation of the internal impeller and elicit blood flow through magnetic coupling to the pump motor [46]. The elimination of a contact bearing system has the potential to significantly improve pump durability [46]. Centrifugal pump design with magnetic and/or hydrodynamic levitation of the impeller enlarges the blood flow path around the impeller to improve pump washing, reducing the thrombosis risks [46]. With reduced friction and a greater degree of blood flow around the impeller, centrifugal pump design increases the sensitivity of pressure-flow relationships, thereby reducing potential suction events and increasing the reliability of the estimated pump flow from pump power and speed parameters [46–48]. A review of centrifugal LVADs currently in use emphasizes the technology and features of these devices.

HVAD

The HeartWare HVAD (Medtronic Inc., Minneapolis, MN) is an implantable, continuous-flow LVAD with centrifugal pump design (Fig. 22.8). The device received CE mark approval for BTT in 2009 and DT in 2012. The FDA approved the HVAD for BTT in 2012 and DT in 2017. In 2018, the FDA approved a less-invasive left thoracotomy implantation technique for the HVAD [49]. The device is relatively small, measuring approximately 6 cm in diameter and 2.8 cm in length. It weighs 145 g with a 50 ml pump volume. The HeartWare HVAD system is composed of the pump, HVAD Controller, HeartWare Monitor, power sources, and batteries.

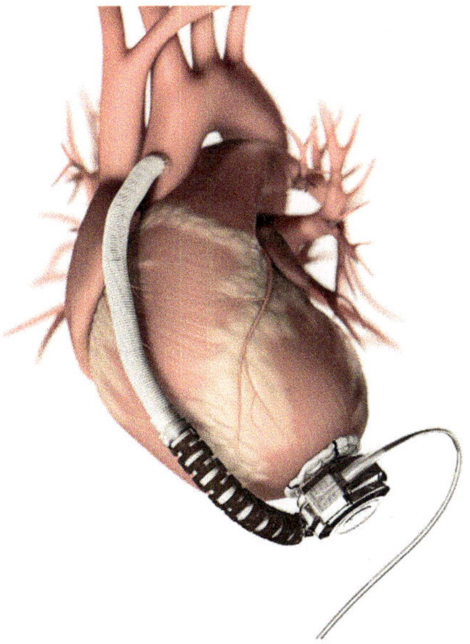

Fig. 22.8 The HVAD is an implantable LVAD with a centrifugal pump design. It received FDA approval for BTT and DT in 2012 and 2017, respectively. It was recently approved by the FDA for a minimally invasive thoracotomy implantation technique. The device measures 6 cm in diameter and 2.8 cm in length. It weighs 145 g with a 50 ml pump volume. (Reproduced with permission of Medtronic Inc.)

The HVAD pump contains several design features unique to centrifugal pumps. The pump consists of an integrated, partially sintered inflow cannula (Fig. 22.9). The inflow cannula and attached pump are positioned directly in the left ventricular apex. The inflow cannula has an outer diameter of 21 mm and a length of 25 mm and contains a circular silicone ring to prevent leakage. The integrated inflow cannula eliminates the bend, inlet stator, and resistance seen in axial-flow devices, helping to improve blood inflow dynamics and miniaturize the pump device size. The size of the device allows for intrapericardial placement of the pump. The pump consists of a small, round front and rear housing assembly made of titanium that contains the impeller, central post, and electromagnetic motor. The pump design uses a hybrid magnetic and hydrodynamic impeller suspension to create frictionless rotation [50]. A wide-bladed impeller contains an outer rotating magnetically levitating bearing (Fig. 22.9). The inner stationary magnetic bear-

Fig. 22.9 The HVAD centrifugal continuous-flow pump features a partially sintered inflow cannula. The pump front and rear housing assembly is made of titanium and contains the impeller, central post, and electromagnetic motor. The pump uses a hybrid magnetic levitation and hydrodynamic impeller suspension. (Reproduced with permission of Medtronic Inc.)

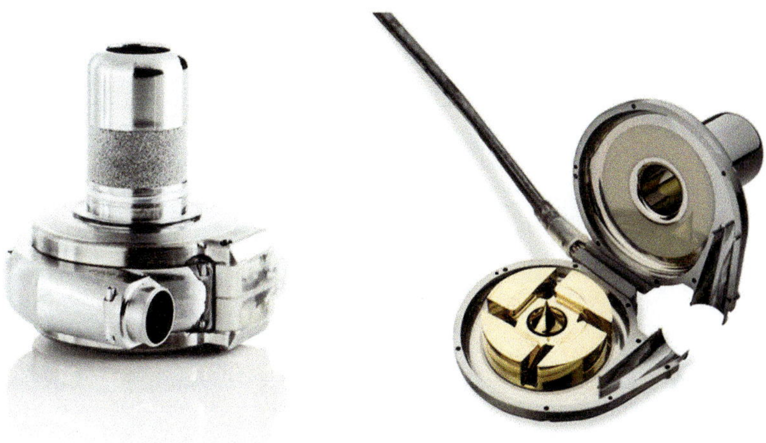

ing is contained within the central post that enters through the center of the impeller. Hydrodynamic bearings are located on the inflow side of the bladed impeller to provide lift force to the impeller. Motor stator electromagnets are located symmetrically in the housing assembly to limit axial force on the impeller. The impeller can be set between 1800 and 4000 rpm and is capable of providing 10 L/min of support. The outflow conduit is located on the rear housing assembly of the pump. The conduit is a 10 mm, gel-impregnated graft with a rigid bend relief system that is anastomosed to the ascending aorta. A flexible driveline exits the pump and is tunneled to an exit point on the abdominal wall.

The driveline connects to a power cord from the HVAD controller. The controller uses a LED display and contains battery indicators, alarm indicators, power indicators, and scroll buttons. The controller settings menu can display battery cycles, alarm settings, hematocrit, pump speed (rpm), and peak and trough flow rates (Fig. 22.10). The controller has two power cord connections, a data cable connection, and a driveline connection. It accepts battery, AC, or DC adapter power and should be connected to two power sources (e.g., two batteries or one battery and an AC adapter) at all times. The controller contains a ventricular suction detection system that continuously tracks the trough flow rates and triggers an alarm at flow rates 40% below the estimated flow baseline [51]. The controller enables this system to be turned on and off. Another unique system that is accessed through the controller is the Lavare Cycle. This is a speed modulation algorithm designed to reduce blood stasis within the left ventricle and prevent thrombosis [51]. The system decreases pump speed 200 rpm below the set speed for 2 seconds and then increases pump speed 200 rpm above the set speed for 1 second [51]. The system cycles every 60 seconds and provides some pulsatility to the device.

The HVAD monitor is a touchscreen device designed to provide a user-friendly interface to monitor and control the HVAD system (Fig. 22.11). The monitor displays pump information, allows adjustment of pump parameters, monitors and reports system errors and alarm conditions, and updates controller software information. It displays power and flow waveforms that can aid in the clinical assessment and management of patients supported by the HVAD [52, 53]. The monitor is designed to use AC power from a wall outlet but contains an internal battery that can support the device during short periods.

The HeartWare HVAD centrifugal pump design features and technology have several clinical implications. The initial HVAD BTT trial in Europe and Australia was a multicenter, prospective, nonrandomized, single-arm study designed to evaluate the safety and efficacy of the device [54]. Fifty patients reached the endpoint of the study, and the device provided safe and effective circulatory support with improvements

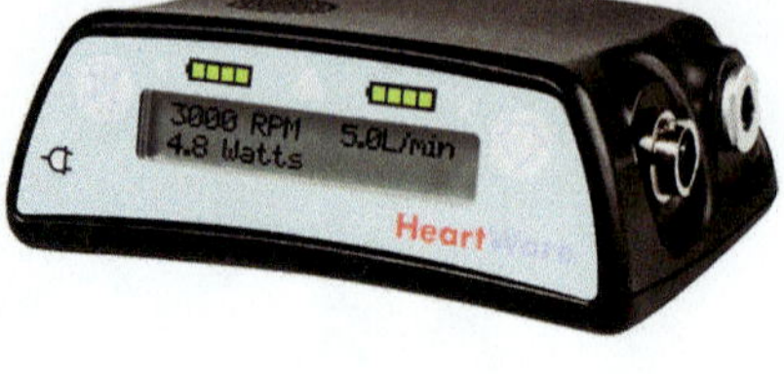

Fig. 22.10 The HVAD controller is small and compact and features a LED display. Controller software enables a Ventricular Suction Detection system and a speed modulation system, the Lavare Cycle. (Reproduced with permission of Medtronic Inc.)

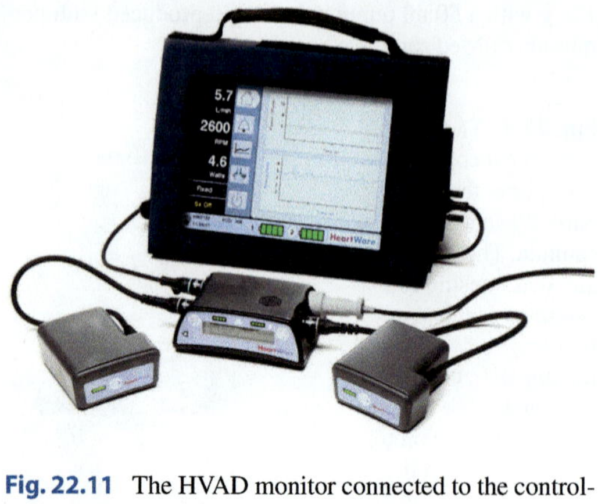

Fig. 22.11 The HVAD monitor connected to the controller and batteries. The touchscreen monitor displays pump information, allows adjustment of pump parameters, and monitors and reports system errors and alarm conditions. It displays power and flow waveforms. (Reproduced with permission of Medtronic Inc.)

in hemodynamics, quality of life, and neurocognitive function compared to medical therapy with 2-year survival similar to heart transplantation [54]. The US BTT trial, the ADVANCE Trial, was a nonrandomized noninferiority study between HVAD and contemporary axial-flow device patients in the INTERMACS registry [55]. The trial and continued access protocol results showed noninferiority and a low adverse event profile similar to historical controls [50, 55]. Survival rates were 91% at 180 days and 84% at 1 year [50]. A multicenter, randomized, noninferiority trial (ENDURANCE) between the HVAD and HeartMate II established HVAD as destination therapy [56]. In this trial, HVAD was found to be noninferior to the axial-flow device with respect to survival, free from disabling stroke or device removal for malfunction or failure [56]. In this trial, the HVAD had fewer device failures relative to the HeartMate II as a result of less pump thrombosis necessitating device exchange but an increased overall stroke rate [56, 57]. Post hoc analysis the ADVANCE and ENDURANCE trials showed an association between hypertension during HVAD support and increased incidence of stroke [57]. The effect of a blood pressure monitoring algorithm in patients supported with HVAD compared to HeartMate II was investigated in the ENDURANCE Supplemental trial [58]. In this trial, blood pressure management was associated with reduced stroke rates in HVAD patients; however, HVAD failed to demonstrate noninferiority versus HeartMate II regarding the pre-specified endpoint of 12-month incidence of transient ischemic attack or stroke [58].

HeartMate 3

The next generation LVAS from the makers of the HeartMate II is the HeartMate 3 LVAS (Abbott, Abbott Park, IL). This pump differs from the axial-flow rotary design of the HeartMate II and uses a magnetically levitated centrifugal continuous-flow pump design. The HeartMate 3 received CE Mark approval in 2015 for BTT and DT. In the United States, it received BTT FDA approval in 2017 and DT approval in 2018. The pump body has a diameter of 5.03 cm and a height of 5.58 cm including the inflow cannula (3.38 cm without inflow cannula). It weighs 200 g with a displaced volume of 80 ml and a priming volume of 21 ml. The HeartMate 3 system consists of the LVAD, system controller, System Monitor, and power options including batteries, Mobile Power Unit, and Power Module (Fig. 22.12).

The magnetically levitated centrifugal continuous-flow pump design of the HeartMate 3 starts with the inflow cannula (Fig. 22.13). The cannula is made of integrated titanium with fused titanium microspheres and is rigidly affixed to the pump cover. The inflow cannula measures 20.5 mm in length and is implanted directly into the left ventricular apex. It is attached to an apical PTFE felt cuff sewn to the epicardium and secured with a slide lock for hemostasis. The pump housing is made of titanium and holds the rotor or impeller, a single stator, and motor. The rotor is fully supported by magnetic levitation

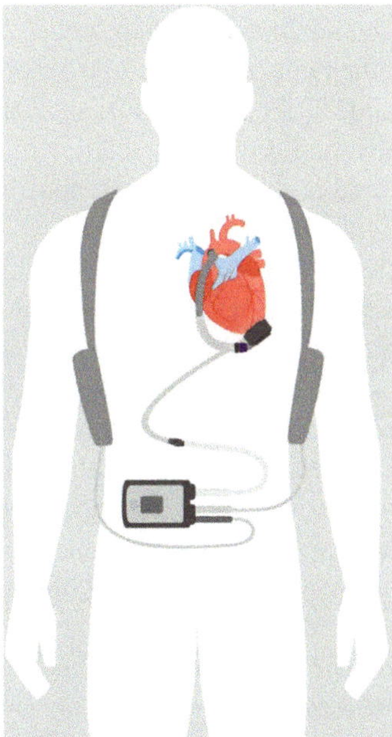

Fig. 22.12 The HeartMate 3 LVAS consists of the LVAD, system controller, System Monitor, and power options including batteries, Mobile Power Unit, and Power Module. (HeartMate II and HeartMate 3 are trademarks of Abbott or its related companies. Reproduced with permission of Abbott, © 2019. All rights reserved)

Fig. 22.13 The HeartMate 3 pump features an integrated titanium inflow cannula with titanium microspheres. The pump housing contains a centrifugal-flow rotor, a single stator, and electromagnetic motor. The rotor is fully supported by magnetic levitation. (HeartMate II and HeartMate 3 are trademarks of Abbott or its related companies. Reproduced with permission of Abbott, © 2019. All rights reserved)

and eliminated the need for mechanical or fluid bearings to increase durability of the pump [59]. The rotor contains a permanent magnet that spins around a central stator. The stator is made of iron pole pieces, a back iron, copper coils, and position sensors. The magnetic interaction between the rotor and iron stator is actively controlled to accomplish drive and levitation of the rotor [59]. The sensors measure the position of the magnet in the rotor, and the current in the coils is adjusted to control the radial position and rotational speed of the rotor [59]. The electronics and software to control the motor drive and levitation of the rotor are integrated in the lower pump housing chamber along with the stator [59].

The pump speed ranges from 3000 to 9000 rpm with an operating voltage between 10 and 17 volts DC and a nominal power consumption of 4 watts. It is capable of generating flows of 10 L/min. Power to the pump is supplied by a two-piece driveline. The pump cable connects to the pump housing and is comprised of a six-conductor silicone sheath. The sheath contains three primary power wires and three backup wires. The driveline is covered with woven polyester to reduce infection and encourage tissue ingrowth at the skin. The pump cable is tunneled through the anterior abdominal wall and connects the modular cable. Blood exits the pump through a sealed, gelatin-impregnated woven polyester outflow graft with proximal bend relief. The outflow graft is anastomosed in an end-to-side construction to the ascending aorta. The pump is implanted directly in the pericardial space due to its small size and centrifugal pump design. Less-invasive, thoracotomy implantation techniques have also been described for the HeartMate 3 [60].

The pump is controlled though the HeartMate 3 System Controller, which acts as the central power and communication center for the LVAS [59] (Fig. 22.14). The Controller constantly monitors system performance, battery charge, and alarm conditions. The Controller receives power from the Power Module, Mobile Power Unit, or lithium-ion batteries and passes it to the pump. Fully charged batteries are capable of 17 hours of system support depending on the activity level of the patient. The Controller also has an integrated emergency backup power supply which can be used for short durations. Similar to the HeartMate II, the HeartMate 3 uses the Power Module with System Monitor to display large-scale system performance and enable clinicians to adjust system settings and operating parameters.

The HeartMate 3 LVAS has several unique features to enhance performance. The system monitors the estimated increase in pump flow caused by ventricular contraction and increase in ventricular pressure during the cardiac cycle. The magnitude of these flow pulses is measured and averaged at 15-second intervals to produce the pulsatility index (PI) [59]. In this system, pump

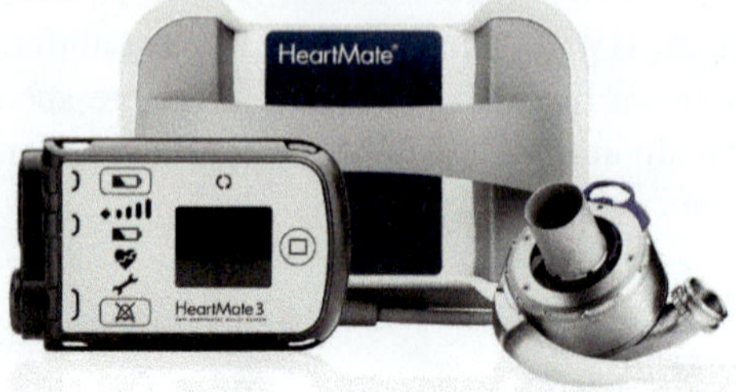

Fig. 22.14 The HeartMate 3 System Controller and Module Power Unit pictured with the pump. The Controller monitors system performance, battery charge, and alarm conditions. Technology features include Pulsatility Index Detection and Artificial Pulse systems. (HeartMate II and HeartMate 3 are trademarks of Abbott or its related companies. Reproduced with permission of Abbott, © 2019. All rights reserved)

flow is estimated by directly measuring pump power. The Pulsatility Index Detection feature recognizes decreases in pulsatility and lowers to the rotor speed to a preset low speed limit to prevent ventricular suction events and potential arrhythmias. The pump will gradually return to its programmed speed. In addition, the pump employs a feature called Artificial Pulse [59]. In this setting, the pump speed will increase 2000 rpm above the set rotor speed for 0.2 seconds and decrease 2000 rpm below the set speed for 0.15 seconds in a 2-second cycle. The effect produces 30 artificial "beats" per minute. The pulsatility prevents stasis and increases washing within the flow path of the pump.

Early clinical experience with the HeartMate 3 has shown that the pump design and technology features translate into improved patient outcomes. The first human implantation took place in Germany in 2015 [61]. Data from the European CE Mark approval trial (ELEVATE) is available with 6-month, 12-month, and 24-month follow-up [62–64]. Survival rates were 82 ± 2%, 81 ± 6%, and 74 ± 6% at 6, 12, and 24 months, respectively [62–64]. At 2 years, no pump malfunctions or pump thromboses occurred [64]. Adverse event rates at 2 years include bleeding requiring surgery (16%), gastrointestinal bleeding (20%), driveline infection (24%), ischemic stroke (16%), hemorrhagic stroke (8%), right heart failure (14%), and outflow graft thrombosis (2%) [64].

In the United States, a nonblinded, randomized trial compared the centrifugal continuous-flow HeartMate 3 with the axial-flow rotary pump HeartMate II in patients with advanced heart failure (MOMENTUM 3) [65]. The primary outcome of the trial was a composite of survival free from disabling stroke or survival free of reoperation to remove or replace the pump [65]. At 6 months, there were no suspected or confirmed pump thromboses with the HeartMate 3 and therefore, a significantly lower rate of reoperation for pump malfunction [65]. There were no significant differences between the two groups in rates of death or disabling stroke [65]. These results were upheld at 2-year follow-up; however, the rate of overall stroke was lower in the HeartMate 3 group compared to the HeartMate II group (10.1% vs 19.2%, hazard ratio, 0.47; 95% CI, 0.27–0.84, $p = 0.02$) [66]. The actuarial all-cause survival of the HeartMate 3 patients at 2 years was 82.8% in the MOMENTUM 3 data [66]. The centrifugal pump design of the HeartMate 3 has decreased pump thrombosis and malfunction with the device when compared to the HeartMate II, likely attributed to a wider blood flow path and improved hemocompatibility [57, 67].

EVAHEART 2

The EVAHEART 2 is the latest iteration of a hydraulically levitated centrifugal LVAS (Sun Medical Technology Research Corporation, Nagano, Japan) (Fig. 22.15). The EVAHEART technology and former models were developed by Sun Medical Technology Research Corporation (Nagano, Japan) in conjunction with Prof. Kenji Yamazaki. Dr. Yamazaki created the concept of the EVAHEART LVAS in collaboration with several universities with the first clinical design arriving in 2002. The first human implantation took place in Japan in 2005. The device received approval by the Japanese PMDA in 2008 after an initial clinical trial [68]. It has received an IDE and approval to launch an EVEHEART 2 bridge to transplant clinical trial in the United States by the FDA. The pump measures 5.1 cm × 6.7 cm and weighs 262 g.

The EVAHEART 2 pump has several novel design features unique to centrifugal continuous blood pump technology. Blood enters the pump through a double-cuff tipless inflow cannula (Fig. 22.16). The inflow design keeps the cannula flush with the endocardium of the left ventricular apex to enhance tissue ingrowth and minimize cannula malposition and thrombus formation. The pump housing and impeller are made of titanium. The blood-contacting surfaces of the pump assembly are coated with MPC, a commercially developed coating to limit thrombosis. An open vane impeller rotates around a central shaft. An interesting feature of the EVAHEART 2 pump is a cooling mechanical seal system for the impeller

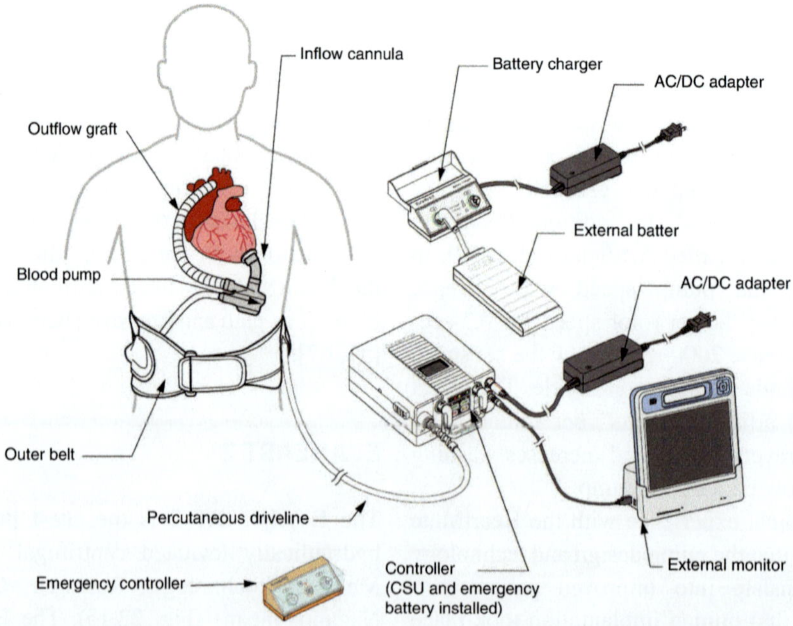

Fig. 22.15 Schematic of the EVAHEART LVAS. (Reproduced with permission from Evaheart, Inc.)

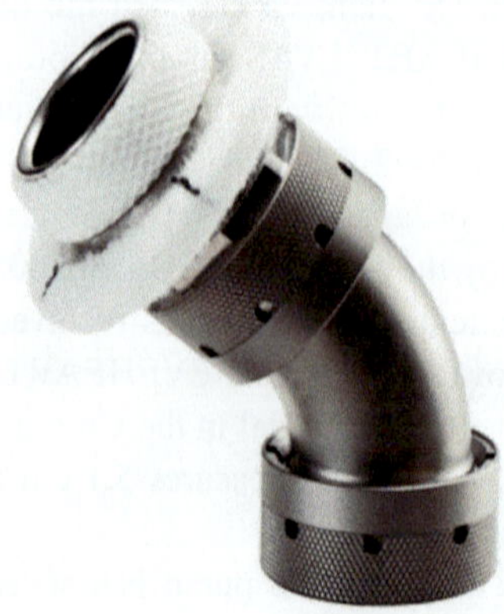

Fig. 22.16 The EVAHEART 2 double-cuff tipless inflow cannula is flush with the endocardium of the left ventricular apex to enhance tissue ingrowth and minimize thrombus formation. (Reproduced with permission from Evaheart, Inc.)

(Fig. 22.17). The impeller is hydraulically levitated by a thin layer of circulating sterile water. The thin water column lubricates the impeller-shaft interaction, thereby reducing friction and heat generation. Blood exits the pump via a 16 mm PTFE outflow graft with polymer spiral reinforcement for bend relief. The outflow graft is anastomosed to the ascending aorta. The pump is capable of generating peak flows up to 14 L/min. With an impeller speed set at 2200 rpm, the pump generated 8 L/min flow against a pressure head of 90 mm Hg.

The EVAHEART 2 LVAS is powered and managed by the Controller. The Controller's display serves at the user interface for the LVAS and provides such information as the pump speed (rpm), power (watts), power sources, batter charge level, and alarm system. Another feature of the EVEHEART LVAS is a Cool Seal Unit (CSU). The CSU circulates sterile water to and from the pump to form the lubricating layer in the hydrodynamic bearing of the pump. The CSU contains an ultrafiltration filter for the sterile water. The water is transported to and from the pump via two tubes contained within the LVAD driveline. Several technology systems have been implemented in the EVEHEART 2 LVAS. These include a rotational speed modulation system and electrocardiography synchronization [69, 70]. These features may improve the hemocompatibility of the device [71] and alter arterial pulsatility and myocardial perfusion [72].

In the initial Japanese approval trial, 18 patients received the EVAHEART device from 2005 to

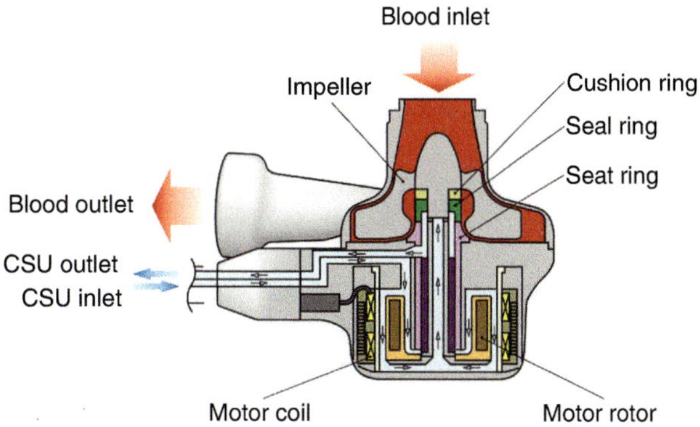

Fig. 22.17 The EVAHEART 2 centrifugal continuous-flow pump features an open vane impeller that is hydraulically levitated with a layer of circulating sterile water that is pumped from the Cool Seal Unit of the Controller. (Reproduced with permission from Evaheart, Inc.)

2008 (Yamazaki). The Kaplan-Meier survival rate at 6 months, 12 months, and 24 months in this small cohort was 89%, 83%, and 70%, respectively [68]. A post-approval study implanted 96 consecutive patients with the device [73]. In this study, the overall patient survival was 87.4% at 1 year [73]. Major adverse events are reported which include ischemic stroke (17.7%), hemorrhagic stroke (13.5%), infection (14.6%), and pump thrombosis (1%) during the 2-year study duration [73].

Conclusion

The pump design and technology features of the continuous-flow LVASs have enabled these devices to become the standard of care for advanced heart failure patients. This review of continuous-flow LVADs illustrates the evolution of pump design and technology which has translated into improved patient outcomes and quality of life. In the current era, >20,000 patients have received an FDA-approved MCS device between 2006 and 2016 in the INTERMACS registry [74]. In fact, over 95% of implants are continuous-flow devices in the registry [74]. The advantage of centrifugal continuous-flow pumps in reducing pump thrombosis and reoperation for pump malfunction in addition to the less-invasive operative options due to the smaller sizes has made these pump design features the current benchmark. Pulsatility systems have been reintroduced into continuous-flow pumps to increase hemocompatibility and limit bleeding complications. However, adverse events including stroke, infection, bleeding, and heart failure remain and continue to limit the success and durability of current LVAD technology [74]. As the technology and pump design of continuous-flow LVADs evolve, patient survival will likely improve, and the focus must remain on limiting morbidity associated with implantable, durable MCS in addition to decreasing the costs associated with these devices [75]. Herein lie the challenges ahead for implantable, continuous-flow circulatory support.

References

1. Everly MJ. Cardiac transplantation in the United States: an analysis of the UNOS registry. Clin Transpl. 2008:35–43.
2. Phillips WM. The artificial heart: history and current status. J Biomech Eng. 1993;115(4b):555–7.
3. Goldstein DJ, Oz MC, Rose EA. Implantable left ventricular assist devices. N Engl J Med. 1998;339(21):1522–33.
4. Rose EA, Gelijns AC, Moskowitz AJ, Heitjan DF, Stevenson LW, Dembitsky W, et al. Long-term use of a left ventricular assist device for end-stage heart failure. N Engl J Med. 2001;345(20):1435–43.
5. Slaughter MS, Pagani FD, Rogers JG, Miller LW, Sun B, Russell SD, et al. Clinical management of continuous-flow left ventricular assist devices in advanced heart failure. J Heart Lung Transplant. 2010;29(4 Suppl):S1–39.

6. Slaughter MS, Rogers JG, Milano CA, Russell SD, Conte JV, Feldman D, et al. Advanced heart failure treated with continuous-flow left ventricular assist device. N Engl J Med. 2009;361(23):2241–51.
7. Griffith BP, Kormos RL, Borovetz HS, Litwak K, Antaki JF, Poirier VL, et al. HeartMate II left ventricular assist system: from concept to first clinical use. Ann Thorac Surg. 2001;71(3 Suppl):S116–20; discussion S4–6.
8. HeartMate II LVAS Operating Manual. Document 103884.F. Thoratec Corporation, Pleasanton, CA.
9. HeartMate II LVAS Intructions for Use. Document 10006960.C. Thoratec Corporation, Pleasanton, CA.
10. Rose EA, Levin HR, Oz MC, Frazier OH, Macmanus Q, Burton NA, et al. Artificial circulatory support with textured interior surfaces. A counterintuitive approach to minimizing thromboembolism. Circulation. 1994;90(5 Pt 2):Ii87–91.
11. Imamura T, Nguyen A, Chung B, Rodgers D, Sarswat N, Kim G, et al. Association of inflow cannula position with left ventricular unloading and clinical outcomes in patients with HeartMate II left ventricular assist device. ASAIO J. 2019;65(4):331–5.
12. Taghavi S, Ward C, Jayarajan SN, Gaughan J, Wilson LM, Mangi AA. Surgical technique influences HeartMate II left ventricular assist device thrombosis. Ann Thorac Surg. 2013;96(4):1259–65.
13. Yuzefpolskaya M, Uriel N, Chow DS, Restaino SW, Mancini DM, Flannery M, et al. Prevalence and timing of bend relief disconnection in patients supported by the late version HeartMate II left ventricular assist device. J Heart Lung Transplant. 2013;32(3):320–5.
14. Waller AH, Dunne R, Stewart GC, Ghosh N, Gosev I, Rybicki FJ, et al. Evaluation of bend relief disconnection in patients supported by a HeartMate II left ventricular assist device. Circ Cardiovasc Imaging. 2014;7(5):844–8.
15. Miller LW, Pagani FD, Russell SD, John R, Boyle AJ, Aaronson KD, et al. Use of a continuous-flow device in patients awaiting heart transplantation. N Engl J Med. 2007;357(9):885–96.
16. Pagani FD, Miller LW, Russell SD, Aaronson KD, John R, Boyle AJ, et al. Extended mechanical circulatory support with a continuous-flow rotary left ventricular assist device. J Am Coll Cardiol. 2009;54(4):312–21.
17. Estep JD, Starling RC, Horstmanshof DA, Milano CA, Selzman CH, Shah KB, et al. Risk assessment and comparative effectiveness of left ventricular assist device and medical management in ambulatory heart failure patients: results from the ROADMAP study. J Am Coll Cardiol. 2015;66(16):1747–61.
18. Starling RC, Estep JD, Horstmanshof DA, Milano CA, Stehlik J, Shah KB, et al. Risk assessment and comparative effectiveness of left ventricular assist device and medical management in ambulatory heart failure patients: the ROADMAP study 2-year results. JACC Heart failure. 2017;5(7):518–27.
19. Heilmann C, Trummer G, Beyersdorf F, Brehm K, Berchtold-Herz M, Schelling J, et al. Acquired Von Willebrand syndrome in patients on long-term support with HeartMate II. Eur J Cardiothorac Surg. 2017;51(3):587–90.
20. Kirklin JK, Naftel DC, Pagani FD, Kormos RL, Myers S, Acker MA, et al. Pump thrombosis in the Thoratec HeartMate II device: an update analysis of the INTERMACS registry. J Heart Lung Transplant. 2015;34(12):1515–26.
21. Klodell CT, Massey HT, Adamson RM, Dean DA, Horstmanshof DA, Ransom JM, et al. Factors related to pump thrombosis with the Heartmate II left ventricular assist device. J Card Surg. 2015;30(10): 775–80.
22. Jarvik R. Jarvik 2000 pump technology and miniaturization. Heart Fail Clin. 2014;10(1 Suppl):S27–38.
23. Westaby S, Frazier OH, Beyersdorf F, Saito S, Siegenthaler MP, Pigott DW, et al. The Jarvik 2000 heart. Clinical validation of the intraventricular position. Eur J Cardiothorac Surg. 2002;22(2):228–32.
24. Selzman CH, Koliopoulou A, Glotzbach JP, McKellar SH. Evolutionary improvements in the Jarvik 2000 left ventricular assist device. ASAIO J. 2018;64(6):827–30.
25. Frazier OH, Shah NA, Myers TJ, Robertson KD, Gregoric ID, Delgado R. Use of the Flowmaker (Jarvik 2000) left ventricular assist device for destination therapy and bridging to transplantation. Cardiology. 2004;101(1–3):111–6.
26. Zeng H, Jarvik R, Catausan G, Moldovan N, Carlisle J. Diamond coated artificial cardiovascular devices. Surface & coatings technology. 2016;302:420–5.
27. Westaby S, Jarvik R, Freeland A, Pigott D, Robson D, Saito S, et al. Postauricular percutaneous power delivery for permanent mechanical circulatory support. J Thorac Cardiovasc Surg. 2002;123(5):977–83.
28. Jarvik R, Westaby S, Katsumata T, Pigott D, Evans RD. LVAD power delivery: a percutaneous approach to avoid infection. Ann Thorac Surg. 1998;65(2):470–3.
29. Frazier OH. Implantation of the Jarvik 2000 left ventricular assist device without the use of cardiopulmonary bypass. Ann Thorac Surg. 2003;75(3):1028–30.
30. Radovancevic B, Gregoric ID, Tamez D, Vrtovec B, Tuzun E, Chee HK, et al. Biventricular support with the Jarvik 2000 axial flow pump: a feasibility study. ASAIO J. 2003;49(5):604–7.
31. Yoshioka D, Toda K, Yoshikawa Y, Sawa Y. Over 1200-day support with dual Jarvik 2000 biventricular assist device. Interact Cardiovasc Thorac Surg. 2014;19(6):1083–4.
32. Westaby S, Siegenthaler M, Beyersdorf F, Massetti M, Pepper J, Khayat A, et al. Destination therapy with a rotary blood pump and novel power delivery. Eur J Cardiothorac Surg. 2010;37(2):350–6.
33. Kohno H, Matsumiya G, Sawa Y, Ono M, Saiki Y, Shiose A, et al. The Jarvik 2000 left ventricular assist device as a bridge to transplantation: Japanese registry for mechanically assisted circulatory support. J Heart Lung Transplant. 2018;37(1):71–8.
34. Selzman CH, Felker E, Sheridan BC, Silvestry S, Daly RC, Anyanwu A, et al. The Jarvik 2000: results

of the United States bridge to transplant trial. J Heart Lung Transplant. 2016;35(4):S38.
35. Noon GP, Loebe M. Current status of the MicroMed DeBakey Noon ventricular assist device. Tex Heart Inst J. 2010;37(6):652–3.
36. Spiliopoulos S, Guersoy D, Koerfer R, Tenderich G. Beneficial aspects of real time flow measurements for the management of acute right ventricular heart failure following continuous flow ventricular assist device implantation. J Cardiothorac Surg. 2012;7:119.
37. Pektok E, Demirozu ZT, Arat N, Yildiz O, Oklu E, Eker D, et al. Remote monitoring of left ventricular assist device parameters after HeartAssist-5 implantation. Artif Organs. 2013;37(9):820–5.
38. Schmitto JD, Hanke JS, Dogan G, Tessmann R, Jeevanandem V, Cohn WE, et al. First implantation of a novel left ventricular assist device: the ReliantHeart aVAD. Ann Thorac Surg. 2017;104(4):e311–e3.
39. Hetzer R, Weng Y, Potapov EV, Pasic M, Drews T, Jurmann M, et al. First experiences with a novel magnetically suspended axial flow left ventricular assist device. Eur J Cardiothorac Surg. 2004;25(6):964–70.
40. Iacovoni A, Centofanti P, Attisani M, Verde A, Terzi A, Senni M, et al. Low incidence of gastrointestinal bleeding and pump thrombosis in patients receiving the INCOR LVAD system in the long-term follow-up. Int J Artif Organs. 2015;38(10):542–7.
41. Goldowsky M. Mini hemoreliable axial flow LVAD with magnetic bearings: part 3: modeling of demo-magnetic bearing and verification. ASAIO J. 2002;48(1):101–5.
42. Goldowsky M. Mini hemoreliable axial flow LVAD with magnetic bearings: part 1: historical overview and concept advantages. ASAIO J. 2002;48(1):96–8.
43. Camboni D, Zerdzitzki M, Hirt S, Tandler R, Weyand M, Schmid C. Reduction of INCOR(R) driveline infection rate with silicone at the driveline exit site. Interact Cardiovasc Thorac Surg. 2017;24(2):222–8.
44. Dean D, Kallel F, Ewald GA, Tatooles A, Sheridan BC, Brewer RJ, et al. Reduction in driveline infection rates: results from the HeartMate II multicenter driveline silicone skin Interface (SSI) registry. J Heart Lung Transplant. 2015;34(6):781–9.
45. Schmid C, Tjan TD, Etz C, Schmidt C, Wenzelburger F, Wilhelm M, et al. First clinical experience with the Incor left ventricular assist device. J Heart Lung Transplant. 2005;24(9):1188–94.
46. Pagani FD. Continuous-flow rotary left ventricular assist devices with "3rd generation" design. Semin Thorac Cardiovasc Surg. 2008;20(3):255–63.
47. Ayre PJ, Vidakovic SS, Tansley GD, Watterson PA, Lovell NH. Sensorless flow and head estimation in the VentrAssist rotary blood pump. Artif Organs. 2000;24(8):585–8.
48. Tsukiya T, Akamatsu T, Nishimura K, Yamada T, Nakazeki T. Use of motor current in flow rate measurement for the magnetically suspended centrifugal blood pump. Artif Organs. 1997;21(5):396–401.
49. Danter MR, McGee EC, Strueber M, Maltais S, Mokadam NA, Weisenthaler GM, et al. A prospective, controlled, un-blinded, multi-center clinical trial to evaluate the thoracotomy implant technique of the HVAD system in patients with advanced heart failure: results of the LATERAL trial. J Heart Lung Transplant. 2017;36(4):S66.
50. Slaughter MS, Pagani FD, McGee EC, Birks EJ, Cotts WG, Gregoric I, et al. HeartWare ventricular assist system for bridge to transplant: combined results of the bridge to transplant and continued access protocol trial. J Heart Lung Transplant. 2013;32(7):675–83.
51. HeartWare HVAD System instructions for use. Part number IFU00375. Medtronic Inc., Minneapolis, MN.
52. Rich JD, Burkhoff D. HVAD flow waveform morphologies: theoretical foundation and implications for clinical practice. ASAIO J. 2017;63(5):526–35.
53. Grinstein J, Rodgers D, Kalantari S, Sayer G, Kim GH, Sarswat N, et al. HVAD waveform analysis as a noninvasive marker of pulmonary capillary wedge pressure: a first step toward the development of a smart left ventricular assist device pump. ASAIO J. 2018;64(1):10–5.
54. Strueber M, O'Driscoll G, Jansz P, Khaghani A, Levy WC, Wieselthaler GM. Multicenter evaluation of an intrapericardial left ventricular assist system. J Am Coll Cardiol. 2011;57(12):1375–82.
55. Aaronson KD, Slaughter MS, Miller LW, McGee EC, Cotts WG, Acker MA, et al. Use of an intrapericardial, continuous-flow, centrifugal pump in patients awaiting heart transplantation. Circulation. 2012;125(25):3191–200.
56. Rogers JG, Pagani FD, Tatooles AJ, Bhat G, Slaughter MS, Birks EJ, et al. Intrapericardial left ventricular assist device for advanced heart failure. N Engl J Med. 2017;376(5):451–60.
57. Schroder JN, Milano CA. A tale of two centrifugal left ventricular assist devices. J Thorac Cardiovasc Surg. 2017;154(3):850–2.
58. Milano CA, Rogers JG, Tatooles AJ, Bhat G, Slaughter MS, Birks EJ, et al. HVAD: the ENDURANCE supplemental trial. JACC Heart failure. 2018;6(9):792–802.
59. HeartMate 3 Left Ventricular Assist System Instructions for Use. Document 100168999.A. Thoratec Corporation, Pleasanton, CA.
60. Schmitto JD, Krabatsch T, Damme L, Netuka I. Less invasive HeartMate 3 left ventricular assist device implantation. J Thorac Dis. 2018;10(Suppl 15):S1692–s5.
61. Schmitto JD, Hanke JS, Rojas SV, Avsar M, Haverich A. First implantation in man of a new magnetically levitated left ventricular assist device (HeartMate III). J Heart Lung Transplant. 2015;34(6):858–60.
62. Gustafsson F, Shaw S, Lavee J, Saeed D, Pya Y, Krabatsch T, et al. Six-month outcomes after treatment of advanced heart failure with a full magnetically levitated continuous flow left ventricular assist device: report from the ELEVATE registry. Eur Heart J. 2018;39(37):3454–60.
63. Krabatsch T, Netuka I, Schmitto JD, Zimpfer D, Garbade J, Rao V, et al. Heartmate 3 fully magnetically

levitated left ventricular assist device for the treatment of advanced heart failure −1 year results from the Ce mark trial. J Cardiothorac Surg. 2017;12(1):23.
64. Schmitto JD, Pya Y, Zimpfer D, Krabatsch T, Garbade J, Rao V, et al. Long-term evaluation of a fully magnetically levitated circulatory support device for advanced heart failure-two-year results from the HeartMate 3 CE mark study. Eur J Heart Fail. 2018;21(1): 90–7.
65. Mehra MR, Naka Y, Uriel N, Goldstein DJ, Cleveland JC Jr, Colombo PC, et al. A fully magnetically levitated circulatory pump for advanced heart failure. N Engl J Med. 2017;376(5):440–50.
66. Mehra MR, Goldstein DJ, Uriel N, Cleveland JC Jr, Yuzefpolskaya M, Salerno C, et al. Two-year outcomes with a magnetically levitated cardiac pump in heart failure. N Engl J Med. 2018;378(15):1386–95.
67. Uriel N, Colombo PC, Cleveland JC, Long JW, Salerno C, Goldstein DJ, et al. Hemocompatibility-related outcomes in the MOMENTUM 3 trial at 6 months: a randomized controlled study of a fully magnetically levitated pump in advanced heart failure. Circulation. 2017;135(21):2003–12.
68. Yamazaki K, Saito S, Nishinaka T, Kurosawa H, Nakatani T, Kobayashi J, et al. 517: Japanese Clinical Trial Results of an Implantable Centrifugal Blood Pump "EVAHEART"2008.
69. Ando M, Nishimura T, Takewa Y, Yamazaki K, Kyo S, Ono M, et al. Electrocardiogram-synchronized rotational speed change mode in rotary pumps could improve pulsatility. Artif Organs. 2011;35(10):941–7.
70. Arakawa M, Nishimura T, Takewa Y, Umeki A, Ando M, Adachi H, et al. Alternation of left ventricular load by a continuous-flow left ventricular assist device with a native heart load control system in a chronic heart failure model. J Thorac Cardiovasc Surg. 2014;148(2):698–704.
71. Naito N, Mizuno T, Nishimura T, Kishimoto S, Takewa Y, Eura Y, et al. Influence of a rotational speed modulation system used with an implantable continuous-flow left ventricular assist device on von Willebrand factor dynamics. Artif Organs. 2016;40(9):877–83.
72. Ando M, Takewa Y, Nishimura T, Yamazaki K, Kyo S, Ono M, et al. A novel counterpulsation mode of rotary left ventricular assist devices can enhance myocardial perfusion. J Artif Organs. 2011;14(3):185–91.
73. Saito S, Yamazaki K, Nishinaka T, Ichihara Y, Ono M, Kyo S, et al. Post-approval study of a highly pulsed, low-shear-rate, continuous-flow, left ventricular assist device, EVAHEART: a Japanese multicenter study using J-MACS. J Heart Lung Transplant. 2014;33(6):599–608.
74. Kirklin JK, Pagani FD, Kormos RL, Stevenson LW, Blume ED, Myers SL, et al. Eighth annual INTERMACS report: special focus on framing the impact of adverse events. J Heart Lung Transplant. 2017;36(10):1080–6.
75. Baras Shreibati J, Goldhaber-Fiebert JD, Banerjee D, Owens DK, Hlatky MA. Cost-effectiveness of left ventricular assist devices in ambulatory patients with advanced heart failure. JACC Heart Fail. 2017;5(2):110–9.

Partial Support with Mechanical Circulatory Support Devices

23

Jordan R. H. Hoffman and Joseph C. Cleveland Jr

Abbreviations

ECLS	Extracorporeal life support
EDP	End-diastolic pressure
EDPVR	End-diastolic pressure-volume relationship
EDV	End-diastolic volume
ESP	End-systolic pressure
ESPVR	End-systolic pressure-volume relation
ESV	End-systolic volume
IABP	Intra-aortic balloon pump
IVC	Inferior vena cava
LV	Left ventricle
LVAD	Left ventricular assist devices
LVEDP	Left ventricular end-diastolic pressure
MAP	Mean arterial pressure
MCS	Mechanical circulatory support
NHLBI	National Heart, Lung, and Blood Institute
NIH	National Institute of Health
PA	Pulmonary artery
PCI	Percutaneous interventions
PCWP	Pulmonary capillary wedge pressure
PVA	Pressure-volume area
PVL	Pressure-volume loop
RA	Right atrium
RV	Right ventricle
SV	Stroke volume
SVR	Systemic vascular resistance
VSD	Ventricular septal defect

Introduction

The use of acute mechanical circulatory support (MCS) closely paralleled the development of cardiopulmonary bypass for open heart surgery in the early 1950s. In fact, as the technology for cardiopulmonary bypass improved, so did the use and development of a multitude of devices for use in the setting of acute cardiac failure.

Currently, many devices indicated for short-term partial support are approved for use for a short period of time, giving the treating team time to formulate a plan for ongoing medical management and additional durable mechanical support. Goals during this period include correction of multisystem organ dysfunction, treatment of metabolic or respiratory acid-base disturbances, and treatment of hemodynamic instability.

Medium-term use of many "short-term" mechanical support devices has been reported

J. R. H. Hoffman, MPH, MD
Vanderbilt University School of Medicine; Section of Surgical Sciences, Department of Cardiac Surgery, Nashville, TN, USA

J. C. Cleveland Jr, MD (✉)
University of Colorado School of Medicine, Division of Cardiothoracic Surgery, Academic Office One, Aurora, CO, USA
e-mail: joseph.cleveland@ucdenver.edu

© Springer Nature Switzerland AG 2020
J. H. Karimov et al. (eds.), *Mechanical Support for Heart Failure*,
https://doi.org/10.1007/978-3-030-47809-4_23

with satisfactory recovery of end-organ perfusion and has therefore led to a redefining of the indications for duration of support in cardiac failure.

The most broad indications for the use of MCS include respiratory failure, cardiac failure, or combined cardiopulmonary failure. Many early trials endorsing the use of acute mechanical support were case reports or case series in patients with post-traumatic cardiopulmonary failure, or "shock lung" [1]. Because of the lack of high-quality evidence and associated cost, the adoption of mechanical cardiopulmonary support was limited. However, in the last few decades, multiple well-designed, randomized trials have continued to demonstrate benefits in terms of improvement in survival, quality of life, and other important outcomes. Because of this, the last decade has seen both an expansion in the indications for and proliferation in use of temporary mechanical cardiac support.

History

The first use of cardiopulmonary bypass occurred in the early 1950s, allowing longer and more complex cardiac operations to be performed. As a result, many patients with postcardiotomy shock due to failed operations were being supported for longer periods of time. Three factors in the evolution of cardiac surgery have encouraged further development of both short- and long-term mechanical support devices: (1) the increasing complexity of cardiac surgical operations in the early 1950s and 1960s, (2) the onset of heart transplantation in the late 1960s, and (3) the realization that donor organs are scarce commodities, thus necessitating "artificial" cardiac support in the 1970s and 1980s.

The earliest use of temporary cardiac support using a balloon pump came in 1962, thus ushering in the era of devices for cardiac support [2]. The first use of temporary postcardiotomy support occurred in 1963 with Liotta's implantation of an "artificial ventricle" [3]. Shortly thereafter, the National Institute of Health (NIH) through the National Heart, Lung, and Blood Institute (NHLBI) formed the Artificial Heart Program, indicating that mechanical support was of national importance. This also represented one of the earliest attempts by the government to influence the trajectory of development of artificial organs. The 1970s and 1980s were dedicated to development "dischargeable" support devices, including the artificial heart and refinement of single ventricle support devices [4].

Understanding of temporary versus full cardiac support was in evolution through the 1990s. Until then, much of the research in cardiac support had focused on short- or long-term *full* cardiac support (government mandates indicating approved devices should be able to support up to 10 liters of flow per minute). However, as ongoing durable ventricular assist device development continued in the early 2000s, the field of acute mechanical support devices providing *partial* cardiac support grew as well [5].

Today, the indications and strategies for partial mechanical circulatory support are numerous and continue to grow. Many "old" ideas have been recycled in an effort to develop new forms of temporary support, a field of study with renewed interest. While durable devices undergo continual refinement, so do the temporary devices that support patients to the point that a durable device is an option. With ongoing excitement about partial support, the landscape that is acute heart failure will certainly change rapidly over the next decade.

Physiologic Basis for Mechanical Support

Heart failure is the inability of the myocardium to pump blood at the appropriate rate to meet the metabolic needs generated by peripheral tissues. Heart failure is broadly classified by ejection fraction, as either being reduced or preserved. Additionally, it can be subcategorized by the ventricle affected (right heart, left heart, or biventricular failure) and the chronicity (acute, chronic, acute-on-chronic) [6]. Some forms of cardiac failure are not due to intrinsic myocardial injury but rather systemic disease, structural dysfunction, or pathology causing external compression.

The heart has evolved several adaptive mechanisms to compensate for ongoing injury including enhanced contractility with rising preload (Frank-Starling law) (Fig. 23.1), structural changes that promote increased contractility (hypertrophy), and initiation of neurohumoral responses (release of catecholamines and activation of the renin-angiotensin-aldosterone system) [7].

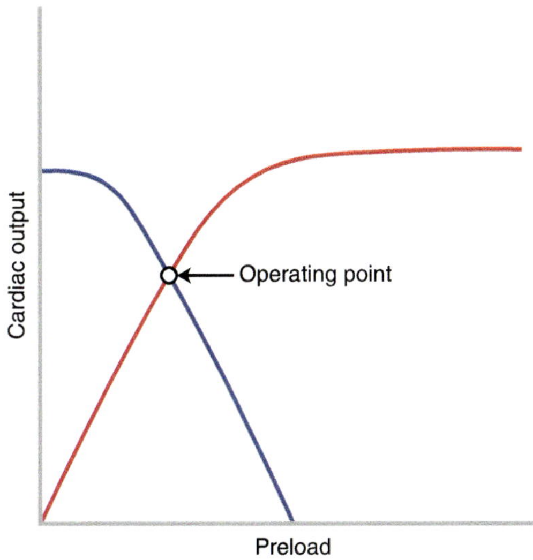

Fig. 23.1 The Frank-Starling Law defining the point at which cardiac output and venous return are equal. The "operating point" establishes optimal loading conditions in both normal and pathologic states. The venous return curve is a result of the combined effects of peripheral resistance, venous capacitance, and overall blood volume. The cardiac output curve reflects changes to contractility (via medication or changes in intrinsic force of contraction) and is indirectly related to changes in peripheral tone

Acute heart failure is caused by conditions in which the heart is abruptly presented with a load that exceeds its capacity to maintain adequate forward flow or in which diastolic ventricular filling is becomes impaired. Chronic heart failure occurs because of a slowly developing intrinsic defect in myocardial contraction [7].

The pathophysiology of heart failure and mechanical support can be understood in terms of pressure-volume loops (PVLs) and the Frank-Starling law, plotting cardiac output or venous return along the y-axis and atrial pressure along the x-axis (Figs. 23.1 and 23.2). The left and right ventricles as well as the atria have distinct P-V loops, representing differences in the underlying physiologic properties of each chamber (Fig. 23.3) [8, 9]. Understanding the components of individual pressure-volume loops allows a clinician to manipulate multiple variables and quickly predict the outcome of an intervention.

Left Ventricle (LV) The standard LV PVL is represented by a smooth edged trapezoid with four distinct phases representing each stage of the cardiac cycle: isovolumic contraction, ejection, isovolumic relaxation, and filling. The location of the LV PVL lies between the end-systolic pressure-volume relationship (ESPVR) and end-diastolic pressure-volume relationship (EDPVR). The position and size of the pressure-

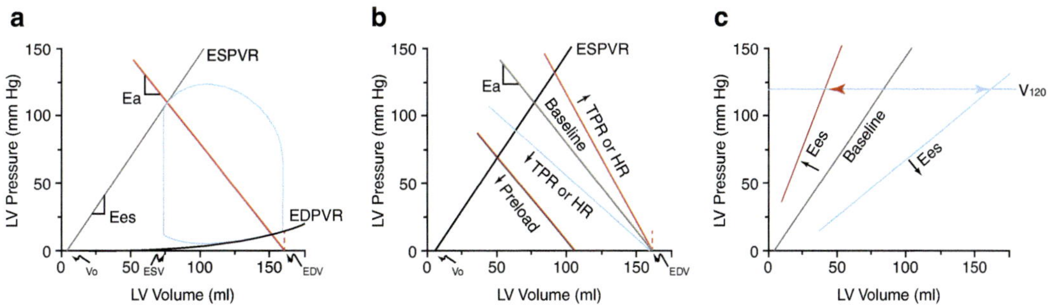

Fig. 23.2 (**a**) A normal pressure-volume loop (PVL) with boundaries as described in the text. The difference of end-diastolic volume (EDV) and end-systolic volume (ESV) is the stroke volume. Stroke work is represented by the area of the PVL. (**b**) Arterial elastance is dependent on peripheral resistance and heart rate. The location on the volume axis is dependent on the preload. (**c**) Changes to the end-systolic pressure-volume relation (ESPVR) are a result of changes in contractility. (Reproduced from Burkhoff et al. [10])

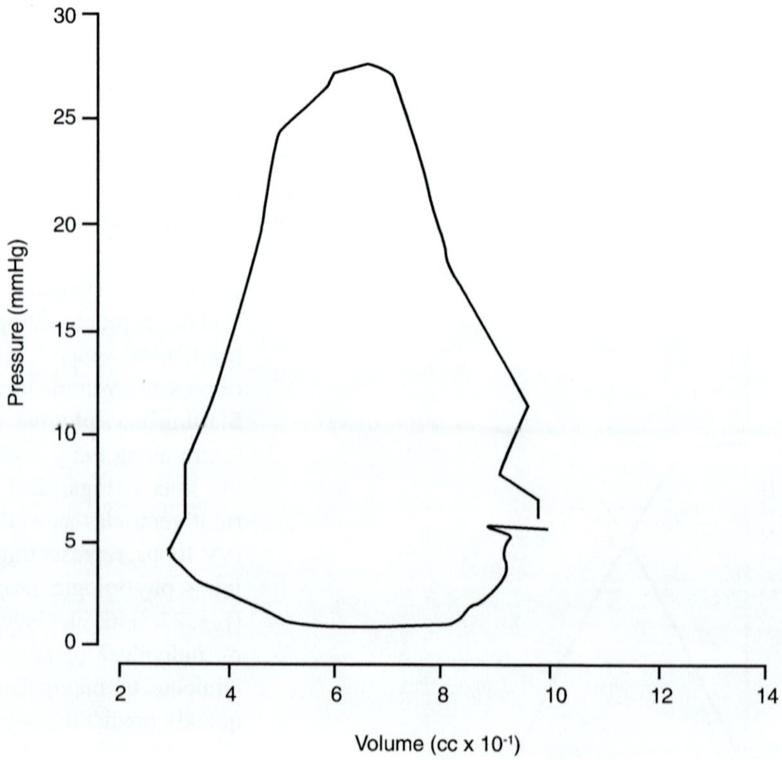

Fig. 23.3 Typical appearance of a right ventricle (RV) pressure-volume loop (PVL). The triangular shape is a result of loss of the isovolumic contraction and relaxation phase owing to near continuous forward flow from the RV to the highly compliant infundibulum and pulmonary vascular bed. (Reproduced from Redington et al. [8], with permission from BMJ Publishing Group Ltd.)

volume loops depend on preload, afterload, and contractility.

Preload is roughly represented by end-diastolic volume or pressure and signifies varying degrees of actin and myosin overlap within the sarcomere. As predicted by the Frank-Starling mechanism, in the non-dilated heart, small changes in end-diastolic pressure (preload) cause large changes in cardiac output (preload dependent). Consequently, if the sarcomere is "overstretched," this unique relationship is lost, and no amount of increased preload will affect contractility (preload independent).

Afterload is represented on the PVL as effective arterial elastance (Ea) which is the slope of the line drawn from the point of end-diastolic volume intersecting with the x-axis through the end-systolic pressure-volume point. Elastance is the tendency for the ventricle to resist change in size when a load is applied and is inversely related to arterial compliance. Arterial elastance is calculated by dividing the end-systolic pressure-volume point by stroke volume, represented by the width of the PVL. Changes to heart rate, vascular resistance, and preload can all cause this line to rotate, expand, or contract, thus changing the shape of the PVL.

Contractility is represented by the slope of the ESPVR. The point on this line intersecting the x-axis is fixed, and therefore changes in contractility will be represented by rotations of the ESPVR. The intersection of the ESPVR with arterial elastance generates a ventricles unique end-diastolic pressure and volume. Increasing contractility will rotate the ESPVR to the left on both the starling curve and the PVL. Decreasing contractility will likewise rotate the ESPVR to the right.

Compliance is the tendency for a ventricle to expand with blood as filling pressure increases (a change in volume for a given change in pressure). If a small change in pressure generates a large change in volume, the ventricle is said to be highly compliant. Compliance is inversely related to ventricular stiffness, which is the slope of the nonlinear end-diastolic pressure-volume relationship (dP/dV). A steeper (leftward shift) ESPVR represents a decrease of compliance, while a flat (rightward shift) ESPVR represents increased compliance.

The boundaries of the pressure-volume loop are dictated by the characteristics of the myocardium, systemic arteries, and systemic veins. Stroke volume (the difference in end-diastolic and end-systolic volume) and stroke work (the area within the PVL) can be calculated from the PVL after manipulation of the "independent" variables above. Taken together, the stroke work and the potential energy of the cardiac cycle (the area between the ESPVR, EDPVR, and diastolic perimeter of the PVL) form the pressure-volume area (PVA), and this is lineraly correlated to myocardial oxygen consumptom. The larger the PVA, the higher the myocardial oxygen consumption. Decreasing myocardial oxygen consumption ("resting the heart") forms the cornerstone for mechanical circulatory support in all forms [10].

Right Ventricle (RV) The right ventricle is responsible for maintaining a low right atrial pressure, thus encouraging venous return to the heart, and supplying continuous, low-pressure flow to the pulmonary vascular bed. Because of the highly compliant nature of the RV myocardium and infundibulum, the RV PVL is distinct in that the loss of isovolumic contraction and isovolumic relaxation leads to a triangular-shaped pressure-volume loop, similar to the continuous flow loops often seen in patients with ventricular assist devices (VAD) [11]. In a healthy patient, continuous flow to the pulmonary vasculature is encouraged by low pulmonary vascular resistance. Contraction of the LV further empties the RV while maintaining a low end-systolic pressure.

Rather than a passive conduit from systemic to pulmonary circulation, the RV is highly complex and sensitive to changes in preload and afterload. In situations where the RV is exposed to increased afterload, the normal triangular-shaped PVL begins to resemble that of an LV, becoming trapezoidal in appearance. For example, the high afterload of pulmonary hypertension requires a period of contraction without ejection to open the pulmonic valve. In response to elevated afterload, the RV will dilate to maintain stroke volume and thus increase myocardial oxygen consumption [9]. The result of these changes is that the normal RV PVL becomes trapezoidal in shape, gaining a distinct isovolumic contraction and relaxation phase.

Ventricular interdependence is the process by which normal or abnormal loading conditions in either ventricle will impact hemodynamics and filling pressures of the contralateral ventricle. This interplay between chambers becomes pronounced in pathologic processes such as intrinsic myocardial disease or constrictive pericardial processes [12]. Ventricular interdependence plays a key role in maintaining forward flow from the RV in patients with temporary or durable mechanical support. Thus, frequent use of echocardiography, specifically targeted to the geometry of the interventricular septum, has been found to be increasingly important in mechanically supported patients. As the LV is unloaded with mechanical support, the geometry of the interventricular septum is changed, ultimately reducing compliance of the RV. This change, coupled with increased preload secondary to now adequate forward flow from the LV, overloads the already dysfunctional RV. This process can be attenuated with attention to septal geometry and adequate RV support during the initial mechanical support period [13].

In the normal heart, cardiac output is controlled by preload, and the myocardium acts as an active pump, ejecting whatever volume is presented to it in the form of systemic venous return. Venous return is affected by three factors, including right atrial pressure, mean systemic filling pressure, and resistance to blood flow return to the right atrium. There are limits to the amount from blood a heart can pump, and in the failing heart, cardiac output is dependent on contractility and afterload as determined by the Frank-Starling law of the heart (Fig. 23.1).

Acute Partial Mechanical Support Devices

The number of devices used for temporary cardiac support in heart failure is increasing. These devices are unique in that they are approved for short duration but also have the ability to provide "full" support in smaller patients. Many centers

Fig. 23.4 Normal hemodynamic values

Measured	Normal value
Systolic blood pressure	100-140 mmHg
Diastolic blood pressure	60-80 mmHg
Mean blood pressure	70-105 mmHg
Cardiac output	4-8 L/min
Central venous pressure	0-6 mmHg 0-8 cmH20
Heart rate	60-100 beats/min
Systolic pulmonary artery pressure	15-25 mmHg
Diastolic pulmonary artery pressure	8-15 mmHg
Mean pulmonary artery pressure	10-15 mmHg
Pulmonary capillary wedge pressure	4-12 mmHg
Right atrial pressure	0-6 mmHg
Right ventricular end-diastolic pressure	0-5 mmHg
Right ventricular end-systolic pressure	15-25 mmHg

now use partial support devices to provide both short- and medium-term cardiac unloading. The various devices are distinct from one another by multiple factors, including the hemodynamic profile, the device size, the technique of insertion, the ability to provide gas exchange, and the distinct physiologic mechanism by which the pump unloads the ventricles of the heart [10].

Echocardiography and pulmonary artery (PA) catheterization form the cornerstone of evaluation in patients undergoing consideration of mechanical support. Additionally, frequent assessment of laboratory values to evaluate tissue perfusion and resolution of end-organ dysfunction is suggested within the first few hours after initiation of mechanical support (Fig. 23.4). After initiation of mechanical support, hemodynamic and laboratory trends may either reveal the need for additional support or indicate that a patient can be successfully weaned from further therapy.

Intra-Aortic Balloon Pump (IABP) Counterpulsation The most widely used partial support device is the intra-aortic balloon pump (IABP). The ease of insertion and the marginal additional benefit it provides make this device one of the most desirable mechanical therapies in patients with refractory heart failure. The counterpulsation mechanism functions such that a cylindrical polyethylene balloon fills rapidly with helium during diastole and deflates during rapidly, prior to systole. The reduction in afterload occurs because of a vacuum effect created during deflation of the balloon. Coronary blood flow is increased during diastole due to the balloon inflation. The mechanisms described combined to increase myocardial oxygen delivery, through increased coronary blood flow during diastole and decrease myocardial oxygen demand while increasing stroke volume, through decreased afterload.

Described physiologically, during IABP therapy, arterial elastance is reduced due to a decreased end-systolic pressure (ESP) and increased stroke volume (SV). Simultaneously, the force required to move a volume of fluid during each contraction, known as stroke work, is unchanged. This is due to the effects of an increased stroke volume with a decreased afterload. The net effect of these physiologic changes is a decrease in the pressure-volume area, a representation of the total mechanical energy generated by the ventricle during contraction [14, 15] (Fig. 23.5). The additional advantage of decreased afterload results in marginal LV unloading, a benefit transmitted to the right ventricle through the pulmonary vascular system. Thus, it can be

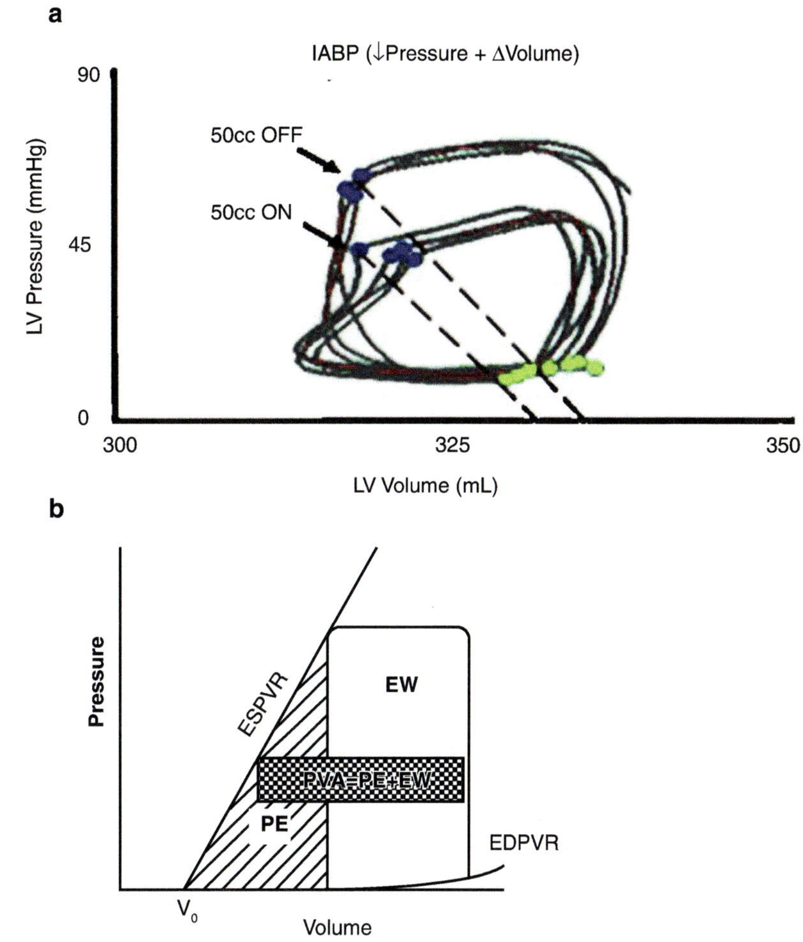

Fig. 23.5 (**a**) Hemodynamic effects during use of the intra-aortic balloon pump. Note the decreased arterial elastance but unchanged stroke work during IABP therapy (Reproduced from Esposito et al. [14]). (**b**) Pressure-volume area (PVA) is the sum total of potential area and stroke work. Myocardial oxygen consumption is linearly correlated with PVA. (Reproduced from Kawaguchi et al. [15])

said that the IABP may also benefit patients with right heart dysfunction.

Clinically, the IABP waveforms can be helpful in determining the amount of support required to maintain adequate hemodynamics as well as the ideal inflation/ deflation period to provide optimal afterload reduction. Balloon inflation should occur in conjunction with the T-wave on electrocardiogram (EKG) or at the dicrotic notch on pressure waveforms. Likewise, balloon deflation should be timed to the R-wave or just before the onset of systole on pressure waveforms (Fig. 23.6) [16]. Interpretation of the balloon pressure waveforms can also provide valuable information as to the status of the balloon pump mechanics or the patient's clinical condition. Important parameters to note from balloon pressure tracings should include the morphology of the waveform which should have a brisk initial rise, followed by a plateau phase, and lastly a fall back to baseline. The filling pressure, the width and height of the pressure waveform, and the regularity and consistency with surrounding waveform tracings should also be noted (Fig. 23.7). Balloon pressure waveforms can indicate conditions such as an improperly sized balloon, changes in systemic vascular resistance (SVR), balloon malposition, abrupt changes in blood pressure, or mechanical failures which include catheter kinking, incomplete balloon emptying, changes in balloon pressure, helium leak, or low helium supply.

IABP therapy can be used in a variety of pathological conditions including cardiogenic shock, intractable angina or ongoing myocardial

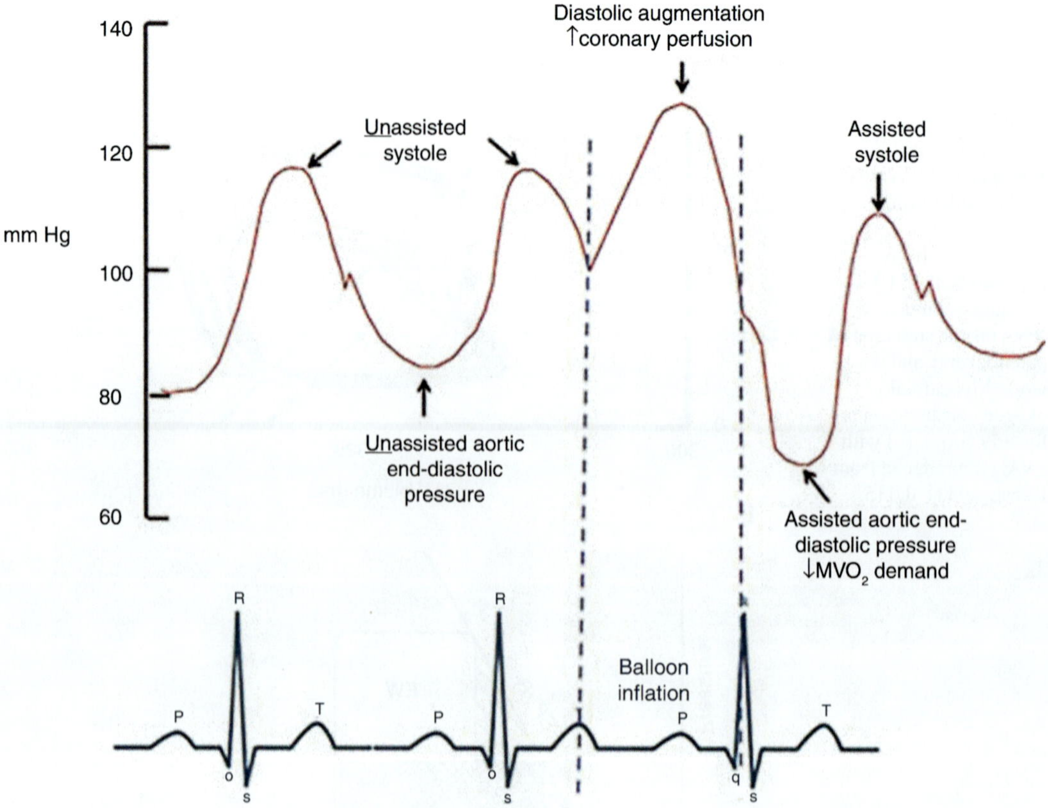

Fig. 23.6 IABP waveforms mapped to EKG tracing. Balloon inflation should occur during diastole and deflation should occur just prior to systole. IABP waveforms should be kept in mind when interpreting hemodynamic changes during support or during troubleshooting. (Reproduced from Patterson et al. [16], with permission from Wolters Kluwer Health, Inc.)

ischemia, postcardiotomy low-cardiac output states, as an adjunct during high-risk percutaneous interventions (PCI), for patients with severe left main coronary artery stenosis, in refractory heart failure as a bridge to further therapy, during intractable ventricular arrhythmias, for mitigation of mechanical complications of myocardial infarction such as acute ventricular septal defect, mitral regurgitation, or free-wall rupture, or critical aortic stenosis awaiting intervention. Contraindications include mild or greater aortic insufficiency, aortic dissection or clinically significant aortic aneurysm, and severe peripheral artery disease. Relative contraindications include uncontrolled sepsis, refractory bleeding condition, aortic endovascular stent grafts, history of aortoiliac, or lower extremity arterial bypass.

IABP placement is unique in that its ease of insertion allows it to be performed at the patient's bedside. Prior to insertion, all radiographic images should be reviewed for anatomy of the aortoiliac system as well as any pathologic conditions that may exist. A balloon pump should be chosen to match the patient's height, and oversizing of the balloon should be avoided as both the additional length and width of a fully inflated, oversized balloon can cause dissection and rupture of the aorta or lower extremity vessels. All equipment needed for IABP insertion is included with the IABP kit. Ultrasound-guided arterial access should be obtained at the level of the common femoral artery. Alternatively, fluoroscopy can also be used to ensure an adequate puncture site. The most common method of IABP insertion is through an introducer sheath; an alterna-

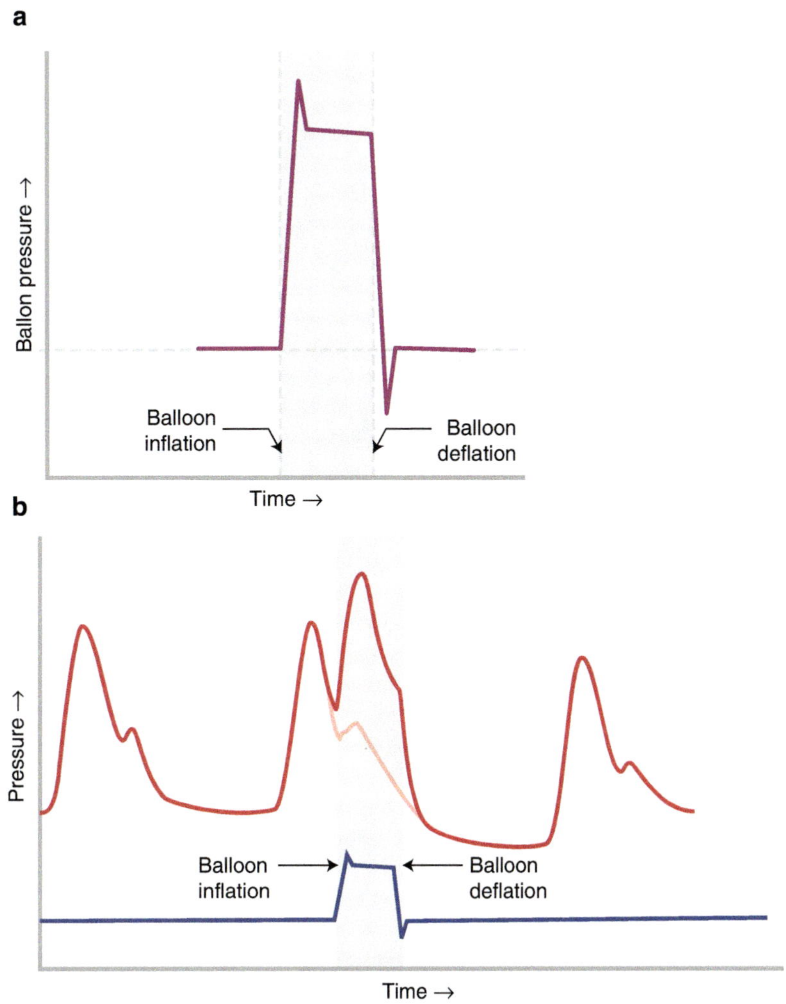

Fig. 23.7 (**a**) Normal IABP balloon filling pressure waveform. (**b**) Normal IABP balloon filling pressure timed with arterial pressure waveforms

tive method allows for "sheathless" insertion which provides the benefit of a small cross-sectional profile for smaller patients or patients otherwise at risk of distal ischemia due small or occluded peripheral vessels. A shallow needle entry can minimize the risk of catheter occlusion from kinking. Ideal balloon position can be confirmed with echocardiography (balloon tip two centimeters distal to the left subclavian artery), fluoroscopy or chest X-ray (balloon tip at the level of the carina), and review of the balloon pressure tracings. Initial therapy should be set to 1:1 or 1:2 IABP augmentation as described below (Fig. 23.8).

Maintenance care during IABP therapy includes twice daily monitoring of the sites securing the IABP to the patient, distal neurovascular examination with confirmation of pulses or signals distal to the balloon pump insertion, assessments of femoral hematoma or puncture site hemorrhage, and at least twice daily assessment of pressure tracing and balloon waveform analysis. IABP tubing should be inspected for the presence of blood, indicating balloon rupture. Daily consideration should be given to balloon pump weaning if a patient's clinical condition improves over the course of IABP therapy. While most patients will require 1:1 balloon pump augmentation, consideration can be given to 1:2 balloon pump augmentation in the setting of need for additional nonpulsatile mechanical support. The balloon pump should be paused during all bal-

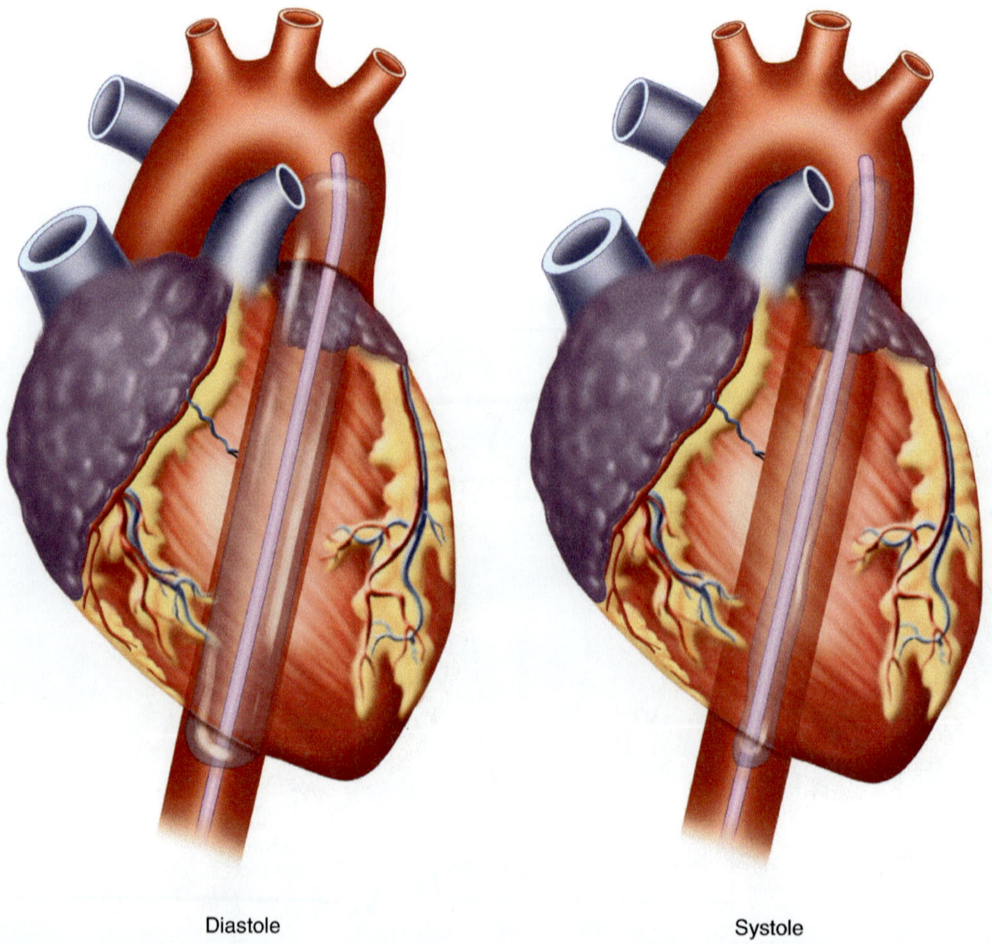

Fig. 23.8 Example of a balloon pump during systole and diastole. Note the ideal IABP position, with the tip located 1–2 cm distal to the left subclavian artery

loon pump position adjustments, and the balloon pump tubing resecured to the patient when accurate position is confirmed. Many centers differ on anticoagulation strategy while on balloon pump therapy, ranging from no anticoagulation to full anticoagulation with heparin or other non-heparin alternatives.

Complications of IABP therapy are uncommon. Close monitoring and early intervention are key to minimizing hemodynamic deterioration during IABP therapy. A class of complications exist related to balloon malposition; occasionally a patient will develop a persistent lactic acidemia due to coverage of visceral vessels by the balloon. In this situation, consideration should be given to insertion of a smaller balloon size or advancement of the balloon superiorly, including coverage of the left subclavian artery if necessary, as a temporizing measure until further stabilization can occur. Another complication related to balloon malposition can be diagnosed by lack of diastolic augmentation, which may be compromised if the balloon has not fully cleared the sheath. The solution to this problem is to advance the balloon further into the aorta. Other complications include bacteremia, groin hematoma or bleeding at the insertion site, ipsilateral limb ischemia due to occlusion of the insertion vessel, or contralateral limb ischemia due to embolization from the balloon, thrombocytopenia or anemia due to hemolysis, aortic dissection, balloon displacement, balloon rupture, or leak.

The ability to wean from balloon pump should be assessed daily. A prerequisite to IABP removal is improvement in clinical condition and hemodynamics with an overall decrease in medical support. If the treatment team feels the patient will tolerate an IABP weaning trial, the balloon pump can be paused to assess changes in hemodynamics. Assessment of weaning parameters should include blood pressure, pulmonary artery and central venous pressures, markers of end-organ perfusion such as vital signs and urine output, and laboratory parameters including arterial blood gasses, lactate, mixed venous oxygen saturations, and creatinine. During a weaning trial, an IABP assist ratio is decreased to 1:3 or 1:4, while the above parameters are monitored. The pump can be paused for up to 30 minutes without anticoagulation. Alternatively, some authors have described weaning balloon volume rather than the assist ratio [17]. Whichever method is used, uniformity in IABP weaning is important to maintain consistency and avoid confusion among providers.

Alternative cannulation strategies exist if a patient is felt to require more than short-term use of a balloon pump. Ideally, alternative cannulation strategies will allow patients to be ambulatory during the period of cardiopulmonary rehabilitation. The most commonly used alternative approach is either a percutaneous or open surgical axillary artery approach. Percutaneous insertion technique is similar to the femoral approach although consideration can be given to using specialized pre-insertion closure devices if short-term use is expected. An open surgical approach requires a surgeon fashioned a graft to either the axillary or subclavian artery. The hemostatic one-way valve is secured to the distal end of the graft which is then tunneled subcutaneously. In both the surgical and percutaneous technique, a combination of fluoroscopy and echo guidance is used to maneuver the aortic arch and correctly position the balloon pump. Pressure triggering is more difficult with the antegrade placement of a balloon pump given the location of the pressure sensing mechanism in the abdominal aorta, and most often, the pump is triggered off of an EKG or intrinsic pacing device in these situations. Maintenance is similar as for that of femoral insertion, with routine hemodynamic and laboratory monitoring. If a graft was used and tunneled subcutaneously, the balloon pump can simply be removed, and the subcutaneous graft left in place with no further operative intervention. Alternatively, the patient can be taken to the OR for graft and balloon pump removal via open surgical exposure. If a percutaneous technique with pre-insertion closure devices used, the balloon pump can be removed and the closure devices tightened; removal should be undertaken in the OR, and surgical exploration should strongly be considered in the case of expanding hematoma or failure of the closure devices. Post-removal monitoring of the balloon pump site with extremity restrictions should be followed for at least 6 hours following removal of a percutaneously inserted device [18, 19].

One of the major challenges with use of a balloon pump is the inability to provide robust cardiac support for patients in extremis. The modest offloading capability of the IABP is offset by its ease of insertion and ability to run without anticoagulation, usually making it the first-line therapy in the armamentarium of mechanical support. While the balloon pump may not completely bridge a patient to recovery, its benefits lie in the temporizing effect it has until a decision for further therapy or de-escalation can be made.

Impella® An Impella® (Abiomed, Inc., Danvers, MA) device is a percutaneous, trans-valvular, axial flow device which uses an impeller to propel blood from the left ventricle to the aorta (Fig. 23.9). It is designed for temporary use under specific circumstances. Similar to durable left ventricular assist devices (LVAD), the pressure-volume loops of the Impella® device are triangular in shape. This reflects diminution of the isovolumic contraction and relaxation phase of the cardiac cycle, as blood is continuously emptied from the LV to the aorta, independent of the cardiac cycle [10]. Advantages to circulatory support with an Impella® come in the form of increased ventricular offloading with increasing

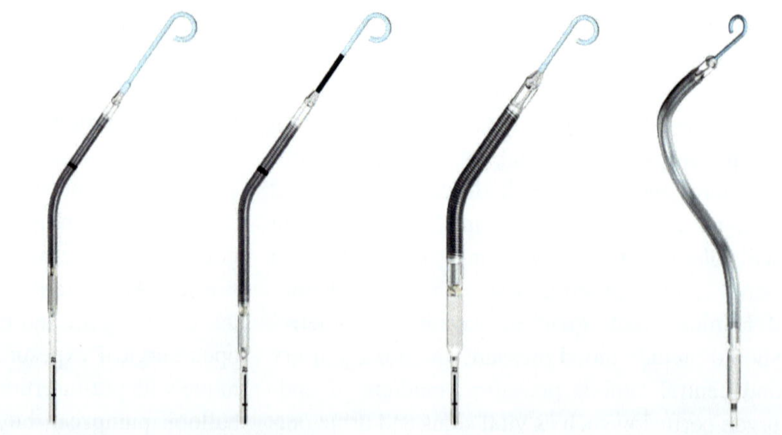

Fig. 23.9 Examples of percutaneously inserted Impella® devices for right- and left-sided support. (Used with permission of Abiomed, Inc. © 2019 Abiomed, Inc.)

levels of support. Progressive ventricular offloading is represented by a leftward shift of the PVL (Fig. 23.10) and a contraction of the pressure-volume area, both corresponding to a decrease in myocardial oxygen requirement while simultaneously increasing oxygen supply through increase coronary blood flow [10, 20]. Additional benefits with use of an LV to aorta pumps include decreased left atrial pressure, pulmonary capillary wedge pressure, and pulmonary congestion. LV offloading is transmitted to the RV, leading to moderately improved RV function.

The Impella® system can support flows up to 5 liters per minute and provide either right-sided or left-sided support depending on the device chosen. It is indicated in cardiogenic shock for short-term use up to 6 days or until myocardial recovery occurs and can be used in most cases of cardiomyopathy to support forward flow from the left ventricle. An Impella RP® is indicated for right-sided support and can be used in isolation or together with left-sided Impella® support for biventricular failure. The right-sided device is designed for use up to 14 days.

A special indication for use of the Impella® system includes indications for use in high-risk percutaneous coronary interventions (PCI) in elective or urgent situations where temporary coronary occlusion is anticipated.

Contraindications include LV thrombus, presence of a mechanical aortic valve, severe aortic stenosis, moderate to severe aortic insufficiency, severe peripheral artery disease precluding placement of a percutaneous device, concomitant respiratory failure, presence of an atrial or ventricular septal defect (VSD), presence of mechanical complications of myocardial infarction including LV rupture, or post-infarct VSD. Contraindications for right-sided Impella® support include thrombus of the iliofemoral venous system, inferior vena cava (IVC) or right ventricular (RV) thrombus, presence of an IVC filter, or severe valve dysfunction.

Placement of right- or left-sided Impella® devices begins with accurate sheath placement under ultrasound and fluoroscopic guidance. During left side placement, a wire is placed into the common femoral artery and advanced up the aorta, through the aortic valve, and should reside in the LV apex, avoiding the mitral valve apparatus. The wire is "pinned" to the drape, and the Impella® is advanced over this wire. After placement is confirmed, tension should be released with gentle traction backward on the device. With right-sided placement, a wire is advanced into the common femoral vein and up the IVC, into the RV and through the pulmonary valve. The wire should come to rest in the left pulmonary artery if possible. Over this wire, the device is advanced while "pinning" the wire to the drape. Inadvertent forward movement of the wire in the pulmonary artery (PA) can cause PA rupture and lethal bleeding. The device should be placed so that the outflow is in the main PA, such that blood is

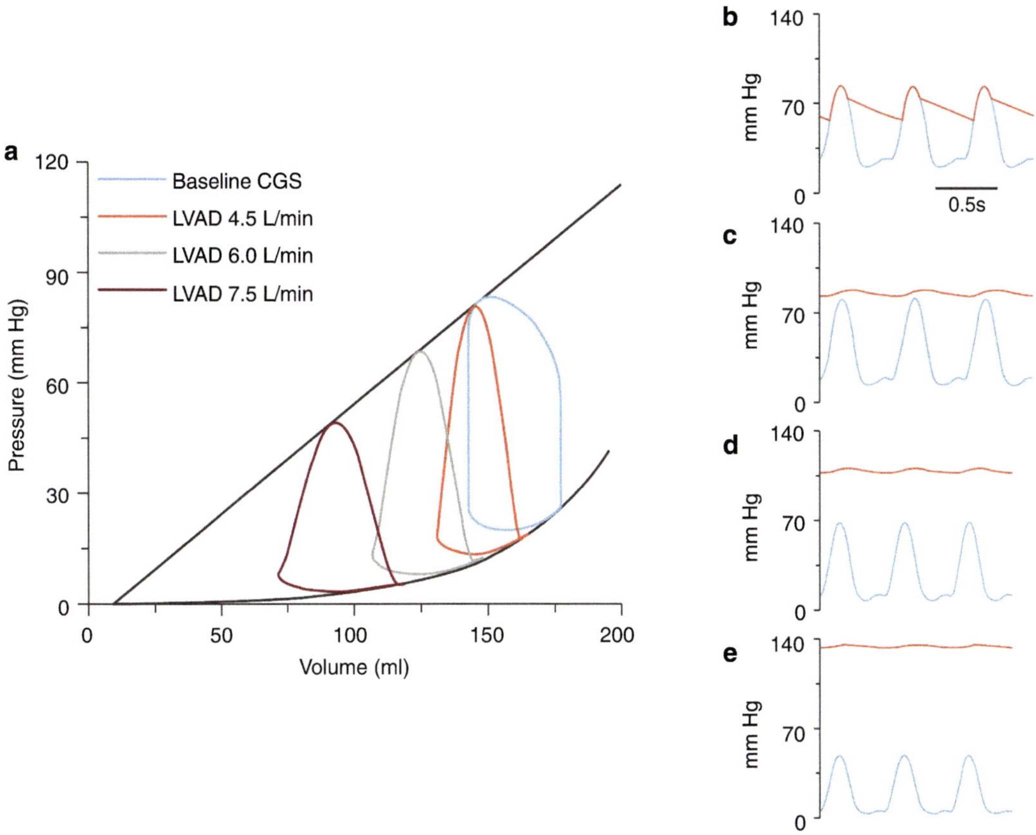

Fig. 23.10 (**a**) Typical appearance of a pressure-volume loop in patients with an Impella device. The triangular shape results from the loss of isovolemic phases of the cardiac cycle, due to continuous forward flow. Note the leftward shift of the PVL with increasing levels of support. (**b**) Increasing support from the Impella device lead to ventriculo-arterial uncoupling, where simultaneous increases in arterial pressure occur with decreases in ventricular pressure. (Reproduced from Burkhoff et al. [10])

distributed equally to both right and left PA. Tension should be removed from the catheter by gentle traction backward; this can easily result in the device being withdrawn into the RV but can be prevented by careful manipulation under fluoroscopic guidance.

Impella® devices can be placed through alternative access sites, the most common being through a graft into the axillary artery. This technique is commonly used for the Impella 5.0® given its larger profile and should also be performed under fluoroscopic guidance with consideration to placement technique, similar to the smaller devices.

A patient with an Impella® device should be assessed daily for the ability to wean mechanical support. Evidence of myocardial recovery on echocardiography, right heart catheterization, or with improving laboratory values (lactate, creatinine, liver function tests) will give clues as to the ability to wean support. Cessation of support should coincide with stabilization of end-organ function, particularly the lungs, kidneys, brain, and gastrointestinal system. Should a patient fail to wean from mechanical support, an additional 24 hours of support is warranted before another weaning trial is attempted. Decannulation of the Impella® device should be performed with surgical exposure of the cannulated artery. The device is removed, and the arteriotomy is repaired.

Complications associated with Impella® use include aortic or mitral valve injury, arrhythmia, myocardial perforation, renal dysfunction, hemolysis, bleeding, stroke, limb ischemia, thrombocytopenia, and vascular injury. Similar complications exist for the right-sided support device but also include liver failure, deep vein thrombosis, tricuspid or pulmonary valve injury, and pulmonary hemorrhage with respiratory failure.

TandemHeart® The TandemHeart (CardiacAssist, Inc., Pittsburgh, PA) device uses a centrifugal pump propel blood drained from the left atrium via a transseptal cannula inserted through the femoral vein back to the systemic circulation via the femoral artery. This mechanism indirectly unloads the LV by decompressing the LA and provides up to 5 liters per minute of blood flow. The TandemHeart® reduces pulmonary capillary wedge pressure (PCWP) and LV end-diastolic pressure (EDP) while maintaining mean arterial pressure (MAP). This allows inotropic medications to be weaned or discontinued, with a net increase in myocardial delivery and decrease in myocardial consumption [21]. Additional benefits to decreased PCWP include increase forward flow of blood from the RV and improvement in levels of oxygenated blood if pulmonary edema is present [10].

Left atrium to femoral artery bypass devices are indicated for up to 6 hours of continuous support, though these devices have been used off label for longer periods. With a maximum speed of 7500 RPM, the TandemHeart® pump can deliver flows up to 4 liters per minute. The indications for use include cardiogenic shock, until recovery occurs or as a bridge to definitive therapy. Like the Impella®, the TandemHeart® can also be used as a temporary support device during high-risk coronary interventions [22–24]. Contraindications are similar to other support devices in patients with advanced heart failure. Additionally, the surgeon placing the device should have comfort with the transseptal approach, as well as the use of fluoroscopy which is necessary for placement of the TandemHeart®.

Placement of the TandemHeart® requires a transseptal puncture. The operator placing the device should be comfortable with this technique before undertaking placement of this device. The TandemHeart® is placed in the cath lab where fluoroscopy is available. Standard femoral venous access is obtained with the assistance of ultrasound or fluoroscopy. A guidewire is advanced from the right atrium (RA) to the LA via the previously made transseptal puncture. The septostomy is serially dilated, and over the guidewire, the 21-french inflow cannula is positioned under fluoroscopic guidance. Next, a 15- or 17- French arterial return cannula is placed in the common femoral artery in the standard fashion. If the patient's femoral arteries are small, smaller return cannulas can be placed bilaterally. The patient should be assessed for the need for an antegrade or retrograde perfusion cannula to the lower extremity of the side in which the return cannula is placed. The system must be cleared of air prior to initiation.

Maintenance while on this device is similar to that of other partial support devices. End-organ function must be monitored daily. Markers of tissue ischemia should be checked at least every 6 hours until stable or down trending. Lastly, clinicians should be aware of the lower risk for hemolysis with this device. Placement can be monitored with daily chest X-rays and echocardiograms documenting cannula position. A devastating complication of TandemHeart® therapy is withdrawal of the transseptal cannula into the right atrium, creating a massive right to left shunt, causing hypoxemia. Therapy should be discontinued immediately if this complication is suspected.

The patient should be assessed for their ability to wean from therapy on a daily basis. Evidence of myocardial recovery will give the clinician clues as to the ability to wean support. Cessation of support should coincide with stabilization of end-organ function. Should a patient fail to wean from mechanical support, an additional 24 hours of support is warranted before another weaning trial is attempted. Decannulation of the TandemHeart® device should be performed with surgical exposure of the cannulated artery. The device is removed and the arteriotomy is repaired. Alternatively, a percutaneous closure device can be used if this has planned prior to TandemHeart® initiation.

Extracorporeal Membrane Oxygenation (ECMO) ECMO, also known as extracorporeal

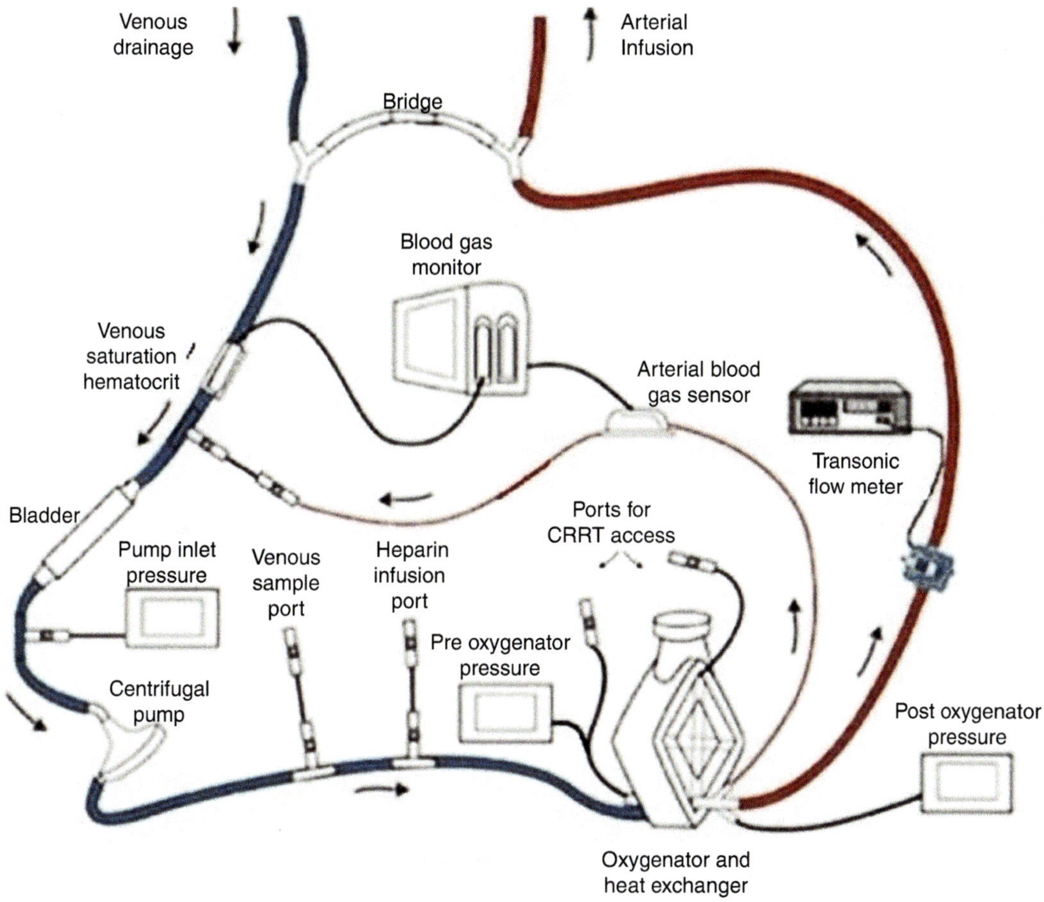

Fig. 23.11 Representation of an ECMO circuit. In its simplest form, an ECMO circuit can consist of only a centrifugal pump, an oxygenator, and tubing. (Reproduced from Lequier et al. [25], with permission from Wolters Kluwer Health, Inc.)

life support (ECLS), has been increasingly used as a rescue therapy for patients presenting with acute cardiopulmonary failure [6]. Venoarterial ECMO removes deoxygenated venous blood from the vena cava and right atrium, circulates the blood through an oxygenator, and returns the blood to the systemic circulation either via a peripheral or central artery. ECMO circuits contain an oxygenator, a gas blender, a heat exchanger, a blood pump with console, and cannulas with a tubing system (Fig. 23.11) [25]. The blood pump and console system vary between hospitals, but the basic setup is similar among all systems. Examples of common pumps in use today include the CentriMag™ Acute Circulatory Support System (Abbott Corporation, Abbott Park, Il), the Revolution® Centrifugal Blood Pump, and Centrifugal Pump Console (SCP® and SCPC®) (LivaNova PLC, London, England, UK), and the CARDIOHELP® System (Maquet, Rastatt, Germany).

Ventricular effects of ECMO include flow-dependent increase in end-diastolic pressure with a concomitant decrease in left ventricular stroke volume. ECMO provides minimal flow-dependent LV unloading compared to other partial mechanical support devices and can increase myocardial work if peripheral resistance remains fixed or if no additional ventricular decompression is provided. The benefits of ECMO include the ability to supplying systemic flow rates while also providing oxygenation and ventilation. Thus, ECMO can limit and even reverse end-

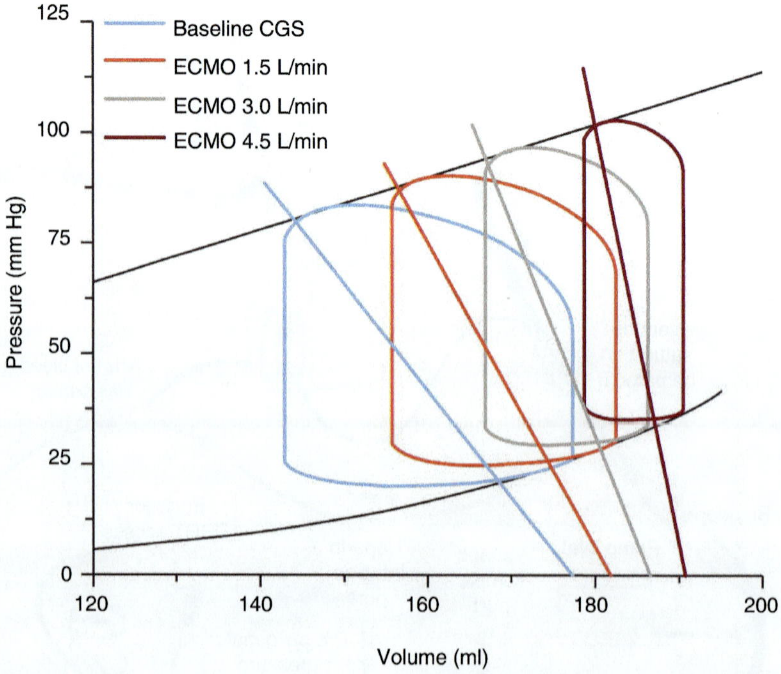

Fig. 23.12 Ventricular effects of ECMO. Note the flow-dependent increase in end-diastolic pressure with a concomitant decrease in left ventricular stroke volume. ECMO provides minimal flow-dependent LV unloading compared to other partial mechanical support devices. (Reproduced from Burkhoff et al. [10])

organ dysfunction by providing an adequate flow of fully oxygenated blood (Fig. 23.12).

Survival after initiation of ECMO is strongly correlated with the indication for which ECLS was initiated; baseline functional status, medical comorbidities, operative history, and duration of ECLS will dictate the prognosis for patients on ECMO. ECMO is indicated for children and adults, in almost all forms of acute or acute-on-chronic cardiogenic shock refractory to medical therapy. When planning ECMO therapy, strong consideration should be given to the goal of care and the long-term plan after the patient has regained stability. Underlying disease processes that are progressive and irreversible are absolute contraindications to ECMO initiation. Other absolute contraindications include permanent neurologic damage, severely depressed functional status, and terminal malignancy. Initiation of ECMO is contraindicated without surgical support.

In most emergency situations, peripheral VA ECMO cannulation allows a physician to rapidly initiate ECMO. Peripheral cannulation involves percutaneous catheter placement into the common femoral artery and common femoral vein with either ultrasound or fluoroscopy. Alternatively, open surgical exposure of these vessels can be used depending on the comfort of the physician. The vessels are serially dilated over a long, stiff wire to accommodate the chosen cannula size. Once placement position is confirmed, the system is manually de-aired, and flow is initiated. Venous cannula placement can be confirmed with transesophageal echo or fluoroscopy. The arterial return cannula is usually positioned with its tip in the distal abdominal aorta and does not need positional confirmation. The cannulas and ECMO tubing are secured to the patient in multiple areas.

Usually, an antegrade perfusion catheter is used to perfuse the extremity ipsilateral to the nearly occlusive arterial cannula. The antegrade catheter consists of a 5-French sheath attached to the return limb of the ECMO circuit and should be placed *before* the arterial cannula, if time permits. Alternatively, a retrograde extremity perfusion catheter can be placed in the posterior tibial artery.

Central VA ECMO cannulation is reserved for the operating room and is most commonly used for postcardiotomy shock after failure to wean from cardiopulmonary bypass. There are numerous cannulation strategies, and a discussion of central VA ECMO is outside the scope of this chapter. Monitoring, maintenance, and weaning are similar to that of peripheral VA ECMO.

Maintenance while on peripheral or central VA ECMO involves attention to labs and end-organ function. Blood gasses and lactates should be monitored often until improvement is noted. Cardiac function should be assessed with a surface echocardiogram and a pulmonary artery catheter. Ventilator management can be difficult in a patient cannulated peripherally or centrally. If lung function is marginal and it is anticipated that long-term support will be needed, conversion to tracheostomy should be discussed with the patient's family. The need for renal replacement should be assessed regularly with assessment of electrolyte and intrinsic renal function, as well as monitoring of urine output. It should be expected that a transient worsening of liver function will occur, and thus liver function should be monitored closely as well. Daily assessment of neurologic function should occur and should be evaluated by weaning sedation and allowing a patient to participate in a neurologic exam if possible. Any change in baseline neurologic status should prompt further investigation with cross-sectional imaging. However, MRI is contraindicated while on ECMO.

Left ventricular venting while on ECMO is important to lower left ventricular end-diastolic pressure (LVEDP), pulmonary capillary wedge pressure (PCWP), and encourage myocardial recover through decompression. Venting can be accomplished with a mini-thoracotomy, through which a cannula can be directly inserted through the LV apex. This cannula can then be attached to the ECMO circuit. An Impella® device can be used percutaneously as an alternative to direct LV puncture for LV venting. A balloon pump provides a marginal amount of LV venting and should be reserved for situations when no other options exist. Pharmacologic measures including inotropes used to increase myocardial contractility offer non-invasive options to LV venting but are unreliable in the failing heart and thus should not be used as the sole mechanism for LV unloading.

The patient should be assessed daily for ability to wean from VA ECMO. Improvement in laboratory values, cardiac function, and decreasing amounts of pharmacological support all indicate that ECMO decannulation can be attempted. Should a patient fail to wean from support, advanced cannulation strategies which would allow a patient to be ambulatory or weaned from a ventilator can be approached. Long-term cannulation strategies are numerous and are outside the scope of this chapter. Nonetheless, long-term cannulation timing and strategies should be tailored to individual patient needs and preferences.

Decannulation from ECMO requires surgical support. An open femoral vessel exposure with proximal and distal control provides the most reliable way to decannulate and repair the femoral vessels. Occasionally, it will be necessary to transiently increase the amount of inotropic support during an ECMO wean, but once decannulation is accomplished, vasoactive and inotropic support can be weaned over multiple days.

A phenomenon known as "mixing cloud" describes the admixture of deoxygenated, stagnant native blood from a contractile myocardium and oxygenated blood delivered peripherally by the ECMO circuit. As contractility recovers and blood is more forcefully ejected from the failing heart, this mixture of oxygenated and deoxygenated blood moves more distally down the aorta. Often, the first sign of recovery of native cardiac function is signaled by a decrease in PaO2 from the right radial arterial line. Because of this, a right radial arterial line is imperative while on VA ECMO. Insufficient monitoring of the arterial mixing cloud may allow stagnant, deoxygenated blood to be ejected to the brain, thus causing a neurologic catastrophe.

Aortic valve opening can be assessed with echocardiography or can be signaled by a pulse pressure and dicrotic notch on the arterial line waveform. Often, with peripheral VA ECMO, the aortic valve fails to open owing to a weak myocardial contraction against an increased afterload caused by aggressive VA ECMO flow. In this situation, stagnant blood in the aortic root can form clot which will then embolize to coronary, head, or visceral vessels. Consequently, the lowest rate of flow that provides oxygenation, end-organ perfusion, and resolution of abnormal labs should be used during extracorporeal support.

Neurologic complications can occur in patients on central or peripheral VA ECMO and are attributable to either low flow states or embolic events. The clinician should maintain a high index of suspicion for neurologic compromise in patients on ECMO. Cross-sectional imaging and electroencephalogram (EEG) should be obtained in all patients with a change in neurologic status, even if transient.

Conclusion

Partial mechanical circulatory support plays an integral role in supporting patients with acute or acute-on-chronic cardiac failure. The indications for temporary support continue to expand, giving clinicians time for treatment, understanding of prognosis, and decision. Current indications include bridge to recovery and bridge to transplant. However, a "bridge to bridge" strategy has emerged as a viable option for patients with cardiac failure who need additional time for complex decision making.

Partial mechanical circulatory support encompasses a wide range of devices that support the failing heart. A thorough understanding of the pathology of heart failure and the physiologic mechanism supporting the use of devices in heart failure patients are important for all clinicians involved in the care of patients with a diagnosis of heart failure. Further research and development will help characterize the optimal device and treatment strategy for different stages of heart failure.

References

1. Hill JD, O'Brien TG, Murray JJ, Dontigny L, Bramson ML, Osborn JJ, Gerbode F. Prolonged extracorporeal oxygenation for acute post-traumatic respiratory failure (shock-lung syndrome). Use of the Bramson membrane lung. N Engl J Med. 1972;286:629–34.
2. Moulopoulos SD, Topaz S, Kolff WJ. Diastolic balloon pumping (with carbon dioxide) in the aorta—a mechanical assistance to the failing circulation. Am Heart J. 1962;63:669–75.
3. Liotta D, William Hall C, Henly WS, Cooley DA, Stanley Crawford E, DeBakey ME. Prolonged assisted circulation during and after cardiac or aortic surgery. Am J Cardiol. 1963;12:399–405.
4. Stewart GC, Givertz MM. Mechanical circulatory support for advanced heart failure: patients and technology in evolution. Circulation. 2012;125:1304–15.
5. Henriques JPS, Remmelink M, Baan J Jr, et al. Safety and feasibility of elective high-risk percutaneous coronary intervention procedures with left ventricular support of the Impella Recover LP 2.5. Am J Cardiol. 2006;97:990–2.
6. Hoffman J. Extracorporeal membrane oxygenation (ECMO). In: Reichman's emergency medicine procedures. 3rd ed; 2018.
7. Kumar V, Abbas AK, Aster JC. Robbins and Cotran pathologic basis of disease. 9th ed: Saunders; 2014. p. 269–76.
8. Redington AN, Gray HH, Hodson ME, Rigby ML, Oldershaw PJ. Characterisation of the normal right ventricular pressure-volume relation by biplane angiography and simultaneous micromanometer pressure measurements. Br Heart J. 1988;59:23–30.
9. Mebazaa A, Karpati P, Renaud E, Algotsson L. Acute right ventricular failure—from pathophysiology to new treatments. Intensive Care Med. 2004;30:185–96.
10. Burkhoff D, Sayer G, Doshi D, Uriel N. Hemodynamics of mechanical circulatory support. J Am Coll Cardiol. 2015;66:2663–74.
11. Stephanazzi J, Guidon-Attali C, Escarment J. Right ventricular function: physiological and pathophysiological features. Annales Françaises d'Anesthésie et de Réanimation. 1997;16:165–86.
12. Visner MC, Arentzen CE, O'Connor MJ, Larson EV, Anderson RW. Alterations in left ventricular three-dimensional dynamic geometry and systolic function during acute right ventricular hypertension in the conscious dog. Circulation. 1983;67:353–65.
13. Santamore WP, Gray LA. Left ventricular contributions to right ventricular systolic function during LVAD support. Ann Thorac Surg. 1996;61:350–6.
14. Esposito M, Bader Y, Pedicini R, Breton C, Mullin A, Kapur NK. The role of acute circulatory support in ST-segment elevation myocardial infarction complicated by cardiogenic shock. Indian Heart J. 2017;69:668–74.
15. Kawaguchi O, Pae WE, Daily BB, Pierce WS. Ventriculoarterial coupling with intra-aortic balloon pump in acute ischemic heart failure. J Thorac Cardiovasc Surg. 1999;117:164–71.
16. Patterson T, Perera D, Redwood SR. Intra-aortic balloon pump for high-risk percutaneous coronary intervention. Circ Cardiovasc Interv. 2014;7:712–20.
17. Gelsomino S, Lozekoot PWJ, Lorusso R, et al. The optimal weaning strategy for intraaortic balloon counterpulsation: volume-based versus rate-based approach in an animal model. Ann Thorac Surg. 2016;101:1485–93.

18. Russo MJ, Jeevanandam V, Stepney J, Merlo A, Johnson EM, Malyala R, Raman J. Intra-aortic balloon pump inserted through the subclavian artery: a minimally invasive approach to mechanical support in the ambulatory end-stage heart failure patient. J Thorac Cardiovasc Surg. 2012;144:951–5.
19. Estep JD, Cordero-Reyes AM, Bhimaraj A, et al. Percutaneous placement of an intra-aortic balloon pump in the left axillary/subclavian position provides safe, ambulatory long-term support as bridge to heart transplantation. JACC: Heart Failure. 2013;1:382–8.
20. Raess DH, Weber DM. Impella 2.5. J Cardiovasc Transl Res. 2009;2:168–72.
21. Burkhoff D, Naidu SS. The science behind percutaneous hemodynamic support: a review and comparison of support strategies. Catheter Cardiovasc Interv. 2012;80:816–29.
22. Nascimbene A, Loyalka P, Gregoric ID, Kar B. Percutaneous coronary intervention with the TandemHeart™ percutaneous left ventricular assist device support: Six years of experience and outcomes. Catheter Cardiovasc Interv. 2015;87:1101–10.
23. Alli OO, Singh IM, Holmes DR, Pulido JN, Park SJ, Rihal CS. Percutaneous left ventricular assist device with TandemHeart for high-risk percutaneous coronary intervention: the Mayo Clinic experience. Catheter Cardiovasc Interv. 2012;80:728–34.
24. Kar B, Adkins LE, Civitello AB, Loyalka P, Palanichamy N, Gemmato CJ, Myers TJ, Gregoric ID, Delgado RM 3rd. Clinical experience with the TandemHeart percutaneous ventricular assist device. Tex Heart Inst J. 2006;33:111–5.
25. Lequier L, Horton SB, McMullan DM, Bartlett RH. Extracorporeal membrane oxygenation circuitry. Pediatr Crit Care Med. 2013;14:S7–12.

24 Percutaneous Mechanical Circulatory Support Technologies

Jerry D. Estep

Introduction

Cardiogenic shock (CS) is a deadly syndrome characterized by hypotension with evidence of tissue hypoperfusion. Historically the number one cause of CS has been acute myocardial infarction (AMI) [1]. More recently, non-AMI-related CS has been reported to be on the rise [2]. The most widely used short-term device placed to treat CS has been the intra-aortic balloon pump (IABP). Utilization, in the United States, however, has been on the decline [3]. Over the past decade, alternative devices, including the Impella devices and venoarterial extracorporeal membrane oxygenation (ECMO), that offer more robust hemodynamic support have been used with increased frequency to treat CS [3]. Despite the availability of these devices, mortality for CS remains high. Based on INTERMACS (Interagency Registry for Mechanically Assisted Circulatory Support) there is a 38% 30-day mortality among patients with the most severe CS, INTERMACS clinical profiles 1 (severe refractory CS) and 2 (CS with progressive decline despite inotrope support) [4]. The purpose of this review is to compare and contrast the available percutaneous mechanical circulatory support (MCS) technologies to treat CS and highlight the available clinical data that supports the use of these devices in this patient population.

J. D. Estep, MD (✉)
Department of Cardiovascular Medicine and Heart and Vascular and Thoracic Institute, Kaufman Center for Heart Failure Treatment and Recovery, Cleveland Clinic, Cleveland, OH, USA
e-mail: estepj@ccf.org

Percutaneous MCS Types and Strategies that Guide Use

Currently available percutaneous MCS devices include the intra-aortic balloon pump (IABP) (Getinge), the micro-axial Impella 2.5, CP, and 5.0 and 5.5 systems (Abiomed Europe, Aachen, Germany), TandemHeart (Cardiac Assist, Inc., Pittsburgh, PA), and ECMO [5] (Fig. 24.1). RV support devices include the Impella RP (Abiomed, Danvers, MA, USA) and the TandemHeart RA-PA (LivaNova, London, UK) with blood delivery from the right atrium or inferior vena cava to the pulmonary artery. Investigational devices include the paracorporeal pulsatile iVAC 2 L (PulseCath BV, Arnhem, the Netherlands) and the HeartMate Percutaneous Heart Pump (Abbott, Lake Bluff, IL, USA). Select indications for percutaneous MCS are listed in Table 24.1.

Temporary MCS devices inserted percutaneously can be used as a bridge to bridge or bridge to recovery strategy (Fig. 24.2). Bridge to bridge, a common strategy, especially in patients with acute on chronic heart failure (HF) complicated by CS, is the case when patients have a temporary device inserted with a provisional

© Springer Nature Switzerland AG 2020
J. H. Karimov et al. (eds.), *Mechanical Support for Heart Failure*,
https://doi.org/10.1007/978-3-030-47809-4_24

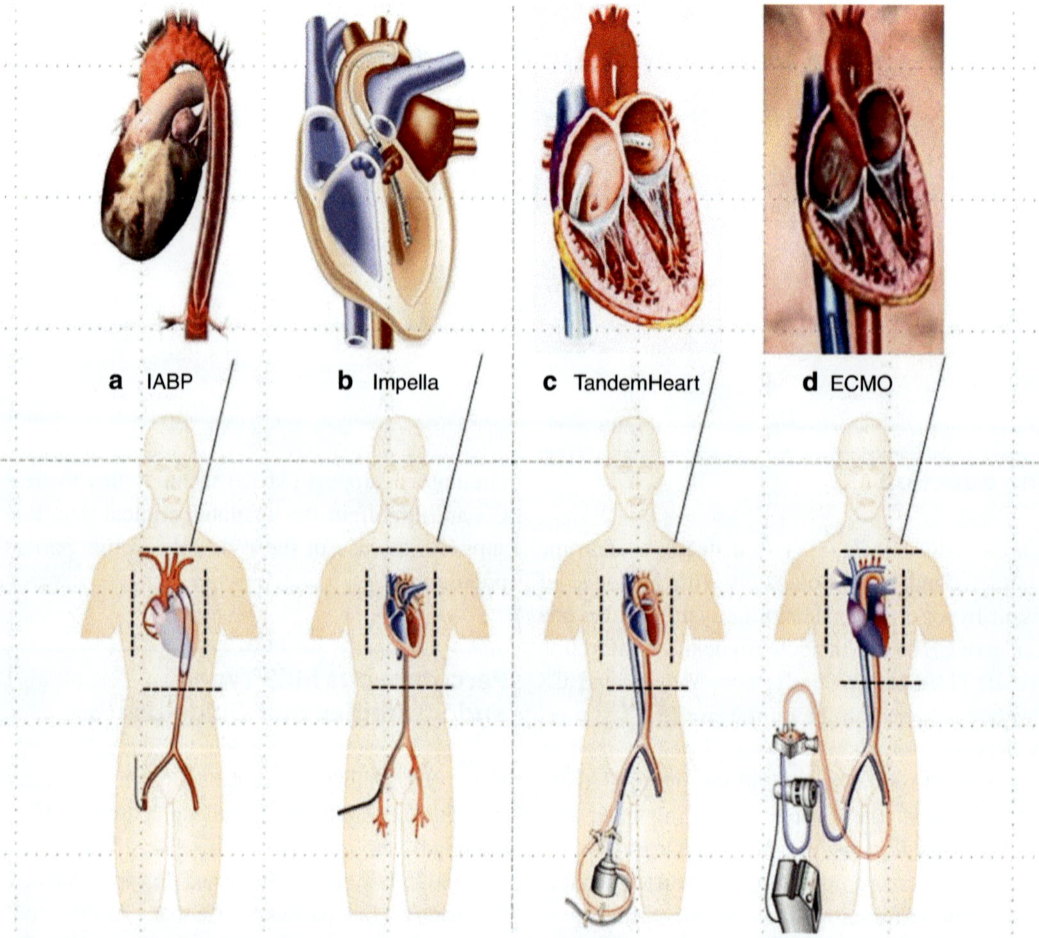

Fig. 24.1 Schematic drawings of current percutaneous mechanical support devices for cardiogenic shock with anatomic features. (**a**) Intra-aortic balloon pump (IABP), (**b**) Impella device, (**c**) TandemHeart, (**d**) VA extracorporeal membrane oxygenation (ECMO), (**d**) intra-aortic balloon pump. (Reproduced from Werdan et al. [5] by permission of Oxford University Press)

plan to transition to a more definitive intervention like a durable left ventricular assist device (LVAD) or as a bridge to transplantation after clinical stabilization. In those with CS, there may be uncertainty about candidacy for a durable MCS or heart transplantation based on the degree of hemodynamic instability, neurological status, psychosocial issues, or concern regarding the reversibility of multisystem organ failure. Temporary MCS can permit hemodynamic optimization and allow the potential reversal of CS-mediated end-organ failure and provide additional time for a more complete social and medical assessment before moving to definitive therapies or palliative care support. Bridge to recovery is the case when the temporary MCS device is placed to support a patient with CS anticipating possible improvement in cardiac function (e.g., acute myocarditis or acute myocardial infarction with successful revascularization) and possible device removal.

To best coordinate and manage patients with varying underlying etiologies and severities of CS, multidisciplinary shock teams of cardiothoracic surgeons, interventional cardiologists, advanced HF and critical care specialists, and other allied health professionals have been created at tertiary centers [6]. A recent observational study suggests that the implementation of a shock team predicated on a multidisciplinary standardized team-based approach emphasizing timely diagnosis, mandatory invasive hemodynamics, and appropriate use of MCS is not only feasible but may result in improved survival in patients with CS [7]. Institutions that manage these types of patients should define the roles and responsibilities of the different team member and establish management and trouble-shooting protocols to minimize bleeding and ischemic complications associated with these different devices. Regional systems for the management of CS have been proposed and consist of a spoke and hub model where tertiary high-volume cardiovascular centers are designated as CS hub centers that facilitate and expedite the transfer of patients with CS from lower-acuity sites [8].

Table 24.1 Select indications for percutaneous mechanical circulatory support

Select indications for percutaneous mechanical circulatory support
Prophylactic use for high-risk percutaneous intervention
Prophylactic use for high-risk ventricular tachycardia ablation
Prophylactic use for high-risk percutaneous valve interventions
Acute myocardial infarction cardiogenic shock
Acute chronic heart failure (ischemic and nonischemic) complicated by cardiogenic shock
Complications of acute myocardial infarction
Refractory ventricular arrhythmias
Postcardiotomy shock or difficulty to wean from cardiopulmonary bypass
Post-heart transplant primary graft failure
Acute cardiac allograft failure complicated by cardiogenic shock
Fulminant myocarditis

Device Types and Specific Device Considerations

IABP

IABP Components and Types

Counterpulsation refers to balloon inflation in diastole and deflation in systole. First conceived in the 1950s, balloon counterpulsation technology was developed to augment coronary blood flow by inflating a balloon in the descending aorta during diastole [9, 10]. The IABP consists of a balloon made of a polyurethane membrane

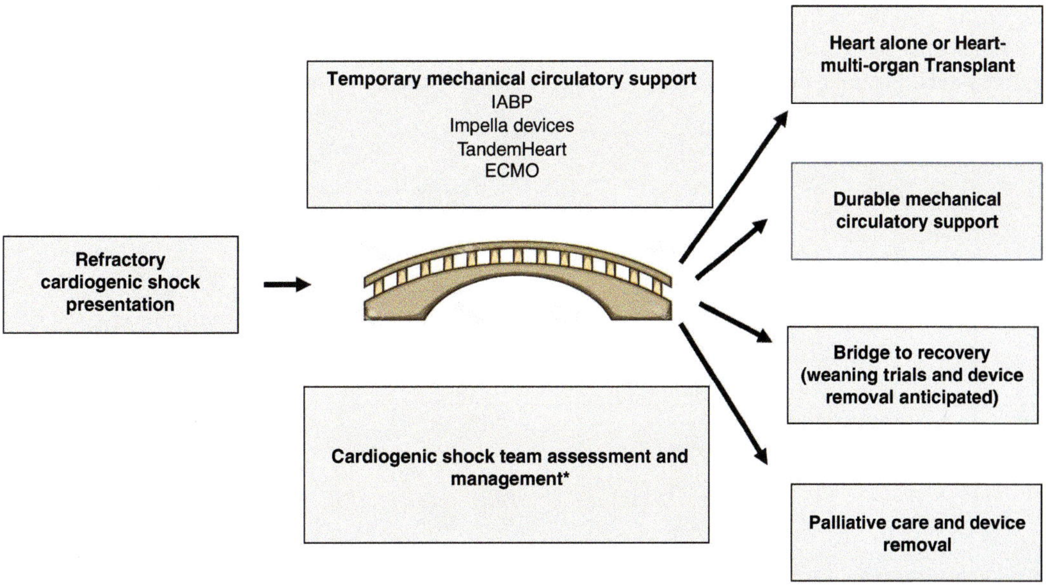

Fig. 24.2 Strategies that guide clinical use of percutaneous MCS device technologies. *includes an advanced heart failure workup to understand patient expectations and durable MCS and/or heart transplant benefits and risks

that is mounted to a flexible double-lumen 7F to 8F catheter that is attached to a pump console. One lumen of the catheter permits guide wire insertion of the IABP into the descending thoracic aorta and serves as access to transduce aortic pressure. The other lumen is attached to a pump console that contains a pump and controls the shuttling of helium gas in and out of the balloon. Helium is used because it can rapidly be transferred in and out of the balloon based on its low viscosity and its ability to be rapidly absorbed in the blood in the case of balloon rupture. The IABP is positioned in the descending thoracic aorta just distal to the left subclavian artery.

Timing of balloon inflation and deflation is based on electrocardiogram (ECG) or pressure triggers. Balloon inflation begins with the onset of diastole, which corresponds with the middle of the T wave on the surface ECG. The peak of the R wave on the ECG is used to time rapid balloon deflation and the onset of left ventricular systole (Fig. 24.3). Diastolic inflation increases diastolic pressure and coronary blood flow, thereby increasing myocardial oxygen supply, and rapid presystolic deflation reduces left ventricular afterload, reducing myocardial oxygen demand [11–15].

An alternative site, albeit off-label use, for percutaneous placement of the IABP is the axillary-subclavian artery. IABP support using this location permits upright sitting and/or ambulation [16].

IABP Hemodynamic Effects

In all types of CS, the hemodynamic condition of the left ventricle (LV) can be illustrated by the pressure-volume (PV) loop. This loop depicts a single cardiac cycle and the four phases and provides information about relaxation and contractile properties, cardiac work, in addition to myocardial oxygen consumption [15, 17–19]. Typically in CS, LV contractility, reflected by Emax, defined as the maximum slope at the end-systolic PV point, and stroke volume and LV stroke work are reduced, and LV end-diastolic pressure (LVEDP) is increased. Specifically with the IABP, counterpulsation reduces both peak LV systolic and diastolic pressures and increases LV stroke volume [17] (Fig. 24.4). Cardiac output enhancement with typical IABP support is mild to modest. The magnitude of systolic unloading and diastolic augmentation provided by the IABP can be directly measured by the tracings obtained from the IABP console (Figs. 24.5 and 24.6). Improvement in mean blood pressure may result, and tissue hypoperfusion signs may improve.

The magnitude of hemodynamic support provided by IABP is related to the magnitude of blood volume displacement. The size of the descending aorta is related to patient weight, height, and age. The greater inflated IABP diam-

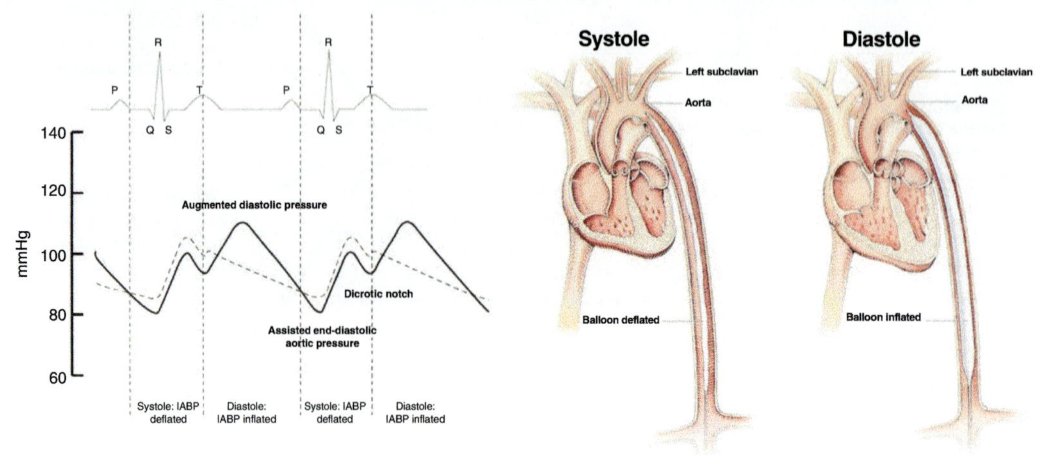

Fig. 24.3 IABP timing. (Reproduced from Abdul et al. [14] with permission from Wolters Kluwer Health, Inc.)

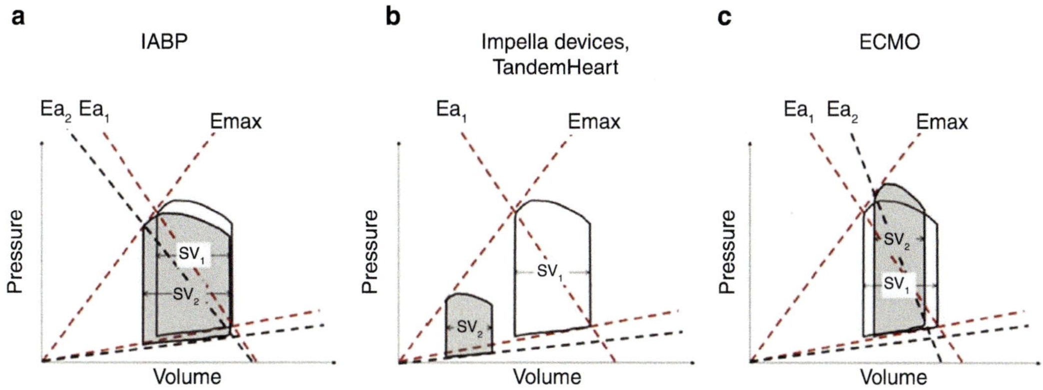

Fig. 24.4 Illustrations of PV loops after activation of device therapy (gray loops). (**a**) Intra-aortic balloon pump (IABP) counterpulsation reduces both peak LV systolic and diastolic pressures and increases LV stroke volume. The net effect is a reduced slope of arterial elastance (Ea2), (**b**) Percutaneous LV assist devices (pLVAD: Impella and TandemHeart) significantly reduce LV pressures, LV volumes, and LV stroke volume. The net effect is a significant reduction in cardiac workload. (**c**) Venoarterial extracorporeal membrane oxygenation (VA-ECMO) without a LV venting strategy increases LV systolic and diastolic pressure while reducing LV stroke volume. The net effect is an increase in arterial elastance (Ea). (Reproduced from Rihal et al. [18] with permission from Elsevier)

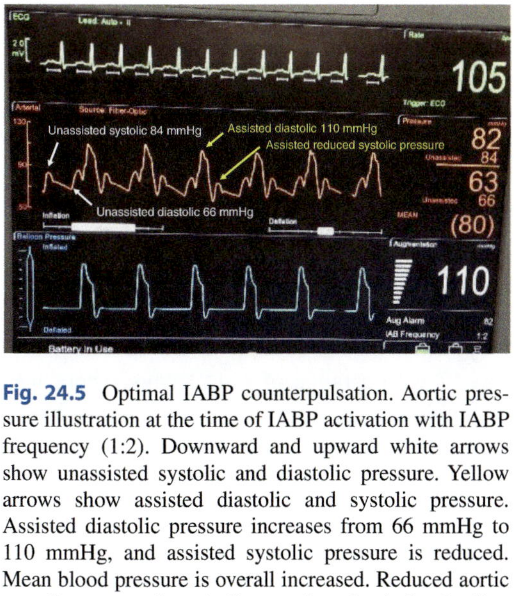

Fig. 24.5 Optimal IABP counterpulsation. Aortic pressure illustration at the time of IABP activation with IABP frequency (1:2). Downward and upward white arrows show unassisted systolic and diastolic pressure. Yellow arrows show assisted diastolic and systolic pressure. Assisted diastolic pressure increases from 66 mmHg to 110 mmHg, and assisted systolic pressure is reduced. Mean blood pressure is overall increased. Reduced aortic systolic pressure is an indicator of mechanical unloading of left ventricular pressure

eter percentage relative to the descending aorta translates to a greater volume of blood displacement. IABP balloon capacity ranges from 34 to 50 cm3 with the 50-cm3 IABP commonly referred to as the MEGA IABP. The larger capacity, 50-cm3 IABPs, provide greater diastolic augmentation and systolic unloading compared with smaller capacity balloon, e.g., the 40-cm3 IABP [20–22]. The hemodynamic impact of the IABP is influenced by several other factors including the frequency and timing of IABP inflation and deflation, its position in the descending aorta, aortic compliance, heart rate, and underlying blood pressure [23, 24]. Erratic balloon inflation can be due to poor ECG quality, electrical interference, and/or cardiac arrhythmias with tachycardia. Fiber-optic-based IABP monitoring is available and can enhance device performance even in the setting of tachycardia [25].

Clinical Data and the IABP

The IABP has been used in patients with myocardial infarction with and without CS, for complications of myocardial infarction (MI), intractable ventricular arrhythmias, high-risk percutaneous coronary interventions, high-risk coronary artery bypass, and as bridge to heart transplantation and to a durable left ventricular assist device [14, 26, 27]. There have been four randomized control trials that have examined the use of IABP support in patients with MI with and without CS and in the setting of high-risk PCI and high-risk CABG [28–31]. The Counterpulsation Reduces Infarct Size Acute Myocardial Infarction (CRISP-AMI)

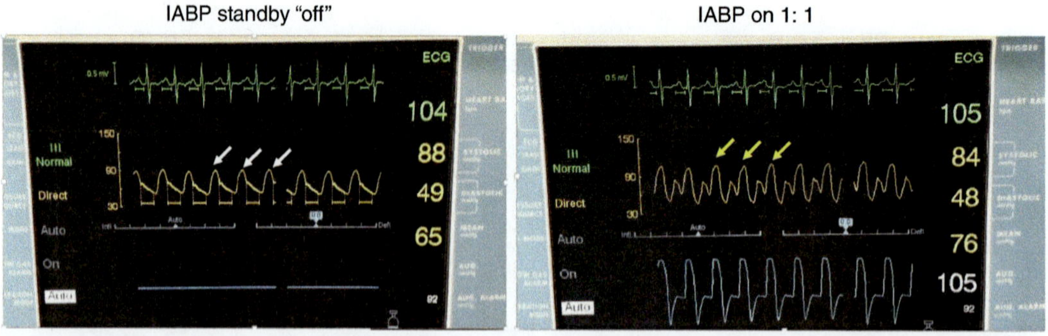

Fig. 24.6 IABP wean study using the standby mode. IABP console illustration of aortic pressure with the IABP on standby (off) and on 1:1 augmentation. In standby mode systolic and diastolic pressures are 88 and 49 mmHg with a mean pressure of 65 mmHg (white arrows). With the IABP turned on with 1:1 augmentation (yellow arrows), diastolic pressure is augmented with an overall increase in mean blood pressure to 76 mmHg

trial showed that IABP implantation immediately before revascularization for an anterior ST-elevation myocardial infarction (MI) did not reduce infarct size or improve short-term survival [28]. The IABP-SHOCK II study was a randomized, open-label, multicenter trial. Patients with CS complicating acute MI who were undergoing early revascularization and optimum medical therapy were randomly assigned (1:1) to IABP versus control (medical therapy). Of 595 patients completing 12-month follow-up, 155 (52%) of 299 patients in the IABP group and 152 (51%) of 296 patients in the control group had died (relative risk [RR] 1·01, 95% CI 0·86–1·18, $p = 0{\cdot}91$) [31]. There were no significant differences in reinfarction (RR 2·60, 95% CI 0·95–7·10, $p = 0{\cdot}05$), recurrent revascularization (0·91, 0·58–1·41, $p = 0{\cdot}77$), or stroke (1·50, 0·25–8·84, $p = 1{\cdot}00$). This trial showed that IABP implantation immediately before revascularization for an anterior ST-elevation MI did not improve short- or long-term survival [28]. In addition, the balloon pump-assisted coronary intervention study (BCIS)-1 also showed short-and long-term mortality with IABP insertion before HR-PCI was not reduced [32]. Overall, these data do not support the general use of IABP in CS complicating acute MI or high-risk PCI treated with primary PCI.

Based on the largest published single center experience with femoral IABP support in patients with acute on chronic HF complicated by CS (N = 132), overall 30-day survival was 84.1%, and 78.0% of patients were successfully bridged to a durable LVAD or heart transplantation or discharge without need for escalation of device support [26]. It is important to note that most of these survivors required a LVAD or heart transplant. This highlights the need for an exit strategy in these patients.

Similarly, based on the largest single center experience with the percutaneously placed axillary IABP, 39 out of 44 patients (89%) were successfully bridged to either heart or heart multiorgan transplant using this novel technique [16]. The median days of IABP in this study was 17 with a range of 6–152 days. There was a significant improvement in creatinine, blood urea nitrogen and total bilirubin, as well as a significant decrease in mean pulmonary artery pressure with slightly lower right atrial pressure suggesting benefit of prolonged use of axillary IABP in a select advanced HF population [16]. Overall outcomes with IABP in acute decompensated chronic HF patients are encouraging, and a femoral IABP is reasonable first-line device for chronic HF patients with CS.

IABP Contraindications

Aortic valve regurgitation of greater than a mild degree has traditionally been considered a contraindication to the IABP as diastolic balloon inflation may worsen the degree of regurgitation. Severe peripheral arterial or aortic disease increases the risk of vascular complications such as thromboembolism to the lower extremities or visceral arteries and is a contraindication to use of the IABP using the femoral artery position [14].

IABP-Associated Complications

Based on registry data, the IABP has a very encouraging safety profile compared to other temporary MCS devices [33]. The majority of complications related to IABPs are vascular in etiology and include limb ischemia, vascular trauma, and stroke. Other complications include thrombocytopenia from platelet deposition on the IABP membrane (or use of heparin), infection, and complications of immobility related to extended IABP when using the femoral approach. Less common, but potentially life-threatening side effects relate to branch aorta or visceral artery ostia trauma including acute kidney injury, bowel ischemia, and atheroembolism. Percutaneously placed transthoracic IABP's carry specific position-related complication risks including device malposition, device kinking, IABP rupture, left upper extremity ischemia, and potential compromise of superior mesenteric artery and bowel ischemia [16].

IABP Regulatory Consideration and Guideline Recommendations

Both the European Society of Cardiology (ESC) and American College of Cardiology (ACC)/ American Heart Association (AHA) guidelines downgraded their recommendation for IABP use in CS in 2012 and 2013 [34]. Based on the most recent guidelines, the 2017 ESC guideline recommendations for the management of CS in ST-elevation MI, state IABP pumping should be considered in patients with hemodynamic instability/CS due to mechanical complications (class IIa, level of evidence C), whereas routine IABP pumping is not indicated in patients with CS and acute MI or acute on chronic HF complicated by CS (class III, level of evidence B) [35, 36]. In contrast, American society-based guidelines do not recommend against IABP in these CS patient populations. Both the ACC/ AHA guidelines (last updated in 2013) and a scientific statement from the AHA on the contemporary management of CS in 2017 describe the IABP as one of the nondurable MCS devices available to help management patients with acute on chronic HF and CS [8, 37].

Impella Devices

Impella Components and Types

Impella devices are catheter-mounted axial-flow pumps that are inserted via a standard catheterization procedure through the femoral artery, into the ascending aorta, across the aortic valve. and into the left ventricle (LV) (Fig. 24.1). The Impella pump is based on an Archimedes screw design and propels the blood from the inlet of the catheter in the LV into the ascending aorta where the outlet is positioned. There are two percutaneously placed available types, the 12-Fr (Impella 2.5) and 14-Fr device (Impella CP) with the Fr size referring to the pump motor diameter both which are mounted to a 9-Fr catheter and can deliver up to 2.5 and ~ 3.7 to 4 l/min of forward flow, respectively [18, 38]. The proximal 9-Fr catheter shaft houses the motor power leads and purge and pressure measurement lumens. The catheter's proximal end consists of the pump motor housing which is in the proximal portion of the ascending aorta. The proximal end also consists of a hub for attachment of a console cable and side arms to permit purge solution and pressure measurements. The tip of the catheter is ideally in the proximity of the LV apex and is a flexible pigtail loop designed to stabilize the device in the LV and minimize perforation. The catheter most proximal end is connected to a controller to monitor catheter performance and alarms and display real-time hemodynamic and catheter position information.

Alternate percutaneous access sites such as the subclavian and axillary artery have been described but are less routinely used [39]. The Impella 5.0 device is the largest of the micro-axial Impella blood pumps that has a 21F pump motor mounted on a 9-Fr catheter and delivers up to 5 l/min of forward flow. The Impella 5.0

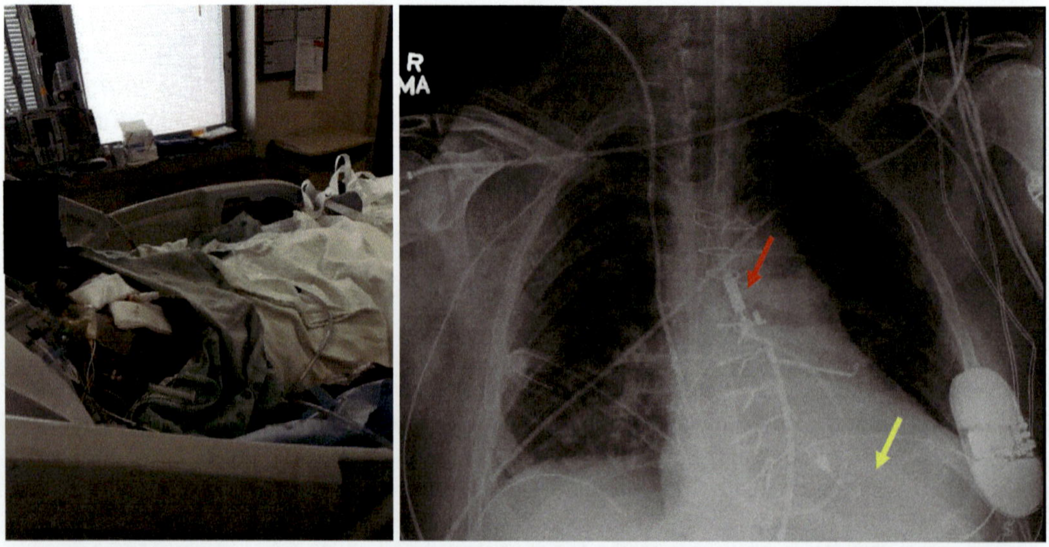

Fig. 24.7 Impella 5.0 support permits upright sitting. Portable plain film shows the Impella device (pigtail) in the left ventricle (yellow arrow) and the housing motor (red arrow) in the proximal ascending aorta

can be inserted surgically into the left ventricle via femoral cutdown or through the axillary artery with the use of a conduit graft and similar to the smaller Impella devices which goes through the ascending aorta, across the valve, and into the LV [18, 38]. An advantage of the axillary surgical approach is the potential for upright sitting and/or ambulation and extended support (Fig. 24.7).

Impella Hemodynamic Effects

The Impella is an LV to aorta assist device that directly unloads the LV and increases forward flow. Unlike the other forms of support, removal of the blood from the LV is not dependent on ejection through the aortic valve. The degree of LV unloading is influenced by the set flow speed (P1 through P9). As the LV becomes increasingly unloaded at a higher flow rate, the pressure-volume loop is shifted leftward, peak LV pressure decreases, and there is a marked decreases in the pressure-volume area. This translates into a reduction in MVO2, decreased pulmonary capillary wedge pressure, and an increase in mean arterial pressure [17, 19].

Several trials have compared the degree of hemodynamic support and/or clinical outcomes in patients treated with an IABP and the Impella or TandemHeart device in the setting of CS and high-risk PCI (Table 24.2). Based on the ISAR-Shock (Efficacy Study of LV Assist Device to Treat Patients With Cardiogenic Shock) study, which randomized a small number of patients presenting with AMI and CS to the Impella 2.5 or IABP, a greater improvement in cardiac index was noted with the Impella 2.5 device compared with IABP (0.49 ± 0.46 L/min/m2 compared with a change of 0.11 ± 0.31 L/min/m2, Impella 2.5 vs IABP; $P < 0.01$) [40]. Similarly based on the PROTECT II trial which examined the use of IABP versus the Impella 2.5 in symptomatic patients with underlying severely depressed LV function undergoing high-risk PCI with complex three vessel disease or unprotected left main coronary artery disease, the Impella device provided superior hemodynamic support in comparison with IABP [41]. The Impella CP compared to the 2.5 offers great hemodynamic support, but comparison of hemodynamic-based trials are lacking.

Clinical Data and the Impella Devices

Large registries define the safety and efficacy of the Impella 2.5 LP device [42–45]. In the USpella registry of patients with AMI-CS treated with Impella devices before PCI, Impella place-

Table 24.2 Selected clinical studies examining cardiogenic shock and percutaneous MCS

Selected device trials	Trial-type arid devices used	Clinical scenario examined	Sample size	Primary end point	Result
ISAR-SHOCK [40]	Randomized Impella 2.5 vs IABP	AMI-cardiogenic shock	$N = 25$	Change of CI from baseline to 30 min	Difference seen (Impella CI change 0.49 ± 0.46 1/min/m^2; IABP change CI = 0.11 ± 0.31 1/min/m^1; $\rho = 0.021$)
PROTECT II trial [41]	Randomized Impella 2.5 vs IABP	High-risk PCI	$N = 452$	30-day incidence of major adverse events	No difference (35.1% for Impella vs 40 1% for IABP; $\rho = 0.22$)
IMPRESS trial [46]	Randomized Impella CP vs IABP	STEMI with cardiogenic shock	$N = 48$	All cause mortality at 30 days	No difference (50% of IABP versus 46% for Impella)
TandemHeart Trial [52]	Randomized TandemHeart vs IABP	AMI with cardiogenic shock	$N = 41$	Cardiac power index within 2 hours after device placement	Difference seen (TandemHeart CPI change from 0.22 to 0.37 W/m^2 when compared to IABP from 0.22 to 0.28 W/m^2; $\rho = 0.004$ intergroup comparison)
TandemHeart vs IABP [53]	Randomized TandemHeart vs IABP	Cardiogenic shock	$N = 42$	PCWP and CI	Difference seen, greater decrease in PCWP and increase in CI and mean BP (TandemHeart vs IABP)

CI cardiac index, *PCWP* pulmonary capillary wedge pressure, *CPI* cardiac power index, *BP* blood pressure

ment resulted in improved survival to hospital discharge, even after adjustment for potential confounding variables [42]. Based on the ISAR-SHOCK (AMI and CS patients), inhospital mortality and 2-year follow-up data showed no significant difference comparing these two devices [40].

The PROTECT II study that randomized patients with heart failure undergoing elective high-risk PCI to the Impella 2.5 or IABP before PCI was terminated early because of expected futility of the 30-day major adverse event composite primary end point [41]. Impella support was, however, associated with more complete revascularization, and at 90 days, there was a trend toward decreased events observed in the intent-to-treat population (40.6% Impella vs 49.3% IABP, $P = 0.066$) [42].

More recently, the "IMPella versus IABP Reduces mortality in STEMI patients treated with primary PCI in Severe cardiogenic SHOCK" (IMPRESS in Severe Shock) compared the Impella CP with IABP in mechanically ventilated patients with AMI-CS in the context of an exploratory assessment of mortality and other safety outcomes [46]. At 30 days, mortality in patients treated with either IABP (N = 24) or Impella CP (N = 24) was similar (50% and 46%, respectively, hazard ratio (HR) with Impella CP, 0.96 (95% confidence interval (CI) 0.42 to 2.18; $p = 0.92$). At 6 months, mortality rates for both Impella CP and IABP were 50% (HR 1.04 (95% CI; 0.47–2.32, $p = 0.923$). In this population, routine treatment with Impella CP was not associated with reduced 30-day mortality compared with IABP [46]. It is important to highlight that this was a markedly underpowered trial to detect a difference in the primary end point in critically ill patient population. Most recently, a matched-pair mortality analysis of 237 Impella-treated vs 237 IABP-treated CS patients demonstrated a lack of mortality benefit with the Impella device (30-day mortality 48.5% vs 46.4%, $P = 0.64$) [55].

There is a paucity of data that has examined the safety and efficacy of Impella 5.0 or 5.5 use in patients with advanced HF. Based on a retrospective evaluation at 3 centers of 58 patients with advanced HF who acutely decompensated and received the Impella 5.0 for bridge to decision, 39 patients survived to next therapy (67%), with most receiving durable MCS (n = 20) or heart transplant (n = 15) [47]. The Impella 5.0 may provide a bridge to bridge strategy for select patients with advanced HF and acute hemodynamic instability; however, further prospective studies are needed.

Contraindications to the Impella Devices

Contraindications for Impella use include moderate to severe aortic regurgitation, aortic valve stenosis/calcification (equivalent to an orifice area of 0.6 cm2 or less), prosthetic aortic valve, aortic dissection, severe aortic or peripheral vascular disease, left ventricular thrombi, bleeding diathesis or inability to tolerate systemic anticoagulation, and left ventricular rupture. A pre-existing ventricular septal defect is also a relative contraindication as direct LV unloading may worsen right to left shunting and hypoxemia [18, 38].

Impella Device-Associated Complications

The most commonly reported complications of Impella placement are vascular injury, bleeding, and limb ischemia [18, 37, 38, 47]. Similar to vascular complications common to all transfemoral procedures, hematoma, arterial-venous fistula, pseudoaneurysm, and retroperitoneal hemorrhage can occur. Hemolysis has been reported within the first 24 h of use in up to 10% of patients and is felt to be provoked by device malposition. Similar to hemolysis in durable continuous-flow LVAD's, persistent hemolysis associated with pigment-induced nephropathy and acute renal dysfunction is an indication for device exchange or removal [37]. Other potential complications include aortic valve injury, cerebral vascular accident/stroke, and thrombocytopenia.

Impella Device Regulatory Consideration and Guideline Recommendations

The FDA label for the Impella 2.5 and CP catheters, in conjunction with the Automated Impella Controller console, states these devices are intended for short-term use (≤4 days) and indicated for the treatment of ongoing CS that occurs immediately (<48 hours) following acute myocardial infarction (AMI) or open-heart surgery as a result of isolated left ventricular failure that is not responsive to optimal medical management and conventional treatment measures with or without an intra-aortic balloon pump. On May 13, 2019, the FDA extended approval of use of the surgically placed Impella 5.0 up to 14 days. Regarding high-risk PCI, the Impella 2.5 and Impella CP are indicated for use (≤6 hours) in elective or urgent, hemodynamically stable patients with severe coronary artery disease, when a heart team, including a cardiac surgeon, has determined high-risk PCI is the appropriate therapeutic option.

The American College of Cardiology, the American Heart Association, and the Society for Cardiovascular Angiography and Intervention have published expert consensus documents and clinical practice guidelines that define hemodynamic support device considerations in the settings of high-risk PCI and STEMI with CS . Elective insertion of an appropriate hemodynamic support device like the Impella device, as an adjunct to PCI, may be reasonable in carefully selected high-risk patients, and alternative LV assist devices for circulatory support may be considered in patients with refractory CS (class IIb, level of evidence C) [48, 49]. Based on a 2015 Consensus statement supported by SCAI/ACC/HFSA/STS the choice of short-term MCS devices is dependent upon vascular anatomy, local expertise, and availability [18]. European-based guidelines are similar. Percutaneous MCS may be considered in selected patients with refractory CS considering patient age, comorbidities, and neurological function without any preference for device selection (IIa C recommendation) [35, 36, 68].

TandemHeart

TandemHeart Components

Different from the Impella devices, the TandemHeart device is an extracorporeal centrifugal flow pump. It is the only available device that percutaneously bypasses the LV by transferring oxygenated blood from the left atrium to the descending aorta [18, 38]. The system includes a 21-Fr transseptal cannula made of wire-reinforced polyurethane with a large end hole and 14-side holes that permits withdrawal of left atrial blood arterial cannulae, a centrifugal blood pump, and a pump console. Blood is pumped from the left atrium into the to the iliofemoral arterial system through 15F, 17F, or 19F femoral arterial cannulae. The TandemHeart pump can provide 3.0 to 5.0 L/min of flow depending on the size of the outflow (arterial) cannula and the set pump speed (3000 to 7500 rpm). The centrifugal blood pump contains a spinning impeller supported by a hydrodynamic bearing. The pump has blood chamber and a motor chamber separated by a polymeric membrane. Heparinized saline flows into one of the chambers to prevent thrombus formation and to provide cooling and lubrication. The external console provides battery backup and controls the pump.

TandemHeart Hemodynamic Effects

Aspiration of the blood from the LA reduces LA and LV filling pressure, LV workload, and myocardial oxygen demand [17, 19, 50]. Due to the parallel configuration relative to the native heart, both the TandemHeart and the native LV may contribute to forward flow. Similar to the Impella devices, the net effect of the reduction in LV pressure, LV volumes, and LV stroke volume is a significant reduction in cardiac workload. In an analysis of 117 patients with severe cardiogenic shock refractory to IABP and/or vasopressor therapy, Kar et al. observed significant improvements in cardiac index, systolic blood pressure, and urine output with TandemHeart support over an average implant time of 6 days. The systolic blood pressure and mixed venous oxygen saturation increased from 75 (IQR: 15) mm Hg to 100 (IQR: 15) mm Hg ($p < 0.001$) and 49 (IQR: 11.5) to 69.3 (IQR: 10) ($p < 0.001$), respectively. The pulmonary capillary wedge pressure also significantly decreased [51]. Thiel et al. and Burkhoff et al. demonstrated in small randomized trials comparing the TandemHeart versus IABP in the setting of CS that the TandemHeart device resulted in a greater increase in cardiac index and decrease in pulmonary capillary wedge pressure [52, 53].

Clinical Data and the TandemHeart

There have been two small multicenter randomized trials comparing the clinical safety and efficacy of the TandemHeart to the IABP in patients with CS [52, 53]. Compared to the IABP, although the TandemHeart device significantly improved hemodynamic parameters, there was no mortality benefit compared with IABP [52, 53]. The largest data single center reported experience with the TandemHeart was reported by Kar and colleagues. In 117 patients with severe refractory CS to IABP and/or vasopressor support, the TandemHeart device improved end-organ perfusion, but despite hemodynamic and laboratory improvements, 30-day mortality remained high at 40% in this critically ill patient population with ~50% of patients requiring CPR during their hospital course [51].

There is limited data that defines the safety and efficacy of the TandemHeart to support patients during high-risk PCI. Alli et al. reported a series of 54 patients deemed high risk for surgery that underwent complex PCI, with left main and multivessel stenting performed in the majority of patients (64%). Procedural success was 97%, and 6-month survival was 87% [54].

TandemHeart Contraindications

Severe peripheral arterial disease, bleeding diathesis, inability to tolerate systemic anticoagulation, and right or left atrial thrombus are contraindications to the use of the TandemHeart [18, 38]. In contrast to the Impella pump, in select cases with peripheral arterial disease, a 5- or 6-Fr sheath can be placed antegrade into the superficial femoral artery and spliced into the arterial outflow cannula to provide limb perfusion either as a prophylactic or treatment measure.

Complications Associated with the TandemHeart

Complications from the device relate to the needed transseptal puncture (cardiac tamponade) with other complications similar to other percutaneous support devices including vascular trauma and limb ischemia. Other possible complications include thrombo- or air embolism and hemolysis. Dislodgement or pullback of the left atrial cannula into the right atrium, particularly during patient movement and/or with transport, will result in massive right to left shunt and refractory hypoxemia requiring the emergent addition of an oxygenator to the system [18, 38, 51]. The cannula may also migrate into a pulmonary vein leading to device malfunction with cannula "chattering" noted on exam secondary to acquired obstruction to flow.

TandemHeart Regulatory Consideration and Guideline Recommendations

Regulatory status includes Federal Drug Administration (FDA) approval to provide extracorporeal circulatory support for up to 6 h and CE mark for use up to 30 days. Unique to the TandemHeart, there is also FDA approval to add an oxygenator to the circuit allowing for concomitant LV unloading and oxygenation. Similar to the Impella devices, the TandemHeart is alternative LV assist device for circulatory support that may be considered in patients with refractory CS or high-risk PCI [18, 48, 49].

VA-ECMO

VA-ECMO Components

Venoarterial extracorporeal membrane oxygenation (VA-ECMO) is a modification of cardiopulmonary bypass that consist of an extracorporeal centrifugal pump like the CentriMag (Thoratec, Pleasanton, California), Rotaflow (Maquet, Rastatt, Germany), Biomedicus, or TandemHeart pumps (TandemLife, Pittsburgh, Pennsylvania), cannulas that displace venous blood from the right atrium into most commonly femoral artery, and a gas exchange unit for normalizing pCO2, pO2, and pH [18, 19, 56]. Blood flows through the membrane oxygenator, and hemoglobin becomes fully saturated with oxygen, and carbon dioxide (CO2) is removed by diffusion. The degree of oxygenation is determined by the flow rate and fraction of inspired oxygen. CO2 removal is primarily controlled by adjusting the sweep speed or the rate of countercurrent gas flow through the oxygenator. This device configuration permits improvement in systemic oxygenation during cardiorespiratory collapse and/or stabilization in biventricular failure. ECMO can be placed at the bedside without fluoroscopic guidance.

VA-ECMO Hemodynamic Effect

Based on the pressure-volume loop, VA-ECMO without a LV venting strategy increases LV systolic and diastolic pressure while reducing both right and left ventricular volumes with a concomitant increase in mean arterial pressure. VA-ECMO can provide complete circulatory support with flows that can exceed 6 L/min depending on cannula sizes [18, 19, 56]. In patients with significantly reduced ejection fractions at baseline, VA-ECMO may cause an increase in wall stress and oxygen demand, which impedes myocardial recovery and may precipitate progressive pulmonary edema acute lung injury and worsen outcomes. VA-ECMO-dependent increases in LV end-diastolic, left atrial, and pulmonary capillary wedge pressures can be mitigated medically by decreases in systemic vascular resistance or by inducing improvements in ventricular contractility or by reducing the VA-ECMO flow rate. Treatment of anticipated or observed increases in LV systolic and end-diastolic pressure includes mechanical venting (e.g., IABP or Impella use) or surgical decompressing the LV [18, 19, 56].

Clinical Data and VA-ECMO

The indications for venoarterial (VA)-ECMO include postcardiotomy shock to multifactorial CS and/or cardiac arrest. There are no large randomized controlled trials with use of ECMO. A recent meta-analysis including only prospective and retrospective cohort studies (four registries of CS and ten registries with cardiac arrest patients undergoing resuscitation) revealed a significant mortality benefit with VA-ECMO use [57]. Nichol et al. reviewed 84 studies of ECMO instituted for CS or cardiac arrest and showed an overall survival of 50% [58]. Analysis of the Extracorporeal Life Support Organization (ELSO) registry for ECMO used in the setting of adult ECPR out-of-hospital cardiac arrest demonstrated an overall survival to hospital discharge 27.6%, and male gender was associated with reduced survival [59]. This observation highlights one of the most important determinants of outcome is time to basic life support ECPR, a commonly used acronym which refers to application of rapid-deployment ECMO to provide circulatory support in patients who fail to achieve a sustained return of spontaneous circulation (ROSC) (defined as 20 consecutive minutes without a need for chest compressions by conventional CPR) [56]. Given the observed increase in LV systolic and end-diastolic pressure, the safety and efficacy of LV unloading during VA-ECMO have also been recently examined. Based on a recent meta-analysis of 17 observational studies, left ventricular unloading was associated with decreased mortality in adult patients with cardiogenic shock treated with VA-ECMO [60]. Aside from a higher rate of hemolysis, there was no additional signal of harm in using a left ventricular unloading strategy, with similar rates of bleeding, multiorgan failure, stroke, or transient ischemic attack, and limb ischemia.

VA-ECMO Contraindications

Contraindications to VA-ECMO include the following: severe irreversible noncardiac organ failure limiting projected survival (e.g., severe anoxic brain injury), aortic dissection, contraindication to anticoagulation or severe coagulopathy, severe peripheral arterial or access limitation, and prolonged asystole as an initial rhythm in the setting of cardiac arrest. Significant aortic insufficiency may worsen with VA-ECMO and may promote increased wall stress without a venting strategy. Central cannulation can be considered for those with significant peripheral artery disease [18, 19, 56, 61].

VA-ECMO Complications

Complications include bleeding, thromboembolic events including limb ischemia and stroke, hemolysis, infection, thrombocytopenia, compartment syndrome, and Harlequin syndrome [56]. Large cannula-related complications include venous thrombosis or distal arterial ischemia [18, 56]. Similar to using the sidearm with the TandemHeart, a second, antegrade, arterial sheath can be inserted into the superficial femoral artery to provide antegrade limb perfusion as a prophylactic or treatment measure. Harlequin syndrome is cyanosis of the upper extremities, while the lower extremities appear pink due to forward flow of deoxygenated blood from failing native lungs mixing unpredictably with retrograde flow from the oxygenator, which can result in inadequate delivery of oxygenated blood into the aortic arch, resulting in upper body and brain hypoxia.

VA-ECMO Regulatory Consideration and Guideline Recommendations

Similar to the Impella and TandemHeart, VA-ECMO is referred to as part of the armamentarium as a support device to treat patients with CS and high-risk PCI in patients with left main, last remaining conduit, or severe multivessel disease with underlying severe LV dysfunction (LVEF <35%) or recent decompensated HF, especially if hypoxemia or RV failure is present [18, 48, 49].

Percutaneous Temporary Right Ventricular MCS

RVF is associated with increased morbidity and mortality. In refractory RVF surgical options include RVAD implantation, VA-ECMO, cardiac transplantation, or a total artificial heart. Historically the IABP was the only available percutaneous device that offered indirect support for patients with right-sided heart failure due to left-sided heart failure. These reported indirect benefits are due to reducing LV afterload and enhancing coronary perfusion. More recent observations suggest that patients with significant RV failure are less likely to respond to counterpulsation. Specifically, more compromised underlying right ventricular functions, as reflected by a lower pulmonary artery pulsatility index, higher central venous pressure, and the presence of biventricular HF in the context of CS, have been deemed features associated with clinical deterioration despite IABP therapy [15, 26, 27, 62, 63].

Percutaneous Right-Sided MCS Device Types

Nondurable MCS options for right ventricular failure include the TandemHeart and the Impella RP catheter. The TandemHeart pump has been used as a right ventricular support system setting of AMI, post-LVAD implantation, severe pulmonary hypertension, and cardiac rejection after orthotopic heart transplantation [38]. This is accomplished by placing a 21-Fr to 28-Fr cannula into the RA and a 21-Fr outflow cannula into the main pulmonary artery using a right internal jugular and femoral venous cannulation configuration or using both femoral veins [38]. The Impella RP (Abiomed Inc.) is an axial-flow catheter with a 22-Fr motor mounted onto an 11-Fr catheter that permits single venous access (e.g., right femoral vein) and propels the blood from the lower right atrium/inferior vena cava to the main pulmonary artery. The Recover Right prospective trial included 30 patients with refractory RV failure to medical treatment [64]. Patients received the Impella RP device at 15 US institutions. The study population included 2 cohorts: 18 patients with RV failure after left ventricular assist device (LVAD) implantation and 12 patients with RVF after cardiotomy or myocardial infarction. The primary end point was survival to 30 days or hospital discharge (whichever was longer). Major secondary end points included indices of safety and efficacy. Hemodynamics improved immediately after initiation of Impella RP support, with an increase in cardiac index from 1.8 ± 0.2 to 3.3 ± 0.23 liters/min/m2 ($p < 0.001$) and a decrease in central venous pressure from 19.2 ± 4 to12.6 ± 1 mmHg ($p < 0.001$). The device was safe and easy to deploy, and patients were supported for an average of 3.0 ± 1.5 days (range, 0.5–7.8 days). Overall survival at 30 days was 73.3%, and all patients discharged were alive at 180 days highlighting the safety and efficacy of this device in this select, critically ill patient population [64].

The more recently available dual-lumen PROTEKDuo (CadiacAssist, Inc. Pittsburgh, PA, USA) cannula can be used as an RVAD with use of a pump like the CentriMag or the TandemHeart centrifugal pump [65, 66]. This cannula contains omnidirectional inflow and outflow ports for simultaneous drainage and reinfusion of the blood to draw the blood from the right atrium and propel it into the pulmonary artery (Fig. 24.8). Placement is guided by fluoroscopy with use of a standard guide wire exchange technique using the right internal jugular and a minimally invasive percutaneous approach. Data regarding the safety and efficacy using this device strategy is limited to case reports and single center observations [65, 66].

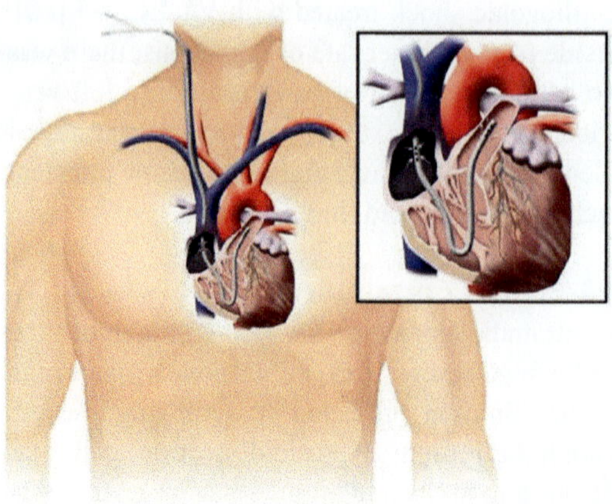

Fig. 24.8 PROTEKDuo cannula for right heart failure support. The PROTEKDuo cannula is a double lumen cannula. The proximal holes drain the blood from right atrium to the pump. The distal holes return the blood to the pulmonary artery. (Reproduced from Kazui et al. [65])

Contraindications to Percutaneous Right-Sided MCS Devices

Contraindications for a right-sided TandemHeart, use of the PROTEKDuo cannula and a centrifugal pump, and the Impella RP include the following: disorders of the pulmonary artery wall or anatomic conditions that would preclude correct positioning or placement of the device. Mechanical tricuspid or pulmonic valves, severe tricuspid or pulmonic valve regurgitation or stenosis, thrombus of the right atrium or inferior vena cava.

Percutaneous Right-Sided MCS Device Complications

Potential adverse effects associated with these systems include the following: bleeding, perforation, cardiac tamponade, device malfunction, hemolysis, hepatic failure, infection, deep venous thrombosis, pulmonary valve and/or tricuspid valve injury, thrombocytopenia, vascular injury, and venous thrombosis.

Percutaneous Right-Sided MCS Device Regulatory Considerations

The Impella RP system is indicated for providing temporary right ventricular support for up to 14 days in patients with a body surface area ≥1.5 m2, who develop acute right heart failure or decompensation following left ventricular assist device implantation, myocardial infarction, heart transplant, or open-heart surgery.

The PROTEKDuo cannula is a 510 K clearance with single cannula use for both venous drainage and reinfusion of the blood via internal jugular vein during extracorporeal life support. There is no FDA approval of ECLS comprising of a pump, oxygenator, and cannula, and, therefore, use of this product in this fashion is "off-label." Similarly, the TandemHeart is "off-label" when placed to provide right-sided support. More recently, providers are using a combination of these right-sided devices with short-term left-sided devices to provide biventricular support as an alternative to using VA-ECMO [67] (Fig. 24.9).

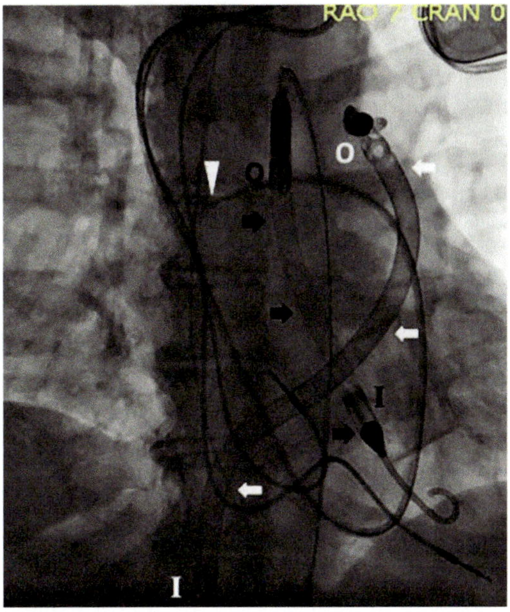

Fig. 24.9 Impella RP and Impella 5.0 biventricular support. Fluoroscopic image showing pulmonary artery catheter (white arrow head), Impella 5.0 (black arrows), and Impella RP (white arrows). The Impella 5.0 has its inlet in the left ventricle (black "I") and outlet in the ascending aorta (black "O"), while the Impella RP has its inlet in the inferior vena cava-right atrial junction (white "I") and outlet in the pulmonary artery (white "O"). (Reproduced from Varian et al. [67] with permission from John Wiley and Sons)

Conclusions

Percutaneously MCS technologies devices have evolved over the past several decades and now include axial and centrifugal flow-based platforms that can be rapidly deployed in patients with CS complicating acute myocardial infarction or acute on chronic HF. Choosing the right percutaneous device for the right patient and clinical scenario is influenced by balancing device-related complications and risk, hemodynamic support needs (LV versus RV or both), operator and institutional experience, and severity of the underlying CS presentation (Table 24.3). Devices with low complication rates may be chosen more liberally in early stages of CS, whereas more robust devices that offer greater hemody-

Table 24.3 Comparison of commercially available devices for short-term mechanical circulatory support

Device	IABP	Impella (2.5; CP; 5.0; RP)	TandemHeart	VA-ECMO
Flow, 1/min	0.5–1	2.5–5	4–6	4–6
Cannula size, F	7–8	12–21	Inflow 21, outflow 15–17	Inflow 18–21, outflow 15–22
Duration of support, FDA approved	9 days	4 days (2.5, CP), 14 days (5.0), 14 days (RP)	21 days	6 h (limited by oxygenator durability)
Ventricles supported	LV	LV or RV	LV or RV	LV and RV
Additional requirements		Surgical cutdown for Impella 5.0	Transseptal puncture	Potential need for LV venting, possible cutdown
Advantages	Easy to place, good safety profile, fewer side effects, especially vascular	Multiple devices to choose from	Highest cardiac output, comparable with VA-ECMO, and no LV distension	Highest cardiac output, complete cardiopulmonary support (including oxygenation and $C0_2$ removal)
Disadvantages	Limited hemodynamic support, contraindicated in severe aortic regurgitation	More invasive and complex to implant than the IABP, unstable position, frequent hemolysis, vascular complications	Need tertiary or quaternary specialized care center, necessitates atrial transseptal puncture with its potential complications, vascular complications, retrograde blood flow	Requires more resources and support staff than other devices, retrograde blood flow with worsening of afterload LV distension), vascular complications, thrombocytopenia

CO_2 carbon dioxide, *FDA* US Food and Drug Administration, *IABP* intra-aortic balloon pump, *LV* left ventricle/ventricular, *RV* right ventricle; *VA-ECMO* venoarterial extracorporeal membrane oxygenation. Modified from Guglin et al. [56]

namic support may be reserved for more severe CS. Beyond device selection, the approach to patients with CS now requires a multidisciplinary understanding of an individual's hemodynamic condition and candidacy for advanced heart failure therapies, including high-risk intervention, durable left ventricular assist devices, or heart transplantation.

References

1. Wayangankar SA, Bangalore S, McCoy LA, et al. Temporal trends and outcomes of patients undergoing percutaneous coronary interventions for cardiogenic shock in the setting of acute myocardial infarction: a report from the Cath-PCI registry. J Am Coll Cardiol Intv. 2016;9:341–51.
2. Puymirat E, Fagon JY, Aegerter P, et al. Cardiogenic shock in intensive care units: evolution of prevalence, patient profile, management and outcomes, 1997–2012. Eur J Heart Fail. 2017;19:192–200.
3. Shah M, Patnaik S, Patel B, et al. Trends in mechanical circulatory support use and hospital mortality among patients with acute myocardial infarction and non-infarction related cardiogenic shock in the United States. Clin Res Cardiol. 2018;107(4):287–303.
4. Alba AC, Rao V, Ivanov J, et al. Usefulness of the INTERMACS scale to predict outcomes after mechanical assist device implantation. J Heart Lung Transplant. 2009;28:827–33.
5. Werdan K, Gielen S, Ebelt H, et al. Mechanical circulatory support in cardiogenic shock. Eur Heart J. 2014;35:156–67.
6. Doll JA, Ohman EM, Patel MR, et al. A team-based approach to patients in cardiogenic shock. Catheter Cardiovasc Interv. 2016;88:424–33.
7. Tehrani BN, Truesdell AG, Sherwood MW, et al. Standardized team-based care for cardiogenic shock. J Am Coll Cardiol. 2019;73(13):1659–69.
8. van Diepen S, Katz JN, Albert NM, et al. Contemporary management of cardiogenic shock a scientific statement from the American heart association. Circulation. 2017;136:e232–68.
9. Kantrowitz A. Experimental augmentation of coronary flow by retardation of the arterial pressure pulse. Surgery. 1953;34:678–87.

10. Kantrowitz A, Tionneland S, Freed PS, et al. Initial clinical experience with intraaortic balloon pumping in cardiogenic shock. JAMA. 1968;203(2):113–8.
11. Williams DO, Korr KS, Gewirtz H, Most AS. The effect of intraaortic balloon counterpulsation on regional myocardial blood flow and oxygen consumption in the presence of coronary artery stenosis in patients with unstable angina. Circulation. 1982;66:593–7.
12. Kern MJ, Aguirre F, Bach R, et al. Augmentation of coronary blood flow by intra-aortic balloon pumping in patients after coronary angioplasty. Circulation. 1993;87:500–11.
13. Schreuder JJ, Maisano F, Donelli A, et al. Beat-to beat effects of intraaortic balloon pump timing on left ventricular performance in patients with low ejection fraction. Ann Thorac Surg. 2005;79(3):872–80.
14. Ihdayhid AR, Chopra S, Rankin J. Intra-aortic balloon pump: indications, efficacy, guidelines and future directions. Curr Opin Cardiol. 2014;29:285–92.
15. Annamalai SK, Buiten L, Esposito ML, et al. Acute hemodynamic effects of intra-aortic balloon counterpulsation pumps in advanced heart failure. J Cardiac Fail. 2017;23:606–14.
16. Estep JD, Cordero-Reyes AM, Bhimaraj A, et al. Percutaneous placement of an intra-aortic balloon pump in the left axillary/subclavian position provides safe, ambulatory long-term support as bridge to heart transplantation. JACC Heart Fail. 2013;1:382–8.
17. Burkhoff D, Mirsky I, Suga H. Assessment of systolic and diastolic ventricular properties via pressure-volume analysis: a guide for clinical, translational, and basic researchers. Am J Physiol Heart Circ Physiol. 2005;289:H501–12.
18. Rihal CS, Naidu SS, Givertz MM, et al. 2015 SCAI/ACC/HFSA/STS clinical expert consensus statement on the use of percutaneous mechanical circulatory support devices in cardiovascular care: endorsed by the American Heart Association, the Cardiological Society of India, and Sociedad Latino Americana de Cardiologia Intervencion; affirmation of value by the Canadian Association of Interventional Cardiology-Association Canadienne de Cardiologie d'intervention. J Am Coll Cardiol. 2015;65:e7–26.
19. Burkhoff D, Sayer G, Doshi D, Uriel N. Hemodynamics of mechanical circulatory support. J Am Coll Cardiol. 2015;66:2663–74.
20. Majithia A, Jumean M, Shih H, et al. The hemodynamic effects of the MEGA intra-aortic balloon counterpulsation pump. J Heart Lung Transplant. 2013;32(4S):S226.
21. Kapur N, Paruchuri V, Majithia A, et al. Hemodynamic effects of standard versus larger-capacity intraaortic balloon counterpulsation pumps. J Invasive Cardiol. 2015;27(4):182–8.
22. Baran DA, Visveswaran GK, Seliem A, DiVita M, Wasty N, Cohen M. Differential responses to larger volume intra-aortic balloon counterpulsation: hemodynamic and clinical outcomes. Catheter Cardiovasc Interv. 2018;92(4):703–10.
23. Papaioannou TG, Stefanadis C. Basic principles of the intra-aortic balloon pump and mechanisms affecting its performance. ASAIO J. 2005;51: 296–300.
24. Rastan AJ, Tillmann E, Subramanian S, Lehmkuhl L, Funkat AK, Leontyev S, Doenst T, Walther T, Gutberlet M, Mohr FW. Visceral arterial compromise during intra-aortic balloon counterpulsation therapy. Circulation. 2010;122(11 Suppl):S92–9.
25. Mulholland J, Yarham G, Clements A, Morris C, Loja D. Mechanical left ventricular support using a 50 cc 8 Fr fibre-optic intra-aortic balloon technology: a case report. Perfusion. 2013;28:109–13.
26. Fried JA, Nair A, Takeda K, et al. Clinical and hemodynamic effects of intra-aortic balloon pump therapy in chronic heart patients with cardiogenic shock. J Heart Lung Transplant. 2018;37:1314–22.
27. Sintek MA, Gdowski M, Lindman BR. Intra-aortic balloon counterpulsation in patients with chronic heart failure and cardiogenic shock: clinical response and predictors of stabilization. J Card Fail. 2015;21(11):868–76.
28. Patel MR, Smalling RW, Thiele H, et al. Intra-aortic balloon counterpulsation and infarct size in patients with acute anterior myocardial infarction without shock: the CRISP AMI randomized trial. JAMA. 2011;306:1329–37.
29. Perera D, Stables R, Clayton T, et al. Long-term mortality data from the balloon pump-assisted coronary intervention study (BCIS-1): a randomized, controlled trial of elective balloon counterpulsation during high-risk percutaneous coronary intervention. Circulation. 2013;127(2):207–12.
30. Thiele H, Zeymer U, Neumann FJ, et al. Intraaortic balloon support for myocardial infarction with cardiogenic shock. N Engl J Med. 2012;367:1287–96.
31. Thiele H, Zeymer U, Neumann FJ, et al. Intra-aortic balloon counterpulsation in acute myocardial infarction complicated by cardiogenic shock (IABP-SHOCK II): final 12 month results of a randomized, open-label trial. Lancet. 2013;382:1638–45.
32. Perera D, Stables R, Thomas M, et al. Elective intra-aortic balloon counterpulsation during high-risk percutaneous coronary intervention: a randomized controlled trial. JAMA. 2010;304(8):867–74.
33. Trost J, Hillis D. Intraaortic balloon counterpulsation. Am J Cardiol. 2006;97(9):1391–8.
34. O'Gara PT, Kushner FG, Ascheim DD, Casey DE Jr, Chung MK, de Lemos JA, Ettinger SM, Fang JC, Fesmire FM, Franklin BA, et al. 2013 ACCF/AHA guideline for the management of ST-elevation myocardial infarction: a report of the American College of Cardiology Foundation/American Heart Association Task Force on Practice Guidelines. J Am Coll Cardiol. 2013;61:e78–e140.
35. Ibanez B, James S, Agewall S, et al. 2017 ESC guidelines for the management of acute myocardial infarction in patients presenting with ST-segment elevation: the task force for the management of acute myocardial infarction in patients presenting with ST-segment

elevation of the European Society of Cardiology (ESC). Eur Heart J. 2018;39:119–77.

36. Ponikowski P, Voors AA, Anker SD, et al. 2016 ESC guidelines for the diagnosis and treatment of acute and chronic heart failure. Eur J. 2016;37:2129–200.
37. Yancy CW, Jessup M, Bozkurt B, et al. 2013 ACCF/AHA guideline for the Management of Heart Failure. A report of the American College of Cardiology Foundation/American Heart Association Task Force on Practice Guidelines. Circulation. 2013;128:e240–327.
38. Kapur NK, Esposito M. Hemodynamic support with percutaneous devices in patients with heart failure. Heart Failure Clin. 2015;11:215–30.
39. Hallak A, Wei L, Berzingi C. Percutaneous axillary access for large-bore arteriotomy: a step-by-step guide. J Card Surg. 2018;33:270–3.
40. Seyfarth M, Sibbing D, Bauer I, et al. A randomized clinical trial to evaluate the safety and efficacy of a percutaneous left ventricular assist device versus intraaortic balloon pumping for treatment of cardiogenic shock caused by myocardial infarction. J Am Coll Cardiol. 2008;52:1584–8.
41. O'Neill WW, Kleiman NS, Moses J, et al. A prospective, randomized clinical trial of hemodynamic support with Impella 2.5 versus intra-aortic balloon pump in patients undergoing high-risk percutaneous coronary intervention: the PROTECT II study. Circulation. 2012;126:1717–27.
42. O'Neill WW, Schreiber T, Wohns DH, Rihal C, Naidu SS, Civitello AB, Dixon SR, Massaro JM, Maini B, Ohman EM. The current use of Impella 2.5 in acute myocardial infarction complicated by cardiogenic shock: results from the USpella registry. J Interv Cardiol. 2014;27:1–11.
43. Maini B, Naidu SS, Mulukutla S, et al. Real-world use of the Impella 2.5 circulatory support system in complex high-risk percutaneous coronary intervention: the USpella Registry. Catheter Cardiovasc Interv. 2012;80(5):717–25.
44. Sjauw KD, Konorza T, Erbel R, et al. Supported high risk percutaneous coronary intervention with the Impella 2.5 device the Europella registry. J Am Coll Cardiol. 2009;54(25):2430–4.
45. Vetrovec GW, Anderson GW, M. The cVAD registry for percutaneous temporary hemodynamic support: a prospective registry of Impella mechanical circulatory support use in high-risk PCI, cardiogenic shock, and decompensated heart failure. Am Heart J. 2018;199:115–21.
46. Ouweneel DM, Eriksen E, Sjauw KD, et al. Impella CP versus intra-aortic balloon pump support in acute myocardial infarction complicated by cardiogenic shock. The IMPRESS in severe shock trial. J Am Coll Card. 2017;69:278–87.
47. Hall SA, Uriel N, Carey SA. Use of a percutaneous temporary circulatory support device as a bridge to decision during acute decompensation of advanced heart failure. J Heart Lung Transplant. 2018;37:100–6.
48. Levine GN, Bates ER, Blankenship JC, et al. 2011 ACCF/AHA/SCAI guideline for percutaneous coronary intervention. A report of the American College of Cardiology Foundation/ American Heart Association Task Force on Practice Guidelines and the Society for Cardiovascular Angiography and Interventions. J Am Coll Cardiol. 2011;58:e44–122.
49. O'Gara PT, Kushner FG, Ascheim DD, et al. 2013 ACCF/AHA guideline for the management of ST-elevation myocardial infarction: a report of the American College of Cardiology Foundation/ American Heart Association Task Force on Practice Guidelines. J Am Coll Cardiol. 2013;61:e78–140.
50. Kapur NK, Paruchuri V, Urbano-Morales JA, Mackey EE, Daly GH, Qiao X, Pandian N, Perides G, Karas RH. Mechanically unloading the left ventricle before coronary reperfusion reduces left ventricular wall stress and myocardial infarct size. Circulation. 2013;128:328–36.
51. Kar B, Gregoric ID, Basra SS, et al. The percutaneous ventricular assist device in severe refractory cardiogenic shock. J Am Coll Cardiol. 2011;57:688–96.
52. Thiele H, Sick P, Boudriot E, et al. Randomized comparison of intra-aortic balloon support versus a percutaneous left ventricular assist device in patients with revascularized acute myocardial infarction complicated by cardiogenic shock. Eur Heart J. 2005;26:1276–83.
53. Burkhoff D, Cohen H, Brunckhorst C, et al. A randomized multicenter clinical study to evaluate the safety and efficacy of the TandemHeart percutaneous ventricular assist device versus conventional therapy with intraaortic balloon pumping for treatment of cardiogenic shock. Am Heart J. 2006;152:e1–8.
54. Alli OO, Singh IM, Holmes DR Jr, et al. Percutaneous left ventricular assist device with Tandem-Heart for high-risk percutaneous coronary intervention: the Mayo Clinic experience. Catheter Cardiovasc Interv. 2012;80(5):728–34.
55. Schrage B, Ibrahim K, Loehn T, et al. Impella support for acute myocardial infarction complicated by cardiogenic shock: a matched-pair IABP-SHOCK II trial 30-day mortality analysis. Circulation. 2019;139:1249–58.
56. Guglin M, Zucker MJ, Bazan VM, et al. The present and future JACC scientific expert panel. Venoarterial ECMO for adults. J Am Coll Cardiol. 2019;73:698–716.
57. Ouweneel DM, Schotborgh JV, Limpens J, Sjauw KD, Engstrom AE, Lagrand WK, Cherpanath TG, Driessen AH, de Mol BA, Henriques JP. Extracorporeal life support during cardiac arrest and cardiogenic shock: a systematic review and metaanalysis. Intensive Care Med. 2016;42:1922–34.
58. Nichol G, Karmy-Jones R, Salerno C, Cantore L, Becker L. Systematic review of percutaneous cardiopulmonary bypass for cardiac arrest or cardiogenic shock states. Resuscitation. 2006;70:381e94.

59. Nasr VG, Raman L, Barbaro RP, et al. Highlights from the extracorporeal life support organization registry: 2006–2017. ASAIO J. 2019;65:537–44.
60. Russo JJ, Aleksova N, Pitcher I, et al. Left ventricular unloading during extracorporeal membrane oxygenation in patients with cardiogenic shock. J Am Coll Cardiol. 2019;73:654–62.
61. Jaski BE, Ortiz B, Alla KR, et al. A 20-year experience with urgent percutaneous cardiopulmonary bypass for salvage of potential survivors of refractory cardiovascular collapse. J Thorac Cardiovasc Surg. 2010;139:753–7.
62. Imamura T, Juricek C, Nguyen A, et al. Predictors of hemodynamic improvement and stabilization following intraaortic balloon pump implantation in patients with advanced heart failure. J Invasive Cardiol. 2018;30:56–61.
63. Krishnamoorthy A, DeVore AD, Sun JL, et al. The impact of a failing right heart in patients supported by intra-aortic balloon counterpulsation. Eur Heart J Acute Cardiovasc Care. 2017;6:709–18.
64. Anderson MB, Goldstein J, Milano C, et al. Benefits of a novel percutaneous ventricular assist device for right heart failure: The prospective RECOVER RIGHT study of the Impella RP device. J Heart Lung Transplant. 2015;34:1549–60.
65. Kazui T, Tran PL, Echeverria A, et al. Minimally invasive approach for percutaneous CentriMag right ventricular assist device support using a single PROTEKDuo cannula. J Cardiothorac Surg. 2016;11:123.
66. Schmack B, Farag M, Kremer J, et al. Results of concomitant groin-free percutaneous temporary RVAD support using a centrifugal pump with a double-lumen jugular venous cannula in LVAD patients. J Thorac Dis. 2019;11(Suppl 6):S913–20.
67. Varian K, Weining D, Weiqin LW, et al. Minimally invasive biventricular mechanical circulatory support with Impella pumps as a bridge to heart transplantation: a first-in-the-world case report. ESC Heart Failure. 2019;6:552–4.
68. Neumann FJ, Sousa Uva M, Ahlsson A, et al. ESC/EACTS Guidelines on myocardial revascularization. 2018.

Computational Fluid Dynamics for Mechanical Circulatory Support Device Development

25

Roland Graefe and Lutz Pauli

CFD Simulation Basics

Simulation Methods and Quality Characteristics

Computational fluid dynamics (CFD) deals with the numerical solution of partial differential equations (PDEs). In particular, when considering blood as a continuum, its general motion can be described by the conservation of mass, momentum, and energy. As we will assume isothermal flows within this chapter, the latter will not be discussed. Instead, we will also present the conservation of a general scalar quantity that is transported with the blood. For a more detailed derivation of the conservation laws, we refer to, e.g., [1].

Conservation of Mass

Consider a closed domain Ω (e.g., the wetted volume of a VAD) in Eulerian frame at a time instance t that contains particles at positions $\boldsymbol{x}(t)$. The velocity of each particle is then defined by

$$\boldsymbol{u}(\boldsymbol{x}(t),t)=\frac{d\boldsymbol{x}(t)}{dt}. \tag{25.1}$$

Mass is conserved in Ω if the total mass remains constant over time, i.e.,

$$\frac{d}{dt}\int_{\Omega}\rho d\Omega=0, \tag{25.2}$$

where ρ is the density of the fluid. If Reynolds transport theorem and divergence theorem are applied, the above equation can be written as

$$\int_{\Omega}\left(\frac{\partial\rho}{\partial t}+\nabla\cdot(\rho\boldsymbol{u})\right)d\Omega=0, \tag{25.3}$$

or equivalently in differential form

$$\frac{\partial\rho}{\partial t}+\nabla\cdot(\rho\boldsymbol{u})=0\,\text{in}\,\Omega. \tag{25.4}$$

The differential form is using the fact that a fluid's control volume/domain can be chosen arbitrarily. It states that all the mass entering the domain Ω has to leave the domain again.

Conservation of Momentum

Momentum can be changed by the action of forces, and its conservation equation is Newton's second law of motion:

$$\frac{d}{dt}\int_{\Omega}\rho\boldsymbol{u}d\Omega=\int_{\Gamma}\sigma\cdot\boldsymbol{n}d\Gamma+\int_{\Omega}\rho\boldsymbol{f}d\Omega, \tag{25.5}$$

where $\boldsymbol{\sigma}$ is a symmetric stress tensor according to Cauchy's stress theorem, $\boldsymbol{n}$ is the outward pointing

R. Graefe, Dr.-Ing (✉)
ReinVAD GmbH, Aachen, Germany
e-mail: graefe@reinvad.de

L. Pauli, Dr.-Ing
MAGMA Gießereitechnologie GmbH,
Aachen, Germany
e-mail: l.pauli@magmasoft.de

© Springer Nature Switzerland AG 2020
J. H. Karimov et al. (eds.), *Mechanical Support for Heart Failure*,
https://doi.org/10.1007/978-3-030-47809-4_25

normal to boundary $\Gamma = \partial\Omega$, and $\boldsymbol{f}$ is the volume force per unit mass. The individual terms can be interpreted as inertial, viscous, and external forces, respectively. Application of Reynolds transport theorem and divergence theorem to Eq. 25.5 gives

$$\frac{\partial}{\partial t}\int_{\Omega}\rho\boldsymbol{u}d\Omega+\int_{\Omega}\nabla\cdot(\rho\boldsymbol{u}\otimes\boldsymbol{u})d\Omega=\int_{\Omega}\nabla\cdot\boldsymbol{\sigma}d\Omega+\int_{\Omega}\rho\boldsymbol{f}d\Omega, \tag{25.6}$$

or in differential form

$$\frac{\partial(\rho\boldsymbol{u})}{\partial t}+\nabla\cdot(\rho\boldsymbol{u}\otimes\boldsymbol{u})-\rho\boldsymbol{f}-\nabla\cdot\boldsymbol{\sigma}=\boldsymbol{0}\text{ in }\Omega, \tag{25.7}$$

where the operator $\otimes$ denotes the tensor product (e.g., in component notation $[\boldsymbol{u}\otimes\boldsymbol{v}]_{ij}=u_iv_j$).

Conservation of a Scalar Quantity

Similar to the change of momentum, a scalar quantity or concentration c inside a domain Ω changes over time due to fluxes $\boldsymbol{j}$ normal to the boundary Γ and sinks or sources $s(c)$ inside of Ω:

$$\frac{d}{dt}\int_{\Omega}cd\Omega+\int_{\Gamma}\boldsymbol{j}\cdot\boldsymbol{n}d\Gamma+\int_{\Omega}s(c)d\Omega=0. \tag{25.8}$$

For concentrations, the fluxes $\boldsymbol{j}$ can be described by Fick's law for mass diffusion, that is, $\boldsymbol{j} = -\nu_d\nabla c$, where ν_d is the diffusion coefficient. Using Fick's law and again Reynolds transport theorem and divergence theorem, the above equation becomes

$$\frac{\partial}{\partial t}\int_{\Omega}cd\Omega+\int_{\Omega}\nabla\cdot(\boldsymbol{u}c)d\Omega=\int_{\Omega}\nabla\cdot(\nu_d\nabla c)d\Omega-\int_{\Omega}s(c)d\Omega, \tag{25.9}$$

or in differential form

$$\frac{\partial c}{\partial t}+\nabla\cdot(\boldsymbol{u}c)-\nabla\cdot(\nu_d\nabla c)+s(c)=0\text{ in }\Omega. \tag{25.10}$$

Discretization Methods

None of the above PDEs can be solved analytically on arbitrary domains. Therefore, in CFD, discretization methods are used to solve the PDEs numerically (using either the differential or the integral form). The most common methods are briefly introduced in the following.

Finite Difference Method

The finite difference method is the historically oldest methods to discretize PDEs. It is also the easiest method to apply to simple geometries.

Let the domain of interest be covered by a structured grid/mesh. At each grid point, the partial derivatives of the PDEs are approximated by finite difference stencils. These stencils are derived by Taylor series expansions or polynomial fitting at the neighboring grid points. This way, a system of simple algebraic equations is obtained per grid node, in which the unknowns appear as variables at that grid node and at a certain number of neighboring grid points [1].

Today, finite difference methods have been mainly replaced by finite volume or finite element methods due to their higher flexibility. However, finite difference stencils are often used to discretize derivatives with respect to time or, in other words, to propagate a numerical quantity in time.

Finite Volume Method

The finite volume method can operate on structured, unstructured, and hybrid grids. The grid defines a finite number of control volumes with associated sampling points, where the sampling points are usually centroids of each control volume. Starting with the integral form of the conservation equations, each flow quantity is discretized at the sampling points. Interpolation is used to express fluxes across control volume boundaries, and integrals are approximated by suitable quadrature formulas.

Since all the approximated terms have a direct physical meaning, the finite volume method is very popular with engineers. Also, most of the commercial CFD tools on the market are based on the finite volume method.

Finite Element Method

The finite element method is similar in the way that it can also operate on structured, unstructured, and hybrid grids. The PDEs are multiplied by a weighting function and integrated over the domain. The flow quantities as well as the weighting terms are discretized by basic functions, e.g., linear shape functions within each element, and integrated using suitable quadrature formulas.

The finite element method is very popular in the field of solid and structural mechanics, but it also has a strong community in the CFD sector. While more difficult to formulate at a first glance, it offers a rich mathematical basis to build upon and is especially advantageous for fluid-structure interaction problems.

Further Discretization Methods

"Classical" finite volume and finite element methods do clearly dominate the software products for CFD simulation of VADs. Nevertheless, there are also other possibilities to compute blood flows. Among others, Lattice Boltzmann methods, smooth particle hydrodynamics, spectral element methods, and discontinuous Galerkin methods are prominent examples. A combination of different methods can be also useful to solve particular problems.

Quality Control

CFD simulation offers a lot of advantages in the design cycle of new or existing VADs. However, the user of CFD software should be always aware that every simulation is only an approximation of the reality. Such an approximation can be extremely good, but also extremely bad. The accuracy of a simulation depends on three kinds of systematic errors: discretization errors, iteration errors, and modeling errors.

Discretization errors are associated with the accuracy of the computational grid. In CFD, we distinguish between surface grids and volume grids. Assuming an accurate digital geometry representation of the device, a surface grid for that device is only able to capture a decent portion of that geometry information. Obviously, a fine grid is more accurate but necessitates more computational resources. In general, the user has to decide whether the grid fineness is sufficient and has to find a compromise between discretization error and the computational burden.

For the volume discretization of the flow domain, it is especially important to capture large gradients in the flow field, which usually occur close to the walls and in regions with flow separation and intensive turbulence. Close to the walls, this leads to boundary layer meshing. For boundary layer meshing, the choice for the height of the first element layer at the walls depends on the *law of the wall* [2]. Since the boundary layer meshing depends a lot on model assumptions, we do not want to provide strict guidelines in this chapter. However, the user should be aware that the boundary layer mesh is crucial for, e.g., a good estimation of the wall shear stress.

Grid refinement studies are useful, and often necessary, to quantify the uncertainty associated with the discretization error. Established procedures are described in, e.g., [3, 4], and involve a comparison of at least three refinement levels of the grid.

Iteration errors arise by solving the system of equations involved. Since direct solvers would be computationally too expensive to apply in CFD, iterative solution methods have to be used. Where a direct method can generate the exact solution of the discretized system starting from scratch, the iterative method is dependent on an initial guess, which is iteratively improved to approximate the solution. For very nonlinear CFD simulations, the initial guess can be important for reaching the correct solution and for reaching that solution fast.

A numerical method is said to be *convergent* if it is *consistent* and *stable*. Consistency means the exact solution is reached if the grid spacing is zero. Stability means the iterative solution method does not diverge, i.e., discretization errors do not amplify. In general, the user has to define a convergence criterion (or stop criterion) for the iterative method and has to make sure it satisfies the quality requirements. In common practice, residuals should drop by at least three orders of magnitude and fluxes like mass flux should be conserved within a range of less than 2–3%.

Modeling errors are by far the most difficult to quantify. For example, the user has to distinguish between Newtonian or non-Newtonian material models, different turbulence models, particular inflow and outflow boundary conditions, different methods for the handling of moving components, single phase or multiphase models, etc. Additionally, the choice of a particular model is associated with certain parameter assumptions. Some common model and parameter choices are discussed in the section *Blood Modeling for VADs*. Some best practices for CFD validation are discussed below. For broader view on the very active research topic of uncertainty quantification, the interested reader is referred to some general introduction given, e.g., in References [5, 6].

Validation

As discussed above and as an important note, it has to be stated that all CFD results are wrong to a certain extend. A validation step is therefore a good option to provide judgement whether the CFD error is acceptably low for the research question at hand. This also means that there cannot be a universal threshold for a specific difference between a CFD derived value and a corresponding experimental result. It is the task of the engineer, researcher, or reader of a publication to judge whether the remaining error is sufficiently low and if the conclusions based on the CFD results can be considered validated.

There are different experiments that are used as validation experiments for CFD results in the field of rotary blood pump design. As gold standard could be considered to provide flow visualization results that are compared to flow fields predicted by CFD. One such study was conducted by Triep et al. as early as 2006 [7]. Figure 25.1 illustrates the experimental setup. The device under investigation is the rotor of a fast-spinning axial blood pump. In the test chamber, the rotor is mounted on a drive shaft that is rotated via an external electric motor. The pump is connected to a hydraulic circuit.

Note that the pump housing is made out of transparent material for optical access. The flow is then illuminated with laser light and via an optical beam guidance system. A high-speed camera is used to take pictures of the flow inside the pump that becomes visible by included tracer particles that are illuminated by the laser light. The pictures are used to calculate the flow field in the test chamber. This measurement technique is referred to as digital particle image velocimetry (DPIV) and is a complex but established flow visualization technique.

Figure 25.2 shows exemplary results from the DPIV measurements and compares them to numerical CFD results with the identical geometry and operational point.

The results confirm that the CFD predictions closely represent the measured flow profiles. Looking at the flow field at the entrance to the rotor and at the leading edge of the blade (Fig. 25.2 left), the results for the velocity profiles predicted by CFD coincide well with the velocity profiles measured by DPIV. This is also true for the region further downstream as shown in Fig. 25.2 on the right. Only in regions of high-velocity gradients, the results show some discrepancies.

Generally, as a validation step of CFD predictions, it is frequent practice to compare experimentally and numerically determined characteristic pump curves. Achieving a numerical setup that accurately predicts the H-Q curve (H, head; Q, pump flow) is then assumed to be an adequate validation. For the second exemplary study, it was decided to additionally check the velocity distributions within the pump, similar to the study by [7]. It was stated by Marseille that a correct prediction of the H-Q curve is a necessary prerequisite, but it would be advisable to ensure a coincidence of flow distributions with the devices as well [8]. It is possible that multiple incorrect predictions of flow phenomena can lead to a correct prediction of the H-Q curve. By ensuring both, a correct prediction of the H-Q curve and velocity distributions, it is very likely that a sufficient validation of the numerical predictions is achieved.

The hydraulic pump components of the 1:1 model are made of transparent polymethylmethacrylate (PMMA) to ensure an optical access for the employed high-speed camera that

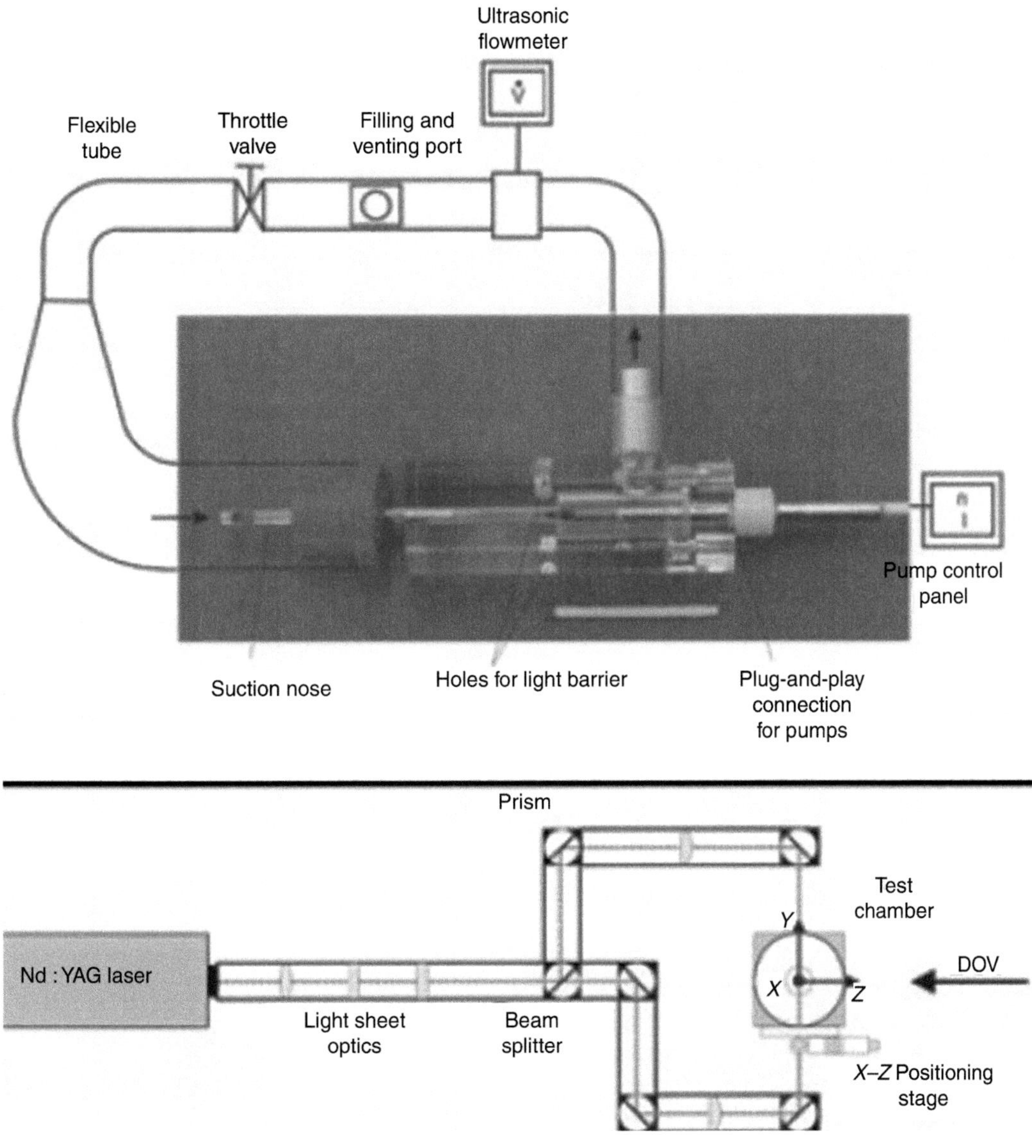

Fig. 25.1 Fluid circuit (above) and optical system (below) for the DPIV measurement of the flow in an axial flow rotary blood pump. (Reprinted and adapted from Siess et al. [7] with permission from John Wiley and Sons)

captures the velocity distribution within the pump. Additionally, a specific Newtonian fluid mixture was used (water-glycerol-sodium iodide, refractive index 1.49) to ensure a matching refractive index with the PMMA. This again improves the optical access, and the resulting velocity distributions are of an improved image quality. As an important note, this procedure requires the introduction of nondimensional pump performance parameters according to the laws of similitude theory because of the increased density of the employed fluid mixture (1740 kg/m3). The nondimensional head coefficient is defined by

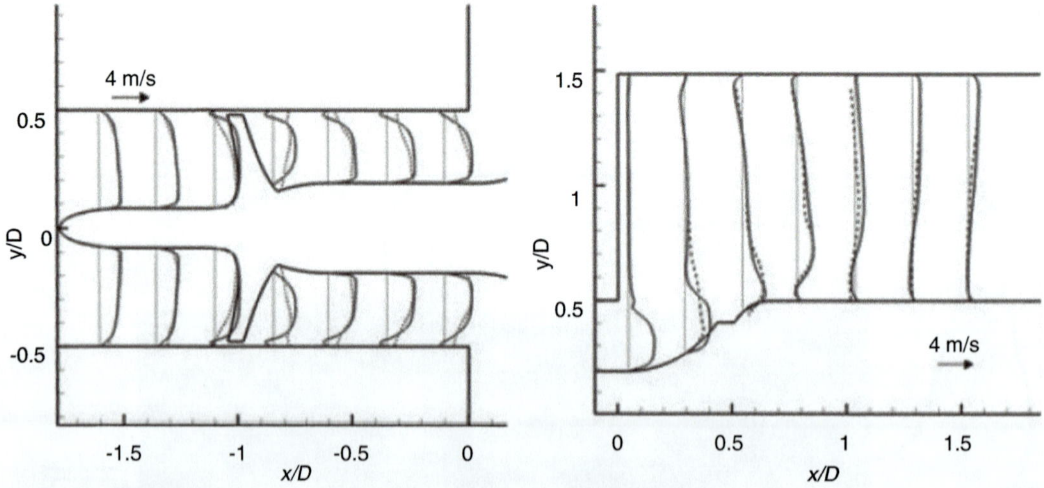

Fig. 25.2 Comparison of the measured velocity profile with DPIV (dashed line) with the results from CFD simulations (solid line). (Reprinted and adapted from Siess [7] with permission from John Wiley and Sons.) Results are shown at the rotor inlet (left) and further downstream (right)

$$\psi = \frac{\Delta p}{\rho \cdot \frac{u_2^2}{2}} \tag{25.11}$$

and the nondimensional flow coefficient is defined by

$$\Phi = \frac{Q}{u_2 \cdot D_2 \cdot b_2 \cdot \pi}. \tag{25.12}$$

According to Spurk et al., it is sufficient to ensure identical pressure and flow coefficients in order to ensure similitude between different fluid flow machines in different operational conditions including size, rotational speed, and fluid properties [9]. A matching pump Reynolds number would additionally increase comparability. Figure 25.3 shows the nondimensional comparison of experimentally and numerically determined characteristic curves. The measured characteristic curve is displayed as an average value including a standard error. The standard error of 8% based on the pressure number at $\Phi =$ 0 represents the variability between different repetitions of the measurements (three repetitions) and due to nonmatching Reynolds numbers.

Both numerically predicted characteristic curves are constantly lower for all flow rates compared to the measured curve except for the highest flow rate. The difference remains within the standard error of 8% and decreases with increasing flow rate. In summary, the numerically predicted H-Q curve represents the measured H-Q curve well. For brevity, for results of the flow visualization refer to [10]. In summary, the numerically predicted velocity distributions represent the measured velocity distributions well since no flow phenomena exist that are capable of changing the characteristic curve that are not captured numerically.

Blood Modeling and Blood Compatibility

Blood Modeling for VADs

Navier-Stokes Equations

In this chapter, blood is considered as an incompressible, isothermal, and single-phase liquid, which means that density is constant over time. As a consequence, Eqs. (25.4) and (25.7) can be simplified and, together with boundary and initial conditions, rewritten as the incompressible Navier-Stokes equations:

$$\rho\left(\frac{\partial \boldsymbol{u}}{\partial t} + \boldsymbol{u} \cdot \nabla \boldsymbol{u} - \boldsymbol{f}\right) - \nabla \cdot \sigma = \boldsymbol{0} \text{ in } \Omega, \tag{25.13}$$

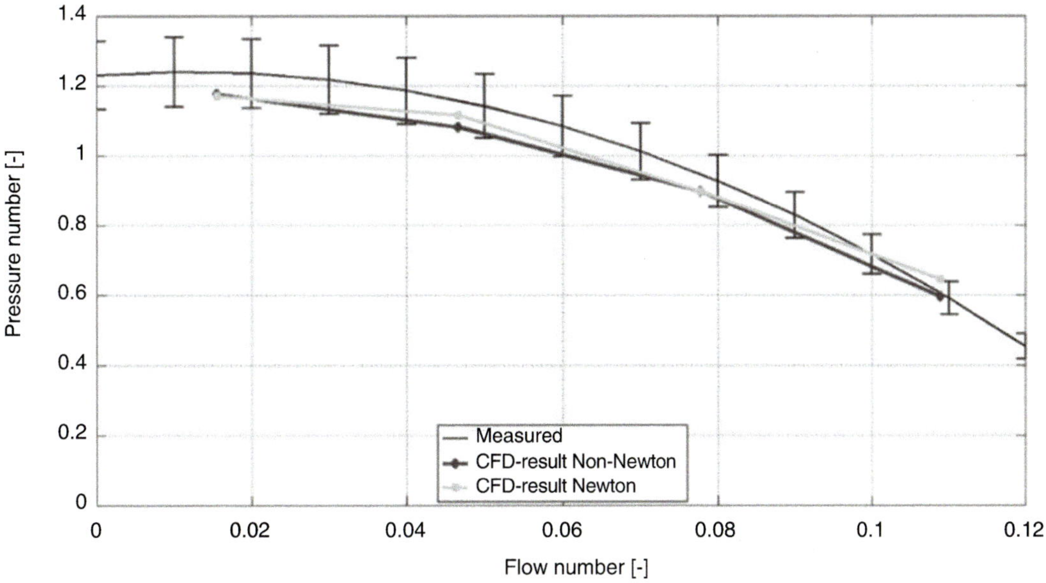

Fig. 25.3 Comparison of H-Q-data represented with dimensionless coefficients from CFD simulation with measurements. (Adapted from Graefe 10] with permission from Shaker Verlag)

$$\nabla \cdot \boldsymbol{u} = 0 \,\text{in}\, \Omega, \quad (25.14)$$

$$\boldsymbol{u} = \boldsymbol{g}_u \,\text{on}\, \Gamma_g, \quad (25.15)$$

$$\boldsymbol{n} \cdot \sigma = \boldsymbol{h} \,\text{on}\, \Gamma_h, \quad (25.16)$$

$$\boldsymbol{u}\big|_{t=0} = \boldsymbol{u}_0. \quad (25.17)$$

In the above equations, Γ_g and Γ_h are complementary subsets of the boundary Γ and $\Gamma_g \neq \varnothing$. The boundary conditions $\boldsymbol{g}_u$ and $\boldsymbol{h}$ are referred as the Dirichlet and Neumann boundary conditions, respectively.

The Navier-Stokes equations are closed by constitutive laws, which specify the physical properties of the material in motion. Some possible equations for blood flow will be presented in a later section. Since such physical properties are always related to the stress tensor $\boldsymbol{\sigma}$, it will be discussed beforehand.

Stress and Strain

Motion, as described by the conservation laws in the section *Simulation Methods and Quality Characteristics*, gives rise to interactions between particles in the fluid. One consequence of these interactions is stress and strain. Stress is the internal response of a particle to external loads. In fluid dynamics, these loads are usually caused by the movement of the fluid itself and can act in normal and in tangential direction. Mathematically, stress can be expressed by a symmetric second-rank tensor, the Cauchy stress tensor $\boldsymbol{\sigma}$, with its components

$$\sigma = \begin{bmatrix} \sigma_{11} & \sigma_{12} & \sigma_{13} \\ \sigma_{12} & \sigma_{22} & \sigma_{23} \\ \sigma_{13} & \sigma_{23} & \sigma_{33} \end{bmatrix} \quad (25.18)$$

For the majority of fluids, the stress tensor depends on the pressure p and the rate-of-strain tensor $\boldsymbol{E}$. The rate-of-strain tensor $\boldsymbol{E}$ is defined by the symmetric part of the velocity gradient

$$\nabla \boldsymbol{u} = \boldsymbol{E} + \boldsymbol{W}, \quad (25.19)$$

$$\boldsymbol{E} = \frac{1}{2}\left(\nabla \boldsymbol{u} + \nabla \boldsymbol{u}^T\right), \quad (25.20)$$

$$\boldsymbol{W} = \frac{1}{2}\left(\nabla \boldsymbol{u} - \nabla \boldsymbol{u}^T\right), \quad (25.21)$$

where $\boldsymbol{W}$ is the antisymmetric part of the velocity gradient and called vorticity tensor.

An important aspect of tensors is its invariant. Invariants are independent from rotations of the coordinate system and therefore, very useful to define constitutive equations. For a symmetric 3× 3 tensor $\boldsymbol{A}$, like stress and rate-of-strain, the three invariants are defined as follows (for a derivation refer to, e.g., Reference [11]):

$$I_A = \mathrm{tr}(\boldsymbol{A}), \tag{25.22}$$

$$II_A = \frac{1}{2}\left(\mathrm{tr}(\boldsymbol{A})^2 - \mathrm{tr}(\boldsymbol{A}^2)\right), \tag{25.23}$$

$$III_A = \det(\boldsymbol{A}), \tag{25.24}$$

where tr($\boldsymbol{A}$) is the trace of $\boldsymbol{A}$, and det($\boldsymbol{A}$) is the determinant of $\boldsymbol{A}$. For the rate-of-strain tensor $\boldsymbol{E}$, the definition of the second invariant can be even simplified. Due to the incompressibility constraint, one has tr($\boldsymbol{E}$) = 0. Thus, we obtain

$$II_E = -\frac{1}{2}\mathrm{tr}(\boldsymbol{E}^2) = -\frac{1}{2}\boldsymbol{E}:\boldsymbol{E}, \tag{25.25}$$

where the colon operator denotes the inner product of two tensors defined by $\boldsymbol{A} : \boldsymbol{B} = \sum_{i,j} a_{ij} \bullet b_{ij}$.

Constitutive Equations

The Navier-Stokes equations (Eqs. 25.13, 25.14, 25.15, 25.16, and 25.17) are closed by the constitutive laws which define (or model) the stress tensor $\boldsymbol{\sigma}$. In general, blood has very complex rheological properties. It shows shear thinning,[1] viscoelastic,[2] and thixotropic[3] behavior. Therefore, every constitutive equation for blood flow is based on certain assumptions. In this chapter, three classes of models will be discussed: Newtonian models, generalized Newtonian (shear thinning) models, and viscoelastic models. For the non-Newtonian classes, a huge number of different models are available in the literature. For the sake of simplicity, we will only discuss two examples for the generalized Newtonian and one example for the viscoelastic models. A collection of different models can be found, for example, in Reference [12]. A general introduction to the modeling of blood rheology is given, e.g., in Reference [13].

Newtonian Models

Newtonian models are the simplest constitutive equations. The rate-of-strain tensor $\boldsymbol{E}$, defined in Eq. (25.20), is linearly related to a shear stress tensor by a constant viscosity μ. Together with the pressure p, the stress tensor $\boldsymbol{\sigma}$ becomes

$$\boldsymbol{\sigma} = -p\boldsymbol{I} + 2\mu\boldsymbol{E}, \tag{25.26}$$

where $\boldsymbol{I}$ is the identity tensor. Blood is usually considered as a Newtonian liquid if the flow operates in shear rate regimes of 1×10^2 s^{-1} to 1×10^4 s^{-1} [14]. Common material parameters are ρ = 1050 kg/m^3 and $\mu = 3.5 \times 10^{-3}$ Pa s for 45% hematocrit and 37°C.

Generalized Newtonian or Shear Thinning Models

If a blood flow is affected by low shear rates, the Newtonian assumption is not sustainable anymore. A popular way to introduce non-Newtonian phenomena into the modeling is to consider only the shear thinning behavior on the constitutive equation; viscoelastic and thixotropic behavior are neglected. Such models are called generalized Newtonian models, where the stress tensor $\boldsymbol{\sigma}$ becomes

$$\boldsymbol{\sigma} = -p\boldsymbol{I} + 2\mu(G_f)\boldsymbol{E}. \tag{25.27}$$

Thus, the modeling assumption is the dependence of viscosity on a scalar shear rate G_f, which is expected to be similar to the behavior in simple shear experiments. The shear rate G_f has to be defined for complex flow fields. With the help of Eqs. (25.23) and (25.25), one can show that for incompressible, viscous fluids one has [15]

$$G_f = \sqrt{-4II_E} = \sqrt{2\boldsymbol{E}:\boldsymbol{E}}. \tag{25.28}$$

We will present two exemplary functions for $\mu(G_f)$. Basically, both of them aim to give a fit of the viscosity function for whole blood as given in Fig. 25.4. The so-called Cross model [16] is given as

$$\mu(G_f) = \mu_\infty + \frac{\mu_0 - \mu_\infty}{1 + (\lambda_C G_f)^{m_C}}, \tag{25.29}$$

where μ_0 and μ_∞ define the limiting viscosities for zero-shear and infinite-shear, respectively.

[1] In a shear thinning, liquid viscosity decreases with increasing shear.

[2] As the name suggests, a viscoelastic liquid shows next to viscous also elastic behavior.

[3] A thixotropic liquid shows time-dependent viscosity changes.

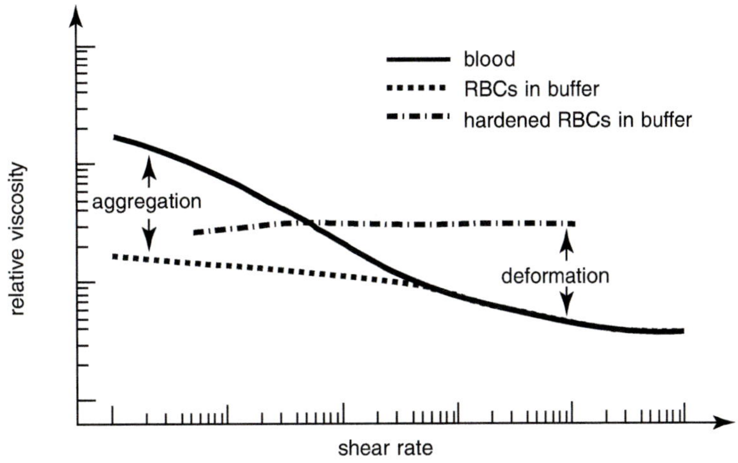

Fig. 25.4 Shear rate-viscosity curves for normal blood (45% hematocrit, temperature of 37 °C), normal RBCs suspended in protein-free buffer (i.e., a medium that prevents RBC aggregation) and hardened RBCs suspended in the buffer. Blood viscosity is shown relative to the viscosity of the solvent (experimental data based on reference [14])

The model parameters λ_C and m_C have to be defined based on the experimental data. The Cross model is a modification of a simple power law to account for the Newtonian flow behavior of blood at high shear rate. The fit to experimental measurements can often be improved by adding a fifth parameter. The enhanced function is termed modified Cross model and given as

$$\mu\left(G_f\right) = \mu_\infty + \frac{\mu_0 - \mu_\infty}{\left(1 + \left(\lambda_C G_f\right)^{m_C}\right)^{a_C}}. \quad (25.30)$$

It should be noted that the modified Cross model can be usually rewritten as the also very popular Carreau-Yasuda model [17]. A possible fit to Fig. 25.4 would be, e.g., given with the parameter set given in Table 25.1.

Viscoelastic and Thixotropic Models

To account for both the viscous and the elastic behavior of blood, there exist two general classes of models: the continuous modeling approach and the modeling of individual cells inside the plasma. The latter has the potential to simulate rheology of blood most accurately, but up to now, it is not computationally feasible for complex domains. Therefore, we will only focus on continuous models for viscoelasticity in blood. Examples on simulation of individual cells can be found in References [18, 19].

Most of the continuous viscoelastic blood models available in the literature belong to the class of generalized Maxwell-type equations. In such models, a new partial differential equation is introduced to predict the temporal and spatial evolution of an extra-stress tensor $\boldsymbol{T}$, that is,

$$\boldsymbol{T} + \lambda_e(\cdot)\check{\boldsymbol{T}} = 2\mu_e(\cdot)\boldsymbol{E} \text{ in } \Omega, \quad (25.31)$$

$$\boldsymbol{T} = \boldsymbol{G}_T \text{ on } \Gamma_g, \quad (25.32)$$

$$\boldsymbol{T}\big|_{t=0} = \boldsymbol{T}_0, \quad (25.33)$$

where $\lambda_e(\cdot)$ and $\mu_e(\cdot)$ are the relaxation time and the "elastic" viscosity, respectively. Both can be functions of other quantities, like the shear rate G_f or an additional degree of freedom. The operator $\check{\boldsymbol{T}}$ is called upper-convected derivative of tensor $\boldsymbol{T}$ and is defined as

$$\check{\boldsymbol{T}} = \frac{\partial \boldsymbol{T}}{\partial t} + \boldsymbol{u} \cdot \nabla \boldsymbol{T} - (\nabla \boldsymbol{u})\boldsymbol{T} + \boldsymbol{T}(\nabla \boldsymbol{u})^T. \quad (25.34)$$

Introducing the plasma viscosity μ_p, Eq. (25.31) can be coupled to the total stress tensor $\boldsymbol{\sigma}$ by

$$\sigma = -p\boldsymbol{I} + 2\mu_p \boldsymbol{E} + \boldsymbol{T}, \quad (25.35)$$

$$\mu = \mu_p + \mu_e. \quad (25.36)$$

In the case where λ_e and μ_e are constant parameters, Eqs. (25.31), (25.35), and (25.36) together with Eqs. (25.13) and (25.14) are called Oldroyd-B model. The Oldroyd-B model is able to represent viscoelastic material behavior, but cannot represent

Table 25.1 An exemplary parameter set for the modified Cross model in Eq. (25.30) that fits the shear rate-viscosity curve in Fig. 25.4

Parameter	Value	Comment
μ_0	0.16	Zero-shear viscosity in pa s
μ_∞	0.0035	Infinite-shear viscosity in pa s
λ_C	8.2	Modified Cross model coefficient
m_C	0.64	Modified Cross model exponent
a_C	1.23	Modified Cross model exponent

shear thinning and thixotropic behavior. A rather simple generalization of the Oldroyd-B model that accounts for both shear thinning and thixotropic behavior is, for example, introduced by Owens [20], where an additional PDE is solved to account for aggregation and fragmentation of RBCs.

It should be noted that, to the authors' best knowledge, no CFD simulation for a VAD has been performed with a fully coupled viscoelastic model so far. However, due to our experience with a viscoelastic hemolysis model (which will be discussed in the section Blood Damage), we believe that modeling such behavior might be particularly important to further understand the flow physics of blood insight a VAD in the future. Especially if VADs operate beyond the design point, the effect of low shear regimes and normal stresses can only be sufficiently represented if viscoelasticity is considered in the model.

Turbulence

Blood can be subject to flow regimes that cause transition from laminar to turbulent flow. In the healthy circulatory system, such flow regimes do usually only occur in the aorta. In medical devices, such as VADs, the existence of turbulent flow is, however, quite common.

The modeling of turbulent flows is a very wide scientific field. We will limit the discussion in this section to a few essential notations. For a deeper introduction into the topic, we refer to the textbook by Pope [2].

Turbulent flows develop if the Reynolds number exceeds a critical limit. In pipe flows, for example, this critical limit occurs at a Reynolds number of $\mathrm{Re} \approx 2300$ [21]. Such a turbulent flow field is characterized by random velocity fluctuations at a wide range of temporal and spatial scales. The Navier-Stokes equations, which are Eqs. (25.13) and (25.14), are able to represent all these scales. However, such a direct numerical simulation (DNS) can become extremely expensive as the entire energy spectrum (down to the so-called Kolmogorov scale) has to be resolved. Two alternatives will be briefly introduced here, based on temporal and spatial filtering of the Navier-Stokes equations.

Temporal Filtering

According to Reynolds [21], a fluctuation of $\boldsymbol{u}$ can be defined by

$$\tilde{\boldsymbol{u}} = \boldsymbol{u} - \langle \boldsymbol{u} \rangle \Leftrightarrow \boldsymbol{u} = \langle \boldsymbol{u} \rangle + \tilde{\boldsymbol{u}}, \tag{25.37}$$

where $\langle \boldsymbol{u} \rangle$ is a temporal mean which may be defined by

$$\langle \boldsymbol{u} \rangle (\boldsymbol{x}) = \frac{1}{\Delta t} \int_t^{t+\Delta t} \boldsymbol{u}(\boldsymbol{x}(t), t)\, dt. \tag{25.38}$$

The decomposition of the current velocity into its temporal mean and its fluctuation is also illustrated in Fig. 25.5.

With the aid of Eq. (25.37), the Navier-Stokes equations (Eqs. (25.13) and (25.14)) can be rewritten for their temporal means. If we assume a Newtonian constitutive equation, we obtain

$$\begin{aligned} &\rho\left(\langle \boldsymbol{u} \rangle \cdot \nabla \langle \boldsymbol{u} \rangle - \langle \boldsymbol{f} \rangle\right) \\ &-\nabla \cdot \left(\langle p \rangle \boldsymbol{I} - \mu\left(\nabla \langle \boldsymbol{u} \rangle + \nabla \langle \boldsymbol{u} \rangle^T\right) + \rho \langle \boldsymbol{u} \otimes \boldsymbol{u} \rangle\right) \\ &= \boldsymbol{0} \text{ in } \Omega, \end{aligned} \tag{25.39}$$

$$\nabla \cdot \langle \boldsymbol{u} \rangle = 0 \text{ in } \Omega, \tag{25.40}$$

which are termed Reynolds equations or Reynolds averaged Navier-Stokes (RANS) equations; for a detailed derivation, we refer to Reference [2]. A comparison of Reynolds equation to Navier-Stokes equations reveals that the only difference is the term $\rho\langle \boldsymbol{u} \otimes \boldsymbol{u} \rangle$, which is referred to as the *Reynolds stress* tensor. Thus, it is only the Reynolds stress tensor that introduced the complexity of random velocity fluctuations to the momentum equation. It is, however, not possible to explicitly compute the Reynolds stresses for the RANS equations, and therefore, several models have been developed to approximate these stresses under certain assumptions.

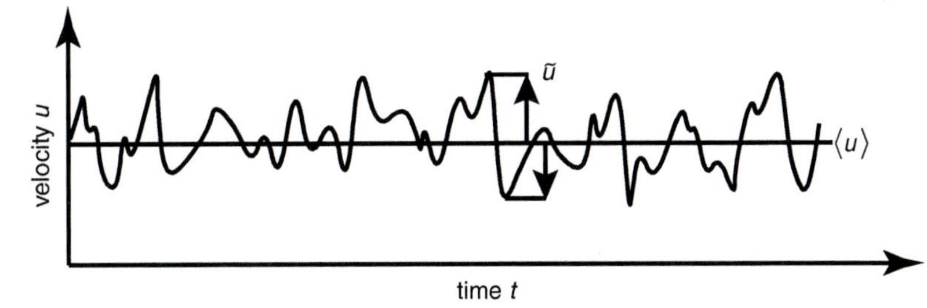

Fig. 25.5 Random fluctuations of a turbulent velocity component at a fixed point over time

Spatial Filtering

Another modeling approach is by means of large eddy simulation (LES). In LES, only the larger unsteady turbulent motion is explicitly computed, whereas the influence of the smaller scales is represented by simple models. Thus, a spatial filtering is applied to the velocity $\boldsymbol{u}$, such that the residual velocity (or subgrid velocity) is defined by

$$\boldsymbol{u}' = \boldsymbol{u} - \bar{\boldsymbol{u}} \Leftrightarrow \boldsymbol{u} = \bar{\boldsymbol{u}} + \boldsymbol{u}'. \quad (25.41)$$

The filtered momentum equation simply becomes

$$\rho\left(\frac{\partial \bar{\boldsymbol{u}}}{\partial t} + \bar{\boldsymbol{u}} \cdot \nabla \bar{\boldsymbol{u}} - \boldsymbol{f}\right) - \nabla \cdot \left(p\boldsymbol{I} - \mu\left(\nabla \bar{\boldsymbol{u}} + \nabla \bar{\boldsymbol{u}}^T\right) + \sigma_r\right) = \boldsymbol{0} \text{ in } \Omega, \quad (25.42)$$

where $\boldsymbol{\sigma}_r$ is a so-called residual stress tensor and thus, similar to the Reynolds stress tensor in Eq. (25.39). A common approach to approximate $\boldsymbol{\sigma}_r$ is an expression with a modeled eddy viscosity ν_e of the form

$$\sigma_r = -\rho\nu_e\left(\nabla \boldsymbol{u} + \nabla \boldsymbol{u}^T\right). \quad (25.43)$$

From a computational point of view, LES is a lot cheaper than DNS but also significantly more expensive than RANS. Compared to RANS model, LES is, however, a lot more accurate in flow regimes with transition to turbulence.

Turbulence in VADs

In common VAD flow simulations, the use of RANS models is currently the standard approach. Due to its ease of use in commercial software and the comparably low computational cost, RANS models offer several advantages. Especially, the SST k-ω model [22] turned out to be the preferred choice in the last couple of years. Nevertheless, more advanced modeling approaches like LES or DES (detached eddy simulation, which is a combination of RANS and LES) are on the advance, thanks to increased computational resources. Such models are able to predict local flow features like flow separation or transition to turbulence more accurately than RANS models [23].

Blood Damage

In devices like VADs, blood is exposed to unphysiological flow conditions or partially bio-incompatible surfaces, and therefore, the blood possibly suffers from damage. Such damage involves various abnormal alterations of the blood, including activation of platelets and leukocytes, damage to proteins, unwanted blood clotting (thrombosis) and thromboembolism, as wells as damage to the erythrocytes (hemolysis).

As hemolysis is up to now the most analyzed mechanism—experimentally as wells as numerically—it will be focused in this chapter. However, we will also present some possibilities to analyze thrombosis.

Hemolysis Estimation

Accurate and reliable modeling of hemolysis in large-scale flow situations like in blood pumps is an open research issue. On the one hand, the microscale effects leading to mechanical hemolysis are, up to now, not fully understood, and, on the other hand, models that resolve the microscale behavior of single RBCs in complex flows would be computationally too expensive. Therefore, a compromise between accuracy and computational efficiency has to be considered already in the modeling. In the following, we will present two approaches to model hemolysis as a macroscale phenomenon. In both cases, the modeled results are given in terms of the index of hemolysis (*IH*), which can be related to hemolysis measurements and gives the percentage of hemolysis. The index of hemolysis is defined by (see Reference [24]) for a derivation)

$$IH = \left(1 - \frac{Hct}{100}\right)\frac{\Delta PHb}{Hb} \cdot 100[\%], \quad (25.44)$$

where *Hct* is the hematocrit in % of the blood sample. The increase in plasma-free hemoglobin ΔPHb and the total hemoglobin content of the blood *Hb* are usually measured in mg/L.

Stress-Based Hemolysis

Motivated from the early findings by Blackshear et al. [25], Giersiepen et al. [26] have been the first to relate shear stress and exposure time to the index of hemolysis by using a simple power law of the form

$$IH = A_{Hb}\sigma_s^{\alpha_{Hb}} t^{\beta_{Hb}}. \quad (25.45)$$

The three parameters in Eq. (25.45) were fitted to the experimental data by Wurzinger et al. [27]. Due to the secondary blood damage effects in Wurzinger's experiments, Giersiepen's parameters are known to overestimate hemolysis. Therefore, there exist several other regressions of the parameters in Eq. (25.45). Those that are frequently used in the literature are listed in Table 25.2. A visualization of the power law with the parameters by Zhang et al. [28] is shown in Fig. 25.6 together with the data points of the experimental measurements.

The power law is a one-dimensional model equation. In order to apply the model to three-dimensional flow problems, the scalar shear stress σ_s is derived from the instantaneous stress tensor $\boldsymbol{\sigma}$ of the Navier-Stokes equations. By using Eq. (25.28), the scalar shear stress becomes

$$\sigma_s = \mu G_f = \mu\sqrt{2\boldsymbol{E} : \boldsymbol{E}}. \quad (25.46)$$

To integrate the power law in time and space, there exist two general approaches: the Eulerian and the Lagrangian approach.

In the Eulerian approach, a convection-diffusion-reaction equation (as introduced in Eq. (25.10)) is used to compute a linearized free plasma hemoglobin ratio f_{Hb} as a field variable in the flow domain. The governing equation with initial and boundary conditions reads

$$\frac{\partial f_{Hb}}{\partial t} + \boldsymbol{u}\cdot\nabla f_{Hb} - \nabla\cdot\left(\nu_{Hb}\nabla f_{Hb}\right) - r_{\Delta_{Hb}} = 0 \text{ in } \Omega, \quad (25.47)$$

$$f_{Hb} = g_{f_{Hb}} \text{ on } \Gamma_g, \quad (25.48)$$

$$\boldsymbol{n}\cdot\left(\nu_{Hb}\nabla f_{Hb}\right) = h_{f_{Hb}} \text{ on } \Gamma_h, \quad (25.49)$$

$$f_{Hb}\big|_{t=0} = f_{Hb}^0, \quad (25.50)$$

where $\nu_{Hb} = 6 \times 10^{-7}\text{cm}^2/\text{s}$ is the self diffusion coefficient for hemoglobin [29]. Following Farinas et al. [30, 31], the reaction or source term can be expressed by

$$r_{\Delta_{Hb}}^a = \left(A_{Hb}\sigma_s^{\alpha_{Hb}}\right)^{1/\beta_{Hb}} \text{(without saturation effect)}, \quad (25.51)$$

or

$$r_{\Delta_{Hb}}^b = \left(A_{Hb}\sigma_s^{\alpha_{Hb}}\right)^{1/\beta_{Hb}}\left(1 - f_{Hb}\right) \text{(with saturation effect)}. \quad (25.52)$$

Thus, the reaction term is the power law (Eq. (25.45)) linearized in time, so that the index of hemolysis can be computed by $IH = f_{Hb}^{\beta_{Hb}}$ or by

$$IH_{out} = \frac{\int_{\Gamma_h}(u\cdot n) f_{Hb}^{\beta_{Hb}}(x)\,d\Gamma}{\int_{\Gamma_h}(u\cdot n)\,d\Gamma}, \quad (25.53)$$

for a velocity-weighted spatial averaging at a Neumann boundary

In the Lagrangian approach, the index of hemolysis is integrated along post processed path lines through the flow domain. Several

Table 25.2 Different parameter regressions for hemolysis modeling with the power law in Eq. (25.45)

A_{Hb}	α_{Hb}	β_{Hb}	Experimental data
3.62×10^{-5}	2.4160	0.7850	Wurzinger et al. [27]
1.8×10^{-6}	1.9910	0.7650	Heuser et al. [72]
1.228×10^{-5}	1.9918	0.6606	Zhang et al. [28]

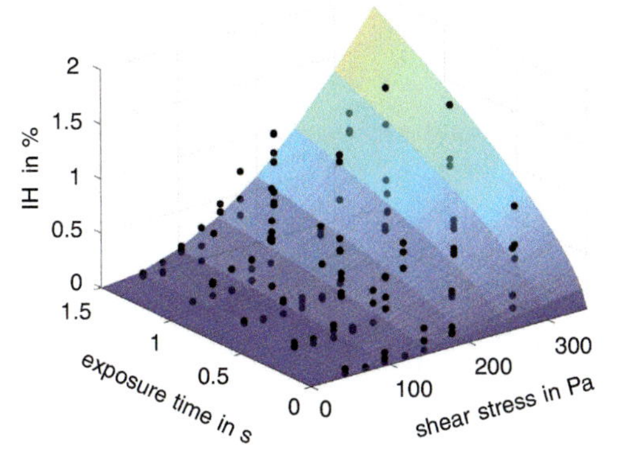

Fig. 25.6 Visualization of the power law with the parameter regression by Zhang et al. [28] (compare Table 25.2) and corresponding data points of the experimental measurements

authors [32–34] write the power law in infinitesimal form, that is,

$$dIH = A_{Hb}\sigma_s^{\alpha_{Hb}} dt^{\beta_{Hb}}, \tag{25.54}$$

and integrate along the path line using a forward Euler scheme. Eq. (25.54) has been criticized by Grigioni et al. [35] and improved for the temporal derivative:

$$dIH = A_{Hb}\beta_{Hb}\sigma_s^{\alpha_{Hb}} t^{\beta_{Hb}-1} dt. \tag{25.55}$$

Later, Grigioni et al. [36] proposed a new formulation that is able to accumulate a mechanical dose along a path line. The new formulation aims to account for the stress history acting on the RBC:

$$dIH = A_{Hb}\beta_{Hb}\left(\int_{t_0}^{t_1}\sigma_s(\varepsilon)^{\alpha_{Hb}/\beta_{Hb}} d\varepsilon\right)^{\beta_{Hb}-1} \sigma_s(t)^{\alpha_{Hb}/\beta_{Hb}} dt. \tag{25.56}$$

However, it should be noted that Goubergrits and Affeld [37] presented a similar idea to include stress history into the path line integration.

From a numerical point of view, the Eulerian approach is beneficial compared to the Lagrangian approach [38]. The Eulerian approach automatically covers the entire flow domain, whereas the Lagrangian approach can cause a biased estimation when the number of tracers is not sufficiently high. Furthermore, some tracers can get "lost" due to stagnation points or vortices in complex flow fields. Due to the nonlinearity of the power law in time, the Lagrangian approach also has the issue that the solution may depend on the choice of the time step size. A benefit compared to the Eulerian approach is, however, the possibility to define damage accumulation along the path line. Nonetheless, the physical relevance of such a stress history is limited as the stress-based hemolysis modeling implicitly assumes an instantaneous deformation of RBCs—a big assumption when considering that an RBC shows very complex viscoelastic deformation under shear flow.

Strain-Based Hemolysis

Due to the limitations of the stress-based hemolysis modeling, several authors worked upon improvements that account for viscoelasticity of the RBC deformation or cell aging. Yeleswarapu et al. [39] presented a scalar damage accumulation that considers cell aging. The used damage function is, however, unknown in complex flow fields. Arwatz and Smits [40] proposed a scalar viscoelastic model as an alternative to the scalar power law; but a connection to multidimensional flows is missing. A viscoelastic threshold model has been proposed by Chen and Sharp [41], which is based on a cell membrane model by Rand [42]. The usage of such a threshold model is also limited, as it is only able to estimate complete rupture of RBCs, but not the occurrence of sublethal damage.

In this section, we will present the strain-based hemolysis model by Arora et al. [43]. They used the model for hemolysis estimations in the GYRO centrifugal blood pump and found good agreement with experimental data [44, 45]. In contrast to Arora, who used a Lagrangian approach, we apply the hemolysis model to the Eulerian approach in this section.

The basic idea of the model is that instead of an instantaneous shear rate G_f (compare Eq. (25.28)) an effective shear rate G_{eff} is computed, based on the viscoelastic deformation (or straining) of an RBC. To compute the RBC deformation on a macroscopic level, it assumed that the RBCs behave similar to droplets in high shear flow. With the aid of a symmetric and positive-definite shape tensor $\boldsymbol{S}$ (also called morphology tensor), Arora et al. formulated the following equation:

$$\begin{aligned}&\frac{\partial \boldsymbol{S}}{\partial t}+\boldsymbol{u}\cdot\nabla \boldsymbol{S}-(\boldsymbol{\Omega S}-\boldsymbol{S\Omega})\\&=-f_1(\boldsymbol{S}-g(\boldsymbol{S})\boldsymbol{I}) && (relaxation)\\&\quad+f_2(\boldsymbol{ES}+\boldsymbol{SE}) && (elongation)\\&+f_3((\boldsymbol{W}-\Omega)\boldsymbol{S}-\boldsymbol{S}(\mathbf{W}-\Omega))\ \text{in}\,\Omega, && (rotation)\end{aligned} \tag{25.57}$$

with boundary and initial conditions

$$\boldsymbol{S}=\boldsymbol{G}_S\ \text{on}\,\Gamma_g, \tag{25.58}$$

$$\boldsymbol{S}\big|_{t=0}=\boldsymbol{S}_0. \tag{25.59}$$

The tensors $\boldsymbol{E}$ and $\boldsymbol{W}$ are the rate-of-strain and vorticity tensors, respectively, as defined in Eqs. (25.20) and (25.21). The tensor $\boldsymbol{\Omega}$ incorporates tank-treading motion, and the function $g(\boldsymbol{S}) = 3III_S/II_S$ ensures that Eq. (25.57) preserves the volume of the droplet, where II_S and III_S are tensor invariants of $\boldsymbol{S}$ as defined in Eqs. (25.23) and (25.24), respectively. The three mechanisms (relaxation, elongation, and rotation) of the droplet model are illustrated in Fig. 25.7. Mechanical properties of an RBC are introduced to Eq. (25.57) by the three parameters f_1, f_2, and f_3, based on experimental findings in the literature [43].

In order to use Eq. (25.57) for hemolysis approximations, the computed shape tensor is coupled to the power law, or more precisely to the reaction term of Eq. (25.47). The distortion of the ellipsoidal droplet can be computed with the following formula:

$$D=\frac{L-B}{L+B}, \tag{25.60}$$

with lengths of the longest and smallest semiaxes of the droplet, L and B (compare Fig. 25.7). Both values can be evaluated by the largest and smallest eigenvalue of tensor **S**, respectively. Having the distortion D an effective shear rate G_{eff} is given by

$$G_{eff}=\frac{2f_1D}{(1-D^2)f_2}. \tag{25.61}$$

In contrast to the instantaneous shear rate G_f, the effective shear rate G_{eff} is able to model the straining that is acting on the RBC itself. By using the definition of the scalar shear stress σ_s, the coupling to Eq. (25.47) is accomplished. Therefore,

$$\sigma_s=\mu G_{eff}=\frac{2\mu f_1 D}{(1-D^2)f_2} \tag{25.62}$$

if the strain-based model is used.

In References [46–48], stress-based and strain-based hemolysis models are compared in benchmark blood pumps. The comparison reveals very significant differences. Stress peaks with short exposure time contribute to the overall hemolysis in the stress-based model, whereas regions with increased shear and long exposure times are responsible for damage in the strain-based model (compare Fig. 25.8).

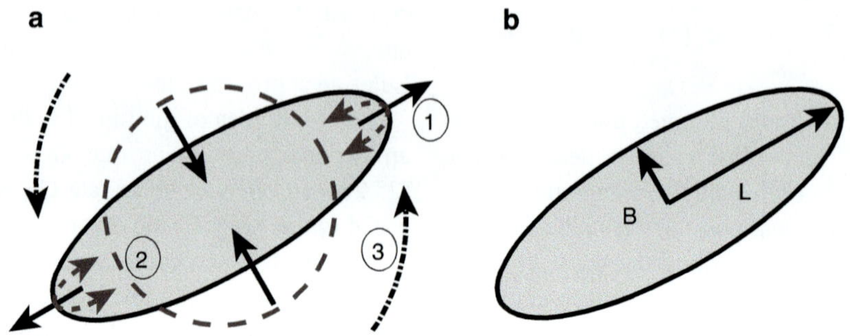

Fig. 25.7 (**a**) A droplet deformation model in analogy to an RBC. The droplet can be elongated due to straining (1). Without tension, the droplet relaxes to a spherical shape (2). Vortices are able to rotate the droplet (3). (**b**) Distortion of a droplet is computed by the longest and smallest semiaxes, L and B

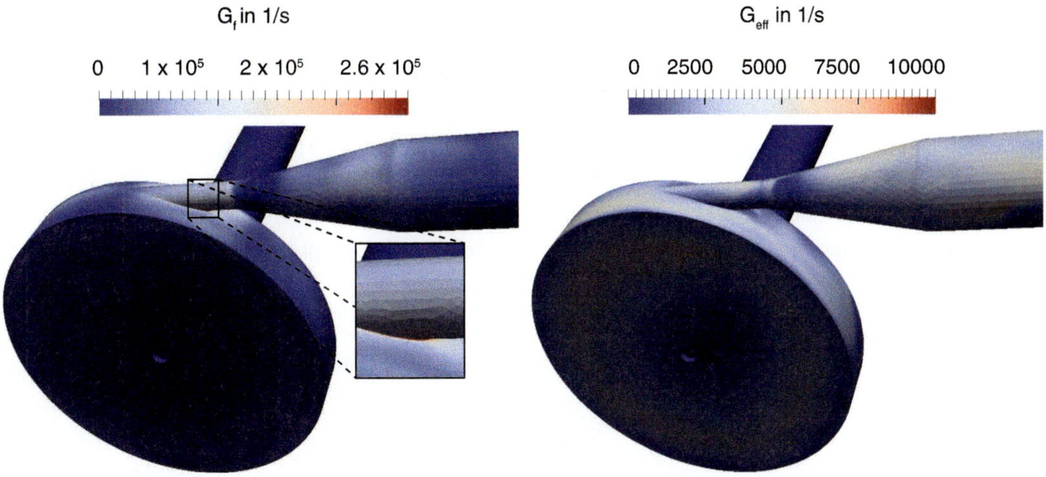

Fig. 25.8 Shear rate distribution at the bottom of a centrifugal blood pump corresponding to a flow rate of 6 L/min, and a rotational speed of 3888 RPM. Comparison of the instantaneous shear rate G_f and the effective shear rate G_{eff}

Thrombosis Estimation

Even though thrombosis estimation is clinically considered more important than hemolysis, successful CFD concepts for VAD design have only been reported in recent years and are still of major concern in the research community. The limited success is mainly based on the fact that in vivo or in vitro thrombosis measurements are extremely challenging and complex.

It is commonly excepted that thrombus formation and deposition are among others driven by a combination of high shear stress and exposure time (similar to hemolysis), but also low shear and stagnation zones. Therefore, several CFD analyses focus on maximizing the washout of the VAD. CFD can be very helpful to find locally stagnant flow and pump volumes with shear stresses below a certain threshold. Lagrangian or preferably Eulerian methods have also been used to analyze the local residence times within VADs [49, 50].

Where full coagulation models or multiscale simulations have shown initial success in simple cardiovascular flow applications, these are still computationally too demanding for full VAD applications. We will briefly present here two examples of successful modeling concepts. The interested reader is also referred to review articles and references therein [51, 52].

Bluestein and coworkers [53] introduced the concept of device thrombogenicity emulation (DTE), which combines in silico numerical studies with in vitro measurements. In the CFD part, a fluid-structure simulation is performed with blood as a Newtonian fluid and platelets as neutrally buoyant solid particles. Using Lagrangian flow trajectories, the stress loading histories experienced by the platelets are calculated by means of a probability density function and evaluated as a thrombogenic footprint for the modeled device. In the experimental part, certain stress histories are emulated in simple blood-shearing devices to identify platelet activation rates. Based on the results, the VAD's design is modified and reanalyzed, leading to an iterative optimization workflow. In Reference [53], the DTE concept is used to design the HeartAssist 5 VAD.

Antaki and coworkers [54, 55] present an Eulerian framework for thrombus deposition. A coupled convection-diffusion-reaction equation system (similar to Eq. (25.10)) is used to incorporate the transport and reactions of ten chemical and biological species involved in the thrombus formation. The processes of thrombus initiation, propagation, and stabilization are represented.

For each species individual reaction rates and boundary conditions are formulated and calibrated with experimental data. For fluid-thrombus interaction, a volume fraction of deposited platelets is computed and coupled to the momentum equation by a resistance force. A list of all equations and parameters is given in Reference [54] including a comparison to two experimental test cases. In Reference [55], the model is also applied to the HeartMate II VAD. Results are compared to the clinical observation of thrombus deposition on the flow straightener of the HeartMate II (compare Fig. 25.9).

Pump Design

In this part, approaches for pump analysis and optimization with CFD are presented. Firstly, methods and exemplary results are illustrated aiming to design hydraulic components for a specific application, explicitly making sure the requirements for pressure head, flow, and rota-

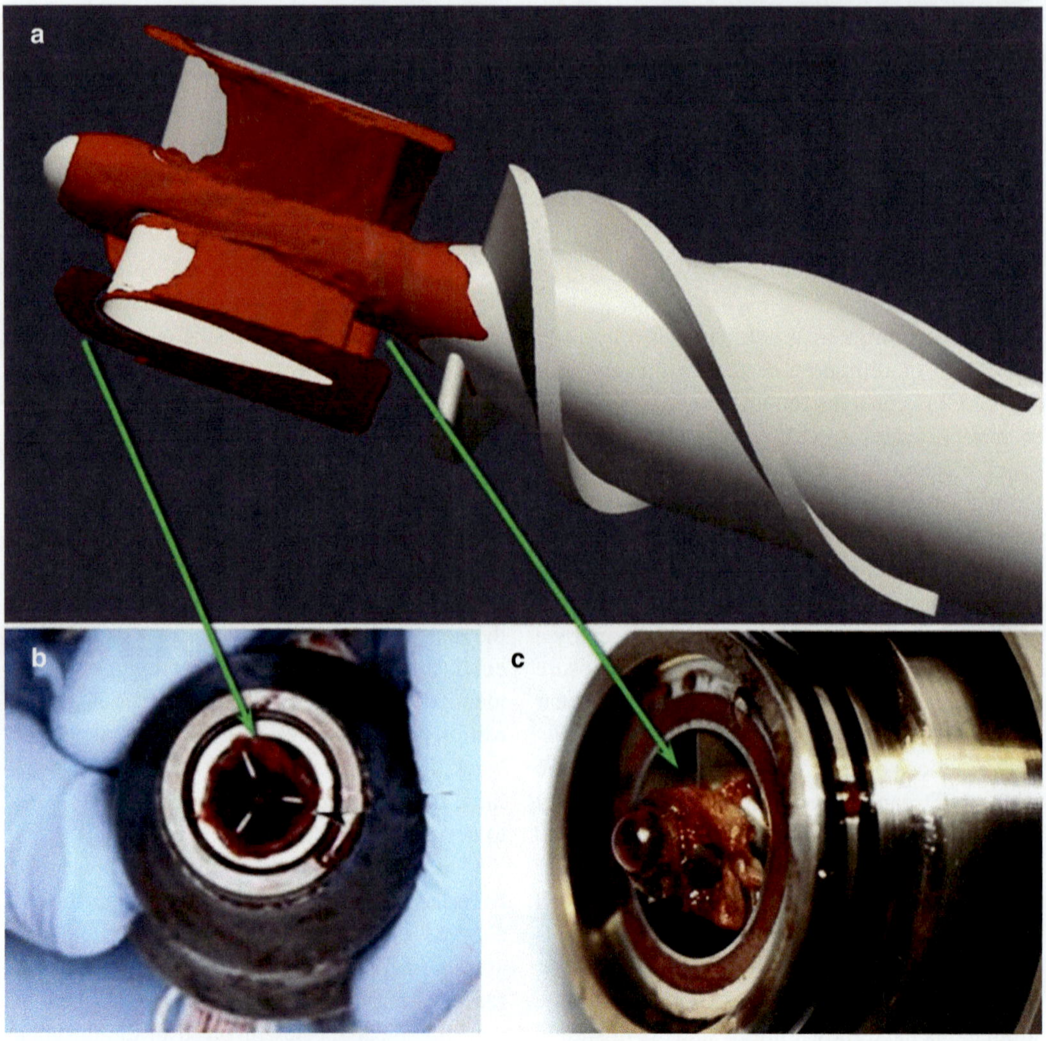

Fig. 25.9 (**a**) Simulated thrombus deposition (in red) after 6.25 hours, corresponding to a flow rate of 4.5 L/min, and a rotational speed of 9000 RPM; (**b**) and (**c**): Clinical observation of thrombus deposition in HeartMate II removed from patients, adapted from [55]. (**b**) Reproduced from Najib, Mohammad Q, Wong, Raymond K. An unusual presentation of left ventricular assist device thrombus. European Heart Journal - Cardiovascular Imaging. 2019:29;(13)6, with permission from Oxford University Press.) (**c**) Reproduced from Mokadam, Nahush A, Andrus, Shauna. Thrombus formation in a HeartMate II. European Journal of Cardio-Thoracic Surgery. 2011;39(3) with permission from Oxford University Press)

tional speed are met within the design limits. The view also covers work on optimization of hydraulic efficiency and rotor torque.

Secondly, methods and results to adapt the characteristic pump curve are shown. The characteristic pump curve determines the hemodynamic interaction with the heart and circulation and, thus, is of elevated interest for rotary pumps that are implanted for mechanical circulatory support.

Lastly, computational fluid dynamics is used to provide input into the rotor-bearing design. This is especially true for devices of the third generation employing a contactless bearing system. This part is expanded to show possibilities to design hydrodynamic bearing structures with CFD.

As early as 2001, Burgreen et al. discussed the use of CFD as a development tool for rotary blood pumps [56]. Figure 25.10 illustrates options in their design approach.

An important aspect of each CFD project is the starting geometry or initial design, for example, of the flow components of a rotary blood pump. This design may be the computer-assisted-design model (CAD-model) of an existing pump or the initial geometry created by experienced pump engineers or, typically, the result of an initial design process using analytical equations. The description of such design procedures is beyond the scope of this chapter. However, the quality of the initial design is crucial for the best approach during a CFD-based optimization project. Weaker designs are best improved by qualitative evaluation, modified with intuition and experience. Geometries can be modified manually. This represents module options "1" in Fig. 25.10. In later stages, when steps for geometry improvements are less evident, quantitative evaluations, formal optimization procedures, and an automatic, software-based geometry modification are indicated (module options "2" in Fig. 25.10).

More recently, especially the design evaluation and geometry modification steps have seen dramatic changes also in the field of CFD-based rotary blood pump design. Fig. 25.11 shows an example with the approach of response-surface-based design space exploration and geometry optimization [57]. These tools are in broad application in industry and available within commercial software packages such as the Ansys package with the flow solvers CFX and Fluent (Ansys, Inc., Canonsburg, USA).

Typically, within a CFD optimization project, constraints exist, for example, due to a desired miniaturization of the pump. This means, geometry parameters can only be allowed to vary within limits. Looking at all geometry parameters subject to optimization, this creates a multidimensional design space or parameter space. At the

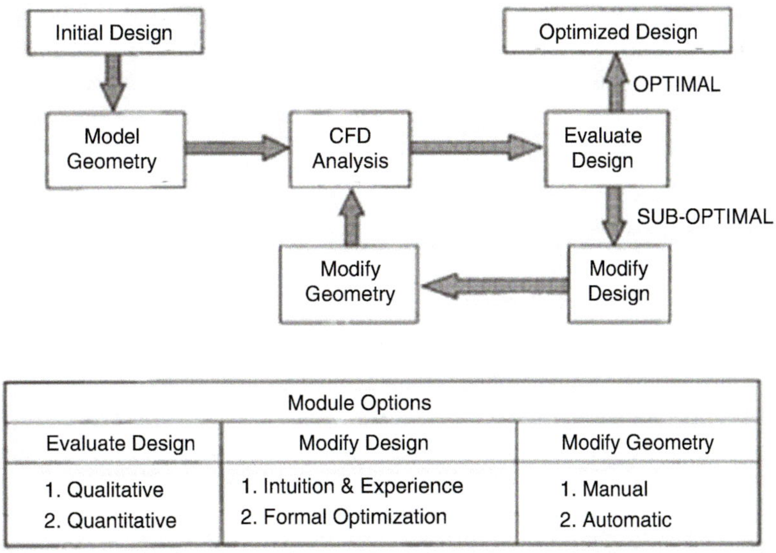

Fig. 25.10 Steps and options in a CFD-based design optimization study (Reprinted and adapted from (Reprinted and adapted from Holmes AJ, Wu ZJ, Antaki JF, et al. Computational fluid dynamics as a development tool for rotary blood Pumps. 2002;25(5), with permission from John Wiley and Sons)

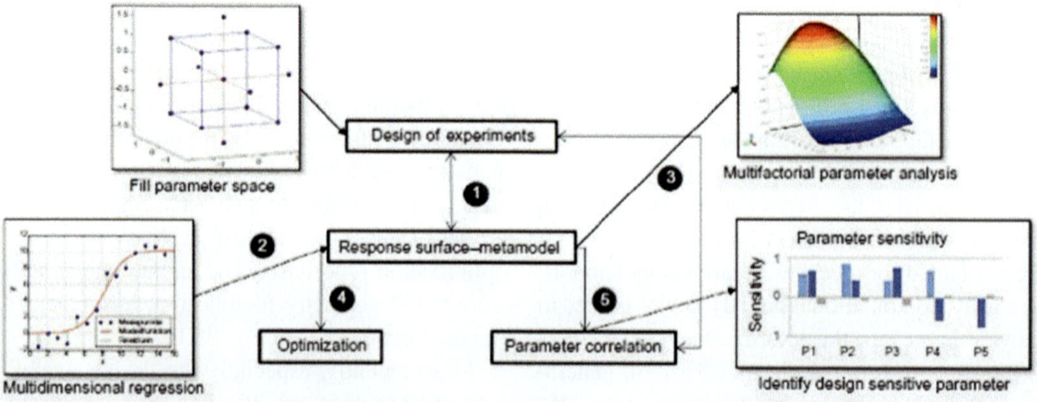

Fig. 25.11 Metamodel-based design exploration and optimization. (Reprinted and adapted from Smith et al. [57] with permission from Elsevier)

beginning, the quantitative effect of a parameter is unknown, and there might be interdependencies between design parameters. Moreover, there might be more than one design goal. This complexity can be dealt with by creating a so-called metamodel that models the effect of design parameters on design goals. If the effects and the interdependence of two design parameters are illustrated on just one design goal parameter, two-dimensional response surface, as illustrated in Fig. 25.11, is created. This metamodel additionally allows for the optimization of the geometry and provides insight into the effect of single design parameters. Hence, the entire process is sometimes referred to as response-surface-based optimization.

In the following, both approaches for CFD-based design optimization, a more manual and intuition driven approach and a metamodel-based approach, are described with an exemplary study.

Performance, Efficiency, and Torque Requirement

The core of each rotary blood pump is the rotor comprising the rotating blades. This is true since this is the component that converts the power provided by the motor from mechanical power into hydraulic power which is of course observable as an increase in flow and pressure generation. Hence, the design of the rotating blades is of crucial importance for numerous design goals such as a high efficiency which in turn is especially important for rotary blood pumps implanted for chronic mechanical circulatory support.

CFD is a traditional development tool for blade design of any turbomachinery, and a classical application to a centrifugal blood pump development project is documented by Steinbrecher [58]. Typically, CFD can be used to model the flow just in one single blade passage.

The figure shows a structured mesh of high quality in the passage between two rotating centrifugal blades and indicates the location of blade inlet angle β_{s1} and blade outlet angle β_{s2}. Both of these design parameters are of very high importance for the later performance of the entire pump. The simplification to investigate just one blade passage allows for fast generation of numerical results. With the aid of specific boundary conditions, the results for an entire rotor can be estimated. Table 25.3 shows an excerpt of the results for the design study conducted in this example. The results are presented for a constant flow rate, and constant rotational speed showing that rotary pump design is in a first step classically done for just one design flow rate and one design rotational speed. The pressure head is defined by

$$H = \frac{\Delta p_t}{\rho \cdot g}, \tag{25.63}$$

and the efficiency is defined by

$$\eta = \frac{\Delta p_t}{T \cdot \omega}. \quad (25.64)$$

The parameters Δp_t, ρ, g, Q, T, and ω denote total pressure difference, density, acceleration due to gravity, volume flow rate, rotor torque, and angular frequency, respectively. As can be seen, the changes of the design parameters blade number, blade angles, and blade thickness affect both design goals, the generation of a specific pressure head H and the efficiency η. In this case, very high hydraulic efficiencies of up to 98.2% were achieved at this particular operational point for the blade passage only. To put these results into perspective, Smith et al. collected hydraulic efficiency data of rotary blood pumps including implantable and extracorporeal devices [59]. The device efficiency of a VAD is a superposition of all component efficiencies like the efficiencies of the motor, electronics, and the hydraulic efficiency itself as denoted in Table 25.3. It was found that best hydraulic efficiencies for most pumps were between 40% and 70% never reaching values >90% which are common for pumps in industrial applications. This shortcoming was attributed to the smaller size and increased friction (viscous) losses [59]. Looking at the results in Table 25.3, it is then the task of the engineering team to make decisions on the choice of these design parameters. Of course, due to the speed of result generation with CFD, more design studies are usually conducted compared to this example.

Table 25.3 Excerpt of quantitative results of an exemplary blade design study at constant flow rate and rotational speed [58]

Number of blades	β_{s1}	β_{s2}	Thickness maximal [mm]	Pressure head H [m]	Efficiency η [%]
6	40.1	49.9	2	1.016	95.5
6	31.2	49.8	2	1.027	95.7
5	40.1	42.1	2	1.025	93.9
7	40.1	42.1	2	1.048	97.3
7	40.1	42.1	1.5	1.064	98.2
6	40.1	42.1	2	1.040	98.1
6	40.1	42.1	1.5	1.053	98.0

Pump Characteristic Curve

A major property of a rotary pump is its characteristic curve, which is often also called *H-Q* curve. It represents the relationship between the pump head H (~ pressure difference) and the pump flow Q at a constant rotational speed (Fig. 25.12). The characteristic curve describes how a specific pump reacts to pressure changes at its inlet and/or its outlet. This capability of changing flow based on pressure changes is generally also known as pressure sensitivity. A flat *H-Q* curve denominates a highly pressure sensitive pump, whereas a steep *H-Q* curve stands for low-pressure sensitivity.

The effect of pressure sensitivity in an LVAD application is not completely understood, but it could have an important influence on patient outcome because, among other aspects, it:

- Provides a means of passive flow control (as discussed below)
- Affects the likelihood of suction events [60]
- Affects the level of arterial pulsatility [60]
- Determines the level of left ventricular unloading and end-diastolic pressure [61]

Figure 25.12 shows the characteristic curves for two different exemplary pumps that are designed differently regarding their flow-guiding structures like the blades. Imagining that both pumps operate at the illustrated point “1,” it is clear that they both provide identical flow and

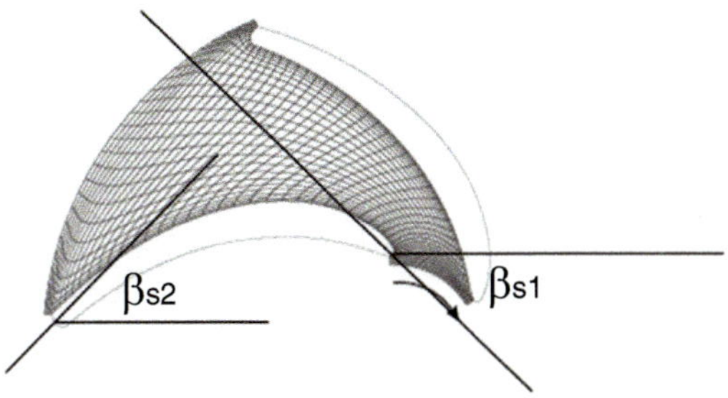

Fig. 25.12 Comparison of the characteristic pump curves of two differently designed rotary blood pumps and illustration of the effect of a pre- and/or afterload change. (Reprinted and adapted from Tenderich et al. [61], with permission from John Wiley and Sons)

pressure generation to the circulation. However, if there is a small change in pump pre-or after-load, in this case called "c" and represented as small decrease either in arterial pressure or small increase in left ventricular pressure, the operational point of the pump with high-pressure sensitivity is able to generate substantially more flow. It then operates in point "2." In case of preload changes, this pump might passively help balancing flow between the pulmonary and systemic circulation in a similar way as the ventricle and analogue to the Frank-Starling mechanism. The pump with low-pressure sensitivity is not able to increase flow to a noticeable extent. It would require an active rotational speed change to move its operational point also to point "2." Such an active physiologic speed control has been proposed by many authors but has not been used as common clinical practice [62, 63].

The characteristic curve shape represents therefore a design goal during the design process of a rotary blood pump used for chronic mechanical circulatory support. In a recent study, the effect of design parameters on the characteristic curve shape was analyzed with CFD [10]. Additionally, among other aspects, the hydraulic efficiency was also considered as a design goal. Therefore, the study represents a multi-objective optimization problem. Figure 25.13 shows the investigated geometry and the design parameters creating the allowable design space. The parameters were chosen based on experience.

Figure 25.13a shows the definition of the blade outlet angle β_2 and a blade wrap factor β_u, which is the blade angle at a relative radius of 0.6. The blade outlet angle affects the theoretical characteristic curve by affecting the blade outlet flow velocity direction. For $\beta_2 = 90°$, the theoretical characteristic curve, also called Euler line, is a horizontal line. In conjunction with the blade wrap factor β_u, the blade wrapping angle θ is defined for a given blade inlet angle. The blade wrap factor β_u is defined and included into this study because it is hypothesized that it affects the amount of fluid friction and therefore the characteristic curve especially at higher flows. Fluid friction is assumed to have a paramount importance in rotary blood pumps because of the highly viscous fluid blood.

Figure 25.13b illustrates the design parameters volute size A_{th}, blade width b_2, and blade tip clearance c_K. Furthermore, it is obvious that the investigated design includes all major pump components like an inlet cannula, a rotor including the blades, a volute, a diffuser leading to the outlet, and also a secondary flow gap which allows for a washout flow back to the inlet region.

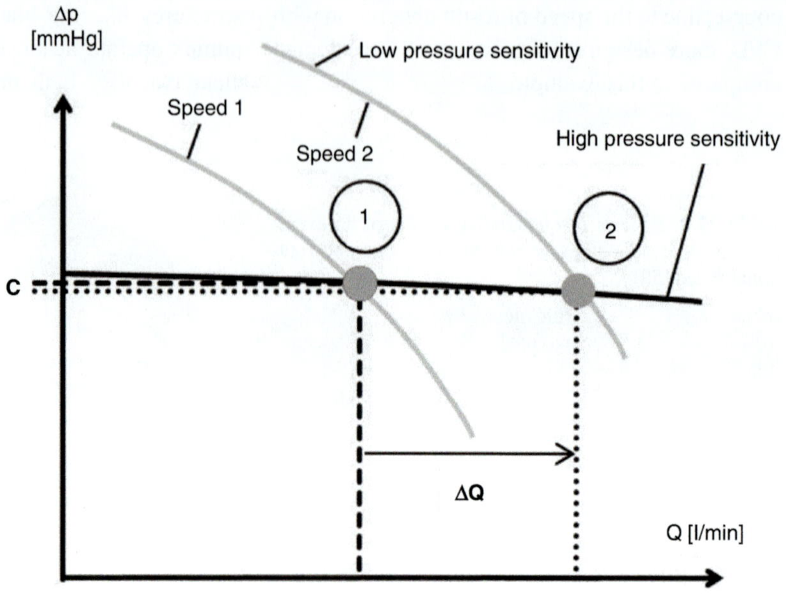

Fig. 25.13 Top view of the investigated centrifugal pump and schematic cross section. (Adapted from Graefe [10] with permission from Shaker Verlag)

Similar to the blade outlet angle, also the blade width affects the theoretical Euler line. The blade width in this study is changed by adapting the outlet blade width b_2, whereas the inlet blade width was kept constant. The blade width was defined as a linear function between the inlet and outlet, and, thus, by changing the outlet blade width, the blade width on the entire radius is affected. By including this parameter into the investigation, an additional supposed effect on friction losses is investigated.

The volute cross-sectional area is included because it was already discovered that it has a significant influence on the location of the best efficiency point of the pump and pressure sensitivity [64, 65]. Gaddum et al. discussed that for an increased volute size, pressure sensitivity increases for a centrifugal flow rotary blood pump under development. The volute size factor A_{th} affects the volute cross-sectional area everywhere around the circumference of the volute. Cross-sectional values that are given in [mm^2] relate to the throat area which is the largest cross-sectional area of the volute at the transition to the diffuser. It is hypothesized that through the volute cross-sectional area, friction and recirculation losses are affected.

Lastly, the blade tip clearance c_K is included in this study. It is known that for a semi-open impeller, which is under investigation here, that the leakage flow over the blade tips is a function of the clearance. This leakage flow disturbs the main flow through the blade passages and therefore leads to undesired turbulences and losses. It is of interest whether this parameter also affects the pressure sensitivity.

In the field of design of experiments and optimization, design parameters are called factors. In order to determine nonlinear effects of individual factors (independent variables) on target variables (dependent variables) and to save resources, a central composite design of experiment on five levels was chosen (Fig. 25.14).

In Fig. 25.14a, it can be seen that around a central data point in the exemplary three-dimensional design space created by factors A, B, and C, there are in total five levels chosen for each factor. It is important to note that the design of the experiment is rotatable and not biased in any direction. Therefore, all data points are eligible to be used to analyze all design factors. Based on best practice procedures, the experiment was reduced to a fractional factorial design of experiment with five factors on five levels. This means that in contrast to full factorial design of experiment like in Figure 25.14a, not all possible data points (factor combinations) are investigated, but a high number is omitted [66].

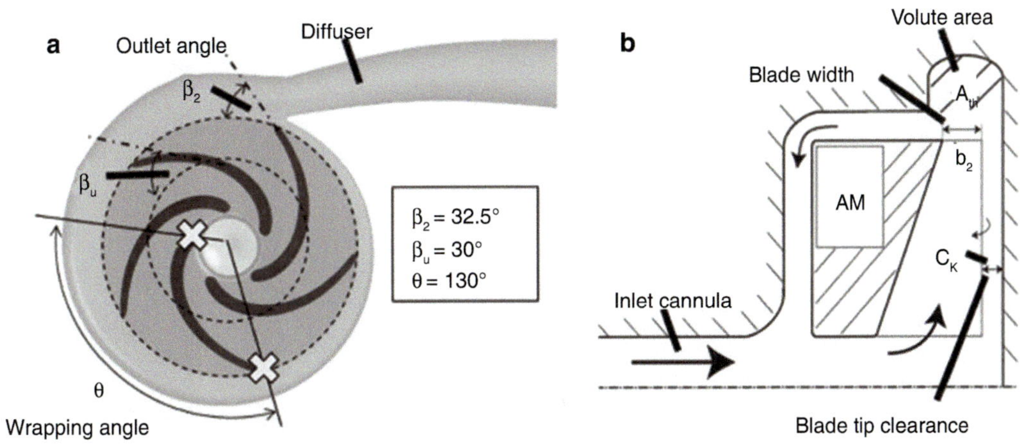

Fig. 25.14 Illustration of a central composite design of experiment with three factors (**a**, **b**, and **c**) and an exemplary response surface formed by data regression based on two factors (Adapted from Graefe [10] with permission from Shaker Verlag)

This led to 27 (instead of 43) factor combinations. Figure 25.14b illustrates an exemplary response surface based on such a design of experiment. The response surface is computed based on the remaining data points and depends on the chosen regression model. In this study, a polynomial regression model or metamodel of second order is used. The quality of the regression is analyzed in [10].

Figure 25.15 depicts the local effects of the design factors affecting the blade geometry for the target variables of pressure sensitivity and efficiency at the design flow at 5 L/min. Individual data points are not shown in order to not overload the graphs. The pressure sensitivity is quantified by averaging the characteristic curve gradient between flow rates of 1 to 8 L/min formed by results in 1 L/min steps. The upper limit is chosen based on the need for a restriction of this study because of limited resources. Local effects denote that when displaying an effect of a factor, the remaining factors are adjusted on their medium level within the design range (e.g., $\beta_2 = 57.5°$). Factor interactions are not considered by displaying a single set of response functions. Because of the low resolution of the design space at the boundaries, these regions are marked "low resolution" for factor values at the thresholds of 20% and 80%.

The response functions for the target variable pressure sensitivity (Fig. 25.15 left column) show a similar behavior for all three design factors. For most of the design space, the pressure sensitivity increases for an increase in blade width (a), blade wrap factor (c), and blade outlet angle (e). The magnitude of the maximal effect on the other hand is different. The biggest increase can be achieved for an increase of blade width from its minimum to almost its maximum value. Looking only at the highly resolved range, the maximal increase of the pressure sensitivity as a function of the blade wrap factor and the blade outlet angle is 27% and 46% based on the effect of the blade width, respectively.

Figure 25.15 (right column) shows the local effect with respect to the efficiency at the design flow at 5 L/min. It increases for most of the design space for an increase of the factors blade width (b) and blade outlet angle (f). There is a local maximum of the efficiency for $\beta_u \approx 55°$ (d).

Figure 25.16 displays the local response functions for the volute size and the blade tip clearance.

Both target variables pressure sensitivity and efficiency increase with an increase of the volute size. A change in blade tip clearance does not show an effect on pressure sensitivity. But for an increase in clearance, the efficiency drops significantly.

Since the response functions given above only show a local effect, the results of an optimization tool such as a genetic global optimization algorithm can be used to further investigate the design

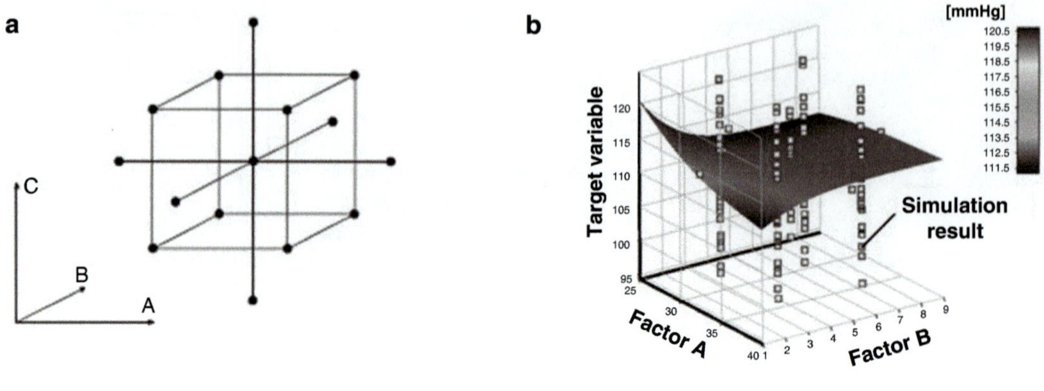

Fig. 25.15 Response functions for the factors affecting the blade geometry (blade width, blade wrap factor, and blade outlet angle) (Adapted from Graefe [10] with permission from Shaker Verlag)

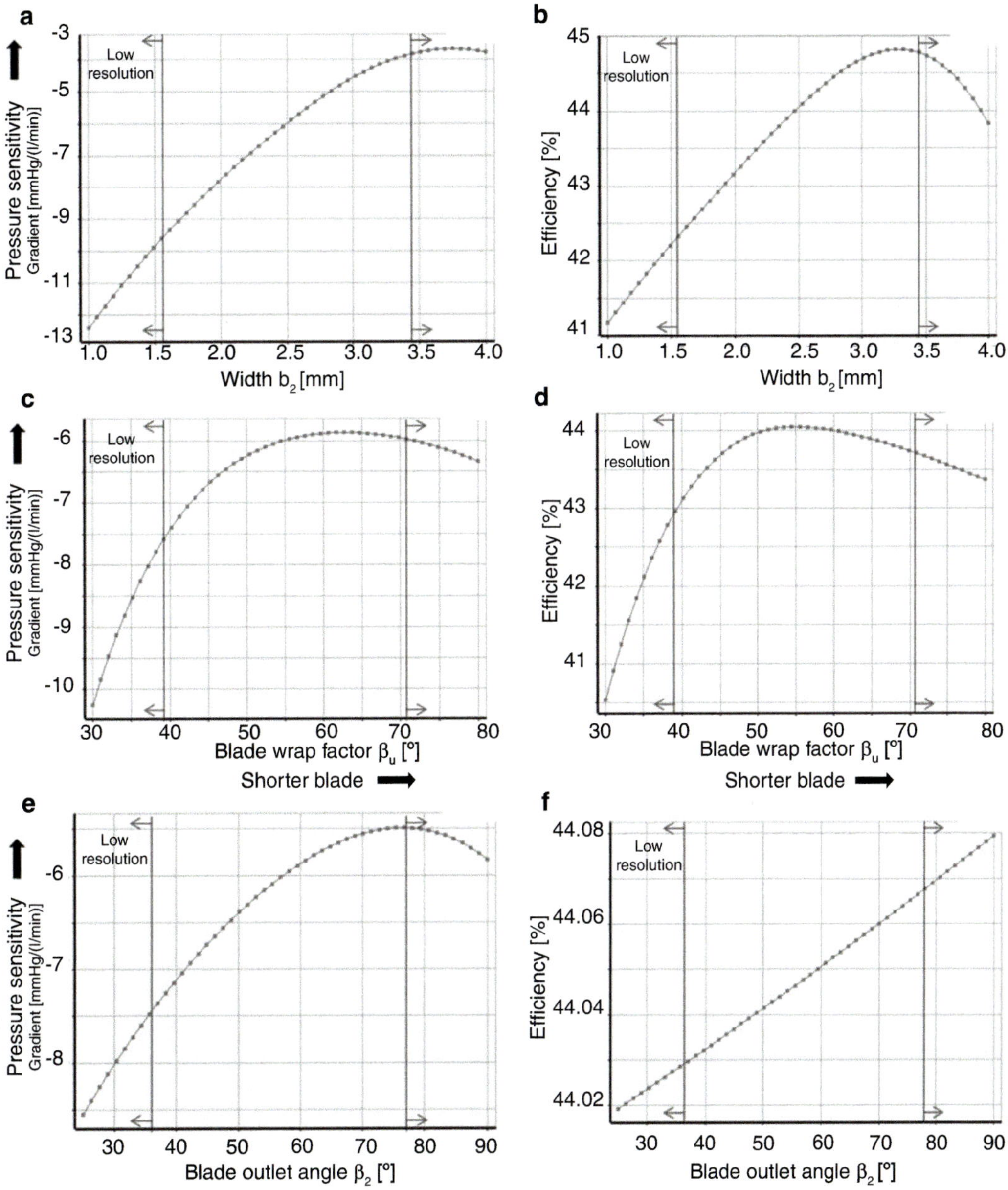

Fig. 25.16 Response functions for the volute size and the blade tip clearance. (Adapted from Graefe [10] with permission from Shaker Verlag)

space. In general, it is very challenging to determine an optimal configuration in a five-dimensional design space without numerical assistance, since factor interactions have to be considered. Since this study aims at finding a favorable compromise between the efficiency and pressure sensitivity, that means two main target variables, a genetic optimization algorithm is employed that is designed to find a global optimum and that can find a favorable trade-off between multiple objectives. The used algorithm is based on the NSGA II algorithm [67, 68]. All necessary evaluations are conducted by computing results based on the metamodel.

Rotor Bearing and Hydrodynamic Bearings

As a third field of application, using CFD as a tool for rotor-bearing design has to be mentioned. For all rotary blood pumps, either extracorporeal or intracorporeal, second or third generation, it is important to know the hydraulic load on the rotor. The load can be an axial or radial force or a tilting moment. For second-generation devices using contact bearings, knowing the load is essential to be able to limit wear of the contacting parts and to ensure failure-free operation. For third generation devices, the hydraulic load must be compensated by bearing means such as passive or active magnetic bearing components or hydrodynamic bearings.

Song et al. conducted a representative study on the analysis of hydraulic loads. The focus of their attention was the development of an axial pump with active and passive magnetic bearing that they called "LEV-VAD" [69]. From a hydraulic point of view, the LEV-VAD was a typical axial flow pump with an inlet flow straightener, one single impeller section and an outlet diffuser. They used commercial CFD software with the solver TASCflow and made use of the implemented special software tools for turbomachinery applications provided by Ansys, Inc. (Canonsburg, USA). Figure 25.17 shows that they investigated the potential operational window of the device in great detail by plotting the instantaneous head-flow point during the time of a heartbeat.

The figure shows that during one single heartbeat, here assumed to be 0.8 s long, the instantaneous operating point of the LVAD changes dramatically. The authors assumed a time-varying boundary condition at the inlet of the pump mimicking the natural change in pressure in the left ventricle during one heartbeat. Such a transient simulation requires higher computational efforts due to the long simulation time in comparison to an investigation of steady-state flows.

Figure 25.17 illustrates that the relation of pressure rise and flow forms a hysteresis that runs counterclockwise. This hysteresis is caused by the fluid mass inertia in the pump, while the gradient of the hysteresis is determined by the gradient of the steady-state characteristic curve of the pump itself. The pressure rise ranges from about 80 to 180 mmHg and from about 3 to 10 L/min instantaneous flow rate illustrating the substantial challenges to design rotary blood pumps for mechanical circulatory support as an LVAD never runs just in one operational point.

Figure 25.18 represents the axial force over time on the rotor for the same transient simulation.

The figure shows that also the axial force on the rotor changes substantially during one single heartbeat. The force ranges from about 3.5 N to about 5.5 N. This is important information for the design process of any bearing technology such as a magnetic bearing. Additionally, Fig. 25.18 illustrates a high-frequency vibration of the axial force that is superimposed. This vibration is caused by the fact that different angular positions around the circumference cause different pressure distributions and consequently time-varying axial forces. With CFD simulations employing sliding mesh interfaces between stationary and rotating volumes and using a high temporal resolution, such important information for the bearing design can be determined.

As mentioned above, once hydraulic or other loads on the rotor are known, bearing components to balance these loads have to be implemented. One approach to completely levitate a rotor is to use hydrodynamic bearings. This approach has successfully proven its applicability with the clinically used examples of the VentrAssist (Ventracor, Australia) and the HVAD (Medtronic, USA). Amaral et al. conducted a study on the design of so-called partially grooved spiral groove bearing (SGB) as a variant of a hydrodynamic bearing and its implementation into the bearing design of a third-generation rotary blood pump [70]. Figure 25.19 shows a schematic of the geometry and the numerical mesh used for a CFD study of the isolated SGB geometry.

In this study, the pump itself is omitted to reduce complexity and enable faster result generation. As a consequence, only the hydrodynamic bearing structure is represented on one plane that

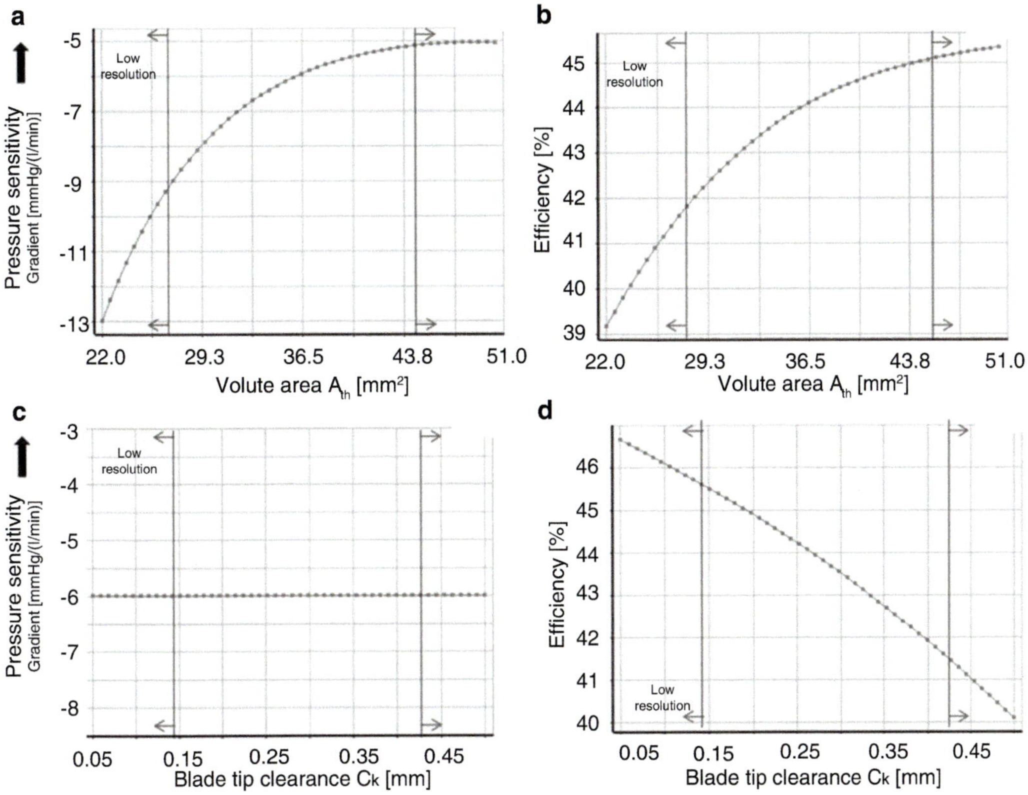

Fig. 25.17 Hysteresis of the instantaneous operational point of an axial rotary blood pump during a heartbeat. (Reprinted and adapted from Song et al. [69], with permission from Wolters Kluwer Health, Inc.)

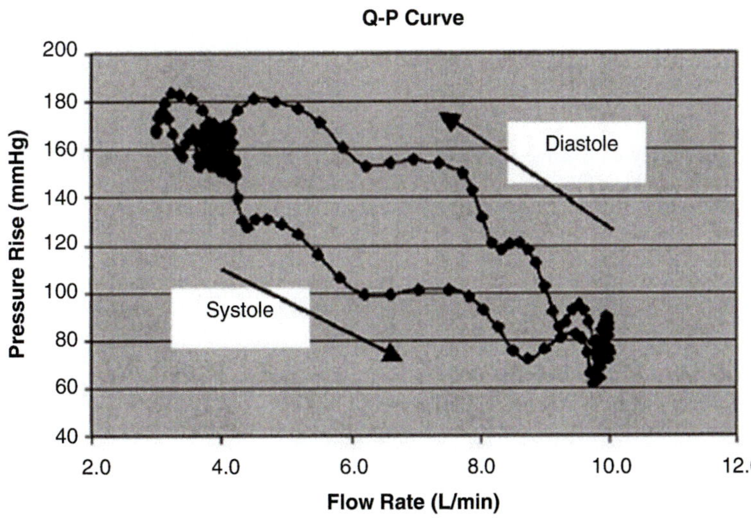

Fig. 25.18 Hydraulic axial force on the rotor of an axial rotary blood pump during a heartbeat. (Reprinted and adapted from Song et al. [69], with permission from Wolters Kluwer Health, Inc.)

is then rotated relative to a second stationary plane. The minimal fluid film thickness between the rotor comprising the hydrodynamic bearing structures and the stationary plane representing the pump casing is called h_2. Additionally, Amaral et al. conducted force measurements as a validation step of the numerical predictions. Figure 25.20 shows the results for the axial force and the wash-

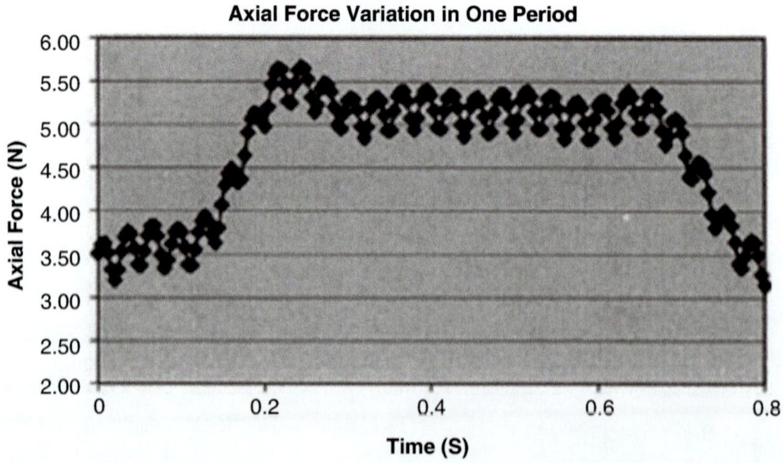

Fig. 25.19 Schematic description of the geometry of a spiral groove hydrodynamic bearing (left) and numerical mesh (right). (Reprinted and adapted from Schmitz-Rode et al. [70], with permission from John Wiley & Sons)

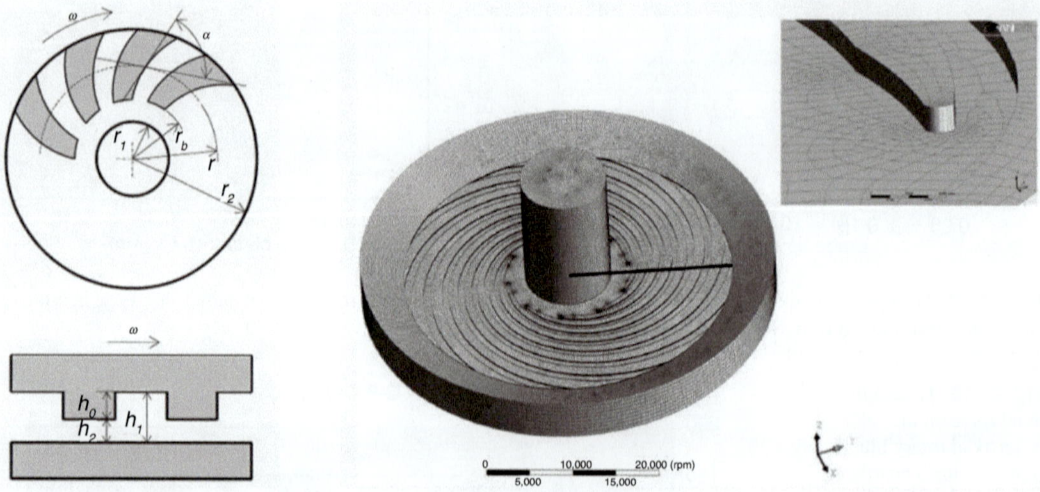

Fig. 25.20 Hydrodynamic axial force created by a spiral groove bearing as a function of fluid film thickness (**a**) and washout of the bearing (**b**). (Reprinted and adapted from Schmitz-Rode et al. [70], with permission from John Wiley & Sons)

out through the so-called secondary flow path that is formed between the SGB and opposing casing wall. Moreover, the results are compared to theoretical predications that are based on Muijderman [71]. The results for the axial force (Fig. 25.20a) show that the SGB is able to generate a substantial axial force, if the film thickness is small. With the help of the CFD results, the initial theoretical approaches could be improved by adding an inertia term and thus rendering the analytical design capabilities of SGBs more effective. Looking at washout (Fig. 25.20b), the CFD results correlated well with the experimental measurements. Through this validation step, the CFD model could then be used to optimize the geometry also of such a delicate geometrical structure as the geometry of a hydrodynamic bearing in a rotary blood pump (Fig. 25.21).

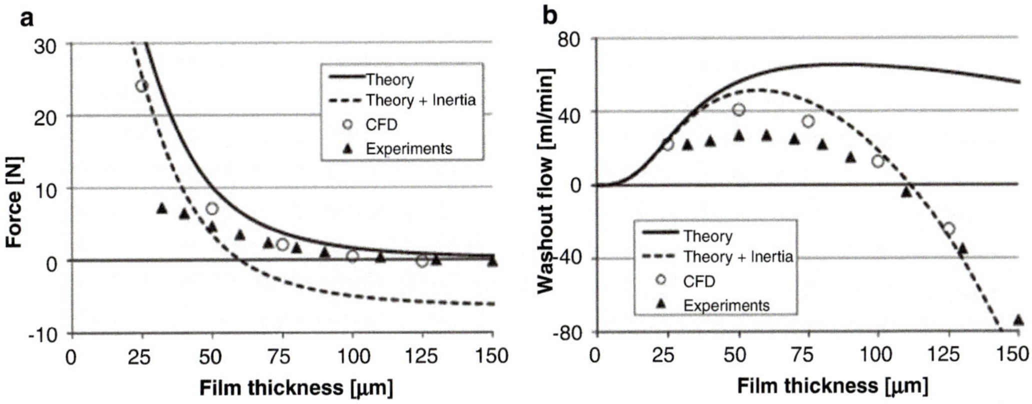

Fig. 25.21 Hydrodynamic axial force created by a spiral groove bearing as a function of fluid film thickness (a) and washout of the bearing (b), reprinted with permission and adapted from [70]

References

1. Ferziger JH, Perić M. Computational methods for fluid dynamics, vol. Bd. 3. Berlin: Springer; 2002.
2. Pope SB. Turbulent Flows. 8th ed: Cambridge University Press; 2011.
3. Celik IB, Ghia U, Roache PJ, Freitas CJ, Coleman H, Raad PE. Procedure for estimation and reporting of uncertainty due to discretization in CFD applications. Journal of Fluids Engineering, Transactions of the ASME. 2008;130:1–4.
4. Eça L, Hoekstra M. A procedure for the estimation of the numerical uncertainty of CFD calculations based on grid refinement studies. Journal of Computational Physics. 2014;262:104–30.
5. Oden JT, Moser R, Ghattas O. Computer predictions with quantified uncertainty, part I. SIAM News. 2010;43:1–3.
6. Oden JT, Moser R, Ghattas O. Computer predictions with quantified uncertainty, part II. SIAM News. 2010;43:1–4.
7. Triep M, Brücker C, Schröder W, Siess T. Computational fluid dynamics and digital particle image velocimetry study of the flow through an optimized micro-axial blood pump. Artif Organs. 2006;30:384–91.
8. Marseille O. Entwicklungs- und Bewertungsverfahren für Rotationsblutpumpen. 2001.
9. Spurk JH. Dimensions analyse: Springer; 1992.
10. Graefe R. Auslegung von Zentrifugalpumpen zur effizienten, drucksensitiven und atraumatischen Herzunterstützung. Band 36, U. D. K. R. u. U. D. m. D. T. S. Univ.-Prof. Dr.-Ing. Dr. med. Steffen Leonhardt, Hrsg., Shaker Verlag. 2016.
11. Holzapfel GA. Nonlinear solid mechanics: a continuum approach for engineering: John Wiley & Sons; 2000.
12. Yilmaz F, Gundogdu MY. A critical review on blood flow in large arteries; relevance to blood rheology, viscosity models, and physiologic conditions. Korea-Australia Rheol J. 2008;20:197–211.
13. Robertson A, Sequeira A, Owens RG. Rheological models for blood. In: Formaggia L, Quarteroni A, Veneziani A, editors. Cardiovascular mathematics: volume 1. Milan, Italy: Springer; 2009.
14. Chien S. Shear dependence of effective cell volume as a determinant of blood viscosity. Science. 1970;168:977–9.
15. Macosko CW. Rheology: principles, measurements, and applications. New York, NY, USA: Wiley-VCH; 1994.
16. Cross MM. Rheology of non-Newtonian fluids: a new flow equation for pseudoplastic systems. J Colloid Sci. 1965;20:417–37.
17. Yasuda K. A multi-mode viscosity model and its applicability to non-newtonian fluids. J Textile Eng. 2006;52:171–3.
18. Freund JB. Numerical simulations of flowing blood cells. Ann Rev Fluid Mec. 2014;46:67–95.
19. Lanotte L, Mauer J, Mendez S, Fedosov D, Fromental J-M, Claveria V, et al. Red cells' dynamic morphologies govern blood shear thinning under microcirculatory flow conditions. Proceedings of the National Academy of Sci United States Am. 2016;13:13289–94.
20. Owens RG. A new microstructure-based constitutive model for human blood. J Non-Newtonian Fluid Mec, 140. 2006:57–70.
21. Reynolds O. On the dynamical theory of incompressible viscous fluids and the determination of the criterion. Papers Mec Phys Sub. 1901;2:535–77.
22. Menter FR. Two-equation eddy-viscosity turbulence models for engineering applications. AIAA J. 1994;32:1598–605.
23. Pauli L, Behr M. On Stabilized Space-Time FEM for Anisotropic Meshes: incompressible Navier-

Stokes equations and applications to blood flow in medical devices. Inter J Numer Methods Fluids. 2017;85:189–209.
24. Paul R, Apel J, Klaus S, Schugner F, Schwindke P, Reul H. Shear stress related blood damage in laminar Couette flow. Artif Organs. 2003;27:517–29.
25. Blackshear PL, Dorman FD, Steinbach JH. Some mechanical effects that influence hemolysis. ASAIO J. 1965;11:112–7.
26. Giersiepen M, Wurzinger LJ, Opitz R, Reul H. Estimation of shear stress-related blood damage in heart valve prostheses - in vitro comparison of 25 aortic valves. Inter J Artif Organs. 1990;13:300–6.
27. Wurzinger LJ, Opitz R, Eckstein H. Mechanical blood trauma: an overview. Angeiologie. 1986;38:81–97.
28. Zhang T, Taskin ME, Fang HB, Pampori A, Jarvik R, Griffith BP, et al. Study of flow-induced hemolysis using novel couette-type blood-shearing devices. Artif Organs. 2011;35:1180–6.
29. Riveros-Moreno V, Wittenberg JB. The self-diffusion coefficients of myoglobin and hemoglobin in concentrated solutions. J Biolog Chemis. 1972;247:895–901.
30. Farinas MI, Garon A. Fast three-dimensional numerical hemolysis approximation. Artif Organs. 2004;28:1016–25.
31. Farinas MI, Garon A, Lacasse D, N'dri D. Asymptotically consistent numerical approximation of hemolysis. J Biomed Eng. 2006;128:688–96.
32. Chan WK, Wong YW, Ding Y, Chua LP, Yu SCM. Numerical investigation of the effect of blade geometry on blood trauma in a centrifugal blood pump. Artif Organs. 2002;26:785–93.
33. Song X, Throckmorton AL, Wood HG, Antaki JF, Olsen DB. Computational fluid dynamics prediction of blood damage in a centrifugal pump. Artif Organs. 2003;27:938–41.
34. Throckmorton AL, Lim DS, McCulloch MA, Jiang W, Song X, Allaire PE, et al. Computational design and experimental performance testing of an axial-flow pediatric ventricular assist device. ASAIO J. 2005;51:629.
35. Grigioni M, Daniele C, Morbiducci U, D'Avenio G, Benedetto GD, Barbaro V. The power-law mathematical model for blood damage prediction: analytical developments and physical inconsistencies. Artif Organs. 2004;28:467–75.
36. Grigioni M, Morbiducci U, D'Avenio G, Benedetto GD, Gaudio CD. A novel formulation for blood trauma prediction by a modified power-law mathematical model. Biomec Modeling Mechanobiol. 2005;4:249–60.
37. Goubergrits L, Affeld K. Numerical estimation of blood damage in artificial organs. Artif Organs. 2004;28:499–507.
38. Hariharan P, D'Souza G, Horner M, Malinauskas RA, Myers MR. Verification benchmarks to assess the implementation of computational fluid dynamics based hemolysis prediction models. J Biomechanical Eng. 2015;137:094501-1–094501-10.
39. Yeleswarapu KK, Antaki JF, Kameneva MV, Rajagopal KR. A mathematical model for shear-induced hemolysis. Artif Organs. 1995;19:576–82.
40. Arwatz G, Smits AJ. A viscoelastic model of shear-induced hemolysis in laminar flow. Biorheology. 2013;50:45–55.
41. Chen Y, Sharp MK. A Strain-Based Flow-Induced Hemolysis Prediction Model Calibrated by In Vitro Erythrocyte Deformation Measurements. Artif Organs. 2011;35:145–56.
42. Rand RP. Mechanical properties of the red cell membrane: II. Viscoelastic breakdown of the membrane. Biophysical J. 1964;4:303–16.
43. Arora D, Behr M, Pasquali M. A tensor-based measure for estimating blood damage. Artif Organs. 2004;28:1002–15.
44. Arora D, Hussain F, Behr M, Pasquali M, Yuri K, Motomura T, et al. Predictions and measurements of mechanical hemolysis in an implantable centrifugal blood pump. ASAIO J. 2005;51:5A.
45. Arora D, Behr M, Pasquali M. Hemolysis estimation in a centrifugal blood pump using a tensor--based measure. Artif Organs. 2006;30:539–47.
46. Pauli L, Nam J, Pasquali M, Behr M. Transient Stress-Based and Strain-Based Hemolysis Estimation in a Simplified Blood Pump. Inter J Numer Methods Biomed Eng. 2013;29:1148–60.
47. Gesenhues L, Pauli L, Behr M. Strain-Based Blood Damage Estimation for Computational Design of Ventricular Assist Devices. Inter J Artif Organs. 2016;39:166–70.
48. Pauli L, Behr M. On the significance of exposure time in computational blood damage estimation. High-Performance Scientific Computing. 2017.
49. Esmaily-Moghadam M, Hsia T-Y, Marsden AL. A non-discrete method for computation of residence time in fluid mechanics simulations. Physics of fluids. 2013;25:110802.
50. Thamsen B, Blümel B, Schaller J, Paschereit CO, Affeld K, Goubergrits L, et al. Numerical analysis of blood damage potential of the heartmate ii and heartware hvad rotary blood pumps. Artif Organs. 2015;39:651–9.
51. Fraser KH, Taskin ME, Griffith BP, Wu ZJ. The use of computational fluid dynamics in the development of ventricular assist devices. Medical Engineering & Physics. 2011;33:263–80.
52. Marsden AL, Bazilevs Y, Long CC, Behr M. Recent Advances in Computational Methodology for Simulation of Mechanical Circulatory Assist Devices. Wiley Interdisciplinary Reviews: Systems Biology and Medicine. 2014;6:169–88.
53. Girdhar G, Xenos M, Alemu Y, Chiu W-C, Lynch BE, Jesty J, et al. Device Thrombogenicity Emulation: A Novel Method for Optimizing Mechanical Circulatory Support Device Thromboresistance. PLoS ONE. 2012;7:–e32463.
54. Wu W-T, Jamiolkowski MA, Wagner WR, Aubry N, Massoudi M, Antaki JF. Multi-Constituent Simulation of Thrombus Deposition. ArXiv. 2016:1–17.

55. Wu W-T, Yang F, Wu J, Aubry N, Massoudi M, Antaki JF. High fidelity computational simulation of thrombus formation in Thoratec HeartMate II continuous flow ventricular assist device. Nature/Scientific Reports 6. 2016;38025.
56. Burgreen GW, Antaki JF, Wu ZJ, Holmes AJ. Computational fluid dynamics as a development tool for rotary blood pumps. Artif Organ. 2001;25:336–40.
57. Smith PA, Wang Y, Groß-Hardt S, Graefe R. Chapter 10 - Hydraulic design. In: Gregory SD, Stevens MC, Fraser JF, editors. Mechanical circulatory and respiratory support: Hrsg., Academic Press; 2018. p. 301–34.
58. Steinbrecher C. Numerische Simulation eines berührungsfrei gelagerten Rotors für eine Blutpumpe. 2004. http://mediatum.ub.tum.de/node?id=601924.
59. Smith WA, Allaire P, Antaki J, Butler KC, Kerkhoffs W, Kink T, et al. Collected nondimensional performance of rotary dynamic blood pumps. ASAIO J. 2004;50:25–32.
60. Moazami N, Fukamachi K, Kobayashi M, Smedira NG, Hoercher KJ, Massiello A, et al. Axial and centrifugal continuous-flow rotary pumps: a translation from pump mechanics to clinical practice. J Heart Lung Transplant. 2013;32:1–11.
61. Graefe R, Beyel C, Henseler A, Körfer R, Steinseifer U, Tenderich G. The Effect of LVAD Pressure Sensitivity on the Assisted Circulation Under Consideration of a Mitral Insufficiency: An In Vitro Study. Artificial Organs. 2018;42:E304–14.
62. Arndt A, Nüsser P, Lampe B. Fully Autonomous Preload - Sensitive Control of Implantable Rotary Blood Pumps. Artificial organs. 2010;34:726–35.
63. AlOmari A-HH, Savkin AV, Stevens M, Mason DG, Timms DL, Salamonsen RF, et al. Developments in control systems for rotary left ventricular assist devices for heart failure patients: a review. Physiol Meas. 2013;34:R1–27.
64. Gülich JF. Kreiselpumpen, vol. 2010. Berlin Heidelberg: Springer; 2010.
65. Gaddum NR, Fraser JF, Timms DL. Increasing the transmitted flow pulse in a rotary left ventricular assist device. Artif Organs. 2012;36:859–67.
66. Kleppmann W. Taschenbuch Versuchsplanung - Produkte und Prozesse optimieren. KG: Carl Hanser Verlag GmbH & Co; 2011.
67. Deb K, Pratap A, Agarwal S, Meyarivan T. A fast and elitist multiobjective genetic algorithm: NSGA-II. Evolutionary Computation, IEEE Transactions on. 2002;6:182–97.
68. Ansys. Help System, Design Exploration User Guid. 2013.
69. Song X, Untaroiu A, Wood HG, Allaire PE, Throckmorton AL, Day SW, et al. Design and transient computational fluid dynamics study of a continuous axial flow ventricular assist device. ASAIO J. 2004;50:215–24.
70. Amaral F, Groß-Hardt S, Timms D, Egger C, Steinseifer U, Schmitz-Rode T. The spiral groove bearing as a mechanism for enhancing the secondary flow in a centrifugal rotary blood pump. Artif Organs. 2013;5
71. Muijderman, Spiral groove bearings. 1964.
72. Heuser G, Opitz R. A Couette viscometer for short time shearing of blood. Biorheology. 1980;17:17–24.

26 Hemodynamic Modelling and Simulations for Mechanical Circulatory Support

Libera Fresiello and Krzysztof Zieliński

Abbreviations

BVAD/LVAD/RVAD	Biventricular/left ventricular/right ventricular assist device
CFD	Computational fluid dynamics
ECMO	Extracorporeal membrane oxygenator
HCS	Hybrid cardiovascular simulator
HI	Hybrid interface
IABP	Intra-aortic balloon pump
LPM	Lumped parameters model
MCL	Mock circulatory loop
MCS	Mechanical circulatory support
Pap	Pulmonary arterial pressure
Pas	Arterial systemic pressure
Pla	Left atrial pressure
Plv	Left ventricular pressure
Pra	Right atrial pressure
TAH	Total artificial heart

L. Fresiello, MSc, PhD (✉)
Department of Cardiovascular Sciences, Cardiac Surgery, Katholieke Universiteit Leuven, Leuven, Belgium

Institute of Clinical Physiology, National Research Council, Pisa, Italy
e-mail: libera.fresiello@gmail.com

K. Zieliński, MSc, PhD
Nalecz Institute of Biocybernetics and Biomedical Engineering, Polish Academy of Sciences, Warsaw, Poland
e-mail: kzielinski@ibib.waw.pl

Models for Mechanical Circulatory Support

Circulatory models provide a valuable platform for testing mechanical circulatory support (MCS) devices interaction with the cardiovascular or (to a larger extent) the cardiorespiratory system. They can serve the purpose of evaluating the design and of assessing the performance of a MCS device in different hemodynamic conditions. Their level of reliability and reproducibility make them a valuable support and in some cases an alternative to animal experiments. It is evident that circulatory models do not replicate the entire complexity of a biological system, but they represent an effective tool for the investigation of patient-device interaction from the hemodynamic point of view.

Cardiovascular Models

In this paragraph, the basic principles and components of computational simulators are described. We mostly focus attention on lumped parameters

© Springer Nature Switzerland AG 2020
J. H. Karimov et al. (eds.), *Mechanical Support for Heart Failure*,
https://doi.org/10.1007/978-3-030-47809-4_26

models (LPMs), a type of models where vessels are lumped into a compartment with compliant and resistive properties. These models are not suitable for complex fluid – structure analysis – or for 3D representation of pressure and flow fields in devices or in patient's specific anatomy. Their advantage is to represent the entire circulation in terms of pressure and flow profiles with a relatively modest computational effort.

LPMs can be used to represent the heart and vasculature and the relative arterio-ventricular and the venous-atrial interactions. Guyton systematically investigated these issues and defined the relationship among cardiac output, stroke volume, and heart rate [1]. His assumption is that in steady state, the cardiac output must correspond to venous return so the working point for a given circulatory condition can be defined as the intersection between the cardiac output and venous return curve.

Sagawa [2] defined the mutual relationship between arterio-ventricular sections and venous-ventricular sections. He provided a graphical approach and an analytical modeling of the interaction among left heart–systemic circulation–right heart–pulmonary circulation that represent to this day the basic theory for cardiovascular modeling. The introduction of these mechanisms into a circulatory model is fundamental for the representations of the effects of a MCS device, from both hemodynamic and ventricular energetic points of view.

Ventricles

The filling of the ventricle is regulated by the venous return, the compliance properties of the ventricular chamber and the state of atrioventricular valves (as regurgitation, stenosis). The relationship between instantaneous ventricular pressure and volume during diastole can be represented as an exponential function:

$$Plv(t) = a \cdot e^{bVlv(t)} + c \tag{26.1}$$

where *Plv* (*Vlv*) is the left ventricular pressure (volume) and *a*, *b*, and *c* are constant parameters.

The empting of the ventricle in systole depends on the filling level reached in diastole. The ventricular behavior is similar to a spring; it generates a tension depending on the stretching level reached in diastole and returns to its unstressed volume at the end of systole. This phenomenon is named Frank-Starling mechanism and states that for constant levels of arterial pressure, the ventricle is capable to adapt its output to the level of preload. The relationship between instantaneous left ventricular pressure and volume in systole can be implemented with a time varying elastance model [2]:

$$Plv(t) = Elvs \cdot vc(t) \cdot (Vlv(t) - Vlv0) \tag{26.2}$$

where *Vlv0* is the zero pressure filling volume, *Elvs* is the end-systolic elastance, and *vc* is the ventricular contraction function. Some models also include a resistive term in Eq. (26.2) to take into account the viscous property of the ventricular muscle during systole [3].

The ventricular contraction function ranges from 0 in diastole to 1 in systole. It can be represented by an exponential [4] or a second-order polynomial [3] or cosine pattern expression [5].

The same equations can be used for the right ventricle.

Atria

As a first approximation, atria can be represented as passive compliances as follows:

$$Cla = \frac{dVla(t)}{dPla(t)} \tag{26.3}$$

where *Cla* represents the elastic properties of the left atrium, and *Vla* and *Pla* represent the left atrial volume and pressure, respectively. This model, although very easily implemented, does not reproduce the left atrial contraction that contributes the 10%–30% to the ventricular filling. To take this into account, a model of active atria can be implemented [5]:

$$Pla(t) = \left[Elam + \frac{(ElaM - Elam)}{2} \cdot ac(t) \right] \cdot (Vla(t) - Vla0) \tag{26.4}$$

where *Pla* is the left atrial pressure, *Vla* is the left atrial volume, *Vla0* is the left atrial volume at zero pressure, *ElaM* (*Elam*) is the maximum and minimum value of atrial elastance, and *ac* is the atrial contraction function ranging from 0 in systole to 2 in diastole.

The same equations can be used to implement the right atrium.

Heart Valves

Heart chambers are connected by valves that can be represented in the simplest way by a diode and a resistance. In case of regurgitation, a reverse diode can be added with different resistance characteristics than the forward one [6]. More complex models include the representation of the blood-leaflet interaction for the description of the opening and closing valve process [5] or the representation of a variable valve area for the study of the transvalvular flow in the presence of specific diseases or prosthetic valve replacement [7].

Vessels

Concerning the modeling of vessels, lumped parameter structure aims at representing the input impedance of the arterial tree in order to reproduce the main pressure-flow relationship at the output of the ventricle [8, 9]. LPMs simulate the vasculature with an equivalent electrical circuit, according to which vessel properties are described by a resistive element reproducing the viscous properties of the vessels, a compliance reproducing the elastic properties of the vessels, and an inertance accounting for the inertial properties of blood motion into the vessels.

Otto Frank first described the arterial tree with a two-element Windkessel model, including resistive and a compliant components [10]. A three-element and a four-element Windkessel model followed with the introduction of an additional resistance [8] and finally of the inertial component [9, 11].

Thanks to the simplicity and to the quick execution, the lumped parameter approach progressively evolved into multicompartment models that provide a more complex and detailed representation of specific peripheral circulatory districts [12–14]. These types of models are suitable for the investigations of blood flow repartition among different circulatory districts in specific conditions, such as during orthostatic stress [12].

An important consideration concerns the venous circulatory section. Its representation in a LPM is fundamental in order to take into account the blood sequestered in the venous bed. Simple constant compliance models can be used for this purpose. In case of particular applications such as hemorrhagic studies, orthostatic studies, or exercise, a nonlinear model of the venous compliance is preferable [12]:

$$\Delta V_i(t) = \frac{2 \cdot \Delta V_{maxi}}{\pi} \arctan\left(\frac{\pi \cdot C_{0i}}{2 \cdot \Delta V_{maxi}} \cdot \Delta P_{trans}(t)\right) \tag{26.5}$$

where ΔV_i (ΔV_{maxi}) is the change (maximal change) of the ith compartment volume, ΔP_{trans} is the transmural pressure, and C_{0i} is the compliance of the ith compartment at the basal transmural pressure.

Autonomic Controls

The next level of complexity comes from the observation that cardiovascular parameters are not constant over time, but change to keep homeostasis in the body. The capability of a model to predict the reaction of the body to an external stimuli affecting flow/pressure (e.g., due to a MCS device) also depends to a certain extent on the inclusion of the main cardiovascular control mechanisms into the model.

Several models of the autonomic controls were developed and coupled with LPMs for the study of complex hemodynamic scenarios such as during tilt tests or hemorrhagic events [12, 13, 15, 16]. These models can help to study the alterations of the sympathetic and parasympathetic nervous system and their effects on the cardiovascular system, such as in case of heart failure, administration of beta-blocker, orthostatic hypotension, chronic hypertension, physical exertion, etc. Baroreflex models are useful for the investigation of the effects of the MCS devices that alter the arterial pressure waveform, thus introducing a "disturbance" in the baroreflex working chain with potential effects on the cardiovascular parameters. Devices such as the intra-aortic balloon pump (IABP) or left ventricular assist device (LVAD) driven in co-counterpulsation were investigated with the LPMs combined with a baroreflex model [17, 18].

Several equations can be found in the literature for baroreflex representation, all sharing a sigmoidal type of shape [12, 13, 19]:

$$Y = \frac{A}{1+e^{B\cdot(P-C)}} + D \tag{26.6}$$

where Y is the output (cardiovascular parameter), P is the input (pressure from the aorta or carotid sinus), A is a constant influencing the Y value at the top plateau, B is the gain coefficient, C is the centering point for which a pressor or depressor stimulus would induce an equal and opposite effect on the cardiovascular parameters, and D is the value of Y at the bottom plateau.

Ursino [13, 20] developed a sophisticated model of the baroreflex control, describing both the afferent and efferent nerve activity and including a model of the baroreceptors stimulation depending on both the intrasinus mean pressure and its pulsatility. The output of the baroreflex function is used in a first-order system control characterized by a time delay and a time constant that permits to reproduce the evolution of the cardiovascular parameters over time.

Cardiorespiratory Interaction

With their simple implementation, LPMs are also particularly suitable for the investigation of the interaction with the respiratory system [21–23]. Cardiorespiratory models are important for the study of the human respiratory control system [21, 22], exercise physiology [24, 25], complex breathing patterns [26], and medical devices such as the extracorporeal membrane oxygenator (ECMO) [27]. We briefly describe here a basic approach for the development of cardiorespiratory models.

Several models of the respiratory systems were developed; some of them reaching high levels of complexity [28]. Their analysis goes beyond the aim of the present chapter. If complex respiratory pathophysiological conditions need to be reproduced, more complex models are needed than the one described here.

A simple way to reproduce the ventilation mechanics is by considering the resistive element of the airways (R) and the elastance of the lungs (E) [29, 30]:

$$Pm - Ppl(t) = R\frac{dVlungs(t)}{dt} + Vlungs(t)\cdot E \tag{26.7}$$

where $Vlungs$ is the lungs volume, Pm is the mouth pressure set equal to the atmospheric pressure, and Ppl is the pleural pressure.

To a first approximation, the ventilation action can be reproduced by a sinusoidal function for Ppl:

$$Ppl(t) = Ppl_0 - \frac{E\cdot TV}{2}\cdot \sin\left(\frac{2\cdot\pi\cdot Freq}{60}\cdot t\right) \tag{26.8}$$

where $Freq$ is the ventilation frequency, TV is the tidal volume, and Ppl_0 is a constant parameter that corresponds to the mean value of Ppl. The representation of the ventilation mechanics is important to describe the effects of ventilation on the vessels in the thorax. From Eq. 26.8, we can obtain the intrathoracic pressure to be used to represent the effect of inspiration and expiration mechanics on the vascular compliances inside the chest.

More rigorous models of ventilation mechanics include the pressure generated by the respiratory muscles instead of Eq. 26.8, usually represented as a ramp or parabolic equation for the inspiration and as decay exponential for the expiration [23, 31].

The ventilation is regulated by central and peripheral chemoreceptors aimed at maintaining arterial oxygen and carbon dioxide (both CO_2 and its derived H^+) constant. A simplified way to implement ventilation control was proposed by Batzel et al. [21]. We report here an adaptation of that method that expresses the ventilation flow (*Vent*) as a function of the oxygen and carbon dioxide tensions (P_{o2} and P_{co2}) sensed in the brain and in the aorta:

$$Vent = G_1 \cdot e^{G_2 \cdot P_{O2}} \cdot \left(P_{CO_2} - P_{CO_2tr}\right) + G_3 \cdot \left(P_{CO_2} - P_{CO_2tr}\right) \quad (26.9)$$

where P_{CO2tr} is the threshold to start a new ventilation cycle, and G_1, G_2, and G_3 are constant parameters that describe the exponential and linear relationship between ventilation and P_{o2} and P_{co2}, respectively [32, 33].

Gas Transport and Exchange

Gas exchange can be modeled as a mass balance equation. Models usually make the assumption that temperature and humidity in the body are constant, as well as atmospheric pressure. Also, some models make the assumptions that O_2 and CO_2 partial pressure are equal in the alveoli and in the surrounding blood (the assumption is valid unless we consider extreme levels of exercise) [34]. The easiest way to represent gas exchange is to implement a mass balance equation relating the average blood flow (cardiac output) with the average ventilation flow (minute ventilation) [35].

The relationship between the partial pressure and the concentration of O_2 and CO_2 leaving the lungs is obtained using the dissociation curves [36, 37].

Concerning the peripheral tissues, a mass balance equation can be implemented as follows:

$$\frac{d\left(C_{O_2iv}(t) \cdot V_i(t)\right)}{dt} = C_{O_2ia}(t) \cdot Q_{ia}(t) - C_{O_2iv}(t) \cdot Q_{iv}(t) - \dot{V}_{O_2i}$$

$$\frac{d\left(C_{CO_2iv}(t) \cdot V_i(t)\right)}{dt} = C_{CO_2ia}(t) \cdot Q_{ia}(t) - C_{CO_2iv}(t) \cdot Q_{iv}(t) + RQ \cdot \dot{V}_{O_2i} \quad (26.10)$$

where Q_{ia} (Q_{iv}) is the arterial (venous) blood flow in the ith circulatory district, C_{O2ia} (C_{O2iv}) is the arterial (venous) oxygen concentration, C_{CO2ia} (C_{CO2iv}) is the arterial (venous) carbon dioxide concentration, V_i is the blood volume in the circulatory district, $\dot{V}_{O2i}$ is the oxygen consumption rate, and RQ is the respiratory quotient. In Eq. (26.10), we assume that the O_2 and CO_2 diffuse quickly enough to consider their concentrations equal in the tissues and in the venous blood.

Peripheral Metabolic Control

Equation 26.10 permits the reproduction of the local O_2 and CO_2 in the peripheral circulation. This equation is the basis for the implementation of the local metabolic control that induces vasodilation if more O_2 supply is needed in a specific vascular region.

$$sf_{RiMet}(t) = Sat\left(\frac{1}{1 + e^{k_{MET} \cdot \left(C_{O2iv}(t) - \frac{C_{O2ivRef}}{2}\right)}}\right) \quad (26.11)$$

where C_{O2iv} is the venous O_2 concentration in a vascular district, $C_{O2ivRef}$ is its reference value, k_{MET} is a constant representing the slope of the static function sf_{RiMet}, and *Sat* is the saturation level of the vasodilation. This function was adapted from [38] and can be fed into a first-order dynamic block to describe the dynamic control of the peripheral resistance of each circulatory district. Equation 26.11 considers only C_{O2iv} as input parameter; more complex controls can also include the effects of other metabolic vasodilator agents [39]. The peripheral metabolic type of control is fundamental to the representation of the exercise physiology [39, 40] (where vasodilation occurs in the exercising vascular regions that require a larger O_2 supply) or of extreme conditions such as during cardiopulmonary bypass [38].

Other Vascular Controls

Other controls can be implemented to increase the fidelity of the physiological model. The myogenic control can be modelled to describe the vasoconstriction of arterioles, depending on local pressure. Such a control is useful for the study of altered flow conditions, as during cardiopulmonary bypass [38].

Also, some vascular districts require specific representations due to their complex hemodynamic and blood flow regulation mechanisms. Examples are the cerebral circulation, renal circulation, and coronary vessels that would require dedicated models for an accurate representation [41, 42]. Their implementation depends on the application assigned to the model and the use envisioned for it.

Congenital Heart Diseases Models

An important application of LPMs is the study of complex hemodynamics as in the case of congenital heart diseases. As an example, LPMs are used to represent the Fontan circulation and to investigate the effects of different therapies. Pure LPMs can be used to compare different therapeutic strategies, such as continuous vs. pulsatile assistance on the cavopulmonary connection and the single ventricle [43]. In other cases, the LPM is interfaced with a 3D computational fluid dynamics (CFD) model of the cavopulmonary connection to assess the hydraulic energy losses across this anatomical tract and to predict the energy gain obtainable by the implantation of a MCS device [44].

Complexity of the Model

The level of complexity of LPMs is limited not by computational restrictions but rather by the difficulty of assigning reliable values to their parameters. Multicompartment models include a large amount of circulatory parameters that can't be estimated directly from hemodynamic measurements on a single subject.

It is evident that the level of complexity of a model needs to be tailored to the specific application. A simple model with a low number of parameters can be characterized from the hemodynamic measurements collected in clinics and can better serve the purpose of representing a patient's specific condition. These types of models can be used to solve basic questions concerning the short-term response of a therapy on a specific patient.

On the other hand, if more sophisticated physiological questions need to be answered, a more complex LPM model has to be used. As an example of complex models, we illustrate the cardiorespiratory model developed by Fresiello et al. [24, 40], used to investigate the exercise response in LVAD patients (Fig. 26.1).

In the case of a complex multicompartment model, the parameters characterization would involve clinical and literature data taken from different sources. The model will not represent a patient's specific condition, but rather a general patient's profile. These types of models are suitable to stratify patients' population in order to identify the best candidates for a therapy or to understand complex phenomena concerning the interaction patient-MCS device. Fresiello et al. tried to synthetize the process of LPM characterization for different profiles of patients by integrating data from different sources and tuning model parameters with an automatic algorithm [45].

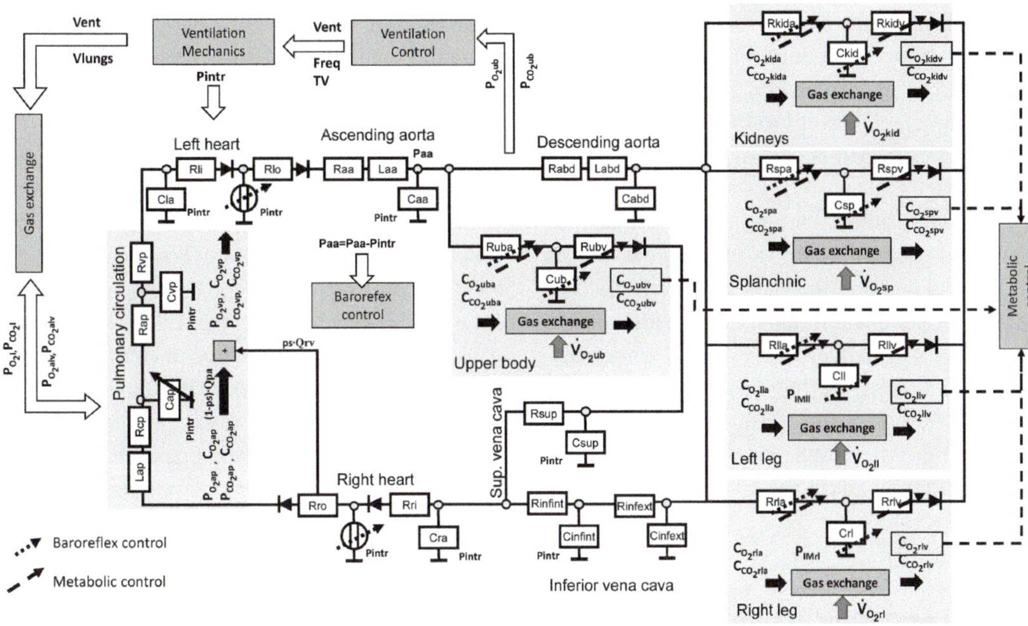

Fig. 26.1 Scheme of a cardiorespiratory model [40]. The heart model includes active atria and ventricles. Compliances, inertances, and resistances are indicated with *C*, *L*, *R* symbols followed by the suffix indicating the vascular regions: arterial/venous pulmonary circulation (*ap/vp*), ascending aorta (*aa*), descending aorta (*abd*), arterial/venous upper body (*uba/ubv*), arterial/venous kidneys (*kida/kidv*), arterial/venous splanchnic circulation (*spa/spv*), arterial/venous left leg (*lla/llv*), arterial/venous right leg (*rla/rlv*), superior vena cava (*sup*), and inferior vena cava (*inf*). With *ps* the pulmonary shunt is indicated. For every ith vascular region, the arterial and venous oxygen concentration (C_{O2ia}, C_{O2iv}) are calculated on the basis of the local flow and the local oxygen consumption ($\dot{V}_{O2i}$). The ventilation control is driven by the oxygen and carbon dioxide partial pressure in the upper body (P_{O2ub}, P_{CO2ub}) and provides the tidal volume (*TV*) and frequency of ventilation (*Freq*). The resulting intrathoracic pressure (*Pintr*) affects all vessels in the chest. A model of baroreflex control regulates peripheral vessels, heart contractility, and heart rate. A model of metabolic peripheral control regulates vasodilation locally, if a low value of C_{O2iv} is detected

Computational Models of MCS devices

Computational models can be used to test the interaction between the cardiovascular system and the MCS devices. In this case, the circulatory model is interfaced with a computational model of the MCS device.

Complex models of volume displacement LVADs have been developed in the past that can represent the filling and emptying cycle and the blood volume stored in the device. These models can also include a simulation of the actuator (pneumatic or electromechanical [46, 47]) that allow investigation of control algorithms and test different optimization driving strategies.

With the introduction of rotary blood pumps, several models were developed in the literature. LVAD hemodynamics can be represented by an equation relating pump flow *Q*, pressure head *H*, and impeller speed *ω* [48–50]:

$$H(t) = \alpha \cdot \omega(t)^2 - a \cdot Q(t) - b \cdot Q(t)^2 - L\frac{dQ(t)}{dt} \tag{26.12}$$

where the quadratic term of the impeller speed is the pressure generated by the device, *a* and *b* are the hydraulic resistance parameters, and the derivative term represents blood inertia effects. Additional resistive and inertial terms can be added to the equation to represent the effects of cannula, including possible turbulent flow phenomena [51], and a nonlinear additional resistance in case of suction [52].

In addition to the hydrodynamic characterization of the pump, some models also include a representation of the motor [49, 50] that relates the motor torque and current to the flow. This equation allows us to estimate LVAD flow from pump-related signals.

The computational model of a LVAD can be connected to a closed-loop LPM, representing the heart and the pulmonary and systemic circulations. Such a model can provide insight into the performance of the LVAD under different conditions, define driving strategies to optimize hemodynamic outcome and/or ventricular energetics, and develop a physiological controller to adapt the LVAD flow to body needs [52–54].

Physical and Hybrid Cardiovascular Simulators for MCS Devices

In this section, we discuss the concept of physical cardiovascular simulators and their application for MCS devices. Physical simulators are usually based on the LPM theory described earlier. They offer the advantage of connecting and testing real MCS devices; in this sense, they overcome the limitation of computational LPMs (where the MCS device has to be modeled, as well). Most ventricular assist device numerical models are based on static pressure-flow characteristics and show some limitations in reproducing dynamics effects like the inertia of fluid or providing an accurate representation of the motor and of its controller. The physical simulators allow to investigate a real MCS device in a more reliable and realistic way.

Physical Cardiovascular Simulators

Physical simulators can be useful tools for both clinicians and engineers. For clinicians, they can be used for training and educational purposes. Simulators can support clinical decision-making, as well as research activities. They can also help to understand physiology or build up practical experience and skills.

Mannequin-like simulators – purely physical simulators – are simpler tools and can be used to train medical personnel on manual skills [55]. More sophisticated mannequin simulators, like the HPS Anesthesia Human Patient Simulator [56], which combines physical user interfaces with some physiological models, are dedicated to more complex training purposes with realistic (e.g., intensive care) scenarios. However, their representation of the circulatory system is rather simple and does not allow reproduction of complex physiological interactions with the MCS devices.

Besides training applications, physical simulators are useful tools in testing new MCS devices at research and development phases. Physical simulators permit to conduct a series of tests on the MCS devices (e.g., optimize their design, test their control algorithms, prove their hemodynamic efficacy and safety with the cardiovascular system). In this sense, physical simulators constitute a valuable test bench that can reduce and potentially replace animal experiments. In addition to the financial advantages, physical simulators also overcome legal and ethical issues related to animal testing. The European Union Directive 2010/63/EU [57] encourages the reduction of animal experiments and the finding of alternative experimental methods.

Many research groups have developed different physical cardiovascular simulators in recent decades [58–66]. These simulators differ mainly in the hardware design and degree of complexity as they are intended to be used for different purposes. Some were developed to understand complex human physiology also in terms of ventricular assist device-heart interaction [60, 63, 64, 66], whereas others were utilized as a test bench (e.g., for medical devices evaluation [59, 62, 65]).

A brief overview of the some selected physical cardiovascular simulators is presented here after, a more complete overview can be found in [4].

A physical cardiovascular simulator for an IABP testing application was presented by Kolyva et al. [64]. It is a mock circulatory loop (MCL), built with polyurethane tubes mimicking the cardiovascular system anatomy (main trunk of the aorta and its branches). The artificial arteries are also ended with syringes mimicking the peripheral vessels compliance and by capillary tubes simulating the peripheral resistance. The left heart is simulated by Abiomed BVS 5000 system (Abiomed Inc., Danvers, MA, USA). This MCL reproduces, to some extent, the anatomy of the systemic circulation and can be used for waveform wave intensity analysis to determine local reflection coefficients for cardiovascular applications [67].

It is worth mentioning that this simulator is open-loop, meaning that only left heart and arterial systemic circulation are simulated. The right heart, venous return, and pulmonary circulation are neglected or simplified to the form of common tubes or tanks to allow fluid to return to the left heart.

For other types of applications, involving the study of preload/afterload and of the interaction heart-MCS devices, closed-loop simulators are needed. Closed-loop simulators are the extension of the open-loop simulators and include the full cardiovascular system representation.

An example of closed-loop physical simulators was developed by Timms et al. [65]. It is a very compact MCL (60 × 60 × 60 cm L × W × H) designed for MCS device testing. Its systemic and pulmonary circulations are based on five-element Windkessel model whose resistance and inertance components are characterized by the dimensions of the pipes (adjustable by proportional control valves); the compliance components are realized by hydraulic chambers (adjustable by regulating the air volume contained). The four heart chambers are modeled by electropneumatic air pressure regulators simulating atrial and ventricular contractility, separated by four swing check valves, mimicking heart valves. Solenoid-controlled connections between the ventricles, as well as between the ventricles and atria, can simulate ventricular septal defects and valve regurgitation, respectively. The simulator also takes the bronchial circulation into account. This MCL has been used, for example, in Starling-like controller development for a dual-rotary total artificial heart (TAH) [68] or for the testing of rotary LVAD as right ventricular assist device (RVAD) for biventricular assistance [69].

Another closed-loop physical simulator was developed by Vukicevic et al. [66] and specifically adapted to reproduce the Fontan circulation with the inclusion of the ventilation mechanics. The systemic circulation is divided into upper and lower body compartments, plus a splanchnic compartment influenced by abdominal respiratory pressure. The pulmonary circulation is divided into left and right lung compartments, taking into account intrapleural changes due to respiration. A patient-specific cavopulmonary connection, based on magnetic resonance imaging data of the patient, is connected to the simulator to mimic the Fontan circulation. An LVAD pulsatile system (EXCOR Berlin Heart, Berlin, Germany) is used to mimic the left ventricle. The whole system was used to simulate and test the influence of the respiration on subdiaphragmatic venous return in patients after the Fontan procedure.

As an alternative to Kolyva's et al. approach [64], in the latter two simulators, the cardiovascular system is reproduced by lumped hydraulic components following the same principles of the Windkessel models. The advantage of using Windkessel components is to have a simulator more compact and adjustable to a specific cardiovascular pathophysiological condition. If more investigations are needed on specific vascular districts, the simulator can be interfaced with 3D anatomical models reconstructed from patients' magnetic resonance imaging data. Despite the advantages of physical simulators, there are also some limitations as: the presence of several physical elements that introduce undesired parasitic components, nonlinearities, low flexibility in simulating different hemodynamic conditions (e.g. to change a resistance a physical element must be substituted, or an actuator is needed).

These limitations cause long-term instabilities and make it time-consuming to update or to modify the hardware of the simulator. Based on these observations, the hybrid simulator concept was introduced.

Hybrid (Hydraulic-Computational) Cardiovascular Simulator

In 1992, Pillon presented in [70] the concept and the prototype of the hybrid cardiovascular simulator. The term "hybrid" means two different simulation environments: the first environment is a software (a domain of computational simulators), and the second is a physical hydraulic structure (a domain of mock simulators). Therefore "hybrid simulator" indicates one simulating system, divided into a physical and a computational part.

The two parts are connected by a hybrid (hydraulic-computational) interface, which provides a bidirectional real-time signal transformation between the hydraulic and numerical components (i.e., for pressure, flow, or volume signals). Since the real-time mutual communication imposes a real-time execution of the computational cardiovascular model, LPMs are better suitable to build up a hybrid cardiovascular simulator. Although there are CFD real-time simulation tools [71], they are usually executed "online" on a server and are less accurate than classic CFD models.

Concerning the hybrid interface, different technical realizations are possible, but they all share the goal of assuring a real-time pressure and flow/volume transformation from the computational domain to the hydraulic domain and vice versa. The pressure, flow, or volume changes in the hydraulic section (e.g., caused by an LVAD) are measured and transferred to the computational section and taken into account in numerical calculations. The pressure, flow, or volume changes in the computational section (e.g., caused by the ventricle) are transferred to the hydraulic side and implemented through the actuators. These latter can be a gear pump, piston, membrane or pusher plate, or another externally controlled component [72].

Hybrid simulators inherit the advantages of computational models, such as simulation accuracy and flexibility and practically cover the same area of application for MCS device testing than MCLs.

Hybrid Cardiovascular Simulators for MCS Device Testing

In the past three decades, many different hybrid cardiovascular simulators were developed for various applications [72–80]. They differ in the level of complexity of the cardiovascular model they embed. Furthermore, their hydraulic and computational parts contribute in different ways to the whole system, ranging from a half-hydraulic and half-numerical simulator to a full hybrid simulator. According to the authors' experience, a full-hybrid approach, whenever possible, represents a better choice to minimize the drawbacks of MCLs and maximize the advantages of computational simulators. More technical details about the hybrid simulators are described in [72].

An example of full hybrid simulator is the one developed at the Nalecz Institute of Biocybernetics and Biomedical Engineering, Polish Academy of Sciences (IBBE-PAS) in cooperation with the Institute of Clinical Physiology of the Italian National Research Council in the frame of the European project SensorART [81]. Now the development of the hybrid cardiovascular simulator (HCS) is continued between IBBE-PAS and the Katholieke Universiteit in Leuven (KUL). This simulator was developed with a novel approach for hybrid interface (HI). The idea was to develop a configurable, modular, and multi-applicable hydraulic-computational HI, suitable for testing different MCS devices.

With the hybrid interface, a term "channel" is involved that relates to one actuator, i.e., one signal transformation and exchange. In the case of LVAD/RVAD systems, two connection points are needed with the cardiovascular simulator. Hence, two channels in the hybrid interface are required – one to communicate with the device input and one to communicate with output graft. It is obvious that for the connection and testing of biventricular assist device (BVAD) systems, a HI with four channels is needed.

Description of the Full Hybrid Cardiovascular Simulator

The HCS used at IBBE-PAS and KUL has a LPM as computational part that includes a representation of atria, ventricles, and pulmonary and systemic circulation. The model embeds most of the features of LPMs described earlier, such as multicompartment circulatory model (upper body, renal, splanchnic, and lower limbs circulations), time-varying elastance model for the ventricles, and baroreflex control. More details are reported in [82].

The hydraulic-computational HI of the HCS (Fig. 26.2) includes four separate channels. Each channel consists of a hydraulic operational chamber connected with a gear pump driven by servo DC motor (the hydraulic actuator). It is also equipped with pressure sensors and a set of safety and operational valves. The HI design was developed to be as universal as possible and configurable for different applications.

To emphasize the HI universal form, an overview of the developed applications for HCS is described in the following paragraphs.

Overview of Hybrid Cardiovascular Simulator Applications

In the following paragraphs, the applications for the presented HCS are described briefly.

LVAD Testing

The HCS can be configured to test pulsatile LVADs in atrio-aortic connections [83]. An example is reported in Fig. 26.3 where the HCS was used to study the hemodynamic interaction between the left ventricle and a pneumatic, membrane based adult MCS device (its pneumatic driver was developed by IBBE-PAS in house). The LPM of the HCS was configured to

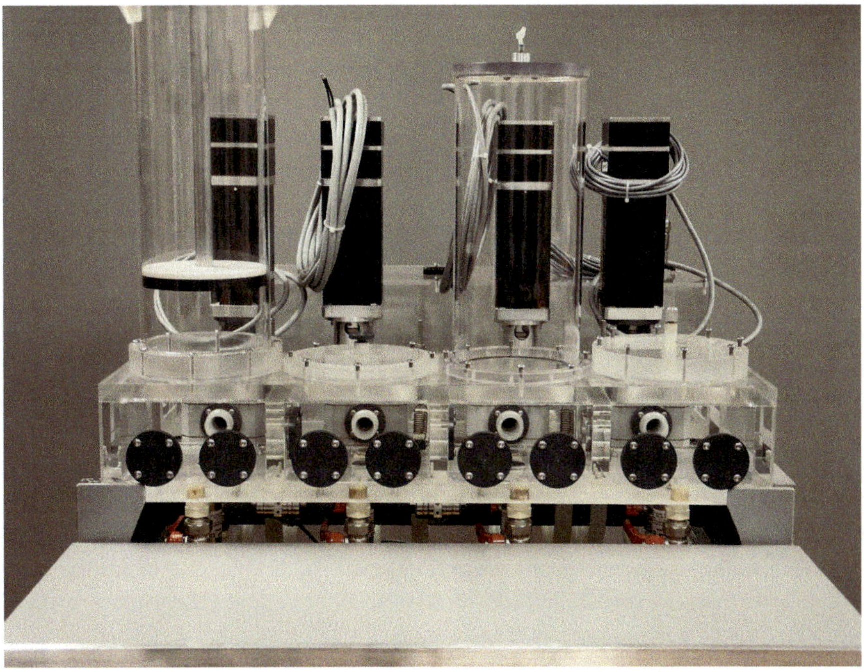

Fig. 26.2 A front view of the hybrid interface, with four channels, each equipped with a DC motor, a gear pump, a pressure sensor, and hydraulic chamber. Dedicated front cups are used to connect external devices (e.g., LVAD). The chambers (channels) are separated by internal caps that can be removed to reduce the number of operational channels or to connect components such as artificial valves. Additional accessories, such as variable or fixed pneumatic compliances, can be mounted on the top of each chamber

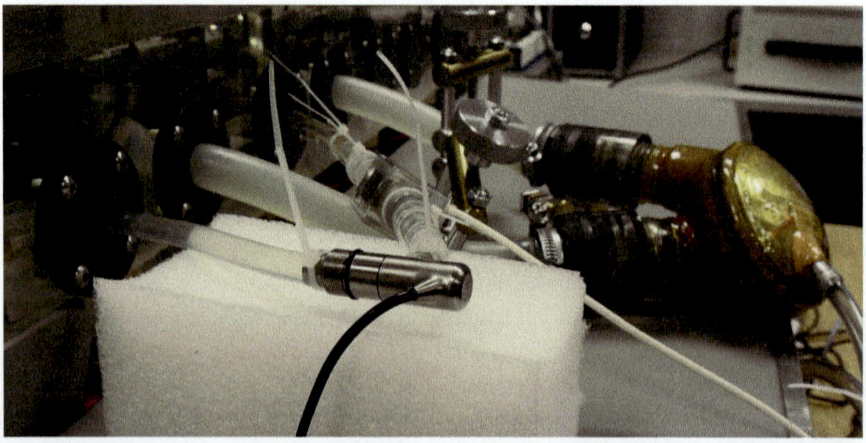

Fig. 26.3 A pulsatile and a rotary LVAD connected to the hybrid interface, between the Pla and Pas channels. Each LVAD can be tested at a time by clamping the other one

provide a digital ECG – similar signal to synchronize the MCS device. The HI configuration, shown in Fig. 26.3, includes an input cannula connected to the first HI channel, where the left atrial pressures (Pla) is reproduced; an output cannula connected to the second channel, where systemic arterial blood pressure (Pas) is simulated.

Using the same HI configuration, a rotary LVAD (CircuLite Synergy Mircropump, HeartWare Inc.) was studied [82]. While in [83] a simple Windkessel model for the arterial and venous systemic circulation was used, here a more sophisticated LPM model, including autonomic controls, was introduced. The hybrid simulator was tuned to reproduce specific animal conditions at baseline (LVAD off) and could simulate the hemodynamic evolution for increasing values of LVAD support (stepwise LVAD speed increase). This study demonstrated that the hybrid simulator approach, thanks to the minimization of hydraulic components and the inclusion of the computational part, assures enough flexibility to reproduce specific hemodynamic conditions. Moreover, the representation of the progressive left ventricular unloading for increasing levels of LVAD support also expanded the use of the HCS to test LVAD physiological control algorithms [84].

The flexibility of the HCS, due to the inclusion of a computational part, allows its applicability for pulsatile and rotary pediatric LVAD testing. The scaling of a computational cardiovascular model to a pediatric size is a much easier process compared to the scaling of the hardware components of a pure MCL.

The HCS was used to test a pediatric, pneumatic LVAD (EXCOR Berlin Heart, Berlin, Germany) [85]. The study investigated the optimal LVADs pulse rate in relation to the native heart rate for asynchronous mechanical support. The HCS configuration is presented in Fig. 26.4. LVAD connections to the HCS were similar to the atrio-aortic assistance described above, but the input cannula was connected to the HI hydraulic chamber mimicking the left ventricular pressure (Plv).

The same configuration was used to investigate an EXCOR Berlin Heart LVAD (Berlin, Germany) membrane rupture observed in clinics [86]. For this application, the HCS was tuned to reproduce the specific pediatric patient's hemodynamic condition in which the rupture occurred. Then the LVAD was connected and tested.

Following the pediatric applications, a new prototype of the rotary LVAD (Jarvik 2015, Jarvik Heart Inc., NY, USA) was tested using the HCS [87]. The HCS permitted the investigation of the feasibility and the potential hemodynamic outcome of the assist device in pediatric patients profiles of different body weights (2–7 Kg).

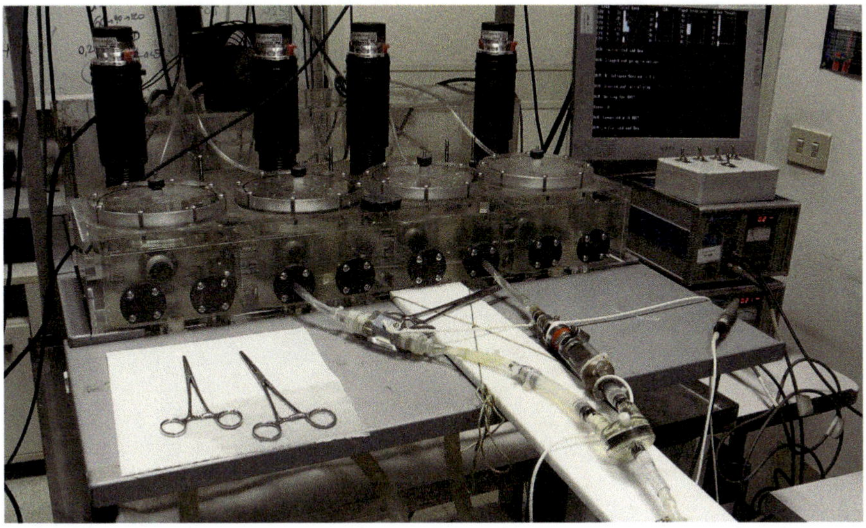

Fig. 26.4 A pneumatic EXCOR Berlin Heart LVAD connected to the hybrid interface, between the Pas and the Plv chambers (second and third chambers from the left)

BVAD and TAH Testing

To configure the HCS for biventricular support, four HI channels are needed (two for the left and two for the right ventricular assistance). The HCS configuration for the BVAD application was reported in [88] and included the representation of the left and right ventricular pressures, and aortic and pulmonary trunk pressures on the four channels.

It is evident that a similar HI configuration can be used to connect a TAH to the HCS, with the difference that in this case, the numerical model of ventricles is removed. The HI configuration for TAH then becomes Pas, Pla, right atrial pressure (Pra) and pulmonary arterial pressure (Pap, see Fig. 26.5). This configuration includes parasitic tube connections between TAH and HCS, which need to be minimized as much as possible.

Artificial Heart Valve Testing

For ventricular assist device and TAH applications, the hybrid interface chambers of the HCS are usually separated by walls in order to work independently. For valve testing, internal caps can be removed from the walls to open a hole between two chambers and to connect an artificial valve accordingly. This solution allows a direct connection of the valve between two chambers to avoid parasitic hydraulic elements between the valve and HCS. The computational model of the HCS is adapted by removing the numerical model of the valve from the LPM. The HCS application for aortic valve testing (Fig. 26.6) is described in more detail in [89]. The HCS configuration can be expanded to test two artificial valves (e.g., mitral and aortic) or aortic biological valves, as well.

Extracorporeal Device Testing

Extracorporeal applications typically require two HI channels, one for the drainage cannula and the second for the infusion cannula. However, when two drainage cannulas are used, a third HI channel is required. This configuration is suitable for extracorporeal LVADs, RVADs, ECMOs, and heart-lung machines. For ECMO, both veno-venous and veno-arterial cannulations can be simulated. However, the HCS allows only investigation of the hydraulic part of the ECMO. Other aspects, such as blood oxygenation, are simulated computationally on the HCS and cannot be combined with the oxygenator of the ECMO physically. More details

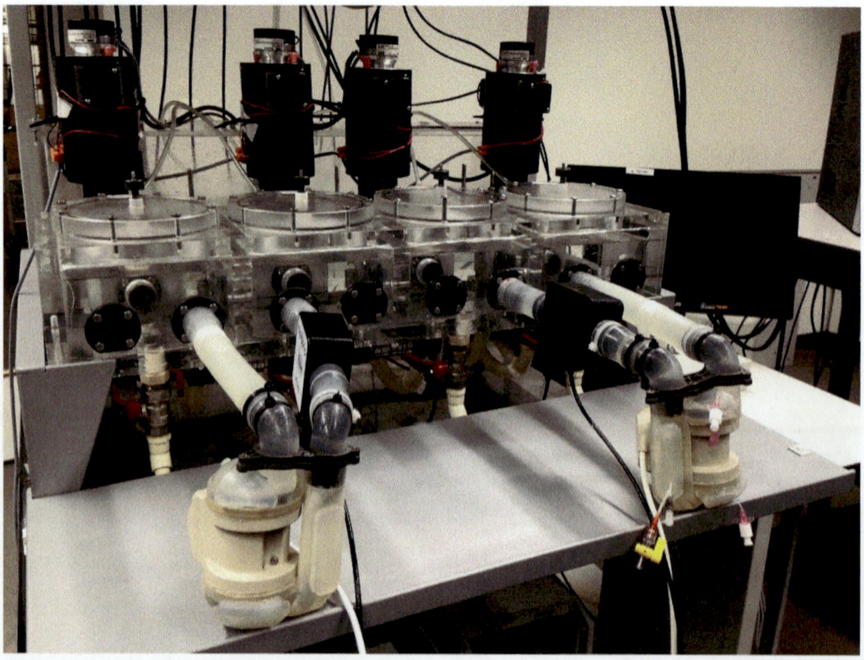

Fig. 26.5 A total artificial heart connected to the hybrid interface. In this case all the four chambers are used to reproduce (from left to right): Pra, Pap, Pas and Pla. The device is a Realheart TAH version 11c (Scandinavian Realheart®, Västerås, Sweden). Silicon pads have been added on the device for this particular experimental version to enable internal pressure monitoring

Fig. 26.6 An artificial aortic valve by St. Jude Medical connected to the hybrid cardiovascular simulator. The first channel reproduces Plv, and the second channel reproduces Pas

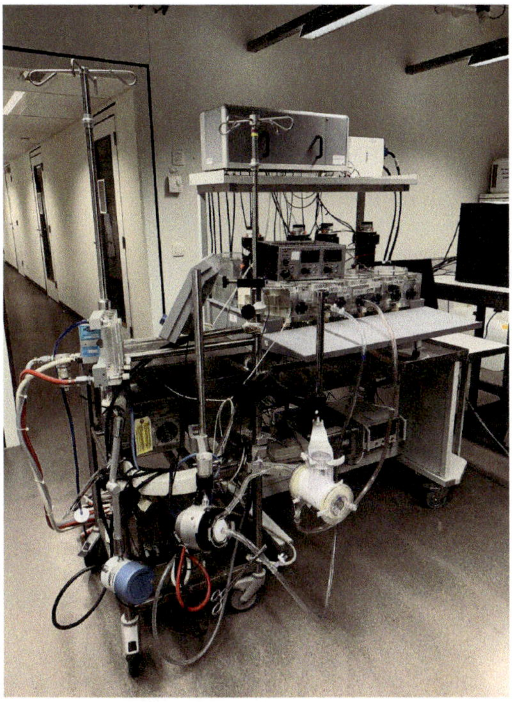

Fig. 26.7 An ECMO connected to the hybrid interface. The venous line is connected to the second HI channel mimicking Pra; the arterial line is connected to the third HI channel mimicking Pas

about the HCS configuration for ECMO in both veno-venous and veno-arterial connections are presented in [90]. An overview of the HCS connected to an ECMO circuit is presented in Fig. 26.7.

Intra-aortic Balloon Pump Testing

The configuration of the HCS for IABP requires an additional hardware component in the HI, specifically, a tube mimicking the aorta, where the balloon can be inserted and connected to the rest of the cardiovascular simulator (as in [64]). Figure 26.8 shows the HI configuration for the IABP, with as additional component a compliant tube inside a hydraulic box. The tube, mimicking the descending aorta, is connected to two HI channels mimicking the ascending aorta upstream and the rest of the descending aorta downstream, respectively. A third HI channel can regulate the pressure inside the box surrounding the tube, in order to mimic different elastic properties of the aorta where the IABP is inserted.

Fig. 26.8 An intra-aortic balloon pump in the tube (mimicking the aorta) connected to the hybrid interface

References

1. Guyton AC, Jones CE, Coleman TG. Circulatory physiology: cardiac output and its regulation. Philadelphia: W.B. Saunders Company; 1973.
2. Sagawa K, Maughan L, Suga H, Sunagawa K. Cardiac contraction and the pressure-volume relationship. New York: Oxford University Press; 1988.
3. Yi Wu, Allaire P, Tao G; Olsen D. Modeling, estimation and control of cardiovascular systems with a left ventricular assist device. EEE Trans Cont Systems Technol 2007;15(4).
4. Ferrari G, Darowski M, Kozarski M. Physical models and components. In: Darowski M, Ferrari G, editors. Comprehensive models of cardiovascular and respiratory systems: their mechanical support and interactions. New York: Nova Science Publishers; 2010.
5. Korakianitis T, Shi Y. A concentrated parameter model for the human cardiovascular system including heart valve dynamics and atrioventricular interaction. Med Eng Phys. 2006;28(7):613–28.
6. Kim YS, Kim EH, Kim HG, Shim EB, Song KS, Lim KM. Mathematical analysis of the effects of valvular regurgitation on the pumping efficacy of continuous and pulsatile left ventricular assist devices. Integr Med Res. 2016 Mar;5(1):22–9.
7. Szabó G, Soans D, Graf AJ, Beller C, Waite L, Hagl S. New computer model of mitral valve hemodynamics during ventricular filling. Eur J Cardiothorac Surg. 2004;26(2):239–47.
8. Westerof N, Elzinga G, Sipkema P. An artificial arterial system for pumping hearts. J Appl Physiol. 1971;31(5):776–81.
9. Stergiopolus N, Berend E, Westerhof N. Total arterial inertance as the fourth element of the Winkessel model. AJP-Heart. 1999;276(1):H81–8.
10. Frank O.. Die Grundform des arteriellen Pulses. Z Biol. 1899;37:483:526.
11. Burkhoff D, Alexander J Jr, Schipke J. Assessment of Windkessel as a model of aortic input impedance. Am J Phys. 1988;255(4 Pt 2):H742–53.
12. Heldt T, Shim EB. Kamm RD and. Mark RG. Computational modeling of cardiovascular response to orthostatic stress. J Appl Physiol. 2002;92:1239–54.
13. Ursino M. Interaction between carotid baroregulation and the pulsating heart: a mathematical model. Am J Physiol Heart Circ Physiol. 1998;275:1733–47.
14. Ferrari G, Kozarski M, Zieliński K, Fresiello L, Di Molfetta A, Górczyńska K, Pałko KJ, Darowski M. A modular computational circulatory model applicable to VAD testing and training. J Artif Organs. 2012;15(1):32–43.
15. Olufsen MS, Ottesen JT, Tran HT. Blood pressure and blood flow variation during postural change from sitting to standing: model development and validation. J Appl Physio. 2005;99:1523–37.
16. Blanco PJ, Trenhago PR, Fernandes L, Gand Feijóo RA. On the integration of the baroreflex control mechanism in a heterogeneous model of the cardiovascular system. Int J Numer MethBiomed Engin. 2012;28:412–33.
17. Fresiello L, Khir AW, Di Molfetta A, Kozarski M, Ferrari G. Effects of intra-aortic balloon pump timing on baroreflex activities in a closed-loop cardiovascular hybrid model. Artif Organs. 2013;37(3): 237–47.
18. Lim KM, Kim IS, Choi SW, Min BG, Won YS, Kim HY, Shim EB. Computational analysis of the effect of the type of LVAD flow on coronary perfusion and ventricular afterload. J Physiol Sci. 2009;59(4):307–16.
19. McDowall LM and Dampney R.A.L. Calculation of threshold and saturation points of sigmoidal baroreflex function curves. Am J Physiol Heart Circ Physiol 2006;291:H2003–H2007.
20. Ursino M, Fiorenzi A, Belardinelli E. The role of pressure pulsatility in the carotid baroreflex control: a computer simulation study. Comput Biol Med. 1996;26(4):297–314.
21. Batzel JJ, Trany HT. Stability of the human respiratory control system. Part I: analysis of a two-dimensional delay state-space model. J Math Biol. 2000;41(1):45–79.
22. Batzel JJ, Trany HT. Stability of the human respiratory control system. Part II: analysis of a two-dimensional delay state-space model. J Math Biol. 2000;41(1):80–102.
23. Van Meurs W. Modeling and simulation in biomedical engineering: applications in cardiorespiratory physiology. New York: McGraw-Hill Professional; 2011.
24. Fresiello L, Rademakers F, Claus P, Ferrari G, Di Molfetta A, Meyns B. Exercise physiology with a left ventricular assist device: analysis of heart-pump interaction with a computational simulator. PLoS One. 2017;12(7).
25. Graefe R, Henseler A, Körfer R, Meyns B, Fresiello L. Influence of left ventricular assist device pressure-flow characteristic on exercise physiology: assessment with a verified numerical model. Int J Artif Organs. 2019;42(9):490–9.
26. Cheng L, Ivanova O, Fan HH, Khoo MC. An integrative model of respiratory and cardiovascular control in sleep-disordered breathing. Respir Physiol Neurobiol. 2010;174(1–2):4–28.
27. Broomé M, Maksuti E, Bjällmark A, Frenckner B, Janerot-Sjöberg B. Closed-loop real-time simulation model of hemodynamics and oxygen transport in the cardiovascular system. Biomed Eng Online. 2013;12:69.
28. Golczewski T, Darowski M. Virtual respiratory system in investigation of CPAP influence on optimal breathing frequency in obstructive lungs disease. Nonlinear Biomed Phys. 2007;1:6.
29. Ben-Tal A. Simplified models for gas exchange in the human lungs. J Theor Biol. 2006;238(2):474–95.
30. Cross TJ, Sabapathy S, Beck KC, Morris NR, Johnson BD. The resistive and elastic work of breathing during exercise in patients with chronic heart failure. Eur Respir J. 2012;39(6):1449–57.

31. Albanese A, Cheng L, Ursino M, Chbat NW. An integrated mathematical model of the human cardiopulmonary system: model development. Am J Physiol Heart Circ Physiol. 2016;310(7):H899–921.
32. Nunn JF. Control of breathing. In: Nunn JF, editor. Applied respiratory physiology with special reference to. Anaesthesia, London: Butterworth & Co Publishers Ltd Press; 1969.
33. Cormack RS, Cunningham DJ, Gee JB. The effect of carbon dioxide on the respiratory response to want of oxygen in man. Q J Exp Physiol Cogn Med Sci. 1957;42(3):303–19.
34. Batzel JJ, Kappel F, Schneditz D, Hien TT. Cardiovascular and respiratory systems, modeling, analysis and control. Philadelphia: Society for Industrial and Applied Mathematics; 2006.
35. Ben-Tal A. Computational models for the study of heart-lung interactions in mammals. Wiley Interdiscip Rev Syst Biol Med. 2012;4(2):163–70.
36. Spencer JL, Firouztale E, Mellins RB. Computational expressions for blood oxygen and carbon dioxide concentrations. Ann Biomed Eng. 1979;7(1):59–66.
37. Golczewski T. Gas exchange in a virtual respiratory system--simulation of ventilation without lung movement. Int J Artif Organs. 2007;30(12):1047–56.
38. Lanzarone E, Liani P, Baselli G, Costantino ML. Model of arterial tree and peripheral control for the study of physiological and assisted circulation. Med Eng Phys. 2007;29(5):542–55.
39. Magosso E, Ursino M. Cardiovascular response to dynamic aerobic exercise: a mathematical model. Med Biol Eng Comput. 2002;40(6):660–74.
40. Fresiello L, Meyns B, Di Molfetta A, Ferrari G. A model of the cardiorespiratory response to aerobic exercise in healthy and heart failure conditions. Front Physiol. 2016;7:189.
41. Ursino M, Di Giammarco P. A mathematical model of the relationship between cerebral blood volume and intracranial pressure changes: the generation of plateau waves. Ann Biomed Eng. 1991;19(1):15–42.
42. Pietrabissa R, Mantero S, Marotta T, Menicanti L. A lumped parameter model to evaluate the fluid dynamics of different coronary bypasses. Med Eng Phys. 1996;18(6):477–84.
43. Di Molfetta A, Ferrari G, Iacobelli R, Filippelli S, Amodeo A. Concurrent use of continuous and pulsatile flow ventricular assist device on a Fontan patient: a simulation study. Artif Organs. 2017;41(1):32–9.
44. Throckmorton AL, Carr JP, Tahir SA, Tate R, Downs EA, Bhavsar SS, Wu Y, Grizzard JD, Moskowitz WB. Mechanical cavopulmonary assistance of a patient-specific Fontan physiology: numerical simulations, lumped parameter modeling, and suction experiments. Artif Organs. 2011;35(11):1036–47.
45. Fresiello L, Ferrari G, Di Molfetta A, Zieliński K, Tzallas A, Jacobs S, Darowski M, Kozarski M, Meyns B, Katertsidis NS, Karvounis EC, Tsipouras MG, Trivella MG. A cardiovascular simulator tailored for training and clinical uses. J Biomed Inform. 2015;57:100–12.
46. Yih-Choung Y, Simaan M, Boston R, Miller PJ, Antaki JF. Estimation of blood pump parameters for cardiovascular system identification. New Mexico: Proceedings of the American Control Conference Albuquerque; 1997.
47. Ferrari G, De Lazzari C, Mimmo R, Tosti G, Ambrosi D. A modular numerical model of the cardiovascular system for studying and training in the field of cardiovascular physiopathology. J Biomed Eng. 1992;14(2):91–107.
48. Moscato F, Danieli GA, Schima H. Dynamic modeling and identification of an axial flow ventricular assist device. Int J Artif Organs. 2009 Jun;32(6):336–43.
49. Choi A, Boston JR, Thomas S, Antaki JF. Modeling and identification of an axial flow blood pump. Alburquerque, New Mexico: Proceeding of the Americal Control Conference; 1997.
50. Kitamura T, Matsushima Y, Tokuyama T, Kono S, Nishimura K, Komeda M, Yanai M, Kijima T, Nojiri C. Physical model-based indirect measurements of blood pressure and flow using a centrifugal pump. Artif Organs. 2000;24(8):589–93.
51. Verkerke GJ, Mihaylov D, Geertsema AA, Lubbers J, Rakhorst G. Numerical simulation of the pulsating catheter pump: a left ventricular assist device. Artif Organs. 1999;23(10):924–31.
52. Chen S, Antaki JF, Simaan MA, Robert Boston JR. Physiological control of left ventricular assist devices based on gradient of flow. Portland, OR, USA: American Control Conference; 2005.
53. Giridharan GA, Skliar M. Physiological control of blood pumps using intrinsic pump parameters: a computer simulation study. Artif Organs. 2006;30: 301–7.
54. Cox LG, Loerakker S, Rutten MC, de Mol BA, van de Vosse FN. A mathematical model to evaluate control strategies for mechanical circulatory support. Artif Organs. 2009;33(8):593–603.
55. Harvey. The Cardiopulmonary Patient Simulator [Internet]. [cited 2020 July 13]. Available from: https://www.laerdal.com/us/item/HARVEY.
56. HPS Anesthesia Human Patient Simulator [Internet]. [cited 2020 July 13]. Available from: https://caehealthcare.com/patient-simulation/hps/.
57. European Union Directive 2010/63/EU [Internet]. [cited 2020 July 13]. Available from: https://eur-lex.europa.eu/eli/dir/2010/63/oj.
58. Kolff WJ. Mock circulation to test pumps designed for permanent replacement of damaged hearts. Cleve Clin Q. 1959;26:223–6.
59. Donovan FM. Design of a hydraulic analog of the circulatory system for evaluating artificial hearts. Biomater Med Devices Artif Organs. 1975;3(4):439–49.
60. Swanson WM, Clark RE. A simple cardiovascular system simulator: design and performance. J Bioeng. 1977;1(2):135–45.
61. Rosenberg G, Phillips WM, Landis DL, Pierce WS. Design and evaluation of the Pennsylvania State University mock circulatory system. ASAIO J. 1981;4(2):41–9.

62. Knierbein B, Reul H, Eilers R, Lange M, Kaufmann R, Rau G. Compact mock loops of the systemic and pulmonary circulation for blood pump testing. Int J Artif Organs. 1992;15(1):40–8.
63. Ferrari G, De Lazzari C, Mimmo R, Ambrosi D, Tosti G. Mock circulatory system for in vitro reproduction of the left ventricle, the arterial tree and their interaction with a left ventricular assist device. J Med Eng Technol. 1994;18(3):87–95.
64. Kolyva C, Biglino G, Pepper JR, Khir AW. A mock circulatory system with physiological distribution of terminal resistance and compliance: application for testing the intra-aortic balloon pump. Artif Organs. 2012;36(3):E62–70.
65. Timms DL, Gregory SD, Greatrex NA, Pearcy MJ, Fraser JF, Steinseifer U. A compact mock circulation loop for the in vitro testing of cardiovascular devices. Artif Organs. 2011;35(4):384–91.
66. Vukicevic M, Conover T, Jaeggli M, Zhou J, Pennati G, Hsia T-Y, et al. Control of respiration-driven retrograde flow in the subdiaphragmatic venous return of the Fontan circulation. ASAIO J. 2014;60(4): 391–9.
67. Li Y, Parker KH, Khir AW. Using wave intensity analysis to determine local reflection coefficient in flexible tubes. J Biomech. 2016;49(13):2709–17.
68. Ng BC, Smith PA, Nestler F, Timms D, Cohn WE, Lim E. Application of adaptive Starling-like controller to Total artificial heart using dual rotary blood pumps. Ann Biomed Eng. 2017;45(3):567–79.
69. Stevens MC, Gregory SD, Nestler F, Thomson B, Choudhary J, Garlick B, et al. In vitro and in vivo characterization of three different modes of pump operation when using a left ventricular assist device as a right ventricular assist device. Artif Organs. 2014;38(11):931–9.
70. Pillon M, Duffour H, Jufer M. In vitro experiments: Circulatory assist device interaction with a virtual cardiovascular system. 1992 14th Annual International Conference of the IEEE Engineering in Medicine and Biology Society. 1992:740–1.
71. ANSYS Discovery Live: Real-Time Simulation Revolution [Internet]. [cited 2020 July 13]. Available from: https://www.ansys.com/about-ansys/advantage-magazine/volume-xi-issue-3-2017/ansys-discovery-live-real-time-simulation-revolution.
72. Zielinski K, Darowski M, Kozarski M, Ferrari G. The need for hybrid modeling in analysis of cardiovascular and respiratory support. Int J Artif Organs. 2016;39(6):265–71.
73. Baloa LA, Boston JR, Antaki JF. Elastance-based control of a mock circulatory system. Ann Biomed Eng. 2001;29(3):244–51.
74. Colacino FM, Moscato F, Piedimonte F, Danieli G, Nicosia S, Arabia M. A modified elastance model to control mock ventricles in real-time: numerical and experimental validation. ASAIO J. 2008;54(6):563–73.
75. Gwak KW, Paden BE, Antaki JF, Ahn IS. Experimental verification of the feasibility of the cardiovascular impedance simulator. IEEE Trans Biomed Eng. 2010;57(5):1176–83.
76. Ferrari G, De Lazzari C, Kozarski M, et al. A hybrid mock circulatory system: testing a prototype under physiologic and pathological conditions. ASAIO J. 2002;48(5):487–94.
77. Cuenca-Navalon E, Finocchiaro T, Laumen M, Fritschi A, Schmitz-Rode T, Steinseifer U. Design and evaluation of a hybrid mock circulatory loop for total artificial heart testing. Int J Artif Organs. 2014;37(1):71–80.
78. Misgeld BJE, Rüschen D, Schwandtner S, Heinke S, Walter M, Leonhardt S. Robust decentralised control of a hydrodynamic human circulatory system simulator. Biomedical Signal Processing and Control. 2015;20:35–44.
79. Nestler F, Bradley AP, Wilson SJ, Timms DL, Frazier OH, Cohn WE. A hybrid mock circulation loop for a total artificial heart. Artif Organs. 2014;38(9):775–82.
80. Ochsner G, Amacher R, Amstutz A, Plass A, Schmid Daners M, Tevaearai H, et al. A novel interface for hybrid mock circulations to evaluate ventricular assist devices. IEEE Trans Biomed Eng. 2013;60(2):507–16.
81. SensorART EU Project. [Internet]. [cited 2020 July 13]. Available from: http://www.sensorart.eu/.
82. Fresiello L, Zieliński K, Jacobs S, Di Molfetta A, Pałko KJ, Bernini F, et al. Reproduction of continuous flow left ventricular assist device experimental data by means of a hybrid cardiovascular model with baroreflex control. Artif Organs. 2014;38(6):456–68.
83. Darowski M, Kozarski M, Ferrari G, Zieliński K, Górczyńska K, Szczepanowski A, et al. A new hybrid (hydro-numerical) model of the circulatory system. Bull Pol Ac Tech. 2013;61(4):993–1003.
84. Tortora G, Fontana R, Fresiello L, Molfetta AD, Silvestri M, Vatteroni M, et al. Experimental integration of Autoregulation Unit for left ventricular assist devices in a cardiovascular hybrid simulator. 2014 36th Annual International Conference of the IEEE Engineering in Medicine and Biology Society. 2014:282–5.
85. Ferrari G, Di Molfetta A, Zieliński K, Fresiello L, Górczyńska K, Pałko KJ, et al. Control of a pediatric pulsatile ventricular assist device: a hybrid cardiovascular model study. Artif Organs. 2017;41(12):1099–108.
86. Di Molfetta A, Filippelli S, Ferrari G, Secinaro A, Zielinski K, Amodeo A. Berlin heart EXCOR ventricular assist device: multilayer membrane rupture in a pediatric patient. Ann Thorac Surg. 2016;102(2):e129–30.
87. Di Molfetta A, Zieliński K, Ferrari G, Kozarski M, Okrzeja P, Iacobelli R, Filipelli S, Perri G, Darowski M, Massetti M, Jarvik R, Amodeo A. Is the new infant Jarvik 2015 suitable for patients<8 kg? In vitro study using a hybrid simulator. Artif Organs. 2019;43(1):E1–8.
88. Kozarski M, Ferrari G, Zieliński K, Górczyńska K, Di Molfetta A, Palko KJ, Fresiello L, Darowski M. Biventricular heart assistance: preliminary tests on

the hybrid (hydro-numerical) circulatory model. Int J Artif Organs. 2010;33(7):450.

89. Kozarski M, Suwalski P, Zieliński K, Górczyńska K, Szafron B, Pałko KJ, et al. A hybrid (hydro-numerical) circulatory model: investigations of mechanical aortic valves and a numerical valve model. Bull Pol Ac: Tech. 2015;63(3):605–12.

90. Zieliński K, Okrzeja P, Stecka A, Kozarski M, Darowski M. A hybrid cardio-pulmonary simulation platform—an application for extracorporeal assist devices. In: Lhotska L, Sukupova L, Lacković I, Ibbott GS, editors. World Congress on Medical Physics and Biomedical Engineering 2018. Singapore: Springer; 2019. p. 703–6. (IFMBE Proceedings).

27 Options for Modeling and Simulations Used in Mechanical Circulatory Support Development

David J. Horvath, Kiyotaka Fukamachi, and Jamshid H. Karimov

Introduction

The study of cardiovascular dynamics now includes the use of a variety of mechanical circulatory support (MCS) blood pumps, such as devices for left, right, and biventricular assist (LVAD, RVAD, and BVAD) and total artificial hearts (TAHs). Characteristics of these devices need to be crafted to accommodate a wide variety of disease states and patient conditions, which present a wide range of hemodynamic environments at the pump ports. Reduction in the number of exploratory *in vivo* experiments in the early stages of development represents an additional advantage, as the simulations can be used to sort through the concepts and protocols before they are enacted.

Understanding the interactions between pump performance and hemodynamic behavior and how these factors relate to pump selection and design is very helpful to those of us trying to conceive, explore, and develop new applications, especially with limited resources. All development programs start with a device that does not yet exist and so cannot be tested on a physical mock loop. Further, bench investigations involving combinations of ventricular and valve disease states are often impractical to duplicate on a basic test loop, thus requiring complete command of a complex physical loop, loop control system, and pumping device, as well as the resources and dedicated personnel needed to set up, operate, and maintain it. A primary objective of simulation is to make it as portable and useful as possible outside the lab environment for people not involved in the creation of its operational software.

Uses and Relative Limitations of Different Approaches

The following simulation approaches represent a progression of tools which can be applied throughout the MCS device development process:

D. J. Horvath, MSME (✉)
R1 Engineering, Euclid, OH, USA

K. Fukamachi, MD, PhD · J. H. Karimov, MD, PhD
Lerner Research Institute, Cleveland Clinic, Cleveland Clinic Lerner College of Medicine of Case Western Reserve University, Cleveland Clinic, Cleveland, OH, USA
e-mail: fukamak@ccf.org; karimoj@ccf.org

© Springer Nature Switzerland AG 2020
J. H. Karimov et al. (eds.), *Mechanical Support for Heart Failure*,
https://doi.org/10.1007/978-3-030-47809-4_27

Numerical simulation →	Basic mock loop →	Hybrid mock loop
Getting started, explore applications, *in vivo* simulation, simulate states beyond practical experimentation	Characterize early MCS devices, verify performance	Explore complex disease states, repetitive testing, controller development

Numerical Simulation *In silico* simulation, as discussed below in this chapter, can overcome some of the practical limitations of a basic mock loop, particularly in the early stages of development. Primarily, with the availability of the software, applications can be explored at low cost and at the same rate that they are considered. The working MCS device and specialized test setup does not have to pre-exist. Complex or combined disease states can be simulated with MCS devices having various operating characteristics for a full range of patient sizes. It can be portable and used outside the laboratory environment in an office or educational setting. Autonomic responses can be included or imposed if they can be described in mathematical terms. Protocols for *in vivo* experiments can be pre-run as a means of structured thought for the election of device and hemodynamic test conditions. The software can also be specialized for educational or diagnostic uses. However, there are primary limitations of using numerical simulations for pump performance:

- The system simulation software requires validation by comparison to a range of actual data. Initially, this can be obtained by analysis of previous experimental studies.
- It cannot be used to verify the device or validate the MCS system.
- It cannot directly address biocompatibility.
- There are a very large number of input parameters, and the user must be competent to select them.
- Ventricular interdependence parameters and bioreceptor reflex responses to contractility, vascular resistance, and beat rate that occur in response to autonomic nervous system inputs require mathematical algorithms to be included.

Although numerical simulation does not directly address biocompatibility, it can be used to compare operating conditions for various patient conditions. Once validated against *in vivo* performance for a particular device, it can be used to compare operating conditions over a range of disease states to inform operating condition input for comparative testing or computational fluid dynamics analysis.

The simulation output is only valid for well-defined patent conditions and disease states. For example, if the simulation inputs are "fine-tuned" to a patient sitting in a chair, it would only be valid while the patient is in the chair (sitting condition). As a result of these limitations, the simulation cannot be considered to be a "virtual patient."

The mathematical cores of the models used to simulate the hemodynamic environment at the pump ports are based primarily on an electrical analogy [1–5]. This distribution of lumped impedances allows sufficient fidelity to model hemodynamic conditions at the pump ports, which is sufficient for device development. A more complex network is needed to resolve the distribution of blood flow throughout the cardiovascular system to individual organs or extremities.

Considering uses beyond device development, a wonderful example of this type of more detailed and advanced simulation, designed for training and clinical uses, is given by Fresiello *et al.* [6]. This model has a broader distribution of impedances to resolve flows to internal organs. The software modules include a preset disease module and modules for baroflex responses, ventricular assist device (VAD) usage, drug infusion, and blood volume control. There is also a module that allows tuning of the simulation inputs to a clinical patient state, to aid in clinical decision-making. Another great example of MCS cardiovascular simulation (CARDIOSIM©) being developed for clinical use is given by Capoccia *et al.* [7], as an informative aid to help guide decision-making with MCS therapy. This software also allows patient-specific modeling (tuning to the patient) and the inclusion of various MCS devices.

Basic Physical Mock Loop Simulation does not replace the need for a physical mock loop in a development program. The performance of the pumping device over the range of required operating conditions must be characterized on a physical mock loop to verify the pump performance and indicate realistic inputs for the simulation. A basic system is a flow loop containing the MCS device, where the impedances are applied and controlled by manual means:

- Resistances – Manual valves are adjusted to discharge pressure loading on the MCS device. These can be simple flow restricting valves, pressure regulating valves, or a combination of both in parallel to allow fine-tuning of the resistance.
- Compliances – Reservoirs that allow volume change in proportion to pressure simulate the lumped compliances of systemic and pulmonary systems. These can be pneumatically pressurized for higher (systemic and pulmonary) pressures, a standpipe open to the atmosphere for lower (atrial) pressures, or a spring-loaded piston.
- Inertance – Tubing of different diameter and length can be used to simulate significant inertial impedances.

Active elements must be added to impose a pulsatile environment upon the tested MCS device, simulating the input of ventricles. A piston-type device or pulsatile VAD is typically used to simulate ventricular input. The four heart valves are simulated by common check valves or artificial clinical valves.

Limitations of the basic physical mock loop include:

- The MCS device and application must pre-exist to use a physical mock loop.
- Modeling of complex or combined disease states is difficult requiring special components and may be beyond practical experimentation.
- Precise repetition of test protocols across MCS device variations or for a series of devices is difficult due to manual operation.
- Biocompatibility cannot be directly demonstrated, although indications for hemolysis can be identified by a specialized loop for blood circulation.
- Autonomic responses such as bioreceptor reflex and ventricular interdependencies must be imposed manually.
- It must be used in a laboratory environment.
- The cost of development and continued maintenance of the test loop and instrumentation system can be significant, depending on the complexity of the system.

Controlled Physical Mock loop (Hybrid Mock Loops) This combines the first two approaches. A hybrid mock loop is created by enhancing the manual controls in a basic mock loop with electrical activation, guided by numerical simulation of the system. This overcomes some of the limitations of both the basic mock loop and numerical simulation. Primarily, the actual hardware can be tested repeatedly with precision over a range of simulated complex or combined disease states and tested in real time with developmental control systems. Related efforts [8–10] use a hybrid system of a physical mock loop coupled to a simulation model-driven control system that can create a wide range of physiologic test conditions, including autoregulatory responses, not otherwise practical to achieve with a basic mock loop. These hybrid systems have the great advantages of using a real MCS device instead of pump simulations and of being an excellent platform for developing real-time physiologic control schemes.

Obvious disadvantages of the hybrid systems are that the user needs total command of all elements, the physical mock loop, instrumentation, control system, and a pre-existing MCS device, and must bear the cost of expert development, maintenance, and use of the system.

A Numerical Simulation Example: The Virtual Mock Loop (VML)

A numerical simulation tool has been developed to study all the MCS applications at our institution [11–14], including LVADs, RVADs, BVADs, and continuous-flow total artificial hearts (CFTAH). A description for the use of this VML, including simulation for an example progression of disease states, including combinations of left systolic, right systolic, and aortic valve regurgitation in interaction with LVAD and BVAD support was reported recently [15]. The VML-based estimation of pump performance and self-regulating capability of the device over wide range of conditions was successfully demonstrated [16] for our CFTAH that was experimentally evaluated using the CFTAH bench mock circulatory loop.

Mathematical Core

This simulation has been developed with particular focus on the hemodynamic environment at the pump ports. The software was coded in MATLAB (MathWorks®, Natick, MA, USA) and simulates the hemodynamics of the human cardiovascular system using a lumped-parameter distribution of impedances for systemic and pulmonary circulation, as well as characteristics of the cardiac chambers, valves, and mechanical support devices. Numerous examples of cardiovascular simulations are available from the literature to build on [1–10]. The mathematical model is commonly described using electrical analogs. Lumped parameters representing the systemic and pulmonary vasculatures are modeled with resistors, capacitors, and inductors. The four heart chambers can be represented with capacitors that continuously vary in capacitance over time in a pattern that approximates beating ventricles and atria. The model for the four valves is based on the comprehensive work described by Korakianitis and Shi [1].

Inputs include values for impedance, systolic and diastolic ventricular and atrial compliances, beat rate, changes in blood volume, and pump characteristic performance. The primary references [1, 2] we used to guide the creation of the cardiovascular impedance core were chosen because of their comprehensive content.

A schematic diagram of the simulation (Fig. 27.1) shows the distribution of impedances in terms of an electrical analogy of resistors (representing vascular resistance), capacitors (compliance), and inductors (inertance). Equation 27.1 governs the fluid acceleration between compliant vessels, as a function of the impedances of resistance, compliance, and inertance.

Fluid acceleration between compliant vessels

$$\frac{dQ}{dt} = \frac{{}^{v_1}\!/_{c_1} - {}^{v_2}\!/_{c_2} - Q \cdot r}{i} \tag{27.1}$$

In Eq. 27.1, Q = flow (mL/s); v_1 and v_2 = volume of the output and input vessel, respectively (mL); c_1 and c_2 = compliance of the output and input vessel, respectively (mL/mm Hg); i = inertance associated with the flow (mm Hg•s^2/mL); and r = resistance associated with the flow (mm Hg•s/mL).

Elastance of the left and right ventricles and atria is modeled by modulating compliance per Equations sets 27.2 and 27.3 in time with the input heart rate.

Ventricular elastance.

$$\text{ventricular elastance}(t) = e_d + (e_s - e_d) \cdot e_v(t)$$

$$e_v(t) = \begin{cases} 1 - \cos\left(\frac{t}{t_a} \cdot \pi\right) t < t_a \\ 1 + \cos\left(\frac{t - t_a}{t_b - t_a} \cdot \pi\right) t_a < t < t_b \\ 0 \, t_b < t \end{cases}$$

$$t_a = \frac{2}{3} \cdot sys \cdot \left(\frac{60}{bpm}\right)$$

$$t_b = sys \cdot \left(\frac{60}{bpm}\right) \tag{27.2}$$

In Equations sets 27.2 and 27.3, bpm = heart rate in beats per minute; sys = percent systole; e_d

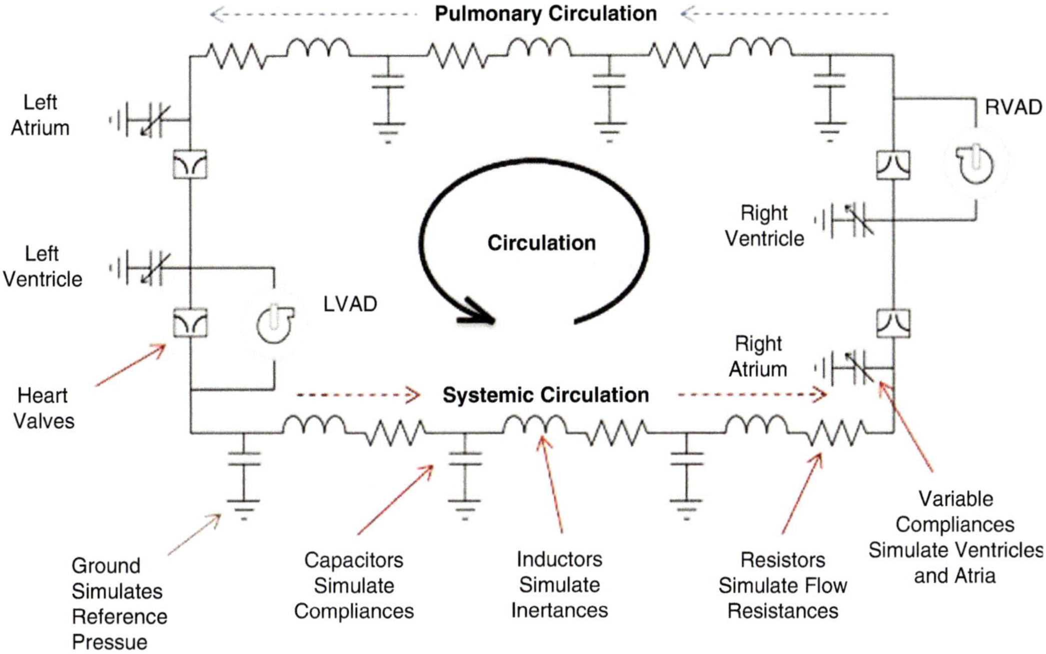

Fig. 27.1 Schematic diagram of simulation circuit using electrical analogies

and e_s = diastolic and systolic elastance, respectively (mm Hg/mL). Ventricular and atrial elastance waveforms are shown (Fig. 27.2) for example input values of systolic and diastolic elastance, beat rate, and percent systole. Note the peak and minimum waveform values for systolic and diastolic elastance are equal to e_s and e_d.

Atrial elastance.

$$atrial\ elastance(t) = e_d + (e_s - e_d)\cdot e_a(t)$$

$$e_a(t) = \begin{cases} 1 - \dfrac{\cos\left(\dfrac{2\pi(t - t_c + 1)}{t_d}\right)}{2} & t < t_c + t_d - 1 \\ 0 & t < t_c \\ 1 - \dfrac{\cos\left(\dfrac{2\pi(t - t_c)}{t_d}\right)}{2} & t \geq t_c \end{cases}$$

$$t_c = 0.92\frac{60}{bpm}$$

$$t_d = 0.09\frac{60}{bpm} \tag{27.3}$$

An increase in ventricular elastance as the end-diastolic volume is approached is imposed by the relationship of Eq. 27.4.

End-diastolic elastance.

$$ventricular\ elastance(t) = ventricular\ elastance(t) + \frac{1}{5}\left(\frac{v}{edvl}\right)^6 \tag{27.4}$$

In Eq. 27.4, v = ventricular volume (mL) and $edvl$ = the end-diastolic volume limit (mL).

Each of the four valves are modeled as swinging check valves with characteristic inputs of orifice loss coefficient, leakage area (for regurgitation), blockage area (stenosis), and dynamic characteristics relating to inertia and damping. Equations 27.5 and 27.6 are the governing equations for valve position and fluid acceleration; the valve static pressure drop as a function of flow and valve position is shown in Eq. 27.7. Regurgitation is modeled by imposing a

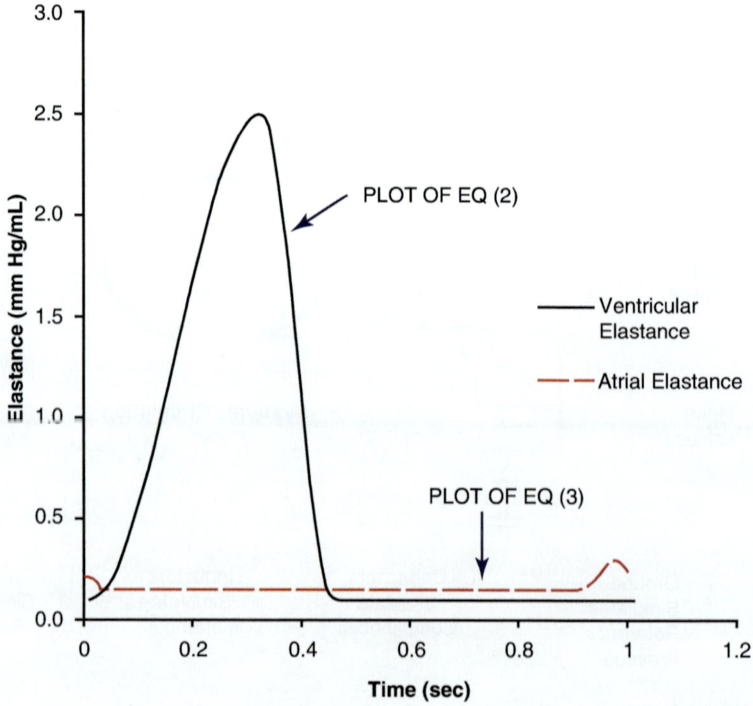

Fig. 27.2 Ventricular and atrial elastance waveforms (from Eqs. 27.2 and 27.3) shown for example inputs of ventricular systolic elastance = 2.5 mm Hg/mL, ventricular diastolic elastance = 0.1 mm Hg/mL, atrial systolic elastance = 0.25 mm Hg/mL, atrial diastolic elastance = 0.15 mm Hg/mL, pulse = 60 bpm, and percent systole = 45

lower limit to valve position to prevent complete closure and to valve stenosis by imposing an upper limiting value to prevent complete opening.

Valve position acceleration.

$$\frac{d^2\theta}{dt^2} = c_p \cdot \Delta P - c_f \cdot \frac{d\theta}{dt} \tag{27.5}$$

Valve fluid acceleration.

$$\frac{dQ}{dt} = \frac{\Delta P - P(Q)}{c_i} \tag{27.6}$$

Valve static pressure drop.

$$P(Q) = \frac{|Q| \cdot Q}{\left(c_v \cdot \left(\frac{1-\cos\theta}{1-\cos 1.3} \right)^2 \right)^2} \tag{27.7}$$

In Eqs. 27.5, 27.6, and 27.7, θ = valve position (range, 0–1.3 radians); Q = flow across the valve (mL/s); $P(Q)$ = static pressure drop as a function of flow and valve position (mm Hg); ΔP = pressure difference across the valve (mm Hg); c_p = valve dynamic constant for pressure (rad/[s^2•mm Hg]); c_f = valve dynamic constant for friction (1/s); c_i = valve dynamic constant for inertia (mL•s^2/mL); and c_v = valve orifice constant (mL/[s•(mm Hg)$^{.5}$]).

The circulatory support pump performance models are based on similitude properties of rotodynamic pumps. A family of head curves (pressure rise vs. flow) at different speeds can be calculated, based on the shape of the head curve and the specified design point in terms of flow, pressure rise, and speed (Fig. 27.3). A generic head curve shape is selected, or the head curve shape for a specific pump can be input as an equation, normalized with the design point ratios (flow/design point flow, pressure rise/design point pressure rise), and mapped to the values of (1,1). Static pump pressure rise for all flows and pump speeds can then be scaled from the normalized head curve. A dynamic head curve (with pulsing flow due to ventricular input) has a hysteresis loop due to inertia of the fluid within the pump. An inertia term is used in Eq. 27.8 to simulate the dynamic transport lag between pressure and flow found in a dynamic head curve to obtain instantaneous pump pressure rise with pulsing flow.

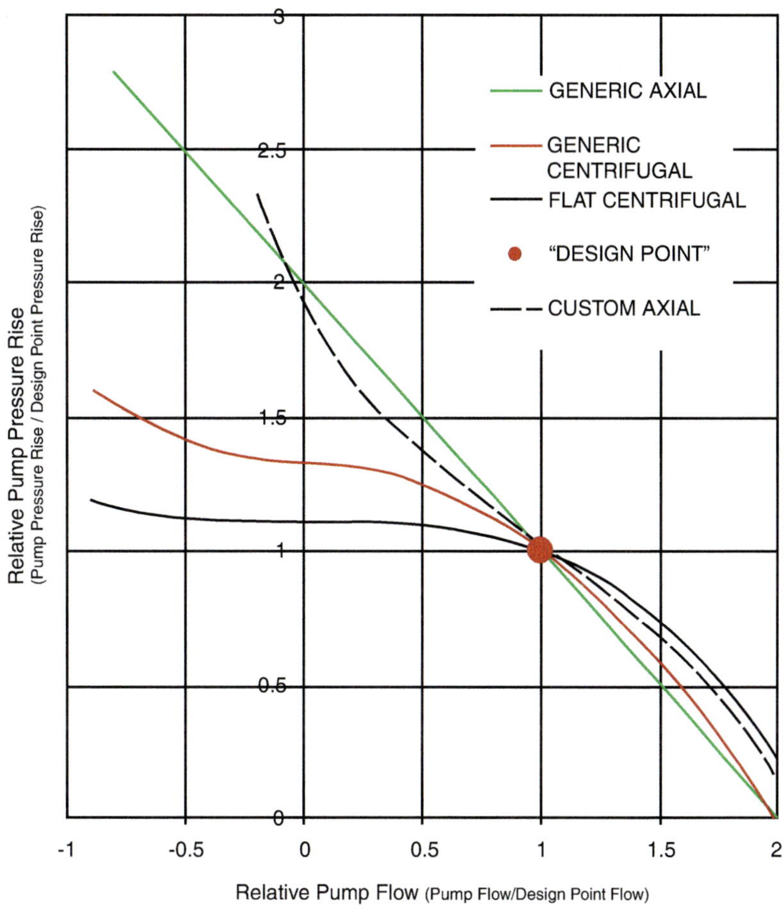

Fig. 27.3 Static head curve shapes for the input of pump performance

Dynamic pressure rise across pump.

$$\Delta P = P(Q) - |Q| \cdot Q \cdot r - \frac{dQ}{dt} \cdot i \quad (27.8)$$

In Eq. 27.8, ΔP = instantaneous pressure rise across the pump (mm Hg); Q = pump flow (mL/s); $P(Q)$ = pressure rise from the static pump head curve (mm Hg); r = pump conduit resistance (combined losses at pump ports) (mm Hg•s^2/mL^2); and i = inertance of fluid in the pump (mL•s^2/mm Hg). The equation also allows for an increase in differential pressure resulting from reverse flow (negative values of Q) through the pump.

The simulation input values for "normal" were taken from selected literature sources, primarily Korakianitis and Shi [1]. A selectable distribution for the inputs was linked to the disease-based input panel, corresponding to the relative severity of disease states. The system's output includes pressure, volume, and flow values throughout the cardiovascular system/pump environment. An input user interface allows selection of a variety of single- or multiple-disease conditions, including stenosis or regurgitation in each of the four valves.

Simulation Inputs

System inputs are made with user interface panels that allow input of simulated patient conditions in either a predefined disease-based mode or by inputting detailed values of impedance distribution.

Disease-Based Input

The disease-based input (Fig. 27.4) allows the user to specify a variety of different conditions that affect the patient. Conditions can be selected individually or in combination. A scaling function can adapt the inputs for subject body weights, with a nominal blood volume of 80 mL/kg. As selections are made, the detailed hemodynamic inputs in the windows described below are populated with default values consistent with the combination of disease conditions and selection of blood volume. All default values can be overwritten by manual entry.

MCS Device Input

Input of left and/or right ventricular assist device (VAD) generic pressure vs. flow and speed characteristics is available under the pump-type menu selection (Fig. 27.5). Characteristic performance for specific pumps can be added. Both the design point and the operating point are input to allow scaling of pump flow vs. pressure characteristics at different speeds. For example, if a "generic centrifugal" pump type is selected with a design point of 78 mm Hg, 4 LPM, and 2800 rpm, the resulting head curves are very similar to those of the HeartWare HVAD (Medtronic, Inc., Fridley, MN, USA) left ventricular assist device (LVAD). BVAD support is simulated by inputting both a left (LVAD) and right ventricular assist device (RVAD). Our continuous-flow total artificial heart (CFTAH) can be simulated using the BVAD inputs with left and right ventricular contractility and also mitral and tricuspid valve pressure drops, all set to zero. To simulate the self-regulating CFTAH, a left/right balancing function is based on the pressures and flows at the four pump ports [16].

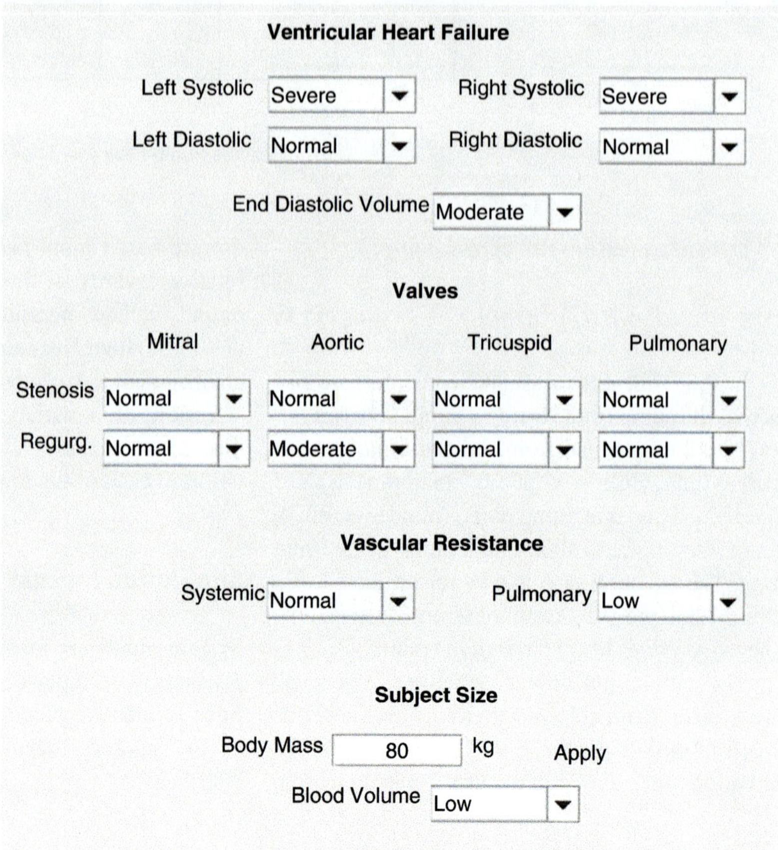

Fig. 27.4 Disease-based inputs for simulations, allowing ranges of disease severity to be specified in combination. Regurg, regurgitation

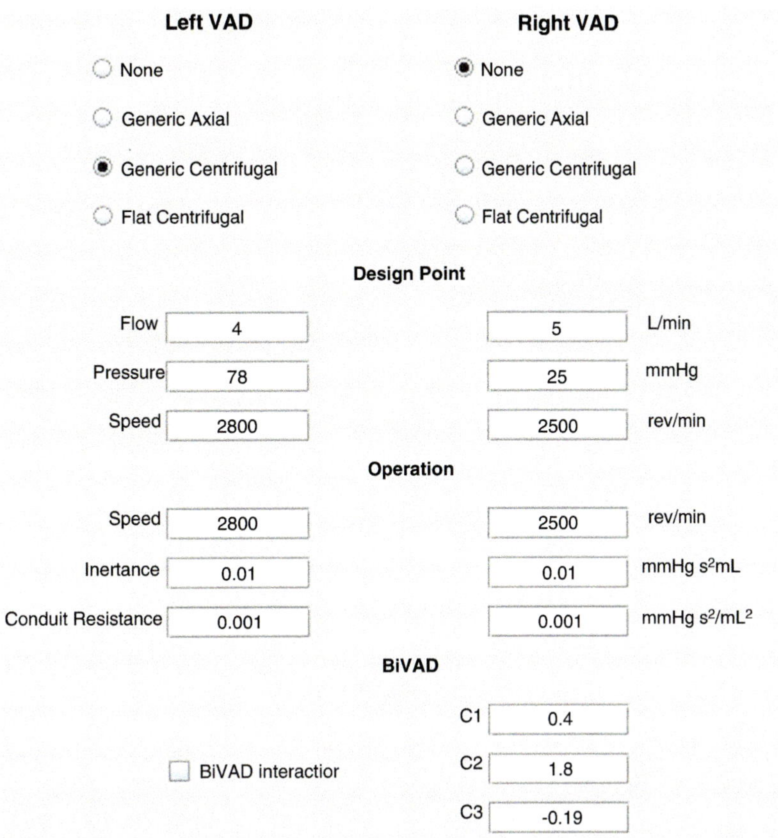

Fig. 27.5 Inputs used to model MCS devices. To model a CFTAH, both left VAD and right VAD are selected, and mitral and tricuspid valve regurgitation set to 100%, and the ventricular compliances set to low values

Valve Input

The variables for Eqs. 27.5, 27.6, and 27.7 comprise the valve inputs. Orifice resistance, percent regurgitation, percent stenosis, and the dynamic constants are input separately for each valve, allowing simulation of combinations of valve disease.

Heart Parameter Input

Compliance inputs are provided for the four heart chambers, end-diastolic volume, beat rate, and percent systole. Systolic and diastolic compliance inputs govern the strength of the atrial and ventricular contractions. Compliance is the inverse of elastance.

Vascular Impedance Input

The distribution of impedance values – input as a series of values for compliance, resistance, and inertance – is lumped for the main arteries, arterioles, and capillaries for the systemic and pulmonary circulations. Values for vascular resistance are linear, meaning pressure drops are proportional to flow rate.

Comparison with Clinical LVAD Flow and Pressure Waveforms

The simulation demonstrates realistic flow and pressure behavior. Because of the availability of recently published clinical experience [17], the HeartWare HVAD was used for the com-

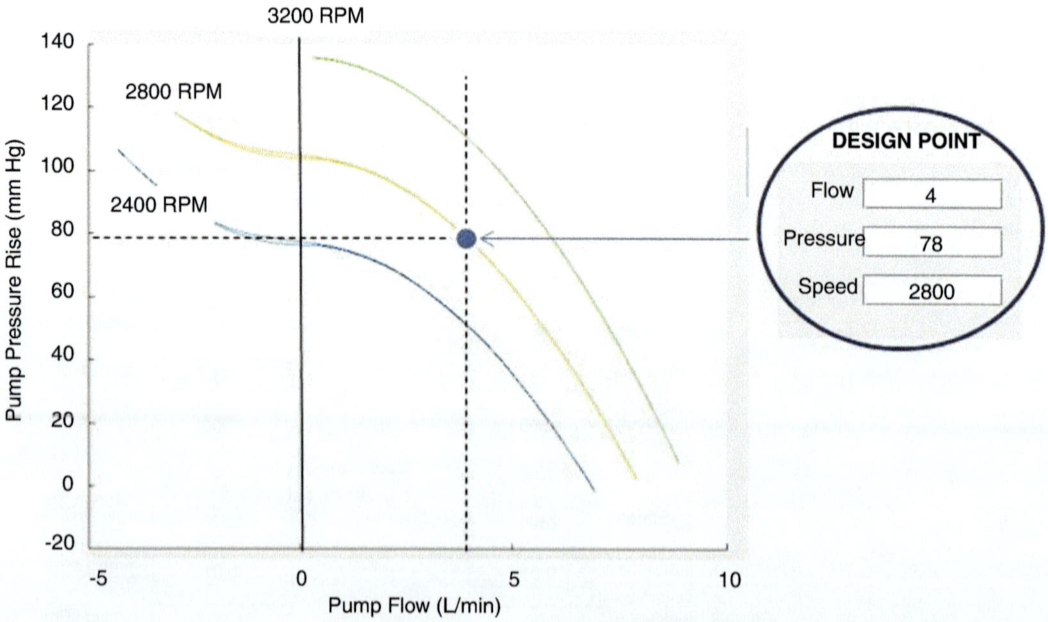

Fig. 27.6 Example simulation inputs used to create HeartWare HVAD static head curves. Example static pump head curves generated by virtual mock loop inputs set to simulate HVAD pump. Pump inertance set to near zero to eliminate hysteresis from curves to show static performance

parisons of simulated vs. actual waveforms for pump flow and ventricular pressure and systemic pressure.

The comparison was made for systolic failure, with LVAD pump inputs set to simulate the HVAD performance static head curves (Fig. 27.6). To create static head curves, the pump inertia input was first set to near zero, so that the hysteresis caused by dynamic conditions would also be near zero. The pump inertial input was then set back to an estimated value of.01 mm Hg•s^2/mL to perform the simulations. Other inputs reflect the baseline normal condition indicated in Table 27.1, except that the systolic compliance was changed from 0.4 to 1.4 mL/mm Hg to simulate a moderate systolic heart failure condition, and the beat rate to 92 bpm. The resulting simulation and actual waveforms for aortic and left ventricular pressures (Fig. 27.7), and the corresponding pump flow waveforms (Fig. 27.8) are shown to be characteristically similar.

An Example Use of the VML: A Progression of Disease States

To illustrate how the model virtual mock loop can be used, example solutions for a hypothetical progression of disease cases leading to severe biventricular disease, with an increasing level of support by MCS devices, are shown in Table 27.1 and Figs. 27.9, 27.10, 27.11, 27.12, 27.13, and 27.14. Following the normal baseline (Case A), the simulated progression starts with the introduction of left systolic failure with an increase in left diastolic volume (Case B). An LVAD is next added (Case C), followed by severe right heart failure (Case D). Severe aortic valve regurgitation is introduced (Case E), followed by an increase in LVAD pump speed (Case F). Next, BVAD support is simulated via the addition of RVAD support (Case G). Finally, the contribution of the BVAD support is shown by removing both the LVAD and RVAD (Case H).

Table 27.1 Virtual mock loop outputs simulating a progression of disease cases with LVAD and BVAD

Case	Disease	VAD	Result	Total flow [L/min]	Pump flow [L/min]	MAP [mm Hg]	PAP [mm Hg]	LAP-RAP [mm Hg]	LAP [mm Hg]	RAP [mm Hg]
A	Nominal "patient" without heart disease	None	Default normal EF 60%	5.5	N/A	99	26	3.1	10.3	7.2
B	Introduce left systolic failure with increased end-diastolic volume	None	Decreased flow; EF 24% Increased LAP	2.9	N/A	54	30	19.4	20.0	6
C	Same as B; LVAD introduced	Centrifugal; nominal speed	Aortic valve closed Increased flow and MAP	4.6	4.9	92	28	11.5	14.0	2.5
D	Introduce severe right heart failure due to LVAD/IVS constraints	Centrifugal; nominal speed	Increased RAP Decreased LAP	4.2	4.4	89	21	−1.2	8.5	9.7
E	Same as D + severe aortic regurgitation	Same as D	Increased pump flow Decreased MAP	3.3	6.2	72	22	0.3	11.1	7.2
F	Increase VAD speed by 20%	Centrifugal; increase speed by 20%	Increased flow and MAP Decreased LAP Increased RAP	4.0	7.5	87	21	−1.8	8.5	9.8
G	Same as F + RVAD (BVAD)	LVAD + RVAD (BVAD) introduced	Increased flow and PAP Increased LAP Decreased RAP	4.4	7.7 L 4.7R	89	28	8.8	12.4	3.6
H	Same as G without BVAD support	None	Severe combined biventricular failure and aortic regurgitation	2.0	N/A	41	20	8.8	14.5	5.7

Note: Beat rate, % systole; and total blood volume inputs were held constant for all example cases
VAD ventricular assist device;,*MAP* mean arterial pressure, *PAP* pulmonary arterial pressure, *LAP* left atrial pressure, *RAP* right atrial pressure, *EF* ejection fraction, *LVAD/RVAD/BVAD* left/right/biventricular assist device, *IVS* intraventricular septum, *L* left, *R* right

A hemodynamic summary for each case is output with a menu of graphical outputs for mean system pressures and flows. In Fig. 27.9, panel A shows the hemodynamic summary, and panel B shows dynamic intraventricular and arterial pressures for the baseline Case A.

Selected pressures and flows can be plotted to compare different simulation cases (Fig. 27.10) comparing normal and BVAD cases. The effects of BVAD support on aortic and pulmonary artery pressures are shown with and without BVAD support (Fig. 27.11). LVAD and RVAD dynamic head curves (Fig. 27.12) indicate pump pressure and flow pulsatility for Case G, BVAD support. Simulations of left intraventricular pressure vs. volume loops (Fig. 27.13) indicate the progression of disease and LVAD support on ventricular loading and aortic valve opening. Simulated VAD dynamic head curves (Fig. 27.14) are plotted in comparison for the disease cases to show the change in pump flow and pressure pulsatility.

Further work in process, based upon the VML core, includes continuing development of this approach for self-regulating CFTAH and BVAD, extracorporeal membrane oxygenation systems, atrial cannulation, and pulsatile devices. It is also being adapted for a CFTAH patient monitoring application.

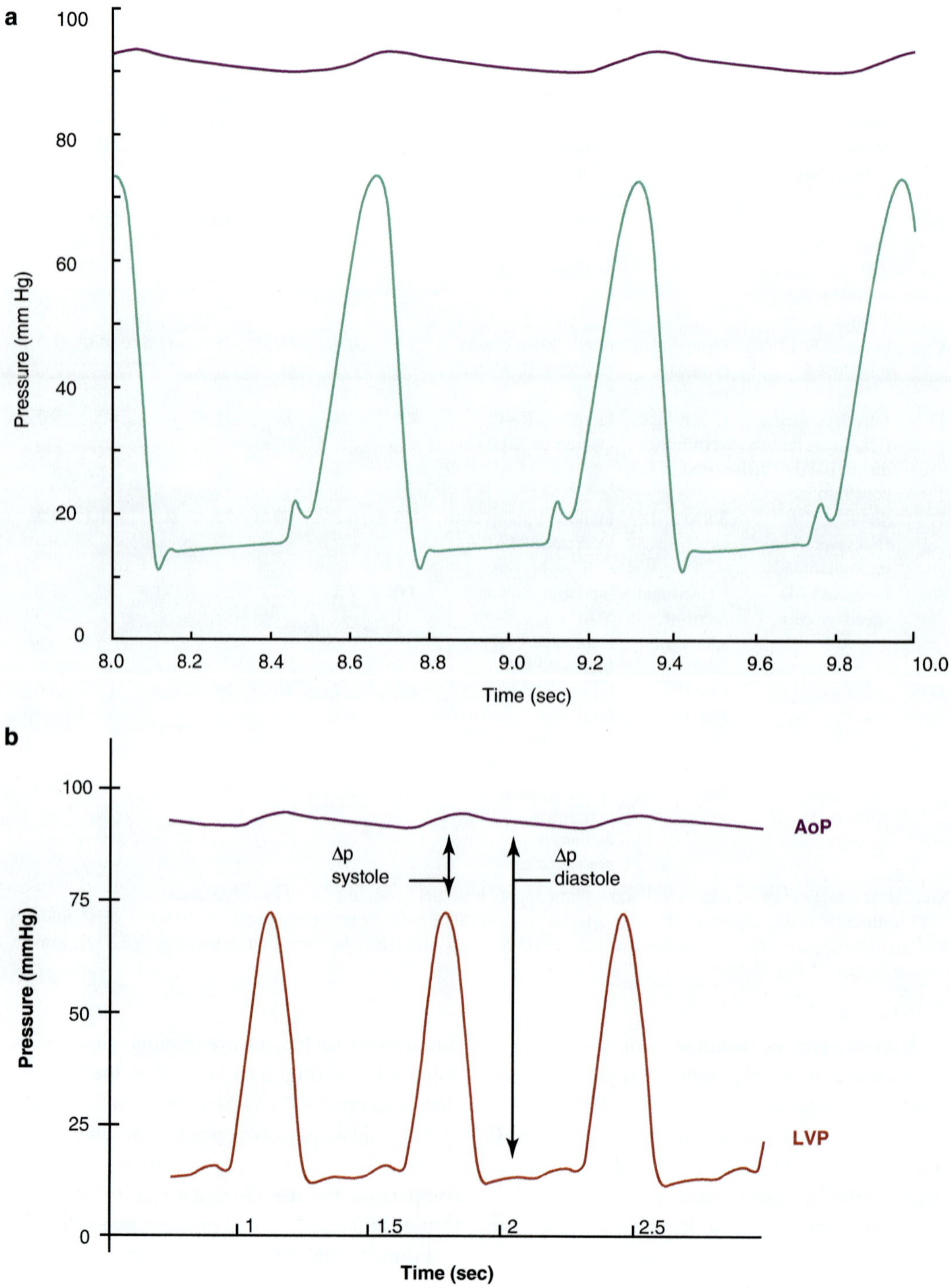

Fig. 27.7 Comparison of simulated aortic and left ventricle wave forms with clinical example. (**a**) – Simulation output (92 BPM, systolic compliance = 1.4 mL/mmHg). (**b**) – Reproduced from Rich JD, Burkhoff et al. [17] with permission from Wolters Kluwer Health, Inc.)

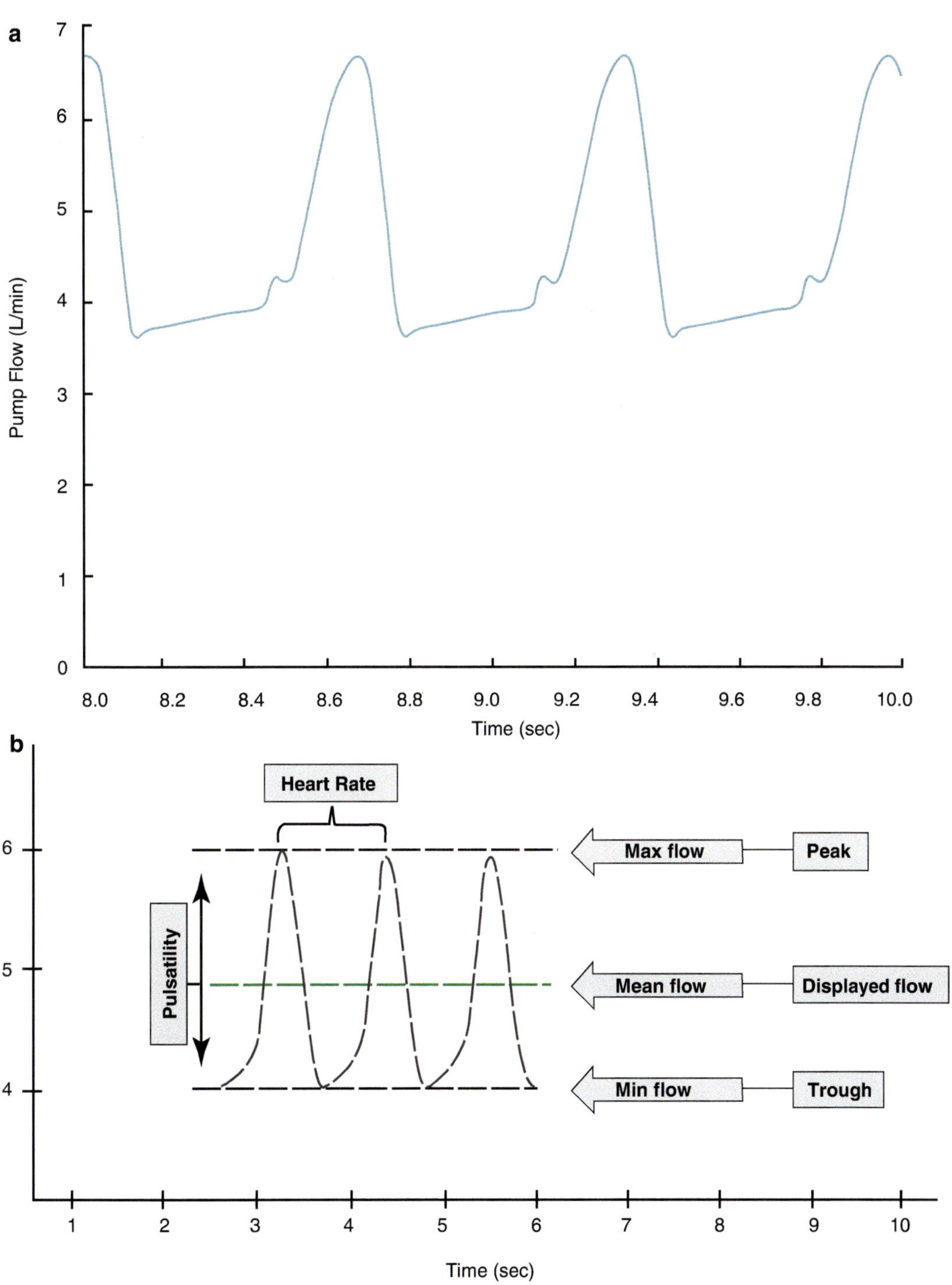

Fig. 27.8 Comparison of simulated clinical pump flow wave form with clinical example. (**a**) – Simulation output (92 BPM, systolic compliance = 1.4 mL/mmHg). (**b**) – Reproduced from Rich et al. [17] with permission from Wolters Kluwer Health, Inc.)

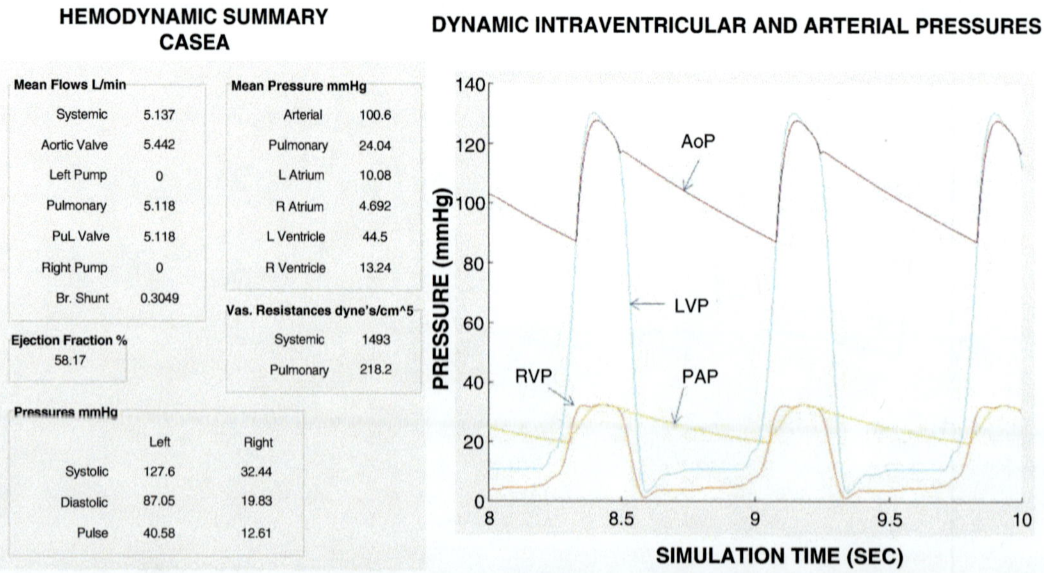

Fig. 27.9 Baseline conditions evaluated. (**a**). Hemodynamic summary for simulated Case A. (**b**). System pressures vs. simulation time for simulated Case A. *Pul. valve* pulmonary valve, *Br. Shunt* bronchial shunt, *L./R. atrium* left/right atrium, *L./R. ventricle* left/right ventricle, *Vas. Resistances* values for vascular resistance, *AoP* aortic pressure, *LVP/RVP* left/right ventricular pressures, *PAP* pulmonary arterial pressure

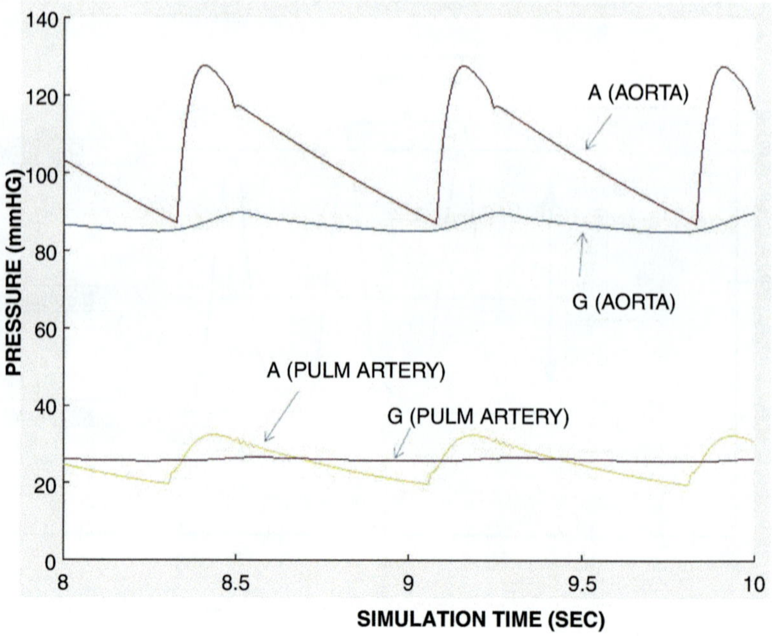

Fig. 27.10 Comparison of systemic and pulmonary pressures for the simulated conditions, normal (A) and diseased with BVAD (G). Aorta, aortic pressure; Pulm artery, pulmonary arterial pressure

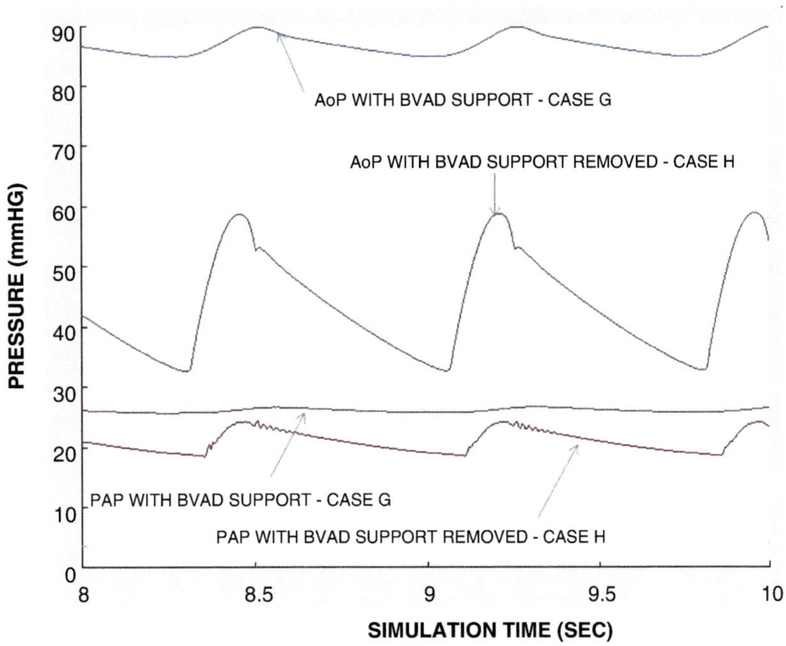

Fig. 27.11 Comparison of simulated Case G (BVAD) system pressures with those of Case H (G with BVAD support removed). AoP, aortic pressure; PAP, pulmonary arterial pressure; BVAD, biventricular assist device

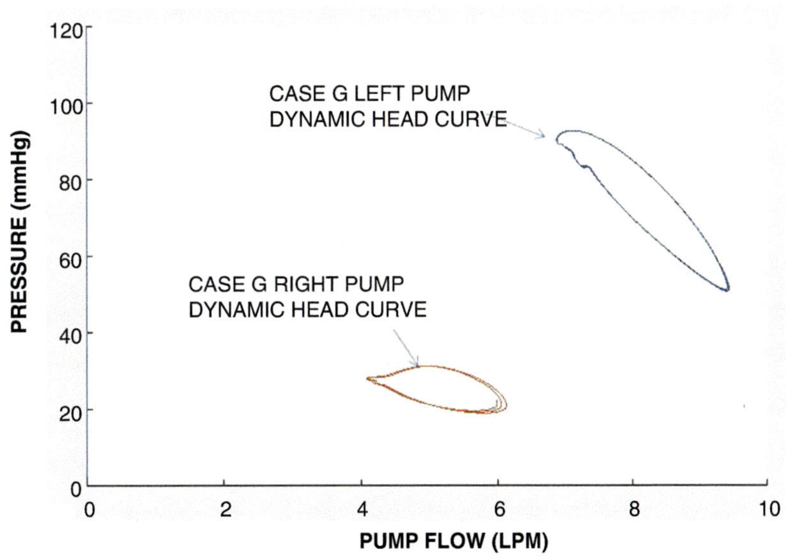

Fig. 27.12 Dynamic head curves for the left and right BVAD pumps of simulated Case G. LPM, liters per minute

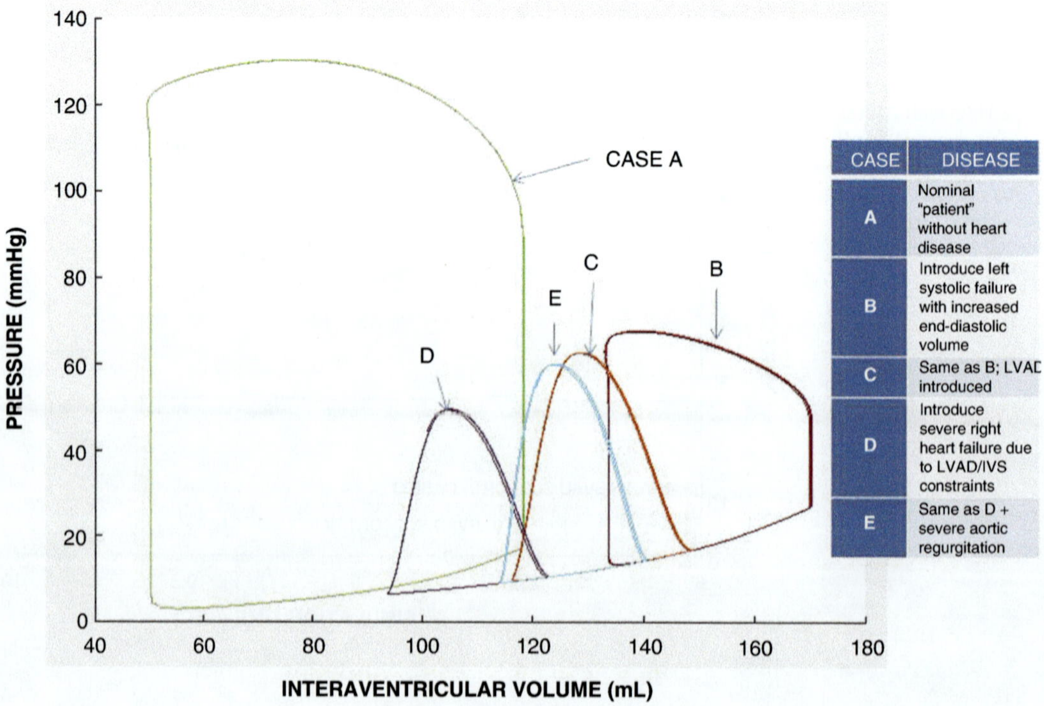

Fig. 27.13 Left ventricle pressure vs. volume loops for simulated Cases A through E. VAD, ventricular assist device; LVAD, left ventricular assist device; IVS, intraventricular septum

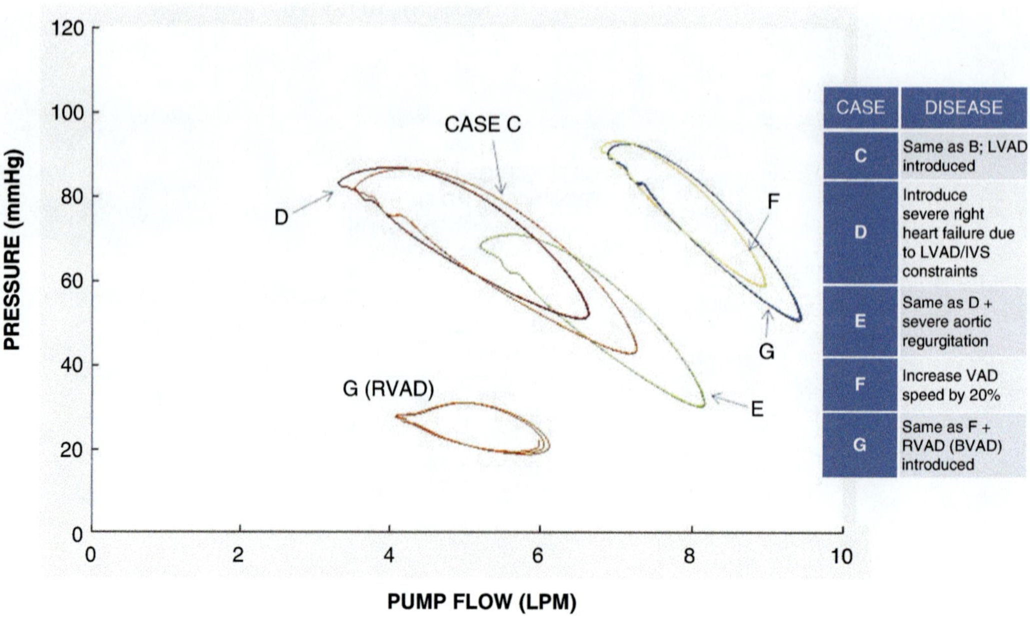

Fig. 27.14 Comparison of VAD pump dynamic head curves for simulated Cases C through G. VAD, ventricular assist device; LVAD/RVAD/BVAD, left/right/biventricular assist device; IVS, intraventricular septum

References

1. Korakianitis T, Shi Y. Numerical simulation of cardiovascular dynamics with healthy and diseased heart valves. J Biomech. 2006;39:1964–82.
2. Shi Y, Korakianitis T, Bowles C. Numerical simulation of cardiovascular dynamics with different types of VAD assistance. J Biomech. 2007;40:2919–33.
3. Hasania K, Navidbkhsh M, Rostami M. Simulation of the cardiovascular system using equivalent electronic system. Biomed Pap Med Fac Univ Palacky Olomouc Czech Repub. 2006;150:105–12.
4. Vollkron S, Schima H, Huber L, Wieselthaler G. Interaction of the cardiovascular system with an implanted rotary assist device: simulation study with a refined computer model. Artif Organs. 2002;26:349–59.
5. Bozkurt S, Safak KK. Evaluating the hemodynamical response of a cardiovascular system under support of a continuous flow left ventricular assist device via numerical modeling and simulations. Comput Math Methods Med. 2013;2013:986430.
6. Frisiello L, Feerrari G, Di Molfetta A, Zielinski H, et al. A cardiovascular simulator tailored for training and clinical use. J Biomed Inform. 2015;57: 100–12.
7. Capoccia M, Singh SA, De Lazzari C. The role of simulation for prospective planning in patients requiring mechanical circulatory support. EMBEC & NBC Proceedings. 2017:559–62.
8. Cuenca-Navalon E, Finocchiaro T, Laumen M, Fritschi A, Schmitz-Rode T, Steinseifer U. Design and evaluation of a hybrid mock circulatory loop for total artificial heart testing. Int J Artif Organs. 2014;37:71–80.
9. Nestler F, Bradley AP, Wilson SJ, Timms DL, Frazier OH, Cohn WE. A hybrid mock circulation loop for a total artificial heart. Artif Organs. 2014;38:775–82.
10. Gregory SD, Greatrex N, Timms D, Pearcy MJ, Fraser JF. Simulation and enhancement of a cardiovascular device test rig. J Simulat. 2010;4:34–41.
11. Horvath D, Byram N, Karimov JH, Kuban B, Sunagawa G, Golding LAR, et al. Mechanism of self-regulation and in vivo performance of the Cleveland Clinic continuous-flow total artificial heart. Artif Organs. 2017;41:411–7.
12. Horvath D, Karimov JH, Byram N, Kuban B, Golding LA, Moazami N, et al. Sensorless suction recognition in the self-regulating Cleveland Clinic continuous-flow total artificial heart. ASAIO J. 2015;61:726–8.
13. Karimov JH, Moazami N, Kobayashi M, Sale S, Such K, Byram N, et al. First report of 90-day support of 2 calves with a continuous-flow total artificial heart. J Thorac Cardiovasc Surg. 2015;150:687–93.e1.
14. Saeed D, Ootaki Y, Ootaki C, Akiyama M, Horai T, Catanese J, et al. Acute in vivo evaluation of an implantable continuous flow biventricular assist system. ASAIO J. 2008;54:20–4.
15. Horvath DJ, Horvath DW, Karimov JH, Byram N, Kuban BD, Miyamoto T, Fukamachi K. Use of a mechanical circulatory support simulation to study pump interactions with the variable hemodynamic environment. Artif Organs. 2018;42:E420–E427.
16. Miyamoto T, Horvath DJ, Horvath DW, Karimov JH, Byram N, Kuban BD, Fukamachi K. Simulated performance of the Cleveland Clinic continuous-flow Total artificial heart using the virtual mock loop. ASAIO J. 2018.
17. Rich JD, Burkhoff D. HVAD flow waveform morphologies: theoretical foundations and implications for clinical practice. ASAIO J. 2017;63:526–35.

Part V

Evolving Technologies and Device Concepts

Emerging Continuous-Flow Blood Pump Technologies

28

Harveen K. Lamba and Jeffrey A. Morgan

Introduction

Although cardiac transplantation is considered the "gold standard" for treatment of advanced heart failure, its epidemiological impact remains low since it is only available to a limited group of patients [1, 2]. In view of the increasing complexity of the donor allocation process, the insertion of a left ventricular assist device (LVAD) may well become the primary treatment for certain patients with advanced heart failure [3]. Since the landmark Randomized Evaluation of Mechanical Assistance for the Treatment of Congestive Heart Failure (REMATCH) trial for inotropic-dependent patients, the evolution from pulsatile to continuous-flow ventricular assist devices (CF-LVAD) has led to an increased survival of patients on prolonged circulatory support [4, 5]. The preliminary results of the Risk Assessment and Comparative Effectiveness of Left Ventricular Assist Device and Medical Management in Ambulatory Heart Failure Patients (ROADMAP) trial have already shown a beneficial effect in implanting CF-LVAD even in ambulatory, non-inotrope-dependent patients.

The challenges facing the field presently are durability issues related to driveline infection, device failure, thrombosis, hemolysis, and the development of a permanent circulatory support system which is anatomically adaptable. Rotary continuous-flow pumps were developed to improve pulsatile pumps or first-generation devices. They are smaller with more favorable durability due to minimal moving parts [6]. Continuous-blood flow generated in a nonphysiological manner avoids the needs for valves and compliance chambers. Pumps with continuous axial flow requiring mechanical bearings and seals in contact with the blood are known as second generation, while rotary pumps with continuous flow based on a magnetically levitated impeller are known as third-generation devices.

H. K. Lamba, MD, MSc (✉)
Division of Cardiothoracic Transplantation and Circulatory Support, Michael E. DeBakey Department of Surgery, Baylor College of Medicine/Texas Heart Institute, Houston, TX, USA
e-mail: Harveen.lamba@bcm.edu

J. A. Morgan, MD
Division of Congestive Heart Failure, Transplant and Mechanical Circulatory Support, Department of Cardiothoracic Surgery, Baylor College of Medicine – Texas Heart Institute, Houston, TX, USA
e-mail: Jeffrey.morgan@bcm.edu

Historical Perspective

Surgical intervention for end-stage heart failure really began in the 1960s with the description of the first clinical use of the intra-aortic balloon pump, the implantation of a ventricular assist device for postcardiotomy shock, and the first successful heart transplantation [7–9]. Initial LVAD designs were intended to simulate the

© Springer Nature Switzerland AG 2020
J. H. Karimov et al. (eds.), *Mechanical Support for Heart Failure*,
https://doi.org/10.1007/978-3-030-47809-4_28

heart's function by creating a pulse with a physiologic 1/3 systole and 2/3 diastole cycle. Limitations of the pulsatile pumps included their large size and weight, but, ultimately, poor durability due to membrane fatigue was the primary practical limitation. Improvements in technology, understanding of biocompatibility, patient selection, surgical technique, and perioperative management have led to significant growth in the field of left ventricular assist devices since their inception. Despite these advancements, issues related to device inconvenience, short battery life, need for a transcutaneous driveline, and device-specific complications have been barriers to a wider adoption of these devices. The most common causes of mortality with these devices are related to infections, bleeding, right heart failure, neurological deficits, and device malfunction [10]. Emerging technologies in continuous-flow mechanical circulatory systems (MCS) are based on past and present issues specifically related to the benefits and weaknesses of previous continuous-flow technology.

Recent Innovations

The new third-generation LVAD devices with noncontact bearings via magnetic levitation have been developed to allow for rotation without friction or wear. Furthermore, these pumps are being designed to allow for remote monitoring capabilities to optimize flow to the body's demand and give clinicians an additional noninvasive tool to monitor device function and optimize treatment. Currently, the only FDA-approved devices employing this design are the HeartWare HVAD and the HeartMate III, while other devices under investigation include the HeartAssist 5.

The HeartWare HVAD is a miniaturized small intrapericardial centrifugal device with magnetic and hydrodynamic levitation. The inflow cannula is cored directly into the left ventricular apex so that the device is entirely above the diaphragm. The impeller rotates centrifugally more slowly than the axial rotary devices, which is thought to result in decreased hemolysis and thrombogenesis and greater mechanical durability. The HVAD was FDA-approved for bridge to transplantation (BTT) in 2012 and has demonstrated a 91% 6-month and 84% 1-year survival [11]. The ENDURANCE trial was designed to evaluate the HVAD as destination therapy (DT), and it demonstrated noninferiority of the HeartWare HVAD as compared to the HeartMate II. However, stroke was more common in HVAD subjects, and post hoc analyses demonstrated increased mean arterial blood pressure as a significant independent risk factor for stroke [12]. The follow-up ENDURANCE supplemental trial was designed to prospectively determine effectiveness of a blood pressure management strategy, and the trial confirmed that blood pressure management resulted in reduced stroke rates in HVAD subjects [13].

The HeartMate III is a third-generation centrifugal intrapericardial VAD with fully magnetically levitated rotors with wider blood flow gaps that can potentially minimize hemolysis, thrombogenesis, and von Willebrand syndrome [14]. Additionally, impeller speeds alternate every 2 seconds to promote "pulsatile" flow. The MOMENTUM 3 clinical trial was designed to evaluate the HeartMate III device for BTT and DT compared to the axial continuous-flow HeartMate II pump [15]. At 2 years, the fully magnetically levitated HeartMate III was superior to the mechanical-bearing axial-flow HeartMate II with regard to survival free of disabling stroke or reoperation to replace or remove a malfunctioning device [16].

In addition to the potential benefits of a smaller, intrapericardial design and decreased device malfunction, new generations of LVADs may allow for noninvasive monitoring of pump function and flow. The HVAD allows for flow waveforms that provide information about HVAD function and patient hemodynamics. The HeartAssist 5 pump is a fifth-generation axial-flow impeller device similar in configuration to the HeartMate II but considerably smaller and is placed in an extrapericardial supradiaphragmatic pocket. It is currently undergoing clinical trial for use as BTT and takes remote monitoring one step further by providing direct flow measurements through a "cellphone system" within the controller – a wireless transmitter that transmits flow, power, and speed data every 15 minutes. These parameters as well as alarm notifications can then be promptly delivered to healthcare providers via text messages or email. In addition to flow data, the HeartAssist 5 tracks speed and electrical current usage by the pump motor providing information about the volume of blood flow and its fluidity. A randomized prospective trial is currently comparing 180-day success among the HeartAssist 5, HeartMate II, and HVAD (ClinicalTrials.gov NCT02205411).

The HeartWare MVAD is a miniaturized (78 g) VAD that is significantly smaller than other devices with a displacement volume of 20 mL. In comparison, the Thoratec HeartMate II and HVAD weigh 290 and 160 g, respectively. The MVAD system combines centrifugal and axial flow into a longitudinal pump. It provides antegrade ejection alongside the native flow that can be varied from full to partial assist by adjusting pump speeds [17]. A platinum alloy impeller is suspended in a ceramic tube by passive magnetic and hydrodynamic forces. A gimbal sewing ring allows modification of device direction and pump depth, with the potential for right ventricular placement and biventricular assist device (BiVAD) support. However, introduction of the MVAD has been delayed due to technical problems including pump thrombosis. The nonrandomized MVAdvantage Trial (ClinicalTrials.gov NCT01831544), designed to study MVAD safety and performance over 24 months, with a primary end point of survival at 6 months, is currently on hold.

Patient Selection

New device designs and future innovations offer great potential to enhance treatment outcomes in advanced heart failure patients. Additionally, appropriate patient selection and timing of LVAD implantation maximizes therapy outcomes. A better understanding of the patient profile that derives the most benefit from LVAD technology has allowed improved survival outcomes. This change is exemplified by the noticeable reduction INTERMACS profile 1 ("crashing and burning") patients at the time of implantation from 44% to 14% in recent years, resulting in 6-month survival rates of 90% or higher [18]. Recently, the ROADMAP study showed that earlier implantation in less sick, nonhospitalized patients results in statistically better survival [19]. Twelve-month survival was greater in the LVAD versus medical management group in this patient group, as was health-related quality of life, although the LVAD group experienced more frequent adverse events and hospitalizations.

On the other hand, the Randomized Evaluation of VAD InterVEntion before Inotropic Therapy (REVIVE-IT) pilot study was recently canceled due to concern over equipoise due to increased risk of device thrombosis. The REVIVE-IT study began in 2009 to investigate the benefits of LVAD implant strategy versus medical management in patients with less advanced heart failure than current LVAD indications to determine if wider application of permanent LVAD use or destination therapy to a less ill population would be associated with improved survival, qualify of life, or functional capacity [20, 21]. The trial was discontinued without ever beginning in 2015 [22]. These studies provide risk-benefit information to assist in patient counseling and physician decision-making for elective LVAD therapy for non-inotrope-dependent patients, but making an informed clinical choice between LVAD and medical therapy remains a challenge.

During the last decade, introduction of individual patient risk profiles have been developed and validated to aid in improved patient selection [23–26]. Because heart failure is a heterogeneous condition, risk stratification needs to be sophisticated and individualized, and although there are published guidelines for patient selection for heart transplantation, there is limited guidance for patient selection for LVAD therapy. Validated tools, such as the seattle heart failure model and the heart failure survival score, tend to underestimate absolute risk and overestimate survival in end-stage HF. New York Heart Association classification, although commonly used in heart failure, is subjective, inconsistent, and susceptible to significant inter-observer variability. The INTERMACS profile of the patient at the time of LVAD implantation allows for prognostic discrimination, with INTERMACS profile 1 patients having the worst outcomes [27]. The HeartMate II risk score (HMRS) was introduced as a measure to predict 90-day and 1-year mortality in patients receiving a continuous-flow HeartMate II (Thoratec, Pleasanton, CA, USA) LVAD [25]. It was derived from and validated within 1122 patients enrolled in the original HeartMate II clinical trials. Although simplistic and easy to calculate, its applicability in routine clinical practice has been quite limited in the diverse population of patients being considered for contemporary LVADs [28]. Machine learning methods using Bayesian analysis can overcome some of the limitations of existing risk scores and are being evaluated for risk stratification for both mortality and major adverse events after LVAD implantation [29]. This risk stratification model is an improvement over HMRS and maintains its accuracy irrespective of device type or device indication. Future integration of this risk tool with electronic health records will allow for easy, rapid risk assessment at the bedside, which will greatly enhance patient selection and clinical decision-making for LVAD therapy.

Surgical Technique

Minimally Invasive Techniques

Due to the continuous process of miniaturization of VADs, the past several years have shown an increased interest in less invasive approaches to LVAD implantation. Potential advantages include reduced trauma, blood loss and coagulopathy, arrhythmogenic complications, as well as reduced length of hospital stay, and safer sternal reentry at time of transplantation [30–34]. An additional advantage of a less invasive approach is minimization of right ventricular dysfunction; the right ventricle remains in its natural position since the pericardium stays intact and the right ventricular function remains sustained.

The most common "less invasive" surgical strategy involves partial sternotomy to access the great vessels with left thoracotomy to access the apex of the left ventricle, using either central or peripheral cannulation for cardiopulmonary bypass. LVAD inflow is obtained via the left ventricular (LV) apex, and outflow is to the ascending aorta. Sternal-sparing techniques have also been described with the outflow graft being sewn to either the ascending aorta (via right mini-thoracotomy) or the axillary artery [35, 36]. Although this approach avoids a full sternotomy, right-sided thoracotomy can induce more postoperative pain and may lead to more difficult access to the ascending aorta, which is crucial in redo or combined cardiac surgery [37].

Most recently the Medtronic HVAD received approval for HVAD implantation via thoracotomy, which was based on data from the LATERAL prospective clinical trial, in which 144 patients, with end-stage heart failure who were eligible for heart transplant, were enrolled at 26 centers in the United States and Canada [38]. The primary endpoint of the trial demonstrated non-inferiority of the HVAD implanted in patients via thoracotomy, where survival at 6 months free from disabling stroke or device explant or exchange due to malfunction was achieved in 88.1 percent of patients. Since the success outcome exceeded the pre-specified performance goal of 77.5 percent, the trial achieved its primary endpoint ($p = 0.0012$). The key secondary endpoint revealed a significant reduction in total length of hospital stay, from an average of 26.1 days down to 18 days ($p < 0.001$). Overall survival among patients receiving an HVAD via the thoracotomy procedure was 88.8 percent at 1 year.

Even with these "less invasive" techniques, a significant thoracotomy with muscle splitting and rib spreading is required because of the profile of even the smallest pumps currently available. Small thoracic incisions may make direct access to the LV apex technically more challenging and result in improper placement of the inflow cannula. Similarly, limited exposure/access to the ascending aorta may encumber emergent cardiopulmonary bypass if needed [30]. The feasibility and safety of minimally invasive LVAD implantation in appropriate patients are fairly established, and outcomes will continue to improve with refinement of technology with newer-generation pumps.

Off Cardiopulmonary Bypass

Implantation of VADs is typically performed with the patient on cardiopulmonary bypass (CPB). However, CPB is associated with harmful effects well-documented in the literature, including activation of inflammatory mediators, increased pulmonary vascular resistance, platelet activation, coagulopathy, and impaired renal function [39]. VAD placement under CPB can exacerbate pre-existing end-organ dysfunction, resulting in postoperative coagulopathy, bleeding, and worsening right heart failure. VAD implantation without the use of CPB could help to minimize these postoperative complications without hemodynamic compromise or excessive bleeding during implantation. Off-pump VAD implantation relies on completion of all surgical steps, including driveline tunneling, outflow graft anastomosis, and de-airing before apical coring. Apical coring is accomplished during rapid ventricular pacing to minimize blood loss and facilitate efficient insertion of the LVAD inflow. The benefits of an off-pump strategy include reduced blood-product transfusion, reoperation for perioperative bleeding, acute kidney injury, and respiratory complications. Minimizing blood product transfusions and reducing exposure to blood antigens decrease the risk of recipient sensitization, thus preserving donor pool availability for bridge to transplantation.

Right Ventricular Support Devices

While right heart failure (RHF) and right ventricular assist device use in LVAD patients has declined due to improved patient selection and perioperative management, RHF remains a serious complication [40]. The incidence of RHF requiring a right ventricular assist device after CF-LVAD implantation has been reported between 5% and 20%, and these patients face a higher mortality than those with durable LVAD support alone [41]. A number of risk scores have been developed to predict right ventricular failure, but prediction of right ventricular failure remains imperfect. Current treatment options for short-term right ventricular support include variety of surgically placed or percutaneous devices: Abbott CentriMag, Abiomed Impella, and TandemHeart.

The CentriMag device consists of an extracorporeal, magnetically levitated blood pump that can provide RV, LV, or extracorporeal membrane oxygenation (ECMO) support, depending on cannulation strategy and the presence or absence of an oxygenator in the circuit. RVAD support with a CentriMag can be configured with transcutaneous cannulas: right atrial venous drainage and pulmonary arterial outflow via sternotomy. Strategies have been developed, with femoral venous drainage and a cannulated graft to the pulmonary artery (PA) that can facilitate discontinuation of RV support without the need for resternotomy. The device can be combined with an oxygenator to provide respiratory ECMO support in conjunction with support of the RV. This setup can be maintained for weeks with low rates of hemolysis and excellent support. Patients can be ambulated but must be maintained in the hospital.

The Abiomed Impella 2.5 is a percutaneous axial-flow device inserted via the femoral vein. It is delivered across the aortic valve via a femoral arterial sheath and sits in the RV outflow tract with the catheter traversing the pulmonic valve and delivering flow to the PA. However, it has a limited output capability of 2.5 liters/min and is marked for only short-term RV support (10 days) [42, 43]. While its benefits include ease of implant, it's limitations include the inability to ambulate or to use an oxygenator with the system as well as positional instability and hemolysis.

The Tandem Life Protek Duo is a specialized cannula designed to facilitate RV support. It can be placed either via the femoral or internal jugular vein and traverses the right heart terminating in the PA under fluoroscopic guidance. It is driven by the Tandem Life extracorporeal pump. A bearing in the pump requires heparinized saline infusion to prevent heat generation.

For more durable, long-term ventricular support, total artificial heart replacement may be considered. Some centers employ off-label use of the Jarvik 2000 or HeartWare HVAD for prolonged right ventricular support, but a major limitation of this approach is that these LVADs are better suited for systemic circulation with higher pressure and resistance. The right ventricle operates in a lower-pressure, less resistant, and more compliant system, and LVADs connected to the right atrium and pulmonary artery may be more prone to suction events.

Emerging durable right ventricular support includes the HeartWare MVAD (see above). Its small size and unique outflow port that originates at a 90° angle from the impeller makes it suitable for right ventricular and biventricular support [44]. The Circulite Synergy Micro-Pump (CircuLite Inc., Saddle Brook, NJ) is the size of an AA battery and has a pumping capacity of 4.25 liters/min and is small enough to be placed in a subcutaneous pocket. The Synergy Micro-Pump is a novel pump that combines axial, centrifugal, and orthogonal flow paths at a 7.5 mL volume and operating speeds of 20,000 to 28,000 rpm. It uses a magnetically and hydrodynamically stabilized rotor and is implanted via a mini right-sided thoracotomy into a subcutaneous pocket. The inflow cannula is placed into the left atrium, and the outflow graft is connected to the right subclavian artery [45]. Although clinical experience is currently limited to left ventricular use, results with recent cadaveric and animal models have been promising [46].

Partial Support

Previous experience with temporary percutaneous mechanical circulatory support such as intra-aortic balloon pumps as a bridge to transplantation has shown utility of partial mechanical support for end-stage heart failure patients. In addition to their potential for right ventricular support, the HeartWare MVAD and Synergy Micro-Pump offer ability to partially support the heart.

The Aortix system is a miniaturized axial-flow pump that represents a radical departure from current percutaneous devices in that it is intended to provide partial circulatory assist in less-sick patients (NYHA III). Passed through the femoral artery, it deploys expandable side anchors that secure the pump in the supradiaphragmatic descending aorta above the renal arteries. It has entraining jets that entrain blood to augment flow ("jet pumping") in series with the circulation, thus decreasing LV afterload while minimizing central thromboembolic risk. In an ischemic heart failure sheep model, the Aortix system not only decreased myocardial oxygen consumption and improved LV ejection fraction and CO but also increased renal blood flow and urine output [47].

The Texas Heart Institute is also developing a miniaturized LVAD that partially supports the left ventricle, the MicroVasc. The MicroVasc compromises a primary impeller, a motor, a washout groove surface, and a magnetic levitation bearing. Unlike other clinically available LVADs, all components of the MicroVasc are aligned in series to minimize pump diameter. Additionally, the MicroVasc is the first LVAD proposed for minimally invasive delivery in a catheterization lab procedure. It will be affixed to the atrial septum and will support the compromised left ventricle in early stage heart failure, thereby avoiding damage to the left ventricular tissues.

Patient Maintenance

An important barrier in the reduction of complications in the foreseeable future will be a consensus across centers on how to best take care of patients with mechanical support devices. The need for a more systematic and organized approach to these patients is highlighted by the Prevention of HeartMate II Pump Thrombosis Through Clinical Management (PREVENT) trial and the long-term balance of anticoagulation versus risk of bleeding as described from the results of the US-TRACE registry [48].

The importance of surgical technique came into recent scrutiny with the reported rise by some centers in the incidence of pump thrombosis following LVAD implantation, particularly the HeartMate II [49, 50]. PREVENT was a prospective, multicenter, single-arm, non-randomized study of 300 patients implanted with a HeartMate II at 24 sites. It recommended structured clinical practices including implantation technique, postoperative anticoagulation strategy, and pump speed management. Adherence to key recommendations of the structured PREVENT recommendations resulted in a significantly lower risk of pump thrombosis, lower composite risk of suspected thrombosis, hemolysis, and ischemic stroke at 6 months [51].

Total Artificial Heart

Severe biventricular failure and or complicated conditions that are not treated well with LVAD implantation are indications for TAH support. Currently, the only device that is available to provide biventricular support is the SynCardia Total Artificial Heart (SynCardia Systems Inc., United States), which is a pulsatile device that functions similar to those of first-generation LVADs. Continuous-flow technology now in widespread use as LVADs may also offer the best answer to a total heart replacement.

However, concerns associated with attenuated pulsatility in LVAD-assisted circulation may play larger role in the context of continuous-flow total artificial hearts, as the deviation of flow patterns from native physiology is more distinct. Most patients with biventricular VAD support retain some degree of cardiac function, and as such maintain some pulsatility. In the case of a continuous-flow total artificial heart implantation, the ventricles are removed and replaced by the device. Therefore, remnant ventricular contractility is omitted, and consequently a continuous-flow TAH generates a completely nonpulsatile outflow.

The most extensive research on continuous-flow total heart perfusion has been performed in a series of in vivo studies at the Texas Heart Institute (Houston, Texas, United States). In preparation for the shift to continuous-flow technology, Texas Heart Institute began investigating dual ventricular assist devices in calves in 2005. The native heart of each animal was excised and replaced with a pair of CF-LVADS to form a continuous-flow total artificial heart (CF-TAH) [52, 53]. Over the course of an 8-year study, CF-TAHs were implanted in 65 calves and were composed of a wide variety of pumps [54–61]. Two of the calves survived to the 90-day milestone, and 29 of them lived at least a week.

These studies lead to the eventual interest in a true continuous-flow total artificial heart. BiVACOR (Houston, TX, USA) and Cleveland Heart (Cleveland, OH, USA) each have developed rotary total artificial hearts that leverage the advantages of centrifugal pumps [62–64]. These rotary pumps lack any mechanical bearings or other sources of mechanical wear and no flexible components or valves, and both utilize a two-sided impeller. These devices are smaller and more energy efficient and have an intrinsic mechanism that allows systemic and pulmonary balance to be adjusted and maintained autonomously.

The Cleveland Heart uses a single, centrally located, rotating assembly that has impellers for both left and right centrifugal pumps. It achieves passive, sensorless autoregulation of pump flows and filling pressures by varying relative left and right pump efficiencies, based on the position of its single moving part, to balance the left and right pump filling pressures and pump outputs. The position of the centrally located rotating assembly is primarily determined by the differential force between the right and left pump filling pressures acting on the impeller. Chronic animal studies are ongoing with the first successful study conducted in 2012, in which 12 calves were implanted and demonstrated good hemodynamic profiles. Recently, one calf was sacrificed after 30 days of support with ideal hemodynamic profile and no pump malfunction. Most recently, two 90-day biocompatibility studies in calves were successfully completed [65].

BiVACOR chronic studies in the calf model over a 7-week period demonstrated normal physiologic parameters for 90 days in a calf [61, 66]. The findings of normal end-organ function and histology after termination fortified the fundamental hypothesis that continuous-flow perfusion can sufficiently support mammalian life. In addition, BiVACOR pulsatile operation in in vivo studies demonstrated its ability to combine the possible physiologic benefits of a pulse with the benefits of a rotary pump [67].

Maintenance of Arterial Pulsatility

Nonpulsatile continuous flow appears to promote the development of arteriovenous malformations throughout the gastrointestinal tract that predispose to recurrent gastrointestinal bleeding [68]. Bleeding is exacerbated by acquired von Willebrand syndrome, which appears to be because of the sheer stress-induced activation of a metalloprotease, ADAMTS-13 that cleaves off the high-molecular-weight multimers essential for platelet activation by von Willebrand factor [69]. Continuous flow with the absence of aortic valve opening predisposes to valve thrombosis. Over time, the aortic valve commissure starts to fuse, and about 30% of all long-term VAD patients develop some degree of aortic regurgitation that may require surgical intervention. A number of strategies have been developed to enhance pulsatility and intermittent aortic valve opening, which may prevent valve or chamber thrombus formation by inhibiting stasis or recirculation zones. These include periodic rpm slowing (Jarvic 2000) or pump speed modulation (HeartWare HVAD and HeartMate III) that generates intrinsic pulsatile flow from the LVAD.

The Jarvik 2000 intermittent low sleep system functions by reducing pump speed for 8 seconds/minute. In vivo studies have suggested that the ILS system allows for aortic valve opening and therefore adequate washout [70]. However, acquired aortic insufficiency has been reported possibly because the duration of speed decrease was too short [71].

Pump speed manipulation may be asynchronous or synchronous, that is, independent of or consistent with the native heart rate. Synchronous pump speed manipulation can maximize LVAD flow during LV systole (copulsation) or diastole (counterpulsation). The latter favors LV unloading, diminished LV pressure work, and the potential for remodeling and recovery. The HeartWare HVAD Lavare cycle functions asynchronously every minute by reducing speed for 2 seconds and then increasing for 1 second before returning to baseline. The intended purpose of the Lavare cycle is to decrease areas of stasis. The HeartMate III uses an artificial pulse by decreasing pump speeds for 1.15 seconds and increasing for 0.2 seconds every 2 seconds. The MOMENTUM 3 study, discussed previously, has demonstrated lower pump thrombosis rates of HeartMate III as compared to HeartMate II. Greater preservation of high-molecular-weight multimers with the HeartMate III in comparison with HeartMate II have also been reported, indicating less shearing [72]. Whether these outcomes were due to the artificial pulse, material biocompatibility, or a combination of these factors is not yet clear.

Transcutaneous Energy Transfer

Current LVADS require a percutaneous driveline to receive power from external batteries or an electrical outlet. Drivelines hinder patient mobility, can be damaged, and often cause infections that may lead to device failure over time. The rate of infection in LVAD patients is prohibitively high when compared with other cardiovascular implants and orthopedic implants [73–79]. Such infections can lead to sepsis, repeated hospitalizations, and lower survival [78, 80–82].

Because driveline infections frequently arise due to trauma to the driveline exit site, the ideal solution is to replace the driveline with a fully implantable energy source that lasts the duration of the patient's lifetime or has the capacity to be recharged without piercing the skin. The development of a safe, effective transcutaneous energy transfer (TET) system for durable ventricular support devices has been in progress for more than three decades [83]. The physical principle of TET is based on an external direct current (DC) power from a control module or battery that is converted into a high-frequency AC current by a power inverter. Next, a primary transmit coil transfers this AC power by inductive coupling through the skin to a secondary receive coil that is implanted subcutaneously. The excitation of a primary coil causes an alternating electromagnetic field which induces a voltage at a secondary coil's terminals. As a result it is possible to supply devices and charge implanted batteries without any electrical conductive connection. TET can supply LVADs with more than enough power, reaching efficiencies of up to 72% for close coil separations [84]. The HeartWare-Dualis system (according to a press release) and coplanar technology (Leviticus Cardio Inc) are examples of technologies based on TET principles [85, 86].

The inductive charging approach would eliminate the driveline and its potential for infection and provide total device implantability and markedly improve patient acceptance and quality of life. However, a number of challenges remain, including the potential for tissue heating due to power losses, infection of implanted material, biocompatibility of the applied materials, and component durability. In addition, TET range and alignment problems limit its applicability [87, 88]. The TET form of energy transmission consumes approximately 20% of the power input, which is less efficient than direct electrical delivery through a percutaneous driveline [89]. In addition, the primary and secondary coils in the TET design must remain close. This proximity restriction requires that the receive coil be implanted just under the skin and that the external transmit coil be secured in a single position on the skin surface. As the distance between the two coils increases, efficiency drops off. Furthermore, TET coils do not tolerate angular misalignment. If the separation between the coils is too great or a misalignment occurs, excess power is transmitted to compensate for the diminished efficiency. For these reasons, alternatives to TET have been explored to allow for greater separation and freedom between the primary and secondary coils.

Discussion

LVADs have revolutionized the treatment options for patients suffering from advanced heart failure. In the past decade, noticeable trends in continuous-flow developments include device miniaturization, minimally invasive and/or percutaneous insertion, and efforts to superimpose pulsatility on continuous flow as well as standardize care. Innovation will certainly accelerate in the years to come. We have already gained considerable gains in the therapy for this disease over the past 50 years, and the future is understandably bright for MCS. As the patient population that can benefit from this technology expands and patient survival continues to improve, progress is needed to minimize morbidity and improve patient quality of life.

References

1. Westaby S. Cardiac transplant or rotary blood pump: contemporary evidence. J Thorac Cardiovasc Surg. 2013;145:24–31.
2. Westaby S, Deng M. Continuous flow blood pumps: the new gold standard for advanced heart failure? Eur J Cardiothorac Surg. 2013;44:4–8.
3. Krabatsch T, Potapov E, Knosalla C, Hetzer R. Ventricular assist devices for all? Eur J Cardiothorac Surg. 2012;42:918–9.
4. Park SJ, Tector A, Piccioni W, Raines E, Gelijns A, Moskowitz A, Rose E, Holman W, Furukawa S, Frazier OH, et al. Left ventricular assist devices as destination therapy: a new look at survival. J Thorac Cardiovasc Surg. 2005;129:9–17.
5. Kirklin JK, Naftel DC, Pagani FD, Kormos RL, Stevenson L, Miller M, Young JB. Long-term mechanical circulatory support (destination therapy): on track to compete with heart transplantation? J Thorac Cardiovasc Surg. 2012;144:584–603.
6. Holman WL, Naftel DC, Eckert CE, Kormos RL, Goldstein DJ, Kirklin JK. Durability of left ventricular assist devices: Interagency Registry for Mechanically Assisted Circulatory Support (INTERMACS) 2006 to 2011. J Thorac Cardiovasc Surg. 2013;146:437–41.
7. Kantrowitz A, Tjonneland S, Freed PS, et al. Initial clinical experience with intraaortic balloon pumping in cardiogenic shock. JAMA. 1968;203:113–8.
8. Kilic A, Ailawadi G. Left ventricular assist devices in heart failure. Expert Rev Cardiovasc Ther. 2012;10:649–56.
9. Barnard CN. The operation. A human cardiac transplant: an interim report of a successful operation performed at Groote Schuur Hospital, Cape Town. S Afr Med J. 1967;41:1271–4.
10. Kirklin JK, Naftel DC, Pagani FD, Kormos RL, Stevenson LW, Blume ED, Miller MA, Baldwin JT, Young JB. Sixth INTERMACS annual report: a 10,000-patient database. J Heart Lung Transplant. 2014;33:555–64.
11. Slaughter MS, Pagani FD, McGee EC, et al. HeartWare Bridge to Transplant ADVANCE Trial

Investigators. HeartWare ventricular assist system for bridge to transplant: combined results of the bridge to transplant and continued access protocol trial. J Heart Lung Transplant. 2013;32:675–83.
12. Rogers JG, Pagani FD, Tatooles AJ, et al. Intrapericardial left ventricular assist device for advanced heart failure. N Engl J Med. 2017;376:451–60. 10.1056.
13. Milano CA, Rogers JG, Tatooles AJ, et al. HVAD: The ENDURANCE Supplemental Trial. JACC HeartFail. 2018;6(9):792–802. https://doi.org/10.1016/j.jchf.2018.05.012.
14. Schmitto JD, Hanke JS, Rojas SV, Avsar M, Haverich A. First implantation in man of a new magnetically levitated left ventricular assist device (HeartMate III). J Heart Lung Transplant. 2015;34:858–60.
15. Mehra MR, Naka Y, Uriel N, et al. A fully magnetically levitated circulatory pump for advanced heart failure. N Engl J Med. 2017;376:440–50.
16. Mehra MR, Goldstein DJ, Uriel N, et al. Two-year outcomes with a magnetically levitated cardiac pump in heart failure. N Engl J Med. 2018;378:1386–95.
17. Tamez D, LaRose JA, Shambaugh C, et al. Early feasibility testing and engineering development of the transapical approach for the HeartWare MVAD ventricular assist system. ASAIO J. 2014;60:170–7.
18. Kirklin JK, Naftel DC, Pagani FD, et al. Seventh INTERMACS annual report: 15,000 patients and counting. J Heart Lung Transplant. 2015;34: 1495–504.
19. Estep JD, et al. Risk assessment and comparative effectiveness of left ventricular assist devices and medical management in ambulatory heart failure patients: results from the ROADMAP study. J Am Coll Cardiol. 2015;66(16):1747–61.
20. Rose EA, Gelijns AC, Moskowitz AJ, et al. Long-term mechanical left ventricular assistance for end-stage heart failure. N Engl J Med. 2001;345:1435–43.
21. Aaronson KD, Pagani FD, Kormos R, et al. Design for the randomized evaluation of VAD Intervention Before Inotropic Therapy (REVIVE-IT) study. (In press). J Heart Lung Transplant. 2016.
22. Pagani FD, Aaronson KD, Kormos R, et al. The NHLBI REVIVE-IT study: understanding its discontinuation in the context of current left ventricular assist device therapy. J Heart Lung Transplant. 2016;35:1277–83.
23. Lietz K, et al. Outcomes of left ventricular assist device implantation in the post-rematch era: implications for patient selection. Circulation. 2007;116(5):497–506.
24. Levy WC, Mozaffarian D, Linker DT, et al. The Seattle Heart Failure Model: prediction of survival in heart failure. Circulation. 2006;113:1424–33.
25. Cowger J, Sundareswaran K, Rogers JG, et al. Predicting survival in patients receiving continuous flow left ventricular assist devices. J Am Coll Cardiol. 2013;61:313–21.
26. Matthews JC, Pagani FD, Haft JW, et al. Model for end-stage liver disease score predicts left ventricular assist device operative transfusion requirements, morbidity, and mortality. Circulation. 2010;121:214–20.
27. Loghmanpour NA, Kormos RL, Kanwar MK, et al. A Bayesian model to predict right ventricular failure following left ventricular assist device therapy. JACC Heart Fail. 2016;4:711–21.
28. Kanwar MK, Lohmueller LC, Kormos RL, et al. Low accuracy of the HeartMate risk score for predicting mortality using the INTERMACS registry data. ASAIO J. 2017;63:251–6.
29. Loghmanpour NA, Kanwar MK, Druzdzel MJ, et al. A new Bayesian network-based risk stratification model for prediction of short-term and long-term LVAD mortality. ASAIO J. 2015;61:313–23.
30. Patil NP, Popov AF, Simon AR. Minimally invasive left ventricular assist device implantation: at the crossroads. J Thorac Dis. 2015;7:564–5.
31. Strueber M, Meyer AL, Feussner M, et al. A minimally invasive off-pump implantation technique for continuous-flow left ventricular assist devices: early experience. J Heart Lung Transplant. 2014;33:851–6.
32. Haberl T, Riebandt J, Mahr S, et al. Viennese approach to minimize the invasiveness of ventricular assist device implantation. Eur J Cardiothorac Surg. 2014;46:991–6.
33. Popov AF, Hosseini MT, Zych B, et al. HeartWare left ventricular assist device implantation through bilateral anterior thoracotomy. Ann Thorac Surg. 2012;93:674–6.
34. Rojas SV, Avsar M, Hanke JS, et al. Minimally invasive ventricular assist device surgery. Artif Organs. 2015;39:473–9.
35. Hill JD, Avery GJ, Egrie G, et al. Less invasive Thoratec LVAD insertion; a surgical technique. Heart Surg Forum. 2000;3:218–23.
36. Gregoric ID, La Francesca S, Myers T, et al. A less invasive approach to axial flow pump insertion. J Heart Lung Transplant. 2008;27:423–6.
37. Samuels LE, Casanova-Ghosh E, Rodriguez R, et al. Left ventricular assist device implantation in high risk destination therapy patients; an alternative surgical approach. J Cardiothorac Surg. 2012;7:21.
38. Strueber M, Cheung A, Mokadam NA, et al. Impact of the thoracotomy implant approach on average length of stay and rehospitalizations in the HVAD LATERAL trial. J Heart Lung Transplant. 2018;37(4):S83–4.
39. Pintar T, Collard CD. The systemic inflammatory response to cardiopulmonary bypass. Anesthesiol Clin North Am. 2003;21:453–64.
40. Kilic A. The future of left ventricular assist devices. J Thorac Dis. 2015;7:2188–93.
41. Kirklin JK, Pagani FD, Kormos RL, et al. Eighth annual INTERMACS report: special focus on framing the impact of adverse events. J Heart Lung Transplant. 2017;36(10):1080–6.
42. Raess DH, Weber DM. Impella 2.5. J Cardiovasc Transl Res. 2009;2(2):168–72.
43. Meyns BP, Simon A, Klotz S, et al. Clinical benefits of partial circulatory support in New York Heart

Association class IIIB and early class IV patients. Eur J Cardiothorac Surg. 2011;39(5):693–8.
44. Pham DT, Paruchuri V, Esposito M, Schraufnagel DP, McSparren D, Civiello D, et al. Hemodynamic Evaluation of HeartWare MVAD in Normal and Infarct Swine Models. 2013;32(4):S184–5. https://doi.org/10.1016/j.healun.2013.01.445.
45. Schmitto JD, Burkhoff D, Avsar M, Fey O, Ziehme P, Buechler G, et al. Two axial-flow synergy micro pumps as a biventricular assist device in an ovine animal model. J Heart Lung Transplant. 2013;31(11):1223–9.
46. Klotz S, Meyns B, Simon A, Wittwer T, Rahmanian P, Schlensak C, et al. Partial mechanical long-term support with the CircuLite synergy pump as bridge-to-transplant in congestive heart failure. Thorac Cardiovasc Surg. 2010;58(Suppl 2):S173–8.
47. Del Rio C, Clifton W, Heuring J, et al. Aortix™, a novel catheter-based intra-vascular assist device, provides cardiorenal support while improving ventriculo-arterial coupling and myocardial demand in sheep with induced chronic ischemic heart failure. J Am Coll Cardiol. 2015;65:10S, A800.
48. Katz JN, Adamson RM, John R, et al. Safety of reduced anti-thrombotic strategies in HeartMate II patients: a one-year analysis of the US-TRACE study. J Heart Lung Transplant. 2015;34:1542–8.
49. Starling RC, Moazami N, Silvestry SC, et al. Unexpected abrupt increase in left ventricular assist device thrombosis. N Engl J Med. 2014;370:33–40.
50. McCarthy FH, Kobrin D, Rame JE, et al. Increasing frequency of left ventricular assist device exchanges in the United States. Ann Thorac Surg. 2015;100: 1660–5.
51. Maltais S, Kilic A, Nathan S, et al. PREVENtion of HeartMate II pump thrombosis through clinical management: the PREVENT multi-center study. J Heart Lung Transplant. 2017;36(1):1–12.
52. Frazier OH, Tuzun E, Cohn W, Tamez D, Kadipasaoglu KA. Total heart replacement with dual centrifugal ventricular assist devices. ASAIO J. 2005;51(3):224–9.
53. Frazier OH, Tuzun E, Cohn W, Conger J, Kadipasaoglu KA. Total heart replacement with dual Intracorporeal continuous-flow pumps in a chronic bovine model: a feasibility study. ASAIO J. 2006;52(2):145–9.
54. Cohn W, Timms D, Frazier OH. Total artificial hearts: past, present, and future. Nat Rev Cardiol. 2015;12:609.
55. Dowling RD, Gray LA, Etoch SW, Laks H, Marelli D, Samuels L, et al. Initial experience with the AbioCor implantable replacement heart system. J Thorac Cardiovasc Surg. 2004;127(1):131–41.
56. Nose Y. FDA approval of totally implantable permanent total artificial heart for humanitarian use. Artif Organs. 2007;31(1):1–3.
57. Haigney P, Hansen P, Forbes J, Sarfaty C. Groundbreaking artificial heart implanted at UMDNJ-Robert Wood Johnson Medical School and Robert Wood Johnson University Hospital. New Jersey: Rutgers; 2009.
58. Golding LR, Jacobs G, Murakami T, Takatani S, Valdes F, Harasaki H, et al. Chronic nonpulsatile blood flow in an alive, awake animal 34-day survival. ASAIO J. 1980;26(1):251–5.
59. Qian KX, Pi KD, Wang YP, Zhao MJ. Toward an implantable impeller total heart. ASAIO J. 1987;33(3):704–7.
60. Slaughter MS, Rogers JG, Milano CA, Russell SD, Conte JV, Feldman D, et al. Advanced heart failure treated with continuous-flow left ventricular assist device. N Engl J Med. 2009;361(23):2241–51.
61. Cohn WE, Handy KM, Parnis SM, Conger JL, Winkler JA, Frazier OH. Eight-year experience with a continuous-flow total artificial heart in calves. ASAIO J. 2014;60(1):25–30.
62. Timms D, Fraser J, Hayne M, et al. The BiVACOR rotary biventricular assist device: concept and in vitro investigation. Artif Organs. 2008;32:816–9.
63. Fukamachi K, Horvath DJ, Massiello AL, et al. An innovative, sensorless, pulsatile, continuous-flow total artificial heart: device design and initial in vitro study. J Heart Lung Transplant. 2010;29:13–20.
64. Fumoto H, Horvath DJ, Rao S, et al. In vivo acute performance of the Cleveland Clinic self-regulating, continuous-flow total artificial heart. J Heart Lung Transplant. 2010;29:21–6.
65. Karimov JH, Fukamachi K, Moazami N, et al. In vivo evaluation of the Cleveland Clinic continuous-flow total artificial heart in calves[abstract]. J Heart Lung Transplant. 2014;3:S165–6.
66. Cohn WE, Winkler JA, Parnis S, Costas GG, Beathard S, Conger J, Frazier OH. Ninety-day survival of a calf implanted with a continuous-flow total artificial heart. ASAIO J. 2014;60(1):25–30.
67. Erkanli K, Kayalar N, Erkanli G, Ercan F, Sener G, Kirali K. Melatonin protects against ischemia/reperfusion injury in skeletal muscle. J Pineal Res. 2005;39:238.
68. Mancini D, Colombo PC. Left ventricular assist devices: a rapidly evolving alternative to transplant. J Am Coll Cardiol. 2015;65:2542–55.
69. Jilma-Stohlawetz P, Quehenberger P, Schima H, et al. Acquired von Willebrand factor deficiency caused by LVAD is ADAMTS-13 and platelet dependent. Thromb Res. 2016;137:196–201.
70. Bruckner BA, Jacob LP, Gregoric ID, et al. Clinical experience with the TandemHeart percutaneous ventricular assist device as a bridge to cardiac transplantation. Tex Heart Inst J. 2008;35:447–50.
71. McIlvennan CK, Allen LA, Nowels C, Brieke A, Cleveland JC, Matlock DD. Decision making for destination therapy left ventricular assist devices: "there was no choice" versus "I thought about it an awful lot". Circ Cardiovasc Qual Outcomes. 2014;7:374–80.
72. Netuka I, Kvasnicka T, Kvasnicka J, et al. Evaluation of von Willebrand factor with a fully magnetically levitated centrifugal continuous-flow left ventricular assist device in advanced heart failure. J Heart Lung Transplant. 2016;35(7):860–7.

73. Monkowski DH, Axelrod P, Fekete T, Hollander T, Furukawa S, Samuel R. Infections associated with ventricular assist devices: epidemiology and effect on prognosis after transplantation. Transpl Infect Dis. 2007;9:114–20.
74. Gordon RJ, Quagliarello B, Lowy FD. Ventricular assist device-related infections. Lancet Infect Dis. 2006;6:426–37.
75. Holman WL, Pamboukian SV, McGiffin DC, Tallaj JA, Cadeiras M, Kirklin JK. Device related infections: are we making progress? J Card Surg. 2010;25:478–83.
76. Raymond AL, Kfoury AG, Bishop CJ, et al. Obesity and left ventricular assist device driveline exit site infection. ASAIO J. 2010;56:57–60.
77. Zierer A, Melby SJ, Voeller RK, et al. Late-onset driveline infections: the Achilles' heel of prolonged left ventricular assist device support. Ann Thorac Surg. 2007;84:515–20.
78. Baddour LM, Epstein AE, Erickson CC, Knight BP. Update on cardiovascular implantable electronic device infections and their management: a scientific statement from the American Heart Association. Circulation. 2010;121:458–77.
79. Kurtz SM, Lau E, Schmier J, Ong KL, Zhao K, Parvizi J. Infection burden for hip and knee arthroplasty in the United States. J Arthroplast. 2008;23:984–91.
80. Goldstein DJ, Naftel D, Holman W, et al. Continuous-flow devices and percutaneous site infections: clinical outcomes. J Heart Lung Transplant. 2012;31: 1151–7.
81. Arnaoutakis GJ, Allen JG, Bonde P, Russell SD, Shah AS, Conte JV. Subcostal exchange of left ventricular assist devices—a novel approach. Circulation. 2010;122:A21162.
82. El-Banayosy A, Arusoglu L, Kizner L, et al. Preliminary experience with the LionHeart left ventricular assist device in patients with end-stage heart failure. Ann Thorac Surg. 2003;75:1469–75.
83. Mussivand T, Miller JA, Santerre PJ, et al. Transcutaneous energy transfer system performance evaluation. Artif Organs. 1993;17:940–7.
84. Zareba KM. The artificial heart—past, present, and future. Med Sci Monit. 2002;8:RA72–7.
85. HeartWare. HeartWare and Dualis MedTech Announce agreement to develop fully implantable ventricular assist system. Available at http://ir.heartware.com/phoenix.zhtml?c¼187755&p¼irol-newsArticle&id¼1616139. Accessed 16 Sept 2013.
86. Kassif Y, Zilbershlag M, Levi M, Plotkin A, Schueler S. A new universal wireless transcutaneous energy transfer (TET) system for implantable LVADs—preliminary in vitro and in vivo results. J Heart Lung Transplant. 2013;32:S140–1.
87. Dissanayake TD, Budgett DM, Hu P, et al. A novel low temperature transcutaneous energy transfer system suitable for high power implantable medical devices: performance and validation in sheep. Artif Organs. 2010;34:E160–7.
88. DUALIS MedTech GmbH. Wireless energy supply sets new standards for medical devices. Available at http://www.dualis-medtech.de. Accessed 6 June 2013.
89. Rintoul TC, Dolgin A. Thoratec transcutaneous energy transformer system: a review and update. ASAIO J. 2004;50:397–400.

29 The Development of Advanced Ventricular Assist Device as a Next Generation Ventricular Assist Device

Takuma Miyamoto, Jamshid H. Karimov, David J. Horvath, Nicole Byram, and Kiyotaka Fukamachi

Introduction

Mechanical circulatory support (MCS) devices have been widely used to treat end-stage heart failure. There are two types of implantable left ventricular assist devices (LVADs): continuous-flow (CF) and pulsatile-flow (PF) LVADs. CF rotary pumps have replaced volume displacement PF pumps in the treatment of end-stage heart failure because of their simplicity, increased mechanical reliability, improved durability, smaller size, and better outcomes [1]. However, existing CF devices have the following clinical limitations: (1) reduced arterial pulsatility; (2) development of aortic regurgitation; (3) threat of pump thrombosis or hemolysis; (4) risk of syncope and death from backflow during pump stoppage; (5) inability to noninvasively evaluate weanability by pump-off test; and (6) narrow operating specifications, meaning there is no universal device without pivot bearings for use as both a left and right ventricular assist device (LVAD and RVAD, respectively).

T. Miyamoto, MD, PhD · N. Byram, BS
Department of Biomedical Engineering, Lerner Research Institute, Cleveland Clinic, Cleveland, OH, USA
e-mail: miyamot@ccf.org; byramn@ccf.org

J. H. Karimov, MD, PhD · K. Fukamachi, MD, PhD (✉)
Lerner Research Institute, Cleveland Clinic, Cleveland Clinic Lerner College of Medicine of Case Western Reserve University, Cleveland Clinic, Cleveland, OH, USA
e-mail: karimoj@ccf.org; fukamak@ccf.org

D. J. Horvath, MSME
Department of Biomedical Engineering, Lerner Research Institute, Cleveland Clinic, Cleveland, OH, USA

R1 Engineering LLC, Euclid, OH, USA

Design Features

We are developing the Advanced ventricular assist device (VAD) (Fig. 29.1) that has many potential new features to address current limitations with CF VADs [2]. The goals of this device development are as follows:

(a) Maintain physiological pulsatility in patients on VAD support. Reinforcing the native pressure and flow pulsatility by dynamically coupling with the native ventricle will create a near-physiological aortic pulse pressure without pump speed modulation. Less flow during diastole will also reduce pressures on the aortic valve during diastole and may reduce development of aortic regurgitation.
(b) Reduce the risk of pump thrombosis or hemolysis by eliminating problematic features (e.g.,

© Springer Nature Switzerland AG 2020
J. H. Karimov et al. (eds.), *Mechanical Support for Heart Failure*,
https://doi.org/10.1007/978-3-030-47809-4_29

inlet/outlet stator vanes and pivot bearings) and offer full passive suspension of the rotor. Passive suspension is much simpler than active magnetic levitation, a complex design with potential for failure.

(c) Prevent backflow by enabling automatic regurgitant flow shutoff in the event of pump stoppage. Enable noninvasive evaluation of pump weanability by regurgitant flow shutoff. Such automatic shutoff will also allow a pump-off test without blood regurgitation or the need to occlude the outflow graft.
(d) Demonstrate wide operating specifications, enabling the use of the Advanced VAD as either an LVAD or RVAD with the same pump and electronic hardware.
(e) Avoid suction events: how to automatically attenuate pump output through automatic aperture closure.

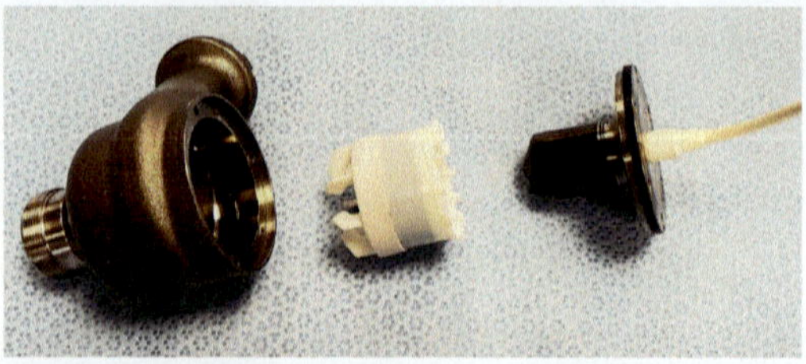

Fig. 29.1 Exploded prototype of the Advanced VAD

The cross section of this pump is shown in Fig. 29.2. The impeller is free to move axially in and out of a housing cavity in response to system pressures, thereby opening and closing an aperture at the impeller discharge. During operation, the axial position modulates along with the cardiac cycle, so that the aperture (a) opens with a surge in output at the start of systole (Fig. 29.2, bottom right) when the left ventricular pressure (inlet pressure) increases and (b) closes during diastole when the left ventricular pressure decreases, attenuating the output (Fig. 29.2, bottom left). In this way, the dynamic coupling of pump capacity is automatically and precisely timed to the action of the ventricle. In CF LVADs, pump flow is determined by the pump speed and the pressure difference between inlet and outflow, so CF LVADs usually pump more flow in systole (low pressure difference) and less flow in diastole (high pressure difference) as long as the left ventricle contracts. The Advanced VAD further increases the flow during systole by opening the aperture and decreases the flow during diastole by closing the aperture, providing much more pulsatility of the pump flow.

A weak magnetic force acts on the rotating assembly in the axial direction to close the aperture, while the hydraulic forces created by rotor

Fig. 29.2 Pump cross section explaining the Advanced VAD function

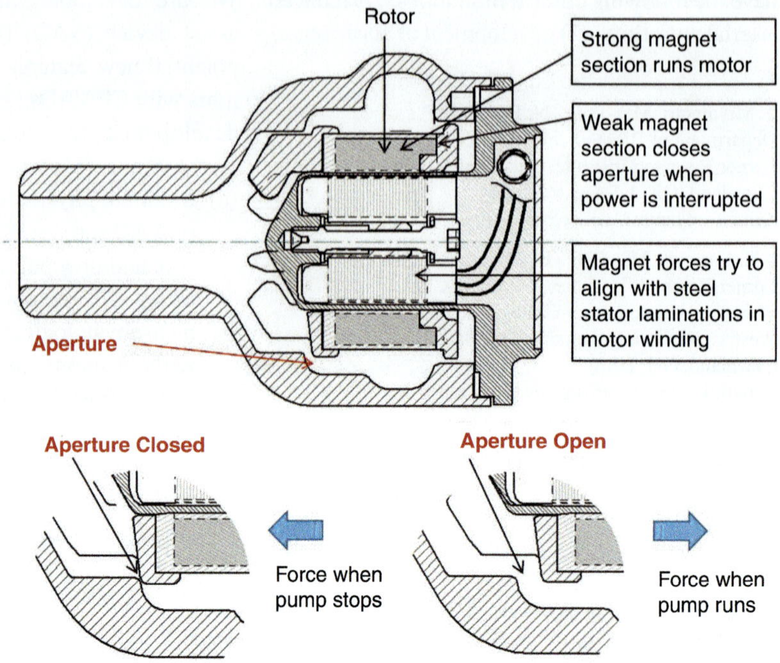

rotation act in the opposite direction to open it. If the rotor stops rotation, the hydraulic forces disappear, and the axial magnetic force causes the impeller discharge aperture to close automatically, inhibiting backflow of blood through the pump (Fig. 29.2, bottom left).

In Vitro Results

The first working prototype of the Advanced VAD (AVα) was bench tested on the static mock loop and compared with HeartMate II (Abbott Inc., Chicago, IL) and HVAD (HeartWare Inc., Framingham, MA). The pressure-flow curves of the Advanced VAD at various pump speeds are shown in Fig. 29.3a, and those of HeartMate II (Fig. 29.3b) and HVAD (Fig. 29.3c) are also shown. Similar to pressure-flow curves for traditional CF pumps, pump flow increases when the pressure rise decreases for a given pump speed, and pump flow also increases as speed is increased for a given pressure rise. However, the normalized pressure-flow curves, in which flow was normalized by dividing by speed and pressure rise was normalized by dividing by speed squared, showed distinct differences from those for the HeartMate II (Fig. 29.3e) and HVAD (Fig. 29.3f). If the pump characteristic remained constant and independent of pump speed, the normalized pump performance curves are not changed by pump speed or pressure differences, so the normalized pressure-flow curves lie along a single curve (Fig. 29.3e, f). However, in the Advanced VAD, the aperture decreases its opening in response to a decrease in pump speed to meet lower clinical requirements for delivery of flow and pressure; therefore, normalized pressure-flow curves do not lie along a single curve but vary based on pump speed (Fig. 29.3d).

Bench testing was also performed to demonstrate the pulsatility on a pulsatile mock loop with a pneumatic mock ventricle (AB5000; Abiomed, Inc., Danvers, MA), modified to have a port on the pumping chamber for the VAD inlet. Figure 29.4 shows the arterial pulse pressure at different pump flows in the Advanced VAD, HeartMate II, and HVAD. At a pump flow of 4 L/min, the Advanced VAD showed the greatest pulse pressure (average 27.8 mm Hg). This pulse pressure was 21% higher than that without the Advanced VAD (23 mm Hg); thus, the pulse amplification factor was 1.21. The pulse amplification factor for

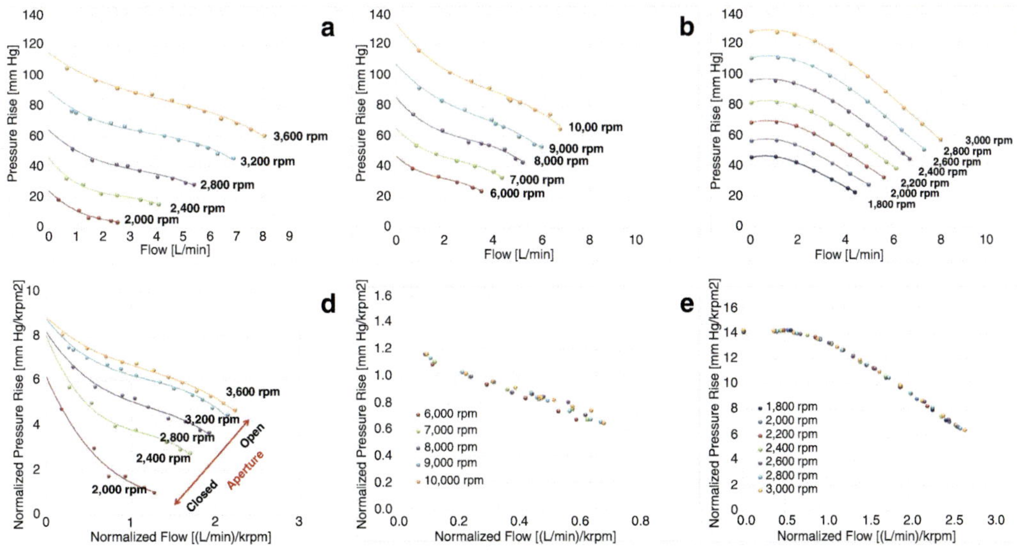

Fig. 29.3 Pressure-flow curves (**a**) and normalized pressure-flow curves (**d**) of the Advanced VAD, HeartMate II (**b**, **e**), and HVAD (**c**, **f**) on the static mock loop

the HeartMate II and HVAD was 0.94 and 0.86, respectively. These results clearly showed that the Advanced VAD made more pulsatile energy than other existing pumps and may prevent some physiologic changes and complications. Figure 29.5a shows representative waveforms of the arterial pressure and pump flow of the Advanced VAD at 3600 rpm on the pulsatile mock loop. The automatic regurgitant flow shut-off feature of the Advanced VAD was also evaluated in bench testing. Each pump was stopped without clamping the outflow conduit to evaluate regurgitant flow through the pump. When each pump was stopped, a marked peak negative flow (regurgitation through the pump) was detected in both the HeartMate II (−4.8 L/min) and HVAD (−5.6 L/min), which disappeared after clamping the outflow conduit (Fig. 29.5b). No such marked levels of negative flow were seen in the Advanced VAD, with or without clamping the outflow conduit (Fig. 29.5b). Therefore, the arterial pressure waveform in the Advanced VAD changed very little when the outflow conduit was clamped, indicating that the pump-off test could be done without clamping the outflow conduit.

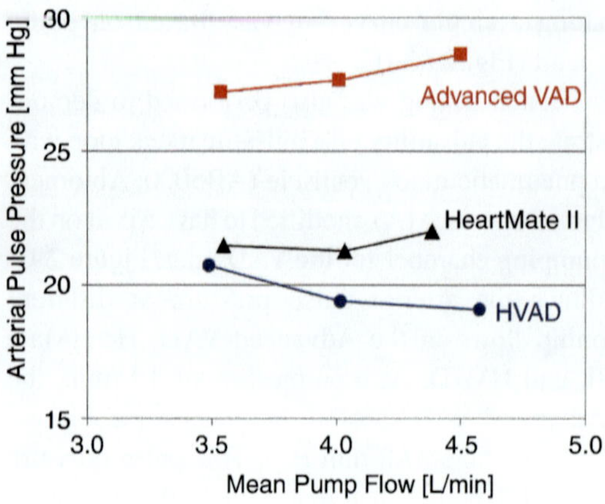

Fig. 29.4 The arterial pulse pressure at different pump flows in the Advanced VAD, HeartMate II, and HVAD

The Development of the Pump Design

To improve the flow pulsatility and the regurgitant shut off features, we have adjusted the spacing of the primary impeller (AV020 3S and AV020 6S) and secondary impeller (AV020 RC) from the second prototype (AV010), which was made with better fabrication and a reduced closing magnet force from AVα. Bench testing of the AV020s was performed on a pulsatile mock loop which can be used

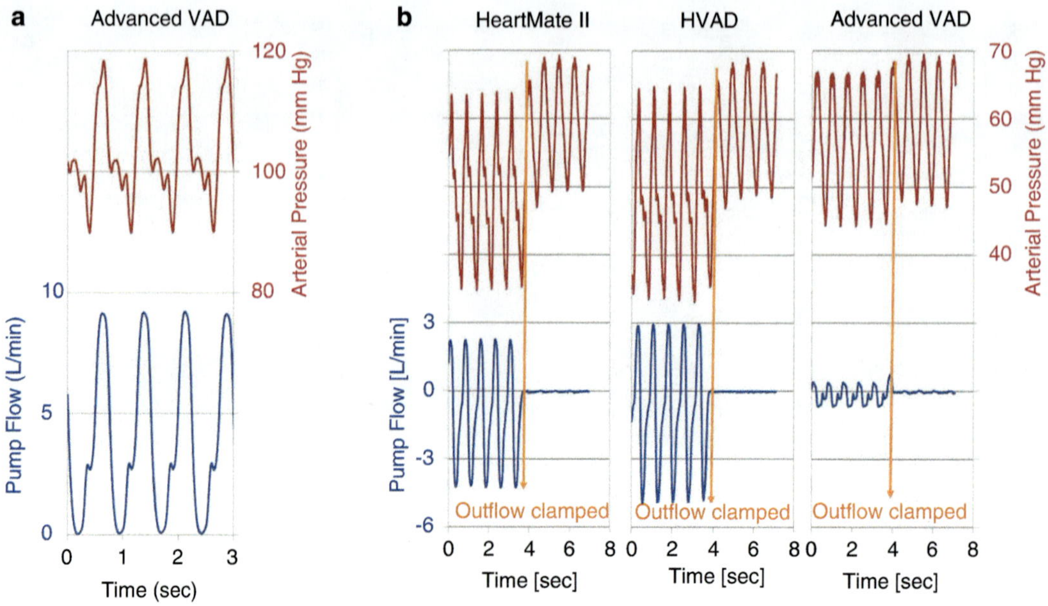

Fig. 29.5 (**a**) Pump flow and arterial pressure with the Advanced VAD running at 3600 rpm on the pulsatile mock loop. (**b**) Arterial pressure and pump flow when the pump was stopped on the pulsatile mock loop with the outflow conduit fully open (the first five beats in each graph) and with the outflow conduit clamped

to obtain static or pulsatile data. The loop uses a pneumatic device to simulate the native ventricle (Abiomed AB5000™). The AV020 RC showed greater pulse pressure (37.3 mm Hg) compared to the AV010 (34.0 mm Hg) (Fig. 29.6a). AV020 3S, while having a slightly higher pulse pressure (39.3 mm Hg), had 40% lower mean pressure output than all other models; AV020 6S ran well but did not outperform the RC version, which demonstrated a pulse augmentation factor of 1.62. When the AV020 pumps were stopped, there was no remarkable regurgitant flow (0.13–0.18 L/min), and regurgitant flow was lower than that of AV010 (Fig. 29.6b). There were no significant changes in the arterial pressure waveform in any of the pumps when the outflow conduit was clamped, indicating that the pump-off test could be done without clamping the outflow conduit. The auto-shutoff feature of this Advanced VAD provides a unique fail-safe mode in the event of power interruption to help reduce hazard to the patient. This study showed improvements of in pulsatility and functionality of the shutoff feature of the new model of Advanced VAD (AV020 RC) over the AV010 and is ready to move into acute in vivo studies.

Computer Fluid Dynamics Analysis

The objective for the proposed modeling effort is to develop and validate a computational fluid dynamics (CFD) model for the Advanced VAD that will help establish a baseline understanding of the pump's hemodynamics, provide insight into experimental results, and guide the next steps in biocompatibility development. The Roache method was used to assess the mesh dependency of the CFD results [3]. ANSYS (Canonsburg, PA) software (versions R18 and R19) DesignModeler, ANSYS Meshing, CFX, and CFD-Post were used for the simulations.

Initially, we focused on creating a three-dimensional CFD model of the Advanced VAD design and compared predicted pump performance with in vitro measurements under steady-state flow conditions. The CFD-predicted hydraulic performance for the Advanced VAD pump agreed very well with in vitro study measurements over the range of intended operating conditions (Fig. 29.7). Leveraging the cylindrical symmetry of the pump housing surrounding the impellers, the multi-frames of reference "fro-

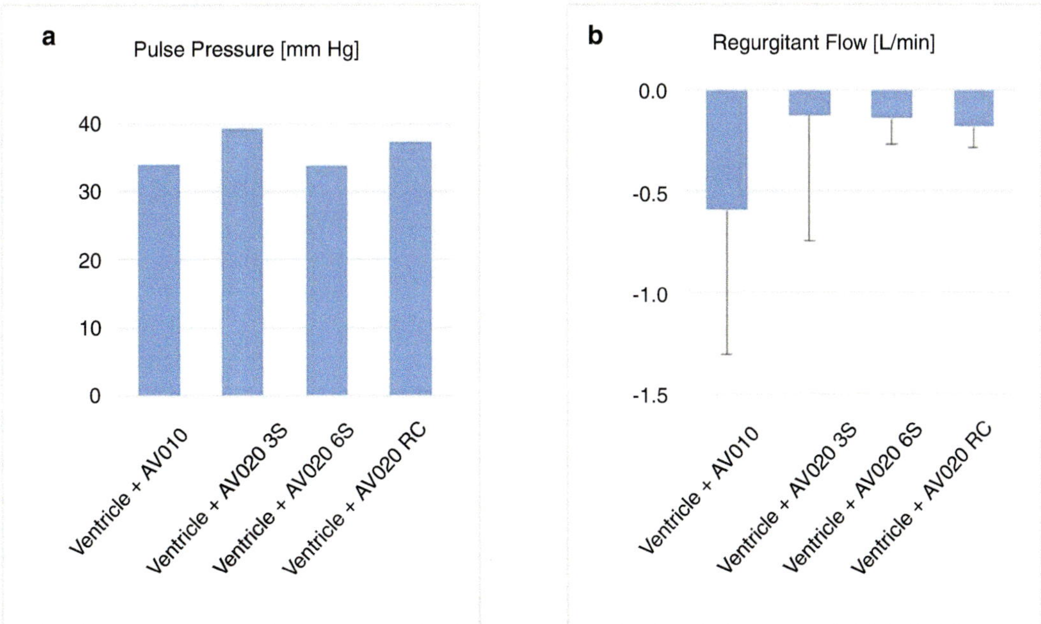

Fig. 29.6 Comparison of pulse pressure (**a**) and regurgitant flow (**b**) when the pumps were stopped

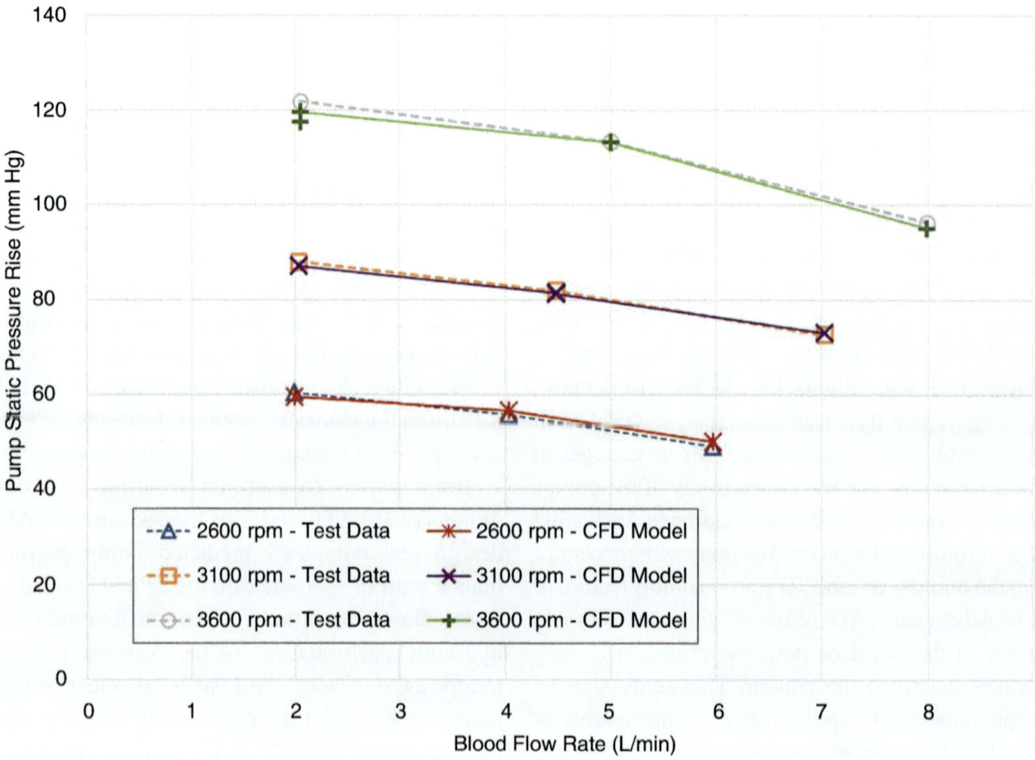

Fig. 29.7 Advanced VAD pump hydraulic performance comparison between in vitro and computer fluid dynamic model

zen rotor" model were applied for these steady-state simulations. A water/glycerin mixture (viscosity = 2.026 cps, density = 1060 kg/m^3) was the fluid used for these comparisons. The effects of turbulent flow within the pump were captured using a Reynolds-averaged Navier-Stokes (RANS) turbulence model (κ/ω shear stress transport). An average inlet static pressure and an outlet mass flow rate were applied as boundary conditions for each of the operating conditions. The values for the average inlet static pressure and mass flow rates were based upon test data values. A constant rotor aperture opening of 0.0220 inches was used for all of the simulations and was based upon the average bench testing value observed under steady-state flow conditions. The measured range of aperture openings ranged from 0.019 inches to 0.025 inches.

After establishing the credibility of the CFD modeling approach, attention was then focused on gaining insight into potential design changes that could improve the biocompatibility of the pump. For these simulations, blood was the fluid used (density = 1060 kg/m^3, viscosity = Cross non-Newtonian blood viscosity model [4, 5]). The nominal flow rate of 4.5 L/min was modeled at two impeller rotational speeds, 3132 rpm to mimic LVAD nominal conditions and 2200 rpm, to simulate RVAD nominal conditions. Relative speed contours and velocity plots, shear stress, residence time plots, and residence time /shear stress plot were analyzed for both solutions. Figure 29.8a shows the results of the relative velocity contours for the LVAD solution, which showed that there is some misalignment of flow approaching the leading edges of impeller blades (flow separation). Figure 29.8b shows fluid shear stress/residence time, plotted on a surface 0.003" from the impeller surface. This image highlights regions with low fluid shear stress and long residence times, which are more susceptible to thrombus formation. The lowest regions of fluid

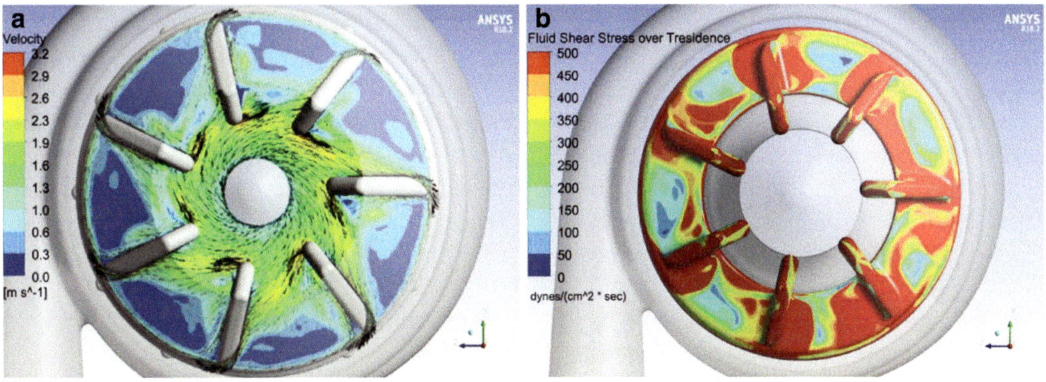

Fig. 29.8 Results of computer fluid dynamic analysis, (**a**) relative velocity contours and (**b**) fluid shear stress/residence time

shear stress/residence time occurred between impeller blades. The CFD solution for the RVAD conditions showed that the lowest regions of fluid shear stress/residence time were also between impeller blades, along the downstream (suction) distal edge of blades. These results were very similar to those observed for the LVAD condition. In addition, there was some misalignment of flow approaching leading edges of impeller blades (flow separation) at the RVAD nominal flow/speed condition. In summary, the CFD analyses indicated that further design optimization could potentially improve the hemocompatibility of the Advanced VAD. Potential areas for refinement include modifying the impeller vane shape (improve alignment to approaching flow to reduce flow separation), the blade height, the number of blades, and the clearance between the top of the blades and the impeller housing. When making these potential design changes, the desired hydraulic performance of the pump will be maintained.

Discussion

Enhancing Pulsatility During LVAD Support

As CF LVADs are now increasingly being used as "destination therapy," and the duration of CF LVAD support is measurably longer, and the importance of pulsatility has recently been reconsidered. How much pulsatile energy was lost during CF LVAD support? Travis et al. [6] evaluated energy loss after CF and PF LVAD support and compared this data to those of patients with normal ventricular function and heart failure. The results showed that at low support levels, PF support restored surplus hemodynamic energy, which is calculated by multiplying the difference between the energy equivalent pressure (defined as the ratio between the area beneath the hemodynamic power curve and the area beneath the pump flow curve) and the mean arterial pressure by 1332 [7], to within 2.5% of normal values, whereas CF support diminished surplus energy by more than 93%. At high support levels, PF support augmented surplus energy by 49% over normal values, whereas CF support further diminished surplus energy by 97%. In the Advanced VAD, the pump "pops its clutch" with every heartbeat, allowing a near physiologic aortic pulse pressure and flow pulse of about 10 L/min (200% flow pulsatility) when in the LVAD mode.

CF support affects the vascular function including endothelial integrity. Amir et al. [8] compared the peripheral artery reactivity of patients supported by VADs (CF LVAD; $n = 10$, PF LVAD; $n = 10$) by measuring flow-mediated dilation. The results showed that the flow-mediated dilation of patients supported with PF was significantly higher than that of patients supported with CF. This means the PF support is associated with a better peripheral vascular reactivity than CF support. Welp et al. [9] examined the effects on plasma renin activity of CF LVAD compared to PF LVAD used as a bridge to cardiac transplantation. Plasma renin activity fell significantly to near normal levels after 21 days of support in both CF- and PF-supported patients. These levels were maintained until 3 weeks of support, when plasma renin activity was significantly lower in patients supported with a PF LVAD. This difference remained until on day 70 of support. The same effect was observed in aldosterone level. As with plasma renin activity, after 3 weeks of support, aldosterone levels were significantly lower in PF-supported patients.

Some CF LVAD-related complications are also considered to be related to reduced pulsatility, such as nonsurgical bleeding and development of aortic insufficiency (AI). Demirozu et al. [10] showed the prevalence of gastrointestinal bleeding and the role of arteriovenous malformations in patients with CF LVADs. Among 172 patients supported with CF LVADs, gastrointestinal bleeding was identified in 32 (19%) patients. Arteriovenous malformations were identified as the source of gastrointestinal bleeding in 10 of 32 (31%) patients. Wever-Pinzon et al. [11] evaluated the impact of the LVAD pulsatility index on the incidence of nonsurgical bleeding among 134 patients supported with a CF LVAD and showed that low pulsatility index was associated with nonsurgical bleeding. Cowger et al. [12] examined the temporal trend of AI following LVAD implant and showed that while AI was similar between CF HeartMate II and PF HeartMate XVE (Thoratec, Inc., Pleasonton, CA) at baseline, the distribution of AI grade was significantly greater in the HeartMate II group compared with the HeartMate XVE group at 1, 3, and 6 months of follow-up. Pak et al. [13] reported the experience with the development of de novo AI during LVAD support in both HeartMate XVE ($n = 93$) and HeartMate II ($n = 73$) patients. They showed the occurrence of AI was higher in HeartMate II group than HeartMate XVE group. Patients in the AI group were more likely to have aortic valves that did not open during echocardiographic assessment. The risk factors of AI development are considered to be multifactorial, such as the loss of pulsatility or the continuous stress to the aortic valve. Our Advanced VAD can't enhance the opening of aortic valve without speed modulation, but unlike other LVADs, it doesn't increase the stress on the aortic valve during diastole when the pump flow is reduced, which may prevent AI development.

Weaning from Ventricular Assist Devices

The introduction of the "destination therapy" also gives us the chance to reconsider weaning from VAD support. Simon et al. [14] sought to define the characteristics of successful bridge to recovery patients and their long-term outcomes. Among 154 adult VAD implants, 10 (6.5%) showed the recovery of ventricular function allowed successful removal of the device without transplantation. The majority of these patients were nonischemic with the etiology in most cases being acute myocarditis or postpartum cardiomyopathy. Birks et al. [15] performed a systematic study combining drug therapy (a combination of angiotensin-converting enzymes, β-blockers, angiotensin II inhibitors, and aldosterone antagonists followed by the β2-agonist clenbuterol) with LVAD support in 20 prospective patients with dilated nonischemic cardiomyopathy eligible for bridge to heart transplantation receiving the HeartMate II. They showed 12 patients (60%) met the explantation criteria. Survival after explantation was 83.3% at 1, 2, and 3 years. Dandel et al. [16] studied 1,038 patients who received a long-term VAD primarily designed as bridge to transplant or permanent therapy. Among them, 96 (9.2%) patients with relevant cardiac improvement underwent VAD explantation. Kaplan-Meier estimates of overall survival after VAD explantation (including post-heart transplant survival for those with heart failure recurrence) revealed probabilities of 72.8 ± 6.6% and 67.0 ± 7.2% for 5- and 10-year survival, respectively. According to the seventh annual report of the Interagency Registry for Mechanically Assisted Circulatory Support [17], the recovery rate from CF LVAD was less than 1%, which is much lower than other reports.

George et al. [18] assessed the safety of speed reduction to determine whether 6,000 rpm is sufficient in assessing the underlying myocardial function in patients with HeartMate II. They showed that speed reduction in patients with the HeartMate II is a safe way to assess underlying LV function. Direct comparison of echocardiographic, hemodynamic, and device flow parameters acquired at 6,000 rpm, 5,000 rpm, and 4,000 rpm revealed no significant difference, suggesting that speed reduction to 6,000 rpm is sufficient to assess the native myocardial function. Dandel et al. [19] also discuss weaning strategy. The complete stops of CF pumps lead to retrograde flow into the LV, which can cause overestimations of LV systolic function. According to them, the misleading retrograde blood flow into the LV during off-pump trials is a major negative factor that can interfere with successful VAD explantation. They suggested that rotor speed reduction to values resulting in close to zero flow in one cardiac cycle was better than complete pump-stop.

We assessed in vitro hemodynamic performance of HeartMate II in low-speed setting to demonstrate the pressure-flow relationship at a pump speed of 6,000 rpm in pulsatile conditions with a mock ventricle [20]. The results showed that the amount of regurgitant flow at 6,000 rpm support depends on the native heart function. Regurgitant flow with the HeartMate II at 6,000 rpm in the normal heart condition was greater than that in the heart failure condition. Despite prior beliefs that at this speed the net flow is zero, our in vitro characterization suggests that the HeartMate II device makes a significant contribution to total flow in the heart failure condition because there is less regurgitant flow at this speed. In the normal heart condition, mean aortic pressure and total flow stayed at the same level. However, in the heart failure condition, mean aortic pressure at 6000 rpm support was higher than with no LVAD support due to significant contribution by the LVAD. Total flow was the same in both conditions, but LVAD flow made up a considerable part of total flow at 6,000 rpm support. We also assessed the HVAD at low speed (1800 rpm) in mock loop [21]. Our results with the HVAD at its lowest pump speed of 1800 rpm mirrored those with the HeartMate II at its lowest pump speed of 6,000 rpm. As previously described, the Advanced VAD has a shut-off feature that prevents backflow when the pump stops, enabling us to evaluate patient weanability more precisely than other LVADs.

Right Heart Failure and Right Ventricular Assist Device

Occurrence and Clinical Results of Right Heart Failure

Right heart failure (RHF) is an unsolved problem after LVAD implantation. We evaluated the incidence of severe RHF after LVAD implantation by retrospective analysis of data from 245 patients who had received PF LVAD at the Cleveland Clinic Foundation and showed RVAD support was required after LVAD insertion for 23 patients (9%) [22]. Dang et al. [23] showed their clinical experience of patients who developed RHF after LVAD implantation for chronic congestive heart failure with 108 patients with LVADs. Of these, 42 (38.9%) developed post-LVAD RHF. Post-LVAD early mortality rate was significantly higher, and bridge to transplantation rate was significantly lower in RHF patients than in non-RHF patients. Kormos et al. [24] evaluated the incidence, risk factors, and effect on outcomes of RHF in 484 patients implanted with the HeartMate II. A total of 98 (20%) of the 484 patients had some form of RHF, including 30 (6%) patients requiring an RVAD. Patients with early RHF had significantly worse outcomes, with those requiring RVADs having the lowest percentage reaching these outcomes. Yoshioka et al. [25] reported the experience with unplanned RVADs using temporary devices early after CF LVAD placement. Of the entire 305 patients, 27 (9%) developed severe RHF and required unplanned RVAD insertion early after LVAD implantation. Seventeen (63%) patients could be weaned from RVAD after median of 14 days. The 1-year survival in the RVAD weaning failure group was 20%, and the RVAD weaning failure group had significantly worse survival compared with the RVAD weaning group ($P < 0.001$) and the isolated LVAD group ($P < 0.001$). They also analyzed late RHF in patients after discharge from the first indexed admission. In the RVAD weaning group, the late RHF-free survival rate at 3, 6, and 12 months was 91%, 91%, and 53% and 96%, 93%, and 90% in the isolated LVAD groups, respectively ($P = 0.002$). They concluded careful attention should be paid to the recurrent or sustained RHF and deterioration of renal function.

Use of Left Ventricular Assist Device as Right Ventricular Assist Device

Since there is no clinically available ventricular assist device specified for RVAD use, LVADs have been used in an RVAD setting. Radovancevic et al. [26] performed acute studies to evaluate the feasibility of using the Jarvik 2000 (Jarvik Heart, Inc., New York, NY) as a biventricular assist device in calves with fibrillating hearts. The right pump was inserted in the right atrium. In this study, Jarvik 2000 showed stable hemodynamic support as a BiVAD setting and showed feasible results. The typical operating range for a CF RVAD pump calls for generating pump flows of 2–6 L/min while raising the pump outlet pressure by 20–50 mm Hg from that at the pump inlet port [27]. The CF RVAD performance requirements are illustrated in Fig. 29.9 by the blue-shaded region within the DexAide CF RVAD pump flow versus differential pressure plot for operating pump speeds between 1800 rpm and 3000 rpm. For comparison, the operating range targeted for the CorAide CF LVAD is also shown (red-shaded region) in Fig. 29.9.

Krabatsch et al. [28] discussed the problem of when HVAD is used as a RVAD. The HVAD LVAD pumps 3–10 L/min with a rotational speed range of 1800–3800 rpm against a pressure head of 50–200 mm Hg. With lower afterload, as in the pulmonary circulation, even with the lowest adjustable speed, the flow would be too high. The inflow cannula of the HVAD is 35 mm in length, and it is too long to insert into the right ventricle. To solve these problems, they reduced the outflow graft diameter stepwise to such a degree that the afterload of the VAD reaches the usual levels of the systemic circulation. To

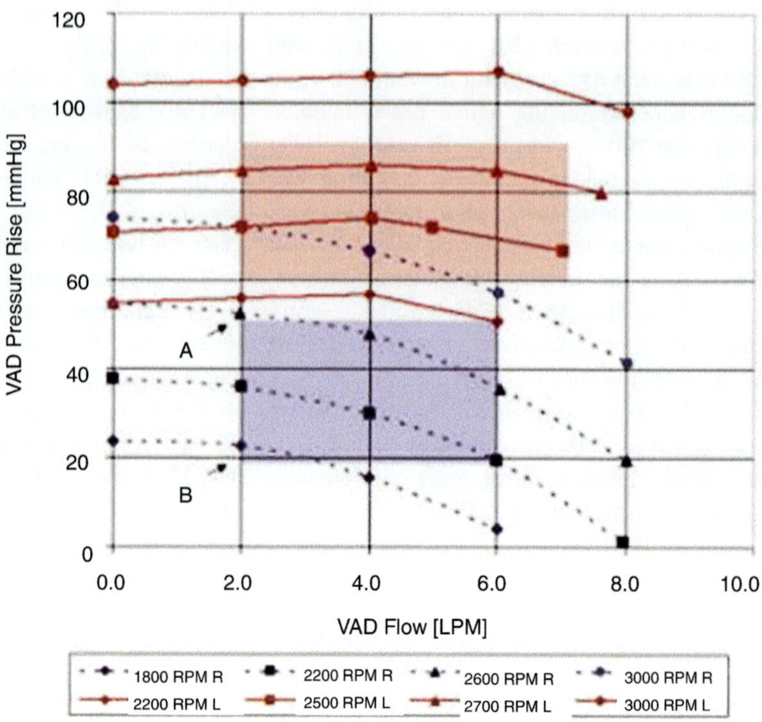

Fig. 29.9 Hydraulic performance of CorAide continuous-flow left ventricular assist device (CF LVAD [red]) versus DexAide continuous-flow right ventricular assist device (CF RVAD [blue]). Red and blue areas represent typical operating requirements for CF LVAD and CF RVAD, respectively. (LPM, liters per minute; RPM, revolutions per minute; VAD, ventricular assist device). (Reproduced from Fukamachi et al. [27], with permissionfrom Elsevier)

reduce the effective length of the inflow cannula, they added two 5 mm silicon suture rings covered with Dacron Velour (DHZB in-house product, made by Berlin Heart GmbH, Berlin, Germany) to the original implantation ring. Marasco et al. [29] implanted the HVAD in the right atrium, suspending the pump through a window in the right lateral pericardium. They also placed spacers over the inflow cannula between the epicardium and the sewing ring, so that only about 15 mm of the inflow cannula passes into the right atrium. To increase the afterload of the device, they crimped the diameter of the outflow graft to about 6 mm with hemaclips over a 3 cm portion of the length. Stevens et al. [30] also discussed the RVAD use of CF LVAD. A RVAD can be operated at a lower speed than LVAD, but this may require operating the pump at a speed lower than recommended by the manufacturer, resulting in potential impeller instability and suboptimal washout within the device. If the RVAD outflow cannula is restricted to a diameter between 6.5 and 8.1 mm, suitable steady-state hemodynamics can be achieved while maintaining impeller stability and optimal device washout. RVAD speed adjustments or outflow diameter changes can accommodate for long-term or transient variations in pulmonary vascular resistance; however, the latter requires the use of an adjustable restriction mechanism.

The Advanced VAD can operate as both an LVAD and RVAD without modification, because at lower RVAD speeds, the aperture opening automatically reduces the area under force of the weak closing magnet, which in turn attenuates hydraulic output for the RVAD application. In our novel Advanced VAD approach, the volute passage, which collects and outputs the flow, is axially offset from the impeller (as in the DexAide RVAD) to even out pressures around the impeller, yielding an axis-symmetrical pressure and flow distribution, which allows the Advanced VAD to operate over a wide range of flows/speeds without driving adverse secondary flow patterns. This broad range of operation allows the Advanced VAD to be used as either an LVAD at high speed (~3300 rpm) or RVAD at low speed (~2300 rpm) without modification, using the same electronic hardware.

Summary

Our initial in vitro study has demonstrated the feasibility of generating high levels of pulsatility and established the functionality of the shutoff feature. The merit of the differential normalized pressure-flow curve depending on the pump speed is that the Advanced VAD can function as either an LVAD or an RVAD using different speed ranges. By introducing the changing aperture, which enables the flow to change automatically with pressure difference between inflow and outflow, the Advanced VAD shows some advantages over existing CF LVADs, including enhanced pulsatility, regurgitant shutoff feature, and wide operating range. Now the pump is ready for the first in vivo study, and while the Advanced VAD requires optimization to improve biocompatibility, this pump could be next generation VAD.

References

1. Slaughter MS, Rogers JG, Milano CA, Russell SD, Conte JV, Feldman D, et al. Advanced heart failure treated with continuous-flow left ventricular assist device. N Engl J Med. 2009;361(23):2241–51.
2. Fukamachi K, Horvath DJ, Byram N, Sunagawa G, Karimov JH, Moazami N. Advanced ventricular assist device with pulse augmentation and automatic regurgitant-flow shut-off. J Heart Lung Transplant. 2016;35(12):1519–21.
3. Roache PJ. Perspective: a method for uniform reporting of grid refinement studies. J Fluids Eng. 1994;116(3):405–13.
4. Banerjee RK, Cho YI, Kensey K, editors. Effect of the non-Newtonian viscosity of blood on steady and pulsatile flow in stenosed arteries: American Society of Mechanical Engineers, Bioengineering Division (Publication) BED; 1991.
5. Cross MM. Rheology of non-Newtonian fluids: a new flow equation for pseudoplastic systems. J Colloid Sci. 1965;20(5):417–37.
6. Travis AR, Giridharan GA, Pantalos GM, Dowling RD, Prabhu SD, Slaughter MS, et al. Vascular pulsatility in patients with a pulsatile- or continuous-flow ventricular assist device. J Thorac Cardiovasc Surg. 2007;133(2):517–24.
7. Undar A, Ji B, Lukic B, Zapanta CM, Kunselman AR, Reibson JD, et al. Quantification of perfusion modes in terms of surplus hemodynamic energy levels in a simulated pediatric CPB model. ASAIO J. 2006;52(6):712–7.

8. Amir O, Radovancevic B, Delgado RM 3rd, Kar B, Radovancevic R, Henderson M, et al. Peripheral vascular reactivity in patients with pulsatile vs axial flow left ventricular assist device support. J Heart Lung Transplant. 2006;25(4):391–4.
9. Welp H, Rukosujew A, Tjan TD, Hoffmeier A, Kosek V, Scheld HH, et al. Effect of pulsatile and non-pulsatile left ventricular assist devices on the renin-angiotensin system in patients with end-stage heart failure. Thorac Cardiovasc Surg. 2010;58(Suppl 2):S185–8.
10. Demirozu ZT, Radovancevic R, Hochman LF, Gregoric ID, Letsou GV, Kar B, et al. Arteriovenous malformation and gastrointestinal bleeding in patients with the HeartMate II left ventricular assist device. J Heart Lung Transplant. 2011;30(8):849–53.
11. Wever-Pinzon O, Selzman CH, Drakos SG, Saidi A, Stoddard GJ, Gilbert EM, et al. Pulsatility and the risk of nonsurgical bleeding in patients supported with the continuous-flow left ventricular assist device HeartMate II. Circ Heart Fail. 2013;6(3):517–26.
12. Cowger J, Pagani FD, Haft JW, Romano MA, Aaronson KD, Kolias TJ. The development of aortic insufficiency in left ventricular assist device-supported patients. Circ Heart Fail. 2010;3(6):668–74.
13. Pak SW, Uriel N, Takayama H, Cappleman S, Song R, Colombo PC, et al. Prevalence of de novo aortic insufficiency during long-term support with left ventricular assist devices. J Heart Lung Transplant. 2010;29(10):1172–6.
14. Simon MA, Kormos RL, Murali S, Nair P, Heffernan M, Gorcsan J, et al. Myocardial recovery using ventricular assist devices: prevalence, clinical characteristics, and outcomes. Circulation. 2005;112(9 Suppl):I32–6.
15. Birks EJ, George RS, Hedger M, Bahrami T, Wilton P, Bowles CT, et al. Reversal of severe heart failure with a continuous-flow left ventricular assist device and pharmacological therapy: a prospective study. Circulation. 2011;123(4):381–90.
16. Dandel M, Weng Y, Siniawski H, Potapov E, Krabatsch T, Lehmkuhl HB, et al. Pre-explant stability of unloading-promoted cardiac improvement predicts outcome after weaning from ventricular assist devices. Circulation. 2012;126(11 Suppl 1):S9–19.
17. Kirklin JK, Naftel DC, Pagani FD, Kormos RL, Stevenson LW, Blume ED, et al. Seventh INTERMACS annual report: 15,000 patients and counting. J Heart Lung Transplant. 2015;34(12):1495–504.
18. George RS, Sabharwal NK, Webb C, Yacoub MH, Bowles CT, Hedger M, et al. Echocardiographic assessment of flow across continuous-flow ventricular assist devices at low speeds. J Heart Lung Transplant. 2010;29(11):1245–52.
19. Dandel M, Hetzer R. Myocardial recovery during mechanical circulatory support: weaning and explantation criteria. Heart Lung Vessel. 2015;7(4):280–8.
20. Sunagawa G, Byram N, Karimov JH, Horvath DJ, Moazami N, Starling RC, et al. In vitro hemodynamic characterization of HeartMate II at 6000 rpm: implications for weaning and recovery. J Thorac Cardiovasc Surg. 2015;150(2):343–8.
21. Sunagawa G, Byram N, Karimov JH, Horvath DJ, Moazami N, Starling RC, et al. The contribution to hemodynamics even at very low pump speeds in the HVAD. Ann Thorac Surg. 2016;101(6):2260–4.
22. Ochiai Y, McCarthy PM, Smedira NG, Banbury MK, Navia JL, Feng J, et al. Predictors of severe right ventricular failure after implantable left ventricular assist device insertion: analysis of 245 patients. Circulation. 2002;106(12 Suppl 1):I198–202.
23. Dang NC, Topkara VK, Mercando M, Kay J, Kruger KH, Aboodi MS, et al. Right heart failure after left ventricular assist device implantation in patients with chronic congestive heart failure. J Heart Lung Transplant. 2006;25(1):1–6.
24. Kormos RL, Teuteberg JJ, Pagani FD, Russell SD, John R, Miller LW, et al. Right ventricular failure in patients with the HeartMate II continuous-flow left ventricular assist device: incidence, risk factors, and effect on outcomes. J Thorac Cardiovasc Surg. 2010;139(5):1316–24.
25. Yoshioka D, Takayama H, Garan RA, Topkara VK, Han J, Kurlansky P, et al. Contemporary outcome of unplanned right ventricular assist device for severe right heart failure after continuous-flow left ventricular assist device insertion. Interact Cardiovasc Thorac Surg. 2017;24(6):828–34.
26. Radovancevic B, Gregoric ID, Tamez D, Vrtovec B, Tuzun E, Chee HK, et al. Biventricular support with the Jarvik 2000 axial flow pump: a feasibility study. ASAIO J. 2003;49(5):604–7.
27. Fukamachi K, Shiose A, Massiello AL, Horvath DJ, Golding LA, Lee S, et al. Implantable continuous-flow right ventricular assist device: lessons learned in the development of a cleveland clinic device. Ann Thorac Surg. 2012;93(5):1746–52.
28. Krabatsch T, Hennig E, Stepanenko A, Schweiger M, Kukucka M, Huebler M, et al. Evaluation of the HeartWare HVAD centrifugal pump for right ventricular assistance in an in vitro model. ASAIO J. 2011;57(3):183–7.
29. Marasco SF, Stornebrink RK, Murphy DA, Bergin PJ, Lo C, McGiffin DC. Long-term right ventricular support with a centrifugal ventricular assist device placed in the right atrium. J Card Surg. 2014;29(6):839–42.
30. Stevens MC, Gregory SD, Nestler F, Thomson B, Choudhary J, Garlick B, et al. In vitro and in vivo characterization of three different modes of pump operation when using a left ventricular assist device as a right ventricular assist device. Artif Organs. 2014;38(11):931–9.

Cleveland Clinic Total Artificial Heart 30

Jamshid H. Karimov, David J. Horvath,
and Kiyotaka Fukamachi

Background

Heart failure (HF) is a serious health issue and a primary contributor to cardiovascular mortality that affects approximately 26 million people worldwide [1]. The incidence of HF is also rising rapidly in developing nations. More than 5.1 million Americans 20 years old and older have HF, and its prevalence has recently been forecast to increase, from 2.42% in 2012 to 2.97% in 2030 (with the number of patients diagnosed with HF to increase by 46%). By 2030, therefore, more than 8 million Americans will be living with HF, and direct medical costs associated with HF are projected to increase from $21 billion to $53 billion [2, 3].

Heart transplantation can provide a remarkable improvement in quality of life and survival in selected patients with end-stage heart disease [4, 5], but a shortage of donor hearts always limits this option [6]. In the United States, an estimated 50,000–100,000 patients are in need of heart transplantation or mechanical circulatory support (MCS), but transplantation is performed in only ~2300 cases per year. Instead, such support is usually provided by left ventricular assist devices (LVADs), the most widely used means for improving the hemodynamic status of both transplant-eligible and non-eligible patients [5].

Initially employed as a bridge to transplantation, MCS was most frequently used to prolong survival while patients were on the waiting list until another therapy could be proposed or a donor organ became available. Today, an increasing number of patients with end-stage HF are being placed on long-term ventricular support as "destination therapy," with better survival and quality of life. During the last decade, the LVAD landscape has moved from volume-displacement pulsatile-flow technology toward continuous-flow (CF) rotary pumps. CF pumps provide greater mechanical reliability, smaller size, improved durability, and better patient outcomes that approach those of heart transplantation. Since its approval by the US Food and Drug Administration (FDA) in April 2008, the HeartMate II (HMII) (Abbott, Lake Bluff, IL), an axial pump, has seen 25% annual growth in the number of LVAD implants throughout the world, exceeding 26,600 total implants as of 2018. The HVAD (Medtronic, Minneapolis, MN), a centrifugal pump, has been implanted in more than 17,000 patients worldwide. The newer HeartMate 3 CF LVAD (Abbott) was compared to its axial-flow predecessor (HMII) in more than 1000 patients in the MOMENTUM trial and demonstrated a lower incidence of pump thrombosis and disabling stroke [7]. According to the eighth

J. H. Karimov, MD, PhD (✉) · K. Fukamachi, MD, PhD
Lerner Research Institute, Cleveland Clinic,
Cleveland Clinic Lerner College of Medicine of Case Western Reserve University, Cleveland Clinic,
Cleveland, OH, USA
e-mail: karimoj@ccf.org; fukamak@ccf.org

D. J. Horvath, MSME
Department of Biomedical Engineering,
Lerner Research Institute, Cleveland Clinic,
Cleveland, OH, USA

R1 Engineering, Euclid, OH, USA

© Springer Nature Switzerland AG 2020
J. H. Karimov et al. (eds.), *Mechanical Support for Heart Failure*,
https://doi.org/10.1007/978-3-030-47809-4_30

annual report of the Interagency Registry for Mechanically Assisted Circulatory Support, more than 95% of implants in the current era are CF devices. This is a pace of more than 2500 implants per year [8], which clearly indicates the future direction of this device therapy.

The major technological advancements have been achieved; however, complications related to MCS device use continue to arise, and they remain the focus of increasingly sophisticated research in device development. Clinicians and engineers continue to expand their understating of the physiological effects of MCS, and device technology is evolving toward meeting the goals of "precision medicine," tailoring therapy to the individual patient.

In this chapter, we report on efforts at Cleveland Clinic to address the demands for end-stage HF treatment using total heart replacement technology, specifically on continuous-flow total artificial heart (CFTAH) platform currently under development in our institution.

Total Artificial Heart Technology

We have made major progress in our understanding of the progression and pathophysiology of HF since the initial conceptualization of a total artificial heart (TAH) using a mechanical device, based on early efforts of Dr. Willem Kolff at Cleveland Clinic in the 1950s. Today, TAHs are used in patients with end-stage HF to replace cardiac function in order to restore adequate viable perfusion. Experience with the available versions of volume-displacement TAHs has been encouraging when the TAH is used as a bridge to transplant in this subset of patients. Clinical experience with the TAH as destination therapy has thus far been limited, but some very promising technologies are currently under development, with some described in this book.

The development of CF rotary pump technology represents an important landmark centered on a novel pump design concept. These newer pumps have almost completely supplanted pulsatile, volume-displacement devices. CF rotary blood pumps are typically designed with an "axial" or "centrifugal" blood flow path, featuring an internal rotor within that path [4]. By eliminating a moving membrane and the multiple valves needed by pulsatile pumps, continuous-flow TAHs have the benefits of a smaller design and potential for superior long-term mechanical dependability.

The TAHs are usually made up of an implantable rotary blood pump, along with external components located outside the patient's body: a wearable controller connected to the pump via a percutaneous cable; wearable rechargeable batteries; battery charger; alternating- and direct-current (AC/DC) power supplies; and a hospital device for system control and monitoring.

If the blood pump is the "heart" of the TAH, the controller is its "brain." The controller drives the pump's electrical motor; controls the pump speed or flow based on user commands or feedback signals; collects, processes, and stores data; performs self-diagnostics; transmits information to and receives data from other system components, i.e., hospital monitor and batteries; and provides various types of user interface (audible, visual, and tactile) [9]. The TAH, in response to its hemodynamic environment, can act as a virtual sensor, providing feedback to indicate and track relative changes in flow and differential pressure. TAHs are connected to their controller units through an exterior driveline. All of the devices' left and right inflows originate from the native atrial chamber to provide blood flow to the pump and outflows connected to the patient's native vessels. The driveline remains outside the body.

Total Artificial Heart Development at Cleveland Clinic

The effort to create an advanced TAH pump in our institution originated with a creative challenge to craft an ultrasimple TAH (Fig. 30.1). Our novel CFTAH concept has undergone several iterations since the device conception as an alpha prototype development, resulting in an *in vivo* laboratory system capable of chronic support and automatic control. Although the system requires no flow or pressure sensors to function, a proposal was made to introduce a rotor position sensor in our research to demonstrate the optional benefits for hemodynamic monitoring.

The CFTAH design was extremely innovative, exploring radical new concepts in TAH design. The CFTAH design comprised a single, valveless CF pump with passive inherent hydraulic flow and pressure regulation to take over the function of the native heart. This "double-ended centrifugal pump" concept had a single, continuously rotating, brushless DC motor and pump assembly with a centrifugal pump on both ends. Pump speed was modulated to create pulsatile flow and pressure. The electrical power response to speed modulation was monitored via controller and used to indicate incipient suction and appropriate speed reduction. This design differentiated itself from current clinical TAHs in that the speed-modulated, single-shaft, dual centrifugal pump allows for a significant reduction in size while eliminating the use of valves and pressure or flow sensors for control, which is believed to improve the overall system reliability.

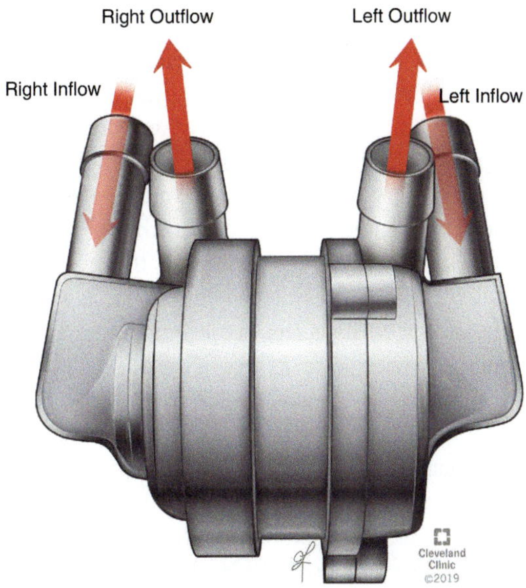

Fig. 30.1 Cleveland Clinic continuous-flow artificial heart. (Reprinted with permission, Cleveland Clinic Center for Medical Art & Photography © 2018–2019. All Rights Reserved)

The core breakthrough of this elegant, innovative pump concept was that the hemodynamic function of balancing systemic and pulmonary flows while minimizing atrial pressure differences is incorporated directly into the pump's hydromechanical design.

This inherent function was engineered as an automatic, immediate, passively regular feature and required no electronic intervention. The CFTAH control scheme was based on this passive self-balancing of pump flows to maintain atrial pressure balance and prevent atrial wall suction without the use of sensors for control feedback.

Preliminary data have shown that these design features do regulate flows and pressures over wide ranges of systemic and pulmonary vascular resistance [10, 11]. Figure 30.2 illustrates how the rotating assembly position affects pump geometry to correct any imbalance in atrial pressure. Note that the rotor magnet is shorter than the steel stator laminations, so that there is a window of axial movement that is relatively free of magnetic restoring forces, allowing rotor movement to respond primarily to pressure forces. Inlet pressures are the most dominant terms in force balancing. If right atrial pressure (RAP) becomes higher than left atrial pressure (LAP), the rotating assembly is pushed to the left (Fig. 30.2a), opening the aperture at the outer diameter of the right pump, increasing right pump output, decreasing RAP, and increasing LAP. Conversely, if LAP is higher than RAP, the rotating assembly is pushed to the right (Fig. 30.2b), closing the aperture of the right pump and reducing right pump output, again driving the system to balance.

This CFTAH self-regulating response also acts to mitigate atrial suction [12–14]. If the left inlet is occluded due to suction, the left inlet pressure drops, pulling the rotating assembly to the left, thereby opening the aperture of the right pump, increasing LAP, and automatically recovering from the suction condition. Similarly, if the right inlet is occluded due to suction, the right inlet pressure drops, pulling the rotating assembly to the right, thereby closing the aperture of the right pump, decreasing the suction on the

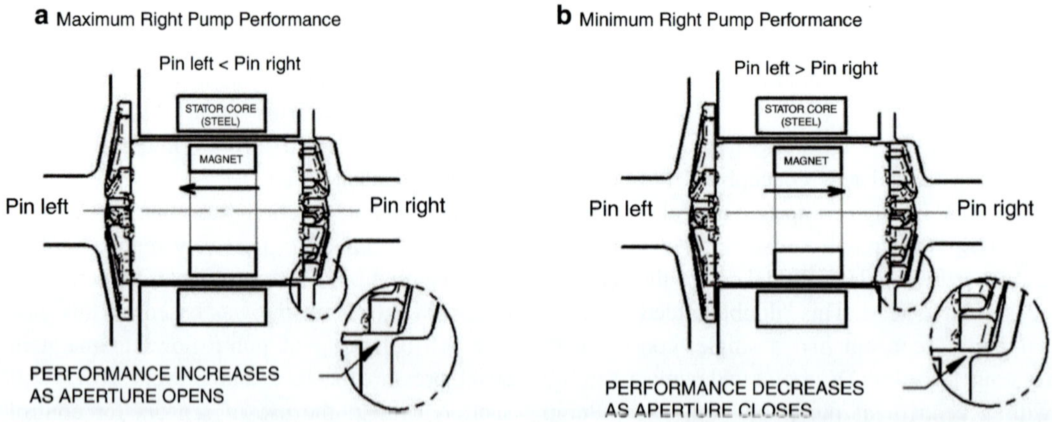

Fig. 30.2 Atrial pressure difference effects on rotor position and right pump performance. (Reproduced from Fukamachi et al. [10], with permission from Elsevier)

right atrium. The stability of the system's dynamic response to enforced perturbations in the hemodynamic environment (simulated suction, extreme vascular resistances, hypovolemia, speed changes, and various levels of speed modulation) has been demonstrated by bench studies and in acute *in vivo* evaluations. Rotor movement was experimentally characterized using a CFTAH pump instrumented with a Hall probe rotor position sensor incorporated seamlessly into the CFTAH assembly.

The relatively high number of complex components in existing pulsatile TAHs, as well as their large size, potential for failure, and limited durability, creates an obvious need for a small, reliable TAH. The availability of a small, simplified alternative with a single moving component will address many problems with existing technologies and provide better management for more patients. Ultimately, such a device would be less costly, as it needs only a single motor controller to balance left and right flows while eliminating valves, compliance chambers, actuator mechanisms, and pressure or flow sensors. The Cleveland Clinic CFTAH accomplishes passive, sensorless autoregulation of pump flows and filling pressures by varying relative left and right pump efficiencies, based on the position of its single moving part, to balance the left and right pump-filling pressures and pump outputs [10, 12].

Our series of *in vivo* single-piece CFTAH experiments has demonstrated desired hemodynamic performance and pump operation within design specifications. From October 2011 to November 2014, the Cleveland Clinic CFTAH was implanted in 17 Jersey calves (weight range, 77.0–93.9 kg) through a median sternotomy (n = 9) or a right thoracotomy (n = 8) [15] for elective chronic implantation of 14, 30, or 90 days. The improved pump design in six animals showed no chronic hemolysis. One animal achieved the intended 30-day duration; two animals achieved intended 90-day duration.

For the first time, two 90-day biocompatibility studies in calves were successfully completed at the planned durations and showed remarkable pump performance and biocompatibility without any postoperative anticoagulation therapy [16]. The CFTAH passively self-balances left and right circulations without electronic intervention. The nominal external dimensions of the latest CFTAH design used in these studies were 98.4 mm (length), 62 mm (diameter), 160 cc (volume displacement), and 486 g (device weight with cable and connector off the scale). All 17 chronic experiments demonstrated good hemodynamic performance of the CFTAH. The postoperative course differed in each animal as described above, in part because of intraoperative complications, or as technical issues that were encountered during this program. All of the longest-surviving cases showed acceptable biocompatibility with no organ thromboembolism. The newer device generation with a revised right impeller is currently being tested *in vivo*.

Pediatric Total Artificial Heart

MCS has become standard therapy for adult patients with end-stage HF; however, in pediatric patients with congenital heart disease, the options for chronic MCS are limited to paracorporeal devices or off-label use of devices intended for implantation in adults. Congenital heart disease and cardiomyopathy often involve both the left and right ventricles; in such cases, heart transplantation or a biventricular assist device (BVAD) or a TAH is needed to adequately sustain both pulmonary and systemic circulations. Heart transplantation in infants and children is an accepted therapy for congenital heart disease, but the availability of suitable donor hearts varies and is never enough to meet the clinical need. MCS is also defined as a standard option, but its use in infants and children is limited due to size and functional range restrictions.

At Cleveland Clinic, we are developing a miniature pediatric version of the CFTAH described above (P-CFTAH; See Chap. 38) with a single motor and single rotating assembly [17] (Fig. 30.3). The Cleveland Clinic team has been the first to report successful implantation of the prototype P-CFTAH in an acute animal study [18]. The results are from a series of implantations of our P-CFTAH in four healthy lambs and recently published [19]. The working model prototype P-CFTAH pump was derived from the adult CFTAH configuration by downsizing it by a scale factor of 0.70 (one-third of its total volume) to enable implantation in the infant chest. The pump flow range (1.5–4.5 L/min) is intended to provide support to patients weighing up to 50 kg (average weight of 14-year-olds) (Fig. 30.4).

The pressure-flow curves demonstrated a broad range of left pump flow. At a delta pressure of 100 mm Hg, the left pump flow range was

Fig. 30.3 Pediatric continuous-flow total artificial heart (P-CFTAH) prototype

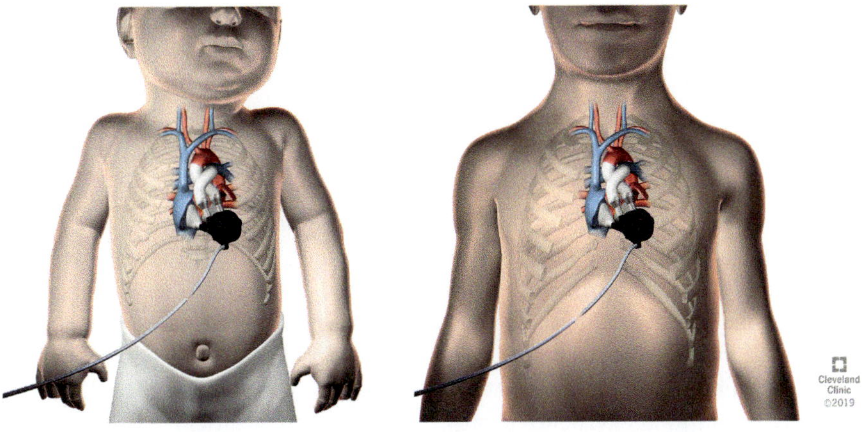

Fig. 30.4 Schematic illustration of P-CFTAH inserted in infant and 4-year-old child's chest. (Reprinted with permission, Cleveland Clinic Center for Medical Art & Photography © 2018–2019. All Rights Reserved)

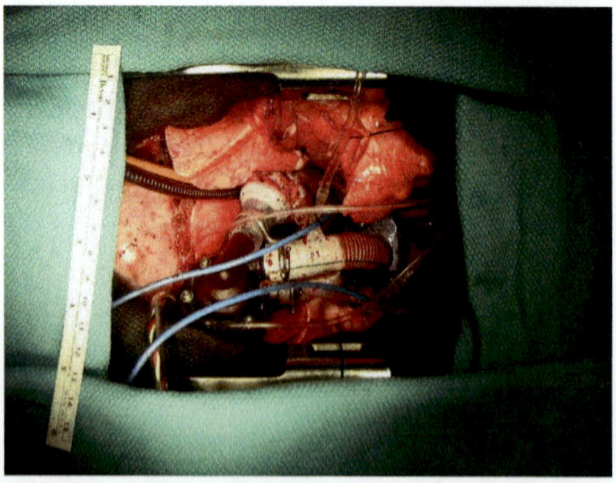

Fig. 30.5 P-CFTAH implantation in animal model

0.4–4.7 L/min with a pump speed of 3500 and 5000 rpm. The difference in left minus right atrial pressures (LAP-RAP) was maintained within ±5 mm Hg with a wide range of systemic to pulmonary vascular resistance ratios (between 1.4 and 35), with and without pump speed modulation. Our findings met the proposed design requirements for pump flow range (1.5–4.5 L/min) and self-regulation (LAP-RAP within −5 to +10 mm Hg). The ongoing *in vivo* experiments will help us evaluate the hydraulic performance and self-regulating mechanical design of the P-CFTAH under selected hemodynamic conditions (Fig. 30.5). Multiple challenges to pediatric MCS devices exist, and these difficulties need to be overcome to advance the field and improve quality of life, safety, and survival of pediatric patients in need of MCS [20].

Infant Total Artificial Heart

In order to explore the possibilities of using the CFTAH platform and designing further downsizing of the pump, we recently tested *in silico* the performance of the initial device model of an infant continuous-flow total artificial heart (I-CFTAH) with one motor and one rotating assembly supported by a hydrodynamic bearing [21]. The I-CFTAH (Fig. 30.6) (flange diameter – 26.9 mm; length – 44 mm) was envisioned to provide hemodynamic support to children ranging from 3 kg to 9 kg, with nominal design point

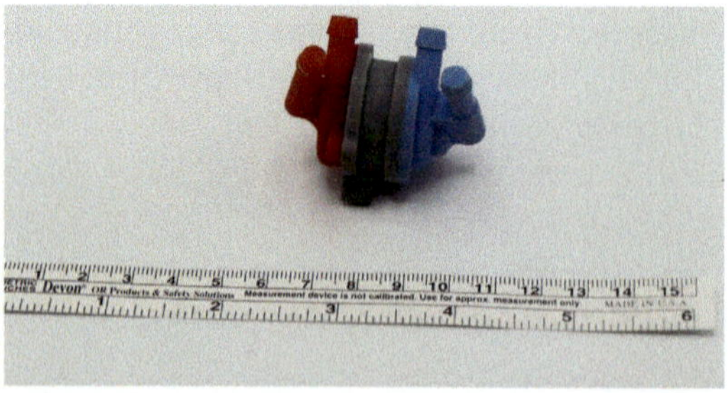

Fig. 30.6 Infant continuous-flow total artificial heart (I-CFTAH) device model

matched to a 6 kg child, requiring 1.0 L/min of flow. I-CFTAH was tested using an *in silico* virtual simulation mock loop (VML; MATLAB; MathWorks®) and simulated the hemodynamics as they surge through the system implanted with a blood pump (run at 0.5 L/min for 3 kg infant).

A lumped-parameter model, including systemic/pulmonary circulation, values for impedance, beat rate, and blood volume, was used. The early assessment during *in silico* experiments demonstrated the intended performance of the scaled-down pump (left and right pump flows – 0.5 L/min at 4380 rpm). The difference between the left and right atrial pressures was maintained within ±5–15 mm Hg. Simulation of I-CFTAH was successfully performed and allowed the creation of arterial pulsation with this device. The development of I-CFTAH is ongoing, and we will be looking to investigate the performance of this device in an animal model.

CFTAH for Biventricular Heart Failure

The use of continuous-flow (non-pulsatile) hemodynamic support using two CF pumps was first performed in our institution as biventricular support in a 75-kg calf supported for 34 days [22]. Today, the efforts are ongoing at Cleveland Clinic, with implantable devices to develop better options for mechanical support of a failing heart using continuous-flow devices [23].

When left ventricular filling remains inadequate despite pharmacotherapy to enhance right ventricular (RV) function and lower pulmonary vascular resistance, a right ventricular assist device (RVAD) may be required [24]. The primary reason VADs have not been designed specifically for RVAD applications is that there is much less clinical need (and therefore a limited market) than those for LVADs [25, 26]. Although clinically approved implantable LVADs are not designed for right-sided support, LVAD implantation is being considered for right heart support simply because an effective, easily implantable RVAD is not yet available. The absence of a dedicated, clinically evaluated, long-term RVAD significantly diminishes the treatment options for patients with isolated and post-LVAD RV failure, as well as for those with biventricular HF requiring two pumps (BVAD) [27, 28]. The inherent CF pump mechanics and translation of clinical features, pump selection, and patient management issues have previously been detailed [12].

The first use of an implantable CF pump for RV support was reported in 2004. A 21-year-old patient with worsened RV failure and elevated peripheral vascular resistance after implantation of a Jarvik 2000 FlowMaker was treated with implantation of a second one [24]. By directing the inlet toward the inferior vena cava-right atrial junction and keeping right atrial pressures at 5–10 cm H_2O, surgeons achieved stable RV unloading and satisfactory blood flow to the left-sided pump. Other groups consecutively reported on limited, off-label clinical use of two LVADs. Saito et al. [28] converted a 6-month survival patient from a Toyobo paracorporeal device (Nipro, Osaka, Japan) to a Jarvik 2000 as RVAD and a DuraHeart (Terumo Heart, Inc., Ann Arbor, MI) for LVAD support.

Successful application of two HVADs (Medtronic, Minneapolis, MN) to treat severe biventricular failure was previously reported [29–31]. A group from the German Heart Institute (Deutsches Herzzentrum) in Berlin reported on ten patients with idiopathic dilated cardiomyopathy who received biventricular HVADs [32]. In three of these patients, physicians were able to turn the RVAD off following long-term BVAD support, having confirmed that the patients had achieved stable hemodynamics and that no thromboembolic events and no regurgitation through the right pump had occurred. Currently, temporary support via RVADs/BVADs is available by CF extracorporeal or paracorporeal circulation support systems [26, 33] such as the CentriMag (Thoratec, Pleasanton, CA), the Bio-Pump (Medtronic, Minneapolis, MN), and the TandemHeart (CardiacAssist, Inc., Pittsburgh, PA). Progress in the development of clinically available rotary pumps now permits physicians to choose a hybrid approach for BVAD placement, using both an implantable LVAD and a temporary extracorporeal RVAD [34–37]. The adaptation of an LVAD to function as an RVAD remains a technical challenge because of both ventricular distinctions and pump characteristics. BVAD support can be effective but continues to be associated with lower survival and a worse quality of life than left ventricular support alone [38].

Despite progress in therapies for left HF, the biventricular failure represents an extremely ill population of end-stage cardiac disease that often involves both ventricles, even if its initial cause was limited to left-sided heart disease. There is a high incidence of adverse events and poor 1-year survival in the BVAD population [39, 40]. With no linkage to any specific pump type, this condition demands timely intervention to sustain deteriorating hemodynamics and to facilitate symptomatic improvement while continuing to limit the long-term success of durable MCS therapies [40]. TAHs are more appropriate than biventricular assist devices (BVADs) for patients with thrombosed ventricles, restrictive cardiomyopathy, cardiac tumors, severe ventricular septal defects, frail ventricles for difficult cannulation, congenital abnormalities, and intractable arrhythmias.

We recently performed a feasibility assessment of MCS using a miniature continuous-flow BVAD based on a pediatric continuous-flow total artificial heart (P-CFTAH) for biventricular support (Figs. 30.7, 30.8, and 30.9). Previously, in its downscaled configuration, the dimensions of the present blood pump have been shown to be implantable for TAH support in patients whose BSA is ≥ 0.3 m^2 [41]. However, the device has never been tested in the presence of the native

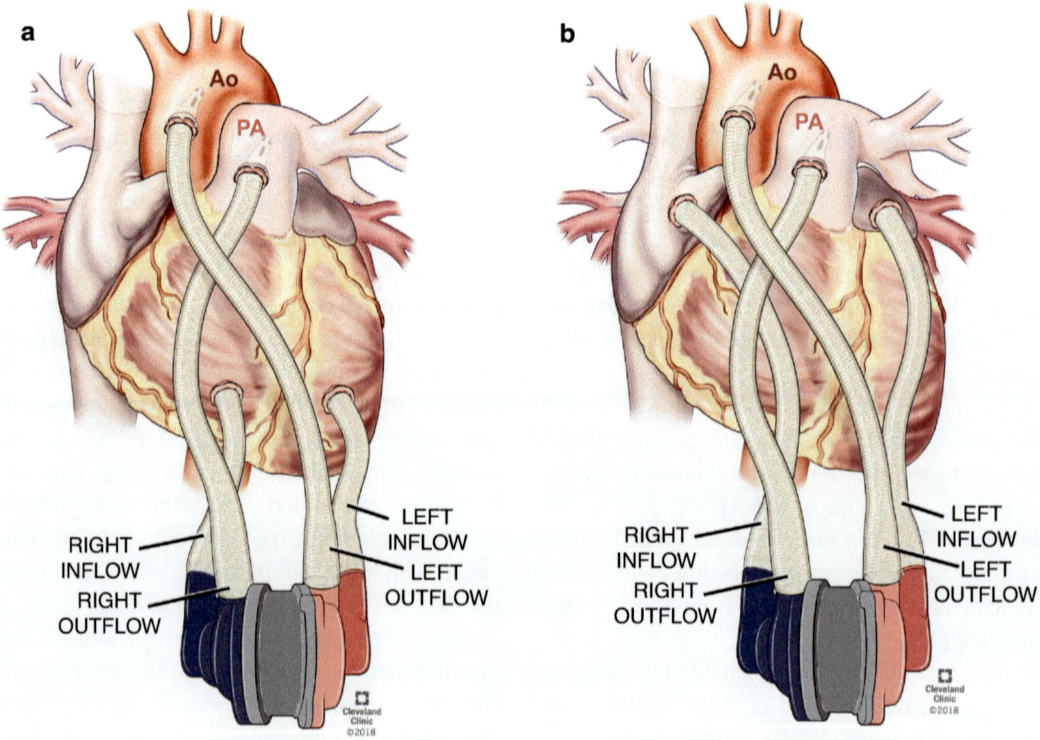

Fig. 30.7 Schematic illustration of biventricular support using P-CFTAH. (**a**) ventricular cannulation; (**b**) atrial cannulation. (Reprinted with permission, Cleveland Clinic Center for Medical Art & Photography © 2018–2019. All Rights Reserved)

heart ventricles. The BVAD used in this study is a single-piece device (P-CFTAH) that is valveless and does not employ flow or pressure sensors for its operation or control [10, 42]. All acute experiments in lambs (n = 4) showed adequate device performance and stability of device operation.

The *in vivo* data results compared favorably with previously collected *in vitro* pump performance. In both cannulation approaches (atrial and ventricular), the BVAD performance was maintained adequately within the 3000–6000 rpm range after implantation. The device self-regulating performance was demonstrated by the left and right atrial pressure difference (LAP-RAP) falling predominantly within the range of –5–10 mm Hg in left and right HF conditions with either atrial or ventricular cannulation types. This early series of acute *in vivo* experiments with the single-piece BVAD support has demonstrated the intended hemodynamic output, device features, and feasibility of BVAD support using one device and one controller unit. More development will be needed to investigate true hemodynamic requirements for BVAD support using single device, degree or self-regulation, and potential advantages or drawbacks in comparison with dual-VAD approach for biventricular HF treatment. This will be the required technology mapping step to indicate the most appropriate patient population for whatever approach will become available in the future.

The adult version of the Cleveland Clinic CFTAH has recently undergone multiple design optimizations (including changes to right impeller and stacked motor configuration) and is currently undergoing animal testing for pump performance and physiologic and hemodynamic parameters (Fig. 30.10). It will also be assessed for its use as BVAD in select HF population where higher hemodynamic output will be required.

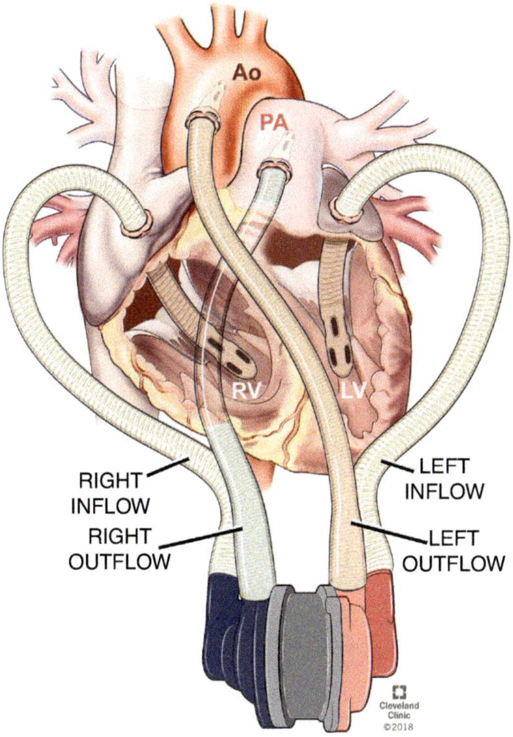

Fig. 30.8 Experimental setup for P-CFTAH implantation as BVAD. (Reprinted with permission, Cleveland Clinic Center for Medical Art & Photography © 2018–2019. All Rights Reserved)

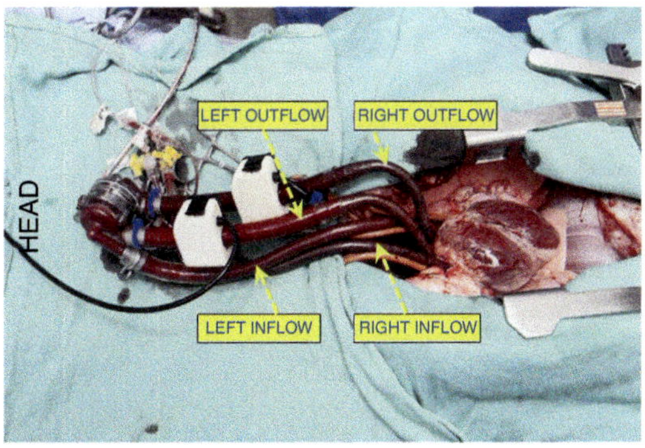

Fig. 30.9 Intraoperative view of biventricular assist device (BVAD) in animal model

Summary

There has been a continuous stream of new devices and modifications to existing technologies in recent years. These therapies have improved outcomes for patients in both morbidity and mortality. Significant technological and therapeutic advances in TAHs have been achieved in recent years, which is a major breakthrough in the field. However, issues with blood pumps continue to be the focus of increasingly sophisticated research in device development. Investigators are sharing insights into the physiological characteristics of mechanical circulation, and newer, more refined device technologies are continually emerging. Biocompatibility of the device components in direct contact with native tissues and blood is improving. However, the biocompatibility of devices and current methods of device control are still far from ideal and merit continuing investigation.

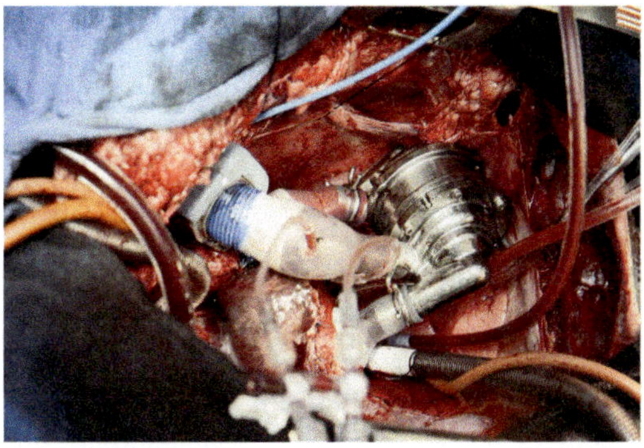

Fig. 30.10 Current version of CFTAH *in vivo*

The theoretical advantages of centrifugal-flow pumps over axial-flow pumps could move the field more toward centrifugal pump design in the future; however, with new device designs and better materials being used, there is certainly much room for improvements in pulsatile blood pump technologies, as well. Centrifugal pumps have better flow characteristics that can minimize hemolysis and perhaps also lead to less need for anticoagulation therapy. In addition, such pumps have a greater degree of "pulsatility" inherent in their design, which may translate into fewer problems with aortic insufficiency, aortic valve fusion, and gastrointestinal bleeding. Finally, the particular pressure/flow characteristics of these pumps and their inherently more accurate flow estimation will likely lead to the development of automated speed control algorithms. These may provide a much safer, more effective way of operating these pumps compared with the fixed-speed mode of operation that is a current design feature.

The studies performed recently at Cleveland Clinic marked a significant milestone in proving that CFTAH technologies can support life for a long duration with a single device. Today, device development and biocompatibility studies at Cleveland Clinic are ongoing under active funding from the National Institutes of Health and institutional support. The implementation of this project and achievement of this important milestone suggests that single-piece, autoregulated, and sensorless devices have significant potential to further development into a valid clinical option in HF patients.

Issues surrounding how best to manage RV failure are still at the crux of every decision that is made for LVAD implantation. Better techniques need to be developed to identify those patients who will or will not benefit from LVAD therapy alone. Developments such as a dedicated, permanently implanted RVAD; safe, effective conduit and pump speed modifications to current LVAD technology; or single-device solutions suitable for BVAD support will greatly expand applications for chronic biventricular failure. Complete cardiac replacement with CFTAHs, while currently in early development and application, will play a key role in addressing a subset of patients with chronic biventricular failure. Smaller, simpler CFTAHs that are proven reliable may even move the field completely in this direction.

It is expected that advances in the field of MCS will expand the use of support devices to the point where they may offer a long-term substitute in terms of quality of life, or at least provide a feasible alternative to heart transplantation for a greater number of terminally ill HF patients [43].

Notably, invasive options to treat end-stage HF continue to be updated and expanded as more innovative devices emerge. The most effective, durable, reliable, and fully autoregulated devices will undoubtedly save lives. However, until the ideal pump has been created and brought into accepted clinical practice, patient survival will continue to be primarily determined by an early decision to implant uni-ventricular or biventricular devices and not by choice of device type.

In addition to early diagnosis and treatment with MCS devices, primary efforts and attention should continue to be provided for better prevention of HF by treating risk factors at an early stage so that the device therapies can be more focused and available when HF is not further manageable by more conservative means. Further research will be needed to boost these endeavors and to benefit the demanding clinical need of the growing HF patient population.

References

1. Ambrosy AP, Fonarow GC, Butler J, Chioncel O, Greene SJ, Vaduganathan M, Nodari S, Lam CS, Sato N, Shah AN, et al. The global health and economic burden of hospitalizations for heart failure: lessons learned from hospitalized heart failure registries. J Am Coll Cardiol. 2014;63(12):1123–33.
2. Go AS, Mozaffarian D, Roger VL, Benjamin EJ, Berry JD, Borden WB, Bravata DM, Dai S, Ford ES, Fox CS, et al. Executive summary: heart disease and stroke statistics–2013 update: a report from the American Heart Association. Circulation. 2013;127(1):143–52.
3. Heidenreich PA, Albert NM, Allen LA, Bluemke DA, Butler J, Fonarow GC, Ikonomidis JS, Khavjou O, Konstam MA, Maddox TM, et al. Forecasting the impact of heart failure in the United States: a policy statement from the American Heart Association. Circ Heart Fail. 2013;6(3):606–19.
4. Garbade J, Bittner HB, Barten MJ, Mohr FW. Current trends in implantable left ventricular assist devices. Cardiol Res Pract. 2011;2011:290561.
5. Wever-Pinzon O, Drakos SG, Kfoury AG, Nativi JN, Gilbert EM, Everitt M, Alharethi R, Brunisholz K, Bader FM, Li DY, et al. Morbidity and mortality in heart transplant candidates supported with mechanical circulatory support: is reappraisal of the current United network for organ sharing thoracic organ allocation policy justified? Circulation. 2013;127(4):452–62.
6. Moazami N, Hoercher KJ, Fukamachi K, Kobayashi M, Smedira NG, Massiello A, Horvath DJ. Mechanical circulatory support for heart failure: past, present and a look at the future. Expert Rev Med Devices. 2013;10(1):55–71.
7. Mehra MR, Uriel N, Naka Y, Cleveland JC Jr, Yuzefpolskaya M, Salerno CT, Walsh MN, Milano CA, Patel CB, Hutchins SW, et al. A fully magnetically levitated left ventricular assist device – final report. N Engl J Med. 2019;380(17):1618–27.

8. Kirklin JK, Pagani FD, Kormos RL, Stevenson LW, Blume ED, Myers SL, Miller MA, Baldwin JT, Young JB, Naftel DC. Eighth annual INTERMACS report: special focus on framing the impact of adverse events. J Heart Lung Transplant. 2017;36(10):1080–6.
9. Medvedev AL, Karimov JH, Kuban BD, Horvath DJ, Moazami N, Fukamachi K. Unlocking the box: basic requirements for an ideal ventricular assist device controller. Expert Rev Med Devices. 2017;14(5):393–400.
10. Fukamachi K, Horvath DJ, Massiello AL, Fumoto H, Horai T, Rao S, Golding LA. An innovative, sensorless, pulsatile, continuous-flow total artificial heart: device design and initial in vitro study. J Heart Lung Transplant. 2010;29(1):13–20.
11. Fumoto H, Horvath DJ, Rao S, Massiello AL, Horai T, Takaseya T, Arakawa Y, Mielke N, Chen JF, Dessoffy R, et al. In vivo acute performance of the Cleveland Clinic self-regulating, continuous-flow total artificial heart. J Heart Lung Transplant. 2010;29(1):21–6.
12. Horvath D, Byram N, Karimov JH, Kuban B, Sunagawa G, Golding LAR, Moazami N, Fukamachi K. Mechanism of self-regulation and in vivo performance of the Cleveland Clinic continuous-flow total artificial heart. Artif Organs. 2017;41(5):411–7.
13. Horvath D, Karimov JH, Byram N, Kuban B, Golding LA, Moazami N, Fukamachi K. Sensorless suction recognition in the self-regulating Cleveland Clinic continuous-flow total artificial heart. ASAIO J. 2015;61(6):726–8.
14. Horvath DJ, Karimov JH, Byram NA, Kuban BD, Sunagawa G, Moazami N, Fukamachi K. Advantages of integrating pressure-regulating devices into mechanical circulatory support pumps. ASAIO J. 2019;65(1):e1–3.
15. Karimov JH, Moazami N, Sunagawa G, Kobayashi M, Byram N, Sale S, Such KA, Horvath DJ, Golding LA, Fukamachi K. Median sternotomy or right thoracotomy techniques for total artificial heart implantation in calves. Artif Organs. 2016;40(10):1022–7.
16. Karimov JH, Moazami N, Kobayashi M, Sale S, Such K, Byram N, Sunagawa G, Horvath D, Gao S, Kuban B, et al. First report of 90-day support of 2 calves with a continuous-flow total artificial heart. J Thorac Cardiovasc Surg. 2015;150(3):687–93. e681.
17. Fukamachi K, Karimov JH, Horvath DJ, Sunagawa G, Byram NA, Kuban BD, Moazami N. Initial in vitro testing of a paediatric continuous-flow total artificial heart. Interact Cardiovasc Thorac Surg. 2018;26(6):897–901.
18. Karimov Jamshid H, Horvath D, Byram N, Sunagawa G, Kuban B, Gao S, Dessoffy R, Fukamachi K. Abstract 16894: initial in vivo results of Cleveland Clinic pediatric continuous-flow total artificial heart implant. Circulation. 2017;136(suppl_1):A16894.
19. Karimov JH, Horvath DJ, Byram N, Sunagawa G, Kuban BD, Gao S, Dessoffy R, Fukamachi K. Early in vivo experience with the pediatric continuous-flow total artificial heart. J Heart Lung Transplant. 2018;37(8):1029–34.
20. Fukamachi K, Karimov JH, Miyamoto T. Challenges in pediatric mechanical circulatory support devices. Artif Organs. 2019;43(5):441–3.
21. Karimov JH, Polakowski A, Horvath D, Byram N, Kado Y, Miyamoto T, Ahmad M, Najm H, Stewart R, Saarel E, et al. Development of continuous-flow total artificial heart for use in infants. J Heart Lung Transplant. 2019;38(4):S482.
22. Golding LR, Jacobs G, Murakami T, Takatani S, Valdes F, Harasaki H, Nose Y. Chronic nonpulsatile blood flow in an alive, awake animal 34-day survival. Transactions. 1980;26:251–5.
23. Karimov JH, Horvath D, Byram N, Polakowski A, Adams J, Kado Y, Miyamoto T, Sale S, Kuban B, Fukamachi K. Mechanical circulatory support for biventricular heart failure using continuous-flow total artificial heart. J Heart Lung Transplant. 2019;38(4):S343.
24. Frazier OH, Myers TJ, Gregoric I. Biventricular assistance with the Jarvik FlowMaker: a case report. J Thorac Cardiovasc Surg. 2004;128(4):625–6.
25. Fukamachi K, Shiose A, Massiello AL, Horvath DJ, Golding LA, Lee S, Starling RC. Implantable continuous-flow right ventricular assist device: lessons learned in the development of a Cleveland Clinic device. Ann Thorac Surg. 2012;93(5):1746–52.
26. Karimov JH, Sunagawa G, Horvath D, Fukamachi K, Starling RC, Moazami N. Limitations to chronic right ventricular assist device support. Ann Thorac Surg. 2016;102(2):651–8.
27. Loforte A, Stepanenko A, Potapov EV, Musumeci F, Dranishnikov N, Schweiger M, Montalto A, Pasic M, Weng Y, Dandel M, et al. Temporary right ventricular mechanical support in high-risk left ventricular assist device recipients versus permanent biventricular or total artificial heart support. Artif Organs. 2013;37(6):523–30.
28. Saito S, Sakaguchi T, Miyagawa S, Yoshikawa Y, Yamauchi T, Ueno T, Kuratani T, Sawa Y. Biventricular support using implantable continuous-flow ventricular assist devices. J Heart Lung Transplant. 2011;30(4):475–8.
29. Strueber M, Meyer AL, Malehsa D, Haverich A. Successful use of the HeartWare HVAD rotary blood pump for biventricular support. J Thorac Cardiovasc Surg. 2010;140(4):936–7.
30. Hetzer R, Krabatsch T, Stepanenko A, Hennig E, Potapov EV. Long-term biventricular support with the HeartWare implantable continuous flow pump. J Heart Lung Transplant. 2010;29(7):822–4.
31. Shehab S, Macdonald PS, Keogh AM, Kotlyar E, Jabbour A, Robson D, Newton PJ, Rao S, Wang L, Allida S, et al. Long-term biventricular HeartWare ventricular assist device support–case series of right atrial and right ventricular implantation outcomes. J Heart Lung Transplant. 2016;35(4):466–73.
32. Potapov E, Schweiger M, Vierecke J, Dandel M, Stepanenko A, Kukucka M, Jurmann B, Hetzer R, Krabatsch T. Discontinuation of HeartWare RVAD

support without device removal in chronic BIVAD patients. ASAIO J. 2012;58(1):15–8.
33. Kaczorowski DJ, Woo YJ. Who needs an RVAD in addition to an LVAD? Cardiol Clin. 2011;29(4):599–605.
34. Chen JM, Levin HR, Rose EA, Addonizio LJ, Landry DW, Sistino JJ, Michler RE, Oz MC. Experience with right ventricular assist devices for perioperative right-sided circulatory failure. Ann Thorac Surg. 1996;61(1):305–10; discussion 311–303.
35. Lazar JF, Swartz MF, Schiralli MP, Schneider M, Pisula B, Hallinan W, Hicks GL Jr, Massey HT. Survival after left ventricular assist device with and without temporary right ventricular support. Ann Thorac Surg. 2013;96(6):2155–9.
36. Meineri M, Van Rensburg AE, Vegas A. Right ventricular failure after LVAD implantation: prevention and treatment. Best Pract Res Clin Anaesthesiol. 2012;26(2):217–29.
37. Raina A, Patarroyo-Aponte M. Prevention and treatment of right ventricular failure during left ventricular assist device therapy. Crit Care Clin. 2018;34(3):439–52.
38. Aissaoui N, Morshuis M, Schoenbrodt M, Hakim Meibodi K, Kizner L, Borgermann J, Gummert J. Temporary right ventricular mechanical circulatory support for the management of right ventricular failure in critically ill patients. J Thorac Cardiovasc Surg. 2013;146(1):186–91.
39. Shah P, Ha R, Singh R, Cotts W, Adler E, Kiernan M, Brambatti M, Meehan K, Phillips S, Kidambi S, et al. Multicenter experience with durable biventricular assist devices. J Heart Lung Transplant. 2018;37(9):1093–101.
40. Vierecke J, Gahl B, de By T, Antretter H, Beyersdorf F, Caliskan K, Krachak V, Loforte A, Potapov E, Schoenrath F, et al. Results of primary biventricular support: an analysis of data from the EUROMACS registry. Eur J Cardiothorac Surg. 2019;56(6):1037–45.
41. Fukamachi K, Karimov JH, Byram NA, Sunagawa G, Dessoffy R, Miyamoto T, Horvath DJ. Anatomical study of the Cleveland Clinic continuous-flow total artificial heart in adult and pediatric configurations. J Artif Organs. 2018;21(3):383–6.
42. Fukamachi K, Horvath DJ, Byram N, Sunagawa G, Karimov JH, Moazami N. Advanced ventricular assist device with pulse augmentation and automatic regurgitant-flow shut-off. J Heart Lung Transplant. 2016;35(12):1519–21.
43. Karimov JH, Polakowski AR, Fukamachi K, Miyamoto T. Progress in mechanical circulatory support: challenges and opportunities. Artif Organs. 2019;43(9):818–20.

31 The ReinVAD LVAD: Smart Technology to Enhance Long-Term Circulatory Support Therapy

Roland Graefe, Erika Minguez, Andreas Henseler, and Reiner Körfer

Pump Architecture and Functionality

Overview and Vision

The ReinVAD left ventricular assist device (LVAD) is an active medical device in preclinical testing. It is being developed by the ReinVAD GmbH in Aachen, Germany. The system is based on the rotary pump principle. It belongs to the newest third-generation pump, and the pump unit is shown in the implanted position in the apex of the left ventricle and connected to the aorta in Fig. 31.1.

Technically speaking, LVADs are fluid flow engines that assist the pumping function of the left ventricle by diverting blood into a bypass circuit connecting the left ventricle to the aorta. A LVAD system usually consists of three main components: the internal component including the pump unit implanted into the left ventricle, a percutaneous driveline, and external components such as a controller and batteries. All such components are further described in this chapter.

A combination of innovative technological developments offers the prerequisite for the implementation of a LVAD system that will enhance long-term circulatory support therapy and thus improve outcome for end-stage heart failure patients. The following list illustrates selected key features of the system, and details on the features are provided in later sections.

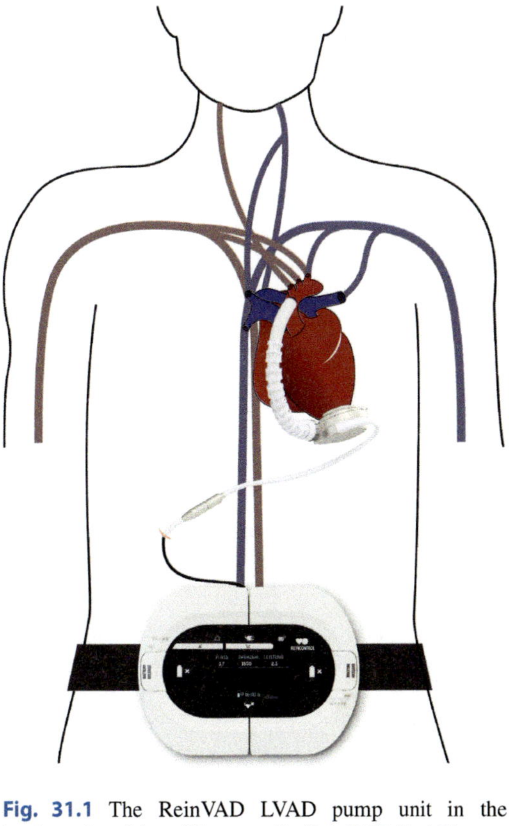

Fig. 31.1 The ReinVAD LVAD pump unit in the implanted position. (Photo credit: ReinVAD GmbH)

R. Graefe, Dr.-Ing (✉) · E. Minguez, MSc
A. Henseler, Dr.-Ing · R. Körfer, Prof. Dr. med. Dr. h. c.
ReinVAD GmbH, Aachen, Germany
e-mail: graefe@reinvad.de; minguez@reinvad.de; henseler@reinvad.de; koerfer@reinvad.de

© Springer Nature Switzerland AG 2020
J. H. Karimov et al. (eds.), *Mechanical Support for Heart Failure*,
https://doi.org/10.1007/978-3-030-47809-4_31

Selected key features of the ReinVAD LVAD *from a technical point of view* are:

- Optimal blood compatibility ensured by a contactless, large-gap hybrid bearing
- Provision of reliable feedback on the therapy to the treating doctor by integration of flow sensor technology
- Increase of survival rates, increase of quality of life of patients, and reduction of costs through avoidance of adverse events by automatic closed-loop rotational speed control
- Avoidance of pump exchange procedures in case of driveline failure or driveline infections by provision of an implantable driveline connector
- Enabling increased mobility of patients by automatic passive adaptation of the support level to the demand and modular, complementary design of the peripherals
- Increase of patient satisfaction with the own artificial heart through modern form and design of the extracorporeal components of daily life

All of these design rationales are further detailed in later sections of this chapter.

Figure 31.2 shows the miniaturized ReinVAD pump unit.

In the following, we present our vision for the new ReinVAD LVAD system: the vision of the company is the development and commercialization of a LVAD system that ensures a long-term, event-free, and life-affirming therapy of advanced left heart failure. Thereby patients waiting for a heart transplant can be treated effectively, and especially also patients without prospect or desire for a transplant can be returned to an active, long-lasting, and meaningful life. The focus during the development of each single component was thus to enhance long-term circulatory support.

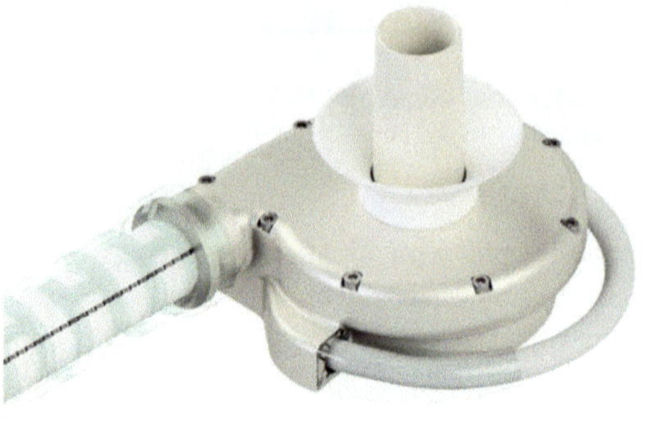

Fig. 31.2 The centrifugal pump unit of the ReinVAD LVAD. (Photo credit: ReinVAD GmbH)

The corresponding design process of the intracorporeal and extracorporeal components of the ReinVAD LVAD pursues the following complementary goals *from a therapy point of view*:

- Increase of survival rates, increase of quality of life of the patients, and reduction of costs through avoidance of therapy-related adverse events
- Provision of reliable feedback on the course of the therapy to the treating doctor through integration of sensor technology
- Enabling of an active life style of patients through automatic adaptation of the support level to the momentary demand and modular, complementary design of the extracorporeal components
- Extension of the patient cohort to smaller patients through miniaturization and intelligent surgical instruments
- Increase of patient satisfaction with the own artificial heart through modern form and design of the extracorporeal components of daily life

Rotor Bearing and CFD-Based Design Approach

Due to the small outer dimensions of its centrifugal pump unit, it can be positioned in the pericardium and connected to the apex of the left ventricle as can be seen in Fig. 31.1. In comparison to the HeartMate III (Abbott, USA), the outer diameter and the critical height of the pump unit have been reduced in order to extend the patient cohort to smaller patients and, additionally, to support forthcoming innovative implantation techniques. Thus, the ReinVAD LVAD can also be implanted in adolescents. Newer implantation techniques employ more but smaller skin incisions in comparison to a classical open-heart surgery with sternotomy. Owing to the small dimensions, different preferences of the surgeons concerning implantation techniques are supported. The aspect of outer dimension specifically concerns the applicability to female patients due to the typically smaller thorax. Here, an implantation might not be feasible at all today in extreme cases, or new implantation techniques are not applicable due to size restrictions, or supplied surgical instruments do not satisfy the specific requirements. Despite the small size of the ReinVAD pump unit, it is exceedingly powerful and can deliver between 2 and more than 10 l/min of blood flow even against elevated arterial pressure.

Figure 31.3 shows the ReinVAD pump unit in a cross-sectional view. Blood enters the pump unit through the inlet cannula from the left ventricle. Powerful blades of the rotor increase blood flow and blood pressure via transfer of angular momentum according to the rotary pump principle. Blood is collected off the rotor in a volute with large and open flow channels and guided towards the tangential outlet of the pump unit. The rotor contains drive magnets that interact magnetically with the motor stator outside of the blood compartment in order to maintain rotation of the rotor.

The ReinVAD LVAD is a third-generation centrifugal pump, which means that the rotor is fully levitated with no contact to the surrounding housing at any point in time during operation. Therefore, there is no potential source of mechanical wear. This contactless bearing system is established by an optimized and patented hydrodynamic axial bearing and a passive magnetic radial bearing as shown in Fig. 31.3. The resulting gap sizes between the rotating rotor and stationary housing are about 30–600 μm and are a result of extensive research to minimize blood trauma due to fluid forces and optimize washout of the bearing area. The lowest clearance gap is found at the axial hydrodynamic bearing, and its size is discussed in more detail below.

Especially for the design of this hydrodynamic bearing, innovative CFD-based tools have been employed in the development phase. The hydrodynamic structure serves as an axial bearing balancing axial thrust forces and also as a tilt restoration mechanism as seen as in Fig. 31.4. Therefore, the axial bearing counteracts all magnetic forces as well as hydraulic thrust forces. As also discussed in Chapter "25 – Computational Fluid Dynamics for Mechanical Circulatory Support Device Development," the design evaluation and geometry modification steps during a CFD project have seen dramatic changes in recent years. Figure 31.5 shows an example with the approach of response-surface-based design space exploration and geometry optimization [1]. These tools are in broad application in industry and available within software packages such as the ANSYS package with the flow solvers CFX and Fluent (ANSYS Inc.,

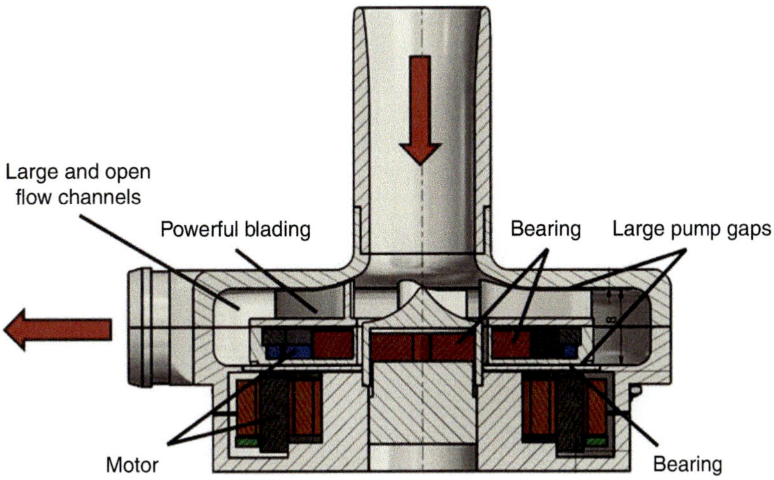

Fig. 31.3 Cross-sectional view of the ReinVAD LVAD pump unit (Photo credit: ReinVAD GmbH)

Canonsburg, USA), and these tools were used during the device development.

Typically, within a CFD optimization project, constraints exist, for example, due to a desired miniaturization of the pump. This means, geometry parameters can only be allowed to vary within limits. Looking at all geometry parameters subject to optimization, this creates a multidimensional design space. At the beginning, the quantitative effect of a parameter is unknown, and there might be interdependencies between design parameters. Moreover, there might be more than one design goal. In the case of the axial hydrodynamic bearing, goals such as a high axial force, a high axial force stiffness, and a high washout were all considered in the development phase. This complexity can be dealt with by creating a so-called metamodel that models the effect of design parameters on design goals. If the effects and the interdependence of two design parameters are thus illustrated on just one design goal parameter, this creates a two-dimensional response surface as illustrated in Fig. 31.5. This metamodel additionally allows for the optimization of the geometry and provides insight into the effect of single design parameters. Hence, the entire process is sometimes referred to as response-surface-based optimization and produced an extremely powerful and well-washed hydrodynamic bearing structure as shown in Fig. 31.4.

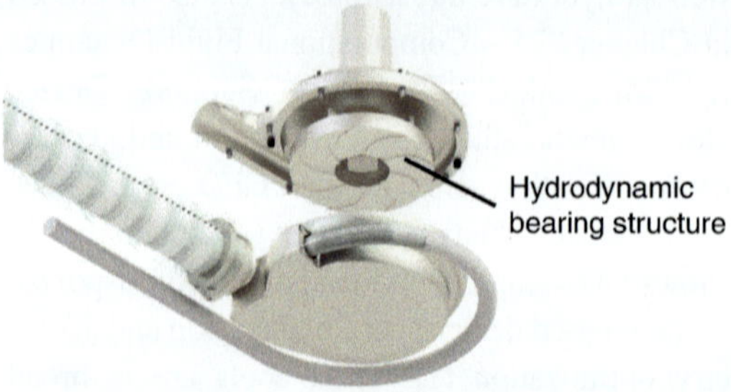

Fig. 31.4 The ReinVAD pump unit and the rotor with the hydrodynamic bearing structures. (Photo credit: ReinVAD GmbH)

Due to the combination of the technical approaches for the entire rotor bearing system, it is called a large-gap *hybrid* bearing. With this contactless bearing design, the ReinVAD pump provides a technology of the latest generation with excellent flow properties to minimize blood trauma. In more detail, Fig. 31.6 illustrates the axial clearance gap between the rotor comprising the hydrodynamic structures and the corresponding pump casing wall as measured with a current pump prototype and with a customized laser position sensor setup.

Figure 31.6 documents that the axial position of the rotor is almost independent of the flow through the device. Consequently, no axial movement is expected during a heartbeat. Moreover, it is shown that for increased rotational speed, the axial clearance increases due to an increased force production of the hydrodynamic bearing. The clearance ranges in the expected operational speed range between about 30 μm and 50 μm. To put results

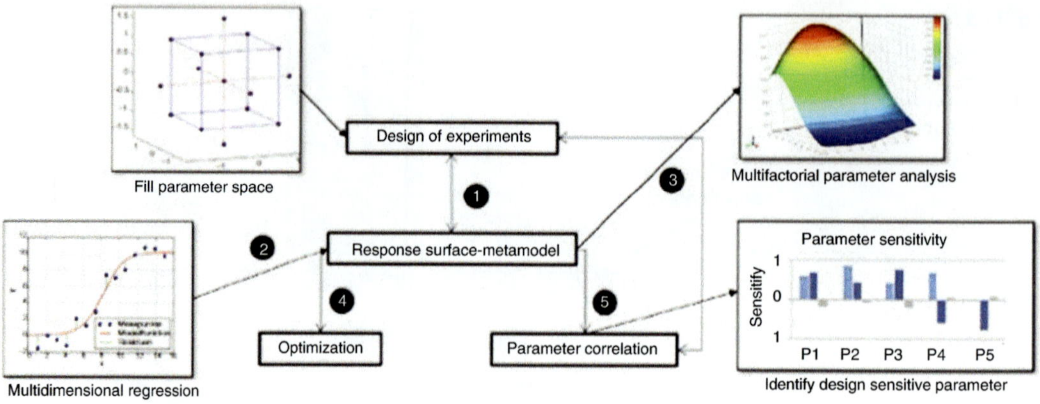

Fig. 31.5 Metamodel-based design exploration and optimization. (Reprinted from Smith et al. [1] with permission from Elsevier)

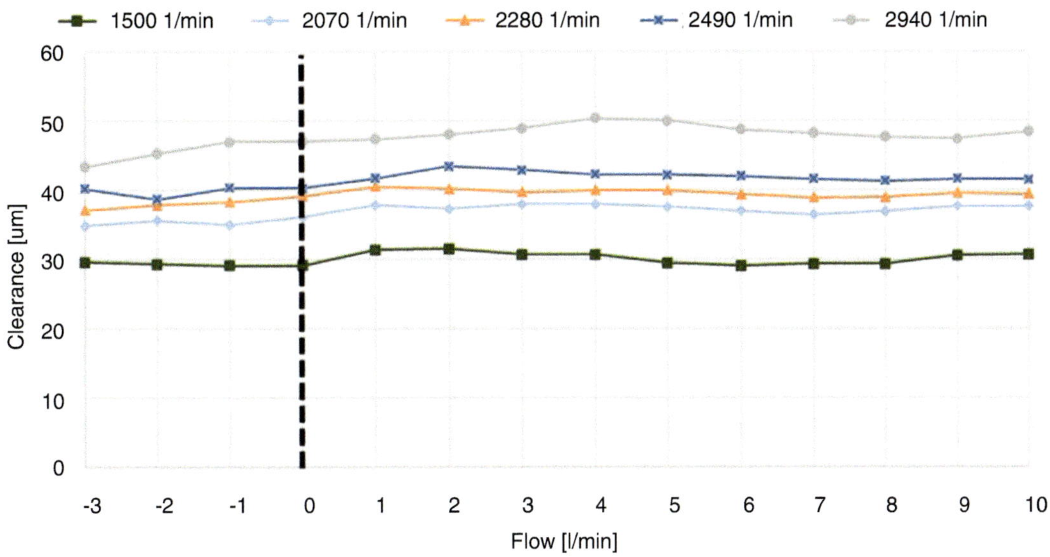

Fig. 31.6 Minimal axial clearance gap between the rotor and the corresponding pump casing wall during operation at rotational speeds between 1500 1/min and 2940 1/min

into perspective, the axial bearing gap of the clinically successful HVAD (Medtronic, USA) is about 22 μm at the main operational point of 2700 1/min and 5 l/min based on measurements by Thamsen et al. [2]. Therefore, the axial clearance gap of the ReinVAD LVAD in the expected main operating range between 2000 and 2500 1/min is almost twice as large.

Pump Performance and Efficiency

One design goal for the ReinVAD LVAD is to provide a pressure-sensitive pumping behavior. More details on the effects when a highly pressure-sensitive pump is implanted in patients are provided in the section Passive Interaction with the Circulation. Nevertheless, the ReinVAD LVAD, as shown in Fig. 31.7, is equipped with a very flat pump characteristic curve in this pressure head-flow diagram and thus provides a high sensitivity to pressure changes.

Figure 31.7 displays that, e.g., for 2490 1/min, the provided pressure generation only drops about 25 mmHg in the flow range between 0 and 10 l/min. This corresponds to an averaged pressure sensitivity of 0.4 (l/min)/mmHg between 0 and 10 l/min. This pressure difference has a minor dependency on the rotational speed, while for higher speeds, the pressure sensitivity tends to increase, and for lower speeds, it tends to decrease. For a full support of patients, the main operational speed range is to be expected between about 2000 and 2500 1/min. Figure 31.7 illustrates that, depending on preload and systemic vascular resistance, the ReinVAD LVAD is thus able to provide substantial flow rates up to 10 l/min and more.

An interesting additional characteristic is displayed in Fig. 31.7. As it can be seen, the generated pressure difference increases considerably for negative flows. This provides an intrinsic barrier to the generation of instantaneous negative flows that, without such a characteristic, might occur during diastole. The effect of negative flows in implanted rotary pumps is yet to be understood, but it seems clear that it would decrease pump efficiency.

The pump system efficiency of the ReinVAD LVAD is illustrated in Fig. 31.8. The efficiency is the ratio of the hydraulic power output in terms of flow and pressure generation to the electrical power input into the system. A high efficiency is desirable due to increased battery lifetime and as a sign of well working pump system.

Efficiency curves of rotary pumps illustrate effectively that they work best in a specific work-

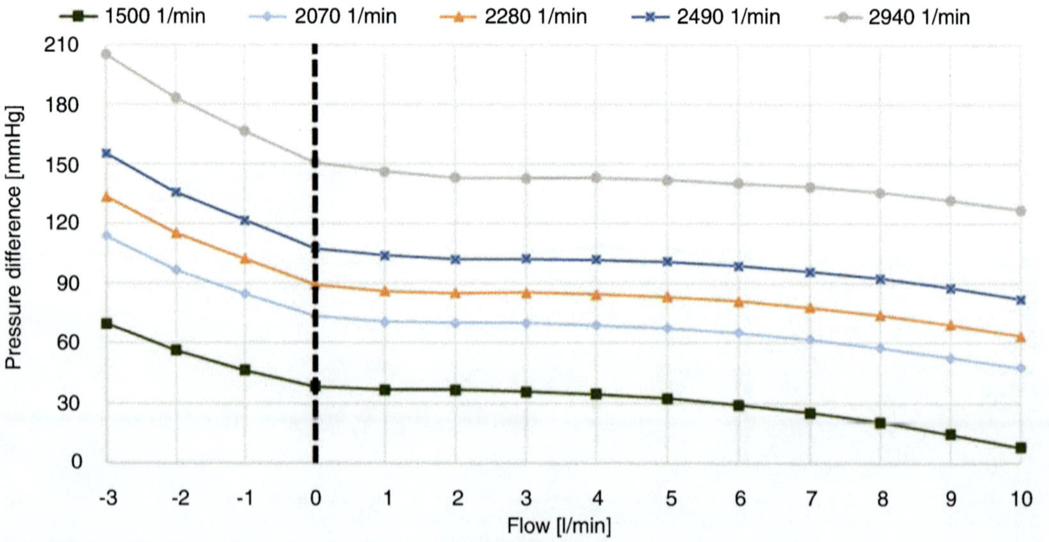

Fig. 31.7 Pump characteristic curves of the ReinVAD LVAD at rotational speeds between 1500 1/min and 2940 1/min

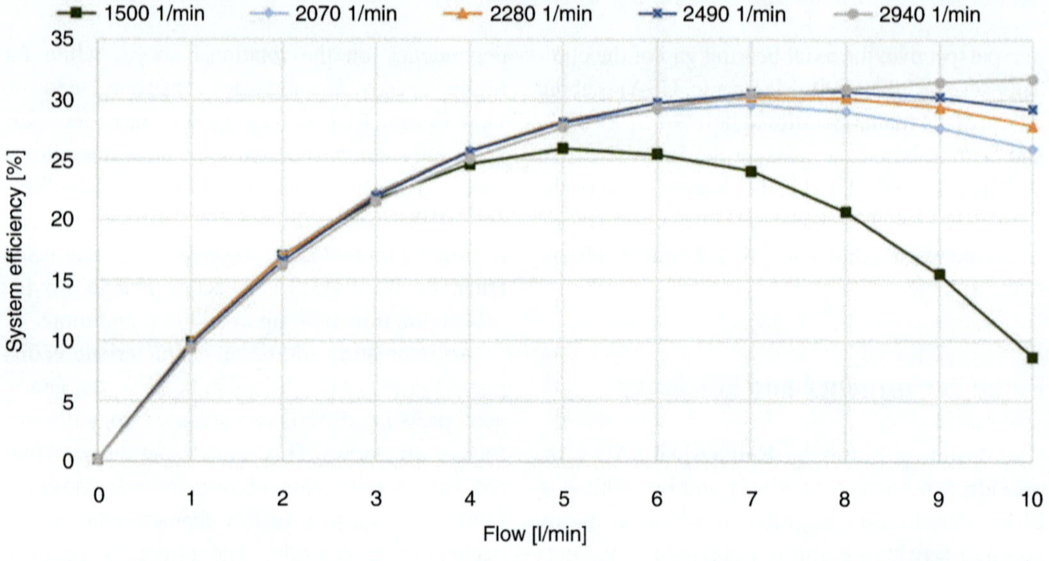

Fig. 31.8 System efficiency of the ReinVAD LVAD at rotational speeds between 1500 1/min and 2940 1/min

ing point, which is the design point. As for all pumps, the efficiency drops to zero if there is no flow through the pump, or the pressure generation decreases to zero. Nevertheless, for the ReinVAD LVAD, measures were taken to provide a system that works well in a broad range of flow rates and rotational speeds. The efficiency curves clearly show a "bulgy" form providing a high system efficiency in a wide flow range. Moreover, the peak system efficiency is located at 5 l/min or above depending on the rotational speed. This design strategy was chosen to provide a high pump pressure sensitivity which requires the peak efficiency to be at elevated flow rates. The peak system efficiency surpasses 30% for increased rotational speeds, which, to our knowledge, has not been achieved by any implantable rotary blood pump.

Figure 31.9 shows the relationship of system power consumption and flow rate.

The relationship differs for the different rotational speeds displayed. However, the curves

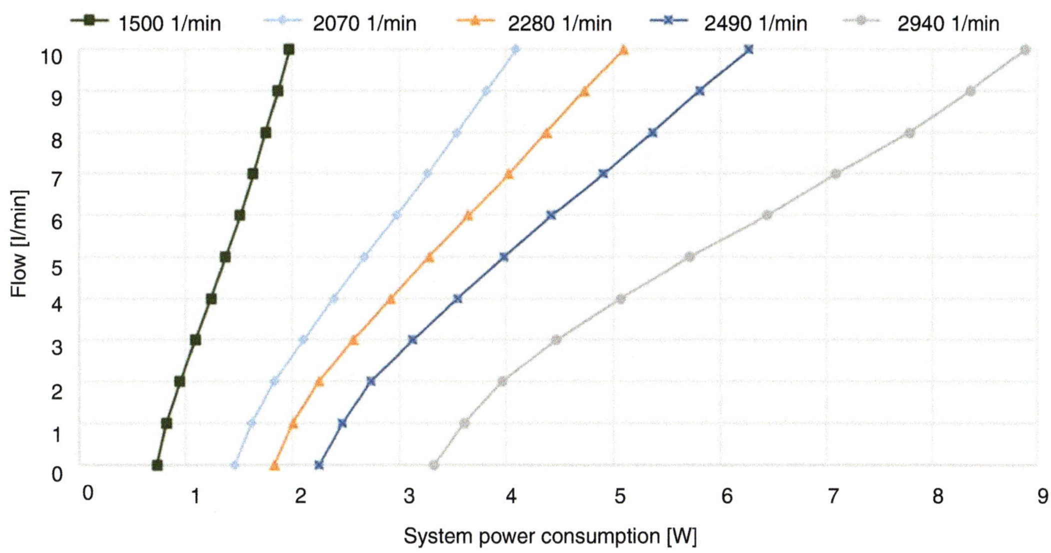

Fig. 31.9 Correlation of power consumption and flow rate for the ReinVAD LVAD at rotational speeds between 1500 1/min and 2940 1/min

show an almost linear relationship between the two variables of power consumption and flow rate, especially for higher flow rates. Moreover and more importantly, the correlation is always a distinct meaning that for a specific rotational speed, a certain power consumption consistently corresponds to a specific flow rate. Consequently, Fig. 31.9 demonstrates that a flow estimation based on system power consumption allows for an accurate flow rate determination.

Technical Data Overview

Table 31.1 provides an overview of the characteristics and features of the ReinVAD LVAD system. The table is not exhaustive and only gives an introduction.

Passive Interaction with the Circulation

Pressure Sensitivity

The ReinVAD LVAD design is optimized to provide a flat H-Q characteristic in the pressure-flow diagram as shown in Fig. 31.7. This is the key in the hydraulic interaction between the LVAD and the human circulation. As one single consequence, it extends potentially provided flow rates to values higher than 10 l/min at typical rotational speeds. This is a unique feature, which cannot be provided by any LVAD pump that is routinely used in the European and US market and that can be placed in the pericardium.

The flat H-Q performance, also quantified by the so-called pressure sensitivity, has major impact on the resulting flow, which is provided by the pump. Pressure sensitivity means that an artificial heart automatically and sensitively reacts to changes in pressure upstream or downstream of the LVAD. This implies that already small changes in pressure lead to significant changes in pump flow rate as illustrated in Fig. 31.10. These changes are a result of the intrinsic pump layout and occur passively at constant rotational speed. This feature is deducted from the natural preload sensitivity of the left ventricle and has various favorable effects concerning the hemodynamic interaction between ventricle, the circulation, and the artificial heart.

Figure 31.10 shows the characteristic curves for two different exemplary pumps that are designed differently regarding their flow guiding structures. One pump has a low-pressure sensitivity, whereas the other pump has a high-pressure sensitivity. Imagining that both pumps operate at

Table 31.1 Overview of the ReinVAD LVAD

Pump type	Centrifugal pump
Planned use	Circulatory support in patients with left ventricular insufficiency as, for example, bridge to transplantation, bridge to recovery, or destination therapy
Application period	Long-term support under continuous operation
Rotor bearing	Contactless, hybrid
Flow range	2–10 l/min and more against physiological arterial pressure
Estimated speed range	1500–2800 1/min
Implantation	Apical within the pericardium, sternotomy, or thoracotomy (less-invasive)
Peak system efficiency	>30%
Battery operating time	16 h or 32 h depending on battery size
Estimated pump weight	138 g
Estimated pump height	22 mm
Estimated pump diameter	55 mm
Special features	Integrated flow sensor Implantable driveline connector Highly pressure-sensitive pump characteristic Automatic safe speed control Innovative one-piece extracorporeal system

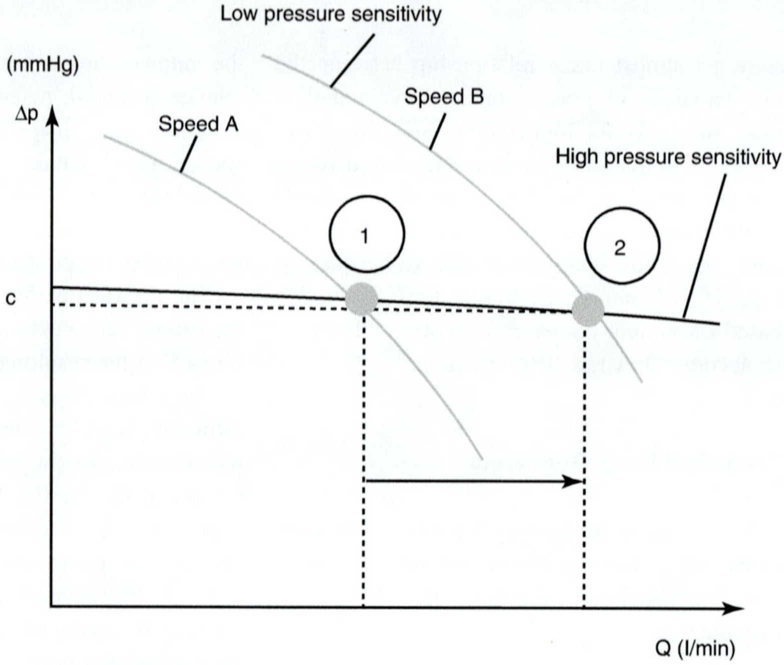

Fig. 31.10 Illustration of the effect of H-Q curve shape (pressure sensitivity) in case of a small decrease "c" in pressure difference over the pump

the illustrated point "1," it is clear that they both provide identical flow and pressure generation to the circulation. However, if there is a small change in pump preload or afterload, in this case called "c" and represented here as small decrease either in arterial pressure or small increase in left ventricular pressure, the operational point of the pump with high-pressure sensitivity is able to generate substantially more flow. It then operates in point "2." In case of preload changes, this pump thus passively helps to balance flow between the pulmonary and systemic circulation in a similar way as the ventricle and analogue to the Frank-Starling mechanism. The pump with low-pressure sensitivity is not able to increase flow to a noticeable extent. It would require an active rotational speed change to move its operational point also to point "2." Such an active physiologic speed control has been proposed by many authors but has not been used as common clinical practice [3–5]. A strong limitation is the provision of a feedback variable as chronic sensors to measure signals such as pressure or flow are not integrated into all systems. Additionally, guaranteeing a safe operation of the control approach might be too much of a challenge as of today. The key advantage of adapting LVAD flow with the implemented LVAD pressure-flow characteristic is the fact that sensors and active control algorithms are not needed.

The effect of high-pressure sensitivity in an LVAD application is not completely understood, but it could have an important influence on patient outcome, because, among other aspects, it has the following effects:

- Provision of a means of passive flow control (as discussed above).
- At low blood volume situations in the ventricle or low venous return, the suction force generated by the pump does not increase due to the flat H-Q characteristic. This is a very effective feature to prevent suction at the pump inlet without any speed change [6].
- The residual arterial pulse is increased, which could have multiple positive effects in comparison to less pulsatile mechanical support [6].
- High-pressure sensitivity increases the level of left ventricular unloading and decreases end-diastolic pressure (see below for details and in [7]).
- The support level is automatically and passively adapted to the momentary flow demand in case of patient exercise (see below for details and in [8]).

This list of effects represents, to some extent, a list of well-grounded speculations, which has to be challenged with clinical data.

Left Ventricular Unloading

The aim of a recent study was to experimentally investigate the effect of LVAD pressure sensitivity on the assisted circulation with a focus on the afterload of the right ventricle (RV) [7].

This study was conducted with an established in vitro model of the human circulation [9]. The model is illustrated in Fig. 31.11. This pneumatically driven model comprises four active chambers and has been widely used, e.g., for testing of a naturally shaped silicone ventricle instead of a cylinder representing the left ventricle employed in this study [10], and investigates mitral regurgitation as a result of inlet cannulation site and speed modulation [11]. Contraction of either the ventricle or atrium is realized by applying a controlled air pressure pulse to the respective chamber. All shunts, except for the mitral valve shunt and the LVAD shunt, were closed for the entire study. Cannulation was limited to ventricular cannulation, and an atrial contraction was omitted for simplicity.

Within this study, the effect of two different pump characteristic curves is investigated under different patient scenarios. The employed characteristic pump curves (illustrated in Fig. 31.12) are established by two custom-built laboratory pumps, so no ReinVAD LVAD prototype was used for this study.

The range of pressure sensitivity of the employed laboratory pumps has been chosen to be roughly representative of available LVADs. Pressure sensitivity is averaged from shut-off at 0 l/min to about 10 l/min to provide a quantitative measure. For further reference, the pump with a mean pressure sensitivity of 0.06 l/min/mmHg is called LowPS, and the pump with a mean pressure sensitivity of 0.35 l/min/mmHg is referred to as HighPS. The local pressure sensitivity for the pump HighPS around 5 l/min is 0.71 l/min/mmHg and 0.06 l/min/mmHg for the pump LowPS as calculated by the gradient between 4 and 6 l/min. To determine the extended pump characteristic curve of the pump LowPS to the full range of 10 l/min, an auxiliary pump had to be added in a test loop, but it was not used in the experiments within this study. The measured data in Fig. 31.12 include inlet and outlet cannulation and, as such, effectively represent the system characteristic curves that interact with the hydraulic model, which represents the patient.

Figure 31.13 shows the main results of the study with the corresponding pressure waveforms of AoP, LVP, and LAP, LVAD flow waveforms,

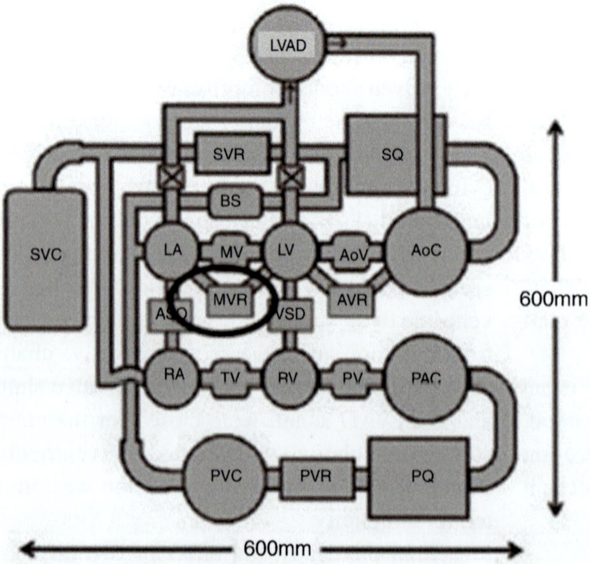

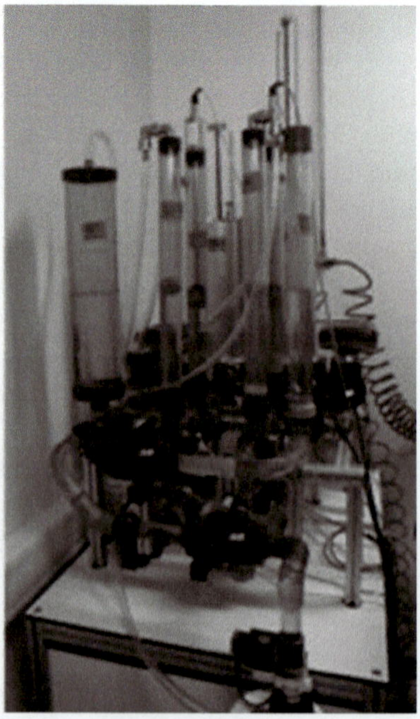

Fig. 31.11 Schematic (left) and image (right) of the hydraulic model of the human circulation. (LA left atrium, MV mitral valve, LV left ventricle, AoV aortic valve, AVR aortic valve regurgitation, AoC aortic compliance, SQ systemic flow sensor, BS bronchial shunt, SVR systemic vascular resistance, SVC systemic vascular compliance, RA right atrium, TV tricuspid valve, RV right ventricle, PV pulmonary valve, PAC pulmonary aorta compliance, PQ pulmonary flow sensor, PVR pulmonary vascular resistance, PVC pulmonary vascular compliance) (Reprinted from Graefe et al. [7], with permission from John Wiley and Sons)

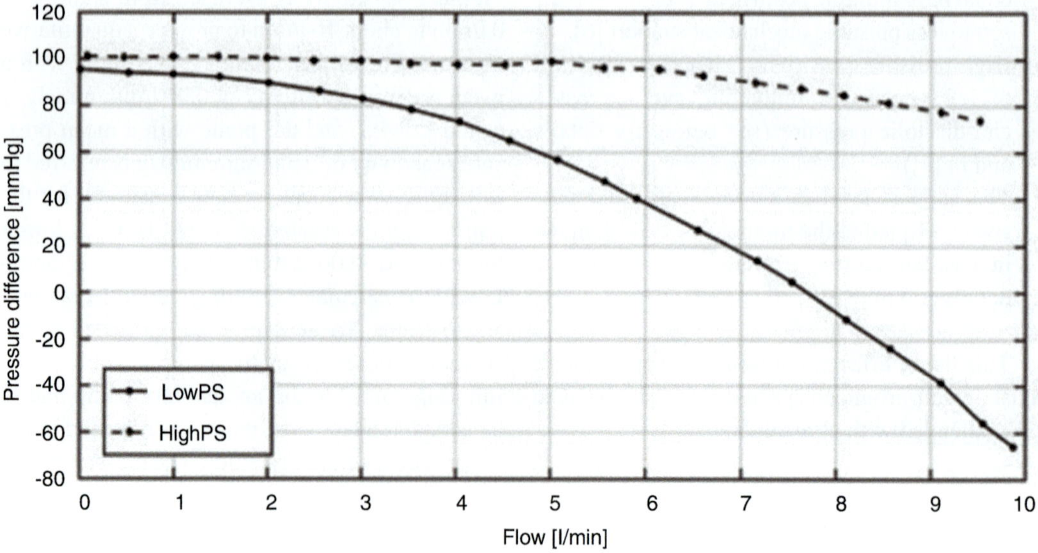

Fig. 31.12 System H-Q curves of the laboratory pumps LowPS (solid line) and HighPS (dashed line) used for the experimental study. (Reprinted from Graefe et al. [7], with permission from John Wiley and Sons)

PV loops of the LV, and dynamic H-Q curves showing the instantaneous operational point of the respective pump. The figure is limited to individual, representative heartbeats of scenarios with decreased ventricular contractility without mitral insufficiency and under full support.

Figure 31.13 (top left) shows that LVP remained below AoP constantly leading to a permanently closed aortic valve. AoP showed a higher pulsatility for support with the pump HighPS, which corresponds well with the higher flow pulsatility (Fig. 31.13, bottom left). Especially during late systole, in this case after about 0.7 s of the presented heartbeats, LVP remained elevated for support with the pump LowPS in comparison to support with the pump HighPS. Instantaneous pump flow rate is high during this period. The LV PV loop is shifted to the left on the EDPVR line for support with the pump HighPS (Fig. 31.13, top right). It is also widened as the pump HighPS quickly draws volume from the LV. The dynamic H-Q curves (Fig. 31.13, bottom right) correspond well with the stationary H-Q curves presented in Fig. 31.12 and with the presented LVAD flow rate waveforms (Fig. 31.13, bottom left). The dynamic H-Q curve of the pump LowPS is steeper, whereas the respective dynamic H-Q curve for the pump HighPS is flatter. A dynamic H-Q curve forms a hysteresis running counterclockwise and represents the actual operating window of the respective pump during one heartbeat. The hysteresis is caused by the inertance of the fluid in the pump and cannulae that are accelerated during systole and decelerated during diastole [12].

The dynamic H-Q curves (Fig. 31.13, bottom right) show that the actual operating windows are shifted towards high flows in comparison to the mean flow rate of about 5 l/min. The shift can be observed for both pumps. At the same time, the pump HighPS reaches more extreme instantaneous pump flow rates, which is also visible in the flow waveforms (Fig. 31.13, bottom left). The shift is mainly caused by the time course of pump pressure difference, which again is mainly influenced by changes in LVP over time. The shift is created for a fixed average pump flow rate, in this case 5 l/min. Time operating at a relatively low

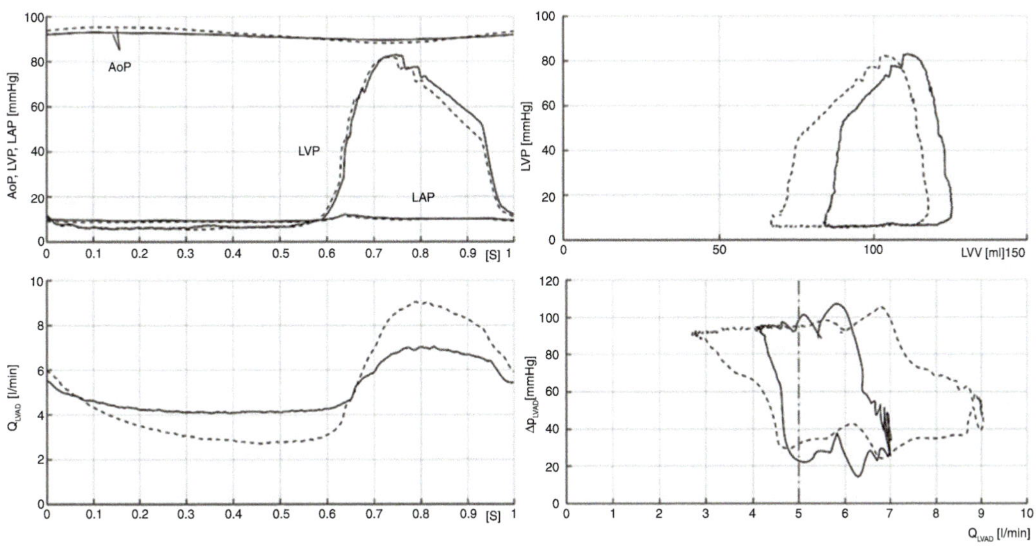

Fig. 31.13 Waveforms of aortic pressure (AoP), left ventricular pressure (LVP) ,and left atrial pressure (LAP, top left), instantaneous flow rates through the pumps (bottom left), left ventricular PV loops (top right), and instantaneous operational points of the pumps (bottom right). Results are shown for a representative heartbeat for the pump LowPS (solid line) and HighPS (dashed line). (Reprinted from Graefe et al. [7], with permission from John Wiley and Sons)

instantaneous flow rate, in this case below 5 l/min and during diastole, is longer than the time operating at higher instantaneous flow rates than 5 l/min in systole. Integrated over time, the average flow rate of 5 l/min is measured.

Looking at the PV loops in Fig. 31.13 (top right), the shift of the PV loops corresponds partially to known changes of inotropy. Figure 31.14 illustrates the effect of a change of inotropy without any LVAD support as taken from [13]. Without LVAD, for an increase in inotropy and because of its effect on ventricular afterload and preload, the PV loops widens, and it is shifted to the left. In this study, however, it is important to note that the setting parameters for LV contractility were maintained identical for support with both pumps. Therefore, the increase in unloading during systole of the pump with higher pressure sensitivity (see Fig. 31.13, lower left) causes the "unit" of the LV and the LVAD to possess a behavior, which is similar to an increased inotropic behavior. In both cases, increased pressure sensitivity and increased inotropy, the slope of the end systolic pressure volume relationship (ESPVR) is increased, and the ESPVR is shifted to the left; the "stroke volume" is increased, the arterial pressure is elevated, and the end-diastolic pressure and volume decrease.

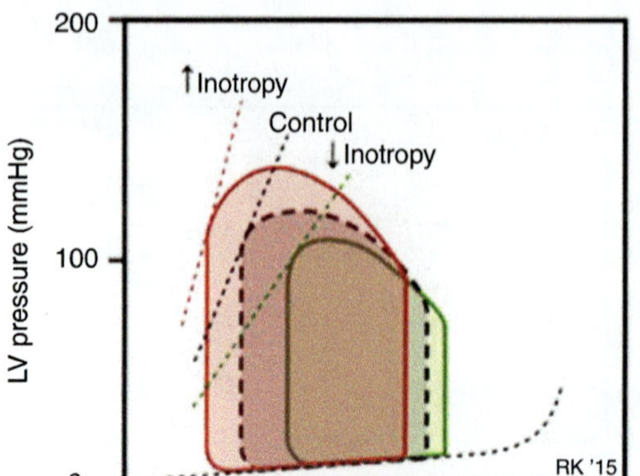

Fig. 31.14 The effect of inotropy on the left ventricular PV loop without any VAD support, reprinted from Cardiovascular Physiology Concepts, cvphysiology.com. (Used with permission from Richard and Klabunde [13])

An important note on the experimental study is the fact that the LV is in the mock circulation loop (MCL) is a rigid cylinder. The inner diameter of the cylinder determines its filling behavior as the height of the water column corresponds to LVP without any contraction force. The correlation is linear. In case of a decrease in LV volume caused by different levels of LVAD pressure sensitivity, the height of the water column will adjust leading to a reduced LVP. The correlation of LVP and LV volume without any contraction force is illustrated, for example, in Fig. 31.13 by means of the filling sections of the illustrated PV loops. It shows that the employed LV is not very stiff in comparison to normal ventricles [14] and that the correlation of volume and pressure is indeed linear. The shift of the PV loops on the other hand occurs on the EDPVR (Fig. 31.13, top right). For stiffer LV and for LV with a typically non-linear EDPVR, the effect of pressure sensitivity might be different.

A limitation of this study originates from the lack of implemented physiological feedback mechanisms in the MCL. Although it has been evaluated for the Frank-Starling mechanism to be implemented in the employed hydraulic MCL [15], this approach was not followed in this study. It is very challenging to adequately represent all relevant feedback mechanisms such as the baroreceptor reflex and metabolic control mechanisms. As a consequence, MCL setting parameters were maintained constant for this study to ease comparison between the investigated pumps.

In conclusion, this study investigated, to the authors' knowledge, for the first time the potential relationship between LVAD pump characteristic in an extended flow range, LV unloading, and RV afterload. Based on the results of this study, pulmonary pressures are reduced for highly pressure-sensitive support in comparison to limited pressure-sensitive support. This seems to be due to increased unloading during systole, and thus, the PV loop of the LV undergoes a leftward and thus somewhat of a downward shift. This behavior is similar to an increase in inotropy without any LVAD support. Consequently, the left ventricle is unloaded to a higher degree at comparable arterial blood pressure and identical cardiac output. Based on the laboratory pumps used, the difference in mean left atrial pressure was about 10.2%.

Exercise Capacity

For LVAD patients, it is known that the device assures a higher cardiac output than seen in chronic heart failure patients at rest, together with a lower wedge pressure, pulmonary pressure, and ventricular volumes. But during exercise, the beneficial effects of the LVAD are less evident as cardiac output increases to values comparable to heart failure patients and below what is observed in healthy subjects [16], the left ventricle is not completely unloaded anymore [17], and the wedge pressure and pulmonary pressure increase abnormally [18]. All these phenomena suggest that current LVADs are designed to properly support patient's hemodynamics at rest but not during exercise.

Salamonsen et al. discuss the shortcoming of LVADs to mimic the preload sensitivity of the natural ventricle enabling a flow balance of systemic and pulmonary circulation at a fixed speed [19]. This was already illustrated in Fig. 31.10. As rotary blood pumps used as LVAD do not differentiate between changes in preload or afterload, the authors conclude that the pressure sensitivity of the investigated LVADs is too low in comparison to the natural preload sensitivity and too high in comparison to the natural afterload sensitivity. Little additional data is known on the hemodynamic changes when applying different pump characteristics as a LVAD, especially in exercise condition.

The aim of this recent work was to investigate the qualitative and especially the quantitative impact of different LVAD pressure sensitivities on the hemodynamic response of the LVAD itself during exercise and ultimately on patient's hemodynamics [8]. To this aim, two LVADs with different pump characteristics were compared in exercise condition by means of a computational cardiorespiratory simulator.

Simulations were run using the cardiorespiratory simulator described in [20] and as shown in Fig. 31.15. Briefly, this is a detailed lumped parameter model implemented in LabVIEW 2016 (National Instrument, Austin, TX, USA). The simulator includes a model for the atria, ventricles, pulmonary circulation, systemic circulation (split into ascending and descending aorta, upper body, kidneys, splanchnic, legs, superior and inferior vena cava), baroreflex, and metabolic peripheral controls, ventilation mechanics, and gas exchange.

The simulator was characterized to reproduce the hemodynamic and ventilation status in end-stage heart failure both at rest condition and specifically also during exercise. The two LVADs were tested at constant rotational speed reproducing an average LVAD patient response to exercise (EXE↑) and a LVAD patient with no chronotropic and inotropic response (EXE→). Given a condition at rest, the simulator reproduces the evolution of the cardiorespiratory system to exercise in terms of chronotropic and inotropic response, peripheral vasodilation, increased cardiac output, and ventilation. More details on the simulator setting parameters are presented in [8].

In this study, the pressure-flow characteristics for two LVADs are implemented, and the effects are compared. LVAD1 represents LVADs with higher pressure sensitivity, whereas LVAD2 represents LVADs with limited pressure sensitivity. In fact, for LVAD1 average pressure sensitivity is 0.21 l/min/mmHg from 0 l/min up to 10 l/min, whereas for LVAD2 average pressure sensitivity is 0.08 l/min/mmHg in the same range of flow. The simulator uses the non-linear H-Q curves of each respective LVAD. For brevity, the stationary H-Q curves are shown together with further results in Fig. 31.18, and it has to be mentioned that the H-Q curve of LVAD1 is a bit less pressure-sensitive than the laboratory pump HighPS shown in Fig. 31.12. Also, LVAD2 has a less steep H-Q curve than the laboratory pump LowPS in Fig. 31.12. Therefore, the difference in H-Q curve shape and thus pressure sensitivity are less than in the previously presented study.

In Fig. 31.16, the main results for the comparison between LVAD1 and LVAD2 are reported for three conditions: REST, EXE↑, and EXE→. It is important to note that the respective speed of

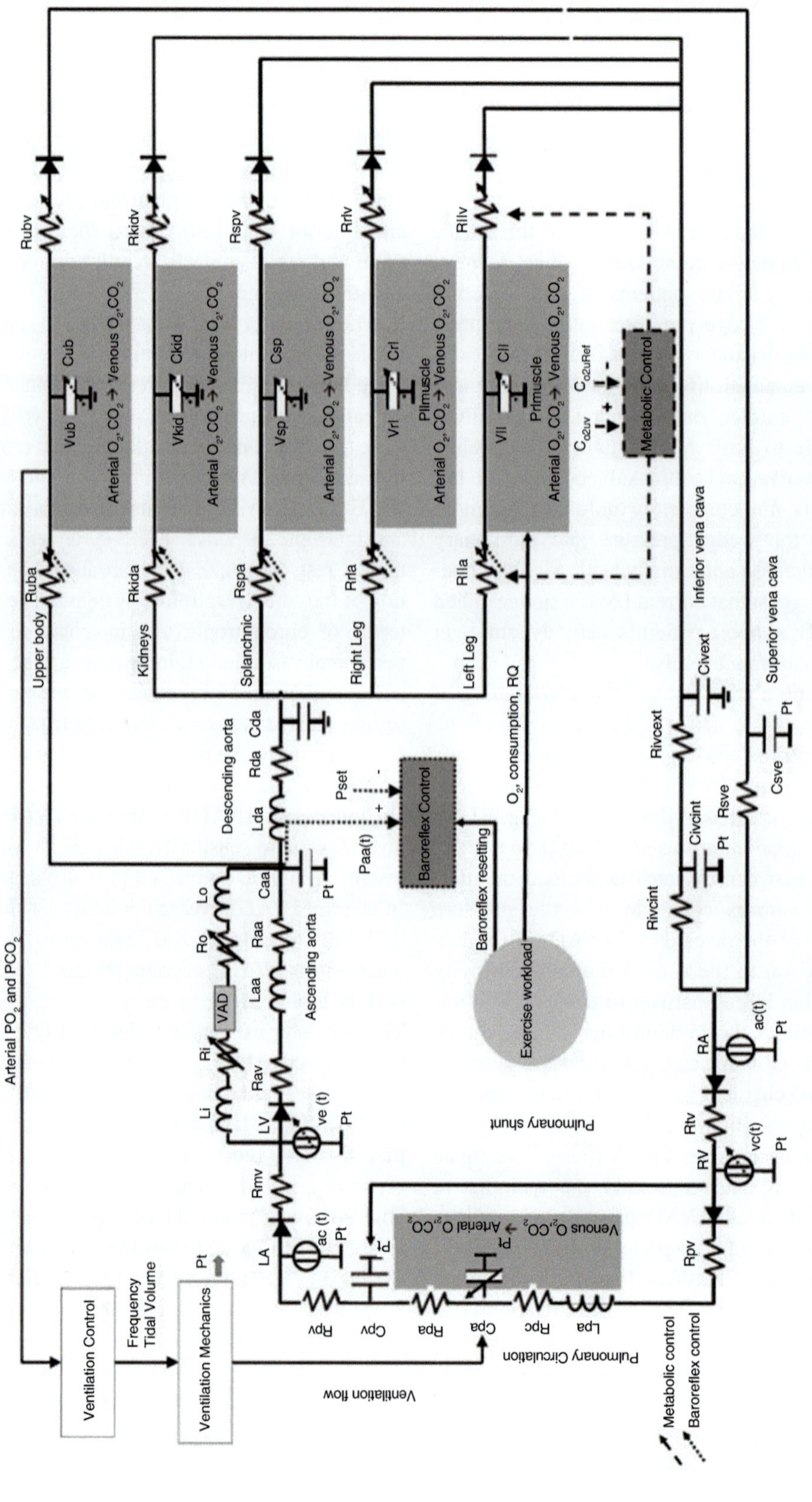

Fig. 31.15 Block diagram of the cardiorespiratory system and of the LVAD, which is connected between LV and the aorta. Further details are described in [20]

each LVAD was adjusted such that the REST condition provided identical baseline hemodynamic results. This was successful as quantitative results shown in the following figures show almost no noticeable difference for support with both pumps at REST. During EXE↑, for both pumps, the heart is allowed to conduct a limited and identical inotropic and chronotropic response to fulfill the increased demand during exercise together with the LVAD. For EXE→, no inotropic or chronotropic response is allowed simulating patients considered as non-responders. In all cases, an exercise of 80 W is simulated, and the rotational speed of each LVAD is, of course, maintained constant for the shift from rest to exercise and throughout the entire study.

Briefly, during exercise with LVAD1, total CO increases from 5.0 l/min at REST l/min to 7.4 l/min at EXE↑ and 6.7 l/min at EXE→. With LVAD2, total CO increases from 4.9 l/min at REST to 7.2 l/min at EXE↑ and 6.1 l/min at EXE→ (Fig. 31.16b). For LVAD2, the CO increase observed at EXE↑ is due to a combined increase in both the native ventricular flow (from 0.0 l/min to 1.6 l/min) and the LVAD flow (from 4.9 l/min to 5.6 l/min); see Fig. 31.16e, f. For LVAD1, CO increase at EXE↑ is mainly sustained by the device (from 5.0 l/min to 6.3 l/min) and in part by the native ventricle (from 0.0 l/min to 1.1 l/min). In case of a heart with no inotropic and chronotropic response, the increase of CO is entirely depending on the device, for both LVAD1 and LVAD2 (Fig. 31.16f).

The results for support with the two LVADs show some differences in terms of aortic pressure (Fig. 31.16a), which is higher with LVAD1 for both EXE↑ and EXE→ (101 and 94 mmHg, respectively) than in LVAD2 (99 and 87 mmHg,

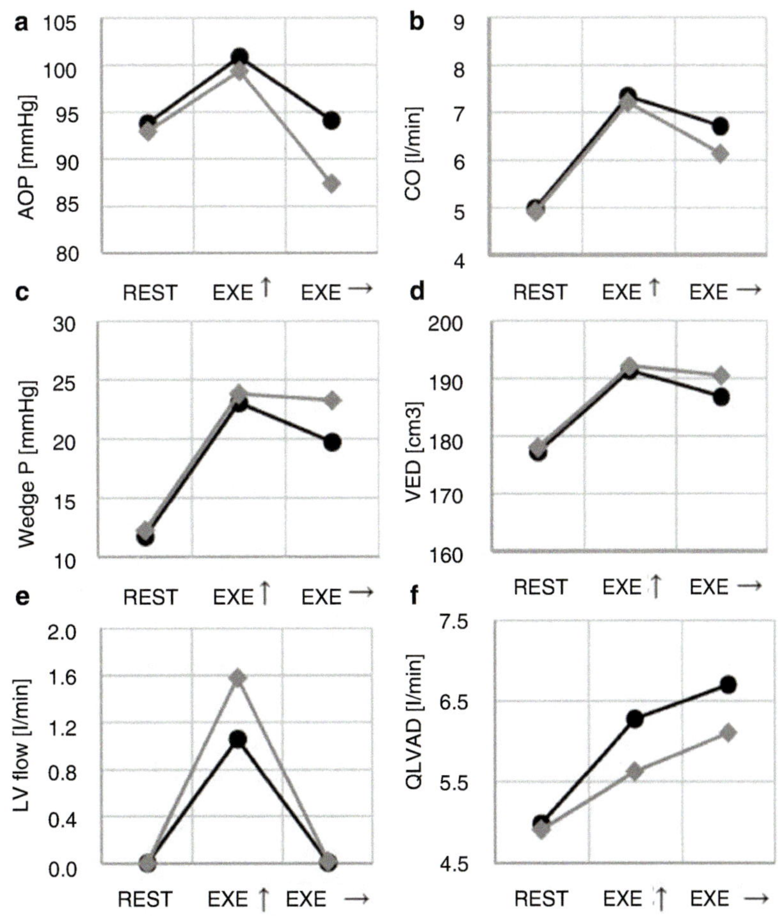

Fig. 31.16 Comparison of simulated hemodynamic data with LVAD1 (black line) and LVAD2 (grey line) at REST, EXE↑, EXE→. Panel A, systemic arterial pressure; panel B, cardiac output; panel C, wedge pressure; panel D, end-diastolic left ventricular volume; panel E, left ventricular flow; panel F, LVAD flow. Further results are shown in [8]

respectively). Left atrial pressure or wedge pressure increases with LVAD1 from 12 mm Hg at REST to 23 mmHg at EXE↑ and 19 mmHg at EXE→. With LVAD2, left atrial pressure increases from 12 mmHg at REST to 24 mmHg at EXE↑ and 23 mmHg at EXE→ (Fig. 31.16c). The end-diastolic volume adjusts correspondingly and is consistently lower for support with the highly pressure-sensitive LVAD1 (Fig. 31.16d).

The left ventricular power delivered to the systemic circulation is similar with the two LVADs in all the three hemodynamic conditions (Fig. 31.17a). The two LVADs provide similar power at REST, but LVAD1 increases the power delivered at EXE↑ and EXE→ (0.67 and 0.82 Watts), while LVAD2 does not increase the power delivered to the circulation, even showing a minimal decrease (0.61 and 0.63 Watts) as reported in Fig. 31.17b.

In Fig. 31.18, the pressure-flow curves for both devices are reported for a number of heartbeats. For both pumps, we observe a right-bottom shift of the working condition, when the patient transitions from rest to exercise. This is true for the averaged working point and also for the hysteresis. For LVAD1, the shift in average flow is from 5.0 l/min to 6.3 l/min and 6.7 l/min and in average pressure generation from 66 mmHg to 58 mmHg and 60 mmHg, respectively (Fig. 31.18a, b). The average working point of LVAD2, on the other hand, shifts in flow from 4.9 l/min at REST to 5.6 l/min at EXE↑ and 6.1 l/min at EXE→ and in pressure generation from 64 mmHg to 55 mmHg and 49 mmHg, respectively (Fig. 31.18c, d).

From the simulations, we can assess that a flatter pressure-flow characteristic, such as the one of LVAD1, can unload the left ventricle better during exercise. Figures 31.16f and 31.18a–c show that LVAD1 can increase its flow at EXE↑ to a larger extent than LVAD2, so that the native ventricle has to pump less flow. This is an important feature in light of previous clinical studies evidencing the incapability of LVADs to assure a full support at exercise so that in some patients, the ventricle starts to eject to accommodate a higher cardiac output [16, 17]. This observation correlates well with Fig. 31.17b showing that less pressure-sensitive LVADs, as implanted in clinical routine, show no increase in power transferred to the circulation or even a decrease in case of patients transitioning to exercise.

Figure 31.18 provides a description of how exercise affects LVAD working conditions. Exercise is associated with a drastic reduction of total peripheral resistance that ultimately provokes a right downshift of the working conditions of the LVAD in the pressure-flow diagram. A flatter pressure-flow characteristic means that especially for lower total peripheral resistance, as in the case of exercise, the LVAD can reach higher flows and

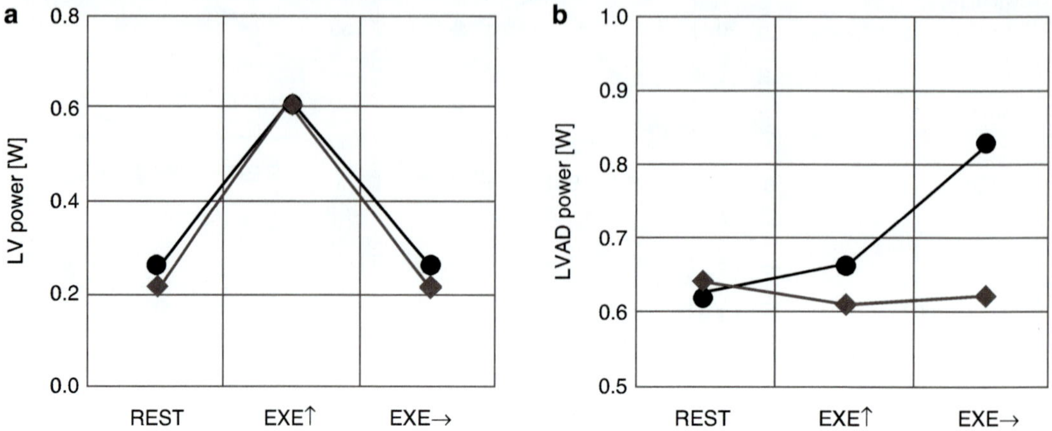

Fig. 31.17 Comparison of simulated hemodynamic data with LVAD1 (black line) and LVAD2 (grey line) at REST, EXE↑, EXE→. Panel A, left ventricular power; panel B, LVAD power. Further results are shown in [8]

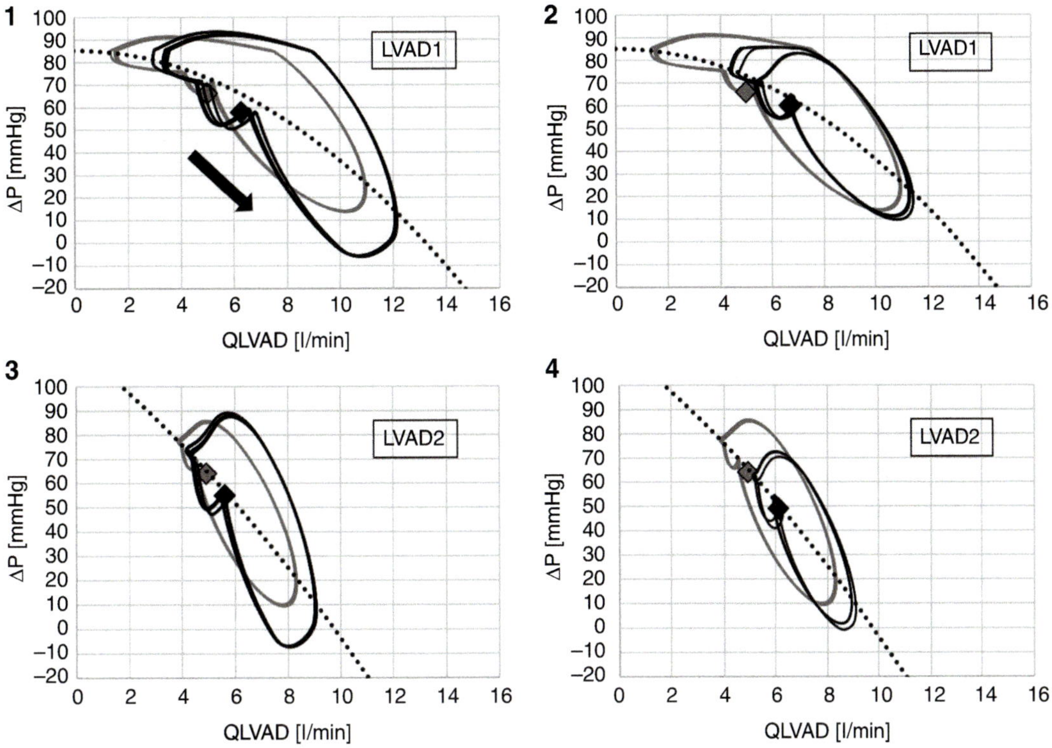

Fig. 31.18 Simulated dynamic pressure-flow curves for LVAD1 (upper panels) and for LVAD2 (bottom panels). Panel 1 and panel 3, data shown at REST (grey) and at EXE↑ (black). Panel 2 and panel 4, data shown at REST (grey) and at EXE→ (black). The shown hysteresis runs counterclockwise and is caused by the inertia in the respective pump and cannulae. The dotted lines illustrate the stationary H-Q curves of the respective LVAD. Further results are shown in [8]

generate a higher pump pressure difference and thus a high pump output power. In addition, the increase of heart rate is associated with a shortening in diastole duration. The native ventricle spends relatively more time in systole, which means that the LVAD spends more time in the lower right area of its dynamic pressure-flow loop. It is also evident that during exercise, LVAD1 pumps a higher flow than LVAD2 almost during the whole cardiac cycle. These two phenomena described above imply that a device such as LVAD1 can naturally increase its flow during exercise to a larger extent than a device with steeper pressure-flow characteristics such as LVAD2.

In summary, the present study investigates the hemodynamic reaction of two LVADs differing in terms of pressure-flow characteristic to exercise conditions at constant rotational speed. Both LVADs increase their flow during exercise, but the one with the flatter characteristic provides a higher flow and pressure difference generation. This results into a higher cardiac output and aortic pressure, lower left atrial pressure, and lower ventricular volumes. In other words and as shown in Fig. 31.17b, highly pressure-sensitive LVADs automatically transfer more hydraulic power to the circulation as soon as the patient transitions into exercise. Looking at EXE↑, the difference was 10%, and at EXE→, the difference in pump output power increased to 32% based on the power at REST. Therefore, especially in patients with poor chronotropic and inotropic response to exercise, for whom the native heart cannot contribute to augment cardiac output, a LVAD with flatter pressure-flow characteristic assures a better hemodynamic support during exercise.

Flow Sensor and Active Speed Control

Integration of a Flow Sensor

One key feature of the ReinVAD LVAD is the provision of a reliable physiological feedback. For this, an ultrasound flow sensor has been implemented in the outflow tract of the ReinVAD LVAD pump unit as shown in Fig. 31.19.

The flow sensor is located downstream of the pump volute and in a short section upstream of the outlet diffuser. In this position, it measures the entire flow that exits the pump unit and enters into the outflow graft and thus into the arterial system of the patient. The sensor works based on the established ultrasound measurement principle. As the measurement accuracy and frequency are adequately high, also flow waveforms during a single heartbeat can be resolved.

Another important aspect of the real flow measurement principle is the independence of the blood viscosity. The sensor provides accurate mean flow and flow waveform measurements, no matter what the current blood viscosity is. This might become especially important during the implantation or in the immediate postoperative time, which typically is characterized by substantial changes in blood viscosity.

Active Closed-Loop Speed Control

Systems currently on the market are set to a fixed rotational speed, which is periodically checked and adjusted manually, if appropriate. The real and exact flow measurement of the ReinVAD system provides numerous opportunities for improvement in this regard. On the one hand, the doctor is now equipped with a reliable feedback on the question whether the specified rotational speed delivers the desired level of cardiac support. If this is not the case, the speed can be adjusted manually. On the other hand, a reliable feedback is available to control the rotational speed of the device automatically and software-based. There is concern that especially the constant speed setting for months and even years employed until today with current LVADs could cause therapy-related complications. This is likely since the patient status can change dramatically in the course of the therapy.

The ReinVAD system provides an automatic speed control feature that works based on the principle of real flow measurement, and the smart control components are illustrated in Fig. 31.20.

The following approaches are being pursued and will be implemented in the control algorithm of the rotational speed of the ReinVAD LVAD.

Fig. 31.19 Position of the flow sensor in the ReinVAD LVAD pump unit. (Photo credit: ReinVAD GmbH)

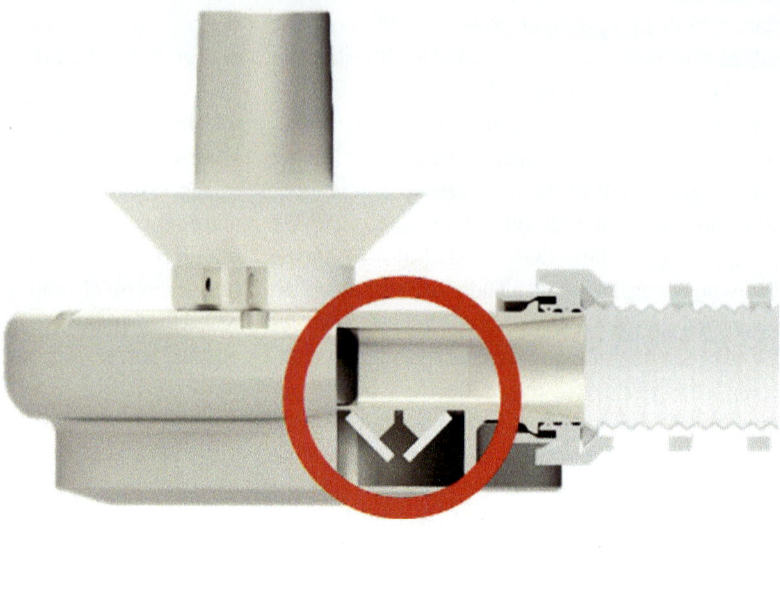

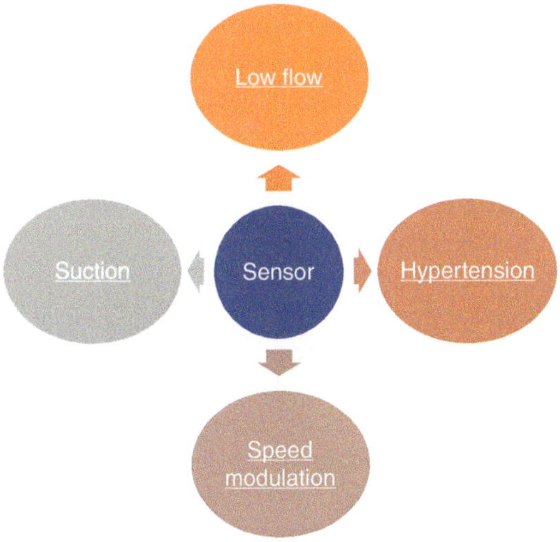

Fig. 31.20 Components of the smart speed control algorithms of the ReinVAD LVAD

Avoidance of low flow operation It is known that low flow operation is potentially positively correlated with higher rates of strokes and higher rates of device malfunction caused by thrombus formation [21, 22]. However, this correlation will be very device specific. Nevertheless, the risk of disturbed flow fields leading to complications at very low flows is a challenge for every blood-guiding device. If, due to changes in patient status, the flow through the device decreases to a critical level, the speed can be increased to increase the flow rate to a predetermined safe level adjustable by the treating doctor. The measured blood flow rate in combination with other intrinsic pump parameters is analyzed in real time by a specific algorithm. Knowing the exact flow condition and controlling the speed of the pump by an intelligent algorithm is the most effective way to provide optimal blood flow and thus to prevent low flow conditions. With this feature, thrombogenicity due to stagnation will be significantly reduced.

Avoidance of arterial hypertension It is commonly known that hypertension, which can be caused by the LVAD itself, causes higher rates of strokes and higher rates of device malfunction caused by device thrombus formation. The rotational speed can be reduced to a predetermined safe level as determined by the treating doctor.

Avoidance of suction events Through excessive rotational speed or corresponding changes in patient status, the left ventricle can be emptied completely, and the persisting suction force at the inlet cannula can cause the ventricular wall to be sucked into the inlet cannula. This causes injuries or even block the entire pump unit. In case of suction, the controller decreases the speed to a safe level to avert this critical flow blockage.

Avoidance of undesired thrombus formation through speed modulation The speed of the rotor of the ReinVAD LVAD will be varied periodically around a mean speed. Thereby, periods of different mean flow rates through the pump unit are established, and the left ventricle is emptied to different levels. Consequently, changing flow conditions are established that break up any blood stagnation regions in the pump unit itself as well as in the left ventricle.

As can be seen, the automatic adaption of the rotational speed is employed for the ReinVAD LVAD following the strategy to avoid device-related adverse events. Solutions to adapt the momentary flow rate to the momentary demand of the patient, e.g., to allow for an increased level of physical activity, are provided by the passive, intrinsic pressure sensitivity as discussed in the section Passive Interaction with the Circulation. The smart speed control algorithm of the ReinVAD LVAD is thus a means to increase safety during the therapy and improve patient outcome.

Driveline Safety

Current state-of-the-art LVADs are powered and controlled via a wired connection (driveline) between the implanted pump unit and the extracorporeal controller. As a result, this driveline needs to permanently penetrate the skin of the patient. Preferably, the spot of penetration is located in the abdominal region to allow short wire length and also safe protection of the wound.

However, driveline infection still is one of the most frequent severe events during LVAD therapy [23]. Besides the improvement of hygiene factors, mechanical movement of the driveline within the wound area is widely accepted to be an important contribution to wound infection. Due to micro-movements, successful growth of the surrounding tissue into the germ barrier of the driveline is hampered.

In order to reduce this kind of wound irritation, it seems beneficial to enhance the flexibility of the driveline. Consequently, especially the bending and twisting forces that act on the driveline are damped and not transmitted to the wound area. Fixation of the driveline on the skin is thus more effective.

Improved quality of life is a desired result of advanced pump technologies that allow increased mobility of the patient. As a side effect, especially the extracorporeal components of the LVAD systems are challenged by the enhanced activity of the patient. Mechanical stress on the extracorporeal part of the driveline is a critical risk for driveline failures, which often leads to pump stop. Therefore, mechanical stability is an important requirement for the driveline. Frequent flexing and kinking should not cause damage of the electrical wires. In addition, the driveline should withstand squeezing and pulling forces that might occur during the handling of the controller. During daily life, the controller is exposed to a number of hazardous situations that can challenge the driveline, i.e., drop of the controller or jamming by a closing door.

Unfortunately, flexibility and mechanical stability are in general complementary properties. The ReinVAD driveline comprises several wires to drive the pump motor and to provide real flow measurement at the outlet of the pump. The wires inside the cable are geometrically distributed in a specific way to improve flexibility. Additionally, the design of the cable ensures that in case of excessive wear, the signal wires for flow measurement are the first to fail before the power lines to the pump. Failing flow measurement can be easily detected by the system to generate a warning before there is severe danger for the patient. By this means, it was possible to eliminate redundant wires, which usually are required from a safety perspective. As a result, the outer diameter of the driveline is reduced to less than 4 mm, which significantly increases the flexibility.

At the controller end of the driveline, a pigtail connector is implemented providing a defined unlocking force. If the controller accidently drops to ground, then the driveline might be stressed by excessive pulling forces. In that case, the pigtail connector unlocks automatically to protect the driveline from being irreversibly damaged.

In the region of skin penetration, the outer surface of the driveline is coated by a non-woven fleece material made of polyurethane. The geometric properties of the fleece material mimic the extracellular matrix in order to optimize cell ingrowth. Thus, wound healing is promoted, which is an important factor to prevent wound infection.

The driveline also includes a separate connector, which is implanted right beneath the skin. This connector is initially factory sealed and can be easily opened during a small surgical intervention in case of a need to exchange the extracorporeal part of driveline. Also, once the future option of a TET system (transcutaneous energy transmission system) will be available, the implanted ReinVAD pump can be upgraded with the corresponding modules minimizing surgical intervention.

Peripherals and Usability

The terms usability engineering and human factors engineering can be considered synonymous [24]. The first challenge is to understand what usability for medical devices means and how to implement the engineering process. The standard IEC 62336-1 defines usability as "characteristic of the user interface that facilitates use and thereby establishes effectiveness, efficiency and user satisfaction in the intended use environment" [25]. Basically, the goal of the usability engineering process is to minimize the hazards related to the normal use of medical devices by the intended users and in the intended environment. The FDA (Food and Drug Administration) has developed a guidance [26], where device types with clear potential for serious harm resulting from use error have been listed. For these types of devices, review of human factors data in premarket submission will help the FDA to evaluate the safety and effectiveness of these devices. This guidance includes VAD systems as a device type for which usability has to be considered.

Figure 31.21 shows the usability engineering process used for the ReinVAD LVAD development. To be able to design safe-to-use user interfaces, the first step is to understand how users use the medical device in question or in case of a new device, the current similar devices on the market. For that, it is necessary to understand also who the users are and where the medical device is going to be used, i.e., the intended environment. To gather this information, we performed a contextual inquiry. A contextual inquiry is a user-centered research method based on a semi-structured interview with the user in his/her environment while observing his/her interaction with the medical device. It is important to plan the contextual inquiry according to the specific user, the scenarios to be performed by the user, and the specific environment where they will be performed. Using this methodology, we gathered detailed information related to the use of the medical device, including the main characteristics about the intended users (Table 31.2) and the use environments (i.e., sterile or non-sterile; hospital, ambulance, in hospital or home use; ambient lighting and noise levels; and other equipment use at the specific environment).

We also collected information about the user actions required with current similar systems for a specific task and the current user needs for a VAD system. Additionally, we conducted research to analyze the reported adverse events and existing recalls during the last years due to poor usability. We used all this information to define the use scenarios and design a breakdown of the user actions necessary to accomplish each task for our ReinVAD system, focusing on the minimization of the number of user actions to facilitate the use of the device (see Fig. 31.21).

To analyze the safety of use for our proposed concept, we mapped each user action to the user profile and the use environment. Then we were able to identify for each specific task possible user errors during normal use due to different causes: perception, cognition, or unexpected/missing user action. The identified use errors were transferred to our risk management process, where they were analyzed and assessed according to ISO 14971, by identifying the hazard situations, defining their risk and necessary risk control measures to avoid or reduce these risks.

As result, the user interface specifications were determined, and we started designing our user interface. This was an iterative process, where a cross-functional group of design experts, development engineers, risk experts, usability experts, and users were involved. Different types of informal formative tests were required such as expert reviews and user's survey/interviews. These activities lead to the development and implementation of our usability concept prototypes. These prototypes were used to conduct a formal formative test. The formative test was divided into different usability tests where frequent and critical tasks were simulated and performed by the participants. During this formative test, the risk of the previously identified use errors was checked, possible not still identified use errors were researched, and the intuitiveness and learnability of the concept was analyzed. Therefore, proven evidence of the concept's advantages was generated, and possible improvements for the concept were identified.

After completion of the formative test, the use of error risk analysis was updated, and we improved the design according to the outcome.

The next steps at the ReinVAD GmbH are the design of the summative evaluation of the user interface based on the hazard-related scenarios to finally verify and validate the product before starting the clinical trials.

The Usability Concept

At the ReinVAD GmbH, we have set our focus in developing a system that is simple, comprehen-

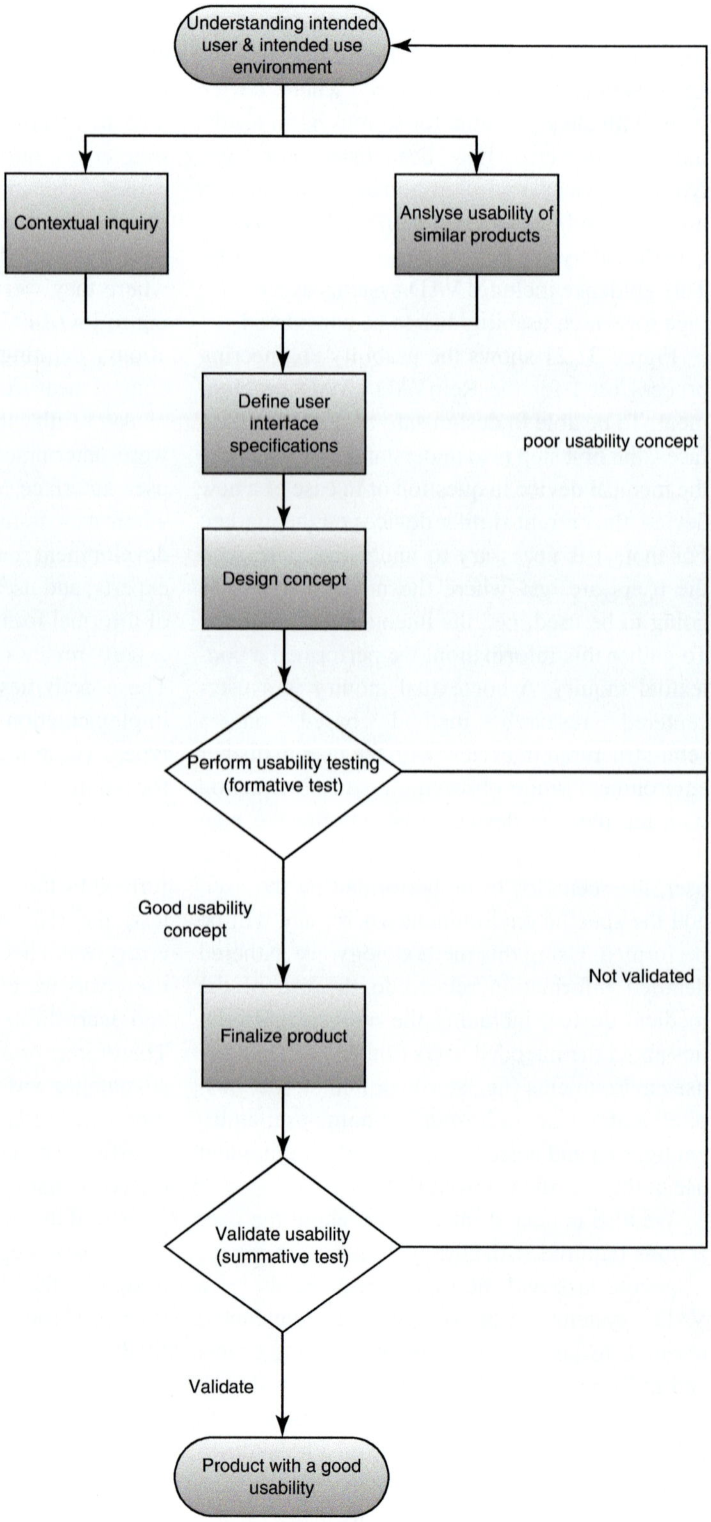

Fig. 31.21 Usability process of the ReinVAD GmbH

sive, and intuitive. Consequently, we can achieve the principles of effectiveness, efficiency and user satisfaction for an adequate usability for our system, which translates also into a minimization and avoidance of known user errors.

The ReinVAD LVAD System

The ReinVAD system consist of different extra-corporeal components (see Fig. 31.22):

Table 31.2 Aspects/characteristics to consider, when defining a user profile

Aspects	What to consider/description
Demographics	Education, age, socioeconomic status, ethnicity, and cultural background
Knowledge and skills	Education, experience level, language, literacy, and health literacy
Limitations	Due to vision, hearing, cognitive, dexterity, and mobility impairments
Performance-shaping factors	Learning style, preference, and tendencies
Wok responsibilities	Task pertinent to the medical device in development

1. The patient controller and batteries to control, monitor, and power the system
2. Clinical monitor to set up the pump parameters, monitor its performance, evaluate and track alarm conditions, and pump performance data saved on the patient controller
3. Charger to charge the batteries, test them, and asses the remaining lifetime
4. AC adapter to power the system, when the patient is indoors
5. Wristband to monitor the system (accessory)

Clinical Monitor

Nowadays, majority of people use computers, tablets, or mobile phones. Accordingly, the clinical monitor of the ReinVAD system is a special, medical grade tablet (see Fig. 31.23), allowing mobility in the hospital due to its size and lightweight, which is less than 1.4 kg. This is thus a reduction of 4.5 times the weight of the clinical monitors currently on the market. This allows easy mobility with the system in the hospital and therefore increases patient safety.

Its graphical user interface is as intuitive and comprehensive as the software used every day and mobile apps. It has been divided in different

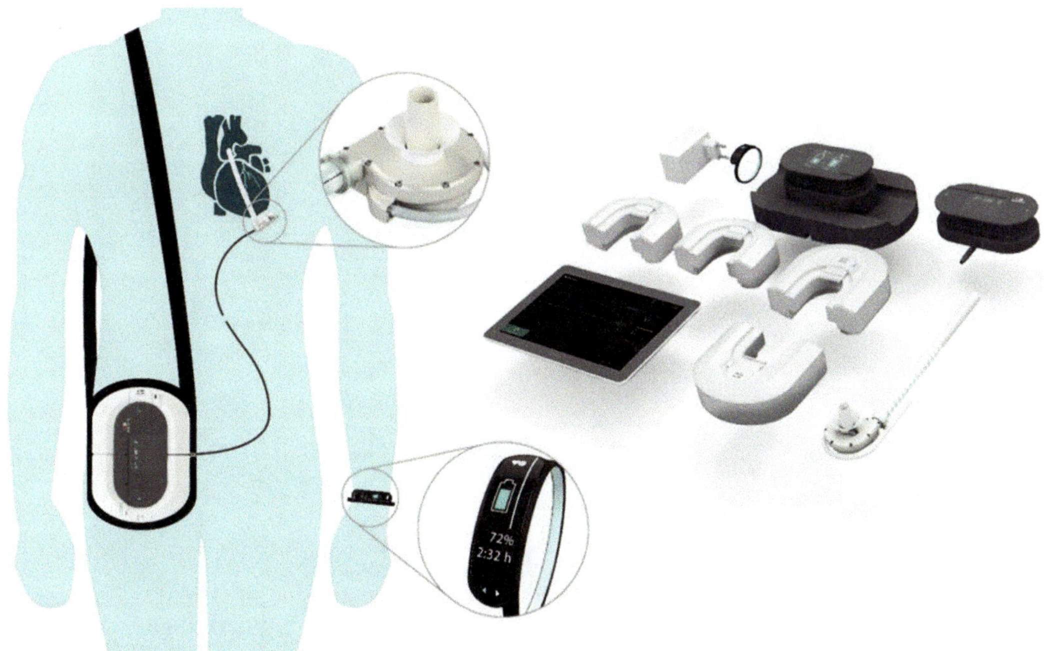

Fig. 31.22 ReinVAD system overview. (Photo credit: ReinVAD GmbH)

sections, where all relevant information can be found at a glance (see Fig. 31.23).

The patient controller and clinical monitor are equipped with a Bluetooth module. Within three clicks, the clinical monitor connects to the patient controller allowing it to setup and monitor the ReinVAD system. This Bluetooth connection and communication use a high-security protocol to ensure safety of the patients and their data.

The wireless feature decreases the risk during the implantation of the system, because no communication cable between patient controller and clinical monitor has to cross between sterile area and non-sterile area in the operation theater (see Fig. 31.24). It increases safety by allowing the patient to be monitored during in-hospital transport and provides freedom to the operator during its use.

One-Piece Concept: Patient Controller and Batteries

Cables are known to be a source of technical problems and use-related problems. Accordingly, we have eliminated some of the cables with the ReinVAD system by designing a one-piece controller. The one-piece concept patient controller integrates controller (see black component in Fig. 31.25) and two batteries connected around it (see white components at both sides of the patient controller in Fig. 31.25).

The benefits of the one-piece controller are:

- Minimization of the number of single components needed to be carried.
- Avoidance of problems with "getting caught," for example, in doors-handles.
- Easy packing of patient controller into the desired carrying system.
- Reduction of the psychological impact in the patient life due to less visible components, cable confusions have been avoided.

Patient Controller

To simplify and improve intuitive use, the graphical user interface (GUI) of the patient controller

Fig. 31.23 GUI layout of the clinical tablet (Photo credit: ReinVAD GmbH)

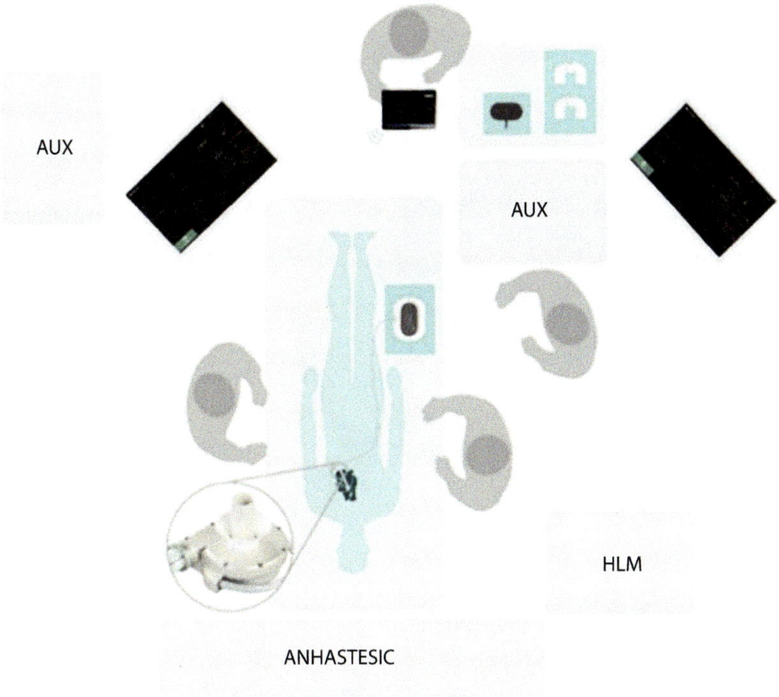

Fig. 31.24 Overview of the ReinVAD system components during implantation showing the VAD pump, a controller and a back-up controller, two batteries, and the clinical monitor transmitting data to two external monitors (Photo credit: ReinVAD GmbH)

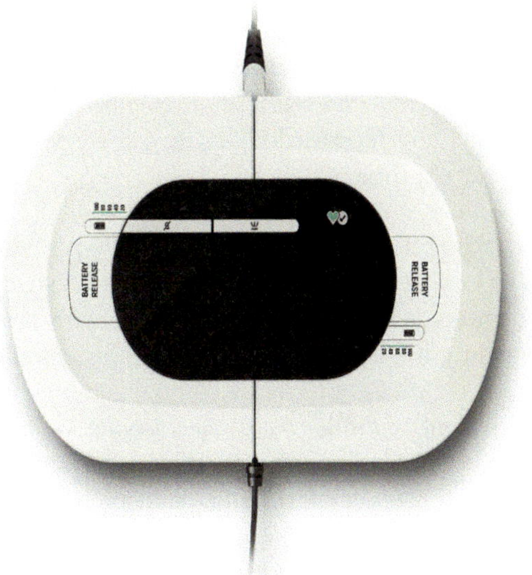

Fig. 31.25 The one-piece controller includes patient controller shown in black and attached batteries around it, shown in white. (Photo credit: ReinVAD GmbH)

incorporates a display 3.5 times bigger than other similar systems (see Fig. 31.26). Its GUI is designed to provide all necessary information for the user at a glance without having to navigate through a cyclic menu. The symbols used in the GUI have been specifically created for the purpose of information desired to communicate. They provide extra information to the user for an intuitive understanding of the situation of the system. Another advantage of the bigger screen is the adequate font size despite the quantity of information, so users with vision deficiencies can also read the displayed information.

The patient controller is ready for the future vision of telemonitoring capabilities due to the built-in wireless feature. Telemonitoring will help the hospital facilities to provide a better care for the patients and reduce the hospitalization cost and cost of care.

Batteries

For the batteries, cable-free connection and an easy-click mechanism have been designed (see Fig. 31.27). By simply sliding the battery towards the patient controller until the user hears a click, the connection is achieved. Then, the batteries are locked to the patient controller. To release a battery, the user needs to press the release button and slide it in opposite direction. With the easy-click

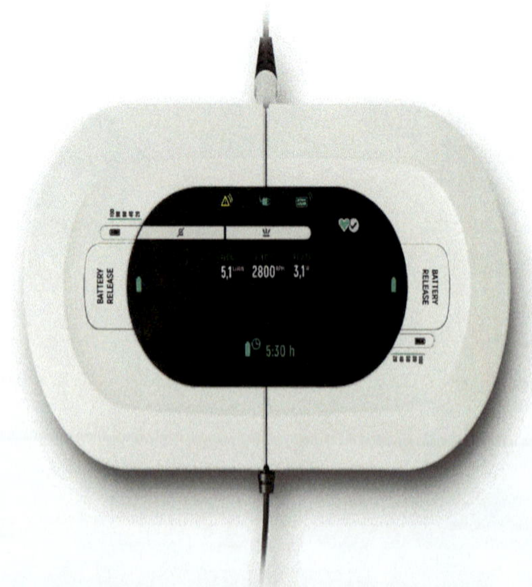

Fig. 31.26 GUI of the patient controller: All information at a glance. (Photo credit: ReinVAD GmbH)

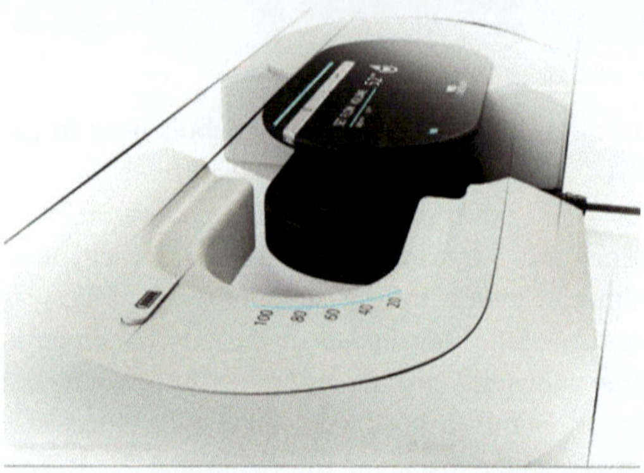

Fig. 31.27 Cable-free connection of the batteries. (Photo credit: ReinVAD GmbH)

mechanism, which does not require alignment of female and male connectors pins or turning, misconnection errors are avoided for users with reduced or limited fine motor skills or vision problems. The time required to connect the batteries is also reduced, which is useful under stress situations.

As explained above, the ReinVAD pump allows for increased mobility and even provides the possibility to perform some sport due to its pressure-sensitive H-Q curve (see Fig. 31.7) and as discussed in the section Exercise Capacity. The extracorporeal components have been designed to support this potential increase of patient mobility.

The daily wearable (i.e., controller and batteries) have been minimized to reduce not only their sizes but also their weight. Also, different battery sizes (see Fig. 31.22) have been developed to provide flexibility to the patient according to his/her necessities and to promote mobility. Using the small batteries, 16 h of support are provided, and the one-piece solution, controller and batteries, weighs 1 kg (see Table 31.3).

Carrying and Protecting System

Another important aspect to support the desired patient mobility is the protecting system, in which the wearable components are carried. The ReinVAD system comprises a versatile carrying and protecting system, which can be used as shoulder bag, rucksack, or even a waist bag (see Fig. 31.28). It is a "all-in-one" system, providing the patient flexibility on how to carry his/her VAD system according to the situation or necessities of the moment. The carrying system is discreet and small to provide privacy for patients and modern looking to keep up with the current time.

Wristband

A wristband (see Fig. 31.22) to monitor the status of the system is another usability feature provided by the ReinVAD system. The patient can check the status of his/her system without taking the patient controller out of the carrying system. The wearable wristband provides privacy, discretion, and comfort to the patients. It is also an easy and modern way to monitor the status of the ReinVAD system.

Charger

The size and weight of the charger has been minimized (see Fig. 31.22). The design has been kept similar to the patient controller, using the same click mechanism to connect the bat-

Table 31.3 Characteristics of the different battery sizes offered by the ReinVAD system

	2× Battery + patient controller	1× Battery + 1× extended battery + patient controller	2× extended batteries + patient controller
Weight	1 kg	1.2 kg	1.4 kg
Runtime	16 h	24 h	32 h

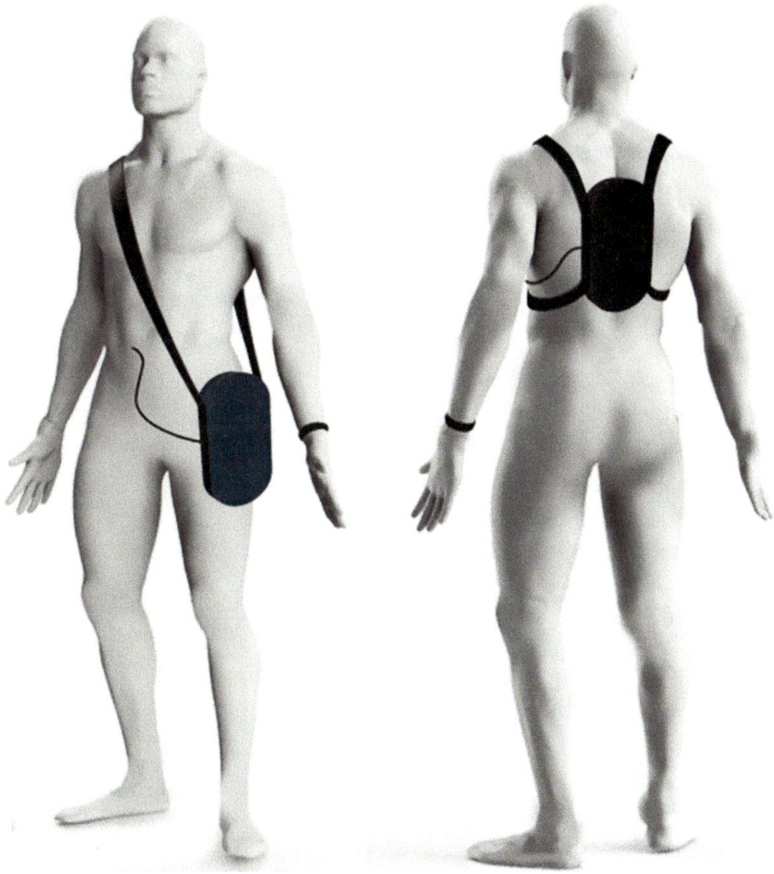

Fig. 31.28 Versatile and discreet carrying system. (Photo credit: ReinVAD GmbH)

teries. Its length is comparable to the patient controller with two connected batteries. It weighs less than 1.5 kg. These characteristics make it easy to use and to transport, also in case of travelling.

AC Adapter

An AC adapter is offered to supply power to the patient controller. Additionally, it offers the possibility to charge the batteries. Its size and weight are like an AC adapter for a laptop. So, for a weekend travel, the charger could be left at home and the AC adapter option can be used, reducing considerably the effort for the patient.

Benefits of the Usability Design

The ReinVAD GmbH has focused its usability concept on simplicity to solve some known usability problems and to adapt to the user needs. This has led to modern looking and discreet extracorporeal components. The number of required cables in the hospital and in daily life of the patient has been minimized by integrating wireless capabilities into the patient controller to communicate with the clinical monitor and a wristband as well as avoiding the use of cables to connect the batteries to the patient controller. The system provides privacy and adapts to the patients' lifestyle thanks to the different sized batteries and the possibilities offered by the carrying system.

Product Development

The ReinVAD system has been developed to improve technical and clinical performance with respect to safety and quality of life. Therefore, a stringent specification and verification of technical requirements have been performed to ensure safe performance of the system. At this point, the ReinVAD LVAD is being verified in a trial using sheep. So far, more than 30 animal experiments have been performed to support the design development process and to show proper functionality of the pump. During the latest 20 experiments, the final design concept of the pump was investigated for a period of up to 90 days. Even with ceasing anticoagulation, the ReinVAD LVAD was able to work without thrombotic complication. Also, no wound infection occurred.

On its way to commercialization, the next steps of the ReinVAD project will be to perform complete product verification including the preclinical trial. These results will demonstrate functional safety of the device and furthermore allow for starting the clinical trial as a final step towards CE certification.

References

1. Smith PA, Wang Y, Groß-Hardt S, Graefe R. Chapter 10 – hydraulic design. In: Gregory SD, Stevens MC, Fraser JF, editors. Mechanical circulatory and respiratory support. London: Academic Press; 2018. p. 301–34.
2. Thamsen B, Plamondon M, Granegger M, Schmid Daners M, Kaufmann R, Neels A, et al. Investigation of the axial gap clearance in a hydrodynamic-passive magnetically levitated rotary blood pump using X-ray radiography. Artif Organs. 2018;42:510–5.
3. Arndt A, Nüsser P, Lampe B. Fully autonomous preload – sensitive control of implantable rotary blood pumps. Artif Organs. 2010;34:726–35.
4. AlOmari A-HH, Savkin AV, Stevens M, Mason DG, Timms DL, Salamonsen RF, et al. Developments in control systems for rotary left ventricular assist devices for heart failure patients: a review. Physiol Meas. 2013;34(1):R1–27.
5. Salamonsen RF, Lim E, Gaddum N, AlOmari A-HH, Gregory SD, Stevens M, et al. Theoretical foundations of a Starling-like controller for rotary blood pumps. Artif Organs. 2012;36(9):787–96.
6. Moazami N, Fukamachi K, Kobayashi M, Smedira NG, Hoercher KJ, Massiello A, et al. Axial and centrifugal continuous-flow rotary pumps: a translation from pump mechanics to clinical practice. J Heart Lung Transplant. 2013;32(1):1–11.
7. Graefe R, Beyel C, Henseler A, Körfer R, Steinseifer U, Tenderich G. The effect of LVAD pressure sensitivity on the assisted circulation under consideration of a mitral insufficiency: an in vitro study. Artif Organs. 2018;42:E304–14.
8. Graefe R, Henseler A, Körfer R, Meyns B, Fresiello L. Influence of LVAD pressure-flow characteristic on exercise physiology: quantification with a validated numerical model. Int J Artif Organs. 2019;42(9):490–9.
9. Timms DL, Gregory SD, Greatrex NA, Pearcy MJ, Fraser JF, Steinseifer U. A compact mock circulation loop for the in vitro testing of cardiovascular devices. Artif Organs. 2010;35(4):384–91.
10. Gregory S, Timms D, Pearcy MJ, Tansley G. A naturally shaped silicone ventricle evaluated in a mock circulation loop: a preliminary study. J Med Eng Technol. 2009;33:185–91.
11. Gregory SD, Stevens MC, Wu EL, Pauls JP, Kleinheyer M, Fraser JF. Mitral valve regurgitation with a rotary left ventricular assist device: the haemodynamic effect of inlet cannulation site and speed modulation. Ann Biomed Eng. 2016;44:2674–82.
12. Pennings KAMA, Martina JR, Rodermans BFM, Lahpor JR, Vosse FN, Mol BAJM, et al. Pump flow estimation from pressure head and power uptake for the HeartAssist5, HeartMate II, and HeartWare VADs. ASAIO J. 2013;59:420–6.
13. Richard P, Klabunde E. Interdependent effects of preload, afterload and inotropy on ventricular pressure-volume loops. 2018. [Online]. Available: http://www.cvphysiology.com/Cardiac%20Function/CF026.
14. Sagawa K, Maughan L, Suga H, Sunagawa K. Cardiac contraction and the pressure-volume relationship. New York: Oxford University Press; 1988.
15. Gregory SD, Stevens M, Timms D, Pearcy M. Replication of the Frank-Starling response in a mock circulation loop. In 2011 Annual International Conference of the IEEE Engineering in Medicine and Biology Society, 2011.
16. Muthiah K, Robson D, Prichard R, Walker R, Gupta S, Keogh AM, et al. Effect of exercise and pump speed modulation on invasive hemodynamics in patients

with centrifugal continuous-flow left ventricular assist devices. J Heart Lung Transplant. 2015;34:522–9.
17. Jacquet L, Vancaenegem O, Pasquet A, Matte P, Poncelet A, Price J, et al. Exercise capacity in patients supported with rotary blood pumps is improved by a spontaneous increase of pump flow at constant pump speed and by a rise in native cardiac output. Artif Organs. 2011;35:682–90.
18. Hayward CS, Fresiello L, Meyns B. Exercise physiology in chronic mechanical circulatory support patients. Curr Opin Cardiol. 2016;31:292–8.
19. Salamonsen RF, Mason DG, Ayre PJ. Response of rotary blood pumps to changes in preload and afterload at a fixed speed setting are unphysiological when compared with the natural heart. Artif Organs. 2011;35:E47–53.
20. Fresiello L, Rademakers F, Claus P, Ferrari G, Molfetta AD, Meyns B. Exercise physiology with a left ventricular assist device: analysis of heart-pump interaction with a computational simulator. PLoS One. 2017;12:e0181879.
21. Starling RC, Moazami N, Silvestry SC, Ewald G, Rogers JG, Milano CA, et al. Unexpected abrupt increase in left ventricular assist device thrombosis. N Engl J Med. 2014;370:33–40.
22. Yang F, Kormos RL, Antaki JF. High-speed visualization of disturbed pathlines in axial flow ventricular assist device under pulsatile conditions. J Thorac Cardiovasc Surg. 2015;150(1):938–44.
23. Kirklin JK, Pagani FD, Kormos RL, Stevenson LW, Blume ED, Myers SL, et al. Eighth annual INTERMACS report: special focus on framing the impact of adverse events. J Heart Lung Transplant. 2017;36:1080–6.
24. Food and Drug Administration (FDA). Applying human factors and usability engineering to medical devices – guidance for industry and food and drug administration staff, 2016.
25. International Organization for Standardization (ISO). IEC 62366-1:2015 medical devices – part 1: application of usability engineering to medical devices, 2015.
26. Food and Drug Administration (FDA). List of highest priority devices for human factors review – draft guidance for industry and food and drug administration staff. 2016.

EVAHEART 2 Left Ventricular Assist System: A Hemocompatible Centrifugal Pump with Physiological Pulsatility

32

Tadashi Motomura

Abbreviations

LVAD	left ventricular assist device
LVAS	left ventricular assist system
cf-LVAD	continuous-flow LVAD
vWF	von Willebrand factor
HRAEs	hemocompatibility-related adverse events

Background

Left ventricular assist devices (LVADs) have become a first-line therapeutic option in patients with heart failure. Continuous-flow LVADs (cf-LVADs) incorporate state-of-the-art technology that has enabled pump miniaturization and long-term durability, yet the clinical outcomes of the current cf-LVADs are not comparable with heart transplantation (5-year survival rate of 90% for heart transplantation compared with <50% for LVADs) because of device-related complications [1]. Design changes in the currently available LVADs have made positive albeit incremental improvements to patient outcomes in terms of cerebrovascular accidents [2]. However, adverse events impact the patient's quality of life, the cost of care [3, 4], and patient survival. Although originally conceived as a means of bridging heart failure patients to transplantation, 46% of LVAD implantations are now performed as destination therapy [5], which requires years (>5 years) of support without heart transplantation. Therefore, post-LVAD adverse events and incremental medical care costs are becoming more serious issues constraining market expansion [6].

T. Motomura, MD, PhD (✉)
Evaheart, Inc., Houston, TX, USA
e-mail: tmotomura@evaheart-usa.com

Limitations of Current Technology

The current pumps on the market focus heavily on miniaturization and intrapericardial positioning [7] (implantation without a pump pocket). As a result, ideal pump performance for full cardiac support may be sacrificed. Adequate flow support to end-stage heart failure patients is important for better organ perfusion and ventricular unloading. Because of their excessive miniaturization, current LVADs are not physiologically responsive and may impair end-organ perfusion by attenuating the native heart pulsatility. Lowered pulsatility has been linked to endothelium dysfunction [8], sympathetic nerve activation [9], abnormality of the renin-angiotensin system [10], and increased risk of arteriovenous malformations [11], which can trigger gastrointestinal bleeding. Also, current cf-LVADs may alter the hemodynamics of blood flow in measurable ways (such as degradation of von Willebrand factor [vWF]

© Springer Nature Switzerland AG 2020
J. H. Karimov et al. (eds.), *Mechanical Support for Heart Failure*,
https://doi.org/10.1007/978-3-030-47809-4_32

multimers) [12], which can lead to serious hemocompatibility-related adverse events such as thrombosis and stroke. In fact, the incidence of pump thrombosis with current cf-LVADs ranges from 2% to 8% and results in 48% mortality [13].

Another limitation of current LVADs is the design of the inflow cannula. In current LVADs, the inflow cannula design has been implicated as a contributing source of thrombus formation through the distortion of blood flow patterns within the ventricular chamber, resulting in nonphysiologic turbulence, shear stress, and stasis (Fig. 32.1). This in turn activates coagulation pathways [13–18]. The inflow cannula may create conditions for early formation of wedge thrombus, and inflow malposition may predispose to the development of right heart failure and cardiac arrhythmia. Optimizing the design of the inflow cannula is therefore crucial in order to minimize these adverse events and improve long-term survival.

Design Advances of the EVAHEART 2 Blood Pump

Rotary pumps generally require a center shaft to suspend the impeller, regardless of pump type (axial flow or centrifugal pump). This center shaft is predisposed to generate blood stagnation, followed by blood clot formation known as pump thrombosis. To avoid pump thrombosis, current cf-LVADs have incorporated a levitated impeller system that completely eliminates the mechanical center shaft by either fluid dynamic force [7] or magnetic force [4]. However, due to either the small diameter or high displacement volume of these impellers, blood delivery performance per impeller rotation is compromised. Therefore, the pump speed of these commercial cf-LVADs is higher (2600–5000 rpm) than that of the EVAHEART pump (1600–2000).

The EVAHEART blood pump was designed to minimize the blood trauma without compromising pump flow capacity and address the risks for hemocompatibility-related adverse events. The EVAHEART pump impeller is levitated by a dynamic fluid (sterile water) suspension system, and potential blood stagnation around the mechanical center shaft area is washed away by a unique wide-gap (open vane) impeller design. Due to this impeller design, the blood flow capacity of the EVAHEART pump is wide enough to allow full cardiac support from small to large patients while running at a lower pump speed compared to other clinically used LVADs. Each centrifugal cf-LVAD has individual performance characteristics that determine how it will interact with the patient. The EVAHEART impeller (Fig. 32.2) minimizes the shear stress [19] to the blood, which plays a main role in inducing hemolysis, platelet activation, and possibly degradation of the vWF high-molecular-weight multimer [20–22]. This impeller design can generate end-

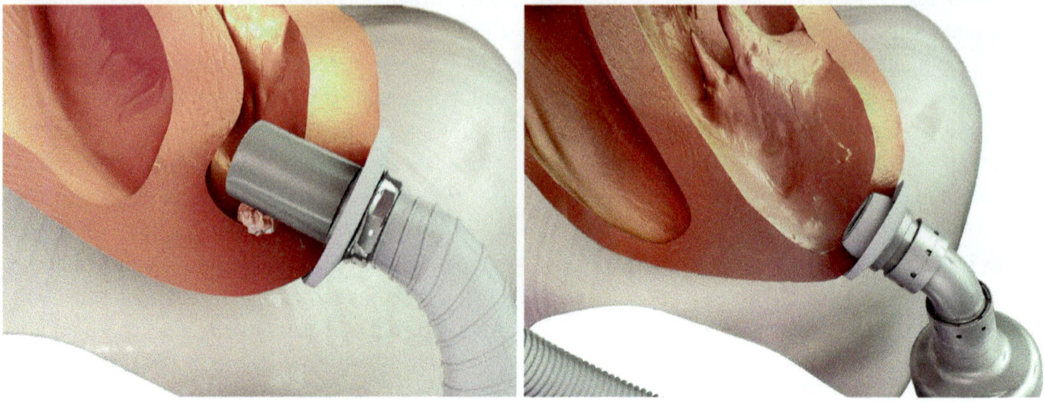

Fig. 32.1 Current left ventricular assist device (LVAD) inflow cannula design (left) compared with the double-cuff tipless (DCT) inflow cannula design (right)

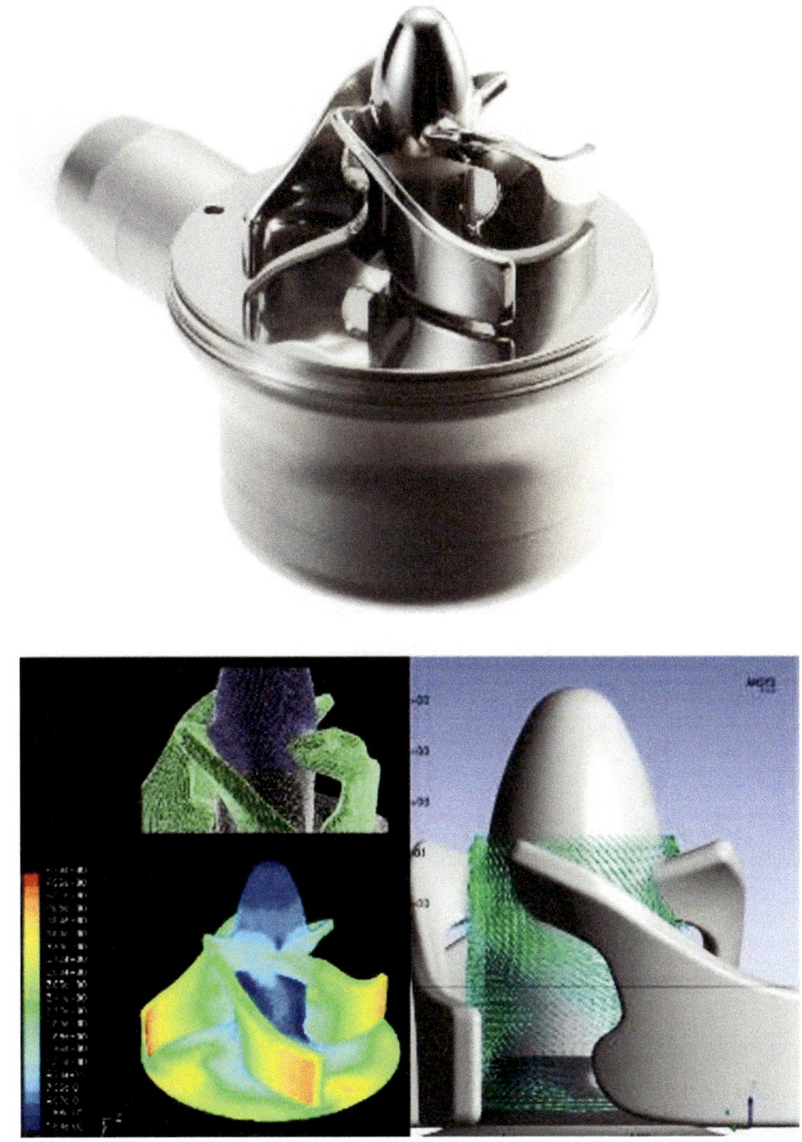

Fig. 32.2 The EVAHEART open vane impeller design and computational fluid dynamics shear stress analysis

systolic high peak flow, which helps to preserve the native aortic pulsatility of the heart [23]. An in vitro mock circulatory loop study demonstrated that the aortic pulse pressure with the EVAHEART blood pump (clinical pump speed range: 1600–2000 rpm) was well preserved in the range of approximately 10–25 mmHg despite a heart failure (cardiac output of 4 L/min) condition (unpublished data) (Fig. 32.3). The flow-head pressure curve (HQ curve) of the EVAHEART is much flatter (more head pressure sensitive) than other clinical LVADs (Fig. 32.4). In other words, as long as the afterload blood pressure is properly managed (i.e., 65–80 mmHg), the EVAHEART pump flow exhibits self-regulating behavior similar to the "Frank-Sterling" law and has minimal pulse attenuation. The minimal pulse attenuation of the EVAHEART pump has been attributed to the slower progression of aortic insufficiency seen in EVAHEART patients compared to patients on other devices [24, 25]. In addition, according to data from the Japanese Registry for Mechanically Assisted Circulatory Support (J-MACS), the incidence of right heart failure is very low (14% at

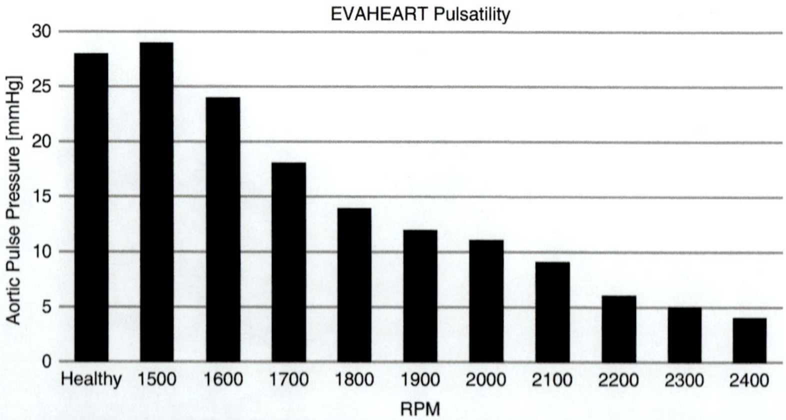

Fig. 32.3 Mock circulatory loop ramp study, performed under heart failure conditions (cardiac output of 4 L/min), for aortic pulsatility assessment for the EVAHEART left ventricular assist system (unpublished data)

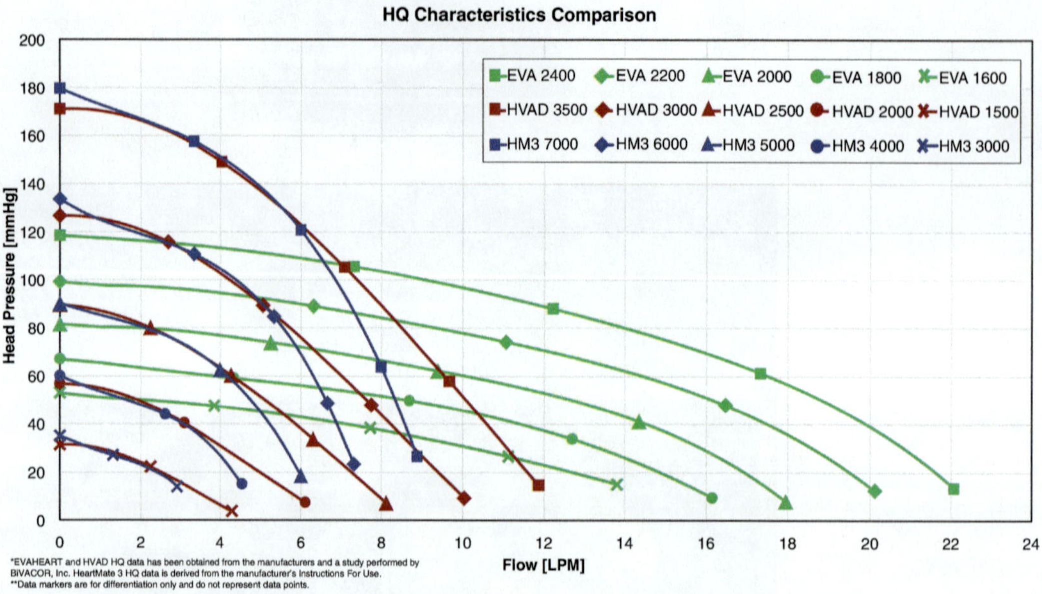

Fig. 32.4 Head pressure-flow (HQ) curve comparison between the EVAHEART (EVA), HeartMate 3 (HM3), and HVAD

2 years for all right heart failure events, with no patients requiring right ventricular assist devices) in patients with the EVAHEART. This may be attributed to the flatter HQ characteristics of the EVAHEART pump, which generates end-systolic high peak flow yet a low amount of blood drainage during the diastolic phase from the left ventricular chamber, thus minimizing the continuous pulling force to the ventricular septal wall. The EVAHEART pump also demonstrates a low incidence of pump thrombosis [26]. This low incidence is also attributed to the open vane design and end-systolic high peak flow, which help to promote washout of the blood around the hydrodynamic bearing. Also contributing is the antithrombogenic 2-methoxyethyloylphosphoryl choline (MPC) coating [27, 28] on the blood-contacting surfaces inside the pump. According to the J-MACS data registry, only 1 case of pump thrombosis (0.5%) was identified out of 190 implantations (75% of patients supported for >1 year, 59% supported for >2 years).

The lower required pump speed and open vane impeller of the EVAHEART blood pump minimize the amount of shear stress that the blood is exposed to. This plays a role in improving the hemocompatibility of the system, such as reducing hemolysis. Thus, the adverse event rate due to hemolysis associated with EVAHEART LVAS implantation is very low [29]. Another hemocompatibility parameter is vWF high-molecular-weight multimers, which may contribute to nonsurgical bleeding, which clinically is represented by gastrointestinal bleeding [30–32]. Post-LVAD gastrointestinal bleeding is a common issue for current commercial LVADs and leads to other thromboembolic events due to the temporary discontinuation of anticoagulation [33]. For example, the MOMENTUM 3 trial investigated clinical outcome measures including post-LVAD adverse event profiles in a comparison of the HeartMate II (HMII) and HeartMate 3 (HM3). Pump thrombosis was significantly improved in the HM3 arm but the incidence of gastrointestinal bleeding remained the same (27.0% in HMII, 27.3% in HM3) [2].

Bartoli et al. compared the EVAHEART LVAS with another commercially available pump using whole human blood in a mock circulatory loop and demonstrated that the EVAHEART LVAS showed significantly less degradation of vWF high-molecular-weight multimers than the HMII (Fig. 32.5) [20]. This result can be attributed to the open vane impeller design and larger flow gaps (700 μm in the EVAHEART compared with 50 μm in the HMII), which creates lower amounts of shear stress (22 Pa of wall shear stress in the EVAHEART compared with 55 Pa of wall shear stress in the HMII). Super physiologic shear stress is the driving factor for adverse events related to blood trauma, so minimizing the levels of shear stress induced on the blood will help to limit the number of adverse events seen clinically. Additionally, according to J-MACS, the incidence of gastrointestinal bleeding with the EVAHEART LVAS seems markedly low (Fig. 32.6).

Recently, Zayat et al. (a group from the RWTH Aachen University) conducted an in vitro mock loop study with human volunteer blood and reported that the EVAHEART pump demonstrated better hemocompatibility than the HM3 in terms of hemolysis parameters, vWF-related parameters, and coagulation/fibrinolytic parameters (Fig. 32.7). This result suggested that a device's hemocompatibility signature is independently related to each specific device regardless of the pump type [22].

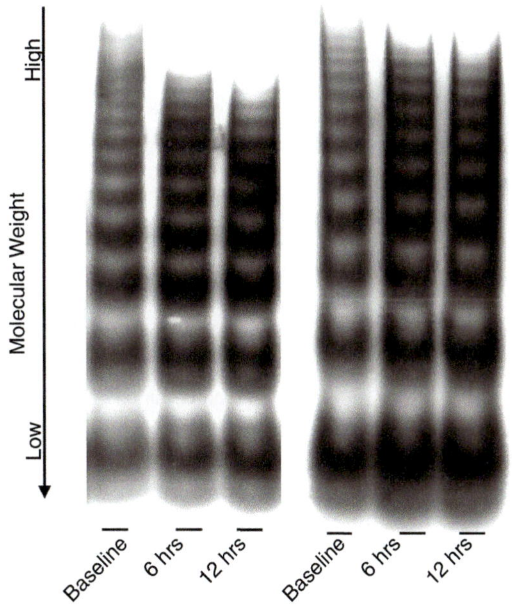

Fig. 32.5 von Willebrand factor large multimer degradation (EVAHEART vs HeartMate II). (Source: Bartoli et al. [20])

Double Cuff Tipless Inflow Cannula

Currently, all clinical LVADs require the insertion of a tubular type inflow cannula into the left ventricular chamber with the inflow tip protruding inside the chamber above the endocardial plane. This creates a hemodynamic obstacle that interferes with blood washout around the space between the cannula wall and the endocardium resulting in thrombus formation, a so-called wedge thrombus. Additionally, none of these inflow cannula designs isolate the healing (clotting) myocardium by securing the endocardium directly adjacent to the endocardium/inflow tip interface. To avoid this issue, inner and/or outer cannula surfaces are texturized to promote tissue ingrowth (endothelialization). The endothelial-

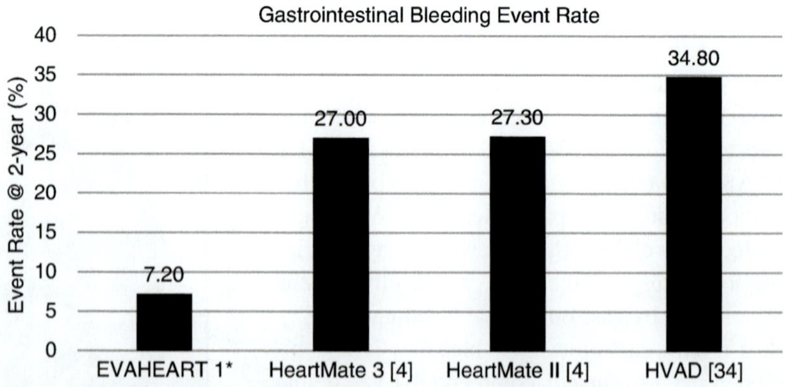

*Data obtained from the Japanese Registry for Mechanically Assisted Circulatory Support (J-MACS)

Fig. 32.6 Gastrointestinal (GI) bleeding profile with EVAHEART, HeartMate II, HeartMate 3, and HVAD

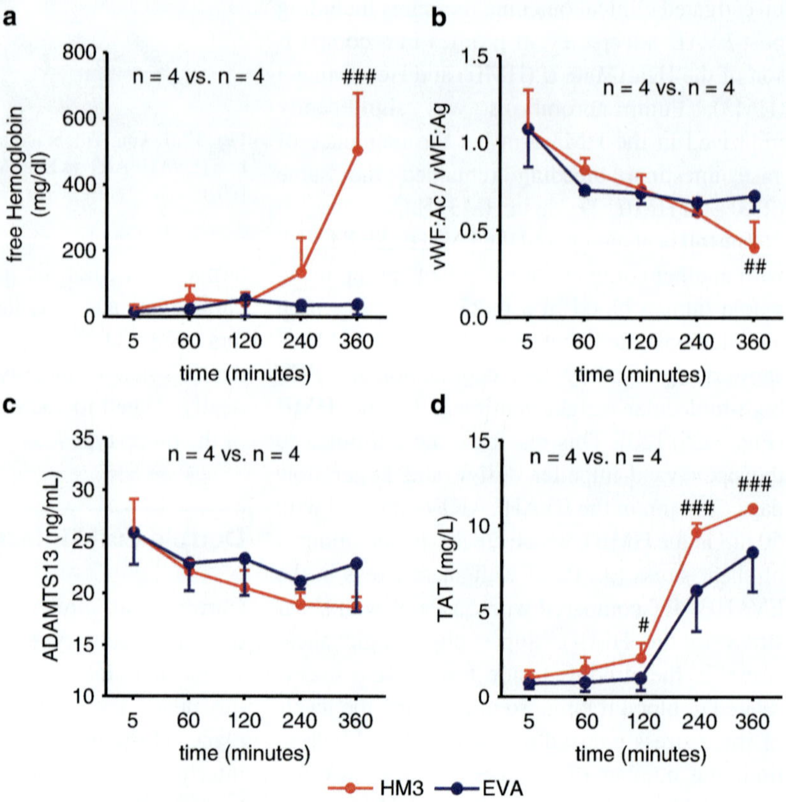

Fig. 32.7 In vitro hemocompatibility study results (EVAHEART vs HeartMate 3) time course of (**a**) free plasma hemoglobin, (**b**) von Willebrand factor (vWF) activity and the vWF antigen ratio (vWF:Ac/vWF:Ag), (**c**) metalloprotease concentration (ADAMTS13), and (**d**) thrombin-antithrombin (TAT) complex under optimal revolutions per minute (RPM) settings. (Source: Zayat et al. [22])

ized foreign surface is more anti-thrombogenic, but in the process of forming complete endothelialization, initial blood-textured surface interaction takes place through the deposition of cells and proteins, such as platelets (white clots), red blood cells (red clots), and fibrin. In the early phase after LVAD implantation (i.e., 14–30 days), these depositions, which anchor onto the textured surface, are friable and can be released into the bloodstream. INTERMACS data suggests that the hazard of neurological events is predominantly increased in the 1–2 months post-LVAD implantation, gradually tapers down, and then plateaus [5].

Malposition of the inflow cannula tip can also occur during LVAD implantation as a result of technical (surgical) failure or migration of the cannula after cardiac remodeling. Increased angulation of the inflow cannula as a result of malposition has been shown in in vitro studies to decrease regional flow velocity, pulsatility, and vortex energy in the left ventricular chamber, which compromises ventricular washout and promotes areas of stagnation that are associated with an increased risk of thrombosis [34]. A malpositioned inflow tip touching the myocardial wall (endocardium in the left ventricle) can trigger thrombus, generation of microemboli, and pannus formation that grows over the inflow tip, which can also contribute to post-LVAD stroke [35–37]. Malposition of the inflow cannula can also result in ventricular wall suction, which increases the risk of arrhythmia and low pump flow, followed by possible pump thrombosis [38]. Recent clinical data indicated a high incidence of pump thrombosis for HMII [39, 40] and a high incidence of stroke for the HVAD [41].

Evaheart, Inc. has developed a double-cuff tipless (DCT) inflow cannula that prevents protrusion into the left ventricular chamber and eliminates myocardial exposure to blood components by moving the hemostatic area to the endocardial surface [42]. The potential technological advantages of the EVAHEART DCT inflow cannula compared with current clinical LVADs (e.g., HMII, HVAD) are presented in Fig. 32.8. The

	HeartMate II Type cannula	HVAD Type cannula	Evaheart DCT* cannula
Appearance of inflow cannula	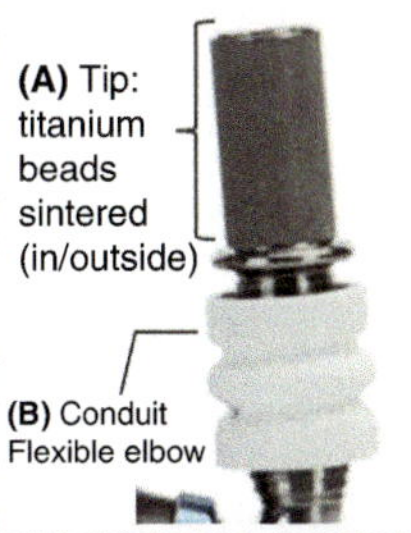	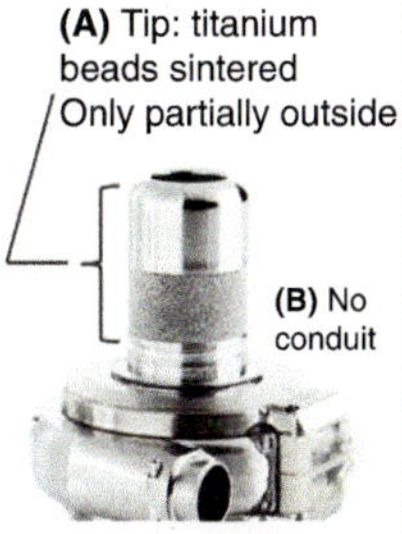	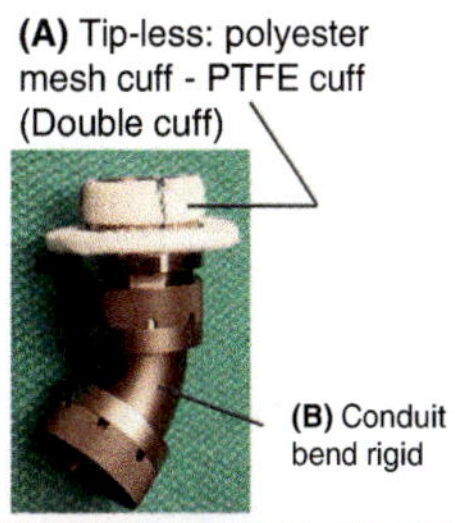
Intra ventricular position	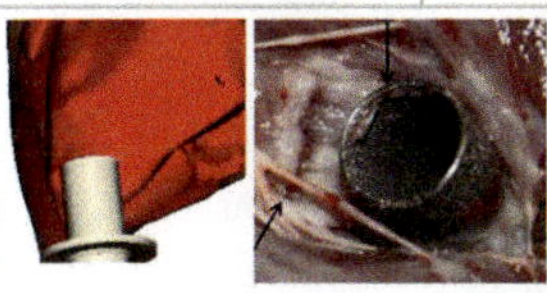	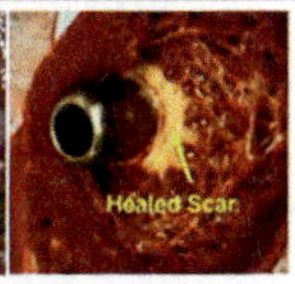	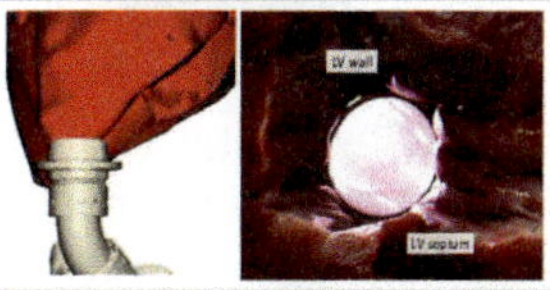
Textured surface (outer surface area)	Titanium beads (5 cm²)	Titanium beads (4.7 cm²)	Knitted polyester mesh (1.9 cm²)
Wedge thrombus risk [34] (no malposition)	+	+	- (no wedge area)
Wedge thrombus risk (with malposition) [34,35]	++	+++	- (no wedge area)
Pannus formation risk [35]	+++	+++	- or minimum
Ventricular wall suction risk [35]	+ (flexible conduit)	++	-
Low flow/pump thrombosis risk	++ (clinical data) [43]	++ (clinical data) [43]	-
*DCT: Double Cuff Tip-less			

Fig. 32.8 Potential technological advantages of the EVAHEART double-cuff tipless (DCT) inflow cannula compared with current left ventricular assist devices

- No cannula tip protrusion
- Proprietary double cuff design
- Lower risk of wedge thrombus
- Quick endothelialization
- Forgiving against inflow malposition
- Lower risk of ventricular wall suction
- **May contribute to lower stroke rate**

Fig. 32.9 The benefits of the EVAHEART double-cuff tipless (DCT) inflow cannula

DCT design consists of a polished and MPC-coated titanium conduit for blood drainage (20-mm outer diameter, 10-mm height), an implantable-grade polyester surgical mesh for approximating the cored apical myocardium, an implantable-grade PTFE suture cuff, and an O-ring sealed connector. The unique double cuff design eliminates protrusion of the inflow tip into the ventricle, which is a possible root cause of wedge thrombus formation and inflow misalignment (Fig. 32.9). In addition, a recent study performed by May-Newman et al. found that eliminating the cannula protrusion into the left ventricle helps to minimize areas of flow stasis and improves intraventricular hemodynamics, including intraventricular pulsatility compared with a protruding cannula [43]. The wide distal suture cuff helps to buffer against a range of myocardial thickness from a thin apex wall to a hypertrophic myocardium. The DCT inflow cannula is implanted by a surgical technique similar to that for placement of a mechanical aortic valve under cardiopulmonary bypass support. After coring the left ventricular apex, 12 mattress sutures are placed, followed by a circumferential running suture to secure the myocardial tissue and suture mesh integrity [44].

EVAHEART Benefits Summary

The EVAHEART 2 LVAS with the DCT inflow cannula has been designed to have remarkable hemocompatibility in order to reduce the incidence of adverse events, especially hemocompatibility-related adverse events, in LVAD patients. The EVAHEART system can provide patients with a more physiological flow by not attenuating their native heart pulsatility, by providing superior hemocompatibility, and by reducing blood trauma, with the goal of improving patient survival and reducing the burden of associated medical care costs.

Acknowledgment We thank Mr. Brian Kelly (Clinical Support Specialist of Evaheart, Inc.) for his assistance with the manuscript preparation and Ms. Jennifer Holmes for reviewing comments that greatly improved the manuscript.

References

1. Kirklin JK, Pagani FD, Kormos RL, Stevenson LW, Blume ED, Myers SL, et al. Eighth annual INTERMACS report: special focus on framing the impact of adverse events. J Heart Lung Transplant. 2017;36(10):1080–6.
2. Mehra MR, Goldstein DJ, Uriel N, Cleveland JC Jr, Yuzefpolskaya M, Salerno C, et al. Two-year outcomes with a magnetically levitated cardiac pump in heart failure. N Engl J Med. 2018;378:1386–95.
3. Baras Shreibati J, Goldhaber-Fiebert JD, Banerjee D, Owens DK, Hlatky MA. Cost-effectiveness of left ventricular assist devices in ambulatory patients with advanced heart failure. JACC Heart Fail. 2017;5(2):110–9.
4. Akhter SA, Badami A, Murray M, Kohmoto T, Lozonschi L, Osaki S, et al. Hospital readmissions after continuous-flow left ventricular assist device implantation: incidence, causes, and cost analysis. Ann Thorac Surg. 2015;100:884–9.
5. Kirklin JK, Naftel DC, Pagani FD, Kormos RL, Stevenson LW, Blume ED, et al. Seventh INTERMACS annual report: 15,000 patients and counting. J Heart Lung Transplant. 2015;34:1495–504.
6. Kwon MH. Cost reduction in left ventricular assist device therapy: an elusive paradigm. J Thorac Cardiovasc Surg. 2018;155:2469–70.
7. Rogers JG, Pagani FD, Tatooles AJ, Bhat G, Slaughter MS, Birks EJ, et al. Intrapericardial left ventricular assist device for advanced heart failure. N Engl J Med. 2017;376(5):451–60.
8. Hutcheson IR, Griffith TM. Release of endothelium-derived relaxing factor is modulated both by frequency and amplitude of pulsatile flow. Am J Physiol. 1991;261:H257–62.
9. Markham DW, Fu Q, Palmer MD, Drazner MH, Meyer DM, Bethea BT, et al. Sympathetic neural and hemodynamic responses to upright tilt in patients with pulsatile and nonpulsatile left ventricular assist devices. Circ Heart Fail. 2013;6(2):293–9.
10. Ootaki C, Yamashita M, Ootaki Y, Kamohara K, Weber S, Klatte R, et al. Reduced pulsatil-

ity induces periarteritis in kidney: role of the local renin-angiotensin system. J Thorac Cardiovasc Surg. 2008;136(1):150–8.

11. Soucy KG, Koenig SC, Giridharan GA, Sobieski MA, Slaughter MS. Defining pulsatility during continuous-flow ventricular assist device support. J Heart Lung Transplant. 2013;32(6):581–7.
12. Miller L. We always need a pulse, or do we?? J Cardiovasc Transl Res. 2012;5:296–301.
13. Kirklin JK, Naftel DC, Pagani FD, Kormos RL, Myers S, Acker MA, et al. Pump thrombosis in the Thoratec HeartMate II device: an update analysis of the INTERMACS Registry. J Heart Lung Transplant. 2015;34(12):1515–26. https://doi.org/10.1016/j.healun.2015.10.024.
14. Kazui T, Zhang A, Greenberg J, Itoh A, Tran PL, Keith AD, et al. Left ventricular assist device inflow angle and pump positional change over time adverse impact on left ventricular assist device function. Ann Thorac Surg. 2016;102:1933–40.
15. Ong C, Dokos S, Chan B, Lim E, Al Abed A, Osman N, et al. Numerical investigation of the effect of cannula placement on thrombosis. Theor Biol Med Model. 2013;10:35.
16. Wong KC, Busen M, Benzinger C, Gang R, Bezema M, Greatrex N, et al. Effect of rotary blood pump pulsatility on potential parameters of blood compatibility and thrombosis in inflow cannula tips. Int J Artif Organs. 2014;37:875–87.
17. Hanke JS, Krabatsch T, Rojas SV, Deniz E, Ismail I, Martens A, et al. In vitro evaluation of inflow cannula fixation techniques in left ventricular assist device surgery. Artif Organs. 2017;41:272–5.
18. Chiu WC, Alemu Y, McLarty AJ, Einav S, Slepian MJ, Bluestein D. Ventricular assist device implantation configurations impact overall mechanical circulatory support system thrombogenic potential. ASAIO J. 2017;63:285–92.
19. Yamane T, Nishida M, Kawamura H, Miyakoshi T, Yamazaki K. Flow visualization for the implantable ventricular assist device EVAHEART. J Artif Organs. 2013;16(1):42–8.
20. Bartoli CR, Kang J, Zhang D, Howard J, Acker M, Alturi P, et al. Left ventricular assist device design reduces von Willebrand factor degradation: a comparative study between the HeartMate II and the EVAHEART left ventricular assist system. Ann Thorac Surg. 2017;103(4):1239–44.
21. Kang J, Zhang DM, Restle DJ, Kallel F, Acker MA, Alturi P, et al. Reduced continuous-flow left ventricular assist device speed does not decrease von Willebrand factor degradation. J Thorac Cardiovasc Surg. 2016;151(6):1747–54.
22. Zayat R, Moza A, Grottke O, Grzanna T, Fechter T, Motomura T, et al. In vitro comparison of the hemocompatibility of two centrifugal left ventricular assist devices. J Thorac Cardiovasc Surg. 2019;157(2):591–9.
23. Yamazaki K, Saito S, Kihara S, Tagusari O, Kurosawa H. Completely pulsatile high flow circulatory support with a constant-speed centrifugal blood pump: mechanisms and early clinical observations. Gen Thorac Cardiovasc Surg. 2007;55(4):158–62.
24. Imamura T, Kinugawa K. Centrifugal pump EVAHEART prevents development of aortic insufficiency preserving pulse pressure. Int Heart J. 2016;57(1):127–8.
25. Fine NM, Park SJ, Stulak JM, Topilsky Y, Daly RC, Pereira NL, et al. Proximal thoracic aorta dimensions after continuous-flow left ventricular assist device implantation: longitudinal changes and relation to aortic valve insufficiency. J Heart Lung Transplant. 2016;35(4):423–32.
26. Saito S, Yamazaki K, Nishinaka T, Ichihara Y, Ono M, Kyo S, et al. Post-approval study of a highly pulsed, low-shear-rate, continuous-flow, left ventricular assist device, EVAHEART: a Japanese multicenter study using J-MACS. J Heart Lung Transplant. 2014;33(6):599–608.
27. Snyder TA, Tsukui H, Kihara S, Akimoto T, Litwak KN, Kameneva MV, et al. Preclinical biocompatibility assessment of the EVAHEART ventricular assist device: coating comparison and platelet activation. J Biomed Mater Res A. 2007;81(1):85–92.
28. Kihara S, Yamazaki K, Litwak KN, Kameneva MV, Ushiyama H, Tokuno T, et al. In vivo evaluation of a MPC polymer coated continuous flow left ventricular assist system. Artif Organs. 2003;27(2):188–92.
29. Ichihara Y, Nishinaka T, Komagamine M, Yamada Y, Yamazaki K. Preservation of von Willebrand factor multimers and function in patients with an EVAHEART centrifugal-type, continuous-flow left ventricular assist device. J Heart Lung Transplant. 2017;36(7):814–7.
30. Aggarwal A, Pant R, Kumar S, Sharma P, Gallagher C, Tatooles AJ, et al. Incidence and management of gastrointestinal bleeding with continuous flow assist devices. Ann Thorac Surg. 2012;93(5):1534–40.
31. Uriel N, Pak SW, Jorde UP, Jude B, Susen S, Vincentelli A, et al. Acquired von Willebrand syndrome after continuous-flow mechanical device support contributes to a high prevalence of bleeding during long-term support and at the time of transplantation. J Am Coll Cardiol. 2010;56(15):1207–13.
32. Letsou GV, Shah N, Gregoric ID, Myers TJ, Delgado R, Frazier OH. Gastrointestinal bleeding from arteriovenous malformations in patients supported by the Jarvik 2000 axial-flow left ventricular assist device. J Heart Lung Transplant. 2005;24(1):105–9.
33. Stulak JM, Lee D, Haft JW, Romano MA, Cowger JA, Park JS, et al. Gastrointestinal bleeding and subsequent risk of thromboembolic events during support with a left ventricular assist device. J Heart Lung Transplant. 2014;33(1):60–4.
34. May-Newman K, Marquez-Maya N, Montes R, Salim S. The effect of inflow cannula angle on the intraventricular flow field of the left ventricular assist device-assisted heart: an in vitro flow visualization study. ASAIO J. 2019 Feb;65(2):139–47.

35. Taghavi S, Ward C, Jayarajan SN, Gaughan J, Wilson LM, Mangi AA. Surgical technique influences HeartMate II left ventricular assist device thrombosis. Ann Thorac Surg. 2013;96(4):1259–65.
36. Adamson RM, Mangi AA, Kormos RL, Farrar DJ, Dembitsky WP. Principles of HeartMate II implantation to avoid pump malposition and migration. J Card Surg. 2015;30(3):296–9.
37. Maltais S, Kilic A, Nathan S, Keebler M, Emani S, Ransom J, et al. PREVENtion of HeartMate II pump thrombosis through clinical management: the PREVENT multi-center study. J Heart Lung Transplant. 2017;36(1):1–12.
38. Milano CA, Simeone AA, Blue LJ, Rogers JG. Presentation and management of left ventricular assist device inflow cannula malposition. J Heart Lung Transplant. 2011;30:838–40.
39. Serious adverse events with implantable left ventricular assist devices (LVADs): FDA safety communication. http://www.fda.gov/MedicalDevices/Safety/AlertsandNotices/ucm457327.htm.
40. Starling RC, Moazami N, Silvestry SC, Ewald G, Rogers JG, Milano CA, et al. Unexpected abrupt increase in left ventricular assist device thrombosis. N Engl J Med. 2014;370:33–40.
41. Pagani FD, Milano CA, Tatooles AJ, Bhat G, Slaughter MS, Birks EJ, et al. HeartWare HVAD for the treatment of patients with advanced heart failure ineligible for cardiac transplantation: results of the ENDURANCE Destination Therapy Trial. J Heart Lung Transplant. 2015;34(4):S9.
42. Motomura T, Tuzun E, Yamazaki K, Tatsumi E, Benkowski R, Yamazaki S. Preclinical evaluation of the EVAHEART 2 centrifugal left ventricular assist device in bovines. ASAIO J. 2018;65(8):845–54. https://doi.org/10.1097/MAT.0000000000000869.
43. May-Newman K, Montes R, Campos J, Marquez-Maya N, Vu V, Zebrowski E, et al. Reducing regional flow stasis and improving intraventricular hemodynamics with a tipless inflow cannula design: An in vitro flow visualization study using the EVAHEART LVAD. Artif Organs. 2019;43(9):834–48.
44. Motomura T, Kelly B. Surgical techniques for implanting the EVAHEART 2 double cuff Tipless inflow cannula. ASAIO J. 2018;65(8):e86–9. https://doi.org/10.1097/MAT.0000000000000914.

33 A Novel Rotary Total Artificial Heart Using a Single Shuttling Impeller

Jeremy Glynn and Richard Wampler

Chapter Summary

There is an unmet need for a small, durable total artificial heart (TAH). Currently, the SynCardia TAH is the only FDA-approved device for cardiac replacement; however, it is large (400 cc displacement volume) and has limited durability. Durable left ventricular assist devices, based on rotary blood pumps, are an accepted therapy for patients with end-stage congestive left heart failure for bridging to cardiac transplantation or destination therapy, but right heart failure is a significant complication for which there is not a dedicated ventricular assist system. Right-sided heart failure can be treated with external pumps such as the CentriMag, and the HeartWare HVAD can be adapted to right heart assistance to provide biventricular assistance with a 1-year survival of 53%.

Leveraging the success of LVADs, based on rotary blood pumps, our group has developed a novel concept for a compact, durable TAH which employs a single shuttling impeller mounted within a hollow shaft to form the pump rotor. Torque is transmitted to the pump rotor by the interaction between current in an external motor stator and magnets embedded in the hollow shaft. Solenoid action causes the rotor to translate axially to act as a shuttle valve, alternately supplying blood flow to the pulmonary and systemic circulation. Rotor speed and dwell time can be independently adjusted to balance flow between the pulmonary and systemic circulation.

The results of in vitro testing of flow control have demonstrated the ability of the system to robustly balance flow. Short-term in vivo experiments have demonstrated good circulatory support, robust flow balance, and low hemolysis. Based on some limitations of the initial design, a second-generation device has been fabricated and undergone in vitro testing. Long-term in vivo tests are anticipated in the near future.

Introduction

The quest to develop a total artificial heart (TAH) was first championed by Dr. Michael DeBakey in the early 1960s and has historically been funded by the National Institute of Health (NIH). Initially, the goal was to completely replace the heart with a TAH, and, in later years, focus shifted to the development of left ventricular assist devices (LVADs). These efforts achieved limited success due, in large part, to a strategy of biomimicry which dictated that mechanical circulatory assist devices (MCAD) should emulate a physiologic pulse. Unfortunately, the engineering challenges

J. Glynn, PhD
Abbott, St. Paul, MN, USA
e-mail: jjglynn@uwalumni.com

R. Wampler, MD (✉)
Oregon Health & Science University, Portland, OR, USA

© Springer Nature Switzerland AG 2020
J. H. Karimov et al. (eds.), *Mechanical Support for Heart Failure*,
https://doi.org/10.1007/978-3-030-47809-4_33

of emulating a pulse were seriously underestimated. Prior to the development of rotary blood pumps, all MCADs leveraged some form of positive displacement pump actuated by pressurized air and, later, with electromechanical systems. Positive displacement pumps also required artificial valves to produce unidirectional flow. Consequently, the pulsatile devices were complicated, large, and not anatomically compatible with most women and smaller men.

Currently, the only FDA-approved TAH is a positive displacement device made by SynCardia which has not enjoyed widespread acceptance and has limited clinical use. This technology is an evolution of the original Jarvik 7 TAH first implanted in Barney Clark in 1982. This clinical trial was short-lived due to very poor clinical outcomes. The basic design concept of the Jarvik 7 TAH has undergone several incarnations which, ultimately, resulted in the current SynCardia. The SynCardia has demonstrated good clinical results in bridging to transplant (BTT) but has limited durability and is cumbersome with bulky peripherals and a notably audible pneumatic drive system. The longest reported time supported by the SynCardia TAH has been 4.6 years [1].

In the development of specific left circulatory support devices, the first FDA-approved LVAD was the Thermo Cardio Systems HeartMate vented electric device which demonstrated significant improvement in survival compared to medical management in the REMATCH trial [2]. However, this device was large, required abdominal placement, and proved to have a limited durability of about 2 years. The cumulative experience with positive displacement pumps, both TAH and LVAD, has demonstrated their utility as BTT devices, but their large size and limited durability proved to be insurmountable barriers to widespread clinical acceptance.

Fortunately, rotary blood pumps proved to be a far superior approach for a mechanical circulatory assist devices. Rotary pumps are significantly smaller than positive displacement pumps, have only one moving part, and have been shown to have lifespans as long as 13 years. In spite of paradigms about the need for pulsatile circulation, mammalian physiology has proven to have the robust ability to adapt to diminished and even absent pulsatility. Despite early concerns of excessive blood trauma from their high rotational speed, these devices can be designed to have very good blood handling capabilities. Though the clinical success of rotary LVADs has been remarkable, there is a significant incidence of concomitant right heart failure. Long-term treatment of biventricular failure could be better treated with a dedicated, compact, durable TAH rather than independent biventricular assist devices.

To date, continuous flow LVADs have been used off-label for biventricular assistance with two separate pumps and, rarely, used to replace the heart. These, some might say heroic, efforts of cardiac replacement have demonstrated limited clinical success, but cardiac replacement with rotary blood technology awaits the development of an integrated TAH. Leveraging the clinical success, mechanical simplicity and durability of rotary pump LVADs, three groups are pursuing development of a TAH based on rotary blood pump technology. Two of these designs, the BiVACOR and Cleveland Clinic SmartHeart, are described elsewhere in this book. In summary, these pumps are similar in that they use two impellers, one for the pulmonary and another for the systemic circulation, which are coupled on the same driveshaft. The primary difference between these designs is the means of suspension. The BiVACOR pump employs complete magnetic suspension, while the SmartHeart utilizes blood lubricated hydrodynamic journal bearings for radial support. Both are small enough to fit in a large percentage of the population, have good blood handling properties, and have been successfully implanted long term in bovines. The fact that the two impellers are on the same shaft means the left and right impeller speeds are necessarily the same. Consequently, there may be limitations in ensuring the TAH can achieve normal flows in all physiologic states, such as severe pulmonary hypertension or other states with significantly abnormal conditions primarily affecting one branch of the circulatory system. In collaboration with the Knight Cardiovascular Institute at Oregon Health & Science University (OHSU), we have developed a novel TAH design which overcomes these limitations [3]. A spinoff company, Orcorazon, Inc., is located in California where a working prototype was developed.

The Orcorazon TAH employs a single moving part that inherently produces physiologically pulsatile flow. This novel design consists of a single impeller on a moving rotor that shuttles to alternately pump blood to the systemic and pulmonary circulations. In comparison to the BiVACOR and SmartHeart with coupled pulmonary and systemic impellers, the Orcorazon TAH is functionally completely uncoupled and capable of providing adequate flow over a wide range of physiologic conditions. This independence has potential to accommodate patients with a very high systemic vascular resistance (peripheral artery disease, hypertension), high pulmonary vascular resistance (pulmonary hypertension), low systemic blood pressure (hypovolemia), and many other physiologic conditions simply by altering TAH settings. In addition, the flow produced is inherently pulsatile. Though the clinical experience with continuous flow LVADs has demonstrated that a pulse is not necessary to sustain life, aberrant tissue remodeling has been observed with these devices, and the lack of a pulse can be disconcerting to the patient and caregivers. In contrast, maintaining pulsatility has no known disadvantage. Following is a description of the TAH design and the results of in vitro testing in a mock circulatory loop [3] and short-term in vivo experiments.

TAH Design

The Orcorazon TAH pump (Fig. 33.1) is comprised of an exterior housing and a single internal rotating assembly (the rotor). The left and right

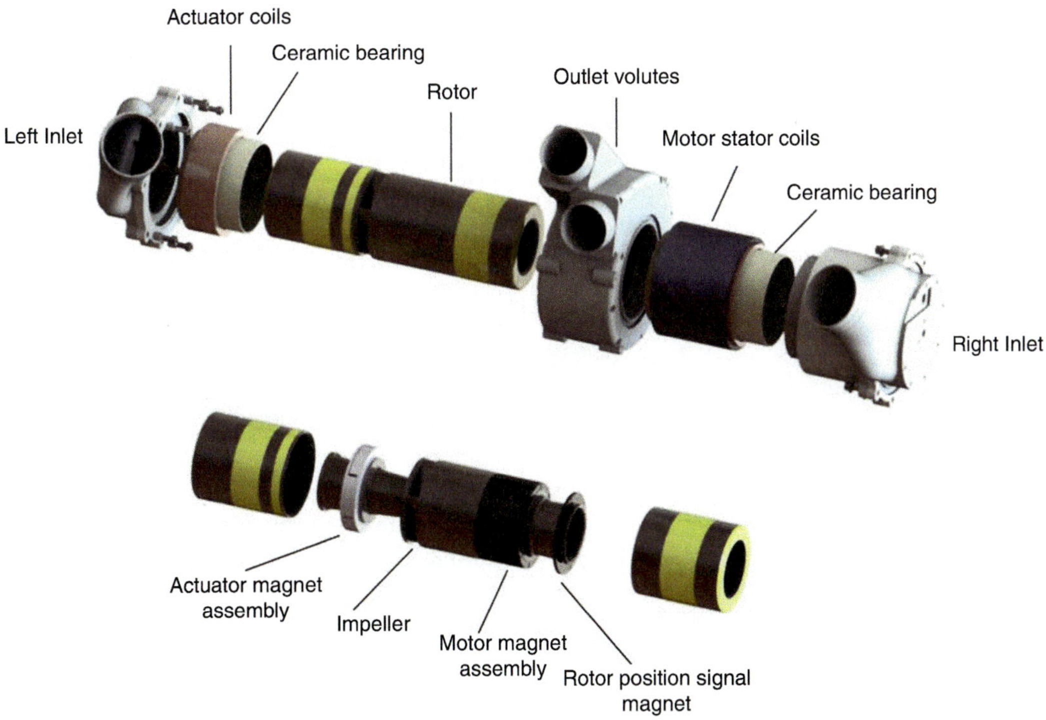

Fig. 33.1 Exploded view of the first-generation Orcorazon total artificial heart (TAH). (TOP) TAH housing. Motor and actuator coils in the TAH housing apply torque and axial force, respectively, to the rotor. Ceramic sleeves line the TAH housing and act as the bearing surfaces. (BOTTOM) TAH rotor. The cylindrical rotor has a single impeller and a hollow bore allowing inflow from either inlet. The bearings on the rotor are coated with titanium nitride (yellow). The actuator and motor magnet assemblies interact with the actuator and motor coils, respectively

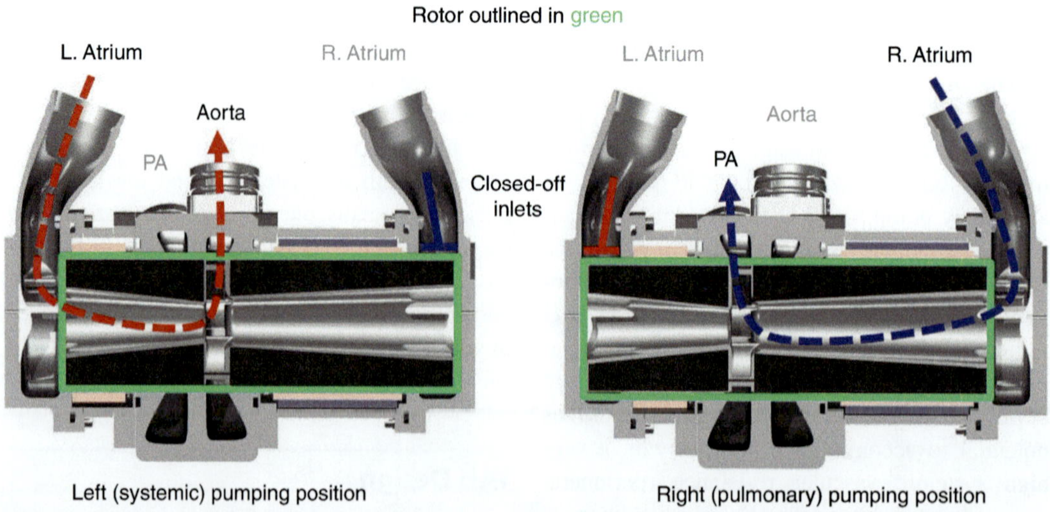

Fig. 33.2 Flow paths in Orcorazon total artificial heart. The rotor is outlined in green within the pump housing, and the primary flow paths are shown with dotted arrows. In the left pumping position, blood flows in the left inlet, through the bore of the rotor, and out to the aorta. Importantly, when the rotor is in this position, flow from the right inlet is blocked by the body of the rotor to exclude inflow from the right atrium. In the right pumping position, the reverse is true. *PA* pulmonary artery

inlets of the TAH housing connect to the left and right atria, respectively, and the outflows connect to the aorta and pulmonary artery via short sections of graft material. The housing contains motor stator coils for applying torque to the rotor, as well as actuator coils to apply axial force to the rotor. The rotor consists of a hollow shaft with centrally located impeller blades and is suspended on blood lubricated hemodynamic bearings. Thus, there is no mechanical contact between the rotor and housing while the TAH is running, resulting in a highly durable design. Because the internal bore is continuous along the full length of the rotor, the impeller can draw blood from either inlet depending on the axial position of the rotor within the housing. To prevent mixing blood between the left and right circulations, the right inlet is closed off by the rotor body when the rotor is in the left pumping position and vice versa. The actuator shuttles the position of the rotor according to a trapezoidal wave function, resulting in brief transit times and longer pumping durations. Shuttling between the left and right pumping positions delivers alternating blood flow to the systemic and pulmonary circulatory systems without requiring valves. The flow paths through the TAH are shown in Fig. 33.2.

The shuttling nature of the Orcorazon TAH provides novel options to control blood flow rates and left/right circulatory balance, as rotor speeds for the left and right pumping positions are independent. The proportion of each pumping cycle that the TAH spends pumping to the left versus the right is also adjustable. Finally, the frequency at which the prototype TAH could shuttle the rotor between the left and right pumping positions can be varied. Extensive bench testing was performed with multiple units of a first generation prototype design. With these prototypes, rotor speeds could be varied from 2000 to 4000 revolutions per minute (RPM) for both left and right pumping positions, and beat rates could be adjusted from 0 (holding on one side) to 70 beats per minute (BPM), where a “beat” is considered as moving the rotor from the left pumping position to the right and back again.

Hydraulic Performance

A prototype Orcorazon TAH (Fig. 33.3) was connected to a bench top loop consisting of two inflow conduits and two outflow conduits that all connected to a shared reservoir filled with a blood analog (40% glycerol in water, viscosity 3.1 cP at 23.8C, Fig. 33.4). Pressure transducers (Utah Medical, Midvale, UT, USA) were inserted into the four inflow and outflow conduits, and flow probes (Transonic, Ithaca, NY, USA) were placed on both outflows. Automated steppers were used to constrict the outflow tubing to adjust outlet pressure. Hydraulic performance was quantified by holding the rotor in either the left or right pumping position, running the TAH at a specific RPM, and recording the resulting pressures and flow rates as the automated stepper gradually constricted the outflow conduit. Hydraulic performance at different rotor speeds for both the left and right flow positions is shown in Fig. 33.5. Because the Orcorazon TAH has a single impeller for both the left and right circulatory systems, hydraulic performance for both flow paths is very similar. Therefore, achieving comparable flow rates at the different physiologic conditions presented by the pulmonary and systemic circulations is primarily achieved by using a higher rotor speed and duty cycle for the higher-pressure systemic circulation.

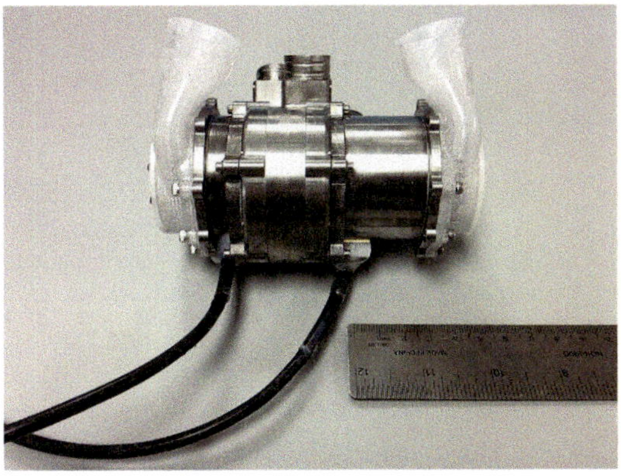

Fig. 33.3 Prototype total artificial heart (TAH) used for mock circulatory loop testing. A prototype TAH was constructed with titanium outlet volutes and housings for the stator and actuator coils. The outlets were 20 mm in diameter. A ceramic sleeve lined the TAH interior housing for the bearing surface. The inlets were 3D-printed polycarbonate with a 25.4 mm orifice. This first-generation prototype is 130 mm long, 90 mm high, with a diameter of 62 mm at its widest point, and a displacement volume of approximately 200 cc

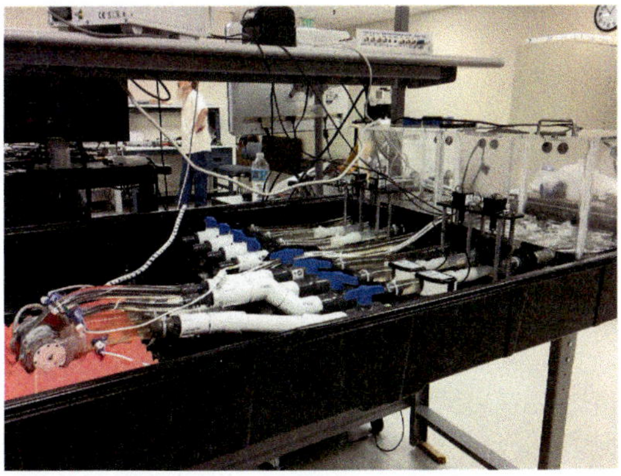

Fig. 33.4 Benchtop testing setup used to evaluate hydraulic performance of the Orcorazon TAH. Automated steppers controlled the resistance of the left and right flow paths. Inflow and outflow rates and pressures were simultaneously recorded to produce pressure-flow (HQ) curves

In Vitro Testing in a Mock Circulatory Loop

A mock circulatory loop was used to evaluate TAH performance across a range of hemodynamic conditions. The loop consisted of two independent circuits for the systemic and pulmonary circulations. Each circuit contained a compliance chamber immediately distal to the pump and an electromechanical cuff to apply resistance. Four reservoirs with variable fluid heights corresponded to the pressures in the left atrium, right atrium, aorta, and pulmonary artery. A glycerol-based blood analogue was used to simulate the viscosity of blood. The resistances of the pulmonary and systemic circuits were varied to simulate a range of normal and pathophysiologic conditions.

Under normal conditions with a mean arterial pressure (MAP) of 92 mm Hg, the TAH provided an average 7.4 L/min systemic blood flow. The peak flow rates to the left and right circulations were 17 and 24 L/min, respectively. Arterial pressures were pulsatile at approximately 60 beats

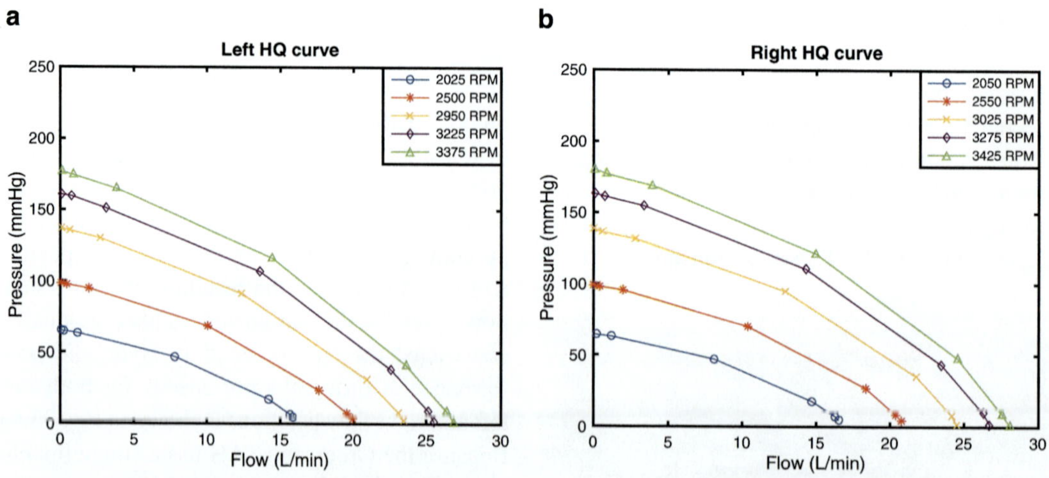

Fig. 33.5 Hydraulic performance of the Orcorazon TAH. Pressure-flow (HQ) curves for the Orcorazon TAH demonstrate flow rates achieved at specific rotor speeds against various pressures. The hydraulic performances for the left (**a**) and right (**b**) circulations are very similar due to symmetric inlet designs and a shared impeller

Table 33.1 Summary of mock circulatory loop studies. Reprinted from Glynn et al. [3]. All values listed are the mean ± standard deviation for each condition

	Normal	High flow	Systemic hypertension	Pulmonary hypertension
Left flow (L/min)	7.4 ± 6.3	9.9 ± 8.7	6.1 ± 5.0	7.4 ± 6.1
Right flow (L/min)	8.3 ± 10.3	11.1 ± 11.8	7.3 ± 7.8	7.5 ± 8.2
AoP (mmHg)	92.2 ± 15.0	99.6 ± 21.4	121.6 ± 12.9	93.5 ± 14.3
Sys/Dia (mmHg)	120/75	130/70	140/100	120/75
PAP (mmHg)	21.8 ± 8.5	24.8 ± 9.8	20.1 ± 7.8	56.2 ± 8.2
Sys/Dia (mmHg)	40/10	40/10	35/10	70/45
LAP (mmHg)	12.9 ± 1.4	11.0 ± 1.4	12.6 ± 1.6	8.9 ± 1.2
RAP (mmHg)	2.7 ± 1.6	2.8 ± 1.6	2.5 ± 1.7	2.4 ± 1.5

AoP aortic pressure, *sys/dia* systolic/diastolic, *PAP* pulmonary artery pressure, *LAP* left atrial pressure, *RAP* right atrial pressure, *RPM* revolutions per minute

per minute, with peak systolic pressure of 120 mmHg and diastolic pressure of 75 mmHg. Pulmonary pressures pulsed between 10 and 40 mmHg. By increasing the TAH rotor speed for both left and right pumping positions, a cardiac output of 10 L/min was achieved against a MAP of 100 mmHg. Pulmonary hypertension was simulated by raising the pulmonary vascular resistance to 7.5 Wood units. By increasing the rotor speed while pumping to the right side, cardiac output of 7.4 L/min was maintained with normal atrial pressures of 8.9 and 2.4 mmHg for the left and right atria, respectively. The ability to use different rotor speeds for the systemic and pulmonary circulations, as well as the ability to adjust the fraction of each cardiac cycle pumping to each circulation, yielded considerable versatility to accommodate a wide range of hemodynamic conditions (Table 33.1).

The mock circulatory loop studies confirmed that the TAH motor design could produce pulsatile flow rates sufficient to support adult physiology across a range of normal and pathophysiologic pressures. Pulse pressures were comparable to those produced by a healthy native heart. Mean atrial pressures were maintained within normal ranges of 2–3 mm Hg for the right atrium and 8–15 mm Hg for the left atrium throughout all test conditions. Generally, mean flow rates measured at the left and right outlets were not identical. This difference is largely due to the secondary flow paths through the TAH, which were intended to provide lubrication to the bearings. The direction of the secondary flow is from high to low pressures, meaning a portion of the blood drawn from the left atrium and intended to be pumped to the aorta will be shunted to the pulmonary artery. This modest left to right shunt is not expected to have physiologic consequences, and atrial pressures were stable despite prolonged periods of unequal left and right flow rates. Thus, even though left and right flow rates were not identical during the mock circulatory loop studies, flows were appropriately "balanced" in that neither circulation became congested or resulted in elevated atrial pressures. These promising studies on the mock circulatory loop demonstrated a versatile TAH design with the anticipated hemodynamic performance and supported further study.

In Vivo Testing in Bovine Calves

Following successful in vitro performance using a mock circulatory loop, a series of animal studies were conducted with Dr. Mark Slaughter and team at the University of Louisville Cardiovascular Innovation Institute, using bovine calves to evaluate in vivo performance. The objective of these initial in vivo studies was to demonstrate proof of concept that the Orcorazon TAH design could provide sufficient blood flow while maintaining appropriate balance between systemic and pulmonary flow in vivo. Secondarily, the studies provided an opportunity to evaluate the pump's hemocompatibility.

These feasibility studies were conducted in three rounds, between which iterative improvements were made to the TAH geometry, implantation technique, and perioperative animal management. The first round of studies identified that the initial inlet geometry and atrial cuff materials were challenging to connect to the juvenile bovine atria, extending the duration of cardiopulmonary bypass and complicating chest closure. After these findings, modification of the TAH inlet geometry and atrial cuff fabrication significantly improved the anatomic fit and facilitated anastomoses to the atria. An additional challenge that emerged during testing was weaning the calves from cardiopulmonary bypass without overdrawing the available atrial blood volume. Because the Orcorazon TAH was alternatively pulsatile, with only a portion of each cycle pumping to either the pulmonary or systemic circulations, inflow rates were necessarily higher than comparable TAH designs with continuous flow. The TAH control software had been designed to produce approximately 60 beats per minute. However, as the TAH actively pulled blood from one atrium while also ejecting blood at a high flow rate, the relatively rapid emptying of the atria contributed to a high rate of atrial collapse events. Atrial collapse events would both choke flow and create a very low pressure at the active inlet, which challenged the actuator to maintain the rotor's position. Atrial collapse and unintended rotor translation caused by the atrial pressure differential were the most significant challenges encountered during the remaining in vivo studies.

Despite this significant challenge, the second round of studies yielded longer durations of support, with two of three animals supported for multiple days with the TAH. During these studies, perioperative management of the calves was also significantly refined, particularly regarding ventilation. The third round of studies were the culmination of early lessons learned regarding TAH geometry, atrial collapse, and perioperative animal management, and successfully supported a calf through to be alert, eating and standing

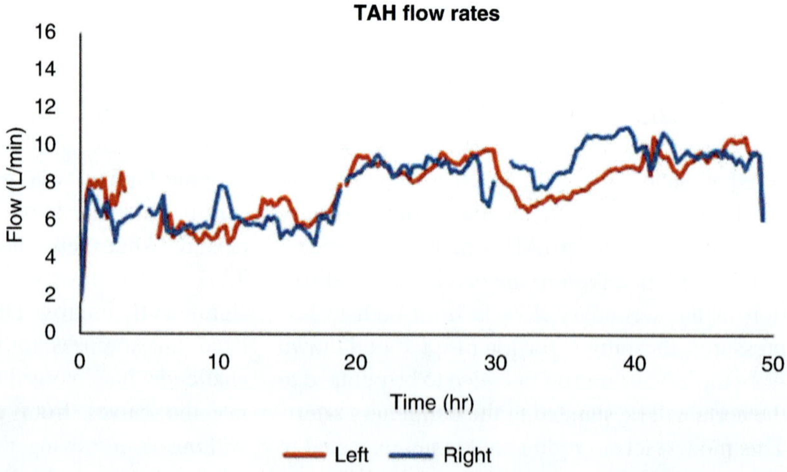

Fig. 33.6 Left and right flow rates throughout a multi-day in vivo study

with assistance. Further in vivo studies are planned but not yet undertaken.

Early lessons learned were critical in achieving postoperative recovery in these in vivo studies. When strictly evaluating the TAH pump performance, the design demonstrated the expected hemodynamic performance without hemocompatibility complications. There was a single instance of thrombus on the rotor, where a thrombus was focally located over the actuator magnets in a case where higher power was experimentally used to maintain rotor position despite the aforementioned atrial collapse events. This higher power expenditure translated into a higher operating temperature of the actuator and likely was responsible for the thrombus deposition. Outside of this instance, no other significant thrombus deposition or hemolysis was observed during the studies. Representative left and right flow rates are shown in Fig. 33.6.

The ability of the TAH to appropriately balance flow between the left and right circulatory systems that was observed in vitro was likewise observed in this in vivo experience. Passive flow balance is a notable achievement of the Orcorazon TAH design and distinct from other TAH designs that require frequent adjustments to maintain balanced left and right flows. The secondary flow paths used to lubricate the hemodynamic bearings, as well as the brief period during shuttling between flow paths, likely help to alleviate pressure imbalances between the atria even if flow rates are not exactly matched between left and right circulations.

In summary, these animal studies demonstrated that the TAH achieved proof-of-concept success criteria by providing sufficient hemodynamic support to support calf recovery post-surgery. During these in vivo studies, the TAH demonstrated flow rates of 7–10 L/min appropriately balanced between left and right circulations with minimal operator adjustment, providing sufficient hemodynamic support for calf recovery. No changes were required to be made to the TAH stator, rotor, or actuator design over the course of these studies, demonstrating a robust TAH design. The most significant challenge experienced during in vivo studies was a propensity for atrial collapse events due to overdrawing the available atrial volume. This scenario is less likely in human subjects with advanced heart failure and pathologically enlarged atria. Nevertheless, subsequent refinements to the TAH control software and hardware were pursued to yield significantly higher shuttle rates. Due to the alternating pulsatile design and short shuttle duration, increasing the beat rate significantly decreases stroke volume with only modest reductions in cardiac output and should directly address the challenges faced during initial in vivo studies.

Second-Generation Pump

Based on the difficulty with atrial collapse and anatomic fit of the pump inlets to atria, significant changes were made to the design. The main goals of the redesign were increase the cycling frequency to at least 200 beats per minute to alleviate atrial collapse events observed weaning from cardiopulmonary bypass and to decrease the axial length of the pump to facilitate anatomically favorable anastomoses with the atria.

The distance between the inlets was significantly reduced though a number of assembly changes to the TAH design. The translation length of the rotor was shortened by decreasing the axial height of the impeller. The overall length of the pump was further reduced by shortening the journal bearing length, incorporating welded junctions between pump components, and designing a shorter motor. Cumulatively, these changes resulted in a decreased axial length from 12.5 to 9.8 cm, which significantly reduced the distance between the inlet orifices to improve anatomic compatibility.

Additionally, shuttle frequency was increased to reduce susceptibility to atrial collapse events. Increasing the shuttling frequency of the pump necessitated a redesign of the position magnets at the end of the rotor and an updated Hall sensor array to produce a cleaner signal for the control algorithm. These changes resulted in the ability to cycle at 300 beats per minute, which was approximately five times faster than the prototypes used in the initial in vivo studies. The effectiveness of shuttle rate to alleviate overdrawing atrial volume was tested on the mock circulatory loop, which was reconfigured to incorporate mock atria to recapitulate the challenges observed in the bovine calf model. Bench testing with variable volume mock atria has demonstrated that this significant increase in shuttle rate is more than sufficient to alleviate atrial collapse issues (Fig. 33.7). Multiple units of the second-generation pump have been

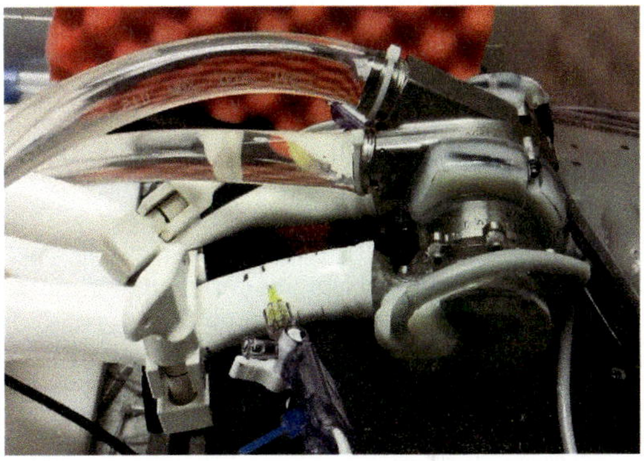

Fig. 33.7 Mock circulatory loop modifications to incorporate atrial volume. Clamps were placed on sections of compliant inflow tubing to simulate atria of various volumes. The right inlet (rear) is being drawn from, as evidenced by the evacuation of volume from the tubing proximal to the inlet

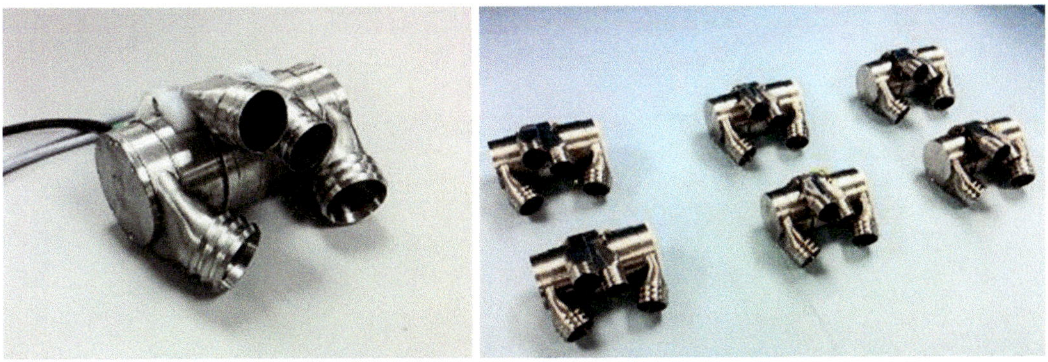

Fig. 33.8 Second-generation OHSU TAH. A single pump has a significantly smaller axial length than the first-generation design. Multiple units of the second-generation device were assembled to facilitate thorough bench and in vivo testing of the refined design

assembled (Fig. 33.8), and in vivo studies are planned to confirm these bench findings in the near future.

Conclusions

The Orcorazon TAH is a novel design that translates the clinically proven technology of rotary blood pumps to an integrated TAH design. Though rotary blood pumps have traditionally been limited to continuous flow devices, the Orcorazon TAH delivers pulsatile flow by alternating pumping between the left and right circulatory systems. As a result of this design, the Orcorazon TAH delivers pulsatile flow without needing to rapidly change the rotor speed.

All TAH designs that have progressed to clinical use have been positive displacement TAHs. In contrast to these devices, the Orcorazon TAH has no artificial valves and uses hemodynamic bearings, resulting in a very quiet, compact, and durable design that can translate to an improved patient experience. Additionally, the small size of the device will enable the TAH to treat smaller patients (adolescents, women) for which there is currently no option for long-term circulatory support. The elegant design of the Orcorazon TAH delivers fully pulsatile flow efficiently and also allows for fine-tuning support to meet physiologic demand by independently controlling the pumping parameters for the left and right circulatory systems. This feature is a unique design advantage that will hopefully lead to better patient outcomes by providing pulsatile blood flow that is optimized for a wide range of individual physiologies.

References

1. Turkish man becomes world's longest supported SynCardia temporary total artificial heart patient | SynCardia. Available at: https://syncardia.com/news/turkish-man-becomes-worlds-longest-supported-syncardia-temporary-total-artificial-heart-patient/. Accessed 13 Jan 2019.
2. Rose EA, et al. Long-term use of a left ventricular assist device for end-stage heart failure. N Engl J Med. 2001;345:1435–43.
3. Glynn J, et al. The OregonHeart Total artificial heart: design and performance on a mock circulatory loop. Artif Organs. 2017;41:904.

New Mechanical Circulatory Device: TORVAD

34

Jeffrey Gohean and Richard Smalling

Introduction

Windmill Cardiovascular Systems, Inc. has developed a new pumping paradigm to deliver synchronous, pulsatile flow with low shear. The TORVAD system consists of an implantable blood pump, an external controller, and a power source. The pump has been designed as a full-support, long-term, implantable VAD for supporting patients with end-stage heart failure and is capable of producing up to 8 L/min of blood flow with a 30 mL stroke volume (Fig. 34.1).

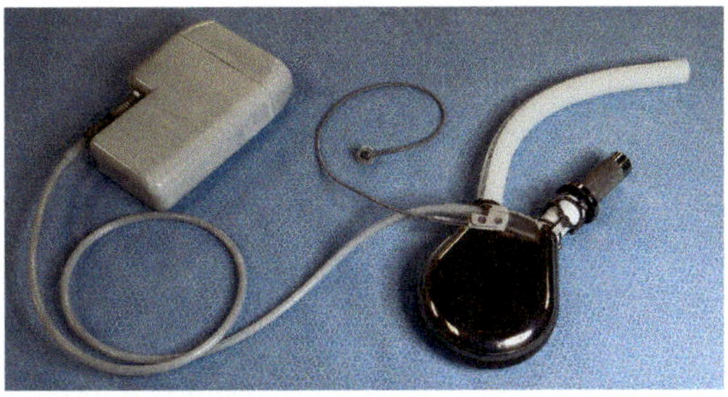

Fig. 34.1 The TORVAD system consists of the pump, magnetic inflow tip/sewing ring coupler, 14 mm outflow graft, integrated ECG sensing lead, and controller with integrated battery connected via a percutaneous driveline

Principle of Operation

The TORVAD pump consists of a toroidal shaped pumping chamber with inlet and outlet ports and two pistons moving within the chamber lumen as illustrated in the schematic in Fig. 34.2. Pumping is achieved by driving one piston around the chamber while holding the other piston between the inlet and outlet ports to act as a virtual valve. Each piston is comprised of rare earth magnets hermetically sealed within a ceramic shell and is independently actuated by a position-controlled motor that rotates a C-shaped magnetic coupling around the outer surface of the pumping chamber. After each pump stroke, the pistons exchange functional roles, achieving positive-displacement, pulsatile flow. This pumping mechanism, which simultaneously fills and ejects without the use of conventional valves, is distinctly different than first-generation sac-type pulsatile VADs, which have separate fill and ejection phases and use conventional check valves.

Various degrees of pulsatility can be achieved by actuating the drive piston at different rates, and surplus hemodynamic energy levels are esti-

J. Gohean, MSME (✉)
Windmill Cardiovascular Systems, Inc., Austin, TX, USA
e-mail: jgohean@windmillcvs.com

R. Smalling, MD, PhD
Windmill Cardiovascular Systems, Inc., Austin, TX, USA

UTHealth/McGovern Medical School, Memorial Hermann Heart and Vascular Institute, Houston, TX, USA
e-mail: rsmalling@windmillcvs.com

© Springer Nature Switzerland AG 2020
J. H. Karimov et al. (eds.), *Mechanical Support for Heart Failure*,
https://doi.org/10.1007/978-3-030-47809-4_34

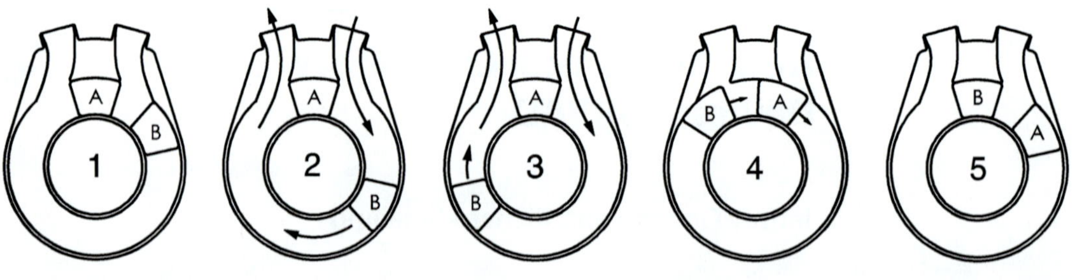

Fig. 34.2 Schematic of the TORVAD, demonstrating how pumping is achieved by independently controlling two pistons (A/B) within a toroidal pumping chamber. From rest (1), one piston (A) is temporarily held stationary between the inflow and outflow ports, while the other piston (B) rotates around the torus (2/3) simultaneously aspirating blood from the left ventricle and ejecting blood into the arterial system. At the end of each pump stroke, the pistons move together (4), exchange roles, and are ready for the next ejection (5)

mated to be 3–4 times the level generated by continuous-flow support, which attenuates the pulsatility of the circulation and has been linked to diminished end-organ perfusion, decreased cerebral oxygen saturation, increased peripheral vascular resistance, and degraded arterial viscoelasticity [1–3].

Pumping blood with this new pumping modality has three advantages: pulsatile flow, synchronous support, and low shear.

Table 34.1 The TORVAD is much smaller than first-generation sac-type pulsatile VADs and comparable in size and weight to existing continuous-flow technologies, which allows for surgical implantation in the pre-peritoneal or pericardial position

	VAD	Weight (g)	Displacement volume (ml)
Pulsatile flow	TORVAD	270	90
	XVE	1150	400
Continuous flow	HeartMate II	290	63
	HVAD	160	50
	HeartMate 3	200	80

System and Surgical Implantation

The pumping mechanism and stroke volume of the TORVAD allows it to be much smaller than first-generation pulsatile VADs (approximately 270 g and 90 cc displacement volume for the TORVAD compared to 1150 g and 400 cc displacement volume for the Thoratec XVE) and comparable to existing continuous-flow devices (290 g and 63 cc displacement volume for the Heartmate II, 200 g and 80 cc for the Heartmate 3, and 160 g and 50 cc displacement volume for the HVAD). This allows the device to be implanted in the pre-peritoneal or pericardial position (Table 34.1).

Cannulation of the ventricular apex is made quickly with a patented magnetic sewing ring coupler and cannula tip and without the use of cardiopulmonary bypass. Blood enters the pump through a reinforced, flexible, inflow cannula and exits via a 14 mm double woven velour graft that is anastomosed to either the ascending or descending aorta. The TORVAD has a standard IS-1 electrocardiogram (ECG) receptacle on the pump to connect a bipolar ECG lead attached to the epicardial surface of the right ventricle to sense ventricular depolarization. Control of the pump is achieved through a 4.7 mm diameter percutaneous driveline connected to a small controller that is worn on a belt or in a pocket.

Sensing Capabilities

The TORVAD has three physiologic sensing capabilities: epicardial ECG, pump flow, and differential pump pressure. Ventricular depolarization is measured using a standard IS-1 ECG lead attached to the epicardial surface. The lead is connected to an integrated receptacle on the pump where the signal is filtered and amplified before being sent to the controller. Controller algorithms are used to detect ventricular depolarization and continuously assess and monitor heart rate and rhythm.

Pump flow rate is known because the TORVAD is a positive displacement pump with a fixed 30 ml stroke volume. Some small leakage flow occurs in the annular gap between the piston and torus walls, but this has been measured to be only about 3%. The differential pressure measurement across the TORVAD is enabled by the dual-piston actuation mechanism of the pump. Because the pump pistons are held stationary during left ventricular contraction, pressure metrics can be determined from sensed motor currents. These metrics can provide important feedback for delivering safe therapy. This information can offer new clinical adjuncts for optimizing heart failure management and efficacy of ventricular assistance. Monitoring arterial blood pressure could also be used to adjust pump operation, providing feedback for physicians as an alert to high or low blood pressure, and for monitoring changes in systemic vascular resistance. Ventricular contractility metrics from the pressure slope (dP/dt) could allow physicians to assess cardiac recovery potential or further progression into heart failure in real time. The pressure measurement can also be used as an additional trigger for synchronization if the ECG signal deteriorates or is lost, similar to the redundant triggering capability of an intra-aortic balloon pump.

Pressure and ECG sensing combined with remote telemetry capabilities may provide clinicians with the ability to remotely monitor patient health status and proactively intervene. For instance, detection of sustained improvement of left ventricular contractility (assessed by ventricular dP/dt increase) could alert the clinician of potential myocardial recovery. This could prompt the first stage in a weaning protocol in which the left ventricular workload is incrementally increased by changing the phase of the TORVAD ejection in the cardiac cycle, by reducing the pump output by skipping beats or by performing half strokes. These options could be used to progressively increase cardiac workload to aid recovery of myocardial function, prevent atrophy, and potentially prompt removal of the device. These weaning functions can be prescribed by the patient's heart failure physician, and the physiologic impact of changing the pump function can be monitored remotely.

Synchronization

Using the integrated ECG sensing lead, the TORVAD synchronizes pump ejections to the cardiac cycle. The default synchronous operation mode is an early diastolic counterpulse. By operating the pump in this mode, the TORVAD briefly ceases to pump blood in systole, allowing the ventricle to eject normally through the aortic valve, which preserves aortic VAD flow and preload sensitivity. This preservation of physiological function is in contrast to CF pumps that pump through the cardiac cycle. The highest VAD flow rate in CF devices is during systole because flow that would normally be ejected by the left ventricle through the aortic valve is instead shunted through the VAD to the point where the aortic valve only opens sporadically or often not at all [4]. As a result, aortic valve commissural fusion and aortic insufficiency (AI) are frequently observed in adult CF VAD recipients [5–8]. Single-center incidence rates for the development of AI has been reported to range between 14.3% and 51% [9–12], and commissural fusion rates exceed 50% in some reports [13–15]. Insufficient or infrequent opening of the aortic valve can also lead to aortic root and left ventricular outflow tract thrombosis [16, 17]. Additionally, when aortic valve flow is diminished with CF support, the inherent normal autoregulation of cardiac output, the so-called Frank-Starling response of cardiac output to changes in preload and afterload, is significantly altered and nonphysiological [18, 19].

Diminished pulsatility and increased low-molecular-weight vWF fragments are also associated with the development of arteriovenous malformations in the GI tract, thereby making patients particularly susceptible to bleeding when combined with systemic anticoagulation therapy. Further, shear-induced AVWD has been causally linked to severe GI bleeding and the need for blood transfusions and surgery in 20–50% of patients [20–25].

By preserving native aortic flow with TORVAD support, patency of the native aortic valve is maintained and the autoregulatory aspect of the heart (Frank-Starling's Law) is preserved, thereby restoring the physiologic preload and afterload sensitivity of the cardiovascular system. The flow conditions during systole and diastole are illustrated in Fig. 34.3, showing the unique ability of the TORVAD to momentarily cease pumping during systole to allow the ventricle to eject and maintain native aortic valve flow. By preserving aortic flow, the TORVAD is capable of providing full hemodynamic support with approximately one-half of the VAD flow of CF devices.

Automatic Response

The system normally operates synchronously with the cardiac cycle by delivering a single 30 mL ejection in early diastole. If needed, the TORVAD can automatically pump asynchronously to deliver up to 8 L/min in the case of cardiac electrical instability.

The TORVAD can be programmed to detect arrhythmias including atrial and ventricular fibrillation and will automatically change its pumping function in response to changes in heart rhythm. In the event of ventricular fibrillation, the pump automatically changes to the asynchronous mode up to 8 L/min to provide full support. The pump controller can evaluate and record events (such as pump-induced premature ventricular contractions or suction events) and automatically adjust its pumping mode to avoid delivering too much flow, thus preventing the known problem of ventricular suction and over-pumping.

Low Shear

Blood shear in the TORVAD is minimized to a level that is orders of magnitude lower than shear induced by CF pump impellers by the combination of low pump speed (mean piston rpm is equal to the heart rate in synchronous support) and a predetermined fixed gap between the piston and torus walls. A fixed gap (approximately 75 μm) is achieved using ceramic micro hydrodynamic bearings. Ceramics have excellent durability and hemocompatibility and have been extensively and successfully used in continuous-flow VADs [26, 27] and artificial joints [28, 29]. Each piston has a chamfered ceramic groove on both the top and bottom faces that interface with chamfered ceramic rails on the top and bottom halves of the torus. The track is angled to provide hydrodynamic forces in both the axial and radial directions. The micro hydrodynamic bearing surface areas are small (approximately 250 μm in width) with gaps of approximately 50 nm. These small areas and gaps are made possible because of an effective viscosity increase of the plasma in this narrow gap that generates a so-called plasma gel layer and provides a thin protective protein buffer between the ceramic surfaces [30–32]. This phenomenon persists even during brief pauses in piston motion, thus maintaining separation and effectively eliminating wear. Dynamic start-stop experiments have confirmed bearing life in excess of 50 years, well beyond the expected life of the device.

There are numerous implications of substantially lower shear and reduced blood trauma compared to currently available devices. In vitro and chronic animal testing has demonstrated that this design has very low hemolysis and does not deplete high-molecular-weight vWF, which could reduce episodes of serious bleeding complications that, as noted earlier, occur in up to 40% of CF VAD patients [33]. Additionally, low-molecular-weight vWF fragments are significantly reduced with the TORVAD compared to CF LVADs, which might reduce the risk of formation of GI arteriovenous malformations and subsequent GI bleeding. Other potential benefits of low shear pumping with the TORVAD include decreased hemolysis, decreased platelet activation with associated pump thrombosis and embolic stroke, and preservation of white blood cell function that could lead to fewer infections.

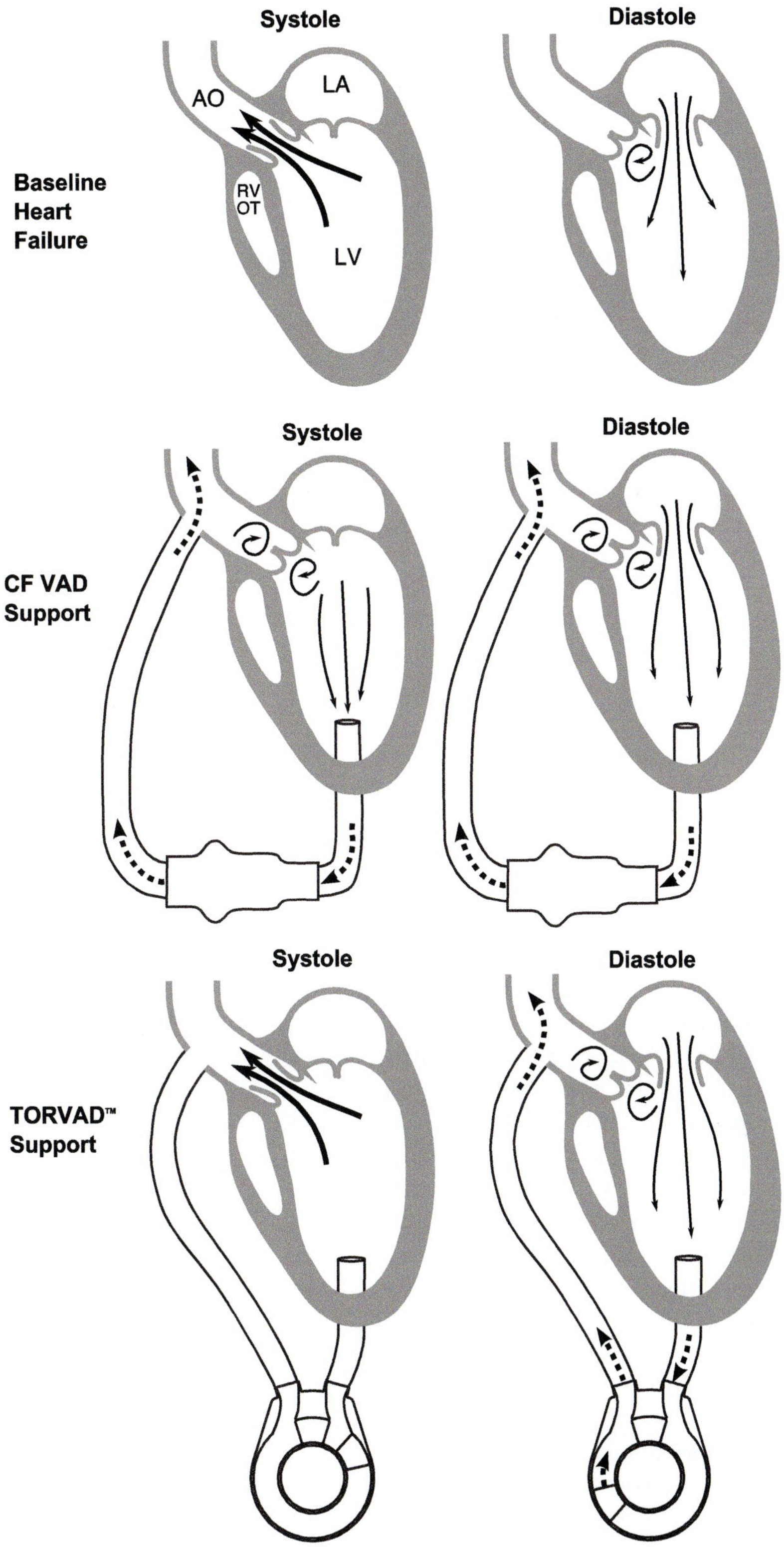

Fig. 34.3 Illustration of continuous flow (CF) and TORVAD support in heart failure. The CF VAD pumps continuously through the cardiac cycle, often eliminating native aortic valve flow during systole. The TORVAD ceases pumping during systole and allows native ejection through the aortic valve, providing full hemodynamic support with only half the VAD flow rate

Pediatric Pump and Platform Technology

The TORVAD system can be scaled for different patient sizes and different patient needs. A 15 ml pediatric TORVAD for children with a body surface area between 0.6 and 1.5 m^2 has been developed. The device was scaled using a variety of computation tools including cardiovascular system modeling, heat transfer analysis, motor and magnetic coupling finite element analysis, and computational fluid dynamics [34]. This analysis provided two important results: low shear is preserved in scaling the TORVAD, and the size of the device scales with the pump stroke volume (e.g., the 15 ml pump has approximately half the displacement volume and weight as a 30 ml pump).

Scaling the TORVAD for pediatrics has established the TORVAD as a platform technology. Pumps with different stroke volumes from 5 to 40 mL can be developed to support different patient sizes (from infants to large adults) as well as different patient needs (end-stage heart failure, partial support for less sick patients, biventricular support, or as a total artificial heart).

Conclusion

The TORVAD is a new pumping paradigm with the potential to significantly improve the treatment of HF using VADs. Synchronization of the pump function with the native heart rhythm can be used to optimize hemodynamic and cardiac mechanics. The TORVAD system has the potential to restore the use of pulsatile assist devices for long-term HF therapy and to increase cardiac recovery rates, which could lead to the use of VADs to treat earlier stages of the disease. The ability to continuously monitor pressure and contractility should also improve patient management and provide the ability to optimize medical therapy remotely. Finally, due to the absence of vWF depletion, maintenance of RV function, and non-thrombogenic pumping surfaces, the TORVAD has the potential to significantly reduce the adverse events (GI bleeding, stroke, and right heart failure) currently associated with all CF LVADs.

References

1. Ji B, Undar A. An evaluation of the benefits of pulsatile versus nonpulsatile perfusion during cardiopulmonary bypass procedures in pediatric and adult cardiac patients. ASAIO J. 2006;52(4):357–61.
2. Undar A, Henderson N, Thurston GB, Masai T, Beyer EA, Frazier OH, et al. The effects of pulsatile versus nonpulsatile perfusion on blood viscoelasticity before and after deep hypothermic circulatory arrest in a neonatal piglet model. Artif Organs. 1999;23(8):717–21.
3. Undar A, Eichstaedt HC, Frazier OH, Fraser CD Jr. Monitoring regional cerebral oxygen saturation using near-infrared spectroscopy during pulsatile hypothermic cardiopulmonary bypass in a neonatal piglet model. ASAIO J. 2000;46(1):103–6.
4. Slaughter MS, Pagani FD, Rogers JG, Miller LW, Sun B, Russell SD, et al. Clinical management of continuous-flow left ventricular assist devices in advanced heart failure. J Heart Lung Transplant. 2010;29(4 Suppl):S1–39.
5. Jorde UP, Uriel N, Nahumi N, Bejar D, Gonzalez-Costello J, Thomas SS, et al. Prevalence, significance, and management of aortic insufficiency in continuous flow left ventricular assist device recipients. Circ Heart Fail. 2014;7(2):310–9.
6. Mudd JO, Cuda JD, Halushka M, Soderlund KA, Conte JV, Russell SD. Fusion of aortic valve commissures in patients supported by a continuous axial flow left ventricular assist device. J Heart Lung Transplant. 2008;27(12):1269–74.
7. Cowger J, Pagani FD, Haft JW, Romano MA, Aaronson KD, Kolias TJ. The development of aortic insufficiency in left ventricular assist device-supported patients. Circ Heart Fail. 2010;3(6):668–74.
8. Hatano M, Kinugawa K, Shiga T, Kato N, Endo M, Hisagi M, et al. Less frequent opening of the aortic valve and a continuous flow pump are risk factors for postoperative onset of aortic insufficiency in patients with a left ventricular assist device. Circ J. 2011;75(5):1147–55.
9. Pak SW, Uriel N, Takayama H, Cappleman S, Song R, Colombo PC, et al. Prevalence of de novo aortic insufficiency during long-term support with left ventricular assist devices. J Heart Lung Transplant. 2010;29(10):1172–6.
10. Toda K, Fujita T, Domae K, Shimahara Y, Kobayashi J, Nakatani T. Late aortic insufficiency related to poor prognosis during left ventricular assist device support. Ann Thorac Surg. 2011;92(3):929–34.
11. Soleimani B, Haouzi A, Manoskey A, Stephenson ER, El-Banayosy A, Pae WE. Development of aortic insufficiency in patients supported with continuous flow left ventricular assist devices. ASAIO J. 2012;58(4):326–9.
12. Aggarwal A, Raghuvir R, Eryazici P, Macaluso G, Sharma P, Blair C, et al. The development of aortic insufficiency in continuous-flow left ventricular

assist device-supported patients. Ann Thorac Surg. 2013;95(2):493–8.
13. Rose AG, Park SJ, Bank AJ, Miller LW. Partial aortic valve fusion induced by left ventricular assist device. Ann Thorac Surg. 2000;70(4):1270–4.
14. Connelly JH, Abrams J, Klima T, Vaughn WK, Frazier OH. Acquired commissural fusion of aortic valves in patients with left ventricular assist devices. J Heart Lung Transplant. 2003;22(12):1291–5.
15. Martina JR, Schipper ME, de Jonge N, Ramjankhan F, de Weger RA, Lahpor JR, et al. Analysis of aortic valve commissural fusion after support with continuous-flow left ventricular assist device. Interact Cardiovasc Thorac Surg. 2013;17(4):616–24.
16. Demirozu ZT, Frazier OH. Aortic valve noncoronary cusp thrombosis after implantation of a nonpulsatile, continuous-flow pump. Tex Heart Inst J. 2012;39(5):618–20.
17. May-Newman K, Wong YK, Adamson R, Hoagland P, Vu V, Dembitsky W. Thromboembolism is linked to intraventricular flow stasis in a patient supported with a left ventricle assist device. ASAIO J. 2013;59(4):452–5.
18. Salamonsen RF, Mason DG, Ayre PJ. Response of rotary blood pumps to changes in preload and afterload at a fixed speed setting are unphysiological when compared with the natural heart. Artif Organs. 2011;35(3):E47–53.
19. Fukamachi K, Shiose A, Massiello A, Horvath DJ, Golding LA, Lee S, et al. Preload sensitivity in cardiac assist devices. Ann Thorac Surg. 2013;95(1):373–80.
20. Demirozu ZT, Radovancevic R, Hochman LF, Gregoric ID, Letsou GV, Kar B, et al. Arteriovenous malformation and gastrointestinal bleeding in patients with the HeartMate II left ventricular assist device. J Heart Lung Transplant. 2011;30(8):849–53.
21. Schaffer JM, Arnaoutakis GJ, Allen JG, Weiss ES, Patel ND, Russell SD, et al. Bleeding complications and blood product utilization with left ventricular assist device implantation. Ann Thorac Surg. 2011;91(3):740–7; discussion 7–9.
22. Stern DR, Kazam J, Edwards P, Maybaum S, Bello RA, D'Alessandro DA, et al. Increased incidence of gastrointestinal bleeding following implantation of the HeartMate II LVAD. J Card Surg. 2010;25(3):352–6.
23. Crow S, John R, Boyle A, Shumway S, Liao K, Colvin-Adams M, et al. Gastrointestinal bleeding rates in recipients of nonpulsatile and pulsatile left ventricular assist devices. J Thorac Cardiovasc Surg. 2009;137(1):208–15.
24. Suarez J, Patel CB, Felker GM, Becker R, Hernandez AF, Rogers JG. Mechanisms of bleeding and approach to patients with axial-flow left ventricular assist devices. Circ Heart Fail. 2011;4(6):779–84.
25. Letsou GV, Shah N, Gregoric ID, Myers TJ, Delgado R, Frazier OH. Gastrointestinal bleeding from arteriovenous malformations in patients supported by the Jarvik 2000 axial-flow left ventricular assist device. J Heart Lung Transplant. 2005;24(1):105–9.
26. Stanfield JR, Selzman CH. In vitro hydrodynamic analysis of pin and cone bearing designs of the Jarvik 2000 adult ventricular assist device. Artif Organs. 2013;37(9):825–33.
27. Sundareswaran KS, Reichenbach SH, Masterson KB, Butler KC, Farrar DJ. Low bearing wear in explanted HeartMate II left ventricular assist devices after chronic clinical support. ASAIO J. 2013;59(1):41–5.
28. Piconi C, Maccauro G. Zirconia as a ceramic biomaterial. Biomaterials. 1999;20(1):1–25.
29. Morita Y, Nakata K, Kim YH, Sekino T, Niihara K, Ikeuchi K. Wear properties of alumina/zirconia composite ceramics for joint prostheses measured with an end-face apparatus. Biomed Mater Eng. 2004;14(3):263–70.
30. Fan J, Myant C, Underwood R, Cann P. Synovial fluid lubrication of artificial joints: protein film formation and composition. Faraday Discuss. 2012;156:69–85; discussion 7–103.
31. Fan J, Myant CW, Underwood R, Cann PM, Hart A. Inlet protein aggregation: a new mechanism for lubricating film formation with model synovial fluids. Proc Inst Mech Eng H. 2011;225(7):696–709.
32. Myant C, Cann P. In contact observation of model synovial fluid lubricating mechanisms. Tribol Int. 2013;63:97–104.
33. Meyer AL, Malehsa D, Budde U, Bara C, Haverich A, Strueber M. Acquired von Willebrand syndrome in patients with a centrifugal or axial continuous flow left ventricular assist device. JACC Heart Fail. 2014;2(2):141–5.
34. Gohean JR, Larson ER, Hsi BH, Kurusz M, Smalling RW, Longoria RG. Scaling the low-shear pulsatile TORVAD for pediatric heart failure. ASAIO J. 2017;63(2):198–206.

35 BiVACOR Total Artificial Heart

Daniel L. Timms and Frank Nestler

Abbreviations

AoP	Aortic pressure
BTT	Bridge to transplant
CPB	Cardiopulmonary bypass
DT	Destination therapy
EFS	Early feasibility study
HMW, LMW	High, Low molecular weight
HQ	Head-flow (e.g., HQ-curve)
HTx	Heart transplantation
LAP	Left atrial pressure
PAP	Pulmonary artery pressure
pfHb	Plasma-free hemoglobin
RAP	Right atrial pressure
RBP	Rotary blood pump
SVC, IVC	Superior, inferior vena cava
TAH	Total artificial heart
VAD	Ventricular assist device
vWF	Von Willebrand Factor
VZP	Virtual zero power (e.g., VZP controller)

D. L. Timms, PhD (✉) · F. Nestler, PhD
BiVACOR, Inc., Houston, TX, USA
e-mail: Daniel.Timms@bivacor.com; Frank.Nestler@bivacor.com

Introduction

Many circulatory support devices are available for patients who require left ventricular support only; however a few options exist for patients who would benefit from biventricular support. Due to limited accessibility of donor organs, the optimal approach to biventricular failure is the replacement of the heart with a total artificial heart (TAH). Traditional approaches have used volume displacement pumps, which have inherent limitations due to their large size, flexing membranes, and valves which conspire to reduce the durability and lifetime of their application. These limitations were also observed throughout the early application of VAD technology, which soon prompted a transition to rotary blood pumps with suspended rotors to reduce size and dramatically increase durability. This has resulted in more than 50,000 patients supported with rotary VADs. The BiVACOR TAH aims to capitalize on the advantages of rotary blood pump technology and apply them to create a TAH that is small and reliable and has inherently balanced blood flows.

The BiVACOR® total artificial heart replaces both ventricles of a failing heart with an implantable rotary biventricular blood pump that utilizes magnetic levitation technology for increased durability (Fig. 35.1). The BiVACOR pump is implanted orthotopically after the diseased ventricles, and valves of the native heart are removed. The patient-carried system consists of the pump,

© Springer Nature Switzerland AG 2020
J. H. Karimov et al. (eds.), *Mechanical Support for Heart Failure*,
https://doi.org/10.1007/978-3-030-47809-4_35

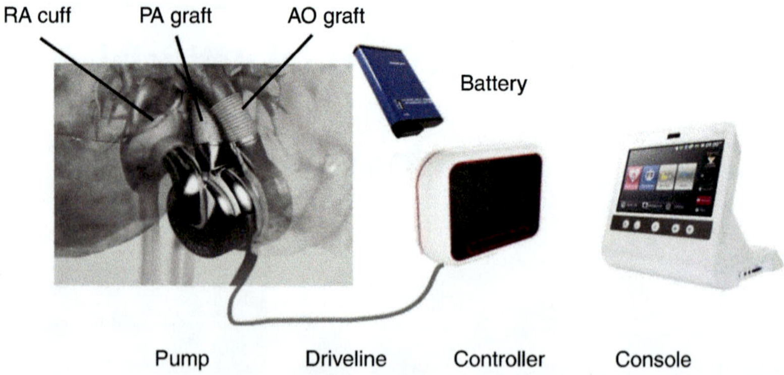

Fig. 35.1 BiVACOR TAH system; AO aorta, PA pulmonary artery, RA right atrium

the driveline, and the battery-powered controller. Trained clinical personnel will have access to advanced device settings and logs via the console. Cuffs form the neo-atria, and grafts connect the TAH to aorta and pulmonary artery.

The device is intended for patients suffering from biventricular failure, requiring heart transplantation (HTx), and is not responsive to optimal medical therapy. It shall be used to bridge the time to transplantation (BTT), supporting recovery to regain HTx-eligibility, or as long-term destination therapy (DT) application for over 10 years and, hence, will be able to be managed in an out-of-hospital setting by the patient and caregiver.

The BiVACOR pump is a single device that is capable of simultaneously supporting both left and right circulatory systems with one dual redundancy motor and one rotor. Over the last years, the BiVACOR program achieved significant development milestones toward a first-article device, and several features were incorporated for a physiological, robust, and biocompatible design (Table 35.1). Open impeller vanes were placed on either side of the rotating disc, one set to pump blood at systemic pressure to the body, the other to pump blood at a reduced pressure to the lungs. Each impeller side can simultaneously pump between 3 and 12 L/min at physiological pressures, enough cardiac output for patients resting or undergoing exercise. Long-term device durability was addressed by suspending and rotating the rotor in a magnetic field, eliminating

Table 35.1 Major user needs and features of the BiVACOR TAH; BTT, bridge to transplant; DT, destination therapy; vWF, von Willebrand factor

Major user needs	Major features
Able to operate for >2 years for BTT, > 10 years for DT	✓Single piece TAH with left and right rotary blood pumps ✓One moving part spinning between 1600–2700 RPM ✓Redundant contact-free suspension design
Suitable for men, women, and some children	✓Small size (<200 mL) and weight (<550 g)
Physiological pressures and flows	✓Inherently balances left and right flow rates ✓Total mean outflow of 3–12 L/min at physiologic pressures ✓Detection and automated response to low atrial volume
Pulsatile outflow	✓Optional speed modulation for pulsatile outflow
Low blood damage	✓Low shear stress on blood with wide clearance gaps ✓Intermitted speed reduction for device washout ✓Biocompatible materials
Redundancy	✓Redundant motor, suspension, and driveline design ✓Replaceable external controller and external driveline
Low incidence of stroke and GI bleed	✓Pulsatile outflow to reduce diastolic pressures ✓Flat pump curves to limit aortic pressure ✓Low platelet activation and low destruction of vWF

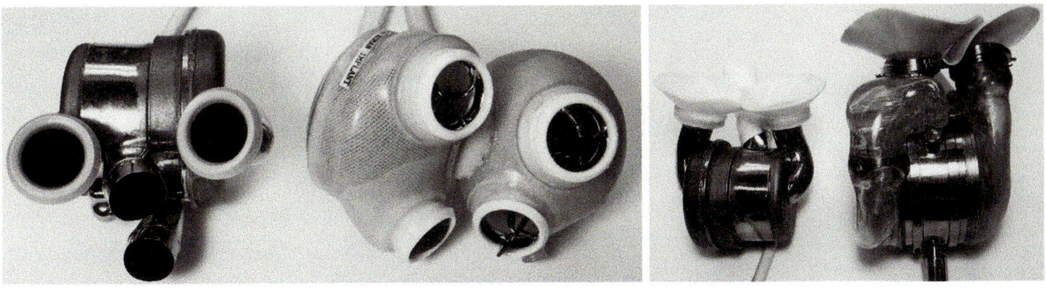

Fig. 35.2 The BiVACOR compared to the SynCardia® TAH [*LEFT*] and compared to the AbioCor® TAH [*RIGHT*]

mechanical wear by maintaining clearances larger than 240 μm in normal operation. This also creates a low shear stress path for delicate blood cells and other blood components. Rotor speed can be modulated to create physiological pressure and flow waveforms while maintaining the blood-friendly operation. The device is considerably smaller than its TAH predecessors that used positive displacement pumps (Fig. 35.2), allowing for implantation into a wider patient population, including women and some children (BSA > 1.4 m^2). An inherent Frank-Starling-like response – achieved through relatively "flat" pump curves – allows for a significant level of passive flow adaptation. This ability not only helps to balance between left and right outflows but also in the adaptation of the total device outflow. The combination of passive flow adaptation, the ability of the rotor to move axially to augmented left/right flow balancing, and an autonomous speed and low inflow volume controller, which does not require blood contacting sensors, allows the BiVACOR device to respond in a manner like the native heart's physiological response and is a key innovation in the field.

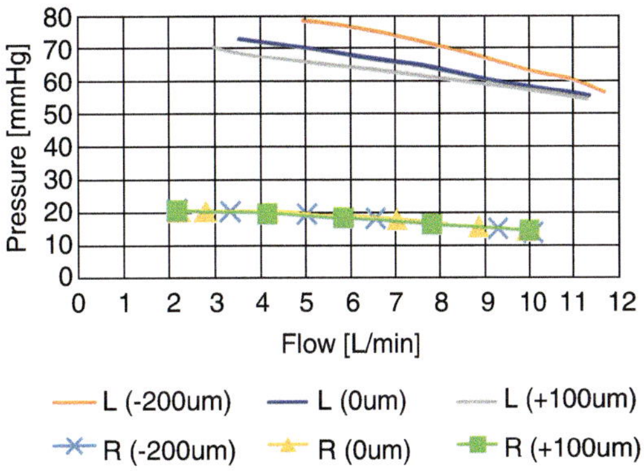

Fig. 35.3 BiVACOR hydraulic performance characteristics showing left (L) and right (R) impeller HQ curves @1975 RPM and varying axial positions (+100, 0 and −200 μm)

Device Description

Hydraulic Design

Left and right hydraulic designs were determined to create a device that exhibits a physiological performance, which can inherently balance pressures and flows over a wide range of vascular conditions. For this purpose, the HQ-curves of both pump sides were matched at the same rotational speed to produce head pressures of approximately 70 mmHg (left) and 20 mmHg (right) at an outflow of 5 L/min (Fig. 35.3). The HQ-curve shapes were then optimized to amplify the inherent Frank-Starling-like response of rotary blood pumps.

In the native ventricles, the Frank-Starling mechanism causes increased ventricular output in response to increased pre-load (atrial pressure). In the BiVACOR TAH, this effect is achieved through selecting pressure sensitive (flat) pump curves, whereby the left pump is less pressure sensitive than the right pump; just as in the native heart.

Fig. 35.4 Oxygenated (red) and deoxygenated (blue) blood flow in the BiVACOR

Flow Paths

The left impeller propels oxygenated blood from the left atrium via the aorta into the systemic circulation, while the right impeller propels deoxygenated blood from the right atrium via the pulmonary artery into the pulmonary circulation (Fig. 35.4). Local stagnation of blood and the occurrence of high shear stresses need to be avoided to ensure biocompatibility of the flow paths inside the device. The former can lead to localized thrombus formation, while the latter can have several effects such as hemolysis, destruction of high molecular weight (HMW) vWF multimers, platelet activation, and leukocyte destruction [1]. Rotary blood pumps often exhibit higher-than-physiological shear rates [2]. However, blood damage can be minimized by designing large gaps within the pump (reducing the shear rates) and by minimizing the time that cells are exposed to high shear regions.

A key design feature achieved through the BiVACOR magnetic levitation system is that all flow paths have clearance gaps of at least 240 μm during normal operation (100 μm under extreme conditions). Large through-flow areas are particularly beneficial in the right side of the device, which may be exposed to emboli from deep vein thrombosis that can pass unobstructed through the pump. Large clearance gaps are also present lateral to the rotor base to ensure sufficient washout of secondary flow paths. The amount of washout flow depends on the left-right pressure gradient and the rotor axial position. Under physiologic conditions, the flow is directed from left to right, like that observed in a ventricular septal defect, and targeted to be just below 1.5 L/min (30% of pump output).

Axial Pressure Capacity

A unique feature of the device is the ability to adjust the relative left and right pump performance (Fig. 35.3) by altering the rotor's axial position within the left and right pumping chambers. A movement toward the left casing reduces the leakage between the impeller vanes and the casing, thus increasing the left pump's efficiency and consequently its output. A movement toward the right casing reduces the left pump efficiency and thus output. The magnitude of left pump pressure variation is referred to as the axial pressure capacity. The left impeller output pressure in the BiVACOR device can vary by up to 12 mmHg at 5 L/min.

Hydrodynamic Backup Bearing

A hydrodynamic thrust bearing, which provides redundancy in the unlikely case of interruption to the magnetic levitation system, has been added to the left impeller blade design. It prevents a permanent touchdown of the rotor on the left side and aims to bridge the time until appropriate measures can be taken: either elective controller/driveline changes or pump replacement/heart transplantation. Each left impeller blade consists of a slope and a horizontal pad area, designed to provide enough lift away from the left side, creating a film capable of producing a force capacity of >20 N, enough to withstand a 10 G shock force.

Motor and Suspension System

The electromagnetic suspension (Maglev) levitates the rotor and is hermetically sealed around the right impeller blades, while the motor drive provides the rotational torque necessary to pump blood and is sealed above the left impeller blades; see Fig. 35.5.

For a redundant drive system, the motor coils are separated into two independent three-phase systems. Each set of windings is driven by a redundant amplifier, driver circuit, and driveline wires, which allow for continued operation in the unlikely event of electrical or mechanical failure within the motor winding, controller electronics, or connection leads. The drive system operates at efficiencies up to $\eta = 80\%$ and provides a torque capacity of up to 150 mNm, which is enough to spin the rotor in both continuous and pulsatile modes at speeds between 1500 to 3500 RPM. Typical motor power consumption under normal flow conditions (5–6 L/min) is 5 W.

The axial attractive forces of the motor and magnetic bearing permanent magnets acting on the rotor are oriented in opposite directions. A movement toward the left side results in an increase of the force toward the motor, while the force toward the magnetic bearing decreases, and vice versa (Fig. 35.6, solid lines). The interaction of the two characteristics results in an unstable but balanced equilibrium point within the axial rotor movement range. Three contact-free displacement sensors are used to determine the axial position and tilt of the rotor within the casing. This position information is processed by the active magnetic bearing controller to determine how much current, and consequently magnetic force is required to maintain the rotor stable in the desired target position.

To achieve a simple and reliable implantable pump, there are no active electrical components in the implantable device, only electromagnetic coils. Consequently, in the unlikely event that electronic issues arise, the active electronics contained in the patient's wearable controller can be replaced without major invasive intervention.

Physiological Interaction

The natural heart continuously balances systemic and pulmonary flow through ventricular interdependence and the Frank-Starling mechanism. When the two failing ventricles are removed for the

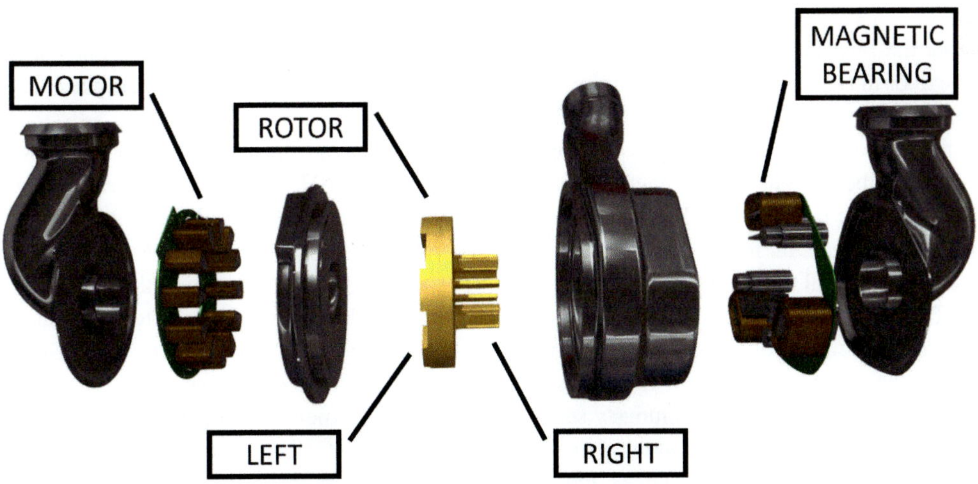

Fig. 35.5 Internal components of the BiVACOR; motor and suspension system

implantation of a TAH, the device must take over this important function and adapt to varying patient conditions and activities. The pressure-sensitive hydraulic designs of the BiVACOR TAH provide an inherent ability to adapt left and right output (flat HQ-curves). To further improve the physiological response, pump speed and axial rotor position can be altered. While in the hospital, those settings could be managed by experienced clinical personnel; however, when the patient leaves the hospital, an autonomous adaptation may allow for an increased patient activity and improved quality of life. With the aim to assist in recovery and restore the patient's quality of life, the BiVACOR TAH provides three features to actively augment the physiological interaction between the TAH and the patient's circulatory system:

- Rotor position control, to balance the relative left and right output flows
- Rotor speed control, to adapt the total mean flow to the vascular demand
- Pulsatility control, to mimic the native heart pressure and flow waveforms

Rotor Position Control

The central element of the implantable pump is the magnetically levitated rotor, which can increase or decrease the left pump's efficiency by axially moving left or right, respectively. The virtual zero power (VZP) controller [3] utilizes this ability to enhance the physiological response by maintaining the rotor in an axial position that requires minimal magnetic bearing power (hence "virtual zero power"). Motor and hydraulic forces act on the rotor toward the left side, while magnetic bearing forces act on the rotor toward the right side (Fig. 35.6). Any change in inlet or outlet pressures translates to an axial hydraulic force applied to the rotor. This force consequently reflects a variation in power consumption of the magnetic bearing to maintain the rotor in the same axial position. Instead, the VZP controller continuously seeks to minimize the bearing power by moving the rotor axially until hydraulic and permanent magnetic forces are in balance. In Fig. 35.6 the position of

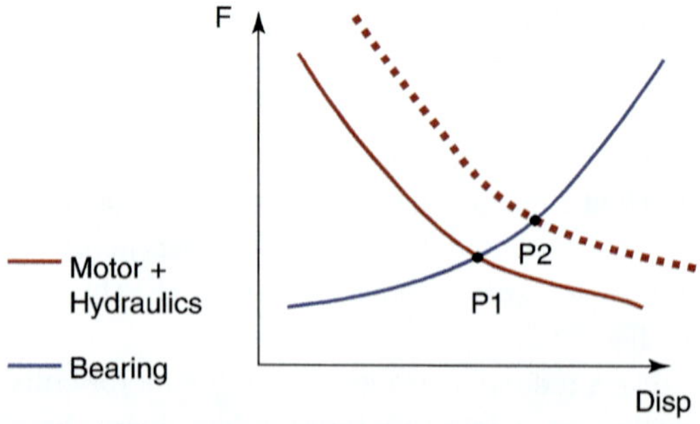

Fig. 35.6 Schematic axial forces (F) on the rotor versus displacement (Disp): Motor and hydraulic forces toward left (red), bearing forces toward right (blue). Dashed: Increased force to left; P1, P2 axial positions of equalized axial forces

equalized forces is marked with "P1." In an example of a decreased left atrial pressure, the total forces toward the left increase, which are displayed by the dashed line. Consequently, the VZP controller autonomously moves the rotor toward further to the right side (position "P2"), and a Frank-Starling-like response is achieved. A decrease in preload (inlet pressure) reduces the relative output of the corresponding pump side, which reduces the risk of over-pumping and helps to maintain left and right outflows balanced.

Rotor Speed Control

In the natural heart, cardiac output changes in response to venous return and the resulting preload due to the Frank-Starling mechanism. This may result in flows increasing beyond 12 L/min during exercise or reducing to less than 5 L/min during sleep.

Due to the pressure-sensitive pump curve, the BiVACOR TAH can inherently track relatively large changes in outflow in response to changes in vascular resistance at a fixed rotational speed. A speed controller can extend this ability further and actively reduce speed to prevent over-pumping and potential atrial collapse or increase speed to improve perfusion.

The implemented mechanism does not require any blood contacting sensors to monitor physiological parameters, as the changes to the physiological system are detected by analyzing the inherent signals of the hermetically sealed magnetic levitation system. Changes in vascular pressures directly translate to hydraulic forces acting on the magnetically levitated rotor. This information is used to identify the onset of atrial collapse, by detecting higher than usual rates of change in magnetic bearing signals. Due to the nonlinear compliant nature of the vascular inflow, the relationship between atrial pressure and its volume is different when the volume is high or low. Namely, when the atrial volume is low, small changes in volume provide larger changes in the atrial pressure in time when compared to when the atrial volume is full. This is demonstrated in Fig. 35.7, where the response to in vivo ventilated breathing are shown with both high (blue) atrial filling pressure and low (red) filling pressures. By monitoring the magnetic bearing current, indications of these forces and the associated hemodynamic state can be identified. In the case of the high filling pressure, there are only minor changes in the magnetic bearing response (bottom graph) to the breathing events indicating that the controller could increase pump speed. However, significant peaks can be seen for the same breathing events when atrial filling is lower, which indicates the need to reduce speed.

Speed Pulsatility

- When a rotary TAH, such as the BiVACOR TAH, is operated at constant speed, the circulation will be perfused with a true continuous flow. There are indications that vasculature and organs may not tolerate the lack of pulsatility, which can lead to adverse events, including non-surgical bleeding, thrombosis, neurological dysfunction, renal and hepatic dysfunction, hypertension, and possibly stroke [4, 5]. Hence, the BiVACOR builds on a powerful motor system that can apply rapid rotor speed modulations creating pulsatile outflows which allows to combine the mechanical reliability of rotary blood pumps and the physiological advantages of pulsatile perfusion [6, 7]. The target characteristics for producing a physiologic pulse like the native heart were to generate peak flow rates up to 15 L/min, aortic pulse pressures >40 mmHg, and arterial dP/dt >500 mmHg/s at beat rates between 30 and 90 BPM. Should future investigations indicate any issues arising from the speed modulation, then the device can still operate in a reduced pulse or continuous-flow mode. To prevent steady-state recirculation areas and to promote the device washout, intermittent speed reductions are also applied.

Anatomical Fitting

The small footprint provides the possibility to implant the device in patients with a reduced chest cavity size, such as children and small adults (BSA > 1.4 m^2). During anatomical tailoring, i.e., optimization of port sizes, shapes, and

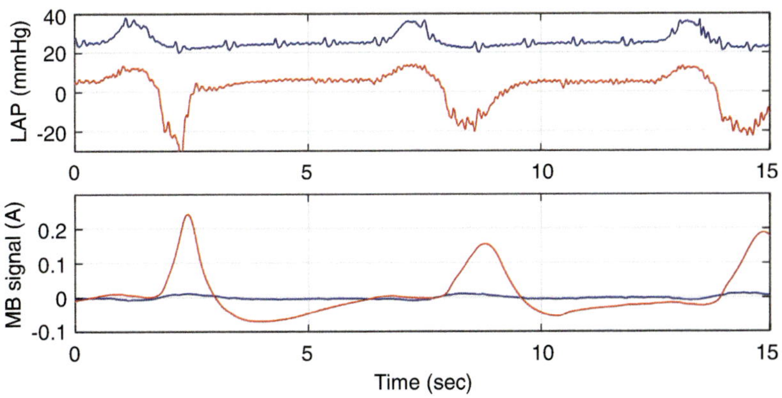

Fig. 35.7 Response of the magnetic bearing (MB) suction detection [BOTTOM] to periodic ventilated breathing when left atrial filling pressure (LAP) is high (blue) and low (red) [TOP]

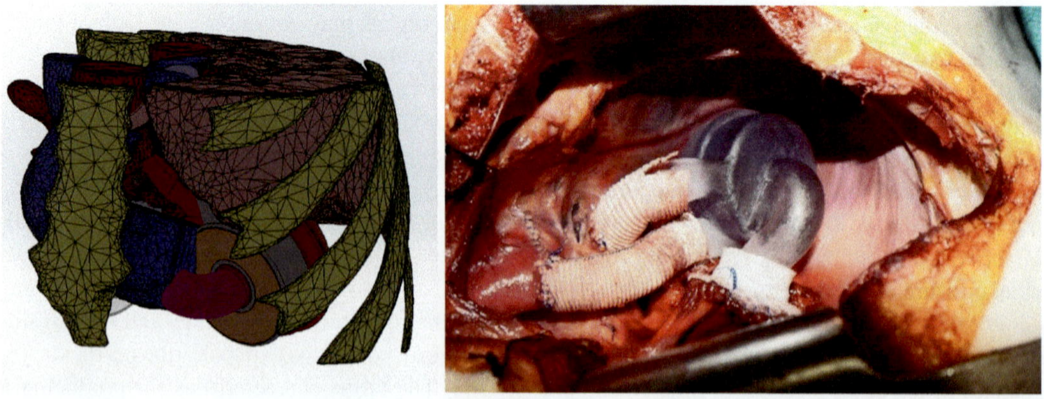

Fig. 35.8 3D human fitting with MIMICS (Materialise NV) [*LEFT*]; cadaver fitting, the BiVACOR is small enough to fit into a small woman (46 kg, BSA 1.44 m^2) [*RIGHT*]

orientations, the design aims were to avoid compressing or obstructing other organs or greater vessels, preserving the tissue necessary to enable future heart transplantation and aligning the outlet ports with great vessels.

Virtual fitting in eight segmented CT scans of various patient sizes, ages, and both genders was initially performed. Twenty variations of the pump, with different port orientations, were drawn and fitted in the computer models of individual patients and in 2 averaged models. One was based on literature and the other a custom-made Statistical Shape Model in MIMICS by Materialise NV (Leuven, Belgium). This enabled the extraction of measurements and coordinates such that both qualitative and quantitative evaluation could be performed (Fig. 35.8). The best fitting pump models were then 3D printed, and implant surgery was simulated in 3 cadavers: a 46 kg female (BSA 1.44 m^2, Fig. 35.8), a 68 kg male (BSA 1.82), and a 91 kg female (BSA 2.06 1.44 m^2).

Biocompatibility

All blood and tissue contacting surfaces of the implantable pump are made from Titanium and undergo either electro- or ultra-polishing. In addition, the rotor is TiN-coated, while the driveline has a silicone jacket. Commercially available vascular polyester grafts are currently used as outflow conduits and inflow conduits are made from Dacron and biocompatible silicone. Most importantly, magnetic rotor levitation eliminates any dynamic mechanical contact to the casing, while all flow paths are designed with blood-friendly gaps >100 µm.

Driveline

The driveline is divided into two sections, implantable device portion and external controller portion, which are connected via an inline connector. The device section of the driveline is terminated at an inline connector near the exit site. The portion of driveline proximal to the connector is sufficiently long to be affixed to the patient's body to reduce cable movement at the exit site and thus minimizing the risk for infection after exit site trauma.

Controller

The portable controller can be placed in a back pack an initially weighs <4 kg with batteries. It is powered by two external (replaceable) batteries, for long-term use (>6 hour run time). The controller has an on-board user interface to provide the patient and caregivers with basic TAH performance metrics such as power consumption and speed but will also display alarms and instructions in case of critical events. For detailed system and event analysis, a console can be connected via a cable to provide access to advanced device data and settings. The controller contains three major electronic control assemblies:

- *Motor Controller* – regulates speed, observes the motor's operational state, and provides emergency alarming in case of a motor fault
- *Magnetic Bearing Controller* – reads the three position sensor signals and computes the required feedback current to allow for stable rotor levitation
- *System Controller* – features system safety and monitoring algorithms, controls the integrated graphical user interface, provides data logging and control of the alarming system, and implements physiological control algorithms

Implantation

The device is implanted in the space previously occupied by the heart through a procedure that is like a traditional heart transplant. To date, the TAH has been successfully implanted in bovine and ovine models. Calves are similar in size to adults while requiring higher resting flow rates representative of human exercising conditions, while sheep are more representative of patient populations smaller in body size and similar resting flow rates to the resting patient.

During in vivo studies, experimental animals undergo intravenous anesthesia, intubation, mechanical ventilation, and instrumentation using our standard procedures. Via left thoracotomy, animals are placed on cardiopulmonary bypass (CPB). Left and right ventricles and appendages are excised at the atrioventricular groove to leave both atria in situ. Quick connect inflow cuffs are sutured onto both left and right atria remnants which are reinforced by felt strips (Fig. 35.9). Outflow grafts are sutured to the pulmonary artery and aorta end-to-end. Fitting tools assist in positioning and trimming of the vascular connections. The anastomoses (four) are tested for leaks with a designed pressure tester. A tunneling tool is used to externalize the driveline percutaneously. The exit site is secured with sutures and surgical dressing.

System Performance

To date the BiVACOR TAH was implanted in 17 chronic experiments, animal models were either Corriente crossbred calves (N = 15) or Sheep (N = 2). In the following the results of the three longest-surviving calves are presented. The durations were 30, 96, and 29 days for the chronic (CH) studies CH06, CH09, and CH14, respectively. All calves were able to stand shortly after the implantation, recovered well, and gained body weight at a normal rate of 0.27 ± 0.05 kg/day (N = 3); the average implantation weight was 81 ± 8 kg. Throughout the experiments the devices were able to maintain stable hemody-

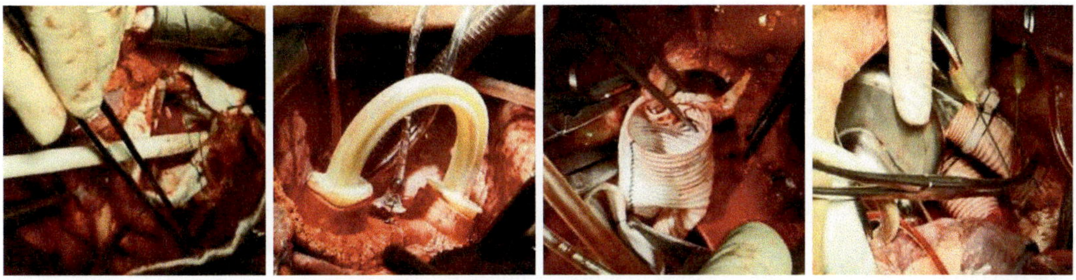

Reinforcing neo-atria rims by felt strips | Locating the cuffs with fitting tool | End-to-end anastomosis of a vascular graft | Device in-situ, cuffs and grafts connected

Fig. 35.9 TAH implantation steps in a bovine model

namics. Aortic pressures, pump flows, and plasma-free hemoglobin (pfHb) levels of the three animals are displayed in Fig. 35.10. In the longest study (CH09), average pump flows were 10.0 ± 1.2 L/min, while aortic pressures were 109 ± 13 mmHg. Plasma-free hemoglobin levels stabilized after implantation and typically remained below 2 mg/dL (study average: 0.9 ± 1.5 mg/dL), while markers for renal and liver health were at normal healthy levels: blood urea nitrogen 13.9 ± 6.3 mg/dL and total bilirubin 0.11 + 0.03 mg/dL, respectively.

Pulsing

Pulse profiles were applied in vivo, and pulse parameters were modified to yield physiological pressure and flow waveforms. Pulsatile hemodynamics, which approach the characteristics of the native cardiovascular pulse, were achieved with a custom speed profile at a rate of 60 BPM and a peak-to-peak amplitude of 1200 RPM. Peak flow rates were above 17 L/min, aortic pulse pressures achieved 40 mmHg, while arterial pressure change rates (dP/dt) were 500 mmHg/s; see Fig. 35.11. During pulsatile operation for more than 2 days, no signs of elevated hemolysis were observed, and pfHb levels remained below 2 mg/dL.

Hydrodynamic Bearing

The hydrodynamic backup bearing functionality was confirmed in-vivo in case CH17 by simulating a magnetic bearing fault. Motor permanent magnet attraction forces pulled the rotor toward the left casing and the rotor continued to spin on a hydrodynamic fluid film without loss of hemodynamic performance. Motor power increased by 12% due in part to greater hemodynamic output as well as increased fluid viscous drag. During the 8-hour operating period, pfHb levels maintained below 2 mg/dL, while hemodynamics, including flow balancing, were stable without mechanical contact between rotor and casing.

Outflow Pressure Sensitivity

Postural change events demonstrate both the effectiveness of the autonomous VZP controller and the device's inherent outflow pressure sensitivity (Fig. 35.12). Due to muscular activation, systemic vasodilation and volume shift during a stand-up of the calf, both AoP and LAP decreased, while RAP increased. While maintaining rotor speed, the pressure-sensitive hydraulic design allowed the device outflow to rise momentarily by more than 3.5 L/min in response to the increased vascular demand. To redistribute the pooled right atrial blood volume, the VZP controller moved the rotor momentarily by more than 50 μm toward the right side, thereby reducing left pump outflow, which allowed the left atrium to refill. Within 4 minutes AoP settled at a new stable level.

Hemocompatibility

To supplement the in vivo results, hemocompatibility was also assessed in terms of hemolysis, platelet activation, and von Willebrand Factor (vWF) degradation. The tests were performed ex vivo in a blood loop, filled with cow blood, in accordance with current standards and publications (ASTM F1841-97, ISO 10993-4, [8–11]). High molecular weight (HMV) vWF only started to degrade at elevated flows of 9 L/min (Fig. 35.13). Plasma-free hemoglobin levels were notably lower when compared to the HeartMate 3 [12]. Altering the rotor position by +/−100 μm had a minor effect on hemocompatibility measures. Overall, hemolysis, platelet activation, and HMW vWF degradation levels of the device are promising and will be evaluated further in the following development phases.

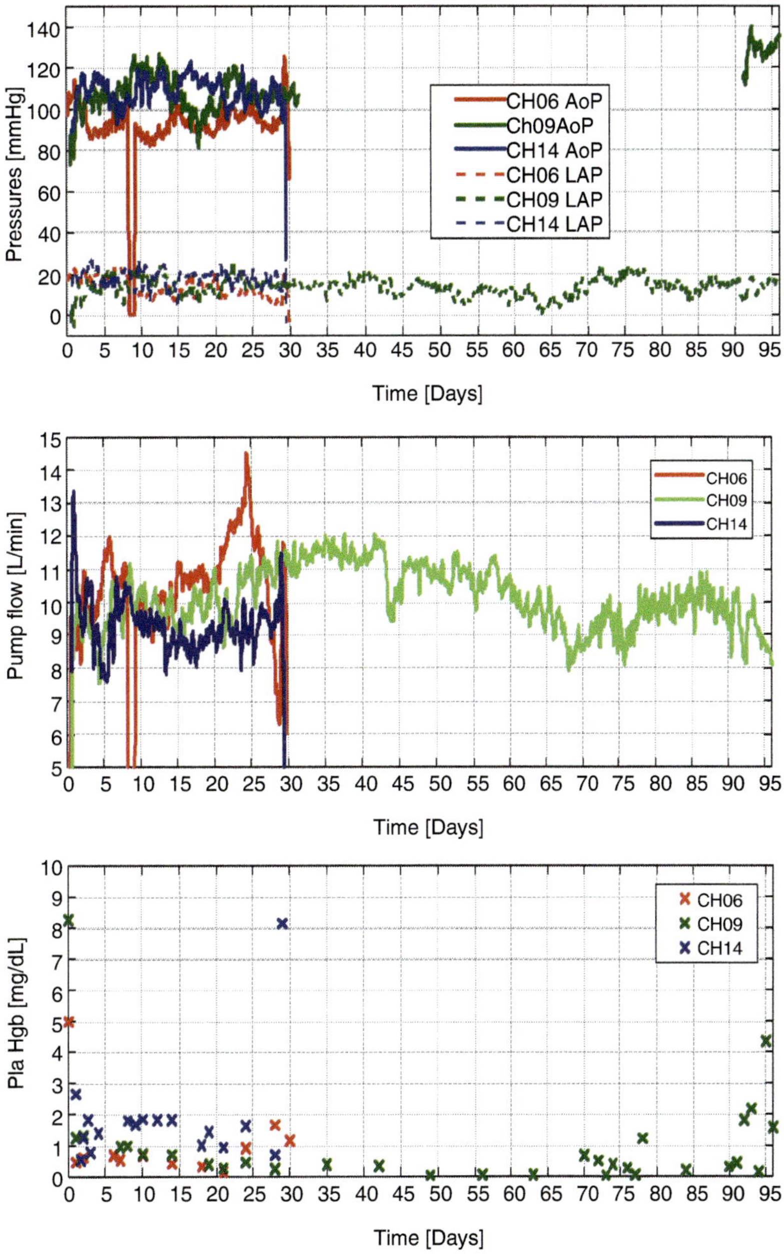

Fig. 35.10 Aortic pressures, pump flow And plasma-free hemoglobin levels of the 3 longest-surviving cases (CH06, CH09, CH14). AoP line of CH09 clotted on POD32 and was reinserted on POD90. AoP aortic pressure, LAP left atrial pressure

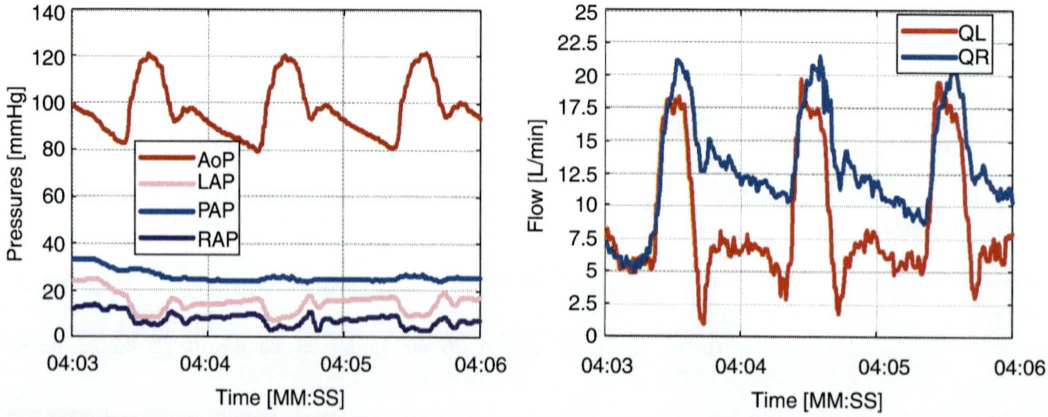

Fig. 35.11 Pulsatile operation applied in vivo (CH17); QL, QR left And right flow, AoP, LAP, PAP, RAP aortic, left atrial, pulmonary artery and right atrial pressure

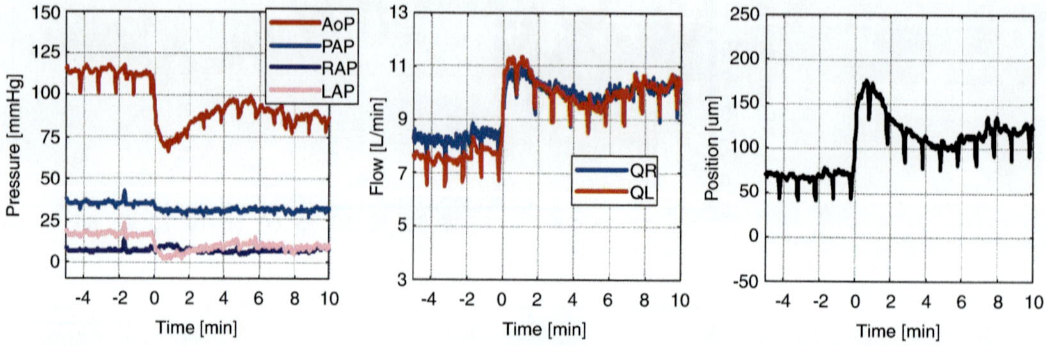

Fig. 35.12 In vivo pressures and flows during stand-up (at 0 min) and the corresponding rotor movement (Study CH12); QL, QR left and right flow; Aop, LAP, PAP, RAP aortic, left atrial, pulmonary artery and right atrial pressure

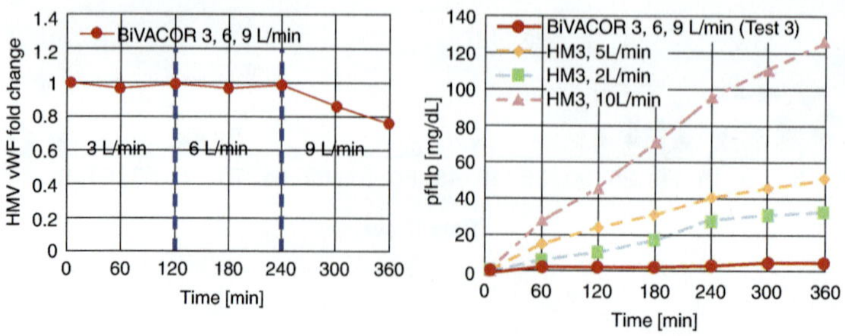

Axial rotor position	NIH [mg/100L]	Platelet activation [%]
Center position	4.02	2.3
+100 μm	3.82	1.8
-100 μm	3.49	1.7

Fig. 35.13 Hemocompatibility measures compared for BiVACOR and HeartMate 3 (HM3). High molecular weight (HMV) Von Willebrand (vWF) factor fold change [*LEFT*], plasma-free hemoglobin (pfHb) levels under different flow conditions: [*MIDDLE*], NIH and platelet activation under in different axial rotor positions [*RIGHT*]

Conclusion

The limited availability of donor hearts motivated the development of mechanical circulatory devices, like the BiVACOR Rotary TAH. The presented data attests that the developed device is small enough to be implanted in many women and children (BSA > 1.4 m^2), and capable of providing enough cardiac output for an adult male undergoing light exercise (>12 L/min). Thereby the highly pressure-sensitive hydraulic designs of both pump sides contribute to the adjustment of total flow and maintaining flow balance without active pump speed management. Further, the utilized Maglev technology used to suspend the wear-free spinning rotor in wide clearances demonstrated to be a key feature to achieve excellent biocompatibility. Equally, rotor speed modulation techniques to produce true physiologic pressure and flow waveforms showed no sign of elevated hemolysis when applied in vivo. Included redundancy features, such as the hydrodynamic dynamic bearing, contribute to device reliability, without impeding hemodynamic performance.

References

1. Ruggeri ZM. Mechanisms initiating platelet thrombus formation. Thromb Haemost. 1997;78(1):611–6.
2. Moazami N, Fukamachi K, Kobayashi M, Smedira NG, Hoercher KJ, Massiello A, et al. Axial and centrifugal continuous-flow rotary pumps: a translation from pump mechanics to clinical practice. J Heart Lung Transplant.. Elsevier. 2013;32(1):1–11.
3. Timms D. Heart pump controller. World: World Intellectual Property Organization; WO2010118476-A1; 2012.
4. Markham DW, Fu Q, Palmer MD, Drazner MH, Meyer DM, Bethea BT, et al. Sympathetic neural and hemodynamic responses to upright tilt in patients with pulsatile and nonpulsatile left ventricular assist devices. Circ Heart Fail. 2013;6(2):293–9.
5. Kirklin JK, Naftel DC, Pagani FD, Kormos RL, Stevenson LW, Blume ED, et al. Seventh INTERMACS annual report: 15,000 patients and counting. J Heart Lung Transplant.. Elsevier. 2015;34(12):1495–504.
6. Kleinheyer M, Timms DL, Greatrex NA, Masuzawa T, Frazier OH, Cohn WE. Pulsatile operation of the BiVACOR TAH - motor design, control and hemodynamics. Conf Proc Annu Int Conf IEEE Eng Med Biol Soc IEEE Eng Med Biol Soc Annu Conf. 2014;2014:5659–62.
7. Imachi K, Chinzei T, Abe Y, Mabuchi K, Imanishi K, Yonezawa T, et al. A new pulsatile total artificial heart using a single centrifugal pump. ASAIO Trans. 1991;37(3):M242–3.
8. Chan CHH, Pieper IL, Hambly R, Radley G, Jones A, Friedmann Y, et al. The CentriMag centrifugal blood pump as a benchmark for in vitro testing of hemocompatibility in implantable ventricular assist devices. Artif Organs. 2015;39(2):93–101.
9. Chan CHH, Hilton A, Foster G, Hawkins K. Reevaluation of the harboe assay as a standardized method of assessment for the hemolytic performance of ventricular assist devices. Artif Organs. 2012;36(8):724–30.
10. Snyder TA, Tsukui H, Kihara S, Akimoto T, Litwak KN, Kameneva MV, et al. Preclinical biocompatibility assessment of the EVAHEART ventricular assist device: coating comparison and platelet activation. Willey Intersci. 2006;81:85–92.
11. Chan CHH, Pieper IL, Fleming S, Friedmann Y, Foster G, Hawkins K, et al. The effect of shear stress on the size, structure, and function of human von Willebrand factor. Artif Organs. 2014;38(9):741–50.
12. Bourque K, Cotter C, Dague C, Harjes D, Dur O, Duhamel J, et al. Design rationale and preclinical evaluation of the HeartMate 3 left ventricular assist system for hemocompatibility. ASAIO J. 2016;62(4):375–83.

36 The Penn State Pediatric Total Artificial Heart

William J. Weiss, Raymond Newswanger, J. Brian Clark, and Jenelle M. Izer

Durable mechanical circulatory support (MCS) in adults has transitioned from an investigational therapy to widespread use. The most common application is a single pump acting as a left ventricular assist device (LVAD). In adults, biventricular support is rarely used, despite the occurrence of right heart failure of moderate severity in 27% of LVAD patients and severe right heart failure in 2.8%. Biventricular VADs (BiVADs) were used in 3.5% of patients requiring mechanical circulatory support (MCS), while the total artificial heart (TAH) was used in 2.1% (Intermacs data from June 2006 to December 2016) [1].

The SynCardia (Tucson, AZ) 70cc TAH is a pneumatically actuated pulsatile pump; it has been implanted in over 1400 patients worldwide [2–4]. In addition to effectively treating right heart failure, the TAH may be preferred over a BiVAD in cases of infiltrative cardiomyopathies (amyloid), restrictive cardiomyopathies, cardiac malignancies, refractory arrhythmias, post infarction VSD, the presence of mechanical heart valve prosthesis, cardiac aneurysm, aortic valve regurgitation, allograft failure, or congenital heart malfunction [5, 6]. Furthermore, the presence of four BiVAD cannulae and difficult positioning of the right cannulae increase the risk of thrombosis; the TAH has been associated with fewer neurologic events and improved survival compared to BiVADs [5]. Finally, direct connection of the TAH pumps to the atria and arteries results in significantly higher flow rates, resulting in more rapid improvement in end-organ function and improved exercise capacity [7].

The SynCardia 70cc requires a body surface area (BSA) of at least 1.7 m^2 and a minimum thoracic size of 10 cm from the sternum to the tenth thoracic vertebra (T10). SynCardia recently introduced a 50cc TAH for use in smaller patients with BSA 1.2–1.79 m^2. The IDE trial was approved in March 2014 and will include ten pediatric patients 10–18 years old.

According to the 2016 Annual Data Report of the US Organ Procurement and Transplantation Network (OPTN) and the Scientific Registry of Transplant Recipients (SRTR) [8], there were 445 pediatric heart transplants performed and 624 listings (age <18 years) in the United States

W. J. Weiss, PhD (✉) · R. Newswanger, PhD
The Pennsylvania State University, College of Medicine, Department of Surgery, Division of Applied Biomedical Engineering, Hershey, PA, USA
e-mail: wweiss@pennstatehealth.psu.edu; rnewswanger@pennstatehealth.psu.edu

J. B. Clark, MD
The Pennsylvania State University, College of Medicine, Department of Pediatrics, Hershey, PA, USA
e-mail: jclark7@pennstatehealth.psu.edu

J. M. Izer, DVM
The Pennsylvania State University, College of Medicine, Department of Comparative Medicine, Hershey, PA, USA
e-mail: jizer@pennstatehealth.psu.edu

© Springer Nature Switzerland AG 2020
J. H. Karimov et al. (eds.), *Mechanical Support for Heart Failure*,
https://doi.org/10.1007/978-3-030-47809-4_36

in 2016. Infants <1 year of age accounted for 27.7% of the waiting list, and children ages 1–5 years accounted for 31.2% of the total. Waitlist mortality and posttransplant survival are poorest in the under 1 year age group (Fig. 36.1a, b). VAD use is increasing in pediatric heart transplantation but remains below the rate used in adults. The percentage of pediatric patients (<18 years of age) on VADs at the time of listing was 6.1% (23.7% at transplant), as compared to 32.1% of adults (52% at transplant).

For infants and small children, bridge to transplant (BTT) is the primary support objective. The Berlin Heart EXCOR (Berlin Heart Inc., Woodlands, TX) is the only viable option for support extending beyond a few weeks. In the Berlin Heart HDE trial, the smallest size 10 ml pump was used in 37.3% of patients. Survival was significantly worse in patients <5 kg, where 2/3 of these patients died [9].

The Need for Pediatric Biventricular Support

Biventricular failure is much more common in pediatrics than in adult MCS. BiVAD support was used in 36.8% of patients in the Berlin Heart trial [9]. The need for biventricular support is related to the disease diagnosis. The most common form of pediatric heart failure leading to heart transplantation (HTx) is cardiomyopathy (CM), which accounted for 43.5% of the HTx primary diagnoses in 2011–2013, of which dilated cardiomyopathy (DCM) makes up nearly 90%. Hypertrophic and restrictive cardiomyopathies are less common but would benefit from the TAH because of difficulty cannulating the small ventricles. In the Berlin Heart trial (204 patients), CM accounted for 62% of LVAD patients and 65% of BiVAD patients.

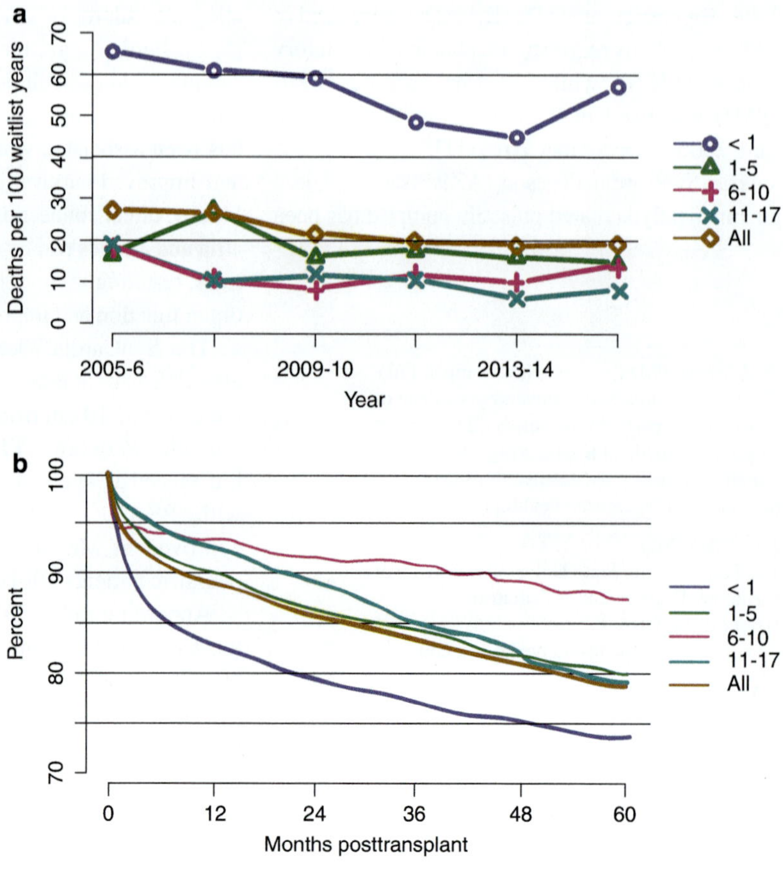

Fig. 36.1 (**a**) Pretransplant mortality rates among pediatrics waitlisted for heart transplant by age (in years). Mortality rates are computed as the number of deaths per 100 patient-years of waiting in the given year. Individual listings are counted separately. Age is determined at the later of listing date or January 1 of the given year. Rates with less than 10 patient-years of exposure are not shown. (From OPTN/SRTR 2016 Annual Data Report: Heart [8]). (**b**) Patient survival among pediatric deceased donor heart transplant recipients, 2004–2011, by age. Recipient survival estimated using unadjusted Kaplan-Meier methods. (From OPTN/SRTR 2016 Annual Data Report: Heart [8])

Single ventricle (SV) anomalies occur in approximately 5 of every 100,000 live births in the United States. Hypoplastic left heart syndrome (HLHS) is the most common form. There is an increasing incidence of HTx for SV patients. Approximately 22% of pediatric HTx patients are SV, though the percentage is increasing, with Voeller et al. [10] reporting 45% of HTx patients as SV, where most had failed single ventricle palliation. The need for BTT support in this group is supported by an analysis by Almond et al. [11] of 750 children supported by ECMO to transplantation, in which 64% had congenital heart disease and 30% were SV. But only 33% of SV patients survived to discharge. In the EXCOR trial, 26 of 255 BiVAD patients were SV, with 15 of these HLHS [12]. Survival in the SV group was 42%, but none of the SV patients weighing <7 kg could be bridged to transplant.

An infant-sized TAH would allow earlier intervention in SV patients. Long-term posttransplant survival is higher in SV patients transplanted as neonates, and circulatory deterioration in SV patients often occurs rapidly. SV patients referred for HTx due to acute cardiac decompensation or Fontan failure have a high mortality [13]. Furthermore, early TAH use as a bridge to transplant may obviate the need for staged surgical palliation with its risk of allosensitization. Allosensitization due to the use of allograft tissue has been associated with a higher rate of sensitization and graft failure [14, 15]. Other potential indications for TAH use in pediatrics are listed in Table 36.1.

The availability of a TAH for infants and small children may eliminate a number of physiologic risk factors that lead to poor HTx survival. The Penn State Pediatric TAH provides full pulsatile cardiac support, which improves end-organ function, which is likely to improve HTx outcomes. Comparing TAH to BiVAD support in adults, the TAH may provide higher cardiac output, because the VAD cannulae add significant resistance and fluid inertance. For the Berlin Heart SV patients [12], multisystem organ failure and pulmonary/respiratory failure accounted for 2/3 of the deaths.

Additionally, pulsatile flow and adequate left preload sensitivity may reduce pulmonary vascular resistance (PVR), e.g., in Fontan patients. High PVR is a risk factor for HTx failure [16]. Availability of a safe TAH would support decisions to use the device earlier, rather than as a last resort. In the Berlin Heart trial, renal failure (abnormal GFR) was an independent predictor of mortality for BiVAD patients [17].

Table 36.1 Potential indications for total artificial heart support in pediatrics

Indication
Biventricular failure
Dysrhythmias
Intraventricular thrombus
Pulmonary hypertension
Single ventricle congenital anomalies
Hypertrophic cardiomyopathies
Restrictive cardiomyopathies
Cardiac malignancies
Ventricular septal defect (VSD)
Mechanical heart valve prosthesis
Ventricular aneurysm
Aortic/pulmonary valve insufficiency
Atrioventricular valve dysfunction
Allograft failure
Fontan failure

Pump Design

The Penn State Pediatric TAH consists of independent left and right pulsatile pumps, as shown in Fig. 36.2. The stroke volume ranges from 12 to 14 ml, depending on preload, afterload, and drive pressures. The pumps are identical except that the right pump outlet is angled to accommodate the pulmonary artery. The pump design is derived from the Penn State Infant VAD, which is based on the adult Thoratec® PVAD® which was designed at Penn State (as the Pierce-Donachy VAD). The Infant VAD was developed under the NHLBI "Pediatric Circulatory Support" contract program (2004–2009) and subsequent NIH grant funding. The device is in the final stages of testing.

The primary objective in the pediatric pump development at Penn State has been the minimization of thrombus formation. Key features include:

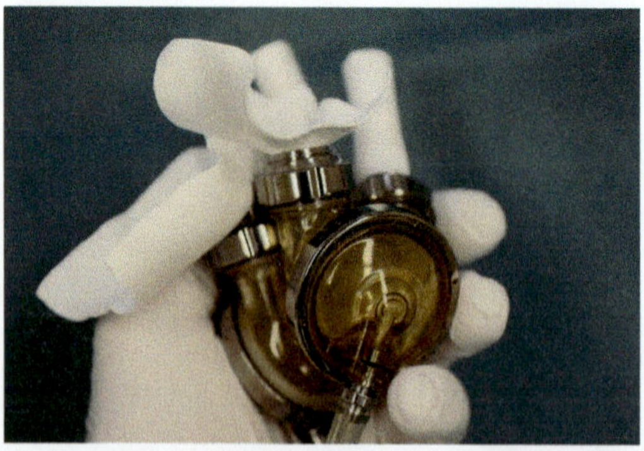

Fig. 36.2 The Penn State pediatric total artificial heart consists of independent left and right pulsatile pumps, with a stroke volume of 12–14 ml. The major diameter of each pump is 50 mm. Atrial and arterial connections are fabricated from ePTFE grafts (expanded polytetrafluoroethylene)

- Custom 17 mm Bjork-Shiley monostrut Delrin disk valves, which provide a large effective orifice area. This results in high inflow velocities [18] in the pump chamber that produce a sustained rotational flow and effective washing of the sac surface, without recirculation regions where thrombus can form. The valves are designed with custom flanges that fit precisely in the pump without steps or gaps. These valves, which are now manufactured in our labs, are of the same design and materials that have been used in the successful Thoratec® adult PVAD™, Arrow LionHeart™ VAD [19, 20], and the Penn State electric total artificial heart [19–24].
- Seamless, highly smooth blood sac, fabricated from segmented polyether urethane urea (SPEUU), which has demonstrated excellent thromboresistance, stability, and durability.
- Pump geometry which promotes rotational flow with effective washout of blood on each cycle.

The VAD may be utilized for left (LVAD), right (RVAD), or biventricular (BiVAD) support. The pump is designed with the option of paracorporeal placement (appropriate for infants and non-ambulatory patients) or implantation (an option in ambulatory children anticipating longer term support). This device is capable of support in excess of 1 year, and durability testing has demonstrated pump operation of 689 days without significant wear. In the most recent series of animal studies, we have demonstrated low thrombogenic risk with no anticoagulation after the initial 2–3 week postoperative acute hypercoagulable phase [25].

Cardiac Output and Anatomic Fit

The required pump stroke volume is driven by the cardiac output requirements and the range of feasible beat rates. For the Penn State Infant VAD, the nominal pump stroke volume of 12 ml was chosen based on a literature search of cardiac output requirements and growth data [26]. For infants aged 0–9 months, normal cardiac output is 0.6–1.45 liters/min, and for a beat rate range of 50–120 per minute, the required stroke volume is 12 ml. (The actual stroke volume depends on preload and afterload pressure and ranges from 12 to 14 ml.)

For larger infants and young children, we plan to develop a 25 ml stroke volume pump which will provide 1.25–3.00 liters/min over a 50–120 beat per minute range, covering an age range of infants to 5–9 years of age.

General fit guidelines for the adult SynCardia TAH include a minimum internal chest dimension from the sternum to the tenth vertebra (T10). An estimate of a similar dimension can be derived from radiological measurements (CT) of intrathoracic anatomical structures. In a study of CPR chest compression depth in infants and children [27], compression depths of up to ½ APD (anterior posterior diameter) are reasonable (APD measured externally). As a starting point, the proposed left pump diameters are ≤½ APD dimensions. In infants, the pumps will likely need to be rotated toward the left chest, which is a technique that has been used in small patients receiving the adult TAH [28]. We have begun modeling anatomic fit using CT or MR image data, tissue segmentation, and 3D modeling software.

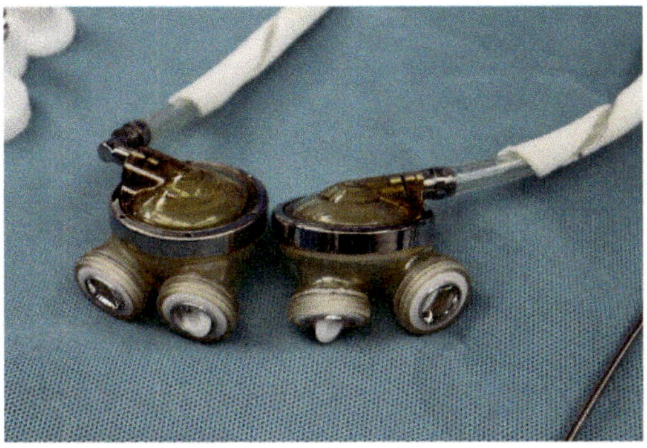

Fig. 36.3 View of the pediatric TAH showing the low-profile tilting disk valves

For orthotopic placement of the TAH, we re-designed the inlet and outlet ports of the VAD to minimize the valve and port length. This is important for preventing compression and occlusion of the arteries and especially the atria. The threaded connectors were re-designed to mate directly to the valves (Fig. 36.3). ePTFE is used for the atria, and an 8 mm ePTFE graft is used for the pulmonary artery and aorta connections. The atrial connectors allow for surgical reconstruction of inflow tracts if required.

The pump case is machined from a translucent polysulfone, which allows visualization of air bubbles in the pump during the de-airing procedure. This material is biocompatible for long term implantable use. Although intended for paracorporeal placement in infants, we have always implanted the pump in animals and have not seen evidence of any adverse tissue response. Polysulfone has also been used in the adult TAH developed at Penn State. It has been observed in our previous animal studies and in human implants of the SynCardia TAH that the pump case motion may lead to pleural effusions. This can be prevented by covering the pumps with a Dacron velour or ePTFE sheet.

We utilize a 17 mm tilting disk valve based on the Bjork-Shiley design, with a special low-profile valve flange that mates precisely with the blood sac and connector. In order to maximize pump output, we are exploring the effect of the disk clearance on backflow, hemolysis, and dynamic stroke volume.

Automatic Control

There are two primary objectives in controlling left and right pump outputs. First (and most important) is to maintain left atrial pressure (LAP) within a range that allows adequate left pump filling but limits LAP to safe levels that prevent pulmonary congestion. In normal physiology, this is achieved by the sensitivities of the ventricles and atria to preload and afterload. In mechanical pumps, right output can exceed left output, e.g., during hypervolemia, leading to a rapid and dangerous rise in left atrial (LA) pressure.

One approach, as used in the SynCardia TAH, is to operate both pumps in a limited fill mode, so that increases in preload to either pump result in an increase in stroke volume (simulating the physiologic Starling response). This response effectively absorbs increasing left atrial volume and pressure at least until full stroke volume is reached. The drawback of this approach is that the limited fill mode does not allow the pump chamber to fully wash each cycle, which may increase the chance of thrombus formation in the pump. In adult pumps with high flow rates, a limited stroke volume may be acceptable. But in an infant-sized pump with lower Reynolds numbers, the dominant viscous effects lead to reduced mixing at the wall, and incomplete filling may increase the risk of thrombus formation.

The philosophy at Penn State has been to always operate the pumps in a *full stroke mode* and to control output by adjusting pump rate. This approach has been used in the PSU pneumatic [29–31] and electric adult TAHs [19, 21–24, 32, 33] and in the VADs [19, 21, 22, 33], including the Infant VAD [34, 35]. This approach requires a measure of pump filling, which serves as the input to an electronic control system to automatically vary pump beat rate.

All pump control functions are achieved by measuring air pressure and airflow in the driver. There are no sensors in the pump. An estimate of the blood volume in the pump is derived by integrating the measured airflow and correcting for air compressibility. Pump filling is detected when

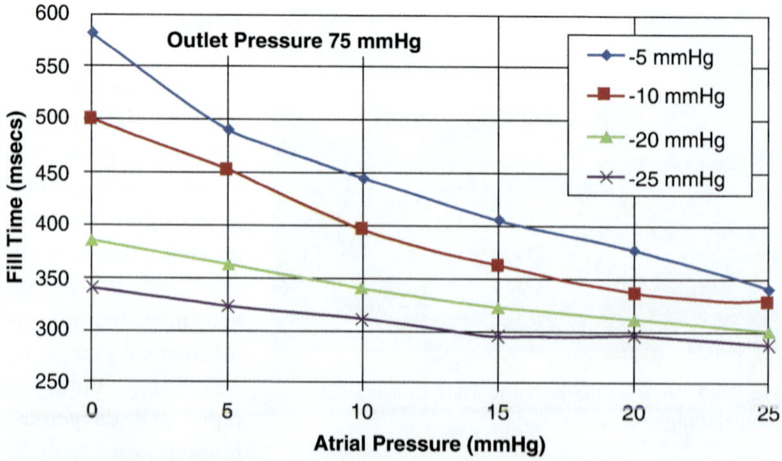

Fig. 36.4 Left pump fill times as a function of atrial pressure and diastolic drive pressure at a mean outlet pressure of 75 mmHg

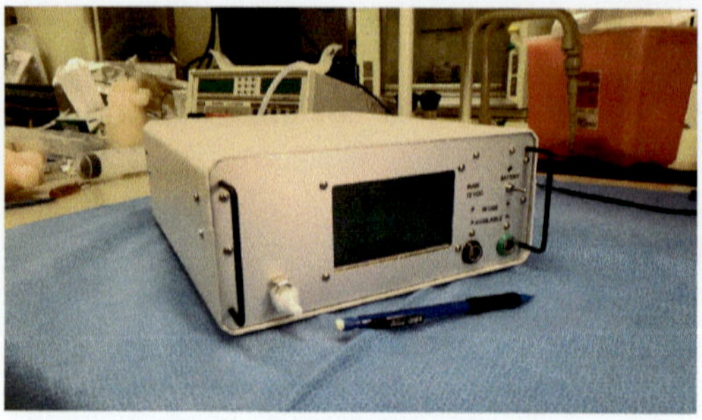

Fig. 36.5 A new portable pneumatic driver has been developed to implement the control functions. The driver weighs 3.5 kg and operates from a 12 volt DC supply

the airflow drops to near zero and the volume estimate has reached a threshold. In vitro data in Fig. 36.4 shows left pump fill times as a function of LAP and diastolic air pressure. With the LAP estimate, the right pump rate can be controlled as a function of LAP (left master/right slave approach).

A new portable pneumatic driver has been developed (Fig. 36.5) to implement the control functions. The driver weighs 3.5 kg and operates from a 12 volt DC supply.

In Vivo Testing

A challenge in the selection of animal model is balancing the required cardiac output versus the ability to fit the device anatomically without compression of the vessels. We are using Dorset-Finn crossbred lambs weighing initially in the 20–25 kg range for the Infant TAH, with the goal of reducing animal size to approximately 15 kg. Study duration will initially be 30 days to limit the effect of growth.

We have performed one acute implant of the Infant TAH in a 26.8 kg Dorset crossbred female lamb (Fig. 36.6).

A thoracotomy was performed through the right fifth intercostal space. The fifth rib was flapped and the retractor placed. The right internal thoracic artery was exposed and controlled and directly cannulated with an arterial catheter. The right lung was packed away with moist sponges and malleable retractors. The pericardium was opened ventral to the phrenic nerve and suspended. The superior and inferior cavae were exposed and mobilized. The aorta was exposed and separated from the PA. Cannulation sutures were placed in the cavae and distal ascending aorta and heparin given. The cavae were looped with tourniquets. Cannulation was performed and cardiopulmonary bypass initiated with cooling toward 32°.

Tourniquets were secured around the cavae and the aortic cross clamp was applied. The heart was incised on the ventricular side of the right atrioventricular groove, and the incision carried across the infundibulum and pulmonary artery at the level of the valve. The aorta was divided at the valve. The ventricles and septum were excised, and excess muscle was trimmed away leaving about 1 cm of myocardium at the AV valves. The coronary sinus was oversewn from within the RA.

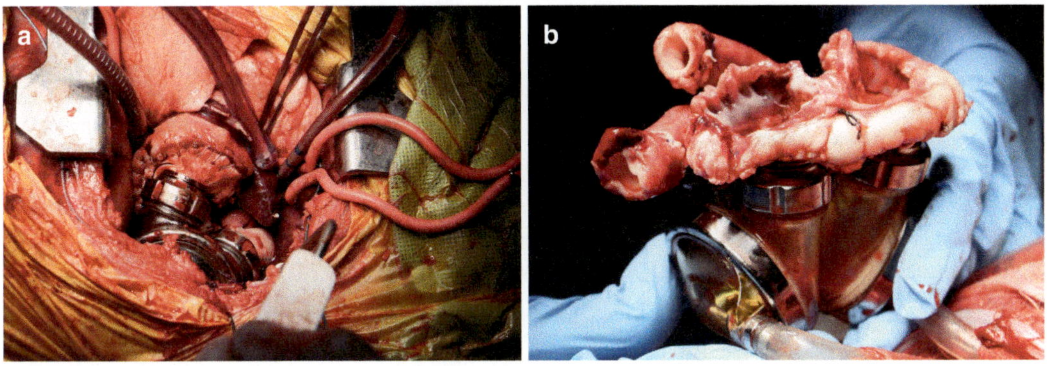

Fig. 36.6 (**a**) Pediatric TAH implanted in a 26.8 kg lamb. The left pump is visible. (**b**) The excised TAH

The left atrial connector cuff was trimmed and sutured to the left atrium with a double layer of running 4-0 Prolene. The process was repeated for the right atrial anastomosis. The grafts for the PA and the aorta were measured, trimmed, and then sutured in place with running 5-0 Prolene. The drivelines were tunneled through the right chest wall. The left and right devices were connected to the inflow and outflow grafts, and deairing maneuvers performed while venting through the arterial suture lines. The arterial suture lines were then secured. The aortic clamp was removed as well as the caval tourniquets.

The pumps were gradually activated and increased as bypass supported was weaned. Space constraints within the limited volume of the mediastinum appeared to compromise inflow to the right pump. The pumps were repositioned which improved filling. Cardiopulmonary bypass was discontinued. Hemodynamic parameters were generally acceptable, although they waxed and waned. Bypass cannulas were removed and cannulation sites secured. Surgical hemostasis was pursued and appeared satisfactory. Chest tubes were placed and the thoracotomy was closed in layers with an intervening catheter for local anesthetic infusion. The skin was closed with staples. Total CPB time was 3 hours 45 minutes. This animal did not recover and was fully heparinized and euthanized postoperatively due to an extremely low hematocrit (10%), with the bleeding source not clearly identified.

Summary

Bridge-to-transplant mechanical circulatory support for infants has been challenging, in part due to the lack of a suitable prosthetic TAH. Penn State is developing an infant-sized pulsatile TAH based on the Penn State pediatric VAD. The valves and connectors have been modified to reduce the pump dimensions and prevent atrial compression. Chronic animal studies are planned, along with anatomic fit studies using human CT and MR data.

Acknowledgments Thanks to Patrick Leibich, Kirby Bletcher, Bradley Doxtater, and Eric Yeager for their outstanding fabrication skills; Gerson Rosenberg, PhD, for design input; Heidi Finicle, Dr. John L. Myers, and Dr. William S. Pierce for their assistance in surgical planning and animal surgery; perfusionists Robert Wise, Karl Woitas, and Gary Irwin; John Reibson and Branka Lukic for instrumentation support; Choon-Sik Jhun, PhD, and Chris Scheib for valve analysis; and surgical technologists Dolly Schepps and Nate Shanaman.

Funding Source This research is supported by NIH grant R01HL131921.

OPTN/SRTR 2016 Annual Data Report is not copyrighted. Readers are free to duplicate and use all or part of the information contained in this publication. Data are not copyrighted and may be used without permission if appropriate citation information is provided.

References

1. Kirklin JK, Pagani FD, Kormos RL, Stevenson LW, Blume ED, Myers SL, et al. Eighth annual INTERMACS report: special focus on framing the impact of adverse events. J Heart Lung Transplant. 2017;36(10):1080–6.
2. Copeland JG. SynCardia total artificial heart: update and future. Tex Heart Inst J. 2013;40(5):587–8.
3. Torregrossa G, Anyanwu A, Zucchetta F, Gerosa G. SynCardia: the total artificial heart. Ann Cardiothorac Surg. 2014;3(6):612–20.
4. Torregrossa G, Morshuis M, Varghese R, Hosseinian L, Vida V, Tarzia V, et al. Results with SynCardia total artificial heart beyond 1 year. ASAIO J. 2014;60(6):626–34.
5. Gerosa G, Scuri S, Iop L, Torregrossa G. Present and future perspectives on total artificial hearts. Ann Cardiothorac Surg. 2014;3(6):595–602.
6. Thanavaro KL, Tang DG, Kasirajan V, Shah KB. Clinical indications for implantation of the total artificial heart. ASAIO J. 2014;60(5):594–6.
7. Copeland J, Copeland H, Nolan P, Gustafson M, Slepian M, Smith R. Results with an anticoagulation protocol in 99 SynCardia total artificial heart recipients. ASAIO J. 2013;59(3):216–20.
8. Colvin M, Smith JM, Hadley N, Skeans MA, Carrico R, Uccellini K, et al. OPTN/SRTR 2016 annual data report: heart. Am J Transplant. 2018;18(Suppl 1):291–362.
9. Almond CS, Morales DL, Blackstone EH, Turrentine MW, Imamura M, Massicotte MP, et al. Berlin Heart EXCOR pediatric ventricular assist device for bridge to heart transplantation in US children. Circulation. 2013;127(16):1702–11.
10. Voeller RK, Epstein DJ, Guthrie TJ, Gandhi SK, Canter CE, Huddleston CB. Trends in the indications and survival in pediatric heart transplants: a 24-year single-center experience in 307 patients. Ann Thorac Surg. 2012;94(3):807–15; discussion 15–6.
11. Almond CS, Singh TP, Gauvreau K, Piercey GE, Fynn-Thompson F, Rycus PT, et al. Extracorporeal membrane oxygenation for bridge to heart transplantation among children in the United States: analysis of data from the Organ Procurement and Transplant Network and Extracorporeal Life Support Organization Registry. Circulation. 2011;123(25):2975–84.
12. Weinstein S, Bello R, Pizarro C, Fynn-Thompson F, Kirklin J, Guleserian K, et al. The use of the Berlin Heart EXCOR in patients with functional single ventricle. J Thorac Cardiovasc Surg. 2014;147(2):697–704; discussion -5.
13. Kirklin JK, Pearce FB, Dabal RJ, Carlo W Jr, McGiffin DC. Cardiac transplantation and mechanical support for functional single ventricle. World J Pediatr Congenit Heart Surg. 2012;3(2):183–93.
14. Ideen C, Albers E, Warner P, Permut L, Kemna M. Effect of initial surgical palliation on allosensitization and post-transplant outcomes in infants with hypoplastic left heart syndrome. J Heart Lung Transplant. 2014;33(11):1178–80.
15. Auerbach SR, Smith JK, Gralla J, Mitchell MB, Campbell DN, Jaggers J, et al. Graft survival is better without prior surgery in cardiac transplantation for functionally univentricular hearts. J Heart Lung Transplant. 2012;31(9):987–95.
16. Kirklin JK, Naftel DC, Kirklin JW, Blackstone EH, White-Williams C, Bourge RC. Pulmonary vascular resistance and the risk of heart transplantation. J Heart Transplant. 1988;7(5):331–6.
17. Zafar F, Jefferies JL, Tjossem CJ, Bryant R 3rd, Jaquiss RD, Wearden PD, et al. Biventricular Berlin Heart EXCOR pediatric use across the United States. Ann Thorac Surg. 2015;99(4):1328–34.
18. Cooper BT, Roszelle BN, Long TC, Deutsch S, Manning KB, Cooper BT, et al. The 12 cc Penn State pulsatile pediatric ventricular assist device: fluid dynamics associated with valve selection. J Biomech Eng. 2008;130(4):041019.
19. Snyder A, Pae WE, Boehmer J, Rosenberg G, Weiss W, Pierce W, et al. First clinical trials of a totally implantable destination therapy ventricular assist system. J Congest Heart Fail Circ Support. 2001;1(4):185–92.
20. Mehta SM, Pae WE Jr, Rosenberg G, Snyder AJ, Weiss WJ, Lewis JP, et al. The LionHeart LVD-2000: a completely implanted left ventricular assist device for chronic circulatory support. Ann Thorac Surg. 2001;71(3 Suppl):S156–61; discussion S83–4.
21. Pierce WS, Rosenberg G, Donachy JH, Landis DL, Weiss W, Hensley F, et al. Postoperative cardiac support with a pulsatile assist pump: techniques and results. Artif Organs. 1987;11(3):247–51.
22. Pierce WS, Snyder AJ, Rosenberg G, Weiss WJ, Pae WE, Waldhausen JA. A long-term ventricular assist system. J Thorac Cardiovasc Surg. 1993;105(3):520–4.
23. Rosenberg G, Snyder A, Weiss W, Landis DL, Geselowitz DB, Pierce WS. A cam-type electric motor-driven left ventricular assist device. J Biomech Eng. 1982;104(3):214–20.
24. Rosenberg G, Snyder AJ, Weiss WJ, Kusagawa H, Rawhouser MA, Prophet GA, et al. Dynamic in vitro and in vivo performance of a permanent total artificial heart. Artif Organs. 1998;22(1):87–94.
25. Lukic B, Clark JB, Izer JM, Cooper TK, Finicle HA, Cysyk J, et al. Chronic ovine studies demonstrate low thromboembolic risk in the Penn State infant VAD. ASAIO J. 2019;65(4):371–9.
26. Connell JM, Khalapyan T, Myers JL, Rosenberg G, Weiss WJ. Anatomic fit assessment for the Penn State pediatric ventricular assist device. ASAIO J. 2007;53(6):687–91.
27. Kao PC, Chiang WC, Yang CW, Chen SJ, Liu YP, Lee CC, et al. What is the correct depth of chest compression for infants and children? A radiological study. Pediatrics. 2009;124(1):49–55.

28. Park SS, Sanders DB, Smith BP, Ryan J, Plasencia J, Osborn MB, et al. Total artificial heart in the pediatric patient with biventricular heart failure. Perfusion. 2014;29(1):82–8.
29. Brickman AD, Landis DL, O'Bannon W, Mortimer VD Jr, Nisley E, Pierce WE. Control system implementation in the air-driven artificial heart. Trans Am Soc Artif Intern Organs. 1974;20(B): 680–4.
30. Landis DL, Pierce WS, Rosenberg G, Donachy JH, Brighton JA. Long-term in vivo automatic electronic control of the artificial heart. Trans Am Soc Artif Intern Organs. 1977;23:519–25.
31. Pierce WS, Landis D, O'Bannon W, Donachy JH, White W, Phillips W, et al. Automatic control of the artificial heart. Trans Am Soc Artif Intern Organs. 1976;22:347–56.
32. Weiss WJ, Rosenberg G, Snyder AJ, Pierce WS, Pae WE, Kuroda H, et al. Steady state hemodynamic and energetic characterization of the Penn State/3M Health Care Total Artificial Heart. ASAIO J. 1999;45(3):189–93.
33. Weiss WJ, Rosenberg G, Snyder AJ, Donachy J Sr, Reibson J, Kawaguchi O, et al. A completely implanted left ventricular assist device. Chronic in vivo testing. ASAIO J. 1993;39(3):M427–32.
34. Weiss WJ. Pulsatile pediatric ventricular assist devices. ASAIO J. 2005;5(15):540–5.
35. Weiss WJ, Carney EL, Clark JB, Peterson R, Cooper TK, Nifong TP, et al. Chronic in vivo testing of the Penn State infant ventricular assist device. ASAIO J. 2012;58(1):65–72.

CorWave LVAD: Insight into Device Concept and Stage of Development

37

Carl Botterbusch, Trevor Snyder, Pier-Paolo Monticone, Louis de Lillers, Alexandra Schmidt, and Charlotte Rasser

Introduction

The rotary pump technology used in all currently marketed left ventricular assist devices (LVADs) dampens the pulsatility of the native heart and damages the blood, which contributes to the onset of severe adverse events: 2 years after LVAD implantation, four out of five patients had a severe adverse event related to the pump [1]. A growing number of cardiac surgeons are demanding more physiologic pressure development from pumps [2, 3].

Based on a disruptive technology that uses an undulating membrane inspired by the swimming movement of marine animals, CorWave is developing a unique cardiac support pump, with the aim of reproducing the heart's pulsatility in a compact size while minimizing blood damage.

C. Botterbusch, MBA (✉) · T. Snyder, PhD
L. de Lillers, Master in Management
A. Schmidt, MSc · C. Rasser, PhD
CorWave SA, Clichy, France
e-mail: carl.botterbusch@corwave.com; trevor.snyder@corwave.com; louis.delillers@corwave.com; alexandra.schmidt@corwave.com; charlotte.rasser@corwave.com

P.-P. Monticone, PhD
Geneva, Switzerland

CorWave Core Technology

CorWave LVAD: A Biomimetic Wave Membrane Pump

CorWave LVAD (see Fig. 37.1) is based on a unique technology that differs from the rotary technologies used in LVADs currently on the market: the wave membrane.

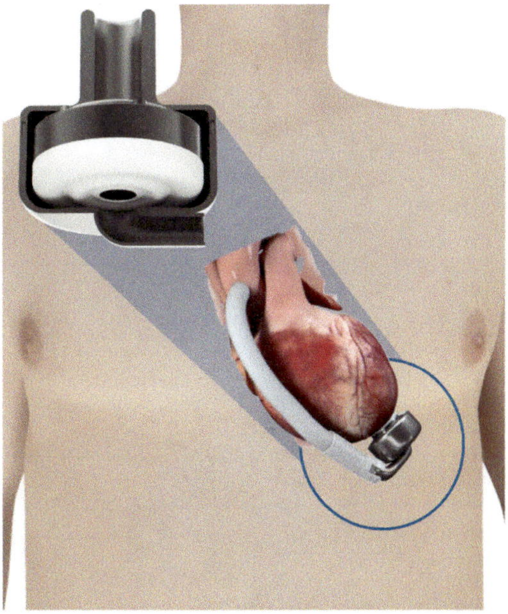

Fig. 37.1 Schematic representation of an implanted CorWave LVAD. The inflow cannula is positioned inside the left ventricle, and the outflow graft is connected to the ascending aorta

© Springer Nature Switzerland AG 2020
J. H. Karimov et al. (eds.), *Mechanical Support for Heart Failure*,
https://doi.org/10.1007/978-3-030-47809-4_37

The biomimetic membrane pump concept was invented by Jean-Baptiste Drevet, a French engineer who conducted fundamental research for nearly 20 years with the Roberval Laboratory of Mechanics, Acoustics and Materials Engineering. The pump concept is inspired from the undulating swimming motion of marine animals. This swimming mode is so effective to move in a fluid that most marine animals, whether fish or marine mammals, use it to propel themselves through water, as described in works by Dr. Fish [4]. In the wave membrane pump, a polymer membrane reproduces the undulations of marine animals, as shown in Fig. 37.2. Since the membrane is fixed at one end and framed by two walls, the undulations create forward flow. Where the marine animal's motion propels it through the fluid, instead the fixed membrane displaces and propels the fluid.

Advantages Derived from This Unique Technology

Physiologic Pulsatility

An important advantage offered by the undulating membrane pump technology is the ability to reproduce physiological pulsatility, defined as all the cyclical variations in pressure, flow, and volume within the vascular system generated by a healthy heart.

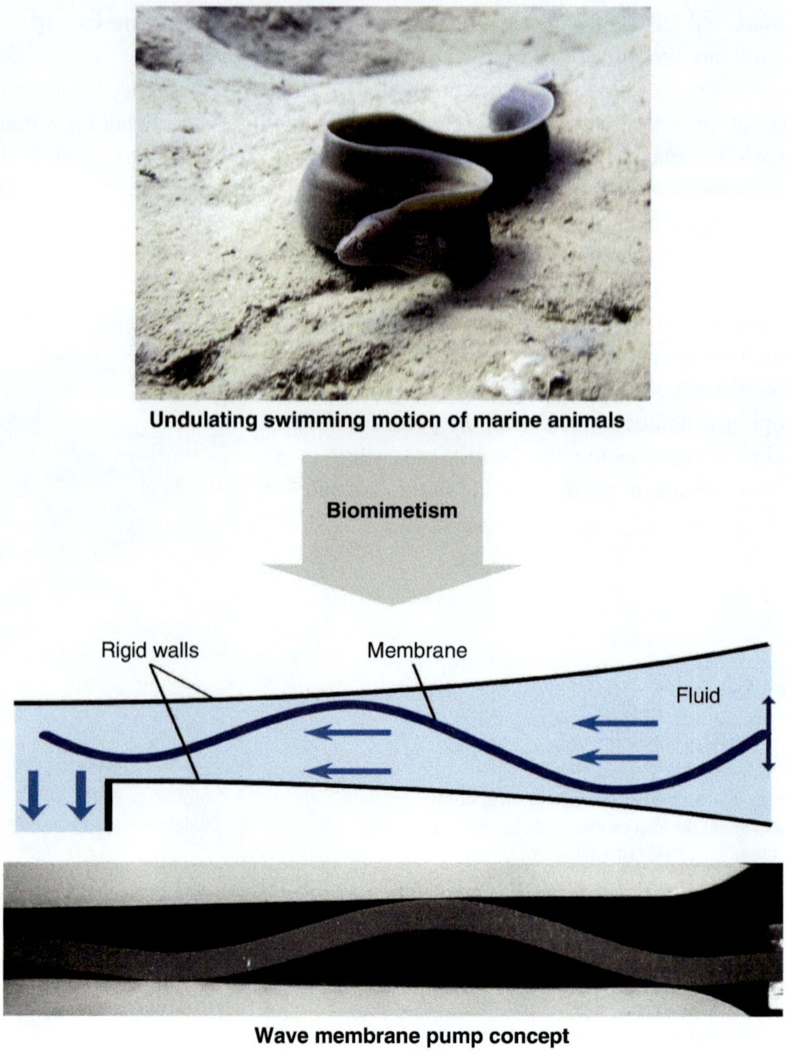

Fig. 37.2 Concept evolution: from the swimming motion of marine animals to CorWave LVAD

Several clinical studies have linked the disturbances in flow caused by rotary pumps to the high incidence of adverse events associated with their use. In particular, Purohit et al. [5] have recently summarized the complications observed in patients supported with continuous-flow LVADs, as shown in Fig. 37.3.

The links between nonphysiologic blood flow and the occurrence of gastrointestinal bleeding are especially well documented. In a meta-analysis, Draper et al. estimated the incidence rate of gastrointestinal bleeding in patients supported with a continuous-flow LVAD at 23%. This incidence rate was 4.5 times lower in patients implanted with a pulsatile-flow LVAD [6]. Furthermore, several teams (from the University of Utah [7], University of Florida [8], and University of Alabama [9]) have reported in independent studies that patients under continuous-flow assist but with significant residual pulsatility presented less gastrointestinal bleeding.

It has also been established that support by continuous-flow LVAD causes acquired von Willebrand factor defect [10–12], which contributes to gastrointestinal bleeding [13]. A recent study [14] suggested that within the scope of flow disorders, abnormal pulsatility is a dominant factor in the onset of von Willebrand factor deficiency.

For these reasons, in recent years pulsatility has become a prominent topic suggested by key opinion leaders in conferences on mechanical circulatory assist devices. Various centers, expert in the implantation of LVADs, such as the Cleveland Clinic (Ohio, USA) and the University of Louisville (Kentucky, USA), have expressed their wish to see technologies developed that enable a more physiologic operation of cardiac pumps, especially by creating a near-normal degree of pulsatility. Dr. Nader Moazami and collaborators thus suggested that "lack of pulsatility may be one factor that has limited the success of continuous flow LVADs and [...] that research needs to focus on methods to generate pulsatility" [3], and clinicians at the University of Louisville concluded that "the ultimate solution will be a return to re-creating our natural physiology" [2].

Pulsatility can be characterized at different places in the vascular system with a wide range of measurements. Variations in pressure in the ascending aorta represent the reference measurement to estimate the pulsatility produced by the heart or an assist pump [15]. Three quantities describe the pressure curve:

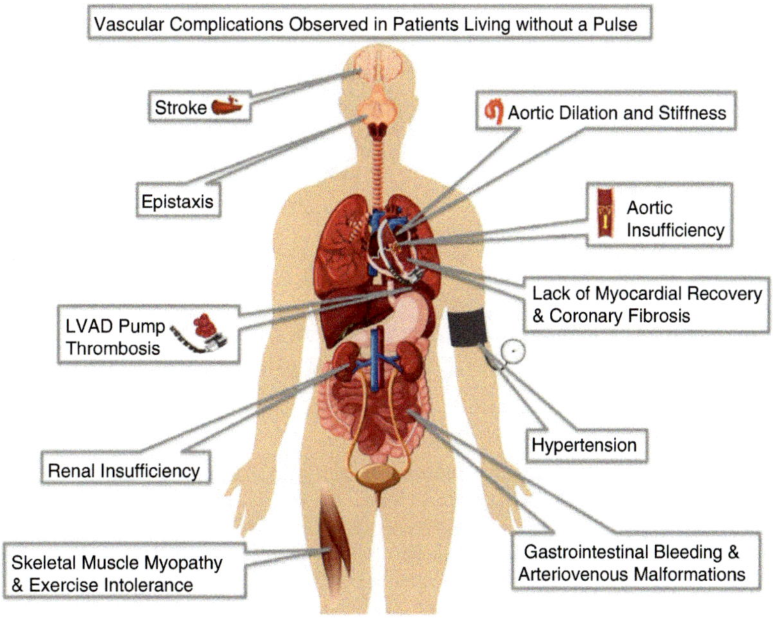

Fig. 37.3 Vascular complications observed in patients living without a pulse. Alterations in vascular properties that result from the unique physiology of blood flow from continuous-flow left ventricular assist *devices* (CF-LVADs) may contribute to the pathogenesis of many common CF-LVAD complications. (Figure reproduced from Purohit et al. [5], with permission of Wolters Kluwer Health, Inc.)

- The pulse pressure (ΔP or PP), the difference between maximum and minimum pressure measured in the aorta during a heart cycle, is commonly used. It quantifies the amplitude of the pulse and degree of expansion of the vascular system. This parameter can only be measured accurately using an arterial catheter or Doppler, contrary to the pulsatility index that some LVADs calculate using electrical data. Pulsatility index is an indirect and noninvasive estimation of the pressure amplitude and especially the residual pulsatility produced by the native heart.
- The maximum pressure curve slope (dP/dt_{max}), the maximum value of the slope of the pressure curve during systole. This value is highly correlated to the power of the pulse wave and, therefore, to the peripheral pulse.
- The heart rate, which is between 60 and 80 beats per minute in healthy adults, and typically higher in heart failure patients.

The typical pulsatility characteristics of a heart failure patient are presented on the arterial pressure curve in Fig. 37.4a. The pulsatility generated by the heart is altered by rotary LVADs currently on the market, which were designed to provide a constant flow between 4 L/min and 6 L/min for total assist. The arterial dP/dt_{max} typically falls below 200 mmHg/s and the pulse pressure below 20 mmHg [16].

In recent years, pump speed modulation algorithms have been developed primarily as attempts to improve pump washing and enable valve opening and to some extent an artificial pulsatility. The principle, summarized in Fig. 37.4b, is to modulate the pump rotational speed to generate flow and pressure variations. However, rotary pumps are not adapted to reproduce fast flow and pressure modulations. The main limitation is rotor inertia, which imposes a significant latency [17] and requires considerable additional energy expenditure to obtain sudden variations in rotational speed [18], thus reducing battery life for patients. By contrast, CorWave wave membrane is lightweight (about 10 g) and only moves a few millimeters by oscillation, which limits its inertia. Pulsatility is created by switching rapidly between levels of output, modulating oscillation frequency and amplitude. A period of high output is generated following detection of systole to augment native cardiac output. Pump flow is reduced during diastole to a low level aimed at preventing back flow through the pump. This does not require additional energy, and the time delay to go from the low output to the high output state is very short – under 80 ms. Finally, the pump is designed to provide a peak flow of up to 15 L/min at physiologic pressures. The creation of a near-physiological pulsatility without a significant increase in electrical consumption is therefore possible, as schematically represented in Fig. 37.4c.

Beyond the three key parameters listed above, other parameters directly or indirectly derived by flow and pressure have been explored, notably the PI (pulsatility index), SHE (surplus hemodynamic energy), and EEP (energy equivalent pressure) [15].

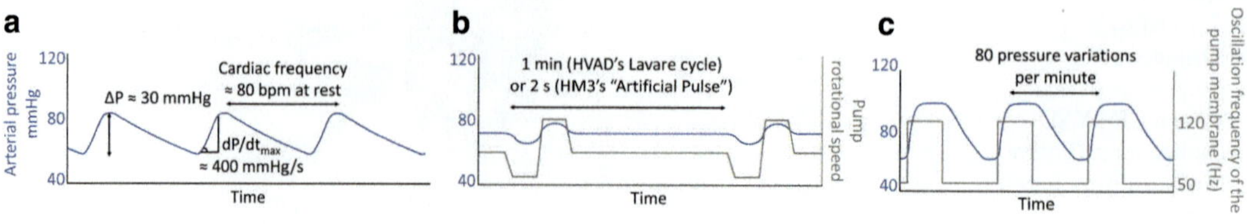

Fig. 37.4 (**a**) Schematic arterial pressure curve in a heart failure patient. (**b**) Principle of algorithms for the modulation of rotary pump speed. A periodic crenel modulation of the pump rotational speed is introduced at a frequency lower than the cardiac rate, so as to re-create a partial ΔP. (**c**) Principle of the algorithm for the modulation of CorWave pump frequency. The oscillation frequency and amplitude of the pump membrane are modulated, synchronously with heartbeats, to reproduce a physiological ΔP. ΔP pulse pressure, bpm beats per minute, HM3 HeartMate 3

Gentle Propulsion of the Blood

The velocity of the blood flow created by LVADs is a key parameter, as it is a predicate of the shear stress that blood is exposed to. In healthy patients, blood is exposed to peak velocities between 1 and 2 m/s when it passes through aortic and mitral valves during systole. In some valvular pathologies, these velocity fields are elevated (higher than 5 m/s [19]) and create mechanical stress on blood. These forces damage blood cells and plasma proteins, leading to hemostasis disorders associated with a high level of gastrointestinal bleeding [20]. Rotary pumps, which operate at several thousand rotations per minute, impose blood velocities estimated to be at least 4 m/s [21], that is, more than twice the flow speeds observed in healthy patients. The membrane in CorWave LVAD oscillates at approximately 100 Hz, which causes the membrane wave to move at an average speed of 1–1.5 m/s and creates equivalent flow speeds in the propelled blood. Therefore, at an equivalent flow (7 L/min), CorWave LVAD exposes less than 1% of the blood volume to speeds higher than 2 m/s, while Selgrade and collaborators [21] estimated that in a centrifugal LVAD approximately 70% of the volume is exposed to flow speeds higher than 2 m/s.

Compact Size

The wave membrane technology has enabled the development of a pump which is compact while providing sufficient flow for full support and the ability to produce the very high peak flows required for full pulsatility.

A Wave Membrane Integrated into a Smart LVAD

Like current rotary LVADs, in addition to the implantable pump itself, the LVAD system developed by CorWave is composed of different components presented in Fig. 37.5.

The pump is designed to be implantable by means of the two main surgical techniques used today for LVADs: the implantation technique based on sternotomy and a less invasive approach, with a left thoracotomy and a hemi-sternotomy or right thoracotomy.

At the heart of CorWave LVAD is the membrane, made of a biocompatible elastomeric material, shown in Fig. 37.6. It is driven by an electromagnetic actuator encapsulated in titanium. As depicted in Fig. 37.7, the electrical current flows in two coils and thus generates electromagnetic fields that move a magnetized ring in an up and down movement of +/– 1 mm. The magnet ring is connected directly to the membrane, thus driving the oscillation of the membrane. A pair of ring springs controls the magnet ring position and provides energy to return the magnet ring to the center point.

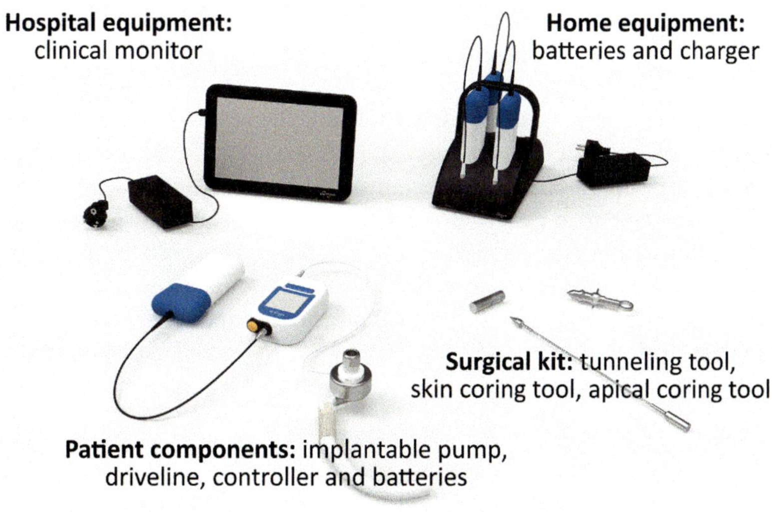

Fig. 37.5 Components of the CorWave LVAD system

Fig. 37.6 Photograph of the main components of CorWave LVAD. The wave membrane (the white circular element with the center hole) is shown in the center element

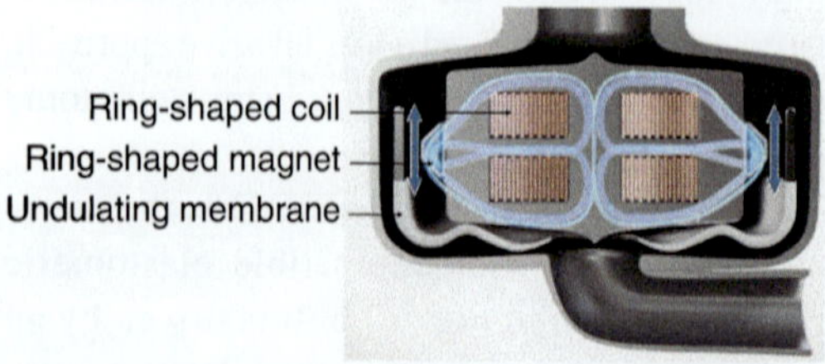

Fig. 37.7 Electromagnetic actuator and transmission of the oscillation to the membrane. The magnetic fields generated in the coils induce the oscillation of the magnet, which drives the oscillation of the wave membrane

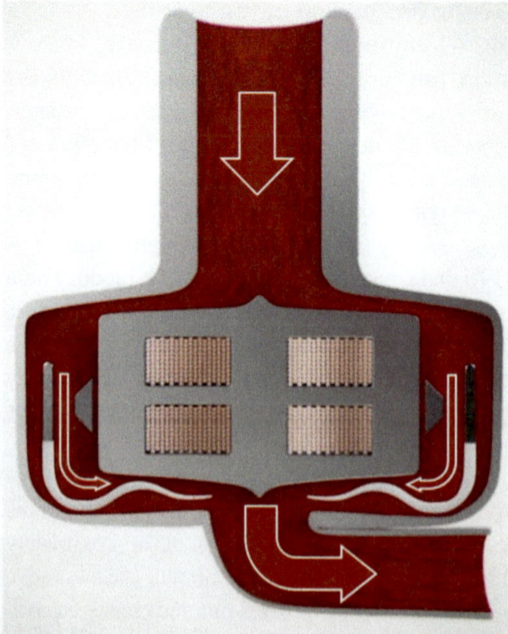

Fig. 37.8 Blood path in the pump. Blood enters the pump through the inlet cannula from the left ventricle, is propelled by the inward undulations of the membrane, and exits through the outlet cannula

The blood flow path through the pump is shown in Fig. 37.8. Blood enters the pump through the inlet cannula and flows around the actuator. It then flows on either side of the flow separator and along both sides of the membrane before being propelled toward the outlet cannula by the inward undulations of the membrane.

In order to regulate the blood flow in real time, the oscillation frequency of the pump is controlled by an algorithm able to synchronize with the heart rate without the use of separate sensors. Indeed, the low inertia of the pump makes it highly sensitive to pressure and flow changes. This allows the implementation of a "smart pulsatility," which adapts to the heart rate at any given time, as well as preventing and correcting possible suction events.

Device Development: From In Silico to In Vivo Testing

In Silico Analyses

A key part of the development of CorWave LVAD is the design of the hydraulic flow path, which includes the geometry of the pump inlet, the body of the pump, and the pump outlet. This design impacts both hydraulic efficiency and blood compatibility, by minimizing hemolysis and vortex formation. The optimization of the hydraulic flow path design requires computational fluid dynamics simulations. Computer simulations of the flow field are routinely used for rotary LVADs. However, the pumping element of CorWave technology is a flexible membrane, which, unlike rigid impellers, is elastically deformed while pumping the blood. Thus, the simulation of fluid behavior in the pump head of CorWave LVAD requires the resolution of a particularly complex fluid structure interaction problem. To tackle this task, CorWave deploys a blend of commercial simulation software and custom solvers developed in collaboration with research centers. An extra challenge is to simulate blood damage and thrombus formation and to validate the simulation with experimental results, for instance, by using optical methods such as digital image correlation, as shown in Fig. 37.9. An example of result from a computational fluid dynamics simulation on a prototype of CorWave LVAD is presented in Fig. 37.10. The fluid velocity is around 1 m/s in most of the fluid volume.

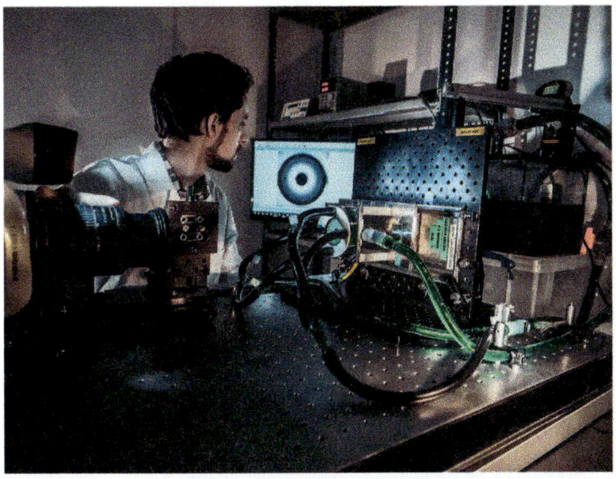

Fig. 37.9 Membrane visualization technique with a high-speed camera. Digital image correlation is used to compare the experimental images with the results from computational fluid dynamics simulations

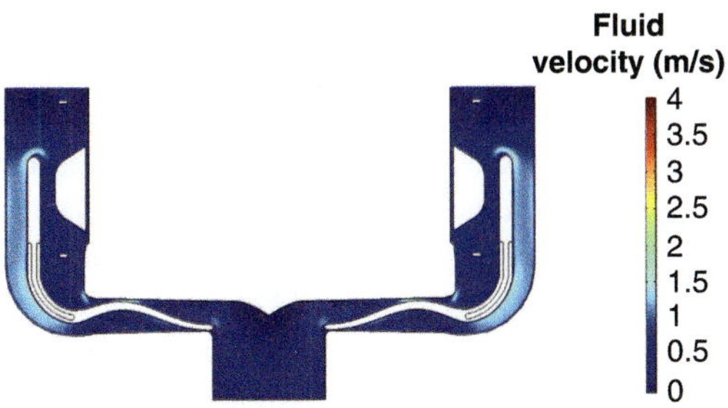

Fig. 37.10 Computational fluid dynamics simulation of CorWave LVAD

Another part of the development process of CorWave LVAD requires a different type of computational simulation. For the optimization of the actuator design, which includes the design of the coils, yoke, and magnet as well as the choice of ferromagnetic material for the three components, 2D and 3D magnetic simulations are utilized. An example from such a simulation is shown in Fig. 37.11.

In Vitro Testing

To validate computational simulations and obtain more data on the hydraulic performances of new prototypes, several test benches have been developed. A mock circulatory loop (see Fig. 37.12) simulates the hydraulic properties of the left heart chambers and valves and of the peripheral circulation, in a healthy adult or a heart failure patient, in order to evaluate the performances of the LVAD in relevant conditions. The hydraulic performances of the current prototype of CorWave LVAD are presented on a pressure-flow curve in Fig. 37.13.

CorWave LVAD operation in pulsatile mode was assessed in a mock circulatory loop and compared with HeartMate II (HM II) and HVAD pumps in the same setup. As can be seen in Fig. 37.14, both HM II and HVAD dampened the pulsatility of the simulated circulatory system of a heart failure patient, resulting in pulse pressure less than 20 mmHg, while CorWave LVAD restored a physiological pulse pressure.

The durability of CorWave LVAD's core component, the polymer membrane, was evaluated using accelerated stress cycling. As shown in Fig. 37.15, a conservative estimation of the minimum membrane durability is 10^{12} cycles, much higher than the maximal expected lifetime of the LVAD.

Blood compatibility of the device is assessed in a blood circulatory loop, according to ASTM F-1841, using both bovine and porcine blood. Assessment of hemolysis has been used as a key

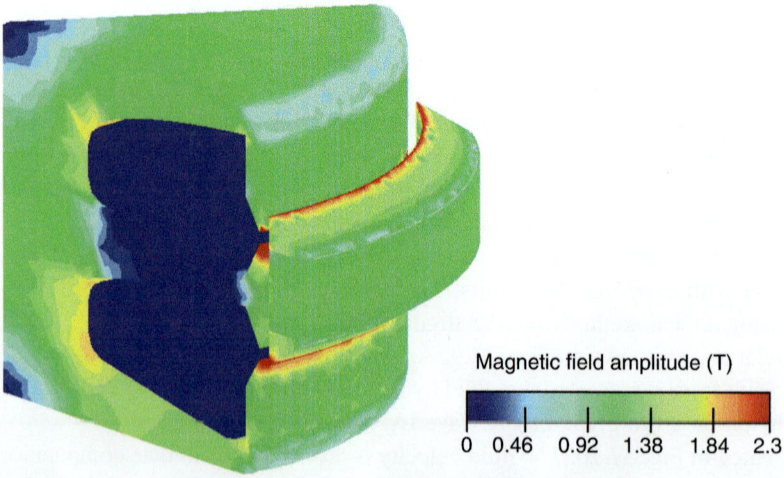

Fig. 37.11 Simulation of the magnetic field in the actuator of CorWave LVAD

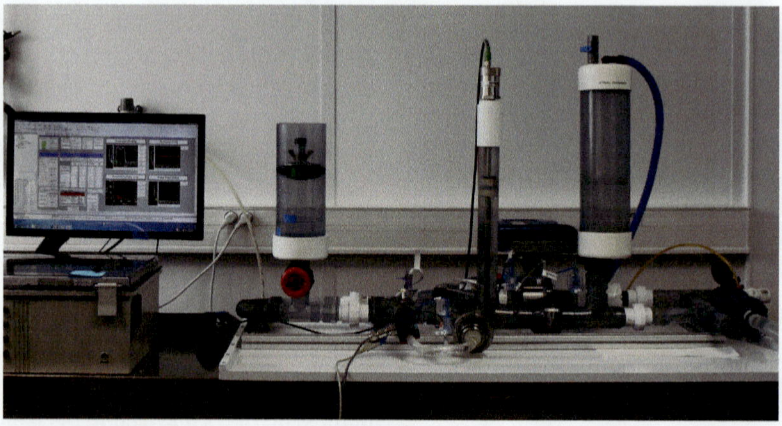

Fig. 37.12 Mock circulatory loop. The left ventricle is modelled by a pump and a compliance chamber. A second compliance chamber represents the aorta and a third compliance chamber the peripheral circulation. Two hydraulic valves play the role of the mitral and aortic valves. The mock loop can be set to simulate the circulation in a healthy individual or in a heart failure patient. An LVAD is then connected to the system, and its effects on pressure and flow are measured

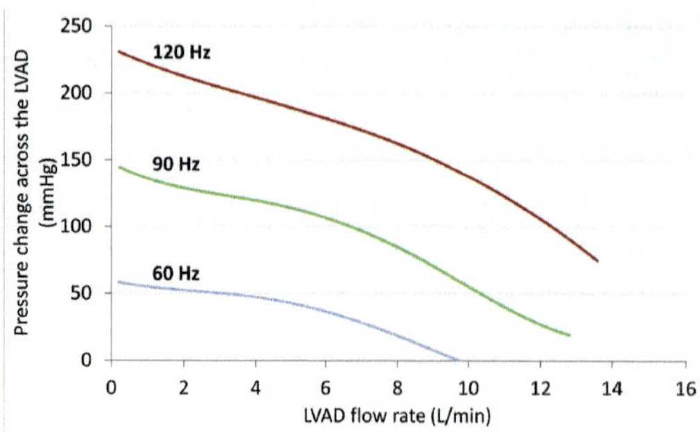

Fig. 37.13 Pressure-flow curve of the current prototype of CorWave LVAD

input to the iterative design process used to optimize the pump performance. To date over 2000 hemolysis tests have been conducted including those on the pump, control devices for normalization, and competitive devices. Blood damage assessment has been conducted over a wide range of operating parameters in continuous flow, as well as in pulsatile operating mode.

The transition of the pump from low to high output in pulsatility mode has been carefully optimized to eliminate any increased blood damage compared to the average of the two in con-

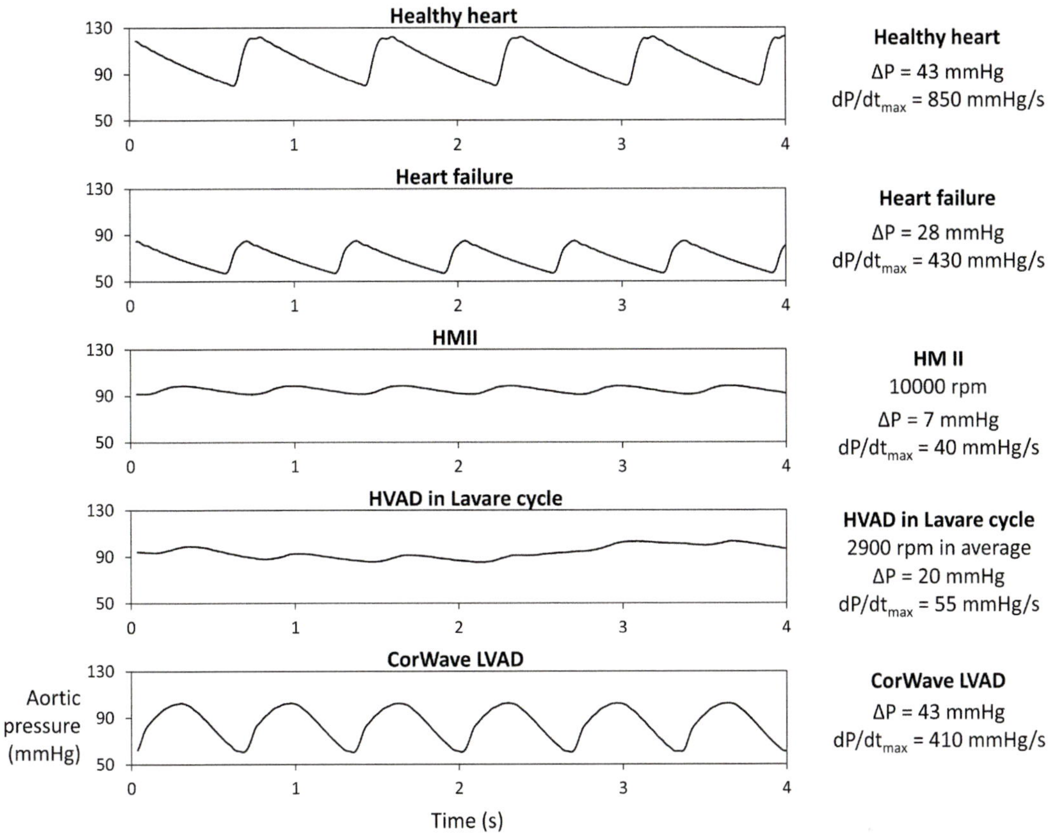

Fig. 37.14 Aortic pressure measured in a mock circulatory loop with different LVADs. The pulse pressure (ΔP) and maximum pressure curve slope (dP/dt_{max}) are much reduced with HeartMate II (HMII) and HVAD, whereas they are maintained or increased with CorWave LVAD. Experiments performed at KU Leuven

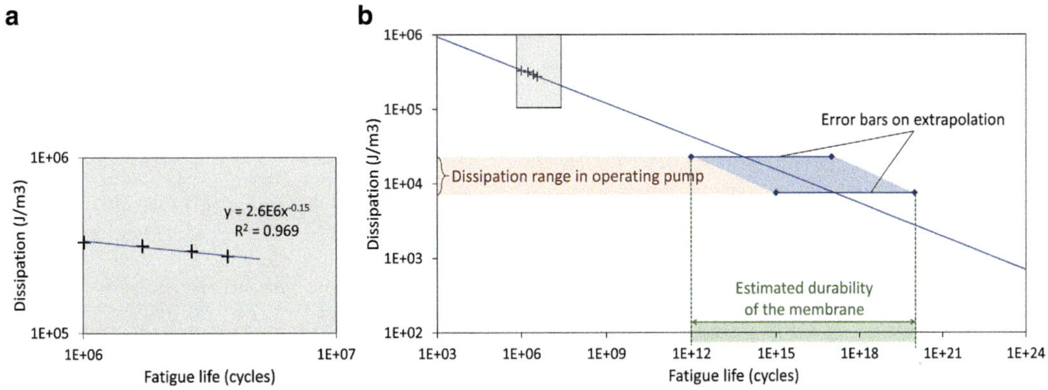

Fig. 37.15 (**a**) A model was constructed from measurements of the membrane durability at high dissipation (i.e., high stress and strain). (**b**) The extrapolation on the measurements based on this model gave a conservative low boundary of the estimated fatigue life of the membrane at 10^{12} cycles, much higher than the maximal expected lifetime of the pump (approximately 3.10^{10} cycles). Experiments performed at Cermel (Tours University)

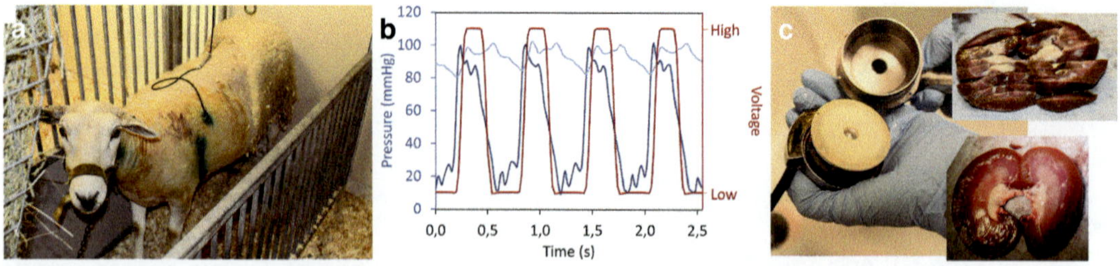

Fig. 37.16 (**a**) Sheep implanted with a CorWave LVAD. (**b**) Synchronization of CorWave LVAD with the animal's heartbeats. Dark blue: Left ventricular pressure. Light blue: Aortic pressure. Red: CorWave LVAD voltage. Work done in collaboration with KU Leuven and Lille University Hospital. (**c**) No significant findings in either the pump or end organs at explant

tinuous mode. The current version of the device meets a clinically acceptable level of hemolysis in both continuous and pulsatile modes.

Assessment of other key blood parameters including von Willebrand factor cleaving and platelet activation is underway.

In Vivo Testing

Animal trials are currently being conducted on CorWave LVAD (see Fig. 37.16a). The synchronization algorithm is being validated in these tests. As shown in Fig. 37.16b, CorWave LVAD is able to operate in a co-pulsatile mode by detecting the animal's heartbeats, using only the data on the instant power consumption of the pump.

Company Development

CorWave was created in Paris in 2011 by MD Start, a medtech incubator. It raised a series A financing in 2013 and a series B financing in 2016. Today, CorWave has over 45 full-time employees, among whom 35 are dedicated to R&D. Its R&D program is supported by the French "Programme d'Investissements d'Avenir." CorWave is currently focusing on achieving long-term implantation in animals, before starting a European clinical trial for CE marking.

References

1. Kirklin JK, Pagani FD, Kormos RL, Stevenson LW, Blume ED, Myers SL, et al. Eighth annual INTERMACS report: special focus on framing the impact of adverse events. J Heart Lung Transplant. 2017;36(10):1080–6.
2. Cheng A, Williamitis CA, Slaughter MS. Comparison of continuous-flow and pulsatile-flow left ventricular assist devices: is there an advantage to pulsatility? Ann Cardiothorac Surg. 2014;3(6):9.
3. Moazami N, Dembitsky WP, Adamson R, Steffen RJ, Soltesz EG, Starling RC, et al. Does pulsatility matter in the era of continuous-flow blood pumps? J Heart Lung Transplant. 2015;34(8):999–1004.
4. Fish FE. Diversity, mechanics and performance of natural aquatic propulsors. In: Liebe R, editor. WIT transactions on state of the art in science and engineering. 1st ed. Southampton: WIT Press; 2006. p. 57–87.
5. Purohit SN, Cornwell WK, Pal JD, Lindenfeld J, Ambardekar AV. Living without a pulse: the vascular implications of continuous-flow left ventricular assist devices. Circ Heart Fail. 2018;11(6):e004670.
6. Draper KV, Huang RJ, Gerson LB. GI bleeding in patients with continuous-flow left ventricular assist devices: a systematic review and meta-analysis. Gastrointest Endosc. 2014;80(3):435–446.e1.
7. Wever-Pinzon O, Selzman CH, Drakos SG, Saidi A, Stoddard GJ, Gilbert EM, et al. Pulsatility and the risk of nonsurgical bleeding in patients supported with the continuous-flow left ventricular assist device HeartMate II. Circ Heart Fail. 2013;6(3):517–26.
8. Jabbar HR, Abbas A, Ahmed M, Klodell CT, Chang M, Dai Y, et al. The incidence, predictors and outcomes of gastrointestinal bleeding in patients with left ventricular assist device (LVAD). Dig Dis Sci. 2015;60(12):3697–706.
9. Edwards AL, Fitzmorris P, Pamboukian SV, George JF, Wilcox CM, Peter S. Association of pulsatility with

gastrointestinal bleeding in a cohort of HeartMate II recipients. ASAIO J. 2018;64:472–9.
10. Crow S, Chen D, Milano C, Thomas W, Joyce L, Piacentino V, et al. Acquired von Willebrand syndrome in continuous-flow ventricular assist device recipients. Ann Thorac Surg. 2010;90(4):1263–9.
11. Meyer AL, Malehsa D, Budde U, Bara C, Haverich A, Strueber M. Acquired von Willebrand syndrome in patients with a centrifugal or axial continuous flow left ventricular assist device. JACC: Heart Fail. 2014;2(2):141–5.
12. Kang J, Hennessy-Strahs S, Kwiatkowski P, Bermudez CA, Acker MA, Atluri P, et al. Continuous-flow LVAD support causes a distinct form of intestinal angiodysplasia – novelty and significance. Circ Res. 2017;121(8):963–9.
13. Uriel N, Pak S-W, Jorde UP, Jude B, Susen S, Vincentelli A, et al. Acquired von Willebrand syndrome after continuous-flow mechanical device support contributes to a high prevalence of bleeding during long-term support and at the time of transplantation. J Am Coll Cardiol. 2010;56(15):1207–13.
14. Vincent F, Rauch A, Loobuyck V, Robin E, Nix C, Vincentelli A, et al. Arterial pulsatility and circulating von Willebrand factor in patients on mechanical circulatory support. J Am Coll Cardiol. 2018;71(19):2106–18.
15. Soucy KG, Koenig SC, Giridharan GA, Sobieski MA, Slaughter MS. Defining pulsatility during continuous-flow ventricular assist device support. J Heart Lung Transplant. 2013;32(6):581–7.
16. Martina JR, Westerhof BE, de Jonge N, van Goudoever J, Westers P, Chamuleau S, et al. Noninvasive arterial blood pressure waveforms in patients with continuous-flow left ventricular assist devices. ASAIO J. 2014;60(2):154–61.
17. Bozkurt S, van de Vosse FN, Rutten MCM. Enhancement of arterial pressure pulsatility by controlling continuous-flow left ventricular assist device flow rate in mock circulatory system. J Med Biol Eng. 2016;36(3):308–15.
18. Soucy KG, Giridharan GA, Choi Y, Sobieski MA, Monreal G, Cheng A, et al. Rotary pump speed modulation for generating pulsatile flow and phasic left ventricular volume unloading in a bovine model of chronic ischemic heart failure. J Heart Lung Transplant. 2015;34(1):122–31.
19. Baumgartner H, Hung J, Bermejo J, Chambers JB, Evangelista A, Griffin BP, et al. Echocardiographic assessment of valve stenosis: EAE/ASE recommendations for clinical practice. J Am Soc Echocardiogr. 2009;22(1):1–23.
20. Vincentelli A, Susen S, Le Tourneau T, Six I, Fabre O, Juthier F, et al. Acquired von Willebrand syndrome in aortic stenosis. N Engl J Med. 2003;349(4):343–9.
21. Selgrade BP, Truskey GA. Computational fluid dynamics analysis to determine shear stresses and rates in a centrifugal left ventricular assist device. Artif Organs. 2012;36(4):E89–96.

Progress on Total Artificial Heart for Pediatric Patients

38

Kiyotaka Fukamachi, Jamshid H. Karimov, and Takuma Miyamoto

Introduction

Heart failure, a major medical problem affecting approximately 5.7 million Americans, is projected to increase by 50% by 2030 [1]. Mechanical circulatory support (MCS) devices, especially ventricular assist devices (VADs), have become an increasingly accepted treatment modality for adult patients with end-stage heart failure as a bridge to transplant and a permanent implant (destination therapy) due to the shortage of available donor hearts [2, 3]. The most recent report of HeartMate 3 (Abbott, Abbott Park, IL) showed 82.8% of actuarial survival at 2 years [4], which is equivalent to that for heart transplant [5].

VADs are also increasingly used recently to support children to heart transplant [6], as nearly 30% of pediatric patients were bridged on some form of MCS device, of which VADs were the primary modality in nearly 60% of all MCS cases [7]. However, patients with congenital heart disease (CHD) are often not mechanically supported because they are at higher risk of VAD complications [8]. Therefore, they are more likely to die while waiting for heart transplant, and CHD is one of the independent predictors of waiting list mortality [8]. As more than 30% of pediatric heart transplant recipients have CHD [7], there is a need for improved support options for CFD patients.

Currently options for MCS modalities are quite limited to external devices: temporary extracorporeal membrane oxygenation (ECMO), a temporary extracorporeal centrifugal VAD such as CentriMag system (Abbott), and the EXCOR VAD (Berlin Heart GmbH, Berlin, Germany). The EXCOR VAD, a paracorporeal pneumatic pump, is currently the only VAD available to provide long-term support to small children awaiting heart transplantation. The Infant Jarvik 2000 (Jarvik Heart, Inc., New York, NY) [9], a downsized version of the adult Jarvik 2000®, was used in a 5.7 kg baby in Italy. In October 2016, the US Food and Drug Administration (FDA) granted conditional approval to conduct an Investigational Device Exemption (IDE) study of the Jarvik 2015, a 40% larger version of the Infant Jarvik 2000. This study (Clinical Trials number = NCT02954497) will compare the implanted, portable electrically powered Jarvik 2015 LVAD System with the Berlin Heart EXCOR Pediatric System.

CHD often involves both ventricles, and in such cases, either a biventricular assist device

K. Fukamachi, MD, PhD (✉) · J. H. Karimov, MD, PhD
Lerner Research Institute, Cleveland Clinic,
Cleveland Clinic Lerner College of Medicine of Case Western Reserve University, Cleveland Clinic,
Cleveland, OH, USA
e-mail: fukamak@ccf.org; karimoj@ccf.org; miyamot@ccf.org

T. Miyamoto, MD, PhD
Department of Biomedical Engineering,
Lerner Research Institute, Cleveland Clinic,
Cleveland, OH, USA
e-mail: fukamak@ccf.org; karimoj@ccf.org; miyamot@ccf.org

© Springer Nature Switzerland AG 2020
J. H. Karimov et al. (eds.), *Mechanical Support for Heart Failure*,
https://doi.org/10.1007/978-3-030-47809-4_38

(BiVAD) or total artificial heart (TAH) is needed. However, no pediatric TAHs exist other than SynCardia 50 cc TAH (SynCardia Systems, Inc., Tucson, AZ), which is under IDE clinical study (Clinical Trials number = NCT02459054). This device, smaller than their regular 70 cc model, is intended only for patients with a body surface area (BSA) of $\geq 1.2\ m^2$. In this chapter, we review clinical needs, challenges, current technologies, devices under development, and future perspective on TAHs for pediatric patients.

Clinical Needs for Pediatric Total Artificial Heart

End-stage heart failure in children remains a significant problem affecting thousands of families each year [10], which is far exceeding the number of pediatric heart transplants (684 reported in 2015) [11]. CHD is a major cause of infant death during the first year and accounts for the largest proportion of mortality (30–50%) resulting from birth defects [12]. From 1999 to 2006, infant mortality accounted for 48.1% (69,252 deaths in the United States) of all mortality from CHD.

Cardiac surgery for neonates and infants with CHD usually addresses complex cardiac anomalies such as hypoplastic left heart syndrome and transposition of the great artery. Complicated surgical procedures (e.g., Norwood procedure for hypoplastic left heart syndrome or complete repair as in the Jatene procedure for transposition of the great artery) can be complicated by severe heart failure postoperatively with failure to thrive. Persistent residual cyanosis and incomplete correction by these palliative procedures often result in unsatisfactory hemodynamic status after surgery.

For pediatric patients with end-stage heart failure, the options for chronic MCS are limited to paracorporeal devices, as no implantable devices are available for patients with a BSA of $<1.2\ m^2$. In addition, VADs may not be optimal for CHD patients because they may not adequately sustain both pulmonary and systemic circulations, and CHD often involves both left and right ventricles. Elevated pulmonary vascular resistance, which is frequently seen in these patients, also limits left ventricular diastolic filling. Anatomic and physiological complexities including both systolic and diastolic dysfunction are other reasons that VADs have not been used in patients with CHD. In such cases, heart transplantation can be an ideal therapy, but donor hearts for infants and small children are very limited [8]. Therefore, either a BiVAD or TAH is needed, but TAHs are more appropriate than BiVADs for patients with a single ventricle, thrombosed ventricles, restrictive cardiomyopathy, cardiac tumors, severe ventricular septal defects, frail ventricles for difficult cannulation, and intractable arrhythmias. The use of the TAH for complex CHD is potentially paradigm shifting for those selected patients by simplifying and optimizing support rather than placing a BiVAD with concurrent cardiac procedures (i.e., closure of a ventricular septal defect, conduit revision, aortic valve replacement) [13]. TAHs have actually been used in patients to bridge complex transposition of the great artery [14, 15] and failed Fontan [16]. The availability of a safe and effective TAH for infants and small children would likely increase survival on the waiting list, as well as posttransplant survival, in this challenging group. The current alternatives, which include surgical palliation, ECMO, and the EXCOR pediatric VAD, have been shown to have significant limitations in smallest patients.

Challenges in Pediatric Total Artificial Heart

According to the most recent Pediatric Interagency Registry for Mechanical Circulatory Support (Pedimacs) database [17], 42 hospitals implanted 432 devices in 364 patients less than 19 years of age between September 2012 and September 2016, but TAH was used in only 8 patients (2%, all were SynCardia 70 cc TAH). This very low usage percentage against the high need described above may be due to many challenges in pediatric TAHs.

SynCardia 50 cc TAH was implanted into a 10-year-old boy in April 2018 who underwent heart transplant in July 2018. This device, smaller than their regular 70 cc model, is intended only for patients with a BSA of ≥1.2 m^2, i.e., that of an average 11-year-old, not an infant. The smallest patient who received SynCardia 50 cc TAH so far had a BSA of 1.16 m^2. Device size is critically important for appropriate fit and sternal chest closure for chronic use of a pump in small children.

Anticoagulation of patients also remains the biggest challenge in the treatment of patients on MCS [18], as there is no clear consensus on the optimal mode and range of anticoagulation of these patients. The anticoagulant and antiplatelet effects in pediatric patients might be different from older adolescents and adults. Also, circulating blood volume is much less than in adults, so a higher fraction of their blood is in contact with the contacts with MCS device. In addition, VADs usually run at a lower pump flow in pediatric patients than in adults, which may contribute to a higher rate of thrombosis. In fact, the rates of thrombosis and bleeding in the pediatric MCS population have been higher than in adults [19].

Prevention of infections is another challenge in pediatric patients. All intracorporeal MCS devices require a relatively large driveline for VADs or pneumatic lines for the SynCardia TAH to provide power to the device, which serves as an easy pathway to the device for bacteria.

A small market size is also a challenge, as no devices can benefit patients unless they are commercialized. Sizable support from the government is essential for device development, preclinical testing, and IDE clinical trial. A good example is the PumpKIN program by the National Heart, Lung, and Blood Institute (NHLBI) that supported development of pediatric VADs [20].

Total Artificial Heart Used in Pediatric Patients

SynCardia 70 cc TAH

Research into the TAH for adult patients began in the United States in 1963 under the impetus of the American Congress. TAHs are now used in patients with end-stage heart failure to replace cardiac function so that adequate viable perfusion can be restored. Experience with the available versions of volume-displacement TAHs has been encouraging when the TAH is used as a bridge to heart transplant in this subset of patients. However, clinical experience with the TAH as a destination therapy (permanent implant) has been limited.

SynCardia 70 cc TAH is the world's first pneumatically driven TAH and has been approved by the FDA since October 2004 as a bridge to heart transplant (Fig. 38.1, right). With more than 1,700 implants worldwide, the 70 cc TAH accounts for more than 600 patient-years of support with the longest support of 1,700 days. A successful bridge to heart transplant rate was 79% [21]. It is currently undergoing an FDA-approved IDE clinical trial for use as destination therapy in adult patients who are not eligible for heart transplantation (Clinical Trials number = NCT02232659). This 70 cc TAH has also been successfully used in small adults and pediatric patients [13–15], but placement within the small thoracic cavity is challenging.

SynCardia 50 cc TAH

To expand the use of the TAH in smaller patients, SynCardia has developed a 50 cc TAH device (Fig. 38.1, left) that is 30% smaller by

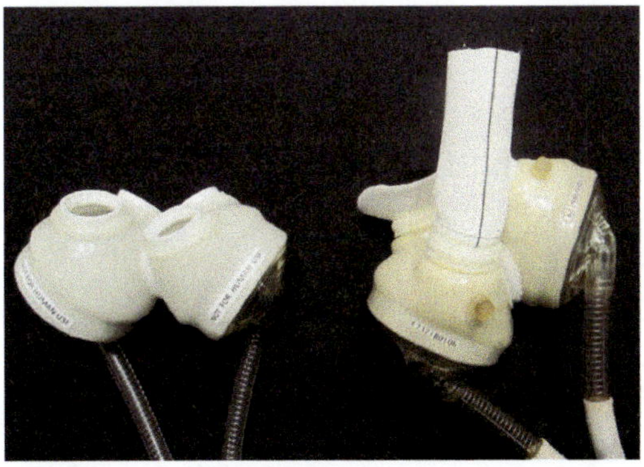

Fig. 38.1 Left: SynCardia 50 cc total artificial heart (TAH) without atrial cuffs or outflow grafts. Right: SynCardia 70 cc TAH with atrial cuffs or outflow grafts

volume than the 70 cc TAH. The SynCardia 50 cc TAH is CE marked as a bridge to transplant in Europe. Recently, FDA cleared the pathway for an IDE trial in the United States, which will open up the option of full cardiac support for patients over 10 years of age with biventricular failure as a bridge to transplant. There are two primary study arms in this IDE study [22]. The pediatric primary arm will include 24 pediatric patients (10–18 years old) and will evaluate the safety and probable benefit of the 50 cc TAH for transplant-eligible pediatric patients to support a Humanitarian Device Exemption application. The adult primary arm will include 24 adult patients (19–75 years old) and will evaluate the safety and efficacy of the 50 cc TAH for transplant-eligible adult patients to support a premarket approval (PMA) application. Primary arm patient enrollment criteria include patients (1) at risk of imminent death from biventricular heart failure; (2) at time of implant, aged 10–18 years (pediatric) or 19–75 years (adults); (3) eligible for donor heart transplant; (4) have two functional atrioventricular valves; (5) on ECMO support ≤3 days and not dialysis-dependent; and (6) have a BSA ≤1.85 m^2 with adequate T10 measurement or adequate room in the chest as determined by 3D imaging assessment or by other standard clinical assessments. Patients who do not meet all the above criteria may still be able to participate in the clinical trial via enrollment in the secondary arm, which will include up to 24 adult and pediatric patients.

Through this IDE clinical trial, patients too small to receive the PMA-approved 70 cc TAH now have access to the smaller 50 cc TAH. Worldwide, 61% of 50 cc TAH implants have been in women, compared to only 11% of 70 cc TAH implants [22]. In addition, 18% of 50 cc TAH implants have been in pediatric patients (<18 years old), compared to only 2% of 70 cc TAH implants. The age range for 50 cc TAH was between 10 and 72 years old, and the smallest BSA was 1.16 m^2.

Pediatric Total Artificial Heart Under Development

Penn State Pediatric Total Artificial Heart

Penn State University has just begun developing a TAH (Fig. 38.2) for infants and small children by using two pulsatile pneumatically driven infant VADs as a bridge to heart transplant, and the work has been supported with federal funding obtained from the NHLBI (Bethesda, Maryland, USA), under grant 1R01HL131921 (PI: Weiss, W). The device is to provide total circulatory support for patients under 1 year of age in severe biventricular failure while awaiting heart transplantation. They are proposing developing a 12 mL stroke volume device for infants (similar to their existing infant VAD) and a 25 mL stroke volume TAH for small children. More details of the device and progress in development are described in Chap. 30 "Cleveland Clinic Total Artificial Heart".

Cleveland Clinic P-CFTAH

We have developed a continuous-flow total artificial heart (CFTAH) for adult patients (Fig. 38.3, left) and have demonstrated its stable hemody-

Fig. 38.2 Illustration of Penn State Pediatric TAH

namics and good biocompatibility without anticoagulation in three long-term calf experiments up to 90 days [23, 24]. The details of the device descriptions and progress in development are described in Chap. 35 "Cleveland Clinic Continuous-Flow Total Artificial Heart," but it is very small (490 g, 160 mL), and it appears that the CFTAH will fit in most patients with BSA ≥1.0 m^2 (corresponding height of ≥130 cm and age of 9 years) [25]. Because it will not fit in infants, we are currently developing a pediatric version (P-CFTAH) by downsizing the adult device by a scale factor of 0.70 (1/3 of total volume) (Fig. 38.3, right). This strategy makes implantation possible in the infant chest with BSA ≥0.3 m^2 (corresponding height of ≥55 cm and age of 1 month) [25].

Similar to the adult CFTAH, the P-CFTAH uses a single-piece rotating assembly that has impellers for both left and right centrifugal pumps (Fig. 38.4). The P-CFTAH is valveless and sensorless, and the position of the centrally

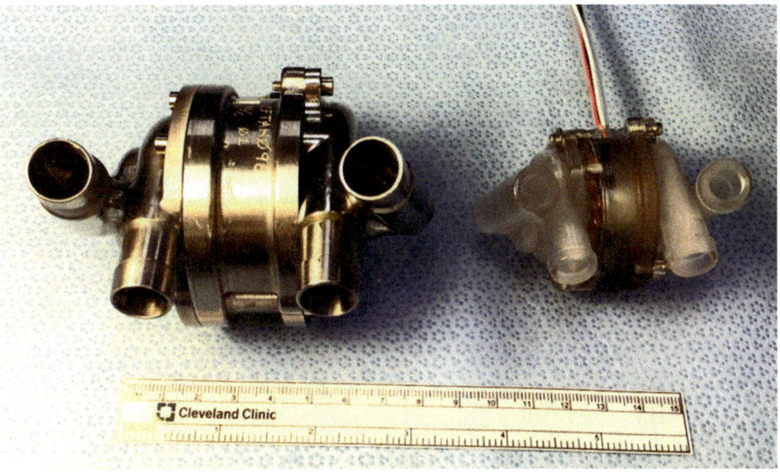

Fig. 38.3 Left: Cleveland Clinic continuous-flow total artificial heart (CFTAH) for adults. Right: pediatric version of Cleveland Clinic CFTAH (P-CFTAH)

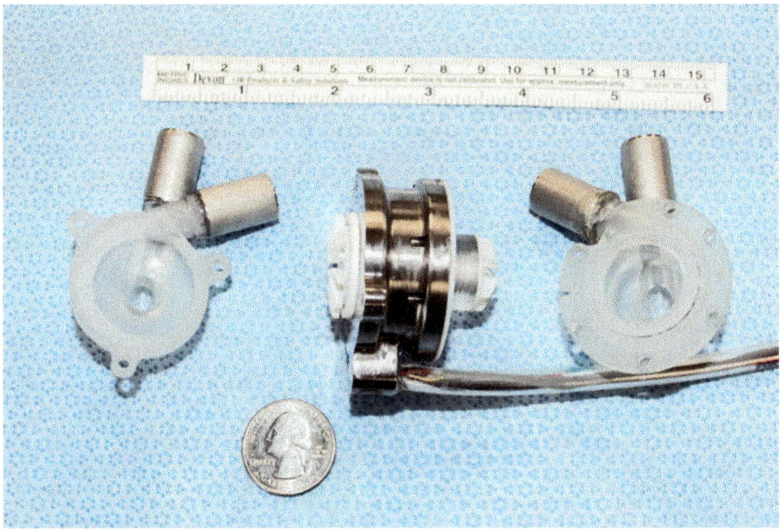

Fig. 38.4 Exploded view of Cleveland Clinic P-CFTAH

located rotating assembly is primarily determined by the differential force between the right and left pump filling pressures acting on this part to balance left and right circulations without electronic intervention. A pulsatile flow can be produced with speed modulation similar to that for CFTAH [26], and an algorithm allows the controller to adjust speed automatically. The nominal external dimensions of the P-CFTAH design are 68.9 mm in length, 43.4 mm in diameter, 54.9 mL in volume displacement, and 167 g in weight.

The pump flow range of the P-CFTAH (1.5–4.5 L/min) is intended to support patients weighing up to 50 kg (average weight of 14-year-olds). The system performance of the initial working prototype of the P-CFTAH was evaluated on the CFTAH mock circulatory loop [27]. The pressure-flow curves demonstrated a broad range of left pump flow (Fig. 38.5). At the delta pressure of 80 mm Hg, the left pump flow ranged from 1 to 4.5 L/min with pump speeds between 3,500 and 5,000 rpm. The atrial pressure balance (the left atrial pressure minus right atrial pressure [LAP-RAP]) was maintained within ±5 mm Hg with a wide range of the systemic to pulmonary vascular resistance (SVR/PVR) ratios (between 1.4 and 35), with and without pump speed modulation (sinusoidal speed change at a frequency of 1 Hz and an amplitude of ±25% of the mean speed) (Fig. 38.6). These data met the proposed design requirements for pump flow range (1.5–4.5 L/min) and self-regulation (LAP-RAP within −5 to +10 mm Hg).

The system performance of the initial working prototype of the P-CFTAH was evaluated in an acute lamb model as well [28]. The P-CFTAH was implanted in four healthy lambs (three males; average weight, 28.7 ± 2.3 kg) through a full median sternotomy in an acute setup without any outflow graft kinking or compressions of atria or vessels (Fig. 38.7). All four acute experiments in lambs showed good anatomical fit and easy implantation of the P-CFTAH, with an average aortic cross-clamp time of 98 ± 18 min. Baseline hemodynamics were stable in all lambs with the pump speed of 3.4 ± 0.2 krpm, pump flow of 2.1 ± 0.9 L/min, mean arterial pressure of 68 ± 10 mm Hg, mean LAP of 6 ± 1 mm Hg, and mean RAP of 6 ± 1 mm Hg. Pump flow and motor current were stable, and no power elevations or mechanical failures occurred during these acute evaluations. *In vivo* left pump performance plots at pump speeds of 3,000–5,000 rpm matched the *in vitro* pressure-flow curves of the respective pump speeds (Fig. 38.5). The P-CFTAH's self-regulating performance was confirmed *in vivo*; the majority of atrial pressure differences (LAP-RAP) remained within ±5 mm Hg, and good left and right atrial balance was maintained in response to manipulations of

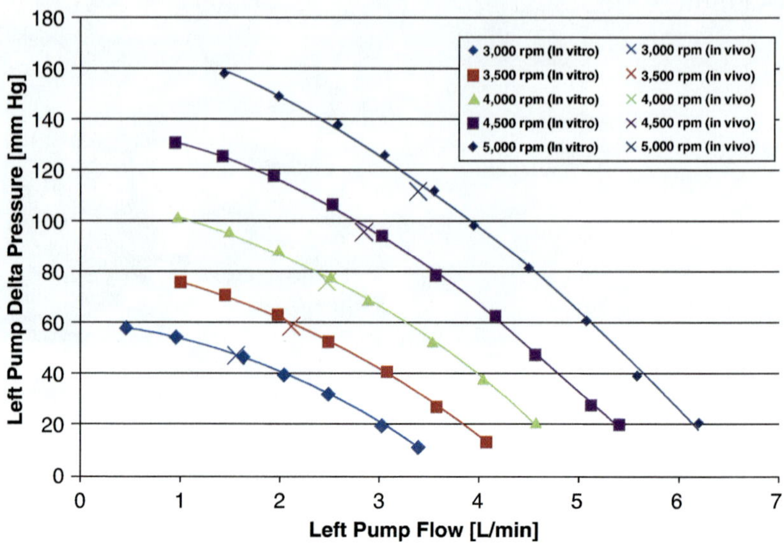

Fig. 38.5 *In vitro* and *in vivo* left pump pressure-flow curves of P-CFTAH

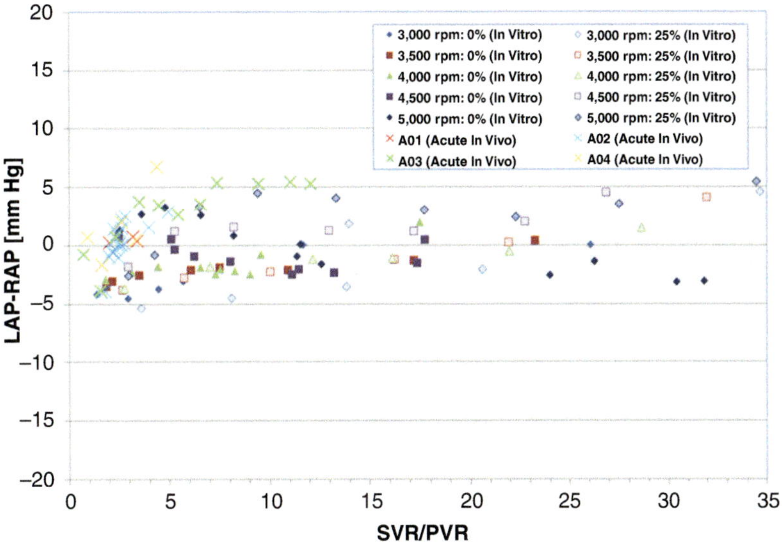

Fig. 38.6 *In vitro* and *in vivo* atrial balance with and without pump speed modulation. LAP left atrial pressure, RAP right atrial pressure, SVR systemic vascular resistance, PVR pulmonary vascular resistance

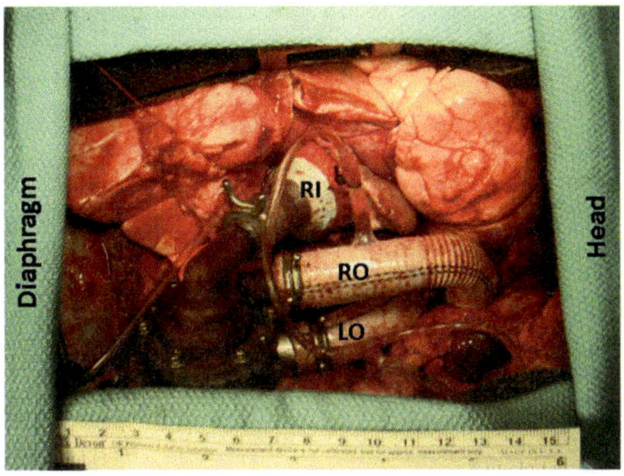

Fig. 38.7 Surgical view of the P-CFTAH's anatomical fit after implantation in a 27.1 kg lamb. P-CFTAH was connected to its inflow and outflow ports. RI right inflow cuff, RO right outflow graft, LO left outflow graft

SVR and PVR (Fig. 38.6). Additionally, the modulation of pump speed was successfully performed to create arterial pulsation (Fig. 38.8).

Downscaling of an adult pump for pediatric applications is in itself not sufficient for optimization and biocompatibility of our pediatric pump because pump flows and pressures differ from those for adult patients with different SVR/PVR ratios, which affect left/right balance self-regulation. However, our initial series of *in vitro* and acute *in vivo* experiments with the P-CFTAH have shown that the key preliminary step of downscaling worked, confirming the desired hemodynamic performance and pump operation within design specifications. The pump flow range (1.5–4.5 L/min) will support patients weighing up to 50 kg (average weight of 14-year-olds). For destination therapy, when the patient grows beyond the requirement of 4.5 L/min, the P-CFTAH could be replaced by the adult version, which has an overlapping pump flow range (3–9 L/min). It is envisioned that such replacement surgery can be done under CPB, by disconnecting the inflow/outflow ports of the P-CFTAH from their respective inflow cuffs and outflow grafts and connecting an adult CFTAH, which has the same port size and configuration.

Future Perspective of Pediatric Total Artificial Heart

Significant technological and therapeutic advances in MCS devices have been achieved in recent years; however, issues with MCS continue to arise, and they remain the focus of increasingly sophisticated research in device development, especially for pediatric patients. In the past 5 years, continuous-flow VADs have totally replaced volume-displacement pulsatile-flow VADs because of their simplicity, greater mechanical reliability, improved durability,

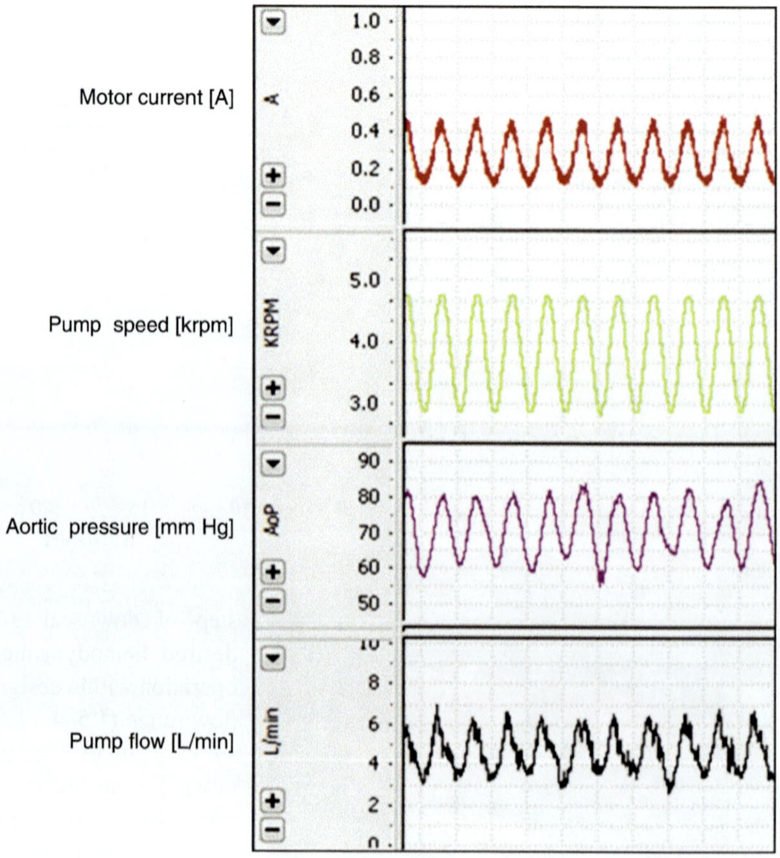

Fig. 38.8 *In vivo* pump speed modulation of ±25% at 100 beats/min

smaller size, and better outcomes [3]. Both SynCardia's and Penn State's TAHs are pneumatically driven pulsatile devices with multiple moving parts (four valves, two diaphragms, and an external pneumatic driver). For the same reasons, we believe that continuous-flow TAHs will soon replace pulsatile TAHs, especially for pediatric application where the size of the devices is one of the key challenges. Complete cardiac replacement with continuous-flow TAHs, while currently in early development and application, will play a key role in addressing a subset of pediatric patients with chronic biventricular failure. Smaller, simpler continuous-flow TAHs that are proven reliable may even move the field completely in this direction and will undoubtedly save many lives. However, until the ideal device has been created and brought into accepted clinical practice, patient survival will continue to be primarily determined by an early decision to implant univentricular or biventricular devices and not by choice of device type.

Conclusions

Currently, SynCardia 50 cc TAH is the only clinically available TAH for pediatric patients, but it is intended only for patients with a BSA of $\geq$1.2 m^2 and cannot be used for infants or small children. Therefore, new TAHs for pediatric patients such as Penn State Pediatric TAH and Cleveland Clinic P-CFTAH are under development to fill this gap. Initial study results on P-CFTAH are encouraging, and further development is warranted.

References

1. Writing Group Members, Mozaffarian D, Benjamin EJ, Go AS, Arnett DK, Blaha MJ, et al. Heart disease and stroke statistics-2016 update: a report from the American Heart Association. Circulation. 2016;133:e38–360.
2. Mancini D, Colombo PC. Left ventricular assist devices: a rapidly evolving alternative to transplant. J Am Coll Cardiol. 2015;65:2542–55.
3. Moazami N, Hoercher KJ, Fukamachi K, Kobayashi M, Smedira NG, Massiello A, et al. Mechanical circulatory support for heart failure: past, present and a look at the future. Expert Rev Med Devices. 2013;10:55–71.
4. Mehra MR, Goldstein DJ, Uriel N, Cleveland JC Jr, Yuzefpolskaya M, Salerno C, et al. Two-year outcomes with a magnetically levitated cardiac pump in heart failure. N Engl J Med. 2018;378:1386–95.
5. Lund LH, Khush KK, Cherikh WS, Goldfarb S, Kucheryavaya AY, Levvey BJ, et al. The registry of the International Society for Heart and Lung Transplantation: thirty-fourth adult heart transplantation report-2017; focus theme: allograft ischemic time. J Heart Lung Transplant. 2017;36:1037–46.
6. Villa CR, Khan MS, Zafar F, Morales DLS, Lorts A. United States trends in pediatric ventricular assist implantation as bridge to transplantation. ASAIO J. 2017;63:470–5.
7. Rossano JW, Dipchand AI, Edwards LB, Goldfarb S, Kucheryavaya AY, Levvey Rn BJ, et al. The registry of the International Society for Heart and Lung Transplantation: nineteenth pediatric heart transplantation report-2016; focus theme: primary diagnostic indications for transplant. J Heart Lung Transplant. 2016;35:1185–95.
8. Zafar F, Castleberry C, Khan MS, Mehta V, Bryant R 3rd, Lorts A, et al. Pediatric heart transplant waiting list mortality in the era of ventricular assist devices. J Heart Lung Transplant. 2015;34:82–8.
9. Kirklin JK. Advances in mechanical assist devices and artificial hearts for children. Curr Opin Pediatr. 2015;27:597–603.
10. Steffen RJ, Miletic KG, Schraufnagel DP, Vargo PR, Fukamachi K, Stewart RD, et al. Mechanical circulatory support in pediatrics. Expert Rev Med Devices. 2016;13:507–14.
11. Rossano JW, Cherikh WS, Chambers DC, Goldfarb S, Khush K, Kucheryavaya AY, et al. The registry of the International Society for Heart and Lung Transplantation: twentieth Pediatric heart transplantation report-2017; focus theme: allograft ischemic time. J Heart Lung Transplant. 2017;36:1060–9.
12. Gilboa SM, Salemi JL, Nembhard WN, Fixler DE, Correa A. Mortality resulting from congenital heart disease among children and adults in the United States, 1999 to 2006. Circulation. 2010;122:2254–63.
13. Villa CR, Morales DLS. The total artificial heart in end-stage congenital heart disease. Front Physiol. 2017;8:131.
14. Morales DL, Khan MS, Gottlieb EA, Krishnamurthy R, Dreyer WJ, Adachi I. Implantation of total artificial heart in congenital heart disease. Semin Thorac Cardiovasc Surg. 2012;24:142–3.
15. Adachi I, Morales DS. Implantation of total artificial heart in congenital heart disease. J Vis Exp. 2014;(89):51569.
16. Rossano JW, Goldberg DJ, Fuller S, Ravishankar C, Montenegro LM, Gaynor JW. Successful use of the total artificial heart in the failing Fontan circulation. Ann Thorac Surg. 2014;97:1438–40.
17. Blume ED, VanderPluym C, Lorts A, Baldwin JT, Rossano JW, Morales DLS, et al. Second annual pediatric interagency registry for mechanical circulatory support (PediMACS) report: pre-implant characteristics and outcomes. J Heart Lung Transplant. 2018;37:38–45.
18. Hetzer R, Potapov EV, Stiller B, Weng Y, Hubler M, Lemmer J, et al. Improvement in survival after mechanical circulatory support with pneumatic pulsatile ventricular assist devices in pediatric patients. Ann Thorac Surg. 2006;82:917–24; discussion 24–5.
19. Reinhartz O, Hill JD, Al-Khaldi A, Pelletier MP, Robbins RC, Farrar DJ. Thoratec ventricular assist devices in pediatric patients: update on clinical results. ASAIO J. 2005;51:501–3.
20. Baldwin JT, Borovetz HS, Duncan BW, Gartner MJ, Jarvik RK, Weiss WJ. The national heart, lung, and blood institute pediatric circulatory support program: a summary of the 5-year experience. Circulation. 2011;123:1233–40.
21. Copeland JG, Smith RG, Arabia FA, Nolan PE, Sethi GK, Tsau PH, et al. CardioWest Total Artificial Heart Investigators. Cardiac replacement with a total artificial heart as a bridge to transplantation. N Engl J Med. 2004;351:859–67.
22. https://syncardia.com/clinicians/clinical-resources/u-s-clinical-trials/. Accessed on 10.21.18.
23. Karimov JH, Moazami N, Kobayashi M, Sale S, Such K, Byram N, et al. First report of 90-day support of 2 calves with a continuous-flow total artificial heart. J Thorac Cardiovasc Surg. 2015;150:687–93.e1. PMCID: PMC4554829.
24. Fukamachi K, Horvath DJ, Massiello AL, Fumoto H, Horai T, Rao S, et al. An innovative, sensorless, pulsatile, continuous-flow total artificial heart: device design and initial in vitro study. J Heart Lung Transplant. 2010;29:13–20. PMCID: PMC2817999.
25. Fukamachi K, Karimov JH, Byram NA, Sunagawa G, Dessoffy R, Miyamoto T, et al. Anatomical study of

the Cleveland Clinic continuous-flow total artificial heart in adult and pediatric configurations. J Artif Organs. 2018;21:383–6.
26. Shiose A, Nowak K, Horvath DJ, Massiello AL, Golding LA, Fukamachi K. Speed modulation of the continuous-flow total artificial heart to simulate a physiologic arterial pressure waveform. ASAIO J. 2010;56:403–9. PMCID: PMC2933186.
27. Fukamachi K, Karimov JH, Horvath DJ, Sunagawa G, Byram NA, Kuban BD, et al. Initial in vitro testing of a paediatric continuous-flow total artificial heart. Interact Cardiovasc Thorac Surg. 2018;26:897–901.
28. Karimov JH, Horvath DJ, Byram N, Sunagawa G, Kuban BD, Gao S, et al. Early in vivo experience with the pediatric continuous-flow total artificial heart. J Heart Lung Transplant. 2018;37:1029–34.

39 Progress on Wireless LVAD and Energy Sources for Mechanical Circulatory Systems

John Valdovinos, Jiheum Park, Joshua Smith, and Pramod Bonde

Introduction

By 2030, the American Heart Association projects that over eight million Americans of the age of 18 will suffer from heart failure [1]. For a portion of this population, mechanical circulatory support (MCS) devices can be used as bridge to transplantation. In the Randomized Evaluation of Mechanical Assistance for the Treatment of Congestive Heart Failure (REMATCH) study, Rose and colleagues demonstrated the superiority of left ventricular assist device (LVAD) therapy over medical drug therapy [2]. Since that milestone study, the design of mechanical circulatory support blood pumps has evolved from paracorporeal pulsatile devices to more compact implantable continuous flow pumps. Continuous flow pumps have increased the 2-year survival of patients compared to patients utilizing pulsatile pumps [3].

Adverse events associated with MCS therapy remain a concern for widespread acceptance among the physician community [4]. On the other hand, quality of life remains a major deterrent for patient acceptance of the current continuous flow technology. One of the most commonly reported adverse events are infections, which constitute an increase in patient morbidity and repeated hospitalization [5]. Driveline infections are the most prevalent source of infection in patients with implanted LVADs [6]. This is because the percutaneous tunnel in which the driveline sits provides a pathway for external pathogens to enter the body [7]. In a retrospective review of patients with LVAD implants, Zierer et al. found that the individual risk that a driveline infection would develop within 1 year was 94% [8]. Eliminating this route for infection is crucial to limiting hospital readmissions and providing heart failure patients with effective and safe MCS therapy [5]. MCS systems that are fully implantable have the potential to eliminate the lingering limitations of current LVAD technology. Specifically, the use of wireless energy transmission systems, like transcutaneous energy transfer

J. Valdovinos, PhD
Formerly Bonde Artificial Heart Lab, Yale School of Medicine, New Haven, CT, USA
e-mail: john.valdovinos@csun.edu

J. Park, PhD
Bonde Artificial Heart Laboratory, Yale School of Medicine, New Haven, CT, USA
e-mail: jiheum.park@yale.edu

J. Smith, PhD
Department of Electrical Engineering, Department of Computer Science and Engineering, University of Washington, Seattle, WA, USA
e-mail: jrs@cs.washington.edu

P. Bonde, MD (✉)
Yale School of Medicine,
New Haven, CT, USA
e-mail: pramod.bonde@yale.edu

© Springer Nature Switzerland AG 2020
J. H. Karimov et al. (eds.), *Mechanical Support for Heart Failure*,
https://doi.org/10.1007/978-3-030-47809-4_39

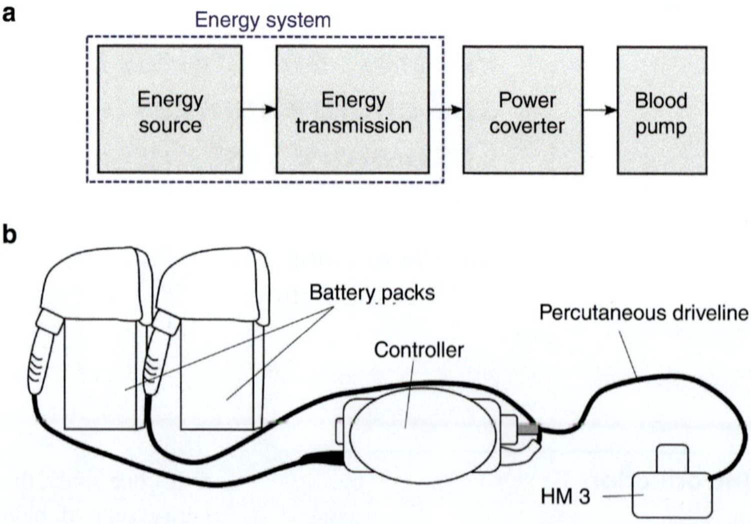

Fig. 39.1 (**a**) Typical scheme and components that make up a mechanical circulatory support system. (**b**) The components of a HeartMate 3 which, like most mechanical circulatory support systems, consists of a power source (lithium-ion battery pack), power transmission (controller and percutaneous driveline), energy converter (motor windings within HM 3), and blood pump (HM 3). HM 3 HeartMate 3

and resonant inductive coupling, to power implantable blood pumps provides a unique approach to achieve total MCS system implantability [9, 10].

This paper provides an overview of the different components that make up the energy system of mechanical circulatory support device and how these components ultimately affect the total implantability of these systems. Typical MCS devices consist of an energy source, method for energy transmission, a power converter, and blood pump (Fig. 39.1a). Current systems, like the HeartMate 3 (Fig. 39.1b), utilize batteries as their energy sources, percutaneous drivelines as the energy transmission, and a slotted DC motor and impeller as the power converter and blood pump. In this review, special attention will be given to the method of energy transmission to power implantable blood pumps, since this is the most appropriate technique to realize a totally implantable support system.

History of Energy Systems in Mechanical Circulatory Support

In 1963, after Gibbons demonstrated the first successful cardiopulmonary bypass in a human, the National Institute of Health (NIH) founded the Artificial Heart Program [11]. The goal of the program was to develop a totally implantable blood pump that could replace the entire hemodynamic function of the heart. Despite the initial enthusiasm, problems with hemolysis, thrombosis, and power supplies led the NIH to fund research on component technology with the goal of achieving safer implantability [12, 13]. One of the main hurdles in developing a totally implantable artificial heart and LVAD was designing a power supply that was both long-lasting and compact. From 1964 to 1977, approximately 31% of the National Heart, Lung, and Blood Institute's (NIH division) research expenditures were allocated for grants related to developing compact and long-lasting energy systems for artificial hearts and LVADs (Fig. 39.2). While significant progress has been made in improving blood pump design and the biocompatibility since 1977, the development of a long-lasting implantable power source has yet to be realized.

Energy Sources

The development of energy sources for MCS devices consists of two basic approaches. The first approach consists of designing an energy source that can provide power to an implantable blood pump for the entirety of the patient's life. The second approach consists of developing an energy source that can be recharged after a few hours use. The first approach inspired the use of

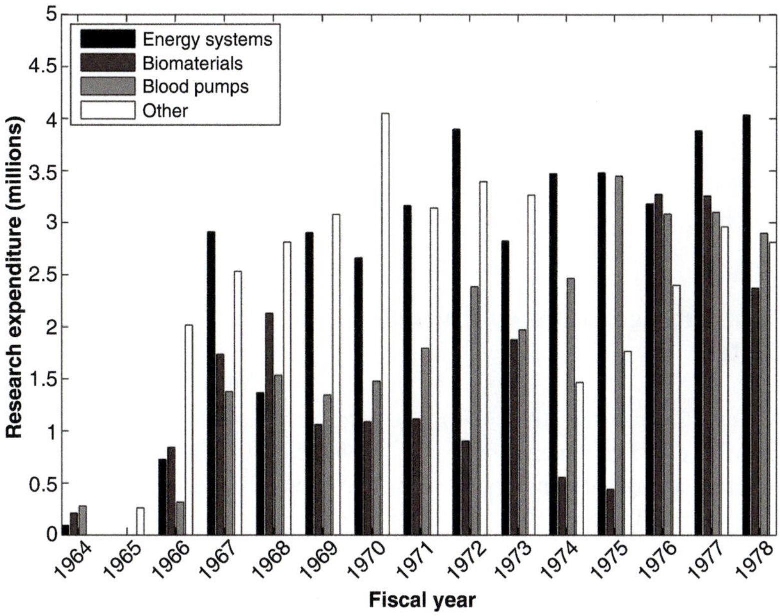

Fig. 39.2 Distribution of NHLBI research expenditures among the various research thrusts between 1964 and 1977. (Data taken from Lubeck [12])

nuclear energy for long-term energy source [14]. In this iteration, plutonium 238, a radioactive isotope, would provide thermal energy during radioactive decay that could be converted to mechanical power via a power converter. Various researchers in the 1970s demonstrated the concept of nuclear-powered blood pumps and predicted a 10-year longevity [15, 16]. While nuclear power is theoretically appealing [17], its practical implementation, cost, and health hazards have limited its utility in the MCS field [18].

The more practical approach, which is currently used in MCS systems, is the use of electrochemical energy sources, like batteries, to power the blood pump for hours until recharge. The use of batteries in medical devices dates back to the use of mercury cell batteries to power implantable cardiac pacemakers [19, 20]. Two main factors affect the duration that batteries can supply charge to MCS devices: energy density and the type of blood pump used. Advancement in battery technology has enabled the use of high-energy-density lithium-ion batteries over nickel cadmium and mercury cell batteries in LVADs [21–23], with their physical size differing in capacity (Fig. 39.3). The HeartMate II (similarly for HeartWare HVAD device), for example, utilizes a 14 V lithium-ion battery with a 4.8 A-h and 78 W-h specifications. Depending on the flow conditions, these batteries can be used approximately 6–10 hours without recharging. The shift from pulsatile pumps to continuous flow devices has enabled for lower current draw (0.1–0.5 A) and lower power requirements (5–10 W). Since the shift to lithium-ion batteries, the progress in energy source technology has not improved substantially for MCS technology. However, the method to transmit this electrical energy to the blood pump has seen rapid progress.

Energy Transmission Methods

As mentioned previously, one of the major limiting factors of current continuous flow pumps is the percutaneous driveline connecting the energy source to the blood pump. Pae et al. implanted wirelessly powered LionHeart LVADs and demonstrated that full implantability of both blood pump and energy system can reduce the number of device-related infections and overall infections when compared to systems that are not totally implantable [24]. Thus, methods to wirelessly power LVADs and MCS devices are crucial in preventing hospital readmissions for device-related infection. Currently, there are two methods to wirelessly power implantable blood pumps that have been demonstrated by various researchers. The first method is via close range inductive coupling, known as transcutaneous energy transfer, and the second method is long-range resonant coupling.

Transcutaneous Energy Transfer System

Systems that use inductive coupling to wirelessly transfer electrical energy across a biological interface are called transcutaneous energy transfer systems (TETS). TETS consist of an energy source, oscillator circuit, primary and secondary coils, rectifier and power conditioning circuit, and the blood pump (Fig. 39.4). The energy source provides the electrical power to be transmitted, while the oscillator converts the direct current electrical power to an alternating current signal in the radio frequency range (3 kHz–300 GHz). This oscillator can be achieved by using a transistor amplifier, LC tank circuit, and positive feedback to create an alternating waveform with amplified voltage to the primary TETS coil [25–28]. The alternating current in the primary coil results in an alternating magnetic field that penetrates the biological interface and induces a voltage across the secondary coil. The induced voltage drives an alternating current through the secondary coil and is rectified to a direct current supply to directly power blood pump or to recharge an internal battery.

The first use of an alternating magnetic field to wirelessly power an implantable blood pump was demonstrated by Kusserow in 1960 [29]. In this iteration, a permanent magnet was attached to an external motor that spun to create an alternating magnetic field through the skin. This external magnet was coupled to an internal magnet that directly transmitted torque to the blood pump. Unlike the direct power transmis-

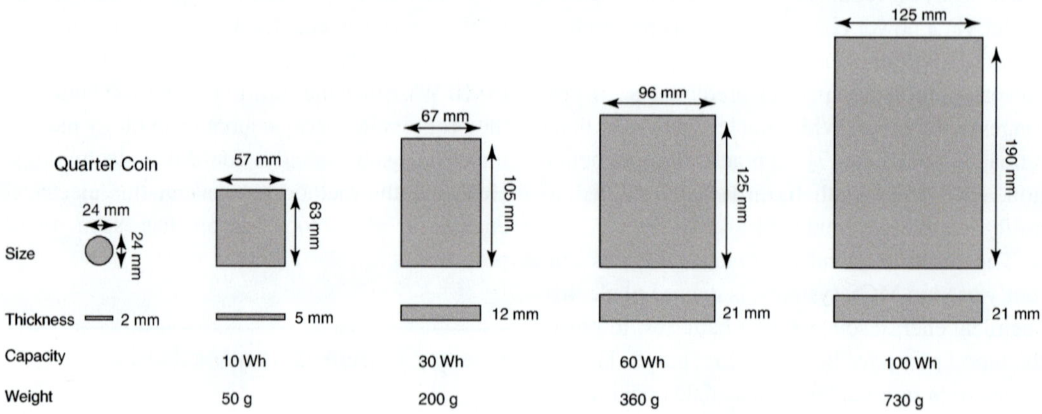

Fig. 39.3 Lithium-ion battery size by capacity

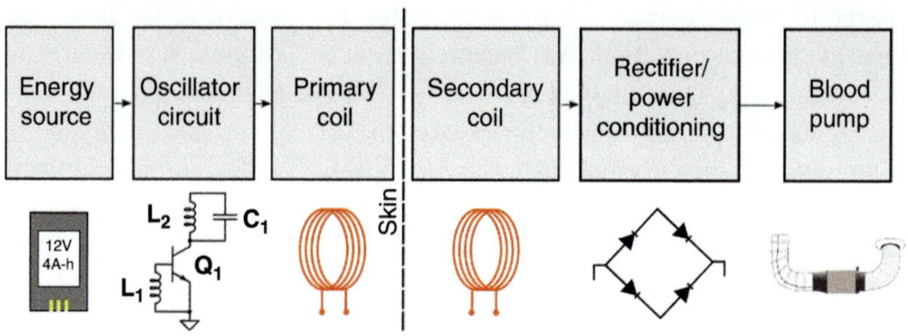

Fig. 39.4 The typical components of a transcutaneous energy transfer system with block diagram on the top and example components on the bottom

sion used by Kusserow, Schuder and colleagues used inductive coils to transmit electrical energy through the closed chest of a dog [25, 30]. A small coil was implanted in the chest of the dog, and power was transmitted to the implanted coil by three orthogonal 2 m square coils. The power needed to transmit power at a relatively large distance (about 1 m) was 1 kW (1% efficiency). In 1981, Sherman et al. showed that the efficiency and bi-directionality of inductive wireless transfer could be improved by placing the secondary and primary coils within a few centimeters from each other across the skin of a pig (Fig. 39.5) [26]. As a result, there are important parameters that contribute to the efficiency and amount of electrical power that can be transmitted to implanted blood pumps when using inductive coupling.

There are two general figures of merit that dictate the efficiency of energy transfer between the primary and secondary coil in TETS: the quality factor (Q-factor) and the coil coupling factor (k). The Q-factor, defined as the ratio of the coil reactance to resistance, is a measure of how much stored energy versus losses there are in the transmitting and receiving coils [31]. Because a coil can be seen as an LRC series circuit with parasitic capacitance (C) and resistance (R), the Q-factor can be defined by Eq. (39.1).

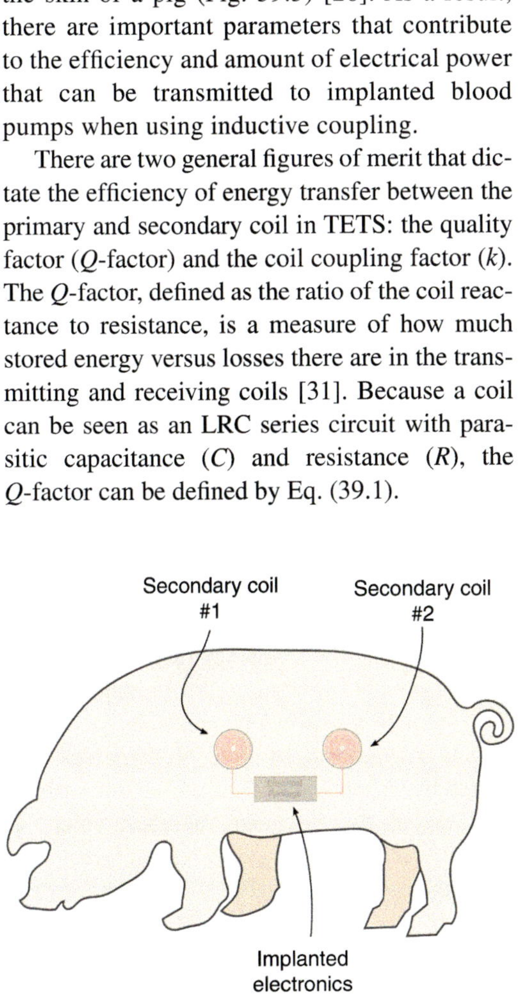

Fig. 39.5 Bi-directional power transfer with two implanted secondary coils in a swine demonstrated by Sherman et al. [26]

$$Q = \frac{2\pi fL}{R} \quad (39.1)$$

where f is the frequency of oscillation, L is the coil inductance, and R is the coil resistance. In order to achieve a high Q-factor, which translates to efficient energy transfer, the resistive losses in the coil need to be minimized. The Q-factor can be changed by altering the number of turns in the coil (affects L and R) and changing the size of the coil.

The coil coupling factor, k, measures the amount of magnetic coupling that exists between two physically close inductors. The coupling factor can be defined by Eq. (39.2).

$$k = \frac{M}{\sqrt{L_1 L_2}} \quad (39.2)$$

where M is the mutual inductance between the two coils, L_1 is the inductance of the primary coil, and L_2 is the inductance of the secondary coil. A common way to increase the coil coupling factor is to increase the mutual inductance of the coils, which is a function of the distance between inductors, alignment, and the permeability of the medium connecting the two coils. Misalignment is a major concern for TET systems. Researchers have gone to great lengths to ensure alignment of the coils to guarantee good coil coupling [32]. For example, Okamoto et al. designed a TET system housed in a casing that allows protrusion of the receiving coil out of the skin for ensured coil alignment [33]. Ozeki et al. developed a TET system with coil malposition sensor that alerts the patient if the coils are misaligned and the system has low efficiency [34]. Some groups have also used ferrite cores, which have large magnetic permeability, to increase coil coupling and reduce the amount of leakage magnetic flux [35–37].

While TETS have shown promise in realizing total implantability of MCS devices, the clinical use of these systems has been limited to two devices. The AbioCor implantable replacement heart (AbioMed Inc., Danvers, MA) was a fully implantable artificial heart that utilized a TET

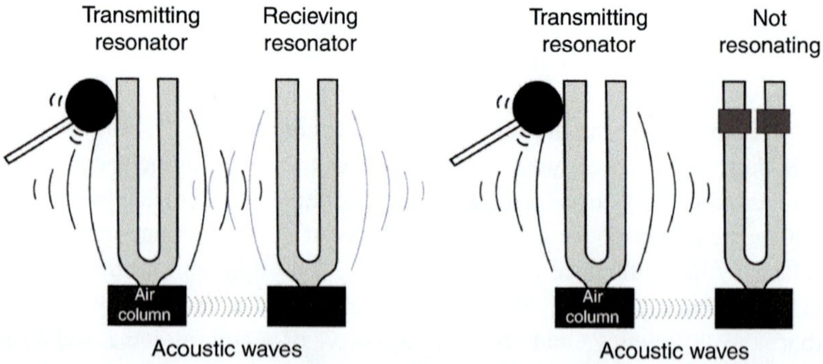

Fig. 39.6 Strongly coupled acoustic system between tuning forks demonstrating resonant coupling

system to power the artificial heart and recharge the internal batteries [38]. The system has been used clinically in few patients, and no significant device-related infections were reported [39, 40]. Similarly, the LionHeart LVAD (Arrow International Inc., Reading, PA) utilizes TETS to wirelessly power the motor of the pulsatile pump and the remaining hardware. In all patients there were no incidences of device-related infections [41]. More recently, a patient with dilated cardiomyopathy was successfully bridged to transplantation with the LionHeart and only reported a problem with the implanted compliance chamber [42]. These clinical experiences demonstrate the utility of TETS in reducing the number of infection-related events but also highlight the limitations of this technology. These include a limited range of efficient power transfer (<20 mm) and susceptibility to interference from other electromagnetic devices [43].

Resonant Coupling Wireless Transmission

The development of wireless transmission technology like TETS has provided the groundwork for MCS device total implantability and reduced rates of device-related infection. Another method to achieve wireless power transmission, particularly at longer ranges, is by using resonantly coupled inductive coils. Unlike traditional inductive coupling, resonantly coupled inductive coils function on the principle of resonant coupling between objects, which have the same natural frequency. This is similar to the tuning fork experiment (Fig. 39.6) in which resonant vibration can be induced in a tuning fork when an identical tuning fork is oscillated at its natural frequency nearby. In particular, two coils with the same resonant frequency will couple strongly at large distance (>10 cm) when one of the coils is perturbed at its resonant condition. There are two distinguishing features between TETS wireless transmission and resonant coupling. First, TETS coils are typically oscillated at frequencies in the 100–500 kHz range, while the frequency of oscillation for resonantly coupled coils is on the order of a few megahertz (depending on the resistance, capacitance, and inductance of the coils). Second, in order for two coils to couple strongly, the quality factor of each coil must be high (~10^3), which is not necessarily true for TETS coils.

Wireless transmission via resonant coupling, like TETs, depends on the *Q*-factor and the magnetic coupling between coils (mutual inductance). However, the rate at which energy is transferred and the frequency response of a resonantly coupled system are more complicated. The interaction between two resonantly coupled coils is analogous to two pendulum masses connected by a spring (Fig. 39.7). Like this system, the there are two frequency modes at which the system oscillates [44], f_1 and f_2, which are defined as,

$$f_1 = \frac{f_a}{\sqrt{1+k}}, f_2 = \frac{f_a}{\sqrt{1-k}} \qquad (39.3)$$

where f_a is the natural frequency at one of the coils and k is the coupling coefficient defined in

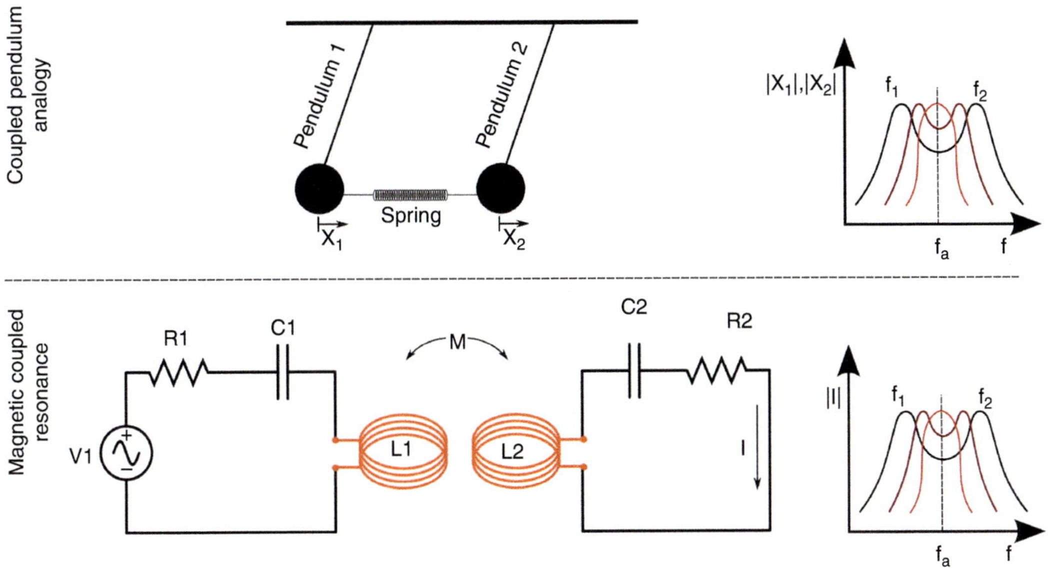

Fig. 39.7 Two pendulums connected by a spring and the resulting frequency response (right) and resonantly coupled coils with frequency response on the right

Eq. (39.2). The first frequency is the frequency at which the pendulum masses are moving in the same direction (in phase). The second frequency is the frequency in which the pendulum masses are moving in opposite directions (out of phase). The amount of coupling defines how close these normal modes are to each other. For systems with larger coupling coefficients, *k*, the normal modes are more separate. For systems with weak coupling, the normal modes approach the natural frequency of one of the coils. This frequency splitting behavior has two implications. First two systems that are coupled can be driven at two frequencies (as opposed to one frequency). More importantly, it also means that the operating frequency of the system is a function of the coupling coefficient. For two resonantly coupled coils, the distance (which affects mutual inductance) strongly influences the frequency at which the system should be operated.

There are two engineering characteristics that distinguish resonant coupled wireless transmission from traditional TETS. Unlike TETS, where the efficiency is directly dictated by the distance between the coupled coils, efficiency in resonantly coupled coils is theoretically not affected by the distance between the coils as long as the system is operated at the correct frequency and the amount of coupling between the coils is sufficient to transfer energy (i.e., spring is stiff enough in pendulum analogy). Thus, the coupling coefficient plays an important role in the amount of energy that is imparted between coils per cycle (or pendulum masses in the analogy) and not the efficiency [45]. The second characteristic is that there does exist a distance at which there isn't sufficient coupling between the coils to extract any useful power from the system. This is known as the critical coupling point. This means that not enough coupling exists between coils (spring not stiff enough in pendulum analogy) to provide any useful energy transfer beyond a particular distance. Practical application of resonantly coupled systems dates as far back as 1914. The technology that has resulted in the MCS field has the potential to revolutionize medical device power transmission.

Nikola Tesla, in 1914, first proposed the use of resonantly coupled inductors for wireless electrical power transmission [46]. Recently, researchers have put this idea into practice. Kurs et al. showed that two large copper coils can be resonantly coupled to achieve electrical power transfer with 50% efficiency across an air gap of

175 cm [47]. Resonant wireless transmission offers advantage of efficient power transfer, minimal interaction with the non-magnetically resonant objects, and relatively low magnetic field for transmitting a few watts [48]. Like the TETS work done by Schuder, Cannon et al. demonstrated that power can be wirelessly transmitted from one resonant transmitting coil to multiple resonant receivers of different sizes [49]. This was achieved by attaching parallel capacitors to tune the resonant frequency of coils of dissimilar sizes. Sample et al. extended the practical use of magnetic resonant coupling by using an adaptive scheme to tune the frequency of operation when the coil separation or orientations are changed [45]. While the preliminary data on resonant coupling has shown the feasibility of medium-range wireless power transmission, the practical use of resonant coupled systems to MCS are limited.

Resonantly coupled wireless transmission not only increases the distance of efficient power transfer, but it also enables dynamic recharging and powering of devices without limitations on mobility. We have utilized the improvements in resonantly coupled wireless transfer to power ventricular assist devices [50]. The system, called Free-range Resonant Electrical Energy Delivery (FREE-D), consists of a transmitter coil, resonator loop, relay coil, receiving vest coil, an implantable receiver coil, and radio frequency to direct current rectifier and regulator (Fig. 39.8a). The FREE-D has been used to power a HeartMate II (Thoratec Inc., Pleasanton, CA) and VentrAssist LVADs in vitro. The LVADs were powered with a 1 m separation from transmitting and receiving coils for a 90-day duration with a total efficiency greater than 50% (Fig. 39.8b). The interference test with several conductive and non-conductive objects was performed to evaluate the potential effects of the interfering objects on wireless power transfer efficiency. The energy transfer efficiency dropped off when conductive material such as coffee mug and cell phone weas introduced in the pathway of the power transmission [9, 10] (Fig. 39.9a), but with adaptive tuning enabled, the system was able to significantly improve the efficiency (Fig. 39.9b) by dynamically changing the optimal frequency (Fig. 39.9c). The in vivo feasibility of the FREE-D system has been evaluated through eight acute animal experiments, four for short (10 cm) and the other four for long (50 cm) separation distance (Fig. 39.10).

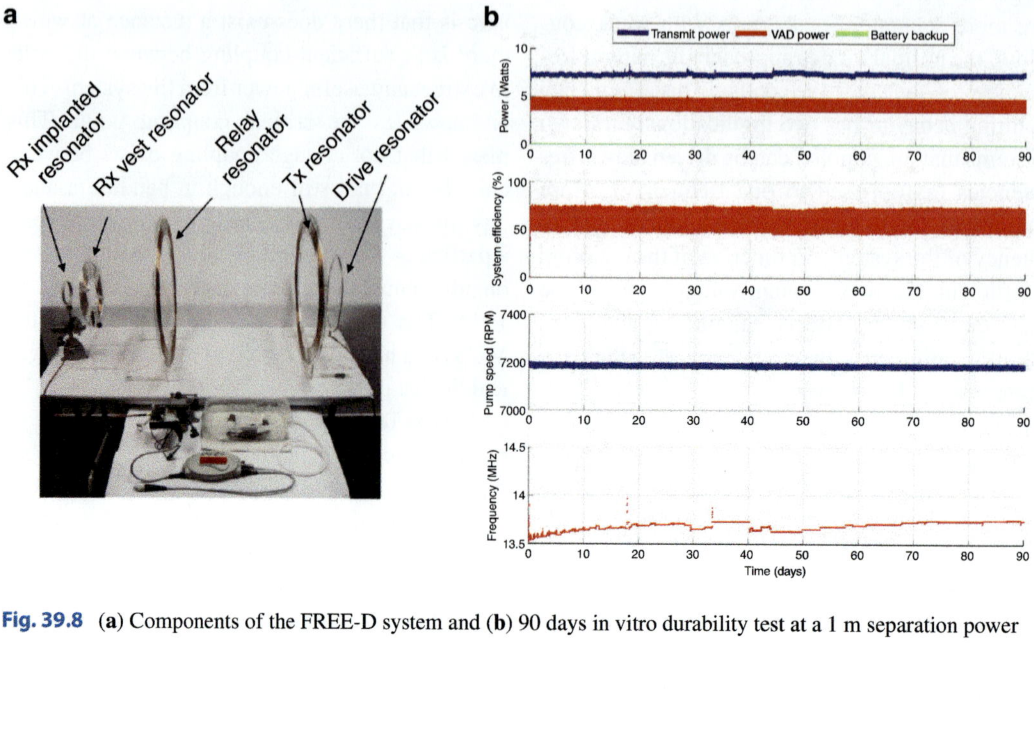

Fig. 39.8 (**a**) Components of the FREE-D system and (**b**) 90 days in vitro durability test at a 1 m separation power

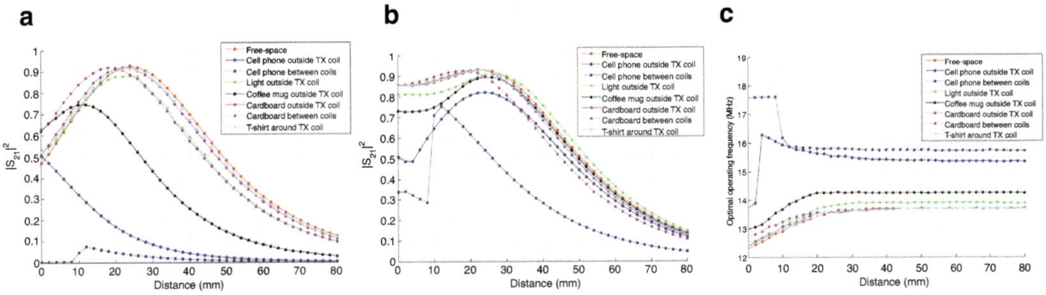

Fig. 39.9 Interference test with adaptive tuning (**a**) disabled and (**b**) enabled. (**c**) The changes in optimal operating frequency when adaptive tuning was enabled

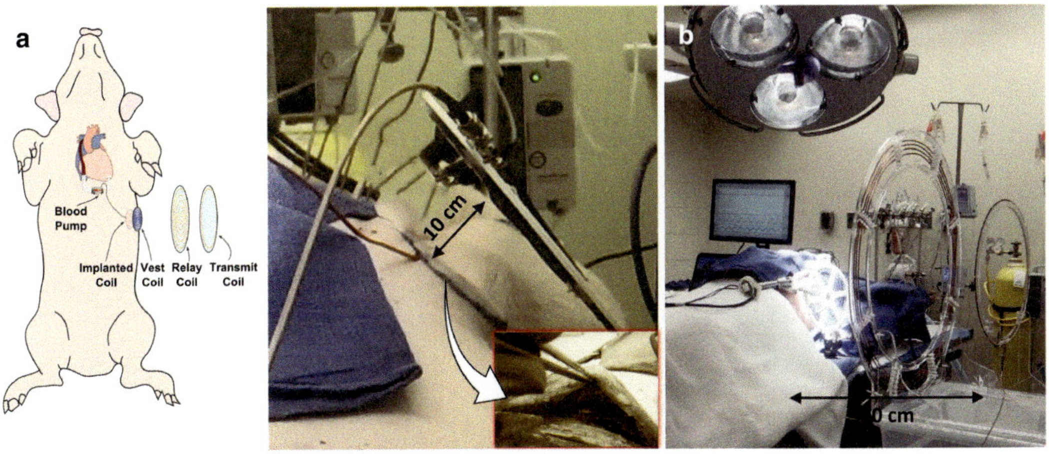

Fig. 39.10 In vivo feasibility of powering an implanted blood pump with the resonant wireless power system (FREE-D system) for (**a**) short range (10 cm) and (**b**) long range (50 cm)

In three-dimensional space, the long-range experiment essentially demonstrates 1-meter working distance as the transmitter can be configured on the other side of the animal body at 50 cm distance. The adaptive tuning techniques allow the system to maintain seamless wireless power delivery with power efficiency of 23% and 17.5% for short and long range, respectively, without any backup battery assist and excessive heat generation (remained below body temperature) (Fig. 39.11) [51].

Future Directions in Wireless Power Transmission

Transcutaneous energy transfer has demonstrated the benefits of total implantability in reducing device-related infections. Resonantly coupled wireless transmission has extended this improvement in MCS technology by enabling better mobility and large distance power transmission. While these modes of wireless energy transfer have improved the overall implantability of MCS systems, there are still significant engineering challenges that must be overcome. For one, misalignment for both TETS and resonantly coupled systems significantly affect system efficiency; in the later this can be addressed by adaptive tuning. In addition, the impedance introduced by biological tissue affects the coupling between resonant coupled inductors. In order to achieve longer distance power transfer, larger coils and potentially more relay coils may need to be used to ensure unconstrained mobility.

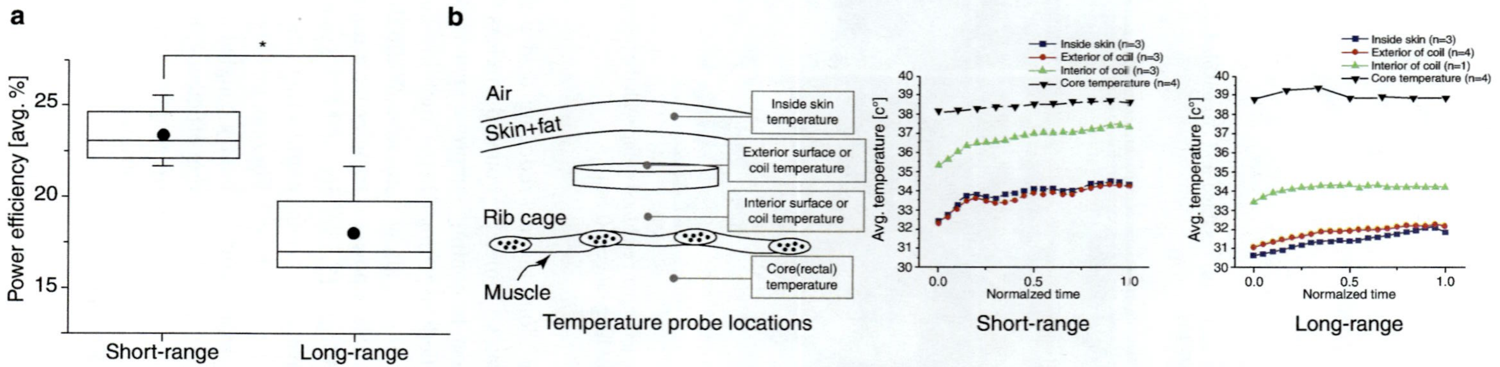

Fig. 39.11 In vivo results of power efficiency and heat generation for (**a**) short- and (**b**) long-range experiments

Conclusion

Since the founding of the NIH Artificial Heart Program, one of the primary goals was to develop a fully implantable support device that could take over the function of the heart. Fifty years later, the improvement in mechanical circulatory support energy systems has brought that goal into realization. Improvement in blood pump technology and battery technology has contributed to the improved outcomes of heart failure patients with implanted LVADs. However, improvement of energy source technology has not improved the overall implantability of MCS systems. The use of TETS and resonantly coupled wireless inductors has made full implantability a reality and allows for further patient mobilization.

References

1. Go AS, Mozaffarian D, Roger VL, Benjamin EJ, Berry JD, Blaha MJ, et al. Heart disease and stroke statistics--2014 update: a report from the American Heart Association. Circulation. 2014;129(3):e28–e292.
2. Rose EA, Gelijns AC, Moskowitz AJ, Heitjan DF, Stevenson LW, Dembitsky W, et al. Long-term use of a left ventricular assist device for end-stage heart failure. N Engl J Med. 2001;345(20):1435–43.
3. Slaughter MS, Rogers JG, Milano CA, Russell SD, Conte JV, Feldman D, et al. Advanced heart failure treated with continuous-flow left ventricular assist device. N Engl J Med. 2009;361(23):2241–51.
4. Bonde P, Ku NC, Genovese EA, Bermudez CA, Bhama JK, Ciarleglio MM, et al. Model for end-stage liver disease score predicts adverse events related to ventricular assist device therapy. Ann Thorac Surg. 2012;93(5):1541–7; discussion 7–8.
5. Bonde P, Dew MA, Meyer D, Tallaj JJ, Martin T, Hollifield KA, et al. 4 National trends in readmission (REA) rates following left ventricular assist device (LVAD) therapy. J Heart Lung Transplant. 2011;30(4):S9.
6. Pereda D, Conte JV. Left ventricular assist device driveline infections. Cardiol Clin. 2011;29(4):515–27.
7. Kormos RL, Bonde P, Bermudez CA, Lockard KL, Genovese EA, Teuteberg JJ, et al. 261: The ventricular assist device (VAD) driveline: what is the price of living with this technology? J Heart Lung Transplant. 2010;29(2):S89.
8. Zierer A, Melby SJ, Voeller RK, Guthrie TJ, Ewald GA, Shelton K, et al. Late-onset driveline infections: the Achilles' heel of prolonged left ventricular assist device support. Ann Thorac Surg. 2007;84(2): 515–20.
9. Waters BH, Sample AP, Bonde P, Smith JR. Powering a ventricular assist device (VAD) with the free-range resonant electrical energy delivery (FREE-D) system. Proc IEEE. 2012;100(1):138–49.
10. Waters BH, Smith JR, Bonde P. Innovative free-range resonant electrical energy delivery system (FREE-D system) for a ventricular assist device using wireless power. ASAIO J. 2014;60(1):31–7.
11. Stewart GC, Givertz MM. Mechanical circulatory support for advanced heart failure patients and technology in evolution. Circulation. 2012;125(10):1304–15.
12. Lubeck DP. The artificial heart. Costs, risks, and benefits--an update. Int J Technol Assess Health Care. 1986;2(3):369–86.
13. Hogness JR, VanAntwerp M. The artificial heart: prototypes, policies, and patients. Washington, DC: National Academies Press; 1991.
14. Mott WE, Cole DW Jr. Development of a nuclear-powered artificial heart. ASAIO J. 1972;18(1):152–6.
15. Norman JC, Molokhia FA, Harmison LT, Whalen RL, Huffman FN. An implantable nuclear-fueled circulatory support system. I. Systems analysis of conception, design, fabrication and initial in vivo testing. Ann Surg. 1972;176(4):492.
16. Whalen R, Molokhia F, Jeffery D, Huffman F, Norman J. Current studies with simulated nuclear-powered left ventricular assist devices. ASAIO J. 1972;18(1):146–51.
17. Tchantchaleishvili V, Bush BS, Swartz MF, Day SW, Massey HT. Plutonium-238: an ideal power source for intracorporeal ventricular assist devices? ASAIO J. 2012;58(6):550–3.
18. Poirier V. Will we see nuclear-powered ventricular assist devices? ASAIO J. 2012;58(6):546–7.
19. Lillehei CW, Gott VL, Hodges PC, Long DM, Bakken EE. Transistor pacemaker for treatment of complete atrioventricular dissociation. JAMA. 1960;172(18):2006–10.
20. Chardack WM, Gage AA, Greatbatch W. A transistorized, self-contained, implantable pacemaker for the long-term correction of complete heart block. Surgery. 1960;48(Oct):643–54.
21. MacLean GK, Aiken PA, Adams WA, Mussivand T. Preliminary evaluation of rechargeable lithium-ion cells for an implantable battery pack. J Power Sources. 1995;56(1):69–74.
22. MacLean GK, Aiken PA, Adams WA, Mussivand T. Comparison of rechargeable lithium and nickel/cadmium battery cells for implantable circulatory support devices. Artif Organs. 1994;18(4):331–4.
23. Okamoto E, Watanabe K, Hashiba K, Inoue T, Iwazawa E, Momoi M, et al. Optimum selection of an implantable secondary battery for an artificial heart by examination of the cycle life test. ASAIO J. 2002;48(5):495–502.
24. Pae WE, Connell JM, Adelowo A, Boehmer JP, Korfer R, El-Banayosy A, et al. Does total implantability reduce infection with the use of a left ventricular assist device? The LionHeart experience in Europe. J Heart Lung Transplant. 2007;26(3):219–29.

25. Schuder JC, Stephenson HE Jr, Townsend JF. Energy transfer into a closed chest by means of stationary coupling coils and a portable high-power oscillator. Trans Am Soc Artif Intern Organs. 1961;7:327–31.
26. Sherman C, Clay W, Dasse K, Daly B. Energy transmission across intact skin for powering artificial internal organs. Trans Am Soc Artif Intern Organs. 1981;27:137–41.
27. Puers R, Vandevoorde G. Recent progress on transcutaneous energy transfer for total artificial heart systems. Artif Organs. 2001;25(5):400–5.
28. Rintoul TC, Dolgin A. Thoratec transcutaneous energy transformer system: a review and update. ASAIO J. 2004;50(4):397–400.
29. Kusserow BK. The use of a magnetic field to remotely power an implantable blood pump. Preliminary report. Trans Am Soc Artif Intern Organs. 1960;6:292–8.
30. Schuder JC, Stephenson HE. Energy transport into the closed chest from a set of very-large mutually orthogonal coils. Trans Am Inst Electr Eng Part I: Commun Electron. 1963;81(6):527–34.
31. Schuder JC. Powering an artificial heart: birth of the inductively coupled-radio frequency system in 1960. Artif Organs. 2002;26(11):909–15.
32. Mehta SM, Pae WE Jr, Rosenberg G, Snyder AJ, Weiss WJ, Lewis JP, et al. The LionHeart LVD-2000: a completely implanted left ventricular assist device for chronic circulatory support. Ann Thorac Surg. 2001;71(3 Suppl):S156–61; discussion S83–4.
33. Okamoto E, Yamamoto Y, Akasaka Y, Motomura T, Mitamura Y, Nose Y. A new transcutaneous energy transmission system with hybrid energy coils for driving an implantable biventricular assist device. Artif Organs. 2009;33(8):622–6.
34. Ozeki T, Chinzei T, Abe Y, Saito I, Isoyama T, Mochizuki S, et al. Functions for detecting malposition of transcutaneous energy transmission coils. ASAIO J. 2003;49(4):469–74.
35. Iwawaki K, Watada M, Takatani S, Um YS, editors. The design of core-type transcutaneous energy transmission systems for artificial heart. Industrial Electronics Society, 2004 IECON 2004 30th annual conference of IEEE; 2–6 November 2004.
36. Miura H, Arai S, Kakubari Y, Sato F, Matsuki H, Sato T. Improvement of the transcutaneous energy transmission system utilizing ferrite cored coils for artificial hearts. IEEE Trans Magn. 2006;42(10): 3578–80.
37. Hassler WL, Dlugos DF, inventors; Google Patents, assignee. Low frequency transcutaneous energy transfer to implanted medical device. United States of America patent US 7599743 B2. 2009.
38. Dowling RD, Gray LA Jr, Etoch SW, Laks H, Marelli D, Samuels L, et al. The AbioCor implantable replacement heart. Ann Thorac Surg. 2003;75(6 Suppl):S93–9.
39. Dowling RD, Gray LA Jr, Etoch SW, Laks H, Marelli D, Samuels L, et al. Initial experience with the AbioCor implantable replacement heart system. J Thorac Cardiovasc Surg. 2004;127(1):131–41.
40. Samuels LE, Dowling R. Total artificial heart: destination therapy. Cardiol Clin. 2003;21(1):115–8.
41. El-Banayosy A, Arusoglu L, Kizner L, Morshuis M, Tenderich G, Pae WE Jr, et al. Preliminary experience with the LionHeart left ventricular assist device in patients with end-stage heart failure. Ann Thorac Surg. 2003;75(5):1469–75.
42. Mehta SM, Silber D, Boehmer JP, Christensen D, Pae WE Jr. Report of the first U.S. patient successfully supported long term with the LionHeart completely implantable left ventricular assist device system. ASAIO J. 2006;52(6):e31–2.
43. Yamamoto T, Koshiji K, Homma A, Tatsumi E, Taenaka Y. Improvement in magnetic field immunity of externally-coupled transcutaneous energy transmission system for a totally implantable artificial heart. J Artif Organs. 2008;11(4):238–40.
44. Fletcher NH, Rossing TD. The physics of musical instruments. 2nd ed. New York: Springer Science & Business Media; 1998. 756 p.
45. Sample AP, Meyer DA, Smith JR. Analysis, experimental results, and range adaptation of magnetically coupled resonators for wireless power transfer. IEEE Trans Ind Electron. 2011;58(2):544–54.
46. Tesla N, inventor; Google Patents, assignee. Apparatus for transmitting electrical energy. United States of America patent US1119732 A. 1914.
47. Kurs A, Karalis A, Moffatt R, Joannopoulos JD, Fisher P, Soljačić M. Wireless power transfer via strongly coupled magnetic resonances. Science. 2007;317(5834):83–6.
48. Karalis A, Joannopoulos JD, Soljačić M. Efficient wireless non-radiative mid-range energy transfer. Ann Phys. 2008;323(1):34–48.
49. Cannon BL, Hoburg JF, Stancil DD, Goldstein SC. Magnetic resonant coupling as a potential means for wireless power transfer to multiple small receivers. IEEE Trans Power Electron. 2009;24(7):1819–25.
50. Waters B, Sample A, Smith J, Bonde P. Toward total implantability using free-range resonant electrical energy delivery system: achieving untethered ventricular assist device operation over large distances. Cardiol Clin. 2011;29(4):609–25.
51. Waters BH, Park J, Bouwmeester JC, Valdovinos J, Geirsson A, Sample AP, et al. Electrical power to run ventricular assist devices using the free-range resonant electrical energy delivery system. J Heart Lung Transplant. 2018;37(12):1467–74.

Part VI

Management Strategies and Limitations of MCS

Risk Factors for Mechanical Circulatory Support Use and Risk Assessment

40

Rajakrishnan Vijayakrishnan and Emma J. Birks

Introduction

Risk factor analysis is a very important component in the evaluation process of advanced heart failure therapies like durable mechanical circulatory support (MCS) and heart transplantation. The use of MCS can be further characterized based on the device strategy into either:

1. Short Term:
   - MCS as short-term therapy like venoarterial extracorporeal membrane oxygenator (VA-ECMO), intra-aortic balloon pump (IABP), impella (right and left sided), and tandem heart

   Or
2. Long term:
   - MCS as destination therapy (DT)
   - VAD as BTT in certain patient population and geographical locations
   - Total artificial heart (TAH) as bridge to transplantation

R. Vijayakrishnan, MD
Advanced Heart Failure Therapeutics and Mechanical Circulatory Support, Norton Thoracic Institute/ Creighton University School of Medicine, St Joseph Hospital and Medical Center, Phoenix, AZ, USA
e-mail: rajakrishnan.vijayakrishnan@dignityhealth.org

E. J. Birks, MD, PhD (✉)
Division of Cardiovascular Medicine, Department of Medicine, University of Kentucky, Gill Heart and Vascular Institute, Louisville, KY, USA
e-mail: emma.birks@louisville.edu

An understanding of the urgency of the need for any form of MCS is very crucial in the success of the therapy. The utilization of the Interagency Registry for Mechanically Assisted Circulatory Support (INTERMACS) database is very crucial in understanding and improving the risk assessment when MCS is considered for advanced heart failure therapy.

Risk Factor Analysis

All patients undergoing durable MCS (VAD: both as DT and BTT) need a comprehensive evaluation prior to implantation. Risk assessment can be broadly divided into at least five categories and are as follows:

1. Severity of systolic heart failure
2. Cardiac factors
3. Noncardiac factors
4. Transplant candidacy
5. VAD operative risk

Severity of Systolic Heart Failure

Historically, the risk analysis of patients with heart failure was derived from New York Heart Association (NYHA) classification of symptoms. Despite good correlation with outcomes in early stages of heart failure, the NYHA class of symp-

© Springer Nature Switzerland AG 2020

J. H. Karimov et al. (eds.), *Mechanical Support for Heart Failure*,
https://doi.org/10.1007/978-3-030-47809-4_40

toms analysis did not further quantify or qualify the risk in advanced Stage D (end stage) systolic heart failure, especially needing surgical interventions. Hence risk factor analysis in end-stage heart failure patients undergoing durable MCS was derived from heart failure risk scores. The Acute Physiology and Chronic Health Evaluation (APACHE) II (1985) and Seattle Heart Failure Model (SHFM, 2006) were used widely to assess and risk stratify chronic heart failure [1, 2]. The COL, Destination Therapy (DT) Risk Score, and Leitz-Miller (LM) scores were developed specifically for LVAD (based on XVE HeartMate I LVAD) for predicting perioperative mortality and 90-day inhospital mortality, respectively [3, 4].

MCS, especially VADs, have also evolved from pulsatile first-generation pumps to current third- and fourth-generation continuous flow pumps and TAH. Based on the eight INTERMACS annual report, up till December 2016, out of a total of 22,866 total durable MCS implants, 957 were pulsatile LVAD, 396 TAH, and the rest continuous flow LVAD (+/−RVAD) [5]. Also, the incidence of DT has increased to ~51% based on the current data.

The outcomes and the risks are affected also by the INTERMACS patient profile (as summarized in Table 40.1); the higher the risk and poorer the outcome, when INTERMACS score is lower, especially profile 1 [6]. From 2008 to 2016, the proportion of VAD patients in INTERMACS profile 1 has been stable at 14–16%. INTERMACS profile 3 (stable but inotrope dependent) now represents approximately 38% of all the implants, whereas the representation of profiles 4–7 has decreased to 12.8% in 2015–2016.

Table 40.1 INTERMACS profile

INTERMACS profile		
Profile	Description	Time frame for intervention
1	*Critical cardiogenic shock* Patients with life-threatening hypotension despite rapidly escalating inotropic support, critical organ hypoperfusion, often confirmed by worsening acidosis and/or lactate levels "Crash and burn"	Definitive intervention needed within hours
2	*Progressive decline* Patient with declining function despite intravenous inotropic support may be manifest by worsening renal function, nutritional depletion, inability to restore volume balance "Sliding on inotropes" also describes declining status in patients unable to tolerate inotropic therapy	Definitive intervention needed within few days
3	*Stable but inotrope dependent* Patient with stable blood pressure, organ function, nutrition, and symptoms on continuous intravenous inotropic support (or a temporary circulatory support device or both), but demonstrating repeated failure to wean from support due to recurrent symptomatic hypotension or renal dysfunction "dependent stability"	Definitive intervention elective over a period of weeks to few months
4	*Resting symptoms* Patient can be stabilized close to normal volume status but experiences daily symptoms of congestion at rest or during ADL. Doses of diuretics generally fluctuate at very high levels. More intensive management and surveillance strategies should be considered, which may in some cases reveal poor compliance that would compromise outcomes with any therapy. Some patients may shuttle between 4 and 5	Definitive intervention elective over period of weeks to few months
5	*Exertion intolerant* Comfortable at rest and with ADL but unable to engage in any other activity, living predominantly within the house. Patients are comfortable at rest without congestive symptoms but may have underlying refractory elevated volume status, often with renal dysfunction. If underlying nutritional status and organ function are marginal, patient may be more at risk than INTERMACS 4 and require definitive intervention	Variable urgency, depends upon maintenance of nutrition, organ function, and activity

Table 40.1 (continued)

INTERMACS profile		
Profile	Description	Time frame for intervention
6	*Exertion limited* Patient without evidence of fluid overload is comfortable at rest and with activities of daily living and minor activities outside the home but fatigues after the first few minutes of any meaningful activity. Attribution to cardiac limitation requires careful measurement of peak oxygen consumption, in some cases with hemodynamic monitoring to confirm severity of cardiac impairment "Walking wounded"	Variable, depends upon maintenance of nutrition, organ function, and activity level
7	*Advanced NYHA III* A placeholder for more precise specification in future, this level includes patients who are without current or recent episodes of unstable fluid balance, living comfortably with meaningful activity limited to mild physical exertion	Transplantation or circulatory support may not currently be indicated
Modifiers for profiles		Possible profiles to modify
TCS	Temporary circulatory support can modify only patients in hospital (other devices would be INTERMACS devices). Includes IABP, ECMO, TandemHeart, Levitronix, BVS 5000 or AB5000, Impella	1,2,3 in hospital
A: Arrhythmia	Recurrent ventricular tachyarrhythmias that have recently contributed substantially to clinical compromise. This includes frequent ICD shock or requirement for external defibrillator, usually more than twice weekly	Any profile
FF: Frequent flyer	Can modify only outpatients, designating a patient requiring frequent emergency visits or hospitalizations for diuretics, ultrafiltration, or temporary intravenous vasoactive therapy	3 if at home, 4,5,6 A frequent flyer would Rarely be profile 7

Reprinted from Stevenson et al. [6, with permission from Elsevier]
ADL, activities of daily living, *ECMO* extracorporeal membrane oxygenator, *IABP* intra-aortic balloon pump, *ICD* implantable cardioverter defibrillator

There are important features of these above-mentioned profiles, which need special mention. In INTERMACS profile 1, use of intra-aortic balloon pump (IABP), associated oliguria, and perhaps with escalating liver function test abnormalities consistent with "shock liver" are important. In describing profile 2, sometimes inotropic infusions cannot be maintained due to tachyarrhythmias, clinical ischemia, or other intolerance. In profile 3, stable mid-level doses of intravenous inotropic agents indicate usually ≤5 μg/kg/min of dobutamine or dopamine, or <0.5 μg /kg/min of milrinone, but may occasionally be stable on higher doses. In the section of modifiers, FF means at least two emergency visits/admissions in the past 3 months or three in the past 6 months. If these admissions were triggered by tachyarrhythmias or ICD shocks, the appropriate modifier would be the arrhythmia modifier "A," rather than "FF."

Even though COL and Leitz-Miller scores were LVAD specific, on a single center multivariate analysis of 86 continuous flow LVADs from June 2000 to May 2009, SHFM score was the best predictor of both 30-day and 90-day mortality on multivariate analysis (*p* 0.08 and *p* 0.09, respectively) [7]. In the same study, 1-year mortality was predicted by three variables, namely, age ≥65 years, serum creatinine ≥1.5 mg/dL, and platelets <148,000/microliter.

Cardiac Factors

Right ventricular systolic function plays a key role in the success of isolated left-sided VAD support, whether it is for BTT or for DT. The anatomic considerations of chest cavity, number of sternotomies (including assessment of adhesion burden), and body size are important considerations prior to any cardiovascular surgery, and hence it is important in durable MCS implantation also. We will discuss nonsurgical/anatomical considerations here and discuss surgical and anatomical factors in the VAD operative risk section.

Right Ventricular Dysfunction

One of the major contributors to increased morbidity (hospital stay, especially ICU stay) and mortality after LVAD implantation has been undiagnosed or rather underdiagnosed right ventricular (RV) systolic failure. The use of inotropes in cardiogenic shock in the preoperative phase prior to implantation of the LVAD masks the presence of RV systolic failure, and it surfaces once the LVAD generates flows more than 5 liters/minute and the patient is without inotropic support and the LVAD pulls the septum to the left. As mentioned earlier, this is important since approximately 38% of the VAD implantation population is in INTERMACS profile 3 [5]. The presence of low output due to LV systolic HF results in decreased preload to RV and hence can mask the RV systolic dysfunction [8].

RV failure can be early as well as late after LVAD implantation. The definitions of both early as well as late RV failure are ill-defined especially since late-RV failure is an emerging concept and difficult to predict. Early RV failure can affect the immediate perioperative period with reduced LV filling and hence LVAD flow, difficulty in weaning cardiopulmonary bypass (CPB), and poor organ/tissue perfusion resulting in multi-organ failure, all of it contributing to increased morbidity and mortality [9]. For standardization purposes, we follow the INTERMACS definition of RV failure (as summarized in Table 40.2) and the severity of RV failure (summarized in Table 40.3).

The LVAD function can distort the normal physiology of LV and subsequently RV contractility (ventricular interdependence) and even tricuspid annular anatomy. This is important since 20–40% of RV output is secondary to LV contraction and systolic interdependence is a result of interventricular septum and diastolic interdependence mediated through the pericardium [11, 12, 13]. In addition, tachyarrhythmias also reduce RV function with supraventricular tachycardia (mostly atrial in origin) resulting in doubling the risk of RVF [14].

Table 40.2 INTERMACS definition of right ventricular failure

INTERMACS definition of right ventricular failure	
Definition	Symptoms or findings of persistent RVF characterized by both of the following: Elevated CVP documented by one of the following: Direct measurement: RHC: RAP >16 mmHg on right heart catheterization Findings of significantly dilated inferior vena cava with absence of inspiratory variation by echocardiography Clinical findings of elevated jugular venous distension at least half way up the neck in an upright patient Manifestations of elevated CVP characterized by one of the following: Peripheral edema (≥2 + either new or unresolved) Presence of ascites or palpable hepatomegaly on physical examination (unmistakable abdominal contour) or by diagnostic imaging Laboratory evidence of worsening hepatic (total bilirubin >2.0 mg/dl) or renal dysfunction (creatinine >2.0 mg/dl)

Interagency Registry for Mechanically Assisted Circulatory Support (INTERMACS) [10]
CVP central venous pressure, *RVF* right ventricular failure, *RHC* right heart catheterization, *RAP* right atrial pressure

Assessment of RVF Prior to LVAD Implantation

Echocardiography

The use of echocardiographic data to assess RV function and size has been very useful in pre-VAD evaluation. Tricuspid annular plane systolic excursion (TAPSE) <7.5 mm has a high specificity (91%) but low sensitivity (46%) and is due to TAPSE being only a marker of RV longitudinal function rather than global motion [15]. RV to LV

Table 40.3 Severity of right ventricular failure (RVF)

INTERMACS right ventricular failure severity scale		
Mild	*VAD implant admission*	Patient meets both criteria for RVF plus: Post-implant inotropes, inhaled nitric oxide, or intravenous vasodilators not continued beyond post-op day 7 after VAD implant AND No inotropes continued beyond post-op day 7 after VAD implant
	Surveillance periods (3 months, 6 months, 12 months, and every 6 months thereafter) following VAD implant	Patient meets both criteria for RVF plus: No readmissions for RVF since last surveillance period AND No inotropes since last surveillance period
Moderate	*VAD implant admission*	Patient meets both criteria for RVF plus: Post-implant inotropes, inhaled nitric oxide, or intravenous vasodilators continued beyond post-op day 7 and up to post-op day 14 following VAD implant
	Surveillance periods (3 months, 6 months, 12 months, and every 6 months thereafter) following VAD implant	Patient meets both criteria for RVF plus: Limited to one readmission for intravenous diuretics/vasodilators to treat RVF since last surveillance period AND No inotropes since last surveillance period
Severe	*VAD implant admission*	Patient meets both criteria for RVF plus: CVP or RAP >16 mmHg AND Prolonged post-implant inotropes, inhaled nitric oxide, or intravenous vasodilators continued beyond post-op day 14 following VAD implant
	Surveillance periods (3 months, 6 months, 12 months, and every 6 months thereafter) following VAD implant	Patient meets both criteria for RVF plus: Need for inotropes at any time since last surveillance period OR Two or more readmissions for intravenous diuretics/vasodilators to treat RVF since last surveillance period OR Requiring RVAD support at any time after hospital discharge OR Death at any time following discharge from the VAD implant hospitalization with RVF as the primary cause
Severe-acute	*VAD implant admission*	Patient meets both criteria for RVF plus: CVP or RAP >16 mmHg AND Need for right ventricular assist device at any time following VAD implant OR Death during the VAD implants hospitalization with RVF as the primary cause

Interagency Registry for Mechanically Assisted Circulatory Support (INTERMACS) [10]
CVP central venous pressure, *RVF* right ventricular failure, *VAD* ventricular assist device, *RAP* right atrial pressure

end diastolic dimension (EDd) ratio has a strong correlation to CVP/PCWP ratio, and both have been shown to have a strong association with RVF [16, 17]. A RVEDd/LVEDd ratio ≥0.75 is related to increased incidence of 30-day RVF (*p* 0.012) as well as composite end point of 30-day death or RVF (*p* 0.03) in continuous flow LVADs. When this is combined with other risk factors/risk scores like the Matthews' score, the strength of risk prediction is increased [18].

Post-LVAD, echocardiography assessment has been useful to predict RVF and smaller

LVEDd; increased left atrial dimension/LVEDd ratio and increased LV ejection fraction are associated with RVF [19]. Strain imaging and speckle tracking echocardiography can be utilized to predict RVF in the pre-LVAD stage, but stain imaging has not been as predictive as expected [20, 21].

Invasive Hemodynamics

Elevated CVP especially in the setting of low PAP is one of the simplest and the best determinants of RVF [22]. Another predictor is the elevated CVP/PCWP ratio >0.63, and this has been proven in the HM II device in bridge-to-transplant population [23]. In this patient population, this ratio was an independent risk for early RVF (*p* 0.009) on multivariate analysis.

RV stroke work index (SWI) is another predictor of RV failure in advanced heart failure. The use of RVSWI in continuous flow LVAD has been studied, and analysis revealed increased incidence of RVF with RVSWI <450 mm Hg·ml/m^2 (*p* 0.012). Despite this result, various multivariate analysis involving RVSWI point toward a nonsignificant predictive trend when compared to biochemical investigations with poorer discrimination scores [18, 24, 25].

Other hemodynamics like elevated transpulmonary gradient, pulmonary vascular resistance, and pulmonary artery pressure show trends toward prediction of RVF but have poor discrimination score [18]. PAPi ([PA systolic – PA diastolic]/RA) is also considered as an independent hemodynamic predictor (on multivariate analysis [*p* 0.0001]) of RVF [26].

Risk Models

Risk models have been suggested based on pulsatile LVAD data, and very few are based on current continuous flow LVADs. The majority of the initial risk models were single center analysis as well as had no clear definition of RVF and hence cannot be used for clinical purposes. The major multicenter trial with continuous-flow LVAD to predict RVF (definition closest to INTERMACS definition) was performed with 484 LVADs and proved that patients with early RVF had 59% survival at 1 year compared to 78% without RVF [22]. The Matthews' score is one such score and is related to very good discrimination score for the prediction of RVF (as summarized in Table 40.4) [18].

Table 40.4 Right ventricular failure risk score

Right ventricular failure risk score and likelihood of RV failure by score strata	
Risk score	Likelihood ratio (95% confidence interval)
≤3.0	0.49 (0.37–0.64)
4.0–5.0	2.8 (1.4–5.9)
≥5.5	7.6 (3.4–17.1)
Presence of a vasopressor requirement: 4 points Aspartate aminotransferase ≥80 IU/l: 2 points Bilirubin ≥2.0 mg/dl: 2.5 points Creatinine ≥2.3 mg/dl: 3 points	

Reprinted from Matthews et al. [Reference 18, with permission from Elsevier]

There are various scoring systems but most of them are single-center studies, and the initial studies were based on pulsatile LVADs with sensitivity and specificity approaching 84 to 90% [18, 27, 28]. The multicenter trial (HM II BTT trial)-derived scoring system revealed the importance of CVP/PCWP (>0.63), ventilatory support, and BUN >39 mg/dL which were directly related to prediction of early RVF and also predicted 1-year survival ($p < 0.001$) in continuous flow LVADs [23].

Risk Factors for Right Ventricular Failure

In conclusion, the risk of RVF (mostly early RVF) is decreased when the following are noted:

- Hemodynamics: CVP ≤ 8 mmHg, CVP/PCWP ≤0.66, RVSWI ≥400 mmHg mL/m^2 and PAPi >4
- Echocardiogram: RVEDd/LVEDd <0.63 and less than moderate tricuspid regurgitation
- End-organ function: Absence of lung, liver, and renal failure (better than stage IIIB chronic kidney disease)

- Normotension
- Normal nutrition
- Virgin chest

The risk factors for RVF following LVAD implantation are summarized in Table 40.5.

Table 40.5 Risk factors for right ventricular failure

Need for mechanical support	1. Requirement of intravenous vasopressor agents 2. Preoperative intra-aortic balloon pump 3. Cardiac arrest at any time in the preoperative period
Gender	**Female**
Preoperative end-organ dysfunction	1. Preoperative ventilator support (five times more likely to develop RVF) [8] 2. Liver dysfunction [18] 3. Renal impairment (or renal replacement therapy preoperatively) [23] 4. Malnutrition [23] 5. Coagulation abnormalities 6. White blood cell count ≥12,200/μL and platelets ≤120,000/μL [29] 7. Elevated pro-BNP and raised C-reactive protein [23, 29]
Echocardiographic features	RVEDd/LVEDd ratio > 0.63 [23]
Hemodynamic features	1. Elevated CVP (>16 mmHg) 2. CVP/PCWP (≥0.63) [23] 3. Low RVSWI <450 mmHg mL/m^2 [18] 4. PAPi <2 [26]
Other risk factors	1. Nonischemic cardiomyopathy 2. Reoperation 3. Presence of severe TR preoperatively [8]

BNP brain natriuretic peptide, *CVP* central venous pressure, *LVEDd* left ventricular end diastolic dimension, *PAPi* pulmonary artery pulsatility index, *PCWP* pulmonary capillary wedge pressure, *RVEDd* right ventricular end diastolic dimension, *RVF* right ventricular failure, *RVSWI* right ventricular stroke work index, *TR* tricuspid regurgitation

Cardiac Arrhythmias

The presence of arrhythmias especially atrial arrhythmias is a hot topic of discussion in the field of heart failure especially end-stage heart failure. The most important predictor of atrial fibrillation (AF) incidence after LVAD insertion is the presence of atrial fibrillation prior to LVAD insertion [30]. Even though paroxysmal AF was not found to have poor prognosis, permanent/persistent AF was associated with increased mortality (0.06) and HF hospitalization ($p < 0.01$) in patients with the HMII LVAD [31]. However, there was no increased risk for thromboembolism or bleeding in this patient population. The studies from Columbia University Medical Center and Mayo Clinic have not shown an increased mortality with the presence of AF in LVAD patients [30, 32]. This could be related to the increased use of beta-blockers (BB) and/or amiodarone in same patient population to control ventricular arrhythmias (VA). Another single-center study from the University of Chicago points toward an increased mortality following the presence of atrial tachycardia (*p* 0. 002) compared to persistent/permanent AF history (*p* 0.03) [33].

The LVAD implantation helps patients tolerate ventricular arrhythmias (sustained ventricular tachycardia (VT) and/or ventricular fibrillation) in end-stage heart failure [34]. In a meta-analysis, the all-cause mortality (at 60 (*p* 0.001), 120 (*p* 0.05), and 180 (*p* 0.05) days) was increased with post-LVAD VA (with preimplant VA as risk factor) [35]. ICD is generally considered to be a good option in high-risk patients since the ICD has been proven to improve survival in systolic heart failure and also in LVAD [36]. Optimal ICD programming has not been defined, but few suggestions for clinical practice are as follows [37]:

- Use of either prolonged detection time
- Higher rate cutoffs (e.g., single zone ≥200 beats per minute)
- VT monitoring zone at lower rates to be used only in the presence of unexplained symptoms

Various mechanisms for VA have been proposed including myocardial scar (independent of insertion site), repolarization abnormalities, myocardial electrolyte shifts, systemic electrolyte shifts, as well as mechanical cause like suction events [38]. LVAD patients who become symptomatic are usually due to the effect of the VAs on the RV systolic function. The above mechanisms may explain the lack of sustained improvement in VA incidence after use of BB in LVAD patient population [34]. Catheter ablation procedures as treatment options are discussed in detail elsewhere.

Despite the decrease in the incidence of early VAs from 47% (2000 to 2007) to <22% from 2008 to 2015, prior cardiac surgery (p 0.023) and pre-continuous flow LVAD VT storm (p 0.003) were found to be independently predictive of early VAs from the preimplantation variables [39]. With the advent of newer CF-LVADs, there needs to be multicenter outcome studies to understand the prognostic importance of ventricular and atrial arrhythmias in the LVAD population.

Noncardiac Factors

Medical Factors

Preexisting chronic obstructive lung disease (COPD) and advanced stages of chronic kidney disease (CKD) are well-established risk factors for death in LVAD patients. COPD is related to higher late-hazard (manifest at approximately 84 months) with a hazard ratio (HR) of 1.27 and p 0.001 [5]. Ventilator-dependent respiratory failure and tracheostomy are known associations with higher post-LVAD mortality and morbidity (HR 4.92; 95% conference interval 1.62–14.93: p 0.005) [40]. Patients with CKD stage IIIB and above (especially end-stage renal disease with dialysis) have very high early (<3 months post-implant) and late hazard for deaths (HR 3.29; p 0.0001 and HR 1.12; $p < 0.0001$, respectively) [5].

Vascular dysfunction in the form of cerebrovascular accident (CVA), peripheral or cerebrovascular diseases, heparin-induced thrombocytopenia (HIT), and chronic coagulopathy imposes major limitation and risk for the implantation of durable mechanical devices like VAD. Pre-LVAD CVA history was associated with a higher incidence of neurological complications when compared to absence of CVA history (27.9% vs 15.5%, respectively, p 0.046). Multiple regression analysis revealed that CVA history (odds ratio 2.37, 95% confidence interval 1.24–5.29; p 0.011) and postoperative infection (odds ratio 2.99, 95% confidence interval 1.16–10.49; p 0.011) were highly associated with the development of neurological complications [41]. Female gender (HR 1.88, p 0.021) and diabetes mellitus (HR 1.99, p 0.009) are also independent predictors of post-LVAD ischemic CVA, whereas age >65 years (HR 1.94, p 0.01) and female gender (HR 1.92, p 0.006) were associated with a higher incidence of hemorrhagic CVA in HMII patients [42]. Hyponatremia and hypoalbuminemia have also been related to post-LVAD neurological complications. Peripheral vascular disease is associated with high late-hazard risk for death with HR 1.28 (p 0.004) [5]. Despite increased risk of thromboembolic events associated with HIT, careful use of alternate agents has resulted in lower complications associated with LVAD but with an increased trend toward bleeding [43]. Chronic coagulopathy has been also associated with increased morbidity but fortunately the incidence is low.

Morbidity is also affected by the increased incidence of gastrointestinal bleeding (GIB) needing transfusion and readmissions requiring multiple endoscopic as well as surgical interventions. Older age (>65 years: HR 1.31, p 0.015), pre-LVAD HCT $\leq$ 31% (HR 1.31, p 0.023), female gender (HR 1.45, p 0.022), and ischemic cardiomyopathy (HR 1.35, p 0.008) are associated with increased incidence of GIB-related events [42].

As mentioned in other sections, malnutrition has a direct impact on LVAD outcomes, both mortality and morbidity. Prognostic nutritional index (PNI) is a simple tool for assessment of nutritional status of patients undergoing LVAD implantation (PNI = [10 $\times$ serum albumin in g/dl] + [0.005 × total lymphocytes 1000/μl]). PNI <30 was associated with 12.2% reduction in postoperative survival (HR: 0.89; 95% confidence interval 0.79–0.99; *p* 0.037) in a single-center study [44]. This increased mortality and morbidity is related to longer ventilatory need, increased infection risk, and poor wound healing. DT Risk Score was the first to emphasize the importance of nutrition in LVAD outcome [4]. Hence patients at risk should be identified by a nutritional specialist and supplemented according to individual caloric and substrate needs prior to elective durable device implantation if time permits.

Uncontrolled diabetes mellitus is a risk for infection, poor wound healing, and CVA, and poorly controlled diabetes is related to increased mortality and morbidity in LVAD patients [45, 46, 47]. The degree of glycemic control (HbA1C < 7 versus >7) and use of insulin in diabetic patients pre-LVAD were not found to influence mortality. A few single-center studies have failed to show a direct relationship of diabetes to post-LVAD mortality [48]. LVAD implantation has improved the glycemic control of diabetes and results in use of lesser medication need for the same control, and this may explain the lack of increased mortality in these single-center studies [47, 49].

Obesity is an established risk factor for cardiovascular disease including heart failure. On retrospective analysis of 3865 LVAD (bridge to transplant) patients (from UNOS database from May 2004 to April 2014), there is no increased risk of death (*p* 0.23) or delisting (*p* 0.92), but the risk of complications like infection and thromboembolism was increased in body mass index (BMI) Class II (BMI 35–39.9) and above (HR 1.48, *p* 0.004) [50]. BMI did not have any noticeable effect on the mortality after LVAD implantation on multivariate analysis. The patients with BMI ≥ 35 have increased postoperative respiratory failure (*p* 0.021) and RV failure (*p* 0.049) but lower reoperative bleeding (*p* 0.047). Despite increased trends toward higher infection and pump thrombosis rates as expected from prior studies, they did not reach statistical significance [51, 52].

Psychosocial Factors

Low socioeconomic status is related to increased mortality and hospitalizations in the heart failure population [53, 54]. When analyzed for LVAD patients who are bridge to transplant, the waitlist mortality was not affected by low socioeconomic status including factors like income, education, insurance, and race [55]. Risk of death has been shown to be reduced when patient can identify a caregiver (social worker determined), who understands the severity of the illness and options available to the patient, and identifies a backup plan and who is able to provide logistical support, compared to patients who live alone (*p* 0.04) [56].

Psychiatric dysfunction including major depression, limited social support, repeated non-adherence to medical management plans, and substance abuse are additional risk factors which need thorough assessment and planning before consideration of VAD implantation and/or heart transplantation. Depression is common in the heart failure population (incidence is 20–25%) and is noted to be present in at least one-third of all the patients undergoing LVAD implantation [57]. Depression results in adverse outcomes in the heart failure population, and a study from Mayo Clinic has revealed severe depression to result in a twofold increased risk of hospitalization and ED visits, a modest increase in outpatient visits, and a fourfold increase in all-cause mortality [58].

Depression has shown to result in poor adherence to treatment plans for heart failure [59]. Depression has been shown to be resulting in higher readmissions post-LVAD destination therapy (HR 1.77; 95% confidence interval 1.40–2.22) after adjustment for other comorbidities and confounders [60]. There was also an increased readmission rate for LVAD complications like hemolysis and drive-line infections and increased mortality in patients with substance abuse (excluding tobacco abuse) [60, 61].

Frailty and Advanced Heart Failure

Frailty is a syndrome characterized by increased vulnerability to acute physiologic stressors including hospitalizations, major surgery, falls, and infection to mention a few [62]. Frailty is an independent predictor of mortality and morbidity especially considering the medical and surgical options available to deal with chronic medical conditions. Despite direct correlation with age (especially elderly), this syndrome affects any age group with chronic medical conditions especially those with end-stage heart failure.

Frailty has been shown to directly and independently predict adverse outcome in heart failure especially considering the invasive surgical options available to prolong quantity and quality of life [63, 64, 65]. Post-cardiac surgery complications were increased ($p < 0.001$) as well as hospital stay and 30-day readmissions which were increased (p 0.026 and 0.014 respectively) in frail patients (independent of the age) compared to those who were pre-frail and non-frail patients [66]. With the advent of VADs (both as BTT and DT), the importance of assessment of frailty has increased. In a meta-analysis, pooled results of 11 eligible studies suggested that patients with frailty compared with non-frail patients significantly had increased the risk of all-cause mortality after undergoing VAD implantation (HR 1.62, 95% confidence interval 1.35–1.94, $p < 0.0001$) [67].

One difficulty we face is how to accurately and consistently define frailty without much variability. The Fried Frailty Index is one of the most widely used scales (developed first as part of the Cardiovascular Health Study [68]) and focuses on five physical components of frailty (Frailty Phenotype [69] as summarized in Table 40.6).

The objective assessment is to be performed by a clinician and includes the following:

Table 40.6 Simplified fried criteria

Simplified fried criteria [70, 71]		
1	Unintentional weight loss	>4.5 kg in last 12 months
2	Walk time	Time to walk 4 meters >6 seconds
3	Grip strength	Assessed by dynamometer
4	Exhaustion	For at least 3 days during the last week "I felt that everything I did was an effort" or "I could not get going"
5	Physical activity	No physical activity, spend most of the time sitting, or rarely a short walk during the last year
Frail: ≥3 criteria, pre-frail 1 or 2 criteria		

Reprinted from Singh et al. [70], with permission from Oxford University Press]

- 5-minute gait speed test
- Hand grip strength using a dynamometer
- Self-reported questionnaire regarding exhaustion/anergia, physical activity, and unintentional weight loss

Currently based on data from the VAD patients, use of single domain of hand grip strength has shown direct correlation with post-operative complications and survival rates as illustrated in the Fig. 40.1 [72]. There are limitations to the use of single domain, but the patients who cannot perform the hand grip strength test are debilitated to start with and have poorer prognosis and are high risk for any surgical procedures.

Despite the availability of modified Fried index like SHARE-FI, there is a limited role in advanced heart failure [73, 74]. The comprehensive geriatric assessment (CGA) scale, deficit accumulation index (DAI: known as Frailty Index), Tilburg Frailty Indicator, Frailty Staging System, and Canadian Study of Health and Aging Clinical Frailty Scale are the other available modified frailty scoring systems, but they are either cumbersome for routine use or are not validated in advanced heart failure and hence have limited utility [75, 76, 77, 78].

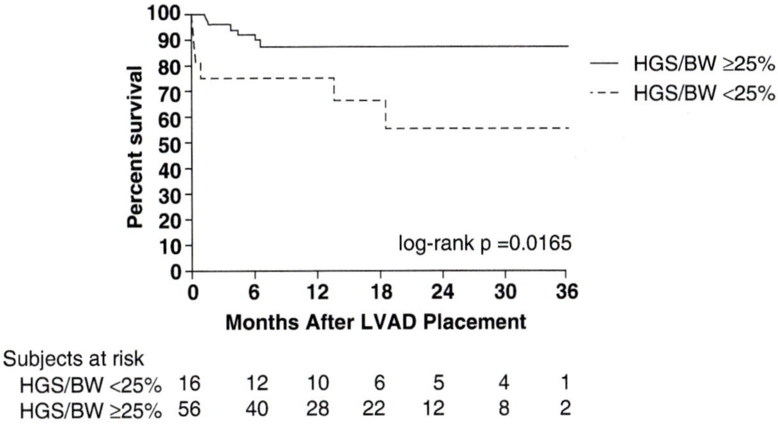

Fig. 40.1 Hand grip strength: correlation with postoperative complications and survival rates. (Reprinted from Chung et al. [72], with permission from Elsevier)

One of the major limitations of frailty assessment for risk for advanced heart failure therapies is that the therapies in question by themselves can improve frailty since frailty is mostly secondary to advanced heart failure-related shock/low output state.

Interventions to Reduce the Risk Factors

If diagnosed at an earlier stage, frailty can be reversible. Exercise-based rehabilitation like cardiac rehabilitation improves the morbidity and mortality of the patients undergoing procedures like TAVR, CABG, coronary interventions, etc. in heart failure and cardiac transplant patients [79, 80].

Dietary counseling and availability of nutritional supplementation with a focus on protein (25–30 grams of high-quality protein per meal) has been associated with improved grip strength and increased muscle mass by reducing sarcopenic muscle loss [81, 82]. Length of hospital stay (especially intensive care unit stay with ventilatory support) is dramatically reduced with the above measures. Vitamin D supplementation (along with calcium) helps to improve muscle function and reduce falls [83], but calcium supplementation may be related with increased risk of myocardial infarction in a women's health initiative study [84].

Transplant Candidacy

Patients receiving an implant as destination therapy (DT) continue to have worse survival compared to patients considered for cardiac transplantation (as illustrated in Fig. 40.2) with a Kaplan-Meier 5-year survival estimate for DT of approximately 35% [5].

This could be explained by the older age group, higher INTERMACs profile, and associated increased comorbidities. Also, younger patients who failed transplant candidacy will also have more frailty and poor psychosocial conditions, contributing to increased risk for morbidity and mortality as described above. The DT strategy is related to an increased risk for deaths with a HR 1.18 (p 0.0003) and an increased late hazard with HR 1.22 ($p < 0.0001$) [5].

VAD Operative Risk

The early mortality after implantation is directly related to operative complications including sepsis, multi-organ failure, bleeding, RV failure, and CVAs. Hence care must be made when patients are selected for elective (or semi-elective) LVAD implantation to avoid high mortality. Unfavorable mediastinal anatomy, thoracic aortic disease, severe malnutrition, liver dysfunction, renal dysfunction, severe lung disease, and obesity contribute to high mortality and morbidity; hence

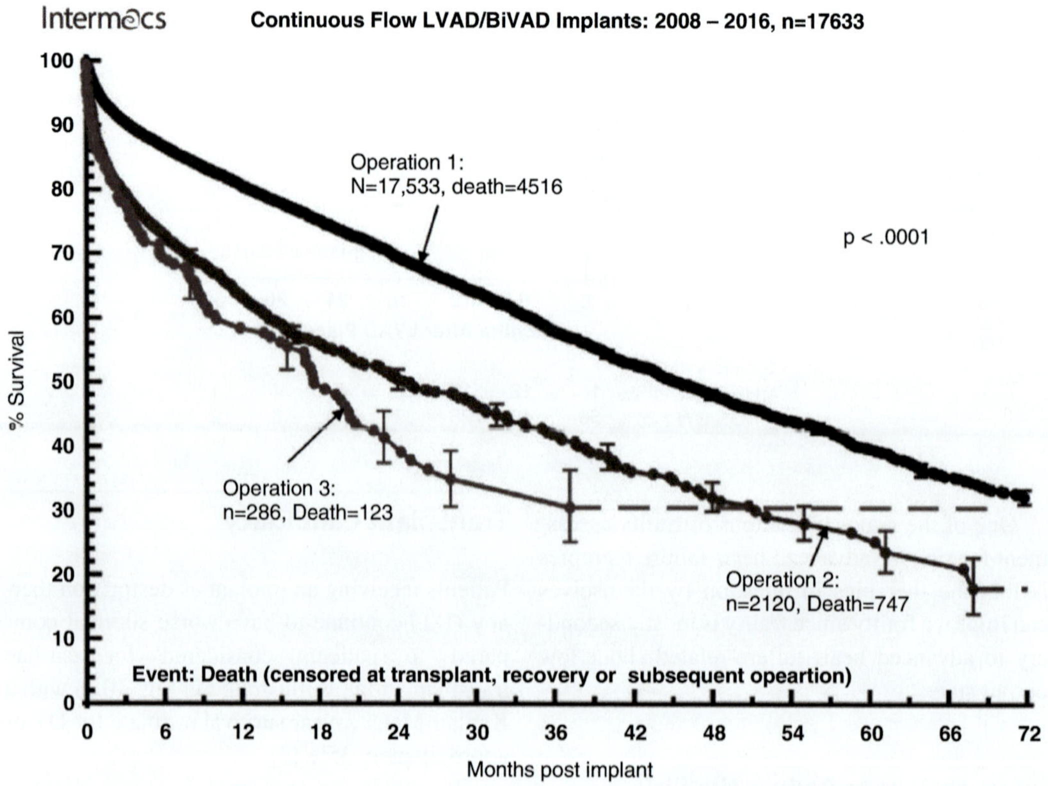

Fig. 40.2 Survival as destination therapy (DT) compared to bridge to transplant (BTT) LVAD implantation. (Reproduced from Kirklin et al. [5], with permission from Elsevier)

preoperative risk assessment must be thorough with a focus on modifying the risks if possible before implantation. The APACHE II score and SHFM are closely related to LVAD outcomes, and the DT Risk Score has pointed out malnutrition as an additional operative risk factor [1, 2, 4].

A high INTERMACS profile and RV failure increase the LVAD operative risk tremendously. Careful understanding of the abovementioned pre-LVAD medical and psychosocial factors helps the multidisciplinary team to accurately assess the VAD operative risk and thus decrease the mortality compared to last decade. The surgeon experience and volume of center will also be contributing to the long-term outcome but have not been established in studies since experienced high-volume centers tend to carry higher INTERMACS and other high-risk patients for implantation thus making it hard to analyze.

A history of prior cardiac surgery affects the short-term outcomes post-LVAD implantation. Early hazard after prior cardiac surgery and CABG is increased and amounts to a hazard ratio of 1.31 (p 0.004) and 1.38 (p 0.001), respectively [5].

Adding concomitant procedures at the time of VAD implantation increases operative time and may adversely affect both short-term and long-term outcomes [85]. In a review of over 2000 patients, repair or replacement of tricuspid valve (for moderate-severe regurgitation) at the time of LVAD did not reduce mortality but was associated with worse early postoperative outcomes [86]. Concomitant cardiac surgery, including valve surgery, results in increased early hazard with HR of 1.53 ($p < 0.0001$) [5]. The need for RVAD or TAH increases the morbidity and mortality when compared to isolated CF-LVAD alone.

On univariate analysis, ventilatory need >7 days, re-intubation, and packed red blood and platelet transfusions were found to be related to increased 90-day all-cause mortality after LVAD, but subsequent multivariate analysis revealed elevated CVP (odds ratio [OR] 1.18; 95% confidence interval 1.014–1.378; *p* 0.033) and higher age (OR 1.14; 95% CI 1.01–1.38; *p* 0.045) as the only independent predictors in BTT patients [87].

Total Artificial Heart

Total artificial heart recipients represent very high-risk advanced stage D heart failure patients with biventricular failure (INTERMACS profile 1 and 2) and multi-organ failure. Hence it is important to understand the risk factors more specifically involved with TAH patients when compared to LVAD alone as implantation strategy. TAHs and BiVADs have different survival profiles and different adverse events, despite having RV failure in common in the decision-making. Studies have shown a trend toward a higher patient survival with TAH when compared to implantable or paracorporeal BiVADs for longer than 90 days [88].

Based on data of TAH implantation between June 23, 2006, and April 30, 2017, risk factors for adverse short-term and long-term outcomes (mortality) were evaluated. Before analyzing the risk factors, it is important to understand the causes of mortality in TAH patients. Multisystem organ failure is the most common cause of death (36.4%), followed by neurologic dysfunction (17.9%) and withdrawal of support (11.7%). Short-term mortality was affected by preimplant renal failure needing dialysis (>60% 6-month mortality: HR 2.5, *p* 0.006) and older age (>60 years versus <40 years: HR 1.6, *p* 0.001) especially in high-volume centers [89]. Long-term mortality was directly related to low albumin (HR 1.9, $p < 0.001$) and high creatinine (HR 1.3, *p* 0.008) at the time of surgery. Low volume (≤10 TAH implants) was also associated with higher mortality with 12-month survival at 36.7% versus 64.8% at centers with an implant volume >10 (HR 3, $p < 0.001$).

Adverse events occurred mostly during the first 3 months of implantation with bleeding (41.3 [rate per 100 patients]), infection (38.8), respiratory failure (21.8), and neurologic dysfunction (14.7) being the major events. Late events (>3 months post-implant) included minor device (controller mostly) malfunction (14.9), infection (12.3), and bleeding (7.1). Hence most of the risk factors for LVAD apply to TAH.

Conclusion

Careful multidisciplinary team evaluation of candidacy for advanced HF therapies like durable MCS (LVAD and TAH) needs to be conducted in all possible circumstances. The analysis of risk scores, echocardiographic, hemodynamic, and laboratory data should be performed in all elective or semi-elective durable MCS implantation. Use of short-term MCS devices is helpful in buying time to complete the assessment before durable MCS is implanted.

Based on available data, the multidisciplinary team needs to thoroughly evaluate the following before every elective LVAD implantation:

1. Severity of heart failure (INTERMACS profile and HF risk scores)
2. Assessment of RV systolic function
3. Frailty testing
4. Assessment of renal function and need for dialysis
5. Noncardiac pulmonary hypertension related to COPD and other restrictive lung pathologies needing prolonged mechanical ventilation/tracheostomy
6. Correction of anemia
7. End-organ function (e.g., liver dysfunction) with potential for reversibility with restoration of cardiac function
8. Nutritional status
9. Psychosocial support
10. Depression
11. Rehabilitation candidacy
12. If age appropriate, continued evaluation for heart (or dual organ) transplant candidacy

Utilization of social worker, nutritionist, and subspecialty physicians (consultants) are helpful for reducing the short-term and long-term morbidity and mortality in durable MCS devices. The availability of a heart transplant program is also helpful in decreasing the morbidity and mortality by the ability to offer heart transplantation as a favorable outcome, especially in TAH patients. Use of palliative care team consultation is very important in making sure that the patient and family knows the nonmedical and nonsurgical options they have in addition to that offered by the medical and surgical teams. The multimodality team approach reduces the anxiety and improves the preparedness for implantation of durable MCS devices and reduces the risk for mechanical circulatory support use.

References

1. Knaus WA, Draper EA, Wagner DP, Zimmerman JE. APACHE II: a severity of disease classification system. Crit Care Med. 1985;13(10):818–29.
2. Levy WC, Mozaffarian D, Linker DT, Sutradhar SC, Anker SD, Cropp AB, et al. The Seattle heart failure model: prediction of survival in heart failure. Circulation. 2006;21(113):1424–33.
3. Rao V, Oz MC, Flannery MA, Catanese KA, Argenziano M, Naka Y. Revised screening scale to predict survival after insertion of a left ventricular assist device. J Thorac Cardiovasc Surg. 2003;125:855–62.
4. Lietz K, Long JW, Kfoury AG, Slaughter MS, Silver MA, Milano CA, et al. Outcomes of left ventricular assist device implantation as destination therapy in the post-REMATCH era: implications for patient selection. Circulation. 2007;116:497–505.
5. Kirklin JK, Pagani FD, Kormos RL, Stevenson LW, Blume ED, Myers SL, et al. Eighth annual INTERMACS report: special focus on framing the impact of adverse events. J Heart Lung Transplant. 2017;36(10):1080–6.
6. Stevenson LW, Pagani FD, Young JB, Jessup M, Miller L, Kormos RL, et al. INTERMACS profiles of advanced heart failure: the current picture. J Heart Lung Transplant. 2009;28(6):535–41.
7. Schaffer JM, Allen JG, Weiss ES, Patel ND, Russell SD, Shah AS, et al. Evaluation of risk indices in continuous-flow left ventricular assist device patients. Ann Thorac Surg. 2009;88:1889–96.
8. Meineri M, Van Rensburg AE, Vegas A. Right ventricular failure after LVAD implantation: prevention and treatment. Best Pract Res Clin Anaesthesiol. 2012;26:217–29.
9. Kukucka M, Potapov E, Stepanenko A, Weller K, Mladenow A, Kuppe H, et al. Acute impact of left ventricular unloading by left ventricular assist device on the right ventricle geometry and function: effect of nitric oxide inhalation. J Thorac Cardiovasc Surg. 2011;141(4):1009–14.
10. Interagency Registry for Mechanically Assisted Circulatory Support (INTERMACS). Appendix A: Adverse event definitions: adult and pediatric patients (2013). Available at http://www.uab.edu/medicine/intermacs/intermacs-documents. Accessed August 14, 2018.
11. Santamore WP, Dell'Italia LJ. Ventricular interdependence: significant left ventricular contributions to right ventricular systolic function. Prog Cardiovasc Dis. 1998;40:289–308.
12. Feneley MP, Gavaghan TP, Baron DW, Branson JA, Roy PR, Morgan JJ. Contribution of left ventricular contraction to the generation of right ventricular systolic pressure in the human heart. Circulation. 1985;71:473–80.
13. Klima UP, Lee MY, Guerrero JL, Laraia PJ, Levine RA, Vlahakes GJ. Determinants of maximal right ventricular function: role of septal shifts. J Thorac Cardiovasc Surg. 2002;123(1):72–80.
14. Brisco M, Sundareswaran KS, Milano CA, Feldman D, Testani JM, Ewald GA, et al; HeartMate II Clinical Investigators. The incidence, risk, and consequences of atrial arrhythmias in patients with continuous-flow left ventricular assist devices. J Card Surg 2014;29:572–580.
15. Puwanant S, Hamilton KK, Klodell CT, Hill JA, Schofield RS, Cleeton TS, et al. Tricuspid annular motion as a predictor of severe right ventricular failure after left ventricular assist device implantation. J Heart Lung Transplant. 2008;27:1102–7.
16. Kukucka M, Stepanenko A, Potapov E, Krabatsch T, Redlin M, Mladenow A, et al. Right-to-left ventricular end-diastolic diameter ratio and prediction of right ventricular failure with continuous-flow left ventricular assist devices. J Heart Lung Transplant. 2011;30:64–9.
17. Vivo RP, Cordero-Reyes AM, Qamar U, Garikipati S, Trevino AR, Aldeiri M, et al. Increased right-to-left ventricle diameter ratio is a strong predictor of right ventricular failure after left ventricular assist device. J Heart Lung Transplant. 2013;32:792–9.
18. Matthews JC, Koelling TM, Pagani FD, Aaronson KD. The right ventricular failure risk score a pre-operative tool for assessing the risk of right ventricular failure in left ventricular assist device candidates. J Am Coll Cardiol. 2008;51(22):2163–72.
19. Kato TS, Farr M, Schulze PC. Usefulness of two-dimensional echocardiographic parameters of the left side of the heart to predict right ventricular failure after left ventricular assist device implantation. Am J Cardiol. 2012;109:246–51.
20. Grant AD, Smedira NG, Starling RC, Marwick TH. Independent and incremental role of quantitative right ventricular evaluation for the prediction of right ventricular failure after left ventricular assist device implantation. J Am Coll Cardiol. 2012;60:521–8.
21. Cameli M, Lisi M, Righini FM, Focardi M, Lunghetti S, Bernazzali S, et al. Speckle tracking echocardiography as a new technique to evaluate right ventricular function in patients with left ventricular assist device therapy. J Heart Lung Transplant. 2013;32:424–30.
22. Dang NC, Topkara VK, Mercando M, Kay J, Kruger KH, Aboodi MS, et al. Right heart failure after left

ventricular assist device implantation in patients with chronic congestive heart failure. J Heart Lung Transplant. 2006;25:1–6.
23. Kormos RL, Teuteberg JJ, Pagani FD, Russell SD, John R, Miller LW, HeartMate II Clinical Investigators, et al. Right ventricular failure in patients with the HeartMate II continuous-flow left ventricular assist device: incidence, risk factors, and effect on outcomes. J Thorac Cardiovasc Surg. 2010;139:1316–24.
24. Kavarana MN, Pessin-Minsley MS, Urtecho J, Catanese KA, Flannery M, Oz MC, et al. Right ventricular dysfunction and organ failure in left ventricular assist device recipients: a continuing problem. Ann Thorac Surg. 2002;73(3):745–50.
25. Fukamachi K, McCarthy PM, Smedira NG, Vargo RL, Starling RC, Young JB. Preoperative risk factors for right ventricular failure after implantable left ventricular assist device insertion. Ann Thorac Surg. 1999;68(6):2181–4.
26. Kang G, Ha R, Banerjee D. Pulmonary artery pulsatility index predicts right ventricular failure after left ventricular assist device implantation. J Heart Lung Transplant. 2016;35:67–73.
27. Fitzpatrick JR 3rd, Frederick JR, Hsu VM, Kozin ED, O'Hara ML, Howell E, et al. Risk score derived from pre-operative data analysis predicts the need for biventricular mechanical circulatory support. J Heart Lung Transplant. 2008;27(12):1286–92.
28. Drakos SG, Janicki L, Horne BD, Kfoury AG, Reid BB, Clayson S, et al. Risk factors predictive of right ventricular failure after left ventricular assist device implantation. Am J Cardiol. 2010;105(7):1030–5.
29. Slaughter MS, Rogers JG, Milano CA, Russell SD, Conte JV, Feldman D, HeartMate II Investigators, et al. Advanced heart failure treated with continuous-flow left ventricular assist device. N Engl J Med. 2009;361:2241–51.
30. Hickey KT, Garan H, Mancini DM, Colombo PC, Naka Y, Sciacca RR, et al. Atrial fibrillation in patients with left ventricular assist devices: incidence, predictors, and clinical outcomes. JACC Clin Electrophysiol. 2016;2(7):793–8.
31. Enriquez AD, Calenda B, Gandhi PU, Nair AP, Anyanwu AC, Pinney SP. Clinical impact of atrial fibrillation in patients with the HeartMate II left ventricular assist device. Am Coll Cardiol. 2014;64:1883–90.
32. Stulak JM, Deo S, Schirger J, Aaronson KD, Park SJ, Joyce LD, et al. Preoperative atrial fibrillation increases risk of thromboembolic events after left ventricular assist device implantation. Ann Thorac Surg. 2013;96:2161–7.
33. Deshmukh A, Kim G, Burke M, Anyanwu E, Jeevanandam V, Uriel N, et al. Atrial arrhythmias and electroanatomical remodeling in patients with left ventricular assist devices. J Am Heart Assoc. 2017;6(3):e005340.
34. Andersen M, Videbaek R, Boesgaard S, Sander K, Hansen PB, Gustafsson F. Incidence of ventricular arrhythmias in patients on long-term support with a continuous-flow assist device (HeartMate II). J Heart Lung Transplant. 2009;28:733–5.
35. Makki N, Mesubi O, Steyers C, Olshansky B, Abraham WT. Meta-analysis of the relation of ventricular arrhythmias to all-cause mortality after implantation of a left ventricular assist device. Am J Cardiol. 2015;116(9):1385–90.
36. Cantillon DJ, Bianco C, Wazni OM, Kanj M, Smedira NG, Wilkoff BL, et al. Electrophysiologic characteristics and catheter ablation of ventricular tachyarrhythmias among patients with heart failure on ventricular assist device support. Heart Rhythm. 2012;9(6):859–64.
37. Nakahara S, Chien C, Gelow J, Dalouk K, Henrikson CA, Mudd J, et al. Ventricular arrhythmias after left ventricular assist device. Circ Arrhythm Electrophysiol. 2013;6(3):648–54.
38. Griffin JM, Katz JN. The burden of ventricular arrhythmias following left ventricular assist device implantation. Arrhythm Electrophysiol Rev. 2014;3(3):145–8.
39. Greet BD, Pujara D, Burkland D, Pollet M, Sudhakar D, Rojas F, et al. Incidence, predictors, and significance of ventricular arrhythmias in patients with continuous-flow left ventricular assist devices: a 15-year institutional experience. JACC Clin Electrophysiol. 2018;4(2):257–64.
40. Tsiouris A, Paone G, Nemeh HW, Borgi J, Williams CT, Lanfear DE, et al. Short and long term outcomes of 200 patients supported by continuous-flow left ventricular assist devices. World J Cardiol. 2015;7(11):792–800.
41. Kato TS, Schulze PC, Yang J, Chan E, Shahzad K, Takayama H, et al. Pre-operative and post-operative risk factors associated with neurologic complications in patients with advanced heart failure supported by a left ventricular assist device. J Heart Lung Transplant. 2012;31:1–8.
42. Boyle AJ, Jorde UP, Sun B, Park SJ, Milano CA, Frazier OH, HeartMate II Clinical Investigators, et al. Pre-operative risk factors of bleeding and stroke during left ventricular assist device support: an analysis of more than 900 HeartMate II outpatients. J Am Coll Cardiol. 2014;63(9):880–8.
43. Schroder JN, Daneshmand MA, Villamizar NR, Petersen RP, Blue LJ, Welsby IJ, et al. Heparin-induced thrombocytopenia in left ventricular assist device bridge-to-transplant patients. Ann Thorac Surg. 2007;84(3):841–5.
44. Yost G, Tatooles A, Bhat G. Preoperative nutritional assessment with the prognostic nutrition index in patients undergoing left ventricular assist device implantation. ASAIO J. 2018;64(1):52–5.
45. Butler J, Howser R, Portner PM, Pierson RN 3rd. Diabetes and outcomes after left ventricular assist device placement. J Card Fail. 2005;11(7):510–5.
46. Usoh CO, Sherazi S, Szepietowska B, Kutyifa V, McNitt S, Papernov A, et al. Influence of diabetes mellitus on outcomes in patients after left ventricular assist device implantation. Ann Thorac Surg. 2018;106(2):555–60.

47. Asleh R, Briasoulis A, Schettle SD, Tchantchaleishvili V, Pereira NL, Edwards BS, et al. Impact of diabetes mellitus on outcomes in patients supported with left ventricular assist devices: a single institutional 9-year experience. Circ Heart Fail. 2017;10(11):e004213.
48. Vest AR, Mistak SM, Hachamovitch R, Mountis MM, Moazami N, Young JB. Outcomes for patients with diabetes after continuous-flow left ventricular assist device implantation. J Card Fail. 2016;22(10):789–96.
49. Mohamedali B, Yost G, Bhat G. Is diabetes mellitus a risk factor for poor outcomes after left ventricular assist device placement? Tex Heart Inst J. 2017;44(2):115–9.
50. Clerkin KJ, Naka Y, Mancini DM, Colombo PC, Topkara VK. The impact of obesity on patients bridged to transplantation with continuous-flow left ventricular assist devices. JACC Heart Fail. 2016;4(10):761–8.
51. Campello E, Zabeo E, Radu CM, Spiezia L, Gavasso S, Fadin M, et al. Hypercoagulability in overweight and obese subjects who are asymptomatic for thrombotic events. Thromb Haemost. 2015;113(1):85–96.
52. Yost G, Coyle L, Gallagher C, Graney N, Siemeck R, Tatooles A, Pappas P, Bhat G. The impact of extreme obesity on outcomes after left ventricular assist device implantation. J Thorac Dis. 2017;9(11):4441–6.
53. Foraker RE, Rose KM, Suchindran CM, Chang PP, McNeill AM, Rosamond WD. Socioeconomic status, Medicaid coverage, clinical comorbidity, and rehospitalization or death after an incident heart failure hospitalization: atherosclerosis risk in communities cohort (1987 to 2004). Circ Heart Fail. 2011;4:308–16.
54. Philbin EF, Dec GW, Jenkins PL, DiSalvo TG. Socioeconomic status as an independent risk factor for hospital readmission for heart failure. Am J Cardiol. 2001;87:1367–71.
55. Clerkin KJ, Garan AR, Wayda B, Givens RC, Yuzefpolskaya M, Nakagawa S, et al. Impact of socioeconomic status on patients supported with a left ventricular assist device: an analysis of the UNOS Database (United Network for Organ Sharing). Circ Heart Fail. 2016;9(10):e003215.
56. Bruce CR, Minard CG, Wilhelms LA, Abraham M, Amione-Guerra J, Pham L, et al. Caregivers of patients with left ventricular assist devices: possible impacts on patients' mortality and interagency registry for mechanically assisted circulatory support-defined morbidity events. Circ Cardiovasc Qual Outcomes. 2017;10(1):e002879.
57. Rutledge T, Reis VA, Linke SE, Greenberg BH, Mills PJ. Depression in heart failure a meta-analytic review of prevalence, intervention effects, and associations with clinical outcomes. J Am Coll Cardiol. 2006;48(8):1527–37.
58. Moraska AR, Chamberlain AM, Shah ND, Vickers KS, Rummans TA, Dunlay SM, et al. Depression, healthcare utilization, and death in heart failure: a community study. Circ Heart Fail. 2013;6(3):387–94.
59. DiMatteo MR, Lepper HS, Croghan TW. Depression is a risk factor for noncompliance with medical treatment: meta-analysis of the effects of anxiety and depression on patient adherence. Arch Intern Med. 2000;160(14):2101–7.
60. Snipelisky D, Stulak JM, Schettle SD, Sharma S, Kushwaha SS, Dunlay SM. Psychosocial characteristics and outcomes in patients with left ventricular assist device implanted as destination therapy. Am Heart J. 2015;170(5):887–94.
61. Cogswell R, Smith E, Hamel A, Bauman L, Herr A, Duval S, et al. Substance abuse at the time of left ventricular assist device implantation is associated with increased mortality. J Heart Lung Transplant. 2014;33(10):1048–55.
62. McDonagh J, Martin L, Ferguson C, Jha SR, Macdonald PS, Davidson PM, et al. Frailty assessment instruments in heart failure: a systematic review. Eur J Cardiovasc Nurs. 2018;17(1):23–35.
63. Cacciatore F, Abete P, Mazzella F, Viati L, Della Morte D, D'Ambrosio D, et al. Frailty predicts long-term mortality in elderly subjects with chronic heart failure. Eur J Clin Investig. 2005;35(12):723–30.
64. Lupón J, González B, Santaeugenia S, Altimir S, Urrutia A, Más D, et al. Prognostic implication of frailty and depressive symptoms in an outpatient population with heart failure. Rev Esp Cardiol. 2008;61(8):835–42.
65. Madan SA, Fida N, Barman P, Sims D, Shin J, Verghese J, et al. Frailty assessment in advanced heart failure. J Card Fail. 2016;22(10):840–4.
66. Robinson TN, Wu DS, Pointer L, Dunn CL, Cleveland JC Jr, Moss M. Simple frailty score predicts postoperative complications across surgical specialties. Am J Surg. 2013;206(4):544–50.
67. Tse G, Gong M, Wong SH, Wu WKK, Bazoukis G, Lampropoulos K, et al. International Health Informatics Study (IHIS) Network. Frailty and clinical outcomes in advanced heart failure patients undergoing left ventricular assist device implantation: a systematic review and meta-analysis. J Am Med Dir Assoc. 2018;19(3):255–61.
68. Fried LP, Tangen CM, Walston J, Newman AB, Hirsch C, Gottdiener J, Cardiovascular Health Study Collaborative Research Group, et al. Frailty in older adults: evidence for a phenotype. J Gerontol A Biol Sci Med Sci. 2001;56(3):M146–56.
69. Fried LP, Borhani NO, Enright P, Furberg CD, Gardin JM, Kronmal RA, et al. The cardiovascular health study: design and rationale. Ann Epidemiol. 1991;1(3):263–76.
70. Singh M, Stewart R, White H. Importance of frailty in patients with cardiovascular disease. Eur Heart J. 2014;35(26):1726–31.
71. Cameron ID, Fairhall N, Langron C, Lockwood K, Monaghan N, Aggar C, et al. A multifactorial interdisciplinary intervention reduces frailty in older people: randomized trial. BMC Med. 2013;11:65.
72. Chung CJ, Wu C, Jones M, Kato TS, Dam TT, Givens RC, et al. Reduced handgrip strength as a marker of frailty predicts clinical outcomes in patients with

heart failure undergoing ventricular assist device placement. J Card Fail. 2014;20(5):310–5.
73. Ferguson C, Inglis SC, Newton PJ, Middleton S, Macdonald PS, Davidson PM. Multi-morbidity, frailty and self-care: important considerations in treatment with anticoagulation drugs. Outcomes of the AFASTER study. Eur J Cardiovasc Nurs. 2017;16(2):113–24.
74. Newton PJ, Davidson PM, Reid CM, Krum H, Hayward C, Sibbritt DW, et al. Acute heart failure admissions in New South Wales and the Australian Capital Territory: the NSW HF Snapshot Study. Med J Aust. 2016;204(3):113.e1–8.
75. Williamson J, Stokoe IH, Gray S, Fisher M, Smith A, Mcghee A, et al. Old people at home their unreported needs. Lancet. 1964;1(7343):1117–20.
76. Pasture O, Onkia O. Canadian study of health and aging: study methods and prevalence of dementia. CMAJ. 1994;150(6):899–913.
77. Gobbens RJ, van Assen MA, Luijkx KG, Schols JM. The predictive validity of the Tilburg Frailty Indicator: disability, health care utilization, and quality of life in a population at risk. Gerontologist. 2012;52(5):619–31.
78. Rockwood K, Song X, MacKnight C, Bergman H, Hogan DB, McDowell I, et al. A global clinical measure of fitness and frailty in elderly people. CMAJ. 2005;173(5):489–95.
79. Suaya JA, Stason WB, Ades PA, Normand SL, Shepard DS. Cardiac rehabilitation and survival in older coronary patients. J Am Coll Cardiol. 2009;54:25–33.
80. Pack QR, Goel K, Lahr BD, Greason KL, Squires RW, Lopez-Jimenez F, Zhang Z, Thomas RJ. Participation in cardiac rehabilitation and survival after coronary artery bypass graft surgery: a community-based study. Circulation. 2013;128:590–7.
81. Paddon-Jones D, Rasmussen BB. Dietary protein recommendations and the prevention of sarcopenia. Curr Opin Clin Nutr Metab Care. 2009;12:86–90.
82. Malafarina V, Uriz-Otano F, Iniesta R, Gil-Guerrero L. Effectiveness of nutritional supplementation on muscle mass in treatment of sarcopenia in old age: a systematic review. J Am Med Direct Assoc. 2013;14:10–7.
83. Murad MH, Elamin KB, Abu Elnour NO, Elamin MB, Alkatib AA, Fatourechi MM, Almandoz JP, Mullan RJ, Lane MA, Liu H, Erwin PJ, Hensrud DD, Montori VM. Clinical review: the effect of vitamin D on falls: a systematic review and meta-analysis. J Clin Endocrinol Metab. 2011;96:2997–3006.
84. Bolland MJ, Grey A, Avenell A, Gamble GD, Reid IR. Calcium supplements with or without vitamin D and risk of cardiovascular events: reanalysis of the Women's health initiative limited access dataset and meta-analysis. BMJ. 2011;342:d2040.
85. Morgan JA, Tsiouris A, Nemeh HW, Hodari A, Karam J, Brewer RJ, et al. Impact of concomitant cardiac procedures performed during implantation of long-term left ventricular assist devices. J Heart Lung Transplant. 2013;32(12):1255–61.
86. Robertson JO, Grau-Sepulveda MV, Okada S, O'Brien SM, Matthew Brennan J, Shah AS, et al. Concomitant tricuspid valve surgery during implantation of continuous-flow left ventricular assist devices: a Society of Thoracic Surgeons database analysis. J Heart Lung Transplant. 2014;33:609–17.
87. Sabashnikov A, Mohite PN, Zych B, García D, Popov AF, Weymann A, et al. Outcomes and predictors of early mortality after continuous-flow left ventricular assist device implantation as a bridge to transplantation. ASAIO J. 2014;60(2):162–9.
88. Kirsch M, Mazzucotelli JP, Roussel JC, Bouchot O, N'loga J, Leprince P, et al. Survival after biventricular circulatory support: does the type of device matter? J Heart Lung Transplant. 2012;31:501–8.
89. Arabía FA, Cantor RS, Koehl DA, Kasirajan V, Gregoric I, Moriguchi JD, Esmailian F, et al. Interagency registry for mechanically assisted circulatory support report on the total artificial heart. J Heart Lung Transplant. 2018;37:1304–12.

41 Gastrointestinal Bleeding in Mechanical Circulatory Support Patients

Katherine M. Klein and Vigneshwar Kasirajan

Introduction

According to the Interagency Registry for Mechanically Assisted Circulatory Support (INTERMACS), a public-private partnership between the National Heart, Lung, and Blood Institute (NHLBI), the US Food and Drug Administration (FDA), Centers for Medicaid and Medicare Services, hospitals, and industry, 22,866 patients have received an FDA-approved mechanical support device between June 23, 2006, and December 31, 2016. With this increased need for advanced mechanical support, the complication of bleeding has become a significant clinical issue in post-implant patients, particularly gastrointestinal bleeding.

Incidence/Prevalence

The incidence of gastrointestinal bleeding in patients with continuous flow left ventricular assist devices (CF-LVADS) is between 18.9% and 22.3% [1]. Despite finding the source of the bleeding in approximately 75% of the patients, the risk for rebleeding is still 35%, making this complication significant for many LVAD patients [1]. The most common location for bleeding is the upper gastrointestinal (GI) tract and most commonly from an arteriovenous malformation (AVM) [1]. Further sites and sources of gastrointestinal bleeding are summarized in Table 41.1. In the ROADMAP study, investigators reported nearly two-thirds of all of their adverse events of patients with VAD support was bleeding, with the majority of bleeding GIB [2].

Clinical Relevance

Beyond the initial hemodynamic concerns with gastrointestinal bleeding, the need for blood transfusions also makes this issue very difficult to manage. As many VAD patients are considered bridge to transplantation status, exposing these patients to further blood transfusions and subsequent higher panel reactive antibodies (PRA) makes transplantation even more difficult. Additionally, the multiple readmissions for bleeding can produce significant burdens on the patient and their families with time and cost of hospitalization. Resource utilization of hospitals can also be affected due to increased intensive care unit (ICU) stays, increased length of stay overall in the hospital, increased readmission rates, multiple procedures with sedation

K. M. Klein, MD · V. Kasirajan, MD (✉)
Pauley Heart Center, Division of Cardiothoracic Surgery, Virginia Commonwealth University, Richmond, VA, USA
e-mail: vigneshwar.kasirajan@vcuhealth.org

© Springer Nature Switzerland AG 2020
J. H. Karimov et al. (eds.), *Mechanical Support for Heart Failure*,
https://doi.org/10.1007/978-3-030-47809-4_41

Table 41.1 Summary of all studies that identify the source of bleeding

	Total incidence	Gastritis	Gastric ulcer	AVM	Divertic-ulitis	Colitis	Colonic polyp	Colonic ulcer	Other	Unknown
Hayes et al.	13.9%			60.0%				20.0%		20.0%
Demirozu et al.	19.0%	31.3%		31.3%	18.8%	3.2%	3.2%		12.5%	
Aggarwal et al.	22.8%	30.4%	8.7%	21.7%			4.3%	13.0%	13.0%	8.7%
Kushnir et al.	34.8%		28.2%	30.8%		5.1%	5.1%			30.8%
Wever-Pinzon et al.	17.2%	8.7%		61.0%	8.7%	4.3%	8.7%		8.7%	

AVM Arteriovenous malformation

(endoscopies, etc.), or even exploratory surgery for bleeding sources. Consequently these patients often also have to have gaps in therapeutic anticoagulation with either stopping their anticoagulation or even full reversal of anticoagulation, leading them to increased risk for thrombosis and subsequent need for pump exchanges. With increased frequency in the hospital with multiple procedures, these patients are also at increased risk for infection leading to even further hospitalizations and procedures.

Pathophysiology

With the amount of concern for bleeding in mechanical device patients, the pathophysiology behind the bleeding has been studied. While all patients with VADS are on anticoagulation, not all bleeding is related solely to their anticoagulation. In reality, most patients who experience GI bleeding have therapeutic or subtherapeutic international normalized ratios (INRs) at the time of presentation [3–5], suggesting a further causative factor related to their bleeding.

Acquired von Willebrand syndrome has been postulated as one potential cause of increased bleeding in the continuous-flow LVAD patients. The lack of pulsatility in the new-generation VADs is thought to produce high shear stress and subsequently induced loss of high-molecular-weight von Willebrand multimers, leading to the further potentiation of bleeding [6–8]. Interestingly, Crow et al. demonstrated patients with continuous-flow LVADs had an absence of high-molecular-weight multimers and once explanted had 100% reversal of the von Willebrand deficiency [9]. Geisen et al. also demonstrated that all patients developed acquired von Willebrand deficiency within 24 hours of VAD implantation and that this deficiency persisted until heart transplantation [10]. Interestingly, the acquired von Willebrand deficiency also disappeared within 24 hours of transplantation [10]. Platelet aggregation abnormalities also are seen in CF-LVAD patients which are also noted to reverse with heart transplantation despite continuation of aspirin therapy in some patients [7]. Further study by Grosman-Rimon et.al showed serum cyclic guanosine monophosphate (cGMP) levels were double in patients with gastrointestinal bleeding when compared to control patients. Additionally, they also showed a continuous rise in serum levels of nitric oxide (NO) and a decrease in platelet-derived growth factor (PDGF) in patients with CF-LVAD support [11]. This is all suggestive of a global change in the coagulation cascade with continuous-flow VAD support and subsequent platelet dysfunction and increased bleeding potential [12]. Geisen et al. in 2018 also studied platelet aggregometry in VAD patients and found severely reduced expression of CD62 (impaired alpha-granule secretion) and impaired CD63 (impaired delta-granule secretion), suggesting further platelet dysfunction with VAD technology [10].

But as previously described, the majority of bleeding in the CF-LVADs has been associated with an arteriovenous malformation (61% of bleeding) [4]. In animal models, the continuous-flow physiology has been shown to lead to microvascular hypoperfusion further propagating localized hypoxia, vascular dilatation, and angiodysplasias [13]. In further research, the continuous flow state leads to smooth muscle dilation, arteriovenous dilation, and again arteriovenous malformations that are at increased risk for bleeding [14].

Diagnosis

With the significant clinical relevance to GI bleeding, multiple strategies have been employed to help diagnose and manage these patients. The first step is a detailed history and physical of each presenting patient. This includes the onset and duration of bleeding, where the bleeding originated (i.e., hemoptysis, hematochezia, melena, or occult bleeding), other associated symptoms, medications including over-the-counter supplements and pain relievers, previous history of any bleeding, recent coagulation studies, any previous endoscopies, and hemodynamic assessment including VAD flow parameters and any event alarms. If the patient is presumed to have an upper gastrointestinal bleed, it is the recommendation of the American College of Gastroenterology to undergo an esophagogastroduodenoscopy (EGD) within 24 hours of presentation, with the caveat that the patient has been stabilized with no further concern for hemodynamic compromise [1]. However, patients with an elevated INR can present a difficult clinical scenario with risk for ongoing bleeding and risk for pump thrombosis if anticoagulation is reversed. Data has suggested that an EGD with an INR 1.3–2.7 does not place the patient at undue risk for bleeding complications [1]. Other means for bleeding assessment include a colonoscopy or a tagged red blood cell imaging scan. A colonoscopy is also an invasive procedure and can present with difficulties in the need for sedation with hemodynamic shifts, visualization of the bleeding site, and continued need for anticoagulation with the ever-present risk of pump thrombosis. A tagged red blood scan also requires access to nuclear medicine facilities, identifying the bleeding when it is brisk enough to be detected by scan, and still often does not yield clinically significant results. Additional strategies for detection and therapy can include capsule endoscopy, mesenteric angiogram, double balloon endoscopy, or even exploratory laparotomy/laparoscopy in severe cases [15, 16].

Management

Upon diagnosis of GIB, management strategies vary with severity of presentation and current VAD settings. The initial treatment should focus on stabilization of the patient with fluid management and acid suppression. Despite the concern for pump thrombosis with reversal of anticoagulation in the bleeding VAD patients, stabilization with reversal has shown to have low evidence for need for pump exchange [1]. An EGD should be performed as soon as possible as most bleeding is associated with an arteriovenous malformation (AVM) and can be both diagnosed and treated endoscopically [16]. Elmunzer et al. did make particular mention that many of these AVMs occur beyond the ligament of Treitz, and therefore, endoscopic management may require double balloon endoscopy to identify and treat the bleeding source [16].

With many GIBs, acid suppression is a key component of bleeding control. Many studies have suggested the use of octreotide, a somatostatin analogue, has reduced the incidence of recurrent bleeding episodes in patients with CF-LVADS [17, 18]. Octreotide is a long-acting somatostatin analogue that helps to reduce splanchnic arterial and portal blood flow limiting the pressure on potential sites of GIB [19]. However, further studies have not been able to show the use of octreotide had reduced the need for transfusion or recurrence of bleeding [20]. Additional studies have described a combination of decreased pump flow speed, adrenaline infusions, and octreotide to successfully treat GIBs, but these show limited success overall [20]. Thalidomide has also been utilized in gastrointestinal bleeding due to its known anti-angiogenic properties, but further adverse events of neuropathy and even death related to sepsis have been reported making its safety and efficacy questionable [20]. Direct use of Wilfactin, a von Willebrand factor concentrate devoid of factor VIII, has also been described in a limited number of patients with severe bleeding noted. However, this subset is very small, and further studies would need to be performed for recommended use [20].

It is of utmost concern to manage the anticoagulation strategy in patients who have presented with an initial bleed or a recurrent gastrointestinal bleed. Many centers have different antithrombotic therapy plans for patients including aspirin and a vitamin K antagonist, aspirin with the addition of dipyridamole/clopidogrel and a vitamin K antagonist, or a vitamin K antagonist solo. Furthermore, many institutions have reported varying target INRs with 1.5–3.5 most commonly used, making standardized protocols difficult. In patients that have presented with GIB, the initial holding of anticoagulation is well tolerated to maintain hemodynamic stability followed by slow reintegration of antiplatelet and vitamin K antagonists. Some studies have suggested utilizing thromboelastography (TEG®) to adjust antiplatelet treatment to obtain a desired clot strength and curve amplitude [21]. Additional monitoring with lactate dehydrogenase levels has also helped to identify when LVAD exchanges should be considered due to thrombosis [21]. Goldstein et.al presented a study where two-thirds of the patient population had an interruption in their antiplatelet or warfarin therapy, and of those patients, only five developed a thrombotic complication after a GIB event, suggesting a limited interruption in anticoagulation therapy is not exceedingly dangerous [22].

Aspirin, i.e., acetylsalicylic acid (ASA), also plays a major role in many centers for VAD patients and can complicate anticoagulation strategies in the bleeding patient. ASA has shown to inhibit platelet function by permanently acetylating cyclooxygenase (COX), and platelets are a major component of the clotting cascade. However, Lorenzo Valerio et al. studied platelet activation with aspirin therapy in VAD patients and found ASA actually had limited ability to modulate shear-mediated platelet activation as seen in VAD patients [23]. This would suggest that aspirin therapy may not need to be stopped in patients with GIB and furthermore studies should be conducted to determine the overall contribution ASA has for VAD patients in preventing thrombosis [23].

Role of Flow

With the advent of LVADs, pump technology has also changed from an axial flow model to a centrifugal pump design. Axial flow VADs utilize a rotating element that functions like a propeller in a pipe that "pushes" blood through the device to the outlet cannula. In contrast, centrifugal pumps utilize a rotating element that spins a disk with blades that "throws" the blood off the blades in a tangential fashion toward the outflow cannula [24]. The degree of pulsatility in the pump, and therefore in the patient, is then dependent on the peak flow during systole or the total swing in flow during a cardiac cycle divided by the average pump flow [24]. The difference in design has led many researchers to question if one pump design leads to more bleeding or hemolysis when directly compared. Moazami, N., et al. [24] described pump hemolysis being directly related to the pump speed, areas of high shear stress, and residence time of blood in the high fluid shear areas of the pumping elements or bearings. With this in mind, axial pumps have higher shear levels when compared to centrifugal pumps, but the blood remains in these high shear areas for a shorter period of time mitigating some of the effects. With the current data, axial pumps may produce more subclinical hemolysis when compared to centrifugal pumps, but this does not translate to any appreciable patient outcome. Furthermore, studies that have monitored free hemoglobin, bilirubin, lactate dehydrogenase, glutamic oxaloacetic transaminase, glutamic pyruvic transaminase, and/or haptoglobin did not show any statistically, or clinically, significant elevation in one particular pump design [24]. Direct comparison of HeartMate II to HeartMate III in a trial designed by Geisen et al. demonstrated less von Willebrand deficiency effect with the HeartMate III and subsequent less bleeding events [10]. Additionally, the HeartMate III patients had less GIB when compared to a similar cohort who received the HeartMate II [10].

The question of pump speed has also been proposed as a method for increasing native pulsatility with speed reduction and potential reduction in von Willebrand factor deficiency. However, Akhter in 2016 published a study that used a mock flow loop of a HeartMate II with varying speeds that did not show a significant impact on vWF degradation [25]. Kang et al. in the same year also tested reduced continuous-flow LVAD speed without significant vWF degradation fragments, further suggesting VAD speed changes should not be utilized to limit or reduce GIB [26].

In addition to pump design, the argument of pulsatility in the VAD population has been of significant interest and concern in the development of gastrointestinal bleeding, stroke, pump thrombosis, and hemolysis. Compared with constant speed operation, pumps that have integrated counterpulsation have shown increases in left ventricle unloading and myocardial supply/demand ratio but decreased pulse pressure. Copulsation devices increase pulse pressure but decrease left ventricle unloading and the myocardial supply/demand ratio by decreasing the diastolic pressure with the increase in systolic pressure [27]. Taking into consideration that with the copulsation mode and counterpulsation mode, the aortic valve does not routinely open, Kishimoto, Y., et.al in 2013 described a novel pump speed algorithm described as delayed copulse mode [28]. They utilized a centrifugal LVAD (EVAHEART®; Sun Medical) that allowed the aortic valve to open with continued high total flow with minimal support for early systole and maximal support shortly after the aortic valve opened. Building on the theory of aortic valve opening to decrease risk for thrombosis in the aortic root and to provide pulsatility, the HeartWare HVAD (HeartWare International, Inc. Framingham, MA) created a Lavare cycle that allows intermittent opening of the aortic valve. The Lavare cycle consists of ±200 rpm variations in 3-second cycles once per minute [27, 29]. However, even with this pulsatility variation, patients continue to develop vascular bed malformations due to high shear stress and resistive indices showing that continuous flow mechanical support is a delicate balance of ventricular unloading, myocardial supply/demand ratio, increased pulse pressure, and intermittent decompression of a compliant vascular system.

Conclusion

Gastrointestinal bleeding still remains one of the most frequent adverse events after VAD implantation. While GIB does not appear to have a major effect on subsequent survival, it does continue to cause significant clinical concern and difficult management decisions [20, 30]. As more patients are diagnosed with heart failure each year and there is a stagnant number of hearts available for transplantation, the increased use of VAD support will likely be utilized. Understanding various management strategies for diagnosis and subsequent treatment of a very common adverse event, gastrointestinal bleeding, will be necessary. With continuous development of VAD support mechanisms and novel anticoagulants, further research will also be necessary to manage these patients successfully.

References

1. Guha A, Eshelbrenner CL, Richards DM, Monsour HP Jr. Gastrointestinal bleeding after continuous-flow left ventricular device implantation: review of pathophysiology and management. Methodist Debakey Cardiovasc J. 2015;11(1):24–7. https://doi.org/10.14797/mdcj-11-1-24.
2. Estep JD, Starling RC, Horstmanshof DA, et al. Risk assessment and comparative effectiveness of left ventricular assist device and medical management in ambulatory heart failure patients: results from the ROADMAP study. JACC. 2015;66(16):1747–61.
3. Kushnir VM, Sharma S, Ewald GA, et al. Evaluation of GI bleeding after implantation of left ventricular assist device. Gastrointest Endosc. 2012;75(5):973–9.
4. Wever-Pinzon O, Selzman CH, Drakos SG, et al. Pulsatility and the risk of nonsurgical bleeding in patients supported with the continuous-flow left ventricular assist device HeartMate II. Circ Heart Fail. 2013;6(3):517–26.

5. Uriel N, Pak SW, Jorde UP, et al. Acquired von Willebrand syndrome after continuous-flow mechanical device support contributes to a high prevalence of bleeding during long-term support and at the time of transplantation. J Am Coll Cardiol. 2010;56(15):1207–13.
6. Vincentelli A, Susen S, Le Tourneau T, et al. Acquired von Willebrand syndrome in aortic stenosis. N Engl J Med. 2003;349(4):343–9.
7. Klovaite J, Gustafsson F, Mortensen SA, Sander K, Nielsen LB. Severely impaired von Willebrand factor-dependent platelet aggregation in patients with a continuous-flow left ventricular assist device (HeartMate II). J Am Coll Cardiol. 2009;53(23):2162–7.
8. Meyer AL, Malehsa D, Budde U, Bara C, Haverich A, Strueber M. Acquired von Willebrand syndrome in patients with a centrifugal or axial continuous flow left ventricular assist device. JACC Heart Fail. 2014;2(2):141–5.
9. Crow S, John R, Boyle A, et al. Gastrointestinal bleeding rates in recipients of nonpulsatile and pulsatile left ventricular assist devices. J Thorac Cardiovasc Surg. 2009;137(1):208–15.
10. Geisen U, Brehm K, Trummer G, et al. Platelet secretion defects and acquired von Willebrand syndrome in patients with ventricular assist devices. J Am Heart Assoc. 2018;7(2):e006519.
11. Grosman-Rimon L, Tumiati LC, Fuks A, Jacobs I, Lalonde SC, Cherney DZ, et al. Increased levels of cGMP in recipients of continuous flow LVADs: implications for gastrointestinal bleeding. J Thorac Cardiovasc Surg. 2016;151:219–27.
12. Crestanello JA. Gastrointestinal bleeding after left ventricular assist device implantation: it is all about the platelets. J Thorac Cardiovasc Surg. 2016;151(1):228–9.
13. Aggarwal A, Pant R, Kumar S, et al. Incidence and management of gastrointestinal bleeding with continuous flow assist devices. Ann Thorac Surg. 2012;93(5):1534–40.
14. Cappell MS, Lebwohl O. Cessation of recurrent bleeding from gastrointestinal angiodysplasias after aortic valve replacement. Ann Intern Med. 1986;105(1):54–7.
15. Islam S, et al. Left ventricular assist devices and gastrointestinal bleeding: a narrative review of case reports and case series. Clin Cardiol. 2013;36(4):190–200.
16. Elmunzer BJ, Padhya KT, Lewis JJ, et al. Endoscopic findings and clinical outcomes in ventricular assist device recipients with gastrointestinal bleeding. Dig Dis Sci. 2011;56(11):3241–6.
17. Shah KB, Gunda S, Emani S, et al. Multicenter evaluation of octreotide as secondary prophylaxis in patients with left ventricular assist devices and gastrointestinal bleeding. Circ Heart Fail. 2017;10(11):e004500.
18. Loyaga-Rendon RY, Hashim T, Tallaj JA, et al. Octreotide in the management of recurrent gastrointestinal bleed in patients supported by continuous flow left ventricular assist devices. ASAIO J. 2015;61(1):107–9.
19. Hayes HM, Dembo LG, Larbalestier R, O'Driscoll G. Management options to treat gastrointestinal bleeding in patients supported on rotary left ventricular assist devices: a single center experience. Artif Organs. 2010;34(9):703–6.
20. Cushing K, Kushnir V. Gastrointestinal bleeding following LVAD placement from top to bottom. Dig Dis Sci. 2016;61(6):1440–7.
21. Baumann Kreuziger LM, Kim B, Wieselthaler GM. Antithrombotic therapy for left ventricular assist devices in adults: a systematic review. J Thromb Haemost. 2015;13(6):946–55.
22. Goldstein DJ, Aaronson KD, et al. Gastrointestinal bleeding in recipients of the HeartWare ventricular assist system. JACC Heart Fail. 2015;3(4):303–13.
23. Valerio L, Tran PL, Sheriff J, et al. Aspirin has limited ability to modulate shear-mediated platelet activation associated with elevated shear stress of ventricular assist devices. Thromb Res. 2016;140:110–7.
24. Moazami N, Fukamachi K, et al. Axial and centrifugal continuous-flow rotary pumps: a translation from pump mechanics to clinical practice. J Heart Lung Transplant. 2013;32(1):1–11.
25. Akhter SA. Continuous-flow left ventricular assist devices and bleeding events: is there a biological role for decreasing pump speed? J Thorac Cardiovasc Surg. 2016;151(6):1755–6.
26. Kang J, Zhang DM, et al. Reduced continuous-flow left ventricular assist device speed does not decrease von Willebrand factor degradation. J Thorac Cardiovasc Surg. 2016;151(6):1747–54.
27. Moazami N, Dembitsky WP, Adamson R, et al. Does pulsatility matter in the era of continuous flow blood pumps? J Heart Lung Transplant. 2015;34(8):999–1004.
28. Kishimoto Y, Takewa Y, Arakawa M, et al. Development of a novel drive mode to prevent aortic insufficiency during continuous-flow LVAD support by synchronizing rotational speed with heartbeat. J Artif Organs. 2013;16(2):129–37.
29. Saeed D, Maxhera B, Sadat N, et al. Gastrointestinal bleeding in patients with HeartWare ventricular assist device: does the activation of the Lavare cycle make a difference? ASAIO J. 2018;64(1):126–8.
30. Kirklin JK, Pagani FD, et al. Eighth annual INTERMACS report: special focus on framing the impact of adverse events. J Heart Lung Transplant. 2017;36(10):1080–6.

42 Postoperative Management Strategies in Mechanical Circulatory Support Patients

Tiffany Buda, Kimberly Miracle, and Marjorie Urban

Abbreviations

AA	Atrial arrhythmia
ACE	Angiotensin-converting enzyme
AE	Adverse event
AF	Atrial fibrillation
AI	Aortic insufficiency
AKI	Acute kidney injury
ARB	Angiotensin II receptor blockers
ARF	Acute rehabilitation facility
AV	Aortic valve
AVF	Arterial venous fistula
BB	Beta-blocker
BMI	Body mass index
BP	Blood pressure
BSI	Bloodstream infection
BTT	Bridge to transplant
cfLVAD	Continuous-flow left ventricular assist device
CI	Cardiac index
CO	Cardiac output
CPR	Cardiopulmonary resuscitation
CT	Computed tomography
CVA	Stroke
CVP	Central venous pressure
DL	Driveline
DLI	Driveline infection
DT	Destination therapy
EKG	Electrocardiogram
EMS	Emergency medical services/system
ESRD	End-stage renal disease
ETCO2	End-tidal carbon dioxide
GIB	Gastrointestinal bleeding
HIT	Heparin-induced thrombocytopenia
HM 3	HeartMate 3
HM2	HeartMate 2
HVAD	HeartWare
I&D	Incision and drainage
ICD	Internal cardiac defibrillator
ICH	Intracerebral hemorrhage
ICU	Intensive care unit
INR	International normalized ration
INTERMACS	Interagency Registry for Mechanically Assisted Circulatory Support
IV	Intravenous
JVD	Jugular vein distension
LAP	Left atrial pressure
LDH	Lactate dehydrogenase
LTAC	Long-term acute care
LV	Left ventricle
LVAD	Left ventricular assist device
LVEDD	Left ventricular end-diastolic volume
MAP	Mean arterial pressure
MCS	Mechanical circulatory support
MR	Mitral regurgitation

T. Buda, MSN, RN, NE (✉) · K. Miracle, MSN, ACNP
M. Urban, MSN, ACNP
Cleveland Clinic, Cleveland, OH, USA
e-mail: budat@ccf.org; miraclk@ccf.org; urbanm1@ccf.org

© Springer Nature Switzerland AG 2020
J. H. Karimov et al. (eds.), *Mechanical Support for Heart Failure*,
https://doi.org/10.1007/978-3-030-47809-4_42

MRSA	Methicillin-resistant *Staphylococcus aureus*
NYHA	New York Heart Association
PAC	Pulmonary artery catheter
PAPI	Pulmonary artery pulsatility index
PCC	Prothrombin complex concentrate
PCWP	Pulmonary capillary wedge pressure
PI	Pulsatility index
POD	Postoperative day
PTT	Partial thromboplastin time
RHC	Right heart catheterization
RPM	Revolutions per minute
RRT	Renal replacement therapy
RV	Right ventricle
RVF	Right ventricular failure
SaO2	Oxygen saturation
SNF	Skilled nursing facility
TAVR	Transcatheter aortic valve replacement
TV	Tricuspid valve
VA	Ventricular arrhythmias
VAC	Vacuum-assisted closure
vWF	von Willebrand factor

Introduction

Long- and short-term mechanical circulatory support (MCS), through the use of left ventricular assist devices (LVADs), has evolved into an increasingly prevalent therapy for the management of advanced, refractory heart failure. In particular, the advent of continuous-flow LVADs (cfLVADs) brought technological advancements which extended life, improved quality of life, reduced adverse events, and allowed successful MCS support outside of the hospital [1–3]. With the transition of cfLVAD management to community settings, it became important to establish a multidisciplinary team of LVAD professionals to globally manage the unique medical, psychological, and emotional needs of these patients throughout their LVAD course. Postoperative management for LVAD patients begins in the acute setting and includes the provision of 24/7/365 support that continues until the cessation of device therapy. Postoperative management additionally includes the enhancement of LVAD-competent and LVAD-aware community and post-acute agencies, and the provision of clinical support for non-LVAD medical personnel who come in contact with LVAD patients. This chapter examines the postoperative management of LVAD patients through a brief discussion of preoperative assessment and optimization, followed by examination of routine critical care management, management of common LVAD adverse events, special considerations associated with LVAD therapy, and the establishment of community-based support for LVAD patients.

Preoperative Assessment and Medical Optimization of LVAD Candidates

Successful postoperative LVAD management begins with astute preoperative examination and medical preparation of LVAD candidates. A thorough medical and psychosocial assessment of LVAD candidates, combined with preoperative clinical optimization, effective patient education, and recurring mutual communication among patients, caregivers, and providers, ensures the best opportunity for achieving successful postoperative goals.

LVAD candidates complete multidisciplinary examinations to determine individual medical risk and compliance, mental health and coping skills, chemical dependency burden and habitual use of opiates/benzodiazepines, caregiver and community support, and financial limitations and insurance coverage [4]. As part of understanding candidacy and surgical risk, LVAD candidates also receive assessments of their New York Heart Association (NYHA) Functional Classification, pulmonary artery pulsatility index (PAPI), and Interagency Registry for Mechanically Assisted Circulatory Support (INTERMACS) classification. Additionally, destination therapy (DT) candidates will meet the Centers for Medicare & Medicaid Services inclusion criteria for destination therapy [4].

Hemodynamic and volume optimization prior to LVAD surgery, through the use of increased inotropic support, aggressive diuresis, and/or the addition of temporary mechanical circulatory support is encouraged, especially in those patients failing maximal medical therapy [4]. End-organ dysfunction and volume status should be corrected, and suboptimal hemodynamics should be enhanced prior to surgery. The use of preoperative pulmonary artery catheter (PAC) guidance and temporary mechanical support might be considered in select patients.

LVAD candidates and their caregivers are fully informed of the expectations, benefits, lifestyle changes, and common adverse events of LVAD therapy during the preoperative period. They receive a multimodal LVAD education through the use of written and audiovisual tools, an LVAD decision aid, interviews with current LVAD patients, and practical education of LVAD equipment. Recurring mutual conversations between patients, caregivers, and heart failure cardiology, cardiac surgery, and MCS team members are encouraged to formulate trust, to provide consistent messaging, and to communicate perioperative and postoperative risks and goals to support shared decision-making [5–8].

Patients and caregivers need to fully understand the implications of LVAD therapy [8]. An LVAD decision aid is used to enhance existing education conducted by the LVAD center to present information in an unbiased, easy to understand format for the purpose of assisting the patient and/or caregiver with making an informed decision versus an emotional decision when faced with few alternatives to prolong life [5–8]. Due to a perceived lack of alternatives, patients are interested in learning how the LVAD will limit and impact their life and seek practical understanding of what is needed to manage the device [5]. An important component of an LVAD decision aid is the sharing of patient and caregiver experience of life on an LVAD.

The DECIDE-LVAD clinical trial demonstrated LVAD decision aids which have been effective in improving patient knowledge [7]. Kostick et al. documented the use of an LVAD knowledge scale to assess patient and caregiver understanding of LVAD therapy [6]. Having the patient and caregiver complete an LVAD-related quiz allows the clinician to assess patient and caregiver understanding, learn of knowledge gaps, and provide further clarification to promote shared decision-making resulting in an informed decision for the patient and their caregiver [6, 8].

Routine Postoperative Care Management

Following LVAD implant, immediate postoperative care takes place in an intensive care unit (ICU) setting, under the management of an LVAD-competent, multidisciplinary team that may include intensivists, anesthesiologists, cardiac surgeons, MCS coordinators, nutritionists, physical therapists, social workers, case managers, and registered nurses [4].

LVAD implantation is associated with an expected postoperative adverse event profile [9], which includes right ventricular (RV) dysfunction, vasoplegia, surgical and non-surgical bleeding, abnormal LVAD flow, cardiac arrhythmias, stroke (CVA), and fluctuating volume status. The focus of early postoperative management centers on hemodynamic stabilization through the control of bleeding, management of arrhythmias, generous right heart support, LVAD speed adjustments, blood pressure (BP) optimization, and adequate end-organ perfusion under echocardiogram and PAC guidance [10].

Hemodynamic Monitoring and Stabilization

From the operating room, an LVAD patient is transferred to an ICU setting, where they will arrive hypothermic, intubated and sedated, requiring inotropic, and either vasopressor or vasodilator support. Postoperative LVAD patients will have invasive monitoring lines that will include a PAC, an arterial line, and possibly a left atrial pressure (LAP) line [4, 11]. During the early postoperative period, continuous hemodynamic monitoring is constant, and the titration of inotropes, vasopressors, vasodilators, epoprostenol therapy, and LVAD speed are used to

Table 42.1 Hemodynamic parameters [4, 12, 13]

Parameter	Goal	Normal range
Cardiac index (CI)	>2.2 L/min/m^2	2.5–5.0 L/min/m^2
Mean arterial pressure (MAP)	65–90 mmHg	65–80 mmHg
Central venous pressure (CVP)	4–14 mmHg	4–12 mmHg
Left atrial pressure (LAP)	6–12 mmHg	6–12 mmHg
Pulmonary capillary wedge pressure (PCWP)	<18 mmHg	6–12 mmHg

achieve hemodynamic goals. (See Table 42.1). The routine use of LAP-guided hemodynamic monitoring is not required in the majority of LVAD patients. However, early recognition and prompt management of pulmonary congestion, through the direct measurement of preload via LAP monitoring, may enhance postoperative recovery and hemodynamic stability in select LVAD patients. Invasive lines are removed from most routine postoperative LVAD patients within 48–72 hours of surgery [4].

Preoperative and Perioperative Infection Risk Assessment and Antibiotic Prophylaxis

Infection prevention strategies begin preoperatively with a focus on candidate selection and screening for nasal methicillin-resistant *Staphylococcus aureus* (MRSA) colonization and perioperatively with surgical technique choice and antibiotic prophylaxis [14, 15]. Prior to surgery, LVAD candidates are evaluated for infection risks such as uncontrolled diabetes, nutritional deficits, and poor dental hygiene. If intranasal MRSA is present, preoperative LVAD candidates will complete 5 days of mupirocin 2% nasal ointment [14–16]. Perioperative antibiotic choice is center-specific and follows standard evidence-based recommendations for gram-negative and gram-positive coverage [4, 16]. The first dose of antibiotic is administered no less than 1 hour from commencement of surgical incision, and the total antibiotic course does not extend beyond 48 hours [4].

Early Right Ventricular Failure (RVF)

Right ventricular dysfunction is commonly present, in varying degrees of severity, in end-stage heart failure patients seeking candidacy for LVAD. Postoperative RV dysfunction is created by high intravascular volume, high venous return to the right heart, insufficient myocardial contractility, and/or pulmonary hypertension. RVF is defined by INTERMACS as persistent signs and symptoms of RV dysfunction, characterized by both the direct measurement of central venous pressure (CVP) >16 mm Hg or jugular venous distension (JVD) at least half way up the neck, combined with indirect findings of peripheral edema ≥2+, the presence of ascites/palpable hepatomegaly/bilirubin >2.0, or renal dysfunction with creatinine >2.0 [17].

Preoperative RV risk stratification and perioperative RV and tricuspid valve (TV) assessment are critical steps toward reducing the risk for early postoperative RVF, through preoperative consideration for inotropic or mechanical RV support and perioperative consideration for surgical tricuspid regurgitation correction, initiation of pulmonary vasodilators, and the addition of temporary mechanical RV support (4). Early postoperative hemodynamic interventions with regard to increased preload, increased afterload, and decreased contractility through PAC guidance will seek to reduce RV workload through inotropic, diuretic, and pulmonary vasodilator titration and the consideration for the addition of ultrafiltration or temporary mechanical RV support [4, 18]. The need for right ventricular assist device (RVAD) at the time of LVAD implant is a predictor for poor LVAD outcome [19].

Postoperative Bleeding: Surgical and Non-surgical

Approximately 30% of all LVAD patients require reoperation for postoperative bleeding, and 50–85% require postoperative blood transfusions [9, 20]. Early postoperative bleeding is a major cause for postoperative risk and is divided into two categories: surgical and non-surgical.

Surgical bleeding results from surgical injury (incisions, anastomoses, cannulation sites, blood vessels), while non-surgical bleeding results from derangements of the coagulation cascade [21]. Sixty-five percent of MCS patients' first bleeding episodes are the result of surgical sources: mediastinum (45%), thoracic space (12%), and chest wall (8%), where hemostasis can only be achieved through surgical correction at the bleeding site [21, 22].

The origins of non-surgical, postoperative bleeding are multifactorial, and each, alone or in combination, creates clotting defects along the coagulation cascade. Protective surgical hypothermia, the presence of residual surgical heparin, and diffuse coagulopathy contribute to postoperative bleeding risk and are easily reversed through the use of postoperative warming protocols, the administration of additional protamine sulfate, and the monitoring and correction of abnormal laboratory parameters (prothrombin time, partial thromboplastin time, platelet count, and fibrinogen). LVAD-acquired coagulopathies begin at the onset of blood pump interaction and result in a reduction in coagulation factors, platelet dysfunction, and impaired von Willebrand factor activity [23]. The presence of LVAD-acquired coagulopathies may contribute to early postoperative bleeding risk; however, the reversal of these coagulopathies is not realized until LVAD therapy is discontinued [23].

Routine LVAD Management

LVAD Speed/Revolutions per Minute (RPM) Choice

LVAD speed choice propels blood flow at a fixed and continuous pace, from the left ventricle (LV), through the pump body, and into the systemic circulation. RPM is the only modifiable parameter on a cfLVAD device and is directly proportional to the amount of blood circulating through the pump. RPM choice attempts to achieve adequate flow through the pump while balancing maximal LV unloading with minimal venous return burden to the right heart [10]. Optimal speed choice takes into consideration LV unloading per left ventricular end-diastolic diameter (LVEDD), midline septum position, minimization of mitral regurgitation (MR) and aortic insufficiency (AI), and minimization of burdensome right heart venous return while also considering the hemodynamic parameters of cardiac output (CO), pulmonary capillary wedge pressure (PCWP), and CVP [4, 24]. In the setting of optimal LVAD speed, clinicians will observe reductions to CVP, PCWP, LVEDD, MR, AI, and NYHA classification, with simultaneous increases to CO, aortic valve (AV) opening, end-organ perfusion, systemic pulsatility, and reverse remodeling [4, 25–27]. RPM choice is individualized and may be adjusted when considering the most recent echocardiographic and PAC findings, hemodynamic evolutions from the immediate postoperative period to the maintenance LVAD period, the development or increase to native contractility, the persistence of left and/or right heart failure symptoms, and desired pulsatility. It cannot be overstated that LVADs are not "gas pedals," and hemodynamic derangements may not be resolved through choice of LVAD speed alone. Observed device-specific speed ranges are displayed in Table 42.2.

Table 42.2 Observed LVAD device speed

LVAD device	Typical speed range (RPMs)
HeartWare (HVAD)	2400–3200
HeartMate II (HM2)	8600–9800
HeartMate 3 (HM3)	4800–6000

(*LVAD* left ventricular assist device, *RPM* revolutions per minute)

Basic Interpretation and Management of LVAD Parameters

Tables 42.3 and 42.4 provide basic interpretation of common LVAD parameters in addition to troubleshooting techniques and management recommendations when abnormal. An LVAD parameter evaluation does not replace the valuable information extracted from physical and laboratory assessment or echocardiography/right heart catheterization guidance. However, when used in

Table 42.3 Understanding LVAD parameters [11, 28–30]

HeartMate parameters			
Pulsatility Index (PI)	Pump Speed (RPM)	Power	Flow
• PI is related to the amount of assistance provided by the LVAD • Lower PI = greater LVAD assist • Higher PI = less LVAD assist • The difference of flow through the LVAD during the cardiac cycle can change with native contraction. Measured over 15 second cycle • Consider decreases in native contractility, low BP, or hypovolemia when PI is low, increases with high PI	• Only modifiable parameter • Determines pump flow • Limited variations in RPM can occur with PI events and suction • Low speed alarm is created when RPM drops below the set low speed	Amount of energy LVAD draws to run pump per set RPM	• Estimated • Directly proportional to RPM • Inversely proportional to pressure within inflow and outflow cannulas • Low flow alarms are created by low RPMs, increases in afterload, or decreases in preload • HM3 flow accuracy depends on current hematocrit as a measure of viscosity
HeartWare parameters			
Flow Pulsatility	Pump Speed (RPM)	Watts	Flow
• Unique feature among cfLVADs • The difference between the peak and trough of the HVAD estimated flow waveform • Affected by LV contractility, RV function, and afterload • Trough should be >2 L/min • Pulsatility should be >2 L/min	• Only modifiable parameter • Determines pump flow	Amount of power LVAD draws to run pump per set RPM	• Estimated • Determined by RPM, current, and blood viscosity • Preload or afterload effect flow • Accuracy of flow is dependent on entering the patient's current hematocrit

PI pulsatility; *BP* blood pressure; *RPM* revolutions per min; *LVAD* left ventricular assist device; *HM3* HeartMate 3; *cfLVAD* continuous-flow LVAD; *HVAD* HeartWare

combination with physical and diagnostic evaluations, the accurate interpretation of LVAD parameters will enhance understanding and guide appropriate treatment options.

Anticoagulation and Antiplatelets

Tailored management of aggressive anticoagulation and antiplatelet strategies in LVAD patients is center-specific and takes into consideration the different thromboembolic risk profiles between LVAD devices and the device-specific recommendations for international normalized ratio (INR) and aspirin dosage [4, 27]. Aggressive postoperative anticoagulation strategies include the early initiation of unfractionated heparin and coumadin once postoperative hemostasis has been achieved (postop day 1 or 2) [4]. Despite a higher risk of bleeding, unfractionated heparin is the most effective anticoagulant with postoperative LVADs [31] and is titrated to center-specifc goal partial thromboplastin time (PTT) ranged between 40 and 80 [4, 10]. Coumadin is titrated to device-specific goal INRs, typically between 2.0 and 3.0 [4, 31]. Table 42.5 summarizes the Cleveland Clinic approach to anticoagulation by LVAD device. Aspirin is initiated at device-specific dosages on postop day 1 or 2 and is titrated based on confirmation of platelet inhibition, observed through arachidonic acid measurement in serial aspirin/clopidogrel panels [4]. In the setting of inadequate platelet response to aspirin, the use of additional antiplatelet agents including clopidogrel and dipyridamole can be proposed, with consideration paid to an individual patients clotting and bleeding profile [4].

Table 42.4 Interpretation and management of abnormal LVAD parameters [28–30]

Abnormal parameter	Etiology	Intervention
High power/ high watts	Hypertension Thrombus AI Driveline fracture	Optimize BP to goal Thrombus/hemolysis evaluation Medical/surgical management of AI Assess waveforms
Low/no power	Obstruction of flow into the pump. Consider thrombus or inflow cannula positioning Inadvertent driveline or battery disconnects Driveline malfunction Malfunction of the peripheral equipment Electrostatic discharge	Immediate imaging assessment and thrombus evaluation May need inotropic support Consider temporizing with volume resuscitation to optimize LV filling if related to cannula positioning May require LVAD exchange May require controller change May require replacement of peripheral equipment
High flow	May be erroneous if power is not also increased Vasodilatation Thrombus	Optimize BP to goal Consider sepsis evaluation Assess waveforms
Low flow	Hypertension Hypotension Hypovolemia LVAD RPM set too high or too low Native recovery Arrhythmia Thrombus Driveline fracture	Optimize BP to goal Correct causes of hypovolemia Optimize RPM choice with imaging guidance Consider reducing low speed alarm limit if patient is hemodynamically stable Antiarrhythmic management Consider weaning LVAD support in select patients Assess history and waveforms
Low speed	Improper low speed setting Driveline/device malfunction Electrostatic discharge Suction events	Assess LVAD history and waveforms immediately Adjust low speed setting Driveline/controller/LVAD exchange if malfunction confirmed Optimize volume/speed
Suction events	Imbalance between RPM choice and LV filling	Decrease RPM with imaging guidance Optimize LV filling with volume resuscitation Optimize preload Assess device history Send waveforms
High pulsatility index	Consider increased native systolic contribution Higher BP Higher preload Arrhythmia	No intervention may be needed if clinically stable and without actionable cause Optimize BP if warranted Optimize antiarrhythmic if warranted PI values of 10 or higher should be discussed with the device company
Low pulsatility index	Consider decreased native systolic contribution Lower BP Hypovolemia Arrhythmia	No intervention may be needed if clinically stable and without actionable cause Optimize antiarrhythmic management if warranted Optimize volume Optimize RPM

AI aortic insufficiency; *LVAD* left ventricular assist device; *RPM* revolutions per minute; *LV* left ventricle; *BP* blood pressure

Management of Common LVAD-Related Adverse Events

Despite incontrovertible evidence that LVADs extend survival and improve heart failure symptoms, functional status, depression, and quality of life [1–3], clinicians continue to identify burdensome, yet common, adverse events (AEs) associated with LVAD therapy, which increase hospital readmissions and

Table 42.5 Current Cleveland Clinic LVAD anticoagulation protocol

Anticoagulation	HeartMate	HeartWare
Aspirin dosage	81 mg[a]	325 mg
Heparin POD #1	500 units/hour (no titration)[b]	500 units/hour (no titration)[b]
Heparin POD #2	Titrate to target PTT 49–67	Titrate to target PTT 49–67
Coumadin (goal INR)	2.0–3.0	2.0–3.0

POD postoperative day; *PTT* partial thromboplastin time; *INR* international normalized ratio; *HIT* heparin-induced thrombocytopenia)

aAspirin dose adjusted based on patient's response [4]

bBivalirudin used in place of heparin if HIT is suspected [4]

patient mortality [2, 12, 32–34]. The occurrence of AEs within a LVAD population is increased based on a combination of multifactorial and patient-specific comorbidities and risks [2, 32, 35]. AEs occur in up to 60% of all LVAD patients within 6 months of therapy and up to 80% of all LVAD patients within 12 months of therapy [33]. The significance, chronicity, and frequency of AEs experienced may influence a LVAD patient's life satisfaction as well as their survival. Prompt identification and treatment of common LVAD AEs are important for successful, long-term management of LVAD-supported patients.

Hemorrhagic and Thrombotic Complications

The continual interface of circulating blood with the synthetic surface of the LVAD exposes vulnerable blood components and endothelial cells to high shear forces [36, 37]. Constant exposure to high shear results in several cellular responses including activation of the fibrinolytic and coagulation pathways, activation of the endothelium, activation of platelets, reduction in von Willebrand (vWF) multimers, and an increase in angiogenesis [32, 36–38]. LVAD-mediated cellular responses contribute to higher incidences of gastrointestinal bleeding (GIB), thrombus, and hemolysis [37]. Modifiable factors such as hypertension, body mass index (BMI), alteration of pulsatility, and dosing of anticoagulation and antiplatelets may also contribute to the prevalence of LVAD-related coagulopathy events [36–40]. Hemorrhagic and thrombotic AEs are not experienced by the majority of LVAD patients, reinforcing our limited understanding of the LVAD's role in the prevalence of these events.

Gastrointestinal Bleeding(GIB)

Occurring in 18–40% of the LVAD population, and contributing to 33% of all LVAD readmissions, GIB events are the most common complication associated with LVAD therapy [2, 36, 38]. Patients typically present with symptomatic anemia and occult blood, and management involves a combined medical and interventional approach including hemodynamic resuscitation through blood transfusions, reversal and/or titration of anticoagulation and antiplatelet therapy, optimization of pulsatility through LVAD speed adjustments, localization of the bleeding source with early endoscopic evaluation [33, 37, 41, 42], and possibly the pharmaceutical inhibition of angiogenesis through the inclusion of alternative therapies such as octreotide, thalidomide, desmopressin, and omega-3 supplementation [43–45]. Pharmaceutical alternative therapies, however, currently represent an area of inconclusive and insufficient research in LVAD GIB prevention.

In the setting of a clinically significant GIB, management strategies depend on presentation. Anticoagulation and antiplatelet use should be suspended, and an elevated INR should be reversed while anemia is corrected and endoscopic diagnostics are pursued [4]. While anticoagulation and antiplatelets are being held, LVAD parameters and hemolysis labs should be trended with greater diligence to detect early signs of looming hemolysis [4]. Management during a GIB should include consultation with a gastroenterologist, where first-line diagnostics include a colonoscopy and/or upper endoscopy, followed by a small bowel evaluation if first-line diagnostics are negative [4]. Patients with GIB recurrence should be considered for repeated endoscopic evaluations, and the intensity or even use of antiplatelets, followed by the intensity or use of anticoagulation, should be considered [4]. Encouragingly, LVAD-related GIB events resolve permanently for those patients where discontinuation of LVAD therapy is an option [36, 38].

Hemolysis

Hemolysis is a common AE after LVAD implant and often foreshadows LVAD thrombosis [39, 46, 47]. INTERMACS defines minor and major hemolysis with the following biomarker limits: serum hemoglobin >20 or lactate dehydrogenase (LDH) >2.5 x the upper limit of normal. Major hemolysis carries the added clinical signs of hemoglobinuria, anemia, bilirubin >2, and pump malfunction/abnormal pump parameters [17, 46]. When the routinely collected biomarker limits of LDH and serum hemoglobin are elevated, management should include medical treatments that attempt to slow progression and minimize the deleterious effects of hemolysis; and a device investigation should be conducted to determine the degree of pump involvement [48, 49]. Medical and practical management of LVAD hemolysis includes hospital admission, intensification of anticoagulation through the addition of intravenous (IV) heparin/bivalirudin (sometimes despite INR), consideration for the intensification of antiplatelet strategies, IV fluid infusions to mitigate heme-induced acute kidney injury (AKI), optimization of AV opening to ensure adequate flow across the AV while simultaneously avoiding low pump speeds, management of hypertension (mean arterial pressure (MAP) >90), optimization of RV function, evaluation for an infectious etiology, and consideration for pump replacement/pump explant/upgrade of transplant status [38, 41, 48–51]. Assessing the level of pump derangement during hemolysis includes device waveform analysis, echocardiography, and contrast imaging with computed tomography (CT) (with careful consideration of renal function under the strain of hemoglobinuria) to evaluate clot deposition within visible areas of the pump circuit and positioning of the pump and graft [52–54]. Twenty percent of LVAD patients presenting with hemolysis will ultimately require device exchange within 6 months of onset [46]; however, hemolysis carries higher risks for stroke, peripheral embolism, abrupt pump malfunction/stoppage, acute kidney injury, and death, leading some clinicians to consider early device replacement as the preferred treatment option over watchful waiting [50, 55–57].

Pump Thrombosis

Pump thrombosis is associated with high morbidity and death and is a feared AE in LVAD therapy [38, 56, 58]. As previously mentioned, a correlation between hemolysis and pump thrombosis exists. LVAD thrombosis may present insidiously as an isolated rise in LDH or flagrantly as cardiogenic shock and pump stoppage. Early investigation and management of elevations in hemolysis biomarkers (mentioned earlier under *hemolysis*) should commence with the specter of thrombosis in mind [50]. Pump thrombosis management begins with prompt identification (hemolysis + heart failure + abnormal pump parameters), anticoagulation intensification, and expedited decision-making regarding surgical intervention (exchange, explant, urgent transplant) [12, 54, 56, 57]. Surgical intervention for pump thrombosis carries a higher success rate for thrombus resolution with reduced mortality and reduced recurrence when compared to medical therapy alone [55]. For this reason, some clinicians advocate for early surgical intervention with device exchange/explant/urgent transplant [50, 55–57]. Unfortunately, significant risks, including renal failure and stroke, can also be associated with device exchange, including a 34% risk of re-thrombosis observed in one post-exchange analysis [56]. Promisingly, a multi-center, prospective investigation (PREVENT) found that the use of a structured surgical implant technique, combined with strict anticoagulation management and choice of pump speed, lowered the incidence of thrombus in HeartMate II (HM2) [48]. When considering the associated risks with surgical exchange, combined with the added risk for thrombus reoccurrence, the prevention and management of LVAD thrombosis and the standardization of surgical implant procedures would benefit from further investigation.

Stroke (Ischemic, Hemorrhagic)

Stroke (CVA) is a devastating AE associated with LVAD therapy and is associated with high mortality [2, 59–61]. Hemorrhagic CVA carries a 30-day mortality of 55–71%, compared to 20–31% with ischemic CVA [60, 61]. A 2017 analysis of INTERMACS stroke outcomes revealed an almost 11% incidence in LVAD patients [61], and Frontera et al. identified higher incidences for CVA at the time of device implant and 9–12 months into LVAD therapy [59]. Predictors for ischemic and hemorrhagic CVA in LVAD patients include female gender, tobacco use, atrial fibrillation, presence of hemolysis/thrombosis, HeartWare (HVAD) pump choice, diagnosis of heparin-induced thrombocytopenia (HIT), use of intra-aortic balloon pump, ischemic etiology, MAP >90, INR <2.0 or >3.0, aspirin therapy <81 milligram/daily, and infection [51, 57, 59, 61–65]. Preventative CVA management should focus on reducing the following modifiable risk factors: physical inactivity, dyslipidemia, dietary risk (excessive sodium, low potassium, suboptimal intake of fruits and vegetables), heavy alcohol consumption, illegal drug abuse, abdominal obesity/BMI >30, sleep-disordered breathing, uncontrolled diabetes, and hemodynamically significant carotid stenosis (>60%) [66].

Acute management of neurological deficits includes prompt brain imaging and consultation with a neurologist. LVAD function and trends are immediately assessed, and hemolysis and thrombosis concerns are investigated [4]. In the setting of hemorrhagic CVA, anticoagulation is completely reversed, and neurosurgery is consulted (4). Anticoagulation reversal is an expedited decision that will include the use of IV vitamin K and/or prothrombin complex concentrate (PCC) while considering that the effective and safe use of PCC in the management of hemorrhagic AEs is rooted in limited information [67]. Following CVA, patients should receive recommended physical, occupational, and speech rehabilitation; and the timing and intensity of anticoagulation and antiplatelets should be coordinated with neurology/neurosurgery.

Acute and Chronic LVAD Infections (Bloodstream Infection (BSI), Driveline Infection (DLI))

Infection carries the second highest cause of death in LVAD patients who survive 6 months and is the second most prevalent AE during the first 3 months of therapy [2, 58]. Risk for infection rises with the duration of LVAD therapy and is experienced in up to 60% of all device patients [68, 69]. DLI is the most common LVAD-related infection, occurring in approximately 19% of patients within the first 12 months of implant [70, 71]. Once prosthetic LVAD components are exposed to infection, a bacterial biofilm matrix forms and adheres to the LVAD components. This biofilm shields the bacteria from the body's immune system and from antibiotic penetration, solidifying the infection's chronicity [72].

Risk for LVAD infections include duration of LVAD therapy, exposed Dacron velour, device choice, showering, surgical LVAD exchange, younger age, driveline (DL) trauma, failure to utilize a driveline immobilization device, negligence in appropriate wound care to the DL exit site, obesity, nutritional status, and diabetes [4, 63, 65, 73–81]. Presentation of LVAD-related infections may include new pain (over the DL exit site, DL tract, or LVAD body), erythema, leukocytosis, drainage from the driveline exit site, fever, rigors, diaphoresis, fatigue, high INR, hypotension, multi-system organ dysfunction, septic embolization/infarcts, and shock. *Staphylococcus* bacteria (*Staphylococcus epidermidis* and *Staphylococcus aureus*) are found in as many as 60% of all LVAD infections, making it the most predominant bacterial species associated with LVAD therapy [71, 81]. Other common bacteria include *Corynebacterium*, *Pseudomonas*, Enterobacteriaceae, and some multi-drug resistance species [73, 81].

In the absence of standardized guidelines for LVAD-related infections, current treatment recommendations originate from expert opinion, prompting early consultation with an infectious disease specialist regarding diagnostics and targeted antibiotic therapy. Diagnostic evaluations of LVAD infections begin with the collection of a DL culture and may additionally include any combination of the following based on patient presentation: serial collection of cultures from appropriate sites (other draining sites and blood), complete blood count, CT, and echocardiography (rule out LVAD or internal cardiac defibrillator (ICD) endocarditis) [4]. Additionally, positron emission tomography (PET) may emerge as a valuable diagnostic tool for identifying the depth or involvement of LVAD-related infections [82, 83]. Initial treatment of early DLIs will include empiric oral antibiotics while awaiting culture results [71]. However infections that include systemic symptoms or concern for underlying abscess will require hospitalization, broad-spectrum IV antibiotics, and evaluation for incision and drainage (I&D) if appropriate. Complete eradication of LVAD-related infections is only accomplished through the removal of all LVAD components [79, 84], which is best achieved in transplant-eligible patients at the time of cardiac transplant. In the absence of complete device removal, surgical I&D will offer palliative control of localized infections of the DL exit site or ascending DL tract and is often combined with negative pressure wound therapy (VAC) or flap coverage [85].

Device Equipment Malfunctions

Proper LVAD function depends on pump component durability and deliberate care of the external peripheral equipment. External LVAD components are constantly exposed to patient and environmental influences. Time, exposure, and frequent manipulation of the external equipment result in expected equipment fails due to wear and tear of LVAD components over the course of treatment.

Controller malfunctions are the most common source for LVAD equipment faults [86]. Controller malfunctions are anxiety-filled events which require a meticulous exchange of the controller, performed ideally at the implanting center to avoid slow or erroneous patient technique. Unfortunately, clinical urgency and/or distance from the hospital may require controller exchanges in the home, at local physician offices, or at local emergency rooms. For obvious reasons of patient safety, it is strongly recommended that patients *never* perform controller exchanges alone.

Driveline malfunction is the most common justification for LVAD replacement [87]. A review of patients who experienced DL malfunction found that 87% had confirmed damage to the external portion of the DL [88]. Additionally, a recent prospective analysis by a single center found that HM2 devices saw more driveline malfunctions than HVAD devices (21% vs 0%) at 3 years [86]. DL malfunction is a clinical emergency because its presentation and pattern for deterioration are unpredictable. A broken DL may manifest with intermittent or sustained pump stoppages, bizarre LVAD parameters, pump vibrations/revving, and critical alarms, or it may present more surreptitiously with vague device alarms/beeps and non-specific data points in the controller history. The DL wire redundancy system allows the pump to continue operating on ungrounded power in the setting of single wire damage. However, complete DL fracture will result in intermittent or sustained pump stoppages, depending on whether electrical connections can still be achieved with external DL position manipulation.

Driveline malfunction is managed according to clinical presentation, device designation (destination therapy (DT) vs bridge to transplant (BTT)), and patient wishes. An evaluation for DL malfunction begins with a thorough visual, tactile, and radiographic inspection of the external DL, with particular focus on areas of obvious damage and waveform analysis (utilizing an ungrounded cable for HM2). Splice repairs to the external DL (whether targeted or blind) are recommended on the advice of device engineering professionals and will be performed by a device technician while the patient remains hospitalized. Clinical situations that include severe pump malfunction, failure of the splice repair, or radiographically confirmed fracture within the internal portion of the DL may warrant urgent device exchange. Select BTT patients may avoid device exchange through urgent transplant listing. For select DT patients, early device exchange may be a reasonable option, when considering the average progression to complete DL failure following splice repair is approximately 10 months [89]. Introduction of the HeartMate 3 (HM3) modular DL now provides HM3 patients with a non-surgical/non-splice option for external DL correction.

Malfunctions in peripheral LVAD equipment (controllers, clips, batteries, battery chargers, cords) will require expected equipment exchanges over the course of LVAD therapy. Typical equipment interventions will include rescue tape application to isolated DL silicone damage, yearly equipment maintenance, and peripheral equipment replacement. Periodically, software and equipment upgrades will require additional equipment maintenance or replacement.

Aortic Insufficiency (AI)

Aortic insufficiency is an AE that has been reported with moderate severity in 30% of LVAD patients at 2 years and may continue to worsen with duration of LVAD therapy [90–92]. Continual LV unloading provided by LVAD therapy reduces LV pressure below aortic pressure and may lead to progressive AV degeneration and commissural fusion [10, 90]. As AI worsens, an increasing portion of blood flow exiting the outflow graft is captured in retrograde flow across the incompetent aortic valve and is returned to the left ventricle (LV)/LVAD circuit. This creates a continuous, ineffective flow of blood from the LV, through the pump/outflow graft, and back into the LV which does not participate in systemic blood flow, resulting in reduced CO, decreased organ perfusion, and heart failure symptoms [90, 93–95]. Risk factors for the development of AI have been reported to include advanced age at time of device implant, pre-existing AI, concurrent MR, use of continuous flow, female gender, low BMI, dilated aortic root, lack of AV opening, and DT designation [93–98]. To decrease the risk of AI progression, pump speed optimization with ramp echocardiogram and PAC guidance is recommended to ensure adequate CO, midline septum, LV unloading, and at least intermittent AV opening [93].

Management choice for AI depends on severity. Medical management focuses on afterload reduction with angiotensin-converting enzyme (ACE) inhibitors or vasodilators to maintain goal MAP <80, maintaining euvolemia with diuretics, and serial pump speed enhancements to reduce transvalvular pressure [4, 10, 93]. Decreased systemic perfusion, despite increased LVAD flow, may indicate worsening AI due to retrograde flow and may require further intervention [94].

Surgical options for AI management should be individualized and may include AV repair, AV replacement with a bioprosthetic valve, or oversewing of the AV. Select LVAD patients, deemed too high risk for surgery, may be candidates for minimally invasive treatment options including percutaneous transcatheter LV outflow graft occlusion or transcatheter AV replacement

(TAVR). These options may relieve the severity of heart failure symptoms; however no clear treatment recommendations for the surgical management of AI exists, and current algorithms require additional study [10, 93, 94].

Late Right Ventricular Failure (RVF)

Left ventricular failure is commonly recognized as the primary cause for *right* ventricular failure; and yet 15–40% of LVAD patients continue to experience RVF despite continuous LV unloading [18, 99]. Late RVF persists when the conditions for preload, afterload, and contractility remain suboptimal [9]. Although a targeted approach to late RV support has not been identified to date, it has been suggested that late RVF is associated with severe preoperative pulmonary hypertension and increased RV preload [100]. A multimodal approach for late right heart support which ensures tailored management of pulmonary resistance, adequate volume management, control of LV afterload, judicious use of inotropic agents, and optimization of LVAD RPM may reduce the prevalence of LVAD patients who experience late RVF [4, 9, 99, 100]. Patients failing maximal drug therapy for RV support should be considered for mechanical right heart support, with progression to urgent transplantation for select transplant-eligible patients [4, 99].

Atrial and Ventricular Arrhythmias and the Use of Internal Cardiac Defibrillators (ICDs)

Atrial and ventricular arrhythmias are prevalent in end-stage heart failure and are generally well tolerated in LVAD patients due to the continuation of systemic circulation throughout the arrhythmia. Nevertheless, a full understanding of the arrhythmia risk profile in an LVAD population is lacking, and consensus regarding best practices for medical and procedural strategies, including ICD programming, is absent [101].

Atrial arrhythmia (AA) prevalence in an LVAD population approaches 50% and typically carries minimal symptomatic effect [102]. However, prolonged or rapid AA may increase risk for hemodynamic compromise and RV dysfunction due to loss of atrial contribution or rapid ventricular response [24, 102]. Management of AA in an LVAD population currently focuses on rhythm suppression and/or rate control with beta-blocker (BB) therapy, digoxin, antiarrhythmics such as amiodarone and dofetilide, and external cardioversion [10, 24, 102]. Refractory AAs with hemodynamic compromise may benefit from catheter ablation; however, permanent pacing post-procedure may be an expected consequence [102]. Discrepant studies exist about thromboembolic risk in LVAD patients experiencing AA [102]. Conveniently, anticoagulation and antiplatelet therapies are utilized in the vast majority of LVAD patients and may reduce the additional thromboembolic risk associated with AA.

Ventricular arrhythmias (VA) are common in an LVAD population, having been reported in 22–59% of LVAD recipients [11, 101–103]. VAs occur more often during the early postoperative period [10, 11, 101–103], with pre-existing VA contributing to higher risk for LVAD-VAs [11, 101–104]. Mechanisms for VAs in LVAD patients are multifactorial and can include post-ischemic VT substrate, LVAD suction events (high pump speed/hypovolemia/native LV recovery), malpositioned inflow cannula, cannula-adjacent scar, RV dysfunction, electrolyte imbalances, and ischemic cardiomyopathy etiology [11, 24, 101, 102, 104]. Similar to AA, cfLVAD patients with VA may not experience symptomatic hemodynamic compromise immediately; however, they may experience recurrent ICD fires or eventual RV failure [10, 11, 101]. Some studies describe increased mortality in LVAD patients with VA [101–104]. Evaluation and management of the underlying mechanisms for VA in LVAD patients will include electrocardiogram (EKG), ICD evaluation, electrolyte replacement, immediate volume resuscitation (until suction event is ruled out), antiarrhythmics, BB therapy (in eligible patients), echocardiogram, RPM reduction, right heart catheterization (RHC), and consultation with an electrophysiology specialist [10, 102,

104]. Review of LVAD history will confirm the presence of suction events, and an echocardiogram will assess suction event risk through assessment of inflow cannula placement and LVEDD, potentially warranting a reduction in pump speed. RHC may be appropriate in patients where decompensated heart failure is suspected of contributing. Medical management with antiarrhythmics (amiodarone, mexiletine, sotalol, and lidocaine) and BBs are initiated or enhanced depending on arrhythmia and patient presentation [10, 11, 24, 101, 102]. Catheter ablation may be warranted for select LVAD patients with refractory VA [24, 102–105], and urgent transplant upgrade can be considered for transplant-eligible patients meeting VT exception criteria.

The association of ICDs with LVAD survival is not well understood, and management opinions are evolving in the era of cfLVADs. The majority of LVAD patients receive ICDs prior to LVAD implant [101, 102], with therapies continued post-LVAD. ICD therapy recommendations for LVAD patients do not currently consider VA burden, symptom burden, heart failure compensation, or preserved systemic circulation. Investigations into the use of ultra-conservative tachy-therapies in select LVAD patients are ongoing [101]. Currently, decisions to pursue new ICD implantations, generator exchanges, and aggressive tachy-therapies in cfLVAD patients, without pre-existing VAs, are controversial and provider-specific [10, 101, 103].

Special Considerations for LVAD Patients

Pulsatility and Blood Pressure (BP) Assessment

Accurate assessment of pulsatility and BP measurement in cfLVADs is necessary for the management of device afterload sensitivity and for the reduction of AEs such as CVA, thrombosis, GIB, AI, and persistent heart failure symptoms [106–108]. The most accurate method for obtaining BP will vary depending on the patient's degree of systemic pulsatility. When considering the combined conditions of preload, afterload, LVAD speed, and native LV contractility, AV opening and pulsatility may be continuous, intermittent, or absent. The degree of AV opening and pulsatility will determine which BP measurement tool will provide the most accurate reading. For patients with a greater degree of pulsatility, traditional BP measurement tools (noninvasive cuff, automated BP) will be accurate and easier to use. However, patients with diminished pulsatility may achieve more accurate BP measurement with Doppler ultrasound. Doppler ultrasound BP collection is the standard of practice in cfLVAD BP management, where accuracy rates are between 94% and 100% [27, 106, 109–111]. The opening Doppler pressure may correlate with either the MAP or the systolic BP depending on the degree of systemic pulsatility. The MAP blood pressure goal for cfLVADs is <80 [4]. Medical management for MAPs >80 will include the use of ACE inhibitors, angiotensin receptor blockers (ARB), BBs, hydralazine, and nitrates [4].

Discharge Preparedness: LVAD Education

Discharge educational preparation for an LVAD patient begins during the preoperative evaluation, where patients and caregivers are first informed of the advantages, limitations, lifestyle changes, complications, medications, and follow-up requirements associated with LVAD therapy. During preoperative discussions, at least one caregiver is identified who will agree to complete VAD education, in addition to the patient, prior to hospital discharge. That caregiver also agrees to facilitate family support for the patient throughout the index LVAD hospitalization and rehabilitation process, 24-hour care at home, and with transportation to outpatient laboratory and LVAD appointments [4, 26]. The information and education provided to patients and caregivers prior to discharge can seem overwhelming in both its volume and significance. The expected psychosocial adjustments and anxiety experienced by new LVAD patients and caregivers can impede educational progression and may require additional counseling and support from MCS social workers to ensure completion of informational goals. Educational objectives are achieved upon successful completion of an individualized, device-specific education, including written, visual, and hands-on techniques, using repetition and demonstration [26, 27, 112].

Discharge Preparedness: Community Readiness

Procedures for readying the community to receive an LVAD patient are necessary and center-specific. MCS team members perform a home safety assessment pre-implant, and again post-implant, to understand the patient's living situation, verify the presence of grounded outlets, verify the presence of a working cell phone and/or landline, and identify potential harms such as switch-controlled AC outlets, extension cords, and generators [4]. Additionally, MCS team members ensure that local community agencies are VAD-prepared prior to hospital discharge. Local branches of the fire department and emergency medical services (EMS), in addition to the closest emergency department and local physician, receive the implanting LVAD center's emergency contact information, basic information regarding LVAD purpose, function, components, alarms, and recommendations for emergency preparedness and cardiopulmonary resuscitation (CPR) [4, 112]. The patient's electric company receives verification of the presence of life-sustaining equipment in the home for inclusion on a priority service restoration list in the event of a power outage [4]. Table 42.6 is an example of a worksheet the patient/caregiver is to complete identifying their community resources and contact information.

Post-acute LVAD Facilities

A portion of medically stable LVAD patients will require a higher level of care at hospital dis-

Table 42.6 Community contact information for LVAD discharge

Community contact information for LVAD Discharge

Fire Department Address: ______________________

Phone Number: ______________________

EMS Address: ______________________

Phone Number: ______________________

Hospital Address: ______________________

Phone Number: ______________________

Physician Address: ______________________

Phone Number: ______________________

Lab for Blood Draws: ______________________

Pharmacy Phone Nmber: ______________________

Electric Company (Please include name of account holder and the account number):

____

Your Address: ______________________

Your Phone Numbers (Land Line and/or Cell): ______________________

(EMS = emergency medical service)

charge, and VAD centers benefit from collaborative relationships with facilities that provide all levels of post-acute care. Depending on the intensity of care the patient requires, VAD-competent acute rehabilitation facilities (ARF), skilled nursing facilities (SNF), long-term acute care hospitals (LTAC), and outpatient dialysis centers are important post-acute care options for LVAD patients. Facilities are selected and maintained based on the standard of their medical and nursing care, their willingness to undergo site visits with MCS team members, their willingness to complete and maintain VAD competency, and their willingness to provide care and communication with the MCS team based on mutually agreed protocols for VAD care and emergency responsiveness. MCS team members provide LVAD training, support, and clinician collaboration to post-acute facilities 24/7/365 and may be required to obtain provider privileges to participate in patient care.

Outpatient Dialysis with LVAD Therapy

LVAD patients with severely impaired renal function, AKI or end-stage renal disease (ESRD) may require temporary or permanent renal replacement therapy (RRT) during their LVAD course [70]. Dialysis support in an LVAD population carries high mortality and infection rates and presents a unique challenge to MCS providers, local nephrologists, and community dialysis agencies [20, 113–115]. Sparse data currently exists in support of a preferred RRT modality or chronic management preferences in an LVAD population. Reluctant outpatient dialysis centers frequently share legitimate concerns regarding inconsistent accuracy of BP measurement, hemodynamically significant fluid shifts, LVAD alarms, and emergency interventions. Despite these concerns, three times weekly hemodialysis sessions in community centers remain the most common outpatient RRT preference for LVAD patients at this time, although peritoneal dialysis is gaining more consideration due to lower risk for bloodstream infection and fluid shifts [115, 116]. Due to the absence, or inconsistent presence, of pulsatile blood flow in cfLVAD therapy, arteriovenous fistula (AVF) maturation was long considered improbable. Promisingly, however, there are limited examples of successful AVF maturation in LVAD patients reported in recent literature [117, 118]. Additional investigations into pre-LVAD predictors for RRT, RRT modality preference, chronic hypotension and anemia management, and AVF maturation techniques need to be conducted to reduce LVAD mortality and facilitate clinical RRT decision-making in an LVAD population.

Emergency Care

Emergency preparedness, rapid response, and 24/7/365 MCS team availability contribute to prompt and appropriate crisis management involving LVAD patients. Despite the fact that LVAD patients, caregivers, and local emergency agencies receive a "VAD-competent" or "VAD-aware" education prior to index hospitalization discharge, they remain inexperienced in their assessment and responsiveness to LVAD emergencies. An unresponsive cfLVAD patient presents a unique challenge to first responders and non-LVAD medical providers. The naturally absent or inconsistent peripheral pulsatility found in most LVAD patients often contributes to inaccurate assessments of BP and oxygen saturation (SaO2) in emergency settings and will also present a challenge during arterial blood gas sampling. Throughout a LVAD crisis situation where BP and SaO2 are questionable, rapid assessment indices for tissue perfusion may be observed through the clinical evaluation of respirations, skin color, skin temperature, and capillary refill, as well as through the evaluation of the laboratory perfusion biomarkers of serum lactate, venous oxygen saturation, and end-tidal carbon dioxide monitoring (ETCO2) [119–124]. Decreases in ETCO2 levels correlate with reductions in BP and CO and elevations in serum lactate and can serve as a beneficial assessment tool in LVAD patients experiencing severe decompensated heart failure or cardiopulmonary shock

[122–124]. Advanced cardiac life support protocols, with chest compressions as a last resort, should be followed for LVAD patients experiencing hypoperfusion [119]. Unresponsive LVAD patients with adequate perfusion, however, should proceed with additional clinical investigation for unresponsiveness, including CVA [119, 120]. An LVAD-specific EMS Field Guide is available and accessible via a smartphone/smart device through the use of a scan code [125]. This guide provides immediate access to LVAD information, including answers to frequently asked LVAD questions, troubleshooting information (alarms, power management, controller changes), and emergency procedures [4, 125]. It cannot be overemphasized that MCS team providers are available around the clock to collaborate and assist non-LVAD personnel during LVAD emergencies.

Discussions with LVAD patients and caregivers regarding responses to severe weather/natural disaster protections, power outages, the use of power generators, the duration of LVAD battery life, the use of backup LVAD equipment, the need for working phones, travel preparations (especially air travel, sea travel, or intercontinental travel), and interactions with non-LVAD medical personnel begin prior to hospital discharge and are reinforced throughout the LVAD course.

Outpatient Management

Following hospital discharge, outpatient interactions are focused on patient and device assessment, prevention and management of VAD-related AEs, and prevention of hospital readmissions [26, 106]. Assessments by MCS team members (cardiologist, surgeon, MCS practitioner/coordinator, social worker) are conducted through regularly scheduled outpatient visits and through routine and urgent phone communications [4, 10, 26]. The frequency of outpatient appointments is center-specific and dependent on clinical stability [4, 10, 26]. In addition to the physical exam, device interrogation and BP assessment, the DL exit site is assessed, and dressing changed, providing an opportunity for ongoing education [4, 10, 26, 106]. Further information regarding pump function is obtained through the downloading of waveforms for evaluation by vendor engineers for routine downloads or with suspicion of pump malfunction [4]. Uploading a picture of the DL exit site into the electronic medical record allows others to observe driveline exit site changes over time, and prevents other care providers from disrupting the intact sterile dressing [26]. During the visit, patient and caregiver coping is assessed with referral to the LVAD social worker for new or worsening psychosocial concerns [26]. Patient and caregivers at risk are followed by the social worker on a routine basis, and referrals to specialists or other supportive agencies as needed.

Routine laboratory tests are completed with each visit and may be required between visits [4, 10, 26, 106]. At set intervals (determined by implanting center) or with concerns of a complication, additional testing is completed. These tests include echocardiogram, RHC, chest X-ray, EKG, 6-minute walk test, and quality of life questionnaires [4, 10, 26, 106].

Understanding follow-up visits is imperative to the prevention of or early detection of AEs; implanting centers may establish shared care sites to provide access to care closer to home for some patients [106, 126]. The success of a shared care site is dependent on both the implanting facility and the shared care site. Shared care sites are developed based on the site's interest in caring for LVAD patients with demonstrated heart failure experience and willingness to obtain necessary device equipment and receive advanced LVAD training [126]. The implanting center shares care protocols to provide consistent care. Together, protocols are established for the frequency of visits between the two sites and define responsibilities for the management of INR, on-call responsibilities, admission options, and emergency responses to avoid fragmentation in care [106, 126].

Whether the patient is being cared for at the implanting center or the shared care site, experienced MCS coordinators or advanced practice providers are available 24/7/365 to address emergencies. Patients and caregivers are encouraged to call the non-urgent office phone to address questions, concerns, and prescription refills among a variety of other needs. The MCS team is responsible for the coordination of care while on LVAD support including accompanying patients to invasive tests, procedures, and surgeries to promote patient safety. The MCS team ensures ongoing education is offered for patients, caregivers, and healthcare providers caring for LVAD patients.

Caregiver Burden

Prior to LVAD implant, patients are required to identify a supportive caregiver to assist with the complexity of LVAD care [4, 127–129]. The social worker evaluates the caregiver's own medical, physical, and emotional concerns that may impact their ability to assist the patient [26, 27]. The caregiver can be a spouse, adult child, extended family member, or close friend [127, 128]. During the LVAD decision-making process, expectations of the caregiver are communicated by the multidisciplinary team [5]. The caregiver must commit to being available post-device implant to learn LVAD care and management and identify an emergency situation and how to respond, how to contact the implanting center, frequency of follow-up visits, and any other additional patient care needs [128, 129]. Being responsible for the patient's care needs solicits feelings of anxiety and fear as the patient transitions to the home environment [27, 128, 129]. Like the patients, caregivers must adapt to lifestyle changes and sacrifices required to accommodate the LVAD patient's care needs [128].

The adaptive process of the caregiver occurs over time and is impacted by the patient's overall recovery and LVAD course [27, 128, 129]. Initially caregivers describe fear, anxiety, and self-doubt regarding their ability to care for an LVAD patient and feeling responsible for their survival [27, 128]. In response to these feelings, caregivers may develop hyper-vigilance interfering with their daily routines [127, 128]. Home care services and frequent follow-up visits are used to help with transition and assess adaptation [27]. For patients living a distance from the implanting center (>2 hours), they are asked to stay in the local area for 1–2 clinic visits, while the multidisciplinary team assesses adaptation to living outside the hospital environment.

The multidisciplinary team must continually assess caregiver burden during routine outpatient visits and with any hospital readmissions [26, 27]. Magid et al. reported the incidence of depression, anxiety, and post-traumatic stress was higher in LVAD caregivers [128]. Support for caregivers include caregiver support groups, one-on-one counseling, or referral to a specialist [26, 128]. Caregivers should be counseled on the importance of maintaining their health and asking for assistance when needed [127].

Conclusion

Mechanical circulatory support, through the use of cfLVADs, is an increasingly prevalent therapy for the management of advanced, refractory heart failure. Continuous-flow LVADs have proven to extend life, improve quality of life, reduce adverse events, and allow successful MCS support outside of the hospital setting. The postoperative management of LVAD patients begins with diligent patient selection, continues through every phase of the LVAD spectrum, involves the multidisciplinary and evidenced-based management of LVAD adverse events, and ends only with the cessation of LVAD therapy. Successful postoperative management requires collaboration between the patient, the caregiver, and the multidisciplinary team.

References

1. Shah KB, Starling RC, Rogers JG, Horstmanshof DA, Long JW, Kasirajan V, et al. Left ventricular assist devices versus medical management in ambulatory heart failure patients: an analysis of INTERMACS profiles 4 and 5 to 7 from the ROADMAP study. J Heart Lung Transplant. 2018;37(6):706–14.
2. Kirklin JK, Pagani FD, Kormos RL, Stevenson LW, Blume ED, Myers SL, et al. Eighth annual INTERMACS report: special focus on framing the

impact of adverse events. J Heart Lung Transplant. 2017;36(10):1080–6.
3. Starling RC, Estep JD, Horstmanshof DA, Milano CA, Stehlik J, Shah KB, et al. Risk assessment and comparative effectiveness of left ventricular assist device and medical management in ambulatory heart failure patients. The ROADMAP study 2- year results. JACC Heart Fail. 2017;5(7): 518–27.
4. Feldman D, Pamboukian SV, Teuteberg JJ, Birks E, Lietz K, Moore SA, et al. The 2013 International Society for Heart and Lung Transplantation guidelines for mechanical circulatory support: executive summary. J Heart Lung Transplant. 2013;32(2):157–87. https://doi.org/10.1016/j.healun.2012.09.013.
5. Blumenthal-Barby JS, Kostick KM, Delgado ED, Volk RJ, Kaplan HM, Wilhelms LA, et al. Assessment of patients' and caregivers' informational and decisional needs for left ventricular assist device placement: implications for informed consent and shared decision making. J Heart Lung Transplant. 2015;34(9):1182–9.
6. Kostick KM, Minard CG, Wilhelms LA, Delgado E, Abraham M, Bruce CR, et al. Development and validation of a patient-centered knowledge scale for left ventricular assist device placement. J Heart Lung Transplant. 2016;35(6):768–76.
7. Allen LA, Mcllvennan CK, Thompson JS, Dunlay SM, LaRue SJ, Lewis EF, et al. Effectiveness of an intervention supporting shared decision making for destination therapy left ventricular assist device. The DECIDE-LVAD randomized clinical trial. JAMA Intern Med. 2018;178(4):520–9.
8. Breathett K, Allen LA, Ambardekar AV. Patient-centered care for left ventricular assist device therapy: current challenges and future direction. Curr Opin Cardiol. 2016;31(3):313–20.
9. Kilic A, Acker MA, Atluri P. Dealing with surgical left ventricular assist device complications. J Thoracic Dis. 2015;7(12):2158–64.
10. DeVore AD, Mentz RJ, Patel CB. Medical management of patients with continuous-flow left ventricular assist devices. Cur Treat Options Cardiovasc Med. 2014;16(2):283. https://doi.org/10.1007/s11936-013-0283-0.
11. Birati EY, Rame JE. Left ventricular assist device management and complications. Crit Care Clin. 2014;30(3):607–27.
12. Imamura T, Jeevanandam V, Kim G, Raikhelkar J, Sarswat N, Kalantari S, et al. Optimal hemodynamics during left ventricular assist device support are associated with reduced readmission rates. Circ Heart Fail. 2019;12(2):e005094.
13. Normal hemodynamic parameters and laboratory values. Edwards Lifesciences LLC; [cited 12 April 2019]. Available from: http://ht.edwards.com/scin/edwards/sitecollectionimages/edwards/products/presep/ar05688_parameters.pdf.
14. Maniar S, Kondareddy S, Topkara VK. Left ventricular assist device-related infections: past, present and future. Expert Rev of Med Devices. 2011;8(5):627–34.
15. Wever-Pinzon O, Stehlik J, Kfoury AG, Terrovitis JV, Diakos NA, Charitos C, et al. Ventricular assist devices: pharmacological aspects of a mechanical therapy. Pharmacol Ther. 2012;134(2):189–99.
16. Acharya MN, Som R, Tsui S. What is the optimum antibiotic prophylaxis in patients undergoing implantation of a left ventricular assist device? Interact Cardiovasc Thorac Surg. 2012;14(2):209–14.
17. Lenderman JC. Appendices [Internet]. UAB - School of Medicine - Interagency Registry for Mechanically Assisted Circulatory Support - Intermacs Appendices. [cited 2018Nov3]. Available from: http://www.uab.edu/medicine/intermacs/intermacs-documents.
18. Fida N, Loebe M, Estep JD, Guha A. Predictors and management of right heart failure after left ventricular assist device implantation. Methodist Debakey Cardiovasc J. 2015;11(1):18–23.
19. Kormos RL, Teuteberg JJ, Pagani FD, Russell SD, John R, Miller LW, et al. Right ventricular failure in patients with the HeartMate II continuous-flow left ventricular assist device: incidence, risk factors, and effect on outcomes. J Thorac Cardiovasc Surg. 2010;139(5):1316–24.
20. Miller LW, Pagani FD, Russell SD, John R, Boyle AJ, Aaronson KD, et al. Use of a continuous-flow device in patients awaiting heart transplantation. N Engl J Med. 2007;357(9):885–96.
21. Gaffey AC, Chen CW, Chung JJ, Han J, Bermudez CA, Wald J, et al. Is there a difference in bleeding after left ventricular assist device implant: centrifugal versus axial? J Cardiothorac Surg. 2018;13(1):22. https://doi.org/10.1186/s13019-018-0703-z.
22. Jessup ML, Goldstein D, Ascheim DD, Teuteberg JJ, Park SJ, Naftel DC, et al. Risk for bleeding after MCSD implant: an analysis of 2358 patients in INTERMACS. J Heart Lung Transplant. 2011;30(4):S9. https://doi.org/10.1016/j.healun.2011.01.012.
23. Nascimbene A, Neelamegham S, Frazier OH, Moake JL, Dong JF. Acquired von Willebrand syndrome associated with left ventricular assist device. Blood. 2016;127(25):3133–41.
24. Pratt AK, Shah NS, Boyce SW. Left ventricular assist device management in the ICU. Crit Care Med. 2014;42(1):158–68.
25. Imamura T, Chung B, Nguyen A, Sayer G, Uriel N. Clinical implications of hemodynamic assessment during left ventricular assist device therapy. J Cardiol. 2018;71(4):352–8.
26. Bellumkonda L, Jacoby D. Hospital to home with mechanical circulatory support. Curr Heart Fail Rep. 2013;10(3):212–8.
27. Smith EM, Franzwa J. Chronic outpatient management of patients with a left ventricular assist device. J Throacic Dis. 2015;7(12):2112–24.
28. HeartWare Ventricular Assist System. [cited: 2 April 2019]. Available from: http://www.heart-

ware.com/sites/default/files/uploads/resources/ifu00001_rev17_hvasinstructionsforuse_us.pdf.

29. Rich J, Burkhoff D. HVAD flow waveform morphologies. ASAIO J. 2017;63(5):526–35.
30. Thoratec Corporation HeartMate 3 Left Ventricular Assist System. [cited: 10April 2019]. Available from: http://www.thoratec.com/_assets/download-tracker/HM3/100168999/100168999.A_HM3.IFU.PMA.Clip.LT.FinalPDF.2018.1012.pdf.
31. Kantorovich A, Fink JM, Militello MA, Bauer SR, Soltesz EG, Moazami N. Comparison of anticoagulation strategies after left ventricular assist device implantation. ASAIO J. 2016;62(2):123–7.
32. Loyaga-Rendon R, Jani M, Fermin D, McDermott J, Vancamp D, Lee S. Prevention and treatment of thrombotic and hemorrhagic complications in patients supported by continuous-flow left ventricular assist devices. Curr Heart Fail Rep. 2017;14(6):465–77.
33. Wever-Pinzon O, Drakos SG, Kfoury AG, Nativi JN, Gilbert EM, Everitt M, et al. Morbidity and mortality in heart transplant candidates supported with mechanical circulatory support: is reappraisal of the current united network for organ sharing thoracic organ allocation policy justified? Circulation. 2013;127(4):452–62.
34. Frazier OH, Rose EA, McCarthy P, Burton NA, Tector A, Levin H, et al. Improved mortality and rehabilitation of transplant candidates treated with a long-term implantable left ventricular assist system. Ann Surg. 1995;222(3):327–38.
35. Kirklin JK, Naftel DC, Kormos RL, Stevenson LW, Pagani FD, Miller MA, et al. Fifth INTERMACS annual report: risk factor analysis from more than 6,000 mechanical circulatory support patients. J Heart Lung Transplant. 2013;32(2):141–56.
36. Shah P, Tantry US, Bliden KP, Gurbel PA. Bleeding and thrombosis associated with ventricular assist device therapy. J Heart Lung Transplant. 2017;36(11):1164–73.
37. Gurvits GE, Fradkov E. Bleeding with the artificial heart: gastrointestinal hemorrhage in CF-LVAD patients. World J Gastroenterol. 2017;23(22):3945–53.
38. Rosenthal JL, Starling RC. Coagulopathy in mechanical circulatory support: a fine balance. Curr Cardiol Rep. 2015;17(12):114. https://doi.org/10.1007/s11886-015-0670-0.
39. Willey JZ, Boehme AK, Castagna F, Yuzefpolskaya M, Garan AR, Topkara V, et al. Hypertension and stroke in patients with left ventricular assist devices (LVADs). Curr Hypertens Rep. 2016;18(2):12. https://doi.org/10.1007/s11906-015-0618-1.
40. Forest SJ, Xie R, Kirklin JK, Cowger J, Xia Y, Dipchand AI, et al. Impact of body mass index on adverse events after implantation of left ventricular assist devices: an IMACS registry analysis. J Heart Lung Transplant. 2018;37(10):1207–17.
41. Harvey L, Holley CT, John R. Gastrointestinal bleed after left ventricular assist device implantation: incidence, management, and prevention. Ann Cardiothorac Surg. 2014;3(5):475–9.
42. Welden CV, Truss W, McGwin G, Weber F, Peter S. Clinical predictors for repeat hospitalizations in left ventricular assist device (LVAD) patients with gastrointestinal bleeding. Gastroenterology Res. 2018;11(2):100–5.
43. Imamura T, Nguyen A, Rodgers D, Kim G, Raikhelkar J, Sarswat N, et al. Omega-3 therapy is associated with reduced gastrointestinal bleeding in patients with continuous-flow left ventricular assist device. Circ Heart Fail. 2018;11(10):e005082. https://doi.org/10.1161/CIRCHEARTFAILURE.118.005082.
44. Seng BJ, Teo LL, Chan LL, Sim DK, Kerk KL, Soon JL, et al. Novel use of low-dose thalidomide in refractory gastrointestinal bleeding in left ventricular assist device patients. The Int J Artif Organs. 2017;40(11):636–40.
45. Juricek C, Imamura T, Nguyen A, Chung B, Rodgers D, Sarswat N, et al. Long-acting Octreotide reduces the recurrence of gastrointestinal bleeding in patients with a continuous-flow left ventricular assist device. J Card Fail. 2018;24(4):249–54.
46. Xia Y, Forest S, Friedmann P, Chou L, Patel S, Jorde U, et al. Factors associated with prolonged survival in left ventricular assist device recipients. Ann Thorac Surg. 2019;107(2):519–26.
47. Katz JN, Jensen BC, Chang PP, Myers SL, Pagani FD, Kirklin JK. A multicenter analysis of clinical hemolysis in patients supported with durable, long-term left ventricular assist device therapy. J Heart Lung Transplant. 2015;34(5):701–9.
48. Maltais S, Kilic A, Nathan S, Keebler M, Emani S, Ransom J, et al. PREVENtion of HeartMate II pump thrombosis through clinical management: the PREVENT multi-center study. J Heart Lung Transplant. 2017;36(1):1–12.
49. Ravichandran A, Parker J, Novak E, Joseph S, Schilling J, Ewald G, Silvestry S. Hemolysis in left ventricular assist device: a retrospective analysis of outcomes. J Heart Lung Transplant. 2014;33(1):44–50.
50. Cowger JA, Romano MA, Shah P, Shah N, Mehta V, Haft JW, et al. Hemolysis: a harbinger of adverse outcome after left ventricular assist device implant. J Heart Lung Transplant. 2014;33(1):35–43.
51. Elmously A, de Biasi AR, Risucci DA, Worku B, Horn EM, Salemi A. Systemic blood pressure trends and antihypertensive utilization following continuous-flow left ventricular assist device implantation: an analysis of the interagency registry for mechanically assisted circulatory support. J Thorac Dis. 2018;10(5):2866–75.
52. Tran BC, Nijjar PS. Role of contrast CT for the diagnosis and the prognosis of suspected LVAD thrombosis. J Card Surg. 2017;32(2):162–5.
53. Estep JD, Vivo RP, Cordero-Reyes AM, Bhimaraj A, Trachtenberg BH, Torre-Amione G, et al. A simplified echocardiographic technique for detect-

ing continuous-flow left ventricular assist device malfunction due to pump thrombosis. J Heart Lung Transplant. 2014;33(6):575–86.
54. Jorde UP, Aaronson KD, Najjar SS, Pagani FD, Hayward C, Zimpfer D, et al. Identification and management of pump thrombus in the HeartWare left ventricular assist device System: a novel approach using log file analysis. JACC Heart Fail. 2015;3(11):849–56.
55. Luc JG, Tchantchaleishvili V, Phan K, Dunlay S, Maltais S, Stulak JM. Medical therapy compared with surgical device exchange for left ventricular assist device thrombosis: a systematic review and meta-analysis. ASAIO J. 2019;65(4):307–17.
56. Hanke JS, Dogan G, Wert L, Ricklefs M, Heimeshoff J, Chatterjee A, et al. Left ventricular assist device exchange for the treatment of HeartMate II pump thrombosis. J Thorac Dis. 2018;10(S15):S1728–36.
57. Levin AP, Saeed O, Willey JZ, Levin CJ, Fried JA, Patel SR, et al. Watchful waiting in continuous-flow left ventricular assist device patients with ongoing hemolysis is associated with an increased risk for cerebrovascular accident or death. Circ Heart Fail. 2016;9(5):e002896. https://doi.org/10.1161/CIRCHEARTFAILURE.115.002896.
58. Kirklin JK, Naftel DC, Kormos RL, Pagani FD, Myers SL, Stevenson LW, et al. Interagency registry for mechanically assisted circulatory support (INTERMACS) analysis of pump thrombosis in the HeartMate II left ventricular assist device. J Heart Lung Transplant. 2014;33(1):12–22.
59. Frontera JA, Starling R, Cho SM, Nowacki AS, Uchino K, Hussain MS, et al. Risk factors, mortality, and timing of ischemic and hemorrhagic stroke with left ventricular assist devices. J Heart Lung Transplant. 2017;36(6):673–83.
60. Cho SM, Moazami N, Frontera JA. Stroke and intracranial hemorrhage in HeartMate II and HeartWare left ventricular assist devices: a systematic review. Neurocrit Care. 2017;27(1):17–25.
61. Acharya D, Loyaga-Rendon R, Morgan CJ, Sands KA, Pamboukian SV, Rajapreyar I, et al. INTERMACS analysis of stroke during support with continuous-flow left ventricular assist devices. JACC Heart Fail. 2017;5(10):703–11.
62. Teuteberg JJ, Slaughter MS, Rogers JG, McGee EC, Pagani FD, Gordon R, et al. The HVAD left ventricular assist device: risk factors for neurological events and risk mitigation strategies. JACC Heart Fail. 2015;3(10):818–28.
63. Stulak JM, Davis ME, Haglund N, Dunlay S, Cowger J, Shah P, et al. Adverse events in contemporary continuous-flow left ventricular assist devices: a multi-institutional comparison shows significant differences. J Thoracic Cardiovasc Surg. 2016;151(1):177–89.
64. Nassif ME, LaRue SJ, Raymer DS, Novak E, Vader JM, Ewald GA, et al. Relationship between anticoagulation intensity and thrombotic or bleeding outcomes among outpatients with continuous-flow left ventricular assist devices. Circ Heart Fail. 2016;9(5):e002680. https://doi.org/10.1161/CIRCHEARTFAILURE.115.002680.
65. Lerman DT, Hamilton KW, Byrne D, Lee DF, Zeitler K, Claridge T, et al. The impact of infection among left ventricular assist device recipients on post-transplantation outcomes: a retrospective review. Transpl Infect Dis. 2018;20(6):e12995. https://doi.org/10.1111/tid.12995.
66. Meschia JF, Bushnell C, Boden-Albala B, Braun LT, Bravata DM, Chaturvedi S, et al. Guidelines for the primary prevention of stroke. A statement for healthcare professionals from the American Heart Association/American Stroke Association. Stroke. 2014;45(12):3754–832.
67. Rimsans J, Levesque A, Lyons E, Sylvester K, Givertz MM, Mehra MR, et al. Four factor prothrombin complex concentrate for warfarin reversal in patients with left ventricular assist devices. J Thromb Thrombolysis. 2018;46(2):180–5.
68. Califano S, Pagani FD, Malani PN. Left ventricular assist device–associated infections. Infect Dis Clin N Am. 2012;26(1):77–87.
69. Nienaber J, Wilhelm MP, Sohail MR. Current concepts in the diagnosis and management of left ventricular assist device infections. Expert Rev Anti-Infect Ther. 2013;11(2):201–10.
70. Kirklin JK, Naftel DC, Pagani FD, Kormos RL, Stevenson LW, Blume ED, et al. Seventh INTERMACS annual report: 15,000 patients and counting. J Heart Lung Transplant. 2015;34(12):1495–504.
71. Nienaber JJC, Kusne S, Riaz T, Walker RC, Baddour LM, Wright AJ, et al. Clinical manifestations and management of left ventricular assist device-associated infections. Clin Infect Dis. 2013;57(10):1438–48.
72. Toba FA, Akashi H, Arrecubieta C, Lowy FD. Role of biofilm in Staphylococcus aureus and Staphylococcus epidermidis ventricular assist device driveline infections. J Thorac Cardiovasc Surg. 2011;141(5):1259–64.
73. Aburjania N, Sherazi S, Tchantchaleishvili V, Alexis J, Hay C. Stopping conventional showering decreases pseudomonas exit site infections in left ventricular assist device patients. J Heart Lung Transplant. 2016;35(4):S363.
74. O'Horo JC, Abu Saleh OM, Stulak JM, Wilhelm MP, Baddour LM, Rizwan SM. Left ventricular assist device infections. A systematic review. ASAIO J. 2018;64(3):287–94.
75. Akhter SA, Russo MJ, Singh A, Valeroso TB, Johnson EM, Anderson AS, et al. 32 modified HeartMate II driveline externalization technique significantly decreases incidence of infection and improves long-term survival. J Heart Lung Transplant. 2012;31(4):S20.
76. Wert L, Hanke JS, Dogan G, Ricklefs M, Fleissner F, Chatterjee A, et al. Reduction of driveline infections

through doubled driveline tunneling of left ventricular assist devices— five-year follow-up. J Thorac Dis. 2018;10(S15):S1703–10.
77. Akay MH, Nathan SS, Radovancevic R, Poglajen G, Jezovnik MK, Candelaria IN, et al. Obesity is associated with driveline infection of left ventricular assist devices. ASAIO J. 2018;1 https://doi.org/10.1097/MAT.0000000000000916.
78. Yu SN, Takayama H, Han J, Garan AR, Topkara VK, Yuzefpolskaya M, et al. Early ald vate outcomes of subcostal exchange of HeartMate II left ventricular assist device: a word of caution. J Heart Lung Transplant. 2018;37(4):S100. https://doi.org/10.1016/j.healun.2018.01.236.
79. Toda K, Sawa Y. Clinical management for complications related to implantable LVAD use. Gen Thorac Cardiovasc Surg. 2015;63(1):1–7.
80. Critsinelis AC, Kurihara C, Kawabori M, Sugiura T, Lee VV, Civitello AB, et al. Predictive value of preoperative serum albumin levels on outcomes in patient undergoing LVAD implantation. J Card Surg. 2018;33(8):469–78.
81. Koval CE, Rakita R. Ventricular assist device related infections and solid organ transplantation. Am J Transplant. 2013;13(s4):348–54.
82. Dell'Aquila AM, Mastrobuoni S, Alles S, Wenning C, Henryk W, Schneider SR, et al. Contributory role of fluorine 18-Fluorodeoxyglucose positron emission tomography/computed tomography in the diagnosis and clinical management of infections in patients supported with a continuous-flow left ventricular assist device. Ann Thorac Surg. 2016;101(1):87–94.
83. Kanapinn P, Burchert W, Körperich H, Körfer J. F-FDG PET/CT-imaging of left ventricular assist device infection: a retrospective quantitative intrapatient analysis. J Nucl Cardiol. 2018;16(1):1–10.
84. Koval CE, Thuita L, Moazami N, Blackstone E. Evolution and impact of drive-line infection in a large cohort of continuous-flow ventricular assist device recipients. J Heart Lung Transplant. 2014;33(11):1164–72.
85. Jacoby A, Stranix JT, Cohen O, Louie E, Balsam LB, Levine JP. Flap coverage for the treatment of exposed left ventricular assist device (LVAD) hardware and intractable LVAD infections. J Card Surg. 2017;32(11):732–7.
86. Kormos RL, Mccall M, Althouse A, Lagazzi L, Schaub R, Kormos MA, et al. Left ventricular assist device malfunctions. Circulation. 2017;136(18):1714–25.
87. Moazami N, Milano CA, John R, Sun B, Adamson RM, Pagani FD, et al. Pump replacement for left ventricular assist device failure can be done safely and is associated with low mortality. Ann Thorac Surg. 2013;95(2):500–5.
88. Kalavrouziotis D, Tong MZ, Starling RC, Massiello A, Soltesz E, Smedira NG, et al. Percutaneous lead dysfunction in the HeartMate II left ventricular assist device. Ann Thorac Surg. 2014;97(4):1373–8.
89. Stulak JM, Schettle S, Haglund N, Dunlay S, Cowger J, Shah P, et al. Percutaneous driveline fracture after implantation of the HeartMate II left ventricular assist device: how durable is driveline repair? ASAIO J. 2017;63(5):542–5.
90. Cowger J. Aortic regurgitation during continuous-flow left ventricular assist device support: an insufficient understanding of an insufficient lesion. J Heart Lung Transplant. 2016;35(8):973–5.
91. Jorde UP, Uriel N, Nahumi N, Bejar D, Gonzalez-Costello J, Thomas SS, et al. Prevalence, significance, and management of aortic insufficiency in continuous flow left ventricular assist device recipients. Circ Heart Fail. 2014;7(2):310–9.
92. Gasparovic H, Kopjar T, Saeed D, Cikes M, Svetina L, Petricevic M, et al. De novo aortic regurgitation after continuous-flow left ventricular assist device implantation. Ann Thorac Surg. 2017;104(2):704–11.
93. Holtz J, Teuteberg J. Management of aortic insufficiency in the continuous flow left ventricular assist device population. Curr Heart Fail Rep. 2014;11(1):103–10.
94. Bouabdallaoui N, El-Hamamsy I, Pham M, Giraldeau G, Parent MC, Carrier M, et al. Aortic regurgitation in patients with a left ventricular assist device: a contemporary review. J Heart Lung Transplant. 2018;37(11):1289–97.
95. Milano CA, Simeone AA. Mechanical circulatory support: devices, outcomes and complications. Heart Fail Review. 2013;18(1):35–53.
96. Deo SV, Sharma V, Cho YH, Shah IK, Park SJ. De novo aortic insufficiency during long-term support on a left ventricular assist device. ASAIO J. 2014;60(2):183–8.
97. Cowger J, Pagani FD, Haft JW, Romano MA, Aaronson KD, Kolias TJ. The development of aortic insufficiency in left ventricular assist device-supported patients. Circ Heart Fail. 2010;3(6):668–74.
98. Aggarwal A, Raghuvir R, Eryazici P, Macaluso G, Sharma P, Blair C, et al. The development of aortic insufficiency in continuous-flow left ventricular assist device-supported patients. Ann Thorac Surg. 2013;95(2):493–8.
99. Voelkel N, Quaife R, Leinwand L, Barst R, McGoon M, Meldrum D, et al. Right ventricular functiuon and failure. Circulation. 2006;114(17):1883–91.
100. Rame JE, Teuteberg JJ, Birks EJ, Rogers JG, Acker MA, Birati EY, et al. An analysis of early versus late right heart failure with an intrapericardial continuous flow LVAD. J Heart Lung Transplant. 2015;34(4):S113–4.

101. Nakahara S, Chien C, Gelow J, Dalouk K, Henrikson CA, Mudd J, et al. Ventricular arrhythmias after left ventricular assist device. Circ Arrhythm Electrophysiol. 2013;6(3):648–54.
102. Kadado AJ, Akar JG, Hummel JP. Arrhythmias after left ventricular assist device implantation: incidence and management. Trends Cardiovasc Med. 2018;28(1):41–50.
103. Lima B, Bansal A, Abraham J, Rich JD, Lee SS, Solemani B, et al. Controversies and challenges of ventricular assist device therapy. Am J Cardiol. 2018;121(10):1219–24.
104. Santangeli P, Rame JE, Birati EY, Marchlinski FE. Management of ventricular arrhythmias in patients with advanced heart failure. J Am Coll Cardiol. 2017;69(14):1842–60.
105. Moss JD, Flatley EE, Beaser AD, Shin JH, Nayak HM, Upadhyay GA, et al. Characterization of ventricular tachycardia after left ventricular assist device implantation as destination therapy. JACC Clin Electrophysiol. 2017;3(12):1412–24.
106. Estep JD, Trachtenberg BH, Loza LP, Bruckner BA. Continuous flow left ventricular assist devices: shared care goals of monitoring and treating patients. Methodist Debakey Cardiovasc J. 2015;11(1):33–44.
107. Milano CA, Rogers JG, Tatooles AJ, Bhat G, Slaughter MS, Birks EJ, et al. HVAD: the ENDURANCE supplemental trial. JACC Heart Fail. 2018;6(9):792–802.
108. Najjar SS, Slaughter MS, Pagani FD, Starling RC, McGee EC, Eckman P, et al. An analysis of pump thrombus events in patients in the HeartWare ADVANCE bridge to transplant and continued access protocol trial. J Heart Lung Transplant. 2014;33(1):23–34.
109. Bennett MK, Roberts CA, Dordunoo D, Shah A, Russell SD. Ideal methodology to assess systemic blood pressure in patients with continuous-flow left ventricular assist devices. J Heart Lung Transplant. 2010;29(5):593–4.
110. Castagna F, Stohr E, Pinsino A, Cockcroft J, Willey J, Baran AR, et al. The unique blood pressure and pulsatility of LVAD patients: current challenges and future opportunities. Curr Hypertens Rep. 2017;19(10):85. https://doi.org/10.1007/s11906-017-0782-6.
111. Lanier GM, Orlanes K, Hayashi Y, Murphy J, Flannery M, Te-Frey R, et al. Validity and reliability of a novel slow cuff-deflation system for noninvasive blood pressure monitoring in patients with continuous-flow left ventricular assist device. Circ Heart Fail. 2013;6(5):1005–12.
112. Widmar SB, Dietrich MS, Minnick AF. How self-care education in ventricular assist device programs is organized and provided: a nation study. Heart Lung. 2014;43(1):25–31.
113. Butler J, Geisberg C, Howser R, Portner PM, Rogers RG, Deng MC, et al. Relationship between renal function and left ventricular assist device use. Ann Thorac Surg. 2006;81(5):1745–51.
114. Bansal N, Hailpern S, Katz R, Hall Y, Tamura M, Kreuter W. Outcomes associated with left ventricular assist devices among recipients with and without end-stage renal disease. JAMA Intern Med. 2018;178(2):204–9.
115. Roehm B, Vest AR, Weiner DE. Left ventricular assist devices, kidney disease, and dialysis. Am J Kidney Dis. 2018;71(2):257–66.
116. Thomas B, Logar C. Anderson A. Renal replacement therapy in congestive heart failure requiring left ventricular assist device augmentation. 2012;32(4):386–92.
117. Sasson T, Wing RE, Foster TH, Kashyap R, Butani D, Waldman DL. Assisted maturation of native fistula in two patients with a continuous flow left ventricular assist device. J Vasc Interv Radiol. 2014;25(5):781–3.
118. Chin AI, Tong K, McVicar JP. Successful hemodialysis arteriovenous fistula creation in a patient with continuous-flow left ventricular assist device support. Am J Kidney Dis. 2017;69(2):314–6.
119. Peberdy MA, Gluck JA, Omato JP, Bemudez CA, Griffin RE, Kasirajan V, et al. Cardiopulmonary resuscitation in adults and children with mechanical circulatory support. A scientific statement from the American Heart Association. Circulation. 2017;135(24):e1115–34.
120. Trinquero P, Pirotte A, Gallagher LP, Iwaki KM, Beach C. Wilcox JE. Left ventricular assist device management in the emergency department. 2018;19(5):834–41.
121. Hasanin A, Mukhtar A, Nassar H. Perfusion indices revisited. J Intensive Care. 2017;5(1):24. https://doi.org/10.1186/s40560-017-0220-5.
122. Aminiahidashti H, Shafiee S, Kiasari AZ, Sazgar M. Applications of end-tidal carbon dioxide (ETCO2) monitoring in emergency department: a narrative review. Emergency (Tehran). 2018;6(1):e5.
123. Kheng C, Rahman N. The use of end-tidal carbon dioxide monitoring in patients with hypotension in the emergency department. Int J Emerg Med. 2012;5:31. https://doi.org/10.1186/1865-1380-5-31.
124. Paiva E, Paxton J, O'Neil B. The use of end-tidal carbon dioxide (ETCO2) measurement to guide management of cardiac arrest: a systematic review. Resuscitation. 2018;123:1–7.
125. EMS Field Guides/MyLVAD [Internet]. MyLVAD.com. 2019 [cited 19 May 2019]. Available from: https://www.mylvad.com/medical-professionals/resource-library/ems-field-guides#.
126. Kierman MS, Joseph SM, Katz JN, Kilic A, Rich JD, Tallman MP, et al. Sharing the care of mechanical circulatory support: collaborative efforts of patients/caregivers, shared care sites, and left ventricular assist device implanting centers. Circ Heart Fail. 2015;8:629–35.

127. Bidwell JT, Lyons KS, Mudd JO, Grady KL, Gelow JM, Hiatt SO, et al. Patient and caregiver determinants of patient quality of life and caregiver strain in left ventricular assist device therapy. J Am Heart Assoc. 2018;7(6):e008080.
128. Magid M, Jones J, Allen LA, Mcllvennan CK, Magid K, Sterling JA, et al. The perceptions of important elements of caregiving for an LVAD patient: a qualitative meta-synthesis. J Cardiovasc Nurs. 2016;31(3):215–25.
129. Adams EE, Wrightson ML. Quality of life with an LVAD: a misunderstood concept. Heart Lung. 2018;47:177–83.

Palliative Care in MCS Patients: Insights into Current Practice and Outcomes

43

Krista Dobbie and Kyle Neale

Abbreviations

ACEi	Angiotensin-converting enzyme inhibitors
ACP	Advance care planning
BTT	Bridge to transplant
CF	Continuous flow
DNR	Do not resuscitate
DT	Destination therapy
DT-LVAD	Left ventricular assist device for destination therapy
GoC	Goals of care
HPM	Hospice and palliative medicine
LVAD	Left ventricular assist device
MCS	Mechanical circulatory support
NYHA	New York Heart Association
PP	Preparedness planning
REMATCH	Randomized Evaluation of Mechanical Assistance for the Treatment of Congestive Heart Failure
TAH	Total artificial heart
VAD	Ventricular assist device

K. Dobbie, MD (✉) · K. Neale, DO
Cleveland Clinic, Cleveland, OH, USA
e-mail: dobbiek@ccf.org; nealek@ccf.org

Historical Context

The history of heart failure, palliative care, and mechanical support begins in the mid-twentieth century. The field benefited from the advent of cardiac catheterization in the 1940s to 1960s which was able to characterize many structural forms of heart disease [1]. In 1964, the first xenotransplantation occurred with a donated chimpanzee heart although the recipient patient died an hour later [2]. Dr. Christiaan Neethling Barnard achieved worldwide notoriety for performing the first human transplant in 1964 to a recipient with advanced cardiomyopathy; the recipient lived for 16 days post-transplant, and the achievement instigated a frenzy of activity with more than 100 transplants performed in 1968. However, interest cooled in subsequent years with only 50 transplants in 1969, 20 in 1970, and less than 10 in 1971 [3].

The 1960s saw parallel investigation into mechanical support. The first left ventricular assistive device was implanted by DeBakey and colleagues in 1966, and in 1969, the first total artificial heart (TAH) was implanted as a bridge to transplant (BTT) [4]. However, TAH carried a high risk of complications and never became a serious alternative to the gold therapy standard of heart transplantation, and instead research in mechanical devices transitioned from achieving cardiac replacement to cardiac support.

© Springer Nature Switzerland AG 2020
J. H. Karimov et al. (eds.), *Mechanical Support for Heart Failure*,
https://doi.org/10.1007/978-3-030-47809-4_43

In 1971, the Framingham study established the first set of criteria for defining congestive heart failure [5]. There was already recognition of the significant symptom burden linked with heart failure, particularly the distressing effects of fatigue and dyspnea. Yet treatment in this decade was largely limited to digitalis, vasodilators, diuretics, and diet restriction. These collective therapies were in a sense all palliative interventions [6]. Framingham showed that 50% of patients diagnosed will die within 5 years, a figure that rivals many aggressive malignancies. Yet, the concept of palliative care as an adjunctive support for terminal disease was in its infancy. Palliation was typically operative in nature and considered when traditional medical options were exhausted [7].

The origins of palliative care grew from the modern hospice movement. Dame Cicely Saunders opened St Christopher's Hospice in the United Kingdom in 1967 but had been involved in end-of-life care for many years [8]. She sparked interest in care of the dying in the United States following a presentation at Yale University in 1963 which later grew with the publication of *On Death and Dying* by Elizabeth Kübler-Ross in 1969 [9]. This movement led to the opening of the first hospice in the United States in 1974. Supportive symptom care for individuals with terminal disease was expanded by Dr. Balfour Mount in 1974. He established the first palliative care service in North America to provide supportive care as co-management with active medical treatment for hospitalized patients and those in the home setting [10]. He coined the term "palliative care" to differentiate his care from hospice [11].

Heart failure treatments through the 1970s – whether mechanical assistive devices or medications to improve hemodynamics and reduce afterload – were aimed primarily at improving cardiovascular function; the neuroendocrine response seen in heart failure was presumed to be a compensatory mechanism. It was not until Dr. Peter Harris published his work on untreated HF patients in India in the mid-1980s that heart failure became understood as a neuroendocrine disease, which gave way to the deployment of angiotensin-converting enzyme inhibitors (ACEi) and beta-blockers which not only decreased the risk of hospitalization and lowered mortality but also dramatically reduced the symptom burden [1].

Aggressive medical heart failure management became more cognizant of symptom burden in the setting of refractory disease and increasing prevalence [12]. Yet, hospice care for people with terminal heart failure remained the exception [13]. A retrospective review found 27 hospice referrals for heart failure to St Christopher's Hospice out of 9920 patients between 1994 and 1999 [14]. Editorials and letters called for greater investigation into quality of life and symptom control beyond the efforts to optimize the underlying disease [15, 16]. This included a recognition of the care pathways in oncology which had palliation as the ultimate goal and arguing for the inclusion of such tracks in heart failure research [17]. This was in no small part due to the recognition of suffering that heart failure patient's experience. Indeed, in a 2001 survey of heart failure patients found that those with Minnesota Living with Heart Failure scores >65, indicative of advanced heart failure, were willing to give up 50% of their remaining life for improved quality of life [18]. The acknowledgement of suffering in end-stage heart failure was part of the larger maturation of palliative care and its increasing recognition by other disciplines [19]. What had once been primarily care of oncology patients receiving cancer-directed therapy was now branching into other life-limiting illnesses, including heart failure, to provide supportive care upstream and "focus on providing relief from the symptoms and stress of a serious illness…[and] to improve quality of life for both the patient and the family" [20].

Meanwhile, advances in mechanical support continued to improve mortality and symptom control. The first successful transplant after utilization of a left ventricular assist device (LVAD) as a bridge to therapy was performed in 1984 [4]. LVADs were formally approved as a bridge to transplant in 1994. The Randomized Evaluation of Mechanical Assistance for the Treatment of Congestive Heart Failure (REMATCH) trial demonstrated decreased mortality and improved quality of life for patients with ventricular assistive devices (VADs) compared to their exclusively medically managed cohorts [21]. This led to its subsequent Federal Drug Administration approval in 2002 as destination therapy (DT), including the HeartMate II and later the HeartWare VAD. There was a progressive increase in VAD implants for DT therapy between 2008 and 2013, and now nearly 46% of VAD implants are designated as destination therapy [22]. Their survival benefits were clear but they also showed significant functional gains. LVADs as destination therapy (DT-LVAD) notably led to improvements in New York Heart Association (NYHA) functional class, a quality of life indicator. At 24 months post-implant, the majority of LVAD patients had improved from NYHA Class IV to Class I or II [23].

This new frontier of heart failure with LVAD therapy was exciting. Patients were given a new lease on life but one with potentially serious complications. The rush to embrace the new mechanical circulatory support (MCS) sometimes led to suboptimal end-of-life care; advance care planning was not always defined [24]. Moreover there was new appreciation for suffering and complications that patients and families experience as a result of this intervention. Palliative care specialists were identified as ideally suited to manage these new care needs, and consequently the joint commission modified their ventricular assist device destination therapy requirements to mandate palliative care as a member of the interdisciplinary specialty team in 2014 [25].

Unfortunately, support services continue to be underutilized [26]. The reasons for this include difficulties with prognostication and the many available treatments available in early-stage disease. The Cleveland Clinic has used the mandated involvement of palliative care in MCS to increase patient access to palliative support along the continuum of their heart failure care, with outpatient and end-of-life support. This enables our team to provide what the World Health Organization defines as "an approach that improves the quality of life of patients and their families facing the problem associated with life-threatening illness, through the prevention and relief of suffering by means of early identification and impeccable assessment and treatment of pain and other problems, physical, psychosocial and spiritual" [27].

Now with nearly half of VAD implants occurring for destination therapy, this translates to a growing population of VAD patients who are older and have multiple comorbidities. This patient population benefits from more comprehensive care through aggressive symptom management, continually reassessing goals of care during potentially life-threatening events, and providing emotional support [28]. Palliative care can also facilitate a timely transition to hospice care which, while improved, remains shockingly low given the nature of the disease. In a recent cohort analysis of 121,990 Medicare beneficiaries in the American Heart Association Get With The Guidelines–Heart Failure registry, hospice discharges increased from 2% in 2005 to 5% in 2014 [29]. In essence, patients with advanced heart failure need greater palliative integration, and an ideal place to make that initial contact is through preparedness planning (PP) in a pre-LVAD consult.

Preparedness Planning

The use of LVADs as DT and not as a BTT poses unique challenges for the medical team, the patient, and their caregivers beginning with the goals of care (GoC) [30]. There will be no hope for curative treatment with a heart transplant in DT, and therefore expectations need to be managed carefully with a focus on quality of life. Doing so may avoid suffering because patients and families admit to feeling overwhelmed with making acute healthcare decisions in times of crisis [31]. PP seeks to mitigate that distress as well as address the other components of advance care planning (ACP) like identifying a surrogate decision-maker, clarifying code status, and promoting understanding of the potential complications (Fig. 43.1).

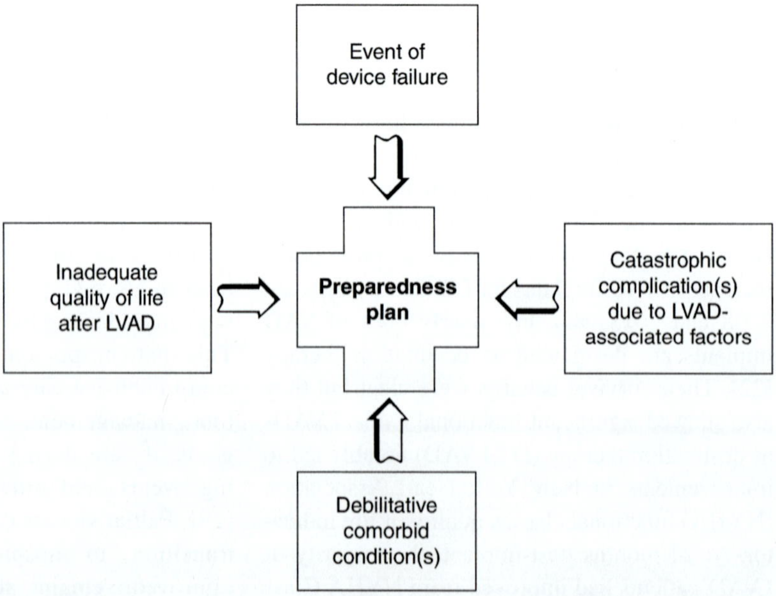

Fig. 43.1 Framework for preparedness planning. (Reprinted from Mayo Clinic Proc, Vol 86, Ed 6, Swetz KM, Freeman MR, Abou Ezzeddine OF, Carter KA, Boilson BA, Ottenberg AL, et al. Palliative medicine consultation for preparedness planning in patients receiving left ventricular assist devices as destination therapy, 495, June 201)

A common misconception still exists among patients in the medical community at large that GoC equates to hospice discussion [33]. On the contrary, GoC discussions ensure patient-centered informed decision-making. PP achieves this through developing a plan with patients and loved ones to allow patients to "live as well as they can, as long as they can, while still accomplishing their overall goals of care" [34]. PP discussions are similar to traditional ACP though with greater weight paid to potential complications. This difference has evolved over time. Traditionally an ACP conversation discussed things like "what does a good day look like for you?" and "what makes life worth living?". These open prompts may then frame a discussion around specific treatment plans. In PP, the discussion should share that sentiment but also include a more narrow focus on frequently seen complications. This has led to increased advance directive completion [35] (Table 43.1).

Table 43.1 Common differences between traditional advance directives and preparedness plans in patients receiving LVAD as destination therapy

Measure to be considered	Advance directive	Preparedness plan
Antibiotics' long-term role	+	++
Artificial nutrition	+	++
Blood transfusions	+	++
Goals and expectations	–	++
Hemodialysis	+	++
Hydration	+	++
Intracranial hemorrhage	–	++
LVAD failure	–	++
LVAD infection	–	++
Organ donation	++	++
Mechanical ventilation	++	++
Postoperative plans for rehabilitation	–	++
Power of attorney appointed	++	++
Psychosocial assessment	–	++
Review of perioperative morbidity and mortality	–	++
Social dynamics reviewed	–	++
Spiritual and/or religious preferences	++	++
Stroke	–	++

– = not generally found in document; + = may be found in document; ++ = often found in document. LVAD = left ventricular assist device

Reprinted from Mayo Clinic Proc, Vol 86, Ed 6, Swetz KM, Freeman MR, Abou Ezzeddine OF, Carter KA, Boilson BA, Ottenberg AL, et al. Palliative medicine consultation for preparedness planning in patients receiving left ventricular assist devices as destination therapy, 495, June 2011

Discussing complications can be difficult and has historically been a reason for avoiding palliative care intervention [34]. The patient has been told that DT-LVAD will improve the quality of their life while extending their life and then is met with a palliative care provider who offers unsettling descriptions of potential adverse effects, like stroke, bleeding, infection, or device failure to name a few. Anecdotally, patients and their loved ones are rarely keen to discuss these matters. However, given the severity of the risks, it is imperative that they have a firm understanding of the more common complications. A "prepare for the worst, hope for the best," communication technique has been used for these types of difficult conversations [36]. Although other authors have found that patients seem most receptive by normalizing the discussion [34]. In either case, it should focus on both device malfunctions and complications and what to do in the event of debilitating comorbid conditions. Overall, the underlying drive should be to elucidate the patient's wishes in the event of a complication which does not improve [34]. Many of these complications will also be brought up by the surgical team as part of their consent procedure. However, it is imperative that palliative care clinicians have a thorough understanding of prognosis, symptom burden, complications, and caregiver needs associated with MCS. Both teams need knowledge and fundamentals of each other's practice as this fosters trust in patients and trust between the team [37].

ACP and PP have traditionally been done through mutual discussion and teach back to ensure informed consent. There are published forms to guide caregivers in these discussions [34]. However, recently there has been recognition that such relationships have an inherent power imbalance which may affect patient decision-making [38]. The DECIDE-LVAD trial recently confirmed this finding in the heart failure population. In a randomized, multicenter trial, the investigators found that the use of a shared decision-making tool – a pamphlet and video – led to "significantly better knowledge and higher concordance between stated values and patient-reported treatment choice" [39].

In short, the PP should help align care of the medical team and the patient's caregivers with the patient and his/her goals of care. Like traditional advance care documents, it should document a medical power of attorney, address organ donation, and address goals of care generally with incorporation of spiritual or religious preferences, but it should also focus on disease-/device-specific complications. The DT-LVAD may provide patients with a new lease on life, but it is important to avoid potential future suffering for patients and families by addressing their goals before implantation.

Complications of LVADs and Opportunities for Palliative Care Involvement

Palliative care teams should know and understand the potential complications of LVADs. The complications must be reviewed with the patient and family during preparedness planning prior to implant, but also for continued care of these patients post-implantation. Fortunately, in the recent VAD era, the overall adverse events are significantly lower, although increasing events have been noted for hemolysis, stoke, renal dysfunction, and respiratory failure [22].

Multisystem organ failure and right heart failure are more prevalent complications and causes of death occurring early after implantation. Later complications and mortality result from infection, multisystem organ failure, and neurological events [22]. While complications can occur, the 1- and 2-year survival with continuous flow LVADS are currently 80% and 70%, respectively [22]. "Longitudinal palliative care" may benefit these patients as they experience repeat hospitalizations, worsening symptom burden, and caregiver distress [40].

LVAD Complications

Bleeding

Bleeding remains the most common adverse-related event in continuous flow (CF) VADS [23, 41]. Bleeding is often characterized as occurring either <30 days (early) or >30 days (late) to differentiate postoperative bleeding from nonsurgical bleeding, and early bleeding is notably associated with decreased survival [42].

Infection

"Device-related infections remain the 'Achilles heel' of MCS" [43]. Infections are one of the leading causes of readmissions for VAD patients [44]. Infection may occur with VAD implantation or develop later and can occur at the driveline site, pump pocket, pump and systemically [45, 46]. VAD-related infections may lead to infective endocarditis and bacteremia [45]. Developing any infection is correlated with decreased survival [46]. Recurrent or life-threatening infections should prompt the care team to engage in ongoing goals of care discussions to ensure that patient preferences are being addressed as device-related infections are a risk factor for hospitalizations and further operative interventions [47]. The authors have found that some patients who are developing multi-drug-resistant driveline infections may opt for hospice care rather than readmission to the hospital when sepsis develops.

Neurological Events

A leading cause of death among patients with continuous flow VADs is hemorrhagic stroke [23]. Other neurologic events may include ischemic stroke and transient ischemic attacks [42]. Neurologic events can occur suddenly with devastating results. The lack of a prodrome and the significant associated morbidity and mortality require that patient's goals of care in the event of a cerebrovascular accident are made clear prior to implantation. Asking patients what quality of life means to them and under what circumstance they would wish to have their VAD deactivated in the event of a disabling stroke should be part of the initial palliative medicine consult.

Device Malfunction

There are various types of device malfunction including pump thrombosis, controller failure, battery failure, driveline fracture, and impeller pump failure [48]. These events can be serious and often require reoperation.

Pump thrombosis represents the most severe and acute complications of VAD device malfunctions [49]. If device replacement is needed, it is usually related to device thrombosis and infection [48]. INTERMACS seventh annual report shows that with each subsequent pump exchange there is a decrease in survival [22]. Patients who encounter more VAD complications feel more dependent on the MCS team and may perceive worse quality of life [50]. Device thrombosis may lead to device exchange, another surgical intervention. The authors have experienced that patients may or may not want to undergo additional surgeries. Patients who have incurred multiple complications may wish to forgo additional operative care or this may not be an option. Palliative care should continue ongoing discussions of goals and patient preferences. However, the palliative care team should always include the MCS team before engaging in additional goals of care to ensure all strategies have been considered when managing complications [40].

Right Heart Failure and Cardiac Arrhythmias

Right ventricular function is imperative for successful outcomes as the right ventricle must provide flow through the pulmonary vasculature to fill the LVAD [51]. Right ventricular dysfunction is associated with higher mortality and higher lengths of ICU and hospital stays [52]. Right ventricular failure could mean extended inotrope therapy and placement of additional right ventricular assist device [23]. Palliative care teams should be aware that right ventricular failure could be a harbinger of increased mortality and remain actively involved with these patients as their families may require more supportive counseling during the patient's ICU stay.

Patients with LVAD implantation can have both supraventricular and ventricular arrhythmias [53]. Studies are mixed on how arrhythmias may affect survival but they have been shown to increase readmission rates [54, 55].

Additional Complications

Less frequently seen complications include hepatic failure, pneumonia, ischemic bowel disease, and GI bleeding [56] (Fig. 43.2).

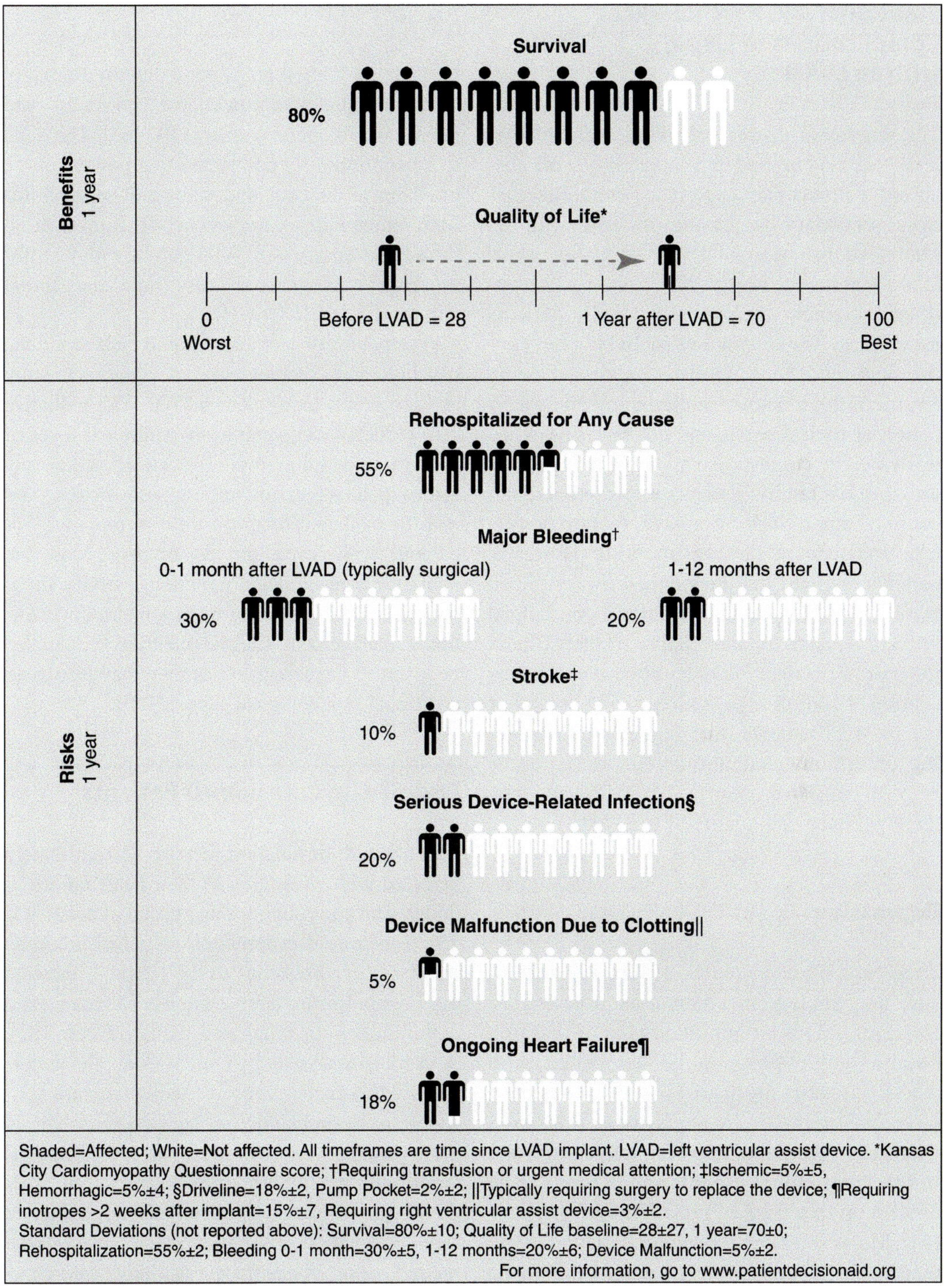

Fig. 43.2 The most common complications for patients with LVADs. (Reprinted from Journal of Pain and Symptom Management, Vol 54, Ed 4, Sara E. Wordingham, Colleen K. McIlvennan, Timothy J. Fendler, Amy L. Behnken, Shannon M. Dunlay, James N. Kirkpatrick, Keith M. Palliative Care Clinicians Caring for Patients Before and After Continuous Flow-Left Ventricular Assist Device, p 604, Oct 2017, with permission from Elsevier)

Emotional and Psychological Complications of Living with an LVAD

The emotional distress of living with MCS is less easily identified but frequently still disabling. Patients may experience emotional distress secondary to having to live with an implanted mechanical device and the uncertainty that device brings [57]. Zambroski identified common themes that VAD patients experience. These emotions included fear, confinement, and hope. Patients expressed being fearful of the unknown, anticipating an exacerbation of their disease, and not knowing when this would occur and also a fear of being a burden to their family. Patients expressed feeling confined due to lack of privacy, feeling powerless, and reduced physical mobility. However, patients reported that they coped by continued hope for the future and that this hope helped them to manage the uncertainty of their illness [58]. Claire Hallas PhD identified that the perception of control is key to improving quality of life in MCS patients. She suggests that reducing uncertainty and normalizing experiences help to improve control in MCS patients, thereby improving quality of life [59].

Depression

There are limited studies evaluating depression and anxiety in continuous flow VADs. However, this is an important area of research because depression can be associated with worse outcomes in heart failure patients and increased risk of readmissions [60, 61]. Clinically significant depression occurs in one in five heart failure patients and is found to be present up to a third of pre-DT-LVAD patients [61]. However, Reynard and colleagues found that independent of psychotropic medications, depression and anxiety scores improve after VAD implantation and appear to remain stable up to 1 year [62].

Anxiety

Patients of VAD therapy have reported anxiety over potential complications, survival, care needs, and financial concerns [32, 63]. The need to be continuously connected to a power source has been a concern and source of anxiety for both patients and caregivers [64]. Bunzel et al. found that spouses of VAD patients may suffer more psychological distress than the actual patient [65].

The palliative care team should understand the psychological complexities of destination therapy and long-term living with a VAD. The palliative care clinician can serve as an additional layer of support, in addition to the MCS team, to actively screen patients for depression and anxiety and help to medically manage these symptoms. For patients with severe anxiety or depression, the palliative care clinician may help identify those patients needing referral to psychiatry for additional management. Careful attention should also be given to the caregivers as they may also need additional psychological support [65].

End-of-Life Care in VAD Patients

Patients with destination therapy will ultimately be faced with worsening of their heart failure or underlying comorbidities and eventual death [21, 23]. The overall progression of patient's illness will bring readmissions to the hospital, increasing symptom burden, caregiver fatigue, and opportunities for integrated palliative care [37]. According to a study by Dunlay et al., the majority of MCS patients receive palliative care consultation pre-implantation, but less than half of these patients were seen in palliative care consultation within 1 month of their death [66]. This data supports that despite mandatory palliative care consultation pre-implant, there appear to be missed opportunities for the palliative care clinicians to address increasing symptom burden and continued goals of care discussions with patients and families.

Illness Trajectories

Understanding the illness trajectories of VAD patients may help the MCS support team and palliative care team identify patients who may benefit from earlier palliative care support. Rizzieri et al. proposed three different disease trajectories for patients nearing end of life with LVADs. The first trajectory is one with very short survival time post-implantation due to postoperative complications; these patients die within 90 days of device implantation [67]. These patients are hospitalized and therefore have easy access to an inpatient palliative care team for symptom management, additional psychosocial support for them and their families, and continued discussions of goals of care as complications arise.

The second trajectory groups patients who survive greater than 90 days after VAD implantation but experience no significant improvement in quality of life or survival compared to their medically managed heart failure cohort. This cohort lacks easily identifiable triggers for palliative care intervention but would likely benefit from ongoing outpatient palliative care follow-up and evaluation of evolving symptom burden.

The last trajectory described by Rizzieri is one of initial improvement in both quality of life and survival but then progression of a gradual decline which occurs as a result of late complications such as infection, sepsis, and neurologic events [67]. Progression of other comorbid conditions such as malignancy or other non-cardiac end-organ dysfunction can lead to this general decline. This subgroup of patients may need to be identified by the MCS team as the team appreciates these changes and refers these patients to re-establish care with the palliative care team.

Dunlay at el. described yet a fourth trajectory for patients. In this trajectory, patients are also experiencing improved quality of life and survival, but an abrupt and unexpected acute event occurs, for example, intracerebral hemorrhage, resulting in an abrupt death [66]. These patients are most likely hospitalized after the event. The crisis of this hospitalization provides an opportunity for the MCS team and palliative care clinician to review and readdress goals of care if necessary. When an abrupt event occurs that dramatically alters quality of life, patients, or their surrogates, may elect to deactivate the LVAD. Palliative care provides an additional layer of support to the multidisciplinary team by offering expert end-of-life symptom control for these patients. Abrupt death can result in complex grief for caregivers and family [68]. Palliative care can support the caregivers or family by referral for grief and bereavement support. Unfortunately, these trajectories do not provide greater prognostication, but contextualizing a patient's care within these trajectories may enable the medical teams to be proactive and anticipate patient and caregiver needs (Fig. 43.3).

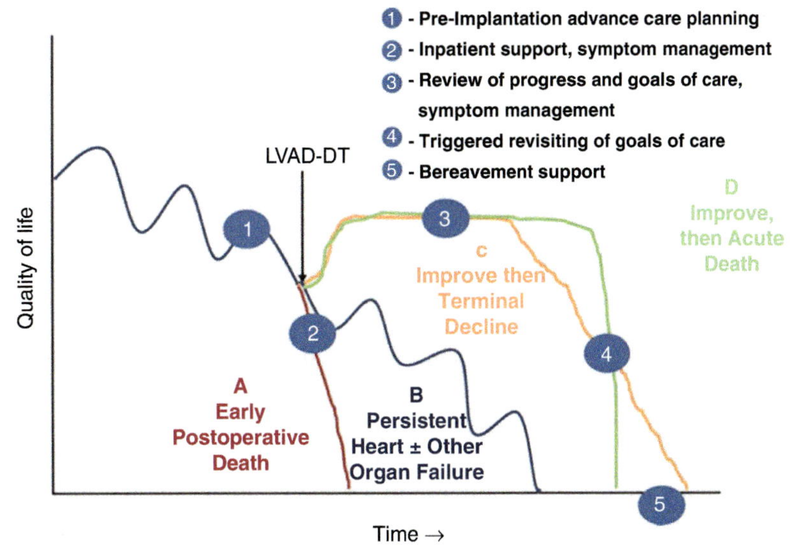

Fig. 43.3 Timeline for role of palliative care providers supporting patients with DT-LVAD. (Reprinted from Journal of Pain and Symptom Management, Vol 54, Ed 4, Sara E. Wordingham, Colleen K. McIlvennan, Timothy J. Fendler, Amy L. Behnken, Shannon M. Dunlay, James N. Kirkpatrick, Keith M. Palliative Care Clinicians Caring for Patients Before and After Continuous Flow-Left Ventricular Assist Device, p 605, Oct 2017, with permission from Elsevier)

Communication and End-of-Life Care

Any clinician delivering care to patients with advanced cardiac technologies must recognize that they have a responsibility to initiate and continue ongoing discussions regarding goals and preferences of the patient or family including device deactivation [69]. This can be a challenge as, unlike cancer, heart failure trajectories are difficult to predict, thereby making prognostication uncertain and physicians reluctant to discuss end-of-life care [70, 71]. However, patients with heart failure and their families prefer open conversations regarding prognosis and want their physicians to initiate end-of-life conversations. Having these conversations are not harmful nor do they destroy hope [72, 73]. Patients are more likely to consider end-of-life issues when they are hospitalized but may not be well enough to fully participate in these conversations or absorb information [74]. Unfortunately, the timing of end-of-life discussions by the healthcare team is often deferred until an emergency when thoughtful decision-making may be impaired [75]. Optimal timing of these discussions is in the ambulatory setting [75]. Yet, due to the unpredictable course of destination therapy, these conversations may not have occurred or may need to be readdressed when acute events occur.

In their article, "Comprehensive Care for Mechanical Circulatory Support: A New Frontier for Synergy with Palliative Care, " published in the July 2011 edition of *Circulation: Heart Failure* [28], Goldstein et al. proposed a conversation guide for clinicians caring for MCS patients to improve communication with patients and their caregivers. Goldstein includes helpful tables to improve communication with mechanical circulator support patients and their caregivers. The authors of this chapter have found this guide very helpful in communicating with VAD patients throughout their illness trajectory. Goldstein's steps include determining what the patients and caregivers understand about their illness and even more importantly what information the patient and caregivers want to know. He then recommends clinicians identify and address any misconceptions and answer specific questions that the patient or family may have. The last several steps include helping the patient express their overall goals of care and working with the patient and family to incorporate goals into a treatment plan that is aligned with medically achievable goals. He encourages the use of the "talk back" approach, having the patient and family reiterate the conversation and plan of care to ensure the patient and family have an accurate understanding. Lastly, he recommends to write this plan down so that it can be reviewed at a later time. This plan of communication can be repeated any time there may be a clinical change in condition and used to continually readdress the patient's goals of care.

Overall, open-ended questions like "What is most important to you now? What are your goals over the next few weeks or months?" serve as conversation starters [76]. By having these conversations, patients are more likely to receive the care that aligns with their own preferences and goals [77, 78]. Additionally, bereaved family members are better able to cope with the loss of their loved one and have less regret, better overall quality of life, and less risk of depression if end-of-life conversations occur [79] (Fig. 43.4a, b).

Deactivation of LVADS and End-of-Life Care

Ethical Considerations

The Patient Self-Determination Act of 1991 established and supports the rights of patients to refuse or withdraw medical treatment [80]. According to the American Heart Association Consensus Statement, patients have the right to request that their VAD be deactivated [75]. While deactivation of a device may seem different from withdrawing of other life-sustaining treatments like mechanical ventilation or dialysis, it has been argued that deactivation is similar and ethically permissible [81]. Swetz et al. surveyed physicians regarding VAD deactivation and found that the majority of physicians perceived VAD support as life-sustaining. Most physicians attrib-

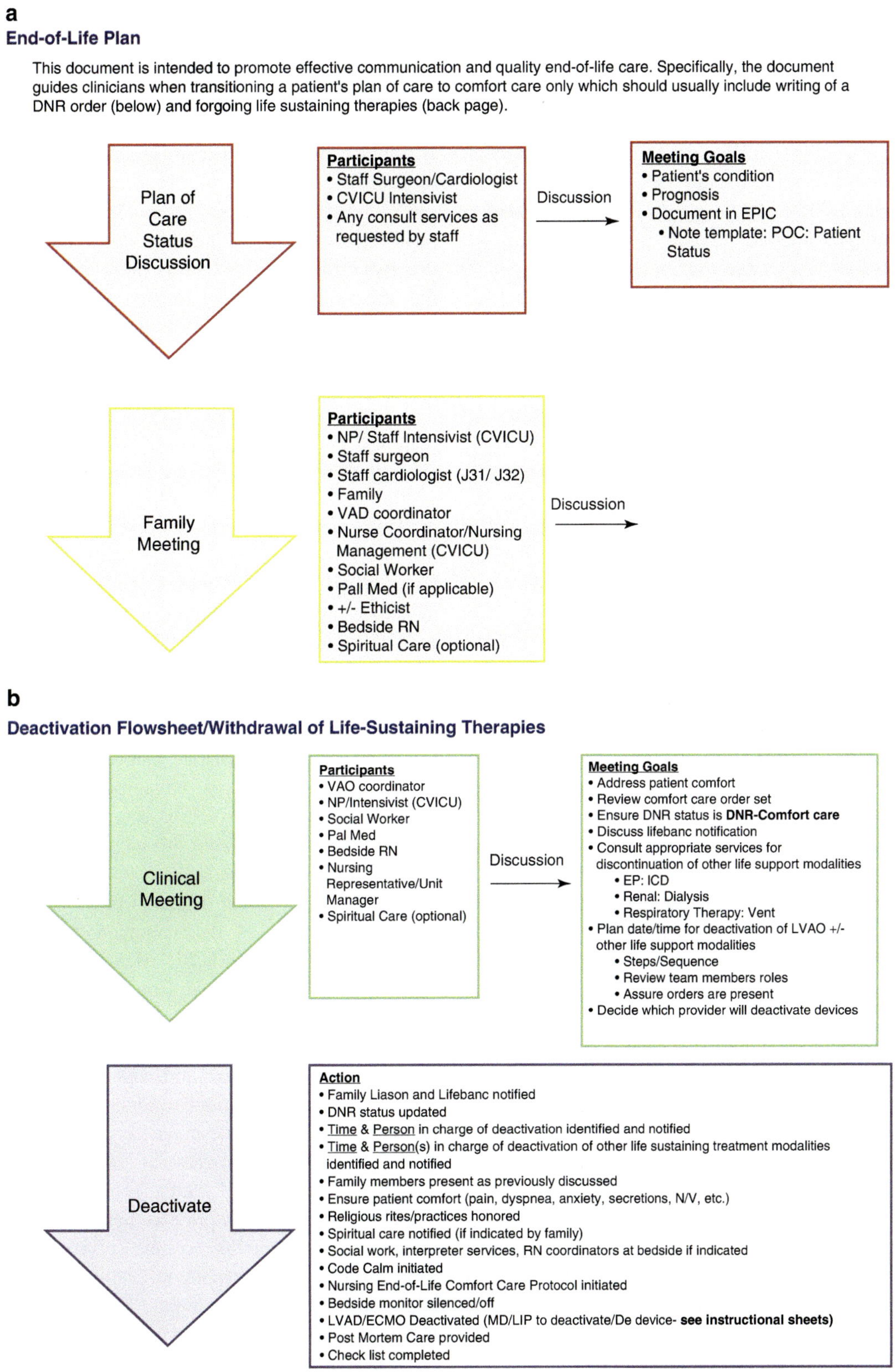

Fig. 43.4 (**a**) End-of-life plan; (**b**) Deactivation flowsheet/withdrawal of life-sustaining therapies

uted the actual death from VAD deactivation as resulting from the underlying disease. However, there are still moral tensions surrounding the management of LVADs in dying patients. Swetz's survey found 13% of physicians equated VAD deactivation with euthanasia [82]. One ethical argument is turning off the device may hasten death as blood pools in the VAD and can lead to pump thrombosis or regurgitant flow into and out of the VAD exacerbating heart failure [83]. However, other ethicists have argued that VAD deactivation is not euthanasia because the intent of deactivation is ending a treatment which is preventing progression of the underlying disease process [24]. Clinicians who object to withdrawing or deactivating devices should not be compelled to do so. Clinicians should inform patients of their objection and if necessary transfer the patients care to other clinicians [84, 85]. LVAD deactivation at end of life has now become an accepted practice in the United States [32].

VAD Deactivation

When is it appropriate to deactivate the LVAD? Currently there are no specific recommendations directing when therapy should be discontinued, but common triggers may include declining quality of life, signs of other organ failure, and irreversible adverse event such as major stroke or hemorrhage [75]. The patient or surrogate may request deactivation when *they* determine the device is no longer a benefit in prolonging their life [63]. Interestingly, when Swetz et al. surveyed clinicians, most believed that psychiatric consultation or ethics consultants should be considered before carrying out a request to deactivate the VAD but only 21% had ever requested a psychiatric consult and only 26% had requested an ethics consult [82].

The lack of guidelines for when to consider LVAD deactivation is reflected in the differing attitudes of cardiologists and palliative physicians. McIlvennan et al. conducted a survey of cardiologists and hospice and palliative medicine (HPM) physicians to gain insight on their perspectives regarding LVAD deactivation. She reported that of the responding physicians, 42% of the cardiologists and 57% of the HPM physicians participated in care of patients during the dying process and requesting deactivation of their devices. She found a majority (92%) of HPM physicians were comfortable with this request, but only a minority of cardiologists (26%) felt comfortable. The majority of HPM physicians agreed that the patient's request to deactivate their VAD should be honored even if they are not actively dying. However only 57% of cardiologists felt that this request should be honored. Cardiologist were more inclined to have the perspective that patients should continue the dying process from their comorbidities rather than deactivate the device. More than half of the cardiologists thought a patient should be imminently dying to deactivate the VAD, but only 2% of HPM physicians shared this perspective. Only 3% of HPM physicians versus 17% of cardiologists reported actually refusing to deactivate a device. Many of the cardiologists believed that a physician should be present to deactivate the VAD, but less than half of the HPM physicians thought an MD should be present. This study highlights the discordance regarding LVAD deactivation between care teams and suggests that "bridging the gaps and engaging in dialog between these two specialties is a critical first step in creating a more cohesive approach to care for LVAD patients" [86].

Dunlay et al. studied deaths in those patients receiving LVADS as destination therapy. 89 patients implanted with DT-LVADS who died at Mayo Clinic were reviewed. Common causes of death included multi-organ failure, hemorrhagic stroke, and heart failure. The majority of patients died in the intensive care unit, and most transitioned to comfort care prior to their death. Most patient's VADs were deactivated, and patients died within 26 hours of deactivation. Unfortunately, only half the patients saw palliative care in consultation prior to their death with very few ever receiving hospice services [66]. This illustrates deficiencies in end-of-life care opportunities for these patients. Despite palliative care involvement, there were low rates of referral to hospice care which means families do not benefit from hospice provided grief and bereavement services after death. The dearth of hospice involvement is unlikely solely due to patient preference, given integration of hospice services in other terminal processes [87]. It is more likely reflective of a need for ongoing goals of care assessments beyond the initial preparedness planning consult.

Actual VAD Deactivation Process and Checklist

LVAD deactivation processes differ between institutions and settings [86]. Successful deactivation of LVADs has been reported in both the home and hospital setting [84, 88, 89]. There have been checklists published to assist with LVAD deactivation [90, 91]. The Cleveland Clinic meeting and device checklist for successful VAD deactivation is outlined below.

Plan of Care Status Discussion

This is a multidisciplinary meeting including members of the patient's healthcare team to discuss prognosis and recommend a plan of care prior to meeting with the patient and family. Participants may include the cardiac surgeon, cardiologist, and relevant consult services such as a nephrologist if patient is on dialysis, infectious disease, neurologist, pulmonologist, etc. It is imperative to include nursing in this meeting including any advanced practice nurses and bedside nurses. This ensures that the entire team is aligned with the plan of care and delivers consistent messaging to the patient and family. A multidisciplinary meeting note is then entered into the electronic medical record.

The Family Meeting

At this meeting prognosis is explained and patient preferences and goals of care are discussed. Advanced directives are reviewed as well as code status to confirm do not resuscitate (DNR) and do not intubate if the patient is not previously intubated. The deactivation process is reviewed as well as withdrawal of other life-sustaining measures such as mechanical ventilation, dialysis, pressors, and implantable cardioverter defibrillator deactivation. Comfort measures are explained and common end-of-life symptoms reviewed. The patient and family are reassured that the team will aggressively manage any symptoms and ensure patient comfort. The patient and family can bring forth any questions or concerns they may have for the team to address. Logistically a time is set for deactivation, it is known who will be present at the time of deactivation, important religious rites are discussed, and organ donation is addressed if requested. During this meeting it is important for families to understand that variability in time to death can occur with VAD deactivation. Specifically, the family and patient should be aware that death may occur in minutes of VAD deactivation or patients may survive a few days. This will help families from becoming upset should the patient die sooner or live longer than expected [88, 91]. A family meeting note is included in the electronic medical record and DNR order placed.

Medical Deactivation

At our institution the device may be deactivated by a physician or an advance practice nurse from our MCS team. A purple heart is placed at the unit desk and outside the patient's room to alert healthcare providers and staff to maintain a sensitive, caring, and respectful attitude. A spiritual care consult should be offered to the patient and family. Comfort care order sets and protocols are initiated to help medically manage symptoms such as dyspnea, agitation, increased secretions, and pain. VAD deactivation can lead to the same end-of-life symptoms as a patient with advanced heart failure such as dyspnea, chest pain, and anxiety [88]. It should be noted that with device deactivation and drop in cardiac output, medications may not circulate well after the device is turned off. Therefore, comfort medications should be given in adequate doses prior to deactivation to ensure comfort [91]. When medications have reached their peak effect, the VAD is deactivated and other life-sustaining support withdrawn.

Time of Patient's Death

At the time of the patient's death, our institution performs *The Pause. The Pause*, developed by Jonathan Bartels, RN, is a 15–30 second period of silence at the time of a patient's death to honor both the patient and healthcare team. See script below.

The Pause Cleveland Clinic Script

Let us take a moment to pause and honor (patient's name or this person.) He/she was someone who loved and was loved; was someone's family member and friend. In our own way and in silence let us take a moment to honor (patient's name). Let us also honor and recognize the care provided by our team.

Where Do LVAD Patients Die?

Both patients implanted with a VAD for either bridge to transplant or destination therapy are most likely to die in the hospital, and the majority of these patients die in the ICU [92]. This is noteworthy as for many patients, their preferred site of death is home [93]. Also troubling is that ICU deaths have been associated with increased risk of psychiatric illness and post-traumatic stress in bereaved caregivers as well as poorer quality of end-of-life care for patients [94].

Hospice care is a team approach for patients with life-limiting illness and focuses on providing care to both the patients and their loved ones. The hospice philosophy is patients should die with dignity and aims to address physical, emotional, and spiritual care. Hospice provides bereavement and counseling for surviving family and friends [95].

However, heart failure patients in general are more likely to be enrolled in hospice care only 3 days prior to their death [96]. Hospice referrals for heart failure patients are more likely to come from acute care hospitals and skilled nursing facilities indicating that the healthcare team is generating this referral as opposed to the patient or family request for hospice services [96]. Late hospice referrals are correlated with unmet needs for quality end-of-life care and families receiving limited information regarding the patient's condition and what to expect with the dying process [97].

Groniger et al. surveyed 23 hospices in the Washington, D.C., Maryland, and Virginia tristate area regarding hospice enrollment for VAD patients. They discovered that most responders felt then when referrals occurred it was too late. Barriers identified to enrolling VAD patients in hospice care were coordination of inotropes, prognostic challenges, and lack of knowledge of MCS. Nearly 90% of responders wanted training from advanced heart failure specialists to improve their clinical knowledge to care for VAD patients.

Caregiver End-of-Life Perceptions

Perhaps the best way to improve end-of-life care for VAD patients is to examine the perceived experience from the bereaved caregiver perspective. McIlvennan et al. conducted in-depth interviews with bereaved caregivers. She discovered that most caregivers felt surprised, anxious, and confused about the dying process and end-of-life medical decisions. Caregivers expressed feeling ill-informed when it came to the discussion of VAD deactivation. For families who met with the palliative care team or transitioned to hospice care, they expressed feeling uncomfortable for they perceived hospice and palliative care to be uneducated with regard to caring for LVAD patients. Sadly, caregivers reported a sense of abandonment when the palliative care or hospice team became their primary team [98].

The Future of Palliative Care for LVAD Patients

Projections show that the prevalence of HF will increase 46% by 2030, resulting in greater than eight million people diagnosed with HF. As this number increases so will the use of mechanical circulatory support. Patients implanted with an LVAD for destination therapy are surviving longer but will eventually die from either their worsening heart failure, device complication, or underlying comorbid conditions. We must improve end-of-life care for these patients by offering them palliative medicine teams and hospice staff that understand care for the MCS patients.

Both cardiologists and palliative care clinicians must become proficient and comfortable with ongoing goals of care discussions, advanced care planning, and discussing patient preferences. Conversations prior to VAD implant for advanced directives and preparedness planning are only the beginning of many conversations. This should be an iterative process occurring when patients are well and readdressed as patients decline. These discussions should include when and under what circumstances the patient would want their device deactivated. Also, these discussions and decisions should be done in conjunction with the patient's medical power of attorney and loved ones to ensure that the patient's wishes are understood by their caregivers.

Teuteberg et al. suggest the MSC team should enhance their knowledge of simple palliative care symptom management and engage in code status and goals of care discussions. If there are refractory symptoms, or family and patient conflicts regarding goals of care, this should trigger referral to a palliative care clinician [99].

In order to accomplish better end of life, palliative medicine clinicians should remain embedded with the MCS team, offer outpatient palliative care visits to manage increasing symptom burden, and remain engaged in goals of care discussions throughout the patient's illness trajectory. By offering this longitudinal care, the palliative medicine and hospice clinician will no longer be a stranger stepping in at the last minute to care for dying patients. This will decrease the feelings of abandonment and confusion that bereaved caregivers experience when loved ones are dying and ensure smoother transitions of care from the VAD team to hospice. If these patients wish to die at home, we need to improve our home hospice support by educating the hospice multidisciplinary team and having the MCS team available to troubleshoot device questions and complications. A comprehensive approach and further research to understand the needs of patients dying on MCS will lead to better end-of-life care for the future generation of MCS patients.

References

1. Ferrari R, Balla C, Fucili A. Heart failure: an historical perspective. [cited 2018 Oct 23]; Available from: https://academic.oup.com/eurheartjsupp/article-abstract/18/suppl_G/G3/2633738.
2. Hardy JD, Kurrus FD, Chavez CM, Neely WA, Eraslan S, Turner MD, et al. Heart transplantation in man. Developmental studies and report of a case. JAMA [Internet]. 1964 [cited 2018 Oct 23];188:1132–40. Available from: http://www.ncbi.nlm.nih.gov/pubmed/14163110.
3. Stolf NAG. History of Heart Transplantation: a Hard and Glorious Journey. Brazilian J Cardiovasc Surg [Internet]. 2017 [cited 2018 Oct 23];32(5):423–7. Available from: http://www.ncbi.nlm.nih.gov/pubmed/29211224.
4. Prinzing A, Herold U, Berkefeld A, Krane M, Lange R, Voss B. Left ventricular assist devices-current state and perspectives. J Thorac Dis [Internet]. 2016 [cited 2018 Oct 15];8(8):E660–6. Available from: http://www.ncbi.nlm.nih.gov/pubmed/27621895.
5. McKee PA, Castelli WP, McNamara PM, Kannel WB. The natural history of congestive heart failure: the Framingham Study. N Engl J Med [Internet]. 1971 [cited 2018 Oct 13];285(26):1441–6. Available from: http://www.ncbi.nlm.nih.gov/pubmed/5122894.
6. Littman D. The natural history of congestive heart failure. N Engl J Med [Internet]. 1971 [cited 2018 Oct 13];285(26):1481–2. Available from: https://www-nejm-org.ccmain.ohionet.org/doi/pdf/10.1056/NEJM197112232852609.
7. Talner NS. Pulmonary arterial banding. Am J Cardiol [Internet]. 1970 [cited 2018 Oct 31];26(4):441–2. Available from: https://www.sciencedirect.com/science/article/pii/0002914970907472.
8. Saunders C. The evolution of palliative care. J R Soc Med [Internet]. 2001 [cited 2018 Nov 1];94(9):430–2. Available from: http://www.ncbi.nlm.nih.gov/pubmed/11535742.
9. Pajka SE. Doctors, death, and denial: the origins of hospice care in 20th century America [Internet]. Vol. 8. 2017 [cited 2018 Oct 31]. Available from: https://elischolar.library.yale.edu/mssa_yale_history; https://elischolar.library.yale.edu/mssa_yale_history/8.
10. Phillips D. Balfour Mount | Palliative Care McGill – McGill University [Internet]. [cited 2018 Oct 30]. Available from: https://www.mcgill.ca/palliativecare/portraits-0/balfour-mount.

11. Lutz S. The history of hospice and palliative care. Curr Probl Cancer [Internet]. 2011 [cited 2018 Oct 30];35(6):304–9. Available from: http://www.ncbi.nlm.nih.gov/pubmed/22136703.
12. Mercadante S, Simonetti MT. Dobutamine as palliative drug in home-care advanced cancer patients. J Pain Symptom Manage [Internet]. 1994 [cited 2018 Oct 15];9(7):480–3. Available from: https://www.sciencedirect.com/science/article/pii/0885392494902054.
13. Beattie JM, Murray RG, Brittle J, Catanheira T. Palliative care in terminal cardiac failure. Small numbers of patients with terminal cardiac failure may make considerable demands on services. BMJ [Internet]. 1995 [cited 2018 Oct 15];310(6991):1411. Available from: http://www.ncbi.nlm.nih.gov/pubmed/7540452.
14. Thorns AR, Gibbs LM, Gibbs JS. Management of severe heart failure by specialist palliative care. Heart [Internet]. 2001 [cited 2018 Oct 31];85(1):93. Available from: http://www.ncbi.nlm.nih.gov/pubmed/11119475.
15. Gibbs JSR, McCoy ASM, Gibbs LME, Rogers AE, Addington-Hall JM. Living with and dying from heart failure: the role of palliative care. Heart [Internet]. 2002 [cited 2018 Oct 15];88 Suppl 2(suppl 2):ii36–9. Available from: http://www.ncbi.nlm.nih.gov/pubmed/12213799.
16. Cleland JGF. Are symptoms the most important target for therapy in chronic heart failure? Prog Cardiovasc Dis [Internet]. 1998 [cited 2018 Oct 15];41(1):59–64. Available from: https://www.sciencedirect.com/science/article/pii/S0033062098800327.
17. Pritzker MR. Chronic heart failure and the quality of life. N Engl J Med [Internet]. 1999 [cited 2018 Oct 31];340(19):1511; author reply 1512. Available from: http://www.nejm.org/doi/abs/10.1056/NEJM199905133401913.
18. Lewis EF, Johnson PA, Johnson W, Collins C, Griffin L, Stevenson LW. Preferences for quality of life or survival expressed by patients with heart failure. J Hear Lung Transplant [Internet]. 2001 [cited 2018 Oct 15];20(9):1016–24. Available from: https://www.sciencedirect.com/science/article/pii/S1053249801002984.
19. Field D, Addington-Hall J. Extending specialist palliative care to all? Soc Sci Med [Internet]. 1999 [cited 2018 Oct 30];48(9):1271–80. Available from: https://www.sciencedirect.com/science/article/abs/pii/S0277953698004304.
20. Palliative Care Definition | What is Palliative Care | CAPC [Internet]. [cited 2018 Oct 30]. Available from: https://www.capc.org/about/palliative-care/.
21. Rose EA, Gelijns AC, Moskowitz AJ, Heitjan DF, Stevenson LW, Dembitsky W, et al. Long-term use of a left ventricular assist device for end-stage heart failure. N Engl J Med [Internet]. 2001 [cited 2018 Oct 28];345(20):1435–43. Available from: http://www.nejm.org/doi/abs/10.1056/NEJMoa012175.
22. Kirklin JK, Naftel DC, Pagani FD, Kormos RL, Stevenson LW, Blume ED, et al. Seventh INTERMACS annual report: 15,000 patients and counting. J Hear Lung Transplant [Internet]. 2015 [cited 2018 Oct 31];34(12):1495–504. Available from: http://www.ncbi.nlm.nih.gov/pubmed/26520247.
23. Slaughter MS, Rogers JG, Milano CA, Russell SD, Conte J V., Feldman D, et al. Advanced heart failure treated with continuous-flow left ventricular assist device. N Engl J Med [Internet]. 2009 [cited 2018 Oct 29];361(23):2241–51. Available from: http://www.nejm.org/doi/10.1056/NEJMoa0909938.
24. Dudzinski DM. Ethics guidelines for destination therapy. Ann Thorac Surg [Internet]. 2006 [cited 2018 Oct 31];81(4):1185–8. Available from: http://linkinghub.elsevier.com/retrieve/pii/S0003497505019612.
25. Nakagawa S, Yuzefpolskaya M, Colombo PC, Naka Y, Blinderman CD. Palliative care interventions before left ventricular assist device implantation in both bridge to transplant and destination therapy. J Palliat Med [Internet]. 2017;20(9):jpm.2016.0568. Available from: http://online.liebertpub.com/doi/10.1089/jpm.2016.0568.
26. Hupcey JE. The state of palliative care and heart failure. Heart Lung [Internet]. 2012 [cited 2018 Oct 31];41(6):529–30. Available from: http://www.ncbi.nlm.nih.gov/pubmed/23121830.
27. WHO | WHO Definition of Palliative Care. WHO [Internet]. 2012 [cited 2018 Oct 30]; Available from: http://www.who.int/cancer/palliative/definition/en/.
28. Goldstein NE, May CW, Meier DE. Comprehensive care for mechanical circulatory support a new frontier for synergy with palliative care. Circ Hear Fail. 2011;4(4):519–27. https://doi.org/10.1161/CIRCHEARTFAILURE.110.957241.
29. Warraich HJ, Xu H, DeVore AD, Matsouaka R, Heidenreich PA, Bhatt DL, et al. Trends in hospice discharge and relative outcomes among medicare patients in the get with the guidelines–heart failure registry. JAMA Cardiol [Internet]. 2018 [cited 2018 Oct 29];3(10):917. Available from: http://cardiology.jamanetwork.com/article.aspx?doi=10.1001/jamacardio.2018.2678.
30. Swetz KM, Freeman MR, AbouEzzeddine OF, Carter KA, Boilson BA, Ottenberg AL, et al. Palliative medicine consultation for preparedness planning in patients receiving left ventricular assist devices as destination therapy. Mayo Clin Proc [Internet]. 2011 [cited 2018 Nov 1];86(6):493–500. Available from: http://www.ncbi.nlm.nih.gov/pubmed/21628614.
31. Matlock DD, Nowels CT, Bekelman DB. Patient perspectives on decision making in heart failure. J Card Fail [Internet]. 2010 [cited 2018 Oct 29];16(10):823–6. Available from: http://www.ncbi.nlm.nih.gov/pubmed/20932464.
32. Swetz KM, Ottenberg AL, Freeman MR, Mueller PS. Palliative care and end-of-life issues in patients treated with left ventricular assist devices as destination therapy. Curr Heart Fail Rep. 2011;8(3):212–8. https://doi.org/10.1007/s11897-011-0060-x.

33. Slomka J, Prince-Paul M, Webel A, Daly BJ. Palliative Care, Hospice, and Advance Care Planning: Views of People Living with HIV and Other Chronic Conditions. J Assoc Nurses AIDS Care [Internet]. 2016 [cited 2018 Oct 29];27(4):476–84. Available from: http://www.ncbi.nlm.nih.gov/pubmed/27053406.
34. Swetz KM, Kamal AH, Matlock DD, Dose AM, Borkenhagen LS, Kimeu AK, et al. Preparedness planning before mechanical circulatory support: A "how-to" guide for palliative medicine clinicians. J Pain Symptom Manage. 2014;47(5):926–35.e6. https://doi.org/10.1016/j.jpainsymman.2013.06.006.
35. Schellinger S, Sidebottom A, Briggs L. Disease specific advance care planning for heart failure patients: implementation in a large health system. J Palliat Med [Internet]. 2011 [cited 2018 Oct 29];14(11):1224–30. Available from: http://www.ncbi.nlm.nih.gov/pubmed/21870958.
36. Back A, Arnold R, Quill T. Hope for the best, and prepare for the worst. Ann Intern Med [Internet]. 2003 [cited 2018 Oct 30];138(5):439–43. Available from: http://annals.org/aim/article-abstract/716134/hope-best-prepare-worst.
37. Wordingham SE, McIlvennan CK, Fendler TJ, Behnken AL, Dunlay SM, Kirkpatrick JN, et al. Palliative care clinicians caring for patients before and after continuous flow-left ventricular assist device. J Pain Symptom Manage [Internet]. 2017;54(4):601–8. Available from: https://doi.org/10.1016/j.jpainsymman.2017.07.007.
38. Fried TR. Shared decision making — finding the sweet spot. N Engl J Med [Internet]. 2016 [cited 2018 Oct 29];374(2):104–6. Available from: http://www.nejm.org/doi/10.1056/NEJMp1510020.
39. Allen LA, McIlvennan CK, Thompson JS, Dunlay SM, LaRue SJ, Lewis EF, et al. Effectiveness of an intervention supporting shared decision making for destination therapy left ventricular assist device: the DECIDE-LVAD randomized clinical trial. JAMA Intern Med [Internet]. 2018 [cited 2018 Oct 31];178(4):520–9. Available from: http://www.ncbi.nlm.nih.gov/pubmed/29482225.
40. Wordingham SE, McIlvennan CK, Fendler TJ, Behnken AL, Dunlay SM, Kirkpatrick JN, et al. Palliative care clinicians caring for patients before and after continuous flow-left ventricular assist device. J Pain Symptom Manage. 2017;54(4):601–8. https://doi.org/10.1016/j.jpainsymman.2017.07.007. Epub 2017 Jul 13.
41. Pagani FD, Miller LW, Russell SD, Aaronson KD, John R, Boyle AJ, et al. Extended mechanical circulatory support with a continuous-flow rotary left ventricular assist device. J Am Coll Cardiol [Internet]. 2009;54(4):312–21. Available from: https://doi.org/10.1016/j.jacc.2009.03.055.
42. McIlvennan CK, Magid KH, Ambardekar A V., Thompson JS, Matlock DD, Allen LA. Clinical outcomes after continuous-flow left ventricular assist device A systematic review. Circ Hear Fail. 2014;7(6):1003–13.
43. Allen JG, Weiss ES, Schaffer JM, Patel ND, Ullrich SL, Russell SD, et al. Quality of life and functional status in patients surviving 12 months after left ventricular assist device implantation. J Hear Lung Transplant [Internet]. 2010;29(3):278–85. Available from: https://doi.org/10.1016/j.healun.2009.07.017.
44. Smedira NG, Hoercher KJ, Lima B, Mountis MM, Starling RC, Thuita L, et al. Unplanned hospital readmissions after heartmate II implantation. Frequency, risk factors, and impact on resource use and survival. JACC Hear Fail [Internet]. 2013;1(1):31–9. Available from: https://doi.org/10.1016/j.jchf.2012.11.001.
45. Trachtenber BH, Cordero-Reyes A, Elias B, Loebe M. A review of infections in patients with left ventricular assist devices: prevention, diagnosis and management. Methodist Debakey Cardiovas J. 2015;11(1):28–32.
46. Goldstein DJ, Naftel D, Holman W, Bellumkonda L, Pamboukian SV, Pagani FD, et al. Continuous-flow devices and percutaneous site infections: clinical outcomes. J Hear Lung Transplant [Internet]. 2012;31(11):1151–7. Available from: https://doi.org/10.1016/j.healun.2012.05.004.
47. Maniar S, Kondareddy S, Topkara VK. Left ventricular assist device-related infections: past, present and future. Expert Rev Med Devices [Internet]. 2011 [cited 2018 Oct 30];8(5):627–34. Available from: http://www.ncbi.nlm.nih.gov/pubmed/22026627.
48. Kormos RL, McCall M, Althouse A, Lagazzi L, Schaub R, Kormos MA, et al. Left ventricular assist device malfunctions: It is more than just the pump. Circulation. 2017;136(18):1714–25.
49. Hanke JS, Dogan G, Wert L, Ricklefs M, Heimeshoff J, Chatterjee A, et al. Left ventricular assist device exchange for the treatment of HeartMate II pump thrombosis. J Thorac Dis. 2018;10(Suppl 15):S1728–36.
50. Modica M, Ferratini M, Torri A, Oliva F, Martinelli L, De Maria R, et al. Quality of life and emotional distress early after left ventricular assist device implant: a mixed-method study. Artif Organs. 2015;39(3):220–7.
51. Kormos RL, Teuteberg JJ, Pagani FD, Russell SD, John R, Miller LW, et al. Right ventricular failure in patients with the HeartMate II continuous-flow left ventricular assist device: Incidence, risk factors, and effect on outcomes. J Thorac Cardiovasc Surg [Internet]. 2010;139(5):1316–24. Available from: https://doi.org/10.1016/j.jtcvs.2009.11.020.
52. Kavarana MN, Pessin-Minsley MS, Urtecho J, Catanese KA, Flannery M, Oz MC, et al. Right ventricular dysfunction and organ failure in left ventricular assist device recipients: a continuing problem. Ann Thorac Surg. 2002;73(3):745–50.
53. Kadado AJ, Akar JG, Hummel JP. Arrhythmias after left ventricular assist device implantation: Incidence and management. Trends Cardiovasc Med [Internet]. 2018 [cited 2018 Oct 30];28(1):41–50. Available from: http://www.ncbi.nlm.nih.gov/pubmed/28734595.

54. Raasch H, Jensen BC, Chang PP, Mounsey JP, Gehi AK, Chung EH, et al. Epidemiology, management, and outcomes of sustained ventricular arrhythmias after continuous-flow left ventricular assist device implantation. Am Heart J [Internet]. 2012;164(3):373–8. Available from: https://doi.org/10.1016/j.ahj.2012.06.018.
55. Brenyo A, Rao M, Koneru S, Hallinan W, Shah S, Massey H, Chen L, Polonsky B, McNitt S, Huang D, Goldenber I AM. Risk of mortality for ventricular arrhythmia in ambulatory LVAD patients. J Cardiovsc Electrophysiol. 2012;23(5):515–20.
56. Slaughter MS, Pagani FD, McGee EC, Birks EJ, Cotts WG, Gregoric I, et al. HeartWare ventricular assist system for bridge to transplant: Combined results of the bridge to transplant and continued access protocol trial. J Hear Lung Transplant [Internet]. 2013 [cited 2018 Nov 1];32(7):675–83. Available from: http://www.ncbi.nlm.nih.gov/pubmed/23796152.
57. MacIver J, Ross HJ. Quality of life and left ventricular assist device support. Circulation. 2012;126(7):866–74.
58. Zambroski CH, Combs P, Cronin SN, Pfeffer C. Edgar Allan Poe, "the pit and the pendulum," and ventricular assist devices. Crit Care Nurse. 2009;29(6):29–39.
59. Hallas C, Banner NR, Wray J. A qualitative study of the psychological experience of patients during and after mechanical cardiac support. J Cardiovasc Nurs. 2009;24(1):31–9.
60. Moraska AR, Chamberlain AM, Shah ND, Vickers KS, Rummans TA, Dunlay SM, et al. Depression, healthcare utilization, and death in heart failure a community study. Circ Hear Fail. 2013;6(3):387–94.
61. Snipelisky D, Stulak JM, Schettle SD, Sharma S, Kushwaha SS, Dunlay SM. Psychosocial characteristics and outcomes in patients with left ventricular assist device implanted as destination therapy. Am Heart J [Internet]. 2015;170(5):887–94. Available from: https://doi.org/10.1016/j.ahj.2015.08.012.
62. Reynard AK, Butler RS, McKee MG, Starling RC, Gorodeski EZ. Frequency of depression and anxiety before and after insertion of a continuous flow left ventricular assist device. Am J Cardiol [Internet]. 2014;114(3):433–40. Available from: https://doi.org/10.1016/j.amjcard.2014.05.015.
63. Byram EK. Upstream palliative care for the patient with a left ventricular assist device as destination therapy. Dimens Crit Care Nurs. 2012;31(1):18–24.
64. Ben Gal T, Jaarsma T. Self-care and communication issues at the end of life of recipients of a left-ventricular assist device as destination therapy. Curr Opin Support Palliat Care. 2013;7(1):29–35.
65. Bunzel B, Laederach-Hofmann K, Wieselthaler G, Roethy W, Wolner E. Mechanical circulatory support as a bridge to heart transplantation: what remains? Long-term emotional sequelae in patients and spouses. J Hear Lung Transplant. 2007;26(4):384–9.
66. Dunlay SM, Strand JJ, Wordingham SE, Stulak JM, Luckhardt AJ, Swetz KM. Dying with a left ventricular assist device as destination therapy. Circ Hear Fail. 2016;9(10):e003096. https://doi.org/10.1161/CIRCHEARTFAILURE.116.003096.
67. Rizzieri AG, Verheijde JL, Rady MY, McGregor JL. Ethical challenges with the left ventricular assist device as a destination therapy. Ethics, Humanit Med [Internet]. 2008 [cited 2018 Oct 5];3:20. Available from: http://www.peh-med.com/content/3/1/20; http://www.peh-med.com/content/3/1/20.
68. Lobb EA, Kristjanson LJ, Aoun SM, Monterosso L, Halkett GKB, Davies A. Predictors of complicated grief: a systematic review of empirical studies. Death Stud [Internet]. 2010 [cited 2018 Oct 30];34(8):673–98. Available from: http://www.tandfonline.com/doi/abs/10.1080/07481187.2010.496686.
69. Matlock DD, Stevenson LW. Life-saving devices reach the end of life with heart failure. Prog Cardiovasc Dis. 2012;55(3):274–81. https://doi.org/10.1016/j.pcad.2012.10.007.
70. Goldstein NE, Lynn J. Trajectory of end-stage heart failure: the influence of technology and implications for policy change. Perspect Biol Med. 2006;49(1):10–8.
71. Hauptman PJ, Swindle J, Hussain Z, Biener L, Burroughs TE. Physician attitudes toward end-stage heart failure: a national survey. Am J Med. 2008;121(2):127–35.
72. Caldwell PH, Arthur HM, Demers C. Preferences of patients with heart failure for prognosis communication. Can J Cardiol. 2007;23(10):791–6. https://doi.org/10.1016/s0828-282x(07)70829-2.
73. Apatira L, Boyd EA, Malvar G, Evans LR, Luce JM, Lo B, et al. Hope, truth, and preparing for death: Perspectives of surrogate decision makers. Ann Intern Med. 2008;149(12):861–8. https://doi.org/10.7326/0003-4819-149-12-200812160-00005.
74. Barclay S., Momen N., Case-Upton S., Kuhn I. and Smith E. End-of-life care conversations with heart failure patients: a systematic literature review and narrative synthesis. Br J Gen Pract. 20011;61(582):e49–62.
75. Allen LA, Stevenson LW, Grady KL, Goldstein NE, Matlock DD, Arnold RM, et al. Decision making in advanced heart failure: a scientific statement from the American Heart Association. Circulation. 2012;125(15):1928–52.
76. Stuart B. Palliative care and hospice in advanced heart failure. J Palliat Med. 2007;10(1):210–28.
77. Silveira MJ, Kim SYH, Langa KM. Advance directives and outcomes of surrogate decision making before death. N Engl J Med. 2010;362(13):1211–8. https://doi.org/10.1056/NEJMsa0907901.
78. Kirchhoff KT, Hammes BJ, Kehl KA, Briggs LA, Brown RL. Effect of a disease-specific advance care planning intervention on end-of-life care. J Am Geriatr Soc. 2012;60(5):946–50.
79. Wright AA, Zhang B, Alaka Ray M, Mack JW, Elizabeth Trice M, Balboni T, et al. Associations between end-of-life discussions, patient mental health, medical care near death, and caregiver bereavement adjustment downloaded from: by a Cleveland Clinic

Foundation User on 10/24/2018 [Internet]. Vol. 300, JAMA. 2008. Available from: www.jama.com.
80. Panetta L. H.R.5835 – 101st Congress (1989–1990): Omnibus Budget Reconciliation Act of 1990. 1990 [cited 2018 Oct 30]; Available from: https://www.congress.gov/bill/101st-congress/house-bill/5835.
81. Trainor J, Caplan A. Left ventricular assist device–destination therapy for symptom management in heart failure. J Hosp Palliat Nurs [Internet]. 2012 [cited 2018 Nov 1];14(4):261–5. Available from: http://content.wkhealth.com/linkback/openurl?sid=WKPTLP:landingpage&an=00129191-201206000-00005.
82. Swetz KM, Cook KE, Ottenberg AL, Chang N, Mueller PS. Clinicians' attitudes regarding withdrawal of left ventricular assist devices in patients approaching the end of life. Eur J Heart Fail. 2013;15(11):1262–6.
83. Bramstedt KA, Wenger NS. When withdrawal of life-sustaining care does more than allow death to take its course: the dilemma of left ventricular assist devices. J Hear Lung Transplant. 2001;20(5):544–8.
84. Mueller PS, Swetz KM, Freeman MR, Carter KA, Crowley ME, Anderson Severson CJ, et al. Ethical analysis of withdrawing ventricular assist device support. Mayo Clin Proc [Internet]. 2010;85(9):791–7. Available from: https://doi.org/10.4065/mcp.2010.0113.
85. Lampert R, Hayes DL, Annas GJ, Farley MA, Goldstein NE, Hamilton RM, et al. HRS Expert Consensus Statement on the Management of Cardiovascular Implantable Electronic Devices (CIEDs) in patients nearing end of life or requesting withdrawal of therapy: this document was developed in collaboration and endorsed by the American College of Cardiology (ACC), the American Geriatrics Society (AGS), the American Academy of Hospice and Palliative Medicine (AAHPM); the American Heart Association (AHA), the European Heart Rhythm Association (EHRA), and the Hospice and Palliative Nur. Hear Rhythm [Internet]. 2010 [cited 2018 Nov 1];7(7):1008–26. Available from: https://www.sciencedirect.com/science/article/pii/S154752711000408X?via%3Dihub.
86. McIlvennan CK, Wordingham SE, Allen LA, Matlock DD, Jones J, Dunlay SM, et al. Deactivation of left ventricular assist devices: differing perspectives of cardiology and hospice/palliative medicine clinicians. J Card Fail. 2017;23(9):708–12. https://doi.org/10.1016/j.cardfail.2016.12.001. Epub 2016 Dec 5.
87. Aldridge MD, Canavan M, Cherlin E, Bradley EH. Has hospice use changed? 2000–2010 utilization patterns. Med Care [Internet]. 2015 [cited 2018 Oct 30];53(1):95–101. Available from: http://www.ncbi.nlm.nih.gov/pubmed/25373406.
88. Brush S, Budge D, Alharethi R, Mccormick AJ, Macpherson JE, Reid BB, et al. End-of-life decision making and implementation in recipients of a destination left ventricular assist device. HEALUN [Internet]. 2010;29:1337–41. Available from: http://www.jhltonline.org.
89. Panke JT, Ruiz G, Elliott T, Pattenden DJ, DeRenzo EG, Rollins EE, et al. Discontinuation of a left ventricular assist device in the home hospice setting. J Pain Symptom Manage [Internet]. 2016;52(2):313–7. Available from: https://doi.org/10.1016/j.jpainsymman.2016.02.010.
90. Schaefer KG, Griffin L, Smith C, May CW, Stevenson LW. An interdisciplinary checklist for left ventricular assist device deactivation. J Palliat Med [Internet]. 2014;17(1):4–5. Available from: http://www.ncbi.nlm.nih.gov/pubmed/24286202.
91. Gafford EF, Luckhardt AJ, Swetz KM. Deactivation of a left ventricular assist device at the end of life #269. J Palliat Med. 2013;16(8):980–2. https://doi.org/10.1089/jpm.2013.9490. Epub 2013 Jun 14.
92. Nakagawa S, Garan AR, Takayama H, Takeda K, Topkara VK, Yuzefpolskaya M, et al. End of life with left ventricular assist device in both bridge to transplant and destination therapy. J Palliat Med [Internet]. 2018;21(9):jpm.2018.0112. Available from: http://www.liebertpub.com/doi/10.1089/jpm.2018.0112.
93. Pritchard RS, Fisher ES, Teno JM, Sharp SM, Reding DJ, Knaus WA, et al. Influence of patient preferences and local health system characteristics on the place of death. J Am Geriatr Soc. 1998;46(10):1242–50.
94. Wright AA, Keating NL, Balboni TA, Matulonis UA, Block SD, Prigerson HG. Place of death: correlations with quality of life of patients with cancer and predictors of bereaved caregivers' mental health. J Clin Oncol. 2010;28(29):4457–64.
95. Preamble and Philosophy | National Hospice and Palliative Care Organization [Internet]. [cited 2018 Oct 31]. Available from: https://www.nhpco.org/ethical-and-position-statements/preamble-and-philosophy#2.
96. Cheung WY, Schaefer K, May CW, Glynn RJ, Curtis LH, Stevenson LW, et al. Enrollment and events of hospice patients with heart failure vs. cancer. J Pain Symptom Manage [Internet]. 2013;45(3):552–60. Available from: https://doi.org/10.1016/j.jpainsymman.2012.03.006.
97. Schockett ER, Teno JM, Miller SC, Stuart B. Late referral to hospice and bereaved family member perception of quality of end-of-life care. J Pain Symp Manag. 2005;30(5):400–7. https://doi.org/10.1016/j.jpainsymman.2005.04.013.
98. McIlvennan CK, Jones J, Allen LA, Swetz KM, Nowels C, Matlock DD. Bereaved caregiver perspectives on the end-of-life experience of patients with a left ventricular assist device. JAMA Intern Med. 2016;176(4):534–9. https://doi.org/10.1001/jamainternmed.2015.8528.
99. Teuteberg W, Maurer M, Neumann A, Death TG. Palliative care throughout the journey of life with a left ventricular assist device is there room for improvement in the care of. Circ Hear Fail. 2016;9(10):e 003564.

Part VII

Future Perspectives

Clinical Demands and Challenges for Future Mechanical Circulatory Support Technologies

44

Adam D. DeVore and Joseph G. Rogers

Introduction

The field of mechanical circulatory support (MCS) has changed rapidly over the past decade. One catalyst of this change was innovation in pump design. The HeartMate II (Abbott) device was first approved in 2008 for bridge-to-transplant therapy. Since that time, MCS devices have become predominantly continuous flow and decreased in size with improved hemocompatibility. In the MOMENTUM 3 trial, survival free of any hemocompatibility-related adverse events at 6 months after randomization was higher for the HeartMate 3 (Abbott) than the HeartMate II (69 ± 4% vs. 55 ± 4%; hazard ratio, 0.62; 95% confidence interval, 0.42 to 0.91) [1]. At the same time, the population of patients living with durable left ventricular assist devices (LVADs) has increased substantially. Presently, >2000 continuous flow LVADs are implanted in adults annually in the United States alone compared with 421 in 2008, and patients are living longer with the device today compared with 2008–2012 [2, 3].

As the field of MCS has matured, new challenges to delivering high-quality care for LVAD patients are emerging (Fig. 44.1). Some of these challenges are related to device evolution and technology, but many of these changes are inevitable growing pains of caring for a large population of aging patients with multiple comorbid conditions. Solutions to these challenges will be important for continuing to improve outcomes and expand the therapy to new populations.

A. D. DeVore, MD, MHS · J. G. Rogers, MD (✉)
Department of Medicine and Duke Clinical Research Institute, Duke University School of Medicine, Durham, NC, USA

Division of Cardiology, Duke University School of Medicine, Durham, NC, USA
e-mail: adam.devore@duke.edu; joseph.rogers@duke.edu

Anticipated Changes to MCS Design

Improvements in hemocompatibility are a remarkable achievement and represent years of biomedical and clinical research. However, significant changes to current devices are still necessary. Infection of the percutaneous driveline and power cord remains a common complication for patients supported on durable LVAD therapy, affecting approximately one in four patients in the first year after implant alone [2, 4]. Totally implantable power sources analogous to current implantable cardioverter defibrillators and pacemakers are possible but will require efficiencies in pump design (i.e., to utilize less energy) and advancements in implantable power sources.

© Springer Nature Switzerland AG 2020
J. H. Karimov et al. (eds.), *Mechanical Support for Heart Failure*,
https://doi.org/10.1007/978-3-030-47809-4_44

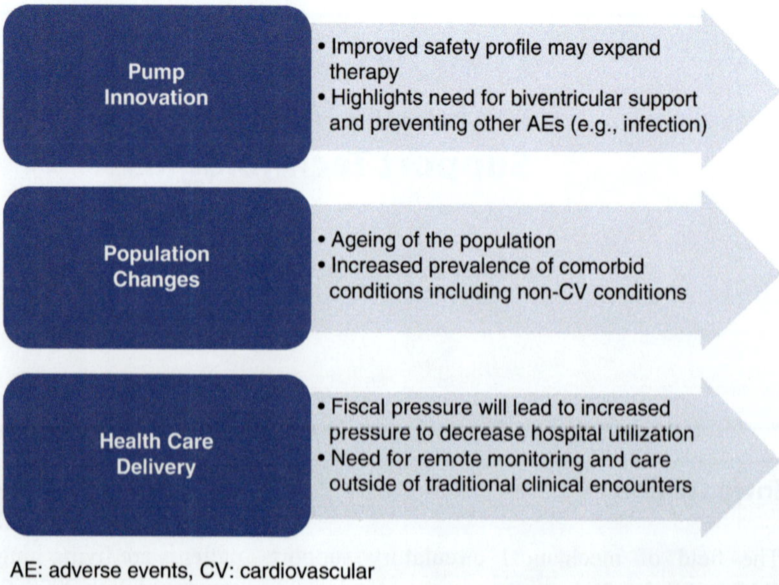

Fig. 44.1 Clinical demands and challenges for future mechanical circulatory support technologies

When fully implantable systems are available, new challenges are likely including determining what patients, if any, should be considered for conversion from external to internal power sources and how to establish sufficient backup systems for patients requiring full hemodynamic support from the device.

Another common complication that has not improved over time is heart failure that recurs late after device implant, typically from progressive right heart failure [5–7]. This complication leads to limitations in health status and is a common cause of rehospitalization. In the HeartMate II destination therapy trial, patients with late right heart failure had worse outcomes at 2 years compared to those without, including worse quality of life (mean Kansas City Cardiomyopathy Questionnaire overall summary score 51 ± 24 vs. 57 ± 26, $P < 0.05$), poorer functional capacity (mean 6-minute walk distance 265 ± 135 meters vs. 373 ± 303 meters, $P < 0.05$), increased rehospitalizations (median 6 [range 2–19] vs. 3 [range 0–27], $P < 0.001$), and decreased survival (58% ± 8 vs. 71% ± 2, $P < 0.001$) [7]. These data highlight the need for improved strategies to reduce the incidence of late right heart failure and for devices that can support the right ventricle, either partially or fully. As durable MCS for the right heart becomes more common, new challenges will emerge in terms of monitoring, including imaging of the supported right ventricle, and for processing large amounts of data for patients on two separate VADs with different controllers, speeds, etc.

Further innovations in MCS design are also likely to lead to a “smart pump” that incorporates physiologic data from intracardiac (e.g., pressure monitors) and extracardiac (e.g., activity monitors) sensors to set pump speeds based on need and estimated/measured filling pressures. While this concept holds great promise, these pumps are also likely to lead to increased complexity in terms of monitoring/calibrating with imaging and hemodynamics and monitoring of data.

Changes to LVAD Healthcare Delivery

In addition to more patients living with durable LVADs, two other factors are significantly impacting MCS healthcare delivery. First, the type of patient implanted has changed over time. Comparing patients implanted in 2012–2017 to those implanted in 2008–2011, patients implanted today are somewhat older (median age 60 years vs. 58 years, $P < 0.0001$), more likely to have severe diabetes (9.5% vs. 3.1%, $P < 0.0001$), more likely to have peripheral vascular disease (4.7% vs. 2.9%, $P < 0.0001$), and more likely to have a history of cancer (5.1% vs. 2.8%, $P < 0.0001$) [2]. The fact that patients with advanced heart failure are older today and with more comorbid conditions is consistent with aging of the population and data on patients hospitalized with acute heart failure [8]. Second, the number of LVAD implant centers has increased over time. In 2016, there were 185 centers enrolled in INTERMACS compared with 88 in 2009 [3, 9]. This has allowed for improved access to care for patients and families and expansion of the therapy to more geographic areas.

Managing LVAD Growth

Similar to other large advanced HF programs, over the last decade, we witnessed significant change in our LVAD volume. In a short amount of time, we began implanting >100 patients with durable LVADs annually and managed a few hundred outpatients living on LVAD support across multiple states. We recognized that different challenges are apparent at different-sized LVAD centers (Table 44.1). For example, a program that implants <25 patients per year requires significant resources to help patients navigate a complex health system, such as coordinating the care of a patient on support that develops a malignancy and some of the oncology team may be less familiar with LVADs. A program with higher volume may have less challenges identifying specialists for common medical problems due to increased exposure of specialists to LVADs in the healthcare system, but other challenges exist such as developing systems of care to manage warfarin for hundreds of outpatients.

Table 44.1 Challenges according to different-sized LVAD program

	LVAD program size	Example of unique challenge
A	<25 implants/year with approximately 50 outpatients on LVAD support	Navigating a complex health system with some clinicians inexperienced in managing patients with an LVAD
B	25–50 implants/year with approximately 100–200 outpatients on LVAD support	Appropriately scaling LVAD clinical team and building an LVAD research program
C	>50 implants/year with >200 outpatients on LVAD support	Managing a controller recall for hundreds of outpatients across a large geographic area

Preparation for New Payment Models in Complex Disease

The rising costs of healthcare in the United States are the focus of the public, payers, and the federal government. Increasingly physicians and health systems are being asked to demonstrate value [10, 11]. The concept of value is best captured considering the cost of achieving high-quality outcomes and service for patients. This simple construct allows one to begin considering opportunities to enhance value. For example, development and utilization of evidence-based practice, increasingly thoughtful use of lab testing and imaging studies, reduction of variation, and reducing adverse events will be critical components of demonstrating value. In LVAD therapy, a large portion of post-implant cost is related to hospitalization utilization [12]. Ultimately cost-effectiveness for MCS will be highly dependent upon streamlining processes and developing innovative care pathways that will provide higher levels of support for this complex patient cohort in an outpatient setting.

Emerging Tools for Managing Patients Outside of Traditional Clinical Encounters

One mechanism to provide MCS patient care outside of traditional clinical encounters is through digital health technology and mobile health (mHealth) devices. Some of the key clinical data obtained through traditional clinical encounters by device interrogation and other testing can be obtained remotely by harnessing existing remote follow-up systems available for other cardiac implantable electronic devices such as pacemakers and implantable cardioverter defibrillators [13, 14]. These systems have the possibility of interacting with other implantable devices (e.g., a CardioMEMS device) to provide remote hemodynamic monitoring of pulmonary artery pressures and potentially reduce complications and hospital stays [15–17]. Innovation in mobile communication technologies (e.g., smartphones) and medical-grade and commercially available wearable devices are also creating unique opportunities for remote monitoring of other physiologic data [18]. This includes routine measurements such as step count, but newer devices hold promise for continuous monitoring of blood pressure despite continuous-flow physiology [19].

These tools are likely to create new challenges for LVAD teams. To efficiently utilize data from multiple implanted devices and wearables, the tools will need to be interconnected and interoperable. For example, many currently available commercial wearable devices are not operable with existing electronic health record platforms. LVAD teams and associated healthcare systems will also need to invest in secure infrastructure and technology solutions in order to safely collate and analyze the data.

The Changing Nature of Clinical Demands

MCS programs will also need to create systems to interpret and respond to large volumes of data from multiple sources. One can imagine that a patient supported on durable MCS in the future may transmit data from an LVAD, an implantable cardioverter defibrillator, a pulmonary artery or left atrial pressure monitor, a wearable with data on activity and blood pressure monitoring, and implanted and wearable devices designed to manage other chronic conditions (e.g., diabetes). Programs will need to create teams and dedicated time to review this information and create treatment pathways for responding to alerts and best practices for LVAD optimization.

Given that the current LVAD population is aging and patients have more comorbid conditions, future MCS teams will also need to rely more on other medical and surgical specialties in the outpatient setting. MCS programs are traditionally multidisciplinary, but we envision more interaction with non-cardiovascular specialties, especially those focused on aging. One specialty that deserves dedicated attention is palliative care.

Palliative Care in the Patient Supported with MCS

As highlighted throughout this text, mechanically assisted circulation has arguably been one of the most impactful cardiovascular treatments ever developed, increasing 2-year survival for patients with advanced heart failure from 10% to nearly 80% [20, 21]. However, device innovation has often overshadowed the potential and real burdens of LVAD therapy resulting in misalignment of goals between patients, families, and their medical team. Elucidating these goals and understanding the implications for advanced therapies is not within the skill sets of many clinicians but remains a critical component of the early and ongoing clinical evaluation.

A detailed description of adverse events experienced by MCS patients has been presented. While some events cause minimal harm, others require heightened, long-term medical care or leave the patient permanently disabled. The cumulative physical and psychological burden on patients and their caregivers has not been measured but certainly plays a significant role in the overall experience of MCS support and influences subsequent decision-making regarding acceptance of the therapy. Further, nearly all MCS patients will require their device to be turned off at the end of life. The complex interweaving of physical and psychological stressors, the challenges of effective communication pertaining to life-sustaining treatments, and end-of-life decision-making are all pertinent reasons to engage palliative care specialists in MCS programs.

By definition, palliative care is interdisciplinary care delivery that focuses on a holistic approach to the patient and his or her family. At its core, palliative care is focused on enhancing quality of life by specifically addressing symptom management; treatment of common comorbidities associated with chronic severe illness; and emotional, psychological, and spiritual concerns. The palliative care team layers incremental support onto standard therapies. Palliative care should be distinguished from hospice. Hospice is an insurance benefit for those with a terminal illness anticipated to survive for less than 6 months. The primary goal of hospice is to provide a high level of supportive care in either an inpatient or outpatient environment that is not focused on curing the underlying condition [22].

Palliative care participation in MCS programs was mandated in the United States by the Joint Commission in 2014 without specific guidance about the role that palliative care teams should serve during each phase of an MCS patient's journey. The literature guiding the role of palliative care in LVAD care is focused primarily on processes rather than outcomes. As a result, the majority of guidance about this integration comes from the perspectives of palliative care specialists and MCS clinicians. In fact, there is little information about the unmet needs expressed by MCS patients and families that would inform our collective palliative care efforts.

A useful framework to consider palliative care interventions throughout the course of the MCS experience is shown in Table 44.2 [23]. To be most impactful, the palliative care clinician should be well-versed in the evaluation of patient and family suitability for MCS support, the typical early and late post-implant course, the likelihood of adverse events, and the long-term benefits and complications of the therapy. Integrating that knowledge with patient and family goals and a clear understanding of the boundaries beyond which a patient would not want to continue VAD support are important data elements for the MCS team. The MCS evaluation should be orchestrated such that each member understands his/her role in obtaining information and providing education. Educational coordination is vital to avoid subsequent confusion and contradictory messages. Patient goals of care should be incorporated into the final decision regarding candidacy. The team also should acknowledge that goals of care may shift during MCS support. A patient's perception of the likelihood and impact of an adverse event may seem only theoretical before implant at a time in which he/she is facing the certain and tangible morbidity and mortality associated with advanced heart failure. Following implant, as singular or cumulative burdens of the therapy become evident, the patient in consultation with his or her family may shift perspectives on tolerability. The MCS team must be receptive to these changes as a natural evolution along the patient journey. Palliative care providers can play a vital role in assisting patients and families through the complex changes that occur on LVAD support.

Table 44.2 Roles for palliative care in MCS therapy

Clinical setting	Goal
Evaluation and preparation for implant	Assure understanding of purpose and limitations of the device; discuss alternative therapies; ensure MCS is aligned with patient goals; symptom management; advanced care planning
Following decision to implant	Support patient and family; manage symptoms
At the time of serious complications	Renew goals of care discussions and likelihood that MCS support is aligned with goals; manage symptoms; support patient and family
End of life	Assess the role of MCS as a therapeutic modality; symptom management when device is discontinued; patient and family support

Adapted from Goldstein et al. [23]

As noted, nearly all MCS patients will require device deactivation at the end of life. In some cases, the patient will be unresponsive, and the care will be similar to withdrawal of life support in other situations. For example, the patient who has experienced a stroke associated with a devastating neurological injury who will not recover or return to a reasonable functional capacity will often be separated from device support. At the other end of the spectrum, however, is the patient who is awake and alert and makes the conscious decision to separate from LVAD support. In this setting, the MCS and palliative care teams are intimately involved in assuring the patient and family understand the implications of this decision, carefully plan device deactivation, and are available to ensure that both the patient and family are supported. Prior study has shown that the average survival in elective device deactivation is less than 20 minutes [24]. In our experience, the time between device deactivation and death is shorter. Ensuring the patient is comfortable and does not experience symptoms such as pronounced pain or dyspnea is of critical importance.

Future Directions

An unmet need for the future of MCS programs will be data to guide treatment pathways. Novel approaches to outpatient management of device complications will need to be rigorously evaluated for safety, efficacy, and cost-effectiveness. These evaluations will ideally be performed in a prospective manner and include randomization to the current standard of care. A relevant and important example is stroke prevention after LVAD implantation. Data from the ADVANCE and ENDURANCE trials demonstrated that blood pressure control was associated with less neurologic events in patients supported on the HeartWare ventricular assist device (Medtronic) [25, 26]. In the ENDURANCE Supplemental Trial, patients were treated with a blood pressure management protocol that included training on and devices for home blood pressure monitoring, close monitoring of home blood pressure diaries, and specified blood pressure goals for clinicians [27]. Patients that received the protocol achieved improved blood pressure control over time and had a numerically lower rate of stroke compared with the ENDURANCE trial (22.3% vs. 16.9%, $p = 0.10$).

The MCS field is also in need of validated quality metrics. As this field has matured, inevitable variability in the medical management of patients on support is apparent [28]. Reducing variation and rapidly implementing best practices will be necessary to continue to improve post-implant outcomes. Some of this work is likely to come from INTERMACS [29], but electronic health records can include more detailed data and are likely to include data from other sources such as device interrogations mentioned above. Current electronic health record data in the United States is focused on billing, but improved data capture is possible through initiatives focused on improving data quality and tools such as natural language processing. With improved data, there is a tremendous opportunity to create learning healthcare systems that continuously measure processes of care and clinical outcomes and their associated links. For the management of heart failure, prior analyses demonstrated that simple process measures such as measurement of left ventricular ejection fraction are not ideally suited for characterizing care quality given an inconsistency associated with clinical outcomes [30]. Instead, the use of multiple measures that include both processes of care and outcomes are needed to fully characterize quality of care [31]. Given the complexity of MCS care, we anticipate that multiple measures will also be necessary to develop useful quality metrics that are tightly linked with improved patient outcomes.

References

1. Uriel N, Colombo PC, Cleveland JC, Long JW, Salerno C, Goldstein DJ, Patel CB, Ewald GA, Tatooles AJ, Silvestry SC, John R, Caldeira C, Jeevanandam V, Boyle AJ, Sundareswaran KS, Sood P, Mehra MR. Hemocompatibility-related outcomes in the MOMENTUM 3 trial at 6 months: a randomized controlled study of a fully magnetically levitated pump in advanced heart failure. Circulation. 2017;135(21):2003–12.
2. Kormos RL, Cowger J, Pagani FD, Teuteberg JJ, Goldstein DJ, Jacobs JP, Higgins RS, Stevenson LW, Stehlik J, Atluri P, Grady KL, Kirklin JK. The Society of Thoracic Surgeons Intermacs database annual report: evolving indications, outcomes, and

scientific partnerships. J Heart Lung Transplant. 2019;38(2):114–26.

3. Kirklin JK, Pagani FD, Kormos RL, Stevenson LW, Blume ED, Myers SL, Miller MA, Baldwin JT, Young JB, Naftel DC. Eight annual INTERMACS report: special focus on framing the impact of adverse events. J Heart Lung Transplant. 2017;36(10):1080–6.
4. Gordon RJ, Weinberg AD, Pagani FD, et al. Prospective, multicenter study of ventricular assist device infections. Circulation. 2013;127:691–702.
5. Takeda K, Takayama H, Colombo PC, et al. Incidence and clinical significance of late right heart failure during continuous-flow left ventricular assist device support. J Heart Lung Transplant. 2015;34:1024–32.
6. Kapelios CJ, Charitos C, Kaldara E, et al. Late-onset right ventricular dysfunction after mechanical support by a continuous-flow left ventricular assist device. J Heart Lung Transplant. 2015;34:1604–10.
7. Rich JD, Gosev I, Patel CB, et al. The incidence, risk factors, and outcomes associated with late right-sided heart failure in patients supported with an axial-flow left ventricular assist device. J Heart Lung Transplant. 2017;36:50–8.
8. Sharma A, Zhao X, Hammill BG, Hernandez AF, Fonarow GC, Felker GM, Yancy CW, Heidenreich PA, Ezekowitz JA, DeVore AD. Trends in noncardiovascular comorbidities among patients hospitalized for heart failure: insights from the Get With The Guidelines-Heart Failure Registry. Circ Heart Fail. 2018;11(6):e004646.
9. Kirklin JK, Naftel DC, Kormos RL, Stevenson LW, Pagani FD, Miller MA, Ulisney KL, Baldwin JT, Young JB. Second INTERMACS annual report: more than 1,000 primary left ventricular assist device implants. J Heart Lung Transplant. 2010;29(1):1–10.
10. Joynt KE. Health policy and cardiovascular medicine: rapid changes, immense opportunities. Circulation. 2015;131:1098–105.
11. Gilstrap LG, Joynt KE. Macra and cardiology—the devil is in the details. JAMA Cardiol. 2017;2:1177–8.
12. Rogers JG, Bostic RR, Tong KB, Adamson R, Russo M, Slaughter MS. Cost effectiveness analysis of continuous flow left ventricular assist devices as destination therapy. Circ Heart Fail. 2012;5:10–6.
13. Landolina M, Perego GB, Lunati M, Curnis A, Guenzati G, Vicentini A, Parati G, Borghi G, Zanaboni P, Valsecchi S, Marzegalli M. Remote monitoring reduces healthcare use and improves quality of care in heart failure patients with implantable defibrillators: the evolution of management strategies of heart failure patients with implantable defibrillators (EVOLVO) study. Circulation. 2012;125(24):2985–92.
14. Zeitler EP, Piccini JP. Remote monitoring of cardiac implantable electronic devices (CIED). Trends Cardiovasc Med. 2016;26(6):568–77.
15. Veenis JF, Manintveld OC, Constantinescu AA, Caliskan K, Birim O, Bekkers JA, van Mieghem NM, den Uil CA, Boersma E, Lenzen MJ, Zijlstra F, Abraham WT, Adamson PB, Brugts JJ. Design and rationale of haemodynamic guidance with CardioMEMS in patients with a left ventricular assist device: the HEMO-VAD pilot study. ESC Heart Fail. 2019;6(1):194–201.
16. Imamura T, Jeevanandam V, Kim G, Raikhelkar J, Sarswat N, Kalantari S, Smith B, Rodgers D, Besser S, Chung B, Nguyen A, Narang N, Ota T, Song T, Juricek C, Mehra M, Costanzo MR, Jorde UP, Burkhoff D, Sayer G, Uriel N. Optimal hemodynamics during left ventricular assist device support are associated with reduced readmission rates. Circ Heart Fail. 2019;12(2):e005094.
17. Imamura T, Nguyen A, Kim G, Raikhelkar J, Sarswat N, Kalantari S, Smith B, Juricek C, Rodgers D, Ota T, Song T, Jeevanandam V, Sayer G, Uriel N. Optimal haemodynamics during left ventricular assist device support are associated with reduced haemocompatibility-related adverse events. Eur J Heart Fail. 2019;21(5):655–62.
18. DeVore AD, Wosik J, Hernandez AF. The future of wearables in heart failure patients. JACC Heart Fail. 2019;7(11):922–32.
19. Sayer G, Kim GH, Rodgers D, Chung B, Nguyen A, Narang N, Raikhelkar J, Kalantari S, Sarswat N, Jeevanandam V, Uriel N. Highly accurate continuous blood pressure measurement in LVAD patients with a non-invasive, non-oscillometric wearable device: a pilot study. J Heart Lung Transplant. 2018;37(4 Supplement):S44.
20. Fang J. Rise of the machines – left ventricular assist devices as permanent therapy for advanced heart failure. N Engl J Med. 2009;361: 2282–5.
21. Mehra MR, Uriel N, Naka Y, Cleveland JC, Yuzefpolskaya M, Salerno CT, Walsh MN, Milano CA, Patel CB, Hutchins SW, Ransom J, Ewald GA, Itoh A, Raval NY, Silvestry SC, Cogswell R, John R, Bhimaraj A, Bruckner BA, Lowes BD, Um JY, Jeevanandam V, Sayer G, Mangi AA, Molina EJ, Sheikh F, Aaronson K, Pagani FD, Cotts WG, Tatooles AJ, Babu A, Chomsky D, Katz JN, Tessmann PB, Dean D, Krishnamoorthy A, Chuang J, Topuria I, Sood P, Goldstein DJ. A fully magnetically levitated left ventricular assist device — final report. N Engl J Med. 2019;380(17):1618–27.
22. Morrison RS, Meier DE. Clinical practice: palliative care. N Engl J Med. 2004;350:2582–90.
23. Goldstein NE, May CW, Meier DE. Comprehensive care for mechanical circulatory support. A new frontier for synergy with palliative care. Circ Heart Fail. 2011;4:519–27.
24. Brush S, Budge D, Alharethi R, McCormick AJ, MacPherson JE, Reid BB, Ledford ID, Smith HK, Stoker S, Clayson SE, Doty JR, Caine WT, Drakos S, Kfoury AG. End-of-life decision making and implementation in recipients of destination left ventricular assist device. J Heart Lung Transplant. 2010;29:1337–41.
25. Teuteberg JJ, Slaughter MS, Rogers JG, et al. The HVAD left ventricular assist device. Risk factors for

neurological events and risk mitigation strategies. JACC Heart Fail. 2015;3(10):818–28.
26. Rogers JG, Pagani FD, Tatooles AJ, Bhat G, Slaughter MS, Birks EJ, Boyce SW, Najjar SS, Jeevanandam V, Anderson AS, Gregoric ID, Mallidi H, Leadley K, Aaronson KD, Frazier OH, Milano CA. Intrapericardial left ventricular assist device for advanced heart failure. N Engl J Med. 2017;376(5):451–60.
27. Milano CA, Rogers JG, Tatooles AJ, Bhat G, Slaughter MS, Birks EJ, Mokadam NA, Mahr C, Miller JS, Markham DW, Jeevanandam V, Uriel N, Aaronson KD, Vassiliades TA. Pagani FD; ENDURANCE investigators. HVAD: the ENDURANCE supplemental trial. JACC Heart Fail. 2018;6(9):792–802.
28. Shreibati JB, Sheng S, Fonarow GC, DeVore AD, Yancy CW, Bhatt DL, Schulte P, Peterson ED, Hernandez A, Heidenreich PA. Heart failure medications prescribed at discharge for patients with left ventricular assist devices. Am Heart J. 2016;179:99–106.
29. STS Intermacs Database. https://www.sts.org/registries-research-center/sts-national-database/sts-intermacs-database. Accessed July 21, 2019.
30. Heidenreich PA, Hernandez AF, Yancy CW, Liang L, Peterson ED, Fonarow GC. Get with the Guidelines program participation, process of care, and outcome for Medicare patients hospitalized with heart failure. Circ Cardiovasc Qual Outcomes. 2012;5:37–43.
31. Hernandez AF, Fonarow GC, Liang L, Heidenreich PA, Yancy C, Peterson ED. The need for multiple measures of hospital quality: results from the Get with the Guidelines-Heart Failure Registry of the American Heart Association. Circulation. 2011;124:712–9.

Smart Mechanical Circulatory Support Devices: Requirements and Perspectives

45

Joshua P. Cysyk and Gerson Rosenberg

In 2017 the INTERMACS report concluded "Despite incremental improvements in survival, the rate of adoption and dissemination of MCS (mechanical circulatory support) therapy using FDA- approved devices has appeared to plateau during the most recent era. Extension of durable devices to a greater proportion of patients with ambulatory heart failure will require more effective neutralization or prevention of major adverse events. Greater emphasis on patient-reported outcomes in addition to traditional assessments of outcomes is needed to understand the totality of the benefit and risk of MCS therapy" [1].

The adverse events include bleeding, stroke, acquired Von Willebrand disease, infection, and device failure. In addition, devices need to have improved response to physiologic need to improve exercise tolerance and quality of life.

In the context of this chapter, the adjective smart refers to devices rationally designed to reduce adverse events and improve quality of life for the patients by both passive and active designs. These devices [heart assist (LVAD) and total artificial heart (TAH)] will need to change operating conditions in response to disturbances and physiologic demand.

The first step in any rational design process is developing a set of requirements usually starting with the customer/user requirements. In this case we will use the patient as the customer or user but keeping in mind that there could be a separate set of customer requirements for the surgeon, cardiologist, or caregiver.

By starting with the highest level general requirements and translating them into more specific requirements followed be detailed specifications a testing plan can be devised to determine if the device design meets the requirements. Here we will define the requirements and be as specific as possible. Whenever possible a design that could potentially satisfy the requirement will be described.

By the nature of the title of this chapter, the requirements and perspectives are in a large part editorial and based upon the authors' experience, suggestions in the current literature, and understanding of current device development.

J. P. Cysyk, PhD · G. Rosenberg, PhD (✉)
The Pennsylvania State University,
College of Medicine, Department of Surgery,
Division of Applied Biomedical Engineering,
Hershey, PA, USA
e-mail: jcysyk@pennstatehealth.psu.edu;
grosenberg@pennstatehealth.psu.edu

© Springer Nature Switzerland AG 2020

J. H. Karimov et al. (eds.), *Mechanical Support for Heart Failure*,
https://doi.org/10.1007/978-3-030-47809-4_45

Device User Requirements

1. The MCS device should provide adequate cardiac output such that the patient is able to perform light to moderate exercise tasks encountered in normal daily living.
2. The device should allow the patient to live a near-normal life style with near-normal quality of life.
3. The device should be easy for the patient to use and be unobtrusive.
4. The device should have few or no adverse events and be a reliable device.
5. The device should be compatible with devices encountered in normal daily activity.

These requirements are, in general, the requirements that have been used for current devices; what is different for future devices is that the specific design requirements and design specifications will be more difficult to meet because adverse events must be reduced.

Specific Device Requirements

These requirements are based on the user requirements and are intended to apply to "smart devices."

1. *The MCS device should provide adequate cardiac output such that the patient is able to perform light to moderate exercise tasks encountered in normal daily living.*

Light to moderate exercise implies playing golf, swimming, climbing stairs, and similar activities. All these activities can be accomplished with a cardiac output of 8 to 10 liters per minute. The device should automatically increase cardiac output, (TAH or LVAD) without user intervention, and not cause adverse events such as infection, arrhythmias, clinically significant hemolysis, or destruction of other formed elements in the blood and arterial venous malformations. It is recognized that in many patients, the cardiac output is right heart limited, and these patients would receive a "smart" artificial heart when available in the future.

2. *The device should allow the patient to live a near-normal life style with near-normal quality of life.*

This requirement implies that a patient should be able to take a shower or take a short bath or engage in activities unencumbered by external batteries and electronics for periods of up to 30 minutes with a 100% safety factor. The device should be small enough so that it does not inhibit patient movement or mobility. With improvements in device efficiency, this can be accomplished with current motor technology and improved batteries. A battery life of 5 years, with replacement possible, with minor surgery after that time, is feasible. This all implies a completely implanted system similar to the Penn State Arrow LionHeart LVAD or the AbioMed AbioCor TAH. Both demonstrated that a completely implanted system is practical and has the potential to reduce device infection. External battery packs should provide for a minimum of 6 hours between charges operating at a resting cardiac output of 5 L/min.

3. *The device should be easy for the patient to use and be as unobtrusive.*

The device should have an easily readable and audible interface that will keep the patient informed of system status. Secure remote monitoring by a physician must be available, and the system must be immune to hacking.

4. *The device should have few or no adverse events and be a reliable device.*

The goal for the maximum rate of serious adverse events should be less than 3% per patient year with service (similar to mechanical heart valves in the 1990s) with serious adverse events defined as any event requiring hospitalization or having a lasting effect. Ultimately, the device reliability goal needs to be 10 years without a critical (life-threating) system failure.

5. *The device should be compatible with devices encountered in normal daily activity.*

The device should not be adversely effected by electromagnetic radiation, typical accelerations encountered, typical temperatures, and humidity and be able to operate satisfactorily, while the patient is in typical positions.

Requirements 2 and 3 are requirements that could be satisfied with current technologies. Implantable electronics are routinely used in devices such as AICDs and Pacemakers and in the past used for a fully implanted LVADs and TAHs. With improvements in electronics and battery technology since the first use of an implanted LVAD, electronics size has been reduced and is limited primarily by battery size. With current lithium batteries, a 5-year battery life is achievable, and with an updated battery/controller, a 10-year mission life is possible. There have been many approaches to automatic control based on indirect measurements of pump flow and head pressure using pump power [7–11]. While these approaches can be useful for detecting over pumping and suction conditions they are unable to directly assess unloading. Thus with sound engineering design and state-of-the-art manufacturing techniques, requirements 2 and 3 can be achieved.

The real challenges to overcome are the requirements in requirement 1. LVAD patients with fixed speed continuous-flow pumps have significantly decreased exercise capacity as compared to heart transplant patients [2] and often are limited by their heart failure in daily routine activities despite the assist device. Patients with residual left ventricular function often can overcome the fixed pump speed and increase native output during exercise; however, the majority of patients with limited LV functionality cannot augment their cardiac output. For these patients increasing pump speed during exercise could increase total pump flow and improve exercise response. Clinical studies that have investigated the role of pump speed during exercise have shown mixed results in increasing peak exercise capacity [3, 4]. Impaired right heart function is often to blame in patients that do not have a demonstrated increased peak exercise capacity while increasing pump speed. For these patients, BiVAD support or total heart replacement would be necessary to improve exercise response. However, pump speed adaptation has been shown to increase submaximal exercise capacity even in patients where peak exercise capacity does not improve. This suggests that routine activities such as walking in the park, playing with grandchildren, or running errands would be possible with automatic speed control, which would greatly increase quality of life in these patients.

The major impediment to an automatic control system is the lack of sensor to assess physiologic demand. The ultimate goal of an automatic control system should be to mimic the Frank-Starling mechanism of the native ventricle [5, 6]. As ventricular loading increases, the LVAD controller should respond by increasing pump speed to increase circulatory support. Likewise, as ventricular loading decreases, the LVAD speed should be reduced in order to minimize the risk of suction. For reliable control, a direct measurement of left ventricular loading is required. There have been many approaches to automatic control based on indirect measurements of pump flow and head pressure using power [7–11]. These approaches are based on using pump power and can be useful for detecting over pumping and suction conditions. However, these characteristics of the pump flow waveform can be useful for detecting suction but are unable to directly assess unloading. The pump power is a function of the flow through the pump and the head pressure across the pump, both of which are functions of both preload and afterload and, therefore, are not ideal for a control system that aims to mimic a response proportional to preload similar to the Frank-Starling mechanism that is independent of afterload.

Pressure or volume sensors in the left ventricle would provide an adequate measure of left ventricular loading. These sensors must be biocompatible and maintain signal integrity for a sufficient time period to allow for control. Pressure sensors have historically been prone to drift that may make long-term measurements difficult.

Baseline pressures drift in excess of 1 mmHg would make the sensors unusable for detecting small changes end diastolic pressure. However, current sensors on the market (The Titan WIHM from ISS, Inc., and the CardioMEMS HF System from St. Jude Medical) have been effective in measuring left atrial filling pressure, which requires the same sensitivity [12–14]. Adapting these sensors to implantation in the left ventricle and providing an output that can be used by LVAD control systems should be possible. Likewise, miniature pressure sensors are available commercially with pressure drift specifications of <1 mmHg/ year (MEAS Switzerland), and some efforts have been made to incorporate these sensors into implantable devices. In addition, periodic calibration using a procedure similar to the Valsalva maneuver could make it possible to extend the lifetime of these devices [15].

Left ventricular volume sensing would provide a direct measure of the flow reserve available. Optimal control and suction detection would be possible through this direct measurement: increase pump support as long as sufficient reserves are available and decrease pump support when the ventricle size decreases and suction is eminent. Control in this manner would be very similar to the current standard-of-care of using echocardiography to determine pump speed [16]. In the clinic setting, pump speed is adjusted in real-time while measuring the size of the ventricle using echocardiography. Typically, the speed set is slightly less than the optimal speed in order to limit the incidents of suction. Using a direct measure of left ventricular volume would allow for a control system to replicate this practice either periodically throughout the day or continuously.

Sonomicrometry ultrasound crystals have been regularly uses in a research setting to measure ventricular volume. Ochsner et al. [17] attached these sensors to a cannula tip and demonstrate that they can be used as an effective sensor for a volume-based control system. More work is required to make a device suitable for long-term implantation. An alternative technique for measuring volume is through measuring the conductance of the blood volume in the ventricle, which is directly proportional to the blood volume [18]. Multi-segmented conductance catheters are commercially available for research purposes (e.g., the Millar Pressure-Volume Loop System from Millar, Inc.). Cysyk et al. [19] incorporated a signal segment of conductance catheter onto the cannula tip of an LVAD and demonstrated that the conductance catheter measurements were linearly proportional to the size of the ventricle measured using echocardiography.

Direct measurement of left ventricle unloading through either pressure or volume measurements, or a combination of both, will be available in the future. LVAD controllers must be able to incorporate these measurements in order to optimize pump speed and reduce the risk of suction. Periodic adjustment of pump speed in the clinic setting is insufficient to achieve the goal of increasing exercise capacity in order to improve quality-of-life in these patients. A real-time control system that can mimic the Frank-Starling mechanisms is necessary to maximize the benefits of mechanical circulatory support.

With a smart control system, pumps must be capable of providing increased flow during periods of increased demand. There is no doubt that a small pump can be designed to pump 10 liters per minute with a head rise of 150 mmHg. The challenge is to do it without creating the stated adverse events at a rate less than 3% per year. The greatest challenge for smart MCS devices is not in making them "smart" but in improving biocompatibility. How can MCS designers address reducing the current rate of adverse events? This is the real challenge to overcome.

If one assumes that pulsatile devices such as the Penn State Arrow LionHeart or a downsized HeartMate XVE pump will not be considered for implantation due to the device size, then the next generation of devices may be either improved rotary pumps or unique pumps such as the TORVAD or CorWave.

The issue of AV malformations is not well understood and has been postulated to be due to lack of or reduced pulsatility; if so this issue potentially could be addressed with by pulsing a continuous-flow device. It is also of interest to note that bleeding and acquired Von Willebrand disease were not associated with pulsatile devices such as the HeartMate XVE or LionHeart.

No matter what the smart device design of the future, it will require an integrated approach of rational design incorporating the latest engineering design tools such as state-of-the-art CFD with large eddy simulation or direct numerical simulation with optimizing tools such as design of experiments (DOE). These tools will require advanced models of hemolysis, thrombosis, and the destruction of VWF. It has been repeatedly demonstrated that excessive fluid stresses will activate platelets, destroy RBCs, and cleave VWF, but the quantification of these stresses is lacking in current models. The other key process that requires improved modeling is the very complex process of thrombosis. We know that excessive stresses can destroy formed elements and that low stresses can create thrombosis, but in order to design improved pumps, we need improved understanding of the thrombosis process and accurate models thereof.

When accurate models of each of these phenomena are developed, then it will be possible to design an improved pump using these models and design techniques. It remains to be seen if in fact rotary pumps can be designed with stresses low enough to not cause hemolysis and clinically significant destruction of VWF.

"Smart" devices can be designed using current models, but a true quantum improvement will require improved understanding and models of thrombosis, hemolysis, and destruction of VWF. All of this requires funding to conduct the required research.

References

1. Kirklin JK, Pagani FD, Kormos RL, Stevenson LW, Blume ED, Myers SL, et al. Eighth annual INTERMACS report: special focus on framing the impact of adverse events. J Heart Lung Transplant. 2017;36(10):1080–6.
2. Kugler C, Malehsa D, Tegtbur U, Guetzlaff E, Meyer AL, Bara C, et al. Health-related quality of life and exercise tolerance in recipients of heart transplants and left ventricular assist devices: a prospective, comparative study. J Heart Lung Transplant. 2011;30(2):204–10.
3. Schmidt T, Bjarnason-Wehrens B, Schulte-Eistrup S, Reiss N. Effects of pump speed changes on exercise capacity in patients supported with a left ventricular assist device-an overview. J Thorac Dis. 2018;10(Suppl 15):S1802–S10.
4. Reiss N, Schmidt T, Workowski A, Willemsen D, Schmitto JD, Haverich A, Bjarnason-Wehrens B. Physical capacity in LVAD patients: hemodynamic principles, diagnostic tools and training control. Int J Artif Organs. 2016;39(9):451–9.
5. Tchantchaleishvili V, Luc JGY, Cohan CM, Phan K, Hübbert L, Day SW, et al. Clinical implications of physiologic flow adjustment in continuous-flow left ventricular assist devices. ASAIO J. 2017;63(3):241–50.
6. Salamonsen RF, Pellegrino V, Fraser JF, Hayes K, Timms D, Lovell NH, et al. Exercise studies in patients with rotary blood pumps: cause, effects, and implications for starling-like control of changes in pump flow. Artif Organs. 2013;37(8):695–703.
7. Arndt A, Nusser P, Graichen K, Muller J, Lampe B. Physiological control of a rotary blood pump with selectable therapeutic options: control of pulsatility gradient. Artif Organs. 2008;32(10): 761–71.
8. Choi S, Boston JR, Antaki JF. Hemodynamic controller for left ventricular assist device based on pulsatility ratio. Artif Organs. 2007;31(2):114–25.
9. Ferreira A, Boston JR, Antaki JF. A control system for rotary blood pumps based on suction detection. IEEE Trans Biomed Eng. 2009;56(3):656–65.
10. Schima H, Vollkron M, Jantsch U, Crevenna R, Roethy W, Benkowski R, et al. First clinical experience with an automatic control system for rotary blood pumps during ergometry and right-heart catheterization. J Heart Lung Transplant. 2006;25(2):167–73.
11. Vollkron M, Schima H, Huber L, Benkowski R, Morello G, Wieselthaler G. Advanced suction detection for an axial flow pump. Artif Organs. 2006;30(9):665–70.
12. Tolia S, Khan Z, Gholkar G, Zughaib M. Validating left ventricular filling pressure measurements in patients with congestive heart failure: CardioMEMS pulmonary arterial diastolic pressure versus left atrial pressure measurement by transthoracic echocardiography. Cardiol Res Pract. 2018;2018:8568356.
13. Hubbert L, Baranowski J, Delshad B, Ahn H. Left atrial pressure monitoring with an implantable wireless pressure sensor after implantation of a left ventricular assist device. ASAIO J. 2017;63(5):e60–e5.
14. Pagani FD. Applications of implantable hemodynamic monitoring in the setting of durable mechanical circulatory support. ASAIO J. 2018;64(3):283–5.
15. Troughton RW, Ritzema J, Eigler NL, Melton IC, Krum H, Adamson PB, et al. Direct left atrial pressure monitoring in severe heart failure: long-term sensor performance. J Cardiovasc Transl Res. 2011;4(1):3–13.
16. Stainback RF, Estep JD, Agler DA, Birks EJ, Bremer M, Hung J, et al. Echocardiography in the manage-

ment of patients with left ventricular assist devices: recommendations from the American Society of Echocardiography. J Am Soc Echocardiogr. 2015;28(8):853–909.
17. Ochsner G, Wilhelm MJ, Amacher R, Petrou A, Cesarovic N, Staufert S, et al. In vivo evaluation of physiological control algorithms for LVADs based on left ventricular volume or pressure. ASAIO J. 2017;63(5):568–77.
18. Baan J, van der Velde ET, de Bruin HG, Smeenk GJ, Koops J, van Dijk AD, et al. Continuous measurement of left ventricular volume in animals and humans by conductance catheter. Circulation. 1984;70(5):812–23.
19. Cysyk J, Newswanger R, Popjes E, Pae W, Jhun CS, Izer J, et al. Cannula tip with integrated volume sensor for rotary blood pump control: early-stage development. ASAIO J. 2018;10. [Epub ahead of print]

Novel Solutions for Patient Monitoring and Mechanical Circulatory Support Device Control

46

Martin Maw, Francesco Moscato, Christoph Gross, Thomas Schlöglhofer, and Heinrich Schima

Introduction

The management of chronic disease necessitates a structured approach to accurately assess the current status of the patient and their disease and a method to decide if and how to act, which results in action. When the human body is understood as a highly interlinked system of sensors and actuators, titled automatons by famous physiologist A.C. Guyton [1], these two main tasks of observation and action could be seen through the lens of monitoring and control, respectively.

If we define the process of monitoring as the observation of the state of the patient, in combination with data processing to ensure the functions of "advising, warning or reminding" as the Latin root, monere would suggest [2] we may state that monitoring indeed needs to be the first step to intervention.

This observation can be done quantitatively by taking regular measurements of system states like body temperature, weight, heart rate, blood pressure, or less easily available parameters, such as blood serum albumin levels. Monitoring can also be understood as qualitative observation, such as patient well-being or psychological status [3].

The purpose of these observations is usually to assess stability of patient health status. If a certain measurement or set of measurements are out of the ordinary, an intervention might have to be implemented in order to change the trajectory of the patient. This intervention is commonly decided by an expert or a panel of experts after taking into account all the acquired data on the patient. So it can be summarized that monitoring means to continually assess whether action has to be taken in order to prevent adverse events or counteract already present unwanted symptoms.

The application of interventions marks the transition from monitoring to control. Certainly, in standard clinical operation, nobody would refer to the tasks of medical therapy as "control." However, in a technical sense, the aforementioned panel of experts constitutes the controller, which takes into account the data, weighs the importance, and then actively takes action, be it via medication, changes to device settings, or even rules or recommendations to the patient, like the intake of fluid, in hypovolemic patients.

With this perspective in mind, this chapter will look at the state of the art of monitoring and control of left ventricular assist device therapy, one of the main pillars of postoperative heart failure patient management. Novel concepts, mainly focusing on automating monitor-

M. Maw, MSc · F. Moscato, PhD · C. Gross, MSc
T. Schlöglhofer, MSc · H. Schima, PhD (✉)
Center for Medical Physics and Biomedical Engineering and Deparment of Cardiac Surgery, Medical University of Vienna, Wien, Austria

Ludwig Boltzmann Institute for Cardiovascular Research, Vienna, Austria
e-mail: francesco.moscato@meduniwien.ac.at; heinrich.schima@meduniwien.ac.at

© Springer Nature Switzerland AG 2020

J. H. Karimov et al. (eds.), *Mechanical Support for Heart Failure*,
https://doi.org/10.1007/978-3-030-47809-4_46

ing and control, are discussed and ultimately the possibility of "closing the loop" of observation and action.

Monitoring in LVAD Therapy

Overview

LVAD therapy is complex and multifaceted. Despite major advancement in recent years, there is still significant risk of adverse events and mortality in LVAD therapy [4]. New York Heart Association IV patients with severe heart failure may experience worsening of their heart failure or comorbidities that may lead to severe consequences if left unchecked. Timely intervention necessitates timely detection; both of which have been linked to improved outcomes in circulatory assistance [5, 6] as well as in cardiac resynchronization therapy, where advanced monitoring has already been established and has been recommended in an international consensus [7].

In this chapter, a brief overview of currently monitored variables is presented, with highlights from novel developments in the field. To structure these developments, monitoring and management tasks in LVAD patients can be roughly framed into three subsections in an outpatient setting.

Clinical Monitoring

This includes any activity involving staff from the implanting center or an affiliated professional institution. The bulk of the workload in clinical monitoring is currently constituted by regular follow-ups to the outpatient clinic, after discharge from the hospital. Clinical monitoring comprises all the monitoring tasks that are done by clinical personnel. Clinical monitoring has access to state-of-the-art equipment and expertise and can thus be considered the most accurate form of monitoring. The frequency of patient visits to the outpatient facilities varies depending on implant centers and patient condition, available resources, and distance to the implanting center [5]. The most common outpatient visit schedule includes initial weekly visits, which transitioned to a monthly schedule once the patient was stable on support for longer durations. Especially large programs, which implant more than 40 pump systems per year, reduce this interval even further to 3 months [8]. While the benefit of a finer mesh of patient contact could be shown, for example, via telephone calls [9], many clinical tests can only be performed in a clinical environment. Currently, a regular visit to the LVAD outpatient clinic includes blood laboratory [10, 11], blood pressure measurement [3], chest x-ray, driveline exit site examination [12], echocardiography [13], review of medical records, as well as online pump parameter readings, logfiles of pump parameters, and patient logs [8].

Patient Self-Monitoring

These are the tasks, which the patient is educated to handle autonomously or with the help of a caregiver. These tasks usually comprise daily monitoring of vital signs and symptoms, as well as monitoring and maintenance of the assist device. As described in the previous chapter, clinical monitoring is the most accurate form of supervision. However, due to limited resources, it cannot be available to all patients, all the time. Thus, some activities are outsourced to the patient or the caregiver. Self-monitoring and self-management are well established among many therapies, with established links between self-management and positive patient outcome in heart failure patients [14].

Self-monitoring and self-management can be described as a "cluster of learned behavior" [15]. So these tasks have to be taught to the patients or caregivers by a VAD professional, with ongoing patient education being highly advised [12]. Those patient tasks include device and system function monitoring, lead and wound monitoring, and monitoring of symptoms and signs of worsening heart failure or side effect/adverse event, psychological status, and blood status in regard to coagulation and arterial blood pressure [16].

It has been reported that patient self-efficacy is correlated with cognitive function of patients, which varies in the heterogeneous LVAD population [17]. Thus many self-monitoring tasks cannot be trusted to be handled consistently by all patients. One of the reported downsides of the outsourcing of the monitoring to the patient or their immediate caregiver is that they can become overwhelmed with the tasks that are demanded of them. Reported compliance rates range from 35% for daily weight monitoring to as low as 9% for symptom reporting. However, accurate self-monitoring and self-management are crucial, as some parameters have to be maintained within very fine target ranges. For example, it is well established, that time in therapeutic range for International Normalized Ratio (INR) measurements is highly correlated with therapy outcome [18]. The crucial task of managing anticoagulation is handled by the patients in about 50% of clinics [8]. Additionally, motivation plays a critical role, as patients are more likely to take the self-monitoring more seriously if they believe that the disease is actually controllable [17]. Since motivation and ability of the patient fluctuate widely between patients and even over the time course of therapy, the crucial tasks of self-monitoring suffer from issues of consistency.

Pump Monitoring

As monitoring is not just the mere recording of data, but also interpretation of said data, there has been significant progress in automating this process. While clinical management holds the greatest expertise regarding decision-making, it is limited by constrained resources. Self-management is performed by the patients themselves and is wrought with problems of varying patient capabilities. Pump monitoring can support the decisions made by patients or clinicians with a great level of expertise, and there is no requirement for allocation of additional resources to activate them, as the used algorithms may run on any readily available hardware. Thus pump monitoring holds great promise for LVAD therapy.

Pump monitoring as it is used here could also be understood as autonomous monitoring, since the data processing is performed according to predefined algorithms, which run on hardware associated with the pump system. Parameters derived from the LVAD system are collected and processed to yield information, which can then be displayed to a monitor screen in real time or tracked in logfiles for retrospective analysis. While the data processing is fully automated, the incorporation into the decision-making process is not, and so the pump monitoring currently only informs other decision-makers, namely, the clinician and the patient. In order to minimize the applied effort of these decision-makers, pump monitoring relies on alerts to signal urgency.

To make autonomous decision based solely on pump monitoring data has also been proposed and has already been implemented in some applications, such as suction detection and reaction in the HMII and HM3 (Abbott Inc., Chicago, IL, USA). These propositions and their implementations will be discussed in the second part of this chapter.

Hierarchy and Interactions Between the Monitoring Modalities

In order to understand the interaction between the monitoring modalities, it is also important to understand the distinction to management blocks. While monitoring consists of information processing and gating, the management blocks maintain and update a set of actions, like fluid intake for self-management, or medication prescription for the clinical management blocks or pump speed control for the pump management block. Monitoring means the preprocessing of information, which is then utilized in the management. Depicting the three monitoring modalities with their respective management task blocks, a schematic of VAD therapy is summarized in Fig. 46.1. Each monitoring block receives input either from the patient via various sources, like sensors or subjective feelings, or from other monitoring blocks that have previously processed or gated information. As we can see with the outer circle, pump monitoring informs self-monitoring via displays of pump information and alarms. The patient via his self-monitoring routine will determine whether this information requires any changes to his self-management, if this information should be forwarded to clinical monitoring, or if it can safely be logged or discarded. Additionally, pump monitoring directly informs clinical monitoring via logfiles, which

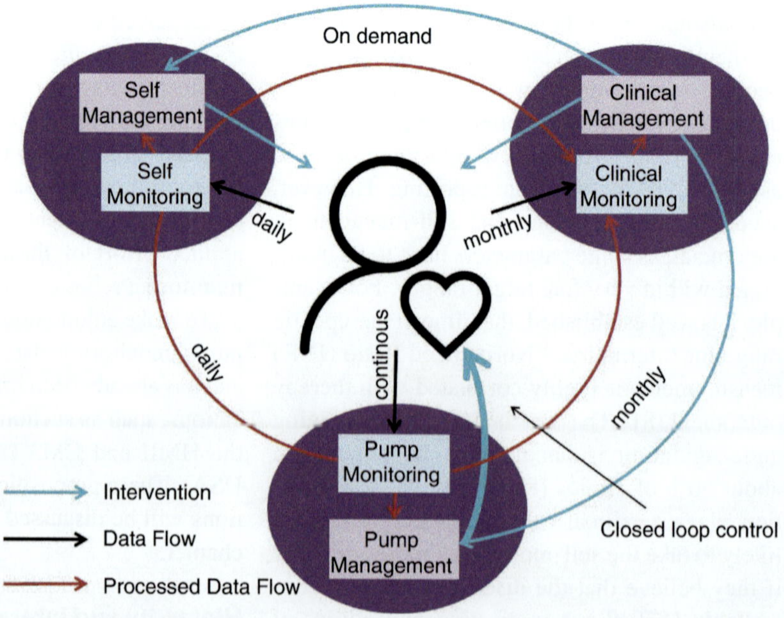

Fig. 46.1 Overview of monitoring entities in a VAD setting depicting clinical monitoring, self-monitoring, and pump monitoring with their respective management blocks. The flow of data and interventions is illustrated by arrows. Additionally, typical time intervals are given

display retrospectively various parameters of the pump over a certain time period.

While all these modalities aim at observing the same entity, namely, the LVAD patient, they differ among several aspects.

Figure 46.2 outlines the differences in the three modalities. Clinical monitoring is usually performed on the basis of every few weeks or months, self-monitoring occurs daily, while pump monitoring may occur up to multiple times per second.

Financial and time burden describes the increased cost of involvement of the modality. While pump monitoring can run without extra cost, self-monitoring necessitates at least the involvement of the patients' time, in the case of blood pressure measurements, but might also utilize one-time use items, such as test strips for INR measurement. Clinical monitoring requires the involvement of at least one professional. This is also reflected in involved people. Whereas clinical monitoring requires a team of technicians and physicians, self-monitoring just requires the patient and sometimes a caregiver, a relative, partner, or professional nurse. Pump monitoring does not require any human activity to process data. However, the interpretation of monitoring output, like alarms or logfiles, still has to be performed at the respective interfaces to these modalities.

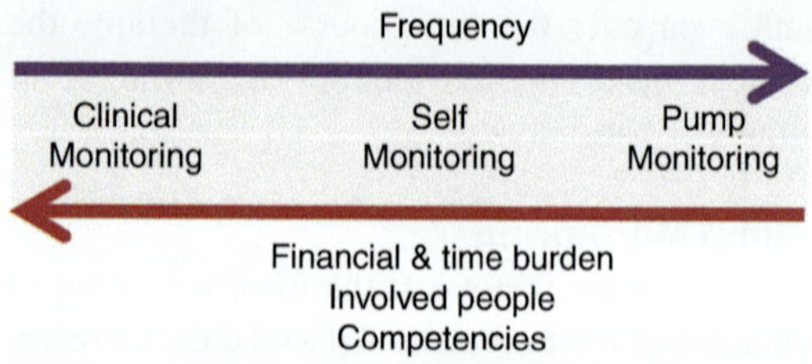

Fig. 46.2 Comparison of monitoring entities in a VAD setting

"Competencies" describe the interface to intervention. In clinical monitoring, the interface to intervention is rather direct. The decision to change pump settings, medical therapy, or even perform surgery can be based on this data.

Novel Monitoring Solutions

While the current approach to monitoring, in combination with improved pumps, is able to provide the best outcomes of VAD therapy since its inception [4], there are still some unsolved issues. Due to the complexity of the treatment, many objectives have to be achieved simultaneously, like anticoagulation, blood pressure, filling status, or driveline exit site maintenance. And even if all objectives are met, adverse events can still burden LVAD therapy. These adverse events may go unnoticed for an extended amount of time, causing patient deterioration [5]. In the following, novel developments in monitoring will be presented.

Telemonitoring

Telemonitoring aims at providing information to the healthcare providers without necessary physical patient contact. This combats the shortcomings of patient self-monitoring, detailed in the previous section. Additional flow of information to the healthcare provider enables earlier detection of potentially dangerous circumstances or enforces compliance with the self-management regimen.

Telephone Monitoring

As recommended in the ISHLT MCS Guidelines [3], between routinely scheduled outpatient visits, monitoring phone calls from the VAD coordinator to the patient or caregiver help proactively identify issues that may adversely affect patient outcomes. This was again supported by additional evidence from a recent publication [6]. In addition to increased cost efficiency of the treatment [19], it was shown to improve 2-year survival in patients that were subjected to a structured telephone interview [5]. In these interventions, open questions and inquiries regarding the patient weight, blood pressure, anticoagulation, pump parameters, or heart failure symptoms were performed. The healthcare providers usually followed a structured algorithm, with set interventions for specific findings. Two-year readmission rates did not differ significantly, while 2-year survival was significantly increased [5]. The potential causes of this significant improvement are the fact that 44.6% of patients died in their home due to undetected problems, in the cohort that did not receive regular phone calls. Treatment of complications also benefited from earlier detection. Pump thrombosis required fewer pump exchanges, and driveline infections could be treated without readmission in the telephone intervention group [5]. Additionally, it was speculated that the closer contact to the healthcare provider lowered the barrier to call the emergency hotline for any types of problem. The increased demand for clinical staff to perform these interviews can be considered a downside to this technology.

Patient Input

Telemonitoring that relies on patient input but not on direct telephone connection was reported to be feasible. In order to support patients in organizing their self-monitoring and self-management, Casida et al. [20] developed a smartphone application. In a first feasibility study, they were able to achieve high scores in usability and acceptability in their initial patient collective. The app assists the patient in structured recording and tracking of their data, such as weight or blood pressure. Additionally, should abnormal data be entered by the patient, the clinicians are contacted automatically. The patient is guided through their daily monitoring regimen including device functioning, symptom reporting, and physiological parameter recording. Additionally, the app provided easy access to resources for patient education, which patients may revisit at any time.

INRtracker.com is a website, which allows patient input of INR values, as well as weight, blood pressure, and other variables. It is linked to the VadLink ® network of ReliantHeart (Reliant Heart Inc., Houston, Tx, USA)

Another application that has been proposed is that of driveline exit site monitoring. This location is prone to infection, which constitutes the most common adverse event in LVAD therapy, occurring in up to 60% of patients [4, 12, 21]. Early signs of infection should be monitored. This is usually done via inspection of the driveline exit site, during the regular changes of the wound dressing. Dressing changes should be performed daily with close inspection of the exit site until fully healed. Once healed and with no drainage present, the frequency of dressing change can be decreased to 1 to 3 times weekly [12]. The patients are advised to notice any redness or swelling at the exit site and report it immediately to the clinical team. Reiss et al. present an approach to transmit images to clinicians in order to assess infection status from image data [22].

This approach can also be used for patients to compare their images to past data.

Telemonitoring of Pump Parameters

Other ways of telemonitoring utilize methods of completely bypassing the patient. The Heart Assist 5 System (Reliant Heart Inc., Houston, TX, USA) allows real-time access to measure pump flowrate by the authorized healthcare provider. The controller establishes an Internet connection via an inserted SIM card. Since time required to continually monitor all patients is prohibitive, these remote monitoring strategies are coupled with decision support systems such as alarms. In a first analysis of the remote monitoring capabilities of the Reliant Heart, the authors reported a high number of alarms especially at the beginning of their experience with the system. However, once the thresholds have been adjusted, they concluded, that remote monitoring was capable of aiding in the "timely recognition of abnormalities" [23]. Remote monitoring was shown to be a valuable tool to reduce cost, patient visits, and time to intervention, as well as increasing psychological well-being of the patients [24]. Its application has not yet been fully embraced by more commonly used pump systems [25]. However, it could be speculated that the recent acquisition of the HeartMate™ pump series (HMII or HM3, Abbott Inc., Chicago, IL, USA) and the HVAD™ (Medtronic, Inc., Fridley, MN, USA) by major medical device companies with established data transfer frameworks, namely, the CareLink™ Network and the Merlin.net™, will extend the monitoring capabilities of widely distributed systems with similar functionality. To fully adapt these solutions, the issue of data analysis will have to be addressed.

Telemonitoring of Additional Sensors

One way for clinicians to obtain valuable information about the patient is to connect them directly to the sensors. Yandrapalli et al. suggest the evaluation of the use of CardioMEMS™ (Abbott Inc., Chicago, IL, USA) in LVAD patients that has proven beneficial in a broader heart failure collective [26]. This system is an implantable pressure sensor and is typically placed in the pulmonary artery. Indeed, Harris et al. report of a case in which CardioMEMS was effectively used to monitor right ventricular function as well as to detect arrhythmias [27]. More recently, Kilic et al. argued for the benefit of pulmonary pressure monitoring in clinical management of LVAD patients [28].

Granegger et al. monitored patient activity as a marker of patient well-being. They correlated irregular patient activity with adverse events. Connection of this sensor data directly to the clinician could prompt the healthcare provider to proactively inquire about the status, if anomalies in activity are detected [29].

Enabling the clinician to access information from an implantable cardiac device (ICD) data can also be seen as an additional sensor [30]. Multiple remote monitoring systems are already in clinical use and can provide clinicians with information regarding ICD status and history as well as providing a measurement for intrathoracic impedance.

Hemodynamic Monitoring

Hemodynamic assessment might be a valuable tool to detect hemodynamic anomalies preceding complications [31]. Additionally, it can be a valuable tool to assess patient status and to tailor therapy to their specific needs.

Blood Pressure Measurement

This hemodynamic measurement is part of the clinical standard evaluation. Hypertension is a risk factor for progressive aortic insufficiency, intracranial hemorrhagic stroke, pump thrombosis, and thromboembolic events [3, 32–34]. Hence, proper blood pressure management is essential. The ISHLT Guidelines [3] recommend a mean arterial pressure (MAP) of <=80mmHg for continuous flow devices. However, autonomous measurement of arterial blood pressure may be difficult or inaccurate due to the decreased arterial pulsatility, with reported success rates of 17–63% [10]. Thus devices operating with the oscillometric method may have difficulties obtaining accurate readings [35]. Devices using

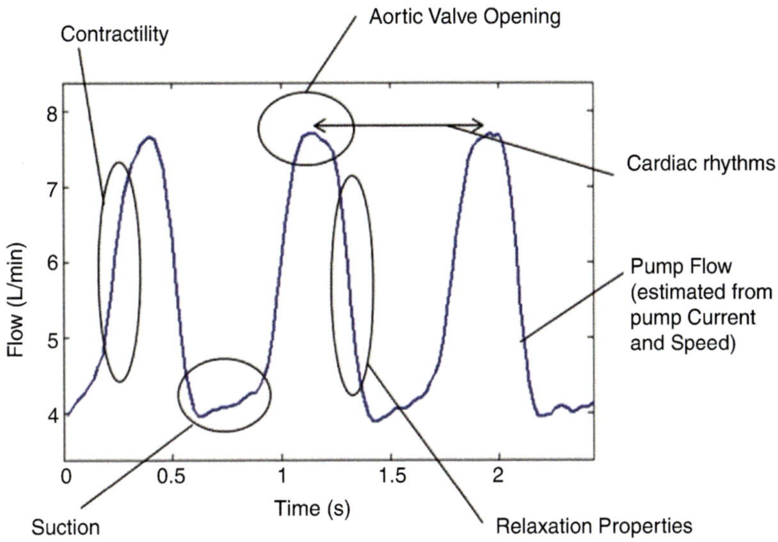

Fig. 46.3 Noninvasive hemodynamic monitoring indices developed by the Vienna Group. Contractility [38], suction [40], aortic valve opening [41, 42, 46], cardiac rhythms [47], and relaxation properties [39] can be derived from estimated pump flow[48]

the ultrasound Doppler method have been evaluated for blood pressure management and have become the clinical standard [36]. These devices tend to work more accurately in patients without a high pulse pressure [37].

Hemodynamic Pump Monitoring

Pump signals have been used to estimate hemodynamic variables noninvasively. These could be used to track patient progress or to detect unwanted pumping states, such as ventricular suction. Variables described in Fig. 46.3, such as contractility [38], relaxation [39], suction [40], aortic valve opening [41, 42], and cardiac rhythms[43], can be extracted. These may aid clinicians in their decisions on the level and strategy of pump support [38]. While this is not currently widely implemented in clinical practice, this type of monitoring holds the promise of deeper insights into patient physiology as well as eventually translation to closed loop control. Recently estimators for ventricular preload were developed [44, 45]. These estimators require nothing more than the data of the pump itself and are thus practically without additional cost.

Alarms

Alarms are an essential feature of pump monitoring. If pump monitoring detects some issue that requires attention, effective alarms make sure proper attention is given. It thereby acts as an interface between pump monitoring and clinical –, as well as patient – monitoring.

As interventions cannot be initiated by the pump itself, except for speed changes, the design of this interface will determine the efficacy of the entire pump monitoring capabilities, including monitoring of basic pump functioning. Thus the choice of auditory and visual alarm strategy [49] and the linguistic details of alarm messages [50] need to be carefully considered [30]. In international use, even the translation of alarm messages and material, as required by EU guidelines [51], can cause misinterpretation with hazardous consequences [50]. With these considerations, a wide range of different alarms can be realized, which should be understood as intuitively as possible. More complex phenomena necessitate a structured approach toward assessing patient overall status, in order to put the alarm into context, such as low-flow alarms [52].

Additionally, a low rate of false alarms is crucial for two reasons. Firstly, every alarm can be considered to bear the risk of the recommended intervention for this specific alarm. In 2015 a voluntarily issued urgent medical device correction letter was published by Thoratec corporation [53]. An unnecessary alarm prompted patients to perform a controller exchange. However, some patients were not sufficiently educated on this task, resulting in two patient deaths and one patient injury. Secondly, alarm fatigue will diminish the perceived urgency of subsequent alarms, if the previous number of false alarms was too high [54].

The management of these prompts by the pump is taught to patients during training. However, alarms that target basic pump functioning, such as whether all cables are connected, are important to be understood by even untrained people [50]. Some events can be safely cleared by the patient, whereas other alarms should signal the patient to contact their healthcare provider immediately. If the alarms are cleared by the patient, alarm logs will still be produced, which can later be reviewed by the clinicians, to establish a pattern and to counteract the cause of those alarms systematically. Currently used alarms either notify the users during danger to pump functioning, such as battery status, or report on anomalies in pump parameters, such as high power consumption or low estimated flow. A previous multicenter study shows room for improvement of currently used alarms as only 37% of LVAD patients felt safe concerning the handling of possible emergency situations [55]. Novel alarms aim at taking an additional step toward being more specific at identifying a certain adverse events. Especially when developing such new alarms, human factors engineering (HFE) must apply to aspects of a device with which a human interacts and to all of the tasks that a human might perform with the device, including all hardware and software interfaces that support all user tasks. HFE-based medical device designs and alarms increase safety, reduce user errors and facilitate recovery from user errors, decrease training time, and increase ease of use [56].

The extra step of interpreting pump parameters in a physiological context has so far been reserved for clinicians and patients. For example, patients have been advised to increase their fluid intake in reaction to low-flow alarms, as this alarm is most commonly caused by hypovolemia. However, by interpreting the pump parameters they are put in a physiological context and the message aims at explicitly stating the cause of the parameter abnormality, the physiologic state of suction, which will receive additional attention in the later sections of this chapter, has been the first such explicitly stated alarm and has already been implemented in the currently used pumps. But additional physiological states, such as pump thrombosis [57], have been proposed. With this different viewpoint, it is then also possible to detect these events even before their symptomatic onset. In a study relating pump parameter-derived factors to pump thrombosis, it was possible to detect pump thrombosis 4 days before current alarms [58].

Thrombus Detection

Device thrombosis is an adverse event with potentially devastating consequences. Even when thrombosis is detected before thromboembolism or pump malfunction occurs, thrombolysis therapy in later stages of thrombus growth is associated with poorer outcome [57]. Thus innovative method detection of pump thrombosis has been developed: Dimitrov et al. were able to correlate observation of microbubbles in the outflow graft in the Doppler echocardiogram with pump thrombosis. Presence of high intensity transient signals was able to detect pump thrombosis in patients before evident clinical symptoms [59].

Another approach to detect thrombosis of the pump by acoustic analysis was presented for the HMII [60] and the HVAD [61]. In this method, a microphone is placed on the thorax of the patient, and the recorded sound is analyzed in regard to the base frequency, which is determined by the rotational speed of the pump and the harmonics. Increased power at a frequency corresponding to the third harmonic was associated with pump thrombosis. A similar approach was reported recently [62].

Summary

Monitoring of VAD patients is a crucial component of successful therapy. Novel developments aim at supporting the patient in their self-monitoring by providing structured regular assistance via personal contact or via specifically developed software. Additionally, the barriers to

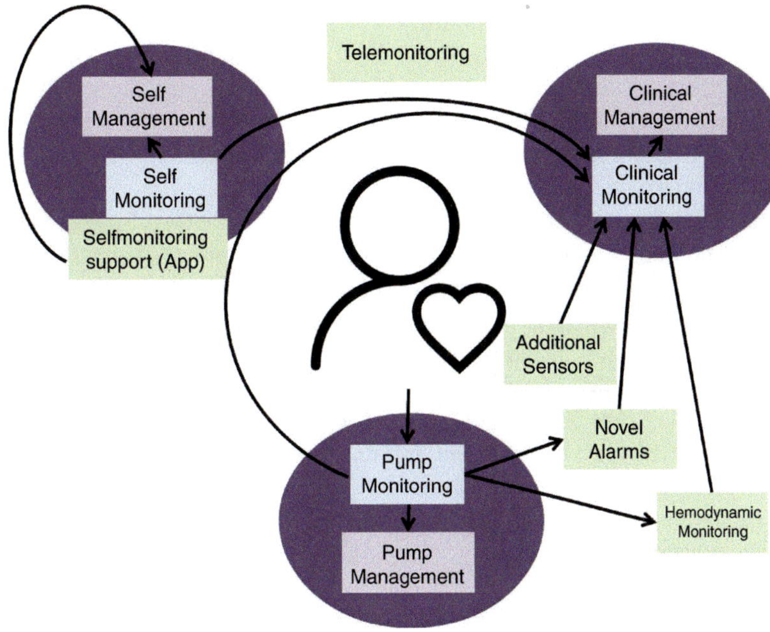

Fig. 46.4 Novel monitoring modalities in a structured overview

relay information to the healthcare provider are minimized (Fig. 46.4).

Clinicians are provided with novel clinical routines for detection of adverse events. Additionally, pump monitoring provides hemodynamic parameters, allowing accurate assessment of hemodynamic status to help in setting the right level of cardiac support. Finally, novel alarms aid in the timely discovery of anomalies.

Control of LVAD

Transition from Monitoring

In the previous section, pump monitoring was discussed as a tool for decision support. It was mainly used as a means to deliver information to the clinician, who may base their decision on whether to increase or decrease the provided support of the LVAD. Currently the level of support is set by adjusting a fixed pump speed, which the internal control of the LVAD will maintain.

However, due to changes in patient state, even throughout the day, be it circadian rhythm changes or changes in activity, different levels of support are warranted. One of the often proposed solutions to this and other problems in VAD therapy is an increased involvement of automatic control of the LVAD.

This automatic control may be achieved, by utilizing the established practices of pump monitoring and bypassing the clinician as an operator, directly altering the support setting, as depicted in Fig. 46.5.

The Clinician as Controller (Clinician in the Loop)

Currently, the hemodynamic interaction between human body and left ventricular assist device is governed by pump design via the pressure-flow characteristic curve. The control objective in clinically available LVADs is to maintain a constant speed, which is set by the physician.

The methodology of determining the optimal set speed is not uniform among the implanting centers [13]; however, there are recommendations, which rely on measurement of MAP, cardiac index and aortic valve opening, as well as echocardiographic evaluation of the septum, left ventricular unloading, and mitral regurgitation [3]. Additionally, depending on the pump type, the manufacturer issues confines for regular pump operation [63]. The PREVENT Trial suggested lower confines on pump speed [64].

When right heart catheterization, in combination with speed ramp tests, was applied to a patient cohort, which had received speed optimization according to ISHLT guidelines, it could be shown that this was the optimal setting in only 42.9% of patients, because information of central venous pressure and pulmonary capillary wedge pressure were not included in the decision-making process [65]. Less invasive techniques using Doppler echocardiography promise to be a sufficiently accurate replacement for the right heart catheterization measurements [66]. Furthermore, a more rigorous application of ramp tests to set speed was suggested [64].

An additional control variable that can be adjusted regularly by the patient is fluid intake. Based on the feedback of the pump, for example, during low-flow scenarios, patients are advised to increase fluid intake to elevate mean circulatory filling pressure and thus also preload. Additionally, in the Jarvik 2000™ (Jarvik Heart, Inc., New York, NY), the patient is allowed to manipulate the speed settings of the pump, thus manually controlling their LVAD [67].

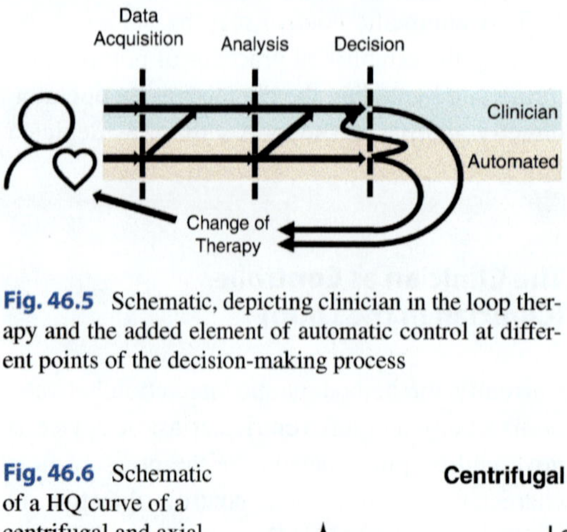

Fig. 46.5 Schematic, depicting clinician in the loop therapy and the added element of automatic control at different points of the decision-making process

Constant Speed Setting

Currently, if a patient requires additional support, e.g., due to physical exercise, the pump will only increase its flowrate (Q) based on the decrease in head pressure (H), which is defined as the pressure difference between pump inlet and outlet. These variables are inversely related via the H (head pressure)-Q (flowrate) curve of the pump, which can be seen in Fig. 46.6. Thus an increase of average left ventricular pressure without an increase in mean arterial pressure results in a decreased average head pressure and an increased average flowrate. However, if mean arterial pressure is increased, without an increase in mean ventricular pressure, flowrate is decreased in constant speed control.

Another drawback of this type of control is that during periods of reduced venous return, the pump might attempt to pump more than what is currently returned to the ventricle and will create unphysiologically low pressures within the ventricle, causing the septum to shift toward the cannula or the collapse of other structures into the pump inlet [40].

Feed Forward Control

Currently employed automatic control is restricted to open loop, meaning not taking any feedback from the patient.

Fig. 46.6 Schematic of a HQ curve of a centrifugal and axial LVAD in constant speed mode

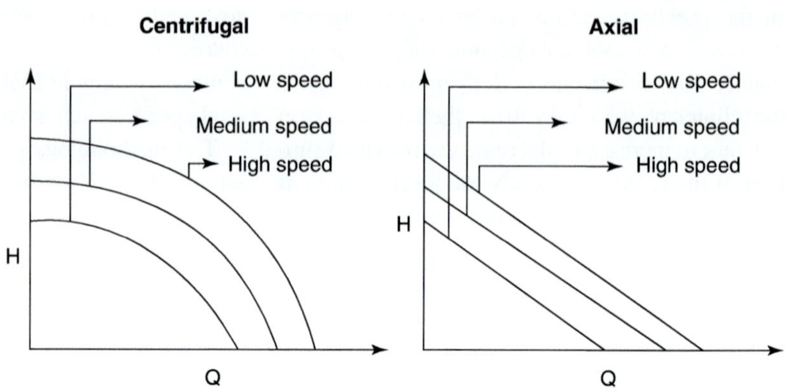

The Lavare Cycle™, which is implemented in the HVAD system, periodically reduces pump speed by 200 rpm for 2 seconds with a subsequent increase of 400 rpm for 1 second before returning to baseline. This is repeated every minute. This improves washout [68], which leads to an improved adverse event profile, but does not change physiological parameters like aortic valve opening [69].

The HMIII features a similar cyclical feed forward control, namely, the artificial pulse. With a 2000 rpm decrease followed by a 4000 rpm increase in a 350ms timeframe, repeated every 2 seconds, this is more severe than the Lavare Cycle™. While it could not be shown that this produces increased washout to regular operation in a simulation study, it was speculated that high rates of change in wall shear stress may dislodge deposits, thus reducing the susceptibility to thrombus growth [70]. The Incor (Berlin Heart, Berlin, Germany) implements a periodic speed reduction in order to generate reverse flow through the pump. The aim of this mechanism is improved washout, AoV opening, and dislodging of small thrombi [71]. And finally the MVAD (Medtronic, Inc., Fridley, MN, USA) utilizes the QPulse™, which also periodically reduces pump speed to allow for AoV opening.

Closed Loop Control

The HeartMate II™ [72] and the HeartMate III™ [73] implemented an algorithm, that senses changes in pulsatility and reacts by temporarily lowering the pump speed. This mechanism is aimed towards detecting and clearing suction events.

Physiologic Control

The design of a physiologic controller requires considerations:

What should be the purpose of the controller?

Based on the answer to this question, it is then possible to answer the more specific design questions:

How should the controller be structured?

What feedback variables should be used?

What is the control strategy?

And finally:

What is the interface to the physician, as the "highest level controller," and what should be provided as an interface?

Control Goals

The goals of the physiological controller should align with the goal of LVAD therapy in general. Clinically three main aims are defined: bridge to transplant, bridge to decision, and destination therapy. Depending on the main aim, further subaims may be weighed according to their importance. Common goals are:

- End organ perfusion
- Myocardial health
- Pulmonary health
- Adverse event prevention
  - Blood compatibility
  - Aortic valve health
- Sensitivity to changing demands

Structure of the Physiological Controller

A physiological controller usually relies on an intermediary control layer in a hierarchical structure (Fig. 46.7). The high level, physiological controller, receives external input, normally from a clinician. The physiological controller then determines a setpoint for the lower level control-

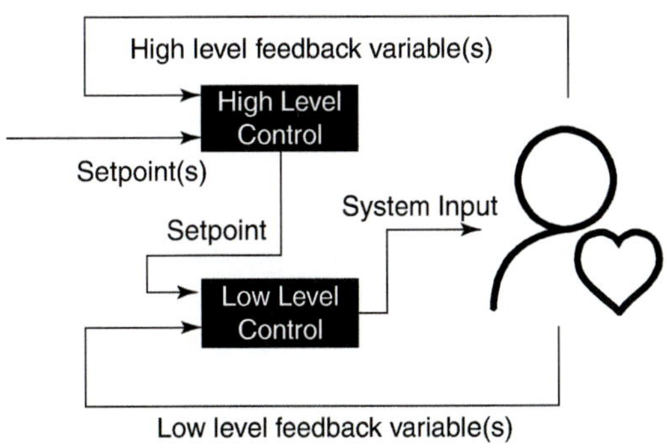

Fig. 46.7 Hierarchy of physiological control. A setpoint is set for a higher order controller by a clinician. This controller will provide a setpoint for the lower level controller, which ultimately acts on the pump

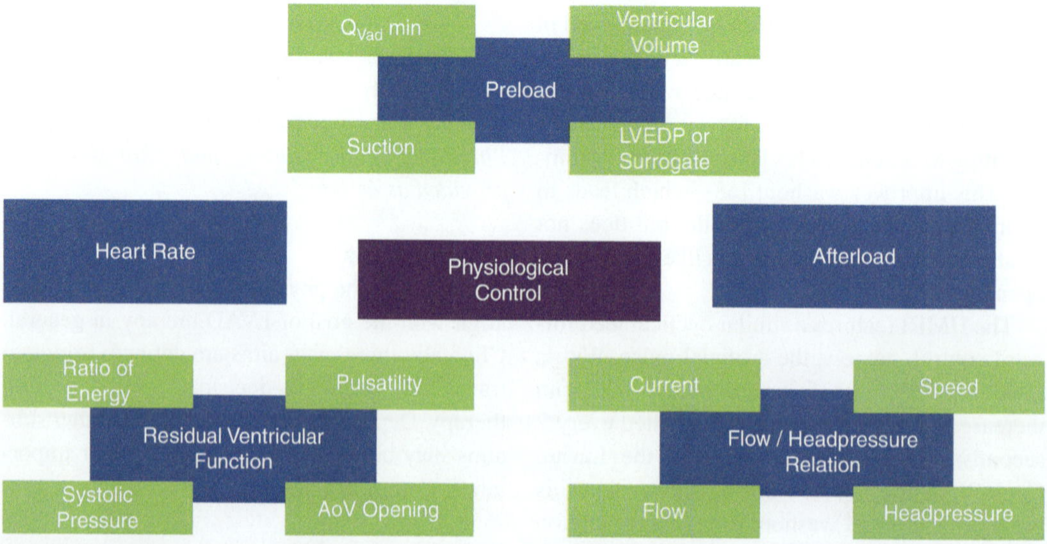

Fig. 46.8 Higher-order control paradigms clustered into five blocks based on conceptual similarity of feedback variable

ler, for example, via a PID Controller. While other low-level control variables, such as pump current, seem feasible and have been studied [74], constant speed control is the main paradigm in low-level control.

Setpoints

The external setpoint in Fig. 46.7 is defined by the clinical team. These setpoints have often been designed to be fixed values to be kept constant. But the controller might also entail some sort of function, such as maximization or minimization of the feedback variable, a linear or nonlinear function, or a combination of multiple feedback variables.

Physiologic Control Variables

The following section describes proposed higher order control circuits. The sorting into categories is based on assuming the primary goal of integration into patient physiology, especially to meet circulatory demand in order to maintain sufficient end organ perfusion. Various parameters have been assessed in previous studies. While some control strategies rely on additional sensors, a majority try to use only input data, which can be derived from readily available feedback variables, like pump speed or pump current. These variables can then be used to estimate hemodynamic variables such as pump flowrate, without additional sensors [48]. Fig. 46.8 displays the commonly used controlled variables. While some papers only utilize one of these, many combine these feedback variables.

Relation of Head Pressure and Flowrate

As described earlier, the LVAD will relate a certain head pressure to a given flowrate, depending on the speed in constant speed mode. This relation is portrayed in the HQ domain (Fig. 46.6). Changes in pump flowrate, even in a constant speed setting, are observed in patients with an increase in average ventricular pressure, if this increase is greater than the increase in aortic pressure. This is observed in exercising individuals and a result of head pressure sensitivity [75]. While the pump and cannula design will determine this relation in the constant speed mode, other methods of shaping this function have been proposed, building upon the paradigm of defining this ratio. The most commonly reported control strategies are constant flow control and constant head pressure control.

Modifications to the HQ Function

The possible relation between these two variables ranges on a spectrum between two extremes. On one end of this spectrum is the paradigm of constant flow control [76–78]. In this case, head pressure is irrelevant and may change freely, as long as pump flowrate is kept constant. The other end of this spectrum is formed by infinite head pressure sensitivity – control. This has been described as "constant head pressure control" [79–81]. In this control strategy, flowrate is allowed to change freely. Via these control strategies, long-term HQ curves could practically be changed to completely flat control curves via constant head pressure control, or they could be made completely vertical, thus indifferent to head pressure by a constant flowrate control. Of course, if the low-level controller continues to be a constant speed controller, these would not change the HQ characteristic within one cardiac cycle. Flat pump characteristic curves would still indicate higher sensitivity to short-term changes of ventricular pressure during systolic contraction, resulting in a greater flowrate variation.

These two extremes and the constant speed paradigm are depicted in Fig. 46.9 for a centrifugal LVAD, which has a nonlinear HQ characteristic and thus different head pressure sensitivities at different working points.

One additional feedback variable, which can be viewed in this domain, is pump current, which has been tested in vivo [82]. As the current is linked to torque via the motor torque constant of the motor, we could also model this as a constant torque control. Torque is mostly influenced by flowrate; thus this control strategy will not be very head pressure sensitive.

Conclusions: Relation of Head Pressure and Flowrate

The inherent limitation of relating a flowrate to a given head pressure is the indifference to the origin of the head pressure change. The controller response to an isolated increase in inlet pressure will be symmetrical to an isolated decrease in afterload. Similarly, it is not possible to differentiate a decrease in preload from an increase in afterload, which might not necessitate the same type of response.

Since the increase of cardiac output is usually the goal of cardiac assistance, designing this HQ function can be put into the perspective of tradeoff between the support flexibility of head pressure sensitivity, with all its drawbacks and the robustness of head pressure insensitivity.

Preload

Preload is arguably the most intuitive parameter to consider, since the Frank-Starling mechanism of the native heart functions in much the same way. Indeed most studies rely on some form of preload to determine the necessary hemodynamic energy to maintain balance between left and right ventricular outputs. However, preload is not uniformly defined and may thus cause some confusion.

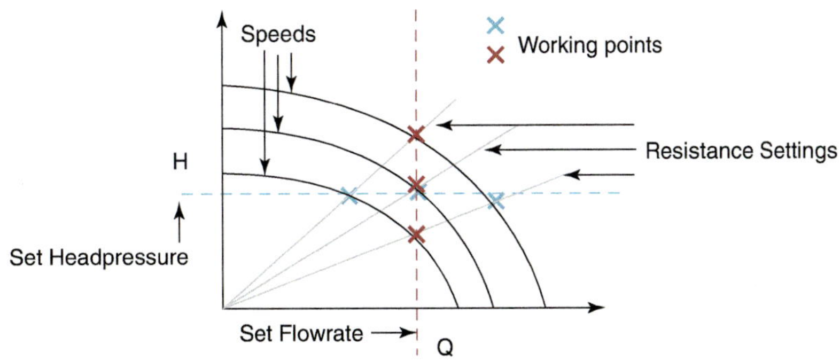

Fig. 46.9 Constant head pressure and flowrate settings and their response to changes in resistance

The terms preload and afterload were originally invented by muscle physiologists to describe the load on the muscle before and during contraction [83]. In cardiac physiology, especially due to the difficult nature of measuring those forces, surrogates have been introduced.

End-diastolic volume has been suggested as a preload surrogate, as it mostly closely correlates with sarcomere length. Linked via the end-diastolic pressure volume relationship is the end-diastolic pressure (LVEDP). The advantages of using pressures are that it might be easier to measure and that it is linked more directly to other interesting hemodynamic variables, such as the left atrial pressure and pulmonary venous pressure. The following is a list of preload surrogates, which have been applied as feedback variables.

Left Ventricular Diastolic Filling Pressure

Bullister et al. [84] utilized sensors to keep the left ventricular diastolic filling pressure, which was defined as the minimum pressure within a certain time window, with a secondary target of keeping the arterial pressure within boundaries. Heart rate was additionally used to toggle the arterial pressure confines. Methods to estimate diastolic ventricular pressure from pump data have also been developed [44, 45]. The estimation has also been used in a control strategy [85].

Left Ventricular End-Diastolic Volume

Ochsner et al. [86] reported on a method to control the LVAD with the ventricular volume as an input variable. In a follow-up paper, they included an ultrasonic volume sensor to continuously measure ventricular volume [87]. The proposed control not only uses ventricular end-diastolic volume but also heart rate to calculate a target pump power.

Q_{Vad}min

A preload surrogate which does not require a sensor is the minimum pump flow. This can be treated as preload surrogate because minimum flow will always occur during diastole. However, it conceptually struggles with the same issues as any pump flowrate, namely, that it is directly related to head pressure. Assuming constant arterial pressure, this means this can be a surrogate for preload, with increasing preload causing an increasing diastolic flowrate. Since aortic pressure is not expected to remain constant, a direct translation to preload is not possible. In order to combat this limitation, Chen et al. argue for a change in controller design [88]. The maximization of the minimum flow corresponds to the controller point that enables maximum volume unloading. A slight increase in speed will lead to suction, which will then again cause the Q_{Vad}min to decrease. The resulting controller should be able to maximize the minimum flowrate even with changing aortic pressure [89–91]. Ventricular properties especially diastolic filling properties can influence [92].

Suction Based

A special case of preload-based control variable is based on the phenomenon of suction. Ventricular suction is often caused by unphysiologically low pressures within the ventricle. Thus, from suction a low pressure in the ventricle can be inferred, specifically the lowest possible pressure, for the given cannula position and ventricular properties. The utilization of this state as a control variable was first proposed in 1991 [93]. To this day, even in control strategies, which use other hemodynamic variables, suction detection algorithms are applied to safely confine pump speed, thereby avoiding overpumping.

An important constraint on suction-based control are presuction conditions such as a shifted septum and concomitant tricuspid valve incompetency. As most suction detection systems aim at detecting partial or complete inflow occlusion, presuction and the geometrical shifts might be tolerated or even encouraged [94, 95].

Another way the LVAD can go into a state of suction is due to inflow cannula malposition. The cannula may face either the ventricular wall or the septum and may be pushed or pulled against those structures, causing suction. Even though this could be an indication of left ventricular volume, it is much less direct and might not be related to any preload metric at all.

Conclusions: Preload

Diastolic pressures, volumes, and flows allow surrogate measurements of ventricular preload. The Frank-Starling mechanism relates cardiac output to preload in the native heart. This mechanism is diminished in heart failure patients. Thus preload-based control could allow to reinstate the Frank-Starling mechanism, underlining the permissive role of the heart [1], which is thereby extended to the heart/LVAD complex. Control based solely on suction detection will theoretically maximally unload the ventricle. However, care has to be taken to not run into presuction conditions. Additionally, chronotropic and inotropic changes can only be picked up indirectly, via their effect on preload.

Heart Rate

Circulatory demand is coupled with heart rate via the sympathetic and parasympathetic nervous system [1]. As long as the patient is chronotropically responsive, heart rate will increase with increased circulatory demand. Since the extent of this response varies from patient to patient, the adequate controller gain usually has to be defined in its magnitude by the clinician. One commonly used method is to apply a linear function to map a target pump flowrate to a heart rate [96, 97].

Karantonis et al. combined heart rate with an accelerometer-derived activity index to get a more accurate reading on patient physiologic demand and to aid in situations that would be a counterindication to heart rate-based control, like chronotropic impairment [98].

Chang et al. utilized a different approach and cite the baroreflex as their intermediator between heart rate and aortic pressure. They attempted to track a reference heart rate in an attempt to stabilize aortic pressure [99].

Heart rate is also often used in combination with other measures to determine a controller setpoint, such as end-diastolic volume [86], end-diastolic pressure [84], or flowrate pulsatility [100].

Conclusions: Heart Rate

Heart rate is a reliable indicator of patient activity. However, chronotropic incompetence and rhythmological problems are expected in a large portion of LVAD patients as evidenced by coinciding ICD implants in more than a third of patients [101] and CRT device implants in an additional subset of patient. While it could be shown that average heart rate could still be reliably detected during periods of arrhythmia, [100] chronotropic response is nonetheless a prerequisite for this type of control

Afterload

Just like preload, the term "afterload" originates from muscle physiologists. As ventricular pressure itself would be determined by too many other factors internal to the ventricle (like contractility), arterial input impedance is the closest surrogate, which is completely external to the ventricle. Arterial input impedance is comprised of static and dynamic components.

Ventricular pressure during the ejection phase can be a surrogate for afterload on the myocardium. However, if the focus is on the heart as a pump and not on the ventricular muscle properties, as is the case in most hemodynamic endeavors for control strategies, pressure afterload as specified by the aortic pressure might be sufficient [83].

There are two reasons why one would like to control afterload. Firstly, controlled muscle loading might aid in recovery and reverse remodeling [102]. Secondly, keeping the aortic pressure constant, via a mechanism akin to the heart rate and inotropy regulating parts of the baroreceptor reflex, the blood demand of the individual tissues could be set by the tissues themselves individually. Thus the resistance could control the cardiac output [103, 104]. Afterload control is often implemented in addition to other control variables in order to confine the aortic pressure and to prevent the previously mentioned downsides of hypertension [105, 106]. The downside of this approach is that of lacking preload sensitivity.

Conclusions: Afterload

Controlling the aortic pressure runs contrary to what the native ventricle does. Thus "circulatory assist device" would be a more accurate description of devices utilizing this type of control than "left ventricular assist device." The healthy left ventricle is afterload insensitive up to the point of failure [1]. Short-term aortic pressure regulation is upheld by some mechanisms partially actuated by the ventricle, such as the baroreceptor reflexes. Many other short-term regulation mechanisms are not influenced by the ventricle. Especially in exercise scenarios, when aortic pressure is increased due to contraction of muscles and sympathetic stimulation, aortic pressure control would respond by decreasing LVAD support. A more practical implementation limitation is the measurement or estimation of AOP, for similar reasons as discussed in the preload control section.

Passive Control

While the other proposed control strategies rely on active electronic control, passive control does not. Gregory et al. evaluated a compliant inflow cannula, which would increase hydraulic resistance when ventricular pressures were low. This was accomplished by a passive contraction, comparable to controlled suction, thus avoiding ventricular suction [107].

Residual Ventricular Function

In this section control strategies rely on measured or estimated function of the ventricle. In contrast to the preload-based controllers, some measure of residual ventricular contraction is used as a surrogate for preload. Higher left atrial pressure will result in stronger contraction due to increase in ventricle preload due to the Frank-Starling mechanism. The contraction of the ventricle is then measured or estimated and can be used to infer preload (Fig. 46.10).

Ratio of Hydraulic Energy

While the blood assist index was developed for an intra-aortic pump, the paradigm of relating pump hydraulic power to total hydraulic power is still applicable to more traditional LVAD approaches [108, 109]. Indeed in [110, 111] two pressure sensors and one flowrate sensor were used to quantify hydraulic power produced by the ventricle. The pump hydraulic power is then set to a fixed ratio to this calculated power, so that a more active ventricle necessitates a higher level of support. This, however, will only accurately assess hydraulic energy in full support, as the aortic ejection cannot be easily quantified, leading to an underestimation of ventricular power.

Pulsatility

Pulsatility is defined as the difference between the maximum and the minimum of a given parameter during a cardiac cycle or time window. These parameters include ventricular pressure, head pressure, or pump flow.

As most of the variation in these variables over the cardiac cycle is produced by the ventricle, it can be seen as a measure of ventricular performance with some caveats. The first caveat is that of nonlinear HQ curve. Different segments of the HQ curve of a centrifugal pump have dif-

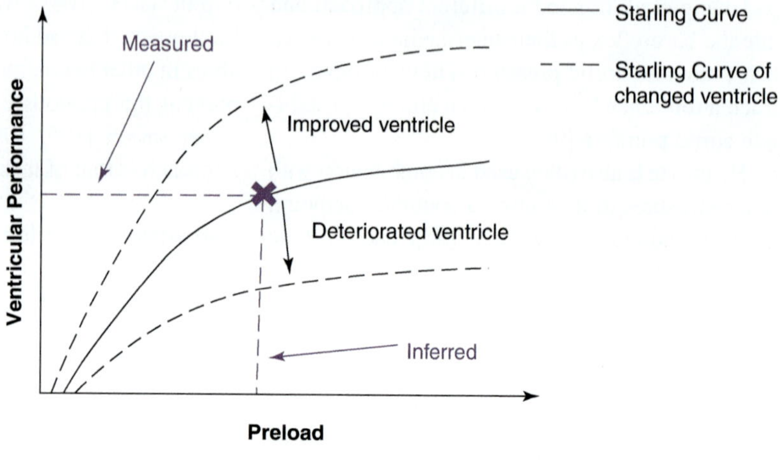

Fig. 46.10 Schematic of the Starling curve showing the relation of ventricular performance and preload. A monotonically increasing function offers possibility of proportional mapping. Changes in inotropy or ventricle health change the slope of the curve

ferent slopes, changing the relation of head pressure and flowrate pulsatility. Since pulsatility is calculated by subtracting the minimum value within a given time window from the maximum, any limiting factors have to be considered as well. The maximum of flow is limited by the x intercept of the HQ curve. Since at 0 head pressure, the aortic valve will open, and the aortic pressure and the ventricular pressure will be approximately the same. The minimum flow is not that strictly bounded and can even become negative if the head pressure becomes sufficiently high for a given pump speed. Another factor that has to be considered is the case of inflow cannula occlusion, which happens during suction. In this case minimum flow will be artificially low.

Cysyk used a pressure sensor at the pump inlet to measure pressure pulsatility. The controller then aimed to maintain a constant pulsatility, with an additional constraint on suction [112]. It was observed that flowrate pulsatility has a minimum at the speed that maximally unloads the ventricle. Thus when a controller aims to maintain a small amount of flow pulsatility, the level of unloading can be adjusted [113, 114].

Systolic Pressure

Petrou et al. [115] investigated a controller based on measured systolic pressure. The setpoint is chosen by the clinician but should remain in full assist. If systolic pressure increases, pump speed will increase as well. Using systolic pressure may have two uses: tracking ventricular function as well as controlling the loading of the heart, similar to the second afterload approach discussed earlier.

Aortic Valve Opening

This support strategy treats the change of full to partial assist as observed variable. Similarly to the suction-based control strategy in the preload section, this is state based. The proposed benefits include aortic valve health and controlled loading of the heart [71, 116]. The objective is to decrease speed of the pump until aortic valve opening is detected. Detection of aortic valve opening was detected by flow waveform analysis [42] or the head pressure pulsatility [116]. The physiologic limitations are that if the ventricle of the patient is not sufficiently strong, partial assist may not be achieved before severe underperfusion, or at all.

Conclusions: Residual Ventricular Function

Control mechanisms based on residual ventricular function aim to amplify the remaining Frank-Starling response of the heart. However, these control algorithms have issues when the heart deteriorates and would all react with a decrease in support if the natural heart progressively deteriorates. This may happen either due to arrhythmia or worsening heart failure symptoms. Thus these control algorithms need to be nested in a cascaded controller, which switches to a safe mode in case of a falling heart. Additionally, estimators of ventricular function become increasingly more difficult in a partial assist support scenario. Changes in inotropy would also change the preload sensitivity of the native ventricle, thus proportionally changing the preload sensitivity of the controller (Fig. 46.10).

Multiobjective Control

A combination of many of the abovementioned principles is often discussed in literature. A recent example is the combination of a measure of residual ventricular function in pump speed pulsatility and constant head pressure control [117]. The most prominent example, and the only control strategy, that has been used in a clinical trial [100, 118] used two setpoints based on heart rate to determine target flowrates. The second part of this control strategy was a venous return index, which was determined by observing changes in the flowrate pulsatility. Additional hierarchical layers were responsible for clearance of suction, utilizing an expert trained suction detection system and setting boundaries of operation, like a minimum acceptable pump flowrate and a maximum power consumption [118]. This multifactor approach allowed the control system to adapt to changes in patient physiology resulting from spiroergometry, Valsalva maneuver, or orthostatic changes. It was shown to safely and reliably increase pump flowrate and venous oxygen saturation during exercise. Additionally, the control algorithm reduced PCWP and PAP compared to a constant speed setting. This study proved the feasibility and the benefits of control directly in patients.

Intrabeat Control

Another type of control is that of short-term speed modulations. These cyclic speed variations usually occur in the timeframe of the native heart rate and can be split into copulse, counterpulse, and asynchronous mode [119]. While the copulse mode increases pump speed during systole, increasing the systemic flowrate pulsatility but also adversely affecting aortic valve opening, and myocardial perfusion, counterpulse mode achieves the opposite. Asynchronous pulsation was shown to produce many of the benefits of synchronized control, without the necessary feedback [120].

Summary

Much effort has been put forward to enable autonomous control of the LVAD. Nonetheless, the main modus operandi in clinically available LVADs is that of constant speed set by the physician.

Recent advances in control applications as well as comparisons of control strategies in a unified test bed [121, 122] aimed at gaining a more complete picture of the potential of this technology. Questions of liability and lack of evidence of long-term stability remain an issue for the widespread application. But isolated mechanisms such as suction detection and clearance, as well as feedforward control mechanisms, have already found their way into clinical practice. Improved monitoring, especially of hemodynamic parameters, could provide the link to control, as the widespread use could improve confidence in the algorithms, which could then eventually be trusted sufficiently to allow autonomous modification of pump support.

Acknowledgement This study was partially supported by the Austrian Science Fund (FWF) KLI357.

References

1. Guyton AC. Textbook of Medical Physiology. 7th ed. London: Saunders; 1986.
2. Aronson JK. Monitoring therapy. Br J Clin Pharmacol. 2005;60:229–30.
3. Feldman D, Pamboukian SV, Teuteberg JJ, et al. The 2013 international society for heart and lung transplantation guidelines for mechanical circulatory support: executive summary. J Heart Lung Transplant. 2013;32:157–87.
4. Kirklin JK, Pagani FD, Kormos RL, Stevenson LW, Blume ED, Myers SL, Miller MA, Baldwin JT, Young JB, Naftel DC. Eighth annual INTERMACS report: special focus on framing the impact of adverse events. J Heart Lung Transplant. 2017;36:1080–6.
5. Schlöglhofer T, Horvat J, Moscato F, et al. A standardized telephone intervention algorithm improves the survival of ventricular assist device outpatients. Artif Organs. 2018; https://doi.org/10.1111/aor.13155.
6. Pamboukian SV, Tallaj JA, Brown RN, et al. Improvement in 2-year survival for ventricular assist device patients after implementation of an intensive surveillance protocol. J Heart Lung Transplant. 2011;30:879–87.
7. Slotwiner D, Varma N, Akar JG, et al. HRS expert consensus statement on remote interrogation and monitoring for cardiovascular implantable electronic devices. Hear Rhythm. 2015;12:e69–e100.
8. Schlöglhofer T, Robson D, Bancroft J, Sørensen G, Kaufmann F, Sweet L, Wrightson N. International coordinator survey results on the outpatient management of patients with the heartware® ventricular assist system. Int J Artif Organs. 2016;39:553–7.
9. Schlöglhofer T, Horvat J, Hartner Z, Riebandt J, Haberl T, Wiedemann D, Dimitrov K, Moscato F, Zimpfer D, Schima H. Standardized telephone intervention algorithm for improved ventricular assist device outpatient management. J Heart Lung Transplant. 2016;35:S388.
10. Estep JD, Trachtenberg BH, Loza LP, Bruckner BA. Continuous flow left ventricular assist devices: shared care goals of monitoring and treating patients. Methodist Debakey Cardiovasc J. 2015;11:33–44.
11. Tchantchaleishvili V, Sagebin F, Ross RE, Hallinan W, Schwarz KQ, Massey HT. Evaluation and treatment of pump thrombosis and hemolysis. Ann Cardiothorac Surg. 2014;3:490–5.
12. Kusne S, Mooney M, Danziger-Isakov L, et al. An ISHLT consensus document for prevention and management strategies for mechanical circulatory support infection. J Heart Lung Transplant. 2017;36:1137–53.
13. Stainback RF, Estep JD, Agler DA, Birks EJ, Bremer M, Hung J, Kirkpatrick JN, Rogers JG, Shah NR. Echocardiography in the management of patients with left ventricular assist devices: recommenda-

tions from the American society of echocardiography. J Am Soc Echocardiogr. 2015;28:853–909.
14. Riegel B, Lee CS, Dickson VV. Self care in patients with chronic heart failure. Nat Rev Cardiol. 2011;8:644–54.
15. Ryan P, Sawin KJ. The individual and family self-management theory: background and perspectives on context, process, and outcomes. Nurs Outlook. 2009;57:217–225.e6.
16. Kato N, Jaarsma T, Ben Gal T. Learning self-care after left ventricular assist device implantation. Curr Heart Fail Rep. 2014;11:290–8.
17. Casida JM, Wu HS, Abshire M, Ghosh B, Yang JJ. Cognition and adherence are self-management factors predicting the quality of life of adults living with a left ventricular assist device. J Heart Lung Transplant. 2017;36:325–30.
18. Halder LC, Richardson LB, Garberich RF, Zimbwa P, Bennett MK. Time in therapeutic range for left ventricular assist device patients anticoagulated with warfarin: a correlation to clinical outcomes. ASAIO J. 2017;63:37–40.
19. Woods A, MacGowan G, Schueler S, Wrightson N, Robinson-Smith N. Reducing readmissions of outpatients with LVADs. Br J Card Nurs. 2015;10.
20. Casida JM, Aikens JE, Craddock H, Aldrich MW, Pagani FD. Development and feasibility of self-management application in left-ventricular assist devices. ASAIO J. 2018;64:159–67.
21. Goldstein DJ, Naftel D, Holman W, Bellumkonda L, Pamboukian SV, Pagani FD, Kirklin J. Continuous-flow devices and percutaneous site infections: clinical outcomes. J Heart Lung Transplant. 2012;31:1151–7.
22. Reiss N, Jansen S, Luneburg N, Feldmann C, Wendl R, Schmitto J, Neubert R. Telemonitoring of driveline exit site for early detection of relevant driveline infections. J Heart Lung Transplant. 2018;37:S135.
23. Hohmann S, Veltmann C, Duncker D, et al. Initial experience with telemonitoring in left ventricular assist device patients. 2018:1–11.
24. Pektok E, Demirozu ZT, Arat N, et al. Remote monitoring of left ventricular assist device parameters after heart assist-5 implantation. Artif Organs. 2013;37:820–5.
25. Reiss N, Schmidt T, Boeckelmann M, Schulte-Eistrup S, Hoffmann JD, Feldmann C, Schmitto JD. Telemonitoring of left-ventricular assist device patients-Current status and future challenges. J Thorac Dis. 2018;10:S1794–801.
26. Yandrapalli S, Raza A, Tariq S, Aronow WS. Ambulatory pulmonary artery pressure monitoring in advanced heart failure patients. World J Cardiol. 2017;9:21.
27. Harris JR, Carlson SK, Wolfson AM, Saxon LA, Doshi RN. Pulmonary arterial pressure sensing in a patient with left ventricular assist device during ventricular arrhythmia. Hear Case Reports. 2017;3:348–51.
28. Kilic A, Katz JN, Joseph SM, Brisco-Bacik MA, Uriel N, Lima B, Agarwal R, Bharmi R, Farrar DJ, Lee S. Changes in pulmonary artery pressure before and after left ventricular assist device implantation in patients utilizing remote haemodynamic monitoring. ESC Hear Fail. 2018;23:S30.
29. Granegger M, Schlöglhofer T, Ober H, Zimpfer D, Schima H, Moscato F. Daily life activity in patients with left ventricular assist devices. Int J Artif Organs. 2016;39:22–7.
30. Glitza JI, Müller-von Aschwege F, Eichelberg M, Reiss N, Schmidt T, Feldmann C, Wendl R, Schmitto JD, Hein A. Advanced telemonitoring of Left Ventricular Assist Device patients for the early detection of thrombosis. J Netw Comput Appl. 2018;118:74–82.
31. Imamura T, Chung B, Nguyen A, Sayer G, Uriel N. Clinical implications of hemodynamic assessment during left ventricular assist device therapy. J Cardiol. 2018;71:352–8.
32. Teuteberg JJ, Slaughter MS, Rogers JG, et al. The HVAD left ventricular assist device: risk factors for neurological events and risk mitigation strategies. JACC Hear Fail. 2015;3:818–28.
33. Najjar SS, Slaughter MS, Pagani FD, et al. An analysis of pump thrombus events in patients in the HeartWare ADVANCE bridge to transplant and continued access protocol trial. J Heart Lung Transplant. 2014;33:23–34.
34. Saeed O, Jermyn R, Kargoli F, et al. Blood pressure and adverse events during continuous flow left ventricular assist device support. Circ Heart Fail. 2015;8:551–6.
35. Bennett MK, Roberts CA, Dordunoo D, Shah A, Russell SD. Ideal methodology to assess systemic blood pressure in patients with continuous-flow left ventricular assist devices. J Heart Lung Transplant. 2010;29:593–4.
36. Lanier GM, Orlanes K, Hayashi Y, et al. Validity and reliability of a novel slow cuff-deflation system for noninvasive blood pressure monitoring in patients with continuous-flow left ventricular assist device. Circ Heart Fail. 2013;6:1005–12.
37. Rangasamy S, Madan S, Saeed O, Goldstein DJ, Jorde UP, Negassa A, Patel SR. Noninvasive measures of pulsatility and blood pressure during continuous-flow left ventricular assist device. ASAIO J. 2018;1.
38. Naiyanetr P, Moscato F, Vollkron M, Zimpfer D, Wieselthaler G, Schima H. Continuous assessment of cardiac function during rotary blood pump support: a contractility index derived from pump flow. J Heart Lung Transplant. 2010;29:37–44.
39. Moscato F, Granegger M, Naiyanetr P, Wieselthaler G, Schima H. Evaluation of left ventricular relaxation in rotary blood pump recipients using the pump flow waveform: a simulation study. Artif Organs. 2012;36:470–8.
40. Vollkron M, Schima H, Huber L, Benkowski R, Morello G, Wieselthaler G. Development of a suction detection system for axial blood pumps. Artif Organs. 2004;28:709–16.

41. Granegger M, Masetti M, Laohasurayodhin R, Schloeglhofer T, Zimpfer D, Schima H, Moscato F. Continuous monitoring of aortic valve opening in rotary blood pump patients. IEEE Trans Biomed Eng. 2016;63:1201–7.
42. Hayward C, Lim CP, Schima H, Macdonald P, Moscato F, Muthiah K, Granegger M. Pump speed waveform analysis to detect aortic valve opening in patients on ventricular assist device support. Artif Organs. 2015;39:704–9.
43. Moscato F, Granegger M, Edelmayer M, Zimpfer D, Schima H. Continuous monitoring of cardiac rhythms in left ventricular assist device patients. Artif Organs. 2014;38:191–8.
44. Grinstein J, Rodgers D, Kalantari S, et al. HVAD waveform analysis as a noninvasive marker of pulmonary capillary wedge pressure: a first step toward the development of a smart left ventricular assist device pump. ASAIO J. 2017:1–6.
45. Lai JV, Muthiah K, Macdonald PS, Jansz P, Hayward CS. Estimation of left ventricular assist device preload using pump flow waveform analysis. J Heart Lung Transplant. 2017;36:240–2.
46. Granegger M, Schima H, Zimpfer D, Moscato F. Assessment of aortic valve opening during rotary blood pump support using pump signals. Artif Organs. 2014;38:290–7.
47. Moscato F, Granegger M, Edelmayer M, Zimpfer D, Schima H. Continuous monitoring of cardiac rhythms in left ventricular assist device patients. Artif Organs. 2014;38:191–8.
48. Granegger M, Moscato F, Casas F, Wieselthaler G, Schima H. Development of a pump flow estimator for rotary blood pumps to enhance monitoring of ventricular function. Artif Organs. 2012;36:691–9.
49. Schlöglhofer T, Schima H. Wearable systems. In: Mech. Circ. Respir. Support. Elsevier; 2018. p. 691–721.
50. Schima H, Schlöglhofer T, Hartner Z, Horvat J, Zimpfer D. Importance of linguistic details in alarm messages of ventricular assist devices. Int J Artif Organs. 2013;36:406–9.
51. Parliament THEE, Council THE, The OF, Union E. 2007. Council Directive 2007/47/EC (Amendment). Off J Eur Union.
52. Trinquero P, Pirotte A, Gallagher LP, Iwaki KM, Beach C, Wilcox JE. Left ventricular assist device management in the emergency department. 2018;19:834–41.
53. Corp. T. 2015. Urgent Medical Device Correction. https://www.hsa.gov.sg/content/dam/HSA/HPRG/Medical_Devices/Updates_and_Safety_reporting/Field_Safety_Corrective_Action/FSN/2015/September/HSA6004101-070-15-19_52 FSN.pdf. Accessed 18 Feb 2019.
54. Borowski M, Görges M, Fried R, Such O, Wrede C, Imhoff M. Medical device alarms. Biomed Tech. 2011;56:73–83.
55. Schima H, Morshuis M, Zimpfer D, Zu Dohna R, Roefe D, Drews T, Schmitto JD, Strüber M, Schlöglhofer T. Usability of ventricular assist devices in daily experience: a multicenter study. Artif Organs. 2015;38:751–60.
56. Association for the Advancement of Medical Instrumentation. Human factor engineering - Design of medical devices. AAMI Guid. 2009:21–5.
57. Jorde UP, Aaronson KD, Najjar SS, et al. Identification and management of pump thrombus in the heartware left ventricular assist device systema novel approach using log file analysis. JACC Hear Fail. 2015;3:849–56.
58. Robesaat JI, Von Aschwege FM, Reiss N, Schmidt T, Feldmann C, Deniz E, Schmitto JD, Hein A. Analysis of LVAD log files for the early detection of pump thrombosis. Proc - IEEE Symp Comput Commun. 2017:236–41.
59. Dimitrov K, Riebandt J, Haberl T, Wiedemann D, Simon P, Laufer G, Schima H, Zimpfer D. High-intensity transient signals in the outflow graft and thrombosis of a heartware left ventricular assist device. Ann Thorac Surg. 2016;101:e83–5.
60. Hubbert L, Sundbom P, Loebe M, Peterzén B, Granfeldt H, Ahn H. Acoustic analysis of a mechanical circulatory support. Artif Organs. 2014;38:593–8.
61. Kaufmann F, Hörmandinger C, Stepanenko A, Kretzschmar A, Soltani S, Krabatsch T, Potapov E, Hetzer R. Acoustic spectral analysis for determining pump thrombosis in rotary blood pumps. ASAIO J. 2014;60:502–7.
62. Feldmann C, Deniz E, Stomps A, et al. An acoustic method for systematic ventricular assist device thrombus evaluation with a novel artificial thrombus model. J Thorac Dis. 2018;10:S1711–9.
63. HeartWare. 2014. HeartWare ® Ventricular Assist System Instructions for Use.
64. Maltais S, Kilic A, Nathan S, et al. PREVENtion of HeartMate II Pump thrombosis through clinical management: the PREVENT multi-center study. J Heart Lung Transplant. 2017;36:1–12.
65. Uriel N, Sayer G, Addetia K, et al. Hemodynamic ramp tests in patients with left ventricular assist devices. JACC Hear Fail. 2016;4:208–17.
66. Mcduff DJ, Estepp JR, Piasecki AM, Blackford EB. A survey of remote optical photoplethysmographic imaging. Methods. 2015:6398–404.
67. Frazier OH, Myers TJ, Gregoric ID, Khan T, Delgado R, Croitoru M, Miller K, Jarvik R, Westaby S. Initial clinical experience with the Jarvik 2000 implantable axial-flow left ventricular assist system. Circulation. 2002;105:2855–60.
68. Zimpfer D, Strueber M, Aigner P, et al. Evaluation of the HeartWare ventricular assist device Lavare cycle in a particle image velocimetry model and in clinical practice. Eur J Cardiothorac Surg. 2016;50:839–48.
69. Clifford R, Kim YD, Robson D, Gross C, Moscato F, Schima H, Macdonald PS, Jansz P, Hayward CS. Effect of the lavare cycle on pump function, aortic valve opening, autonomic function and activity

outcomes in continuous flow lvad patients. J Heart Lung Transplant. 2017;36:S82.

70. Wiegmann L, Thamsen B, de Zélicourt D, Granegger M, Boës S, Schmid Daners M, Meboldt M, Kurtcuoglu V. Fluid dynamics in the HeartMate 3: influence of the artificial pulse feature and residual cardiac pulsation. Artif Organs. 2018:0–3.
71. Arndt A, Nüsser P, Lampe B. Fully autonomous preload-sensitive control of implantable rotary blood pumps. Artif Organs. 2010;34:726–35.
72. Corp. T. 2009. HeartMateII Instructions for Use. https://www.accessdata.fda.gov/cdrh_docs/pdf6/P060040S005c.pdf. Accessed 14 Nov 2018.
73. Corp. T. 2017. HeartMateIII Instructions for Use. https://www.accessdata.fda.gov/cdrh_docs/pdf16/P160054C.pdf. Accessed 14 Nov 2018.
74. Nishimura K, Tsukiya T, Akamatsu T, Kono S, Yamada T, Ban T (1997) Nishimura1997_Control_of_the_Pressure_Flow_Relationship.pdf.
75. Gross C, Marko C, Mikl J, Altenberger J, Schlöglhofer T, Schima H, Zimpfer D, Moscato F (2018) LVAD pump flow does not adequately increase with exercise. Artif Organs 0:1–7
76. Casas F, Ahmed N, Reeves A. Minimal sensor count approach to fuzzy logic rotary blood pump flow control. ASAIO J. 2007;53:140–6.
77. Song Z, Gu K, Gao B, Wan F, Chang Y, Zeng Y. Hemodynamic effects of various support modes of continuous flow LVADs on the cardiovascular system: a numerical study. Med Sci Monit. 2014;20:733–41.
78. Verbeni A, Fontana R, Silvestri M, Tortora G, Vatteroni M, Trivella MG, Dario P. An innovative adaptive control strategy for sensorized left ventricular assist devices. IEEE Trans Biomed Circuits Syst. 2014;8:660–8.
79. Waters T, Allaire P, Tao G, Adams M, Bearnson G, Wei N, Hilton E, Baloh M, Olsen D, Khanwilkar P. Motor feedback physiological control for a continuous flow ventricular assist device. Artif Organs. 1999;23:480–6.
80. Giridharan GA, Skliar M, Olsen DB, Pantalos GM. Modeling and control of a brushless dc axial flow ventricular assist device. ASAIO J. 2002;48:272–89.
81. Giridharan GA, Skliar M. Nonlinear controller for ventricular assist devices. Artif Organs. 2002;26:980–4.
82. Nishimura K, Tsukiya T, Akamatsu T, Kono S, Yamada T, Ban T. Control of the pressure flow relationship with a magnetically suspended centrifugal pump in a chronic animal experiment. ASAIO J. 1997;43:M553–6.
83. Sagawa K, Maughan L, Hiroyuki Suga KS. Cardiac contraction and the pressure-volume relationship: Oxford University Press; 1988.
84. Bullister E, Reich S, Sluetz J. Physiologic control algorithms for rotary blood pumps using pressure sensor input. Artif Organs. 2002;26:931–8.
85. Alomari AH, Savkin AV, Ayre PJ, Lim E, Mason DG, Salamonsen RF, Fraser JF, Lovell NH. Non-invasive estimation and control of inlet pressure in an implantable rotary blood pump for heart failure patients. Physiol Meas. 2011; https://doi.org/10.1088/0967-3334/32/8/004.
86. Ochsner G, Amacher R, Wilhelm MJ, Vandenberghe S, Tevaearai H, Plass A, Amstutz A, Falk V, Schmid Daners M. A physiological controller for turbodynamic ventricular assist devices based on a measurement of the left ventricular volume. Artif Organs. 2014;38:527–38.
87. Ochsner G, Wilhelm MJ, Amacher R, et al. In vivo evaluation of physiologic control algorithms for left ventricular assist devices based on left ventricular volume or pressure. ASAIO J. 2017;63:568–77.
88. Chen SCS, Antaki JF, Simaan MA, Boston JR. Physiological control of left ventricular assist devices based on gradient of flow. Proc Am Control Conf. 2005, 2005; https://doi.org/10.1109/ACC.2005.1470571.
89. Gwak K-W. Application of extremum seeking control to turbodynamic blood pumps. ASAIO J. 2007;53:403–9.
90. Gwak KW, Antaki JF, Paden BE, Kang B. Safety-enhanced optimal control of turbodynamic blood pumps. Artif Organs. 2011;35:725–32.
91. Simaan MA, Ferreira A, Member S, Chen S, Antaki JF, Galati DG. A dynamical state space representation and performance analysis of a feedback-controlled rotary left ventricular assist device. IEEE Trans Control Syst Technol. 2009;17:15–28.
92. Moscato F, Granegger M, Naiyanetr P, Wieselthaler G, Schima H. Evaluation of left ventricular relaxation in rotary blood pump recipients using the pump flow waveform: a simulation study. Artif Organs. 2012;36:470–8.
93. Schima H, Trubel W, Moritz A, Wieselthaler G, Stohr HG, Thoma H, Losert U, Wolner E. Noninvasive monitoring of rotary blood pumps: necessity, Possibilities, and Limitations. 1992;16:195–202.
94. Ferreira A, Boston JR, Antaki JF. A control system for rotary blood pumps based on suction detection. IEEE Trans Biomed Eng. 2009;56:656–65.
95. Ferreira A, Boston JR, Antaki JF. A rule-based controller based on suction detection for rotary blood pumps. Annu Int Conf IEEE Eng Med Biol – Proc. 2007:3978–81.
96. Fu M, Xu L. Computer simulation of sensorless fuzzy control of a rotary blood pump to assure normal physiology. ASAIO J. 2000;46:273–8.
97. Ohuchi K, Kikugawa D, Takahashi K, Uemura M, Nakamura M, Murakami T, Sakamoto T, Takatani S. Control strategy for rotary blood pumps. Artif Organs. 2001;25:366–70.
98. Karantonis DM, Lim E, Mason DG, Salamonsen RF, Ayre PJ, Lovell NH. Noninvasive activity-based control of an implantable rotary blood pump: comparative software simulation study. Artif Organs. 2010;34:34–45.

99. Chang Y, Gao B, Gu K. A model-free adaptive control to a blood pump based on heart rate. ASAIO J. 2011;57:262–7.
100. Vollkron M, Schima H, Huber L, Benkowski R, Morello G, Wieselthaler G. Development of a reliable automatic speed control system for rotary blood pumps. J Heart Lung Transplant. 2005;24:1878–85.
101. Vakil K, Kazmirczak F, Sathnur N, Adabag S, Cantillon DJ, Kiehl EL, Koene R, Cogswell R, Anand I, Roukoz H. Implantable cardioverter-defibrillator use in patients with left ventricular assist devices. JACC Hear Fail. 2016;4:772–9.
102. Moscato F, Arabia M, Colacino FM, Naiyanetr P, Danieli GA, Schima H. Left ventricle afterload impedance control by an axial flow ventricular assist device: a potential tool for ventricular recovery. Artif Organs. 2010;34:736–44.
103. Wu Y, Allaire P, Tao G, Wood H, Olsen D, Tribble C. An advanced physiological controller design for a left ventricular assist device to prevent left ventricular collapse. Artif Organs. 2003;27:926–30.
104. Faragallah G, Wang Y, Divo E, Simaan M. A new control system for left ventricular assist devices based on patient-specific physiological demand. Inverse Probl Sci Eng. 2012;20:721–34.
105. Antaki JF, Boston JR, Simaan MA. Control of Heart Assist Devices. Proc 42nd IEEE Conf Decis Control. 2003. 4084–9.
106. He P, Bai J, Xia DD. Optimum control of the Hemopump as a left-ventricular assist device. Med Biol Eng Comput. 2005;43:136–41.
107. Gregory SD, Stevens MC, Pauls JP, et al. In vivo evaluation of active and passive physiological control systems for rotary left and right ventricular assist devices. Artif Organs. 2016;40:894–903.
108. Gu K, Gao B, Chang Y, Zeng Y. The effect of captopril on the performance of the control strategies of BJUT-II VAD. Biomed Eng Online. 2016;15:399–410.
109. Gao B, Gu K, Zeng Y, Liu Y, Chang Y. A blood assist index control by intraaorta pump: a control strategy for ventricular recovery. ASAIO J. 2011;57:358–62.
110. Schrodel F, Schindler D, Claver A, Hein M, Ketelhut M, Abel D. A physiological control strategy for continuous-flow left ventricular assist devices: the power ratio controller. In: 2016 Eur. IEEE: Control Conf; 2016. p. 2416–22.
111. Habigt M, Ketelhut M, Gesenhues J, Schrödel F, Hein M, Mechelinck M, Schmitz-Rode T, Abel D, Rossaint R. Comparison of novel physiological load-adaptive control strategies for ventricular assist devices. Biomed Tech (Berl). 2017;62:149–60.
112. Cysyk J, Jhun C-S, Newswanger R, Weiss W, Rosenberg G. Rotary blood pump control using integrated inlet pressure sensor. Conf Proc IEEE Eng Med Biol Soc. 2011;2011:373–6.
113. Choi S, Antaki JE, Boston R, Thomas D. A sensorless approach to control of a turbodynamic left ventricular assist system. IEEE Trans Control Syst Technol. 2001;9:473–82.
114. Gaddum NR, Stevens M, Lim E, Fraser J, Lovell N, Mason D, Timms D, Salamonsen R. Starling-like flow control of a left ventricular assist device: in vitro validation. Artif Organs. 2014; https://doi.org/10.1111/aor.12221.
115. Petrou A, Ochsner G, Amacher R, Pergantis P, Rebholz M, Meboldt M, Schmid Daners M. A physiological controller for turbodynamic ventricular assist devices based on left ventricular systolic pressure. Artif Organs. 2016;40:842–55.
116. Arndt A, Nüsser P, Graichen K, Müller J, Lampe B. Physiological control of a rotary blood pump with selectable therapeutic options: control of pulsatility gradient. Artif Organs. 2008;32:761–71.
117. Wang Y, Koenig SC, Slaughter MS, Giridharan GA. Suction prevention and physiologic control of continuous flow left ventricular assist devices using intrinsic pump parameters. ASAIO J. 2015;61:170–7.
118. Schima H, Vollkron M, Jantsch U, Crevenna R, Roethy W, Benkowski R, Morello G, Quittan M, Hiesmayr M, Wieselthaler G. First clinical experience with an automatic control system for rotary blood pumps during ergometry and right-heart catheterization. J Heart Lung Transplant. 2006;25:167–73.
119. Tchantchaleishvili V, Luc JGY, Cohan CM, Phan K, Hübbert L, Day SW, Massey HT. Clinical implications of physiologic flow adjustment in continuous-flow left ventricular assist devices. ASAIO J. 2017;63:241–50.
120. Soucy KG, Giridharan GA, Choi Y, Sobieski MA, Monreal G, Cheng A, Schumer E, Slaughter MS, Koenig SC. Rotary pump speed modulation for generating pulsatile flow and phasic left ventricular volume unloading in a bovine model of chronic ischemic heart failure. J Heart Lung Transplant. 2015;34:122–31.
121. Pauls JP, Stevens MC, Bartnikowski N, Fraser JF, Gregory SD, Tansley G. Evaluation of physiological control systems for rotary left ventricular assist devices: an in-vitro study. Ann Biomed Eng. 2016;44:1–11.
122. Petrou A, Lee J, Dual S, Ochsner G, Meboldt M, Schmid Daners M. Standardized comparison of selected physiological controllers for rotary blood pumps: in vitro study. Artif Organs. 2018;42:E29–42.

Index

A
Abbott PediMag Blood Pump, 85
AbioCor implantable replacement heart, 9, 84, 193–196, 200, 613
AbioMed AbioCor TAH, 702
Abiomed Impella 2.5, 473
Acetylsalicylic acid (ASA), 644
Acid suppression, 643
Active blood pressure management, 318
Active closed-loop speed control, 522
  arterial hypertension avoidance, 523
  low flow operation avoidance, 523
  suction events avoidance, 523
  undesired thrombus formation avoidance, 523
Acute myocardial infarction (AMI), 141
Acute Physiology and Chronic Health Evaluation (APACHE) II, 624
Acute rehabilitation facilities (ARF), 662
ADAMTS-13, 476
Adaptive tuning techniques, 617
Adult respiratory distress syndrome (ARDS), 41, 209
Advanced device therapy, 151
Advanced heart failure, 632, 633
Advanced "smart pump" techniques, 126
Advanced ventricular assist device
  computer fluid dynamics analysis, 485–487
  exploded prototype, 482
  goals, 481, 482
  pressure-flow curves, 483
  pulsatility during LVAD support, 487, 488
  pulse pressure comparison, 484
  pump cross-section, 482, 483
  pump flow and arterial pressure, 483, 484
  pump hydraulic performance, *in vitro vs.* computer fluid dynamics model, 485
  RHF and RVAD
    LVADs use, 490, 491
    occurrence and clinical results, 489, 490
  weaning from, 488, 489
Adverse events (AEs), 653
Aggressive medical heart failure management, 672
Akutsu III total artificial heart, 7
American Society of Echocardiography (ASE), 156
Anesthetic induction, 156
Angiotensin converting enzyme inhibitors (ACEi), 135, 153, 658, 672
Angiotensin receptor blocker (ARB), 153, 660
Angiotensin receptor-neprilysin inhibitor (ARNI), 135
ANSYS Meshing, 485
Anticoagulation strategy, 644
Antiplatelet and anticoagulant therapies, 231
Aortic insufficiency (AI), 171, 658
Aortic pressures, 572, 573
Aortic stenosis (AS), 171
Aortix system, 474
Apical coring device, 174
Arrhythmias, 167
Arteriovenous fistula (AVF) maturation, 662
Artificial blood pumps, 318
Artificial heart, 4
Assist devices, 111
Atrial arrhythmia (AA), 659
Atrial elastance, 453
AV020 pumps, 485
Axial pressure capacity, 566
Axipump, 28
Azygos flow concept, 39

B
Berlin heart GmbH, 45
Berlin Heart INCOR
  axial-flow design, 348
  commercial availability, 348
  control unit, 348
  design features, 347, 348
  inflow cannula, 347
  levitation system, 348
  pump assembly, 347
  suction protection, 348
  survival rates, 348
β-blockade, 135
Biocompatability, 570
Biocompatible elastomeric material, 591
Bio-Medicus pump, 23
Biomimetic membrane pump, 587, 588
BiVACOR continuous-flow pump, 200, 201, 475, 476

© Springer Nature Switzerland AG 2020
J. H. Karimov et al. (eds.), *Mechanical Support for Heart Failure*,
https://doi.org/10.1007/978-3-030-47809-4

BiVACOR total artificial heart
  anatomical fitting, 569, 570
  biocompatibility, 570
  controller, 570, 571
  driveline, 570
  hydraulic designs
    axial pressure capacity, 566
    characteristics, 565
    flow paths, 566
    Frank-Starling mechanism, 565
    hydrodynamic backup bearing, 566
  implantable rotary biventricular blood pump, 563
  implantation, 571
  long term device durability, 564
  major user needs and features, 564
  motor and suspension system, 567, 568
  patient-carried system, 563
  physiologic interaction
    rotor position control, 568
    rotor speed control, 568, 569
    speed pulsatility, 569
  rotary blood pump technology, 563
  SynCardia® TAH vs. AbioCor® TAH, 565
  system performance
    hemocompatibility, 572, 574
    hydrodynamic bearing, 572
    outflow pressure sensitivity, 572, 574
    pulse profiles, 572, 574
Biventricular assist device (BVAD), 497, 499–502
Biventricular failure
  crista supraventricularis, 178
  echocardiographic parameters, 179
  epidemiology, 178
  evaluate RV failure, 181
  hemodynamic parameters, 180
  isovolumic contraction and relaxation phases, 178
  non-dilated restrictive cardiomyopathies, 177
  post-operative optimization
    MCS options, 185
    medical management, 184–185
    surveillance, 184
  preoperative optimization, 182
  right ventricular failure, 177
  risk stratification
    echocardiographic parameters, 179
    invasive hemodynamic data, 180
    myocardial deformation/strain, 179
    risk scores, 181, 182
  surgical and intraoperative management strategies
    filling pressures and PVR monitoring, 184
    functional MR, 183
    iNO, 184
    intraoperative transesophageal echocardiogram, 183
    planned and unplanned BiVADs, 182
    right ventricular dysfunction, 183
    sternotomy *vs*. lateral thoracotomy, 183
    temporary paracorporeal RVADs, 182
    tricuspid regurgitation, 183
  transverse muscle fibers, 178
Bjork-Shiley design, 581
Bladeless Tesla shear pump, 23
Blalock-Taussig operation, 38
Blood measure measurement, 712, 713
Blood pump circulation
  aortic and ventricular pressures, 65
  axial and centrifugal flow pump ramp tests, 70
  Cf-LVADs, 64
  clinical LVAD physiology, 76
  HeartMate II axial flow pump, 65
  HeartMate II instantaneous HQ, 68
  HVAD waveform, 70
  hypertension, 76, 78
  impact of LVAD speed, 73
  LVAD flow and total flow
    HeartMate II echocardiographic ramp test, 71
    HVAD echocardiographic ramp test, 71
    impact on CVP and PCWP, 73–75
    impact on left ventricle, 71
    impact on total body flow, 71, 72
    LV unloading, 72
    ramp test, 69–71
  LVAD hemodynamics, 63
  pathologic conditions and VAD responses, 77
  pressure-flow relationship, 64
  pressure-volume loops, 67
  pulmonary hypertension, 76
  PVA, 66
  RV failure with LVAD, 75–76
  simulation model, 69
  VAD-exercise physiology
    chronotropic incompetence and preload, 78
    effect of pump speed adjustment, 79
    hemodynamic factors, 78
    hemodynamic support and exercise, 78
  ventricular assist device, 63–64
  ventricular mechanics and energetics
    artificial flow pulsatility, 67–69
    cf-LVADs impacts, 66
    instantaneous pressure-flow relationships, 66, 67
    myocardial energetics, 66
    pump power and flow relationship, 67
Bridge to candidacy (BTC), 94
Bridge to cardiac transplantation (BTT), 28
Bridge to decision (BTD), 131
Bridge to recovery (BTR), 63, 95, 488
Bridge to transplantation (BTT), 63, 94, 95, 131, 203, 232, 248, 658

C

Candesartan in Heart Failure: Assessment of Reduction in Mortality and morbidity (CHARM) trials, 135
Cardiac catheterization, 671
Cardiac factors
  arrhythmias, 629, 630
  right ventricular failure
    echocardiography, 626, 627
    invasive hemodynamics, 628

risk factors, 628, 629
risk models, 628
Cardiac remission, 126
Cardiac transplantation and destination therapy, 33
Cardiogenic shock (CS), 141
causes, 379
IABP
ACC/AHA guidelines, 385
clinical data, 383, 384
complications, 385
components and types, 381, 382
contraindications, 384
ESC and AHA guidelines, 385
hemodynamic effects, 382–384
Impella devices
clinical data, 386–388
complications, 388
components and types, 385, 386
consensus documents and clinical practice guidelines, 388
contraindications, 388
FDA approval, 388
hemodynamic effects, 386, 387
SCAI/ACC/HFSA/STS statement, 388
indications, 379, 381
multidisciplinary standardized team-based approach, 381
percutaneous temporary right ventricular MCS
complications, 393
contraindications, 393
device types, 392
regulatory considerations, 393
short-term mechanical circulatory support, 394
TandemHeart
clinical data, 389
complications, 390
components, 388, 389
contraindications, 389
FDA approval, 390
hemodynamic effects, 389
temporary MCS device, 379–381
venoarterial extracorporeal membrane oxygenation
clinical data, 390, 391
complications, 391
components, 390
contraindications, 391
hemodynamic effect, 390
CARDIOHELP® System, 373
CardioMEMS, 330
Cardiopulmonary bypass (CPB), 91, 232, 473
Cardiopulmonary resuscitation (CPR), 208, 661
Cardiorespiratory model, 434, 435
CARDIOSIM©, 450
CARMAT bioprosthetic artificial heart (C-TAH), 201, 202
CARMAT Total Artificial Heart, 201
Carotid stenosis, 99
Carreau-Yasuda model, 407
Central venous pressure (CVP), 75, 171
Centrifugal LVADs, 348
Centrifugal pump inflow cannula, 160
CentriMag device, 45, 85, 86, 473, 499
CentriMag™ Acute Circulatory Support System, 373
Cerebrovascular accidents (CVA), 194
CF rotary blood pumps, 494
Chronic obstructive pulmonary disease (COPD), 209
Chronic venous thromboembolic disease, 98
Cirulite Synergy Micro-Pump, 474
Class III Medical Devices, 87
Cleveland Clinic SmartHeart CFTAH, 200
Cleveland Heart, 475
Clinical requirements, MCS
end-stage heart failure, 92–93
evaluation of candidates
abnormal coagulation, 98, 99
clinical requirements, MCS
evaluation of candidates
clinical examination, 96
hemodynamic factors, 96
for infection, 99
gastrointestinal bleeding, 98
hemodynamic evaluation, 96
hepatic function, 98
laboratory studies, 96
of nutrition, 99
peripheral vascular disease, 99
psychosocial assessment, 96
psychosocial evaluation, 100
pulmonary function, 98
renal function, 96–98
screening for malignancy, 99
goals of
coronary perfusion, 94
maintain end-organ perfusion, 93
unloading failing ventricle, 93
heart failure pathophysiology, 92
HeartMate II, 91
ideal pump characteristics, 102
indications for
BTC, 94
BTR, 95
BTT, 94
destination therapy, 95
Laplace's Law, 94
long term MCS, 92, 93
MVAD, 104
non-hemodynamic factors, 101, 102
patient selection and eligibility, 100
pulsatile pumps, 91
recipient selection, 101
right ventricular failure, 97
right ventricular support, 102–104
schematic of HVAD, 103
temporary MCS, 91, 92
Closed loop control, 717
Coil coupling factor (k), 613
Competencies, 710

Computational fluid dynamics (CFD), 485–487, 593
characteristic curves, 404, 405
conservation of mass, 399
conservation of momentum, 399, 400
DPIV measurements, 402, 404
experimental setup, 402, 403
finite difference method, 400
finite element method, 400, 401
finite volume method, 400
H-Q curve, 402, 404
hydraulic pump components, 402
Newtonian fluid mixture, 403
non-dimensional flow coefficient, 404
non-dimensional head coefficient, 403
pressure and flow coefficients, 404
quality control, 401, 402
scalar quantity, 400
validation step, 402
Congenital heart disease (CHD), 203, 599
Congestive heart failure (CHF), 33
Conservation of mass, 399
Conservation of momentum, 399, 400
Contactless bearing system, 507
Contemporary rotary CF-LVAD pumps, 151
Contextual inquiry, 525
Continuous flow devices(CFD), 158, 318
Continuous flow left ventricular assist devices (Cf-LVADs), 64, 151, 469, 473, 475, 641, 648
advantages, 337
adverse events, 281
aortic pulse pressure, 281
on aortic valve structure and function, 287, 288
arterial adaptation, 284
arterial pressure tracings, 279, 280
arterial pulsatility, 284
ascending aortic anastomosis, 287
axial-flow and centrifugal rotary pump designs, 338
balance point approach, 284
centrifugal LVADs, 348
on cerebrovascular system, 291, 292
clinical limitations, 481
clinical relevance of pulsatility, 280
complications, 279
for congestive heart failure, 279
degree of mechanical circulatory support, 280
endothelial shear stress and circumferential wall stress, 284
EVAHEART 2, 353–355
ex-vivo beating heart study, 286
FDA approved devices, 281
on gastrointestinal (GI) bleeding, 292, 293
HeartAssist5, 346, 347
HeartMate 3, 351–353
HeartMate II, 339–342
HeartWare HVAD, 349–351
on hematological parameters, 288, 289
INCOR, 347, 348
inflow cannula, 346
intramyocardial arteriovenous shunts, 287
Jarvik 2000, 342–344, 346
long-term effects, 280, 284
myocardial arteriovenous shunts, 287
myocardial oxygen demand, 286
myocardial perfusion, 286
nonpulsatile circulation, 280
outflow-graft anastomosis, 287, 346
patient specific simulations, 280
percutaneously implantable CFVADs, 281
postoperative right heart failure, 287
preload and afterload sensitive pumps, 281
pressure/volume (PV) curves, 281, 283
pump differential pressure, 281, 282
regional myocardial perfusion, 286
on renal, hepatic and pulmonary system, 289, 290
survival rates, 279
on systemic inflammatory response and endocrine system, 290, 291
tissue CO_2 production (TCO2), 286
two-chamber ex-vivo mock loop study, 287
Continuous flow pumps, 112, 113, 317
Continuous-flow rotary blood pumps, 91
Continuous-flow total artificial heart (CFTAH), 199, 602, 603
atrial pressure, in LAP and RAP, 495, 496
biventricular heart failure
BVAD, 499–501
HVAD, 499
LVAD, 499
P-CFTAH, 500, 501
RVAD, 499
centrifugal-flow pumps *vs*. axial-flow pumps, 501
Cleveland Clinic, 494–496, 502
current version of, *in vivo*, 500, 501
design, 495
double-ended centrifugal pump, 495
I-CFTAH, 498
P-CFTAH, 497, 498
Controller malfunctions, 657
Convection-diffusion-reaction equation, 410
Cooley, D., 38
Copulsation devices, 645
CorAide continuous-flow left ventricular assist device, 490
CorAide (Dr. Leonard Golding), 44
Coronary blood flow (CBF), 66
CorWave LVAD
advantages
compact size, 591
gentle propulsion of blood, 591
physiologic pulsatility, 588–590
biocompatible elastomeric material, 591
biomimetic membrane pump, 587, 588
blood flow path, 592
company development, 596
components, 591
device development
in silico analysis, 592–594
in vitro testing, 593–596
in vivo testing, 596
electromagnetic actuator and transmission, 591, 592
membrane visualization technique, 593
surgical techniques, 591

Cross-circulation technique, 40
Cross model, 406, 408
Cytokines, 93

D
DeBakey, M., 39
DeBakey's ventricular assist pump, 5, 6
DECIDE-LVAD clinical trial, 649
De-escalation therapy, 143
Degree of pulsatility, 121
Dennis's machine, 40
Design Failure Mode and Effects Analysis (dFMEA), 123
DesignModeler, 485
Design of experiments (DOE), 705
Destination therapy (DT), 63, 95, 131, 152, 488, 493, 494, 648, 658, 673
Device malfunction, 676
DexAide continuous-flow right ventricular assist device, 490
Disease-based input, 456
Double-ended centrifugal pump, 495
Driveline malfunction, 657
Driveline safety, 523, 524

E
Early right ventricular failure (RVF), 650
Earnshaw's Theorem, 114
ECMO console, 57
ECMO system, 209
Extracorporeal life support (ECLS), 112
El Bayad. Life, 23
Electrical circuit analogy, 110
Emergency medical services (EMS), 661
End-diastolic elastance, 453
End-diastolic pressure (EDP), 255
End-diastolic pressure-volume relation (EDPVR), 255
End-diastolic volume (EDV), 255
End-stage renal disease (ESRD), 136
ENDURANCE Supplemental Trial, 470
Energy equivalent pressure (EEP), 237
Energy sources, 610–612
Energy systems, history of, 610, 611
Energy transmission methods
  resonant-coupling wireless transmission, 614–618
  transcutaneous energy transfer systems, 612–614
Engineering MCS devices
  engineering requirements, 85, 87
  magnetically-levitated bearings, 85
  next generation of MCS devices, 85
  polycarbonate housings, 85
  polymer blood sacs, 83, 85
  risk analysis to, 87–89
EUROMACS risk score, 155
EVAHEART 2 left ventricular assist system, 353, 355
  advantages, 542
  axial/centrifugal pump, 536
  double cuff tipless inflow cannula, 539–542
  flow-head pressure curve, 537, 538
  gastrointestinal bleeding profile, 539, 540
  impeller design, 536–539
  in vitro hemocompatibility study, 539, 540
  in vitro mock circulatory loop study, 537
  limitations, 535, 536
  von Willebrand factor large multimer degradation, 539
EVAHEART LVAS, 354, 355
EXCOR BerlinHeart LVAD, 440, 441
Extracorporeal cardiopulmonary resuscitation (ECPR), 207, 217
Extracorporeal centrifugal membrane oxygenation (ECMO), 57, 58
Extracorporeal circulation, 39
Extracorporeal life support (ECLS), 207
Extracorporeal Life Support Organization (ELSO), 216, 217
Extracorporeal membrane oxygenation (ECMO) techniques, 10, 92, 237, 238, 373–376, 473
  acute cardiopulmonary failure, 217
  cardiopulmonary bypass technique, 207
  complications of
    bleeding, 212
    Harlequin syndrome, 214
    infection, 213
    left ventricular overload, 213–214
    limb ischemia, 213
    recirculation, 214
    stroke, 212
  2009 H1N1 influenza epidemic, 207
  high risk and complex therapy, 207
  indications and contraindications, 208, 209
  LVAD, 207
  management of
    anticoagulation, 212
    blood gases, 212
    initiation of support, 211
    monitoring, 211–212
    open chest cannulation, 209–211
    ventilator management, 212
  program organization, 215–216
  refractory cardiac or pulmonary failure, 217
  system, 209
  transport guidelines, 216, 217
  VA-ECMO, 207, 208
  VAV-ECMO, 207
  VV-ECMO, 207
  weaning, 214, 215

F
Failure mode and effect analysis (FMEA), 125
Feed forward control, 716, 717
Finite difference method, 400
Finite element method, 400, 401
Finite volume method, 400
Flow sensor integration, 522
Frailty, 632, 633
Framingham Heart Study, 3
Frank-Starling law, 254–256, 261, 361, 418, 430, 558, 565, 567, 704

Frazier, O.H., 43
Frozen rotor model, 486
Fully-implanted left-ventricular assist system (FILVAS), 126

G

Gastrointestinal (GIB), 654, 655
- axial flow model, 644, 645
- clinical relevance, 641
- diagnosis, 643
- incidence/prevalence, 641
- management strategies, 643, 644
- pathophysiology, 642, 643
- sources, 642

Glass perfusion pump, 4

H

Harlequin syndrome, 211, 214
HeartAssist5, 470, 471
- CE Mark approval, 347
- design and features, 346, 347
- HeartAttendant, 347
- IDE clinical trial approval, 346
- implantation, 346
- of remote monitoring, 347
- pump assembly, 346
- US BTT IDE trial, 347

Heart failure-related circulatory dysfunction, 109
Heart Failure Survival Score, 471
Heart Incor™ VADs (Robert Kroslowitz), 45
Heart-lung machine, 4, 39
HeartMate II echocardiographic ramp test, 71
HeartMate II instantaneous HQ, 68
HeartMate II LVAS, 12, 45, 58, 83, 132, 151, 199, 339, 493, 693
- adverse events, 342
- alarm system, 341
- anticoagulation therapy, 341
- BTT trials, 341
- CE Mark approval, 339
- complication, 341
- description, 339
- design and features, 340
- FDA approval, 339
- FDA destination therapy approval, 339
- inflow positioning, 340
- outflow graft, 339
- pump rotor apparatus, 339
- risk factor analysis, 341
- system controller, 339–341
- system monitor, 341
- woven polyester graft, 340

HeartMate 3 device (HM3), 58, 59, 83, 84, 227, 258, 351–353, 470, 476, 610
HeartMate LVAD, 242, 243
HeartMate Risk Score (HMRS), 136, 472
HeartMate XVE, 83, 232
HeartQuest™, 45
HeartWare HVAD, 13, 349–351, 456–458, 470, 474, 656
Heartware MVAD, 104, 471
Hematocrit post-implantation, 246
Hemocompatibility, 120, 572, 574
Hemodynamic monitoring
- alarms, 713, 714
- blood pressure measurement, 712, 713
- pump, 713
- thrombus detection, 714

Hemodynamic pulsatility, 301
Hemodyne pump system, 22, 24
Hemolysis, 22, 655
Hemopump system, 24, 25, 27, 83
Heparin-induced thrombocytopenia (HIT), 212, 656
Hermetically sealed magnetic levitation system, 569
Highly-accelerated life tests (HALT), 125
HM II and HM III displays, 320, 322
Homeostasis, 110
Human factors and usability engineering (HFUE) techniques, 125
HVAD echocardiographic ramp test, 71
Hybrid cardiovascular simulator (HCS)
- artificial valve testing, 441–443
- BVAD and TAH testing, 441, 442
- configuration, 440, 441
- flexibility, 440
- IABP, 443
- LPM, 439, 440
- pneumatic LVAD, 440
- Windkessel model, 440

Hybrid mock circulation (HMC) maps, 264, 265
Hybrid mock loop, 451
Hydraulic designs
- axial pressure capacity, 566
- characteristics, 565
- flow paths, 566
- Frank-Starling mechanism, 565
- hydrodynamic backup bearing, 566

Hydrodynamic bearings, 113
Hypoplastic left heart syndrome (HLHS), 579
Hypovolemia, 167

I

Impella pump catheter, 53–55, 369–372
Implantable electronics, 703
Implanted LVAD device, 59
In silico analysis, 592–594
In vitro mock circulatory loop study, 537
In vitro testing, 593–596
In vivo testing, 596
Inductive wireless transfer, 613
Infant continuous-flow total artificial heart (I-CFTAH), 498
Inhaled nitric oxide (iNO), 180
Input impedance (Zart), 321, 322
Interagency Registry for Mechanically Assisted Circulatory Support (INTERMACS), 12, 102, 132, 133, 159, 178, 198, 471, 472, 488, 493–494, 624–625, 648, 701

International Society for Heart and Lung Transplantation (ISHLT), 91, 178
Intra-aortic balloon pump (IABP), 43, 52–53, 112, 142, 213, 238, 239, 364–369, 443
Intrabeat control, 724
Intravascular devices, 111
Ischemic myocardium, 255
ISO Standard for Circulatory Support Devices, 88

J
Jarvik 2000 FlowMaker, 499
Jarvik 2000 ventricular assist system, 474, 476
  adverse events, 346
  CE Mark certification, 342, 344
  description, 342
  design and technology features, 342–344
  development, 342
  FlowMaker controller, 343, 344
  implantation techniques, 343
  Intermittent Low Speed mode, 343, 345
  Investigational Device Exemption, 342
  lithium-ion batteries, 344
  PMDA approval, 344
  PumpKIN clinical trial, 344
  RELIVE trial, 342
Jarvik-7 total artificial heart, 8
Jugular venous distension (JVD), 650

K
Kantrowitz, A., 42

L
Laplace's Law, 94
Large-gap hybrid bearing, 508
Late right ventricular failure (RVF), 659
LATERAL prospective clinical trial, 472
Lavare cycle, 258, 324, 325
Left atrial pressure monitoring (LAP), 166–167
Left heart catheterization, 171
Left ventricular assist device (LVAD), 33, 55, 63, 111, 131, 191, 493, 499, 502, 545, 577, 609, 648, 671–673, 693
  anesthetic induction, 156
  anticoagulation status, 154
  aortic anastomosis, 225, 226
  Aortix system, 474
  apical sewing ring placement, 222–224
  arrhythmias, 167
  arterial pulsatility maintenance, 476
  color flow doppler, 162
  CW doppler, 162
  deairing, 226
  destination therapy, 152
  device strategy @implantation, 152
  driveline tunneling, 223, 225
  electrocautery, 225
  EUROMACS risk score, 155
  first generation devices, 469
  fluid management, 165, 166
  HeartAssist 5 pump, 470, 471
  HeartMate III device, 227, 470
  HeartWare HVAD, 470
  Heartware MVAD, 471
  hemodynamic factors
    contractility and loading conditions, 157
    PAPi, 157–159
    ventriculo-arterial coupling, 157
  hepatic dysfunction, 153
  historical perspective, 469, 470
  HVAD, 151
  hypertension, 167
  incision approach, 221, 222
  INTERMACS profile, 152
  intraoperative echocardiography
    LVAD implantation, 156
    RV tissue Doppler velocities, 156
  intraoperative management, 155–156
  LV apex exposure, 222
  median sternotomy, 221
  MicroVasc, 474
  minimally invasive technique, 221, 472, 473
  off cardiopulmonary bypass, 473
  operating room and monitoring lines
    chest closure, 176
    coring, 173
    implanting LVAD, 173–175
    outflow graft, 175, 176
    pump implantation, 173
    sewing cuff, 173
    surgical access, 172
    temporary RVAD, 176
    tunneling of driveline, 172
  outflow graft tunneling, 225
  patient maintenance, 474, 475
  patient positioning, 221
  patient selection, 471, 472
  perioperative factors, 154
  pharmacological support, 165
  pre-operative anesthetic considerations, 151–153
  pre-operative CT and CXR, 221
  preoperative optimization, 171, 172
  preoperative RV risk stratification, 154–156
  post CPB and ICU management
    pharmacological support, 164, 165
    RV failure, 163, 164
    suction events, 162
  postoperative bleeding, 166, 167
  postoperative management, 176
  post-operative RVF risk stratification, 155
  prior cardiac surgeries, 154
  renal dysfunction, 153
  right heart failure, 155
    Abiomed Impella 2.5, 473
    Centrimag device, 473
    Cirulite Synergy Micro-Pump, 474
    incidence, 473
    Synergy Micro-pump, 474

Left ventricular assist device (LVAD) *(con't)*
Tandem Life Protek Duo, 473
second generation devices, 469
separating from CPB, 159–161
single lumen endotracheal intubation, 221
sternal-sparing approaches, 221
stratifying RV support, 164
suction event, 163
total artificial heart, 475, 476
transcutaneous energy transfer, 476, 477
TEE spatial and directional guidance, 160
TTE examination, 157
ventriculotomy, 226
Left ventricular diastolic filling pressure, 720
Left ventricular end diastolic dimension (LVEDD), 71, 651
Left ventricular end diastolic pressure (LVEDP), 66
Left ventricular end diastolic volume, 720
Left ventricular end systolic dimension (LVESD), 71
Levitronix LLC, 45
Lillehei-DeWall oxygenator, 39
LiMAx test, 146
LionHeart LVAD, 84, 614
Liotta-Cooley artificial heart, 7
Long and short-term mechanical circulatory support (MCS), 92, 93, 648
Long-term acute care hospitals (LTAC), 662
Lumped parameters models (LPMs), 430
arterio-ventricular sections and venous-ventricular sections, 430
atria, 430, 431
autonomic controls, 431, 432
cardiorespiratory interaction, 432, 433
congenital heart diseases models, 434
gas transport and exchange, 433
heart chambers, 431
level of complexity, 434
modeling of vessels, 431
peripheral metabolic control, 433, 434
vascular controls, 434
ventricle filling, 430
LVAD pump speed modulation therapy, 319
LVAD therapy, 153
clinical monitoring, 708
closed loop control, 717
constant speed setting, 716
feed forward control, 716, 717
hemodynamic monitoring
alarms, 713, 714
blood pressure measurement, 712, 713
pump, 713
thrombus detection, 714
monitoring modalities, 709, 710
optimal set speed, 715
patient self monitoring, 708, 709
physiologic control (*see* Physiologic control)
pump monitoring, 709
telemonitoring, 711
CardioMEMS™, 712
patient input, 711
pump parameters, 712
telephone-monitoring, 711
transition from monitoring, 715
variable, 716
LV unloading, 72

M

Magnetic bearings, 113
MATLAB, 452
Maximum pressure curve slope (dP/dt_{max}), 590
MCS, engineering perspectives
advantageously alter therapy, 126
aortic and ventricular pressure, 115
cardiac remission, 126
centrifugal pump H-Q characteristic curve, 120
confident diligence, 126
continuous flow pumps, 113
convenient conversion factor, 114
development, verification, and validation
design control principles, 124
design reviews, 124
game-changing, 123
HALT, 125
HFUE, 125
input requirement, 124
long-term reliability and durability, 125
new product development, 124
NPD programs, 123
product sustaining efforts, 123
real-world conditions, 124
risk management, 125
simulated real-world conditions, 125
sustaining of existing products, 124
VAD systems, 125
eclipse earlier-generation products, 126
fundamentals, first principles, and design
afterload sensitive, 116
arterial pressure, 116
cardiac systole, 114
continuous flow pumps, 114
efficiency, 116
mechanical (artificial) circulation, 111–112
native circulation, 110–111
preload sensitive, 116
pressure head, 114
pump design, 116–122
pump types, 112–114
volume flow rate, 114, 116
H-Q characteristic, 115
human circulatory system, 110
implant location influencing pump, 119
life expectancy and quality of life, 109
LVAD system, 118
MCS-related adverse events, 126
power-Q characteristic, 117
pulsatility, 121
sensor-less flow estimation, 121
simplified negative feedback model, 111
suction detection algorithm, 122
surviving cold weather, 109
TETS technology, 126

typical LVAD (bypass) configuration, 112
Mean arterial pressure (MAP), 93, 302
Mechanical bearings, 113
Mechanical circulatory assist devices (MCAD), 19, 545
Mechanical circulatory support (MCS), 19, 37, 493, 494, 497–499, 502, 599
acute on chronic congestive heart failure, 51
advantages, 360
adverse events, 609
anticoagulation therapy, 51
blood pump, 610
cardiovascular physical simulators, 436, 437
centrifugal pumps, 309, 310
circulatory models, 429
clinical demands and challenges
ADVANCE and ENDURANCE trials, 698
anticipated changes, 693–694
changing nature of, 696
delivering high-quality care, 693
different sized LVAD program, 695
electronic health record data, 698
LVAD health care delivery changes, 694–695
managing LVAD growth, 695
managing patients outside, 695–696
new payment models in complex disease, 695
palliative care, 696–698
processes of care and outcomes, 698
clinical requirements (*see* Clinical requirements, MCS)
computational models, 435, 436
continuous flow pumps, 310, 609
DeBakey's ventricular assist pump, 5, 6
degree of pulsatility, 302
diminished pulsatility, 302
durable, implantable, 52
echocardiography and pulmonary artery catheterization, 364
ECMO, 57, 58, 372–376
energy equivalent pressure, 306, 307
energy sources, 610–612
energy transmission methods
resonant-coupling wireless transmission, 614–618
transcutaneous energy transfer systems, 612–614
engineering perspectives (*see* MCS, engineering perspectives)
evolution of cardiac surgery, 360
extracorporeal life support, 301
extracorporeal membrane oxygenation, 301
full hybrid cardiovascular simulator, 439
gastrointestinal bleeding
axial flow model, 644, 645
clinical relevance, 641
diagnosis, 643
incidence/prevalence, 641
management strategies, 643, 644
pathophysiology, 642, 643
sources, 642
glass perfusion pump, 4
goal of permanent replacement, 8–10
heart-lung machine, 4
HeartMate 3, 58, 59, 610
history of energy systems, 610, 611
hybrid cardiovascular simulators, 438
hydraulic-computater cardiovascular simulator, 438
IABP counterpulsation, 52, 53, 364–369
i-cor system, 310–313
Impella®, 53–55, 369–372
indications, 360
intra-aortic balloon pumps, 301, 309
ischemic/non-ischemic acute decompensated ventricular failure, 51
Liotta's implantation, 360
long-term MCS implantation
cardiorenal syndrome, 146
ECLS or short-term devices, 144
ENDURANCE Supplemental trial, 148
failing ventricle and maintain sufficient end-organ perfusion, 144
HeartMate II device, 147
HeartMate II risk score, 146
INTERMACS registry, 145, 146
intra- and early postoperative decision making, 147
LiMAx test, 146
longterm mortality, 147
LVAD implantation, 145
PENN-Columbia risk score, 146
peripheral vascular disease, 146
renal dysfunction, 146
renine-angiotensine-aldosterone system, 146
RV/LV diameter ratio, 147
RVAD implantation, 147
RVF, 147
transcutaneous energy transfer, 144
MAP, 302
maximum rate of change of pressure, 305
mechanisms of action, 301
method for energy transmission, 610
mortality rates, 3
National Heart Act, 3
non-durable, percutaneous, 51
non-pulsatile and pulsatile MCS, 302
palliative care (*see* Palliative care)
patient selection
HeartMate II trial, 132
HeartMate III, 132
INTERMACS profiles, 132–134
intraaortic balloon pump, 132
MOMENTUM 3 trial, 132
REMATCH trial, 132
ROADMAP trial, 134
percutaneous or implantable ventricular assist devices, 301
physiologic basis, 360–363
physiological models, heart chambers, 431 (*see* Lumped parameters models (LPMs))
pneumatic pumps, 309
power converter, 610
pulsatile MCS devices, 302
pulsatility index, 304
pulse power index, 305, 306
pulse pressure, 303, 304
recognition of clinical deterioration

Mechanical circulatory support (MCS) *(con't)*
exercise cardiopulmonary stress testing, 135
hemodynamic criteria, 135
inotrope dependence, 135–136
medication intolerance, 135
renal and hepatic dysfunction, 136
repeat hospitalization, 135
risk factor analysis (*see* Risk factor analysis)
roller pumps, 309
rotary pumps, 309
short-term, 141–144
surplus hemodynamic energy, 307, 308
symptomatic heart failure, 3
systolic pressure, diastolic pressure, 302
TAH, 1950s-1960s, 6, 7
TandemHeart®, 55, 56, 372
timing for LVAD implantation, 136, 137
TORVAD, 310
total artificial hearts and FDA regulatory status, 10
total hemodynamic energy, 307
U.S. Artificial Heart Program, 4
VAD, 10–13
ventricular assist devices and FDA regulatory status, 14
wireless energy transmission systems, 609
Medical Management in Ambulatory Heart Failure Patients (ROADMAP) trial, 469
Membrane visualization technique, 593
Methicillin-resistant staphylococcus aureus (MRSA), 650
MicroMed DeBakey (Dr. Michael DeBakey), 44, 45
MicroVasc, 474
Microvascular circulation, 233, 235
Milrinone, 180
Mitral stenosis (MS), 171
Mitral valve surgery (MVS), 183
Mock circulation loop (MCL), 516, 593, 594
Modern MCS devices, 111
MOMENTUM 3 trial, 132, 470, 476, 493, 693
Motor and suspension system, 567, 568
Multiobjective control, 723
Multi-objective physiologic controller, 262
Multi-segmented conductance catheters, 704
Multisystem organ failure, 675

N
National Heart Act, 3
National Heart, Lung, and Blood Institute (NHLBI), 318–319
National Institutes of Health (NIH), 43
Navier-Stokes equations, 404, 405, 408
New product development (NPD), 123
Newtonian models, 406
New York Heart Association classification, 471
New York Heart Association Functional Classification (NYHA), 648
Nitroprusside, 180
Non-cardiac factors
advanced heart failure, 632, 633
frailty, 632, 633
interventions, 633
medical factors, 630, 631
psychosocial factors, 631
Non-pulsatile pumps, 232
Nonrandomized MVAdvantage Trial, 471
Novacor left ventricular assist system, 11, 241, 242
Numerically controlled physical mock loop, 451
Numerical simulation
application, 450
atrial elastance, 453
circuit using electrical analogies, 452, 453
clinical uses, 450
disease-based input, 456
end-diastolic elastance, 453
heart parameters input, 457
limitations for pump performance, 450
MCS device input, 456, 457
output, 450
pressure rise *vs.* flow, 454, 455
simulation inputs, 455
valve fluid acceleration, 454
valve input, 457
valve position acceleration, 454
valve static pressure drop, 454
vascular impedance input, 457
ventricular elastance, 452
VML (*see* Virtual mock loop (VML))

O
Oxygenator, 40

P
Palliative care
actual VAD deactivation process and checklist
family meeting, 683
medical deactivation, 683
plan of care, 683
time of patient's death, 684
advanced directives, 685
caregiver end of life perceptions, 684
emotional and psychological complications
anxiety, 678
depression, 678
end of life care, 685
and communication, 680
and LVAD deactivation, 682, 683
in VAD patients, 678
ethical considerations, 680, 682
HF, prevalence of, 684
history, 671–673
hospice care, 684
illness trajectories, 679
LVAD complications
bleeding, 675
device malfunction, 676
infection, 676
neurologic events, 676

right heart failure, 676
multisystem organ failure, 675
preparedness planning, 673–675, 685
right heart failure, 675
Patent foramen ovale (PFO), 158
Patient Self Determination Act, 680
Pediatric continuous-flow total artificial heart (P-CFTAH), 497, 603–605
Pediatric mechanical circulatory devices
balloon atrial septostomy, 40
cardiac ultrasound, 40
cardiopulmonary bypass machine, 38, 40
color-flow Doppler, 40
congenital cardiac anomalies, 38
congenital cardiac surgical procedure, 39
creation of angiography, 40
cross-circulation procedure, 39
DeWall-Lillehei helix bubble oxygenator, 40
hypothermia, 39
magnetic resonance imaging, 40
open-heart surgery, 40
pediatric ECMO, 41, 42
pediatric mechanical circulatory devices
Blalock-Taussig operation, 38
hypothermia, 39
pediatric ventricular assist devices
Archimedes screw-type pump, 44
Berlin heart GmbH, 45
CentriMag® pump, 45
extracorporeal Biomedicus BP-80 device, 43
extracorporeal centrifugal device, 45
extracorporeal VAD, 43
first centrifugal pump, 43
HeartMate Vented Electric assist device, 45
HeartMate XVE LVAS, 44
Hemopump, 44
Levitronix LLC, 45
long-term mechanical circulatory support, 42
MEDOS/HIA DeltaStream, 43
NHLBI artificial heart program, 42
Novacor LVAD, 43
pediatric VAD and ECMO device, 46
pediatric-size EXCOR VAD, 45
PediMag® pump, 45
Pierce-Donachy VAD, 43
pulsatile *vs.* non-pulsatile pumps, 44
PumpKIN program, 46
second-generation pumps, 44
TCI, 43, 44
third generation centrifugal devices, 45
Toyobo and Zeon pumps, 44
viscous impeller pump, 46
world's first heart transplant, 42
perfusion machine, 37
Pediatric total artificial heart
Cleveland Clinic P-CFTAH, 602–606
clinical challenges, 600, 601
clinical needs, 600
continuous-flow VADs, 605
MCS devices, 605
Penn State University, 602
SynCardia 50cc, 601, 602
SynCardia 70cc, 601
PediMag blood pump, 45, 85–87
Penn State Arrow LionHeart LVAD, 702, 704
Penn state pediatric total artificial heart
automatic control, 581, 582
biventricular support, 578, 579
cardiac output and anatomic fit, 580, 581
durable MCS, 577
in vivo testing, 582, 583
indications, 579
left ventricular assist device, 577
pump design, 579, 580
SynCardia 70cc, 577
waitlist mortality and post-transplant survival, 578
Peripheral grafting, 238
Pharmaceuticals and Medical Devices Agency (PMDA) approval, 344
Physical mock loop, 451
Physiological pulsatility, 588–590
Physiologic control
control goals, 717
headpressure and flowrate, 718
heart-rate, 721
HQ-curves, modifications, 719
intrabeat control, 724
multiobjective control, 723
passive, 722
preload based control
left ventricular diastolic filling pressure, 720
left ventricular end diastolic volume, 720
QVadmin, 720
suction based, 720
residual ventricular function
aortic valve opening, 723
hydraulic energy, 722
pulsatility, 722, 723
systolic pressure, 723
setpoint, 718
structure, 717
variables, 718
Pierce-Donachy (Thoratec) pumps, 239
Plasma-free hemoglobin (pfHb) levels, 19, 572, 573
Pneumatically driven ventricular assist device (VAD), 42
Pneumatic pulsatile pump, 243
Positive displacement pumps, 112
Positive end-expiratory pressure (PEEP), 212
Postoperative management strategies
caregiver burden, 664
cfLVAD, 648
community contact information, 661
critical care management
early RVF, 650
hemodynamic monitoring and stabilization, 649–650
LVAD implantation, 649
postoperative bleeding, 651
risk assessment and antibiotic prophylaxis, 650
discharge educational preparation, 660, 661

Postoperative management strategies *(con't)*
- discharge preparedness, 661
- emergency care, 662–663
- heartmate parameters, 652
- hemodynamic parameters, 650
- LVAD device speed, 651
- LVAD related adverse events
  - acute and chronic LVAD infections, 656, 657
  - aortic insufficiency, 658
  - atrial and ventricular arrhythmias, 659, 660
  - device equipment malfunctions, 657–658
  - GIB, 654, 655
  - hemolysis, 655
  - hemorrhagic and thrombotic complications, 654
  - late RVF, 659
  - pump thrombosis, 655, 656
  - stroke, 656
- medical preparation of LVAD candidates, 648, 649
- outpatient dialysis, 662
- outpatient management, 663–664
- post-acute LVAD facilities, 661–662
- pulsatility and BP measurement, 660
- routine LVAD management
  - anticoagulation and antiplatelets, 652–653
  - basic interpretation and management, 651–652
  - LVAD speed/RPM choice, 651

Preload based control
- left ventricular diastolic filling pressure, 720
- left ventricular end diastolic volume, 720

Preload-responsive speed (PRS) controller, 261, 262
Premarket approval (PMA), 44
Preparedness planning (PP), 673–675
Pressure or volume sensors, 703, 704
Pressure sensitive hydraulic designs, 568
Pressure sensitivity, 511–513
Pressure–volume area (PVA), 66
Prevention of HeartMate II Pump Thrombosis Through Clinical Management (PREVENT) trial, 474, 475
Prognostic nutritional index (PNI), 630
Prophylactic intravenous, 172
PROTECT II trial, 54
PROTEKDuo cannula, 392, 393
Pulmonary artery catheter (PAC), 649
Pulmonary artery diastolic pressure (PADP), 76
Pulmonary artery pulsatility index (PAPi), 52, 97, 157–159, 180
Pulmonary capillary wedge pressure (PCWP), 75, 180
Pulmonary hypertension (PH), 76
Pulmonary vascular resistance (PVR), 180
Pulmonary wedge pressure (PWP), 25
Pulsatile devices
- cardiopulmonary bypass pump, 238
- ECMO, 237, 238
- extracorporeal pulsatile ventricular assist devices
  - biventricular configuration, 240
  - in children, 240, 241
  - connection options, 239
  - direct ventricular drainage, 240
  - Pierce-Donachy (Thoratec) pumps, 239
  - thoratec VADs, 240
  - univentricular/biventricular configuration, 240
- HeartMate LVAD, 242, 243
- IABP, 239
- implantable pulsatile LVADs, 241
- intra-aortic balloon pumps, 238
- microvascular circulation, 233, 234
- Novacor left ventricular assist system, 241, 242
- peripheral grafting, 238
- risk factor analysis, 247
- Rotoflow centrifugal continuous flow pump, 238
- selection and management protocols, 247
- VentriFlow pulsatile pump, 238

Pulsatile flow (PF), 318
Pulsatile perfusion, 235–237
Pulsatile pumps, 91
Pulsatile volume-displacement pumps, 91
Pulsatility and Blood Pressure (BP) assessment, 660
Pulsatility index (PI), 58, 237, 319, 320
Pulse pressure, 319, 321, 590
Pump flows, 572, 573
Pump monitoring, 709
Pump shutoff, 118
Pump speed modulation algorithms, 318, 319
Pump thrombosis, 655, 656, 676, 714
PumpKIN program, 46
Pump-oxygenator, 231

Q

Quality factor (Q-factor), 613

R

Ramp test, 69–71
Randomized Evaluation of Mechanical Assistance for the Treatment of Congestive Heart Failure (REMATCH) trial, 469
ReinHeart® TAH, 202, 203
ReinVAD left ventricular assist device
- active closed-loop speed control, 522
  - hypertension avoidance, 523
  - low flow operation avoidance, 523
  - suction events avoidance, 523
  - undesired thrombus formation avoidance, 523
- aspects/characteristics, 525, 527
- causes, 525
- characteristics and features, 511, 512
- contextual inquiry, 525
- driveline safety, 523, 524
- flow sensor integration, 522
- human factors process, 524
- informal formative tests, 525
- passive interaction with circulation
  - excercise capacity, 517, 519–521
  - left ventricular unloading, 513, 515, 516
  - pressure sensitivity, 511–513
- product development, 532
- pump architecture and functionality
  - ANSYS package, 507

axial clearance gap, 508
centrifugal pump unit, 506, 507
components, 505
contactless bearing system, 507
features, 506
geometry parameters, 508
hydrodynamic structure, 507
innovative technological developments, 505
intracorporeal and extracorporeal components, 506
large-gap hybrid bearing, 508
lowest clearance gap, 507
metamodel, 508
miniaturized ReinVAD pump unit, 506
response-surface-based design space exploration and geometry optimization, 507, 508
rotary pump principle, 505
rotor bearing, 507
pump performance and efficiency, 509, 510
usability, 525
AC adapter, 531
advantages, 531
batteries, 529–531
carrying and protecting system, 530
charger, 530
clinical monitor, 527, 528
patient controller, 528, 529
wristband, 530
user interface specifications, 525
REMATCH trial, 132, 135, 279
Renin-angiotensin-aldosterone system (RAAS), 238, 291
Replacement devices, 111
Residual ventricular function
aortic valve opening, 723
hydraulic energy, 722
pulsatility, 722, 723
systolic pressure, 723
Resonant-coupling wireless transmission, 614–618
Resonantly coupled wireless transmission, 617
Revolution® Centrifugal Blood pump, 373
Reynolds Averaged Navier Stokes (RANS) turbulence model, 486
Reynolds number, 118
Right heart catheterization (RHC), 180
Right heart failure (RHF), 154
Right ventricular assist device (RVAD), 165, 499, 502, 650
Right ventricular failure (RVF), 147, 156, 177, 649
echocardiography, 626, 627
invasive hemodynamics, 628
risk factors, 628, 629
risk models, 628
Right ventricular stroke work index (RVSWI), 137, 157, 180
Right ventricular systolic pressure (RVSP), 71
Risk Assessment and Comparative Effectiveness Management in Ambulatory Heart Failure Patients (ROADMAP) study, 317
Risk factor analysis
cardiac factors
arrhythmias, 629, 630
right ventricular dysfunction, 626, 627
RVF (*see* Right ventricular failure (RVF))
INTERMACS patient profile, 624–625
long term therpy, 623
non-cardiac factors
advanced heart failure, 632, 633
frailty, 632, 633
interventions, 633
medical factors, 630, 631
psychosocial factors, 631
short-term therapy, 623
systolic heart failure, 623–625
total artificial heart, 635
transplant candidacy, 633, 634
VAD operative risk, 633, 634
Roache method, 485
Roller pump, 39
Rolling screen oxygenator, 40
Rotary blood pumps
accumulation of thrombus, 20
Archimedes screw pump, 23
arterio-venous malformations, 31
axial flux gap motor, 21
Bio-Medicus pump, 23
bladeless Tesla shear pump, 23
BUN or urine hemoglobin spillage, 22
cardiopulmonary bypass, 19
centrifugal pumps, 21
commercial centrifugal pumps, 21
complete magnetic suspension, 31
CorAide LVAD, 34
CorAide LVAS, 29
denatured protein, 20
El Bayad. Life, 23
epiphany, 20
fixed wing device, 20
flexible inflow cannula, 25
gastrointestinal bleeding, 31, 32
Heartmate II, 34
HeartWare HVAD, 29–31, 34
Hemodyne pump, 22, 24
hemolysis, 22
Hemopump system, 24, 25
human implantation
circulatory physiology react, 25
DeBakey LVAD, 29
electromagnetic torque coupling, 27
HeartMate II, 28, 29
Hemopump hydraulics, 27
integrated motor pump design, 27
Kitty Hawk, 27
multicenter clinical trial, 26
'new' physiology and clinical management, 26
OKT3, 26
outcomes, 28
patient selection, 25–26
safety and efficacy, 27
silicone rubber plug, 26
vasopressor infusions, 26

Rotary blood pumps *(con't)*
  Kranich's recovery, 26
  left ventricular assist devices, 21
  lost wax casting, 21
  LVADs, 21, 24
  motor magnets, 27
  percutaneous site infection, 32–33
  pump flow, 33
  pump intelligence, 33–34
  pump speed, 34
  rate of excretion, 22
  sensorless commutation, 24
  serendipity, 20
  stroke, 32
  submersible pumps, 23
  TAH built during resideny, 21
  tolerance to hemolysis, 22
  Viaderm, 33
Rotary blood pump speed modulation
  artificial pulse mode algorithm, 323, 324
  asynchronous timing, 327, 329
  axial and centrifugal continuous-flow rotary pumps, 325, 326
  benefits and tradeoffs, 329
  clinical best practices, 322
  design criteria, 323
  engineering parameters, 322
  engineering tradeoffs, 329, 330
  HM III, and HVAD speed modulation algorithms, 324
  in vitro and in silico testing, 330, 331
  in vivo (large animal models) testing, 330
  intrinsic pump parameters, 322
  physiology of continuous-flow LVADs, 326
  pulsatility and volume unloading, 327
  pump configuration, 324, 325
  synchronous co-pulsation, 327
  synchronous counter-pulsation, 327
  washing, 323, 324
Rotary blood pump technology, 563
Rotary continuous flow pumps, 469
Rotoflow centrifugal continuous flow pump, 238
RV stroke work index (SWI), 628

S

Seattle Heart Failure Model (SHFM), 136, 471, 624
Self-balancing flow system, 200
Sensor-less flow estimation, 121
Sensors, 703, 704
Serene stupidity, 20
Sewing cuff, 173
Short-term MCS, 141–143
Significant mitral regurgitation (MR), 183
Simulation
  clinical LVAD flow *vs.* pressure wave forms, 457, 458, 460, 461
  HeartWare HVAD static head curves, 458
  hybrid mock loop, 451
  numerical (*see* Numerical simulation)
  objective, 449
  physical mock loop, 451
Skilled nursing facilities (SNF), 662
Sleep-disordered breathing, 98
Smart mechanical circulatory support devices
  automatic control system, 703
  biocompatibility, 704
  challenges, 703
  design of experiments, 705
  device user requirements, 702
  implantable electronics, 703
  left ventricular volume sensing, 704
  physiologic need, 701
  pressure or volume sensors in left ventricle, 703
  pulsatile devices, 704
  sonomicrometry ultrasound crystals, 704
  specific device requirements
    compatiblity with other devices, 703
    easy usability, 702
    hacking immunity, 702
    light to moderate exercise, 702
    no adverse events and reliability, 702
    normal life style with near normal quality of life, 702
Smart pump, 126
Sonomicrometry ultrasound crystals, 704
Speed pulsatility, 569
Starling mechanism, 243
Statistical Shape Model, 570
Strain-based hemolysis, 411–413
Stratifying RV support, 164
Stress-based hemolysis, 410, 411
Stroke (CVA), 656
Suction event, 162, 163, 176
Supportive symptom care, 672
Surplus hemodynamic energy (SHE), 237, 318–321
Swan-Ganz catheter, 172
SynCardia-CardioWest TAH, 9
Syncardia TAH, 9, 10, 85, 185, 191, 196, 200, 243, 244, 475, 546, 601, 602
  bridge-to-transplantation device, 195
  clinical implementation of, 197
  evolution of, 195
  external drive console, 196
  four-layer polyurethane diaphragm, 195
  Freedom Driver, 196
  patient characteristics and outcomes, 198–199
  smaller 50cc model, 197
Synergy Micro-pump, 474
Syn-Hall valves, 243
Systolic heart failure, 623–625
Systolic-pressure (SP) controller, 262

T

TandemHeart®, 45, 55, 56, 372, 499
Tandem Life Protek Duo, 473
Teichholz formula, 254
Telemonitoring, 711
  CardioMEMS™, 712

patient input, 711
pump parameters, 712
telephone-monitoring, 711
Temporary and durable devices, 100, 101
Terumo DuraHeartTM (Dr. Chisato Nojiri), 44
TET communication system, 9
Thermo Cardio Systems Heartmate vented electric device, 546
Thermo cardiosystems incorporated (TCI), 43, 44
Thoratec, 45
Thoratec HeartMate XVE™, 84
Thoratec Paracorporeal Ventricular Assist Device, 83
Thoratec PVAD system, 11, 83
Thromboelastography (TEG), 212
Tilt table testing, 78
TORVAD system, 704
automatic response, 558
low shear, 558
pediatric pump and platform technology, 560
physiologic sensing capabilities, 556, 557
principle of operation, 555, 556
synchronization, 557–559
system and surgical implantation, 556
Total artificial heart (TAH), 6, 7, 111, 243, 245, 246, 635
AbioCor®, 193–195
BiVACOR pump, 200, 201, 546
biventricular assistance, 546
CF rotary blood pumps, 494
CFTAH (*see* Continuous-flow total artificial heart (CFTAH))
Cleveland Clinic SmartHeart CFTAH, 200
Continuous Flow TAH, 199–200
C-TAH, 201, 202
escalating medical urgency, 191
expanding patient populations, 203–204
history, 545
hydraulic performance, 549, 550
in vitro testing
bovine calves, 551, 552
mock circulatory loop, 549–551
Jarvik 7 TAH, 546
importance of pulsatility, 204
left ventricular assist devices, 545
mechanical circulatory assist devices, 545
national research program, 193
Orcorazon TAH, 547, 548
positive displacement pumps, 546
ReinHeart® TAH, 202, 203
rotary blood pumps, 546
second generation pump, 553
SmartHeart, 546
Syncardia temporary TAH, 191, 546
A70cc model, 197
bridge-to-transplantation device, 195
clinical implementation of, 197
evolution of, 195
external drive console, 196
four-layer polyurethane diaphragm, 195
Freedom Driver, 196
patient characteristics and outcomes, 198–199
smaller 50cc model, 197
Thermo Cardio Systems Heartmate vented electric device, 546
two-ventricle system, 191
Transcatheter AV replacement (TAVR), 658–659
Transcutaneous energy transfer, 617
Transcutaneous energy transfer system(TETS), 9, 84, 126, 476, 477, 612–614
Transesophageal echocardiographic (TEE), 157
Tricuspid annular plane systolic excursion (TAPSE), 137
Tricuspid regurgitation (TR), 183
Tricuspid valve (TV) assessment, 650
Tricuspid valve surgery (TVS), 183

U

Usability engineering process, 524, 525
U.S. Artificial Heart Program, 4

V

Vadovations, 46
Veno-arterial ECMO (VA-ECMO), 207, 210
Veno-arterial-veno ECMO (VAV-ECMO), 207
Veno-venous ECMO (VV-ECMO), 207, 210
VentrAssistTM, 45
Ventricular arrhythmias (VA), 167, 659
Ventricular assist devices (VADs), 10–13, 83, 232, 253, 254, 599
blood damage, 409
constant-pump-speed operation, 270
constant-speed operation, 264
constitutive equations, 406
ejection fraction, 254
electronics of, 264
generalized Newtonian/shear thinning models, 406, 407
hemocompatibility, 270
hemodynamic testing conditions, 267
hemolysis estimation, 409, 410
HMC maps, 264, 265
multi-objective physiologic controller, 269, 270
Navier-Stokes equations, 404, 405
Newtonian models, 406
optimal control strategy, 269
pathological hemodynamics, 273
pathophysiologic determinants
afterload and contractility changes, 257
afterload measures, 256, 257
drift, 258
ejection fraction, 254
elastance and contractility of heart, 255
elevated physical strain, 257
feedback-based physiologic control system, 257
hematocrit, 257
ischemic event, 255
mitral valve insufficiency, 254
non-pulsatile VAD flow, 257
physiologic controller, 257
preload measures, 255, 256

Ventricular assist devices (VADs) *(con't)*
pressure sensors, 257
pump data, 256
pump speed setting, 254
stiff vasculature, 256
superimposed pulsations, 257
Teichholz formula, 254
performance, efficiency and torque requirement, 416, 417
physiologic pump controllers, 259, 264
feedback-based physiologic controllers, 260, 261
fixed-speed pumps, 258
Frank-Starling mechanism, 261
multi-objective physiologic controller, 262, 263
non-physiologic and non-pulsatile arterial flow, 258
optimal control strategy, 262, 263
over- and underpumping events, 258
physiologic controllers, 258
PRS controller, 261, 262
pump speed modulations, 258, 260
SP controller, 262
underpumping situation, 258
preload-based physiologic controllers, 267
PRS controller, 267, 268, 271, 272
pump characteristic curve, 417–421
pump design, 414–416
pump speed changes, 264
rotor bearing and hydrodynamic bearings, 422–424
safety algorithms, 269
spatial filtering, 409
SP controller, 268, 269, 271
strain-based hemolysis, 410–413
stress and strain, 405, 406
temporal filtering, 408, 409
thrombosis estimation, 413, 414
turbulence in, 408, 409
viscoelastic and thixotropic models, 407, 408
Ventricular elastance, 452
Ventricular interdependence, 567
Ventricular suction, 122
VentriFlow pulsatile pump, 238
VentriFlow True Pulse Pump, 238
Virchow's endothelial injury, 121
Virtual mock loop (VML), 452
applications, 452
baseline conditions, 458
hemodynamic summary, 459, 462
left intraventricular pressure *vs.* volume loops, 459, 464
LVAD and BVAD, 458, 459
LVAD and RVAD dynamic head curves, 459, 463
system pressures *vs.* simulation time, 459, 462
systemic *vs.* pulmonary pressures, 459, 462
VAD pump dynamic head curves comparison, 459, 464
with and without BVAD support, 459, 463
Virtual Zero Power (VZP) controller, 568
Viscous impeller pump (VIP), 46
Visual analogue score (VAS), 159
Volume flow rate, 114
Volumetric efficiency, 120
von Willebrand (vWF) multimers, 654
von Willebrand factor deficiency, 645
Von Willebrand syndrome, 476

W
WHO II pulmonary hypertension, 184
Windkessel effect, 236, 440
Wireless energy transmission systems, 609

X
Xenotransplantation, 671